Problem-Solving Strategies and Model Boxes

Note for users of the three-volume edition:
Volume 1 (pp. 1–600) includes Chapters 1–21.
Volume 2 (pp. 601–1062) includes Chapters 22–36.
Volume 3 (pp. 1021–1240) includes Chapters 36–42.

Chapters 37–42 are not in the Standard Edition.

Brief Contents

PHYSICS

FOR SCIENTISTS AND ENGINEERS **A STRATEGIC APPROACH** 4/E

VOLUME ONE

RANDALL D. KNIGHT

California Polytechnic State University
San Luis Obispo

PEARSON

Editor-in-Chief:	Jeanne Zalesky
Acquisitions Editor:	Darien Estes
Project Manager:	Martha Steele
Program Manager:	Katie Conley
Senior Development Editor:	Alice Houston, Ph.D.
Art Development Editors:	Alice Houston, Kim Brucker, and Margot Otway
Development Manager:	Cathy Murphy
Program and Project Management Team Lead:	Kristen Flathman
Production Management:	Rose Kernan
Compositor:	Cenveo® Publisher Services
Design Manager:	Mark Ong
Cover Designer:	John Walker
Illustrators:	Rolin Graphics
Rights & Permissions Project Manager:	Maya Gomez
Rights & Permissions Management:	Rachel Youdelman
Photo Researcher:	Eric Schrader
Manufacturing Buyer:	Maura Zaldivar-Garcia
Executive Marketing Manager:	Christy Lesko
Cover Photo Credit:	Thomas Vogel/Getty Images

Library of Congress Cataloging-in-Publication Data On File

ISBN 10: 0-134-11068-4; ISBN 13: 978-0-134-11068-4 (Volume 1)

www.pearsonhighered.com   1 2 3 4 5 6 7 8 9 10—**V011**—18 17 16 15

About the Author

Randy Knight taught introductory physics for 32 years at Ohio State University and California Polytechnic State University, where he is Professor Emeritus of Physics. Professor Knight received a Ph.D. in physics from the University of California, Berkeley and was a post-doctoral fellow at the Harvard-Smithsonian Center for Astrophysics before joining the faculty at Ohio State University. It was at Ohio State that he began to learn about the research in physics education that, many years later, led to *Five Easy Lessons: Strategies for Successful Physics Teaching* and this book, as well as *College Physics: A Strategic Approach,* co-authored with Brian Jones and Stuart Field. Professor Knight's research interests are in the fields of laser spectroscopy and environmental science. When he's not in front of a computer, you can find Randy hiking, sea kayaking, playing the piano, or spending time with his wife Sally and their five cats.

A research-driven approach, fine-tuned for even greater ease-of-use and student success

REVISED COVERAGE AND ORGANIZATION GIVE INSTRUCTORS GREATER CHOICE AND FLEXIBILITY

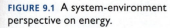
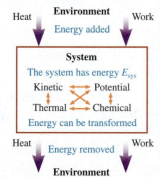

FIGURE 9.1 A system-environment perspective on energy.

NEW! CHAPTER ORGANIZATION allows instructors to more easily present material as needed to complement labs, course schedules, and different teaching styles. Work and energy are now covered before momentum, oscillations are grouped with mechanical waves, and optics appears after electricity and magnetism. Unchanged is Knight's unique approach of working from concrete to abstract, using multiple representations, balancing qualitative with quantitative, and addressing misconceptions.

NEW! ADVANCED TOPICS as optional sections add even more flexibility for instructors' individual courses. Topics include rocket propulsion, gyroscopes and precession, the wave equation (including for electromagnetic waves), the speed of sound in gases, and more details on the interference of light.

11.6 ADVANCED TOPIC Rocket Propulsion

Newton's second law $\vec{F} = m\vec{a}$ applies to objects whose mass does not change. That's an excellent assumption for balls and bicycles, but what about something like a rocket that loses a significant amount of mass as its fuel is burned? Problems of varying mass are solved with momentum rather than acceleration. We'll look at one important example.

FIGURE 11.29 shows a rocket being propelled by the thrust of burning fuel but *not* influenced by gravity or drag. Perhaps it is a rocket in deep space where gravity is very weak in comparison to the rocket's thrust. This may not be highly realistic, but ignoring gravity allows us to understand the essentials of rocket propulsion without making the mathematics too complicated. Rocket propulsion with gravity is a Challenge Problem in the end-of-chapter problems.

The system rocket + exhaust gases is an isolated system, so its total momentum is conserved. The basic idea is simple: As exhaust gases are shot out the back, the rocket "recoils" in the opposite direction. Putting this idea on a mathematical footing is fairly straightforward—it's basically the same as analyzing an explosion—but we have to be extremely careful with signs.

We'll use a before-and-after approach, as we do with all momentum problems. Th...

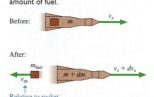

FIGURE 11.29 A before-and-after pictorial representation of a rocket burning a small amount of fuel.

60. ‖ A clever engineer designs a "sprong" that obeys the force law
CALC $F_x = -q(x - x_{eq})^3$, where x_{eq} is the equilibrium position of the end of the sprong and q is the sprong constant. For simplicity, we'll let $x_{eq} = 0$ m. Then $F_x = -qx^3$.
   a. What are the units of q?
   b. Find an expression for the potential energy of a stretched or compressed sprong.
   c. A sprong-loaded toy gun shoots a 20 g plastic ball. What is the launch speed if the sprong constant is 40,000, with the units you found in part a, and the sprong is compressed 10 cm? Assume the barrel is frictionless.

NEW! MORE CALCULUS-BASED PROBLEMS have been added, along with an icon to make these easy to identify. The significantly revised end-of-chapter problem sets, extensively class-tested and both calibrated and improved using MasteringPhysics® data, expand the range of physics and math skills students will use to solve problems.

Built from the ground up on physics education research and crafted using key ideas from learning theory, Knight has set the standard for effective and accessible pedagogical materials in physics. In this fourth edition, Knight continues to refine and expand the instructional techniques to take students further.

NEW AND UPDATED LEARNING TOOLS PROMOTE DEEPER AND BETTER-CONNECTED UNDERSTANDING

NEW! MODEL BOXES enhance the text's emphasis on modeling—analyzing a complex, real-world situation in terms of simple but reasonable idealizations that can be applied over and over in solving problems. These fundamental simplifications are developed in the text and then deployed more explicitly in the worked examples, helping students to recognize when and how to use recurring models, a key critical-thinking skill.

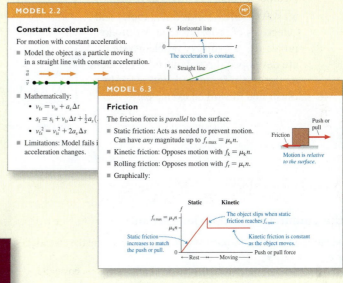

MODEL 2.2

Constant acceleration
For motion with constant acceleration.
- Model the object as a particle moving in a straight line with constant acceleration.
- Mathematically:
  - $v_{fs} = v_{is} + a_s \Delta t$
  - $s_f = s_i + v_{is} \Delta t + \frac{1}{2} a_s (\ldots$
  - $v_{fs}^2 = v_{is}^2 + 2a_s \Delta s$
- Limitations: Model fails i...
  acceleration changes.

MODEL 6.3

Friction
The friction force is *parallel* to the surface.
- Static friction: Acts as needed to prevent motion. Can have *any* magnitude up to $f_{s\,max} = \mu_s n$.
- Kinetic friction: Opposes motion with $f_k = \mu_k n$.
- Rolling friction: Opposes motion with $f_r = \mu_r n$.
- Graphically:

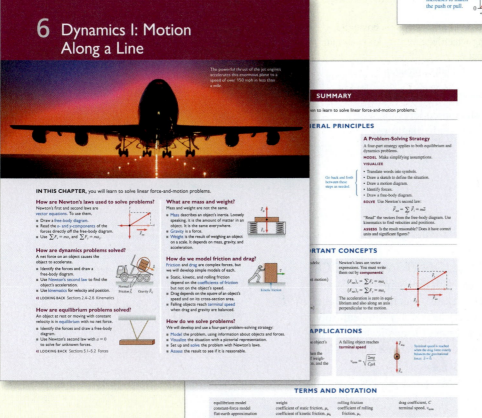

6 Dynamics I: Motion Along a Line

The powerful thrust of the jet engines accelerates this enormous plane to a speed of over 150 mph in less than a mile.

IN THIS CHAPTER, you will learn to solve linear force-and-motion problems.

How are Newton's laws used to solve problems?
Newton's first and second laws are vector equations. To use them,
- Draw a free-body diagram.
- Read the x- and y-components of the forces directly off the free-body diagram.
- Use $\sum F_x = ma_x$ and $\sum F_y = ma_y$.

How are dynamics problems solved?
A net force on an object causes the object to accelerate.
- Identify the forces and draw a free-body diagram.
- Use Newton's second law to find the object's acceleration.
- Use kinematics for velocity and position.
  « LOOKING BACK Sections 2.4–2.6 Kinematics

How are equilibrium problems solved?
An object at rest or moving with constant velocity is in equilibrium with no net force.
- Identify the forces and draw a free-body diagram.
- Use Newton's second law with $a = 0$ to solve for unknown forces.
  « LOOKING BACK Sections 5.1–5.2 Forces

What are mass and weight?
Mass and weight are not the same.
- Mass describes an object's inertia. Loosely speaking, it is the amount of matter in an object. It is the same everywhere.
- Weight is the result of weighing an object on a scale. It depends on mass, gravity, and acceleration.

How do we model friction and drag?
Friction and drag are complex forces, but we will develop simple models of each.
- Static, kinetic, and rolling friction depend on the coefficients of friction but not on the object's speed.
- Drag depends on the square of an object's speed and on its cross-section area.
- Falling objects reach terminal speed when drag and gravity are balanced.

How do we solve problems?
We will develop and use a four-part problem-solving strategy:
- Model the problem, using information about objects and forces.
- Visualize the situation with a pictorial representation.
- Set up and solve the problem with Newton's laws.
- Assess the result to see if it is reasonable.

SUMMARY

...en to learn to solve linear force-and-motion problems.

...IERAL PRINCIPLES

A Problem-Solving Strategy
A four-part strategy applies to both equilibrium and dynamics problems.
MODEL Make simplifying assumptions.
VISUALIZE
- Translate words into symbols.
- Draw a sketch to define the situation.
- Draw a motion diagram.
- Identify forces.
- Draw a free-body diagram.
SOLVE Use Newton's second law:
$$\vec{F}_{net} = \sum_i \vec{F}_i = m\vec{a}$$
"Read" the vectors from the free-body diagram. Use kinematics to find velocities and positions.
ASSESS Is the result reasonable? Does it have correct units and significant figures?

...RTANT CONCEPTS

Newton's laws are vector expressions. You must write them out by components:
$(F_{net})_x = \sum F_x = ma_x$
$(F_{net})_y = \sum F_y = ma_y$
The acceleration is zero in equilibrium and also along an axis perpendicular to the motion.

...APPLICATIONS

A falling object reaches terminal speed
$$v_{term} = \sqrt{\frac{2mg}{C\rho A}}$$

TERMS AND NOTATION

equilibrium model
constant-force model
flat-earth approximation
weight
coefficient of static friction, μ_s
coefficient of kinetic friction, μ_k
rolling friction
coefficient of rolling friction, μ_r
drag coefficient, C
terminal speed, v_{term}

REVISED! ENHANCED CHAPTER PREVIEWS, based on the educational psychology concept of an "advance organizer," have been reconceived to address the questions students are most likely to ask themselves while studying the material for the first time. Questions cover the important ideas, and provide a big-picture overview of the chapter's key principles. Each chapter concludes with the visual Chapter Summary, consolidating and structuring understanding.

A STRUCTURED AND CONSISTENT APPROACH BUILDS PROBLEM-SOLVING SKILLS AND CONFIDENCE

With a research-based 4-step problem-solving framework used throughout the text, students learn the importance of making assumptions (in the MODEL step) and gathering information and making sketches (in the VISUALIZE step) before treating the problem mathematically (SOLVE) and then analyzing their results (ASSESS).

Detailed **PROBLEM-SOLVING STRATEGIES** for different topics and categories of problems (circular-motion problems, calorimetry problems, etc.) are developed throughout, each one built on the 4-step framework and carefully illustrated in worked examples.

TACTICS BOXES give step-by-step procedures for developing specific skills (drawing free-body diagrams, using ray tracing, etc.).

The **REVISED STUDENT WORKBOOK** is tightly integrated with the main text–allowing students to practice skills from the text's Tactics Boxes, work through the steps of Problem-Solving Strategies, and assess the applicability of the Models. The workbook is referenced throughout the text with the icon .

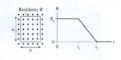
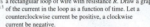

MasteringPhysics

THE ULTIMATE RESOURCE BEFORE, DURING, AND AFTER CLASS

BEFORE CLASS

NEW! INTERACTIVE PRELECTURE VIDEOS address the rapidly growing movement toward pre-lecture teaching and flipped classrooms. These whiteboard-style animations provide an introduction to key topics with embedded assessment to help students prepare and professors identify student misconceptions before lecture.

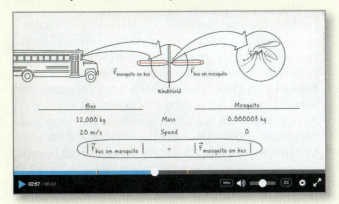

NEW! DYNAMIC STUDY MODULES (DSMs) continuously assess students' performance in real time to provide personalized question and explanation content until students master the module with confidence. The DSMs cover basic math skills and key definitions and relationships for topics across all of mechanics and electricity and magnetism.

DURING CLASS

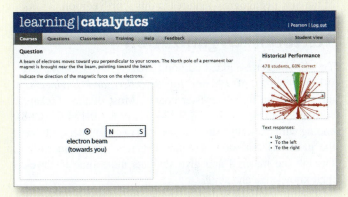

NEW! LEARNING CATALYTICS™ is an interactive classroom tool that uses students' devices to engage them in more sophisticated tasks and thinking. Learning Catalytics enables instructors to generate classroom discussion and promote peer-to-peer learning to help students develop critical-thinking skills. Instructors can take advantage of real-time analytics to find out where students are struggling and adjust their instructional strategy.

AFTER CLASS

NEW! ENHANCED END-OF-CHAPTER QUESTIONS offer students instructional support when and where they need it, including links to the eText, math remediation, and wrong-answer feedback for homework assignments.

ADAPTIVE FOLLOW-UPS are personalized assignments that pair Mastering's powerful content with Knewton's adaptive learning engine to provide individualized help to students before misconceptions take hold. These adaptive follow-ups address topics students struggled with on assigned homework, including core prerequisite topics.

Preface to the Instructor

This fourth edition of *Physics for Scientists and Engineers: A Strategic Approach* continues to build on the research-driven instructional techniques introduced in the first edition and the extensive feedback from thousands of users. From the beginning, the objectives have been:

- To produce a textbook that is more focused and coherent, less encyclopedic.
- To move key results from physics education research into the classroom in a way that allows instructors to use a range of teaching styles.
- To provide a balance of quantitative reasoning and conceptual understanding, with special attention to concepts known to cause student difficulties.
- To develop students' problem-solving skills in a systematic manner.

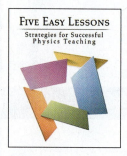

These goals and the rationale behind them are discussed at length in the *Instructor's Guide* and in my small paperback book, *Five Easy Lessons: Strategies for Successful Physics Teaching*. Please request a copy from your local Pearson sales representative if it is of interest to you (ISBN 978-0-805-38702-5).

What's New to This Edition

For this fourth edition, we continue to apply the best results from educational research and to tailor them for this course and its students. At the same time, the extensive feedback we've received from both instructors and students has led to many changes and improvements to the text, the figures, and the end-of-chapter problems. These include:

- **Chapter ordering changes** allow instructors to more easily organize content as needed to accommodate labs, schedules, and different teaching styles. Work and energy are now covered before momentum, oscillations are grouped with mechanical waves, and optics appears after electricity and magnetism.
- **Addition of advanced topics** as optional sections further expands instructors' options. Topics include rocket propulsion, gyroscopes, the wave equation (for mechanical and electromagnetic waves), the speed of sound in gases, and more details on the interference of light.
- **Model boxes** enhance the text's emphasis on modeling— analyzing a complex, real-world situation in terms of simple but reasonable idealizations that can be applied over and over in solving problems. These fundamental simplifications

are developed in the text and then deployed more explicitly in the worked examples, helping students to recognize when and how to use recurring models.

- **Enhanced chapter previews** have been redesigned, with student input, to address the questions students are most likely to ask themselves while studying the material for the first time. The previews provide a big-picture overview of the chapter's key principles.
- **Looking Back pointers** enable students to look back at a previous chapter when it's important to review concepts. Pointers provide the specific section to consult at the exact point in the text where they need to use this material.
- **Focused Part Overviews and Knowledge Structures** consolidate understanding of groups of chapters and give a tighter structure to the book as a whole. Reworked Knowledge Structures provide more targeted detail on overarching themes.
- **Updated visual program** that has been enhanced by revising over 500 pieces of art to increase the focus on key ideas.
- **Significantly revised end-of-chapter problem sets** include more challenging problems to expand the range of physics and math skills students will use to solve problems. A new icon for calculus-based problems has been added.

At the front of this book, you'll find an illustrated walkthrough of the new pedagogical features in this fourth edition.

Textbook Organization

The 42-chapter extended edition (ISBN 978-0-133-94265-1 / 0-133-94265-1) of *Physics for Scientists and Engineers* is intended for a three-semester course. Most of the 36-chapter standard edition (ISBN 978-0-134-08149-6 / 0-134-08149-8), ending with relativity, can be covered in two semesters, although the judicious omission of a few chapters will avoid rushing through the material and give students more time to develop their knowledge and skills.

The full textbook is divided into eight parts: Part I: *Newton's Laws,* Part II: *Conservation Laws,* Part III: *Applications of Newtonian Mechanics,* Part IV: *Oscillations and Waves,* Part V: *Thermodynamics,* Part VI: *Electricity and Magnetism,* Part VII: *Optics,* and Part VIII: *Relativity and Quantum Physics.* Note that covering the parts in this order is by no means essential. Each topic is self-contained, and Parts III–VII can be rearranged to suit an instructor's needs. Part VII: *Optics* does need to follow Part IV: *Oscillations and Waves,* but optics can be taught either before or after electricity and magnetism.

There's a growing sentiment that quantum physics is quickly becoming the province of engineers, not just scientists, and that even a two-semester course should include a reasonable introduction to quantum ideas. The *Instructor's Guide* outlines a couple of routes through the book that allow most of the quantum physics chapters to be included in a two-semester course. I've written the book with the hope that an increasing number of instructors will choose one of these routes.

- **Extended edition,** with modern physics (ISBN 978-0-133-94265-1 / 0-133-94265-1): Chapters 1–42.
- **Standard edition** (ISBN 978-0-134-08149-6 / 0-134-08149-8): Chapters 1-36.
- **Volume 1** (ISBN 978-0-134-11068-4 / 0-134-11068-4) covers mechanics, waves, and thermodynamics: Chapters 1–21.
- **Volume 2** (ISBN 978-0-134-11066-0 / 0-134-11066-8) covers electricity and magnetism and optics, plus relativity: Chapters 22–36.
- **Volume 3** (ISBN 978-0-134-11065-3 / 0-134-11065-X) covers relativity and quantum physics: Chapters 36–42.

The Student Workbook

A key component of *Physics for Scientists and Engineers: A Strategic Approach* is the accompanying *Student Workbook.* The workbook bridges the gap between textbook and home-

work problems by providing students the opportunity to learn and practice skills prior to using those skills in quantitative end-of-chapter problems, much as a musician practices technique separately from performance pieces. The workbook exercises, which are keyed to each section of the textbook, focus on developing specific skills, ranging from identifying forces and drawing free-body diagrams to interpreting wave functions.

The workbook exercises, which are generally qualitative and/or graphical, draw heavily upon the physics education research literature. The exercises deal with issues known to cause student difficulties and employ techniques that have proven to be effective at overcoming those difficulties. **New to the fourth edition workbook** are exercises that provide guided practice for the textbook's Model boxes. The workbook exercises can be used in class as part of an active-learning teaching strategy, in recitation sections, or as assigned homework. More information about effective use of the *Student Workbook* can be found in the *Instructor's Guide.*

Instructional Package

Physics for Scientists and Engineers: A Strategic Approach, fourth edition, provides an integrated teaching and learning package of support material for students and instructors. **NOTE** For convenience, most instructor supplements can be downloaded from the "Instructor Resources" area of MasteringPhysics® and the Instructor Resource Center (www.pearsonhighered.com/educator).

Name of Supplement	Print	Online	Instructor or Student Supplement	Description
MasteringPhysics with Pearson eText ISBN 0-134-08313-X		✓	Instructor and Student Supplement	This product features all of the resources of MasteringPhysics in addition to the new Pearson eText 2.0. Now available on smartphones and tablets, Pearson eText 2.0 comprises the full text, including videos and other rich media. Students can configure reading settings, including resizeable type and night-reading mode, take notes, and highlight, bookmark, and search the text.
Instructor's Solutions Manual ISBN 0-134-09246-5		✓	Instructor Supplement	This comprehensive solutions manual contains complete solutions to all end-of-chapter questions and problems. All problem solutions follow the Model/Visualize/Solve/Assess problem-solving strategy used in the text.
Instructor's Guide ISBN 0-134-09248-1		✓	Instructor Supplement	Written by Randy Knight, this resource provides chapter-by-chapter creative ideas and teaching tips for use in your class. It also contains an extensive review of results of what has been learned from physics education research and provides guidelines for using active-learning techniques in your classroom.
TestGen Test Bank ISBN 0-134-09245-7		✓	Instructor Supplement	The Test Bank contains over 2,000 high-quality conceptual and multiple-choice questions. Test files are provided in both TestGen® and Word format.
Instructor's Resource DVD ISBN 0-134-09247-3	✓	✓	Instructor Supplement	This cross-platform DVD includes an Image Library; editable content for Key Equations, Problem-Solving Strategies, Math Relationship Boxes, Model Boxes, and Tactic Boxes; PowerPoint Lecture Slides and Clicker Questions; Instructor's Guide, Instructor's Solutions Manual; Solutions to Student Workbook exercises; and PhET simulations.
Student Workbook Extended (Ch 1–42) ISBN 0-134-08316-4 Standard (Ch 1–36) ISBN 0-134-08315-6 Volume 1 (Ch 1–21) ISBN 0-134-11064-1 Volume 2 (Ch 22–36) ISBN 0-134-11063-3 Volume 3 (Ch 36–42) ISBN 0-134-11060-9	✓		Student Supplement	For a more detailed description of the *Student Workbook*, see page v.

Acknowledgments

I have relied upon conversations with and, especially, the written publications of many members of the physics education research community. Those whose influence can be seen in these pages include Wendy Adams, the late Arnold Arons, Stuart Field, Uri Ganiel, Ibrahim Halloun, Richard Hake, Ken Heller, Paula Heron, David Hestenes, Brian Jones, the late Leonard Jossem, Jill Larkin, Priscilla Laws, John Mallinckrodt, Kandiah Manivannan, Richard Mayer, Lillian McDermott and members of the Physics Education Research Group at the University of Washington, David Meltzer, Edward "Joe" Redish, Fred Reif, Jeffery Saul, Rachel Scherr, Bruce Sherwood, Josip Slisko, David Sokoloff, Richard Steinberg, Ronald Thornton, Sheila Tobias, Alan Van Heuleven, Carl Wieman, and Michael Wittmann. John Rigden, founder and director of the Introductory University Physics Project, provided the impetus that got me started down this path. Early development of the materials was supported by the National Science Foundation as the Physics for the Year 2000 project; their support is gratefully acknowledged.

I especially want to thank my editors, Jeanne Zalesky and Becky Ruden, development editor Alice Houston, project manager Martha Steele, art development editors Kim Brucker and Margot Otway, and all the other staff at Pearson for their enthusiasm and hard work on this project. Rose Kernan and the team at Cenveo along with photo researcher Eric Schrader get a good deal of the credit for making this complex project all come together. Larry Smith, Larry Stookey, and Michael Ottinger have done an outstanding job of checking the solutions to every end-of-chapter problem and updating the Instructor's Solutions Manual. John Filaseta must be thanked for carefully writing out the solutions to the Student Workbook exercises, and Jason Harlow for putting together the Lecture Slides. In addition to the reviewers and classroom testers listed below, who gave invaluable feedback, I am particularly grateful to Charlie Hibbard for his close scrutiny of every word and figure.

Finally, I am endlessly grateful to my wife Sally for her love, encouragement, and patience, and to our many cats, past and present, who are always ready to suggest "Dinner time?" when I'm in need of a phrase.

Randy Knight, September 2015
rknight@calpoly.edu

Reviewers and Classroom Testers

Gary B. Adams, *Arizona State University*
Ed Adelson, *Ohio State University*
Kyle Altmann, *Elon University*
Wayne R. Anderson, *Sacramento City College*
James H. Andrews, *Youngstown State University*
Kevin Ankoviak, *Las Positas College*
David Balogh, *Fresno City College*
Dewayne Beery, *Buffalo State College*

Joseph Bellina, *Saint Mary's College*
James R. Benbrook, *University of Houston*
David Besson, *University of Kansas*
Matthew Block, *California State University, Sacramento*
Randy Bohn, *University of Toledo*
Richard A. Bone, *Florida International University*
Gregory Boutis, *York College*
Art Braundmeier, *University of Southern Illinois, Edwardsville*
Carl Bromberg, *Michigan State University*
Meade Brooks, *Collin College*
Douglas Brown, *Cabrillo College*
Ronald Brown, *California Polytechnic State University, San Luis Obispo*
Mike Broyles, *Collin County Community College*
Debra Burris, *University of Central Arkansas*
James Carolan, *University of British Columbia*
Michael Chapman, *Georgia Tech University*
Norbert Chencinski, *College of Staten Island*
Tonya Coffey, *Appalachian State University*
Kristi Concannon, *King's College*
Desmond Cook, *Old Dominion University*
Sean Cordry, *Northwestern College of Iowa*
Robert L. Corey, *South Dakota School of Mines*
Michael Crescimanno, *Youngstown State University*
Dennis Crossley, *University of Wisconsin–Sheboygan*
Wei Cui, *Purdue University*
Robert J. Culbertson, *Arizona State University*
Danielle Dalafave, *The College of New Jersey*
Purna C. Das, *Purdue University North Central*
Chad Davies, *Gordon College*
William DeGraffenreid, *California State University–Sacramento*
Dwain Desbien, *Estrella Mountain Community College*
John F. Devlin, *University of Michigan, Dearborn*
John DiBartolo, *Polytechnic University*
Alex Dickison, *Seminole Community College*
Chaden Djalali, *University of South Carolina*
Margaret Dobrowolska, *University of Notre Dame*
Sandra Doty, *Denison University*
Miles J. Dresser, *Washington State University*
Taner Edis, *Truman State University*
Charlotte Elster, *Ohio University*
Robert J. Endorf, *University of Cincinnati*
Tilahun Eneyew, *Embry-Riddle Aeronautical University*
F. Paul Esposito, *University of Cincinnati*
John Evans, *Lee University*
Harold T. Evensen, *University of Wisconsin–Platteville*
Michael R. Falvo, *University of North Carolina*
Abbas Faridi, *Orange Coast College*
Nail Fazleev, *University of Texas–Arlington*
Stuart Field, *Colorado State University*
Daniel Finley, *University of New Mexico*
Jane D. Flood, *Muhlenberg College*
Michael Franklin, *Northwestern Michigan College*

Jonathan Friedman, *Amherst College*
Thomas Furtak, *Colorado School of Mines*
Alina Gabryszewska-Kukawa, *Delta State University*
Lev Gasparov, *University of North Florida*
Richard Gass, *University of Cincinnati*
Delena Gatch, *Georgia Southern University*
J. David Gavenda, *University of Texas, Austin*
Stuart Gazes, *University of Chicago*
Katherine M. Gietzen, *Southwest Missouri State University*
Robert Glosser, *University of Texas, Dallas*
William Golightly, *University of California, Berkeley*
Paul Gresser, *University of Maryland*
C. Frank Griffin, *University of Akron*
John B. Gruber, *San Jose State University*
Thomas D. Gutierrez, *California Polytechnic State University, San Luis Obispo*
Stephen Haas, *University of Southern California*
John Hamilton, *University of Hawaii at Hilo*
Jason Harlow, *University of Toronto*
Randy Harris, *University of California, Davis*
Nathan Harshman, *American University*
J. E. Hasbun, *University of West Georgia*
Nicole Herbots, *Arizona State University*
Jim Hetrick, *University of Michigan–Dearborn*
Scott Hildreth, *Chabot College*
David Hobbs, *South Plains College*
Laurent Hodges, *Iowa State University*
Mark Hollabaugh, *Normandale Community College*
Steven Hubbard, *Lorain County Community College*
John L. Hubisz, *North Carolina State University*
Shane Hutson, *Vanderbilt University*
George Igo, *University of California, Los Angeles*
David C. Ingram, *Ohio University*
Bob Jacobsen, *University of California, Berkeley*
Rong-Sheng Jin, *Florida Institute of Technology*
Marty Johnston, *University of St. Thomas*
Stanley T. Jones, *University of Alabama*
Darrell Judge, *University of Southern California*
Pawan Kahol, *Missouri State University*
Teruki Kamon, *Texas A&M University*
Richard Karas, *California State University, San Marcos*
Deborah Katz, *U.S. Naval Academy*
Miron Kaufman, *Cleveland State University*
Katherine Keilty, *Kingwood College*
Roman Kezerashvili, *New York City College of Technology*
Peter Kjeer, *Bethany Lutheran College*
M. Kotlarchyk, *Rochester Institute of Technology*
Fred Krauss, *Delta College*
Cagliyan Kurdak, *University of Michigan*
Fred Kuttner, *University of California, Santa Cruz*
H. Sarma Lakkaraju, *San Jose State University*
Darrell R. Lamm, *Georgia Institute of Technology*
Robert LaMontagne, *Providence College*
Eric T. Lane, *University of Tennessee–Chattanooga*

Alessandra Lanzara, *University of California, Berkeley*
Lee H. LaRue, *Paris Junior College*
Sen-Ben Liao, *Massachusetts Institute of Technology*
Dean Livelybrooks, *University of Oregon*
Chun-Min Lo, *University of South Florida*
Olga Lobban, *Saint Mary's University*
Ramon Lopez, *Florida Institute of Technology*
Vaman M. Naik, *University of Michigan, Dearborn*
Kevin Mackay, *Grove City College*
Carl Maes, *University of Arizona*
Rizwan Mahmood, *Slippery Rock University*
Mani Manivannan, *Missouri State University*
Mark E. Mattson, *James Madison University*
Richard McCorkle, *University of Rhode Island*
James McDonald, *University of Hartford*
James McGuire, *Tulane University*
Stephen R. McNeil, *Brigham Young University–Idaho*
Theresa Moreau, *Amherst College*
Gary Morris, *Rice University*
Michael A. Morrison, *University of Oklahoma*
Richard Mowat, *North Carolina State University*
Eric Murray, *Georgia Institute of Technology*
Michael Murray, *University of Kansas*
Taha Mzoughi, *Mississippi State University*
Scott Nutter, *Northern Kentucky University*
Craig Ogilvie, *Iowa State University*
Benedict Y. Oh, *University of Wisconsin*
Martin Okafor, *Georgia Perimeter College*
Halina Opyrchal, *New Jersey Institute of Technology*
Derek Padilla, *Santa Rosa Junior College*
Yibin Pan, *University of Wisconsin–Madison*
Georgia Papaefthymiou, *Villanova University*
Peggy Perozzo, *Mary Baldwin College*
Brian K. Pickett, *Purdue University, Calumet*
Joe Pifer, *Rutgers University*
Dale Pleticha, *Gordon College*
Marie Plumb, *Jamestown Community College*
Robert Pompi, *SUNY-Binghamton*
David Potter, *Austin Community College–Rio Grande Campus*
Chandra Prayaga, *University of West Florida*
Kenneth M. Purcell, *University of Southern Indiana*
Didarul Qadir, *Central Michigan University*
Steve Quon, *Ventura College*
Michael Read, *College of the Siskiyous*
Lawrence Rees, *Brigham Young University*
Richard J. Reimann, *Boise State University*
Michael Rodman, *Spokane Falls Community College*
Sharon Rosell, *Central Washington University*
Anthony Russo, *Northwest Florida State College*
Freddie Salsbury, *Wake Forest University*
Otto F. Sankey, *Arizona State University*
Jeff Sanny, *Loyola Marymount University*
Rachel E. Scherr, *University of Maryland*
Carl Schneider, *U.S. Naval Academy*

Bruce Schumm, *University of California, Santa Cruz*
Bartlett M. Sheinberg, *Houston Community College*
Douglas Sherman, *San Jose State University*
Elizabeth H. Simmons, *Boston University*
Marlina Slamet, *Sacred Heart University*
Alan Slavin, *Trent College*
Alexander Raymond Small, *California State Polytechnic University, Pomona*
Larry Smith, *Snow College*
William S. Smith, *Boise State University*
Paul Sokol, *Pennsylvania State University*
LTC Bryndol Sones, *United States Military Academy*
Chris Sorensen, *Kansas State University*
Brian Steinkamp, *University of Southern Indiana*
Anna and Ivan Stern, *AW Tutor Center*
Gay B. Stewart, *University of Arkansas*
Michael Strauss, *University of Oklahoma*
Chin-Che Tin, *Auburn University*
Christos Valiotis, *Antelope Valley College*
Andrew Vanture, *Everett Community College*
Arthur Viescas, *Pennsylvania State University*
Ernst D. Von Meerwall, *University of Akron*
Chris Vuille, *Embry-Riddle Aeronautical University*
Jerry Wagner, *Rochester Institute of Technology*
Robert Webb, *Texas A&M University*
Zodiac Webster, *California State University, San Bernardino*
Robert Weidman, *Michigan Technical University*
Fred Weitfeldt, *Tulane University*
Gary Williams, *University of California, Los Angeles*
Lynda Williams, *Santa Rosa Junior College*
Jeff Allen Winger, *Mississippi State University*
Carey Witkov, *Broward Community College*
Ronald Zammit, *California Polytechnic State University, San Luis Obispo*
Darin T. Zimmerman, *Pennsylvania State University, Altoona*
Fredy Zypman, *Yeshiva University*

Student Focus Groups

California Polytechnic State University, San Luis Obispo
Matthew Bailey
James Caudill
Andres Gonzalez
Mytch Johnson

California State University, Sacramento
Logan Allen
Andrew Fujikawa
Sagar Gupta
Marlene Juarez
Craig Kovac
Alissa McKown
Kenneth Mercado
Douglas Ostheimer
Ian Tabbada
James Womack

Santa Rosa Junior College
Kyle Doughty
Tacho Gardiner
Erik Gonzalez
Joseph Goodwin
Chelsea Irmer
Vatsal Pandya
Parth Parikh
Andrew Prosser
David Reynolds
Brian Swain
Grace Woods

Stanford University
Montserrat Cordero
Rylan Edlin
Michael Goodemote II
Stewart Isaacs
David LaFehr
Sergio Rebeles
Jack Takahashi

Preface to the Student

From Me to You

The most incomprehensible thing about the universe is that it is comprehensible.
　—Albert Einstein

The day I went into physics class it was death.
　—Sylvia Plath, *The Bell Jar*

Let's have a little chat before we start. A rather one-sided chat, admittedly, because you can't respond, but that's OK. I've talked with many of your fellow students over the years, so I have a pretty good idea of what's on your mind.

What's your reaction to taking physics? Fear and loathing? Uncertainty? Excitement? All the above? Let's face it, physics has a bit of an image problem on campus. You've probably heard that it's difficult, maybe impossible unless you're an Einstein. Things that you've heard, your experiences in other science courses, and many other factors all color your *expectations* about what this course is going to be like.

It's true that there are many new ideas to be learned in physics and that the course, like college courses in general, is going to be much faster paced than science courses you had in high school. I think it's fair to say that it will be an *intense* course. But we can avoid many potential problems and difficulties if we can establish, here at the beginning, what this course is about and what is expected of you—and of me!

Just what is physics, anyway? Physics is a way of thinking about the physical aspects of nature. Physics is not better than art or biology or poetry or religion, which are also ways to think about nature; it's simply different. One of the things this course will emphasize is that physics is a human endeavor. The ideas presented in this book were not found in a cave or conveyed to us by aliens; they were discovered and developed by real people engaged in a struggle with real issues.

You might be surprised to hear that physics is not about "facts." Oh, not that facts are unimportant, but physics is far more focused on discovering *relationships* and *patterns* than on learning facts for their own sake.

 For example, the colors of the rainbow appear both when white light passes through a prism and—as in this photo—when white light reflects from a thin film of oil on water. What does this pattern tell us about the nature of light?

Our emphasis on relationships and patterns means that there's not a lot of memorization when you study physics. Some—there are still definitions and equations to learn—but less than in many other courses. Our emphasis, instead, will be on thinking and reasoning. This is important to factor into your expectations for the course.

Perhaps most important of all, *physics is not math!* Physics is much broader. We're going to look for patterns and relationships in nature, develop the logic that relates different ideas, and search for the reasons *why* things happen as they do. In doing so, we're going to stress qualitative reasoning, pictorial and graphical reasoning, and reasoning by analogy. And yes, we will use math, but it's just one tool among many.

It will save you much frustration if you're aware of this physics–math distinction up front. Many of you, I know, want to find a formula and plug numbers into it—that is, to do a math problem. Maybe that worked in high school science courses, but it is *not* what this course expects of you. We'll certainly do many calculations, but the specific numbers are usually the last and least important step in the analysis.

As you study, you'll sometimes be baffled, puzzled, and confused. That's perfectly normal and to be expected. Making mistakes is OK too *if* you're willing to learn from the experience. No one is born knowing how to do physics any more than he or she is born knowing how to play the piano or shoot basketballs. The ability to do physics comes from practice, repetition, and struggling with the ideas until you "own" them and can apply them yourself in new situations. There's no way to make learning effortless, at least for anything worth learning, so expect to have some difficult moments ahead. But also expect to have some moments of excitement at the joy of discovery. There will be instants at which the pieces suddenly click into place and you *know* that you understand a powerful idea. There will be times when you'll surprise yourself by successfully working a difficult problem that you didn't think you could solve. My hope, as an author, is that the excitement and sense of adventure will far outweigh the difficulties and frustrations.

Getting the Most Out of Your Course

Many of you, I suspect, would like to know the "best" way to study for this course. There is no best way. People are different, and what works for one student is less effective for another. But I do want to stress that *reading the text* is vitally important. The basic knowledge for this course is written down on these pages, and your instructor's *number-one expectation* is that you will read carefully to find and learn that knowledge.

Despite there being no best way to study, I will suggest *one* way that is successful for many students.

1. **Read each chapter *before* it is discussed in class.** I cannot stress too strongly how important this step is. Class attendance is much more effective if you are prepared. When you first read a chapter, focus on learning new vocabulary, definitions, and notation. There's a list of terms and notations at the end of each chapter. Learn them! You won't understand

what's being discussed or how the ideas are being used if you don't know what the terms and symbols mean.

2. **Participate actively in class.** Take notes, ask and answer questions, and participate in discussion groups. There is ample scientific evidence that *active participation* is much more effective for learning science than passive listening.

3. **After class, go back for a careful re-reading of the chapter.** In your second reading, pay closer attention to the details and the worked examples. Look for the *logic* behind each example (I've highlighted this to make it clear), not just at what formula is being used. And use the textbook tools that are designed to help your learning, such as the problem-solving strategies, the chapter summaries, and the exercises in the *Student Workbook*.

4. **Finally, apply what you have learned to the homework problems at the end of each chapter.** I strongly encourage you to form a study group with two or three classmates. There's good evidence that students who study regularly with a group do better than the rugged individualists who try to go it alone.

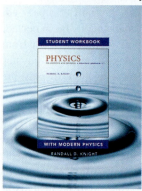

Did someone mention a workbook? The companion *Student Workbook* is a vital part of the course. Its questions and exercises ask you to reason *qualitatively*, to use graphical information, and to give explanations. It is through these exercises that you will learn what the concepts mean and will practice the reasoning skills appropriate to the chapter. You will then have acquired the baseline knowledge and confidence you need *before* turning to the end-of-chapter homework problems. In sports or in music, you would never think of performing before you practice, so why would you want to do so in physics? The workbook is where you practice and work on basic skills.

Many of you, I know, will be tempted to go straight to the homework problems and then thumb through the text looking for a formula that seems like it will work. That approach will not succeed in this course, and it's guaranteed to make you frustrated and discouraged. Very few homework problems are of the "plug and chug" variety where you simply put numbers into a formula. To work the homework problems successfully, you need a better study strategy—either the one outlined above or your own—that helps you learn the concepts and the relationships between the ideas.

Getting the Most Out of Your Textbook

Your textbook provides many features designed to help you learn the concepts of physics and solve problems more effectively.

■ **TACTICS BOXES** give step-by-step procedures for particular skills, such as interpreting graphs or drawing special

diagrams. Tactics Box steps are explicitly illustrated in subsequent worked examples, and these are often the starting point of a full *Problem-Solving Strategy*.

■ **PROBLEM-SOLVING STRATEGIES** are provided for each broad class of problems—problems characteristic of a chapter or group of chapters. The strategies follow a consistent four-step approach to help you develop confidence and proficient problem-solving skills: **MODEL, VISUALIZE, SOLVE, ASSESS**.

■ Worked **EXAMPLES** illustrate good problem-solving practices through the consistent use of the four-step problem-solving approach The worked examples are often very detailed and carefully lead you through the *reasoning* behind the solution as well as the numerical calculations.

■ **STOP TO THINK** questions embedded in the chapter allow you to quickly assess whether you've understood the main idea of a section. A correct answer will give you confidence to move on to the next section. An incorrect answer will alert you to re-read the previous section.

■ Blue annotations on figures help you better understand what the figure is showing. They will help you to interpret graphs; translate between graphs, math, and pictures; grasp difficult concepts through a visual analogy; and develop many other important skills.

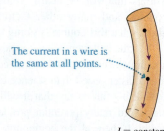

The current in a wire is the same at all points.

I = constant

■ Schematic *Chapter Summaries* help you organize what you have learned into a hierarchy, from general principles (top) to applications (bottom). Side-by-side pictorial, graphical, textual, and mathematical representations are used to help you translate between these key representations.

■ Each part of the book ends with a **KNOWLEDGE STRUCTURE** designed to help you see the forest rather than just the trees.

Now that you know more about what is expected of you, what can you expect of me? That's a little trickier because the book is already written! Nonetheless, the book was prepared on the basis of what I think my students throughout the years have expected—and wanted—from their physics textbook. Further, I've listened to the extensive feedback I have received from thousands of students like you, and their instructors, who used the first three editions of this book.

You should know that these course materials—the text and the workbook—are based on extensive research about how students learn physics and the challenges they face. The effectiveness of many of the exercises has been demonstrated through extensive class testing. I've written the book in an informal style that I hope you will find appealing and that will encourage you to do the reading. And, finally, I have endeavored to make clear not only that physics, as a technical body of knowledge, is relevant to your profession but also that physics is an exciting adventure of the human mind.

I hope you'll enjoy the time we're going to spend together.

Detailed Contents

Volume 1 contains chapters 1–21; Volume 2 contains chapters 22–36; Volume 3 contains chapters 36–42.

Part V Thermodynamics

Part VI Electricity and Magnetism

Newton's Laws

OVERVIEW

Why Things Move

Each of the seven parts of this book opens with an overview to give you a look ahead, a glimpse at where your journey will take you in the next few chapters. It's easy to lose sight of the big picture while you're busy negotiating the terrain of each chapter. In Part I, the big picture, in a word, is *motion*.

There are two big questions we must tackle:

■ **How do we describe motion?** It is easy to say that an object moves, but it's not obvious how we should measure or characterize the motion if we want to analyze it mathematically. The mathematical description of motion is called *kinematics,* and it is the subject matter of Chapters 1 through 4.

■ **How do we explain motion?** Why do objects have the particular motion they do? Why, when you toss a ball upward, does it go up and then come back down rather than keep going up? Are there "laws of nature" that allow us to predict an object's motion? The explanation of motion in terms of its causes is called *dynamics,* and it is the topic of Chapters 5 through 8.

Two key ideas for answering these questions are *force* (the "cause") and *acceleration* (the "effect"). A variety of pictorial and graphical tools will be developed in Chapters 1 through 5 to help you develop an *intuition* for the connection between force and acceleration. You'll then put this knowledge to use in Chapters 5 through 8 as you analyze motion of increasing complexity.

Another important tool will be the use of *models.* Reality is extremely complicated. We would never be able to develop a science if we had to keep track of every little detail of every situation. A model is a simplified description of reality— much as a model airplane is a simplified version of a real airplane—used to reduce the complexity of a problem to the point where it can be analyzed and understood. We will introduce several important models of motion, paying close attention, especially in these earlier chapters, to where simplifying assumptions are being made, and why.

The "laws of motion" were discovered by Isaac Newton roughly 350 years ago, so the study of motion is hardly cutting-edge science. Nonetheless, it is still extremely important. Mechanics—the science of motion—is the basis for much of engineering and applied science, and many of the ideas introduced here will be needed later to understand things like the motion of waves and the motion of electrons through circuits. Newton's mechanics is the foundation of much of contemporary science, thus we will start at the beginning.

Motion can be slow and steady, or fast and sudden. This rocket, with its rapid acceleration, is responding to forces exerted on it by thrust, gravity, and the air.

1 Concepts of Motion

Motion takes many forms. The ski jumper seen here is an example of translational motion.

IN THIS CHAPTER, you will learn the fundamental concepts of motion.

What is a chapter preview?

Each chapter starts with an overview. Think of it as a roadmap to help you get oriented and make the most of your studying.

« LOOKING BACK A Looking Back reference tells you what material from previous chapters is especially important for understanding the new topics. A quick review will help your learning. You will find additional Looking Back references within the chapter, right at the point they're needed.

What is motion?

Before solving motion problems, we must learn to *describe* motion. We will use

- Motion diagrams
- Graphs
- Pictures

Motion concepts introduced in this chapter include position, velocity, and acceleration.

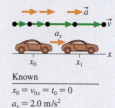

Why do we need vectors?

Many of the quantities used to describe motion, such as velocity, have both a size and a direction. We use vectors to represent these quantities. This chapter introduces graphical techniques to add and subtract vectors. Chapter 3 will explore vectors in more detail.

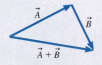

Why are units and significant figures important?

Scientists and engineers must communicate their ideas to others. To do so, we have to agree about the *units* in which quantities are measured. In physics we use metric units, called SI units. We also need rules for telling others how accurately a quantity is known. You will learn the rules for using significant figures correctly.

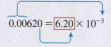

$$0.00620 = \boxed{6.20} \times 10^{-3}$$

Why is motion important?

The universe is in motion, from the smallest scale of electrons and atoms to the largest scale of entire galaxies. We'll start with the motion of everyday objects, such as cars and balls and people. Later we'll study the motions of waves, of atoms in gases, and of electrons in circuits. Motion is the one theme that will be with us from the first chapter to the last.

1.1 Motion Diagrams

Motion is a theme that will appear in one form or another throughout this entire book. Although we all have intuition about motion, based on our experiences, some of the important aspects of motion turn out to be rather subtle. So rather than jumping immediately into a lot of mathematics and calculations, this first chapter focuses on *visualizing* motion and becoming familiar with the *concepts* needed to describe a moving object. Our goal is to lay the foundations for understanding motion.

FIGURE 1.1 Four basic types of motion.

Linear motion

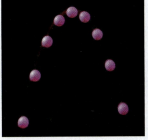
Circular motion

Projectile motion

Rotational motion

To begin, let's define **motion** as the change of an object's position with time. **FIGURE 1.1** shows four basic types of motion that we will study in this book. The first three—linear, circular, and projectile motion—in which the object moves through space are called **translational motion.** The path along which the object moves, whether straight or curved, is called the object's **trajectory.** Rotational motion is somewhat different because there's movement but the object as a whole doesn't change position. We'll defer rotational motion until later and, for now, focus on translational motion.

Making a Motion Diagram

An easy way to study motion is to make a video of a moving object. A video camera, as you probably know, takes images at a fixed rate, typically 30 every second. Each separate image is called a *frame.* As an example, **FIGURE 1.2** shows four frames from a video of a car going past. Not surprisingly, the car is in a somewhat different position in each frame.

Suppose we edit the video by layering the frames on top of each other, creating the composite image shown in **FIGURE 1.3**. This edited image, showing an object's position at several *equally spaced instants of time,* is called a **motion diagram.** As the examples below show, we can define concepts such as constant speed, speeding up, and slowing down in terms of how an object appears in a motion diagram.

NOTE It's important to keep the camera in a *fixed position* as the object moves by. Don't "pan" it to track the moving object.

FIGURE 1.2 Four frames from a video.

FIGURE 1.3 A motion diagram of the car shows all the frames simultaneously.

The same amount of time elapses between each image and the next.

Examples of motion diagrams

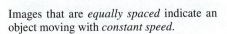
Images that are *equally spaced* indicate an object moving with *constant speed.*

An *increasing distance* between the images shows that the object is *speeding up.*

A *decreasing distance* between the images shows that the object is *slowing down.*

STOP TO THINK 1.1 Which car is going faster, A or B? Assume there are equal intervals of time between the frames of both videos.

Car A   Car B

NOTE Each chapter will have several *Stop to Think* questions. These questions are designed to see if you've understood the basic ideas that have been presented. The answers are given at the end of the book, but you should make a serious effort to think about these questions before turning to the answers.

1.2 Models and Modeling

The real world is messy and complicated. Our goal in physics is to brush aside many of the real-world details in order to discern patterns that occur over and over. For example, a swinging pendulum, a vibrating guitar string, a sound wave, and jiggling atoms in a crystal are all very different—yet perhaps not so different. Each is an example of a system moving back and forth around an equilibrium position. If we focus on understanding a very simple oscillating system, such as a mass on a spring, we'll automatically understand quite a bit about the many real-world manifestations of oscillations.

Stripping away the details to focus on essential features is a process called *modeling*. A **model** is a highly simplified picture of reality, but one that still captures the essence of what we want to study. Thus "mass on a spring" is a simple but realistic model of almost all oscillating systems.

Models allow us to make sense of complex situations by providing a framework for thinking about them. One could go so far as to say that developing and testing models is at the heart of the scientific process. Albert Einstein once said, "Physics should be as simple as possible—but not simpler." We want to find the simplest model that allows us to understand the phenomenon we're studying, but we can't make the model so simple that key aspects of the phenomenon get lost.

We'll develop and use many models throughout this textbook; they'll be one of our most important thinking tools. These models will be of two types:

- *Descriptive models:* What are the essential characteristics and properties of a phenomenon? How do we describe it in the simplest possible terms? For example, the mass-on-a-spring model of an oscillating system is a descriptive model.
- *Explanatory models:* Why do things happen as they do? Explanatory models, based on the laws of physics, have predictive power, allowing us to test—against experimental data—whether a model provides an adequate explanation of our observations.

The Particle Model

For many types of motion, such as that of balls, cars, and rockets, the motion of the object *as a whole* is not influenced by the details of the object's size and shape. All we really need to keep track of is the motion of a single point on the object, so we can treat the object *as if* all its mass were concentrated into this single point. An object that can be represented as a mass at a single point in space is called a **particle.** A particle has no size, no shape, and no distinction between top and bottom or between front and back.

If we model an object as a particle, we can represent the object in each frame of a motion diagram as a simple dot rather than having to draw a full picture. **FIGURE 1.4** shows how much simpler motion diagrams appear when the object is represented as a particle. Note that the dots have been numbered 0, 1, 2, . . . to tell the sequence in which the frames were exposed.

We can model an airplane's takeoff as a particle (a descriptive model) undergoing constant acceleration (a descriptive model) in response to constant forces (an explanatory model).

FIGURE 1.4 Motion diagrams in which the object is modeled as a particle.

(a) Motion diagram of a rocket launch

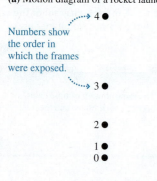

Numbers show the order in which the frames were exposed.

(b) Motion diagram of a car stopping

The same amount of time elapses between each image and the next.

Treating an object as a particle is, of course, a simplification of reality—but that's what modeling is all about. The **particle model** of motion is a simplification in which we treat a moving object as if all of its mass were concentrated at a single point. The particle model is an excellent approximation of reality for the translational motion of cars, planes, rockets, and similar objects.

Of course, not everything can be modeled as a particle; models have their limits. Consider, for example, a rotating gear. The center doesn't move at all while each tooth is moving in a different direction. We'll need to develop new models when we get to new types of motion, but the particle model will serve us well throughout Part I of this book.

STOP TO THINK 1.2 Three motion diagrams are shown. Which is a dust particle settling to the floor at constant speed, which is a ball dropped from the roof of a building, and which is a descending rocket slowing to make a soft landing on Mars?

(a) 0 ●
1 ●
2 ●
3 ●
4 ●
5 ●

(b) 0 ●
1 ●
2 ●
3 ●
4 ●
5 ●

(c) 0 ●
1 ●
2 ●
3 ●
4 ●
5 ●

1.3 Position, Time, and Displacement

To use a motion diagram, you would like to know *where* the object is (i.e., its *position*) and *when* the object was at that position (i.e., the *time*). Position measurements can be made by laying a coordinate-system grid over a motion diagram. You can then measure the (x, y) coordinates of each point in the motion diagram. Of course, the world does not come with a coordinate system attached. A coordinate system is an artificial grid that *you* place over a problem in order to analyze the motion. You place the origin of your coordinate system wherever you wish, and different observers of a moving object might all choose to use different origins.

Time, in a sense, is also a coordinate system, although you may never have thought of time this way. You can pick an arbitrary point in the motion and label it "$t = 0$ seconds." This is simply the instant you decide to start your clock or stopwatch, so it is the origin of your time coordinate. Different observers might choose to start their clocks at different moments. A video frame labeled "$t = 4$ seconds" was taken 4 seconds after you started your clock.

We typically choose $t = 0$ to represent the "beginning" of a problem, but the object may have been moving before then. Those earlier instants would be measured as negative times, just as objects on the x-axis to the left of the origin have negative values of position. Negative numbers are not to be avoided; they simply locate an event in space or time *relative to an origin*.

To illustrate, **FIGURE 1.5a** shows a sled sliding down a snow-covered hill. **FIGURE 1.5b** is a motion diagram for the sled, over which we've drawn an xy-coordinate system. You can see that the sled's position is $(x_3, y_3) = (15 \text{ m}, 15 \text{ m})$ at time $t_3 = 3$ s. Notice how we've used subscripts to indicate the time and the object's position in a specific frame of the motion diagram.

NOTE The frame at $t = 0$ is frame 0. That is why the fourth frame is labeled 3.

Another way to locate the sled is to draw its **position vector:** an arrow from the origin to the point representing the sled. The position vector is given the symbol $\vec{r}$. Figure 1.5b shows the position vector $\vec{r}_3 = (21 \text{ m}, 45°)$. The position vector $\vec{r}$ does not tell us anything different than the coordinates (x, y). It simply provides the information in an alternative form.

FIGURE 1.5 Motion diagram of a sled with frames made every 1 s.

(a)

(b)
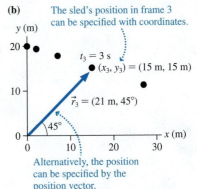

The sled's position in frame 3 can be specified with coordinates.

$t_3 = 3$ s
$(x_3, y_3) = (15 \text{ m}, 15 \text{ m})$

$\vec{r}_3 = (21 \text{ m}, 45°)$

45°

Alternatively, the position can be specified by the position vector.

Scalars and Vectors

Some physical quantities, such as time, mass, and temperature, can be described completely by a single number with a unit. For example, the mass of an object is 6 kg and its temperature is 30°C. A single number (with a unit) that describes a physical quantity is called a **scalar.** A scalar can be positive, negative, or zero.

Many other quantities, however, have a directional aspect and cannot be described by a single number. To describe the motion of a car, for example, you must specify not only how fast it is moving, but also the *direction* in which it is moving. A quantity having both a *size* (the "How far?" or "How fast?") and a *direction* (the "Which way?") is called a **vector.** The size or length of a vector is called its *magnitude*. Vectors will be studied thoroughly in Chapter 3, so all we need for now is a little basic information.

We indicate a vector by drawing an arrow over the letter that represents the quantity. Thus $\vec{r}$ and $\vec{A}$ are symbols for vectors, whereas r and A, without the arrows, are symbols for scalars. In handwritten work you must draw arrows over all symbols that represent vectors. This may seem strange until you get used to it, but it is very important because we will often use both r and $\vec{r}$, or both A and $\vec{A}$, in the same problem, and they mean different things! Note that the arrow over the symbol always points to the right, regardless of which direction the actual vector points. Thus we write $\vec{r}$ or $\vec{A}$, never $\overleftarrow{r}$ or $\overleftarrow{A}$.

Displacement

We said that motion is the change in an object's position with time, but how do we show a change of position? A motion diagram is the perfect tool. **FIGURE 1.6** is the motion diagram of a sled sliding down a snow-covered hill. To show how the sled's position changes between, say, $t_3 = 3$ s and $t_4 = 4$ s, we draw a vector arrow between the two dots of the motion diagram. This vector is the sled's **displacement,** which is given the symbol $\Delta\vec{r}$. The Greek letter delta (Δ) is used in math and science to indicate the *change* in a quantity. In this case, as we'll show, the displacement $\Delta\vec{r}$ is the change in an object's position.

FIGURE 1.6 The sled undergoes a displacement $\Delta\vec{r}$ from position $\vec{r}_3$ to position $\vec{r}_4$.

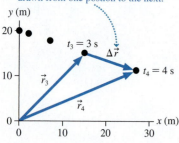

The sled's displacement between $t_3 = 3$ s and $t_4 = 4$ s is the vector drawn from one postion to the next.

| NOTE $\Delta\vec{r}$ is a *single* symbol. You cannot cancel out or remove the Δ.

Notice how the sled's position vector $\vec{r}_4$ is a combination of its early position $\vec{r}_3$ with the displacement vector $\Delta\vec{r}$. In fact, $\vec{r}_4$ is the *vector sum* of the vectors $\vec{r}_3$ and $\Delta\vec{r}$. This is written

$$\vec{r}_4 = \vec{r}_3 + \Delta\vec{r} \tag{1.1}$$

Here we're adding vector quantities, not numbers, and vector addition differs from "regular" addition. We'll explore vector addition more thoroughly in Chapter 3, but for now you can add two vectors $\vec{A}$ and $\vec{B}$ with the three-step procedure shown in Tactics Box 1.1.

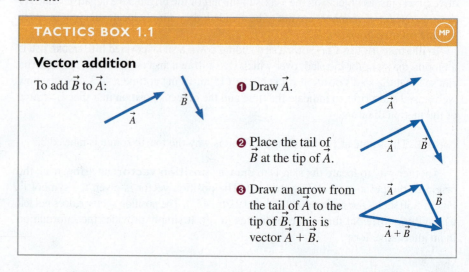

TACTICS BOX 1.1

Vector addition

To add $\vec{B}$ to $\vec{A}$:

❶ Draw $\vec{A}$.

❷ Place the tail of $\vec{B}$ at the tip of $\vec{A}$.

❸ Draw an arrow from the tail of $\vec{A}$ to the tip of $\vec{B}$. This is vector $\vec{A} + \vec{B}$.

If you examine Figure 1.6, you'll see that the steps of Tactics Box 1.1 are exactly how $\vec{r}_3$ and $\Delta \vec{r}$ are added to give $\vec{r}_4$.

NOTE A vector is not tied to a particular location on the page. You can move a vector around as long as you don't change its length or the direction it points. Vector $\vec{B}$ is not changed by sliding it to where its tail is at the tip of $\vec{A}$.

Equation 1.1 told us that $\vec{r}_4 = \vec{r}_3 + \Delta \vec{r}$. This is easily rearranged to give a more precise definition of displacement: **The displacement $\Delta \vec{r}$ of an object as it moves from an initial position $\vec{r}_i$ to a final position $\vec{r}_f$ is**

$$\Delta \vec{r} = \vec{r}_f - \vec{r}_i \qquad (1.2)$$

Graphically, $\Delta \vec{r}$ is a vector arrow drawn from position $\vec{r}_i$ to position $\vec{r}_f$.

NOTE To be more general, we've written Equation 1.2 in terms of an *initial position* and a *final position,* indicated by subscripts i and f. We'll frequently use i and f when writing general equations, then use specific numbers or values, such as 3 and 4, when working a problem.

This definition of $\Delta \vec{r}$ involves *vector subtraction.* With numbers, subtraction is the same as the addition of a negative number. That is, $5 - 3$ is the same as $5 + (-3)$. Similarly, we can use the rules for vector addition to find $\vec{A} - \vec{B} = \vec{A} + (-\vec{B})$ if we first define what we mean by $-\vec{B}$. As **FIGURE 1.7** shows, the negative of vector $\vec{B}$ is a vector with the same length but pointing in the opposite direction. This makes sense because $\vec{B} - \vec{B} = \vec{B} + (-\vec{B}) = \vec{0}$, where $\vec{0}$, a vector with zero length, is called the **zero vector.**

FIGURE 1.7 The negative of a vector.

Vector $-\vec{B}$ has the same length as $\vec{B}$ but points in the opposite direction.

$\vec{B} + (-\vec{B}) = \vec{0}$ because the sum returns to the starting point.

The zero vector $\vec{0}$ has no length.

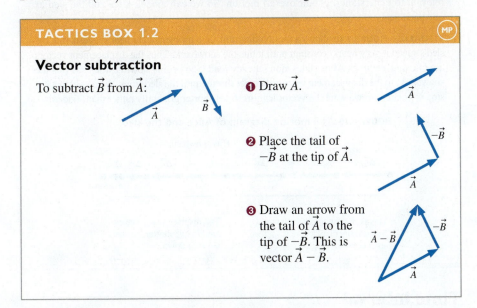

TACTICS BOX 1.2 (MP)

Vector subtraction

To subtract $\vec{B}$ from $\vec{A}$:

❶ Draw $\vec{A}$.

❷ Place the tail of $-\vec{B}$ at the tip of $\vec{A}$.

❸ Draw an arrow from the tail of $\vec{A}$ to the tip of $-\vec{B}$. This is vector $\vec{A} - \vec{B}$.

FIGURE 1.8 uses the vector subtraction rules of Tactics Box 1.2 to prove that the displacement $\Delta \vec{r}$ is simply the vector connecting the dots of a motion diagram.

▼ **FIGURE 1.8** Using vector subtraction to find $\Delta \vec{r} = \vec{r}_f - \vec{r}_i$.

(a) Initial and final position vectors

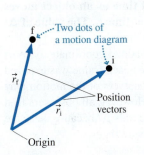

Two dots of a motion diagram

Position vectors

Origin

(b) Procedure for finding the particle's displacement vector $\Delta \vec{r}$

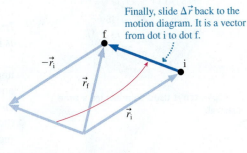

❶ Draw $\vec{r}_f$.

❷ Draw $-\vec{r}_i$ at the tip of $\vec{r}_f$.

❸ Draw $\vec{r}_f - \vec{r}_i$. This is $\Delta \vec{r}$.

Finally, slide $\Delta \vec{r}$ back to the motion diagram. It is a vector from dot i to dot f.

FIGURE 1.9 Motion diagrams with the displacement vectors.

(a) Rocket launch

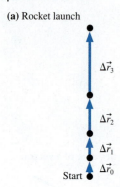

(b) Car stopping

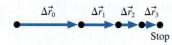

Motion Diagrams with Displacement Vectors

The first step in analyzing a motion diagram is to determine all of the displacement vectors. As Figure 1.8 shows, the displacement vectors are simply the arrows connecting each dot to the next. Label each arrow with a *vector* symbol $\Delta \vec{r}_n$, starting with $n = 0$. **FIGURE 1.9** shows the motion diagrams of Figure 1.4 redrawn to include the displacement vectors. You do not need to show the position vectors.

NOTE When an object either starts from rest or ends at rest, the initial or final dots are *as close together* as you can draw the displacement vector arrow connecting them. In addition, just to be clear, you should write "Start" or "Stop" beside the initial or final dot. It is important to distinguish stopping from merely slowing down.

Now we can conclude, more precisely than before, that, as time proceeds:

- An object is speeding up if its displacement vectors are increasing in length.
- An object is slowing down if its displacement vectors are decreasing in length.

EXAMPLE 1.1 | **Headfirst into the snow**

Alice is sliding along a smooth, icy road on her sled when she suddenly runs headfirst into a large, very soft snowbank that gradually brings her to a halt. Draw a motion diagram for Alice. Show and label all displacement vectors.

MODEL The details of Alice and the sled—their size, shape, color, and so on—are not relevant to understanding their overall motion. So we can model Alice and the sled as one particle.

VISUALIZE **FIGURE 1.10** shows a motion diagram. The problem statement suggests that the sled's speed is very nearly constant until it hits the snowbank. Thus the displacement vectors are of equal length as Alice slides along the icy road. She begins slowing when she hits the snowbank, so the displacement vectors then get shorter until the sled stops. We're told that her stop is gradual, so we want the vector lengths to get shorter gradually rather than suddenly.

FIGURE 1.10 The motion diagram of Alice and the sled.

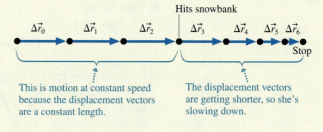

Time Interval

It's also useful to consider a *change* in time. For example, the clock readings of two frames of film might be t_1 and t_2. The specific values are arbitrary because they are timed relative to an arbitrary instant that you chose to call $t = 0$. But the **time interval** $\Delta t = t_2 - t_1$ is *not* arbitrary. It represents the elapsed time for the object to move from one position to the next.

The time interval $\Delta t = t_f - t_i$ measures the elapsed time as an object moves from an initial position $\vec{r}_i$ at time t_i to a final position $\vec{r}_f$ at time t_f. The value of Δt is independent of the specific clock used to measure the times.

To summarize the main idea of this section, we have added coordinate systems and clocks to our motion diagrams in order to measure *when* each frame was exposed and *where* the object was located at that time. Different observers of the motion may choose different coordinate systems and different clocks. However, all observers find the *same* values for the displacements $\Delta \vec{r}$ and the time intervals Δt because these are independent of the specific coordinate system used to measure them.

A stopwatch is used to measure a time interval.

1.4 Velocity

It's no surprise that, during a given time interval, a speeding bullet travels farther than a speeding snail. To extend our study of motion so that we can compare the bullet to the snail, we need a way to measure how fast or how slowly an object moves.

One quantity that measures an object's fastness or slowness is its **average speed,** defined as the ratio

$$\text{average speed} = \frac{\text{distance traveled}}{\text{time interval spent traveling}} = \frac{d}{\Delta t} \qquad (1.3)$$

If you drive 15 miles (mi) in 30 minutes ($\frac{1}{2}$ h), your average speed is

$$\text{average speed} = \frac{15 \text{ mi}}{\frac{1}{2} \text{ h}} = 30 \text{ mph} \qquad (1.4)$$

Although the concept of speed is widely used in our day-to-day lives, it is not a sufficient basis for a science of motion. To see why, imagine you're trying to land a jet plane on an aircraft carrier. It matters a great deal to you whether the aircraft carrier is moving at 20 mph (miles per hour) to the north or 20 mph to the east. Simply knowing that the boat's speed is 20 mph is not enough information!

It's the displacement $\Delta \vec{r}$, a vector quantity, that tells us not only the distance traveled by a moving object, but also the *direction* of motion. Consequently, a more useful ratio than $d/\Delta t$ is the ratio $\Delta \vec{r}/\Delta t$. In addition to measuring how fast an object moves, this ratio is a vector that points in the direction of motion.

It is convenient to give this ratio a name. We call it the **average velocity,** and it has the symbol $\vec{v}_{\text{avg}}$. **The average velocity of an object during the time interval Δt, in which the object undergoes a displacement $\Delta \vec{r}$, is the vector**

$$\vec{v}_{\text{avg}} = \frac{\Delta \vec{r}}{\Delta t} \qquad (1.5)$$

An object's average velocity vector points in the same direction as the displacement vector $\Delta \vec{r}$. This is the direction of motion.

> **NOTE** In everyday language we do not make a distinction between speed and velocity, but in physics *the distinction is very important.* In particular, speed is simply "How fast?" whereas velocity is "How fast, and in which direction?" As we go along we will be giving other words more precise meanings in physics than they have in everyday language.

As an example, **FIGURE 1.11a** shows two ships that move 5 miles in 15 minutes. Using Equation 1.5 with $\Delta t = 0.25$ h, we find

$$\vec{v}_{\text{avg A}} = (20 \text{ mph, north})$$
$$\vec{v}_{\text{avg B}} = (20 \text{ mph, east}) \qquad (1.6)$$

Both ships have a speed of 20 mph, but their velocities are different. Notice how the velocity *vectors* in **FIGURE 1.11b** point in the direction of motion.

> **NOTE** Our goal in this chapter is to *visualize* motion with motion diagrams. Strictly speaking, the vector we have defined in Equation 1.5, and the vector we will show on motion diagrams, is the *average* velocity $\vec{v}_{\text{avg}}$. But to allow the motion diagram to be a useful tool, we will drop the subscript and refer to the average velocity as simply $\vec{v}$. Our definitions and symbols, which somewhat blur the distinction between average and instantaneous quantities, are adequate for visualization purposes, but they're not the final word. We will refine these definitions in Chapter 2, where our goal will be to develop the mathematics of motion.

The victory goes to the runner with the highest average speed.

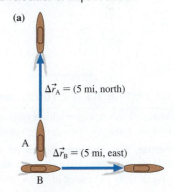
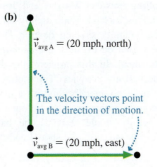

FIGURE 1.11 The displacement vectors and velocities of ships A and B.

(a)

$\Delta \vec{r}_A = (5 \text{ mi, north})$

A

$\Delta \vec{r}_B = (5 \text{ mi, east})$

B

(b)

$\vec{v}_{\text{avg A}} = (20 \text{ mph, north})$

The velocity vectors point in the direction of motion.

$\vec{v}_{\text{avg B}} = (20 \text{ mph, east})$

Motion Diagrams with Velocity Vectors

FIGURE 1.12 Motion diagram of the tortoise racing the hare.

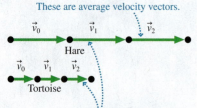

These are average velocity vectors.

Hare

Tortoise

The length of each arrow represents the average speed. The hare moves faster than the tortoise.

The velocity vector points in the same direction as the displacement $\Delta \vec{r}$, and the length of $\vec{v}$ is directly proportional to the length of $\Delta \vec{r}$. Consequently, the vectors connecting each dot of a motion diagram to the next, which we previously labeled as displacements, could equally well be identified as velocity vectors.

This idea is illustrated in **FIGURE 1.12**, which shows four frames from the motion diagram of a tortoise racing a hare. The vectors connecting the dots are now labeled as velocity vectors $\vec{v}$. **The length of a velocity vector represents the average speed with which the object moves between the two points.** Longer velocity vectors indicate faster motion. You can see that the hare moves faster than the tortoise.

Notice that the hare's velocity vectors do not change; each has the same length and direction. We say the hare is moving with *constant velocity*. The tortoise is also moving with its own constant velocity.

EXAMPLE 1.2 | **Accelerating up a hill**

The light turns green and a car accelerates, starting from rest, up a 20° hill. Draw a motion diagram showing the car's velocity.

MODEL Use the particle model to represent the car as a dot.

VISUALIZE The car's motion takes place along a straight line, but the line is neither horizontal nor vertical. A motion diagram should show the object moving with the correct orientation—in this case, at an angle of 20°. **FIGURE 1.13** shows several frames of the motion diagram, where we see the car speeding up. The car starts from rest, so the first arrow is drawn as short as possible and the first dot is labeled "Start." The displacement vectors have been drawn from each dot to the next, but then they are identified and labeled as average velocity vectors $\vec{v}$.

FIGURE 1.13 Motion diagram of a car accelerating up a hill.

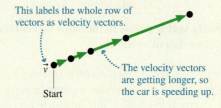

This labels the whole row of vectors as velocity vectors.

$\vec{v}$

The velocity vectors are getting longer, so the car is speeding up.

Start

EXAMPLE 1.3 | **A rolling soccer ball**

Marcos kicks a soccer ball. It rolls along the ground until stopped by Jose. Draw a motion diagram of the ball.

MODEL This example is typical of how many problems in science and engineering are worded. The problem does not give a clear statement of where the motion begins or ends. Are we interested in the motion of the ball just during the time it is rolling between Marcos and Jose? What about the motion *as* Marcos kicks it (ball rapidly speeding up) or *as* Jose stops it (ball rapidly slowing down)? The point is that *you* will often be called on to make a *reasonable interpretation* of a problem statement. In this problem, the details of kicking and stopping the ball are complex. The motion of the ball across the ground is easier to describe, and it's a motion you might expect to learn about in a physics class. So our *interpretation* is that the motion diagram should start as the ball leaves Marcos's foot (ball already moving) and should end the instant it touches Jose's foot

(ball still moving). In between, the ball will slow down a little. We will model the ball as a particle.

VISUALIZE With this interpretation in mind, **FIGURE 1.14** shows the motion diagram of the ball. Notice how, in contrast to the car of Figure 1.13, the ball is already moving as the motion diagram video begins. As before, the average velocity vectors are found by connecting the dots. You can see that the average velocity vectors get shorter as the ball slows. Each $\vec{v}$ is different, so this is *not* constant-velocity motion.

FIGURE 1.14 Motion diagram of a soccer ball rolling from Marcos to Jose.

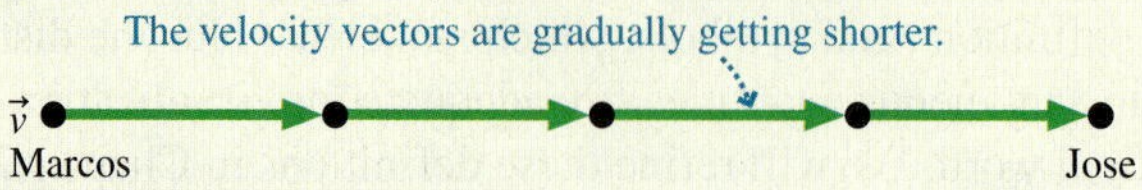

The velocity vectors are gradually getting shorter.

$\vec{v}$
Marcos   Jose

STOP TO THINK 1.3 A particle moves from position 1 to position 2 during the time interval Δt. Which vector shows the particle's average velocity?

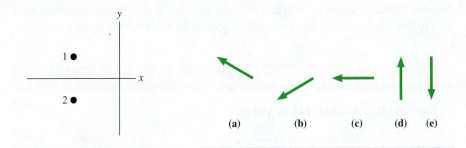

1.5 Linear Acceleration

Position, time, and velocity are important concepts, and at first glance they might appear to be sufficient to describe motion. But that is not the case. Sometimes an object's velocity is constant, as it was in Figure 1.12. More often, an object's velocity changes as it moves, as in Figures 1.13 and 1.14. We need one more motion concept to describe a *change* in the velocity.

Because velocity is a vector, it can change in two possible ways:

1. The magnitude can change, indicating a change in speed; or
2. The direction can change, indicating that the object has changed direction.

We will concentrate for now on the first case, a change in speed. The car accelerating up a hill in Figure 1.13 was an example in which the magnitude of the velocity vector changed but not the direction. We'll return to the second case in Chapter 4.

When we wanted to measure changes in position, the ratio $\Delta \vec{r}/\Delta t$ was useful. This ratio is the *rate of change of position*. By analogy, consider an object whose velocity changes from an initial $\vec{v}_i$ to a final $\vec{v}_f$ during the time interval Δt. Just as $\Delta \vec{r} = \vec{r}_f - \vec{r}_i$ is the change of position, the quantity $\Delta \vec{v} = \vec{v}_f - \vec{v}_i$ is the change of velocity. The ratio $\Delta \vec{v}/\Delta t$ is then the *rate of change of velocity*. It has a large magnitude for objects that speed up quickly and a small magnitude for objects that speed up slowly.

The ratio $\Delta \vec{v}/\Delta t$ is called the **average acceleration,** and its symbol is $\vec{a}_{avg}$. **The average acceleration of an object during the time interval Δt, in which the object's velocity changes by $\Delta \vec{v}$, is the vector**

$$\vec{a}_{avg} = \frac{\Delta \vec{v}}{\Delta t} \qquad (1.7)$$

The average acceleration vector points in the same direction as the vector $\Delta \vec{v}$.

Acceleration is a fairly abstract concept. Yet it is essential to develop a good intuition about acceleration because it will be a key concept for understanding why objects move as they do. Motion diagrams will be an important tool for developing that intuition.

The Audi TT accelerates from 0 to 60 mph in 6 s.

NOTE As we did with velocity, we will drop the subscript and refer to the average acceleration as simply $\vec{a}$. This is adequate for visualization purposes, but not the final word. We will refine the definition of acceleration in Chapter 2.

Finding the Acceleration Vectors on a Motion Diagram

Let's look at how we can determine the average acceleration vector $\vec{a}$ from a motion diagram. From its definition, Equation 1.7, we see that $\vec{a}$ **points in the same direction as $\Delta\vec{v}$,** the change of velocity. This critical idea is the basis for a technique to find $\vec{a}$.

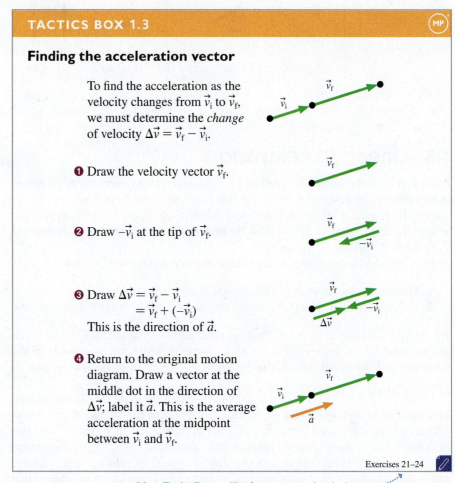

TACTICS BOX 1.3

Finding the acceleration vector

To find the acceleration as the velocity changes from $\vec{v}_i$ to $\vec{v}_f$, we must determine the *change* of velocity $\Delta\vec{v} = \vec{v}_f - \vec{v}_i$.

❶ Draw the velocity vector $\vec{v}_f$.

❷ Draw $-\vec{v}_i$ at the tip of $\vec{v}_f$.

❸ Draw $\Delta\vec{v} = \vec{v}_f - \vec{v}_i$
$= \vec{v}_f + (-\vec{v}_i)$
This is the direction of $\vec{a}$.

❹ Return to the original motion diagram. Draw a vector at the middle dot in the direction of $\Delta\vec{v}$; label it $\vec{a}$. This is the average acceleration at the midpoint between $\vec{v}_i$ and $\vec{v}_f$.

Exercises 21–24

Many Tactics Boxes will refer you to exercises in the *Student Workbook* where you can practice the new skill.

Notice that the acceleration vector goes beside the middle dot, not beside the velocity vectors. This is because each acceleration vector is determined by the *difference* between the *two* velocity vectors on either side of a dot. The length of $\vec{a}$ does not have to be the exact length of $\Delta\vec{v}$; it is the direction of $\vec{a}$ that is most important.

The procedure of Tactics Box 1.3 can be repeated to find $\vec{a}$ at each point in the motion diagram. Note that we cannot determine $\vec{a}$ at the first and last points because we have only one velocity vector and can't find $\Delta\vec{v}$.

The Complete Motion Diagram

You've now seen several *Tactics Boxes* that help you accomplish specific tasks. Tactics Boxes will appear in nearly every chapter in this book. We'll also, where appropriate, provide *Problem-Solving Strategies*.

PROBLEM-SOLVING STRATEGY 1.1

Motion diagrams

MODEL Determine whether it is appropriate to model the moving object as a particle. Make simplifying assumptions when interpreting the problem statement.

VISUALIZE A complete motion diagram consists of:

- The position of the object in each frame of the video, shown as a dot. Use five or six dots to make the motion clear but without overcrowding the picture. More complex motions may need more dots.
- The average velocity vectors, found by connecting each dot in the motion diagram to the next with a vector arrow. There is *one* velocity vector linking each *two* position dots. Label the row of velocity vectors $\vec{v}$.
- The average acceleration vectors, found using Tactics Box 1.3. There is *one* acceleration vector linking each *two* velocity vectors. Each acceleration vector is drawn at the dot between the two velocity vectors it links. Use $\vec{0}$ to indicate a point at which the acceleration is zero. Label the row of acceleration vectors $\vec{a}$.

STOP TO THINK 1.4 A particle undergoes acceleration $\vec{a}$ while moving from point 1 to point 2. Which of the choices shows the most likely velocity vector $\vec{v}_2$ as the particle leaves point 2?

   (a)   (b)   (c)   (d)

Examples of Motion Diagrams

Let's look at some examples of the full strategy for drawing motion diagrams.

EXAMPLE 1.4 | **The first astronauts land on Mars**

A spaceship carrying the first astronauts to Mars descends safely to the surface. Draw a motion diagram for the last few seconds of the descent.

MODEL The spaceship is small in comparison with the distance traveled, and the spaceship does not change size or shape, so it's reasonable to model the spaceship as a particle. We'll assume that its motion in the last few seconds is straight down. The problem ends as the spacecraft touches the surface.

VISUALIZE FIGURE 1.15 shows a complete motion diagram as the spaceship descends and slows, using its rockets, until it comes to rest on the surface. Notice how the dots get closer together as it slows. The inset uses the steps of Tactics Box 1.3 (numbered circles) to show how the acceleration vector $\vec{a}$ is determined at one point. All the other acceleration vectors will be similar because for each pair of velocity vectors the earlier one is longer than the later one.

FIGURE 1.15 Motion diagram of a spaceship landing on Mars.

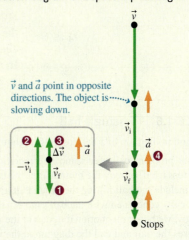

EXAMPLE 1.5 | **Skiing through the woods**

A skier glides along smooth, horizontal snow at constant speed, then speeds up going down a hill. Draw the skier' motion diagram.

MODEL Model the skier as a particle. It's reasonable to assume that the downhill slope is a straight line. Although the motion as a whole is not linear, we can treat the skier's motion as two separate linear motions.

VISUALIZE **FIGURE 1.16** shows a complete motion diagram of the skier. The dots are equally spaced for the horizontal motion, indicating constant speed; then the dots get farther apart as the skier speeds up going down the hill. The insets show how the average acceleration vector $\vec{a}$ is determined for the horizontal motion and along the slope. All the other acceleration vectors along the slope will be similar to the one shown because each velocity vector is longer than the preceding one. Notice that we've explicitly written $\vec{0}$ for the acceleration beside the dots where the velocity is constant. The acceleration at the point where the direction changes will be considered in Chapter 4.

FIGURE 1.16 Motion diagram of a skier.

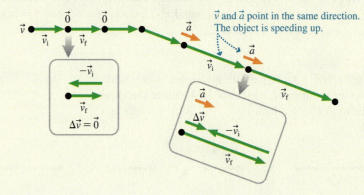

Notice something interesting in Figures 1.15 and 1.16. Where the object is speeding up, the acceleration and velocity vectors point in the *same direction*. Where the object is slowing down, the acceleration and velocity vectors point in *opposite directions*. These results are always true for motion in a straight line. **For motion along a line:**

- **An object is speeding up if and only if $\vec{v}$ and $\vec{a}$ point in the same direction.**
- **An object is slowing down if and only if $\vec{v}$ and $\vec{a}$ point in opposite directions.**
- **An object's velocity is constant if and only if $\vec{a} = \vec{0}$.**

NOTE In everyday language, we use the word *accelerate* to mean "speed up" and the word *decelerate* to mean "slow down." But speeding up and slowing down are both changes in the velocity and consequently, by our definition, *both* are accelerations. In physics, *acceleration* refers to changing the velocity, no matter what the change is, and not just to speeding up.

EXAMPLE 1.6 | **Tossing a ball**

Draw the motion diagram of a ball tossed straight up in the air.

MODEL This problem calls for some interpretation. Should we include the toss itself, or only the motion after the ball is released? Should we include the ball hitting the ground? It appears that this problem is really concerned with the ball's motion through the air. Consequently, we begin the motion diagram at the instant that the tosser releases the ball and end the diagram at the instant the ball hits the ground. We will consider neither the toss nor the impact. And, of course, we will model the ball as a particle.

VISUALIZE We have a slight difficulty here because the ball retraces its route as it falls. A literal motion diagram would show the upward motion and downward motion on top of each other, leading to confusion. We can avoid this difficulty by horizontally separating the upward motion and downward motion diagrams. This will not affect our conclusions because it does not change any of the vectors. **FIGURE 1.17** shows the motion diagram drawn this way. Notice that the very top dot is shown twice—as the end point of the upward motion and the beginning point of the downward motion.

The ball slows down as it rises. You've learned that the acceleration vectors point opposite the velocity vectors for an object that is slowing down along a line, and they are shown accordingly. Similarly, $\vec{a}$ and $\vec{v}$ point in the same direction as the falling ball speeds up. Notice something interesting: The acceleration vectors point downward both while the ball is rising *and* while it is falling. Both "speeding up" and "slowing down" occur with the *same* acceleration vector. This is an important conclusion, one worth pausing to think about.

Now let's look at the top point on the ball's trajectory. The velocity vectors point upward but are getting shorter as the ball approaches the top. As the ball starts to fall, the velocity vectors point downward and are getting longer. There must be a moment—just an instant as $\vec{v}$ switches from pointing up to pointing down—when the velocity is zero. Indeed, the ball's velocity *is* zero for an instant at the precise top of the motion!

But what about the acceleration at the top? The inset shows how the average acceleration is determined from the last upward velocity before the top point and the first downward velocity. We find that the acceleration at the top is pointing downward, just as it does elsewhere in the motion.

Many people expect the acceleration to be zero at the highest point. But the velocity at the top point *is* changing—from up to down. If the velocity is changing, there *must* be an acceleration. A downward-pointing acceleration vector is needed to turn the velocity vector from up to down. Another way to think about this is to note that zero acceleration would mean no change of velocity. When the ball reached zero velocity at the top, it would hang there and not fall if the acceleration were also zero!

FIGURE 1.17 Motion diagram of a ball tossed straight up in the air.

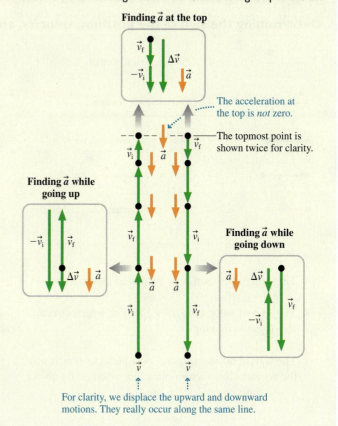

Finding $\vec{a}$ at the top

The acceleration at the top is *not* zero.

The topmost point is shown twice for clarity.

Finding $\vec{a}$ while going up

Finding $\vec{a}$ while going down

For clarity, we displace the upward and downward motions. They really occur along the same line.

1.6 Motion in One Dimension

An object's motion can be described in terms of three fundamental quantities: its position $\vec{r}$, velocity $\vec{v}$, and acceleration $\vec{a}$. These are vectors, but for motion in one dimension, the vectors are restricted to point only "forward" or "backward." Consequently, we can describe one-dimensional motion with the simpler quantities x, v_x, and a_x (or y, v_y, and a_y). However, we need to give each of these quantities an explicit *sign*, positive or negative, to indicate whether the position, velocity, or acceleration vector points forward or backward.

Determining the Signs of Position, Velocity, and Acceleration

Position, velocity, and acceleration are measured with respect to a coordinate system, a grid or axis that *you* impose on a problem to analyze the motion. We will find it convenient to use an x-axis to describe both horizontal motion and motion along an inclined plane. A y-axis will be used for vertical motion. A coordinate axis has two essential features:

1. An origin to define zero; and
2. An x or y label (with units) to indicate the positive end of the axis.

In this textbook, we will follow the convention that **the positive end of an x-axis is to the right and the positive end of a y-axis is up.** The signs of position, velocity, and acceleration are based on this convention.

TACTICS BOX 1.4   ⓂⓅ

Determining the sign of the position, velocity, and acceleration

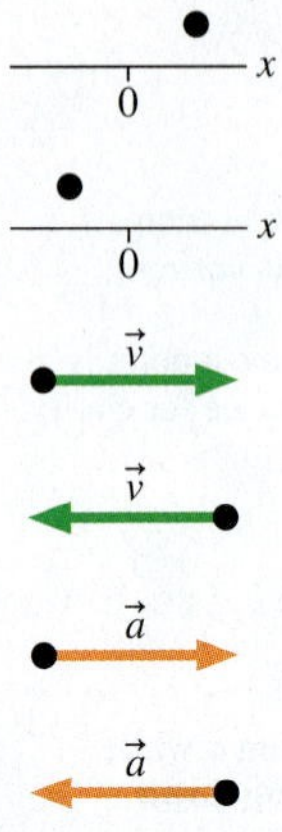

$x > 0$ Position to right of origin.

$x < 0$ Position to left of origin.

$v_x > 0$ Direction of motion is to the right.

$v_x < 0$ Direction of motion is to the left.

$a_x > 0$ Acceleration vector points to the right.

$a_x < 0$ Acceleration vector points to the left.

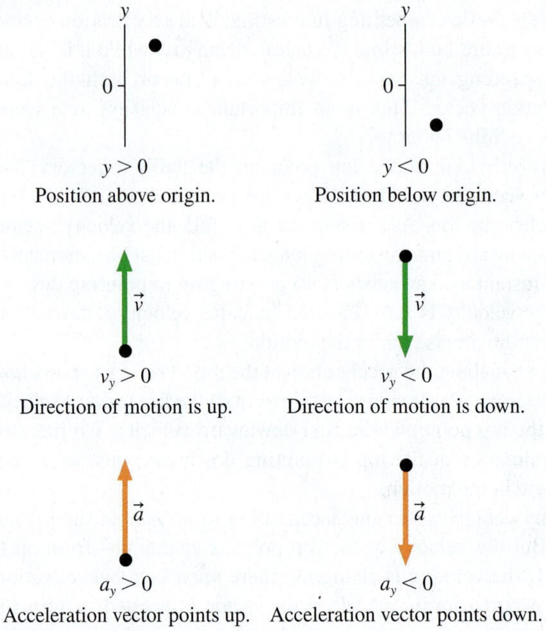

■ The sign of position (x or y) tells us *where* an object is.

■ The sign of velocity (v_x or v_y) tells us *which direction* the object is moving.

■ The sign of acceleration (a_x or a_y) tells us which way the acceleration vector points, *not* whether the object is speeding up or slowing down.

$y > 0$
Position above origin.

$y < 0$
Position below origin.

$v_y > 0$
Direction of motion is up.

$v_y < 0$
Direction of motion is down.

$a_y > 0$
Acceleration vector points up.

$a_y < 0$
Acceleration vector points down.

Exercises 30–31 ✎

Acceleration is where things get a bit tricky. A natural tendency is to think that a positive value of a_x or a_y describes an object that is speeding up while a negative value describes an object that is slowing down (decelerating). However, this interpretation *does not work*.

Acceleration is defined as $\vec{a}_{avg} = \Delta\vec{v}/\Delta t$. The direction of $\vec{a}$ can be determined by using a motion diagram to find the direction of $\Delta\vec{v}$. The one-dimensional acceleration a_x (or a_y) is then positive if the vector $\vec{a}$ points to the right (or up), negative if $\vec{a}$ points to the left (or down).

FIGURE 1.18 shows that this method for determining the sign of a does not conform to the simple idea of speeding up and slowing down. The object in Figure 1.18a has a positive acceleration ($a_x > 0$) not because it is speeding up but because the vector $\vec{a}$ points in the positive direction. Compare this with the motion diagram of Figure 1.18b. Here the object is slowing down, but it still has a positive acceleration ($a_x > 0$) because $\vec{a}$ points to the right.

In the previous section, we found that an object is speeding up if $\vec{v}$ and $\vec{a}$ point in the same direction, slowing down if they point in opposite directions. For one-dimensional motion this rule becomes:

■ An object is speeding up if and only if v_x and a_x have the same sign.
■ An object is slowing down if and only if v_x and a_x have opposite signs.
■ An object's velocity is constant if and only if $a_x = 0$.

Notice how the first two of these rules are at work in Figure 1.18.

FIGURE 1.18 One of these objects is speeding up, the other slowing down, but they both have a positive acceleration a_x.

(a) Speeding up to the right

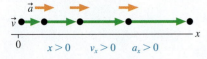

$x > 0$ $v_x > 0$ $a_x > 0$

(b) Slowing down to the left

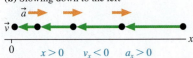

$x > 0$ $v_x < 0$ $a_x > 0$

Position-versus-Time Graphs

FIGURE 1.19 is a motion diagram, made at 1 frame per minute, of a student walking to school. You can see that she leaves home at a time we choose to call $t = 0$ min and makes steady progress for a while. Beginning at $t = 3$ min there is a period where the distance traveled during each time interval becomes less—perhaps she slowed down to speak with a friend. Then she picks up the pace, and the distances within each interval are longer.

FIGURE 1.19 The motion diagram of a student walking to school and a coordinate axis for making measurements.

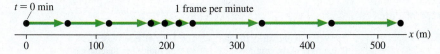

Figure 1.19 includes a coordinate axis, and you can see that every dot in a motion diagram occurs at a specific position. **TABLE 1.1** shows the student's positions at different times as measured along this axis. For example, she is at position $x = 120$ m at $t = 2$ min.

The motion diagram is one way to represent the student's motion. Another is to make a graph of the measurements in Table 1.1. **FIGURE 1.20a** is a graph of x versus t for the student. The motion diagram tells us only where the student is at a few discrete points of time, so this graph of the data shows only points, no lines.

NOTE A graph of "a versus b" means that a is graphed on the vertical axis and b on the horizontal axis. Saying "graph a versus b" is really a shorthand way of saying "graph a as a function of b."

TABLE 1.1 Measured positions of a student walking to school

Time t (min)	Position x (m)	Time t (min)	Position x (m)
0	0	5	220
1	60	6	240
2	120	7	340
3	180	8	440
4	200	9	540

FIGURE 1.20 Position graphs of the student's motion.

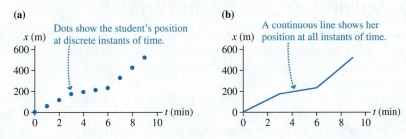

However, common sense tells us the following. First, the student was *somewhere specific* at all times. That is, there was never a time when she failed to have a well-defined position, nor could she occupy two positions at one time. (As reasonable as this belief appears to be, it will be severely questioned and found not entirely accurate when we get to quantum physics!) Second, the student moved *continuously* through all intervening points of space. She could not go from $x = 100$ m to $x = 200$ m without passing through every point in between. It is thus quite reasonable to believe that her motion can be shown as a continuous line passing through the measured points, as shown in **FIGURE 1.20b**. A continuous line or curve showing an object's position as a function of time is called a **position-versus-time graph** or, sometimes, just a *position graph*.

NOTE A graph is *not* a "picture" of the motion. The student is walking along a straight line, but the graph itself is not a straight line. Further, we've graphed her position on the vertical axis even though her motion is horizontal. Graphs are *abstract representations* of motion. We will place significant emphasis on the process of interpreting graphs, and many of the exercises and problems will give you a chance to practice these skills.

EXAMPLE 1.7 | **Interpreting a position graph**

The graph in **FIGURE 1.21a** represents the motion of a car along a straight road. Describe the motion of the car.

MODEL We'll model the car as a particle with a precise position at each instant.

VISUALIZE As **FIGURE 1.21b** shows, the graph represents a car that travels to the left for 30 minutes, stops for 10 minutes, then travels back to the right for 40 minutes.

FIGURE 1.21 Position-versus-time graph of a car.

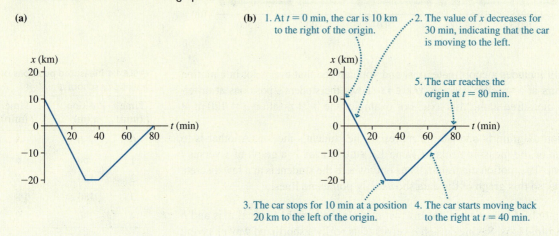

(a)

(b) 1. At $t = 0$ min, the car is 10 km to the right of the origin. 2. The value of x decreases for 30 min, indicating that the car is moving to the left.

5. The car reaches the origin at $t = 80$ min.

3. The car stops for 10 min at a position 20 km to the left of the origin. 4. The car starts moving back to the right at $t = 40$ min.

1.7 Solving Problems in Physics

Physics is not mathematics. Math problems are clearly stated, such as "What is 2 + 2?" Physics is about the world around us, and to describe that world we must use language. Now, language is wonderful—we couldn't communicate without it—but language can sometimes be imprecise or ambiguous.

The challenge when reading a physics problem is to translate the words into symbols that can be manipulated, calculated, and graphed. **The translation from words to symbols is the heart of problem solving in physics.** This is the point where ambiguous words and phrases must be clarified, where the imprecise must be made precise, and where you arrive at an understanding of exactly what the question is asking.

Using Symbols

Symbols are a language that allows us to talk with precision about the relationships in a problem. As with any language, we all need to agree to use words or symbols in the same way if we want to communicate with each other. Many of the ways we use symbols in science and engineering are somewhat arbitrary, often reflecting historical roots. Nonetheless, practicing scientists and engineers have come to agree on how to use the language of symbols. Learning this language is part of learning physics.

We will use subscripts on symbols, such as x_3, to designate a particular point in the problem. Scientists usually label the starting point of the problem with the subscript "0," not the subscript "1" that you might expect. When using subscripts, make sure that all symbols referring to the same point in the problem have the *same numerical subscript*. To have the same point in a problem characterized by position x_1 but velocity v_{2x} is guaranteed to lead to confusion!

Drawing Pictures

You may have been told that the first step in solving a physics problem is to "draw a picture," but perhaps you didn't know why, or what to draw. The purpose of drawing a picture is to aid you in the words-to-symbols translation. Complex problems have far more information than you can keep in your head at one time. Think of a picture as a "memory extension," helping you organize and keep track of vital information.

Although any picture is better than none, there really is a *method* for drawing pictures that will help you be a better problem solver. It is called the **pictorial representation** of the problem. We'll add other pictorial representations as we go along, but the following procedure is appropriate for motion problems.

TACTICS BOX 1.5

Drawing a pictorial representation

❶ **Draw a motion diagram.** The motion diagram develops your intuition for the motion.

❷ **Establish a coordinate system.** Select your axes and origin to match the motion. For one-dimensional motion, you want either the x-axis or the y-axis parallel to the motion. The coordinate system determines whether the signs of v and a are positive or negative.

❸ **Sketch the situation.** Not just any sketch. Show the object at the *beginning* of the motion, at the *end,* and at any point where the character of the motion changes. Show the object, not just a dot, but very simple drawings are adequate.

❹ **Define symbols.** Use the sketch to define symbols representing quantities such as position, velocity, acceleration, and time. *Every* variable used later in the mathematical solution should be defined on the sketch. Some will have known values, others are initially unknown, but all should be given symbolic names.

❺ **List known information.** Make a table of the quantities whose values you can determine from the problem statement or that can be found quickly with simple geometry or unit conversions. Some quantities are implied by the problem, rather than explicitly given. Others are determined by your choice of coordinate system.

❻ **Identify the desired unknowns.** What quantity or quantities will allow you to answer the question? These should have been defined as symbols in step 4. Don't list every unknown, only the one or two needed to answer the question.

It's not an overstatement to say that a well-done pictorial representation of the problem will take you halfway to the solution. The following example illustrates how to construct a pictorial representation for a problem that is typical of problems you will see in the next few chapters.

EXAMPLE 1.8 | Drawing a pictorial representation

Draw a pictorial representation for the following problem: A rocket sled accelerates horizontally at 50 m/s² for 5.0 s, then coasts for 3.0 s. What is the total distance traveled?

VISUALIZE FIGURE 1.22, on the next page, is the pictorial representation. The motion diagram shows an acceleration phase followed by a coasting phase. Because the motion is horizontal, the appropriate coordinate system is an x-axis. We've chosen to place the origin at the starting point. The motion has a beginning, an end, and a point where the motion changes from accelerating to coasting, and these are the three sled positions sketched in the figure. The quantities x, v_x, and t are needed at each of three *points,* so these

have been defined on the sketch and distinguished by subscripts. Accelerations are associated with *intervals* between the points, so only two accelerations are defined. Values for three quantities are given in the problem statement, although we need to use the motion diagram, where $\vec{a}$ points to the right, and our choice of coordinate system to know that $a_{0x} = +50$ m/s² rather than -50 m/s². The values $x_0 = 0$ m and $t_0 = 0$ s are choices we made when setting up the coordinate system. The value $v_{0x} = 0$ m/s is part of our *interpretation* of the problem. Finally, we identify x_2 as the quantity that will answer the question. We now understand quite a bit about the problem and would be ready to start a quantitative analysis.

Continued

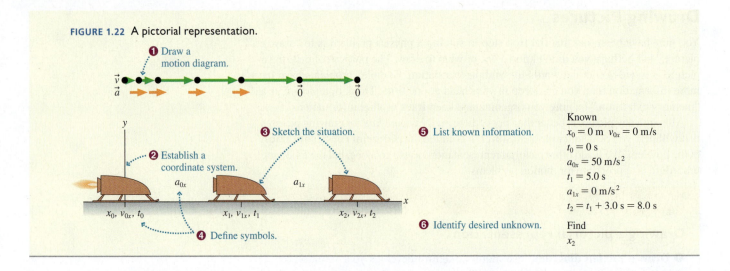

FIGURE 1.22 A pictorial representation.

① Draw a motion diagram.

② Establish a coordinate system.

③ Sketch the situation.

④ Define symbols.

⑤ List known information.

⑥ Identify desired unknown.

Known
$x_0 = 0$ m $v_{0x} = 0$ m/s
$t_0 = 0$ s
$a_{0x} = 50$ m/s^2
$t_1 = 5.0$ s
$a_{1x} = 0$ m/s^2
$t_2 = t_1 + 3.0$ s $= 8.0$ s

Find
x_2

We didn't *solve* the problem; that is not the purpose of the pictorial representation. The pictorial representation is a systematic way to go about interpreting a problem and getting ready for a mathematical solution. Although this is a simple problem, and you probably know how to solve it if you've taken physics before, you will soon be faced with much more challenging problems. Learning good problem-solving skills at the beginning, while the problems are easy, will make them second nature later when you really need them.

Representations

A picture is one way to *represent* your knowledge of a situation. You could also represent your knowledge using words, graphs, or equations. Each **representation of knowledge** gives us a different perspective on the problem. The more tools you have for thinking about a complex problem, the more likely you are to solve it.

There are four representations of knowledge that we will use over and over:

1. The *verbal* representation. A problem statement, in words, is a verbal representation of knowledge. So is an explanation that you write.
2. The *pictorial* representation. The pictorial representation, which we've just presented, is the most literal depiction of the situation.
3. The *graphical* representation. We will make extensive use of graphs.
4. The *mathematical* representation. Equations that can be used to find the numerical values of specific quantities are the mathematical representation.

NOTE The mathematical representation is only one of many. Much of physics is more about thinking and reasoning than it is about solving equations.

A Problem-Solving Strategy

One of the goals of this textbook is to help you learn a *strategy* for solving physics problems. The purpose of a strategy is to guide you in the right direction with minimal wasted effort. The four-part problem-solving strategy shown on the next page—**Model, Visualize, Solve, Assess**—is based on using different representations of knowledge. You will see this problem-solving strategy used consistently in the worked examples throughout this textbook, and you should endeavor to apply it to your own problem solving.

Throughout this textbook we will emphasize the first two steps. They are the *physics* of the problem, as opposed to the mathematics of solving the resulting equations. This is not to say that those mathematical operations are always easy—in many cases they are not. But our primary goal is to understand the physics.

A new building requires careful planning. The architect's visualization and drawings have to be complete before the detailed procedures of construction get under way. The same is true for solving problems in physics.

GENERAL PROBLEM-SOLVING STRATEGY

MODEL It's impossible to treat every detail of a situation. Simplify the situation with a model that captures the essential features. For example, the object in a mechanics problem is often represented as a particle.

VISUALIZE This is where expert problem solvers put most of their effort.

- Draw a *pictorial representation*. This helps you visualize important aspects of the physics and assess the information you are given. It starts the process of translating the problem into symbols.
- Use a *graphical representation* if it is appropriate for the problem.
- Go back and forth between these representations; they need not be done in any particular order.

SOLVE Only after modeling and visualizing are complete is it time to develop a *mathematical representation* with specific equations that must be solved. All symbols used here should have been defined in the pictorial representation.

ASSESS Is your result believable? Does it have proper units? Does it make sense?

Textbook illustrations are obviously more sophisticated than what you would draw on your own paper. To show you a figure very much like what *you* should draw, the final example of this section is in a "pencil sketch" style. We will include one or more pencil-sketch examples in nearly every chapter to illustrate exactly what a good problem solver would draw.

EXAMPLE 1.9 | Launching a weather rocket

Use the first two steps of the problem-solving strategy to analyze the following problem: A small rocket, such as those used for meteorological measurements of the atmosphere, is launched vertically with an acceleration of 30 m/s². It runs out of fuel after 30 s. What is its maximum altitude?

MODEL We need to do some interpretation. Common sense tells us that the rocket does not stop the instant it runs out of fuel. Instead, it continues upward, while slowing, until it reaches its maximum altitude. This second half of the motion, after running out of fuel, is like the ball that was tossed upward in the first half of Example 1.6. Because the problem does not ask about the rocket's descent, we conclude that the problem ends at the point of maximum altitude. We'll model the rocket as a particle.

VISUALIZE FIGURE 1.23 shows the pictorial representation in pencil-sketch style. The rocket is speeding up during the first half of the motion, so $\vec{a}_0$ points upward, in the positive y-direction. Thus the initial acceleration is $a_{0y} = 30$ m/s². During the second half, as the rocket slows, $\vec{a}_1$ points downward. Thus a_{1y} is a negative number.

This information is included with the known information. Although the velocity v_{2y} wasn't given in the problem statement, it must—just like for the ball in Example 1.6—be zero at the very top of the trajectory. Last, we have identified y_2 as the desired unknown. This, of course, is not the only unknown in the problem, but it is the one we are specifically asked to find.

ASSESS If you've had a previous physics class, you may be tempted to assign a_{1y} the value -9.8 m/s², the free-fall acceleration.

FIGURE 1.23 Pictorial representation for the rocket.

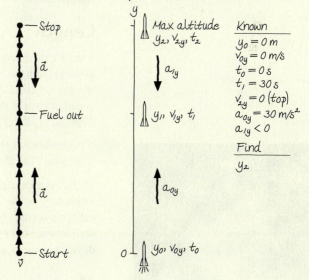

However, that would be true only if there is no air resistance on the rocket. We will need to consider the *forces* acting on the rocket during the second half of its motion before we can determine a value for a_{1y}. For now, all that we can safely conclude is that a_{1y} is negative.

Our task in this section is not to *solve* problems—all that in due time—but to focus on what is happening in a problem. In other words, to make the translation from words to symbols in preparation for subsequent mathematical analysis. Modeling and the pictorial representation will be our most important tools.

1.8 Units and Significant Figures

TABLE 1.2 The basic SI units

Quantity	Unit	Abbreviation
time	second	s
length	meter	m
mass	kilogram	kg

An atomic clock at the National Institute of Standards and Technology is the primary standard of time.

Science is based upon experimental measurements, and measurements require *units*. The system of units used in science is called *le Système Internationale d'Unités*. These are commonly referred to as **SI units.** In casual speaking we often refer to *metric units*.

All of the quantities needed to understand motion can be expressed in terms of the three basic SI units shown in **TABLE 1.2**. Other quantities can be expressed as a combination of these basic units. Velocity, expressed in meters per second or m/s, is a ratio of the length unit to the time unit.

Time

The standard of time prior to 1960 was based on the *mean solar day*. As time-keeping accuracy and astronomical observations improved, it became apparent that the earth's rotation is not perfectly steady. Meanwhile, physicists had been developing a device called an *atomic clock*. This instrument is able to measure, with incredibly high precision, the frequency of radio waves absorbed by atoms as they move between two closely spaced energy levels. This frequency can be reproduced with great accuracy at many laboratories around the world. Consequently, the SI unit of time—the second—was redefined in 1967 as follows:

> One *second* is the time required for 9,192,631,770 oscillations of the radio wave absorbed by the cesium-133 atom. The abbreviation for second is the letter s.

Several radio stations around the world broadcast a signal whose frequency is linked directly to the atomic clocks. This signal is the time standard, and any time-measuring equipment you use was calibrated from this time standard.

Length

The SI unit of length—the meter—was originally defined as one ten-millionth of the distance from the north pole to the equator along a line passing through Paris. There are obvious practical difficulties with implementing this definition, and it was later abandoned in favor of the distance between two scratches on a platinum-iridium bar stored in a special vault in Paris. The present definition, agreed to in 1983, is as follows:

> One *meter* is the distance traveled by light in vacuum during 1/299,792,458 of a second. The abbreviation for meter is the letter m.

This is equivalent to defining the speed of light to be exactly 299,792,458 m/s. Laser technology is used in various national laboratories to implement this definition and to calibrate secondary standards that are easier to use. These standards ultimately make their way to your ruler or to a meter stick. It is worth keeping in mind that any measuring device you use is only as accurate as the care with which it was calibrated.

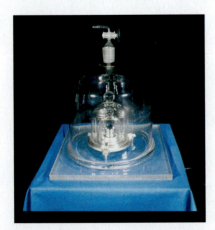

By international agreement, this metal cylinder, stored in Paris, is the definition of the kilogram.

Mass

The original unit of mass, the gram, was defined as the mass of 1 cubic centimeter of water. That is why you know the density of water as 1 g/cm^3. This definition proved to be impractical when scientists needed to make very accurate measurements. The SI unit of mass—the kilogram—was redefined in 1889 as:

> One *kilogram* is the mass of the international standard kilogram, a polished platinum-iridium cylinder stored in Paris. The abbreviation for kilogram is kg.

The kilogram is the only SI unit still defined by a manufactured object. Despite the prefix *kilo,* it is the kilogram, not the gram, that is the SI unit.

Using Prefixes

We will have many occasions to use lengths, times, and masses that are either much less or much greater than the standards of 1 meter, 1 second, and 1 kilogram. We will do so by using *prefixes* to denote various powers of 10. TABLE 1.3 lists the common prefixes that will be used frequently throughout this book. Memorize it! Few things in science are learned by rote memory, but this list is one of them. A more extensive list of prefixes is shown inside the front cover of the book.

Although prefixes make it easier to talk about quantities, the SI units are meters, seconds, and kilograms. Quantities given with prefixed units must be converted to SI units before any calculations are done. Unit conversions are best done at the very beginning of a problem, as part of the pictorial representation.

TABLE 1.3 Common prefixes

Prefix	Power of 10	Abbreviation
giga-	10^9	G
mega-	10^6	M
kilo-	10^3	k
centi-	10^{-2}	c
milli-	10^{-3}	m
micro-	10^{-6}	μ
nano-	10^{-9}	n

Unit Conversions

Although SI units are our standard, we cannot entirely forget that the United States still uses English units. Thus it remains important to be able to convert back and forth between SI units and English units. TABLE 1.4 shows several frequently used conversions, and these are worth memorizing if you do not already know them. While the English system was originally based on the length of the king's foot, it is interesting to note that today the conversion 1 in = 2.54 cm is the *definition* of the inch. In other words, the English system for lengths is now based on the meter!

There are various techniques for doing unit conversions. One effective method is to write the conversion factor as a ratio equal to one. For example, using information in Tables 1.3 and 1.4, we have

TABLE 1.4 Useful unit conversions

1 in = 2.54 cm
1 mi = 1.609 km
1 mph = 0.447 m/s
1 m = 39.37 in
1 km = 0.621 mi
1 m/s = 2.24 mph

$$\frac{10^{-6}\,\text{m}}{1\,\mu\text{m}} = 1 \qquad \text{and} \qquad \frac{2.54\,\text{cm}}{1\,\text{in}} = 1$$

Because multiplying any expression by 1 does not change its value, these ratios are easily used for conversions. To convert 3.5 μm to meters we compute

$$3.5\,\mu\text{m} \times \frac{10^{-6}\,\text{m}}{1\,\mu\text{m}} = 3.5 \times 10^{-6}\,\text{m}$$

Similarly, the conversion of 2 feet to meters is

$$2.00\,\text{ft} \times \frac{12\,\text{in}}{1\,\text{ft}} \times \frac{2.54\,\text{cm}}{1\,\text{in}} \times \frac{10^{-2}\,\text{m}}{1\,\text{cm}} = 0.610\,\text{m}$$

Notice how units in the numerator and in the denominator cancel until only the desired units remain at the end. You can continue this process of multiplying by 1 as many times as necessary to complete all the conversions.

Assessment

As we get further into problem solving, you will need to decide whether or not the answer to a problem "makes sense." To determine this, at least until you have more experience with SI units, you may need to convert from SI units back to the English units in which you think. But this conversion does not need to be very accurate. For example, if you are working a problem about automobile speeds and reach an answer of 35 m/s, all you really want to know is whether or not this is a realistic speed for a car. That requires a "quick and dirty" conversion, not a conversion of great accuracy.

TABLE 1.5 Approximate conversion factors. Use these for assessment, not in problem solving.

$1 \text{ cm} \approx \frac{1}{2} \text{ in}$
$10 \text{ cm} \approx 4 \text{ in}$
$1 \text{ m} \approx 1 \text{ yard}$
$1 \text{ m} \approx 3 \text{ feet}$
$1 \text{ km} \approx 0.6 \text{ mile}$
$1 \text{ m/s} \approx 2 \text{ mph}$

TABLE 1.5 shows several approximate conversion factors that can be used to assess the answer to a problem. Using 1 m/s ≈ 2 mph, you find that 35 m/s is roughly 70 mph, a reasonable speed for a car. But an answer of 350 m/s, which you might get after making a calculation error, would be an unreasonable 700 mph. Practice with these will allow you to develop intuition for metric units.

NOTE These approximate conversion factors are accurate to only one significant figure. This is sufficient to assess the answer to a problem, but do *not* use the conversion factors from Table 1.5 for converting English units to SI units at the start of a problem. Use Table 1.4.

Significant Figures

It is necessary to say a few words about a perennial source of difficulty: significant figures. Mathematics is a subject where numbers and relationships can be as precise as desired, but physics deals with a real world of ambiguity. It is important in science and engineering to state clearly what you know about a situation—no less and, especially, no more. Numbers provide one way to specify your knowledge.

If you report that a length has a value of 6.2 m, the implication is that the actual value falls between 6.15 m and 6.25 m and thus rounds to 6.2 m. If that is the case, then reporting a value of simply 6 m is saying less than you know; you are withholding information. On the other hand, to report the number as 6.213 m is wrong. Any person reviewing your work—perhaps a client who hired you—would interpret the number 6.213 m as meaning that the actual length falls between 6.2125 m and 6.2135 m, thus rounding to 6.213 m. In this case, you are claiming to have knowledge and information that you do not really possess.

The way to state your knowledge precisely is through the proper use of **significant figures.** You can think of a significant figure as being a digit that is reliably known. A number such as 6.2 m has *two* significant figures because the next decimal place—the one-hundredths—is not reliably known. As **FIGURE 1.24** shows, the best way to determine how many significant figures a number has is to write it in scientific notation.

FIGURE 1.24 Determining significant figures.

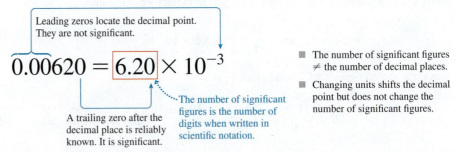

Leading zeros locate the decimal point. They are not significant.

$$0.00620 = \boxed{6.20} \times 10^{-3}$$

A trailing zero after the decimal place is reliably known. It is significant.

The number of significant figures is the number of digits when written in scientific notation.

- The number of significant figures ≠ the number of decimal places.
- Changing units shifts the decimal point but does not change the number of significant figures.

What about numbers like 320 m and 20 kg? Whole numbers with trailing zeros are ambiguous unless written in scientific notation. Even so, writing 2.0×10^1 kg is tedious, and few practicing scientists or engineers would do so. In this textbook, we'll adopt the rule that *whole numbers always have at least two significant figures,* even if one of those is a trailing zero. By this rule, 320 m, 20 kg, and 8000 s each have two significant figures, but 8050 s would have three.

Calculations with numbers follow the "weakest link" rule. The saying, which you probably know, is that "a chain is only as strong as its weakest link." If nine out of ten links in a chain can support a 1000 pound weight, that strength is meaningless if the tenth link can support only 200 pounds. Nine out of the ten numbers used in a calculation might be known with a precision of 0.01%; but if the tenth number is poorly known, with a precision of only 10%, then the result of the calculation cannot possibly be more precise than 10%.

TACTICS BOX 1.6

Using significant figures

❶ When multiplying or dividing several numbers, or taking roots, the number of significant figures in the answer should match the number of significant figures of the *least* precisely known number used in the calculation.

❷ When adding or subtracting several numbers, the number of decimal places in the answer should match the *smallest* number of decimal places of any number used in the calculation.

❸ Exact numbers are perfectly known and do not affect the number of significant figures an answer should have. Examples of exact numbers are the 2 and the π in the formula $C = 2\pi r$ for the circumference of a circle.

❹ It is acceptable to keep one or two extra digits during intermediate steps of a calculation, to minimize rounding error, as long as the final answer is reported with the proper number of significant figures.

❺ Examples and problems in this textbook will normally provide data to either two or three significant figures, as is appropriate to the situation. **The appropriate number of significant figures for the answer is determined by the data provided.**

Exercises 38–39

NOTE Be careful! Many calculators have a default setting that shows two decimal places, such as 5.23. This is dangerous. If you need to calculate 5.23/58.5, your calculator will show 0.09 and it is all too easy to write that down as an answer. By doing so, you have reduced a calculation of two numbers having three significant figures to an answer with only one significant figure. The proper result of this division is 0.0894 or 8.94×10^{-2}. You will avoid this error if you keep your calculator set to display numbers in *scientific notation* with two decimal places.

EXAMPLE 1.10 | **Using significant figures**

An object consists of two pieces. The mass of one piece has been measured to be 6.47 kg. The volume of the second piece, which is made of aluminum, has been measured to be 4.44×10^{-4} m^3. A handbook lists the density of aluminum as 2.7×10^3 kg/m^3. What is the total mass of the object?

SOLVE First, calculate the mass of the second piece:

$$m = (4.44 \times 10^{-4}\, \text{m}^3)(2.7 \times 10^3\, \text{kg/m}^3)$$
$$= 1.199\, \text{kg} = 1.2\, \text{kg}$$

The number of significant figures of a product must match that of the *least* precisely known number, which is the two-significant-figure density of aluminum. Now add the two masses:

$$\begin{array}{r} 6.47\ \text{kg} \\ +\ 1.2\ \ \text{kg} \\ \hline 7.7\ \ \text{kg} \end{array}$$

The sum is 7.67 kg, but the hundredths place is not reliable because the second mass has no reliable information about this digit. Thus we must round to the one decimal place of the 1.2 kg. The best we can say, with reliability, is that the total mass is 7.7 kg.

Proper use of significant figures is part of the "culture" of science and engineering. We will frequently emphasize these "cultural issues" because you must learn to speak the same language as the natives if you wish to communicate effectively. Most students know the rules of significant figures, having learned them in high school, but many fail to apply them. It is important to understand the reasons for significant figures and to get in the habit of using them properly.

TABLE 1.6 Some approximate lengths

	Length (m)
Altitude of jet planes	10,000
Distance across campus	1000
Length of a football field	100
Length of a classroom	10
Length of your arm	1
Width of a textbook	0.1
Length of a fingernail	0.01

TABLE 1.7 Some approximate masses

	Mass (kg)
Small car	1000
Large human	100
Medium-size dog	10
Science textbook	1
Apple	0.1
Pencil	0.01
Raisin	0.001

Orders of Magnitude and Estimating

Precise calculations are appropriate when we have precise data, but there are many times when a very rough estimate is sufficient. Suppose you see a rock fall off a cliff and would like to know how fast it was going when it hit the ground. By doing a mental comparison with the speeds of familiar objects, such as cars and bicycles, you might judge that the rock was traveling at "about" 20 mph.

This is a one-significant-figure estimate. With some luck, you can distinguish 20 mph from either 10 mph or 30 mph, but you certainly cannot distinguish 20 mph from 21 mph. A one-significant-figure estimate or calculation, such as this, is called an **order-of-magnitude estimate.** An order-of-magnitude estimate is indicated by the symbol ~, which indicates even less precision than the "approximately equal" symbol ≈. You would say that the speed of the rock is $v \sim 20$ mph.

A useful skill is to make reliable estimates on the basis of known information, simple reasoning, and common sense. This is a skill that is acquired by practice. Many chapters in this book will have homework problems that ask you to make order-of-magnitude estimates. The following example is a typical estimation problem.

TABLES 1.6 and **1.7** have information that will be useful for doing estimates.

EXAMPLE 1.11 | **Estimating a sprinter's speed**

Estimate the speed with which an Olympic sprinter crosses the finish line of the 100 m dash.

SOLVE We do need one piece of information, but it is a widely known piece of sports trivia. That is, world-class sprinters run the 100 m dash in about 10 s. Their *average* speed is $v_{avg} \approx (100 \text{ m})/(10 \text{ s}) \approx 10$ m/s. But that's only average. They go slower than average at the beginning, and they cross the finish line at a speed faster than average. How much faster? Twice as fast, 20 m/s, would be ≈40 mph. Sprinters don't seem like they're running as fast as a 40 mph car, so this probably is too fast. Let's *estimate* that their final speed is 50% faster than the average. Thus they cross the finish line at $v \sim 15$ m/s.

STOP TO THINK 1.5 Rank in order, from the most to the least, the number of significant figures in the following numbers. For example, if b has more than c, c has the same number as a, and a has more than d, you could give your answer as $b > c = a > d$.

a. 82   b. 0.0052   c. 0.430   d. 4.321×10^{-10}

SUMMARY

The goal of Chapter 1 has been to learn the fundamental concepts of motion.

GENERAL STRATEGY

Problem Solving

MODEL Make simplifying assumptions.

VISUALIZE Use:

- **Pictorial representation**
- **Graphical representation**

SOLVE Use a **mathematical representation** to find numerical answers.

ASSESS Does the answer have the proper units and correct significant figures? Does it make sense?

Motion Diagrams

- Help visualize motion.
- Provide a tool for finding acceleration vectors.

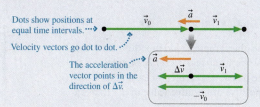

Dots show positions at equal time intervals.

Velocity vectors go dot to dot.

The acceleration vector points in the direction of $\Delta\vec{v}$.

▶ These are the *average* velocity and acceleration vectors.

IMPORTANT CONCEPTS

The **particle model** represents a moving object as if all its mass were concentrated at a single point.

Position locates an object with respect to a chosen coordinate system. Change in position is called **displacement.**

Velocity is the rate of change of the position vector $\vec{r}$.

Acceleration is the rate of change of the velocity vector $\vec{v}$.

An object has an acceleration if it

- Changes speed and/or
- Changes direction.

Pictorial Representation

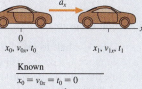

❶ Draw a motion diagram.

❷ Establish coordinates.

❸ Sketch the situation.

❹ Define symbols.

❺ List knowns.

❻ Identify desired unknown.

Known
$x_0 = v_{0x} = t_0 = 0$
$a_x = 2.0 \text{ m/s}^2 \quad t_1 = 2.0 \text{ s}$
Find
x_1

APPLICATIONS

For **motion along a line:**

- Speeding up: $\vec{v}$ and $\vec{a}$ point in the same direction, v_x and a_x have the same sign.
- Slowing down: $\vec{v}$ and $\vec{a}$ point in opposite directions, v_x and a_x have opposite signs.
- Constant speed: $\vec{a} = \vec{0}$, $a_x = 0$.

Acceleration a_x is positive if $\vec{a}$ points right, negative if $\vec{a}$ points left. The sign of a_x does *not* imply speeding up or slowing down.

Significant figures are reliably known digits. The number of significant figures for:

- **Multiplication, division, powers** is set by the value with the fewest significant figures.
- **Addition, subtraction** is set by the value with the smallest number of decimal places.

The appropriate number of significant figures in a calculation is determined by the data provided.

TERMS AND NOTATION

motion	particle model	time interval, Δt	representation of knowledge
translational motion	position vector, $\vec{r}$	average speed	SI units
trajectory	scalar	average velocity, $\vec{v}$	significant figures
motion diagram	vector	average acceleration, $\vec{a}$	order-of-magnitude estimate
model	displacement, $\Delta\vec{r}$	position-versus-time graph	
particle	zero vector, $\vec{0}$	pictorial representation	

CONCEPTUAL QUESTIONS

1. How many significant figures does each of the following numbers have?
   a. 0.73   b. 7.30   c. 73   d. 0.073
2. How many significant figures does each of the following numbers have?
   a. 290   b. 2.90×10^4 c. 0.0029   d. 2.90
3. Is the particle in **FIGURE Q1.3** speeding up? Slowing down? Or can you tell? Explain.

   FIGURE Q1.3

4. Does the object represented in **FIGURE Q1.4** have a positive or negative value of a_x? Explain.
5. Does the object represented in **FIGURE Q1.5** have a positive or negative value of a_y? Explain.

   FIGURE Q1.4   **FIGURE Q1.5**

6. Determine the signs (positive, negative, or zero) of the position, velocity, and acceleration for the particle in **FIGURE Q1.6**.

   FIGURE Q1.6

7. Determine the signs (positive, negative, or zero) of the position, velocity, and acceleration for the particle in **FIGURE Q1.7**.

   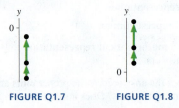
   FIGURE Q1.7   **FIGURE Q1.8**

8. Determine the signs (positive, negative, or zero) of the position, velocity, and acceleration for the particle in **FIGURE Q1.8**.

EXERCISES AND PROBLEMS

Exercises

Section 1.1 Motion Diagrams

1. | A car skids to a halt to avoid hitting an object in the road. Draw a basic motion diagram, using the images from the video, from the time the skid begins until the car is stopped.
2. | A rocket is launched straight up. Draw a basic motion diagram, using the images from the video, from the moment of liftoff until the rocket is at an altitude of 500 m.
3. | You are watching a jet ski race. A racer speeds up from rest to 70 mph in just a few seconds, then continues at a constant speed. Draw a basic motion diagram of the jet ski, using images from the video, from 10 s before reaching top speed until 10 s afterward.

Section 1.2 Models and Modeling

4. | a. Write a paragraph describing the particle model. What is it, and why is it important?
   b. Give two examples of situations, different from those described in the text, for which the particle model is appropriate.
   c. Give an example of a situation, different from those described in the text, for which it would be inappropriate.

Section 1.3 Position, Time, and Displacement

Section 1.4 Velocity

5. | You drop a soccer ball from your third-story balcony. Use the particle model to draw a motion diagram showing the ball's position and average velocity vectors from the time you release the ball until the instant it touches the ground.

6. | A baseball player starts running to the left to catch the ball as soon as the hit is made. Use the particle model to draw a motion diagram showing the position and average velocity vectors of the player during the first few seconds of the run.
7. | A softball player slides into second base. Use the particle model to draw a motion diagram showing his position and his average velocity vectors from the time he begins to slide until he reaches the base.

Section 1.5 Linear Acceleration

8. | a. **FIGURE EX1.8** shows the first three points of a motion diagram. Is the object's average speed between points 1 and 2 greater than, less than, or equal to its average speed between points 0 and 1? Explain how you can tell.
   b. Use Tactics Box 1.3 to find the average acceleration vector at point 1. Draw the completed motion diagram, showing the velocity vectors and acceleration vector.

   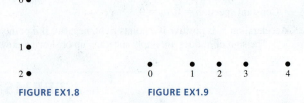
   FIGURE EX1.8   **FIGURE EX1.9**

9. | **FIGURE EX1.9** shows five points of a motion diagram. Use Tactics Box 1.3 to find the average acceleration vectors at points 1, 2, and 3. Draw the completed motion diagram showing velocity vectors and acceleration vectors.

10. ‖ **FIGURE EX1.10** shows two dots of a motion diagram and vector $\vec{v}_1$. Copy this figure, then add dot 3 and the next velocity vector $\vec{v}_2$ if the acceleration vector $\vec{a}$ at dot 2 (a) points up and (b) points down.

FIGURE EX1.10   **FIGURE EX1.11**

11. ‖ **FIGURE EX1.11** shows two dots of a motion diagram and vector $\vec{v}_2$. Copy this figure, then add dot 4 and the next velocity vector $\vec{v}_3$ if the acceleration vector $\vec{a}$ at dot 3 (a) points right and (b) points left.

12. | A speed skater accelerates from rest and then keeps skating at a constant speed. Draw a complete motion diagram of the skater.

13. | A car travels to the left at a steady speed for a few seconds, then brakes for a stop sign. Draw a complete motion diagram of the car.

14. | A goose flies toward a pond. It lands on the water and slides for some distance before it comes to a stop. Draw the motion diagram of the goose, starting shortly before it hits the water and assuming the motion is entirely horizontal.

15. | You use a long rubber band to launch a paper wad straight up. Draw a complete motion diagram of the paper wad from the moment you release the stretched rubber band until the paper wad reaches its highest point.

16. | A roof tile falls straight down from a two-story building. It lands in a swimming pool and settles gently to the bottom. Draw a complete motion diagram of the tile.

17. | Your roommate drops a tennis ball from a third-story balcony. It hits the sidewalk and bounces as high as the second story. Draw a complete motion diagram of the tennis ball from the time it is released until it reaches the maximum height on its bounce. Be sure to determine and show the acceleration at the lowest point.

Section 1.6 Motion in One Dimension

18. ‖ **FIGURE EX1.18** shows the motion diagram of a drag racer. The camera took one frame every 2 s.

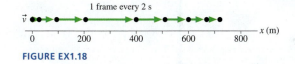

FIGURE EX1.18

a. Measure the x-value of the racer at each dot. List your data in a table similar to Table 1.1, showing each position and the time at which it occurred.

b. Make a position-versus-time graph for the drag racer. Because you have data only at certain instants, your graph should consist of dots that are not connected together.

19. | Write a short description of the motion of a real object for which **FIGURE EX1.19** would be a realistic position-versus-time graph.

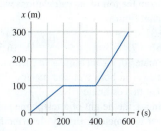

FIGURE EX1.19

20. | Write a short description of the motion of a real object for which **FIGURE EX1.20** would be a realistic position-versus-time graph.

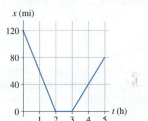

FIGURE EX1.20

Section 1.7 Solving Problems in Physics

21. ‖ Draw a pictorial representation for the following problem. Do *not* solve the problem. The light turns green, and a bicyclist starts forward with an acceleration of 1.5 m/s². How far must she travel to reach a speed of 7.5 m/s?

22. ‖ Draw a pictorial representation for the following problem. Do *not* solve the problem. What acceleration does a rocket need to reach a speed of 200 m/s at a height of 1.0 km?

Section 1.8 Units and Significant Figures

23. | How many significant figures are there in the following values?
   a. 0.05×10^{-4} b. 0.00340
   c. 7.2×10^4 d. 103.00

24. ‖ Convert the following to SI units:
   a. 8.0 in   b. 66 ft/s
   c. 60 mph   d. 14 in²

25. | Convert the following to SI units:
   a. 75 in   b. 3.45×10^6 yr
   c. 62 ft/day   d. 2.2×10^4 mi²

26. ‖ Using the approximate conversion factors in Table 1.5, convert the following SI units to English units *without* using your calculator.
   a. 30 cm   b. 25 m/s
   c. 5 km   d. 0.5 cm

27. | Using the approximate conversion factors in Table 1.5, convert the following to SI units *without* using your calculator.
   a. 20 ft   b. 60 mi
   c. 60 mph   d. 8 in

28. | Compute the following numbers, applying the significant figure rules adopted in this textbook.
    a. 33.3×25.4
    b. $33.3 - 25.4$
    c. $\sqrt{33.3}$
    d. $333.3 \div 25.4$

29. | Perform the following calculations with the correct number of significant figures.
    a. 159.31×204.6
    b. $5.1125 + 0.67 + 3.2$
    c. $7.662 - 7.425$
    d. $16.5/3.45$

30. | Estimate (don't measure!) the length of a typical car. Give your answer in both feet and meters. Briefly describe how you arrived at this estimate.

31. | Estimate the height of a telephone pole. Give your answer in both feet and meters. Briefly describe how you arrived at this estimate.

32. | Estimate the average speed with which the hair on your head
    BIO grows. Give your answer in both m/s and μm/hour. Briefly describe how you arrived at this estimate.

33. | Motor neurons in mammals transmit signals from the brain
    BIO to skeletal muscles at approximately 25 m/s. Estimate how long in ms it takes a signal to get from your brain to your hand.

Problems

For Problems 34 through 43, draw a complete pictorial representation. **Do *not* solve these problems or do any mathematics.**

34. | A Porsche accelerates from a stoplight at 5.0 m/s^2 for five seconds, then coasts for three more seconds. How far has it traveled?

35. | A jet plane is cruising at 300 m/s when suddenly the pilot turns the engines up to full throttle. After traveling 4.0 km, the jet is moving with a speed of 400 m/s. What is the jet's acceleration as it speeds up?

36. | Sam is recklessly driving 60 mph in a 30 mph speed zone when he suddenly sees the police. He steps on the brakes and slows to 30 mph in three seconds, looking nonchalant as he passes the officer. How far does he travel while braking?

37. | You would like to stick a wet spit wad on the ceiling, so you toss it straight up with a speed of 10 m/s. How long does it take to reach the ceiling, 3.0 m above?

38. | A speed skater moving across frictionless ice at 8.0 m/s hits a 5.0-m-wide patch of rough ice. She slows steadily, then continues on at 6.0 m/s. What is her acceleration on the rough ice?

39. | Santa loses his footing and slides down a frictionless, snowy roof that is tilted at an angle of 30°. If Santa slides 10 m before reaching the edge, what is his speed as he leaves the roof?

40. | A motorist is traveling at 20 m/s. He is 60 m from a stoplight when he sees it turn yellow. His reaction time, before stepping on the brake, is 0.50 s. What steady deceleration while braking will bring him to a stop right at the light?

41. | A car traveling at 30 m/s runs out of gas while traveling up a 10° slope. How far up the hill will the car coast before starting to roll back down?

42. ‖ Ice hockey star Bruce Blades is 5.0 m from the blue line and gliding toward it at a speed of 4.0 m/s. You are 20 m from the blue line, directly behind Bruce. You want to pass the puck to Bruce. With what speed should you shoot the puck down the ice so that it reaches Bruce exactly as he crosses the blue line?

43. ‖ David is driving a steady 30 m/s when he passes Tina, who is sitting in her car at rest. Tina begins to accelerate at a steady 2.0 m/s^2 at the instant when David passes. How far does Tina drive before passing David?

Problems 44 through 48 show a motion diagram. For each of these problems, write a one or two sentence "story" about a *real object* that has this motion diagram. Your stories should talk about people or objects by name and say what they are doing. Problems 34 through 43 are examples of motion short stories.

44. |

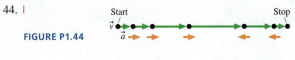

FIGURE P1.44

45. |

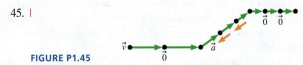

FIGURE P1.45

46. |

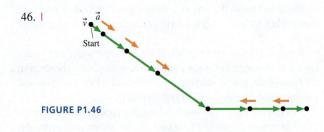

FIGURE P1.46

47. |

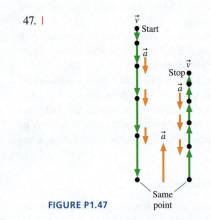

FIGURE P1.47

48. |

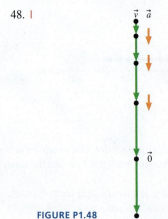

FIGURE P1.48

Problems 49 through 52 show a partial motion diagram. For each:

a. Complete the motion diagram by adding acceleration vectors.

b. Write a physics *problem* for which this is the correct motion diagram. Be imaginative! Don't forget to include enough information to make the problem complete and to state clearly what is to be found.

c. Draw a pictorial representation for your problem.

49.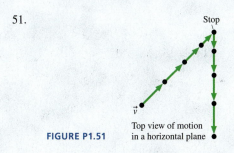

    FIGURE P1.49

50.

    FIGURE P1.50

51.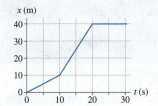

    FIGURE P1.51

52.

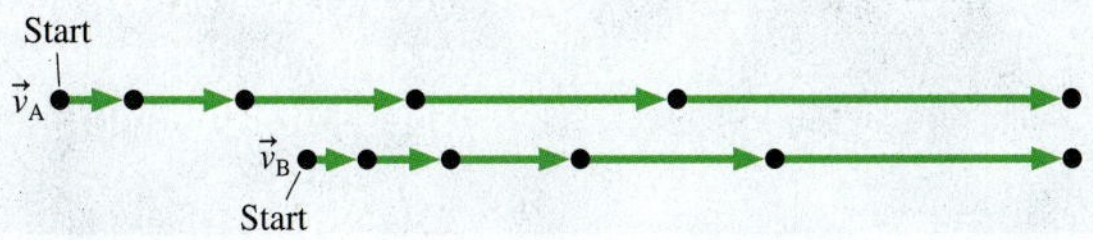

    FIGURE P1.52

53. | A regulation soccer field for international play is a rectangle with a length between 100 m and 110 m and a width between 64 m and 75 m. What are the smallest and largest areas that the field could be?

54. | As an architect, you are designing a new house. A window has a height between 140 cm and 150 cm and a width between 74 cm and 70 cm. What are the smallest and largest areas that the window could be?

55. || A 5.4-cm-diameter cylinder has a length of 12.5 cm. What is the cylinder's volume in SI units?

56. | An intravenous saline drip has 9.0 g of sodium chloride per
BIO liter of water. By definition, 1 mL = 1 cm^3. Express the salt concentration in kg/m^3.

57. || The quantity called *mass density* is the mass per unit volume of a substance. What are the mass densities in SI units of the following objects?

a. A 215 cm^3 solid with a mass of 0.0179 kg.

b. 95 cm^3 of a liquid with a mass of 77 g.

58. | **FIGURE P1.58** shows a motion diagram of a car traveling down a street. The camera took one frame every 10 s. A distance scale is provided.

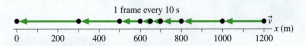

FIGURE P1.58

a. Measure the *x*-value of the car at each dot. Place your data in a table, similar to Table 1.1, showing each position and the instant of time at which it occurred.

b. Make a position-versus-time graph for the car. Because you have data only at certain instants of time, your graph should consist of dots that are not connected together.

59. | Write a short description of a real object for which **FIGURE P1.59** would be a realistic position-versus-time graph.

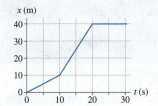

FIGURE P1.59

60. | Write a short description of a real object for which **FIGURE P1.60** would be a realistic position-versus-time graph.

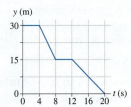

FIGURE P1.60

2 Kinematics in One Dimension

This Japanese "bullet train" accelerates slowly but steadily until reaching a speed of 300 km/h.

IN THIS CHAPTER, you will learn to solve problems about motion along a straight line.

What is kinematics?

Kinematics is the mathematical description of motion. We begin with motion along a straight line. Our primary tools will be an object's position, velocity, and acceleration.

« LOOKING BACK Sections 1.4–1.6 Velocity, acceleration, and Tactics Box 1.4 about signs

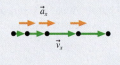

What are models?

A model is a simplified description of a situation that focuses on essential features while ignoring many details. Models allow us to make sense of complex situations by seeing them as variations on a common theme, all with the same underlying physics.

> **MODEL 2.1**
> Look for model boxes like this throughout the book.
> ■ Key figures
> ■ Key equations
> ■ Model limitations

How are graphs used in kinematics?

Graphs are a very important visual representation of motion, and learning to "think graphically" is one of our goals. We'll work with graphs showing how position, velocity, and acceleration change with time. These graphs are related to each other:

■ Velocity is the slope of the position graph.
■ Acceleration is the slope of the velocity graph.

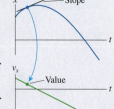

What is free fall?

Free fall is motion under the influence of gravity only. Free fall is not literally "falling" because it also applies to objects thrown straight up and to projectiles. Surprisingly, all objects in free fall, *regardless of their mass*, have the same acceleration. Motion on a frictionless inclined plane is closely related to free-fall motion.

How is calculus used in kinematics?

Motion is change, and calculus is the mathematical tool for describing a quantity's rate of change. We'll find that

■ Velocity is the time derivative of position.
■ Acceleration is the time derivative of velocity.

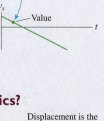

How will I use kinematics?

The equations of motion that you learn in this chapter will be used throughout the entire book. In Part I, we'll see how an object's motion is related to forces acting on the object. We'll later apply these kinematic equations to the motion of waves and to the motion of charged particles in electric and magnetic fields.

2.1 Uniform Motion

The simplest possible motion is motion along a straight line at a constant, unvarying speed. We call this **uniform motion.** Because velocity is the combination of speed and direction, **uniform motion is motion with constant velocity.**

FIGURE 2.1 shows the motion diagram of an object in uniform motion. For example, this might be you riding your bicycle along a straight line at a perfectly steady 5 m/s (≈ 10 mph). Notice how all the displacements are exactly the same; this is a characteristic of uniform motion.

If we make a position-versus-time graph—remember that position is graphed on the *vertical* axis—it's a straight line. In fact, an alternative definition is that **an object's motion is uniform if and only if its position-versus-time graph is a straight line.**

« Section 1.4 defined an object's **average velocity** as $\Delta \vec{r}/\Delta t$. For one-dimensional motion, this is simply $\Delta x/\Delta t$ (for horizontal motion) or $\Delta y/\Delta t$ (for vertical motion). You can see in Figure 2.1 that Δx and Δt are, respectively, the "rise" and "run" of the position graph. Because rise over run is the slope of a line,

$$v_{avg} \equiv \frac{\Delta x}{\Delta t} \quad \text{or} \quad \frac{\Delta y}{\Delta t} = \text{slope of the position-versus-time graph} \qquad (2.1)$$

That is, **the average velocity is the slope of the position-versus-time graph.** Velocity has units of "length per time," such as "miles per hour." The SI units of velocity are meters per second, abbreviated m/s.

> **NOTE** The symbol $\equiv$ in Equation 2.1 stands for "is defined as." This is a stronger statement than the two sides simply being equal.

The constant slope of a straight-line graph is another way to see that the velocity is constant for uniform motion. There's no real need to specify "average" for a velocity that doesn't change, so we will drop the subscript and refer to the average velocity as v_x or v_y.

An object's **speed** v is how fast it's going, independent of direction. This is simply $v = |v_x|$ or $v = |v_y|$, the magnitude or absolute value of the object's velocity. Although we will use speed from time to time, our mathematical analysis of motion is based on velocity, not speed. The subscript in v_x or v_y is an essential part of the notation, reminding us that, even in one dimension, the velocity is a vector.

FIGURE 2.1 Motion diagram and position graph for uniform motion.

The displacements between successive frames are the same.

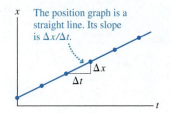

The position graph is a straight line. Its slope is $\Delta x/\Delta t$.

EXAMPLE 2.1 | Relating a velocity graph to a position graph

FIGURE 2.2 is the position-versus-time graph of a car.

a. Draw the car's velocity-versus-time graph.

b. Describe the car's motion.

MODEL Model the car as a particle, with a well-defined position at each instant of time.

VISUALIZE Figure 2.2 is the graphical representation.

SOLVE a. The car's position-versus-time graph is a sequence of three straight lines. Each of these straight lines represents uniform motion at a constant velocity. We can determine the car's velocity during each interval of time by measuring the slope of the line.

The position graph starts out sloping downward—a negative slope. Although the car moves a distance of 4.0 m during the first 2.0 s, its *displacement* is

$$\Delta x = x_{\text{at } 2.0\,s} - x_{\text{at } 0.0\,s} = -4.0\,\text{m} - 0.0\,\text{m} = -4.0\,\text{m}$$

The time interval for this displacement is $\Delta t = 2.0$ s, so the velocity during this interval is

$$v_x = \frac{\Delta x}{\Delta t} = \frac{-4.0\,\text{m}}{2.0\,\text{s}} = -2.0\,\text{m/s}$$

The car's position does not change from $t = 2$ s to $t = 4$ s ($\Delta x = 0$), so $v_x = 0$. Finally, the displacement between $t = 4$ s and $t = 6$ s is $\Delta x = 10.0$ m. Thus the velocity during this interval is

$$v_x = \frac{10.0\,\text{m}}{2.0\,\text{s}} = 5.0\,\text{m/s}$$

These velocities are shown on the velocity-versus-time graph of FIGURE 2.3.

b. The car backs up for 2 s at 2.0 m/s, sits at rest for 2 s, then drives forward at 5.0 m/s for at least 2 s. We can't tell from the graph what happens for $t > 6$ s.

ASSESS The velocity graph and the position graph look completely different. The *value* of the velocity graph at any instant of time equals the *slope* of the position graph.

Continued

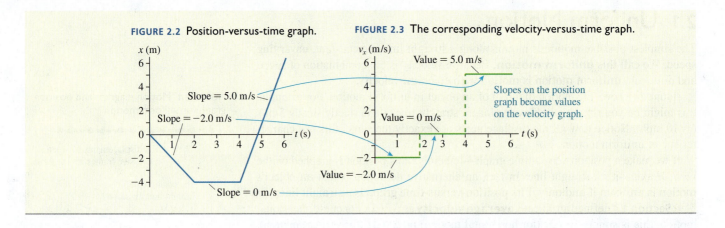

FIGURE 2.2 Position-versus-time graph.

FIGURE 2.3 The corresponding velocity-versus-time graph.

Example 2.1 brought out several points that are worth emphasizing.

TACTICS BOX 2.1 (MP)

Interpreting position-versus-time graphs

❶ Steeper slopes correspond to faster speeds.
❷ Negative slopes correspond to negative velocities and, hence, to motion to the left (or down).
❸ The slope is a ratio of intervals, $\Delta x / \Delta t$, not a ratio of coordinates. That is, the slope is *not* simply x / t.

Exercises 1–3

NOTE We are distinguishing between the *actual* slope and the *physically mean-ingful* slope. If you were to use a ruler to measure the rise and the run of the graph, you could compute the actual slope of the line as drawn on the page. That is not the slope to which we are referring when we equate the velocity with the slope of the line. Instead, we find the *physically meaningful* slope by measuring the rise and run using the scales along the axes. The "rise" Δx is some number of meters; the "run" Δt is some number of seconds. The physically meaningful rise and run include units, and the ratio of these units gives the units of the slope.

The Mathematics of Uniform Motion

The physics of the motion is the same regardless of whether an object moves along the x-axis, the y-axis, or any other straight line. Consequently, it will be convenient to write equations for a "generic axis" that we will call the s-axis. The position of an object will be represented by the symbol s and its velocity by v_s.

NOTE In a specific problem you should use either x or y rather than s.

Consider an object in uniform motion along the s-axis with the linear position-versus-time graph shown in FIGURE 2.4. The object's **initial position** is s_i at time t_i. The term *initial position* refers to the starting point of our analysis or the starting point in a problem; the object may or may not have been in motion prior to t_i. At a later time t_f, the ending point of our analysis, the object's **final position** is s_f.

The object's velocity v_s along the s-axis can be determined by finding the slope of the graph:

$$v_s = \frac{\text{rise}}{\text{run}} = \frac{\Delta s}{\Delta t} = \frac{s_f - s_i}{t_f - t_i} \tag{2.2}$$

FIGURE 2.4 The velocity is found from the slope of the position-versus-time graph.

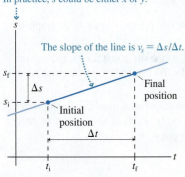

We will use s as a generic label for position. In practice, s could be either x or y.

Equation 2.2 is easily rearranged to give

$$s_f = s_i + v_s \Delta t \qquad \text{(uniform motion)} \qquad (2.3)$$

Equation 2.3 tells us that the object's position increases linearly as the elapsed time Δt increases—exactly as we see in the straight-line position graph.

The Uniform-Motion Model

Chapter 1 introduced a *model* as a simplified picture of reality, but one that still captures the essence of what we want to study. When it comes to motion, few real objects move with a precisely constant velocity. Even so, there are many cases in which it is quite reasonable to model their motion as being uniform. That is, uniform motion is a very good approximation of their actual, but more complex, motion. The **uniform-motion model** is a coherent set of representations—words, pictures, graphs, and equations—that allows us to explain an object's motion and to predict where the object will be at a future instant of time.

MODEL 2.1

Uniform motion

For motion with constant velocity.

- Model the object as a particle moving in a straight line at constant speed:

- Mathematically:
  - $v_s = \Delta s / \Delta t$
  - $s_f = s_i + v_s \Delta t$

- Limitations: Model fails if the particle has a significant change of speed or direction.

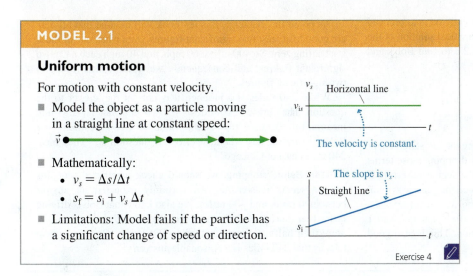

The velocity is constant.

The slope is v_s.

Straight line

Exercise 4

EXAMPLE 2.2 | **Lunch in Cleveland?**

Bob leaves home in Chicago at 9:00 A.M. and drives east at 60 mph. Susan, 400 miles to the east in Pittsburgh, leaves at the same time and travels west at 40 mph. Where will they meet for lunch?

MODEL Here is a problem where, for the first time, we can really put all four aspects of our problem-solving strategy into play. To begin, we'll model Bob's and Susan's cars as being in uniform

motion. Their real motion is certainly more complex, but over a long drive it's reasonable to approximate their motion as constant speed along a straight line.

VISUALIZE FIGURE 2.5 shows the pictorial representation. The equal spacings of the dots in the motion diagram indicate that the motion is uniform. In evaluating the given information, we

FIGURE 2.5 Pictorial representation for Example 2.2.

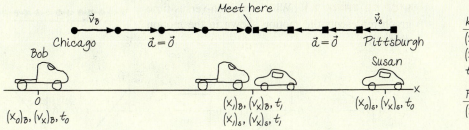

Continued

recognize that the starting time of 9:00 A.M. is not relevant to the problem. Consequently, the initial time is chosen as simply $t_0 = 0$ h. Bob and Susan are traveling in opposite directions, hence one of the velocities must be a negative number. We have chosen a coordinate system in which Bob starts at the origin and moves to the right (east) while Susan is moving to the left (west). Thus Susan has the negative velocity. Notice how we've assigned position, velocity, and time symbols to each point in the motion. Pay special attention to how subscripts are used to distinguish different points in the problem and to distinguish Bob's symbols from Susan's.

One purpose of the pictorial representation is to establish what we need to find. Bob and Susan meet when they have the same position at the same time t_1. Thus we want to find $(x_1)_B$ at the time when $(x_1)_B = (x_1)_S$. Notice that $(x_1)_B$ and $(x_1)_S$ are Bob's and Susan's *positions,* which are equal when they meet, not the distances they have traveled.

SOLVE The goal of the mathematical representation is to proceed from the pictorial representation to a mathematical solution of the problem. We can begin by using Equation 2.3 to find Bob's and Susan's positions at time t_1 when they meet:

$$(x_1)_B = (x_0)_B + (v_x)_B(t_1 - t_0) = (v_x)_B t_1$$
$$(x_1)_S = (x_0)_S + (v_x)_S(t_1 - t_0) = (x_0)_S + (v_x)_S t_1$$

Notice two things. First, we started by writing the *full* statement of Equation 2.3. Only then did we simplify by dropping those terms known to be zero. You're less likely to make accidental errors if you follow this procedure. Second, we replaced the generic symbol s with the specific horizontal-position symbol x, and we replaced the generic subscripts i and f with the specific symbols 0 and 1 that we defined in the pictorial representation. This is also good problem-solving technique.

The condition that Bob and Susan meet is

$$(x_1)_B = (x_1)_S$$

By equating the right-hand sides of the above equations, we get

$$(v_x)_B t_1 = (x_0)_S + (v_x)_S t_1$$

Solving for t_1 we find that they meet at time

$$t_1 = \frac{(x_0)_S}{(v_x)_B - (v_x)_S} = \frac{400 \text{ miles}}{60 \text{ mph} - (-40) \text{ mph}} = 4.0 \text{ hours}$$

Finally, inserting this time back into the equation for $(x_1)_B$ gives

$$(x_1)_B = \left(60 \, \frac{\text{miles}}{\text{hour}}\right) \times (4.0 \text{ hours}) = 240 \text{ miles}$$

As noted in Chapter 1, this textbook will assume that all data are good to at least two significant figures, even when one of those is a trailing zero. So 400 miles, 60 mph, and 40 mph each have two significant figures, and consequently we've calculated results to two significant figures.

While 240 miles is a number, it is not yet the answer to the question. The phrase "240 miles" by itself does not say anything meaningful. Because this is the value of Bob's *position,* and Bob was driving east, the answer to the question is, "They meet 240 miles east of Chicago."

ASSESS Before stopping, we should check whether or not this answer seems reasonable. We certainly expected an answer between 0 miles and 400 miles. We also know that Bob is driving faster than Susan, so we expect that their meeting point will be *more* than halfway from Chicago to Pittsburgh. Our assessment tells us that 240 miles is a reasonable answer.

FIGURE 2.6 Position-versus-time graphs for Bob and Susan.

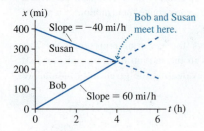

It is instructive to look at this example from a graphical perspective. **FIGURE 2.6** shows position-versus-time graphs for Bob and Susan. Notice the negative slope for Susan's graph, indicating her negative velocity. The point of interest is the intersection of the two lines; this is where Bob and Susan have the same position at the same time. Our method of solution, in which we equated $(x_1)_B$ and $(x_1)_S$, is really just solving the mathematical problem of finding the intersection of two lines. This procedure is useful for many problems in which there are two moving objects.

STOP TO THINK 2.1 Which position-versus-time graph represents the motion shown in the motion diagram?

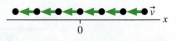

(a)   (b)   (c)   (d)   (e)

2.2 Instantaneous Velocity

Uniform motion is simple, but objects rarely travel for long with a constant velocity. Far more common is a velocity that changes with time. For example, **FIGURE 2.7** shows the motion diagram and position graph of a car speeding up after the light turns green. Notice how the velocity vectors increase in length, causing the graph to curve upward as the car's displacements get larger and larger.

If you were to watch the car's speedometer, you would see it increase from 0 mph to 10 mph to 20 mph and so on. At any instant of time, the speedometer tells you how fast the car is going *at that instant*. If we include directional information, we can define an object's **instantaneous velocity**—speed and direction—as its velocity at a single instant of time.

For uniform motion, the slope of the straight-line position graph is the object's velocity. **FIGURE 2.8** shows that there's a similar connection between instantaneous velocity and the slope of a curved position graph.

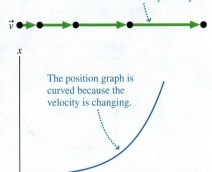

FIGURE 2.7 Motion diagram and position graph of a car speeding up.

The spacing between the dots increases as the car speeds up.

The position graph is curved because the velocity is changing.

FIGURE 2.8 Instantaneous velocity at time t is the slope of the tangent to the curve at that instant.

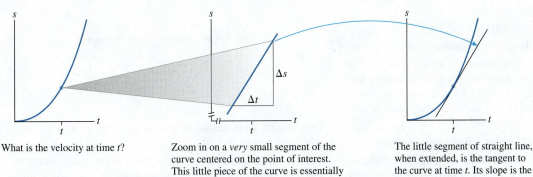

What is the velocity at time t?

Zoom in on a *very* small segment of the curve centered on the point of interest. This little piece of the curve is essentially a straight line. Its slope $\Delta s/\Delta t$ is the average velocity during the interval Δt.

The little segment of straight line, when extended, is the tangent to the curve at time t. Its slope is the instantaneous velocity at time t.

What we see graphically is that the average velocity $v_{avg} = \Delta s/\Delta t$ becomes a better and better approximation to the instantaneous velocity v_s as the time interval Δt over which the average is taken gets smaller and smaller. We can state this idea mathematically in terms of the limit $\Delta t \rightarrow 0$:

$$v_s \equiv \lim_{\Delta t \to 0} \frac{\Delta s}{\Delta t} = \frac{ds}{dt} \qquad \text{(instantaneous velocity)} \qquad (2.4)$$

As Δt continues to get smaller, the average velocity $v_{avg} = \Delta s/\Delta t$ reaches a constant or *limiting* value. That is, **the instantaneous velocity at time t is the average velocity during a time interval Δt, centered on t, as Δt approaches zero.** In calculus, this limit is called *the derivative of s with respect to t*, and it is denoted ds/dt.

Graphically, $\Delta s/\Delta t$ is the slope of a straight line. As Δt gets smaller (i.e., more and more magnification), the straight line becomes a better and better approximation of the curve *at that one point*. In the limit $\Delta t \rightarrow 0$, the straight line is tangent to the curve. As Figure 2.8 shows, **the instantaneous velocity at time t is the slope of the line that is tangent to the position-versus-time graph at time t.** That is,

$$v_s = \text{slope of the position-versus-time graph at time } t \qquad (2.5)$$

The steeper the slope, the larger the magnitude of the velocity.

EXAMPLE 2.3 | **Finding velocity from position graphically**

FIGURE 2.9 shows the position-versus-time graph of an elevator.

a. At which labeled point or points does the elevator have the least velocity?

b. At which point or points does the elevator have maximum velocity?

c. Sketch an approximate velocity-versus-time graph for the elevator.

FIGURE 2.9 Position-versus-time graph.

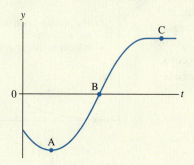

MODEL Model the elevator as a particle.

VISUALIZE Figure 2.9 is the graphical representation.

SOLVE a. At any instant, an object's velocity is the slope of its position graph. **FIGURE 2.10a** shows that the elevator has the least velocity—no velocity at all!—at points A and C where the slope is zero. At point A, the velocity is only instantaneously zero. At point C, the elevator has actually stopped and remains at rest.

b. The elevator has maximum velocity at B, the point of steepest slope.

c. Although we cannot find an exact velocity-versus-time graph, we can see that the slope, and hence v_y, is initially negative, becomes zero at point A, rises to a maximum value at point B, decreases back to zero a little before point C, then remains at zero thereafter.

FIGURE 2.10 The velocity-versus-time graph is found from the slope of the position graph.

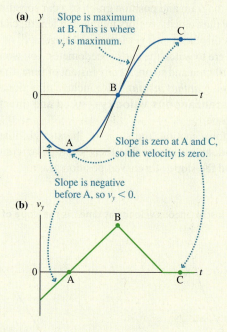

Thus **FIGURE 2.10b** shows, at least approximately, the elevator's velocity-versus-time graph.

ASSESS Once again, the shape of the velocity graph bears no resemblance to the shape of the position graph. You must transfer *slope* information from the position graph to *value* information on the velocity graph.

A Little Calculus: Derivatives

Calculus—invented simultaneously in England by Newton and in Germany by Leibniz—is designed to deal with instantaneous quantities. In other words, it provides us with the tools for evaluating limits such as the one in Equation 2.4.

The notation ds/dt is called *the derivative of s with respect to t*, and Equation 2.4 defines it as the limiting value of a ratio. As Figure 2.8 showed, ds/dt can be interpreted graphically as the slope of the line that is tangent to the position graph.

The most common functions we will use in Parts I and II of this book are powers and polynomials. Consider the function $u(t) = ct^n$, where c and n are constants. The symbol u is a "dummy name" to represent any function of time, such as $x(t)$ or $y(t)$. The following result is proven in calculus:

Scientists and engineers must use calculus to calculate the orbits of satellites.

$$\text{The derivative of } u = ct^n \text{ is } \frac{du}{dt} = nct^{n-1} \tag{2.6}$$

For example, suppose the position of a particle as a function of time is $s(t) = 2t^2$ m, where t is in s. We can find the particle's velocity $v_s = ds/dt$ by using Equation 2.6 with $c = 2$ and $n = 2$ to calculate

$$v_s = \frac{ds}{dt} = 2 \cdot 2t^{2-1} = 4t$$

This is an expression for the particle's velocity as a function of time.

FIGURE 2.11 shows the particle's position and velocity graphs. It is critically important to understand the relationship between these two graphs. The *value* of the velocity graph at any instant of time, which we can read directly off the vertical axis, is the *slope* of the position graph at that same time. This is illustrated at $t = 3$ s.

A value that doesn't change with time, such as the position of an object at rest, can be represented by the function $u = c = $ constant. That is, the exponent of t^n is $n = 0$. You can see from Equation 2.6 that the derivative of a constant is zero. That is,

$$\frac{du}{dt} = 0 \text{ if } u = c = \text{constant} \tag{2.7}$$

This makes sense. The graph of the function $u = c$ is simply a horizontal line. The slope of a horizontal line—which is what the derivative du/dt measures—is zero.

The only other information we need about derivatives for now is how to evaluate the derivative of the sum of two functions. Let u and w be two separate functions of time. You will learn in calculus that

$$\frac{d}{dt}(u + w) = \frac{du}{dt} + \frac{dw}{dt} \tag{2.8}$$

That is, the derivative of a sum is the sum of the derivatives.

NOTE You may have learned in calculus to take the derivative dy/dx, where y is a function of x. The derivatives we use in physics are the same; only the notation is different. We're interested in how quantities change with time, so our derivatives are with respect to t instead of x.

FIGURE 2.11 Position-versus-time graph and the corresponding velocity-versus-time graph.

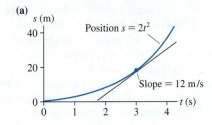

(a)

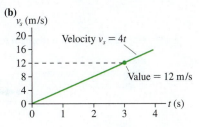

(b)

EXAMPLE 2.4 | **Using calculus to find the velocity**

A particle's position is given by the function $x(t) = (-t^3 + 3t)$ m, where t is in s.

a. What are the particle's position and velocity at $t = 2$ s?

b. Draw graphs of x and v_x during the interval -3 s $\leq t \leq 3$ s.

c. Draw a motion diagram to illustrate this motion.

SOLVE

a. We can compute the position directly from the function x:

$$x(\text{at } t = 2 \text{ s}) = -(2)^3 + (3)(2) = -8 + 6 = -2 \text{ m}$$

The velocity is $v_x = dx/dt$. The function for x is the sum of two polynomials, so

$$v_x = \frac{dx}{dt} = \frac{d}{dt}(-t^3 + 3t) = \frac{d}{dt}(-t^3) + \frac{d}{dt}(3t)$$

The first derivative is a power with $c = -1$ and $n = 3$; the second has $c = 3$ and $n = 1$. Using Equation 2.6, we have

$$v_x = (-3t^2 + 3) \text{ m/s}$$

where t is in s. Evaluating the velocity at $t = 2$ s gives

$$v_x(\text{at } t = 2 \text{ s}) = -3(2)^2 + 3 = -9 \text{ m/s}$$

The negative sign indicates that the particle, at this instant of time, is moving to the *left* at a speed of 9 m/s.

b. **FIGURE 2.12** shows the position graph and the velocity graph. You can make graphs like these with a graphing calculator or graphing software. The slope of the position-versus-time graph at $t = 2$ s is -9 m/s; this becomes the *value* that is graphed for the velocity at $t = 2$ s.

FIGURE 2.12 Position and velocity graphs.

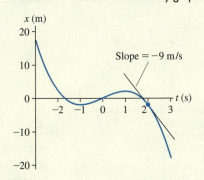

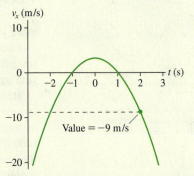

Continued

c. Finally, we can interpret the graphs in Figure 2.12 to draw the motion diagram shown in **FIGURE 2.13**.

- The particle is initially to the right of the origin ($x > 0$ at $t = -3$ s) but moving to the left ($v_x < 0$). Its *speed* is slowing ($v = |v_x|$ is decreasing), so the velocity vector arrows are getting shorter.
- The particle passes the origin $x = 0$ m at $t \approx -1.5$ s, but it is still moving to the left.
- The position reaches a minimum at $t = -1$ s; the particle is as far left as it is going. The velocity is *instantaneously $v_x = 0$ m/s* as the particle reverses direction.

- The particle moves back to the right between $t = -1$ s and $t = 1$ s ($v_x > 0$).
- The particle turns around again at $t = 1$ s and begins moving back to the left ($v_x < 0$). It keeps speeding up, then disappears off to the left.

A point in the motion where a particle reverses direction is called a **turning point.** It is a point where the velocity is instantaneously zero while the position is a maximum or minimum. This particle has two turning points, at $t = -1$ s and again at $t = +1$ s. We will see many other examples of turning points.

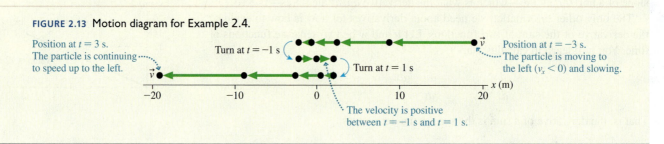

FIGURE 2.13 Motion diagram for Example 2.4.

Position at $t = 3$ s. The particle is continuing to speed up to the left.

Turn at $t = -1$ s

Turn at $t = 1$ s

Position at $t = -3$ s. The particle is moving to the left ($v_x < 0$) and slowing.

The velocity is positive between $t = -1$ s and $t = 1$ s.

STOP TO THINK 2.2 Which velocity-versus-time graph goes with the position-versus-time graph on the left?

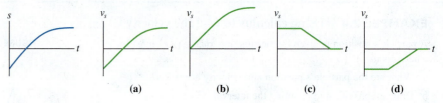

(a) (b) (c) (d)

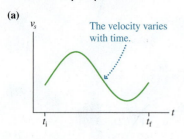

FIGURE 2.14 Approximating a velocity-versus-time graph with a series of constant-velocity steps.

(a)

The velocity varies with time.

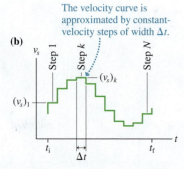

The velocity curve is approximated by constant-velocity steps of width Δt.

(b)

2.3 Finding Position from Velocity

Equation 2.4 allows us to find the instantaneous velocity v_s if we know the position s as a function of time. But what about the reverse problem? Can we use the object's velocity to calculate its position at some future time t? Equation 2.3, $s_f = s_i + v_s \Delta t$, does this for the case of uniform motion with a constant velocity. We need to find a more general expression that is valid when v_s is not constant.

FIGURE 2.14a is a velocity-versus-time graph for an object whose velocity varies with time. Suppose we know the object's position to be s_i at an initial time t_i. Our goal is to find its position s_f at a later time t_f.

Because we know how to handle constant velocities, using Equation 2.3, let's *approximate* the velocity function of Figure 2.14a as a series of constant-velocity steps of width Δt. This is illustrated in **FIGURE 2.14b**. During the first step, from time t_i to time $t_i + \Delta t$, the velocity has the constant value $(v_s)_1$. The velocity during step k has the constant value $(v_s)_k$. Although the approximation shown in the figure is rather rough, with only 11 steps, we can easily imagine that it could be made as accurate as desired by having more and more ever-narrower steps.

The velocity during each step is constant (uniform motion), so we can apply Equation 2.3 to each step. The object's displacement Δs_1 during the first step is simply $\Delta s_1 = (v_s)_1 \Delta t$. The displacement during the second step $\Delta s_2 = (v_s)_2 \Delta t$, and during step k the displacement is $\Delta s_k = (v_s)_k \Delta t$.

The total displacement of the object between t_i and t_f can be approximated as the sum of all the individual displacements during each of the N constant-velocity steps. That is,

$$\Delta s = s_f - s_i \approx \Delta s_1 + \Delta s_2 + \cdots + \Delta s_N = \sum_{k=1}^{N} (v_s)_k \Delta t \qquad (2.9)$$

where Σ (Greek sigma) is the symbol for summation. With a simple rearrangement, the particle's final position is

$$s_f \approx s_i + \sum_{k=1}^{N} (v_s)_k \Delta t \qquad (2.10)$$

Our goal was to use the object's velocity to find its final position s_f. Equation 2.10 nearly reaches that goal, but Equation 2.10 is only approximate because the constant-velocity steps are only an approximation of the true velocity graph. But if we now let $\Delta t \to 0$, each step's width approaches zero while the total number of steps N approaches infinity. In this limit, the series of steps becomes a perfect replica of the velocity-versus-time graph and Equation 2.10 becomes exact. Thus

$$s_f = s_i + \lim_{\Delta t \to 0} \sum_{k=1}^{N} (v_s)_k \Delta t = s_i + \int_{t_i}^{t_f} v_s \, dt \qquad (2.11)$$

The expression on the right is read, "the integral of $v_s \, dt$ from t_i to t_f." Equation 2.11 is the result that we were seeking. It allows us to predict an object's position s_f at a future time t_f.

We can give Equation 2.11 an important geometric interpretation. **FIGURE 2.15** shows step k in the approximation of the velocity graph as a tall, thin rectangle of height $(v_s)_k$ and width Δt. The product $\Delta s_k = (v_s)_k \Delta t$ is the area (base × height) of this small rectangle. The sum in Equation 2.11 adds up all of these rectangular areas to give the total area enclosed between the t-axis and the tops of the steps. The limit of this sum as $\Delta t \to 0$ is the total area enclosed between the t-axis and the velocity curve. This is called the "area under the curve." Thus a graphical interpretation of Equation 2.11 is

$$s_f = s_i + \text{area under the velocity curve } v_s \text{ between } t_i \text{ and } t_f \qquad (2.12)$$

NOTE Wait a minute! The displacement $\Delta s = s_f - s_i$ is a length. How can a length equal an area? Recall earlier, when we found that the velocity is the slope of the position graph, we made a distinction between the *actual* slope and the *physically meaningful* slope? The same distinction applies here. We need to measure the quantities we are using, v_s and Δt, by referring to the scales on the axes. Δt is some number of seconds while v_s is some number of meters per second. When these are multiplied together, the *physically meaningful* area has units of meters.

FIGURE 2.15 The total displacement Δs is the "area under the curve."

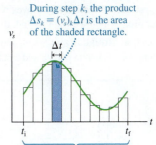

During step k, the product $\Delta s_k = (v_s)_k \Delta t$ is the area of the shaded rectangle.

During the interval t_i to t_f, the total displacement Δs is the "area under the curve."

EXAMPLE 2.5 | The displacement during a drag race

FIGURE 2.16 shows the velocity-versus-time graph of a drag racer. How far does the racer move during the first 3.0 s?

FIGURE 2.16 Velocity-versus-time graph for Example 2.5.

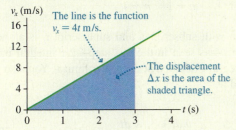

The line is the function $v_x = 4t$ m/s.

The displacement Δx is the area of the shaded triangle.

MODEL Model the drag racer as a particle with a well-defined position at all times.

VISUALIZE Figure 2.16 is the graphical representation.

SOLVE The question "How far?" indicates that we need to find a displacement Δx rather than a position x. According to Equation 2.12, the car's displacement $\Delta x = x_f - x_i$ between $t = 0$ s and $t = 3$ s is the area under the curve from $t = 0$ s to $t = 3$ s. The curve in this case is an angled line, so the area is that of a triangle:

$$\Delta x = \text{area of triangle between } t = 0 \text{ s and } t = 3 \text{ s}$$
$$= \tfrac{1}{2} \times \text{base} \times \text{height}$$
$$= \tfrac{1}{2} \times 3 \text{ s} \times 12 \text{ m/s} = 18 \text{ m}$$

The drag racer moves 18 m during the first 3 seconds.

ASSESS The "area" is a product of s with m/s, so Δx has the proper units of m.

EXAMPLE 2.6 | **Finding the turning point**

FIGURE 2.17 is the velocity graph for a particle that starts at $x_i = 30$ m at time $t_i = 0$ s.

a. Draw a motion diagram for the particle.

b. Where is the particle's turning point?

c. At what time does the particle reach the origin?

FIGURE 2.17 Velocity-versus-time graph for the particle of Example 2.6.

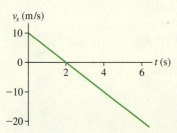

VISUALIZE The particle is initially 30 m to the right of the origin and moving *to the right* ($v_x > 0$) with a speed of 10 m/s. But v_x is decreasing, so the particle is slowing down. At $t = 2$ s the velocity, just for an instant, is zero before becoming negative. This is the turning point. The velocity is negative for $t > 2$ s, so the particle has reversed direction and moves back toward the origin. At some later time, which we want to find, the particle will pass $x = 0$ m.

SOLVE a. **FIGURE 2.18** shows the motion diagram. The distance scale will be established in parts b and c but is shown here for convenience.

FIGURE 2.18 Motion diagram for the particle whose velocity graph was shown in Figure 2.17.

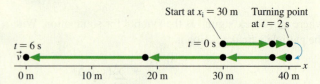

b. The particle reaches the turning point at $t = 2$ s. To learn *where* it is at that time we need to find the displacement during the first two seconds. We can do this by finding the area under the curve between $t = 0$ s and $t = 2$ s:

$$x(\text{at } t = 2 \text{ s}) = x_i + \text{area under the curve between 0 s and 2 s}$$
$$= 30 \text{ m} + \tfrac{1}{2}(2 \text{ s} - 0 \text{ s})(10 \text{ m/s} - 0 \text{ m/s})$$
$$= 40 \text{ m}$$

The turning point is at $x = 40$ m.

c. The particle needs to move $\Delta x = -40$ m to get from the turning point to the origin. That is, the area under the curve from $t = 2$ s to the desired time t needs to be -40 m. Because the curve is below the axis, with negative values of v_x, the area to the right of $t = 2$ s is a *negative* area. With a bit of geometry, you will find that the triangle with a base extending from $t = 2$ s to $t = 6$ s has an area of -40 m. Thus the particle reaches the origin at $t = 6$ s.

A Little More Calculus: Integrals

Taking the derivative of a function is equivalent to finding the slope of a graph of the function. Similarly, evaluating an integral is equivalent to finding the area under a graph of the function. The graphical method is very important for building intuition about motion but is limited in its practical application. Just as derivatives of standard functions can be evaluated and tabulated, so can integrals.

The integral in Equation 2.11 is called a *definite integral* because there are two definite boundaries to the area we want to find. These boundaries are called the lower (t_i) and upper (t_f) *limits of integration*. For the important function $u(t) = ct^n$, the essential result from calculus is that

$$\int_{t_i}^{t_f} u \, dt = \int_{t_i}^{t_f} ct^n \, dt = \frac{ct^{n+1}}{n+1} \Big|_{t_i}^{t_f} = \frac{ct_f^{n+1}}{n+1} - \frac{ct_i^{n+1}}{n+1} \quad (n \neq -1) \quad (2.13)$$

The vertical bar in the third step with subscript t_i and superscript t_f is a shorthand notation from calculus that means—as seen in the last step—the integral evaluated at the upper limit t_f *minus* the integral evaluated at the lower limit t_i. You also need to know that for two functions u and w,

$$\int_{t_i}^{t_f} (u + w) \, dt = \int_{t_i}^{t_f} u \, dt + \int_{t_i}^{t_f} w \, dt \quad (2.14)$$

That is, the integral of a sum is equal to the sum of the integrals.

EXAMPLE 2.7 | Using calculus to find the position

Use calculus to solve Example 2.6.

SOLVE Figure 2.17 is a linear graph. Its "y-intercept" is seen to be 10 m/s and its slope is −5 (m/s)/s. Thus the velocity can be described by the equation

$$v_x = (10 - 5t) \text{ m/s}$$

where t is in s. We can find the position x at time t by using Equation 2.11:

$$x = x_i + \int_0^t v_x \, dt = 30 \text{ m} + \int_0^t (10 - 5t) \, dt$$

$$= 30 \text{ m} + \int_0^t 10 \, dt - \int_0^t 5t \, dt$$

We used Equation 2.14 for the integral of a sum to get the final expression. The first integral is a function of the form $u = ct^n$ with $c = 10$ and $n = 0$; the second is of the form $u = ct^n$ with $c = 5$ and $n = 1$. Using Equation 2.13, we have

$$\int_0^t 10 \, dt = 10t \Big|_0^t = 10 \cdot t - 10 \cdot 0 = 10t \text{ m}$$

and

$$\int_0^t 5t \, dt = \tfrac{5}{2} t^2 \Big|_0^t = \tfrac{5}{2} \cdot t^2 - \tfrac{5}{2} \cdot 0^2 = \tfrac{5}{2} t^2 \text{ m}$$

Combining the pieces gives

$$x = \left(30 + 10t - \tfrac{5}{2} t^2\right) \text{ m}$$

This is a general result for the position at *any* time t.

The particle's turning point occurs at $t = 2$ s, and its position at that time is

$$x(\text{at } t = 2 \text{ s}) = 30 + (10)(2) - \tfrac{5}{2}(2)^2 = 40 \text{ m}$$

The time at which the particle reaches the origin is found by setting $x = 0$ m:

$$30 + 10t - \tfrac{5}{2} t^2 = 0$$

This quadratic equation has two solutions: $t = -2$ s or $t = 6$ s.

When we solve a quadratic equation, we cannot just arbitrarily select the root we want. Instead, we must decide which is the *meaningful* root. Here the negative root refers to a time before the problem began, so the meaningful one is the positive root, $t = 6$ s.

ASSESS The results agree with the answers we found previously from a graphical solution.

STOP TO THINK 2.3 Which position-versus-time graph goes with the velocity-versus-time graph on the left? The particle's position at $t_i = 0$ s is $x_i = -10$ m.

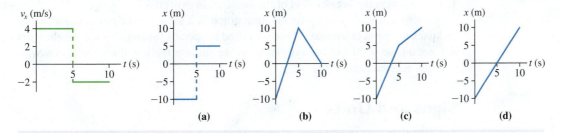

(a)   (b)   (c)   (d)

2.4 Motion with Constant Acceleration

We need one more major concept to describe one-dimensional motion: acceleration. Acceleration, as we noted in Chapter 1, is a rather abstract concept. Nonetheless, acceleration is the linchpin of mechanics. We will see very shortly that Newton's laws relate the acceleration of an object to the forces that are exerted on it.

Let's conduct a race between a Volkswagen Beetle and a Porsche to see which can achieve a velocity of 30 m/s ($\approx$ 60 mph) in the shortest time. Both cars are equipped with computers that will record the speedometer reading 10 times each second. This gives a nearly continuous record of the *instantaneous* velocity of each car. **TABLE 2.1** shows some of the data. The velocity-versus-time graphs, based on these data, are shown in **FIGURE 2.19** on the next page.

How can we describe the difference in performance of the two cars? It is not that one has a different velocity from the other; both achieve every velocity between 0 and 30 m/s. The distinction is how long it took each to *change* its velocity from 0 to 30 m/s. The Porsche changed velocity quickly, in 6.0 s, while the VW needed 15 s to make

TABLE 2.1 Velocities of a Porsche and a Volkswagen Beetle

t (s)	v_{Porsche} (m/s)	v_{VW} (m/s)
0.0	0.0	0.0
0.1	0.5	0.2
0.2	1.0	0.4
0.3	1.5	0.6
$\vdots$	$\vdots$	$\vdots$

FIGURE 2.19 Velocity-versus-time graphs for the Porsche and the VW Beetle.

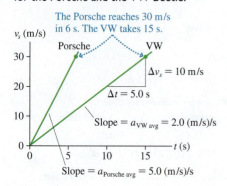

the same velocity change. Because the Porsche had a velocity change $\Delta v_s = 30$ m/s during a time interval $\Delta t = 6.0$ s, the *rate* at which its velocity changed was

$$\text{rate of velocity change} = \frac{\Delta v_s}{\Delta t} = \frac{30 \text{ m/s}}{6.0 \text{ s}} = 5.0 \text{ (m/s)/s} \qquad (2.15)$$

Notice the units. They are units of "velocity per second." A rate of velocity change of 5.0 "meters per second per second" means that the velocity increases by 5.0 m/s during the first second, by another 5.0 m/s during the next second, and so on. In fact, the velocity will increase by 5.0 m/s during any second in which it is changing at the rate of 5.0 (m/s)/s.

Chapter 1 introduced *acceleration* as "the rate of change of velocity." That is, acceleration measures how quickly or slowly an object's velocity changes. In parallel with our treatment of velocity, let's define the **average acceleration** a_{avg} during the time interval Δt to be

$$a_{\text{avg}} \equiv \frac{\Delta v_s}{\Delta t} \qquad \text{(average acceleration)} \qquad (2.16)$$

Equations 2.15 and 2.16 show that the Porsche had the rather large acceleration of 5.0 (m/s)/s.

Because Δv_s and Δt are the "rise" and "run" of a velocity-versus-time graph, we see that a_{avg} can be interpreted graphically as the *slope* of a straight-line velocity-versus-time graph. In other words,

$$a_{\text{avg}} = \text{slope of the velocity-versus-time graph} \qquad (2.17)$$

Figure 2.19 uses this idea to show that the VW's average acceleration is

$$a_{\text{VW avg}} = \frac{\Delta v_s}{\Delta t} = \frac{10 \text{ m/s}}{5.0 \text{ s}} = 2.0 \text{ (m/s)/s}$$

This is less than the acceleration of the Porsche, as expected.

An object whose velocity-versus-time graph is a straight-line graph has a steady and unchanging acceleration. There's no need to specify "average" if the acceleration is constant, so we'll use the symbol a_s as we discuss motion along the *s*-axis with constant acceleration.

Signs and Units

An important aspect of acceleration is its *sign*. Acceleration $\vec{a}$, like position $\vec{r}$ and velocity $\vec{v}$, is a vector. For motion in one dimension, the sign of a_x (or a_y) is positive if the vector $\vec{a}$ points to the right (or up), negative if it points to the left (or down). This was illustrated in « Figure 1.18 and the very important « Tactics Box 1.4, which you may wish to review. It's particularly important to emphasize that positive and negative values of a_s do *not* correspond to "speeding up" and "slowing down."

EXAMPLE 2.8 | **Relating acceleration to velocity**

a. A bicyclist has a velocity of 6 m/s and a constant acceleration of 2 (m/s)/s. What is her velocity 1 s later? 2 s later?

b. A bicyclist has a velocity of −6 m/s and a constant acceleration of 2 (m/s)/s. What is his velocity 1 s later? 2 s later?

SOLVE

a. An acceleration of 2 (m/s)/s *means* that the velocity increases by 2 m/s every 1 s. If the bicyclist's initial velocity is 6 m/s, then 1 s later her velocity will be 8 m/s. After 2 s, which is 1

additional second later, it will increase by another 2 m/s to 10 m/s. After 3 s it will be 12 m/s. Here a positive a_x is causing the bicyclist to speed up.

b. If the bicyclist's initial velocity is a *negative* −6 m/s but the acceleration is a positive +2 (m/s)/s, then 1 s later his velocity will be −4 m/s. After 2 s it will be −2 m/s, and so on. In this case, a positive a_x is causing the object to *slow down* (decreasing speed v). This agrees with the rule from Tactics Box 1.4: An object is slowing down if and only if v_x and a_x have opposite signs.

NOTE It is customary to abbreviate the acceleration units (m/s)/s as m/s^2. For example, the bicyclists in Example 2.8 had an acceleration of 2 m/s^2. We will use this notation, but keep in mind the *meaning* of the notation as "(meters per second) per second."

EXAMPLE 2.9 | **Running the court**

A basketball player starts at the left end of the court and moves with the velocity shown in **FIGURE 2.20**. Draw a motion diagram and an acceleration-versus-time graph for the basketball player.

FIGURE 2.20 Velocity-versus-time graph for the basketball player of Example 2.9.

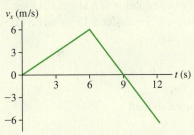

VISUALIZE The velocity is positive (motion to the right) and increasing for the first 6 s, so the velocity arrows in the motion diagram are to the right and getting longer. From $t = 6$ s to 9 s the motion is still to the right (v_x is still positive), but the arrows are getting shorter because v_x is decreasing. There's a turning point at $t = 9$ s, when $v_x = 0$, and after that the motion is to the left (v_x is negative) and getting faster. The motion diagram of **FIGURE 2.21a** shows the velocity and the acceleration vectors.

SOLVE Acceleration is the slope of the velocity graph. For the first 6 s, the slope has the constant value

$$a_x = \frac{\Delta v_x}{\Delta t} = \frac{6.0 \text{ m/s}}{6.0 \text{ s}} = 1.0 \text{ m/s}^2$$

The velocity then decreases by 12 m/s during the 6 s interval from $t = 6$ s to $t = 12$ s, so

$$a_x = \frac{\Delta v_x}{\Delta t} = \frac{-12 \text{ m/s}}{6.0 \text{ s}} = -2.0 \text{ m/s}^2$$

The acceleration graph for these 12 s is shown in **FIGURE 2.21b**. Notice that there is no change in the acceleration at $t = 9$ s, the turning point.

FIGURE 2.21 Motion diagram and acceleration graph for Example 2.9.

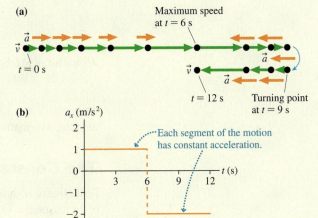

ASSESS The *sign* of a_x does *not* tell us whether the object is speeding up or slowing down. The basketball player is slowing down from $t = 6$ s to $t = 9$ s, then speeding up from $t = 9$ s to $t = 12$ s. Nonetheless, his acceleration is negative during this entire interval because his acceleration vector, as seen in the motion diagram, always points to the left.

The Kinematic Equations of Constant Acceleration

Consider an object whose acceleration a_s remains constant during the time interval $\Delta t = t_f - t_i$. At the beginning of this interval, at time t_i, the object has initial velocity v_{is} and initial position s_i. Note that t_i is often zero, but it does not have to be. We would like to predict the object's final position s_f and final velocity v_{fs} at time t_f.

The object's velocity is changing because the object is accelerating. **FIGURE 2.22a** shows the acceleration-versus-time graph, a horizontal line between t_i and t_f. It is not hard to find the object's velocity v_{fs} at a later time t_f. By definition,

$$a_s = \frac{\Delta v_s}{\Delta t} = \frac{v_{fs} - v_{is}}{\Delta t} \qquad (2.18)$$

which is easily rearranged to give

$$v_{fs} = v_{is} + a_s \Delta t \qquad (2.19)$$

The velocity-versus-time graph, shown in **FIGURE 2.22b**, is a straight line that starts at v_{is} and has slope a_s.

FIGURE 2.22 Acceleration and velocity graphs for constant acceleration.

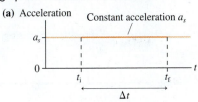

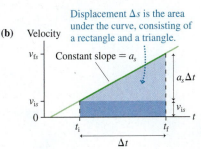

As you learned in the last section, the object's final position is

$$s_f = s_i + \text{area under the velocity curve } v_s \text{ between } t_i \text{ and } t_f \quad (2.20)$$

The shaded area in Figure 2.22b can be subdivided into a rectangle of area $v_{is}\,\Delta t$ and a triangle of area $\frac{1}{2}(a_s\Delta t)(\Delta t) = \frac{1}{2}a_s(\Delta t)^2$. Adding these gives

$$s_f = s_i + v_{is}\,\Delta t + \tfrac{1}{2}a_s(\Delta t)^2 \quad (2.21)$$

where $\Delta t = t_f - t_i$ is the elapsed time. The quadratic dependence on Δt causes the position-versus-time graph for constant-acceleration motion to have a parabolic shape, as shown in Model 2.2.

Equations 2.19 and 2.21 are two of the basic kinematic equations for motion with *constant* acceleration. They allow us to predict an object's position and velocity at a future instant of time. We need one more equation to complete our set, a direct relation between position and velocity. First use Equation 2.19 to write $\Delta t = (v_{fs} - v_{is})/a_s$. Substitute this into Equation 2.21, giving

$$s_f = s_i + v_{is}\left(\frac{v_{fs} - v_{is}}{a_s}\right) + \tfrac{1}{2}a_s\left(\frac{v_{fs} - v_{is}}{a_s}\right)^2 \quad (2.22)$$

With a bit of algebra, this is rearranged to read

$$v_{fs}^2 = v_{is}^2 + 2a_s\,\Delta s \quad (2.23)$$

where $\Delta s = s_f - s_i$ is the *displacement* (not the distance!). Equation 2.23 is the last of the three kinematic equations for motion with constant acceleration.

The Constant-Acceleration Model

Few objects with changing velocity have a perfectly constant acceleration, but it is often reasonable to model their acceleration as being constant. We do so by utilizing the **constant-acceleration model.** Once again, a model is a set of words, pictures, graphs, and equations that allows us to explain and predict an object's motion.

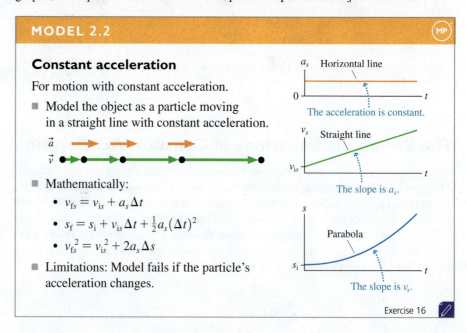

MODEL 2.2

Constant acceleration

For motion with constant acceleration.

■ Model the object as a particle moving in a straight line with constant acceleration.

$\vec{a}$

$\vec{v}$

■ Mathematically:
  • $v_{fs} = v_{is} + a_s\,\Delta t$
  • $s_f = s_i + v_{is}\,\Delta t + \tfrac{1}{2}a_s(\Delta t)^2$
  • $v_{fs}^2 = v_{is}^2 + 2a_s\,\Delta s$

■ Limitations: Model fails if the particle's acceleration changes.

a_s Horizontal line
0
The acceleration is constant.

v_s Straight line
v_{is}
The slope is a_s.

s
Parabola
s_i
The slope is v_s.

Exercise 16

In this text, we'll usually model runners, cars, planes, and rockets as having constant acceleration. Their actual acceleration is often more complicated (for example, a car's acceleration gradually decreases rather than remaining constant until full speed is reached), but the mathematical complexity of dealing with realistic accelerations would detract from the physics we're trying to learn.

The constant-acceleration model is the basis for a problem-solving strategy.

PROBLEM-SOLVING STRATEGY 2.1

Kinematics with constant acceleration

MODEL Model the object as having constant acceleration.

VISUALIZE Use different representations of the information in the problem.

- Draw a *pictorial representation*. This helps you assess the information you are given and starts the process of translating the problem into symbols.
- Use a *graphical representation* if it is appropriate for the problem.
- Go back and forth between these two representations as needed.

SOLVE The mathematical representation is based on the three kinematic equations:

$$v_{fs} = v_{is} + a_s \Delta t$$
$$s_f = s_i + v_{is} \Delta t + \tfrac{1}{2} a_s (\Delta t)^2$$
$$v_{fs}^2 = v_{is}^2 + 2a_s \Delta s$$

- Use x or y, as appropriate to the problem, rather than the generic s.
- Replace i and f with numerical subscripts defined in the pictorial representation.

ASSESS Check that your result has the correct units and significant figures, is reasonable, and answers the question.

NOTE You are strongly encouraged to solve problems on the Dynamics Worksheets found at the back of the Student Workbook. These worksheets will help you use the Problem-Solving Strategy and develop good problem-solving skills.

EXAMPLE 2.10 | The motion of a rocket sled

A rocket sled's engines fire for 5.0 s, boosting the sled to a speed of 250 m/s. The sled then deploys a braking parachute, slowing by 3.0 m/s per second until it stops. What is the total distance traveled?

MODEL We're not given the sled's initial acceleration, while the rockets are firing, but rocket sleds are aerodynamically shaped to minimize air resistance and so it seems reasonable to model the sled as a particle undergoing constant acceleration.

VISUALIZE FIGURE 2.23 shows the pictorial representation. We've made the reasonable assumptions that the sled starts from rest and that the braking parachute is deployed just as the rocket burn ends. There are three points of interest in this problem: the start, the change from propulsion to braking, and the stop. Each of these points has been assigned a position, velocity, and time. Notice that we've replaced the generic subscripts i and f of the kinematic equations with the numerical subscripts 0, 1, and 2. Accelerations are associated not with specific points in the motion but with the

intervals between the points, so acceleration a_{0x} is the acceleration between points 0 and 1 while acceleration a_{1x} is the acceleration between points 1 and 2. The acceleration vector $\vec{a}_1$ points to the left, so a_{1x} is negative. The sled stops at the end point, so $v_{2x} = 0$ m/s.

SOLVE We know how long the rocket burn lasts and the velocity at the end of the burn. Because we're modeling the sled as having uniform acceleration, we can use the first kinematic equation of Problem-Solving Strategy 2.1 to write

$$v_{1x} = v_{0x} + a_{0x}(t_1 - t_0) = a_{0x}t_1$$

We started with the complete equation, then simplified by noting which terms were zero. Solving for the boost-phase acceleration, we have

$$a_{0x} = \frac{v_{1x}}{t_1} = \frac{250 \text{ m/s}}{5.0 \text{ s}} = 50 \text{ m/s}^2$$

Notice that we worked algebraically until the last step—a hallmark of good problem-solving technique that minimizes the chances of

---

FIGURE 2.23 Pictorial representation of the rocket sled.

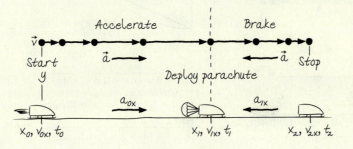

Continued

calculation errors. Also, in accord with the significant figure rules of Chapter 1, 50 m/s² is considered to have two significant figures.

Now we have enough information to find out how far the sled travels while the rockets are firing. The second kinematic equation of Problem-Solving Strategy 2.1 is

$$x_1 = x_0 + v_{0x}(t_1 - t_0) + \tfrac{1}{2}a_{0x}(t_1 - t_0)^2 = \tfrac{1}{2}a_{0x}t_1^2$$
$$= \tfrac{1}{2}(50 \text{ m/s}^2)(5.0 \text{ s})^2 = 625 \text{ m}$$

The braking phase is a little different because we don't know how long it lasts. But we do know both the initial and final velocities, so we can use the third kinematic equation of Problem-Solving Strategy 2.1:

$$v_{2x}^2 = v_{1x}^2 + 2a_{1x}\Delta x = v_{1x}^2 + 2a_{1x}(x_2 - x_1)$$

Notice that Δx is *not* x_2; it's the displacement $(x_2 - x_1)$ during the braking phase. We can now solve for x_2:

$$x_2 = x_1 + \frac{v_{2x}^2 - v_{1x}^2}{2a_{1x}}$$
$$= 625 \text{ m} + \frac{0 - (250 \text{ m/s})^2}{2(-3.0 \text{ m/s}^2)} = 11{,}000 \text{ m}$$

We kept three significant figures for x_1 at an intermediate stage of the calculation but rounded to two significant figures at the end.

ASSESS The total distance is 11 km ≈ 7 mi. That's large but believable. Using the approximate conversion factor 1 m/s ≈ 2 mph from Table 1.5, we see that the top speed is ≈ 500 mph. It will take a long distance for the sled to gradually stop from such a high speed.

EXAMPLE 2.11 | A two-car race

Fred is driving his Volkswagen Beetle at a steady 20 m/s when he passes Betty sitting at rest in her Porsche. Betty instantly begins accelerating at 5.0 m/s². How far does Betty have to drive to overtake Fred?

MODEL Model the VW as a particle in uniform motion and the Porsche as a particle with constant acceleration.

VISUALIZE FIGURE 2.24 is the pictorial representation. Fred's motion diagram is one of uniform motion, while Betty's shows uniform acceleration. Fred is ahead in frames 1, 2, and 3, but Betty catches up with him in frame 4. The coordinate system shows the cars with the same position at the start and at the end—but with the important difference that Betty's Porsche has an acceleration while Fred's VW does not.

SOLVE This problem is similar to Example 2.2, in which Bob and Susan met for lunch. As we did there, we want to find Betty's position $(x_1)_B$ at the instant t_1 when $(x_1)_B = (x_1)_F$. We know, from the models of uniform motion and uniform acceleration, that Fred's position graph is a straight line but Betty's is a parabola. The position graphs in Figure 2.24 show that we're solving for the intersection point of the line and the parabola.

Fred's and Betty's positions at t_1 are

$$(x_1)_F = (x_0)_F + (v_{0x})_F(t_1 - t_0) = (v_{0x})_F t_1$$
$$(x_1)_B = (x_0)_B + (v_{0x})_B(t_1 - t_0) + \tfrac{1}{2}(a_{0x})_B(t_1 - t_0)^2 = \tfrac{1}{2}(a_{0x})_B t_1^2$$

By equating these,

$$(v_{0x})_F t_1 = \tfrac{1}{2}(a_{0x})_B t_1^2$$

we can solve for the time when Betty passes Fred:

$$t_1\left[\tfrac{1}{2}(a_{0x})_B t_1 - (v_{0x})_F\right] = 0$$
$$t_1 = \begin{cases} 0 \text{ s} \\ 2(v_{0x})_F/(a_{0x})_B = 8.0 \text{ s} \end{cases}$$

Interestingly, there are two solutions. That's not surprising, when you think about it, because the line and the parabola of the position graphs have *two* intersection points: when Fred first passes Betty, and 8.0 s later when Betty passes Fred. We're interested in only the second of these points. We can now use either of the distance equations to find $(x_1)_B = (x_1)_F = 160 \text{ m}$. Betty has to drive 160 m to overtake Fred.

ASSESS 160 m ≈ 160 yards. Because Betty starts from rest while Fred is moving at 20 m/s ≈ 40 mph, needing 160 yards to catch him seems reasonable.

NOTE The purpose of the Assess step is not to prove that an answer must be right but to rule out answers that, with a little thought, are clearly wrong.

FIGURE 2.24 Pictorial representation for Example 2.11.

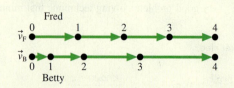

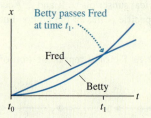

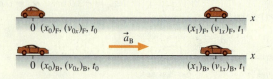

Known
$(x_0)_F = 0 \text{ m}$ $(x_0)_B = 0 \text{ m}$ $t_0 = 0 \text{ s}$
$(v_{0x})_F = 20 \text{ m/s}$ $(v_{0x})_B = 0 \text{ m/s}$
$(a_{0x})_B = 5.0 \text{ m/s}^2$ $(v_{1x})_F = 20 \text{ m/s}$

Find
$(x_1)_B$ at t_1 when $(x_1)_B = (x_1)_F$

STOP TO THINK 2.4 Which velocity-versus-time graph or graphs go with the acceleration-versus-time graph on the left? The particle is initially moving to the right.

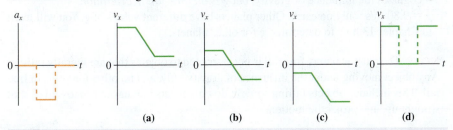

(a)   (b)   (c)   (d)

2.5 Free Fall

The motion of an object moving under the influence of gravity only, and no other forces, is called **free fall.** Strictly speaking, free fall occurs only in a vacuum, where there is no air resistance. Fortunately, the effect of air resistance is small for "heavy objects," so we'll make only a very slight error in treating these objects *as if* they were in free fall. For very light objects, such as a feather, or for objects that fall through very large distances and gain very high speeds, the effect of air resistance is *not* negligible. Motion with air resistance is a problem we will study in Chapter 6. Until then, we will restrict our attention to "heavy objects" and will make the reasonable assumption that falling objects are in free fall.

Galileo, in the 17th century, was the first to make detailed measurements of falling objects. The story of Galileo dropping different weights from the leaning bell tower at the cathedral in Pisa is well known, although historians cannot confirm its truth. Based on his measurements, wherever they took place, Galileo developed a *model* for motion in the absence of air resistance:

- Two objects dropped from the same height will, if air resistance can be neglected, hit the ground at the same time and with the same speed.
- Consequently, **any two objects in free fall, regardless of their mass, have the same acceleration $\vec{a}_{\text{free fall}}$.**

FIGURE 2.25a shows the motion diagram of an object that was released from rest and falls freely. **FIGURE 2.25b** shows the object's velocity graph. The motion diagram and graph are identical for a falling pebble and a falling boulder. The fact that the velocity graph is a straight line tells us the motion is one of constant acceleration, and $a_{\text{free fall}}$ is found from the slope of the graph. Careful measurements show that the value of $a_{\text{free fall}}$ varies ever so slightly at different places on the earth, due to the slightly nonspherical shape of the earth and to the fact that the earth is rotating. A global average, at sea level, is

$$\vec{a}_{\text{free fall}} = (9.80 \text{ m/s}^2, \text{ vertically downward}) \tag{2.24}$$

Vertically downward means along a line toward the center of the earth.

The length, or magnitude, of $\vec{a}_{\text{free fall}}$ is known as the **free-fall acceleration,** and it has the special symbol g:

$$g = 9.80 \text{ m/s}^2 \text{ (free-fall acceleration)}$$

Several points about free fall are worthy of note:

- g, by definition, is *always* positive. **There will never be a problem that will use a negative value for g.** But, you say, objects fall when you release them rather than rise, so how can g be positive?
- g is *not* the acceleration $a_{\text{free fall}}$, but simply its magnitude. Because we've chosen the y-axis to point vertically upward, the downward acceleration vector $\vec{a}_{\text{free fall}}$ has the one-dimensional acceleration

$$a_y = a_{\text{free fall}} = -g \tag{2.25}$$

It is a_y that is negative, not g.

In a vacuum, the apple and feather fall at the same rate and hit the ground at the same time.

FIGURE 2.25 Motion of an object in free fall.

(a)

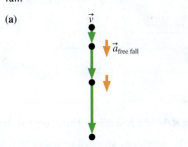

(b)

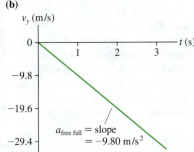

- We can model free fall as motion with constant acceleration, with $a_y = -g$.
- g is not called "gravity." Gravity is a force, not an acceleration. The symbol g recognizes the influence of gravity, but g is *the free-fall acceleration.*
- $g = 9.80$ m/s^2 only on earth. Other planets have different values of g. You will learn in Chapter 13 how to determine g for other planets.

NOTE Despite the name, free fall is not restricted to objects that are literally falling. Any object moving under the influence of gravity only, and no other forces, is in free fall. This includes objects falling straight down, objects that have been tossed or shot straight up, and projectile motion.

EXAMPLE 2.12 | **A falling rock**

A rock is dropped from the top of a 20-m-tall building. What is its impact velocity?

MODEL A rock is fairly heavy, and air resistance is probably not a serious concern in a fall of only 20 m. It seems reasonable to model the rock's motion as free fall: constant acceleration with $a_y = a_{\text{free fall}} = -g$.

VISUALIZE FIGURE 2.26 shows the pictorial representation. We have placed the origin at the ground, which makes $y_0 = 20$ m. Although the rock falls 20 m, it is important to notice that the *displacement* is $\Delta y = y_1 - y_0 = -20$ m.

FIGURE 2.26 Pictorial representation of a falling rock.

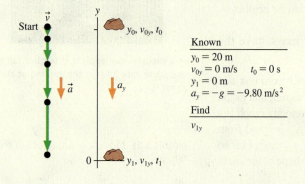

Known
$y_0 = 20$ m
$v_{0y} = 0$ m/s $t_0 = 0$ s
$y_1 = 0$ m
$a_y = -g = -9.80$ m/s^2

Find
v_{1y}

SOLVE In this problem we know the displacement but not the time, which suggests that we use the third kinematic equation from Problem-Solving Strategy 2.1:

$$v_{1y}{}^2 = v_{0y}{}^2 + 2a_y\,\Delta y = -2g\,\Delta y$$

We started by writing the general equation, then noted that $v_{0y} = 0$ m/s and substituted $a_y = -g$. Solving for v_{1y}:

$$v_{1y} = \sqrt{-2g\Delta y} = \sqrt{-2(9.8 \text{ m/s}^2)(-20 \text{ m})} = \pm 20 \text{ m/s}$$

A common error would be to say, "The rock fell 20 m, so $\Delta y = 20$ m." That would have you trying to take the square root of a negative number. As noted above, Δy is a *displacement,* not a distance, and in this case $\Delta y = -20$ m.

The $\pm$ sign indicates that there are two mathematical solutions; therefore, we have to use physical reasoning to choose between them. The rock does hit with a *speed* of 20 m/s, but the question asks for the impact *velocity.* The velocity vector points down, so the sign of v_{1y} is negative. Thus the impact velocity is -20 m/s.

ASSESS Is the answer reasonable? Well, 20 m is about 60 feet, or about the height of a five- or six-story building. Using 1 m/s $\approx$ 2 mph, we see that 20 m/s $\approx$ 40 mph. That seems quite reasonable for the speed of an object after falling five or six stories. If we had misplaced a decimal point, though, and found 2.0 m/s, we would be suspicious that this was much too small after converting it to $\approx$ 4 mph.

EXAMPLE 2.13 | **Finding the height of a leap**

The springbok, an antelope found in Africa, gets its name from its remarkable jumping ability. When startled, a springbok will leap straight up into the air—a maneuver called a "pronk." A springbok goes into a crouch to perform a pronk. It then extends its legs forcefully, accelerating at 35 m/s^2 for 0.70 m as

its legs straighten. Legs fully extended, it leaves the ground and rises into the air. How high does it go?

MODEL The springbok is changing shape as it leaps, so can we reasonably model it as a particle? We can if we focus on the *body* of the springbok, treating the expanding legs like external springs. Initially, the body of the springbok is driven upward by its legs. We'll model this as a particle—the body—undergoing constant acceleration. Once the springbok's feet leave the ground, we'll model the motion of the springbok's body as a particle in free fall.

VISUALIZE FIGURE 2.27 shows the pictorial representation. This is a problem with a beginning point, an end point, and a point in between where the nature of the motion changes. We've identified these points with subscripts 0, 1, and 2. The motion from 0 to 1 is a rapid upward acceleration until the springbok's feet leave the ground at 1. Even though the springbok is moving upward from 1 to 2, this is free-fall motion because the springbok is now moving under the influence of gravity only.

FIGURE 2.27 Pictorial representation of a startled springbok.

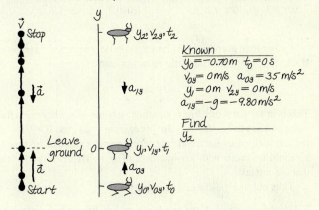

How do we put "How high?" into symbols? The clue is that the very top point of the trajectory is a *turning point,* and we've seen that the instantaneous velocity at a turning point is $v_{2y} = 0$.

This was not explicitly stated but is part of our interpretation of the problem.

SOLVE For the first part of the motion, pushing off, we know a displacement but not a time interval. We can use

$$v_{1y}^2 = v_{0y}^2 + 2a_{0y}\,\Delta y = 2(35\ \text{m/s}^2)(0.70\ \text{m}) = 49\ \text{m}^2/\text{s}^2$$

$$v_{1y} = \sqrt{49\ \text{m}^2/\text{s}^2} = 7.0\ \text{m/s}$$

The springbok leaves the ground with a velocity of 7.0 m/s. This is the starting point for the problem of a projectile launched straight up from the ground. One possible solution is to use the velocity equation to find how long it takes to reach maximum height, then the position equation to calculate the maximum height. But that takes two separate calculations. It is easier to make another use of the velocity-displacement equation:

$$v_{2y}^2 = 0 = v_{1y}^2 + 2a_{1y}\,\Delta y = v_{1y}^2 - 2g(y_2 - y_1)$$

where now the acceleration is $a_{1y} = -g$. Using $y_1 = 0$, we can solve for y_2, the height of the leap:

$$y_2 = \frac{v_{1y}^2}{2g} = \frac{(7.0\ \text{m/s})^2}{2(9.80\ \text{m/s}^2)} = 2.5\ \text{m}$$

ASSESS 2.5 m is a bit over 8 feet, a remarkable vertical jump. But these animals are known for their jumping ability, so the answer seems reasonable. Note that it is especially important in a multipart problem like this to use numerical subscripts to distinguish different points in the motion.

2.6 Motion on an Inclined Plane

FIGURE 2.28a shows a problem closely related to free fall: that of motion down a straight, but frictionless, inclined plane, such as a skier going down a slope on frictionless snow. What is the object's acceleration? Although we're not yet prepared to give a rigorous derivation, we can deduce the acceleration with a plausibility argument.

FIGURE 2.28b shows the free-fall acceleration $\vec{a}_{\text{free fall}}$ the object would have if the incline suddenly vanished. The free-fall acceleration points straight down. This vector can be broken into two pieces: a vector $\vec{a}_\parallel$ that is parallel to the incline and a vector $\vec{a}_\perp$ that is perpendicular to the incline. The surface of the incline somehow "blocks" $\vec{a}_\perp$, through a process we will examine in Chapter 6, but $\vec{a}_\parallel$ is unhindered. It is this piece of $\vec{a}_{\text{free fall}}$, parallel to the incline, that accelerates the object.

By definition, the length, or magnitude, of $\vec{a}_{\text{free fall}}$ is g. Vector $\vec{a}_\parallel$ is opposite angle θ (Greek *theta*), so the length, or magnitude, of $\vec{a}_\parallel$ must be $g \sin\theta$. Consequently, the one-dimensional acceleration along the incline is

$$a_s = \pm g \sin\theta \tag{2.26}$$

The correct sign depends on the direction in which the ramp is tilted. Examples will illustrate.

Equation 2.26 makes sense. Suppose the plane is perfectly horizontal. If you place an object on a horizontal surface, you expect it to stay at rest with no acceleration. Equation 2.26 gives $a_s = 0$ when $\theta = 0°$, in agreement with our expectations. Now suppose you tilt the plane until it becomes vertical, at $\theta = 90°$. Without friction, an object would simply fall, in free fall, parallel to the vertical surface. Equation 2.26 gives $a_s = -g = a_{\text{free fall}}$ when $\theta = 90°$, again in agreement with our expectations. Equation 2.26 gives the correct result in these *limiting cases*.

FIGURE 2.28 Acceleration on an inclined plane.

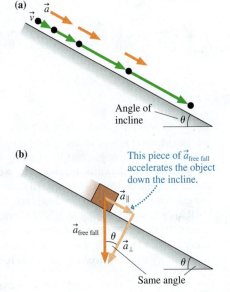

(a)

Angle of incline θ

(b)

This piece of $\vec{a}_{\text{free fall}}$ accelerates the object down the incline.

$\vec{a}_\parallel$

$\vec{a}_{\text{free fall}}$ θ $\vec{a}_\perp$

θ

Same angle

EXAMPLE 2.14 | Measuring acceleration

In the laboratory, a 2.00-m-long track has been inclined as shown in **FIGURE 2.29**. Your task is to measure the acceleration of a cart on the ramp and to compare your result with what you might have expected. You have available five "photogates" that measure the cart's speed as it passes through. You place a gate every 30 cm from a line you mark near the top of the track as the starting line. One run generates the data shown in the table. The first entry isn't a photogate, but it is a valid data point because you know the cart's speed is zero at the point where you release it.

FIGURE 2.29 The experimental setup.

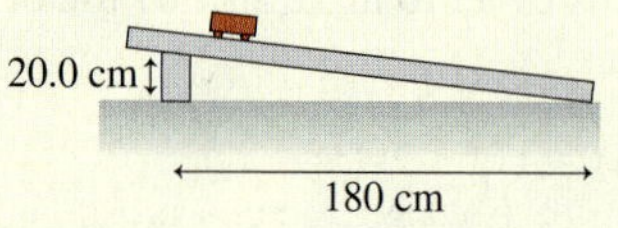

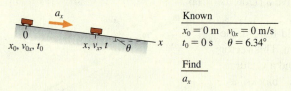

Distance (cm)	Speed (m/s)
0	0.00
30	0.75
60	1.15
90	1.38
120	1.56
150	1.76

NOTE Physics is an experimental science. Our knowledge of the universe is grounded in observations and measurements. Consequently, some examples and homework problems throughout this book will be based on data. Data-based homework problems require the use of a spreadsheet, graphing software, or a graphing calculator in which you can "fit" data with a straight line.

MODEL Model the cart as a particle.

VISUALIZE FIGURE 2.30 shows the pictorial representation. The track and axis are tilted at angle $\theta = \tan^{-1}(20.0\,\text{cm}/180\,\text{cm}) = 6.34°$. This is motion on an inclined plane, so you might expect the cart's acceleration to be $a_x = g\sin\theta = 1.08\,\text{m/s}^2$.

FIGURE 2.30 The pictorial representation of the cart on the track.

	Known
	$x_0 = 0\,\text{m}$ $v_{0x} = 0\,\text{m/s}$
	$t_0 = 0\,\text{s}$ $\theta = 6.34°$
	Find
	a_x

SOLVE In analyzing data, we want to use *all* the data. Further, we almost always want to use graphs when we have a series of measurements. We might start by graphing speed versus distance traveled. This is shown in **FIGURE 2.31a**, where we've converted distances to meters. As expected, speed increases with distance, but the graph isn't linear and that makes it hard to analyze.

Rather than proceeding by trial and error, let's be guided by theory. *If* the cart has constant acceleration—which we don't yet know and need to confirm—the third kinematic equation tells us that velocity and displacement should be related by

$$v_x^2 = v_{0x}^2 + 2a_x\Delta x = 2a_x x$$

The last step was based on starting from rest ($v_{0x} = 0$) at the origin ($\Delta x = x - x_0 = x$).

FIGURE 2.31 Graphs of velocity and of velocity squared.

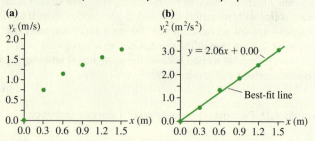

Rather than graphing v_x versus x, suppose we graph v_x^2 versus x. If we let $y = v_x^2$, the kinematic equation reads

$$y = 2a_x x$$

This is in the form of a linear equation: $y = mx + b$, where m is the slope and b is the y-intercept. In this case, $m = 2a_x$ and $b = 0$. So if the cart really does have constant acceleration, a graph of v_x^2 versus x should be linear with a y-intercept of zero. This is a prediction that we can test.

Thus our analysis has three steps:

1. Graph v_x^2 versus x. If the graph is a straight line with a y-intercept of zero (or very close to zero), then we can conclude that the cart has constant acceleration on the ramp. If not, the acceleration is *not* constant and we cannot use the kinematic equations for constant acceleration.
2. If the graph has the correct shape, we can determine its slope m.
3. Because kinematics predicts $m = 2a_x$, the acceleration must be $a_x = m/2$.

FIGURE 2.31b is the graph of v_x^2 versus x. It does turn out to be a straight line with a y-intercept of zero, and this is the evidence we need that the cart has a constant acceleration on the ramp. To proceed, we want to determine the slope by finding the straight line that is the "best fit" to the data. This is a statistical technique, justified in a statistics class, but one that is implemented in spreadsheets and graphing calculators. The solid line in Figure 2.31b is the best-fit line for this data, and its equation is shown. We see that the slope is $m = 2.06\,\text{m/s}^2$. **Slopes have units,** and the units come not from the fitting procedure but by looking at the axes of the graph. Here the vertical axis is velocity squared, with units of $(\text{m/s})^2$, while the horizontal axis is position, measured in m. Thus the slope, rise over run, has units of m/s^2.

Finally, we can determine that the cart's acceleration was

$$a_x = \frac{m}{2} = 1.03\,\text{m/s}^2$$

This is about 5% less than the $1.08\,\text{m/s}^2$ we expected. Two possibilities come to mind. Perhaps the distances used to find the tilt angle weren't measured accurately. Or, more likely, the cart rolls with a small bit of friction. The predicted acceleration $a_x = g\sin\theta$ is for a *frictionless* inclined plane; any friction would decrease the acceleration.

ASSESS The acceleration is just slightly less than predicted for a frictionless incline, so the result is reasonable.

Thinking Graphically

A good way to solidify your intuitive understanding of motion is to consider the problem of a hard, smooth ball rolling on a smooth track. The track is made up of several straight segments connected together. Each segment may be either horizontal or inclined. Your task is to analyze the ball's motion graphically.

There are a small number of rules to follow:

1. Assume that the ball passes smoothly from one segment of the track to the next, with no abrupt change of speed and without ever leaving the track.
2. The graphs have no numbers, but they should show the correct *relationships*. For example, the position graph should be steeper in regions of higher speed.
3. The position s is the position measured *along* the track. Similarly, v_s and a_s are the velocity and acceleration parallel to the track.

EXAMPLE 2.15 | From track to graphs

Draw position, velocity, and acceleration graphs for the ball on the smooth track of **FIGURE 2.32**.

FIGURE 2.32 A ball rolling along a track.

$v_{0s} > 0$

VISUALIZE It is often easiest to begin with the velocity. There is no acceleration on the horizontal surface ($a_s = 0$ if $\theta = 0°$), so the velocity remains constant at v_{0s} until the ball reaches the slope. The slope is an inclined plane where the ball has constant acceleration. The velocity increases linearly with time during constant-acceleration motion. The ball returns to constant-velocity motion after reaching the bottom horizontal segment. The middle graph of **FIGURE 2.33** shows the velocity.

We can easily draw the acceleration graph. The acceleration is zero while the ball is on the horizontal segments and has a constant positive value on the slope. These accelerations are consistent with the slope of the velocity graph: zero slope, then positive slope, then a return to zero. The acceleration cannot

really change instantly from zero to a nonzero value, but the change can be so quick that we do not see it on the time scale of the graph. That is what the vertical dotted lines imply.

Finally, we need to find the position-versus-time graph. The position increases linearly with time during the first segment at constant velocity. It also does so during the third segment of motion, but with a steeper slope to indicate a faster velocity. In between, while the acceleration is nonzero but constant, the position graph has a *parabolic* shape. Notice that the parabolic section blends *smoothly* into the straight lines on either side. An abrupt change of slope (a "kink") would indicate an abrupt change in velocity and would violate rule 1.

FIGURE 2.33 Motion graphs for the ball in Example 2.15.

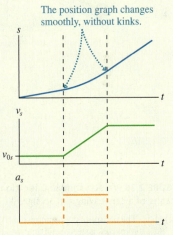

The position graph changes smoothly, without kinks.

EXAMPLE 2.16 | From graphs to track

FIGURE 2.34 shows a set of motion graphs for a ball moving on a track. Draw a picture of the track and describe the ball's initial condition. Each segment of the track is *straight,* but the segments may be tilted.

VISUALIZE The ball starts with initial velocity $v_{0s} > 0$ and maintains this velocity for awhile; there's no acceleration. Thus the ball must start out rolling to the right on a horizontal track. At the end of the motion, the ball is again rolling on a horizontal track (no acceleration, constant velocity), but it's rolling to the *left* because v_s is negative. Further, the final speed ($|v_s|$) is greater than the initial speed. The middle section of the graph shows us what happens. The ball starts slowing with constant acceleration (rolling uphill), reaches a turning point (s is maximum, $v_s = 0$), then speeds up in the opposite direction (rolling downhill). This is still a negative acceleration because the ball is speeding up in the negative

FIGURE 2.34 Motion graphs of a ball rolling on a track of unknown shape.

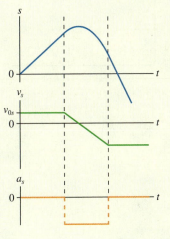

Continued

s-direction. It must roll farther downhill than it had rolled uphill before reaching a horizontal section of track. **FIGURE 2.35** shows the track and the initial conditions that are responsible for the graphs of Figure 2.34.

FIGURE 2.35 Track responsible for the motion graphs of Figure 2.34.

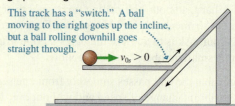

This track has a "switch." A ball moving to the right goes up the incline, but a ball rolling downhill goes straight through.

$v_{0s} > 0$

STOP TO THINK 2.5 The ball rolls up the ramp, then back down. Which is the correct acceleration graph?

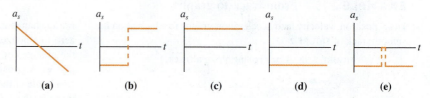

2.7 ADVANCED TOPIC Instantaneous Acceleration

FIGURE 2.36 Velocity and acceleration graphs of a car leaving a stop sign.

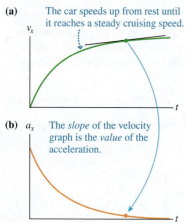

(a) The car speeds up from rest until it reaches a steady cruising speed.

v_x

(b) a_x The *slope* of the velocity graph is the *value* of the acceleration.

Although the constant-acceleration model is very useful, real moving objects only rarely have constant acceleration. For example, **FIGURE 2.36a** is a realistic velocity-versus-time graph for a car leaving a stop sign. The graph is not a straight line, so this is *not* motion with constant acceleration.

We can define an instantaneous acceleration much as we defined the instantaneous velocity. The instantaneous velocity at time t is the slope of the position-versus-time graph at that time or, mathematically, the derivative of the position with respect to time. By analogy: **The instantaneous acceleration a_s is the slope of the line that is tangent to the velocity-versus-time curve at time t.** Mathematically, this is

$$a_s = \frac{dv_s}{dt} = \text{slope of the velocity-versus-time graph at time } t \qquad (2.27)$$

FIGURE 2.36b applies this idea by showing the car's acceleration graph. At each instant of time, the *value* of the car's acceleration is the *slope* of its velocity graph. The initially steep slope indicates a large initial acceleration. The acceleration decreases to zero as the car reaches cruising speed.

The reverse problem—to find the velocity v_s if we know the acceleration a_s at all instants of time—is also important. Again, with analogy to velocity and position, we have

$$v_{fs} = v_{is} + \int_{t_i}^{t_f} a_s \, dt \qquad (2.28)$$

The graphical interpretation of Equation 2.28 is

$$v_{fs} = v_{is} + \text{area under the acceleration curve } a_s \text{ between } t_i \text{ and } t_f \qquad (2.29)$$

EXAMPLE 2.17 | Finding velocity from acceleration

FIGURE 2.37 shows the acceleration graph for a particle with an initial velocity of 10 m/s. What is the particle's velocity at $t = 8$ s?

FIGURE 2.37 Acceleration graph for Example 2.17.

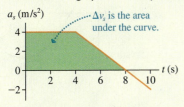

MODEL We're told this is the motion of a particle.

VISUALIZE Figure 2.37 is a graphical representation of the motion.

SOLVE The change in velocity is found as the area under the acceleration curve:

$$v_{fs} = v_{is} + \text{area under the acceleration curve } a_s \text{ between } t_i \text{ and } t_f$$

The area under the curve between $t_i = 0$ s and $t_f = 8$ s can be subdivided into a rectangle (0 s $\leq t \leq 4$ s) and a triangle (4 s $\leq t \leq 8$ s). These areas are easily computed. Thus

$$
\begin{aligned}
v_s(\text{at } t = 8 \text{ s}) &= 10 \text{ m/s} + (4 \text{ (m/s)/s})(4 \text{ s}) \\
&\quad + \tfrac{1}{2}(4 \text{ (m/s)/s})(4 \text{ s}) \\
&= 34 \text{ m/s}
\end{aligned}
$$

EXAMPLE 2.18 | A realistic car acceleration

Starting from rest, a car takes T seconds to reach its cruising speed v_{max}. A plausible expression for the velocity as a function of time is

$$
v_x(t) = \begin{cases} v_{max}\left(\dfrac{2t}{T} - \dfrac{t^2}{T^2}\right) & t \leq T \\[2mm] v_{max} & t \geq T \end{cases}
$$

a. Demonstrate that this is a plausible function by drawing velocity and acceleration graphs.

b. Find an expression for the distance traveled at time T in terms of T and the maximum acceleration a_{max}.

c. What are the maximum acceleration and the distance traveled for a car that reaches a cruising speed of 15 m/s in 8.0 s?

MODEL Model the car as a particle.

VISUALIZE **FIGURE 2.38a** shows the velocity graph. It's an inverted parabola that reaches v_{max} at time T and then holds that value. From the slope, we see that the acceleration should start at a maximum value a_{max}, steadily decrease until T, and be zero for $t > T$.

SOLVE

a. We can find an expression for a_x by taking the derivative of v_x. Starting with $t \leq T$, and using Equation 2.6 for the derivatives of polynomials, we find

$$a_x = \frac{dv_x}{dt} = v_{max}\left(\frac{2}{T} - \frac{2t}{T^2}\right) = \frac{2v_{max}}{T}\left(1 - \frac{t}{T}\right) = a_{max}\left(1 - \frac{t}{T}\right)$$

where $a_{max} = 2v_{max}/T$. For $t \geq T$, $a_x = 0$. Altogether,

$$
a_x(t) = \begin{cases} a_{max}\left(1 - \dfrac{t}{T}\right) & t \leq T \\[2mm] 0 & t \geq T \end{cases}
$$

This expression for the acceleration is graphed in **FIGURE 2.38b**. The acceleration decreases linearly from a_{max} to 0 as the car accelerates from rest to its cruising speed.

b. To find the position as a function of time, we need to integrate the velocity (Equation 2.11) using Equation 2.13 for the integrals of polynomials. At time T, when cruising speed is reached,

FIGURE 2.38 Velocity and acceleration graphs for Example 2.18.

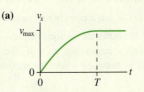

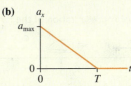

$$x_T = x_0 + \int_0^T v_x\, dt = 0 + \frac{2v_{max}}{T}\int_0^T t\, dt - \frac{v_{max}}{T^2}\int_0^T t^2\, dt$$

$$= \frac{2v_{max}}{T}\frac{t^2}{2}\bigg|_0^T - \frac{v_{max}}{T^2}\frac{t^3}{3}\bigg|_0^T$$

$$= v_{max}T - \tfrac{1}{3}v_{max}T = \tfrac{2}{3}v_{max}T$$

Recalling that $a_{max} = 2v_{max}/T$, we can write the distance traveled as

$$x_T = \tfrac{2}{3}v_{max}T = \tfrac{1}{3}\left(\frac{2v_{max}}{T}\right)T^2 = \tfrac{1}{3}a_{max}T^2$$

If the acceleration stayed constant, the distance would be $\tfrac{1}{2}aT^2$. We have found a similar expression but, because the acceleration is steadily decreasing, a smaller fraction in front.

c. With $v_{max} = 15$ m/s and $T = 8.0$ s, realistic values for city driving, we find

$$a_{max} = \frac{2v_{max}}{T} = \frac{2(15 \text{ m/s})}{8.0 \text{ s}} = 3.75 \text{ m/s}^2$$

$$x_T = \tfrac{1}{3}a_{max}T^2 = \tfrac{1}{3}(3.75 \text{ m/s}^2)(8.0 \text{ s})^2 = 80 \text{ m}$$

ASSESS 80 m in 8.0 s to reach a cruising speed of 15 m/s $\approx$ 30 mph is very reasonable. This gives us good reason to believe that a car's initial acceleration is $\approx \tfrac{1}{3}g$.

STOP TO THINK 2.6   Rank in order, from most positive to least positive, the accelerations at points A to C.

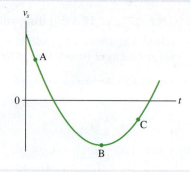

a. $a_A > a_B > a_C$
b. $a_C > a_A > a_B$
c. $a_C > a_B > a_A$
d. $a_B > a_A > a_C$

CHALLENGE EXAMPLE 2.19 | Rocketing along

A rocket sled accelerates along a long, horizontal rail. Starting from rest, two rockets burn for 10 s, providing a constant acceleration. One rocket then burns out, halving the acceleration, but the other burns for an additional 5 s to boost the sled's speed to 625 m/s. How far has the sled traveled when the second rocket burns out?

MODEL Model the rocket sled as a particle with constant acceleration.

VISUALIZE FIGURE 2.39 shows the pictorial representation. This is a two-part problem with a beginning, an end (the second rocket burns out), and a point in between where the motion changes (the first rocket burns out).

FIGURE 2.39 The pictorial representation of the rocket sled.

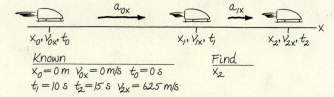

Known	Find
$x_0 = 0$ m $v_{0x} = 0$ m/s $t_0 = 0$ s	x_2
$t_1 = 10$ s $t_2 = 15$ s $v_{2x} = 625$ m/s	

SOLVE The difficulty with this problem is that there's not enough information to completely analyze either the first or the second part of the motion. A successful solution will require combining information about both parts of the motion, and that can be done only by working algebraically, not worrying about numbers until the end of the problem. A well-drawn pictorial representation and clearly defined symbols are essential.

The first part of the motion, with both rockets firing, has acceleration a_{0x}. The sled's position and velocity when the first rocket burns out are

$$x_1 = x_0 + v_{0x}\,\Delta t + \tfrac{1}{2}a_{0x}(\Delta t)^2 = \tfrac{1}{2}a_{0x}t_1^2$$
$$v_{1x} = v_{0x} + a_{0x}\,\Delta t = a_{0x}t_1$$

where we simplified as much as possible by knowing that the sled started from rest at the origin at $t_0 = 0$ s. We can't compute numerical values, but these are valid algebraic expressions that we can carry over to the second part of the motion.

From t_1 to t_2, the acceleration is a smaller a_{1x}. The velocity when the second rocket burns out is

$$v_{2x} = v_{1x} + a_{1x}\,\Delta t = a_{0x}t_1 + a_{1x}(t_2 - t_1)$$

where for v_{1x} we used the algebraic result from the first part of the motion. Now we have enough information to complete the solution. We know that the acceleration is halved when the first rocket burns out, so $a_{1x} = \tfrac{1}{2}a_{0x}$. Thus

$$v_{2x} = 625 \text{ m/s} = a_{0x}(10 \text{ s}) + \tfrac{1}{2}a_{0x}(5 \text{ s}) = (12.5 \text{ s})a_{0x}$$

Solving, we find $a_{0x} = 50$ m/s^2.

With the acceleration now known, we can calculate the position and velocity when the first rocket burns out:

$$x_1 = \tfrac{1}{2}a_{0x}t_1^2 = \tfrac{1}{2}(50 \text{ m/s}^2)(10 \text{ s})^2 = 2500 \text{ m}$$
$$v_{1x} = a_{0x}t_1 = (50 \text{ m/s}^2)(10 \text{ s}) = 500 \text{ m/s}$$

Finally, the position when the second rocket burns out is

$$x_2 = x_1 + v_{1x}\,\Delta t + \tfrac{1}{2}a_{1x}(\Delta t)^2$$
$$= 2500 \text{ m} + (500 \text{ m/s})(5 \text{ s}) + \tfrac{1}{2}(25 \text{ m/s}^2)(5 \text{ s})^2 = 5300 \text{ m}$$

The sled has traveled 5300 m when it reaches 625 m/s at the burnout of the second rocket.

ASSESS 5300 m is 5.3 km, or roughly 3 miles. That's a long way to travel in 15 s! But the sled reaches incredibly high speeds. At the final speed of 625 m/s, over 1200 mph, the sled would travel nearly 10 km in 15 s. So 5.3 km in 15 s for the accelerating sled seems reasonable.

SUMMARY

The goal of Chapter 2 has been to learn to solve problems about motion along a straight line.

GENERAL PRINCIPLES

Kinematics describes motion in terms of position, velocity, and acceleration.

General kinematic relationships are given **mathematically** by:

Instantaneous velocity $v_s = ds/dt =$ slope of position graph

Instantaneous acceleration $a_s = dv_s/dt =$ slope of velocity graph

Final position $s_f = s_i + \int_{t_i}^{t_f} v_s \, dt = s_i + \begin{cases} \text{area under the velocity} \\ \text{curve from } t_i \text{ to } t_f \end{cases}$

Final velocity $v_{fs} = v_{is} + \int_{t_i}^{t_f} a_s \, dt = v_{is} + \begin{cases} \text{area under the acceleration} \\ \text{curve from } t_i \text{ to } t_f \end{cases}$

Solving Kinematics Problems

MODEL Uniform motion or constant acceleration.

VISUALIZE Draw a pictorial representation.

SOLVE

- Uniform motion $s_f = s_i + v_s \Delta t$
- Constant acceleration $v_{fs} = v_{is} + a_s \Delta t$
  $$s_f = s_i + v_s \Delta t + \tfrac{1}{2} a_s (\Delta t)^2$$
  $$v_{fs}^2 = v_{is}^2 + 2 a_s \Delta s$$

ASSESS Is the result reasonable?

IMPORTANT CONCEPTS

Position, velocity, and acceleration are related graphically.

- The slope of the position-versus-time graph is the value on the velocity graph.

- The slope of the velocity graph is the value on the acceleration graph.

- s is a maximum or minimum at a turning point, and $v_s = 0$.

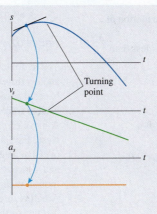

Turning point

- Displacement is the area under the velocity curve.

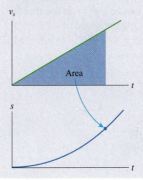

Area

APPLICATIONS

The **sign of** v_s indicates the direction of motion.

- $v_s > 0$ is motion to the right or up.
- $v_s < 0$ is motion to the left or down.

The **sign of** a_s indicates which way $\vec{a}$ points, *not* whether the object is speeding up or slowing down.

- $a_s > 0$ if $\vec{a}$ points to the right or up.
- $a_s < 0$ if $\vec{a}$ points to the left or down.
- The direction of $\vec{a}$ is found with a motion diagram.

An object is **speeding up** if and only if v_s and a_s have the same sign.

An object is **slowing down** if and only if v_s and a_s have opposite signs.

Free fall is constant-acceleration motion with

$$a_y = -g = -9.80 \text{ m/s}^2$$

Motion on an inclined plane has $a_s = \pm g \sin \theta$. The sign depends on the direction of the tilt.

TERMS AND NOTATION

kinematics
uniform motion
average velocity, v_{avg}
speed, v

initial position, s_i
final position, s_f
uniform-motion model
instantaneous velocity, v_s

turning point
average acceleration, a_{avg}
constant-acceleration model
free fall

free-fall acceleration, g
instantaneous acceleration, a_s

CONCEPTUAL QUESTIONS

For Questions 1 through 3, interpret the position graph given in each figure by writing a very short "story" of what is happening. Be creative! Have characters and situations! Simply saying that "a car moves 100 meters to the right" doesn't qualify as a story. Your stories should make *specific reference* to information you obtain from the graph, such as distance moved or time elapsed.

1.
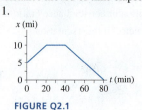

FIGURE Q2.1

2.
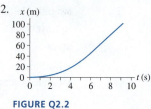

FIGURE Q2.2

3.

FIGURE Q2.3

4. **FIGURE Q2.4** shows a position-versus-time graph for the motion of objects A and B as they move along the same axis.
   a. At the instant $t = 1$ s, is the speed of A greater than, less than, or equal to the speed of B? Explain.
   b. Do objects A and B ever have the *same* speed? If so, at what time or times? Explain.

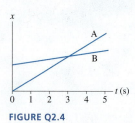

FIGURE Q2.4

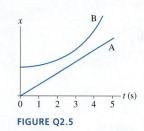

FIGURE Q2.5

5. **FIGURE Q2.5** shows a position-versus-time graph for the motion of objects A and B as they move along the same axis.
   a. At the instant $t = 1$ s, is the speed of A greater than, less than, or equal to the speed of B? Explain.
   b. Do objects A and B ever have the *same* speed? If so, at what time or times? Explain.

6. **FIGURE Q2.6** shows the position-versus-time graph for a moving object. At which lettered point or points:
   a. Is the object *moving* the slowest?
   b. Is the object moving the fastest?
   c. Is the object at rest?
   d. Is the object moving to the left?

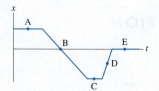

FIGURE Q2.6

7. **FIGURE Q2.7** shows the position-versus-time graph for a moving object. At which lettered point or points:
   a. Is the object moving the fastest?
   b. Is the object moving to the left?
   c. Is the object speeding up?
   d. Is the object turning around?

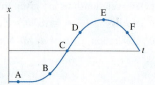

FIGURE Q2.7

8. **FIGURE Q2.8** shows six frames from the motion diagrams of two moving cars, A and B.
   a. Do the two cars ever have the same position at one instant of time? If so, in which frame number (or numbers)?
   b. Do the two cars ever have the same velocity at one instant of time? If so, between which two frames?

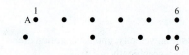

FIGURE Q2.8

9. You're driving along the highway at a steady speed of 60 mph when another driver decides to pass you. At the moment when the front of his car is exactly even with the front of your car, and you turn your head to smile at him, do the two cars have equal velocities? Explain.

10. A bicycle is traveling east. Can its acceleration vector ever point west? Explain.

11. (a) Give an example of a vertical motion with a positive velocity and a negative acceleration. (b) Give an example of a vertical motion with a negative velocity and a negative acceleration.

12. A ball is thrown straight up into the air. At each of the following instants, is the magnitude of the ball's acceleration greater than g, equal to g, less than g, or 0? Explain.
    a. Just after leaving your hand.
    b. At the very top (maximum height).
    c. Just before hitting the ground.

13. A rock is *thrown* (not dropped) straight down from a bridge into the river below. At each of the following instants, is the magnitude of the rock's acceleration greater than g, equal to g, less than g, or 0? Explain.
    a. Immediately after being released.
    b. Just before hitting the water.

14. **FIGURE Q2.14** shows the velocity-versus-time graph for a moving object. At which lettered point or points:
    a. Is the object speeding up?
    b. Is the object slowing down?
    c. Is the object moving to the left?
    d. Is the object moving to the right?

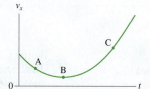

FIGURE Q2.14

EXERCISES AND PROBLEMS

Exercises

Section 2.1 Uniform Motion

1. ‖ Alan leaves Los Angeles at 8:00 A.M. to drive to San Francisco, 400 mi away. He travels at a steady 50 mph. Beth leaves Los Angeles at 9:00 A.M. and drives a steady 60 mph.
   a. Who gets to San Francisco first?
   b. How long does the first to arrive have to wait for the second?

2. ‖ Julie drives 100 mi to Grandmother's house. On the way to Grandmother's, Julie drives half the distance at 40 mph and half the distance at 60 mph. On her return trip, she drives half the time at 40 mph and half the time at 60 mph.
   a. What is Julie's average speed on the way to Grandmother's house?
   b. What is her average speed on the return trip?

3. ‖ Larry leaves home at 9:05 and runs at constant speed to the lamppost seen in **FIGURE EX2.3**. He reaches the lamppost at 9:07, immediately turns, and runs to the tree. Larry arrives at the tree at 9:10.
   a. What is Larry's average velocity, in m/min, during each of these two intervals?
   b. What is Larry's average velocity for the entire run?

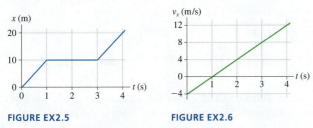

FIGURE EX2.3

4. ‖ **FIGURE EX2.4** is the position-versus-time graph of a jogger. What is the jogger's velocity at $t = 10$ s, at $t = 25$ s, and at $t = 35$ s?

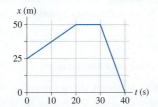

FIGURE EX2.4

Section 2.2 Instantaneous Velocity

Section 2.3 Finding Position from Velocity

5. | **FIGURE EX2.5** shows the position graph of a particle.
   a. Draw the particle's velocity graph for the interval $0 \text{ s} \leq t \leq 4 \text{ s}$.
   b. Does this particle have a turning point or points? If so, at what time or times?

FIGURE EX2.5   **FIGURE EX2.6**

6. ‖ A particle starts from $x_0 = 10$ m at $t_0 = 0$ s and moves with the velocity graph shown in **FIGURE EX2.6**.
   a. Does this particle have a turning point? If so, at what time?
   b. What is the object's position at $t = 2$ s and 4 s?

7. ‖ **FIGURE EX2.7** is a somewhat idealized graph of the velocity
   BIO of blood in the ascending aorta during one beat of the heart. Approximately how far, in cm, does the blood move during one beat?

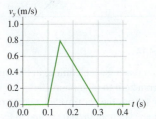

FIGURE EX2.7

8. | **FIGURE EX2.8** shows the velocity graph for a particle having initial position $x_0 = 0$ m at $t_0 = 0$ s. At what time or times is the particle found at $x = 35$ m?

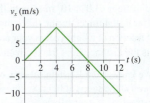

FIGURE EX2.8

Section 2.4 Motion with Constant Acceleration

9. ‖ **FIGURE EX2.9** shows the velocity graph of a particle. Draw the particle's acceleration graph for the interval $0 \text{ s} \leq t \leq 4 \text{ s}$.

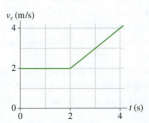

FIGURE EX2.9

10. ‖ **FIGURE EX2.7** showed the velocity graph of blood in the aorta.
    BIO What is the blood's acceleration during each phase of the motion, speeding up and slowing down?

11. ‖ **FIGURE EX2.11** shows the velocity graph of a particle moving along the x-axis. Its initial position is $x_0 = 2.0$ m at $t_0 = 0$ s. At $t = 2.0$ s, what are the particle's (a) position, (b) velocity, and (c) acceleration?

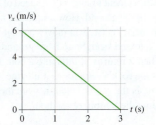

FIGURE EX2.11

12. ‖ **FIGURE EX2.12** shows the velocity-versus-time graph for a particle moving along the x-axis. Its initial position is $x_0 = 2.0$ m at $t_0 = 0$ s.
    a. What are the particle's position, velocity, and acceleration at $t = 1.0$ s?
    b. What are the particle's position, velocity, and acceleration at $t = 3.0$ s?

FIGURE EX2.12

13. ‖ a. What constant acceleration, in SI units, must a car have to go from zero to 60 mph in 10 s?
    b. How far has the car traveled when it reaches 60 mph? Give your answer both in SI units and in feet.
14. ‖ A jet plane is cruising at 300 m/s when suddenly the pilot turns the engines up to full throttle. After traveling 4.0 km, the jet is moving with a speed of 400 m/s. What is the jet's acceleration, assuming it to be a constant acceleration?
15. ‖ a. How many days will it take a spaceship to accelerate to the speed of light $(3.0 \times 10^8$ m/s$)$ with the acceleration g?
    b. How far will it travel during this interval?
    c. What fraction of a light year is your answer to part b? A light year is the distance light travels in one year.

    NOTE We know, from Einstein's theory of relativity, that no object can travel at the speed of light. So this problem, while interesting and instructive, is not realistic.

16. ‖ When you sneeze, the air in your lungs accelerates from rest
    BIO to 150 km/h in approximately 0.50 s. What is the acceleration of the air in m/s^2?
17. ‖ A speed skater moving to the left across frictionless ice at 8.0 m/s hits a 5.0-m-wide patch of rough ice. She slows steadily, then continues on at 6.0 m/s. What is her acceleration on the rough ice?
18. ‖ A Porsche challenges a Honda to a 400 m race. Because the Porsche's acceleration of 3.5 m/s^2 is larger than the Honda's 3.0 m/s^2, the Honda gets a 1.0 s head start. Who wins? By how many seconds?
19. ‖ A car starts from rest at a stop sign. It accelerates at 4.0 m/s^2 for 6.0 s, coasts for 2.0 s, and then slows down at a rate of 3.0 m/s^2 for the next stop sign. How far apart are the stop signs?

Section 2.5 Free Fall

20. ‖ Ball bearings are made by letting spherical drops of molten metal fall inside a tall tower—called a *shot tower*—and solidify as they fall.
    a. If a bearing needs 4.0 s to solidify enough for impact, how high must the tower be?
    b. What is the bearing's impact velocity?
21. ‖ A student standing on the ground throws a ball straight up. The ball leaves the student's hand with a speed of 15 m/s when the hand is 2.0 m above the ground. How long is the ball in the air before it hits the ground? (The student moves her hand out of the way.)
22. ‖ A rock is tossed straight up from ground level with a speed of 20 m/s. When it returns, it falls into a hole 10 m deep.
    a. What is the rock's velocity as it hits the bottom of the hole?
    b. How long is the rock in the air, from the instant it is released until it hits the bottom of the hole?

23. ‖ When jumping, a flea accelerates at an astounding 1000 m/s^2,
    BIO but over only the very short distance of 0.50 mm. If a flea jumps straight up, and if air resistance is neglected (a rather poor approximation in this situation), how high does the flea go?
24. ‖ As a science project, you drop a watermelon off the top of the Empire State Building, 320 m above the sidewalk. It so happens that Superman flies by at the instant you release the watermelon. Superman is headed straight down with a speed of 35 m/s. How fast is the watermelon going when it passes Superman?
25. ‖‖ A rock is dropped from the top of a tall building. The rock's displacement in the last second before it hits the ground is 45% of the entire distance it falls. How tall is the building?

Section 2.6 Motion on an Inclined Plane

26. ‖ A skier is gliding along at 3.0 m/s on horizontal, frictionless snow. He suddenly starts down a 10° incline. His speed at the bottom is 15 m/s.
    a. What is the length of the incline?
    b. How long does it take him to reach the bottom?
27. ‖ A car traveling at 30 m/s runs out of gas while traveling up a 10° slope. How far up the hill will it coast before starting to roll back down?
28. ‖ Santa loses his footing and slides down a frictionless, snowy roof that is tilted at an angle of 30°. If Santa slides 10 m before reaching the edge, what is his speed as he leaves the roof?
29. ‖ A snowboarder glides down a 50-m-long, 15° hill. She then glides horizontally for 10 m before reaching a 25° upward slope. Assume the snow is frictionless.
    a. What is her velocity at the bottom of the hill?
    b. How far can she travel up the 25° slope?
30. ‖ A small child gives a plastic frog a big push at the bottom of a slippery 2.0-m-long, 1.0-m-high ramp, starting it with a speed of 5.0 m/s. What is the frog's speed as it flies off the top of the ramp?

Section 2.7 Instantaneous Acceleration

31. ‖ **FIGURE EX2.31** shows the acceleration-versus-time graph of a particle moving along the x-axis. Its initial velocity is $v_{0x} = 8.0$ m/s at $t_0 = 0$ s. What is the particle's velocity at $t = 4.0$ s?

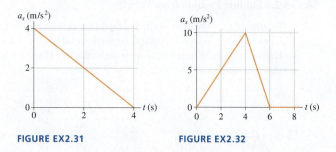

FIGURE EX2.31   **FIGURE EX2.32**

32. ‖ **FIGURE EX2.32** shows the acceleration graph for a particle that starts from rest at $t = 0$ s. What is the particle's velocity at $t = 6$ s?
33. ‖ A particle moving along the x-axis has its position described by
    CALC the function $x = (2t^3 + 2t + 1)$ m, where t is in s. At $t = 2$ s what are the particle's (a) position, (b) velocity, and (c) acceleration?

34. ‖ A particle moving along the x-axis has its velocity described
CALC by the function $v_x = 2t^2$ m/s, where t is in s. Its initial position is
$x_0 = 1$ m at $t_0 = 0$ s. At $t = 1$ s what are the particle's (a) position,
(b) velocity, and (c) acceleration?

35. ‖ The position of a particle is given by the function
CALC $x = (2t^3 - 9t^2 + 12)$ m, where t is in s.
   a. At what time or times is $v_x = 0$ m/s?
   b. What are the particle's position and its acceleration at this time(s)?

36. ‖ The position of a particle is given by the function
CALC $x = (2t^3 - 6t^2 + 12)$ m, where t is in s.
   a. At what time does the particle reach its minimum velocity?
      What is $(v_x)_{min}$?
   b. At what time is the acceleration zero?

Problems

37. ‖ Particles A, B, and C move along the x-axis. Particle C has
an initial velocity of 10 m/s. In **FIGURE P2.37**, the graph for A is a
position-versus-time graph; the graph for B is a velocity-versus-
time graph; the graph for C is an acceleration-versus-time graph.
Find each particle's velocity at $t = 7.0$ s.

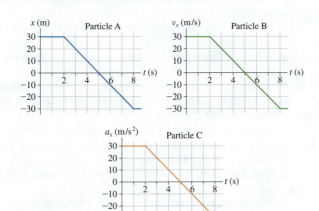

FIGURE P2.37

38. ‖ A block is suspended from a spring, pulled down, and released.
The block's position-versus-time graph is shown in **FIGURE P2.38**.
   a. At what times is the velocity zero? At what times is the veloc-
      ity most positive? Most negative?
   b. Draw a reasonable velocity-versus-time graph.

FIGURE P2.38

39. ‖ A particle's velocity is described by the function
CALC $v_x = (t^2 - 7t + 10)$ m/s, where t is in s.
   a. At what times does the particle reach its turning points?
   b. What is the particle's acceleration at each of the turning points?

40. ‖‖ A particle's velocity is described by the function $v_x = kt^2$ m/s,
CALC where k is a constant and t is in s. The particle's position at $t_0 = 0$ s
is $x_0 = -9.0$ m. At $t_1 = 3.0$ s, the particle is at $x_1 = 9.0$ m.
Determine the value of the constant k. Be sure to include the
proper units.

41. ‖ A particle's acceleration is described by the function
CALC $a_x = (10 - t)$ m/s², where t is in s. Its initial conditions are
$x_0 = 0$ m and $v_{0x} = 0$ m/s at $t = 0$ s.
   a. At what time is the velocity again zero?
   b. What is the particle's position at that time?

42. ‖ A particle's velocity is given by the function
CALC $v_x = (2.0$ m/s$)\sin(\pi t)$, where t is in s.
   a. What is the first time after $t = 0$ s when the particle reaches
      a turning point?
   b. What is the particle's acceleration at that time?

43. ‖ A ball rolls along the smooth track shown in **FIGURE P2.43**.
Each segment of the track is straight, and the ball passes smoothly
from one segment to the next without changing speed or leaving
the track. Draw three vertically stacked graphs showing position, velocity, and acceleration versus time. Each graph should
have the same time axis, and the proportions of the graph should
be qualitatively correct. Assume that the ball has enough speed
to reach the top.

FIGURE P2.43

44. ‖ Draw position, velocity, and acceleration graphs for the ball
shown in **FIGURE P2.44**. See Problem 43 for more information.

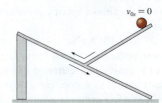

FIGURE P2.44

45. ‖ **FIGURE P2.45** shows a set of kinematic graphs for a ball rolling
on a track. All segments of the track are straight lines, but some
may be tilted. Draw a picture of the track and also indicate the
ball's initial condition.

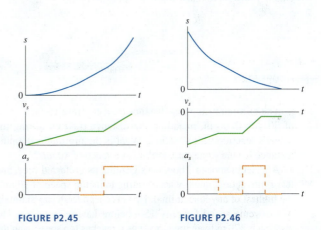

FIGURE P2.45   **FIGURE P2.46**

46. ‖ **FIGURE P2.46** shows a set of kinematic graphs for a ball rolling
on a track. All segments of the track are straight lines, but some
may be tilted. Draw a picture of the track and also indicate the
ball's initial condition.

47. ‖ The takeoff speed for an Airbus A320 jetliner is 80 m/s. Velocity data measured during takeoff are as shown.

t (s)	v_x (m/s)
0	0
10	23
20	46
30	69

   a. Is the jetliner's acceleration constant during takeoff? Explain.
   b. At what time do the wheels leave the ground?
   c. For safety reasons, in case of an aborted takeoff, the runway must be three times the takeoff distance. Can an A320 take off safely on a 2.5-mi-long runway?

48. ‖ You are driving to the grocery store at 20 m/s. You are 110 m from an intersection when the traffic light turns red. Assume that your reaction time is 0.50 s and that your car brakes with constant acceleration. What magnitude braking acceleration will bring you to a stop exactly at the intersection?

49. ‖ You're driving down the highway late one night at 20 m/s when a deer steps onto the road 35 m in front of you. Your reaction time before stepping on the brakes is 0.50 s, and the maximum deceleration of your car is 10 m/s².
   a. How much distance is between you and the deer when you come to a stop?
   b. What is the maximum speed you could have and still not hit the deer?

50. ‖ Two cars are driving at the same constant speed on a straight road, with car 1 in front of car 2. Car 1 suddenly starts to brake with constant acceleration and stops in 10 m. At the instant car 1 comes to a stop, car 2 begins to brake with the same acceleration. It comes to a halt just as it reaches the back of car 1. What was the separation between the cars before they starting braking?

51. ‖ You are playing miniature golf at the golf course shown in **FIGURE P2.51**. Due to the fake plastic grass, the ball decelerates at 1.0 m/s² when rolling horizontally and at 6.0 m/s² on the slope. What is the slowest speed with which the ball can leave your golf club if you wish to make a hole in one?

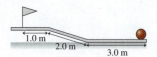

FIGURE P2.51

52. ‖ The minimum stopping distance for a car traveling at a speed of 30 m/s is 60 m, including the distance traveled during the driver's reaction time of 0.50 s. What is the minimum stopping distance for the same car traveling at a speed of 40 m/s?

53. ‖
   BIO A cheetah spots a Thomson's gazelle, its preferred prey, and leaps into action, quickly accelerating to its top speed of 30 m/s, the highest of any land animal. However, a cheetah can maintain this extreme speed for only 15 s before having to let up. The cheetah is 170 m from the gazelle as it reaches top speed, and the gazelle sees the cheetah at just this instant. With negligible reaction time, the gazelle heads directly away from the cheetah, accelerating at 4.6 m/s² for 5.0 s, then running at constant speed. Does the gazelle escape? If so, by what distance is the gazelle in front when the cheetah gives up?

54. ‖ You are at a train station, standing next to the train at the front of the first car. The train starts moving with constant acceleration, and 5.0 s later the back of the first car passes you. How long does it take after the train starts moving until the back of the seventh car passes you? All cars are the same length.

55. ‖‖ A 200 kg weather rocket is loaded with 100 kg of fuel and fired straight up. It accelerates upward at 30 m/s² for 30 s, then runs out of fuel. Ignore any air resistance effects.
   a. What is the rocket's maximum altitude?
   b. How long is the rocket in the air before hitting the ground?

56. ‖‖ A 1000 kg weather rocket is launched straight up. The rocket motor provides a constant acceleration for 16 s, then the motor stops. The rocket altitude 20 s after launch is 5100 m. You can ignore any effects of air resistance. What was the rocket's acceleration during the first 16 s?

57. ‖‖ A lead ball is dropped into a lake from a diving board 5.0 m above the water. After entering the water, it sinks to the bottom with a constant velocity equal to the velocity with which it hit the water. The ball reaches the bottom 3.0 s after it is released. How deep is the lake?

58. ‖ A hotel elevator ascends 200 m with a maximum speed of 5.0 m/s. Its acceleration and deceleration both have a magnitude of 1.0 m/s².
   a. How far does the elevator move while accelerating to full speed from rest?
   b. How long does it take to make the complete trip from bottom to top?

59. ‖ A basketball player can jump to a height of 55 cm. How far above the floor can he jump in an elevator that is descending at a constant 1.0 m/s?

60. ‖ You are 9.0 m from the door of your bus, behind the bus, when it pulls away with an acceleration of 1.0 m/s². You instantly start running toward the still-open door at 4.5 m/s.
   a. How long does it take for you to reach the open door and jump in?
   b. What is the maximum time you can wait before starting to run and still catch the bus?

61. ‖ Ann and Carol are driving their cars along the same straight road. Carol is located at $x = 2.4$ mi at $t = 0$ h and drives at a steady 36 mph. Ann, who is traveling in the same direction, is located at $x = 0.0$ mi at $t = 0.50$ h and drives at a steady 50 mph.
   a. At what time does Ann overtake Carol?
   b. What is their position at this instant?
   c. Draw a position-versus-time graph showing the motion of both Ann and Carol.

62. ‖ Amir starts riding his bike up a 200-m-long slope at a speed of 18 km/h, decelerating at 0.20 m/s² as he goes up. At the same instant, Becky starts down from the top at a speed of 6.0 km/h, accelerating at 0.40 m/s² as she goes down. How far has Amir ridden when they pass?

63. ‖ A very slippery block of ice slides down a smooth ramp tilted at angle θ. The ice is released from rest at vertical height h above the bottom of the ramp. Find an expression for the speed of the ice at the bottom.

64. ‖ Bob is driving the getaway car after the big bank robbery. He's going 50 m/s when his headlights suddenly reveal a nail strip that the cops have placed across the road 150 m in front of him. If Bob can stop in time, he can throw the car into reverse and escape. But if he crosses the nail strip, all his tires will go flat and he will be caught. Bob's reaction time before he can hit the brakes is 0.60 s, and his car's maximum deceleration is 10 m/s². Does Bob stop before or after the nail strip? By what distance?

65. ‖ One game at the amusement park has you push a puck up a long, frictionless ramp. You win a stuffed animal if the puck, at its highest point, comes to within 10 cm of the end of the ramp without going off. You give the puck a push, releasing it with a speed of 5.0 m/s when it is 8.5 m from the end of the ramp. The puck's speed after traveling 3.0 m is 4.0 m/s. How far is it from the end when it stops?

66. ‖ A motorist is driving at 20 m/s when she sees that a traffic light 200 m ahead has just turned red. She knows that this light stays red for 15 s, and she wants to reach the light just as it turns green again. It takes her 1.0 s to step on the brakes and begin slowing. What is her speed as she reaches the light at the instant it turns green?

67. ‖ Nicole throws a ball straight up. Chad watches the ball from a window 5.0 m above the point where Nicole released it. The ball passes Chad on the way up, and it has a speed of 10 m/s as it passes him on the way back down. How fast did Nicole throw the ball?

68. ‖ David is driving a steady 30 m/s when he passes Tina, who is sitting in her car at rest. Tina begins to accelerate at a steady 2.0 m/s^2 at the instant when David passes.
   a. How far does Tina drive before passing David?
   b. What is her speed as she passes him?

69. ‖ A cat is sleeping on the floor in the middle of a 3.0-m-wide room when a barking dog enters with a speed of 1.50 m/s. As the dog enters, the cat (as only cats can do) immediately accelerates at 0.85 m/s^2 toward an open window on the opposite side of the room. The dog (all bark and no bite) is a bit startled by the cat and begins to slow down at 0.10 m/s^2 as soon as it enters the room. How far is the cat in front of the dog as it leaps through the window?

70. ‖‖ Water drops fall from the edge of a roof at a steady rate. A fifth drop starts to fall just as the first drop hits the ground. At this instant, the second and third drops are exactly at the bottom and top edges of a 1.00-m-tall window. How high is the edge of the roof?

71. ‖‖ I was driving along at 20 m/s, trying to change a CD and not watching where I was going. When I looked up, I found myself 45 m from a railroad crossing. And wouldn't you know it, a train moving at 30 m/s was only 60 m from the crossing. In a split second, I realized that the train was going to beat me to the crossing and that I didn't have enough distance to stop. My only hope was to accelerate enough to cross the tracks before the train arrived. If my reaction time before starting to accelerate was 0.50 s, what minimum acceleration did my car need for me to be here today writing these words?

72. ‖ As an astronaut visiting Planet X, you're assigned to measure the free-fall acceleration. Getting out your meter stick and stop watch, you time the fall of a heavy ball from several heights. Your data are as follows:

Height (m)	Fall time (s)
0.0	0.00
1.0	0.54
2.0	0.72
3.0	0.91
4.0	1.01
5.0	1.17

Analyze these data to determine the free-fall acceleration on Planet X. Your analysis method should involve fitting a straight line to an appropriate graph, similar to the analysis in Example 2.14.

73. ‖ Your goal in laboratory is to launch a ball of mass m straight up so that it reaches exactly height h above the top of the launching tube. You and your lab partners will earn fewer points if the ball goes too high or too low. The launch tube uses compressed air to accelerate the ball over a distance d, and you have a table of data telling you how to set the air compressor to achieve a desired acceleration. Find an expression for the acceleration that will earn you maximum points.

74. ‖ When a 1984 Alfa Romeo Spider sports car accelerates at
CALC the maximum possible rate, its motion during the first 20 s is extremely well modeled by the simple equation

$$v_x^2 = \frac{2P}{m} t$$

where $P = 3.6 \times 10^4$ watts is the car's power output, $m = 1200$ kg is its mass, and v_x is in m/s. That is, the square of the car's velocity increases linearly with time.
   a. Find an algebraic expression in terms of P, m, and t for the car's acceleration at time t.
   b. What is the car's speed at $t = 2$ s and $t = 10$ s?
   c. Evaluate the acceleration at $t = 2$ s and $t = 10$ s.

75. ‖ The two masses in **FIGURE P2.75** slide on frictionless wires. They are connected by a pivoting rigid rod of length L. Prove that $v_{2x} = -v_{1y} \tan\theta$.

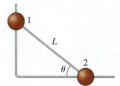

FIGURE P2.75

In Problems 76 through 79, you are given the kinematic equation or equations that are used to solve a problem. For each of these, you are to:
   a. Write a *realistic* problem for which this is the correct equation(s). Be sure that the answer your problem requests is consistent with the equation(s) given.
   b. Draw the pictorial representation for your problem.
   c. Finish the solution of the problem.

76. $64 \text{ m} = 0 \text{ m} + (32 \text{ m/s})(4 \text{ s} - 0 \text{ s}) + \frac{1}{2} a_x (4 \text{ s} - 0 \text{ s})^2$

77. $(10 \text{ m/s})^2 = v_{0y}^2 - 2(9.8 \text{ m/s}^2)(10 \text{ m} - 0 \text{ m})$

78. $(0 \text{ m/s})^2 = (5 \text{ m/s})^2 - 2(9.8 \text{ m/s}^2)(\sin 10°)(x_1 - 0 \text{ m})$

79. $v_{1x} = 0 \text{ m/s} + (20 \text{ m/s}^2)(5 \text{ s} - 0 \text{ s})$
   $x_1 = 0 \text{ m} + (0 \text{ m/s})(5 \text{ s} - 0 \text{ s}) + \frac{1}{2}(20 \text{ m/s}^2)(5 \text{ s} - 0 \text{ s})^2$
   $x_2 = x_1 + v_{1x}(10 \text{ s} - 5 \text{ s})$

Challenge Problems

80. ‖‖ A rocket is launched straight up with constant acceleration. Four seconds after liftoff, a bolt falls off the side of the rocket. The bolt hits the ground 6.0 s later. What was the rocket's acceleration?

81. ‖‖ Careful measurements have been made of Olympic sprinters in the 100 meter dash. A simple but reasonably accurate model is that a sprinter accelerates at 3.6 m/s^2 for $3\frac{1}{3}$ s, then runs at constant velocity to the finish line.
   a. What is the race time for a sprinter who follows this model?
   b. A sprinter could run a faster race by accelerating faster at the beginning, thus reaching top speed sooner. If a sprinter's top speed is the same as in part a, what acceleration would he need to run the 100 meter dash in 9.9 s?
   c. By what percent did the sprinter need to increase his acceleration in order to decrease his time by 1%?

82. ||| Careful measurements have been made of Olympic sprinters
CALC in the 100 meter dash. A quite realistic model is that the sprinter's velocity is given by

$$v_x = a(1 - e^{-bt})$$

where t is in s, v_x is in m/s, and the constants a and b are characteristic of the sprinter. Sprinter Carl Lewis's run at the 1987 World Championships is modeled with $a = 11.81$ m/s and $b = 0.6887$ s^{-1}.

a. What was Lewis's acceleration at $t = 0$ s, 2.00 s, and 4.00 s?

b. Find an expression for the distance traveled at time t.

c. Your expression from part b is a transcendental equation, meaning that you can't solve it for t. However, it's not hard to use trial and error to find the time needed to travel a specific distance. To the nearest 0.01 s, find the time Lewis needed to sprint 100.0 m. His official time was 0.01 s more than your answer, showing that this model is very good, but not perfect.

83. ||| A sprinter can accelerate with constant acceleration for 4.0 s before reaching top speed. He can run the 100 meter dash in 10.0 s. What is his speed as he crosses the finish line?

84. ||| A rubber ball is shot straight up from the ground with speed v_0. Simultaneously, a second rubber ball at height h directly above the first ball is dropped from rest.

a. At what height above the ground do the balls collide? Your answer will be an *algebraic expression* in terms of h, v_0, and g.

b. What is the maximum value of h for which a collision occurs before the first ball falls back to the ground?

c. For what value of h does the collision occur at the instant when the first ball is at its highest point?

85. ||| The Starship Enterprise returns from warp drive to ordinary space with a forward speed of 50 km/s. To the crew's great surprise, a Klingon ship is 100 km directly ahead, traveling in the same direction at a mere 20 km/s. Without evasive action, the Enterprise will overtake and collide with the Klingons in just slightly over 3.0 s. The Enterprise's computers react instantly to brake the ship. What magnitude acceleration does the Enterprise need to just barely avoid a collision with the Klingon ship? Assume the acceleration is constant.

Hint: Draw a position-versus-time graph showing the motions of both the Enterprise and the Klingon ship. Let $x_0 = 0$ km be the location of the Enterprise as it returns from warp drive. How do you show graphically the situation in which the collision is "barely avoided"? Once you decide what it looks like graphically, express that situation mathematically.

3 Vectors and Coordinate Systems

Wind has both a speed and a direction, hence the motion of the wind is described by a vector.

IN THIS CHAPTER, you will learn how vectors are represented and used.

What is a vector?

A vector is a quantity with both a size—its magnitude—and a direction. Vectors you'll meet in the next few chapters include position, displacement, velocity, acceleration, force, and momentum.

« LOOKING BACK Tactics Boxes 1.1 and 1.2 Vector addition and subtraction

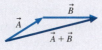

How are vectors added and subtracted?

Vectors are added "tip to tail." The order of addition does not matter. To subtract vectors, turn the subtraction into addition by writing $\vec{A} - \vec{B} = \vec{A} + (-\vec{B})$. The vector $-\vec{B}$ is the same length as $\vec{B}$ but points in the opposite direction.

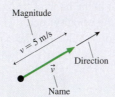

What are unit vectors?

Unit vectors define what we *mean* by the $+x$- and $+y$-directions in space.

- A unit vector has magnitude 1.
- A unit vector has no units.

Unit vectors simply point.

What are components?

Components of vectors are the pieces of vectors parallel to the coordinate axes—in the directions of the unit vectors. We write

$$\vec{E} = E_x\hat{\imath} + E_y\hat{\jmath}$$

Components simplify vector math.

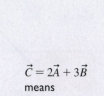

How are components used?

Components let us do vector math with algebra, which is easier and more precise than adding and subtracting vectors using geometry and trigonometry. Multiplying a vector by a number simply multiplies all of the vector's components by that number.

$$\vec{C} = 2\vec{A} + 3\vec{B}$$

means

$$\begin{cases} C_x = 2A_x + 3B_x \\ C_y = 2A_y + 3B_y \end{cases}$$

How will I use vectors?

Vectors appear everywhere in physics and engineering—from velocities to electric fields and from forces to fluid flows. The tools and techniques you learn in this chapter will be used throughout your studies and your professional career.

3.1 Scalars and Vectors

A quantity that is fully described by a single number (with units) is called a **scalar.** Mass, temperature, volume and energy are all scalars. We will often use an algebraic symbol to represent a scalar quantity. Thus m will represent mass, T temperature, V volume, E energy, and so on.

Our universe has three dimensions, so some quantities also need a direction for a full description. If you ask someone for directions to the post office, the reply "Go three blocks" will not be very helpful. A full description might be, "Go three blocks south." A quantity having both a size and a direction is called a **vector.**

The mathematical term for the length, or size, of a vector is **magnitude,** so we can also say that **a vector is a quantity having a magnitude and a direction.**

FIGURE 3.1 shows that the *geometric representation* of a vector is an arrow, with the tail of the arrow (not its tip!) placed at the point where the measurement is made. An arrow makes a natural representation of a vector because it inherently has both a length and a direction. As you've already seen, we label vectors by drawing a small arrow over the letter that represents the vector: $\vec{r}$ for position, $\vec{v}$ for velocity, $\vec{a}$ for acceleration.

FIGURE 3.1 The velocity vector $\vec{v}$ has both a magnitude and a direction.

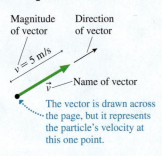

Magnitude of vector   Direction of vector

$v = 5$ m/s

$\vec{v}$ — Name of vector

The vector is drawn across the page, but it represents the particle's velocity at this one point.

NOTE Although the vector arrow is drawn across the page, from its tail to its tip, this does *not* indicate that the vector "stretches" across this distance. Instead, the vector arrow tells us the value of the vector quantity only at the one point where the tail of the vector is placed. ◄

The magnitude of a vector can be written using absolute value signs or, more frequently, as the letter without the arrow. For example, the magnitude of the velocity vector in Figure 3.1 is $v = |\vec{v}| = 5$ m/s. This is the object's *speed.* The magnitude of the acceleration vector $\vec{a}$ is written a. **The magnitude of a vector is a scalar.** Note that magnitude of a vector cannot be a negative number; it must be positive or zero, with appropriate units.

It is important to get in the habit of using the arrow symbol for vectors. If you omit the vector arrow from the velocity vector $\vec{v}$ and write only v, then you're referring only to the object's speed, not its velocity. The symbols $\vec{r}$ and r, or $\vec{v}$ and v, do *not* represent the same thing.

FIGURE 3.2 Displacement vectors.

(a)

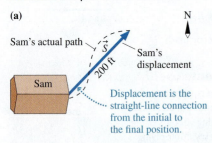

N

Sam's actual path

$\vec{S}$

200 ft

Sam's displacement

Displacement is the straight-line connection from the initial to the final position.

Sam

(b)

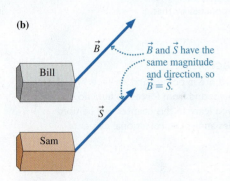

$\vec{B}$

Bill

$\vec{B}$ and $\vec{S}$ have the same magnitude and direction, so $\vec{B} = \vec{S}$.

$\vec{S}$

Sam

3.2 Using Vectors

Suppose Sam starts from his front door, walks across the street, and ends up 200 ft to the northeast of where he started. Sam's displacement, which we will label $\vec{S}$, is shown in **FIGURE 3.2a**. The displacement vector is a *straight-line connection* from his initial to his final position, not necessarily his actual path.

To describe a vector we must specify both its magnitude and its direction. We can write Sam's displacement as $\vec{S} = (200 \text{ ft, northeast})$. The magnitude of Sam's displacement is $S = |\vec{S}| = 200$ ft, the distance between his initial and final points.

Sam's next-door neighbor Bill also walks 200 ft to the northeast, starting from his own front door. Bill's displacement $\vec{B} = (200 \text{ ft, northeast})$ has the same magnitude and direction as Sam's displacement $\vec{S}$. Because vectors are defined by their magnitude and direction, **two vectors are equal if they have the same magnitude and direction.** Thus the two displacements in **FIGURE 3.2b** are equal to each other, and we can write $\vec{B} = \vec{S}$.

NOTE A vector is unchanged if you move it to a different point on the page as long as you don't change its length or the direction it points. ◄

Vector Addition

If you earn $50 on Saturday and $60 on Sunday, your *net* income for the weekend is the sum of $50 and $60. With numbers, the word *net* implies addition. The same is true with vectors. For example, **FIGURE 3.3** shows the displacement of a hiker who first hikes 4 miles to the east, then 3 miles to the north. The first leg of the hike is described by the displacement $\vec{A} = (4\text{ mi, east})$. The second leg of the hike has displacement $\vec{B} = (3\text{ mi, north})$. Vector $\vec{C}$ is the *net displacement* because it describes the net result of the hiker's first having displacement $\vec{A}$, then displacement $\vec{B}$.

The net displacement $\vec{C}$ is an initial displacement $\vec{A}$ *plus* a second displacement $\vec{B}$, or

$$\vec{C} = \vec{A} + \vec{B} \tag{3.1}$$

The sum of two vectors is called the **resultant vector.** It's not hard to show that vector addition is commutative: $\vec{A} + \vec{B} = \vec{B} + \vec{A}$. That is, you can add vectors in any order you wish.

≪ Tactics Box 1.1 on page 6 showed the three-step procedure for adding two vectors, and it's highly recommended that you turn back for a quick review. This tip-to-tail method for adding vectors, which is used to find $\vec{C} = \vec{A} + \vec{B}$ in Figure 3.3, is called **graphical addition.** Any two vectors of the same type—two velocity vectors or two force vectors—can be added in exactly the same way.

The graphical method for adding vectors is straightforward, but we need to do a little geometry to come up with a complete description of the resultant vector $\vec{C}$. Vector $\vec{C}$ of Figure 3.3 is defined by its magnitude C and by its direction. Because the three vectors $\vec{A}$, $\vec{B}$, and $\vec{C}$ form a right triangle, the magnitude, or length, of $\vec{C}$ is given by the Pythagorean theorem:

$$C = \sqrt{A^2 + B^2} = \sqrt{(4\text{ mi})^2 + (3\text{ mi})^2} = 5\text{ mi} \tag{3.2}$$

Notice that Equation 3.2 uses the magnitudes A and B of the vectors $\vec{A}$ and $\vec{B}$. The angle θ, which is used in Figure 3.3 to describe the direction of $\vec{C}$, is easily found for a right triangle:

$$\theta = \tan^{-1}\left(\frac{B}{A}\right) = \tan^{-1}\left(\frac{3\text{ mi}}{4\text{ mi}}\right) = 37° \tag{3.3}$$

Altogether, the hiker's net displacement is $\vec{C} = \vec{A} + \vec{B} = (5\text{ mi, } 37°\text{ north of east})$.

NOTE Vector mathematics makes extensive use of geometry and trigonometry. Appendix A, at the end of this book, contains a brief review of these topics.

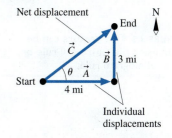

FIGURE 3.3 The net displacement $\vec{C}$ resulting from two displacements $\vec{A}$ and $\vec{B}$.

EXAMPLE 3.1 | Using graphical addition to find a displacement

A bird flies 100 m due east from a tree, then 50 m northwest (that is, 45° north of west). What is the bird's net displacement?

VISUALIZE FIGURE 3.4 shows the two individual displacements, which we've called $\vec{A}$ and $\vec{B}$. The net displacement is the vector sum $\vec{C} = \vec{A} + \vec{B}$, which is found graphically.

FIGURE 3.4 The bird's net displacement is $\vec{C} = \vec{A} + \vec{B}$.

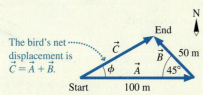

SOLVE The two displacements are $\vec{A} = (100\text{ m, east})$ and $\vec{B} = (50\text{ m, northwest})$. The net displacement $\vec{C} = \vec{A} + \vec{B}$ is found by drawing a vector from the initial to the final position. But

describing $\vec{C}$ is a bit trickier than the example of the hiker because $\vec{A}$ and $\vec{B}$ are not at right angles. First, we can find the magnitude of $\vec{C}$ by using the law of cosines from trigonometry:

$$C^2 = A^2 + B^2 - 2AB\cos 45°$$
$$= (100\text{ m})^2 + (50\text{ m})^2 - 2(100\text{ m})(50\text{ m})\cos 45°$$
$$= 5430\text{ m}^2$$

Thus $C = \sqrt{5430\text{ m}^2} = 74\text{ m}$. Then a second use of the law of cosines can determine angle ϕ (the Greek letter phi):

$$B^2 = A^2 + C^2 - 2AC\cos\phi$$

$$\phi = \cos^{-1}\left[\frac{A^2 + C^2 - B^2}{2AC}\right] = 29°$$

The bird's net displacement is

$$\vec{C} = (74\text{ m, } 29°\text{ north of east})$$

It is often convenient to draw two vectors with their tails together, as shown in **FIGURE 3.5a**. To evaluate $\vec{D} + \vec{E}$, you could move vector $\vec{E}$ over to where its tail is on the tip of $\vec{D}$, then use the tip-to-tail rule of graphical addition. That gives vector $F = \vec{D} + \vec{E}$ in **FIGURE 3.5b**. Alternatively, **FIGURE 3.5c** shows that the vector sum $\vec{D} + \vec{E}$ can be found as the diagonal of the parallelogram defined by $\vec{D}$ and $\vec{E}$. This method for vector addition is called the *parallelogram rule* of vector addition.

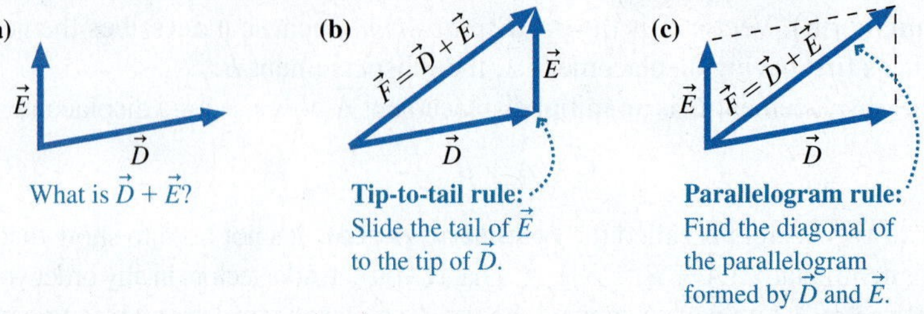

▶ **FIGURE 3.5** Two vectors can be added using the tip-to-tail rule or the parallelogram rule.

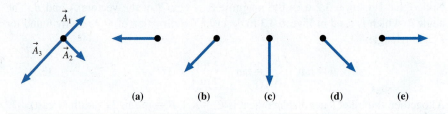

(a)
$\vec{E}$
$\vec{D}$
What is $\vec{D} + \vec{E}$?

(b)
$\vec{F} = \vec{D} + \vec{E}$
$\vec{E}$
$\vec{D}$
Tip-to-tail rule:
Slide the tail of $\vec{E}$
to the tip of $\vec{D}$.

(c)
$\vec{E}$ $\vec{F} = \vec{D} + \vec{E}$
$\vec{D}$
Parallelogram rule:
Find the diagonal of
the parallelogram
formed by $\vec{D}$ and $\vec{E}$.

FIGURE 3.6 The net displacement after four individual displacements.

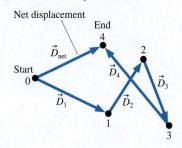

Net displacement
End
4
$\vec{D}_{net}$
Start
0
$\vec{D}_4$
2
$\vec{D}_3$
$\vec{D}_1$
$\vec{D}_2$
1
3

Vector addition is easily extended to more than two vectors. **FIGURE 3.6** shows the path of a hiker moving from initial position 0 to position 1, then position 2, then position 3, and finally arriving at position 4. These four segments are described by displacement vectors $\vec{D}_1$, $\vec{D}_2$, $\vec{D}_3$, and $\vec{D}_4$. The hiker's *net* displacement, an arrow from position 0 to position 4, is the vector $\vec{D}_{net}$. In this case,

$$\vec{D}_{net} = \vec{D}_1 + \vec{D}_2 + \vec{D}_3 + \vec{D}_4 \tag{3.4}$$

The vector sum is found by using the tip-to-tail method three times in succession.

STOP TO THINK 3.1   Which figure shows $\vec{A}_1 + \vec{A}_2 + \vec{A}_3$?

$\vec{A}_1$
$\vec{A}_3$ $\vec{A}_2$

(a)   (b)   (c)   (d)   (e)

More Vector Mathematics

In addition to adding vectors, we will need to subtract vectors (« Tactics Box 1.2 on page 7), multiply vectors by scalars, and understand how to interpret the negative of a vector. These operations are illustrated in **FIGURE 3.7**.

FIGURE 3.7 Working with vectors.

The length of $\vec{B}$ is "stretched" by the factor c. That is, $B = cA$.
$\vec{A} = (A, \theta)$
θ
θ
$\vec{B} = c\vec{A} = (cA, \theta)$
$\vec{B}$ points in the same direction as $\vec{A}$.
Multiplication by a scalar

$\vec{A} + (-\vec{A}) = \vec{0}$. The tip of $-\vec{A}$ returns to the starting point.
$\vec{A}$
$-\vec{A}$
Vector $-\vec{A}$ is equal in magnitude but opposite in direction to $\vec{A}$.
The **zero vector** $\vec{0}$ has zero length
The negative of a vector

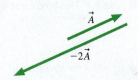

$\vec{A}$
$-2\vec{A}$
Multiplication by a negative scalar

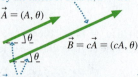

$\vec{A}$
$\vec{C}$
Vector subtraction: What is $\vec{A} - \vec{C}$?
Write it as $\vec{A} + (-\vec{C})$ and add!

$-\vec{C}$
$\vec{A} - \vec{C}$
$\vec{A}$
Tip-to-tail subtraction using $-\vec{C}$

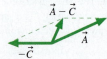

$\vec{A} - \vec{C}$
$\vec{A}$
$-\vec{C}$
Parallelogram subtraction using $-\vec{C}$

EXAMPLE 3.2 | Velocity and displacement

Carolyn drives her car north at 30 km/h for 1 hour, east at 60 km/h for 2 hours, then north at 50 km/h for 1 hour. What is Carolyn's net displacement?

SOLVE Chapter 1 defined average velocity as

$$\vec{v} = \frac{\Delta \vec{r}}{\Delta t}$$

so the displacement $\Delta \vec{r}$ during the time interval Δt is $\Delta \vec{r} = (\Delta t)\vec{v}$. This is multiplication of the vector $\vec{v}$ by the scalar Δt. Carolyn's velocity during the first hour is $\vec{v}_1 = (30 \text{ km/h, north})$, so her displacement during this interval is

$$\Delta \vec{r}_1 = (1 \text{ hour})(30 \text{ km/h, north}) = (30 \text{ km, north})$$

Similarly,

$$\Delta \vec{r}_2 = (2 \text{ hours})(60 \text{ km/h, east}) = (120 \text{ km, east})$$

$$\Delta \vec{r}_3 = (1 \text{ hour})(50 \text{ km/h, north}) = (50 \text{ km, north})$$

In this case, multiplication by a scalar changes not only the length of the vector but also its units, from km/h to km. The direction, however, is unchanged. Carolyn's net displacement is

$$\Delta \vec{r}_{\text{net}} = \Delta \vec{r}_1 + \Delta \vec{r}_2 + \Delta \vec{r}_3$$

FIGURE 3.8 The net displacement is the vector sum $\Delta \vec{r}_{\text{net}} = \Delta \vec{r}_1 + \Delta \vec{r}_2 + \Delta \vec{r}_3$.

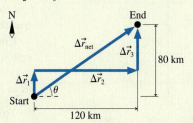

This addition of the three vectors is shown in **FIGURE 3.8**, using the tip-to-tail method. $\Delta \vec{r}_{\text{net}}$ stretches from Carolyn's initial position to her final position. The magnitude of her net displacement is found using the Pythagorean theorem:

$$r_{\text{net}} = \sqrt{(120 \text{ km})^2 + (80 \text{ km})^2} = 144 \text{ km}$$

The direction of $\Delta \vec{r}_{\text{net}}$ is described by angle θ, which is

$$\theta = \tan^{-1}\left(\frac{80 \text{ km}}{120 \text{ km}}\right) = 34°$$

Thus Carolyn's net displacement is $\Delta \vec{r}_{\text{net}} = (144 \text{ km}, 34°$ north of east).

STOP TO THINK 3.2   Which figure shows $2\vec{A} - \vec{B}$?

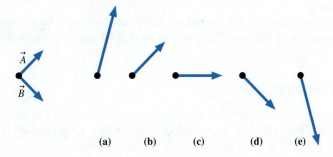

(a)   (b)   (c)   (d)   (e)

3.3 Coordinate Systems and Vector Components

Vectors do not require a coordinate system. We can add and subtract vectors graphically, and we will do so frequently to clarify our understanding of a situation. But the graphical addition of vectors is not an especially good way to find quantitative results. In this section we will introduce a *coordinate representation* of vectors that will be the basis of an easier method for doing vector calculations.

Coordinate Systems

The world does not come with a coordinate system attached to it. A coordinate system is an artificially imposed grid that you place on a problem in order to make quantitative measurements. You are free to choose:

- Where to place the origin, and
- How to orient the axes.

Different problem solvers may choose to use different coordinate systems; that is perfectly acceptable. However, some coordinate systems will make a problem easier

A GPS uses satellite signals to find your position in the earth's coordinate system with amazing accuracy.

FIGURE 3.9 A conventional xy-coordinate system and the quadrants of the xy-plane.

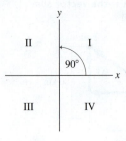

to solve. Part of our goal is to learn how to choose an appropriate coordinate system for each problem.

FIGURE 3.9 shows the xy-coordinate system we will use in this book. The placement of the axes is not entirely arbitrary. By convention, the positive y-axis is located 90° *counterclockwise* (ccw) from the positive x-axis. Figure 3.9 also identifies the four **quadrants** of the coordinate system, I through IV.

Coordinate axes have a positive end and a negative end, separated by zero at the origin where the two axes cross. When you draw a coordinate system, it is important to label the axes. This is done by placing x and y labels at the *positive* ends of the axes, as in Figure 3.9. The purpose of the labels is twofold:

- To identify which axis is which, and
- To identify the positive ends of the axes.

This will be important when you need to determine whether the quantities in a problem should be assigned positive or negative values.

Component Vectors

FIGURE 3.10 Component vectors $\vec{A}_x$ and $\vec{A}_y$ are drawn parallel to the coordinate axes such that $\vec{A} = \vec{A}_x + \vec{A}_y$.

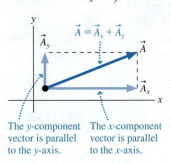

The y-component vector is parallel to the y-axis.

The x-component vector is parallel to the x-axis.

FIGURE 3.10 shows a vector $\vec{A}$ and an xy-coordinate system that we've chosen. Once the directions of the axes are known, we can define two new vectors *parallel to the axes* that we call the **component vectors** of $\vec{A}$. You can see, using the parallelogram rule, that $\vec{A}$ is the vector sum of the two component vectors:

$$\vec{A} = \vec{A}_x + \vec{A}_y \tag{3.5}$$

In essence, we have broken vector $\vec{A}$ into two perpendicular vectors that are parallel to the coordinate axes. This process is called the **decomposition** of vector $\vec{A}$ into its component vectors.

> **NOTE** It is not necessary for the tail of $\vec{A}$ to be at the origin. All we need to know is the *orientation* of the coordinate system so that we can draw $\vec{A}_x$ and $\vec{A}_y$ parallel to the axes.

Components

You learned in Chapters 1 and 2 to give the kinematic variable v_x a positive sign if the velocity vector $\vec{v}$ points toward the positive end of the x-axis, a negative sign if $\vec{v}$ points in the negative x-direction. We need to extend this idea to vectors in general.

Suppose vector $\vec{A}$ has been decomposed into component vectors $\vec{A}_x$ and $\vec{A}_y$ parallel to the coordinate axes. We can describe each component vector with a single number called the **component**. The x-*component* and y-*component* of vector $\vec{A}$, denoted A_x and A_y, are determined as follows:

TACTICS BOX 3.1

Determining the components of a vector

❶ The absolute value $|A_x|$ of the x-component A_x is the magnitude of the component vector $\vec{A}_x$.

❷ The sign of A_x is positive if $\vec{A}_x$ points in the positive x-direction (right), negative if $\vec{A}_x$ points in the negative x-direction (left).

❸ The y-component A_y is determined similarly.

Exercises 10–18

In other words, the component A_x tells us two things: how big $\vec{A}_x$ is and, with its sign, which end of the axis $\vec{A}_x$ points toward. **FIGURE 3.11** shows three examples of determining the components of a vector.

FIGURE 3.11 Determining the components of a vector.

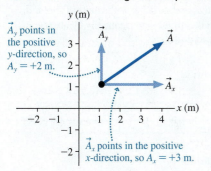
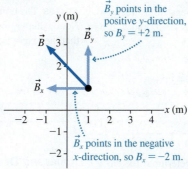
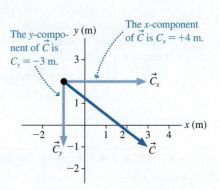

NOTE Beware of the somewhat confusing terminology. $\vec{A}_x$ and $\vec{A}_y$ are called *component vectors,* whereas A_x and A_y are simply called *components.* The components A_x and A_y are just numbers (with units), so make sure you do *not* put arrow symbols over the components.

We will frequently need to decompose a vector into its components. We will also need to "reassemble" a vector from its components. In other words, we need to move back and forth between the geometric and the component representations of a vector. **FIGURE 3.12** shows how this is done.

The magnitude and direction of $\vec{A}$ are found from the components. In this example,
$$A = \sqrt{A_x^2 + A_y^2} \qquad \theta = \tan^{-1}(A_y/A_x)$$

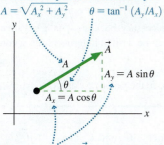

The components of $\vec{A}$ are found from the magnitude and direction.

The angle is defined differently. In this example, the magnitude and direction are
$$B = \sqrt{B_x^2 + B_y^2} \qquad \phi = \tan^{-1}(B_x/|B_y|)$$

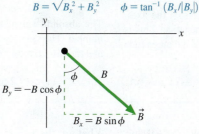

Minus signs must be inserted manually, depending on the vector's direction.

◀ **FIGURE 3.12** Moving between the geometric representation and the component representation.

Each decomposition requires that you pay close attention to the direction in which the vector points and the angles that are defined.

- If a component vector points left (or down), you must *manually* insert a minus sign in front of the component, as was done for B_y in Figure 3.12.
- The role of sines and cosines can be reversed, depending upon which angle is used to define the direction. Compare A_x and B_x.
- The angle used to define direction is almost always between 0° and 90°, so you must take the inverse tangent of a positive number. Use absolute values of the components, as was done to find angle ϕ (Greek phi) in Figure 3.12.

EXAMPLE 3.3 | **Finding the components of an acceleration vector**

Seen from above, a hummingbird's acceleration is (6.0 m/s², 30° south of west). Find the x- and y-components of the acceleration vector $\vec{a}$.

VISUALIZE It's important to *draw* vectors. **FIGURE 3.13** establishes a map-like coordinate system with the x-axis pointing east and the y-axis north. Vector $\vec{a}$ is then decomposed into components parallel to the axes. Notice that the axes are "acceleration axes" with units of acceleration, not xy-axes, because we're measuring an acceleration vector.

▶ **FIGURE 3.13** Decomposition of $\vec{a}$.

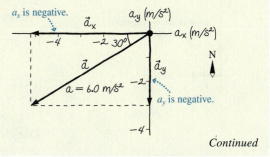

Continued

SOLVE The acceleration vector points to the left (negative x-direction) and down (negative y-direction), so the components a_x and a_y are both negative:

$$a_x = -a\cos 30° = -(6.0\ \text{m/s}^2)\cos 30° = -5.2\ \text{m/s}^2$$
$$a_y = -a\sin 30° = -(6.0\ \text{m/s}^2)\sin 30° = -3.0\ \text{m/s}^2$$

ASSESS The units of a_x and a_y are the same as the units of vector $\vec{a}$. Notice that we had to insert the minus signs manually by observing that the vector points left and down.

EXAMPLE 3.4 | **Finding the direction of motion**

FIGURE 3.14 shows a car's velocity vector $\vec{v}$. Determine the car's speed and direction of motion.

FIGURE 3.14 The velocity vector $\vec{v}$ of Example 3.4.

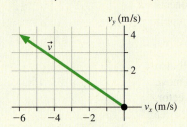

FIGURE 3.15 Decomposition of $\vec{v}$.

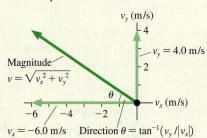

VISUALIZE FIGURE 3.15 shows the components v_x and v_y and defines an angle θ with which we can specify the direction of motion.

SOLVE We can read the components of $\vec{v}$ directly from the axes: $v_x = -6.0\ \text{m/s}$ and $v_y = 4.0\ \text{m/s}$. Notice that v_x is negative. This is enough information to find the car's speed v, which is the magnitude of $\vec{v}$:

$$v = \sqrt{v_x^2 + v_y^2} = \sqrt{(-6.0\ \text{m/s})^2 + (4.0\ \text{m/s})^2} = 7.2\ \text{m/s}$$

From trigonometry, angle θ is

$$\theta = \tan^{-1}\left(\frac{v_y}{|v_x|}\right) = \tan^{-1}\left(\frac{4.0\ \text{m/s}}{6.0\ \text{m/s}}\right) = 34°$$

The absolute value signs are necessary because v_x is a negative number. The velocity vector $\vec{v}$ can be written in terms of the speed and the direction of motion as

$$\vec{v} = (7.2\ \text{m/s}, 34° \text{ above the negative } x\text{-axis})$$

STOP TO THINK 3.3 What are the x- and y-components C_x and C_y of vector $\vec{C}$?

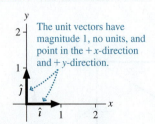

3.4 Unit Vectors and Vector Algebra

FIGURE 3.16 The unit vectors $\hat{\imath}$ and $\hat{\jmath}$.

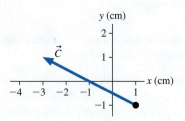

The vectors (1, $+x$-direction) and (1, $+y$-direction), shown in **FIGURE 3.16**, have some interesting and useful properties. Each has a magnitude of 1, has no units, and is parallel to a coordinate axis. A vector with these properties is called a **unit vector.** These unit vectors have the special symbols

$$\hat{\imath} \equiv (1, \text{ positive } x\text{-direction})$$
$$\hat{\jmath} \equiv (1, \text{ positive } y\text{-direction})$$

The notation $\hat{\imath}$ (read "i hat") and $\hat{\jmath}$ (read "j hat") indicates a unit vector with a magnitude of 1. Recall that the symbol $\equiv$ means "is defined as."

Unit vectors establish the directions of the positive axes of the coordinate system. Our choice of a coordinate system may be arbitrary, but once we decide to place a coordinate system on a problem we need something to tell us "That direction is the positive x-direction." This is what the unit vectors do.

The unit vectors provide a useful way to write component vectors. The component vector $\vec{A}_x$ is the piece of vector $\vec{A}$ that is parallel to the x-axis. Similarly, $\vec{A}_y$ is parallel to the y-axis. Because, by definition, the vector $\hat{\imath}$ points along the x-axis and $\hat{\jmath}$ points along the y-axis, we can write

$$\vec{A}_x = A_x \hat{\imath}$$
$$\vec{A}_y = A_y \hat{\jmath}$$

(3.6)

Equations 3.6 separate each component vector into a length and a direction. The full decomposition of vector $\vec{A}$ can then be written

$$\vec{A} = \vec{A}_x + \vec{A}_y = A_x \hat{\imath} + A_y \hat{\jmath}$$

(3.7)

FIGURE 3.17 shows how the unit vectors and the components fit together to form vector $\vec{A}$.

NOTE In three dimensions, the unit vector along the $+z$-direction is called $\hat{k}$, and to describe vector $\vec{A}$ we would include an additional component vector $\vec{A}_z = A_z \hat{k}$.

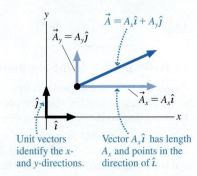

FIGURE 3.17 The decomposition of vector $\vec{A}$ is $A_x\hat{\imath} + A_y\hat{\jmath}$.

Unit vectors identify the x- and y-directions.

Vector $A_x\hat{\imath}$ has length A_x and points in the direction of $\hat{\imath}$.

EXAMPLE 3.5 | Run rabbit run!

A rabbit, escaping a fox, runs 40.0° north of west at 10.0 m/s. A coordinate system is established with the positive x-axis to the east and the positive y-axis to the north. Write the rabbit's velocity in terms of components and unit vectors.

VISUALIZE FIGURE 3.18 shows the rabbit's velocity vector and the coordinate axes. We're showing a velocity vector, so the axes are labeled v_x and v_y rather than x and y.

FIGURE 3.18 The velocity vector $\vec{v}$ is decomposed into components v_x and v_y.

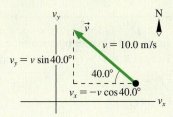

SOLVE 10.0 m/s is the rabbit's *speed,* not its velocity. The velocity, which includes directional information, is

$$\vec{v} = (10.0 \text{ m/s}, 40.0° \text{ north of west})$$

Vector $\vec{v}$ points to the left and up, so the components v_x and v_y are negative and positive, respectively. The components are

$$v_x = -(10.0 \text{ m/s}) \cos 40.0° = -7.66 \text{ m/s}$$
$$v_y = +(10.0 \text{ m/s}) \sin 40.0° = 6.43 \text{ m/s}$$

With v_x and v_y now known, the rabbit's velocity vector is

$$\vec{v} = v_x \hat{\imath} + v_y \hat{\jmath} = (-7.66\hat{\imath} + 6.43\hat{\jmath}) \text{ m/s}$$

Notice that we've pulled the units to the end, rather than writing them with each component.

ASSESS Notice that the minus sign for v_x was inserted manually. **Signs don't occur automatically; you have to set them after checking the vector's direction.**

Vector Math

You learned in Section 3.2 how to add vectors graphically, but it is a tedious problem in geometry and trigonometry to find precise values for the magnitude and direction of the resultant. The addition and subtraction of vectors become much easier if we use components and unit vectors.

To see this, let's evaluate the vector sum $\vec{D} = \vec{A} + \vec{B} + \vec{C}$. To begin, write this sum in terms of the components of each vector:

$$\vec{D} = D_x\hat{\imath} + D_y\hat{\jmath} = \vec{A} + \vec{B} + \vec{C}$$

$$= (A_x\hat{\imath} + A_y\hat{\jmath}) + (B_x\hat{\imath} + B_y\hat{\jmath}) + (C_x\hat{\imath} + C_y\hat{\jmath})$$

(3.8)

We can group together all the x-components and all the y-components on the right side, in which case Equation 3.8 is

$$(D_x)\hat{\imath} + (D_y)\hat{\jmath} = (A_x + B_x + C_x)\hat{\imath} + (A_y + B_y + C_y)\hat{\jmath}$$

(3.9)

Comparing the x- and y-components on the left and right sides of Equation 3.9, we find:

$$D_x = A_x + B_x + C_x$$
$$D_y = A_y + B_y + C_y$$

(3.10)

Stated in words, Equation 3.10 says that we can perform vector addition by adding the x-components of the individual vectors to give the x-component of the resultant and by adding the y-components of the individual vectors to give the y-component of the resultant. This method of vector addition is called **algebraic addition.**

EXAMPLE 3.6 | **Using algebraic addition to find a displacement**

Example 3.1 was about a bird that flew 100 m to the east, then 50 m to the northwest. Use the algebraic addition of vectors to find the bird's net displacement.

VISUALIZE FIGURE 3.19 shows displacement vectors $\vec{A} = (100$ m, east$)$ and $\vec{B} = (50$ m, northwest$)$. We draw vectors tip-to-tail to add them graphically, but it's usually easier to draw them all from the origin if we are going to use algebraic addition.

FIGURE 3.19 The net displacement is $\vec{C} = \vec{A} + \vec{B}$.

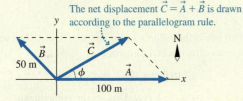

The net displacement $\vec{C} = \vec{A} + \vec{B}$ is drawn according to the parallelogram rule.

SOLVE To add the vectors algebraically we must know their components. From the figure these are seen to be

$$\vec{A} = 100\,\hat{\imath}\ \text{m}$$

$$\vec{B} = (-50\cos 45°\,\hat{\imath} + 50\sin 45°\,\hat{\jmath})\ \text{m} = (-35.3\,\hat{\imath} + 35.3\,\hat{\jmath})\ \text{m}$$

Notice that vector quantities must include units. Also notice, as you would expect from the figure, that $\vec{B}$ has a negative x-component. Adding $\vec{A}$ and $\vec{B}$ by components gives

$$\vec{C} = \vec{A} + \vec{B} = 100\,\hat{\imath}\ \text{m} + (-35.3\,\hat{\imath} + 35.3\,\hat{\jmath})\ \text{m}$$
$$= (100\ \text{m} - 35.3\ \text{m})\hat{\imath} + (35.3\ \text{m})\hat{\jmath} = (64.7\hat{\imath} + 35.3\hat{\jmath})\ \text{m}$$

This would be a perfectly acceptable answer for many purposes. However, we need to calculate the magnitude and direction of $\vec{C}$ if we want to compare this result to our earlier answer. The magnitude of $\vec{C}$ is

$$C = \sqrt{C_x^2 + C_y^2} = \sqrt{(64.7\ \text{m})^2 + (35.3\ \text{m})^2} = 74\ \text{m}$$

The angle ϕ, as defined in Figure 3.19, is

$$\phi = \tan^{-1}\left(\frac{C_y}{|C_x|}\right) = \tan^{-1}\left(\frac{35.3\ \text{m}}{64.7\ \text{m}}\right) = 29°$$

Thus $\vec{C} = (74$ m, 29° north of west$)$, in perfect agreement with Example 3.1.

Vector subtraction and the multiplication of a vector by a scalar, using components, are very much like vector addition. To find $\vec{R} = \vec{P} - \vec{Q}$ we would compute

$$R_x = P_x - Q_x$$
$$R_y = P_y - Q_y$$

(3.11)

Similarly, $\vec{T} = c\vec{S}$ would be

$$T_x = cS_x$$
$$T_y = cS_y$$

(3.12)

In other words, a vector equation is interpreted as meaning: Equate the x-components on both sides of the equals sign, then equate the y-components, and then the z-components. Vector notation allows us to write these three equations in a much more compact form.

Tilted Axes and Arbitrary Directions

FIGURE 3.20 A coordinate system with tilted axes.

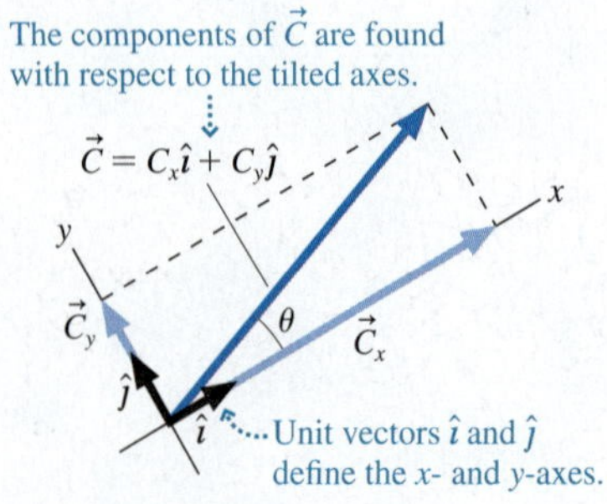

The components of $\vec{C}$ are found with respect to the tilted axes.

$$\vec{C} = C_x\hat{\imath} + C_y\hat{\jmath}$$

Unit vectors $\hat{\imath}$ and $\hat{\jmath}$ define the x- and y-axes.

As we've noted, the coordinate system is entirely your choice. It is a grid that you impose on the problem in a manner that will make the problem easiest to solve. As you've already seen in Chapter 2, it is often convenient to tilt the axes of the coordinate system, such as those shown in FIGURE 3.20. The axes are perpendicular, and the y-axis is oriented correctly with respect to the x-axis, so this is a legitimate coordinate system. There is no requirement that the x-axis has to be horizontal.

Finding components with tilted axes is no harder than what we have done so far. Vector $\vec{C}$ in Figure 3.20 can be decomposed into $\vec{C} = C_x\hat{\imath} + C_y\hat{\jmath}$, where $C_x = C\cos\theta$ and $C_y = C\sin\theta$. Note that the unit vectors $\hat{\imath}$ and $\hat{\jmath}$ correspond to the *axes*, not to "horizontal" and "vertical," so they are also tilted.

Tilted axes are useful if you need to determine component vectors "parallel to" and "perpendicular to" an arbitrary line or surface. This is illustrated in the following example.

EXAMPLE 3.7 | Muscle and bone

The deltoid—the rounded muscle across the top of your upper arm—allows you to lift your arm away from your side. It does so by pulling on an attachment point on the humerus, the upper arm bone, at an angle of 15° with respect to the humerus. If you hold your arm at an angle 30° below horizontal, the deltoid must pull with a force of 720 N to support the weight of your arm, as shown in FIGURE 3.21a. (You'll learn in Chapter 5 that force is a vector

FIGURE 3.21 Finding the components of force parallel and perpendicular to the humerus.

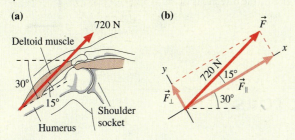

(a)

(b)

quantity measured in units of *newtons*, abbreviated N.) What are the components of the muscle force parallel to and perpendicular to the bone?

VISUALIZE FIGURE 3.21b shows a tilted coordinate system with the x-axis parallel to the humerus. The force $\vec{F}$ is shown 15° from the x-axis. The component of force parallel to the bone, which we can denote $F_\parallel$, is equivalent to the x-component: $F_\parallel = F_x$. Similarly, the component of force perpendicular to the bone is $F_\perp = F_y$.

SOLVE From the geometry of Figure 3.21b, we see that

$$F_\parallel = F \cos 15° = (720 \text{ N}) \cos 15° = 695 \text{ N}$$

$$F_\perp = F \sin 15° = (720 \text{ N}) \sin 15° = 186 \text{ N}$$

ASSESS The muscle pulls nearly parallel to the bone, so we expected $F_\parallel \approx 720$ N and $F_\perp \ll F_\parallel$. Thus our results seem reasonable.

STOP TO THINK 3.4 Angle ϕ that specifies the direction of $\vec{C}$ is given by

a. $\tan^{-1}(|C_x|/C_y)$

b. $\tan^{-1}(C_x/|C_y|)$

c. $\tan^{-1}(|C_x|/|C_y|)$

d. $\tan^{-1}(|C_y|/C_x)$

e. $\tan^{-1}(C_y/|C_x|)$

f. $\tan^{-1}(|C_y|/|C_x|)$

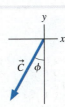

CHALLENGE EXAMPLE 3.8 | Finding the net force

FIGURE 3.22 shows three forces acting at one point. What is the net force $\vec{F}_{net} = \vec{F}_1 + \vec{F}_2 + \vec{F}_3$?

VISUALIZE Figure 3.22 shows the forces and a tilted coordinate system.

SOLVE The vector equation $\vec{F}_{net} = \vec{F}_1 + \vec{F}_2 + \vec{F}_3$ is really two simultaneous equations:

$$(F_{net})_x = F_{1x} + F_{2x} + F_{3x}$$

$$(F_{net})_y = F_{1y} + F_{2y} + F_{3y}$$

The components of the forces are determined with respect to the axes. Thus

$$F_{1x} = F_1 \cos 45° = (50 \text{ N}) \cos 45° = 35 \text{ N}$$

$$F_{1y} = F_1 \sin 45° = (50 \text{ N}) \sin 45° = 35 \text{ N}$$

$\vec{F}_2$ is easier. It is pointing along the y-axis, so $F_{2x} = 0$ N and $F_{2y} = 20$ N. To find the components of $\vec{F}_3$, we need to recognize—because $\vec{F}_3$ points straight down—that the angle between $\vec{F}_3$ and the x-axis is 75°. Thus

$$F_{3x} = F_3 \cos 75° = (57 \text{ N}) \cos 75° = 15 \text{ N}$$

$$F_{3y} = -F_3 \sin 75° = -(57 \text{ N}) \sin 75° = -55 \text{ N}$$

FIGURE 3.22 Three forces.

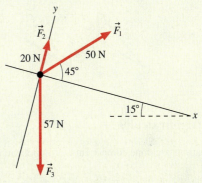

The minus sign in F_{3y} is critical, and it appears not from some formula but because we recognized—from the figure—that the y-component of $\vec{F}_3$, points in the −y-direction. Combining the pieces, we have

$$(F_{net})_x = 35 \text{ N} + 0 \text{ N} + 15 \text{ N} = 50 \text{ N}$$

$$(F_{net})_y = 35 \text{ N} + 20 \text{ N} + (-55 \text{ N}) = 0 \text{ N}$$

Thus the net force is $\vec{F}_{net} = 50\hat{\imath}$ N. It points along the x-axis of the tilted coordinate system.

ASSESS Notice that all work was done with reference to the axes of the coordinate system, not with respect to vertical or horizontal.

SUMMARY

The goals of Chapter 3 have been to learn how vectors are represented and used.

IMPORTANT CONCEPTS

A vector is a quantity described by both a magnitude and a direction.

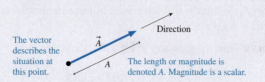

The vector describes the situation at this point.

Direction

The length or magnitude is denoted A. Magnitude is a scalar.

Unit Vectors

Unit vectors have magnitude 1 and no units. Unit vectors $\hat{\imath}$ and $\hat{\jmath}$ define the directions of the x- and y-axes.

USING VECTORS

Components

The component vectors are parallel to the x- and y-axes:

$$\vec{A} = \vec{A}_x + \vec{A}_y = A_x \hat{\imath} + A_y \hat{\jmath}$$

In the figure at the right, for example:

$$A_x = A\cos\theta \qquad A = \sqrt{A_x^2 + A_y^2}$$
$$A_y = A\sin\theta \qquad \theta = \tan^{-1}(A_y/A_x)$$

▶ Minus signs need to be included if the vector points down or left.

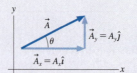

$$
\begin{array}{c|c}
A_x < 0 & A_x > 0 \\
A_y > 0 & A_y > 0 \\
\hline
A_x < 0 & A_x > 0 \\
A_y < 0 & A_y < 0
\end{array}
$$

The components A_x and A_y are the magnitudes of the component vectors $\vec{A}_x$ and $\vec{A}_y$ *and* a plus or minus sign to show whether the component vector points toward the positive end or the negative end of the axis.

Working Graphically

Addition 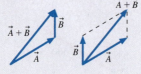   Negative   Subtraction   Multiplication

Working Algebraically

Vector calculations are done component by component: $\vec{C} = 2\vec{A} + \vec{B}$ means $\begin{cases} C_x = 2A_x + B_x \\ C_y = 2A_y + B_y \end{cases}$

The magnitude of $\vec{C}$ is then $C = \sqrt{C_x^2 + C_y^2}$ and its direction is found using $\tan^{-1}$.

TERMS AND NOTATION

scalar	resultant vector	quadrants	component
vector	graphical addition	component vector	unit vector, $\hat{\imath}$ or $\hat{\jmath}$
magnitude	zero vector, $\vec{0}$	decomposition	algebraic addition

CONCEPTUAL QUESTIONS

1. Can the magnitude of the displacement vector be more than the distance traveled? Less than the distance traveled? Explain.
2. If $\vec{C} = \vec{A} + \vec{B}$, can $C = A + B$? Can $C > A + B$? For each, show how or explain why not.
3. If $\vec{C} = \vec{A} + \vec{B}$, can $C = 0$? Can $C < 0$? For each, show how or explain why not.
4. Is it possible to add a scalar to a vector? If so, demonstrate. If not, explain why not.
5. How would you define the *zero vector* $\vec{0}$?
6. Can a vector have a component equal to zero and still have nonzero magnitude? Explain.

7. Can a vector have zero magnitude if one of its components is nonzero? Explain.
8. Suppose two vectors have unequal magnitudes. Can their sum be zero? Explain.
9. Are the following statements true or false? Explain your answer.
   a. The magnitude of a vector can be different in different coordinate systems.
   b. The direction of a vector can be different in different coordinate systems.
   c. The components of a vector can be different in different coordinate systems.

EXERCISES AND PROBLEMS

Exercises

Section 3.1 Scalars and Vectors

Section 3.2 Using Vectors

1. | Trace the vectors in **FIGURE EX3.1** onto your paper. Then find (a) $\vec{A} + \vec{B}$ and (b) $\vec{A} - \vec{B}$.

FIGURE EX3.1   **FIGURE EX3.2**

2. | Trace the vectors in **FIGURE EX3.2** onto your paper. Then find (a) $\vec{A} + \vec{B}$ and (b) $\vec{A} - \vec{B}$.

Section 3.3 Coordinate Systems and Vector Components

3. | a. What are the x- and y-components of vector $\vec{E}$ shown in **FIGURE EX3.3** in terms of the angle θ and the magnitude E?
   b. For the same vector, what are the x- and y-components in terms of the angle ϕ and the magnitude E?

FIGURE EX3.3

4. ‖ A velocity vector 40° below the positive x-axis has a y-component of -10 m/s. What is the value of its x-component?
5. | A position vector in the first quadrant has an x-component of 6 m and a magnitude of 10 m. What is the value of its y-component?
6. | Draw each of the following vectors. Then find its x- and y-components.
   a. $\vec{a} = (3.5 \text{ m/s}^2, \text{ negative } x\text{-direction})$
   b. $\vec{v} = (440 \text{ m/s}, 30° \text{ below the positive } x\text{-axis})$
   c. $\vec{r} = (12 \text{ m}, 40° \text{ above the positive } x\text{-axis})$
7. ‖ Draw each of the following vectors. Then find its x- and y-components.
   a. $\vec{v} = (7.5 \text{ m/s}, 30° \text{ clockwise from the positive } y\text{-axis})$
   b. $\vec{a} = (1.5 \text{ m/s}^2, 30° \text{ above the negative } x\text{-axis})$
   c. $\vec{F} = (50.0 \text{ N}, 36.9° \text{ counterclockwise from the positive } y\text{-axis})$

8. | Let $\vec{C} = (3.15 \text{ m}, 15° \text{ above the negative } x\text{-axis})$ and $\vec{D} = (25.6 \text{ m}, 30° \text{ to the right of the negative } y\text{-axis})$. Find the x- and y-components of each vector.
9. | A runner is training for an upcoming marathon by running around a 100-m-diameter circular track at constant speed. Let a coordinate system have its origin at the center of the circle with the x-axis pointing east and the y-axis north. The runner starts at $(x, y) = (50 \text{ m}, 0 \text{ m})$ and runs 2.5 times around the track in a clockwise direction. What is his displacement vector? Give your answer as a magnitude and direction.

Section 3.4 Unit Vectors and Vector Algebra

10. ‖ Draw each of the following vectors, label an angle that specifies the vector's direction, then find its magnitude and direction.
   a. $\vec{B} = -4.0\hat{\imath} + 4.0\hat{\jmath}$
   b. $\vec{r} = (-2.0\hat{\imath} - 1.0\hat{\jmath}) \text{ cm}$
   c. $\vec{v} = (-10\hat{\imath} - 100\hat{\jmath}) \text{ m/s}$
   d. $\vec{a} = (20\hat{\imath} + 10\hat{\jmath}) \text{ m/s}^2$
11. ‖ Draw each of the following vectors, label an angle that specifies the vector's direction, and then find the vector's magnitude and direction.
   a. $\vec{A} = 3.0\hat{\imath} + 7.0\hat{\jmath}$
   b. $\vec{a} = (-2.0\hat{\imath} + 4.5\hat{\jmath}) \text{ m/s}^2$
   c. $\vec{v} = (14\hat{\imath} - 11\hat{\jmath}) \text{ m/s}$
   d. $\vec{r} = (-2.2\hat{\imath} - 3.3\hat{\jmath}) \text{ m}$
12. ‖ Let $\vec{A} = 2\hat{\imath} + 3\hat{\jmath}$, $\vec{B} = 2\hat{\imath} - 4\hat{\jmath}$, and $\vec{C} = \vec{A} + \vec{B}$.
   a. Write vector $\vec{C}$ in component form.
   b. Draw a coordinate system and on it show vectors $\vec{A}$, $\vec{B}$, and $\vec{C}$.
   c. What are the magnitude and direction of vector $\vec{C}$?
13. | Let $\vec{A} = 4\hat{\imath} - 2\hat{\jmath}$, $\vec{B} = -3\hat{\imath} + 5\hat{\jmath}$, and $\vec{C} = \vec{A} + \vec{B}$.
   a. Write vector $\vec{C}$ in component form.
   b. Draw a coordinate system and on it show vectors $\vec{A}$, $\vec{B}$, and $\vec{C}$.
   c. What are the magnitude and direction of vector $\vec{C}$?
14. | Let $\vec{A} = 4\hat{\imath} - 2\hat{\jmath}$, $\vec{B} = -3\hat{\imath} + 5\hat{\jmath}$, and $\vec{D} = \vec{A} - \vec{B}$.
   a. Write vector $\vec{D}$ in component form.
   b. Draw a coordinate system and on it show vectors $\vec{A}$, $\vec{B}$, and $\vec{D}$.
   c. What are the magnitude and direction of vector $\vec{D}$?
15. | Let $\vec{A} = 4\hat{\imath} - 2\hat{\jmath}$, $\vec{B} = -3\hat{\imath} + 5\hat{\jmath}$, and $\vec{E} = 2\vec{A} + 3\vec{B}$.
   a. Write vector $\vec{E}$ in component form.
   b. Draw a coordinate system and on it show vectors $\vec{A}$, $\vec{B}$, and $\vec{E}$.
   c. What are the magnitude and direction of vector $\vec{E}$?

16. | Let $\vec{A} = 4\hat{i} - 2\hat{j}$, $\vec{B} = -3\hat{i} + 5\hat{j}$, and $\vec{F} = \vec{A} - 4\vec{B}$.
    a. Write vector $\vec{F}$ in component form.
    b. Draw a coordinate system and on it show vectors $\vec{A}$, $\vec{B}$, and $\vec{F}$.
    c. What are the magnitude and direction of vector $\vec{F}$?

17. | Let $\vec{E} = 2\hat{i} + 3\hat{j}$ and $\vec{F} = 2\hat{i} - 2\hat{j}$. Find the magnitude of
    a. $\vec{E}$ and $\vec{F}$ b. $\vec{E} + \vec{F}$ c. $-\vec{E} - 2\vec{F}$

18. | Let $\vec{B} = (5.0$ m, $30°$ counterclockwise from vertical). Find the x- and y-components of $\vec{B}$ in each of the two coordinate systems shown in **FIGURE EX3.18**.

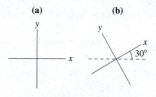

FIGURE EX3.18

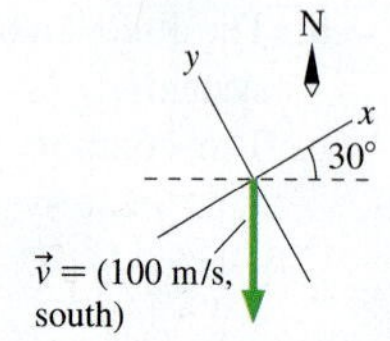

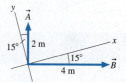

$\vec{v} = (100$ m/s, south)

FIGURE EX3.19

19. | What are the x- and y-components of the velocity vector shown in **FIGURE EX3.19**?

20. || For the three vectors shown in **FIGURE EX3.20**, $\vec{A} + \vec{B} + \vec{C} = 1\hat{j}$. What is vector $\vec{B}$?
    a. Write $\vec{B}$ in component form.
    b. Write $\vec{B}$ as a magnitude and a direction.

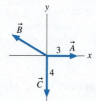

FIGURE EX3.20

21. | The *magnetic field* inside an instrument is $\vec{B} = (2.0\hat{i} - 1.0\hat{j})$ T where $\vec{B}$ represents the magnetic field vector and T stands for *tesla,* the unit of the magnetic field. What are the magnitude and direction of the magnetic field?

Problems

22. || Let $\vec{A} = (3.0$ m, $20°$ south of east), $\vec{B} = (2.0$ m, north), and $\vec{C} = (5.0$ m, $70°$ south of west).
    a. Draw and label $\vec{A}$, $\vec{B}$, and $\vec{C}$ with their tails at the origin. Use a coordinate system with the x-axis to the east.
    b. Write $\vec{A}$, $\vec{B}$, and $\vec{C}$ in component form, using unit vectors.
    c. Find the magnitude and the direction of $\vec{D} = \vec{A} + \vec{B} + \vec{C}$.

23. || The position of a particle as a function of time is given by
    CALC $\vec{r} = (5.0\hat{i} + 4.0\hat{j})t^2$ m, where t is in seconds.
    a. What is the particle's distance from the origin at t = 0, 2, and 5 s?
    b. Find an expression for the particle's velocity $\vec{v}$ as a function of time.
    c. What is the particle's speed at t = 0, 2, and 5 s?

24. || a. What is the angle ϕ between vectors $\vec{E}$ and $\vec{F}$ in **FIGURE P3.24**?
    b. Use geometry and trigonometry to determine the magnitude and direction of $\vec{G} = \vec{E} + \vec{F}$.
    c. Use components to determine the magnitude and direction of $\vec{G} = \vec{E} + \vec{F}$.

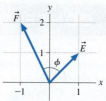

FIGURE P3.24

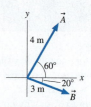

FIGURE P3.25

25. || **FIGURE P3.25** shows vectors $\vec{A}$ and $\vec{B}$. Find vector $\vec{C}$ such that $\vec{A} + \vec{B} + \vec{C} = \vec{0}$. Write your answer in component form.

26. ||| **FIGURE P3.26** shows vectors $\vec{A}$ and $\vec{B}$. Find $\vec{D} = 2\vec{A} + \vec{B}$. Write your answer in component form.

FIGURE P3.26

27. || Find a vector that points in the same direction as the vector $(\hat{i} + \hat{j})$ and whose magnitude is 1.

28. || While vacationing in the mountains you do some hiking. In the morning, your displacement is $\vec{S}_{\text{morning}} = (2000$ m, east) + $(3000$ m, north) + $(200$ m, vertical). After lunch, your displacement is $\vec{S}_{\text{afternoon}} = (1500$ m, west) + $(2000$ m, north) − $(300$ m, vertical).
    a. At the end of the hike, how much higher or lower are you compared to your starting point?
    b. What is the magnitude of your net displacement for the day?

29. || The minute hand on a watch is 2.0 cm in length. What is the displacement vector of the tip of the minute hand
    a. From 8:00 to 8:20 A.M.?
    b. From 8:00 to 9:00 A.M.?

30. || You go to an amusement park with your friend Betty, who wants to ride the 30-m-diameter Ferris wheel. She starts the ride at the lowest point of a wheel that, as you face it, rotates counterclockwise. What is her displacement vector when the wheel has rotated by an angle of $60°$? Give your answer as a magnitude and direction.

31. || Ruth sets out to visit her friend Ward, who lives 50 mi north and 100 mi east of her. She starts by driving east, but after 30 mi she comes to a detour that takes her 15 mi south before going east again. She then drives east for 8 mi and runs out of gas, so Ward flies there in his small plane to get her. What is Ward's displacement vector? Give your answer (a) in component form, using a coordinate system in which the y-axis points north, and (b) as a magnitude and direction.

32. | A cannon tilted upward at $30°$ fires a cannonball with a speed of 100 m/s. What is the component of the cannonball's velocity parallel to the ground?

33. | You are fixing the roof of your house when a hammer breaks loose and slides down. The roof makes an angle of $35°$ with the horizontal, and the hammer is moving at 4.5 m/s when it reaches the edge. What are the horizontal and vertical components of the hammer's velocity just as it leaves the roof?

34. | Jack and Jill ran up the hill at 3.0 m/s. The horizontal component of Jill's velocity vector was 2.5 m/s.
    a. What was the angle of the hill?
    b. What was the vertical component of Jill's velocity?

35. | A pine cone falls straight down from a pine tree growing on a 20° slope. The pine cone hits the ground with a speed of 10 m/s. What is the component of the pine cone's impact velocity (a) parallel to the ground and (b) perpendicular to the ground?

36. | Kami is walking through the airport with her two-wheeled suitcase. The suitcase handle is tilted 40° from vertical, and Kami pulls parallel to the handle with a force of 120 N. (Force is measured in *newtons*, abbreviated N.) What are the horizontal and vertical components of her applied force?

37. ‖ Dee is on a swing in the playground. The chains are 2.5 m long, and the tension in each chain is 450 N when Dee is 55 cm above the lowest point of her swing. Tension is a vector directed along the chain, measured in *newtons*, abbreviated N. What are the horizontal and vertical components of the tension at this point in the swing?

38. ‖ Your neighbor Paul has rented a truck with a loading ramp. The ramp is tilted upward at 25°, and Paul is pulling a large crate up the ramp with a rope that angles 10° above the ramp. If Paul pulls with a force of 550 N, what are the horizontal and vertical components of his force? (Force is measured in *newtons*, abbreviated N.)

39. ‖ Tom is climbing a 3.0-m-long ladder that leans against a vertical wall, contacting the wall 2.5 m above the ground. His weight of 680 N is a vector pointing vertically downward. (Weight is measured in *newtons*, abbreviated N.) What are the components of Tom's weight parallel and perpendicular to the ladder?

40. ‖ The treasure map in **FIGURE P3.40** gives the following directions to the buried treasure: "Start at the old oak tree, walk due north for 500 paces, then due east for 100 paces. Dig." But when you arrive, you find an angry dragon just north of the tree. To avoid the dragon, you set off along the yellow brick road at an angle 60° east of north. After walking 300 paces you see an opening through the woods. Which direction should you go, and how far, to reach the treasure?

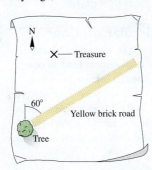

FIGURE P3.40

41. ‖‖ The bacterium *E. coli* is a single-cell organism that lives in the
BIO gut of healthy animals, including humans. When grown in a uniform medium in the laboratory, these bacteria swim along zig-zag paths at a constant speed of 20 μm/s. **FIGURE P3.41** shows the trajectory of an *E. coli* as it moves from point A to point E. What are the magnitude and direction of the bacterium's average velocity for the entire trip?

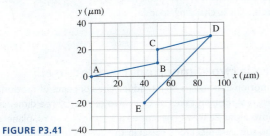

FIGURE P3.41

42. ‖ A flock of ducks is trying to migrate south for the winter, but they keep being blown off course by a wind blowing from the west at 6.0 m/s. A wise elder duck finally realizes that the solution is to fly at an angle to the wind. If the ducks can fly at 8.0 m/s relative to the air, what direction should they head in order to move directly south?

43. ‖ **FIGURE P3.43** shows three ropes tied together in a knot. One of your friends pulls on a rope with 3.0 units of force and another pulls on a second rope with 5.0 units of force. How hard and in what direction must you pull on the third rope to keep the knot from moving?

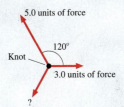

FIGURE P3.43

44. ‖ Four forces are exerted on the object shown in **FIGURE P3.44**. (Forces are measured in *newtons*, abbreviated N.) The *net force* on the object is $\vec{F}_{net} = \vec{F}_1 + \vec{F}_2 + \vec{F}_3 + \vec{F}_4 = 4.0\hat{\imath}$ N. What are (a) $\vec{F}_3$ and (b) $\vec{F}_4$? Give your answers in component form.

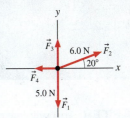

FIGURE P3.44

45. ‖ **FIGURE P3.45** shows four electric charges located at the corners of a rectangle. Like charges, you will recall, repel each other while opposite charges attract. Charge B exerts a repulsive force (directly *away from* B) on charge A of 3.0 N. Charge C exerts an attractive force (directly *toward* C) on charge A of 6.0 N. Finally, charge D exerts an attractive force of 2.0 N on charge A. Assuming that forces are vectors, what are the magnitude and direction of the net force $\vec{F}_{net}$ exerted on charge A?

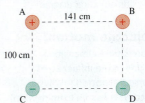

FIGURE P3.45

4 Kinematics in Two Dimensions

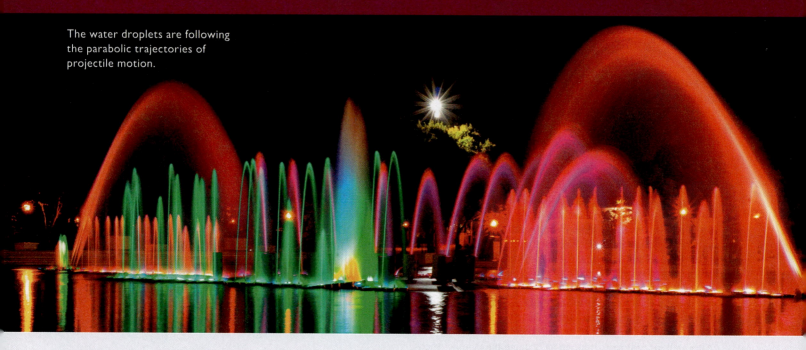

The water droplets are following the parabolic trajectories of projectile motion.

IN THIS CHAPTER, you will learn how to solve problems about motion in a plane.

How do objects accelerate in two dimensions?

An object accelerates when it changes velocity. In two dimensions, velocity can change by changing magnitude (speed) or by changing direction. These are represented by acceleration components tangent to and perpendicular to an object's trajectory.

« LOOKING BACK Section 1.5 Finding acceleration vectors on a motion diagram

What is projectile motion?

Projectile motion is two-dimensional free-fall motion under the influence of only gravity. Projectile motion follows a parabolic trajectory. It has uniform motion in the horizontal direction and $a_y = -g$ in the vertical direction.

« LOOKING BACK Section 2.5 Free fall

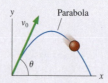

What is relative motion?

Coordinate systems that move relative to each other are called reference frames. If object C has velocity $\vec{v}_{CA}$ relative to a reference frame A, and if A moves with velocity $\vec{v}_{AB}$ relative to another reference frame B, then the velocity of C in reference frame B is $\vec{v}_{CB} = \vec{v}_{CA} + \vec{v}_{AB}$.

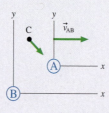

What is circular motion?

An object moving in a circle (or rotating) has an *angular displacement* instead of a linear displacement. Circular motion is described by angular velocity ω (analogous to velocity v_s) and angular acceleration α (analogous to acceleration a_s). We'll study both uniform and accelerated circular motion.

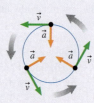

What is centripetal acceleration?

An object in circular motion is always changing direction. The acceleration of changing direction—called centripetal acceleration—points to the center of the circle. All circular motion has a centripetal acceleration. An object also has a *tangential acceleration* if it is changing speed.

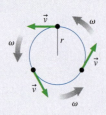

Where is two-dimensional motion used?

Linear motion allowed us to introduce the concepts of motion, but most real motion takes place in two or even three dimensions. Balls move along curved trajectories, cars turn corners, planets orbit the sun, and electrons spiral in the earth's magnetic field. Where is two-dimensional motion used? Everywhere!

4.1 Motion in Two Dimensions

Motion diagrams are an important tool for visualizing motion, and we'll continue to use them, but we also need to develop a mathematical description of motion in two dimensions. For convenience, we'll say that any two-dimensional motion is in the xy-plane regardless of whether the plane of motion is horizontal or vertical.

FIGURE 4.1 shows a particle moving along a curved path—its *trajectory*—in the xy-plane. We can locate the particle in terms of its position vector $\vec{r} = x\hat{\imath} + y\hat{\jmath}$.

NOTE In Chapter 2 we made extensive use of position-versus-time graphs, either x versus t or y versus t. Figure 4.1, like many of the graphs we'll use in this chapter, is a graph of y versus x. In other words, it's an actual *picture* of the trajectory, not an abstract representation of the motion.

FIGURE 4.2a shows the particle moving from position $\vec{r}_1$ at time t_1 to position $\vec{r}_2$ at a later time t_2. The average velocity—pointing in the direction of the displacement $\Delta \vec{r}$—is

$$\vec{v}_{\text{avg}} = \frac{\Delta \vec{r}}{\Delta t} = \frac{\Delta x}{\Delta t}\hat{\imath} + \frac{\Delta y}{\Delta t}\hat{\jmath} \qquad (4.1)$$

You learned in Chapter 2 that the instantaneous velocity is the limit of $\vec{v}_{\text{avg}}$ as $\Delta t \to 0$. As Δt decreases, point 2 moves closer to point 1 until, as **FIGURE 4.2b** shows, the displacement vector becomes tangent to the curve. Consequently, **the instantaneous velocity vector $\vec{v}$ is tangent to the trajectory.**

Mathematically, the limit of Equation 4.1 gives

$$\vec{v} = \lim_{\Delta t \to 0} \frac{\Delta \vec{r}}{\Delta t} = \frac{d\vec{r}}{dt} = \frac{dx}{dt}\hat{\imath} + \frac{dy}{dt}\hat{\jmath} \qquad (4.2)$$

We can also write the velocity vector in terms of its x- and y-components as

$$\vec{v} = v_x\hat{\imath} + v_y\hat{\jmath} \qquad (4.3)$$

Comparing Equations 4.2 and 4.3, you can see that the velocity vector $\vec{v}$ has x- and y-components

$$v_x = \frac{dx}{dt} \qquad \text{and} \qquad v_y = \frac{dy}{dt} \qquad (4.4)$$

That is, the x-component v_x of the velocity vector is the rate dx/dt at which the particle's x-coordinate is changing. The y-component is similar.

FIGURE 4.2c illustrates another important feature of the velocity vector. If the vector's angle θ is measured from the positive x-direction, the velocity vector components are

$$\begin{aligned} v_x &= v\cos\theta \\ v_y &= v\sin\theta \end{aligned} \qquad (4.5)$$

where

$$v = \sqrt{v_x^2 + v_y^2} \qquad (4.6)$$

is the particle's *speed* at that point. Speed is always a positive number (or zero), whereas the components are *signed* quantities (i.e., they can be positive or negative) to convey information about the direction of the velocity vector. Conversely, we can use the two velocity components to determine the direction of motion:

$$\theta = \tan^{-1}\left(\frac{v_y}{v_x}\right) \qquad (4.7)$$

NOTE In Chapter 2, you learned that the *value* of the velocity is the *slope* of the position-versus-time graph. Now we see that the *direction* of the velocity vector $\vec{v}$ is the *tangent* to the y-versus-x graph of the trajectory. **FIGURE 4.3**, on the next page, reminds you that these two graphs use different interpretations of the tangent lines. The tangent to the trajectory does not tell us anything about how fast the particle is moving.

FIGURE 4.1 A particle moving along a trajectory in the xy-plane.

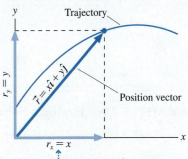

The x- and y-components of $\vec{r}$ are simply x and y.

FIGURE 4.2 The instantaneous velocity vector is tangent to the trajectory.

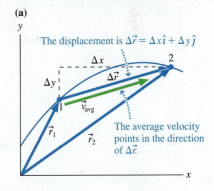

(a)

The displacement is $\Delta\vec{r} = \Delta x\hat{\imath} + \Delta y\hat{\jmath}$

The average velocity points in the direction of $\Delta\vec{r}$.

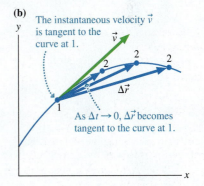

(b)

The instantaneous velocity $\vec{v}$ is tangent to the curve at 1.

As $\Delta t \to 0$, $\Delta\vec{r}$ becomes tangent to the curve at 1.

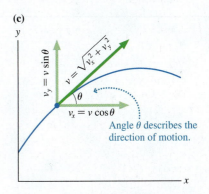

(c)

Angle θ describes the direction of motion.

FIGURE 4.3 Two different uses of tangent lines.

Position-versus-time graph

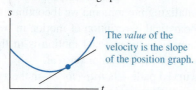

The *value* of the velocity is the slope of the position graph.

Trajectory

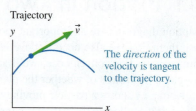

The *direction* of the velocity is tangent to the trajectory.

EXAMPLE 4.1 | **Finding velocity**

A sports car's position on a winding road is given by

$$\vec{r} = (6.0t - 0.10t^2 + 0.00048t^3)\hat{\imath} + (8.0t + 0.060t^2 - 0.00095t^3)\hat{\jmath}$$

where the y-axis points north, t is in s, and r is in m. What are the car's speed and direction at $t = 120$ s?

MODEL Model the car as a particle.

SOLVE Velocity is the derivative of position, so

$$v_x = \frac{dx}{dt} = 6.0 - 2(0.10t) + 3(0.00048t^2)$$

$$v_y = \frac{dy}{dt} = 8.0 + 2(0.060t) - 3(0.00095t^2)$$

Written as a vector, the velocity is

$$\vec{v} = (6.0 - 0.20t + 0.00144t^2)\hat{\imath} + (8.0 + 0.120t - 0.00285t^2)\hat{\jmath}$$

where t is in s and v is in m/s. At $t = 120$ s, we can calculate $\vec{v} = (2.7\hat{\imath} - 18.6\hat{\jmath})$ m/s. The car's speed at this instant is

$$v = \sqrt{v_x^2 + v_y^2} = \sqrt{(2.7 \text{ m/s})^2 + (-18.6 \text{ m/s})^2} = 19 \text{ m/s}$$

The velocity vector has a negative y-component, so the direction of motion is to the right (east) and down (south). The angle below the x-axis is

$$\theta = \tan^{-1}\left(\frac{|-18.6 \text{ m/s}|}{2.7 \text{ m/s}}\right) = 82°$$

So, at this instant, the car is headed 82° south of east at a speed of 19 m/s.

STOP TO THINK 4.1 During which time interval or intervals is the particle described by these position graphs at rest? More than one may be correct.

a. 0–1 s

b. 1–2 s

c. 2–3 s

d. 3–4 s

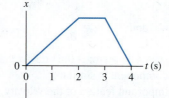

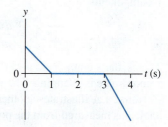

Acceleration Graphically

In « Section 1.5 we defined the *average acceleration* $\vec{a}_{\text{avg}}$ of a moving object to be

$$\vec{a}_{\text{avg}} = \frac{\Delta\vec{v}}{\Delta t} \quad (4.8)$$

From its definition, we see that $\vec{a}$ **points in the same direction as** $\Delta\vec{v}$, the change of velocity. As an object moves, its velocity vector can change in two possible ways:

1. The magnitude of $\vec{v}$ can change, indicating a change in speed, or
2. The direction of $\vec{v}$ can change, indicating that the object has changed direction.

The kinematics of Chapter 2 considered only the acceleration due to changing speed. Now it's time to look at the acceleration associated with changing direction. Tactics Box 4.1 shows how we can use the velocity vectors on a motion diagram to determine the direction of the average acceleration vector. This is an extension of Tactics Box 1.3, which showed how to find $\vec{a}$ for one-dimensional motion.

TACTICS BOX 4.1

Finding the acceleration vector

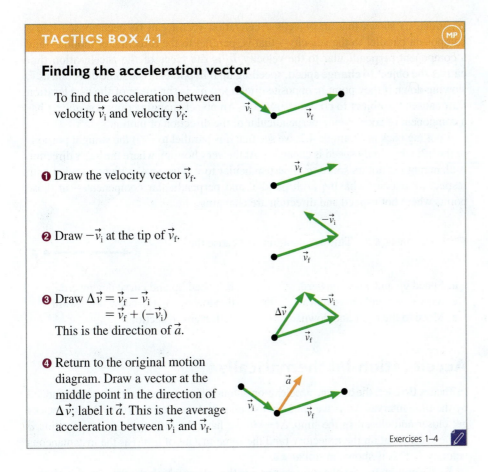

To find the acceleration between velocity $\vec{v}_i$ and velocity $\vec{v}_f$:

❶ Draw the velocity vector $\vec{v}_f$.

❷ Draw $-\vec{v}_i$ at the tip of $\vec{v}_f$.

❸ Draw $\Delta \vec{v} = \vec{v}_f - \vec{v}_i$
$\qquad = \vec{v}_f + (-\vec{v}_i)$
This is the direction of $\vec{a}$.

❹ Return to the original motion diagram. Draw a vector at the middle point in the direction of $\Delta \vec{v}$; label it $\vec{a}$. This is the average acceleration between $\vec{v}_i$ and $\vec{v}_f$.

Exercises 1–4

Our everyday use of the word "accelerate" means "speed up." The mathematical definition of acceleration—the rate of change of velocity—also includes slowing down, as you learned in Chapter 2, as well as changing direction. All these are motions that change the velocity.

EXAMPLE 4.2 | **Through the valley**

A ball rolls down a long hill, through the valley, and back up the other side. Draw a complete motion diagram of the ball.

MODEL Model the ball as a particle.

VISUALIZE FIGURE 4.4 is the motion diagram. Where the particle moves along a *straight line,* it speeds up if $\vec{a}$ and $\vec{v}$ point in the same direction and slows down if $\vec{a}$ and $\vec{v}$ point in opposite

directions. This important idea was the basis for the one-dimensional kinematics we developed in Chapter 2. When the direction of $\vec{v}$ changes, as it does when the ball goes through the valley, we need to use vector subtraction to find the direction of $\Delta \vec{v}$ and thus of $\vec{a}$. The procedure is shown at two points in the motion diagram.

FIGURE 4.4 The motion diagram of the ball of Example 4.2.

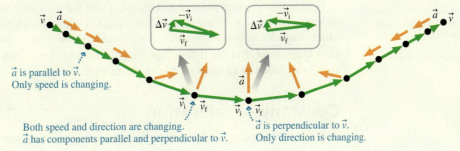

$\vec{a}$ is parallel to $\vec{v}$.
Only speed is changing.

Both speed and direction are changing.
$\vec{a}$ has components parallel and perpendicular to $\vec{v}$.

$\vec{a}$ is perpendicular to $\vec{v}$.
Only direction is changing.

FIGURE 4.5 Analyzing the acceleration vector.

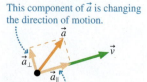

This component of $\vec{a}$ is changing the direction of motion.

This component of $\vec{a}$ is changing the speed of the motion.

FIGURE 4.5 shows that an object's acceleration vector can be decomposed into a component parallel to the velocity—that is, parallel to the direction of motion—and a component perpendicular to the velocity. $\vec{a}_\parallel$ **is the piece of the acceleration that causes the object to change speed,** speeding up if $\vec{a}_\parallel$ points in the same direction as $\vec{v}$, slowing down if they point in opposite directions. $\vec{a}_\perp$ **is the piece of the acceleration that causes the object to change direction.** An object changing direction *always* has a component of acceleration perpendicular to the direction of motion.

Looking back at Example 4.2, we see that $\vec{a}$ is parallel to $\vec{v}$ on the straight portions of the hill where only speed is changing. At the very bottom, where the ball's direction is changing but not its speed, $\vec{a}$ is perpendicular to $\vec{v}$. The acceleration is angled with respect to velocity—having both parallel and perpendicular components—at those points where both speed and direction are changing.

STOP TO THINK 4.2 This acceleration will cause the particle to

a. Speed up and curve upward.
b. Speed up and curve downward.
c. Slow down and curve upward.
d. Slow down and curve downward.
e. Move to the right and down.
f. Reverse direction.

Acceleration Mathematically

FIGURE 4.6 The instantaneous acceleration $\vec{a}$.

(a) The parallel component is associated with a change of speed.

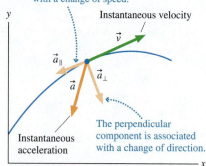

Instantaneous velocity

Instantaneous acceleration

The perpendicular component is associated with a change of direction.

(b)

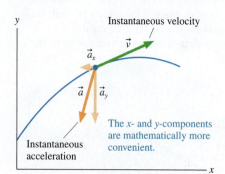

Instantaneous velocity

Instantaneous acceleration

The x- and y-components are mathematically more convenient.

In Tactics Box 4.1, the average acceleration is found from two velocity vectors separated by the time interval Δt. If we let Δt get smaller and smaller, the two velocity vectors get closer and closer. In the limit $\Delta t \rightarrow 0$, we have the instantaneous acceleration $\vec{a}$ at the same point on the trajectory (and the same instant of time) as the instantaneous velocity $\vec{v}$. This is shown in **FIGURE 4.6**.

By definition, the acceleration vector $\vec{a}$ is the rate at which the velocity $\vec{v}$ is changing at that instant. To show this, Figure 4.6a decomposes $\vec{a}$ into components $\vec{a}_\parallel$ and $\vec{a}_\perp$ that are parallel and perpendicular to the trajectory. As we just showed, $\vec{a}_\parallel$ is associated with a change of speed, and $\vec{a}_\perp$ is associated with a change of direction. Both kinds of changes are accelerations. Notice that $\vec{a}_\perp$ always points toward the "inside" of the curve because that is the direction in which $\vec{v}$ is changing.

Although the parallel and perpendicular components of $\vec{a}$ convey important ideas about acceleration, it's often more practical to write $\vec{a}$ in terms of the x- and y-components shown in Figure 4.6b. Because $\vec{v} = v_x \hat{\imath} + v_y \hat{\jmath}$, we find

$$\vec{a} = a_x \hat{\imath} + a_y \hat{\jmath} = \frac{d\vec{v}}{dt} = \frac{dv_x}{dt} \hat{\imath} + \frac{dv_y}{dt} \hat{\jmath} \tag{4.9}$$

from which we see that

$$a_x = \frac{dv_x}{dt} \quad \text{and} \quad a_y = \frac{dv_y}{dt} \tag{4.10}$$

That is, the x-component of $\vec{a}$ is the rate dv_x/dt at which the x-component of velocity is changing.

Notice that Figures 4.6a and 4.6b show the *same* acceleration vector; all that differs is how we've chosen to decompose it. For motion with constant acceleration, which includes projectile motion, the decomposition into x- and y-components is most convenient. But we'll find that the parallel and perpendicular components are especially suited to an analysis of circular motion.

Constant Acceleration

If the acceleration $\vec{a} = a_x \hat{\imath} + a_y \hat{\jmath}$ is constant, then the two components a_x and a_y are both constant. In this case, everything you learned about constant-acceleration kinematics in « Section 2.4 carries over to two-dimensional motion.

Consider a particle that moves with constant acceleration from an initial position $\vec{r}_i = x_i\hat{\imath} + y_i\hat{\jmath}$, starting with initial velocity $\vec{v}_i = v_{ix}\hat{\imath} + v_{iy}\hat{\jmath}$. Its position and velocity at a final point f are

$$x_f = x_i + v_{ix}\,\Delta t + \tfrac{1}{2}a_x(\Delta t)^2 \qquad y_f = y_i + v_{iy}\,\Delta t + \tfrac{1}{2}a_y(\Delta t)^2$$
$$v_{fx} = v_{ix} + a_x\,\Delta t \qquad\qquad v_{fy} = v_{iy} + a_y\,\Delta t \qquad\qquad (4.11)$$

There are *many* quantities to keep track of in two-dimensional kinematics, making the pictorial representation all the more important as a problem-solving tool.

NOTE For constant acceleration, the x-component of the motion and the y-component of the motion are independent of each other. However, they remain connected through the fact that Δt must be the same for both.

EXAMPLE 4.3 | Plotting a spacecraft trajectory

In the distant future, a small spacecraft is drifting "north" through the galaxy at 680 m/s when it receives a command to return to the starship. The pilot rotates the spacecraft until the nose is pointed 25° north of east, then engages the ion engine. The spacecraft accelerates at 75 m/s². Plot the spacecraft's trajectory for the first 20 s.

MODEL Model the spacecraft as a particle with constant acceleration.

VISUALIZE FIGURE 4.7 shows a pictorial representation in which the y-axis points north and the spacecraft starts at the origin. Notice that each point in the motion is labeled with *two* positions (x and y), *two* velocity components (v_x and v_y), and the time t. This will be our standard labeling scheme for trajectory problems.

SOLVE The acceleration vector has both x- and y-components; their values have been calculated in the pictorial representation. But it is a *constant* acceleration, so we can write

$$x_1 = x_0 + v_{0x}(t_1 - t_0) + \tfrac{1}{2}a_x(t_1 - t_0)^2$$
$$= 34.0t_1^2 \text{ m}$$
$$y_1 = y_0 + v_{0y}(t_1 - t_0) + \tfrac{1}{2}a_y(t_1 - t_0)^2$$
$$= 680t_1 + 15.8t_1^2 \text{ m}$$

where t_1 is in s. Graphing software produces the trajectory shown in FIGURE 4.8. The trajectory is a parabola, which is characteristic of two-dimensional motion with constant acceleration.

FIGURE 4.7 Pictorial representation of the spacecraft.

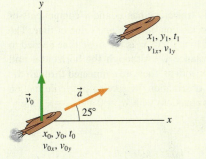

Known
$x_0 = y_0 = 0$ m $v_{0x} = 0$ m/s $v_{0y} = 680$ m/s
$a_x = (75 \text{ m/s}^2)\cos 25° = 68.0 \text{ m/s}^2$
$a_y = (75 \text{ m/s}^2)\sin 25° = 31.6 \text{ m/s}^2$
$t_0 = 0$ s $t_1 = 0$ s to 20 s

Find
x_1 and y_1

FIGURE 4.8 The spacecraft trajectory.

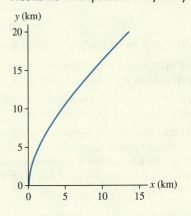

4.2 Projectile Motion

Baseballs and tennis balls flying through the air, Olympic divers, and daredevils shot from cannons all exhibit what we call *projectile motion*. A **projectile** is **an object that moves in two dimensions under the influence of only gravity.** Projectile motion is an extension of the free-fall motion we studied in Chapter 2. We will continue to neglect the influence of air resistance, leading to results that are a good approximation of reality for relatively heavy objects moving relatively slowly over relatively short distances. As we'll see, projectiles in two dimensions follow a *parabolic trajectory* like the one seen in FIGURE 4.9.

The start of a projectile's motion, be it thrown by hand or shot from a gun, is called the *launch*, and the angle θ of the initial velocity $\vec{v}_0$ above the horizontal (i.e., above

FIGURE 4.9 A parabolic trajectory.

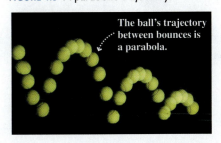

The ball's trajectory between bounces is a parabola.

FIGURE 4.10 A projectile launched with initial velocity $\vec{v}_0$.

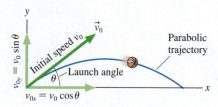

the x-axis) is called the **launch angle**. **FIGURE 4.10** illustrates the relationship between the initial velocity vector $\vec{v}_0$ and the initial values of the components v_{0x} and v_{0y}. You can see that

$$v_{0x} = v_0 \cos\theta$$
$$v_{0y} = v_0 \sin\theta \qquad (4.12)$$

where v_0 is the initial speed.

NOTE A projectile launched at an angle *below* the horizontal (such as a ball thrown downward from the roof of a building) has *negative* values for θ and v_{0y}. However, the *speed* v_0 is always positive.

Gravity acts downward, and we know that objects released from rest fall straight down, not sideways. Hence a projectile has no horizontal acceleration, while its vertical acceleration is simply that of free fall. Thus

$$a_x = 0$$
$$a_y = -g \qquad \text{(projectile motion)} \qquad (4.13)$$

FIGURE 4.11 The velocity and acceleration vectors of a projectile.

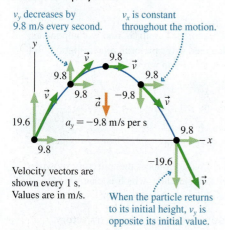

v_y decreases by 9.8 m/s every second.

v_x is constant throughout the motion.

$a_y = -9.8$ m/s per s

Velocity vectors are shown every 1 s. Values are in m/s.

When the particle returns to its initial height, v_y is opposite its initial value.

In other words, **the vertical component of acceleration a_y is just the familiar $-g$ of free fall, while the horizontal component a_x is zero. Projectiles are in free fall.**

To see how these conditions influence the motion, **FIGURE 4.11** shows a projectile launched from $(x_0, y_0) = (0 \text{ m}, 0 \text{ m})$ with an initial velocity $\vec{v}_0 = (9.8\hat{\imath} + 19.6\hat{\jmath})$ m/s. The value of v_x never changes because there's no horizontal acceleration, but v_y decreases by 9.8 m/s every second. This is what it *means* to accelerate at $a_y = -9.8 \text{ m/s}^2 = (-9.8 \text{ m/s})$ per second.

You can see from Figure 4.11 that **projectile motion is made up of two independent motions:** uniform motion at constant velocity in the horizontal direction and free-fall motion in the vertical direction. The kinematic equations that describe these two motions are simply Equations 4.11 with $a_x = 0$ and $a_y = -g$.

EXAMPLE 4.4 | Don't try this at home!

A stunt man drives a car off a 10.0-m-high cliff at a speed of 20.0 m/s. How far does the car land from the base of the cliff?

MODEL Model the car as a particle in free fall. Assume that the car is moving horizontally as it leaves the cliff.

VISUALIZE The pictorial representation, shown in **FIGURE 4.12**, is *very* important because the number of quantities to keep track of is quite large. We have chosen to put the origin at the base of the cliff. The assumption that the car is moving horizontally as it leaves the cliff leads to $v_{0x} = v_0$ and $v_{0y} = 0$ m/s.

FIGURE 4.12 Pictorial representation for the car of Example 4.4.

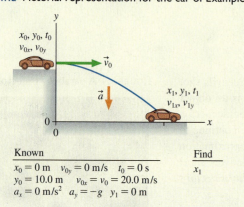

Known	Find
$x_0 = 0$ m $v_{0y} = 0$ m/s $t_0 = 0$ s	x_1
$y_0 = 10.0$ m $v_{0x} = v_0 = 20.0$ m/s	
$a_x = 0$ m/s^2 $a_y = -g$ $y_1 = 0$ m	

SOLVE Each point on the trajectory has x- and y-components of position, velocity, and acceleration but only *one* value of time. The time needed to move horizontally to x_1 is the *same* time needed to fall vertically through distance y_0. **Although the horizontal and vertical motions are independent, they are connected through the time t.** This is a critical observation for solving projectile motion problems. The kinematics equations with $a_x = 0$ and $a_y = -g$ are

$$x_1 = x_0 + v_{0x}(t_1 - t_0) = v_0 t_1$$
$$y_1 = 0 = y_0 + v_{0y}(t_1 - t_0) - \tfrac{1}{2}g(t_1 - t_0)^2 = y_0 - \tfrac{1}{2}g t_1^2$$

We can use the vertical equation to determine the time t_1 needed to fall distance y_0:

$$t_1 = \sqrt{\frac{2y_0}{g}} = \sqrt{\frac{2(10.0 \text{ m})}{9.80 \text{ m/s}^2}} = 1.43 \text{ s}$$

We then insert this expression for t into the horizontal equation to find the distance traveled:

$$x_1 = v_0 t_1 = (20.0 \text{ m/s})(1.43 \text{ s}) = 28.6 \text{ m}$$

ASSESS The cliff height is ≈ 33 ft and the initial speed is $v_0 \approx 40$ mph. Traveling $x_1 = 29$ m ≈ 95 ft before hitting the ground seems reasonable.

The x- and y-equations of Example 4.4 are parametric equations. It's not hard to eliminate t and write an expression for y as a function of x. From the x_1 equation, $t_1 = x_1/v_0$. Substituting this into the y_1 equation, we find

$$y = y_0 - \frac{g}{2v_0^2}x^2 \qquad (4.14)$$

The graph of $y = cx^2$ is a parabola, so Equation 4.14 represents an inverted parabola that starts from height y_0. This proves, as we asserted previously, that a projectile follows a parabolic trajectory.

Reasoning About Projectile Motion

Suppose a heavy ball is launched exactly horizontally at height h above a horizontal field. At the exact instant that the ball is launched, a second ball is simply dropped from height h. Which ball hits the ground first?

It may seem hard to believe, but—if air resistance is neglected—the balls hit the ground *simultaneously*. They do so because the horizontal and vertical components of projectile motion are independent of each other. The initial horizontal velocity of the first ball has *no* influence over its vertical motion. Neither ball has any initial motion in the vertical direction, so both fall distance h in the same amount of time. You can see this in **FIGURE 4.13**.

FIGURE 4.14a shows a useful way to think about the trajectory of a projectile. Without gravity, a projectile would follow a straight line. Because of gravity, the particle at time t has "fallen" a distance $\frac{1}{2}gt^2$ below this line. The separation grows as $\frac{1}{2}gt^2$, giving the trajectory its parabolic shape.

Use this idea to think about the following "classic" problem in physics:

A hungry bow-and-arrow hunter in the jungle wants to shoot down a coconut that is hanging from the branch of a tree. He points his arrow directly at the coconut, but as luck would have it, the coconut falls from the branch at the *exact* instant the hunter releases the string. Does the arrow hit the coconut?

You might think that the arrow will miss the falling coconut, but it doesn't. Although the arrow travels very fast, it follows a slightly curved parabolic trajectory, not a straight line. Had the coconut stayed on the tree, the arrow would have curved under its target as gravity caused it to fall a distance $\frac{1}{2}gt^2$ below the straight line. But $\frac{1}{2}gt^2$ is also the distance the coconut falls while the arrow is in flight. Thus, as **FIGURE 4.14b** shows, the arrow and the coconut fall the same distance and meet at the same point!

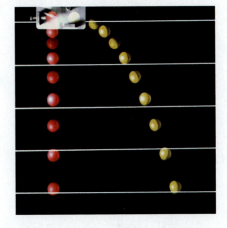

FIGURE 4.13 A projectile launched horizontally falls in the same time as a projectile that is released from rest.

FIGURE 4.14 A projectile follows a parabolic trajectory because it "falls" a distance $\frac{1}{2}gt^2$ below a straight-line trajectory.

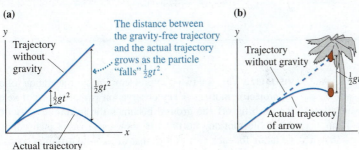

(a)

Trajectory without gravity

The distance between the gravity-free trajectory and the actual trajectory grows as the particle "falls" $\frac{1}{2}gt^2$.

$\frac{1}{2}gt^2$

Actual trajectory

(b)

Trajectory without gravity

$\frac{1}{2}gt^2$

Actual trajectory of arrow

The Projectile Motion Model

Projectile motion is an ideal that's rarely achieved by real objects. Nonetheless, the **projectile motion model** is another important simplification of reality that we can add to our growing list of models.

MODEL 4.1

Projectile motion

For motion under the influence of only gravity.

- ■ Model the object as a particle launched with speed v_0 at angle θ:
- ■ Mathematically:
  - **Uniform motion** in the horizontal direction with $v_x = v_0 \cos\theta$.
  - **Constant acceleration** in the vertical direction with $a_y = -g$.
  - Same Δt for both motions.
- ■ Limitations: Model fails if air resistance is significant.

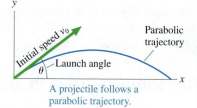

Initial speed v_0

Launch angle θ

Parabolic trajectory

A projectile follows a parabolic trajectory.

Exercise 9

PROBLEM-SOLVING STRATEGY 4.1

Projectile motion problems

MODEL Is it reasonable to ignore air resistance? If so, use the projectile motion model.

VISUALIZE Establish a coordinate system with the x-axis horizontal and the y-axis vertical. Define symbols and identify what the problem is trying to find. For a launch at angle θ, the initial velocity components are $v_{ix} = v_0 \cos\theta$ and $v_{iy} = v_0 \sin\theta$.

SOLVE The acceleration is known: $a_x = 0$ and $a_y = -g$. Thus the problem is one of two-dimensional kinematics. The kinematic equations are

Horizontal	Vertical
$x_f = x_i + v_{ix}\,\Delta t$	$y_f = y_i + v_{iy}\,\Delta t - \frac{1}{2}g(\Delta t)^2$
$v_{fx} = v_{ix} = \text{constant}$	$v_{fy} = v_{iy} - g\,\Delta t$

Δt is the same for the horizontal and vertical components of the motion. Find Δt from one component, then use that value for the other component.

ASSESS Check that your result has correct units and significant figures, is reasonable, and answers the question.

EXAMPLE 4.5 | **Jumping frog contest**

Frogs, with their long, strong legs, are excellent jumpers. And thanks to the good folks of Calaveras County, California, who have a jumping frog contest every year in honor of a Mark Twain story, we have very good data on how far a determined frog can jump.

High-speed cameras show that a good jumper goes into a crouch, then rapidly extends his legs by typically 15 cm during a 65 ms push off, leaving the ground at a 30° angle. How far does this frog leap?

MODEL Model the push off as linear motion with uniform acceleration. A bullfrog is fairly heavy and dense, so ignore air resistance and model the leap as projectile motion.

VISUALIZE This is a two-part problem: linear acceleration followed by projectile motion. A key observation is that **the final velocity for pushing off the ground becomes the initial velocity of the projectile motion.** FIGURE 4.15 shows a separate pictorial representation for each part. Notice that we've used different coordinate systems for the two parts; coordinate systems are our choice, and for each part of the motion we've chosen the coordinate system that makes the problem easiest to solve.

SOLVE While pushing off, the frog travels 15 cm = 0.15 m in 65 ms = 0.065 s. We could find his speed at the end of pushing off if we knew the acceleration. Because the initial velocity is zero,

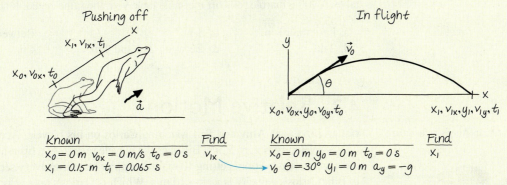

FIGURE 4.15 Pictorial representations of the jumping frog.

we can find the acceleration from the position-acceleration-time kinematic equation:

$$x_1 = x_0 + v_{0x}\,\Delta t + \tfrac{1}{2}a_x(\Delta t)^2 = \tfrac{1}{2}a_x(\Delta t)^2$$

$$a_x = \frac{2x_1}{(\Delta t)^2} = \frac{2(0.15\text{ m})}{(0.065\text{ s})^2} = 71\text{ m/s}^2$$

This is a substantial acceleration, but it doesn't last long. At the end of the 65 ms push off, the frog's velocity is

$$v_{1x} = v_{0x} + a_x\,\Delta t = (71\text{ m/s}^2)(0.065\text{ s}) = 4.62\text{ m/s}$$

We'll keep an extra significant figure here to avoid round-off error in the second half of the problem.

The end of the push off is the beginning of the projectile motion, so the second part of the problem is to find the distance of a projectile launched with velocity $\vec{v}_0 = (4.62\text{ m/s}, 30°)$. The initial x- and y-components of the launch velocity are

$$v_{0x} = v_0\cos\theta \qquad v_{0y} = v_0\sin\theta$$

The kinematic equations of projectile motion, with $a_x = 0$ and $a_y = -g$, are

$$x_1 = x_0 + v_{0x}\,\Delta t \qquad y_1 = y_0 + v_{0y}\,\Delta t - \tfrac{1}{2}g(\Delta t)^2$$
$$= (v_0\cos\theta)\,\Delta t \qquad\quad = (v_0\sin\theta)\,\Delta t - \tfrac{1}{2}g(\Delta t)^2$$

We can find the time of flight from the vertical equation by setting $y_1 = 0$:

$$0 = (v_0\sin\theta)\,\Delta t - \tfrac{1}{2}g(\Delta t)^2 = (v_0\sin\theta - \tfrac{1}{2}g\,\Delta t)\,\Delta t$$

and thus

$$\Delta t = 0 \qquad\text{or}\qquad \Delta t = \frac{2v_0\sin\theta}{g}$$

Both are legitimate solutions. The first corresponds to the instant when $y = 0$ at the launch, the second to when $y = 0$ as the frog hits the ground. Clearly, we want the second solution. Substituting this expression for Δt into the equation for x_1 gives

$$x_1 = (v_0\cos\theta)\frac{2v_0\sin\theta}{g} = \frac{2v_0^2\sin\theta\cos\theta}{g}$$

We can simplify this result with the trigonometric identity $2\sin\theta\cos\theta = \sin(2\theta)$. Thus the distance traveled by the frog is

$$x_1 = \frac{v_0^2\sin(2\theta)}{g}$$

Using $v_0 = 4.62\text{ m/s}$ and $\theta = 30°$, we find that the frog leaps a distance of 1.9 m.

ASSESS 1.9 m is about 6 feet, or about 10 times the frog's body length. That's pretty amazing, but true. Jumps of 2.2 m have been recorded in the lab. And the Calaveras County record holder, Rosie the Ribeter, covered 6.5 m—21 feet—in three jumps!

The distance a projectile travels is called its *range*. As Example 4.5 found, a projectile that lands at the same elevation from which it was launched has

$$\text{range} = \frac{v_0^2\sin(2\theta)}{g} \tag{4.15}$$

The maximum range occurs for $\theta = 45°$, where $\sin(2\theta) = 1$. But there's more that we can learn from this equation. Because $\sin(180° - x) = \sin x$, it follows that $\sin(2(90° - \theta)) = \sin(2\theta)$. Consequently, a projectile launched either at angle θ *or* at angle $(90° - \theta)$ will travel the same distance *over level ground*. **FIGURE 4.16** shows several trajectories of projectiles launched with the same initial speed.

NOTE Equation 4.15 is *not* a general result. It applies *only* in situations where the projectile lands at the same elevation from which it was fired.

FIGURE 4.16 Trajectories of a projectile launched at different angles with a speed of 99 m/s.

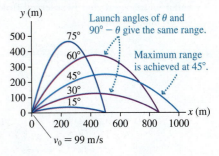

STOP TO THINK 4.3 A 50 g marble rolls off a table and hits 2 m from the base of the table. A 100 g marble rolls off the same table with the same speed. It lands at distance

a. Less than 1 m.   b. 1 m.   c. Between 1 m and 2 m.
d. 2 m.   e. Between 2 m and 4 m.   f. 4 m.

4.3 Relative Motion

FIGURE 4.17 Velocities in Amy's reference frame.

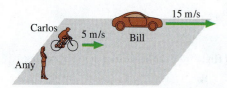

FIGURE 4.17 shows Amy and Bill watching Carlos on his bicycle. According to Amy, Carlos's velocity is $v_x = 5$ m/s. Bill sees the bicycle receding in his rearview mirror, in the *negative x-*direction, getting 10 m farther away from him every second. According to Bill, Carlos's velocity is $v_x = -10$ m/s. Which is Carlos's *true* velocity?

Velocity is not a concept that can be true or false. Carlos's velocity *relative to Amy* is $(v_x)_{CA} = 5$ m/s, where the subscript notation means "C relative to A." Similarly, Carlos's velocity *relative to Bill* is $(v_x)_{CB} = -10$ m/s. These are both valid descriptions of Carlos's motion.

It's not hard to see how to combine the velocities for one-dimensional motion:

The first subscript is the same on both sides.   The last subscript is the same on both sides.

$$(v_x)_{CB} = (v_x)_{CA} + (v_x)_{AB} \qquad (4.16)$$

The inner subscripts "cancel."

We'll justify this relationship later in this section and then extend it to two-dimensional motion.

Equation 4.16 tells us that the velocity of C relative to B is the velocity of C relative to A *plus* the velocity of A relative to B. Note that

$$(v_x)_{AB} = -(v_x)_{BA} \qquad (4.17)$$

because if B is moving to the right relative to A, then A is moving to the left relative to B. In Figure 4.17, Bill is moving to the right relative to Amy with $(v_x)_{BA} = 15$ m/s, so $(v_x)_{AB} = -15$ m/s. Knowing that Carlos's velocity relative to Amy is 5 m/s, we find that Carlos's velocity relative to Bill is, as expected, $(v_x)_{CB} = (v_x)_{CA} + (v_x)_{AB} = 5$ m/s $+ (-15)$ m/s $= -10$ m/s.

EXAMPLE 4.6 | **A speeding bullet**

The police are chasing a bank robber. While driving at 50 m/s, they fire a bullet to shoot out a tire of his car. The police gun shoots bullets at 300 m/s. What is the bullet's speed as measured by a TV camera crew parked beside the road?

MODEL Assume that all motion is in the positive *x-*direction. The bullet is the object that is observed from both the police car and the ground.

SOLVE The bullet B's velocity relative to the gun G is $(v_x)_{BG} = 300$ m/s. The gun, inside the car, is traveling relative to the TV crew C at $(v_x)_{GC} = 50$ m/s. We can combine these values to find that the bullet's velocity relative to the TV crew on the ground is

$$(v_x)_{BC} = (v_x)_{BG} + (v_x)_{GC} = 300 \text{ m/s} + 50 \text{ m/s} = 350 \text{ m/s}$$

ASSESS It should be no surprise in this simple situation that we simply add the velocities.

Reference Frames

A coordinate system in which an experimenter (possibly with the assistance of helpers) makes position and time measurements of physical events is called a **reference frame**. In Figure 4.17, Amy and Bill each had their own reference frame (where they were at rest) in which they measured Carlos's velocity.

More generally, **FIGURE 4.18** shows two reference frames, A and B, and an object C. It is assumed that the reference frames are moving with respect to each other. At this instant of time, the position vector of C in reference frame A is $\vec{r}_{CA}$, meaning "the position of C relative to the origin of frame A." Similarly, $\vec{r}_{CB}$ is the position vector of C in reference frame B. Using vector addition, you can see that

$$\vec{r}_{CB} = \vec{r}_{CA} + \vec{r}_{AB} \tag{4.18}$$

where $\vec{r}_{AB}$ locates the origin of A relative to the origin of B.

In general, object C is moving relative to both reference frames. To find its velocity in each reference frame, take the time derivative of Equation 4.18:

$$\frac{d\vec{r}_{CB}}{dt} = \frac{d\vec{r}_{CA}}{dt} + \frac{d\vec{r}_{AB}}{dt} \tag{4.19}$$

By definition, $d\vec{r}/dt$ is a velocity. The first derivative is $\vec{v}_{CB}$, the velocity of C relative to B. Similarly, the second derivative is the velocity of C relative to A, $\vec{v}_{CA}$. The last derivative is slightly different because it doesn't refer to object C. Instead, this is the velocity $\vec{v}_{AB}$ of reference frame A relative to reference frame B. As we noted in one dimension, $\vec{v}_{AB} = -\vec{v}_{BA}$.

Writing Equation 4.19 in terms of velocities, we have

$$\vec{v}_{CB} = \vec{v}_{CA} + \vec{v}_{AB} \tag{4.20}$$

This relationship between velocities in different reference frames was recognized by Galileo in his pioneering studies of motion, hence it is known as the **Galilean transformation of velocity.** If you know an object's velocity in one reference frame, you can *transform* it into the velocity that would be measured in a different reference frame. Just as in one dimension, the velocity of C relative to B is the velocity of C relative to A plus the velocity of A relative to B, *but* you must add the velocities as vectors for two-dimensional motion.

As we've seen, the Galilean velocity transformation is pretty much common sense for one-dimensional motion. The real usefulness appears when an object travels in a *medium* moving with respect to the earth. For example, a boat moves relative to the water. What is the boat's net motion if the water is a flowing river? Airplanes fly relative to the air, but the air at high altitudes often flows at high speed. Navigation of boats and planes requires knowing both the motion of the vessel in the medium and the motion of the medium relative to the earth.

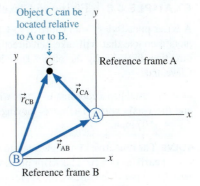

FIGURE 4.18 Two reference frames.

EXAMPLE 4.7 | Flying to Cleveland I

Cleveland is 300 miles east of Chicago. A plane leaves Chicago flying due east at 500 mph. The pilot forgot to check the weather and doesn't know that the wind is blowing to the south at 50 mph. What is the plane's ground speed? Where is the plane 0.60 h later, when the pilot expects to land in Cleveland?

MODEL Establish a coordinate system with the x-axis pointing east and the y-axis north. The plane P flies in the air, so its velocity relative to the air A is $\vec{v}_{PA} = 500\hat{\imath}$ mph. Meanwhile, the air is moving relative to the ground G at $\vec{v}_{AG} = -50\hat{\jmath}$ mph.

FIGURE 4.19 The wind causes a plane flying due east in the air to move to the southeast relative to the ground.

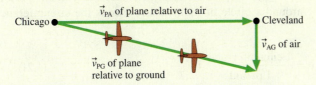

SOLVE The velocity equation $\vec{v}_{PG} = \vec{v}_{PA} + \vec{v}_{AG}$ is a vector-addition equation. **FIGURE 4.19** shows graphically what happens. Although the nose of the plane points east, the wind carries the plane in a direction somewhat south of east. The plane's velocity relative to the ground is

$$\vec{v}_{PG} = \vec{v}_{PA} + \vec{v}_{AG} = (500\hat{\imath} - 50\hat{\jmath})\ \text{mph}$$

The plane's ground speed is

$$v = \sqrt{(v_x)_{PG}^2 + (v_y)_{PG}^2} = 502\ \text{mph}$$

After flying for 0.60 h at this velocity, the plane's location (relative to Chicago) is

$$x = (v_x)_{PG}t = (500\ \text{mph})(0.60\ \text{h}) = 300\ \text{mi}$$
$$y = (v_y)_{PG}t = (-50\ \text{mph})(0.60\ \text{h}) = -30\ \text{mi}$$

The plane is 30 mi due south of Cleveland! Although the pilot thought he was flying to the east, his actual heading has been $\tan^{-1}(50\ \text{mph}/500\ \text{mph}) = \tan^{-1}(0.10) = 5.71°$ south of east.

EXAMPLE 4.8 | Flying to Cleveland II

A wiser pilot flying from Chicago to Cleveland on the same day plots a course that will take her directly to Cleveland. In which direction does she fly the plane? How long does it take to reach Cleveland?

MODEL Establish a coordinate system with the x-axis pointing east and the y-axis north. The air is moving relative to the ground at $\vec{v}_{AG} = -50\hat{j}$ mph.

SOLVE The objective of navigation is to move between two points on the earth's surface. The wiser pilot, who knows that the wind will affect her plane, draws the vector picture of **FIGURE 4.20**. She sees that she'll need $(v_y)_{PG} = 0$, in order to fly due east to Cleveland. This will require turning the nose of the plane at an angle θ north of east, making $\vec{v}_{PA} = (500\cos\theta\,\hat{i} + 500\sin\theta\,\hat{j})$ mph.

The velocity equation is $\vec{v}_{PG} = \vec{v}_{PA} + \vec{v}_{AG}$. The desired heading is found from setting the y-component of this equation to zero:

$$(v_y)_{PG} = (v_y)_{PA} + (v_y)_{AG} = (500\sin\theta - 50)\text{ mph} = 0\text{ mph}$$

$$\theta = \sin^{-1}\left(\frac{50\text{ mph}}{500\text{ mph}}\right) = 5.74°$$

FIGURE 4.20 To travel due east in a south wind, a pilot has to point the plane somewhat to the northeast.

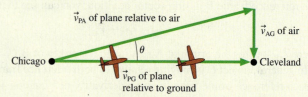

The plane's velocity relative to the ground is then $\vec{v}_{PG} = (500\text{ mph}) \times \cos 5.74°\,\hat{i} = 497\hat{i}$ mph. This is slightly slower than the speed relative to the air. The time needed to fly to Cleveland at this speed is

$$t = \frac{300\text{ mi}}{497\text{ mph}} = 0.604\text{ h}$$

It takes 0.004 h = 14 s longer to reach Cleveland than it would on a day without wind.

ASSESS A boat crossing a river or an ocean current faces the same difficulties. These are exactly the kinds of calculations performed by pilots of boats and planes as part of navigation.

STOP TO THINK 4.4 A plane traveling horizontally to the right at 100 m/s flies past a helicopter that is going straight up at 20 m/s. From the helicopter's perspective, the plane's direction and speed are

a. Right and up, less than 100 m/s.
b. Right and up, 100 m/s.
c. Right and up, more than 100 m/s.
d. Right and down, less than 100 m/s.
e. Right and down, 100 m/s.
f. Right and down, more than 100 m/s.

4.4 Uniform Circular Motion

FIGURE 4.21 A particle in uniform circular motion.

The velocity is tangent to the circle.
The velocity vectors are all the same length.

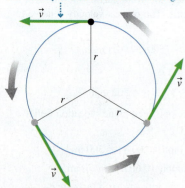

FIGURE 4.21 shows a particle moving around a circle of radius r. The particle might be a satellite in an orbit, a ball on the end of a string, or even just a dot painted on the side of a rotating wheel. Circular motion is another example of motion in a plane, but it is quite different from projectile motion.

To begin the study of circular motion, consider a particle that moves at *constant speed* around a circle of radius r. This is called **uniform circular motion.** Regardless of what the particle represents, its velocity vector $\vec{v}$ is always tangent to the circle. The particle's speed v is constant, so vector $\vec{v}$ is always the same length.

The time interval it takes the particle to go around the circle once, completing one revolution (abbreviated rev), is called the **period** of the motion. Period is represented by the symbol T. It's easy to relate the particle's period T to its speed v. For a particle moving with constant speed, speed is simply distance/time. In one period, the particle moves once around a circle of radius r and travels the circumference $2\pi r$. Thus

$$v = \frac{1\text{ circumference}}{1\text{ period}} = \frac{2\pi r}{T} \tag{4.21}$$

EXAMPLE 4.9 | **A rotating crankshaft**

A 4.0-cm-diameter crankshaft turns at 2400 rpm (revolutions per minute). What is the speed of a point on the surface of the crankshaft?

SOLVE We need to determine the time it takes the crankshaft to make 1 rev. First, we convert 2400 rpm to revolutions per second:

$$\frac{2400 \text{ rev}}{1 \text{ min}} \times \frac{1 \text{ min}}{60 \text{ s}} = 40 \text{ rev/s}$$

If the crankshaft turns 40 times in 1 s, the time for 1 rev is

$$T = \frac{1}{40} \text{ s} = 0.025 \text{ s}$$

Thus the speed of a point on the surface, where $r = 2.0$ cm $= 0.020$ m, is

$$v = \frac{2\pi r}{T} = \frac{2\pi(0.020 \text{ m})}{0.025 \text{ s}} = 5.0 \text{ m/s}$$

Angular Position

Rather than using xy-coordinates, it will be more convenient to describe the position of a particle in circular motion by its distance r from the center of the circle and its angle θ from the positive x-axis. This is shown in **FIGURE 4.22**. The angle θ is the **angular position** of the particle.

We can distinguish a position above the x-axis from a position that is an equal angle below the x-axis by *defining* θ to be positive when measured *counterclockwise* (ccw) from the positive x-axis. An angle measured clockwise (cw) from the positive x-axis has a negative value. "Clockwise" and "counterclockwise" in circular motion are analogous, respectively, to "left of the origin" and "right of the origin" in linear motion, which we associated with negative and positive values of x. A particle $30°$ below the positive x-axis is equally well described by either $\theta = -30°$ or $\theta = +330°$. We could also describe this particle by $\theta = \frac{11}{12}$ rev, where *revolutions* are another way to measure the angle.

Although degrees and revolutions are widely used measures of angle, mathematicians and scientists usually find it more useful to measure the angle θ in Figure 4.22 by using the **arc length** s that the particle travels along the edge of a circle of radius r. We define the angular unit of **radians** such that

$$\theta(\text{radians}) \equiv \frac{s}{r} \tag{4.22}$$

The radian, which is abbreviated rad, is the SI unit of angle. An angle of 1 rad has an arc length s exactly equal to the radius r.

The arc length completely around a circle is the circle's circumference $2\pi r$. Thus the angle of a full circle is

$$\theta_{\text{full circle}} = \frac{2\pi r}{r} = 2\pi \text{ rad}$$

This relationship is the basis for the well-known conversion factors

$$1 \text{ rev} = 360° = 2\pi \text{ rad}$$

As a simple example of converting between radians and degrees, let's convert an angle of 1 rad to degrees:

$$1 \text{ rad} = 1 \text{ rad} \times \frac{360°}{2\pi \text{ rad}} = 57.3°$$

FIGURE 4.22 A particle's position is described by distance r and angle θ.

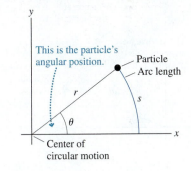

This is the particle's angular position.

Particle
Arc length
r
s
θ
Center of circular motion

Circular motion is one of the most common types of motion.

Thus a rough approximation is $1 \text{ rad} \approx 60°$. We will often specify angles in degrees, but keep in mind that the SI unit is the radian.

An important consequence of Equation 4.22 is that the arc length spanning angle θ is

$$s = r\theta \qquad (\text{with } \theta \text{ in rad}) \qquad (4.23)$$

This is a result that we will use often, but it is valid *only* if θ is measured in radians and not in degrees. This very simple relationship between angle and arc length is one of the primary motivations for using radians.

> **NOTE** Units of angle are often troublesome. Unlike the kilogram or the second, for which we have standards, the radian is a *defined* unit. It's really just a *name* to remind us that we're dealing with an angle. Consequently, the radian unit sometimes appears or disappears without warning. This seems rather mysterious until you get used to it. This textbook will call your attention to such behavior the first few times it occurs. With a little practice, you'll soon learn when the rad unit is needed and when it's not.

Angular Velocity

FIGURE 4.23 shows a particle moving in a circle from an initial angular position θ_i at time t_i to a final angular position θ_f at a later time t_f. The change $\Delta\theta = \theta_f - \theta_i$ is called the **angular displacement.** We can measure the particle's circular motion in terms of the rate of change of θ, just as we measured the particle's linear motion in terms of the rate of change of its position s.

In analogy with linear motion, let's define the *average angular velocity* to be

$$\text{average angular velocity} \equiv \frac{\Delta\theta}{\Delta t} \qquad (4.24)$$

As the time interval Δt becomes very small, $\Delta t \to 0$, we arrive at the definition of the instantaneous **angular velocity:**

$$\omega \equiv \lim_{\Delta t \to 0} \frac{\Delta\theta}{\Delta t} = \frac{d\theta}{dt} \qquad (\text{angular velocity}) \qquad (4.25)$$

The symbol ω is a lowercase Greek omega, *not* an ordinary w. The SI unit of angular velocity is rad/s, but °/s, rev/s, and rev/min are also common units. Revolutions per minute is abbreviated rpm.

Angular velocity is the *rate* at which a particle's angular position is changing as it moves around a circle. A particle that starts from $\theta = 0$ rad with an angular velocity of 0.5 rad/s will be at angle $\theta = 0.5$ rad after 1 s, at $\theta = 1.0$ rad after 2 s, at $\theta = 1.5$ rad after 3 s, and so on. Its angular position is increasing at the *rate* of 0.5 radian per second. **A particle moves with uniform circular motion if and only if its angular velocity ω is constant and unchanging**.

Angular velocity, like the velocity v_s of one-dimensional motion, can be positive or negative. The signs shown in **FIGURE 4.24** are based on the fact that θ was defined to be positive for a counterclockwise rotation. Because the definition $\omega = d\theta/dt$ for circular motion parallels the definition $v_s = ds/dt$ for linear motion, the graphical relationships we found between v_s and s in Chapter 2 apply equally well to ω and θ:

$$\omega = \text{slope of the } \theta\text{-versus-}t \text{ graph at time } t$$
$$\theta_f = \theta_i + \text{area under the } \omega\text{-versus-}t \text{ curve between } t_i \text{ and } t_f$$
$$= \theta_i + \omega\Delta t \qquad (4.26)$$

You will see many more instances where circular motion is analogous to linear motion with angular variables replacing linear variables. Thus much of what you learned about linear kinematics carries over to circular motion.

FIGURE 4.23 A particle moves with angular velocity ω.

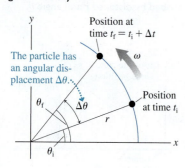

FIGURE 4.24 Positive and negative angular velocities.

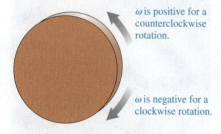

ω is positive for a counterclockwise rotation.

ω is negative for a clockwise rotation.

EXAMPLE 4.10 | A graphical representation of circular motion

FIGURE 4.25 shows the angular position of a painted dot on the edge of a rotating wheel. Describe the wheel's motion and draw an ω-versus-t graph.

FIGURE 4.25 Angular position graph for the wheel of Example 4.10.

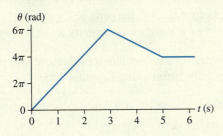

SOLVE Although circular motion seems to "start over" every revolution (every 2π rad), the angular position θ continues to increase. $\theta = 6\pi$ rad corresponds to three revolutions. This wheel makes 3 ccw rev (because θ is getting more positive) in 3 s, immediately reverses direction and makes 1 cw rev in 2 s, then stops at $t = 5$ s

and holds the position $\theta = 4\pi$ rad. The angular velocity is found by measuring the slope of the graph:

$t = 0\text{–}3$ s	slope $= \Delta\theta/\Delta t = 6\pi$ rad/3 s $= 2\pi$ rad/s
$t = 3\text{–}5$ s	slope $= \Delta\theta/\Delta t = -2\pi$ rad/2 s $= -\pi$ rad/s
$t > 5$ s	slope $= \Delta\theta/\Delta t = 0$ rad/s

These results are shown as an ω-versus-t graph in **FIGURE 4.26**. For the first 3 s, the motion is uniform circular motion with $\omega = 2\pi$ rad/s. The wheel then changes to a different uniform circular motion with $\omega = -\pi$ rad/s for 2 s, then stops.

FIGURE 4.26 ω-versus-t graph for the wheel of Example 4.10.

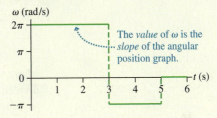

The *value* of ω is the *slope* of the angular position graph.

NOTE In physics, we nearly always want to give results as numerical values. Example 4.9 had a π in the equation, but we used its numerical value to compute $v = 5.0$ m/s. However, angles in radians are an exception to this rule. It's okay to leave a π in the value of θ or ω, and we have done so in Example 4.10.

Not surprisingly, the angular velocity ω is closely related to the period and speed of the motion. As a particle goes around a circle one time, its angular displacement is $\Delta\theta = 2\pi$ rad during the interval $\Delta t = T$. Thus, using the definition of angular velocity, we find

$$|\omega| = \frac{2\pi \text{ rad}}{T} \quad \text{or} \quad T = \frac{2\pi \text{ rad}}{|\omega|} \quad\quad (4.27)$$

The period alone gives only the absolute value of $|\omega|$. You need to know the direction of motion to determine the sign of ω.

EXAMPLE 4.11 | At the roulette wheel

A small steel roulette ball rolls ccw around the inside of a 30-cm-diameter roulette wheel. The ball completes 2.0 rev in 1.20 s.

a. What is the ball's angular velocity?

b. What is the ball's position at $t = 2.0$ s? Assume $\theta_i = 0$.

MODEL Model the ball as a particle in uniform circular motion.

SOLVE a. The period of the ball's motion, the time for 1 rev, is $T = 0.60$ s. Angular velocity is positive for ccw motion, so

$$\omega = \frac{2\pi \text{ rad}}{T} = \frac{2\pi \text{ rad}}{0.60 \text{ s}} = 10.47 \text{ rad/s}$$

b. The ball starts at $\theta_i = 0$ rad. After $\Delta t = 2.0$ s, its position is

$$\theta_f = 0 \text{ rad} + (10.47 \text{ rad/s})(2.0 \text{ s}) = 20.94 \text{ rad}$$

where we've kept an extra significant figure to avoid round-off error. Although this is a mathematically acceptable answer, an observer would say that the ball is always located somewhere between 0° and 360°. Thus it is common practice to subtract an integer number of 2π rad, representing the completed revolutions. Because $20.94/2\pi = 3.333$, we can write

$$\begin{aligned}\theta_f &= 20.94 \text{ rad} = 3.333 \times 2\pi \text{ rad} \\ &= 3 \times 2\pi \text{ rad} + 0.333 \times 2\pi \text{ rad} \\ &= 3 \times 2\pi \text{ rad} + 2.09 \text{ rad}\end{aligned}$$

In other words, at $t = 2.0$ s the ball has completed 3 rev and is 2.09 rad $= 120°$ into its fourth revolution. An observer would say that the ball's position is $\theta_f = 120°$.

As Figure 4.21 showed, the velocity vector $\vec{v}$ is always tangent to the circle. In other words, the velocity vector has only a *tangential component,* which we will designate v_t. The tangential velocity is positive for ccw motion, negative for cw motion.

Combining $v = 2\pi r/T$ for the speed with $\omega = 2\pi/T$ for the angular velocity—but keeping the sign of ω to indicate the direction of motion—we see that the tangential velocity and the angular velocity are related by

$$v_t = \omega r \qquad \text{(with } \omega \text{ in rad/s)} \qquad (4.28)$$

Because v_t is the only nonzero component of $\vec{v}$, the particle's speed is $v = |v_t| = |\omega|r$. We'll sometimes write this as $v = \omega r$ if there's no ambiguity about the sign of ω.

> **NOTE** While it may be convenient in some problems to measure ω in rev/s or rpm, you must convert to SI units of rad/s before using Equation 4.28.

As a simple example, a particle moving cw at 2.0 m/s in a circle of radius 40 cm has angular velocity

$$\omega = \frac{v_t}{r} = \frac{-2.0 \text{ m/s}}{0.40 \text{ m}} = -5.0 \text{ rad/s}$$

where v_t and ω are negative because the motion is clockwise. Notice the units. Velocity divided by distance has units of s^{-1}. But because the division, in this case, gives us an angular quantity, we've inserted the *dimensionless* unit rad to give ω the appropriate units of rad/s.

STOP TO THINK 4.5 A particle moves cw around a circle at constant speed for 2.0 s. It then reverses direction and moves ccw at half the original speed until it has traveled through the same angle. Which is the particle's angle-versus-time graph?

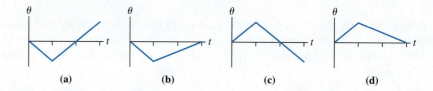

    (a)   (b)   (c)   (d)

FIGURE 4.27 Using Tactics Box 4.1 to find Maria's acceleration on the Ferris wheel.

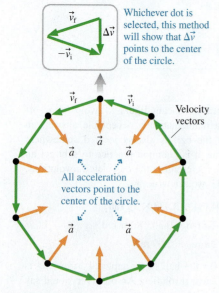

Maria's acceleration is an acceleration of changing direction, not of changing speed.

4.5 Centripetal Acceleration

FIGURE 4.27 shows a motion diagram of Maria riding a Ferris wheel at the amusement park. Maria has constant speed but *not* constant velocity because her velocity vector is changing direction. She may not be speeding up, but Maria *is* accelerating because her velocity is changing. The inset to Figure 4.27 applies the rules of Tactics Box 4.1 to find that—at every point—**Maria's acceleration vector points toward the center of the circle.** This is an acceleration due to changing direction rather than changing speed. Because the instantaneous velocity is tangent to the circle, $\vec{v}$ and $\vec{a}$ are perpendicular to each other at all points on the circle.

The acceleration of uniform circular motion is called **centripetal acceleration,** a term from a Greek root meaning "center seeking." Centripetal acceleration is not a new type of acceleration; all we are doing is *naming* an acceleration that corresponds to a particular type of motion. The magnitude of the centripetal acceleration is constant because each successive $\Delta\vec{v}$ in the motion diagram has the same length.

The motion diagram tells us the direction of $\vec{a}$, but it doesn't give us a value for a. To complete our description of uniform circular motion, we need to find a quantitative relationship between a and the particle's speed v. **FIGURE 4.28** shows the velocity $\vec{v}_i$

at one instant of motion and the velocity $\vec{v}_f$ an infinitesimal amount of time dt later. During this small interval of time, the particle has moved through the infinitesimal angle $d\theta$ and traveled distance $ds = r\,d\theta$.

By definition, the acceleration is $\vec{a} = d\vec{v}/dt$. We can see from the inset to Figure 4.28 that $d\vec{v}$ points toward the center of the circle—that is, $\vec{a}$ is a centripetal acceleration. To find the magnitude of $\vec{a}$, we can see from the isosceles triangle of velocity vectors that, if $d\theta$ is in radians,

$$dv = |d\vec{v}| = v\,d\theta \qquad (4.29)$$

For uniform circular motion at constant speed, $v = ds/dt = r\,d\theta/dt$ and thus the time to rotate through angle $d\theta$ is

$$dt = \frac{r\,d\theta}{v} \qquad (4.30)$$

Combining Equations 4.29 and 4.30, we see that the acceleration has magnitude

$$a = |\vec{a}| = \frac{|d\vec{v}|}{dt} = \frac{v\,d\theta}{r\,d\theta/v} = \frac{v^2}{r}$$

In vector notation, we can write

$$\vec{a} = \left(\frac{v^2}{r},\ \text{toward center of circle}\right) \qquad \text{(centripetal acceleration)} \qquad (4.31)$$

Using Equation 4.28, $v = \omega r$, we can also express the magnitude of the centripetal acceleration in terms of the angular velocity ω as

$$a = \omega^2 r \qquad (4.32)$$

NOTE Centripetal acceleration is not a constant acceleration. The magnitude of the centripetal acceleration is constant during uniform circular motion, but the direction of $\vec{a}$ is constantly changing. **Thus the constant-acceleration kinematics equations of Chapter 2 do *not* apply to circular motion.**

The Uniform Circular Motion Model

The **uniform circular motion model** is especially important because it applies not only to particles moving in circles but also to the uniform rotation of solid objects.

MODEL 4.2

Uniform circular motion

For motion with constant angular velocity ω.

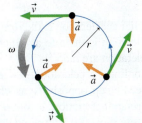

- Applies to a particle moving along a circular trajectory at constant speed or to points on a solid object rotating at a steady rate.

- Mathematically:
  - The tangential velocity is $v_t = \omega r$.
  - The centripetal acceleration is v^2/r or $\omega^2 r$.
  - ω and v_t are positive for ccw rotation, negative for cw rotation.

The velocity is tangent to the circle.
The acceleration points to the center.

- Limitations: Model fails if rotation isn't steady.

Exercise 20

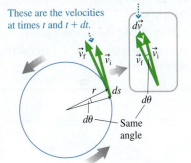

FIGURE 4.28 Finding the acceleration of circular motion.

$d\vec{v}$ is the arc of a circle with arc length $dv = v\,d\theta$.

These are the velocities at times t and $t + dt$.

Same angle

EXAMPLE 4.12 | **The acceleration of a Ferris wheel**

A typical carnival Ferris wheel has a radius of 9.0 m and rotates 4.0 times per minute. What speed and acceleration do the riders experience?

MODEL Model the rider as a particle in uniform circular motion.

SOLVE The period is $T = \frac{1}{4}$ min = 15 s. From Equation 4.21, a rider's speed is

$$v = \frac{2\pi r}{T} = \frac{2\pi(9.0 \text{ m})}{15 \text{ s}} = 3.77 \text{ m/s}$$

Consequently, the centripetal acceleration has magnitude

$$a = \frac{v^2}{r} = \frac{(3.77 \text{ m/s})^2}{9.0 \text{ m}} = 1.6 \text{ m/s}^2$$

ASSESS This was not intended to be a profound problem, merely to illustrate how centripetal acceleration is computed. The acceleration is enough to be noticed and make the ride interesting, but not enough to be scary.

STOP TO THINK 4.6 Rank in order, from largest to smallest, the centripetal accelerations a_a to a_e of particles a to e.

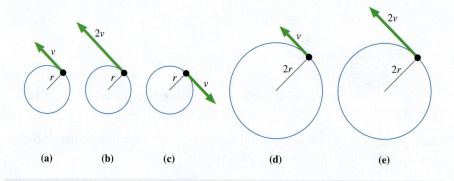

(a)　(b)　(c)　(d)　(e)

4.6 Nonuniform Circular Motion

A roller coaster car doing a loop-the-loop slows down as it goes up one side, speeds up as it comes back down the other. The ball in a roulette wheel gradually slows until it stops. Circular motion with a changing speed is called **nonuniform circular motion.** As you'll see, nonuniform circular motion is analogous to accelerated linear motion.

FIGURE 4.29 shows a point speeding up as it moves around a circle. This might be a car speeding up around a curve or simply a point on a solid object that is rotating faster and faster. The key feature of the motion is a *changing angular velocity.* For linear motion, we defined acceleration as $a_x = dv_x/dt$. By analogy, let's define the **angular acceleration** α (Greek alpha) of a rotating object, or a point on the object, to be

$$\alpha \equiv \frac{d\omega}{dt} \quad \text{(angular acceleration)} \tag{4.33}$$

Angular acceleration is the *rate* at which the angular velocity ω changes, just as linear acceleration is the rate at which the linear velocity v_x changes. The units of angular acceleration are rad/s^2.

For linear acceleration, you learned that a_x and v_x have the same sign when an object is speeding up, opposite signs when it is slowing down. The same rule applies to circular and rotational motion: ω and α have the same sign when the rotation is speeding up, opposite signs if it is slowing down. These ideas are illustrated in **FIGURE 4.30.**

> **NOTE** Be careful with the sign of α. You learned in Chapter 2 that positive and negative values of the acceleration can't be interpreted as simply "speeding up" and "slowing down." Similarly, positive and negative values of angular acceleration can't be interpreted as a rotation that is speeding up or slowing down.

FIGURE 4.29 Circular motion with a changing angular velocity.

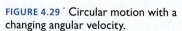

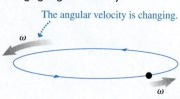

FIGURE 4.30 The signs of angular velocity and acceleration. The rotation is speeding up if ω and α have the same sign, slowing down if they have opposite signs.

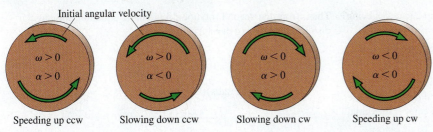

Initial angular velocity

| $\omega > 0$ $\alpha > 0$ | $\omega > 0$ $\alpha < 0$ | $\omega < 0$ $\alpha > 0$ | $\omega < 0$ $\alpha < 0$ |
| Speeding up ccw | Slowing down ccw | Slowing down cw | Speeding up cw |

Angular position, angular velocity, and angular acceleration are defined exactly the same as linear position, velocity, and acceleration—simply starting with an angular rather than a linear measurement of position. Consequently, **the graphical interpretation and the kinematic equations of circular/rotational motion with constant angular acceleration are exactly the same as for linear motion with constant acceleration**. This is shown in the **constant angular acceleration model** below. All the problem-solving techniques you learned in Chapter 2 for linear motion carry over to circular and rotational motion.

MODEL 4.3

Constant angular acceleration

For motion with constant angular acceleration α.

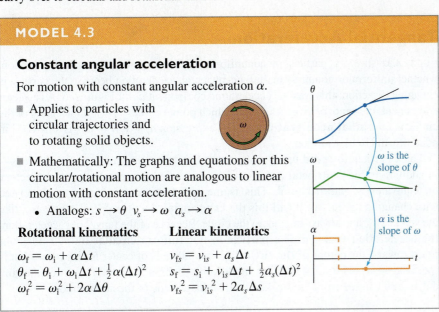

- Applies to particles with circular trajectories and to rotating solid objects.

- Mathematically: The graphs and equations for this circular/rotational motion are analogous to linear motion with constant acceleration.
  - Analogs: $s \rightarrow \theta \quad v_s \rightarrow \omega \quad a_s \rightarrow \alpha$

ω is the slope of θ

α is the slope of ω

Rotational kinematics	Linear kinematics
$\omega_f = \omega_i + \alpha \Delta t$	$v_{fs} = v_{is} + a_s \Delta t$
$\theta_f = \theta_i + \omega_i \Delta t + \frac{1}{2}\alpha(\Delta t)^2$	$s_f = s_i + v_{is} \Delta t + \frac{1}{2}a_s(\Delta t)^2$
$\omega_f^2 = \omega_i^2 + 2\alpha \Delta\theta$	$v_{fs}^2 = v_{is}^2 + 2a_s \Delta s$

EXAMPLE 4.13 | A rotating wheel

FIGURE 4.31a is a graph of angular velocity versus time for a rotating wheel. Describe the motion and draw a graph of angular acceleration versus time.

SOLVE This is a wheel that starts from rest, gradually speeds up *counterclockwise* until reaching top speed at t_1, maintains a constant angular velocity until t_2, then gradually slows down until stopping at t_3. The motion is always ccw because ω is always positive. The angular acceleration graph of **FIGURE 4.32b** is based on the fact that α is the slope of the ω-versus-t graph.

Conversely, the initial linear increase of ω can be seen as the increasing area under the α-versus-t graph as t increases from 0 to t_1. The angular velocity doesn't change from t_1 to t_2 when the area under the α-versus-t is zero.

▶ **FIGURE 4.31** ω-versus-t graph and the corresponding α-versus-t graph for a rotating wheel.

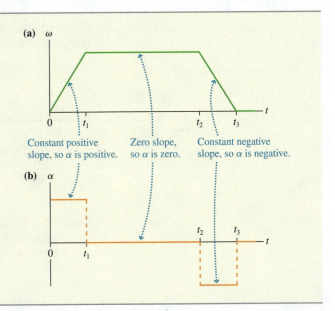

(a) ω

Constant positive slope, so α is positive.   Zero slope, so α is zero.   Constant negative slope, so α is negative.

(b) α

EXAMPLE 4.14 | **A slowing fan**

A ceiling fan spinning at 60 rpm coasts to a stop 25 s after being turned off. How many revolutions does it make while stopping?

MODEL Model the fan as a rotating object with constant angular acceleration.

SOLVE We don't know which direction the fan is rotating, but the fact that the rotation is slowing tells us that ω and α have opposite signs. We'll assume that ω is positive. We need to convert the initial angular velocity to SI units:

$$\omega_i = 60 \, \frac{\text{rev}}{\text{min}} \times \frac{1 \, \text{min}}{60 \, \text{s}} \times \frac{2\pi \, \text{rad}}{1 \, \text{rev}} = 6.28 \, \text{rad/s}$$

We can use the first rotational kinematics equation in Model 4.3 to find the angular acceleration:

$$\alpha = \frac{\omega_f - \omega_i}{\Delta t} = \frac{0 \, \text{rad/s} - 6.28 \, \text{rad/s}}{25 \, \text{s}} = -0.25 \, \text{rad/s}^2$$

Then, from the second rotational kinematic equation, the angular displacement during these 25 s is

$$\Delta\theta = \omega_i \Delta t + \tfrac{1}{2}\alpha(\Delta t)^2$$

$$= (6.28 \, \text{rad/s})(25 \, \text{s}) + \tfrac{1}{2}(-0.25 \, \text{rad/s}^2)(25 \, \text{s})^2$$

$$= 78.9 \, \text{rad} \times \frac{1 \, \text{rev}}{2\pi \, \text{rad}} = 13 \, \text{rev}$$

The kinematic equation returns an angle in rad, but the question asks for revolutions, so the last step was a unit conversion.

ASSESS Turning through 13 rev in 25 s while stopping seems reasonable. Notice that the problem is solved just like the linear kinematics problems you learned to solve in Chapter 2.

Tangential Acceleration

FIGURE 4.32 Acceleration in nonuniform circular motion.

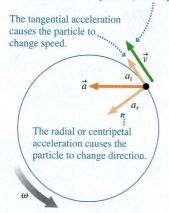

The velocity is always tangent to the circle, so the radial component v_r is always zero.

The tangential acceleration causes the particle to change speed.

The radial or centripetal acceleration causes the particle to change direction.

FIGURE 4.32 shows a particle in nonuniform circular motion. *Any* circular motion, whether uniform or nonuniform, has a centripetal acceleration because the particle is changing direction; this was the acceleration component $\vec{a}_\perp$ of Figure 4.6. As a vector component, the centripetal acceleration, which points radially toward the center of the circle, is the **radial acceleration** a_r. The expression $a_r = v_t^2/r = \omega^2 r$ is still valid in nonuniform circular motion.

For a particle to speed up or slow down as it moves around a circle, it needs—in addition to the centripetal acceleration—an acceleration parallel to the trajectory or, equivalently, parallel to $\vec{v}$. This is the acceleration component $\vec{a}_\parallel$ associated with changing speed. We'll call this the **tangential acceleration** a_t because, like the velocity v_t, it is always tangent to the circle. Because of the tangential acceleration, **the acceleration vector $\vec{a}$ of a particle in nonuniform circular motion does not point toward the center of the circle.** It points "ahead" of center for a particle that is speeding up, as in Figure 4.32, but it would point "behind" center for a particle slowing down. You can see from Figure 4.32 that the magnitude of the acceleration is

$$a = \sqrt{a_r^2 + a_t^2} \tag{4.34}$$

If a_t is constant, then the arc length s traveled by the particle around the circle and the tangential velocity v_t are found from constant-acceleration kinematics:

$$s_f = s_i + v_{it}\Delta t + \tfrac{1}{2}a_t(\Delta t)^2$$
$$v_{ft} = v_{it} + a_t\Delta t \tag{4.35}$$

Because tangential acceleration is the rate at which the tangential velocity changes, $a_t = dv_t/dt$, and we already know that the tangential velocity is related to the angular velocity by $v_t = \omega r$, it follows that

$$a_t = \frac{dv_t}{dt} = \frac{d(\omega r)}{dt} = \frac{d\omega}{dt}r = \alpha r \tag{4.36}$$

Thus $v_t = \omega r$ and $a_t = \alpha r$ are analogous equations for the tangential velocity and acceleration. In Example 4.14, where we found the fan to have angular acceleration $\alpha = -0.25 \, \text{rad/s}^2$, a blade tip 65 cm from the center would have tangential acceleration

$$a_t = \alpha r = (-0.25 \, \text{rad/s}^2)(0.65 \, \text{m}) = -0.16 \, \text{m/s}^2$$

EXAMPLE 4.15 | Analyzing rotational data

You've been assigned the task of measuring the start-up character-istics of a large industrial motor. After several seconds, when the motor has reached full speed, you know that the angular acceleration will be zero, but you hypothesize that the angular acceleration may be constant during the first couple of seconds as the motor speed increases. To find out, you attach a shaft encoder to the 3.0-cm-diameter axle. A shaft encoder is a device that converts the angular position of a shaft or axle to a signal that can be read by a computer. After setting the computer program to read four values a second, you start the motor and acquire the following data:

Time (s)	Angle (°)	Time (s)	Angle (°)
0.00	0	1.00	267
0.25	16	1.25	428
0.50	69	1.50	620
0.75	161		

a. Do the data support your hypothesis of a constant angular acceleration? If so, what is the angular acceleration? If not, is the angular acceleration increasing or decreasing with time?
b. A 76-cm-diameter blade is attached to the motor shaft. At what time does the acceleration of the tip of the blade reach 10 m/s²?

MODEL The axle is rotating with nonuniform circular motion. Model the tip of the blade as a particle.

VISUALIZE FIGURE 4.33 shows that the blade tip has both a tangential and a radial acceleration.

FIGURE 4.33 Pictorial representation of the axle and blade.

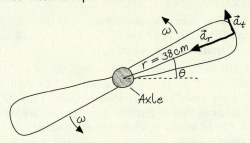

SOLVE a. *If* the motor starts up with constant angular acceleration, with $\theta_i = 0$ and $\omega_i = 0$ rad/s, the angle-time equation of rotation-al kinematics is $\theta = \frac{1}{2}\alpha t^2$. This can be written as a linear equation $y = mx + b$ if we let $\theta = y$ and $t^2 = x$. That is, constant angular acceleration predicts that a graph of θ versus t^2 should be a straight line with slope $m = \frac{1}{2}\alpha$ and y-intercept $b = 0$. We can test this.

FIGURE 4.34 is the graph of θ versus t^2, and it confirms our hypoth-esis that the motor starts up with constant angular acceleration. The

FIGURE 4.34 Graph of θ versus t^2 for the motor shaft.

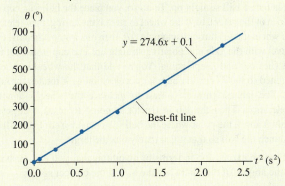

best-fit line, found using a spreadsheet, gives a slope of 274.6°/s². The units come not from the spreadsheet but by looking at the units of rise (°) over run (s² because we're graphing t^2 on the x-axis). Thus the angular acceleration is

$$\alpha = 2m = 549.2°/s^2 \times \frac{\pi \text{ rad}}{180°} = 9.6 \text{ rad/s}^2$$

where we used $180° = \pi$ rad to convert to SI units of rad/s².
c. The magnitude of the linear acceleration is

$$a = \sqrt{a_r^2 + a_t^2}$$

The tangential acceleration of the blade tip is

$$a_t = \alpha r = (9.6 \text{ rad/s}^2)(0.38 \text{ m}) = 3.65 \text{ m/s}^2$$

We were careful to use the blade's radius, not its diameter, and we kept an extra significant figure to avoid round-off error. The radial (centripetal) acceleration increases as the rotation speed increases, and the total acceleration reaches 10 m/s² when

$$a_r = \sqrt{a^2 - a_t^2} = \sqrt{(10 \text{ m/s}^2)^2 - (3.65 \text{ m/s}^2)^2} = 9.31 \text{ m/s}^2$$

Radial acceleration is $a_r = \omega^2 r$, so the corresponding angular velocity is

$$\omega = \sqrt{\frac{a_r}{r}} = \sqrt{\frac{9.31 \text{ m/s}^2}{0.38 \text{ m}}} = 4.95 \text{ rad/s}$$

For constant angular acceleration, $\omega = \alpha t$, so this angular velocity is achieved at

$$t = \frac{\omega}{\alpha} = \frac{4.95 \text{ rad/s}}{9.6 \text{ rad/s}^2} = 0.52 \text{ s}$$

Thus it takes 0.52 s for the acceleration of the blade tip to reach 10 m/s².

ASSESS The acceleration at the tip of a long blade is likely to be large. It seems plausible that the acceleration would reach 10 m/s² in ≈ 0.5 s.

STOP TO THINK 4.7 The fan blade is slowing down. What are the signs of ω and α?

a. ω is positive and α is positive.
b. ω is positive and α is negative.
c. ω is negative and α is positive.
d. ω is negative and α is negative.

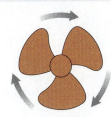

CHALLENGE EXAMPLE 4.16 | Hit the target!

One day when you come into lab, you see a spring-loaded wheel that can launch a ball straight up. To do so, you place the ball in a cup on the rim of the wheel, turn the wheel to stretch the spring, then release. The wheel rotates through an angle $\Delta\theta$, then hits a stop when the cup is level with the axle and pointing straight up. The cup stops, but the ball flies out and keeps going. You're told that the wheel has been designed to have constant angular acceleration as it rotates through $\Delta\theta$. The lab assignment is first to measure the wheel's angular acceleration. Then the lab instructor is going to place a target at height h above the point where the ball is launched. Your task will be to launch the ball so that it just barely hits the target.

a. Find an expression in terms of quantities that you can measure for the angle $\Delta\theta$ that launches the ball at the correct speed.

b. Evaluate $\Delta\theta$ if the wheel's diameter is 62 cm, you've determined that its angular acceleration is 200 rad/s², the mass of the ball is 25 g, and the instructor places the target 190 cm above the launch point.

MODEL Model the ball as a particle. It first undergoes circular motion that we'll model as having constant angular acceleration. We'll then ignore air resistance and model the vertical motion as free fall.

VISUALIZE FIGURE 4.35 is a pictorial representation. This is a two-part problem, with the speed at the end of the angular acceleration being the launch speed for the vertical motion. We've chosen to call the wheel radius R and the target height h. These and the angular

FIGURE 4.35 Pictorial representation of the ball launcher.

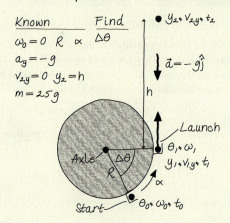

acceleration α are considered "known" because we will measure them, but we don't have numerical values at this time.

SOLVE

a. The circular motion problem and the vertical motion problem are connected through the ball's speed: The final speed of the angular acceleration is the launch speed of the vertical motion. We don't know anything about time intervals, which suggests using the kinematic equations that relate distance and acceleration (for the vertical motion) and angle and angular acceleration (for the circular motion). For the angular acceleration, with $\omega_0 = 0$ rad/s,

$$\omega_1^2 = \omega_0^2 + 2\alpha\,\Delta\theta = 2\alpha\,\Delta\theta$$

The final speed of the ball and cup, when the wheel hits the stop, is

$$v_1 = \omega_1 R = R\sqrt{2\alpha\,\Delta\theta}$$

Thus the vertical-motion problem begins with the ball being shot upward with velocity $v_{1y} = R\sqrt{2\alpha\,\Delta\theta}$. How high does it go? The highest point is the point where $v_{2y} = 0$, so the free-fall equation is

$$v_{2y}^2 = 0 = v_{1y}^2 - 2g\,\Delta y = R^2 \cdot 2\alpha\,\Delta\theta - 2gh$$

Rather than solve for height h, we need to solve for the angle that produces a given height. This is

$$\Delta\theta = \frac{gh}{\alpha R^2}$$

Once we've determined the properties of the wheel and then measured the height at which our instructor places the target, we'll quickly be able to calculate the angle through which we should pull back the wheel to launch the ball.

b. For the values given in the problem statement, $\Delta\theta = 0.969$ rad = 56°. Don't forget that equations involving angles need values in radians and return values in radians.

ASSESS The angle needed to be less than 90° or else the ball would fall out of the cup before launch. And an angle of only a few degrees would seem suspiciously small. Thus 56° seems to be reasonable. Notice that the mass was not needed in this problem. Part of becoming a better problem solver is evaluating the information you have to see what is relevant. Some homework problems will help you develop this skill by providing information that isn't necessary.

SUMMARY

The goal of Chapter 4 has been to learn how to solve problems about motion in a plane.

GENERAL PRINCIPLES

The **instantaneous velocity**

$$\vec{v} = d\vec{r}/dt$$

is a vector tangent to the trajectory.

The **instantaneous acceleration** is

$$\vec{a} = d\vec{v}/dt$$

$\vec{a}_\parallel$, the component of $\vec{a}$ parallel to $\vec{v}$, is responsible for change of *speed*. $\vec{a}_\perp$, the component of $\vec{a}$ perpendicular to $\vec{v}$, is responsible for change of *direction*.

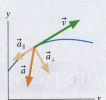

Relative Motion

If object C moves relative to reference frame A with velocity $\vec{v}_{CA}$, then it moves relative to a different reference frame B with velocity

$$\vec{v}_{CB} = \vec{v}_{CA} + \vec{v}_{AB}$$

where $\vec{v}_{AB}$ is the velocity of A relative to B. This is the Galilean transformation of velocity.

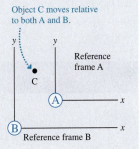

Object C moves relative to both A and B.

Reference frame A

Reference frame B

IMPORTANT CONCEPTS

Uniform Circular Motion

Angular velocity $\omega = d\theta/dt$.

v_t and ω are constant:

$$v_t = \omega r$$

The centripetal acceleration points toward the center of the circle:

$$a = \frac{v^2}{r} = \omega^2 r$$

It changes the particle's direction but not its speed.

Nonuniform Circular Motion

Angular acceleration $\alpha = d\omega/dt$.

The radial acceleration

$$a_r = \frac{v^2}{r} = \omega^2 r$$

changes the particle's direction. The tangential component

$$a_t = \alpha r$$

changes the particle's speed.

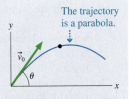

APPLICATIONS

Kinematics in two dimensions

If $\vec{a}$ is constant, then the x- and y-components of motion are independent of each other.

$$x_f = x_i + v_{ix}\,\Delta t + \tfrac{1}{2}a_x(\Delta t)^2$$
$$y_f = y_i + v_{iy}\,\Delta t + \tfrac{1}{2}a_y(\Delta t)^2$$
$$v_{fx} = v_{ix} + a_x\,\Delta t$$
$$v_{fy} = v_{iy} + a_y\,\Delta t$$

Projectile motion is motion under the influence of only gravity.

MODEL Model as a particle launched with speed v_0 at angle θ.

VISUALIZE Use coordinates with the x-axis horizontal and the y-axis vertical.

SOLVE The horizontal motion is uniform with $v_x = v_0\cos\theta$. The vertical motion is free fall with $a_y = -g$. The x and y kinematic equations have the *same* value for Δt.

The trajectory is a parabola.

Circular motion kinematics

Period $T = \dfrac{2\pi r}{v} = \dfrac{2\pi}{\omega}$

Angular position $\theta = \dfrac{s}{r}$

Constant angular acceleration

$$\omega_f = \omega_i + \alpha\,\Delta t$$
$$\theta_f = \theta_i + \omega_i\,\Delta t + \tfrac{1}{2}\alpha(\Delta t)^2$$
$$\omega_f^2 = \omega_i^2 + 2\alpha\,\Delta\theta$$

Circular motion graphs and kinematics are analogous to linear motion with constant acceleration.

Angle, angular velocity, and angular acceleration are related graphically.

- The angular velocity is the slope of the angular position graph.
- The angular acceleration is the slope of the angular velocity graph.

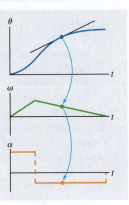

TERMS AND NOTATION

projectile	uniform circular motion	angular velocity, ω	angular acceleration, α
launch angle, θ	period, T	centripetal acceleration	constant angular acceleration
projectile motion model	angular position, θ	uniform circular motion	model
reference frame	arc length, s	model	radial acceleration, a_r
Galilean transformation	radians	nonuniform circular	tangential acceleration, a_t
of velocity	angular displacement, $\Delta\theta$	motion	

CONCEPTUAL QUESTIONS

1. a. At this instant, is the particle in **FIGURE Q4.1** speeding up, slowing down, or traveling at constant speed?
    b. Is this particle curving to the right, curving to the left, or traveling straight?

FIGURE Q4.1   **FIGURE Q4.2**

2. a. At this instant, is the particle in **FIGURE Q4.2** speeding up, slowing down, or traveling at constant speed?
    b. Is this particle curving upward, curving downward, or traveling straight?
3. Tarzan swings through the jungle by hanging from a vine.
    a. Immediately after stepping off a branch to swing over to another tree, is Tarzan's acceleration $\vec{a}$ zero or not zero? If not zero, which way does it point? Explain.
    b. Answer the same question at the lowest point in Tarzan's swing.
4. A projectile is launched at an angle of 30°.
    a. Is there any point on the trajectory where $\vec{v}$ and $\vec{a}$ are parallel to each other? If so, where?
    b. Is there any point where $\vec{v}$ and $\vec{a}$ are perpendicular to each other? If so, where?
5. For a projectile, which of the following quantities are constant during the flight: $x, y, r, v_x, v_y, v, a_x, a_y$? Which of these quantities are zero throughout the flight?
6. A cart that is rolling at constant velocity on a level table fires a ball straight up.
    a. When the ball comes back down, will it land in front of the launching tube, behind the launching tube, or directly in the tube? Explain.
    b. Will your answer change if the cart is accelerating in the forward direction? If so, how?
7. A rock is thrown from a bridge at an angle 30° below horizontal. Immediately after the rock is released, is the magnitude of its acceleration greater than, less than, or equal to g? Explain.
8. Anita is running to the right at 5 m/s in **FIGURE Q4.8**. Balls 1 and 2 are thrown toward her by friends standing on the ground. According to Anita, both balls are approaching her at 10 m/s.

Which ball was thrown at a faster speed? Or were they thrown with the same speed? Explain.

FIGURE Q4.8

9. An electromagnet on the ceiling of an airplane holds a steel ball. When a button is pushed, the magnet releases the ball. The experiment is first done while the plane is parked on the ground, and the point where the ball hits the floor is marked with an X. Then the experiment is repeated while the plane is flying level at a steady 500 mph. Does the ball land slightly in front of the X (toward the nose of the plane), on the X, or slightly behind the X (toward the tail of the plane)? Explain.
10. Zack is driving past his house in **FIGURE Q4.10**. He wants to toss his physics book out the window and have it land in his driveway. If he lets go of the book exactly as he passes the end of the driveway, should he direct his throw outward and toward the front of the car (throw 1), straight outward (throw 2), or outward and toward the back of the car (throw 3)? Explain.

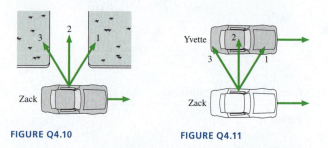

FIGURE Q4.10   **FIGURE Q4.11**

11. In **FIGURE Q4.11**, Yvette and Zack are driving down the freeway side by side with their windows down. Zack wants to toss his physics book out the window and have it land in Yvette's front seat. Ignoring air resistance, should he direct his throw outward and toward the front of the car (throw 1), straight outward (throw 2), or outward and toward the back of the car (throw 3)? Explain.

12. In uniform circular motion, which of the following quantities are constant: speed, instantaneous velocity, the tangential component of velocity, the radial component of acceleration, the tangential component of acceleration? Which of these quantities are zero throughout the motion?

13. **FIGURE Q4.13** shows three points on a steadily rotating wheel.
    a. Rank in order, from largest to smallest, the angular velocities ω_1, ω_2, and ω_3 of these points. Explain.
    b. Rank in order, from largest to smallest, the speeds v_1, v_2, and v_3 of these points. Explain.

FIGURE Q4.13

14. **FIGURE Q4.14** shows four rotating wheels. For each, determine the signs (+ or −) of ω and α.

(a)	(b)	(c)	(d)
Speeding up	Slowing down	Slowing down	Speeding up

FIGURE Q4.14

15. **FIGURE Q4.15** shows a pendulum at one end point of its arc.
    a. At this point, is ω positive, negative, or zero? Explain.
    b. At this point, is α positive, negative, or zero? Explain.

FIGURE Q4.15

EXERCISES AND PROBLEMS

Exercises

Section 4.1 Motion in Two Dimensions

Problems 1 and 2 show a partial motion diagram. For each:
a. Complete the motion diagram by adding acceleration vectors.
b. Write a physics *problem* for which this is the correct motion diagram. Be imaginative! Don't forget to include enough information to make the problem complete and to state clearly what is to be found.

1. |

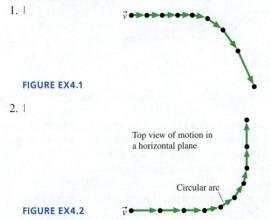

FIGURE EX4.1

2. |

Top view of motion in a horizontal plane

Circular arc

FIGURE EX4.2

Answer Problems 3 through 5 by choosing one of the eight labeled acceleration vectors or selecting option I: $\vec{a} = \vec{0}$.

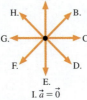

3. ‖ At this instant, the particle has steady speed and is curving to the right. What is the direction of its acceleration?

FIGURE EX4.3

4. ‖ At this instant, the particle is speeding up and curving upward. What is the direction of its acceleration?

FIGURE EX4.4

5. ‖ At this instant, the particle is speeding up and curving downward. What is the direction of its acceleration?

FIGURE EX4.5

6. ‖ A rocket-powered hockey puck moves on a horizontal frictionless table. **FIGURE EX4.6** shows graphs of v_x and v_y, the x- and y-components of the puck's velocity. The puck starts at the origin.
    a. In which direction is the puck moving at $t = 2$ s? Give your answer as an angle from the x-axis.
    b. How far from the origin is the puck at $t = 5$ s?

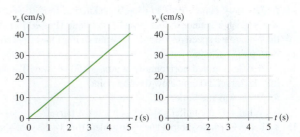

FIGURE EX4.6

7. ‖ A rocket-powered hockey puck moves on a horizontal frictionless table. **FIGURE EX4.7** shows graphs of v_x and v_y, the x- and y-components of the puck's velocity. The puck starts at the origin. What is the magnitude of the puck's acceleration at $t = 5$ s?

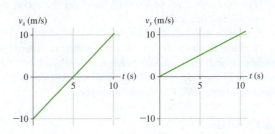

FIGURE EX4.7

8. ‖ A particle's trajectory is described by $x = \left(\frac{1}{2}t^3 - 2t^2\right)$ m and
CALC $y = \left(\frac{1}{2}t^2 - 2t\right)$ m, where t is in s.
   a. What are the particle's position and speed at $t = 0$ s and $t = 4$ s?
   b. What is the particle's direction of motion, measured as an angle from the x-axis, at $t = 0$ s and $t = 4$ s?

9. ‖ A particle moving in the xy-plane has velocity $\vec{v} = (2t\hat{i} + (3 - t^2)\hat{j})$ m/s, where t is in s. What is the particle's acceleration vector at $t = 4$ s?
CALC

10. ‖ You have a remote-controlled car that has been programmed to
CALC have velocity $\vec{v} = (-3t\hat{i} + 2t^2\hat{j})$ m/s, where t is in s. At $t = 0$ s, the car is at $\vec{r}_0 = (3.0\hat{i} + 2.0\hat{j})$ m. What are the car's (a) position vector and (b) acceleration vector at $t = 2.0$ s?

Section 4.2 Projectile Motion

11. ‖ A ball thrown horizontally at 25 m/s travels a horizontal distance of 50 m before hitting the ground. From what height was the ball thrown?

12. ‖ A physics student on Planet Exidor throws a ball, and it follows the parabolic trajectory shown in FIGURE EX4.12. The ball's position is shown at 1 s intervals until $t = 3$ s. At $t = 1$ s, the ball's velocity is $\vec{v} = (2.0\hat{i} + 2.0\hat{j})$ m/s.
   a. Determine the ball's velocity at $t = 0$ s, 2 s, and 3 s.
   b. What is the value of g on Planet Exidor?
   c. What was the ball's launch angle?

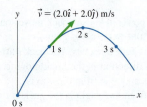

FIGURE EX4.12

13. ‖ A supply plane needs to drop a package of food to scientists working on a glacier in Greenland. The plane flies 100 m above the glacier at a speed of 150 m/s. How far short of the target should it drop the package?

14. ‖ A rifle is aimed horizontally at a target 50 m away. The bullet hits the target 2.0 cm below the aim point.
   a. What was the bullet's flight time?
   b. What was the bullet's speed as it left the barrel?

15. ‖ In the Olympic shotput event, an athlete throws the shot with an initial speed of 12.0 m/s at a 40.0° angle from the horizontal. The shot leaves her hand at a height of 1.80 m above the ground. How far does the shot travel?

16. ‖ On the Apollo 14 mission to the moon, astronaut Alan Shepard hit a golf ball with a 6 iron. The free-fall acceleration on the moon is 1/6 of its value on earth. Suppose he hit the ball with a speed of 25 m/s at an angle 30° above the horizontal.
   a. How much farther did the ball travel on the moon than it would have on earth?
   b. For how much more time was the ball in flight?

17. ‖ A baseball player friend of yours wants to determine his pitching speed. You have him stand on a ledge and throw the ball horizontally from an elevation 4.0 m above the ground. The ball lands 25 m away. What is his pitching speed?

Section 4.3 Relative Motion

18. ‖ A boat takes 3.0 hours to travel 30 km down a river, then 5.0 hours to return. How fast is the river flowing?

19. ‖ When the moving sidewalk at the airport is broken, as it often seems to be, it takes you 50 s to walk from your gate to baggage claim. When it is working and you stand on the moving sidewalk the entire way, without walking, it takes 75 s to travel the same distance. How long will it take you to travel from the gate to baggage claim if you walk while riding on the moving sidewalk?

20. ‖ Mary needs to row her boat across a 100-m-wide river that is flowing to the east at a speed of 1.0 m/s. Mary can row with a speed of 2.0 m/s.
   a. If Mary points her boat due north, how far from her intended landing spot will she be when she reaches the opposite shore?
   b. What is her speed with respect to the shore?

21. ‖ A kayaker needs to paddle north across a 100-m-wide harbor. The tide is going out, creating a tidal current that flows to the east at 2.0 m/s. The kayaker can paddle with a speed of 3.0 m/s.
   a. In which direction should he paddle in order to travel straight across the harbor?
   b. How long will it take him to cross?

22. ‖ Susan, driving north at 60 mph, and Trent, driving east at 45 mph, are approaching an intersection. What is Trent's speed relative to Susan's reference frame?

Section 4.4 Uniform Circular Motion

23. ‖ FIGURE EX4.23 shows the angular-velocity-versus-time graph for a particle moving in a circle. How many revolutions does the object make during the first 4 s?

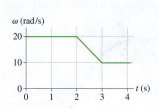

FIGURE EX4.23

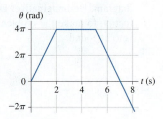

FIGURE EX4.24

24. ‖ FIGURE EX4.24 shows the angular-position-versus-time graph for a particle moving in a circle. What is the particle's angular velocity at (a) $t = 1$ s, (b) $t = 4$ s, and (c) $t = 7$ s?

25. ‖ FIGURE EX4.25 shows the angular-velocity-versus-time graph for a particle moving in a circle, starting from $\theta_0 = 0$ rad at $t = 0$ s. Draw the angular-position-versus-time graph. Include an appropriate scale on both axes.

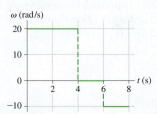

FIGURE EX4.25

26. ‖ The earth's radius is about 4000 miles. Kampala, the capital of Uganda, and Singapore are both nearly on the equator. The distance between them is 5000 miles. The flight from Kampala to Singapore takes 9.0 hours. What is the plane's angular velocity with respect to the earth's surface? Give your answer in °/h.

27. ‖ An old-fashioned single-play vinyl record rotates on a turntable at 45 rpm. What are (a) the angular velocity in rad/s and (b) the period of the motion?

28. ‖ As the earth rotates, what is the speed of (a) a physics student in Miami, Florida, at latitude 26°, and (b) a physics student in Fairbanks, Alaska, at latitude 65°? Ignore the revolution of the earth around the sun. The radius of the earth is 6400 km.

29. | How fast must a plane fly along the earth's equator so that the sun stands still relative to the passengers? In which direction must the plane fly, east to west or west to east? Give your answer in both km/h and mph. The earth's radius is 6400 km.

30. ‖ A 3000-m-high mountain is located on the equator. How much faster does a climber on top of the mountain move than a surfer at a nearby beach? The earth's radius is 6400 km.

Section 4.5 Centripetal Acceleration

31. | Peregrine falcons are known for their maneuvering ability. In a
BIO tight circular turn, a falcon can attain a centripetal acceleration 1.5 times the free-fall acceleration. What is the radius of the turn if the falcon is flying at 25 m/s?

32. | To withstand "g-forces" of up to 10 g's, caused by suddenly
BIO pulling out of a steep dive, fighter jet pilots train on a "human centrifuge." 10 g's is an acceleration of 98 m/s². If the length of the centrifuge arm is 12 m, at what speed is the rider moving when she experiences 10 g's?

33. ‖ The radius of the earth's very nearly circular orbit around the sun is 1.5×10^{11} m. Find the magnitude of the earth's (a) velocity, (b) angular velocity, and (c) centripetal acceleration as it travels around the sun. Assume a year of 365 days.

34. ‖ A speck of dust on a spinning DVD has a centripetal acceleration of 20 m/s².
   a. What is the acceleration of a different speck of dust that is twice as far from the center of the disk?
   b. What would be the acceleration of the first speck of dust if the disk's angular velocity was doubled?

35. ‖ Your roommate is working on his bicycle and has the bike upside down. He spins the 60-cm-diameter wheel, and you notice that a pebble stuck in the tread goes by three times every second. What are the pebble's speed and acceleration?

Section 4.6 Nonuniform Circular Motion

36. | FIGURE EX4.36 shows the angular velocity graph of the crankshaft in a car. What is the crankshaft's angular acceleration at (a) $t = 1$ s, (b) $t = 3$ s, and (c) $t = 5$ s?

FIGURE EX4.36

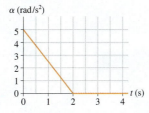

FIGURE EX4.37

37. ‖ FIGURE EX4.37 shows the angular acceleration graph of a turntable that starts from rest. What is the turntable's angular velocity at (a) $t = 1$ s, (b) $t = 2$ s, and (c) $t = 3$ s?

38. ‖ FIGURE EX4.38 shows the angular-velocity-versus-time graph for a particle moving in a circle. How many revolutions does the object make during the first 4 s?

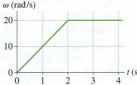

FIGURE EX4.38

39. ‖ A wheel initially rotating at 60 rpm experiences the angular acceleration shown in FIGURE EX4.39. What is the wheel's angular velocity, in rpm, at $t = 3.0$ s?

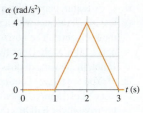

FIGURE EX4.39

40. ‖ A 5.0-m-diameter merry-go-round is initially turning with a 4.0 s period. It slows down and stops in 20 s.
   a. Before slowing, what is the speed of a child on the rim?
   b. How many revolutions does the merry-go-round make as it stops?

41. ‖ An electric fan goes from rest to 1800 rpm in 4.0 s. What is its angular acceleration?

42. ‖ A bicycle wheel is rotating at 50 rpm when the cyclist begins to pedal harder, giving the wheel a constant angular acceleration of 0.50 rad/s².
   a. What is the wheel's angular velocity, in rpm, 10 s later?
   b. How many revolutions does the wheel make during this time?

43. ‖ Starting from rest, a DVD steadily accelerates to 500 rpm in 1.0 s, rotates at this angular speed for 3.0 s, then steadily decelerates to a halt in 2.0 s. How many revolutions does it make?

Problems

44. ‖‖ A spaceship maneuvering near Planet Zeta is located at $\vec{r} = (600\hat{i} - 400\hat{j} + 200\hat{k}) \times 10^3$ km, relative to the planet, and traveling at $\vec{v} = 9500\hat{i}$ m/s. It turns on its thruster engine and accelerates with $\vec{a} = (40\hat{i} - 20\hat{k})$ m/s² for 35 min. What is the spaceship's position when the engine shuts off? Give your answer as a position vector measured in km.

45. ‖‖ A particle moving in the xy-plane has velocity $\vec{v}_0 = v_{0x}\hat{i} + v_{0y}\hat{j}$
CALC at $t = 0$. It undergoes acceleration $\vec{a} = bt\hat{i} - cv_y\hat{j}$, where b and c are constants. Find an expression for the particle's velocity at a later time t.

46. ‖ A projectile's horizontal range over level ground is $v_0^2 \sin 2\theta/g$. At what launch angle or angles will the projectile land at half of its maximum possible range?

47. ‖ a. A projectile is launched with speed v_0 and angle θ. Derive an expression for the projectile's maximum height h.
   b. A baseball is hit with a speed of 33.6 m/s. Calculate its height and the distance traveled if it is hit at angles of 30.0°, 45.0°, and 60.0°.

48. ‖‖ A projectile is launched from ground level at angle θ and speed
CALC v_0 into a headwind that causes a constant horizontal acceleration of magnitude a opposite the direction of motion.
   a. Find an expression in terms of a and g for the launch angle that gives maximum range.
   b. What is the angle for maximum range if a is 10% of g?

49. ‖‖ A gray kangaroo can bound across level ground with each jump
BIO carrying it 10 m from the takeoff point. Typically the kangaroo leaves the ground at a 20° angle. If this is so:
   a. What is its takeoff speed?
   b. What is its maximum height above the ground?

50. ‖ A ball is thrown toward a cliff of height h with a speed of 30 m/s and an angle of 60° above horizontal. It lands on the edge of the cliff 4.0 s later.
   a. How high is the cliff?
   b. What was the maximum height of the ball?
   c. What is the ball's impact speed?

51. ‖ A tennis player hits a ball 2.0 m above the ground. The ball leaves his racquet with a speed of 20.0 m/s at an angle 5.0° above the horizontal. The horizontal distance to the net is 7.0 m, and the net is 1.0 m high. Does the ball clear the net? If so, by how much? If not, by how much does it miss?

52. ‖ You are target shooting using a toy gun that fires a small ball at a speed of 15 m/s. When the gun is fired at an angle of 30° above horizontal, the ball hits the bull's-eye of a target at the same height as the gun. Then the target distance is halved. At what angle must you aim the gun to hit the bull's-eye in its new position? (Mathematically there are two solutions to this problem; the physically reasonable answer is the smaller of the two.)

53. ‖ A 35 g steel ball is held by a ceiling-mounted electromagnet 3.5 m above the floor. A compressed-air cannon sits on the floor, 4.0 m to one side of the point directly under the ball. When a button is pressed, the ball drops and, simultaneously, the cannon fires a 25 g plastic ball. The two balls collide 1.0 m above the floor. What was the launch speed of the plastic ball?

54. ‖ You are watching an archery tournament when you start wondering how fast an arrow is shot from the bow. Remembering your physics, you ask one of the archers to shoot an arrow parallel to the ground. You find the arrow stuck in the ground 60 m away, making a 3.0° angle with the ground. How fast was the arrow shot?

55. ‖ You're 6.0 m from one wall of the house seen in **FIGURE P4.55**. You want to toss a ball to your friend who is 6.0 m from the opposite wall. The throw and catch each occur 1.0 m above the ground.
    a. What minimum speed will allow the ball to clear the roof?
    b. At what angle should you toss the ball?

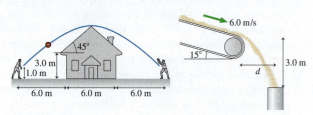

FIGURE P4.55   **FIGURE P4.56**

56. ‖ Sand moves without slipping at 6.0 m/s down a conveyer that is tilted at 15°. The sand enters a pipe 3.0 m below the end of the conveyer belt, as shown in **FIGURE P4.56**. What is the horizontal distance d between the conveyer belt and the pipe?

57. ‖ A stunt man drives a car at a speed of 20 m/s off a 30-m-high cliff. The road leading to the cliff is inclined upward at an angle of 20°.
    a. How far from the base of the cliff does the car land?
    b. What is the car's impact speed?

58. ‖ A javelin thrower standing at rest holds the center of the javelin
BIO behind her head, then accelerates it through a distance of 70 cm as she throws. She releases the javelin 2.0 m above the ground traveling at an angle of 30° above the horizontal. Top-rated javelin throwers do throw at about a 30° angle, not the 45° you might have expected, because the biomechanics of the arm allow them to throw the javelin much faster at 30° than they would be able to at 45°. In this throw, the javelin hits the ground 62 m away. What was the acceleration of the javelin during the throw? Assume that it has a constant acceleration.

59. ‖ A rubber ball is dropped onto a ramp that is tilted at 20°, as shown in **FIGURE P4.59**. A bouncing ball obeys the "law of reflection," which says that the ball leaves the surface at the same angle it approached the surface. The ball's next bounce is 3.0 m to the right of its first bounce. What is the ball's rebound speed on its first bounce?

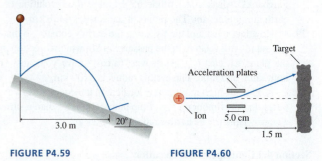

FIGURE P4.59   **FIGURE P4.60**

60. ‖‖‖ You are asked to consult for the city's research hospital, where
BIO a group of doctors is investigating the bombardment of cancer tumors with high-energy ions. As **FIGURE P4.60** shows, ions are fired directly toward the center of the tumor at speeds of 5.0×10^6 m/s. To cover the entire tumor area, the ions are deflected sideways by passing them between two charged metal plates that accelerate the ions perpendicular to the direction of their initial motion. The acceleration region is 5.0 cm long, and the ends of the acceleration plates are 1.5 m from the target. What sideways acceleration is required to deflect an ion 2.0 cm to one side?

61. ‖ Ships A and B leave port together. For the next two hours, ship A travels at 20 mph in a direction 30° west of north while ship B travels 20° east of north at 25 mph.
    a. What is the distance between the two ships two hours after they depart?
    b. What is the speed of ship A as seen by ship B?

62. ‖ While driving north at 25 m/s during a rainstorm you notice that the rain makes an angle of 38° with the vertical. While driving back home moments later at the same speed but in the opposite direction, you see that the rain is falling straight down. From these observations, determine the speed and angle of the raindrops relative to the ground.

63. ‖ You've been assigned the task of using a shaft encoder—a device that measures the angle of a shaft or axle and provides a signal to a computer—to analyze the rotation of an engine crankshaft under certain conditions. The table lists the crankshaft's angles over a 0.6 s interval.

Time (s)	Angle (rad)
0.0	0.0
0.1	2.0
0.2	3.2
0.3	4.3
0.4	5.3
0.5	6.1
0.6	7.0

Is the crankshaft rotating with uniform circular motion? If so, what is its angular velocity in rpm? If not, is the angular acceleration positive or negative?

64. ‖ A circular track has several concentric rings where people can run at their leisure. Phil runs on the outermost track with radius r_P while Annie runs on an inner track with radius $r_A = 0.80r_P$. The runners start side by side, along a radial line, and run at the same speed in a counterclockwise direction. How many revolutions has Annie made when Annie's and Phil's velocity vectors point in opposite directions for the first time?

65. ‖‖‖ A typical laboratory centrifuge rotates at 4000 rpm. Test tubes have to be placed into a centrifuge very carefully because of the very large accelerations.
    a. What is the acceleration at the end of a test tube that is 10 cm from the axis of rotation?
    b. For comparison, what is the magnitude of the acceleration a test tube would experience if dropped from a height of 1.0 m and stopped in a 1.0-ms-long encounter with a hard floor?

66. ‖ Astronauts use a centrifuge to simulate the acceleration of a
    BIO rocket launch. The centrifuge takes 30 s to speed up from rest to its top speed of 1 rotation every 1.3 s. The astronaut is strapped into a seat 6.0 m from the axis.
    a. What is the astronaut's tangential acceleration during the first 30 s?
    b. How many g's of acceleration does the astronaut experience when the device is rotating at top speed? Each 9.8 m/s² of acceleration is 1 g.

67. ‖ Communications satellites are placed in a circular orbit where they stay directly over a fixed point on the equator as the earth rotates. These are called *geosynchronous orbits*. The radius of the earth is 6.37×10^6 m, and the altitude of a geosynchronous orbit is 3.58×10^7 m ($\approx$22,000 miles). What are (a) the speed and (b) the magnitude of the acceleration of a satellite in a geosynchronous orbit?

68. ‖ A computer hard disk 8.0 cm in diameter is initially at rest. A small dot is painted on the edge of the disk. The disk accelerates at 600 rad/s² for $\frac{1}{2}$ s, then coasts at a steady angular velocity for another $\frac{1}{2}$ s.
    a. What is the speed of the dot at $t = 1.0$ s?
    b. Through how many revolutions has the disk turned?

69. ‖ A high-speed drill rotating ccw at 2400 rpm comes to a halt in 2.5 s.
    a. What is the magnitude of the drill's angular acceleration?
    b. How many revolutions does it make as it stops?

70. ‖ A turbine is spinning at 3800 rpm. Friction in the bearings is so low that it takes 10 min to coast to a stop. How many revolutions does the turbine make while stopping?

71. ‖ Your 64-cm-diameter car tire is rotating at 3.5 rev/s when suddenly you press down hard on the accelerator. After traveling 200 m, the tire's rotation has increased to 6.0 rev/s. What was the tire's angular acceleration? Give your answer in rad/s².

72. ‖ The angular velocity of a process control motor is
    CALC $\omega = \left(20 - \frac{1}{2}t^2\right)$ rad/s, where t is in seconds.
    a. At what time does the motor reverse direction?
    b. Through what angle does the motor turn between $t = 0$ s and the instant at which it reverses direction?

73. ‖ A Ferris wheel of radius R speeds up with angular acceleration α starting from rest. Find an expression for the (a) velocity and (b) centripetal acceleration of a rider after the Ferris wheel has rotated through angle $\Delta\theta$.

74. ‖ A 6.0-cm-diameter gear rotates with angular velocity $\omega =$
    CALC $\left(2.0 + \frac{1}{2}t^2\right)$ rad/s, where t is in seconds. At $t = 4.0$ s, what are:
    a. The gear's angular acceleration?
    b. The tangential acceleration of a tooth on the gear?

75. ‖ A painted tooth on a spinning gear has angular acceleration
    CALC $\alpha = (20 - t)$ rad/s², where t is in s. Its initial angular velocity, at $t = 0$ s, is 300 rpm. What is the tooth's angular velocity in rpm at $t = 20$ s?

76. ‖‖‖ A car starts from rest on a curve with a radius of 120 m and accelerates tangentially at 1.0 m/s². Through what angle will the car have traveled when the magnitude of its total acceleration is 2.0 m/s²?

77. ‖‖‖ A long string is wrapped around a 6.0-cm-diameter cylinder, initially at rest, that is free to rotate on an axle. The string is then pulled with a constant acceleration of 1.5 m/s² until 1.0 m of string has been unwound. If the string unwinds without slipping, what is the cylinder's angular speed, in rpm, at this time?

In Problems 78 through 80 you are given the equations that are used to solve a problem. For each of these, you are to
   a. Write a realistic problem for which these are the correct equations. Be sure that the answer your problem requests is consistent with the equations given.
   b. Finish the solution of the problem, including a pictorial representation.

78. $100 \text{ m} = 0 \text{ m} + (50\cos\theta \text{ m/s})t_1$
    $0 \text{ m} = 0 \text{ m} + (50\sin\theta \text{ m/s})t_1 - \frac{1}{2}(9.80 \text{ m/s}^2)t_1^2$

79. $v_x = -(6.0\cos 45°) \text{ m/s} + 3.0 \text{ m/s}$
    $v_y = (6.0 \sin 45°) \text{ m/s} + 0 \text{ m/s}$
    $100 \text{ m} = v_y t_1, \ x_1 = v_x t_1$

80. $2.5 \text{ rad} = 0 \text{ rad} + \omega_i(10 \text{ s}) + \left((1.5 \text{ m/s}^2)/2(50 \text{ m})\right)(10 \text{ s})^2$
    $\omega_f = \omega_i + \left((1.5 \text{ m/s}^2)/(50 \text{ m})\right)(10 \text{ s})$

Challenge Problems

81. ‖‖‖ In one contest at the county fair, seen in **FIGURE CP4.81**, a spring-loaded plunger launches a ball at a speed of 3.0 m/s from one corner of a smooth, flat board that is tilted up at a 20° angle. To win, you must make the ball hit a small target at the adjacent corner, 2.50 m away. At what angle θ should you tilt the ball launcher?

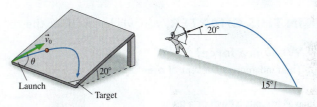

FIGURE CP4.81　　　　　　　**FIGURE CP4.82**

82. ‖‖‖ An archer standing on a 15° slope shoots an arrow 20° above the horizontal, as shown in **FIGURE CP4.82**. How far down the slope does the arrow hit if it is shot with a speed of 50 m/s from 1.75 m above the ground?

83. ‖‖‖ A skateboarder starts up a 1.0-m-high, 30° ramp at a speed of 7.0 m/s. The skateboard wheels roll without friction. At the top she leaves the ramp and sails through the air. How far from the end of the ramp does the skateboarder touch down?

84. ‖‖‖ A cannon on a train car fires a projectile to the right with speed
    CALC v_0, relative to the train, from a barrel elevated at angle θ. The cannon fires just as the train, which had been cruising to the right along a level track with speed v_{train}, begins to accelerate with acceleration a, which can be either positive (speeding up) or negative (slowing down). Find an expression for the angle at which the projectile should be fired so that it lands as far as possible from the cannon. You can ignore the small height of the cannon above the track.

85. ‖‖‖ A child in danger of drowning in a river is being carried downstream by a current that flows uniformly with a speed of 2.0 m/s. The child is 200 m from the shore and 1500 m upstream of the boat dock from which the rescue team sets out. If their boat speed is 8.0 m/s with respect to the water, at what angle from the shore should the pilot leave the shore to go directly to the child?

5 Force and Motion

These ice boats are a memorable example of the connection between force and motion.

IN THIS CHAPTER, you will learn about the connection between force and motion.

What is a force?
The fundamental concept of **mechanics** is force.

- A force is a push or a pull.
- A force acts on an object.
- A force requires an agent.
- A force is a vector.

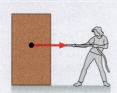

How do we identify forces?
A force can be a contact force or a long-range force.

- Contact forces occur at points where the environment touches the object.
- Contact forces disappear the instant contact is lost. Forces have no memory.
- Long-range forces include gravity and magnetism.

Thrust force $\vec{F}_{\text{thrust}}$

Gravity $\vec{F}_G$ Normal force $\vec{n}$

How do we show forces?
Forces can be displayed on a free-body diagram. You'll draw all forces—both pushes and pulls—as vectors with their tails on the particle. A well-drawn free-body diagram is an essential step in solving problems, as you'll see in the next chapter.

What do forces do?
A net force causes an object to *accelerate* with an acceleration directly proportional to the size of the force. This is Newton's second law, the most important statement in mechanics. For a particle of mass m,

$$\vec{a} = \frac{1}{m}\vec{F}_{\text{net}}$$

« LOOKING BACK Sections 1.4, 2.4, and 3.2 Acceleration and vector addition

What is Newton's first law?
Newton's first law—an object at rest stays at rest and an object in motion continues moving at constant speed in a straight line if and only if the net force on the object is zero—helps us define what a force *is*. It is also the basis for identifying the reference frames—called inertial reference frames—in which Newton's laws are valid.

What good are forces?
Kinematics describes *how* an object moves. For the more important tasks of knowing *why* an object moves and being able to predict its position and orientation at a future time, we have to know the forces acting on the object. Relating force to motion is the subject of **dynamics,** and it is one of the most important underpinnings of all science and engineering.

5.1 Force

The two major issues that this chapter will examine are:

- What is a force?
- What is the connection between force and motion?

We begin with the first of these questions in the table below.

What is a force?

A force is a push or a pull.

Our commonsense idea of a **force** is that it is a *push* or a *pull*. We will refine this idea as we go along, but it is an adequate starting point. Notice our careful choice of words: We refer to "*a force*," rather than simply "force." We want to think of a force as a very specific *action*, so that we can talk about a single force or perhaps about two or three individual forces that we can clearly distinguish. Hence the concrete idea of "a force" acting on an object.

A force acts on an object.

Implicit in our concept of force is that a **force acts on an object.** In other words, pushes and pulls are applied *to* something—an object. From the object's perspective, it has a force *exerted* on it. Forces do not exist in isolation from the object that experiences them.

A force requires an agent.

Every force has an **agent,** something that acts or exerts power. That is, a force has a specific, identifiable *cause.* As you throw a ball, it is your hand, while in contact with the ball, that is the agent or the cause of the force exerted on the ball. *If* a force is being exerted on an object, you must be able to identify a specific cause (i.e., the agent) of that force. Conversely, a force is not exerted on an object *unless* you can identify a specific cause or agent. Although this idea may seem to be stating the obvious, you will find it to be a powerful tool for avoiding some common misconceptions about what is and is not a force.

A force is a vector.

If you push an object, you can push either gently or very hard. Similarly, you can push either left or right, up or down. To quantify a push, we need to specify both a magnitude *and* a direction. It should thus come as no surprise that force is a vector. The general symbol for a force is the vector symbol $\vec{F}$. The size or strength of a force is its magnitude F.

A force can be either a contact force …

There are two basic classes of forces, depending on whether the agent touches the object or not. **Contact forces** are forces that act on an object by touching it at a point of contact. The bat must touch the ball to hit it. A string must be tied to an object to pull it. The majority of forces that we will examine are contact forces.

… or a long-range force.

Long-range forces are forces that act on an object without physical contact. Magnetism is an example of a long-range force. You have undoubtedly held a magnet over a paper clip and seen the paper clip leap up to the magnet. A coffee cup released from your hand is pulled to the earth by the long-range force of gravity.

NOTE In the particle model, objects cannot exert forces on themselves. A force on an object will always have an agent or cause external to the object. Now, there are certainly objects that have internal forces (think of all the forces inside the engine of your car!), but the particle model is not valid if you need to consider those internal forces. If you are going to treat your car as a particle and look only at the overall motion of the car as a whole, that motion will be a consequence of external forces acting on the car.

Force Vectors

We can use a simple diagram to visualize how forces are exerted on objects.

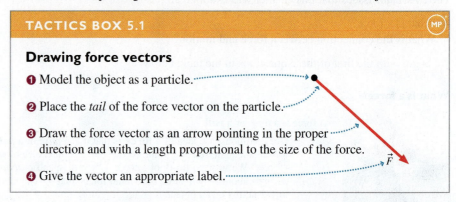

TACTICS BOX 5.1

Drawing force vectors

❶ Model the object as a particle.

❷ Place the *tail* of the force vector on the particle.

❸ Draw the force vector as an arrow pointing in the proper direction and with a length proportional to the size of the force.

❹ Give the vector an appropriate label.

$\vec{F}$

Step 2 may seem contrary to what a "push" should do, but recall that moving a vector does not change it as long as the length and angle do not change. The vector $\vec{F}$ is the same regardless of whether the tail or the tip is placed on the particle. **FIGURE 5.1** shows three examples of force vectors.

FIGURE 5.1 Three examples of forces and their vector representations.

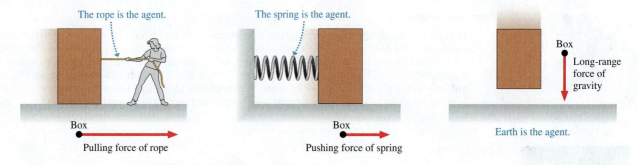

The rope is the agent.

The spring is the agent.

Box

Long-range force of gravity

Earth is the agent.

Box

Pulling force of rope

Box

Pushing force of spring

Combining Forces

FIGURE 5.2 Two forces applied to a box.

(a)

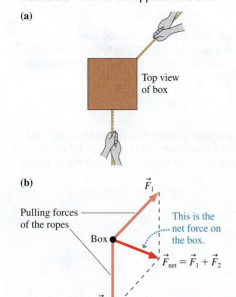

Top view of box

(b)

Pulling forces of the ropes

Box

$\vec{F}_1$

This is the net force on the box.

$\vec{F}_{net} = \vec{F}_1 + \vec{F}_2$

$\vec{F}_2$

FIGURE 5.2a shows a box being pulled by two ropes, each exerting a force on the box. How will the box respond? Experimentally, we find that when several forces $\vec{F}_1$, $\vec{F}_2$, $\vec{F}_3$, ... are exerted on an object, they combine to form a **net force** given by the *vector sum* of *all* the forces:

$$\vec{F}_{net} \equiv \sum_{i=1}^{N} \vec{F}_i = \vec{F}_1 + \vec{F}_2 + \cdots + \vec{F}_N \tag{5.1}$$

Recall that $\equiv$ is the symbol meaning "is defined as." Mathematically, this summation is called a **superposition of forces**. **FIGURE 5.2b** shows the net force on the box.

STOP TO THINK 5.1 Two of the three forces exerted on an object are shown. The net force points to the left. Which is the missing third force?

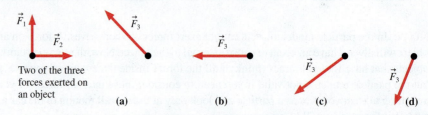

$\vec{F}_1$

$\vec{F}_2$

$\vec{F}_3$

$\vec{F}_3$

$\vec{F}_3$

$\vec{F}_3$

Two of the three forces exerted on an object

(a)　　(b)　　(c)　　(d)

5.2 A Short Catalog of Forces

There are many forces we will deal with over and over. This section will introduce you to some of them. Many of these forces have special symbols. As you learn the major forces, be sure to learn the symbol for each.

Gravity

Gravity—the only long-range force we will encounter in the next few chapters—keeps you in your chair and the planets in their orbits around the sun. We'll have a thorough look at gravity in Chapter 13. For now we'll concentrate on objects on or near the surface of the earth (or other planet).

The pull of a planet on an object on or near the surface is called the **gravitational force.** The agent for the gravitational force is the *entire planet*. Gravity acts on *all* objects, whether moving or at rest. The symbol for gravitational force is $\vec{F}_G$. **The gravitational force vector always points vertically downward,** as shown in **FIGURE 5.3**.

> **NOTE** We often refer to "the weight" of an object. For an object at rest on the surface of a planet, its weight is simply the magnitude F_G of the gravitational force. However, weight and gravitational force are not the same thing, nor is weight the same as mass. We will briefly examine mass later in the chapter, and we'll explore the rather subtle connections among gravity, weight, and mass in Chapter 6.

Spring Force

Springs exert one of the most common contact forces. **A spring can either push (when compressed) or pull (when stretched). FIGURE 5.4** shows the **spring force,** for which we use the symbol $\vec{F}_{Sp}$. In both cases, pushing and pulling, the tail of the force vector is placed on the particle in the force diagram.

Although you may think of a spring as a metal coil that can be stretched or compressed, this is only one type of spring. Hold a ruler, or any other thin piece of wood or metal, by the ends and bend it slightly. It flexes. When you let go, it "springs" back to its original shape. This is just as much a spring as is a metal coil.

Tension Force

When a string or rope or wire pulls on an object, it exerts a contact force that we call the **tension force,** represented by a capital $\vec{T}$. **The direction of the tension force is always along the direction of the string or rope,** as you can see in **FIGURE 5.5**. The commonplace reference to "the tension" in a string is an informal expression for T, the size or magnitude of the tension force.

> **NOTE** Tension is represented by the symbol T. This is logical, but there's a risk of confusing the tension T with the identical symbol T for the period of a particle in circular motion. The number of symbols used in science and engineering is so large that some letters are used several times to represent different quantities. The use of T is the first time we've run into this problem, but it won't be the last. You must be alert to the *context* of a symbol's use to deduce its meaning.

We can obtain a deeper understanding of some forces and interactions with a picture of what's happening at the atomic level. You'll recall from chemistry that matter consists of *atoms* that are attracted to each other by *molecular bonds.* Although the details are complex, governed by quantum physics, we can often use a simple **ball-and-spring model** of a solid to get an idea of what's happening at the atomic level.

FIGURE 5.3 Gravity.

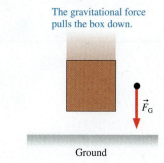

The gravitational force pulls the box down.

$\vec{F}_G$

Ground

FIGURE 5.4 The spring force.

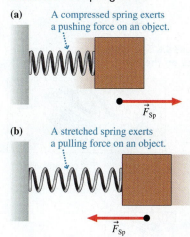

(a) A compressed spring exerts a pushing force on an object.

$\vec{F}_{Sp}$

(b) A stretched spring exerts a pulling force on an object.

$\vec{F}_{Sp}$

FIGURE 5.5 Tension.

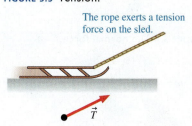

The rope exerts a tension force on the sled.

$\vec{T}$

MODEL 5.1

Ball-and-spring model of solids

Solids consist of atoms held together by molecular bonds.

■ Represent the solid as **an array of balls connected by springs.**

■ Pulling on or pushing on a solid causes the bonds to be stretched or compressed. **Stretched or compressed bonds exert spring forces.**

■ There are an immense number of bonds. The force of one bond is very tiny, but the combined force of all bonds can be very large.

■ Limitations: Model fails for liquids and gases.

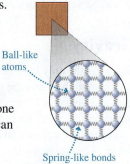

Ball-like atoms

Spring-like bonds

In the case of tension, pulling on the ends of a string or rope stretches the spring-like molecular bonds ever so slightly. What we call "tension" is then the net spring force being exerted by trillions and trillions of microscopic springs.

Normal Force

If you sit on a bed, the springs in the mattress compress and, as a consequence of the compression, exert an upward force on you. Stiffer springs would show less compression but still exert an upward force. The compression of extremely stiff springs might be measurable only by sensitive instruments. Nonetheless, the springs would compress ever so slightly and exert an upward spring force on you.

FIGURE 5.6 shows an object resting on top of a sturdy table. The table may not visibly flex or sag, but—just as you do to the bed—the object compresses the spring-like molecular bonds in the table. The size of the compression is very small, but it is not zero. As a consequence, the compressed "molecular springs" *push upward* on the object. We say that "the table" exerts the upward force, but it is important to understand that the pushing is *really* done by molecular bonds.

We can extend this idea. Suppose you place your hand on a wall and lean against it. Does the wall exert a force on your hand? As you lean, you compress the molecular bonds in the wall and, as a consequence, they push outward against your hand. So the answer is yes, the wall does exert a force on you.

The force the table surface exerts is vertical; the force the wall exerts is horizontal. In all cases, the force exerted on an object that is pressing against a surface is in a direction *perpendicular* to the surface. Mathematicians refer to a line that is perpendicular to a surface as being *normal* to the surface. In keeping with this terminology, we define the **normal force** as **the force exerted perpendicular to a surface (the agent) against an object that is pressing against the surface.** The symbol for the normal force is $\vec{n}$.

We're not using the word *normal* to imply that the force is an "ordinary" force or to distinguish it from an "abnormal force." A surface exerts a force *perpendicular* (i.e., normal) to itself as the molecular springs press *outward*. **FIGURE 5.7** shows an object on an inclined surface, a common situation.

In essence, the normal force is just a spring force, but one exerted by a vast number of microscopic springs acting at once. The normal force is responsible for the "solidness" of solids. It is what prevents you from passing right through the chair you are sitting in and what causes the pain and the lump if you bang your head into a door.

FIGURE 5.6 The table exerts an upward force on the book.

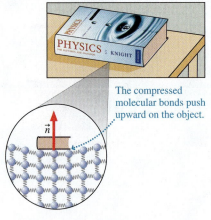

The compressed molecular bonds push upward on the object.

FIGURE 5.7 The normal force.

$\vec{n}$

The surface pushes outward against the bottom of the frog.

Friction

Friction, like the normal force, is exerted by a surface. But whereas the normal force is perpendicular to the surface, **the friction force is always *parallel* to the surface.** It is useful to distinguish between two kinds of friction:

■ *Kinetic friction,* denoted $\vec{f}_k$, appears as an object slides across a surface. This is a force that "opposes the motion," meaning that the friction force vector $\vec{f}_k$ points in a direction opposite the velocity vector $\vec{v}$ (i.e., "the motion").

■ *Static friction,* denoted $\vec{f}_s$, is the force that keeps an object "stuck" on a surface and prevents its motion. Finding the direction of $\vec{f}_s$ is a little trickier than finding it for $\vec{f}_k$. Static friction points opposite the direction in which the object *would* move if there were no friction. That is, it points in the direction necessary to *prevent* motion.

FIGURE 5.8 shows examples of kinetic and static friction.

> **NOTE** A surface exerts a kinetic friction force when an object moves *relative to* the surface. A package on a conveyor belt is in motion, but it does not experience a kinetic friction force because it is not moving relative to the belt. So to be precise, we should say that the kinetic friction force points opposite to an object's motion *relative to* a surface.

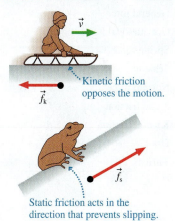

FIGURE 5.8 Kinetic and static friction.

Kinetic friction opposes the motion.

Static friction acts in the direction that prevents slipping.

Drag

Friction at a surface is one example of a *resistive force,* a force that opposes or resists motion. Resistive forces are also experienced by objects moving through fluids—gases and liquids. The resistive force of a fluid is called **drag,** with symbol $\vec{F}_{drag}$. **Drag, like kinetic friction, points opposite the direction of motion.** **FIGURE 5.9** shows an example.

Drag can be a significant force for objects moving at high speeds or in dense fluids. Hold your arm out the window as you ride in a car and feel how the air resistance against it increases rapidly as the car's speed increases. Drop a lightweight object into a beaker of water and watch how slowly it settles to the bottom.

For objects that are heavy and compact, that move in air, and whose speed is not too great, the drag force of air resistance is fairly small. To keep things as simple as possible, **you can neglect air resistance in all problems unless a problem explicitly asks you to include it.**

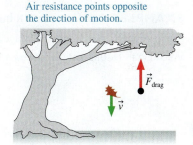

FIGURE 5.9 Air resistance is an example of drag.

Air resistance points opposite the direction of motion.

Thrust

A jet airplane obviously has a force that propels it forward during takeoff. Likewise for the rocket being launched in **FIGURE 5.10**. This force, called **thrust,** occurs when a jet or rocket engine expels gas molecules at high speed. Thrust is a contact force, with the exhaust gas being the agent that pushes on the engine. The process by which thrust is generated is rather subtle, and we will postpone a full discussion until we study Newton's third law in Chapter 7. For now, we will treat thrust as a force opposite the direction in which the exhaust gas is expelled. There's no special symbol for thrust, so we will call it $\vec{F}_{thrust}$.

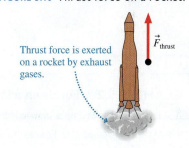

FIGURE 5.10 Thrust force on a rocket.

Thrust force is exerted on a rocket by exhaust gases.

Electric and Magnetic Forces

Electricity and magnetism, like gravity, exert long-range forces. We will study electric and magnetic forces in detail in Part VI. For now, it is worth noting that the forces holding molecules together—the molecular bonds—are not actually tiny springs. Atoms and molecules are made of charged particles—electrons and protons—and what we call a molecular bond is really an electric force between these particles. So when we say that the normal force and the tension force are due to "molecular springs," or that friction is due to atoms running into each other, what we're really saying is that these forces, at the most fundamental level, are actually electric forces between the charged particles in the atoms.

5.3 Identifying Forces

A typical physics problem describes an object that is being pushed and pulled in various directions. Some forces are given explicitly; others are only implied. In order to proceed, it is necessary to determine all the forces that act on the object. The procedure for identifying forces will become part of the *pictorial representation* of the problem.

Force	Notation
General force	$\vec{F}$
Gravitational force	$\vec{F}_G$
Spring force	$\vec{F}_{Sp}$
Tension	$\vec{T}$
Normal force	$\vec{n}$
Static friction	$\vec{f}_s$
Kinetic friction	$\vec{f}_k$
Drag	$\vec{F}_{drag}$
Thrust	$\vec{F}_{thrust}$

TACTICS BOX 5.2

Identifying forces

❶ **Identify the object of interest.** This is the object you wish to study.
❷ **Draw a picture of the situation.** Show the object of interest and all other objects—such as ropes, springs, or surfaces—that touch it.
❸ **Draw a closed curve around the object.** Only the object of interest is inside the curve; everything else is outside.
❹ **Locate every point on the boundary of this curve where other objects touch the object of interest.** These are the points where *contact forces* are exerted on the object.
❺ **Name and label each contact force acting on the object.** There is at least one force at each point of contact; there may be more than one. When necessary, use subscripts to distinguish forces of the same type.
❻ **Name and label each long-range force acting on the object.** For now, the only long-range force is the gravitational force.

Exercises 3–8

EXAMPLE 5.1 | **Forces on a bungee jumper**

A bungee jumper has leapt off a bridge and is nearing the bottom of her fall. What forces are being exerted on the jumper?

VISUALIZE **FIGURE 5.11** Forces on a bungee jumper.

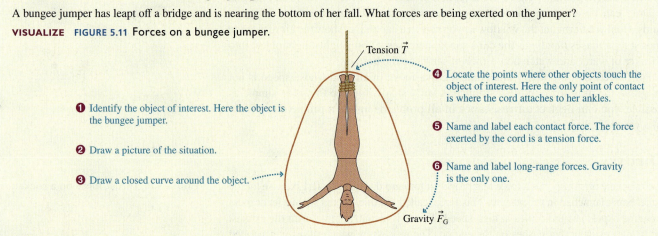

Tension $\vec{T}$

❶ Identify the object of interest. Here the object is the bungee jumper.

❷ Draw a picture of the situation.

❸ Draw a closed curve around the object.

❹ Locate the points where other objects touch the object of interest. Here the only point of contact is where the cord attaches to her ankles.

❺ Name and label each contact force. The force exerted by the cord is a tension force.

❻ Name and label long-range forces. Gravity is the only one.

Gravity $\vec{F}_G$

EXAMPLE 5.2 | **Forces on a skier**

A skier is being towed up a snow-covered hill by a tow rope. What forces are being exerted on the skier?

VISUALIZE **FIGURE 5.12** Forces on a skier.

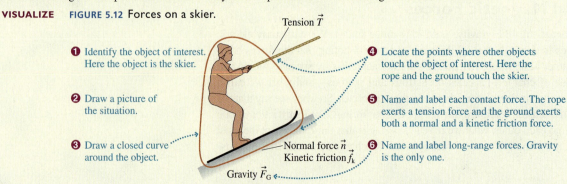

Tension $\vec{T}$

❶ Identify the object of interest. Here the object is the skier.

❷ Draw a picture of the situation.

❸ Draw a closed curve around the object.

❹ Locate the points where other objects touch the object of interest. Here the rope and the ground touch the skier.

❺ Name and label each contact force. The rope exerts a tension force and the ground exerts both a normal and a kinetic friction force.

❻ Name and label long-range forces. Gravity is the only one.

Normal force $\vec{n}$
Kinetic friction $\vec{f}_k$
Gravity $\vec{F}_G$

NOTE You might have expected two friction forces and two normal forces in Example 5.2, one on each ski. Keep in mind, however, that we're working within the particle model, which represents the skier by a single point. A particle has only one contact with the ground, so there is one normal force and one friction force.

EXAMPLE 5.3 | **Forces on a rocket**

A rocket is being launched to place a new satellite in orbit. Air resistance is not negligible. What forces are being exerted on the rocket?

VISUALIZE This drawing is much more like the sketch you would make when identifying forces as part of solving a problem.

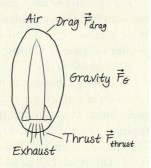

▶ **FIGURE 5.13** Forces on a rocket.

STOP TO THINK 5.2 You've just kicked a rock, and it is now sliding across the ground 2 m in front of you. Which of these forces act on the rock? List all that apply.

a. Gravity, acting downward.
b. The normal force, acting upward.
c. The force of the kick, acting in the direction of motion.
d. Friction, acting opposite the direction of motion.

5.4 What Do Forces Do?

Having learned to identify forces, we ask the next question: How does an object move when a force is exerted on it? The only way to answer this question is to do experiments. Let's conduct a "virtual experiment," one you can easily visualize. Imagine using your fingers to stretch a rubber band to a certain length—say 10 centimeters—that you can measure with a ruler, as shown in **FIGURE 5.14**. You know that a stretched rubber band exerts a force—a spring force—because your fingers *feel* the pull. Furthermore, this is a reproducible force; the rubber band exerts the same force every time you stretch it to this length. We'll call this the *standard force F*. Not surprisingly, two identical rubber bands exert twice the pull of one rubber band, and N side-by-side rubber bands exert N times the standard force: $F_{net} = NF$.

Now attach one rubber band to a 1 kg block and stretch it to the standard length. The object experiences the same force F as did your finger. The rubber band gives us a way of applying a known and reproducible force to an object. Then imagine using the rubber band to pull the block across a horizontal, frictionless table. (We can imagine a frictionless table since this is a virtual experiment, but in practice you could nearly eliminate friction by supporting the object on a cushion of air.)

If you stretch the rubber band and then release the object, the object moves toward your hand. But as it does so, the rubber band gets shorter and the pulling force decreases. To keep the pulling force constant, you must *move your hand* at just the right speed to keep the length of the rubber band from changing! **FIGURE 5.15a** shows the experiment being carried out. Once the motion is complete, you can use motion diagrams and kinematics to analyze the object's motion.

FIGURE 5.14 A reproducible force.

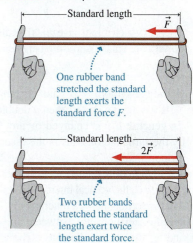

One rubber band stretched the standard length exerts the standard force F.

Two rubber bands stretched the standard length exert twice the standard force.

FIGURE 5.15 Measuring the motion of a 1 kg block that is pulled with a constant force.

(a)

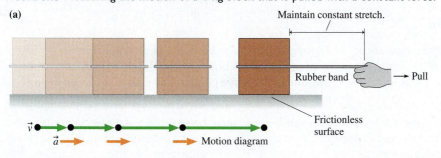

Maintain constant stretch.

Rubber band → Pull

Frictionless surface

$\vec{v}$

$\vec{a}$ → Motion diagram

(b)

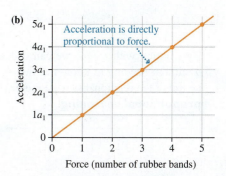

Acceleration is directly proportional to force.

Acceleration

$5a_1$
$4a_1$
$3a_1$
$2a_1$
$1a_1$
0

0 1 2 3 4 5

Force (number of rubber bands)

The first important finding of this experiment is that **an object pulled with a constant force moves with a constant acceleration.** That is, the answer to the question What does a force do? is: A force causes an object to accelerate, and a constant force produces a constant acceleration.

What happens if you increase the force by using several rubber bands? To find out, use two rubber bands, then three rubber bands, then four, and so on. With N rubber bands, the force on the block is NF. FIGURE 5.15b shows the results of this experiment. You can see that doubling the force causes twice the acceleration, tripling the force causes three times the acceleration, and so on. The graph reveals our second important finding: **The acceleration is directly proportional to the force.** This result can be written as

$$a = cF \tag{5.2}$$

where c, called the *proportionality constant,* is the slope of the graph.

MATHEMATICAL ASIDE

Proportionality and proportional reasoning

The concept of **proportionality** arises frequently in physics. A quantity symbolized by u is *proportional* to another quantity symbolized by v if

$$u = cv$$

where c (which might have units) is called the **proportionality constant**. This relationship between u and v is often written

$$u \propto v$$

where the symbol $\propto$ means "is proportional to."

If v is doubled to $2v$, then u doubles to $c(2v) = 2(cv) = 2u$. In general, if v is changed by any factor f, then u changes by the same factor. This is the essence of what we *mean* by proportionality.

A graph of u versus v is a straight line *passing through the origin* (i.e., the vertical intercept is zero) with slope $= c$. Notice that proportionality is a much more specific relationship between u and v than mere linearity. The linear equation $u = cv + b$ has a straight-line graph, but it doesn't pass through the origin (unless b happens to be zero) and doubling v does not double u.

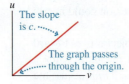

u is proportional to v.

If $u \propto v$, then $u_1 = cv_1$ and $u_2 = cv_2$. Dividing the second equation by the first, we find

$$\frac{u_2}{u_1} = \frac{v_2}{v_1}$$

By working with *ratios*, we can deduce information about u without needing to know the value of c. (This would not be true if the relationship were merely linear.) This is called **proportional reasoning.**

Proportionality is not limited to being linearly proportional. The graph on the left shows that u is clearly not proportional to w. But a graph of u versus $1/w^2$ *is* a straight line passing through the origin; thus, in this case, u is proportional to $1/w^2$, or $u \propto 1/w^2$. We would say that "u is proportional to the inverse square of w."

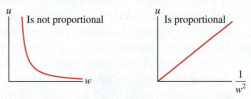

u is proportional to the inverse square of w.

EXAMPLE u is proportional to the inverse square of w. By what factor does u change if w is tripled?

SOLUTION This is an opportunity for proportional reasoning; we don't need to know the proportionality constant. If u is proportional to $1/w^2$, then

$$\frac{u_2}{u_1} = \frac{1/w_2^2}{1/w_1^2} = \frac{w_1^2}{w_2^2} = \left(\frac{w_1}{w_2}\right)^2$$

Tripling w, with $w_2/w_1 = 3$, and thus $w_1/w_2 = \frac{1}{3}$, changes u to

$$u_2 = \left(\frac{w_1}{w_2}\right)^2 u_1 = \left(\frac{1}{3}\right)^2 u_1 = \frac{1}{9} u_1$$

Tripling w causes u to become $\frac{1}{9}$ of its original value.

Many *Student Workbook* and end-of-chapter homework questions will require proportional reasoning. It's an important skill to learn.

The final question for our virtual experiment is: How does the acceleration depend on the mass of the object being pulled? To find out, apply the *same force*—for example, the standard force of one rubber band—to a 2 kg block, then a 3 kg block, and so on, and for each measure the acceleration. Doing so gives you the results shown in **FIGURE 5.16**. An object with twice the mass of the original block has only half the acceleration when both are subjected to the same force.

Mathematically, the graph of Figure 5.16 is one of *inverse proportionality*. That is, **the acceleration is inversely proportional to the object's mass.** We can combine these results—that the acceleration is directly proportional to the force applied and inversely proportional to the object's mass—into the single statement

$$a = \frac{F}{m} \tag{5.3}$$

if we define the basic unit of force as the force that causes a 1 kg mass to accelerate at 1 m/s^2. That is,

$$1 \text{ basic unit of force} \equiv 1 \text{ kg} \times 1 \frac{\text{m}}{\text{s}^2} = 1 \frac{\text{kg m}}{\text{s}^2}$$

This basic unit of force is called a newton:

> One **newton** is the force that causes a 1 kg mass to accelerate at 1 m/s^2. The abbreviation for newton is N. Mathematically, $1 \text{ N} = 1 \text{ kg m/s}^2$.

TABLE 5.1 lists some typical forces. As you can see, "typical" forces on "typical" objects are likely to be in the range 0.01–10,000 N.

Mass

We've been using the term *mass* without a clear definition. As we learned in Chapter 1, the SI unit of mass, the kilogram, is based on a particular metal block kept in a vault in Paris. This suggests that *mass* is the amount of matter an object contains, and that is certainly our everyday concept of mass. Now we see that a more precise way of defining an object's mass is in terms of its acceleration in response to a force. Figure 5.16 shows that an object with twice the amount of matter accelerates only half as much in response to the same force. The more matter an object has, the more it *resists* accelerating in response to a force. You're familiar with this idea: Your car is much harder to push than your bicycle. The tendency of an object to resist a *change* in its velocity (i.e., to resist acceleration) is called **inertia.** Consequently, the mass used in Equation 5.3, a measure of an object's resistance to changing its motion, is called **inertial mass.** We'll meet a different concept of mass, *gravitational mass,* when we study Newton's law of gravity in Chapter 13.

STOP TO THINK 5.3 Two rubber bands stretched to the standard length cause an object to accelerate at 2 m/s^2. Suppose another object with twice the mass is pulled by four rubber bands stretched to the standard length. The acceleration of this second object is

a. 1 m/s^2
b. 2 m/s^2
c. 4 m/s^2
d. 8 m/s^2
e. 16 m/s^2

Hint: Use proportional reasoning.

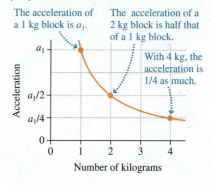

FIGURE 5.16 Acceleration is inversely proportional to mass.

The acceleration of a 1 kg block is a_1.

The acceleration of a 2 kg block is half that of a 1 kg block.

With 4 kg, the acceleration is 1/4 as much.

TABLE 5.1 Approximate magnitude of some typical forces

Force	Approximate magnitude (newtons)
Weight of a U.S. quarter	0.05
Weight of 1/4 cup sugar	0.5
Weight of a 1 pound object	5
Weight of a house cat	50
Weight of a 110 pound person	500
Propulsion force of a car	5,000
Thrust force of a small jet engine	50,000

5.5 Newton's Second Law

Equation 5.3 is an important finding, but our experiment was limited to looking at an object's response to a single applied force. Realistically, an object is likely to be subjected to several distinct forces $\vec{F}_1$, $\vec{F}_2$, $\vec{F}_3$, ... that may point in different directions. What happens then? In that case, it is found experimentally that the acceleration is determined by the *net* force.

Newton was the first to recognize the connection between force and motion. This relationship is known today as **Newton's second law**.

> **Newton's second law** An object of mass m subjected to forces $\vec{F}_1$, $\vec{F}_2$, $\vec{F}_3$, ... will undergo an acceleration $\vec{a}$ given by
>
> $$\vec{a} = \frac{\vec{F}_{net}}{m} \qquad (5.4)$$
>
> where the net force $\vec{F}_{net} = \vec{F}_1 + \vec{F}_2 + \vec{F}_3 + \cdots$ is the vector sum of all forces acting on the object. The acceleration vector $\vec{a}$ points in the same direction as the net force vector $\vec{F}_{net}$.

The significance of Newton's second law cannot be overstated. There was no reason to suspect that there should be any simple relationship between force and acceleration. Yet there it is, a simple but exceedingly powerful equation relating the two. The critical idea is that **an object accelerates in the direction of the net force vector $\vec{F}_{net}$**.

We can rewrite Newton's second law in the form

$$\vec{F}_{net} = m\vec{a} \qquad (5.5)$$

which is how you'll see it presented in many textbooks. Equations 5.4 and 5.5 are mathematically equivalent, but Equation 5.4 better describes the central idea of Newtonian mechanics: A force applied to an object causes the object to accelerate.

It's also worth noting that **the object responds only to the forces acting on it at *this instant*.** The object has no memory of forces that may have been exerted at earlier times. This idea is sometimes called **Newton's zeroth law.**

> **NOTE** Be careful not to think that one force "overcomes" the others to determine the motion. Forces are not in competition with each other! It is $\vec{F}_{net}$, the sum of *all* the forces, that determines the acceleration $\vec{a}$.

As an example, **FIGURE 5.17a** shows a box being pulled by two ropes. The ropes exert tension forces $\vec{T}_1$ and $\vec{T}_2$ on the box. **FIGURE 5.17b** represents the box as a particle, shows the forces acting on the box, and adds them graphically to find the net force $\vec{F}_{net}$. The box will accelerate in the direction of $\vec{F}_{net}$ with acceleration

$$\vec{a} = \frac{\vec{F}_{net}}{m} = \frac{\vec{T}_1 + \vec{T}_2}{m}$$

> **NOTE** The acceleration is *not* $(T_1 + T_2)/m$. You must add the forces as *vectors,* not merely add their magnitudes as scalars.

Forces Are Interactions

There's one more important aspect of forces. If you push against a door (the object) to close it, the door pushes back against your hand (the agent). If a tow rope pulls on a car (the object), the car pulls back on the rope (the agent). In general, if an agent exerts a force on an object, the object exerts a force on the agent. We really need to think of a force as an *interaction* between two objects. This idea is captured in **Newton's third law**—that for every action there is an equal but opposite reaction.

FIGURE 5.17 Acceleration of a pulled box.

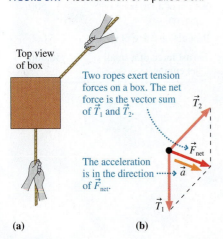

Top view
of box

Two ropes exert tension forces on a box. The net force is the vector sum of $\vec{T}_1$ and $\vec{T}_2$.

$\vec{T}_2$

$\vec{F}_{net}$

The acceleration is in the direction of $\vec{F}_{net}$.

$\vec{a}$

$\vec{T}_1$

(a)   (b)

Although the interaction perspective is a more exact way to view forces, it adds complications that we would like to avoid for now. Our approach will be to start by focusing on how a single object responds to forces exerted on it. Then, in Chapter 7, we'll return to Newton's third law and the larger issue of how two or more objects interact with each other.

STOP TO THINK 5.4   Three forces act on an object. In which direction does the object accelerate?

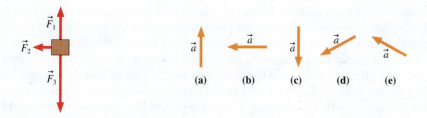

5.6 Newton's First Law

For 2000 years, scientists and philosophers thought that the "natural state" of an object is to be at rest. An object at rest requires no explanation. A moving object, though, is not in its natural state and thus requires an explanation: Why is this object moving? What keeps it going?

Galileo, in around 1600, was one of the first scientists to carry out controlled experiments. Many careful measurements in which he minimized the influence of friction led Galileo to conclude that in the absence of friction or air resistance, a moving object would continue to move along a straight line forever with no loss of speed. In other words, the natural state of an object—its behavior if free of external influences—is not rest but is *uniform motion* with constant velocity! "At rest" has no special significance in Galileo's view of motion; it is simply uniform motion that happens to have $\vec{v} = \vec{0}$.

It was left to Newton to generalize this result, and today we call it **Newton's first law** of motion.

FIGURE 5.18 Two examples of mechanical equilibrium.

> **Newton's first law**   An object that is at rest will remain at rest, or an object that is moving will continue to move in a straight line with constant velocity, if and only if the net force acting on the object is zero.

Newton's first law is also known as the *law of inertia*. If an object is at rest, it has a tendency to stay at rest. If it is moving, it has a tendency to continue moving with the *same velocity*.

NOTE The first law refers to *net* force. An object can remain at rest, or can move in a straight line with constant velocity, even though forces are exerted on it as long as the *net* force is zero.

Notice the "if and only if" aspect of Newton's first law. If an object is at rest or moves with constant velocity, then we can conclude that there is no net force acting on it. Conversely, if no net force is acting on it, we can conclude that the object will have constant velocity, not just constant speed. The direction remains constant, too!

An object on which the net force is zero—and thus is either at rest or moving in a straight line with constant velocity—is said to be in **mechanical equilibrium**. As **FIGURE 5.18** shows, objects in mechanical equilibrium have no acceleration: $\vec{a} = \vec{0}$.

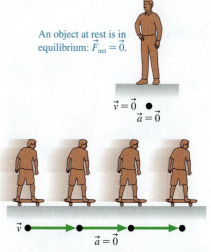

An object at rest is in equilibrium: $\vec{F}_{net} = \vec{0}$.

$\vec{v} = \vec{0}$
$\vec{a} = \vec{0}$

$\vec{v}$

$\vec{a} = \vec{0}$

An object moving in a straight line at constant velocity is also in equilibrium: $\vec{F}_{net} = \vec{0}$.

What Good Is Newton's First Law?

So what causes an object to move? Newton's first law says **no cause is needed for an object to move!** Uniform motion is the object's natural state. Nothing at all is required for it to remain in that state. The proper question, according to Newton, is: What causes an object to *change* its velocity? Newton, with Galileo's help, also gave us the answer. **A *force* is what causes an object to change its velocity.**

The preceding paragraph contains the essence of Newtonian mechanics. This new perspective on motion, however, is often contrary to our common experience. We all know perfectly well that you must keep pushing an object—exerting a force on it—to keep it moving. Newton is asking us to change our point of view and to consider motion *from the object's perspective* rather than from our personal perspective. As far as the object is concerned, our push is just one of several forces acting on it. Others might include friction, air resistance, or gravity. Only by knowing the *net* force can we determine the object's motion.

Newton's first law may seem to be merely a special case of Newton's second law. After all, the equation $\vec{F}_{net} = m\vec{a}$ tells us that an object moving with constant velocity ($\vec{a} = \vec{0}$) has $\vec{F}_{net} = \vec{0}$. The difficulty is that the second law assumes that we already know what force is. The purpose of the first law is to *identify* a force as something that disturbs a state of equilibrium. The second law then describes how the object responds to this force. Thus from a *logical* perspective, the first law really is a separate statement that must precede the second law. But this is a rather formal distinction. From a pedagogical perspective it is better—as we have done—to use a commonsense understanding of force and start with Newton's second law.

Inertial Reference Frames

If a car stops suddenly, you may be "thrown" into the windshield if you're not wearing your seat belt. You have a very real forward acceleration *relative to the car,* but is there a force pushing you forward? A force is a push or a pull caused by an identifiable agent in contact with the object. Although you *seem* to be pushed forward, there's no agent to do the pushing.

The difficulty—an acceleration without an apparent force—comes from using an inappropriate reference frame. Your acceleration measured in a reference frame attached to the car is not the same as your acceleration measured in a reference frame attached to the ground. Newton's second law says $\vec{F}_{net} = m\vec{a}$. But which $\vec{a}$? Measured in which reference frame?

We define an **inertial reference frame** as a reference frame in which Newton's first law is valid. If $\vec{a} = \vec{0}$ (an object is at rest or moving with constant velocity) only when $\vec{F}_{net} = \vec{0}$, then the reference frame in which $\vec{a}$ is measured is an inertial reference frame.

Not all reference frames are inertial reference frames. **FIGURE 5.19a** shows a physics student cruising at constant velocity in an airplane. If the student places a ball on the floor, it stays there. There are no horizontal forces, and the ball remains at rest relative to the airplane. That is, $\vec{a} = \vec{0}$ in the airplane's reference frame when $\vec{F}_{net} = \vec{0}$. Newton's first law is satisfied, so this airplane is an inertial reference frame.

The physics student in **FIGURE 5.19b** conducts the same experiment during takeoff. He carefully places the ball on the floor just as the airplane starts to accelerate down the runway. You can imagine what happens. The ball rolls to the back of the plane as the passengers are being pressed back into their seats. Nothing exerts a horizontal contact force on the ball, yet the ball accelerates *in the plane's reference frame.* This violates Newton's first law, so the plane is *not* an inertial reference frame during takeoff.

In the first example, the plane is traveling with constant velocity. In the second, the plane is accelerating. **Accelerating reference frames are not inertial reference frames.** Consequently, Newton's laws are not valid in an accelerating reference frame.

The earth is not exactly an inertial reference frame because the earth rotates on its axis and orbits the sun. However, the earth's acceleration is so small that violations of Newton's laws can be measured only in very careful experiments. We will treat the earth and laboratories attached to the earth as inertial reference frames, an approximation that is exceedingly well justified.

This guy thinks there's a force hurling him into the windshield. What a dummy!

FIGURE 5.19 Reference frames.

(a) Cruising at constant speed.

The ball stays in place; the airplane is an inertial reference frame.

(b) Accelerating during takeoff.

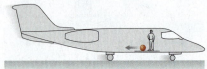

The ball accelerates toward the back even though there are no horizontal forces; the airplane is *not* an inertial reference frame.

To understand the motion of the passengers in a braking car, you need to measure velocities and accelerations *relative to the ground*. From the perspective of an observer on the ground, the body of a passenger in a braking car tries to continue moving forward with constant velocity, exactly as we would expect on the basis of Newton's first law, while his immediate surroundings are decelerating. The passenger is not "thrown" into the windshield. Instead, the windshield runs into the passenger!

Thinking About Force

It is important to identify correctly all the forces acting on an object. It is equally important not to include forces that do not really exist. We have established a number of criteria for identifying forces; the three critical ones are:

- A force has an agent. Something tangible and identifiable causes the force.
- Forces exist at the point of contact between the agent and the object experiencing the force (except for the few special cases of long-range forces).
- Forces exist due to interactions happening *now*, not due to what happened in the past.

Consider a bowling ball rolling along on a smooth floor. It is very tempting to think that a horizontal "force of motion" keeps it moving in the forward direction. But *nothing contacts the ball* except the floor. No agent is giving the ball a forward push. According to our definition, then, there is *no* forward "force of motion" acting on the ball. So what keeps it going? Recall our discussion of the first law: *No* cause is needed to keep an object moving at constant velocity. It continues to move forward simply because of its inertia.

A related problem occurs if you throw a ball. A pushing force was indeed required to accelerate the ball *as it was thrown*. But that force disappears the instant the ball loses contact with your hand. The force does not stick with the ball as the ball travels through the air. Once the ball has acquired a velocity, *nothing* is needed to keep it moving with that velocity.

There's no "force of motion" or any other forward force on this arrow. It continues to move because of inertia.

5.7 Free-Body Diagrams

Having discussed at length what is and is not a force, we are ready to assemble our knowledge about force and motion into a single diagram called a *free-body diagram*. You will learn in the next chapter how to write the equations of motion directly from the free-body diagram. Solution of the equations is a mathematical exercise—possibly a difficult one, but nonetheless an exercise that could be done by a computer. The *physics* of the problem, as distinct from the purely calculational aspects, are the steps that lead to the free-body diagram.

A **free-body diagram,** part of the *pictorial representation* of a problem, represents the object as a particle and shows *all* of the forces acting on the object.

TACTICS BOX 5.3   (MP)

Drawing a free-body diagram

❶ **Identify all forces acting on the object.** This step was described in Tactics Box 5.2.

❷ **Draw a coordinate system.** Use the axes defined in your pictorial representation.

❸ **Represent the object as a dot at the origin of the coordinate axes.** This is the particle model.

❹ **Draw vectors representing each of the identified forces.** This was described in Tactics Box 5.1. Be sure to label each force vector.

❺ **Draw and label the *net force* vector $\vec{F}_{\text{net}}$.** Draw this vector beside the diagram, not on the particle. Or, if appropriate, write $\vec{F}_{\text{net}} = \vec{0}$. Then check that $\vec{F}_{\text{net}}$ points in the same direction as the acceleration vector $\vec{a}$ on your motion diagram.

Exercises 24–29

EXAMPLE 5.4 | **An elevator accelerates upward**

An elevator, suspended by a cable, speeds up as it moves upward from the ground floor. Identify the forces and draw a free-body diagram of the elevator.

MODEL Model the elevator as a particle.

VISUALIZE

FIGURE 5.20 Free-body diagram of an elevator accelerating upward.

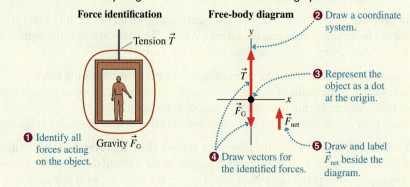

ASSESS The coordinate axes, with a vertical y-axis, are the ones we would use in a pictorial representation of the motion. The elevator is accelerating upward, so $\vec{F}_{net}$ must point upward. For this to be true, the magnitude of $\vec{T}$ must be larger than the magnitude of $\vec{F}_G$. The diagram has been drawn accordingly.

EXAMPLE 5.5 | **An ice block shoots across a frozen lake**

Bobby straps a small model rocket to a block of ice and shoots it across the smooth surface of a frozen lake. Friction is negligible. Draw a pictorial representation of the block of ice.

MODEL Model the block of ice as a particle. The pictorial representation consists of a motion diagram to determine $\vec{a}$, a force-identification picture, and a free-body diagram. The statement of the situation implies that friction is negligible.

VISUALIZE

FIGURE 5.21 Pictorial representation for a block of ice shooting across a frictionless frozen lake.

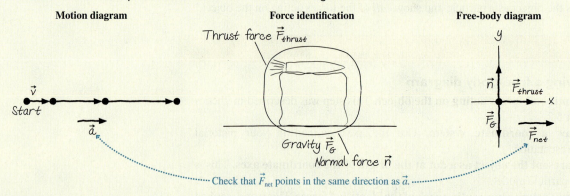

ASSESS The motion diagram tells us that the acceleration is in the positive x-direction. According to the rules of vector addition, this can be true only if the upward-pointing $\vec{n}$ and the downward-pointing $\vec{F}_G$ are equal in magnitude and thus cancel each other. The vectors have been drawn accordingly, and this leaves the net force vector pointing toward the right, in agreement with $\vec{a}$ from the motion diagram.

EXAMPLE 5.6 | A skier is pulled up a hill

A tow rope pulls a skier up a snow-covered hill at a constant speed. Draw a pictorial representation of the skier.

MODEL This is Example 5.2 again with the additional information that the skier is moving at constant speed. The skier will be modeled as a particle in *mechanical equilibrium*. If we were doing a kinematics problem, the pictorial representation would use a tilted coordinate system with the *x*-axis parallel to the slope, so we use these same tilted coordinate axes for the free-body diagram.

VISUALIZE

FIGURE 5.22 Pictorial representation for a skier being towed at a constant speed.

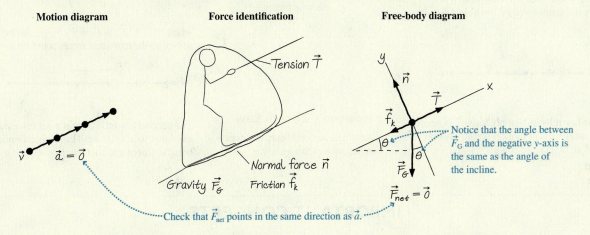

ASSESS We have shown $\vec{T}$ pulling parallel to the slope and $\vec{f}_k$, which opposes the direction of motion, pointing down the slope. $\vec{n}$ is perpendicular to the surface and thus along the *y*-axis. Finally, and this is important, the gravitational force $\vec{F}_G$ is *vertically* downward, *not* along the negative *y*-axis. In fact, you should convince yourself from the geometry that the angle θ between the $\vec{F}_G$ vector and the negative *y*-axis is the same as the angle θ of the incline above the horizontal. The skier moves in a straight line with constant speed, so $\vec{a} = \vec{0}$ and, from Newton's first law, $\vec{F}_{net} = \vec{0}$. Thus we have drawn the vectors such that the *y*-component of $\vec{F}_G$ is equal in magnitude to $\vec{n}$. Similarly, $\vec{T}$ must be large enough to match the negative *x*-components of both $\vec{f}_k$ and $\vec{F}_G$.

Free-body diagrams will be our major tool for the next several chapters. Careful practice with the workbook exercises and homework in this chapter will pay immediate benefits in the next chapter. Indeed, it is not too much to assert that a problem is half solved, or even more, when you complete the free-body diagram.

STOP TO THINK 5.5 An elevator suspended by a cable is moving upward and slowing to a stop. Which free-body diagram is correct?

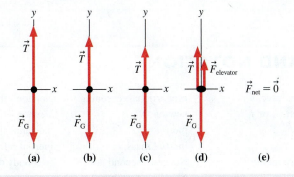

SUMMARY

The goal of Chapter 5 has been to learn about the connection between force and motion.

GENERAL PRINCIPLES

Newton's First Law

An object at rest will remain at rest, or an object that is moving will continue to move in a straight line with constant velocity, if and only if the net force on the object is zero.

Newton's laws are valid only in inertial reference frames.

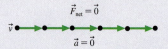

$$\vec{F}_{net} = \vec{0}$$
$$\vec{a} = \vec{0}$$

The first law tells us that no "cause" is needed for motion. Uniform motion is the "natural state" of an object.

Newton's Zeroth Law

An object responds only to forces acting on it *at this instant*.

Newton's Second Law

An object with mass m has acceleration

$$\vec{a} = \frac{1}{m}\vec{F}_{net}$$

where $\vec{F}_{net} = \vec{F}_1 + \vec{F}_2 + \vec{F}_3 + \cdots$ is the vector sum of all the individual forces acting on the object.

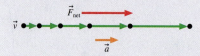

The second law tells us that a net force causes an object to accelerate. This is the connection between force and motion.

IMPORTANT CONCEPTS

Acceleration is the link to kinematics.

From $\vec{F}_{net}$, find $\vec{a}$.

From a, find v and x.

$\vec{a} = \vec{0}$ is the condition for **equilibrium**.

An object **at rest** is in equilibrium.

So is an object with **constant velocity**.

Equilibrium occurs if and only if $\vec{F}_{net} = \vec{0}$.

Mass is the resistance of an object to acceleration. It is an intrinsic property of an object.

Mass is the inverse of the slope. Larger mass, smaller slope.

Force is a push or a pull on an object.

- Force is a vector, with a magnitude and a direction.
- Force requires an agent.
- Force is either a contact force or a long-range force.

KEY SKILLS

Identifying Forces

Forces are identified by locating the points where other objects touch the object of interest. These are points where contact forces are exerted. In addition, objects with mass feel a long-range gravitational force.

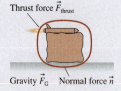

Thrust force $\vec{F}_{thrust}$

Gravity $\vec{F}_G$ Normal force $\vec{n}$

Free-Body Diagrams

A free-body diagram represents the object as a particle at the origin of a coordinate system. Force vectors are drawn with their tails on the particle. The net force vector is drawn beside the diagram.

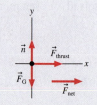

TERMS AND NOTATION

mechanics	net force, $\vec{F}_{net}$	normal force, $\vec{n}$	proportional reasoning	Newton's third law
dynamics	superposition of forces	friction, $\vec{f}_k$ or $\vec{f}_s$	newton, N	Newton's first law
force, $\vec{F}$	gravitational force, $\vec{F}_G$	drag, $\vec{F}_{drag}$	inertia	mechanical equilibrium
agent	spring force, $\vec{F}_{Sp}$	thrust, $\vec{F}_{thrust}$	inertial mass, m	inertial reference frame
contact force	tension force, $\vec{T}$	proportionality	Newton's second law	free-body diagram
long-range force	ball-and-spring model	proportionality constant	Newton's zeroth law	

CONCEPTUAL QUESTIONS

1. An elevator suspended by a cable is descending at constant velocity. How many force vectors would be shown on a free-body diagram? Name them.

2. A compressed spring is pushing a block across a rough horizontal table. How many force vectors would be shown on a free-body diagram? Name them.

3. A brick is falling from the roof of a three-story building. How many force vectors would be shown on a free-body diagram? Name them.

4. In **FIGURE Q5.4**, block B is falling and dragging block A across a table. How many force vectors would be shown on a free-body diagram of block A? Name them.

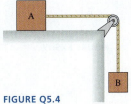

FIGURE Q5.4

5. You toss a ball straight up in the air. Immediately after you let go of it, what force or forces are acting on the ball? For each force you name, (a) state whether it is a contact force or a long-range force and (b) identify the agent of the force.

6. A constant force applied to A causes A to accelerate at 5 m/s². The same force applied to B causes an acceleration of 3 m/s². Applied to C, it causes an acceleration of 8 m/s².
   a. Which object has the largest mass? Explain.
   b. Which object has the smallest mass?
   c. What is the ratio m_A/m_B of the mass of A to the mass of B?

7. An object experiencing a constant force accelerates at 10 m/s². What will the acceleration of this object be if
   a. The force is doubled? Explain.
   b. The mass is doubled?
   c. The force is doubled *and* the mass is doubled?

8. An object experiencing a constant force accelerates at 8 m/s². What will the acceleration of this object be if
   a. The force is halved? Explain.
   b. The mass is halved?
   c. The force is halved *and* the mass is halved?

9. If an object is at rest, can you conclude that there are no forces acting on it? Explain.

10. If a force is exerted on an object, is it possible for that object to be moving with constant velocity? Explain.

11. Is the statement "An object always moves in the direction of the net force acting on it" true or false? Explain.

12. Newton's second law says $\vec{F}_{net} = m\vec{a}$. So is $m\vec{a}$ a force? Explain.

13. Is it possible for the friction force on an object to be in the direction of motion? If so, give an example. If not, why not?

14. Suppose you press your physics book against a wall hard enough to keep it from moving. Does the friction force on the book point (a) into the wall, (b) out of the wall, (c) up, (d) down, or (e) is there no friction force? Explain.

15. **FIGURE Q5.15** shows a hollow tube forming three-quarters of a circle. It is lying flat on a table. A ball is shot through the tube at high speed. As the ball emerges from the other end, does it follow path A, path B, or path C? Explain.

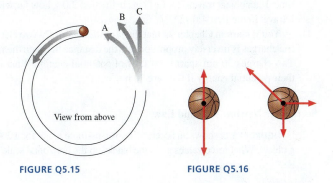

FIGURE Q5.15   FIGURE Q5.16

16. Which, if either, of the basketballs in **FIGURE Q5.16** are in equilibrium? Explain.

17. Which of the following are inertial reference frames? Explain.
   a. A car driving at steady speed on a straight and level road.
   b. A car driving at steady speed up a 10° incline.
   c. A car speeding up after leaving a stop sign.
   d. A car driving at steady speed around a curve.

EXERCISES AND PROBLEMS

Exercises

Section 5.3 Identifying Forces

1. | A car is parked on a steep hill. Identify the forces on the car.
2. | A chandelier hangs from a chain in the middle of a dining room. Identify the forces on the chandelier.
3. | A baseball player is sliding into second base. Identify the forces on the baseball player.
4. ‖ A jet plane is speeding down the runway during takeoff. Air resistance is not negligible. Identify the forces on the jet.
5. ‖ An arrow has just been shot from a bow and is now traveling horizontally. Air resistance is not negligible. Identify the forces on the arrow.

Section 5.4 What Do Forces Do?

6. | Two rubber bands cause an object to accelerate with acceleration *a*. How many rubber bands are needed to cause an object with half the mass to accelerate three times as quickly?
7. | Two rubber bands pulling on an object cause it to accelerate at 1.2 m/s².
   a. What will be the object's acceleration if it is pulled by four rubber bands?
   b. What will be the acceleration of two of these objects glued together if they are pulled by two rubber bands?

8. ‖ **FIGURE EX5.8** shows acceleration-versus-force graphs for two objects pulled by rubber bands. What is the mass ratio m_1/m_2?

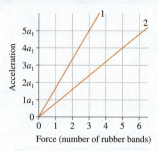

FIGURE EX5.8

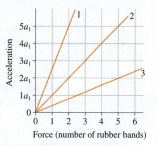

FIGURE EX5.9

9. ‖ **FIGURE EX5.9** shows an acceleration-versus-force graph for three objects pulled by rubber bands. The mass of object 2 is 0.20 kg. What are the masses of objects 1 and 3? Explain your reasoning.

10. ‖ For an object starting from rest and accelerating with constant acceleration, distance traveled is proportional to the square of the time. If an object travels 2.0 furlongs in the first 2.0 s, how far will it travel in the first 4.0 s?

11. ‖ You'll learn in Chapter 25 that the *potential energy* of two electric charges is inversely proportional to the distance between them. Two charges 30 nm apart have 1.0 J of potential energy. What is their potential energy if they are 20 nm apart?

Section 5.5 Newton's Second Law

12. | **FIGURE EX5.12** shows an acceleration-versus-force graph for a 200 g object. What force values go in the blanks on the horizontal scale?

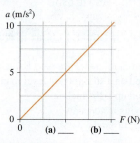

FIGURE EX5.12

FIGURE EX5.13

13. | **FIGURE EX5.13** shows an acceleration-versus-force graph for a 500 g object. What acceleration values go in the blanks on the vertical scale?

14. ‖ **FIGURE EX5.14** shows the acceleration of objects of different mass that experience the same force. What is the magnitude of the force?

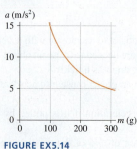

FIGURE EX5.14

FIGURE EX5.15

15. | **FIGURE EX5.15** shows an object's acceleration-versus-force graph. What is the object's mass?

16. | Based on the information in Table 5.1, *estimate*
   a. The weight of a laptop computer.
   b. The propulsion force of a bicycle.

Section 5.6 Newton's First Law

Exercises 17 through 19 show two of the three forces acting on an object in equilibrium. Redraw the diagram, showing all three forces. Label the third force $\vec{F}_3$.

17. ‖

FIGURE EX5.17

18. ‖

FIGURE EX5.18

19. ‖

FIGURE EX5.19

Section 5.7 Free-Body Diagrams

Exercises 20 through 22 show a free-body diagram. For each, write a short description of a real object for which this would be the correct free-body diagram. Use Examples 5.4, 5.5, and 5.6 as examples of what a description should be like.

20. | 21. | 22. |

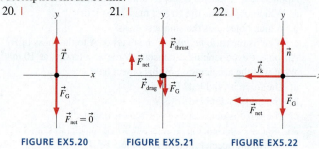

FIGURE EX5.20   **FIGURE EX5.21**   **FIGURE EX5.22**

Exercises 23 through 27 describe a situation. For each, identify all forces acting on the object and draw a free-body diagram of the object.

23. | A cat is sitting on a window sill.

24. | An ice hockey puck glides across frictionless ice.

25. | Your physics textbook is sliding across the table.

26. | A steel beam, suspended by a single cable, is being lowered by a crane at a steadily decreasing speed.

27. | A jet plane is accelerating down the runway during takeoff. Friction is negligible, but air resistance is not.

Problems

28. | Redraw the two motion diagrams shown in **FIGURE P5.28**, then draw a vector beside each one to show the direction of the net force acting on the object. Explain your reasoning.

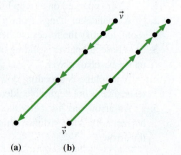

FIGURE P5.28   **(a)**   **(b)**

29. | A single force with x-component F_x acts on a 2.0 kg object as it moves along the x-axis. The object's acceleration graph (a_x versus t) is shown in **FIGURE P5.29**. Draw a graph of F_x versus t.

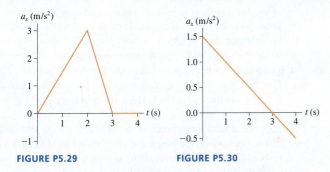

FIGURE P5.29   **FIGURE P5.30**

30. || A single force with x-component F_x acts on a 500 g object as it moves along the x-axis. The object's acceleration graph (a_x versus t) is shown in **FIGURE P5.30**. Draw a graph of F_x versus t.

31. | A single force with x-component F_x acts on a 2.0 kg object as it moves along the x-axis. A graph of F_x versus t is shown in **FIGURE P5.31**. Draw an acceleration graph (a_x versus t) for this object.

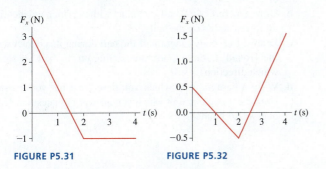

FIGURE P5.31   **FIGURE P5.32**

32. || A single force with x-component F_x acts on a 500 g object as it moves along the x-axis. A graph of F_x versus t is shown in **FIGURE P5.32**. Draw an acceleration graph (a_x versus t) for this object.

33. | A constant force is applied to an object, causing the object to accelerate at 8.0 m/s². What will the acceleration be if
   a. The force is doubled?
   b. The object's mass is doubled?
   c. The force and the object's mass are both doubled?
   d. The force is doubled and the object's mass is halved?

34. | A constant force is applied to an object, causing the object to accelerate at 10 m/s². What will the acceleration be if
   a. The force is halved?
   b. The object's mass is halved?
   c. The force and the object's mass are both halved?
   d. The force is halved and the object's mass is doubled?

Problems 35 through 40 show a free-body diagram. For each:
   a. Identify the direction of the acceleration vector $\vec{a}$ and show it as a vector next to your diagram. Or, if appropriate, write $\vec{a} = \vec{0}$.
   b. If possible, identify the direction of the velocity vector $\vec{v}$ and show it as a labeled vector.
   c. Write a short description of a real object for which this is the correct free-body diagram. Use Examples 5.4, 5.5, and 5.6 as models of what a description should be like.

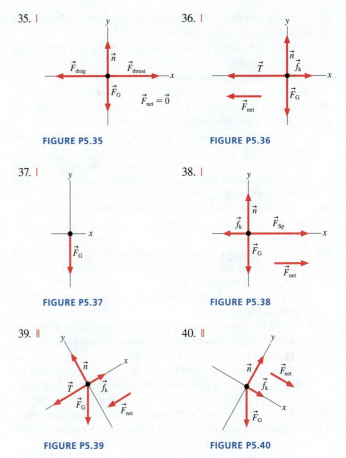

FIGURE P5.35   **FIGURE P5.36**

FIGURE P5.37   **FIGURE P5.38**

FIGURE P5.39   **FIGURE P5.40**

41. || In lab, you propel a cart with four known forces while using an ultrasonic motion detector to measure the cart's acceleration. Your data are as follows:

Force (N)	Acceleration (m/s²)
0.25	0.5
0.50	0.8
0.75	1.3
1.00	1.8

   a. How should you graph these data so as to determine the mass of the cart from the slope of the line? That is, what values should you graph on the horizontal axis and what on the vertical axis?
   b. Is there another data point that would be reasonable to add, even though you made no measurements? If so, what is it?
   c. What is your best determination of the cart's mass?

Problems 42 through 52 describe a situation. For each, draw a motion diagram, a force-identification diagram, and a free-body diagram.
42. | An elevator, suspended by a single cable, has just left the tenth floor and is speeding up as it descends toward the ground floor.
43. || A rocket is being launched straight up. Air resistance is not negligible.
44. | A Styrofoam ball has just been shot straight up. Air resistance is not negligible.
45. | You are a rock climber going upward at a steady pace on a vertical wall.

46. ‖ You've slammed on the brakes and your car is skidding to a stop while going down a 20° hill.

47. | You've just kicked a rock on the sidewalk and it is now sliding along the concrete.

48. ‖ You've jumped down from a platform. Your feet are touching the ground and your knees are flexing as you stop.

49. ‖ You are bungee jumping from a high bridge. You are moving downward while the bungee cord is stretching.

50. ‖ Your friend went for a loop-the-loop ride at the amusement park. Her car is upside down at the top of the loop.

51. ‖ A spring-loaded gun shoots a plastic ball. The trigger has just been pulled and the ball is starting to move down the barrel. The barrel is horizontal.

52. ‖ A person on a bridge throws a rock straight down toward the water. The rock has just been released.

53. ‖ The leaf hopper, champion jumper of the insect world, can
BIO jump straight up at 4 m/s^2. The jump itself lasts a mere 1 ms before the insect is clear of the ground.
   a. Draw a free-body diagram of this mighty leaper while the jump is taking place.
   b. While the jump is taking place, is the force of the ground on the leaf hopper greater than, less than, or equal to the force of gravity on the leaf hopper? Explain.

54. ‖ A bag of groceries is on the seat of your car as you stop for a stop light. The bag does not slide. Draw a motion diagram, a force-identification diagram, and a free-body diagram for the bag.

55. ‖ A heavy box is in the back of a truck. The truck is accelerating to the right. Draw a motion diagram, a force-identification diagram, and a free-body diagram for the box.

56. ‖ If a car stops suddenly, you feel "thrown forward." We'd like to understand what happens to the passengers as a car stops.

Imagine yourself sitting on a *very* slippery bench inside a car. This bench has no friction, no seat back, and there's nothing for you to hold onto.
   a. Draw a picture and identify all of the forces acting on you as the car travels at a perfectly steady speed on level ground.
   b. Draw your free-body diagram. Is there a net force on you? If so, in which direction?
   c. Repeat parts a and b with the car slowing down.
   d. Describe what happens to you as the car slows down.
   e. Use Newton's laws to explain why you seem to be "thrown forward" as the car stops. Is there really a force pushing you forward?
   f. Suppose now that the bench is not slippery. As the car slows down, you stay on the bench and don't slide off. What force is responsible for your deceleration? In which direction does this force point? Include a free-body diagram as part of your answer.

57. ‖ A rubber ball bounces. We'd like to understand *how* the ball bounces.
   a. A rubber ball has been dropped and is bouncing off the floor. Draw a motion diagram of the ball during the brief time interval that it is in contact with the floor. Show 4 or 5 frames as the ball compresses, then another 4 or 5 frames as it expands. What is the direction of $\vec{a}$ during each of these parts of the motion?
   b. Draw a picture of the ball in contact with the floor and identify all forces acting on the ball.
   c. Draw a free-body diagram of the ball during its contact with the ground. Is there a net force acting on the ball? If so, in which direction?
   d. Write a paragraph in which you describe what you learned from parts a to c and in which you answer the question: How does a ball bounce?

6 Dynamics I: Motion Along a Line

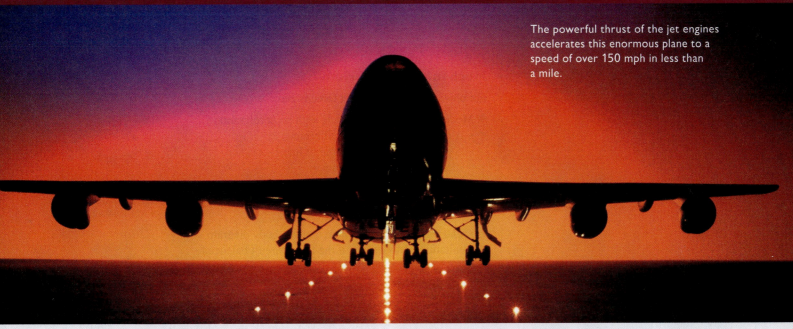

The powerful thrust of the jet engines accelerates this enormous plane to a speed of over 150 mph in less than a mile.

IN THIS CHAPTER, you will learn to solve linear force-and-motion problems.

How are Newton's laws used to solve problems?

Newton's first and second laws are vector equations. To use them,

- Draw a free-body diagram.
- Read the x- and y-components of the forces directly off the free-body diagram.
- Use $\sum F_x = ma_x$ and $\sum F_y = ma_y$.

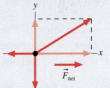

How are dynamics problems solved?

A net force on an object causes the object to accelerate.

- Identify the forces and draw a free-body diagram.
- Use Newton's second law to find the object's acceleration.
- Use kinematics for velocity and position.

Normal $\vec{n}$
Friction $\vec{f}_s$ Gravity $\vec{F}_G$

« LOOKING BACK Sections 2.4–2.6 Kinematics

How are equilibrium problems solved?

An object at rest or moving with constant velocity is in equilibrium with no net force.

- Identify the forces and draw a free-body diagram.
- Use Newton's second law with $a = 0$ to solve for unknown forces.

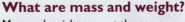

« LOOKING BACK Sections 5.1–5.2 Forces

What are mass and weight?

Mass and weight are not the same.

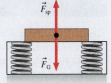

- Mass describes an object's inertia. Loosely speaking, it is the amount of matter in an object. It is the same everywhere.
- Gravity is a force.
- Weight is the result of weighing an object on a scale. It depends on mass, gravity, and acceleration.

How do we model friction and drag?

Friction and drag are complex forces, but we will develop simple models of each.

Kinetic friction

- Static, kinetic, and rolling friction depend on the coefficients of friction but not on the object's speed.
- Drag depends on the *square* of an object's speed and on its cross-section area.
- Falling objects reach terminal speed when drag and gravity are balanced.

How do we solve problems?

We will develop and use a four-part problem-solving strategy:

- Model the problem, using information about objects and forces.
- Visualize the situation with a pictorial representation.
- Set up and solve the problem with Newton's laws.
- Assess the result to see if it is reasonable.

6.1 The Equilibrium Model

Kinematics is a description of *how* an object moves. But our goal is deeper: We would like an explanation for *why* an object moves as it does. Galileo and Newton discovered that motion is determined by forces. In the absence of a net force, an object is at rest or moves with constant velocity. **Its acceleration is zero,** and this is the basis for our first explanatory model: the **equilibrium model.**

MODEL 6.1

Mechanical equilibrium

For objects on which the net force is zero.

- Model the object as a particle with no acceleration.
  - A particle at rest is in equilibrium.
  - A particle moving in a straight line at constant speed is also in equilibrium.
- Mathematically: $\vec{a} = \vec{0}$ in equilibrium; thus
  - **Newton's second law** is $\vec{F}_{net} = \sum_i \vec{F}_i = \vec{0}$.
  - The forces are "read" from the free-body diagram.
- Limitations: Model fails if the forces aren't balanced.

$\vec{F}_{net} = \vec{0}$
$\vec{a} = \vec{0}$

The object is at rest or moves with constant velocity.

Newton's laws are *vector equations*. The requirement for equilibrium, $\vec{F}_{net} = \vec{0}$ and thus $\vec{a} = \vec{0}$, is a shorthand way of writing two simultaneous equations:

$$(F_{net})_x = \sum_i (F_i)_x = 0 \quad \text{and} \quad (F_{net})_y = \sum_i (F_i)_y = 0 \tag{6.1}$$

In other words, the sum of all *x*-components and the sum of all *y*-components must simultaneously be zero. Although real-world situations often have forces pointing in three dimensions, thus requiring a third equation for the *z*-component of $\vec{F}_{net}$, we will restrict ourselves for now to problems that can be analyzed in two dimensions.

NOTE The equilibrium condition of Equations 6.1 applies only to particles, which cannot rotate. Equilibrium of an extended object, which can rotate, requires an additional condition that we will study in Chapter 12.

Equilibrium problems occur frequently. Let's look at a couple of examples.

The concept of equilibrium is essential for the engineering analysis of stationary objects such as bridges.

EXAMPLE 6.1 | **Finding the force on the kneecap**

Your kneecap (patella) is attached by a tendon to your quadriceps muscle. This tendon pulls at a 10° angle relative to the femur, the bone of your upper leg. The patella is also attached to your lower leg (tibia) by a tendon that pulls parallel to the leg. To balance these forces, the end of your femur pushes outward on the patella. Bending your knee increases the tension in the tendons, and both have a tension of 60 N when the knee is bent to make a 70° angle between the upper and lower leg. What force does the femur exert on the kneecap in this position?

MODEL Model the kneecap as a particle in equilibrium.

FIGURE 6.1 Pictorial representation of the kneecap in equilibrium.

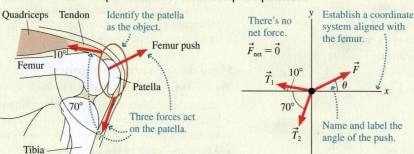
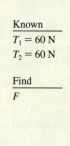

Known
$T_1 = 60$ N
$T_2 = 60$ N

Find
F

VISUALIZE FIGURE 6.1 shows how to draw a pictorial representation. We've chosen to align the x-axis with the femur. The three forces—shown on the free-body diagram—are labeled $\vec{T}_1$ and $\vec{T}_2$ for the tensions and $\vec{F}$ for the femur's push. Notice that we've *defined* angle θ to indicate the direction of the femur's force on the kneecap.

SOLVE This is an equilibrium problem, with three forces on the kneecap that must sum to zero. For $\vec{a} = \vec{0}$, Newton's second law, written in component form, is

$$(F_{net})_x = \sum_i (F_i)_x = T_{1x} + T_{2x} + F_x = 0$$

$$(F_{net})_y = \sum_i (F_i)_y = T_{1y} + T_{2y} + F_y = 0$$

NOTE You might have been tempted to write $-T_{1x}$ in the equation since $\vec{T}_1$ points to the left. But the net force, by definition, is the *sum* of all the individual forces. The fact that $\vec{T}_1$ points to the left will be taken into account when we *evaluate* the components.

The components of the force vectors can be evaluated directly from the free-body diagram:

$$T_{1x} = -T_1 \cos 10° \qquad T_{1y} = T_1 \sin 10°$$

$$T_{2x} = -T_2 \cos 70° \qquad T_{2y} = -T_2 \sin 70°$$

$$F_x = F \cos \theta \qquad F_y = F \sin \theta$$

This is where signs enter, with T_{1x} being assigned a negative value because $\vec{T}_1$ points to the left. Similarly, $\vec{T}_2$ points both to the left and down, so both T_{2x} and T_{2y} are negative. With these components, Newton's second law becomes

$$-T_1 \cos 10° - T_2 \cos 70° + F \cos \theta = 0$$

$$T_1 \sin 10° - T_2 \sin 70° + F \sin \theta = 0$$

These are two simultaneous equations for the two unknowns F and θ. We will encounter equations of this form on many occasions,

so make a note of the method of solution. First, rewrite the two equations as

$$F \cos \theta = T_1 \cos 10° + T_2 \cos 70°$$

$$F \sin \theta = -T_1 \sin 10° + T_2 \sin 70°$$

Next, divide the second equation by the first to eliminate F:

$$\frac{F \sin \theta}{F \cos \theta} = \tan \theta = \frac{-T_1 \sin 10° + T_2 \sin 70°}{T_1 \cos 10° + T_2 \cos 70°}$$

Then solve for θ:

$$\theta = \tan^{-1}\left(\frac{-T_1 \sin 10° + T_2 \sin 70°}{T_1 \cos 10° + T_2 \cos 70°}\right)$$

$$= \tan^{-1}\left(\frac{-(60 \text{ N}) \sin 10° + (60 \text{ N}) \sin 70°}{(60 \text{ N}) \cos 10° + (60 \text{ N}) \cos 70°}\right) = 30°$$

Finally, use θ to find F:

$$F = \frac{T_1 \cos 10° + T_2 \cos 70°}{\cos \theta}$$

$$= \frac{(60 \text{ N}) \cos 10° + (60 \text{ N}) \cos 70°}{\cos 30°} = 92 \text{ N}$$

The question asked What force? and force is a vector, so we must specify both the magnitude and the direction. With the knee in this position, the femur exerts a force $\vec{F} = (92 \text{ N}, 30° \text{ above the femur})$ on the kneecap.

ASSESS The magnitude of the force would be 0 N if the leg were straight, 120 N if the knee could be bent 180° so that the two tendons pull in parallel. The knee is closer to fully bent than to straight, so we would expect a femur force between 60 N and 120 N. Thus the calculated magnitude of 92 N seems reasonable.

EXAMPLE 6.2 | Towing a car up a hill

A car with a weight of 15,000 N is being towed up a 20° slope at constant velocity. Friction is negligible. The tow rope is rated at 6000 N maximum tension. Will it break?

MODEL Model the car as a particle in equilibrium.

VISUALIZE Part of our analysis of the problem statement is to determine which quantity or quantities allow us to answer the yes-or-no

question. In this case, we need to calculate the tension in the rope. FIGURE 6.2 shows the pictorial representation. Note the similarities to Examples 5.2 and 5.6 in Chapter 5, which you may want to review.

We noted in Chapter 5 that the weight of an object at rest is the magnitude F_G of the gravitational force acting on it, and that information has been listed as known.

FIGURE 6.2 Pictorial representation of a car being towed up a hill.

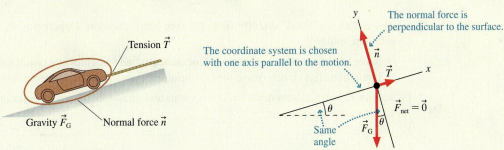

Continued

SOLVE The free-body diagram shows forces $\vec{T}$, $\vec{n}$, and $\vec{F}_G$ acting on the car. Newton's second law with $\vec{a} = \vec{0}$ is

$$(F_{net})_x = \sum F_x = T_x + n_x + (F_G)_x = 0$$
$$(F_{net})_y = \sum F_y = T_y + n_y + (F_G)_y = 0$$

From here on, we'll use $\sum F_x$ and $\sum F_y$, without the label i, as a simple shorthand notation to indicate that we're adding all the x-components and all the y-components of the forces.

We can find the components directly from the free-body diagram:

$$T_x = T \qquad\qquad T_y = 0$$
$$n_x = 0 \qquad\qquad n_y = n$$
$$(F_G)_x = -F_G \sin\theta \qquad (F_G)_y = -F_G \cos\theta$$

NOTE The gravitational force has both x- and y-components in this coordinate system, both of which are negative due to the direction of the vector $\vec{F}_G$. You'll see this situation often, so be sure you understand where $(F_G)_x$ and $(F_G)_y$ come from.

With these components, the second law becomes

$$T - F_G \sin\theta = 0$$
$$n - F_G \cos\theta = 0$$

The first of these can be rewritten as

$$T = F_G \sin\theta = (15,000 \text{ N}) \sin 20° = 5100 \text{ N}$$

Because $T < 6000$ N, we conclude that the rope will *not* break. It turned out that we did not need the y-component equation in this problem.

ASSESS Because there's no friction, it would not take *any* tension force to keep the car rolling along a horizontal surface ($\theta = 0°$). At the other extreme, $\theta = 90°$, the tension force would need to equal the car's weight ($T = 15,000$ N) to lift the car straight up at constant velocity. The tension force for a 20° slope should be somewhere in between, and 5100 N is a little less than half the weight of the car. That our result is reasonable doesn't prove it's right, but we have at least ruled out careless errors that give unreasonable results.

6.2 Using Newton's Second Law

The essence of Newtonian mechanics, introduced in « Section 5.4, can be expressed in two steps:

- The forces acting on an object determine its acceleration $\vec{a} = \vec{F}_{net}/m$.
- The object's trajectory can be determined by using $\vec{a}$ in the equations of kinematics.

These two ideas are the basis of a problem-solving strategy.

PROBLEM-SOLVING STRATEGY 6.1   (MP)

Newtonian mechanics

MODEL Model the object as a particle. Make other simplifications depending on what kinds of forces are acting.

VISUALIZE Draw a **pictorial representation**.

- Show important points in the motion with a sketch, establish a coordinate system, define symbols, and identify what the problem is trying to find.
- Use a motion diagram to determine the object's acceleration vector $\vec{a}$. The acceleration is zero for an object in equilibrium.
- Identify all forces acting on the object *at this instant* and show them on a free-body diagram.
- It's OK to go back and forth between these steps as you visualize the situation.

SOLVE The mathematical representation is based on Newton's second law:

$$\vec{F}_{net} = \sum_i \vec{F}_i = m\vec{a}$$

The forces are "read" directly from the free-body diagram. Depending on the problem, either

- Solve for the acceleration, then use kinematics to find velocities and positions; or
- Use kinematics to determine the acceleration, then solve for unknown forces.

ASSESS Check that your result has correct units and significant figures, is reasonable, and answers the question.

Exercise 23

Newton's second law is a vector equation. To apply the step labeled Solve, you must write the second law as two simultaneous equations:

$$(F_{net})_x = \sum F_x = ma_x$$
$$(F_{net})_y = \sum F_y = ma_y$$

(6.2)

The primary goal of this chapter is to illustrate the use of this strategy.

EXAMPLE 6.3 | **Speed of a towed car**

A 1500 kg car is pulled by a tow truck. The tension in the tow rope is 2500 N, and a 200 N friction force opposes the motion. If the car starts from rest, what is its speed after 5.0 seconds?

MODEL Model the car as an accelerating particle. We'll assume, as part of our *interpretation* of the problem, that the road is horizontal and that the direction of motion is to the right.

VISUALIZE FIGURE 6.3 shows the pictorial representation. We've established a coordinate system and defined symbols to represent kinematic quantities. We've identified the speed v_1, rather than the velocity v_{1x}, as what we're trying to find.

SOLVE We begin with Newton's second law:

$$(F_{net})_x = \sum F_x = T_x + f_x + n_x + (F_G)_x = ma_x$$
$$(F_{net})_y = \sum F_y = T_y + f_y + n_y + (F_G)_y = ma_y$$

All four forces acting on the car have been included in the vector sum. The equations are perfectly general, with + signs everywhere, because the four vectors are *added* to give $\vec{F}_{net}$. We can now "read" the vector components from the free-body diagram:

$$T_x = +T \quad T_y = 0 \quad n_x = 0 \quad n_y = +n$$
$$f_x = -f \quad f_y = 0 \quad (F_G)_x = 0 \quad (F_G)_y = -F_G$$

The signs, which we had to insert by hand, depend on which way the vectors point. Substituting these into the second-law equations and dividing by m give

$$a_x = \frac{1}{m}(T - f)$$
$$= \frac{1}{1500 \text{ kg}}(2500 \text{ N} - 200 \text{ N}) = 1.53 \text{ m/s}^2$$
$$a_y = \frac{1}{m}(n - F_G)$$

NOTE Newton's second law has allowed us to determine a_x exactly but has given only an algebraic expression for a_y. However, we know *from the motion diagram* that $a_y = 0$! That is, the motion is purely along the x-axis, so there is *no* acceleration along the y-axis. The requirement $a_y = 0$ allows us to conclude that $n = F_G$.

Because a_x is a constant 1.53 m/s², we can finish by using constant-acceleration kinematics to find the velocity:

$$v_{1x} = v_{0x} + a_x \Delta t$$
$$= 0 + (1.53 \text{ m/s}^2)(5.0 \text{ s}) = 7.7 \text{ m/s}$$

The problem asked for the *speed* after 5.0 s, which is $v_1 = 7.7$ m/s.

ASSESS 7.7 m/s $\approx$ 15 mph, a quite reasonable speed after 5 s of acceleration.

FIGURE 6.3 Pictorial representation of a car being towed.

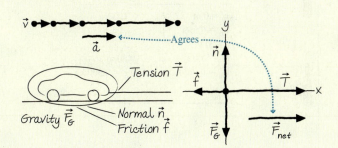

If all the forces acting on an object are constant, as in the last example, then the object moves with constant acceleration and we can deploy the uniform-acceleration model of kinetics. Now not all forces are constant—you will later meet forces that vary with position or time—but in many situations it is reasonable to model the motion as being due to constant forces. The **constant-force model** will be our most important dynamics model for the next several chapters.

MODEL 6.2

Constant force

For objects on which the net force is constant.

- Model the object as a particle with uniform acceleration.
  - The particle accelerates in the direction of the net force.
- Mathematically:
  - **Newton's second law** is $\vec{F}_{net} = \sum_i \vec{F}_i = m\vec{a}$.
  - Use the kinematics of constant acceleration.
- Limitations: Model fails if the forces aren't constant.

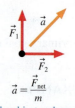

$$\vec{a} = \frac{\vec{F}_{net}}{m}$$

The object undergoes uniform acceleration.

EXAMPLE 6.4 | Altitude of a rocket

A 500 g model rocket with a weight of 4.90 N is launched straight up. The small rocket motor burns for 5.00 s and has a steady thrust of 20.0 N. What maximum altitude does the rocket reach?

MODEL We'll model the rocket as a particle acted on by constant forces by neglecting the velocity-dependent air resistance (rockets have very aerodynamic shapes) and neglecting the mass loss of the burned fuel.

VISUALIZE The pictorial representation of **FIGURE 6.4** finds that this is a two-part problem. First, the rocket accelerates straight up. Second, the rocket continues going up as it slows down, a free-fall situation. The maximum altitude is at the end of the second part of the motion.

SOLVE We now know what the problem is asking, have established relevant symbols and coordinates, and know what the forces are.

We begin the mathematical representation by writing Newton's second law, in component form, as the rocket accelerates upward. The free-body diagram shows two forces, so

$$(F_{net})_x = \sum F_x = (F_{thrust})_x + (F_G)_x = ma_{0x}$$
$$(F_{net})_y = \sum F_y = (F_{thrust})_y + (F_G)_y = ma_{0y}$$

The fact that vector $\vec{F}_G$ points downward—and which might have tempted you to use a minus sign in the y-equation—will be taken into account when we *evaluate* the components. None of the vectors in this problem has an x-component, so only the y-component of the second law is needed. We can use the free-body diagram to see that

$$(F_{thrust})_y = +F_{thrust}$$
$$(F_G)_y = -F_G$$

FIGURE 6.4 Pictorial representation of a rocket launch.

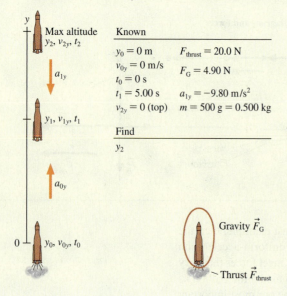

Known
$y_0 = 0$ m $F_{thrust} = 20.0$ N
$v_{0y} = 0$ m/s $F_G = 4.90$ N
$t_0 = 0$ s
$t_1 = 5.00$ s $a_{1y} = -9.80$ m/s^2
$v_{2y} = 0$ (top) $m = 500$ g $= 0.500$ kg

Find
y_2

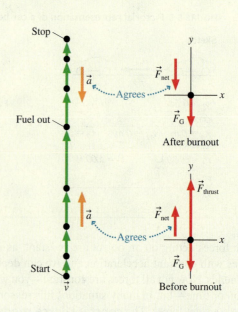

This is the point at which the directional information about the force vectors enters. The y-component of the second law is then

$$a_{0y} = \frac{1}{m}(F_{\text{thrust}} - F_G)$$

$$= \frac{20.0 \text{ N} - 4.90 \text{ N}}{0.500 \text{ kg}} = 30.2 \text{ m/s}^2$$

Notice that we converted the mass to SI units of kilograms before doing any calculations and that, because of the definition of the newton, the division of newtons by kilograms automatically gives the correct SI units of acceleration.

The acceleration of the rocket is constant until it runs out of fuel, so we can use constant-acceleration kinematics to find the altitude and velocity at burnout ($\Delta t = t_1 = 5.00$ s):

$$y_1 = y_0 + v_{0y}\,\Delta t + \tfrac{1}{2}a_{0y}(\Delta t)^2$$

$$= \tfrac{1}{2}a_{0y}(\Delta t)^2 = 377 \text{ m}$$

$$v_{1y} = v_{0y} + a_{0y}\,\Delta t = a_{0y}\,\Delta t = 151 \text{ m/s}$$

The only force on the rocket after burnout is gravity, so the second part of the motion is free fall. We do not know how long it takes to reach the top, but we do know that the final velocity is $v_{2y} = 0$. Constant-acceleration kinematics with $a_{1y} = -g$ gives

$$v_{2y}^2 = 0 = v_{1y}^2 - 2g\,\Delta y = v_{1y}^2 - 2g(y_2 - y_1)$$

which we can solve to find

$$y_2 = y_1 + \frac{v_{1y}^2}{2g} = 377 \text{ m} + \frac{(151 \text{ m/s})^2}{2(9.80 \text{ m/s}^2)}$$

$$= 1540 \text{ m} = 1.54 \text{ km}$$

ASSESS The maximum altitude reached by this rocket is 1.54 km, or just slightly under one mile. While this does not seem unreasonable for a high-acceleration rocket, the neglect of air resistance was probably not a terribly realistic assumption.

STOP TO THINK 6.1 A Martian lander is approaching the surface. It is slowing its descent by firing its rocket motor. Which is the correct free-body diagram?

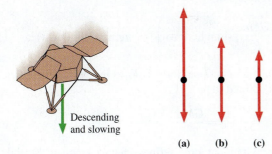

Descending and slowing

(a)   (b)   (c)   (d)   (e)

6.3 Mass, Weight, and Gravity

Ordinary language does not make a large distinction between mass and weight. However, these are separate and distinct concepts in science and engineering. We need to understand how they differ, and how they're related to gravity, if we're going to think clearly about force and motion.

Mass: An Intrinsic Property

Mass, you'll recall from **«** Section 5.4, is a scalar quantity that describes an object's inertia. Loosely speaking, it also describes the amount of matter in an object. **Mass is an intrinsic property of an object.** It tells us something about the object, regardless of where the object is, what it's doing, or whatever forces may be acting on it.

A *pan balance*, shown in **FIGURE 6.5**, is a device for measuring mass. Although a pan balance requires gravity to function, it does not depend on the strength of gravity. Consequently, the pan balance would give the same result on another planet.

FIGURE 6.5 A pan balance measures mass.

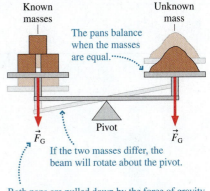

Known masses

Unknown mass

The pans balance when the masses are equal.

$\vec{F}_G$

Pivot

$\vec{F}_G$

If the two masses differ, the beam will rotate about the pivot.

Both pans are pulled down by the force of gravity.

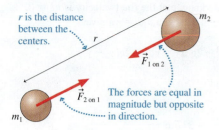

FIGURE 6.6 Newton's law of gravity.

r is the distance between the centers.

$\vec{F}_{1 \text{ on } 2}$

$\vec{F}_{2 \text{ on } 1}$ The forces are equal in magnitude but opposite in direction.

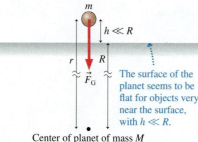

FIGURE 6.7 Gravity near the surface of a planet.

$h \ll R$

$\vec{F}_G$ The surface of the planet seems to be flat for objects very near the surface, with $h \ll R$.

Center of planet of mass M

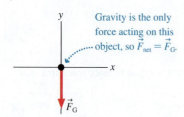

FIGURE 6.8 The free-body diagram of an object in free fall.

Gravity is the only force acting on this object, so $\vec{F}_{\text{net}} = \vec{F}_G$.

$\vec{F}_G$

Gravity: A Force

The idea of gravity has a long and interesting history intertwined with our evolving ideas about the solar system. It was Newton who—along with discovering his three laws of motion—first recognized that **gravity is an attractive, long-range force between _any_ two objects.**

FIGURE 6.6 shows two objects with masses m_1 and m_2 separated by distance r. Each object pulls on the other with a force given by _Newton's law of gravity:_

$$F_{1 \text{ on } 2} = F_{2 \text{ on } 1} = \frac{Gm_1 m_2}{r^2} \qquad \text{(Newton's law of gravity)} \qquad (6.3)$$

where $G = 6.67 \times 10^{-11}$ N m^2/kg^2, called the _gravitational constant,_ is one of the basic constants of nature. Notice that gravity is _not_ a constant force—the force gets weaker as the distance between the objects increases.

The gravitational force between two human-sized objects is minuscule, completely insignificant in comparison with other forces. That's why you're not aware of being tugged toward everything around you. Only when one or both objects are planet-sized or larger does gravity become an important force. Indeed, Chapter 13 will explore in detail the application of Newton's law of gravity to the orbits of satellites and planets.

For objects moving near the surface of the earth (or other planet), things like balls and cars and planes that we'll be studying in the next few chapters, we can make the **flat-earth approximation** shown in **FIGURE 6.7**. That is, if the object's height above the surface is very small in comparison with the size of the planet, then the curvature of the surface is not noticeable and there's virtually no difference between r and the planet's radius R. Consequently, a very good approximation for the gravitational force of the planet on mass m is simply

$$\vec{F}_G = \vec{F}_{\text{planet on } m} = \left(\frac{GMm}{R^2}, \text{straight down} \right) = (mg, \text{straight down}) \qquad (6.4)$$

The magnitude or size of the gravitational force is $F_G = mg$, where the quantity g—a property of the planet—is defined to be

$$g = \frac{GM}{R^2} \qquad (6.5)$$

Also, the direction of the gravitational force defines what we _mean_ by "straight down."

But why did we choose to call it g, a symbol we've already used for free-fall acceleration? To see the connection, recall that free fall is motion under the influence of gravity only. **FIGURE 6.8** shows the free-body diagram of an object in free fall near the surface of a planet. With $\vec{F}_{\text{net}} = \vec{F}_G$, Newton's second law predicts the acceleration to be

$$\vec{a}_{\text{free fall}} = \frac{\vec{F}_{\text{net}}}{m} = \frac{\vec{F}_G}{m} = (g, \text{straight down}) \qquad (6.6)$$

Because g is a property of the planet, independent of the object, **all objects on the same planet, regardless of mass, have the same free-fall acceleration.** We introduced this idea in Chapter 2 as an experimental discovery of Galileo, but now we see that the mass independence of $\vec{a}_{\text{free fall}}$ is a prediction of Newton's law of gravity.

But does Newton's law predict the correct value, which we know from experiment to be $g = |a_{\text{free fall}}| = 9.80$ m/s^2? We can use the average radius ($R_{\text{earth}} = 6.37 \times 10^6$ m) and mass ($M_{\text{earth}} = 5.98 \times 10^{24}$ kg) of the earth to calculate

$$g_{\text{earth}} = \frac{GM_{\text{earth}}}{(R_{\text{earth}})^2} = \frac{(6.67 \times 10^{-11} \text{ N m}^2/\text{kg}^2)(5.98 \times 10^{24} \text{ kg})}{(6.37 \times 10^6 \text{ m})^2} = 9.83 \text{ N/kg}$$

You should convince yourself that N/kg is equivalent to m/s^2, so $g_{\text{earth}} = 9.83$ m/s^2.

NOTE Astronomical data are provided inside the back cover of the book.

Newton's prediction is very close, but it's not quite right. The free-fall acceleration *would* be 9.83 m/s^2 on a stationary earth, but, in reality, the earth is rotating on its axis. The "missing" 0.03 m/s^2 is due to the earth's rotation, a claim we'll justify when we study circular motion in Chapter 8. Because we're on the outside of a rotating sphere, rather like being on the outside edge of a merry-go-round, the effect of rotation is to "weaken" gravity.

Strictly speaking, Newton's laws of motion are not valid in an earth-based reference frame because it is rotating and thus is not an inertial reference frame. Fortunately, we can use Newton's laws to analyze motion near the earth's surface, and we can use $F_G = mg$ for the gravitational force *if* we use $g = |a_{\text{free fall}}| = 9.80$ m/s^2 rather than $g = g_{\text{earth}}$. (This assertion is proved in more advanced classes.) In our rotating reference frame, $\vec{F}_G$ is the *effective gravitational force*, the true gravitational force given by Newton's law of gravity plus a small correction due to our rotation. This is the force to show on free-body diagrams and use in calculations.

Weight: A Measurement

When you weigh yourself, you stand on a *spring scale* and compress a spring. The reading of a spring scale, such as the one shown in **FIGURE 6.9**, is F_{Sp}, the magnitude of the upward force the spring is exerting.

With that in mind, let's define the **weight** of an object to be the reading F_{Sp} of a calibrated spring scale when the object is at rest relative to the scale. That is, **weight is a measurement, the result of "weighing" an object.** Because F_{Sp} is a force, weight is measured in newtons.

If the object and scale in Figure 6.9 are stationary, then the object being weighed is in equilibrium. $\vec{F}_{\text{net}} = \vec{0}$ only if the upward spring force exactly balances the downward gravitational force of magnitude mg:

$$F_{\text{Sp}} = F_G = mg \qquad (6.7)$$

Because we defined weight as the reading F_{Sp} of a spring scale, the weight of a stationary object is

$$w = mg \qquad \text{(weight of a stationary object)} \qquad (6.8)$$

Note that the scale does not "know" the weight of the object. All it can do is to measure how much its spring is compressed. On earth, a student with a mass of 70 kg has weight $w = (70 \text{ kg})(9.80 \text{ m/s}^2) = 686$ N *because* he compresses a spring until the spring pushes upward with 686 N. On a different planet, with a different value for g, the compression of the spring would be different and the student's weight would be different.

> **NOTE** Mass and weight are not the same thing. Mass, in kg, is an intrinsic property of an object; its value is unique and always the same. Weight, in N, depends on the object's mass, but it also depends on the situation—the strength of gravity and, as we will see, whether or not the object is accelerating. Weight is *not* a property of the object, and thus weight does not have a unique value.

Surprisingly, you cannot directly feel or sense gravity. Your *sensation*—how heavy you feel—is due to contact forces pressing against you, forces that touch you and activate nerve endings in your skin. As you read this, your sensation of weight is due to the normal force exerted on you by the chair in which you are sitting. When you stand, you feel the contact force of the floor pushing against your feet.

But recall the sensations you feel while accelerating. You feel "heavy" when an elevator suddenly accelerates upward, but this sensation vanishes as soon as the elevator reaches a steady speed. Your stomach seems to rise a little and you feel lighter than normal as the upward-moving elevator brakes to a halt or a roller coaster goes over the top. Has your weight actually changed?

FIGURE 6.9 A spring scale measures weight.

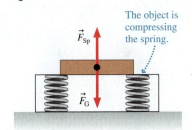

The object is compressing the spring.
$\vec{F}_{\text{Sp}}$
$\vec{F}_G$

FIGURE 6.10 A man weighing himself in an accelerating elevator.

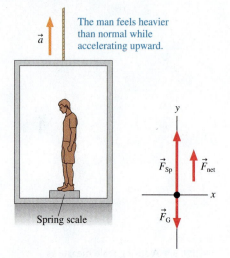

The man feels heavier than normal while accelerating upward.

Spring scale

To answer this question, **FIGURE 6.10** shows a man weighing himself on a spring scale in an accelerating elevator. The only forces acting on the man are the upward spring force of the scale and the downward gravitational force. This seems to be the same situation as Figure 6.9, but there's one big difference: The man is accelerating, hence there must be a net force on the man in the direction of $\vec{a}$.

For the net force $\vec{F}_{net}$ to point upward, the magnitude of the spring force must be *greater* than the magnitude of the gravitational force. That is, $F_{Sp} > mg$. Looking at the free-body diagram in Figure 6.10, we see that the y-component of Newton's second law is

$$(F_{net})_y = (F_{Sp})_y + (F_G)_y = F_{Sp} - mg = ma_y \qquad (6.9)$$

where m is the man's mass.

We defined weight as the reading F_{sp} of a calibrated spring scale *when the object is at rest relative to the scale*. That is the case here as the scale and man accelerate upward together. Thus the man's weight as he accelerates vertically is

$$w = \text{scale reading } F_{Sp} = mg + ma_y = mg\left(1 + \frac{a_y}{g}\right) \qquad (6.10)$$

If an object is either at rest or moving with constant velocity, then $a_y = 0$ and $w = mg$. That is, the weight of a stationary object is the magnitude of the (effective) gravitational force acting on it. But its weight differs if it has a vertical acceleration.

You *do* weigh more when accelerating upward ($a_y > 0$) because the reading of a scale—a weighing—increases. Similarly, your weight is less when the acceleration vector $\vec{a}$ points downward ($a_y < 0$) because the scale reading goes down. Weight, as we've defined it, corresponds to your sensation of heaviness or lightness.*

We found Equation 6.10 by considering a person in an accelerating elevator, but it applies to any object with a vertical acceleration. Further, an object doesn't really have to be on a scale to have a weight; an object's weight is the magnitude of the contact force supporting it. It makes no difference whether this is the spring force of the scale or simply the normal force of the floor.

NOTE Informally, we sometimes say "This object weighs such and such" or "The weight of this object is . . . ." We'll interpret these expressions as meaning mg, the weight of an object of mass m at rest ($a_y = 0$) on the surface of the earth or some other astronomical body.

Astronauts are weightless as they orbit the earth.

Weightlessness

Suppose the elevator cable breaks and the elevator, along with the man and his scale, plunges straight down in free fall! What will the scale read? When the free-fall acceleration $a_y = -g$ is used in Equation 6.10, we find $w = 0$. In other words, *the man has no weight!*

Suppose, as the elevator falls, the man inside releases a ball from his hand. In the absence of air resistance, as Galileo discovered, both the man and the ball would fall at the same rate. From the man's perspective, the ball would appear to "float" beside him. Similarly, the scale would float beneath him and not press against his feet. He is what we call *weightless*. Gravity is still pulling down on him—that's why he's falling—but he has no *sensation* of weight as everything floats around him in free fall.

But isn't this exactly what happens to astronauts orbiting the earth? If an astronaut tries to stand on a scale, it does not exert any force against her feet and reads zero. She is said to be weightless. But if the criterion to be weightless is to be in free fall, and if astronauts orbiting the earth are weightless, does this mean that they are in free fall? This is a very interesting question to which we shall return in Chapter 8.

---

* Surprisingly, there is no universally agreed-upon definition of *weight*. Some textbooks define weight as the gravitational force on an object, $\vec{w} = (mg, \text{down})$. In that case, the scale reading of an accelerating object, and your sensation of weight, is often called *apparent weight*. This textbook prefers the definition of *weight* as being what a scale reads, the result of a weighing measurement.

An elevator that has descended from the 50th floor is coming to a halt at the 1st floor. As it does, your weight is

a. More than mg.   b. Less than mg.   c. Equal to mg.   d. Zero.

6.4 Friction

Friction is absolutely essential for many things we do. Without friction you could not walk, drive, or even sit down (you would slide right off the chair!). Although friction is a complicated force, many aspects of friction can be described with a simple model.

Static Friction

« Section 5.2 defined *static friction* $\vec{f}_s$ as the force on an object that keeps it from slipping. **FIGURE 6.11** shows a rope pulling on a box that, due to static friction, isn't moving. The box is in equilibrium, so the static friction force must exactly balance the tension force:

$$f_s = T \qquad (6.11)$$

To determine the direction of $\vec{f}_s$, decide which way the object would move if there were no friction. The static friction force $\vec{f}_s$ points in the *opposite* direction to prevent the motion.

Unlike the gravitational force, which has the precise and unambiguous magnitude $F_G = mg$, the size of the static friction force depends on how hard you push or pull. The harder the rope in Figure 6.11 pulls, the harder the floor pulls back. Reduce the tension, and the static friction force will automatically be reduced to match. Static friction acts in *response* to an applied force. **FIGURE 6.12** illustrates this idea.

But there's clearly a limit to how big f_s can get. If you pull hard enough, the object slips and starts to move. In other words, the static friction force has a *maximum* possible size $f_{s\,max}$.

- An object remains at rest as long as $f_s < f_{s\,max}$.
- The object slips when $f_s = f_{s\,max}$.
- A static friction force $f_s > f_{s\,max}$ is not physically possible.

Experiments with friction show that $f_{s\,max}$ is proportional to the magnitude of the normal force. That is,

$$f_{s\,max} = \mu_s n \qquad (6.12)$$

where the proportionality constant μ_s is called the **coefficient of static friction.** The coefficient is a dimensionless number that depends on the materials of which the object and the surface are made. **TABLE 6.1** on the next page shows some typical coefficients of friction. It is to be emphasized that these are only approximate; the exact value of the coefficient depends on the roughness, cleanliness, and dryness of the surfaces.

> **NOTE** The static friction force is *not* given by Equation 6.12; this equation is simply the maximum possible static friction. The static friction force is not found with an equation but by determining how much force is needed to maintain equilibrium.

Kinetic Friction

Once the box starts to slide, as in **FIGURE 6.13**, the static friction force is replaced by a kinetic friction force $\vec{f}_k$. Experiments show that kinetic friction, unlike static friction, has a nearly *constant* magnitude. Furthermore, the size of the kinetic friction force

FIGURE 6.11 Static friction keeps an object from slipping.

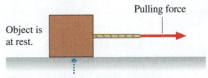

The direction of static friction is opposite to the pull, preventing motion.

FIGURE 6.12 Static friction acts in response to an applied force.

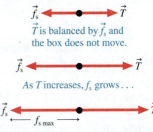

FIGURE 6.13 The kinetic friction force is opposite the direction of motion.

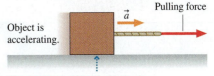

The direction of kinetic friction is opposite to the motion.

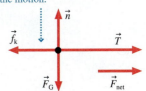

is *less* than the maximum static friction, $f_k < f_{s\,max}$, which explains why it is easier to keep the box moving than it was to start it moving. The direction of $\vec{f}_k$ is always opposite to the direction in which an object slides across the surface.

The kinetic friction force is also proportional to the magnitude of the normal force:

$$f_k = \mu_k n \tag{6.13}$$

where μ_k is called the **coefficient of kinetic friction.** Table 6.1 includes typical values of μ_k. You can see that $\mu_k < \mu_s$, causing the kinetic friction to be less than the maximum static friction.

Rolling Friction

If you slam on the brakes hard enough, your car tires slide against the road surface and leave skid marks. This is kinetic friction. A wheel *rolling* on a surface also experiences friction, but not kinetic friction. As **FIGURE 6.14** shows, the portion of the wheel that contacts the surface is stationary with respect to the surface, not sliding. The interaction of this contact area with the surface causes **rolling friction.** The force of rolling friction can be calculated in terms of a **coefficient of rolling friction** μ_r:

$$f_r = \mu_r n \tag{6.14}$$

Rolling friction acts very much like kinetic friction, but values of μ_r (see Table 6.1) are much lower than values of μ_k. This is why it is easier to roll an object on wheels than to slide it.

A Model of Friction

The friction equations are not "laws of nature" on a level with Newton's laws. Instead, they provide a reasonably accurate, but not perfect, description of how friction forces act. That is, they are a *model* of friction. And because we characterize friction in terms of constant forces, this model of friction meshes nicely with our model of dynamics with constant force.

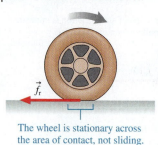

FIGURE 6.14 Rolling friction is also opposite the direction of motion

The wheel is stationary across the area of contact, not sliding.

TABLE 6.1 Coefficients of friction

Materials	Static μ_s	Kinetic μ_k	Rolling μ_r
Rubber on dry concrete	1.00	0.80	0.02
Rubber on wet concrete	0.30	0.25	0.02
Steel on steel (dry)	0.80	0.60	0.002
Steel on steel (lubricated)	0.10	0.05	
Wood on wood	0.50	0.20	
Wood on snow	0.12	0.06	
Ice on ice	0.10	0.03	

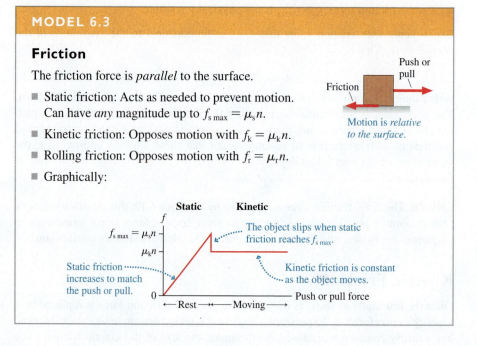

MODEL 6.3

Friction

The friction force is *parallel* to the surface.

- Static friction: Acts as needed to prevent motion. Can have *any* magnitude up to $f_{s\,max} = \mu_s n$.
- Kinetic friction: Opposes motion with $f_k = \mu_k n$.
- Rolling friction: Opposes motion with $f_r = \mu_r n$.
- Graphically:

Push or pull

Friction

Motion is *relative* to the surface.

Static   Kinetic

f

$f_{s\,max} = \mu_s n$

$\mu_k n$

The object slips when static friction reaches $f_{s\,max}$.

Static friction increases to match the push or pull.

Kinetic friction is constant as the object moves.

Push or pull force

0

← Rest → ← Moving →

STOP TO THINK 6.3 Rank in order, from largest to smallest, the sizes of the friction forces $\vec{f}_a$ to $\vec{f}_e$ in these 5 different situations. The box and the floor are made of the same materials in all situations.

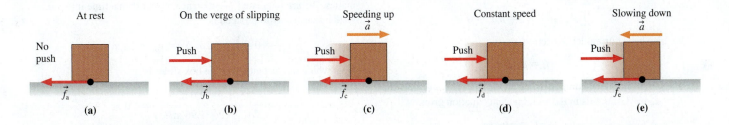

| At rest | On the verge of slipping | Speeding up | Constant speed | Slowing down |

(a)   (b)   (c)   (d)   (e)

EXAMPLE 6.5 | How far does a box slide?

Carol pushes a 25 kg wood box across a wood floor at a steady speed of 2.0 m/s. How much force does Carol exert on the box? If she stops pushing, how far will the box slide before coming to rest?

MODEL This situation can be modeled as dynamics with constant force—one of the forces being friction. Notice that this is a two-part problem: first while Carol is pushing the box, then as it slides after she releases it.

VISUALIZE This is a fairly complex situation, one that calls for careful visualization. **FIGURE 6.15** shows the pictorial representation both while Carol pushes, when $\vec{a} = \vec{0}$, and after she stops. We've placed $x = 0$ at the point where she stops pushing because this is the point where the kinematics calculation for How far? will begin. Notice that each part of the motion needs its own free-body diagram. The box is moving until the very instant that the problem ends, so only kinetic friction is relevant.

SOLVE We'll start by finding how hard Carol has to push to keep the box moving at a steady speed. The box is in equilibrium (constant velocity, $\vec{a} = \vec{0}$), and Newton's second law is

$$\sum F_x = F_{push} - f_k = 0$$
$$\sum F_y = n - F_G = n - mg = 0$$

where we've used $F_G = mg$ for the gravitational force. The negative sign occurs in the first equation because $\vec{f}_k$ points to the left and thus the *component* is negative: $(f_k)_x = -f_k$. Similarly, $(F_G)_y = -F_G$ because the gravitational force vector—with magnitude mg—points down. In addition to Newton's laws, we also have our model of kinetic friction:

$$f_k = \mu_k n$$

Altogether we have three simultaneous equations in the three unknowns F_{push}, f_k, and n. Fortunately, these equations are easy to solve. The y-component of Newton's second law tells us that $n = mg$. We can then find the friction force to be

$$f_k = \mu_k mg$$

We substitute this into the x-component of the second law, giving

$$F_{push} = f_k = \mu_k mg$$
$$= (0.20)(25 \text{ kg})(9.80 \text{ m/s}^2) = 49 \text{ N}$$

Carol pushes this hard to keep the box moving at a steady speed.

The box is not in equilibrium after Carol stops pushing it. Our strategy for the second half of the problem is to use Newton's second law to find the acceleration, then use constant-acceleration kinematics to find how far the box moves before stopping. We know

FIGURE 6.15 Pictorial representation of a box sliding across a floor.

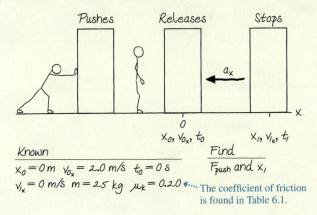

Known
$x_0 = 0$ m $v_{0x} = 2.0$ m/s $t_0 = 0$ s
$v_{1x} = 0$ m/s $m = 25$ kg $\mu_k = 0.20$ ◁···· The coefficient of friction is found in Table 6.1.

Find
F_{push} and x_1

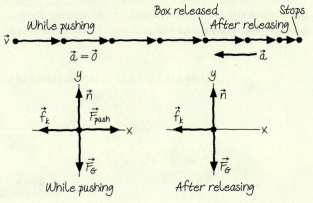

Continued

from the motion diagram that $a_y = 0$. Newton's second law, applied to the second free-body diagram of Figure 6.15, is

$$\sum F_x = -f_k = ma_x$$
$$\sum F_y = n - mg = ma_y = 0$$

We also have our model of friction,

$$f_k = \mu_k n$$

We see from the y-component equation that $n = mg$, and thus $f_k = \mu_k mg$. Using this in the x-component equation gives

$$ma_x = -f_k = -\mu_k mg$$

This is easily solved to find the box's acceleration:

$$a_x = -\mu_k g = -(0.20)(9.80 \text{ m/s}^2) = -1.96 \text{ m/s}^2$$

The acceleration component a_x is negative because the acceleration vector $\vec{a}$ points to the left, as we see from the motion diagram.

Now we are left with a problem of constant-acceleration kinematics. We are interested in a distance, rather than a time interval, so the easiest way to proceed is

$$v_{1x}^2 = 0 = v_{0x}^2 + 2a_x \Delta x = v_{0x}^2 + 2a_x x_1$$

from which the distance that the box slides is

$$x_1 = \frac{-v_{0x}^2}{2a_x} = \frac{-(2.0 \text{ m/s})^2}{2(-1.96 \text{ m/s}^2)} = 1.0 \text{ m}$$

ASSESS Carol was pushing at 2 m/s $\approx$ 4 mph, which is fairly fast. The box slides 1.0 m, which is slightly over 3 feet. That sounds reasonable.

NOTE Example 6.5 needed both the horizontal and the vertical components of the second law even though the motion was entirely horizontal. This need is typical when friction is involved because we must find the normal force before we can evaluate the friction force.

EXAMPLE 6.6 | Dumping a file cabinet

A 50 kg steel file cabinet is in the back of a dump truck. The truck's bed, also made of steel, is slowly tilted. What is the size of the static friction force on the cabinet when the bed is tilted 20°? At what angle will the file cabinet begin to slide?

MODEL Model the file cabinet as a particle in equilibrium. We'll also use the model of static friction. The file cabinet will slip when the static friction force reaches its maximum value $f_{s\,max}$.

VISUALIZE FIGURE 6.16 shows the pictorial representation when the truck bed is tilted at angle θ. We can make the analysis easier if we tilt the coordinate system to match the bed of the truck.

SOLVE The file cabinet is in equilibrium. Newton's second law is

$$(F_{net})_x = \sum F_x = n_x + (F_G)_x + (f_s)_x = 0$$
$$(F_{net})_y = \sum F_y = n_y + (F_G)_y + (f_s)_y = 0$$

From the free-body diagram we see that f_s has only a *negative* x-component and that n has only a positive y-component. The gravitational force vector can be written $\vec{F}_G = +F_G \sin\theta\,\hat{\imath} - F_G \cos\theta\,\hat{\jmath}$,

so $\vec{F}_G$ has both x- and y-components in this coordinate system. Thus the second law becomes

$$\sum F_x = F_G \sin\theta - f_s = mg \sin\theta - f_s = 0$$
$$\sum F_y = n - F_G \cos\theta = n - mg \cos\theta = 0$$

where we've used $F_G = mg$.

You might be tempted to solve the y-component equation for n, then to use Equation 6.12 to calculate the static friction force as $\mu_s n$. **But Equation 6.12 does *not* say $f_s = \mu_s n$.** Equation 6.12 gives only the maximum possible static friction force $f_{s\,max}$, the point at which the object slips. In nearly all situations, the actual static friction force is less than $f_{s\,max}$. In this problem, we can use the x-component equation—which tells us that static friction has to exactly balance the component of the gravitational force along the incline—to find the size of the static friction force:

$$f_s = mg \sin\theta = (50 \text{ kg})(9.80 \text{ m/s}^2) \sin 20°$$
$$= 170 \text{ N}$$

FIGURE 6.16 The pictorial representation of a file cabinet in a tilted dump truck.

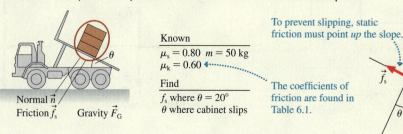

Known
$\mu_s = 0.80$ $m = 50$ kg
$\mu_k = 0.60$

Find
f_s where $\theta = 20°$
θ where cabinet slips

Normal $\vec{n}$
Friction $\vec{f}_s$ Gravity $\vec{F}_G$

To prevent slipping, static friction must point *up* the slope.

The coefficients of friction are found in Table 6.1.

Slipping occurs when the static friction reaches its maximum value

$$f_s = f_{s\,max} = \mu_s n$$

From the y-component of Newton's law we see that $n = mg\cos\theta$. Consequently,

$$f_{s\,max} = \mu_s mg\cos\theta$$

Substituting this into the x-component of the first law gives

$$mg\sin\theta - \mu_s mg\cos\theta = 0$$

The mg in both terms cancels, and we find

$$\frac{\sin\theta}{\cos\theta} = \tan\theta = \mu_s$$

$$\theta = \tan^{-1}\mu_s = \tan^{-1}(0.80) = 39°$$

ASSESS Steel doesn't slide all that well on unlubricated steel, so a fairly large angle is not surprising. The answer seems reasonable.

NOTE A common error is to use simply $n = mg$. Be sure to evaluate the normal force within the context of each specific problem. In this example, $n = mg\cos\theta$.

Causes of Friction

It is worth a brief pause to look at the *causes* of friction. All surfaces, even those quite smooth to the touch, are very rough on a microscopic scale. When two objects are placed in contact, they do not make a smooth fit. Instead, as **FIGURE 6.17** shows, the high points on one surface become jammed against the high points on the other surface, while the low points are not in contact at all. The amount of contact depends on how hard the surfaces are pushed together, which is why friction forces are proportional to n.

At the points of actual contact, the atoms in the two materials are pressed closely together and molecular bonds are established between them. These bonds are the "cause" of the static friction force. For an object to slip, you must push it hard enough to break these molecular bonds between the surfaces. Once they are broken, and the two surfaces are sliding against each other, there are still attractive forces between the atoms on the opposing surfaces as the high points of the materials push past each other. However, the atoms move past each other so quickly that they do not have time to establish the tight bonds of static friction. That is why the kinetic friction force is smaller. Friction can be minimized with lubrication, a very thin film of liquid between the surfaces that allows them to "float" past each other with many fewer points in actual contact.

FIGURE 6.17 An atomic-level view of friction.

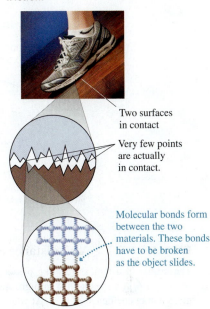

Two surfaces in contact

Very few points are actually in contact.

Molecular bonds form between the two materials. These bonds have to be broken as the object slides.

6.5 Drag

The air exerts a drag force on objects as they move through the air. You experience drag forces every day as you jog, bicycle, ski, or drive your car. The drag force $\vec{F}_{drag}$

- Is opposite in direction to $\vec{v}$.
- Increases in magnitude as the object's speed increases.

FIGURE 6.18 illustrates the drag force.

Drag is a more complex force than ordinary friction because drag depends on the object's speed. Drag also depends on the object's shape and on the density of the medium through which it moves. Fortunately, we can use a fairly simple *model* of drag if the following three conditions are met:

- The object is moving through the air near the earth's surface.
- The object's size (diameter) is between a few millimeters and a few meters.
- The object's speed is less than a few hundred meters per second.

These conditions are usually satisfied for balls, people, cars, and many other objects in our everyday world. Under these conditions, the drag force on an object moving with speed v can be written

$$\vec{F}_{drag} = \left(\tfrac{1}{2}C\rho Av^2, \text{ direction opposite the motion}\right) \qquad (6.15)$$

FIGURE 6.18 The drag force on a high-speed motorcyclist is significant.

$\vec{F}_{drag}$

The symbols in Equation 6.15 are:

- A is the *cross-section area* of the object as it "faces into the wind," as illustrated in **FIGURE 6.19**.
- ρ is the density of the air, which is 1.3 kg/m³ at atmospheric pressure and 0°C, a common reference point of pressure and temperature.
- C is the **drag coefficient.** It is smaller for aerodynamically shaped objects, larger for objects presenting a flat face to the wind. Figure 6.19 gives approximate values for a sphere and two cylinders. C is dimensionless; it has no units.

Notice that the drag force is proportional to the *square* of the object's speed. So drag is *not* a constant force (unless v is constant) and you cannot use constant-acceleration kinematics.

This model of drag fails for objects that are very small (such as dust particles), very fast (such as bullets), or that move in liquids (such as water). Motion in a liquid will be considered in Challenge Problems 6.76 and 6.77, but otherwise we'll leave these situations to more advanced textbooks.

FIGURE 6.19 Cross-section areas for objects of different shape.

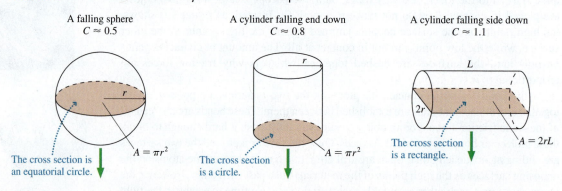

A falling sphere
$C \approx 0.5$

A cylinder falling end down
$C \approx 0.8$

A cylinder falling side down
$C \approx 1.1$

The cross section is an equatorial circle.
$A = \pi r^2$

The cross section is a circle.
$A = \pi r^2$

The cross section is a rectangle.
$A = 2rL$

EXAMPLE 6.7 | Air resistance compared to rolling friction

The profile of a typical 1500 kg passenger car, as seen from the front, is 1.6 m wide and 1.4 m high. Aerodynamic body shaping gives a drag coefficient of 0.35. At what speed does the magnitude of the drag equal the magnitude of the rolling friction?

MODEL Model the car as a particle. Use the models of rolling friction and drag. Note that this is *not* a constant-force situation.

VISUALIZE **FIGURE 6.20** shows the car and a free-body diagram. A full pictorial representation is not needed because we won't be doing any kinematics calculations.

FIGURE 6.20 A car experiences both rolling friction and drag.

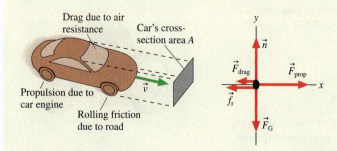

Drag due to air resistance

Car's cross-section area A

Propulsion due to car engine

Rolling friction due to road

SOLVE Drag is less than friction at low speeds, where air resistance is negligible. But drag increases as v increases, so there will be a speed at which the two forces are equal in size. Above this speed, drag is more important than rolling friction.

There's no motion and no acceleration in the vertical direction, so we can see from the free-body diagram that $n = F_G = mg$. Thus $f_r = \mu_r mg$. Equating friction and drag, we have

$$\tfrac{1}{2} C\rho A v^2 = \mu_r mg$$

Solving for v, we find

$$v = \sqrt{\frac{2\mu_r mg}{C\rho A}} = \sqrt{\frac{2(0.02)(1500\ \text{kg})(9.80\ \text{m/s}^2)}{(0.35)(1.3\ \text{kg/m}^3)\ (1.4\ \text{m} \times 1.6\ \text{m})}} = 24\ \text{m/s}$$

where the value of μ_r for rubber on concrete was taken from Table 6.1.

ASSESS 24 m/s is approximately 50 mph, a reasonable result. This calculation shows that we can reasonably ignore air resistance for car speeds less than 30 or 40 mph. Calculations that neglect drag will be increasingly inaccurate as speeds go above 50 mph.

Terminal Speed

The drag force increases as an object falls and gains speed. If the object falls far enough, it will eventually reach a speed, shown in **FIGURE 6.21**, at which $F_{\text{drag}} = F_G$. That is, the drag force will be equal and opposite to the gravitational force. The net force at this speed is $\vec{F}_{\text{net}} = \vec{0}$, so there is no further acceleration and the object falls with a *constant* speed. The speed at which the exact balance between the upward drag force and the downward gravitational force causes an object to fall without acceleration is called the **terminal speed** v_{term}. Once an object has reached terminal speed, it will continue falling at that speed until it hits the ground.

It's not hard to compute the terminal speed. It is the speed, by definition, at which $F_{\text{drag}} = F_G$ or, equivalently, $\frac{1}{2}C\rho Av^2 = mg$. This speed is

$$v_{\text{term}} = \sqrt{\frac{2mg}{C\rho A}} \tag{6.16}$$

A more massive object has a larger terminal speed than a less massive object of equal size and shape. A 10-cm-diameter lead ball, with a mass of 6 kg, has a terminal speed of 160 m/s, while a 10-cm-diameter Styrofoam ball, with a mass of 50 g, has a terminal speed of only 15 m/s.

A popular use of Equation 6.16 is to find the terminal speed of a skydiver. A skydiver is rather like the cylinder of Figure 6.19 falling "side down," for which we see that $C \approx 1.1$. A typical skydiver is 1.8 m long and 0.40 m wide ($A = 0.72$ m^2) and has a mass of 75 kg. His terminal speed is

$$v_{\text{term}} = \sqrt{\frac{2mg}{C\rho A}} = \sqrt{\frac{2(75\text{ kg})(9.8\text{ m/s}^2)}{(1.1)(1.3\text{ kg/m}^3)(0.72\text{ m}^2)}} = 38\text{ m/s}$$

This is roughly 90 mph. A higher speed can be reached by falling feet first or head first, which reduces the area A and the drag coefficient.

Although we've focused our analysis on objects moving vertically, the same ideas apply to objects moving horizontally. If an object is thrown or shot horizontally, $\vec{F}_{\text{drag}}$ causes the object to slow down. An airplane reaches its maximum speed, which is analogous to the terminal speed, when the drag is equal and opposite to the thrust: $F_{\text{drag}} = F_{\text{thrust}}$. The net force is then zero and the plane cannot go any faster. The maximum speed of a passenger jet is about 550 mph.

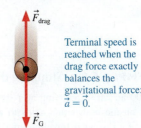

FIGURE 6.21 An object falling at terminal speed.

Terminal speed is reached when the drag force exactly balances the gravitational force: $\vec{a} = \vec{0}$.

STOP TO THINK 6.4 The terminal speed of a Styrofoam ball is 15 m/s. Suppose a Styrofoam ball is shot straight down from a high tower with an initial speed of 30 m/s. Which velocity graph is correct?

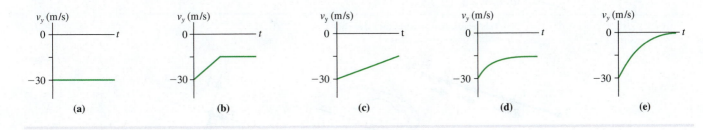

6.6 More Examples of Newton's Second Law

We will finish this chapter with four additional examples in which we use the problem-solving strategy in more complex scenarios.

EXAMPLE 6.8 | **Stopping distances**

A 1500 kg car is traveling at a speed of 30 m/s when the driver slams on the brakes and skids to a halt. Determine the stopping distance if the car is traveling up a 10° slope, down a 10° slope, or on a level road.

MODEL Model the car's motion as dynamics with constant force and use the model of kinetic friction. We want to solve the problem only once, not three separate times, so we'll leave the slope angle θ unspecified until the end.

VISUALIZE FIGURE 6.22 shows the pictorial representation. We've shown the car sliding uphill, but these representations work equally well for a level or downhill slide if we let θ be zero or negative, respectively. We've used a tilted coordinate system so that the motion is along one of the axes. We've *assumed* that the car is traveling to the right, although the problem didn't state this. You could equally well make the opposite assumption, but you would have to be careful with negative values of x and v_x. The car *skids* to a halt, so we've taken the coefficient of *kinetic* friction for rubber on concrete from Table 6.1.

SOLVE Newton's second law and the model of kinetic friction are

$$\sum F_x = n_x + (F_G)_x + (f_k)_x$$
$$= -mg\sin\theta - f_k = ma_x$$
$$\sum F_y = n_y + (F_G)_y + (f_k)_y$$
$$= n - mg\cos\theta = ma_y = 0$$
$$f_k = \mu_k n$$

We've written these equations by "reading" the motion diagram and the free-body diagram. Notice that both components of the gravitational force vector $\vec{F}_G$ are negative. $a_y = 0$ because the motion is entirely along the x-axis.

The second equation gives $n = mg\cos\theta$. Using this in the friction model, we find $f_k = \mu_k mg\cos\theta$. Inserting this result back into the first equation then gives

$$ma_x = -mg\sin\theta - \mu_k mg\cos\theta$$
$$= -mg(\sin\theta + \mu_k\cos\theta)$$
$$a_x = -g(\sin\theta + \mu_k\cos\theta)$$

This is a constant acceleration. Constant-acceleration kinematics gives

$$v_{1x}^2 = 0 = v_{0x}^2 + 2a_x(x_1 - x_0) = v_{0x}^2 + 2a_x x_1$$

which we can solve for the stopping distance x_1:

$$x_1 = -\frac{v_{0x}^2}{2a_x} = \frac{v_{0x}^2}{2g(\sin\theta + \mu_k\cos\theta)}$$

Notice how the minus sign in the expression for a_x canceled the minus sign in the expression for x_1. Evaluating our result at the three different angles gives the stopping distances:

$$x_1 = \begin{cases} 48 \text{ m} & \theta = 10° & \text{uphill} \\ 57 \text{ m} & \theta = 0° & \text{level} \\ 75 \text{ m} & \theta = -10° & \text{downhill} \end{cases}$$

The implications are clear about the danger of driving downhill too fast!

ASSESS 30 m/s $\approx$ 60 mph and 57 m $\approx$ 180 feet on a level surface. This is similar to the stopping distances you learned when you got your driver's license, so the results seem reasonable. Additional confirmation comes from noting that the expression for a_x becomes $-g\sin\theta$ if $\mu_k = 0$. This is what you learned in Chapter 2 for the acceleration on a frictionless inclined plane.

FIGURE 6.22 Pictorial representation of a skidding car.

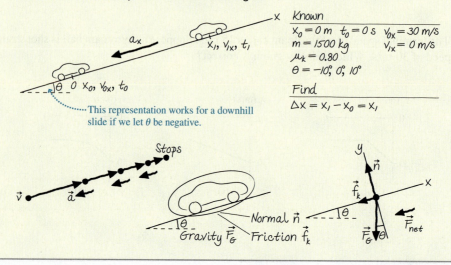

EXAMPLE 6.9 | Measuring the tension pulling a cart

Your instructor has set up a lecture demonstration in which a 250 g cart can roll along a level, 2.00-m-long track while its velocity is measured with a motion detector. First, the instructor simply gives the cart a push and measures its velocity as it rolls down the track. The data below show that the cart slows slightly before reaching the end of the track. Then, as **FIGURE 6.23** shows, the instructor attaches a string to the cart and uses a falling weight to pull the cart. She then asks you to determine the tension in the string. For extra credit, find the coefficient of rolling friction.

Time (s)	Rolled velocity (m/s)	Pulled velocity (m/s)
0.00	1.20	0.00
0.25	1.17	0.36
0.50	1.15	0.80
0.75	1.12	1.21
1.00	1.08	1.52
1.25	1.04	1.93
1.50	1.02	2.33

FIGURE 6.23 The experimental arrangement.

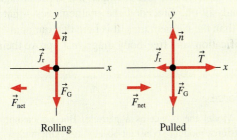

MODEL Model the cart as a particle acted on by constant forces.

VISUALIZE The cart changes velocity—it accelerates—when both pulled and rolled. Consequently, there must be a net force for both motions. For rolling, force identification finds that the only horizontal force is rolling friction, a force that opposes the motion and slows the cart. There is no "force of motion" or "force of the hand" because the hand is no longer in contact with the cart. (Recall Newton's "zeroth law": The cart responds only to forces applied *at this instant*.) Pulling adds a tension force in the direction of motion. The two free-body diagrams are shown in **FIGURE 6.24**.

FIGURE 6.24 Pictorial representations of the cart.

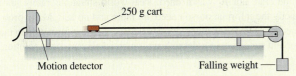

SOLVE The cart's acceleration when pulled, which we can find from the velocity data, will allow us to find the net force. Isolating the tension force will require knowing the friction force, but we can determine that from the rolling motion. For the rolling motion, Newton's second law can be written by "reading" the free-body diagram on the left:

$$\sum F_x = (f_r)_x = -f_r = ma_x = ma_{roll}$$
$$\sum F_y = n_y + (F_G)_y = n - mg = 0$$

Make sure you understand where the signs come from and how we used our knowledge that $\vec{a}$ has only an x-component, which we called a_{roll}. The magnitude of the friction force, which is all we'll need to determine the tension, is found from the x-component equation:

$$f_r = -ma_{roll} = -m \times \text{slope of the rolling-velocity graph}$$

But we'll need to do a bit more analysis to get the coefficient of rolling friction. The y-component equation tells us that $n = mg$. Using this in the model of rolling friction, $f_r = \mu_r n = \mu_r mg$, we see that the coefficient of rolling friction is

$$\mu_r = \frac{f_r}{mg}$$

The x-component equation of Newton's second law when the cart is pulled is

$$\sum F_x = T + (f_r)_x = T - f_r = ma_x = ma_{pulled}$$

Thus the tension that we seek is

$$T = f_r + ma_{pulled} = f_r + m \times \text{slope of the pulled-velocity graph}$$

FIGURE 6.25 shows the graphs of the velocity data. The accelerations are the slopes of these lines, and from the equations of the best-fit lines we find $a_{roll} = -0.124 \text{ m/s}^2$ and $a_{pulled} = 1.55 \text{ m/s}^2$. Thus the friction force is

$$f_r = -ma_{roll} = -(0.25 \text{ kg})(-0.124 \text{ m/s}^2) = 0.031 \text{ N}$$

Knowing this, we find that the string tension pulling the cart is

$$T = f_r + ma_{pulled} = 0.031 \text{ N} + (0.25 \text{ kg})(1.55 \text{ m/s}^2) = 0.42 \text{ N}$$

and the coefficient of rolling friction is

$$\mu_r = \frac{f_r}{mg} = \frac{0.031 \text{ N}}{(0.25 \text{ kg})(9.80 \text{ m/s}^2)} = 0.013$$

FIGURE 6.25 The velocity graphs of the rolling and pulled motion. The slopes of these graphs are the cart's acceleration.

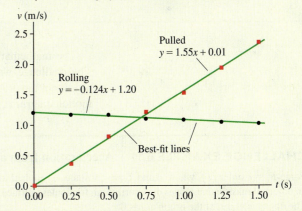

ASSESS The coefficient of rolling friction is very small, but it's similar to the values in Table 6.1 and thus believable. That gives us confidence that our value for the tension is also correct. It's reasonable that the tension needed to accelerate the cart is small because the cart is light and there's very little friction.

EXAMPLE 6.10 | **Make sure the cargo doesn't slide**

A 100 kg box of dimensions 50 cm × 50 cm × 50 cm is in the back of a flatbed truck. The coefficients of friction between the box and the bed of the truck are $\mu_s = 0.40$ and $\mu_k = 0.20$. What is the maximum acceleration the truck can have without the box slipping?

MODEL This is a somewhat different problem from any we have looked at thus far. Let the box, which we'll model as a particle, be the object of interest. It contacts other objects only where it touches the truck bed, so only the truck can exert contact forces on the box. If the box does *not* slip, then there is no motion of the box *relative to the truck* and the box must accelerate *with the truck*: $a_{box} = a_{truck}$. As the box accelerates, it must, according to Newton's second law, have a net force acting on it. But from what?

Imagine, for a moment, that the truck bed is frictionless. The box would slide backward (as seen in the truck's reference frame) as the truck accelerates. The force that prevents sliding is *static friction*, so the truck must exert a static friction force on the box to "pull" the box along with it and prevent the box from sliding *relative to the truck*.

VISUALIZE This situation is shown in **FIGURE 6.26**. There is only one horizontal force on the box, $\vec{f}_s$, and it points in the *forward* direction to accelerate the box. Notice that we're solving the problem with the ground as our reference frame. Newton's laws are not valid in the accelerating truck because it is not an inertial reference frame.

SOLVE Newton's second law, which we can "read" from the free-body diagram, is

$$\sum F_x = f_s = ma_x$$
$$\sum F_y = n - F_G = n - mg = ma_y = 0$$

Now, static friction, you will recall, can be *any* value between 0 and $f_{s\,max}$. If the truck accelerates slowly, so that the box doesn't slip, then $f_s < f_{s\,max}$. However, we're interested in the acceleration a_{max} at which the box begins to slip. This is the acceleration at which f_s reaches its maximum possible value

$$f_s = f_{s\,max} = \mu_s n$$

The y-equation of the second law and the friction model combine to give $f_{s\,max} = \mu_s mg$. Substituting this into the x-equation, and noting that a_x is now a_{max}, we find

$$a_{max} = \frac{f_{s\,max}}{m} = \mu_s g = 3.9 \text{ m/s}^2$$

The truck must keep its acceleration less than 3.9 m/s² if slipping is to be avoided.

ASSESS 3.9 m/s² is about one-third of g. You may have noticed that items in a car or truck are likely to *tip over* when you start or stop, but they slide only if you really floor it and accelerate very quickly. So this answer seems reasonable. Notice that neither the dimensions of the crate nor μ_k was needed. Real-world situations rarely have exactly the information you need, no more and no less. Many problems in this textbook will require you to assess the information in the problem statement in order to learn which is relevant to the solution.

FIGURE 6.26 Pictorial representation for the box in a flatbed truck.

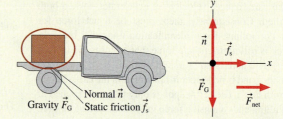

Known
m = 100 kg
Box dimensions 50 cm × 50 cm × 50 cm
$\mu_s = 0.40$ $\mu_k = 0.20$

Find
Acceleration at which box slips

Gravity $\vec{F}_G$ Normal $\vec{n}$ Static friction $\vec{f}_s$

The mathematical representation of this last example was quite straightforward. The challenge was in the analysis that preceded the mathematics—that is, in the *physics* of the problem rather than the mathematics. It is here that our analysis tools—motion diagrams, force identification, and free-body diagrams—prove their value.

CHALLENGE EXAMPLE 6.11 | **Acceleration from a variable force**

Force $F_x = c \sin(\pi t/T)$, where c and T are constants, is applied to an object of mass m that moves on a horizontal, frictionless surface. The object is at rest at the origin at $t = 0$.

a. Find an expression for the object's velocity. Graph your result for $0 \le t \le T$.

b. What is the maximum velocity of a 500 g object if $c = 2.5$ N and $T = 1.0$ s?

MODEL Model the object as a particle. But we cannot use the constant-force model or constant-acceleration kinematics.

VISUALIZE The sine function is 0 at $t = 0$ and again at $t = T$, when the value of the argument is π rad. Over the interval $0 \le t \le T$, the force grows from 0 to c and then returns to 0, always pointing in the positive x-direction. **FIGURE 6.27** shows a graph of the force and a pictorial representation.

FIGURE 6.27 Pictorial representation for a variable force.

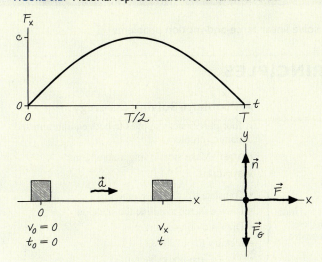

Then we integrate both sides from the initial conditions ($v_x = v_{0x} = 0$ at $t = t_0 = 0$) to the final conditions (v_x at the later time t):

$$\int_0^{v_x} dv_x = \frac{c}{m} \int_0^t \sin\left(\frac{\pi t}{T}\right) dt$$

The fraction c/m is a constant that we could take outside the integral. The integral on the right side is of the form

$$\int \sin(bx)\, dx = -\frac{1}{b} \cos(bx)$$

Using this, and integrating both sides of the equation, we find

$$v_x \Big|_0^{v_x} = v_x - 0 = -\frac{cT}{\pi m} \cos\left(\frac{\pi t}{T}\right)\Big|_0^t = -\frac{cT}{\pi m}\left(\cos\left(\frac{\pi t}{T}\right) - 1\right)$$

Simplifying, we find the object's velocity at time t is

$$v_x = \frac{cT}{\pi m}\left(1 - \cos\left(\frac{\pi t}{T}\right)\right)$$

This expression is graphed in **FIGURE 6.28**, where we see that, as predicted, maximum velocity is reached at $t = T$.

FIGURE 6.28 The object's velocity as a function of time.

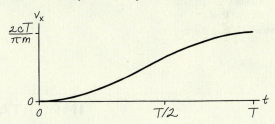

SOLVE The object's acceleration increases between 0 and $T/2$ as the force increases. You might expect the object to slow down between $T/2$ and T as the force decreases. However, *there's still a net force in the positive x-direction, so there must be an acceleration in the positive x-direction.* The object continues to speed up, only more slowly as the acceleration decreases. Maximum velocity is reached at $t = T$.

a. This is not constant-acceleration motion, so we cannot use the familiar equations of constant-acceleration kinematics. Instead, we must use the definition of acceleration as the rate of change—the time derivative—of velocity. With no friction, we need only the x-component equation of Newton's second law:

$$a_x = \frac{dv_x}{dt} = \frac{F_{net}}{m} = \frac{c}{m}\sin\left(\frac{\pi t}{T}\right)$$

First we rewrite this as

$$dv_x = \frac{c}{m}\sin\left(\frac{\pi t}{T}\right) dt$$

b. Maximum velocity, at $t = T$, is

$$v_{max} = \frac{cT}{\pi m}(1 - \cos\pi) = \frac{2cT}{\pi m} = \frac{2(2.5\ \text{N})(1.0\ \text{s})}{\pi(0.50\ \text{kg})} = 3.2\ \text{m/s}$$

ASSESS A steady 2.5 N force would cause a 0.5 kg object to accelerate at 5 m/s^2 and reach a speed of 5 m/s in 1 s. A variable force with a maximum of 2.5 N will produce less acceleration, so a top speed of 3.2 m/s seems reasonable.

SUMMARY

The goal of Chapter 6 has been to learn to solve linear force-and-motion problems.

GENERAL PRINCIPLES

Two Explanatory Models

An object on which there is no net force is in **mechanical equilibrium**.

- Objects at rest.
- Objects moving with constant velocity.
- Newton's second law applies with $\vec{a} = \vec{0}$.

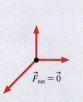

An object on which the net force is constant undergoes **dynamics with constant force**.

- The object accelerates.
- The kinematic model is that of constant acceleration.
- Newton's second law applies.

A Problem-Solving Strategy

A four-part strategy applies to both equilibrium and dynamics problems.

MODEL Make simplifying assumptions.

VISUALIZE

Go back and forth between these steps as needed.

- Translate words into symbols.
- Draw a sketch to define the situation.
- Draw a motion diagram.
- Identify forces.
- Draw a free-body diagram.

SOLVE Use Newton's second law:

$$\vec{F}_{net} = \sum_i \vec{F}_i = m\vec{a}$$

"Read" the vectors from the free-body diagram. Use kinematics to find velocities and positions.

ASSESS Is the result reasonable? Does it have correct units and significant figures?

IMPORTANT CONCEPTS

Specific information about three important descriptive models:

Gravity　$\vec{F}_G = (mg, \text{ downward})$

Friction　$\vec{f}_s = (0 \text{ to } \mu_s n, \text{ direction as necessary to prevent motion})$

$\vec{f}_k = (\mu_k n, \text{ direction opposite the motion})$

$\vec{f}_r = (\mu_r n, \text{ direction opposite the motion})$

Drag　$\vec{F}_{drag} = \left(\frac{1}{2}C\rho A v^2, \text{ direction opposite the motion}\right)$

Newton's laws are vector expressions. You must write them out by **components**:

$$(F_{net})_x = \sum F_x = ma_x$$

$$(F_{net})_y = \sum F_y = ma_y$$

The acceleration is zero in equilibrium and also along an axis perpendicular to the motion.

APPLICATIONS

Mass is an intrinsic property of an object that describes the object's inertia and, loosely speaking, its quantity of matter.

The **weight** of an object is the reading of a spring scale when the object is at rest relative to the scale. Weight is the result of weighing. An object's weight depends on its mass, its acceleration, and the strength of gravity. An object in free fall is weightless.

A falling object reaches **terminal speed**

$$v_{term} = \sqrt{\frac{2mg}{C\rho A}}$$

Terminal speed is reached when the drag force exactly balances the gravitational force: $\vec{a} = \vec{0}$.

TERMS AND NOTATION

CONCEPTUAL QUESTIONS

1. Are the objects described here in equilibrium while at rest, in equilibrium while in motion, or not in equilibrium at all? Explain.
   a. A 200 pound barbell is held over your head.
   b. A girder is lifted at constant speed by a crane.
   c. A girder is being lowered into place. It is slowing down.
   d. A jet plane has reached its cruising speed and altitude.
   e. A box in the back of a truck doesn't slide as the truck stops.

2. A ball tossed straight up has $v = 0$ at its highest point. Is it in equilibrium? Explain.

3. Kat, Matt, and Nat are arguing about why a physics book on a table doesn't fall. According to Kat, "Gravity pulls down on it, but the table is in the way so it can't fall." "Nonsense," says Matt. "An upward force simply overcomes the downward force to prevent it from falling." "But what about Newton's first law?" counters Nat. "It's not moving, so there can't be any forces acting on it." None of the statements is exactly correct. Who comes closest, and how would you change his or her statement to make it correct?

4. If you know all of the forces acting on a moving object, can you tell the direction the object is moving? If yes, explain how. If no, give an example.

5. An elevator, hanging from a single cable, moves upward at constant speed. Friction and air resistance are negligible. Is the tension in the cable greater than, less than, or equal to the gravitational force on the elevator? Explain. Include a free-body diagram as part of your explanation.

6. An elevator, hanging from a single cable, moves downward and is slowing. Friction and air resistance are negligible. Is the tension in the cable greater than, less than, or equal to the gravitational force on the elevator? Explain. Include a free-body diagram as part of your explanation.

7. Are the following statements true or false? Explain.
   a. The mass of an object depends on its location.
   b. The weight of an object depends on its location.
   c. Mass and weight describe the same thing in different units.

8. An astronaut takes his bathroom scale to the moon and then stands on it. Is the reading of the scale his weight? Explain.

9. The four balls in FIGURE Q6.9 have been thrown straight up. They have the same size, but different masses. Air resistance is negligible. Rank in order, from largest to smallest, the magnitude of the net force acting on each ball. Some may be equal. Give your answer in the form a $>$ b $>$ c $=$ d and explain your ranking.

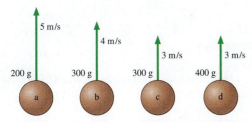

FIGURE Q6.9

10. Suppose you attempt to pour out 100 g of salt, using a pan balance for measurements, while in a rocket accelerating upward. Will the quantity of salt be too much, too little, or the correct amount? Explain.

11. An astronaut orbiting the earth is handed two balls that have identical outward appearances. However, one is hollow while the other is filled with lead. How can the astronaut determine which is which? Cutting or altering the balls is not allowed.

12. A hand presses down on the book in FIGURE Q6.12. Is the normal force of the table on the book larger than, smaller than, or equal to mg?

FIGURE Q6.12　　Book of mass m

13. Boxes A and B in FIGURE Q6.13 both remain at rest. Is the friction force on A larger than, smaller than, or equal to the friction force on B? Explain.

FIGURE Q6.13

14. Suppose you push a hockey puck of mass m across frictionless ice for 1.0 s, starting from rest, giving the puck speed v after traveling distance d. If you repeat the experiment with a puck of mass $2m$, pushing with the same force,
   a. How long will you have to push for the puck to reach the same speed v?
   b. How long will you have to push for the puck to travel the same distance d?

15. A block pushed along the floor with velocity v_{0x} slides a distance d after the pushing force is removed.
   a. If the mass of the block is doubled but its initial velocity is not changed, what distance does the block slide before stopping?
   b. If the initial velocity is doubled to $2v_{0x}$ but the mass is not changed, what distance does the block slide before stopping?

16. A crate of fragile dishes is in the back of a pickup truck. The truck accelerates north from a stop sign, and the crate moves without slipping. Does the friction force on the crate point north or south? Or is the friction force zero? Explain.

17. Five balls move through the air as shown in FIGURE Q6.17. All five have the same size and shape. Air resistance is not negligible. Rank in order, from largest to smallest, the magnitudes of the accelerations a_a to a_e. Some may be equal. Give your answer in the form a $>$ b $=$ c $>$ d $>$ e and explain your ranking.

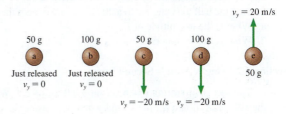

FIGURE Q6.17

EXERCISES AND PROBLEMS

Exercises

Section 6.1 The Equilibrium Model

1. ‖ The three ropes in **FIGURE EX6.1** are tied to a small, very light ring. Two of these ropes are anchored to walls at right angles with the tensions shown in the figure. What are the magnitude and direction of the tension $\vec{T}_3$ in the third rope?

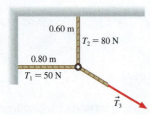

FIGURE EX6.1

2. | The three ropes in **FIGURE EX6.2** are tied to a small, very light ring. Two of the ropes are anchored to walls at right angles, and the third rope pulls as shown. What are T_1 and T_2, the magnitudes of the tension forces in the first two ropes?

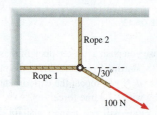

FIGURE EX6.2

3. | A football coach sits on a sled while two of his players build their strength by dragging the sled across the field with ropes. The friction force on the sled is 1000 N, the players have equal pulls, and the angle between the two ropes is 20°. How hard must each player pull to drag the coach at a steady 2.0 m/s?

4. ‖ A 20 kg loudspeaker is suspended 2.0 m below the ceiling by two 3.0-m-long cables that angle outward at equal angles. What is the tension in the cables?

5. | A 65 kg gymnast wedges himself between two closely spaced vertical walls by pressing his hands and feet against the walls. What is the magnitude of the friction force on each hand and foot? Assume they are all equal.

6. ‖ A construction worker with a weight of 850 N stands on a roof that is sloped at 20°. What is the magnitude of the normal force of the roof on the worker?

7. ‖ In an electricity experiment, a 1.0 g plastic ball is suspended on a 60-cm-long string and given an electric charge. A charged rod brought near the ball exerts a horizontal electrical force $\vec{F}_{\text{elec}}$ on it, causing the ball to swing out to a 20° angle and remain there.
   a. What is the magnitude of $\vec{F}_{\text{elec}}$?
   b. What is the tension in the string?

Section 6.2 Using Newton's Second Law

8. | The forces in **FIGURE EX6.8** act on a 2.0 kg object. What are the values of a_x and a_y, the x- and y-components of the object's acceleration?

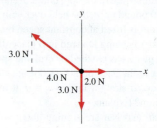

FIGURE EX6.8

9. | The forces in **FIGURE EX6.9** act on a 2.0 kg object. What are the values of a_x and a_y, the x- and y-components of the object's acceleration?

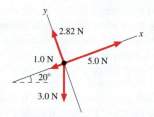

FIGURE EX6.9

10. | **FIGURE EX6.10** shows the velocity graph of a 2.0 kg object as it moves along the x-axis. What is the net force acting on this object at $t = 1$ s? At 4 s? At 7 s?

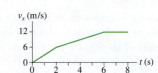

FIGURE EX6.10

11. ‖ **FIGURE EX6.11** shows the force acting on a 2.0 kg object as it moves along the x-axis. The object is at rest at the origin at $t = 0$ s. What are its acceleration and velocity at $t = 6$ s?

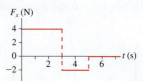

FIGURE EX6.11

12. | A horizontal rope is tied to a 50 kg box on frictionless ice. What is the tension in the rope if:
   a. The box is at rest?
   b. The box moves at a steady 5.0 m/s?
   c. The box has $v_x = 5.0$ m/s and $a_x = 5.0$ m/s²?

13. | A 50 kg box hangs from a rope. What is the tension in the rope if:
   a. The box is at rest?
   b. The box moves up at a steady 5.0 m/s?
   c. The box has $v_y = 5.0$ m/s and is speeding up at 5.0 m/s²?
   d. The box has $v_y = 5.0$ m/s and is slowing down at 5.0 m/s²?

14. | A 2.0×10^7 kg train applies its brakes with the intent of slowing down at a 1.2 m/s^2 rate. What magnitude force must its brakes provide?

15. || A 8.0×10^4 kg spaceship is at rest in deep space. Its thrusters provide a force of 1200 kN. The spaceship fires its thrusters for 20 s, then coasts for 12 km. How long does it take the spaceship to coast this distance?

16. || The position of a 2.0 kg mass is given by $x = (2t^3 - 3t^2)$ m,
CALC where t is in seconds. What is the net horizontal force on the mass at (a) $t = 0$ s and (b) $t = 1$ s?

Section 6.3 Mass, Weight, and Gravity

17. | A woman has a mass of 55 kg.
   a. What is her weight while standing on earth?
   b. What are her mass and her weight on Mars, where $g = 3.76$ m/s^2?

18. | It takes the elevator in a skyscraper 4.0 s to reach its cruising speed of 10 m/s. A 60 kg passenger gets aboard on the ground floor. What is the passenger's weight
   a. Before the elevator starts moving?
   b. While the elevator is speeding up?
   c. After the elevator reaches its cruising speed?

19. || Zach, whose mass is 80 kg, is in an elevator descending at 10 m/s. The elevator takes 3.0 s to brake to a stop at the first floor.
   a. What is Zach's weight before the elevator starts braking?
   b. What is Zach's weight while the elevator is braking?

20. || **FIGURE EX6.20** shows the velocity graph of a 75 kg passenger in an elevator. What is the passenger's weight at $t = 1$ s? At 5 s? At 9 s?

FIGURE EX6.20

21. | What thrust does a 200 g model rocket need in order to have a vertical acceleration of 10 m/s^2
   a. On earth?
   b. On the moon, where $g = 1.62$ m/s^2?

22. || A 20,000 kg rocket has a rocket motor that generates 3.0×10^5 N of thrust. Assume no air resistance.
   a. What is the rocket's initial upward acceleration?
   b. At an altitude of 5000 m the rocket's acceleration has increased to 6.0 m/s^2. What mass of fuel has it burned?

23. || The earth is 1.50×10^{11} m from the sun. The earth's mass is 5.98×10^{24} kg, while the mass of the sun is 1.99×10^{30} kg. What is earth's acceleration toward the sun?

Section 6.4 Friction

24. | Bonnie and Clyde are sliding a 300 kg bank safe across the floor to their getaway car. The safe slides with a constant speed if Clyde pushes from behind with 385 N of force while Bonnie pulls forward on a rope with 350 N of force. What is the safe's coefficient of kinetic friction on the bank floor?

25. | A stubborn, 120 kg mule sits down and refuses to move. To drag the mule to the barn, the exasperated farmer ties a rope around the mule and pulls with his maximum force of 800 N. The coefficients of friction between the mule and the ground are $\mu_s = 0.8$ and $\mu_k = 0.5$. Is the farmer able to move the mule?

26. || A 10 kg crate is placed on a horizontal conveyor belt. The materials are such that $\mu_s = 0.5$ and $\mu_k = 0.3$.
   a. Draw a free-body diagram showing all the forces on the crate if the conveyer belt runs at constant speed.
   b. Draw a free-body diagram showing all the forces on the crate if the conveyer belt is speeding up.
   c. What is the maximum acceleration the belt can have without the crate slipping?

27. || Bob is pulling a 30 kg filing cabinet with a force of 200 N, but the filing cabinet refuses to move. The coefficient of static friction between the filing cabinet and the floor is 0.80. What is the magnitude of the friction force on the filing cabinet?

28. || A rubber-wheeled 50 kg cart rolls down a 15° concrete incline. What is the cart's acceleration if rolling friction is (a) neglected and (b) included?

29. || A 4000 kg truck is parked on a 15° slope. How big is the friction force on the truck? The coefficient of static friction between the tires and the road is 0.90.

30. | A 1500 kg car skids to a halt on a wet road where $\mu_k = 0.50$. How fast was the car traveling if it leaves 65-m-long skid marks?

31. || A 50,000 kg locomotive is traveling at 10 m/s when its engine and brakes both fail. How far will the locomotive roll before it comes to a stop? Assume the track is level.

32. || You and your friend Peter are putting new shingles on a roof pitched at 25°. You're sitting on the very top of the roof when Peter, who is at the edge of the roof directly below you, 5.0 m away, asks you for the box of nails. Rather than carry the 2.5 kg box of nails down to Peter, you decide to give the box a push and have it slide down to him. If the coefficient of kinetic friction between the box and the roof is 0.55, with what speed should you push the box to have it gently come to rest right at the edge of the roof?

33. || An Airbus A320 jetliner has a takeoff mass of 75,000 kg. It reaches its takeoff speed of 82 m/s (180 mph) in 35 s. What is the thrust of the engines? You can neglect air resistance but not rolling friction.

Section 6.5 Drag

34. || A medium-sized jet has a 3.8-m-diameter fuselage and a loaded mass of 85,000 kg. The drag on an airplane is primarily due to the cylindrical fuselage, and aerodynamic shaping gives it a drag coefficient of 0.37. How much thrust must the jet's engines provide to cruise at 230 m/s at an altitude where the air density is 1.0 kg/m^3?

35. ||| A 75 kg skydiver can be modeled as a rectangular "box" with dimensions 20 cm × 40 cm × 180 cm. What is his terminal speed if he falls feet first? Use 0.8 for the drag coefficient.

36. ||| A 6.5-cm-diameter ball has a terminal speed of 26 m/s. What is the ball's mass?

Problems

37. || A 2.0 kg object initially at rest at the origin is subjected to the time-varying force shown in **FIGURE P6.37**. What is the object's velocity at $t = 4$ s?

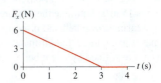

FIGURE P6.37

38. ‖ A 5.0 kg object initially at rest at the origin is subjected to the time-varying force shown in **FIGURE P6.38**. What is the object's velocity at $t = 6$ s?

FIGURE P6.38

39. ‖ The 1000 kg steel beam in **FIGURE P6.39** is supported by two ropes. What is the tension in each?

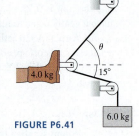

FIGURE P6.39

40. ‖ Henry, whose mass is 95 kg, stands on a bathroom scale in an elevator. The scale reads 830 N for the first 3.0 s after the elevator starts moving, then 930 N for the next 3.0 s. What is the elevator's velocity 6.0 s after starting?

41. ‖
BIO An accident victim with a broken leg is being placed in traction. The patient wears a special boot with a pulley attached to the sole. The foot and boot together have a mass of 4.0 kg, and the doctor has decided to hang a 6.0 kg mass from the rope. The boot is held suspended by the ropes, as shown in **FIGURE P6.41**, and does not touch the bed.

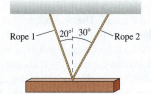

FIGURE P6.41

   a. Determine the amount of tension in the rope by using Newton's laws to analyze the hanging mass.

   b. The net traction force needs to pull straight out on the leg. What is the proper angle θ for the upper rope?

   c. What is the net traction force pulling on the leg?

   Hint: If the pulleys are frictionless, which we will assume, the tension in the rope is constant from one end to the other.

42. ‖
BIO Seat belts and air bags save lives by reducing the forces exerted on the driver and passengers in an automobile collision. Cars are designed with a "crumple zone" in the front of the car. In the event of an impact, the passenger compartment decelerates over a distance of about 1 m as the front of the car crumples. An occupant restrained by seat belts and air bags decelerates with the car. By contrast, an unrestrained occupant keeps moving forward with no loss of speed (Newton's first law!) until hitting the dashboard or windshield. These are unyielding surfaces, and the unfortunate occupant then decelerates over a distance of only about 5 mm.

   a. A 60 kg person is in a head-on collision. The car's speed at impact is 15 m/s. Estimate the net force on the person if he or she is wearing a seat belt and if the air bag deploys.

   b. Estimate the net force that ultimately stops the person if he or she is not restrained by a seat belt or air bag.

43. ‖ The piston of a machine exerts a constant force on a ball as it moves horizontally through a distance of 15 cm. You use a motion detector to measure the speed of five different balls as they come off the piston; the data are shown in the table. Use theory to find two quantities that, when graphed, should give a straight line. Then use the graph to find the size of the piston's force.

Mass (g)	Speed (m/s)
200	9.4
400	6.3
600	5.2
800	4.9
1000	4.0

44. ‖ Compressed air is used to fire a 50 g ball vertically upward from a 1.0-m-tall tube. The air exerts an upward force of 2.0 N on the ball as long as it is in the tube. How high does the ball go above the top of the tube?

45. ‖ a. A rocket of mass m is launched straight up with thrust $\vec{F}_{thrust}$. Find an expression for the rocket's speed at height h if air resistance is neglected.

   b. The motor of a 350 g model rocket generates 9.5 N thrust. If air resistance can be neglected, what will be the rocket's speed as it reaches a height of 85 m?

46. ‖ A rifle with a barrel length of 60 cm fires a 10 g bullet with a horizontal speed of 400 m/s. The bullet strikes a block of wood and penetrates to a depth of 12 cm.

   a. What resistive force (assumed to be constant) does the wood exert on the bullet?

   b. How long does it take the bullet to come to rest?

47. ‖ A truck with a heavy load has a total mass of 7500 kg. It is climbing a 15° incline at a steady 15 m/s when, unfortunately, the poorly secured load falls off! Immediately after losing the load, the truck begins to accelerate at 1.5 m/s². What was the mass of the load? Ignore rolling friction.

48. ‖ An object of mass m is at rest at the top of a smooth slope of height h and length L. The coefficient of kinetic friction between the object and the surface, μ_k, is small enough that the object will slide down the slope after being given a very small push to get it started. Find an expression for the object's speed at the bottom of the slope.

49. ‖ Sam, whose mass is 75 kg, takes off across level snow on his jet-powered skis. The skis have a thrust of 200 N and a coefficient of kinetic friction on snow of 0.10. Unfortunately, the skis run out of fuel after only 10 s.

   a. What is Sam's top speed?

   b. How far has Sam traveled when he finally coasts to a stop?

50. ‖ A baggage handler drops your 10 kg suitcase onto a conveyor belt running at 2.0 m/s. The materials are such that $\mu_s = 0.50$ and $\mu_k = 0.30$. How far is your suitcase dragged before it is riding smoothly on the belt?

51. ‖ A 2.0 kg wood block is launched up a wooden ramp that is inclined at a 30° angle. The block's initial speed is 10 m/s.

   a. What vertical height does the block reach above its starting point?

   b. What speed does it have when it slides back down to its starting point?

52. ‖ It's a snowy day and you're pulling a friend along a level road on a sled. You've both been taking physics, so she asks what you think the coefficient of friction between the sled and the snow is. You've been walking at a steady 1.5 m/s, and the rope pulls up on the sled at a 30° angle. You estimate that the mass of the sled, with your friend on it, is 60 kg and that you're pulling with a force of 75 N. What answer will you give?

53. ‖ A large box of mass M is pulled across a horizontal, frictionless surface by a horizontal rope with tension T. A small box of mass m sits on top of the large box. The coefficients of static and kinetic friction between the two boxes are μ_s and μ_k, respectively. Find an expression for the maximum tension T_{max} for which the small box rides on top of the large box without slipping.

54. ‖ A large box of mass M is moving on a horizontal surface at speed v_0. A small box of mass m sits on top of the large box. The coefficients of static and kinetic friction between the two boxes are μ_s and μ_k, respectively. Find an expression for the shortest distance d_{min} in which the large box can stop without the small box slipping.

55. ‖ You're driving along at 25 m/s with your aunt's valuable antiques in the back of your pickup truck when suddenly you see a giant hole in the road 55 m ahead of you. Fortunately, your foot is right beside the brake and your reaction time is zero!
   a. Can you stop the truck before it falls into the hole?
   b. If your answer to part a is yes, can you stop without the antiques sliding and being damaged? Their coefficients of friction are $\mu_s = 0.60$ and $\mu_k = 0.30$.
   Hint: You're not trying to stop in the shortest possible distance. What's your best strategy for avoiding damage to the antiques?

56. ‖ The 2.0 kg wood box in **FIGURE P6.56** slides down a vertical wood wall while you push on it at a 45° angle. What magnitude of force should you apply to cause the box to slide down at a constant speed?

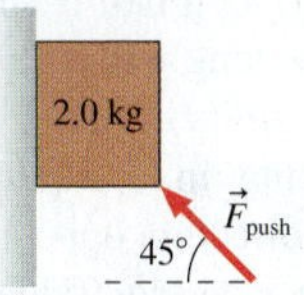

2.0 kg

$\vec{F}_{push}$

45°

FIGURE P6.56

57. ‖ A 1.0 kg wood block is pressed against a vertical wood wall by the 12 N force shown in **FIGURE P6.57**. If the block is initially at rest, will it move upward, move downward, or stay at rest?

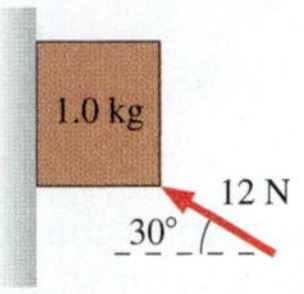

1.0 kg

12 N

30°

FIGURE P6.57

58. ‖ A person with compromised pinch strength in his fingers can exert a force of only 6.0 N to either side of a pinch-held object, such as the book shown in **FIGURE P6.58**. What is the heaviest book he can hold vertically before it slips out of his fingers? The coefficient of static friction between his fingers and the book cover is 0.80.

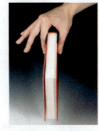

FIGURE P6.58

59. ‖ A ball is shot from a compressed-air gun at twice its terminal speed.
   a. What is the ball's initial acceleration, as a multiple of g, if it is shot straight up?
   b. What is the ball's initial acceleration, as a multiple of g, if it is shot straight down?

60. ‖‖ Starting from rest, a 2500 kg helicopter accelerates straight up at a constant 2.0 m/s². What is the helicopter's height at the moment its blades are providing an upward force of 26 kN? The helicopter can be modeled as a 2.6-m-diameter sphere.

61. ‖ Astronauts in space "weigh" themselves by oscillating on a **CALC** spring. Suppose the position of an oscillating 75 kg astronaut is given by $x = (0.30 \text{ m}) \sin((\pi \text{ rad/s}) \times t)$, where t is in s. What force does the spring exert on the astronaut at (a) $t = 1.0$ s and (b) 1.5 s? Note that the angle of the sine function is in radians.

62. ‖ A particle of mass m moving along the x-axis experiences the **CALC** net force $F_x = ct$, where c is a constant. The particle has velocity v_{0x} at $t = 0$. Find an algebraic expression for the particle's velocity v_x at a later time t.

63. ‖ At $t = 0$, an object of mass m is at rest at $x = 0$ on a horizontal, **CALC** frictionless surface. A horizontal force $F_x = F_0(1 - t/T)$, which decreases from F_0 at $t = 0$ to zero at $t = T$, is exerted on the object. Find an expression for the object's (a) velocity and (b) position at time T.

64. ‖ At $t = 0$, an object of mass m is at rest at $x = 0$ on a horizontal, frictionless surface. Starting at $t = 0$, a horizontal force $F_x = F_0 e^{-t/T}$ is exerted on the object.
   a. Find and graph an expression for the object's velocity at an arbitrary later time t.
   b. What is the object's velocity after a very long time has elapsed?

65. ‖‖ Large objects have inertia and tend to keep moving—Newton's **BIO** first law. Life is very different for small microorganisms that swim through water. For them, drag forces are so large that they instantly stop, without coasting, if they cease their swimming motion. To swim at constant speed, they must exert a constant propulsion force by rotating corkscrew-like flagella or beating hair-like cilia. The quadratic model of drag of Equation 6.15 fails for very small particles. Instead, a small object moving in a liquid experiences a *linear* drag force, $\vec{F}_{drag} = (bv, \text{ direction opposite the motion})$, where b is a constant. For a sphere of radius R, the drag constant can be shown to be $b = 6\pi\eta R$, where η is the *viscosity* of the liquid. Water at 20°C has viscosity 1.0×10^{-3} N s/m².
   a. A *paramecium* is about 100 μm long. If it's modeled as a sphere, how much propulsion force must it exert to swim at a typical speed of 1.0 mm/s? How about the propulsion force of a 2.0-μm-diameter *E. coli* bacterium swimming at 30 μm/s?
   b. The propulsion forces are very small, but so are the organisms. To judge whether the propulsion force is large or small *relative to the organism*, compute the acceleration that the propulsion force could give each organism if there were no drag. The density of both organisms is the same as that of water, 1000 kg/m³.

66. ‖‖ A 60 kg skater is gliding across frictionless ice at 4.0 m/s. Air **CALC** resistance is not negligible. You can model the skater as a 170-cm-tall, 36-cm-diameter cylinder. What is the skater's speed 2.0 s later?

67. ‖‖ Very small objects, such as dust particles, experience a *linear* drag force, $\vec{F}_{drag} = (bv, \text{ direction opposite the motion})$, where b is a constant. That is, the quadratic model of drag of Equation 6.15 fails for very small particles. For a sphere of radius R, the drag constant can be shown to be $b = 6\pi\eta R$, where η is the *viscosity* of the gas.
   a. Find an expression for the terminal speed v_{term} of a spherical particle of radius R and mass m falling through a gas of viscosity η.
   b. Suppose a gust of wind has carried a 50-μm-diameter dust particle to a height of 300 m. If the wind suddenly stops, how long will it take the dust particle to settle back to the ground? Dust has a density of 2700 kg/m³, the viscosity of 25°C air is 2.0×10^{-5} N s/m², and you can assume that the falling dust particle reaches terminal speed almost instantly.

Problems 68 and 69 show a free-body diagram. For each:

a. Write a realistic dynamics problem for which this is the correct free-body diagram. Your problem should ask a question that can be answered with a value of position or velocity (such as "How far?" or "How fast?"), and should give sufficient information to allow a solution.

b. Solve your problem!

68.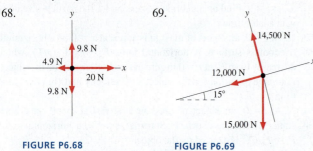

FIGURE P6.68

FIGURE P6.69

In Problems 70 through 72 you are given the dynamics equations that are used to solve a problem. For each of these, you are to

a. Write a realistic problem for which these are the correct equations.

b. Draw the free-body diagram and the pictorial representation for your problem.

c. Finish the solution of the problem.

70. $-0.80n = (1500 \text{ kg})a_x$

$n - (1500 \text{ kg})(9.80 \text{ m/s}^2) = 0$

71. $T - 0.20n - (20 \text{ kg})(9.80 \text{ m/s}^2)\sin 20°$

$= (20 \text{ kg})(2.0 \text{ m/s}^2)$

$n - (20 \text{ kg})(9.80 \text{ m/s}^2)\cos 20° = 0$

72. $(100 \text{ N})\cos 30° - f_k = (20 \text{ kg})a_x$

$n + (100 \text{ N})\sin 30° - (20 \text{ kg})(9.80 \text{ m/s}^2) = 0$

$f_k = 0.20n$

Challenge Problems

73. ⦀ A block of mass m is at rest at the origin at $t = 0$. It is pushed
CALC with constant force F_0 from $x = 0$ to $x = L$ across a horizontal surface whose coefficient of kinetic friction is $\mu_k = \mu_0(1 - x/L)$. That is, the coefficient of friction decreases from μ_0 at $x = 0$ to zero at $x = L$.

a. Use what you've learned in calculus to prove that

$$a_x = v_x \frac{dv_x}{dx}$$

b. Find an expression for the block's speed as it reaches position L.

74. ⦀ A spring-loaded toy gun exerts a variable force on a plastic ball as
CALC the spring expands. Consider a horizontal spring and a ball of mass m whose position when barely touching a fully expanded spring is $x = 0$. The ball is pushed to the left, compressing the spring. You'll learn in Chapter 9 that the spring force on the ball, when the ball is at position x (which is negative), can be written as $(F_{Sp})_x = -kx$, where k is called the *spring constant*. The minus sign is needed to make the x-component of the force positive. Suppose the ball is initially pushed to $x_0 = -L$, then released and shot to the right.

a. Use what you've learned in calculus to prove that

$$a_x = v_x \frac{dv_x}{dx}$$

b. Find an expression, in terms of m, k, and L, for the speed of the ball as it comes off the spring at $x = 0$.

75. ⦀ **FIGURE CP6.75** shows an
CALC *accelerometer,* a device for measuring the horizontal acceleration of cars and airplanes. A ball is free to roll on a parabolic track described by the equation $y = x^2$, where both x and y are in meters. A scale along the bottom is used to measure the ball's horizontal position x.

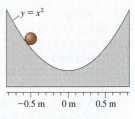

FIGURE CP6.75

a. Find an expression that allows you to use a measured position x (in m) to compute the acceleration a_x (in m/s^2). (For example, $a_x = 3x$ is a possible expression.)

b. What is the acceleration if $x = 20$ cm?

76. ⦀ An object moving in a liquid experiences a *linear* drag
CALC force: $\vec{F}_{\text{drag}} = (bv$, direction opposite the motion), where b is a constant called the *drag coefficient*. For a sphere of radius R, the drag constant can be computed as $b = 6\pi\eta R$, where η is the *viscosity* of the liquid.

a. Find an algebraic expression for $v_x(t)$, the x-component of velocity as a function of time, for a spherical particle of radius R and mass m that is shot horizontally with initial speed v_0 through a liquid of viscosity η.

b. Water at 20°C has viscosity $\eta = 1.0 \times 10^{-3}$ N s/m^2. Suppose a 4.0-cm-diameter, 33 g ball is shot horizontally into a tank of 20°C water. How long will it take for the horizontal speed to decrease to 50% of its initial value?

77. ⦀ An object moving in a liquid experiences a *linear* drag
CALC force: $\vec{F}_{\text{drag}} = (bv$, direction opposite the motion), where b is a constant called the *drag coefficient*. For a sphere of radius R, the drag constant can be computed as $b = 6\pi\eta R$, where η is the *viscosity* of the liquid.

a. Use what you've learned in calculus to prove that

$$a_x = v_x \frac{dv_x}{dx}$$

b. Find an algebraic expression for $v_x(x)$, the x-component of velocity as a function of distance traveled, for a spherical particle of radius R and mass m that is shot horizontally with initial speed v_0 through a liquid of viscosity η.

c. Water at 20°C has viscosity $\eta = 1.0 \times 10^{-3}$ N s/m^2. Suppose a 1.0-cm-diameter, 1.0 g marble is shot horizontally into a tank of 20°C water at 10 cm/s. How far will it travel before stopping?

78. ⦀ An object with cross section A is shot horizontally across
CALC frictionless ice. Its initial velocity is v_{0x} at $t_0 = 0$ s. Air resistance is not negligible.

a. Show that the velocity at time t is given by the expression

$$v_x = \frac{v_{0x}}{1 + C\rho A v_{0x} t/2m}$$

b. A 1.6-m-wide, 1.4-m-high, 1500 kg car with a drag coefficient of 0.35 hits a very slick patch of ice while going 20 m/s. If friction is neglected, how long will it take until the car's speed drops to 10 m/s? To 5 m/s?

c. Assess whether or not it is reasonable to neglect kinetic friction.

7 Newton's Third Law

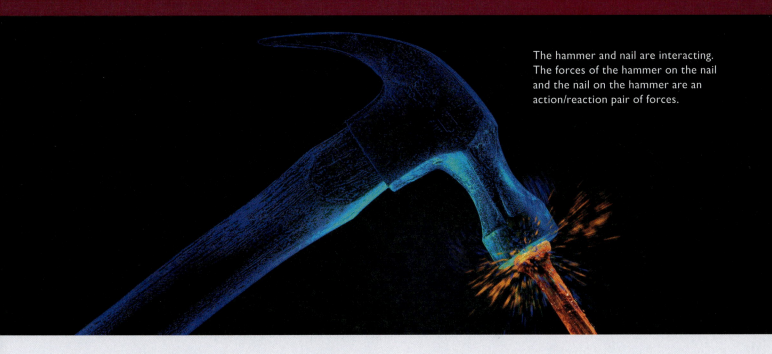

The hammer and nail are interacting. The forces of the hammer on the nail and the nail on the hammer are an action/reaction pair of forces.

IN THIS CHAPTER, you will use Newton's third law to understand how objects interact.

What is an interaction?

All forces are interactions in which objects exert forces on each other. If A pushes on B, then B pushes back on A. These two forces form an action/reaction pair of forces. One can't exist without the other.

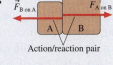

Action/reaction pair

« LOOKING BACK Section 5.5 Forces, interactions, and Newton's second law

What is an interaction diagram?

We will often analyze a problem by defining a system—the objects of interest—and the larger environment that acts on the system. An interaction diagram is a key visual tool for identifying action/reaction forces of interaction *inside* the system and external forces from agents in the environment.

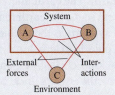

How do we model ropes and pulleys?

A common way that two objects interact is to be connected via a rope or cable or string. Pulleys change the direction of the tension forces. We will often model

- Ropes and strings as massless;
- Pulleys as massless and frictionless.

The objects' accelerations are constrained to have the same magnitude.

What is Newton's third law?

Newton's third law governs interactions:

- Every force is a member of an action/reaction pair.
- The two members of a pair act on different objects.
- The two members of a pair are equal in magnitude but opposite in direction.

How is Newton's third law used?

The dynamics problem-solving strategy of Chapter 6 is still our primary tool.

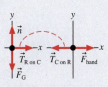

- Draw a free-body diagram for *each object*.
- Identify and show action/reaction pairs.
- Use Newton's second law for *each object*.
- Relate forces with Newton's third law.

« LOOKING BACK Section 6.2 Problem-Solving Strategy 6.1

Why is Newton's third law important?

We started our study of dynamics with only the first two of Newton's laws in order to practice identifying and using forces. But objects in the real world don't exist in isolation—they *interact* with each other. Newton's third law gives us a much more complete view of mechanics. The third law is also an essential tool in the practical application of physics to problems in engineering and technology.

7.1 Interacting Objects

FIGURE 7.1 The hammer and nail are interacting with each other.

The force of the nail on the hammer

The force of the hammer on the nail

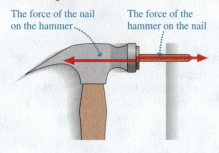

FIGURE 7.2 An action/reaction pair of forces.

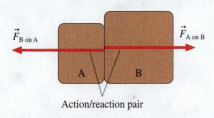

$\vec{F}_{\text{B on A}}$ $\vec{F}_{\text{A on B}}$

A   B

Action/reaction pair

FIGURE 7.1 shows a hammer hitting a nail. The hammer exerts a force on the nail as it drives the nail forward. At the same time, the nail exerts a force on the hammer. If you're not sure that it does, imagine hitting the nail with a glass hammer. It's the force of the nail on the hammer that would cause the glass to shatter.

In fact, any time an object A pushes or pulls on another object B, B pushes or pulls back on A. When you pull someone with a rope in a tug-of-war, that person pulls back on you. Your chair pushes up on you (the normal force) as you push down on the chair. These are examples of an **interaction**, the mutual influence of two objects on each other.

To be more specific, if object A exerts a force $\vec{F}_{\text{A on B}}$ on object B, then object B exerts a force $\vec{F}_{\text{B on A}}$ on object A. This pair of forces, shown in **FIGURE 7.2**, is called an **action/reaction pair.** Two objects interact by exerting an action/reaction pair of forces on each other. Notice the very explicit subscripts on the force vectors. The first letter is the *agent;* the second letter is the object on which the force acts. $\vec{F}_{\text{A on B}}$ is a force exerted *by* A *on* B.

NOTE The name "action/reaction pair" is somewhat misleading. The forces occur simultaneously, and we cannot say which is the "action" and which the "reaction." **An action/reaction pair of forces exists as a pair, or not at all.**

The hammer and nail interact through contact forces. Does the same idea hold true for long-range forces such as gravity? Newton was the first to realize that it does. His evidence was the tides. Astronomers had known since antiquity that the tides depend on the phase of the moon, but Newton was the first to understand that tides are the ocean's response to the gravitational pull of the moon on the earth.

Objects, Systems, and the Environment

Chapters 5 and 6 considered forces acting on a single object that we modeled as a particle. **FIGURE 7.3a** shows a diagrammatic representation of single-particle dynamics. We can use Newton's second law, $\vec{a} = \vec{F}_{\text{net}}/m$, to determine the particle's acceleration.

We now want to extend the particle model to situations in which two or more objects, each represented as a particle, interact with each other. For example, **FIGURE 7.3b** shows three objects interacting via action/reaction pairs of forces. The forces can be given labels such as $\vec{F}_{\text{A on B}}$ and $\vec{F}_{\text{B on A}}$. How do these particles move?

We will often be interested in the motion of some of the objects, say objects A and B, but not of others. For example, objects A and B might be the hammer and the nail, while object C is the earth. The earth interacts with both the hammer and the nail via gravity, but in a practical sense the earth remains "at rest" while the hammer and nail move. Let's define the **system** as those objects whose motion we want to analyze and the **environment** as objects external to the system.

FIGURE 7.3c is a new kind of diagram, an **interaction diagram,** in which we've enclosed the objects of the system in a box and represented interactions as lines

FIGURE 7.3 Single-particle dynamics and a model of interacting objects.

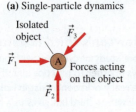

(a) Single-particle dynamics

Isolated object

$\vec{F}_3$

$\vec{F}_1$ A   Forces acting on the object

$\vec{F}_2$

This is a force diagram.

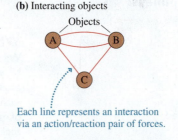

(b) Interacting objects

Objects

A   B

C

Each line represents an interaction via an action/reaction pair of forces.

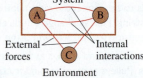

(c) System and environment

System

A   B

External forces   Internal interactions

C

Environment

This is an *interaction diagram.*

connecting objects. This is a rather abstract, schematic diagram, but it captures the essence of the interactions. Notice that interactions with objects in the environment are called **external forces.** For the hammer and nail, the gravitational force on each—an interaction with the earth—is an external force.

NOTE *Every* force is one member of an action/reaction pair, so there is no such thing as a true "external force." What we call an external force is simply an interaction between an object of interest, one we've chosen to place inside the system, and an object whose motion is not of interest.

The bat and the ball are interacting with each other.

7.2 Analyzing Interacting Objects

TACTICS BOX 7.1

Analyzing interacting objects

❶ **Represent each object as a circle** with a name and label. Place each in the correct position relative to other objects. The surface of the earth (label S; contact forces) and the entire earth (label EE; long-range forces) should be considered separate objects.

❷ **Identify interactions.** Draw *one* connecting line between relevant circles to represent each interaction.

  ■ Every interaction line connects two and only two objects.
  ■ A surface can have two interactions: friction (parallel to the surface) and a normal force (perpendicular to the surface).
  ■ The entire earth interacts only by the long-range gravitational force.

❸ **Identify the system.** Identify the objects of interest; draw and label a box enclosing them. This completes the interaction diagram.

❹ **Draw a free-body diagram for each object in the system.** Include only the forces acting *on* each object, not forces exerted by the object.

  ■ Every interaction line crossing the system boundary is one external force acting on an object. The usual symbols, such as $\vec{n}$ and $\vec{T}$ can be used.
  ■ Every interaction line within the system represents an action/reaction pair of forces. There is one force vector on *each* of the objects, and these forces point in opposite directions. Use labels like $\vec{F}_{\text{A on B}}$ and $\vec{F}_{\text{B on A}}$.
  ■ Connect the two action/reaction forces—which must be on *different* free-body diagrams—with a dashed line.

Exercises 1–7

We'll illustrate these ideas with two concrete examples. The first example will be much longer than usual because we'll go carefully through all the steps in the reasoning.

EXAMPLE 7.1 | **Pushing a crate**

FIGURE 7.4 shows a person pushing a large crate across a rough surface. Identify all interactions, show them on an interaction diagram, then draw free-body diagrams of the person and the crate.

FIGURE 7.4 A person pushes a crate across a rough floor.

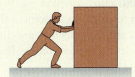

VISUALIZE The interaction diagram of **FIGURE 7.5** on the next page starts by representing every object as a circle in the correct position but separated from all other objects. The person and the crate are obvious objects. The earth is also an object that both exerts and experiences forces, but it's necessary to distinguish between the surface, which exerts contact forces, and the entire earth, which exerts the long-range gravitational force.

   Figure 7.5 also identifies the various interactions. Some, like the pushing interaction between the person and the crate, are fairly

Continued

FIGURE 7.5 The interaction diagram.

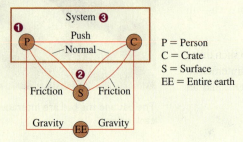

System ❸
Push
Normal
❷
Friction Friction
Gravity Gravity

P = Person
C = Crate
S = Surface
EE = Entire earth

FIGURE 7.6 Free-body diagrams of the person and the crate.

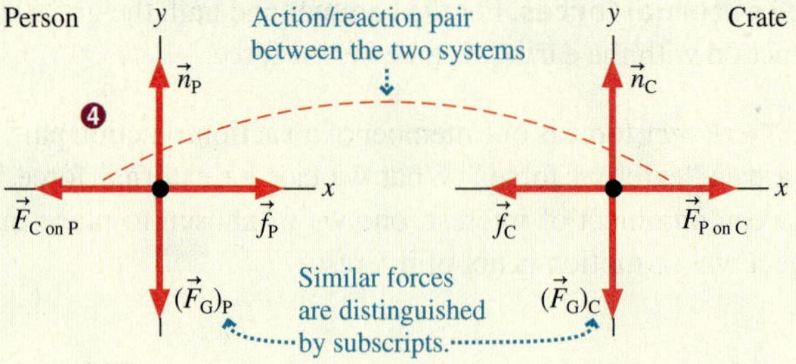

Person

Action/reaction pair
between the two systems

Crate

$\vec{n}_P$

$\vec{n}_C$

$\vec{F}_{C\,on\,P}$ $\vec{f}_P$ $\vec{f}_C$ $\vec{F}_{P\,on\,C}$

$(\vec{F}_G)_P$ Similar forces
are distinguished
by subscripts. $(\vec{F}_G)_C$

obvious. The interactions with the earth are a little trickier. Gravity, a long-range force, is an interaction between each object and the earth as a whole. Friction forces and normal forces are contact interactions between each object and the earth's surface. These are two different interactions, so two interaction lines connect the crate to the surface and the person to the surface. Finally, we've enclosed the person and crate in a box labeled System. These are the objects whose motion we wish to analyze.

NOTE Interactions are between two *different* objects. None of the interactions are between an object and itself.

We can now draw free-body diagrams for the objects in the system, the crate and the person. **FIGURE 7.6** correctly locates the crate's free-body diagram to the right of the person's free-body diagram. For each, three interaction lines cross the system boundary and thus represent external forces. These are the gravitational force from the entire earth, the upward normal force from the surface, and a friction force from the surface. We can use familiar labels such as $\vec{n}_P$ and $\vec{f}_C$, but **it's very important to distinguish different forces with subscripts.** There's now more than one normal force. If you call both simply $\vec{n}$, you're almost certain to make mistakes when you start writing out the second-law equations.

The directions of the normal forces and the gravitational forces are clear, but we have to be careful with friction. Friction force $\vec{f}_C$ is kinetic friction of the crate sliding across the surface, so it

points left, opposite the motion. But what about friction between the person and the surface? It is tempting to draw force $\vec{f}_P$ pointing to the left. After all, friction forces are supposed to be in the direction opposite the motion. But if we did so, the person would have two forces to the left, $\vec{F}_{C\,on\,P}$ and $\vec{f}_P$, and none to the right, causing the person to accelerate *backward!* That is clearly not what happens, so what is wrong?

Imagine pushing a crate to the right across loose sand. Each time you take a step, you tend to kick the sand to the *left*, behind you. Thus friction force $\vec{f}_{P\,on\,S}$, the force of the person pushing against the earth's surface, is to the *left*. In reaction, the force of the earth's surface against the person is a friction force to the *right*. It is force $\vec{f}_{S\,on\,P}$, which we've shortened to $\vec{f}_P$, that causes the person to accelerate in the forward direction. Further, as we'll discuss more below, this is a *static* friction force; your foot is planted on the ground, not sliding across the surface.

Finally, we have one internal interaction. The crate is pushed with force $\vec{F}_{P\,on\,C}$. If A pushes or pulls on B, then B pushes or pulls back on A, so the reaction to force $\vec{F}_{P\,on\,C}$ is $\vec{F}_{C\,on\,P}$, the crate pushing back against the person's hands. Force $\vec{F}_{P\,on\,C}$ is a force exerted on the crate, so it's shown on the crate's free-body diagram. Force $\vec{F}_{C\,on\,P}$ is exerted on the person, so it is drawn on the person's free-body diagram. **The two forces of an action/reaction pair never occur on the same free-body diagram.** We've connected forces $\vec{F}_{P\,on\,C}$ and $\vec{F}_{C\,on\,P}$ with a dashed line to show that they are an action/reaction pair.

Propulsion

The friction force $\vec{f}_P$ (force of surface on person) is an example of **propulsion.** It is the force that a system with an internal source of energy uses to drive itself forward. Propulsion is an important feature not only of walking or running but also of the forward motion of cars, jets, and rockets. Propulsion is somewhat counterintuitive, so it is worth a closer look.

If you try to walk across a frictionless floor, your foot slips and slides *backward.* In order for you to walk, the floor needs to have friction so that your foot *sticks* to the floor as you straighten your leg, moving your body forward. The friction that prevents slipping is *static* friction. Static friction, you will recall, acts in the direction that prevents slipping. The static friction force $\vec{f}_P$ has to point in the *forward* direction to prevent your foot from slipping backward. It is this forward-directed static friction force that propels you forward! The force of your foot on the floor, the other half of the action/reaction pair, is in the opposite direction.

The distinction between you and the crate is that you have an *internal source of energy* that allows you to straighten your leg by pushing backward against the surface.

In essence, you walk by pushing the earth away from you. The earth's surface responds by pushing you forward. These are static friction forces. In contrast, all the crate can do is slide, so *kinetic* friction opposes the motion of the crate.

FIGURE 7.7 shows how propulsion works. A car uses its motor to spin the tires, causing the tires to push backward against the ground. This is why dirt and gravel are kicked backward, not forward. The earth's surface responds by pushing the car forward. These are also *static* friction forces. The tire is rolling, but the bottom of the tire, where it contacts the road, is instantaneously at rest. If it weren't, you would leave one giant skid mark as you drove and would burn off the tread within a few miles.

FIGURE 7.7 Examples of propulsion.

The person pushes backward against the earth.
The earth pushes forward on the person.

Static friction

The car pushes backward against the earth.
The earth pushes forward on the car.

Static friction

The rocket pushes the hot gases backward.
The gases push the rocket forward.

Thrust

EXAMPLE 7.2 | Towing a car

A tow truck uses a rope to pull a car along a horizontal road, as shown in **FIGURE 7.8**. Identify all interactions, show them on an interaction diagram, then draw free-body diagrams of each object in the system.

FIGURE 7.8 A truck towing a car.

FIGURE 7.9 The interaction diagram.

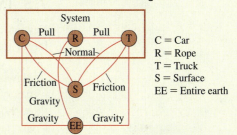

C = Car
R = Rope
T = Truck
S = Surface
EE = Entire earth

VISUALIZE The interaction diagram of **FIGURE 7.9** represents the objects as separate circles, but with the correct relative positions. The rope is shown as a separate object. Many of the interactions are identical to those in Example 7.1. The system—the objects in motion—consists of the truck, the rope, and the car.

The three objects in the system require three free-body diagrams, shown in **FIGURE 7.10**. Gravity, friction, and normal forces at the surface are all interactions that cross the system boundary and are shown as external forces. The car is an

inert object rolling along. It would slow and stop if the rope were cut, so the surface must exert a rolling friction force $\vec{f}_C$ to the left. The truck, however, has an internal source of energy. The truck's drive wheels push the ground to the left with force $\vec{f}_{T \text{ on } S}$. In reaction, the ground propels the truck forward, to the right, with force $\vec{f}_T$.

We next need to identify the forces between the car, the truck, and the rope. The rope pulls on the car with a tension force $\vec{T}_{R \text{ on } C}$. You might be tempted to put the reaction force on the truck because we say that "the truck pulls the car," but the truck is not in contact

FIGURE 7.10 Free-body diagrams of Example 7.2.

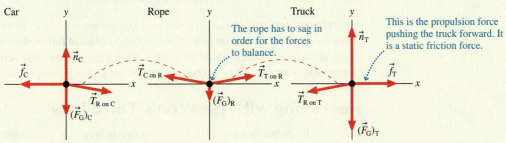

The rope has to sag in order for the forces to balance.

This is the propulsion force pushing the truck forward. It is a static friction force.

Continued

with the car. The truck pulls on the rope, then the rope pulls on the car. Thus the reaction to $\vec{T}_{\text{R on C}}$ is a force on the *rope:* $\vec{T}_{\text{C on R}}$. These are an action/reaction pair. At the other end, $\vec{T}_{\text{T on R}}$ and $\vec{T}_{\text{R on T}}$ are also an action/reaction pair.

> **NOTE** Drawing an interaction diagram helps you avoid mistakes because it shows very clearly what is interacting with what.

Notice that the tension forces of the rope *cannot* be horizontal. If they were, the rope's free-body diagram would show a net downward force, because of its weight, and the rope would accelerate downward.

The tension forces $\vec{T}_{\text{T on R}}$ and $\vec{T}_{\text{C on R}}$ have to angle slightly upward to balance the gravitational force, so any real rope has to sag at least a little in the center.

ASSESS Make sure you avoid the common error of considering $\vec{n}$ and $\vec{F}_{\text{G}}$ to be an action/reaction pair. These are both forces on the *same* object, whereas the two forces of an action/reaction pair are always on two *different* objects that are interacting with each other. The normal and gravitational forces are often equal in magnitude, as they are in this example, but that doesn't make them an action/reaction pair of forces.

STOP TO THINK 7.1 A rope of negligible mass pulls a crate across the floor. The rope and crate are the system; the hand pulling the rope is part of the environment. What, if anything, is wrong with the free-body diagrams?

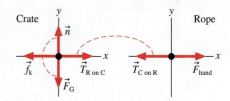

7.3 Newton's Third Law

Newton was the first to recognize how the two members of an action/reaction pair of forces are related to each other. Today we know this as **Newton's third law:**

> **Newton's third law** Every force occurs as one member of an action/reaction pair of forces.
>
> - The two members of an action/reaction pair act on two *different* objects.
> - The two members of an action/reaction pair are equal in magnitude but opposite in direction: $\vec{F}_{\text{A on B}} = -\vec{F}_{\text{B on A}}$.

We deduced most of the third law in Section 7.2. There we found that the two members of an action/reaction pair are always opposite in direction (see Figures 7.6 and 7.10). According to the third law, this will always be true. But the most significant portion of the third law, which is by no means obvious, is that the two members of an action/reaction pair have *equal* magnitudes. That is, $F_{\text{A on B}} = F_{\text{B on A}}$. This is the quantitative relationship that will allow you to solve problems of interacting objects.

Newton's third law is frequently stated as "For every action there is an equal but opposite reaction." While this is indeed a catchy phrase, it lacks the preciseness of our preferred version. In particular, it fails to capture an essential feature of action/reaction pairs—that they each act on a *different* object.

> **NOTE** Newton's third law extends and completes our concept of *force*. We can now recognize force as an *interaction* between objects rather than as some "thing" with an independent existence of its own. The concept of an interaction will become increasingly important as we begin to study the laws of energy and momentum.

Reasoning with Newton's Third Law

Newton's third law is easy to state but harder to grasp. For example, consider what happens when you release a ball. Not surprisingly, it falls down. But if the ball and the earth exert equal and opposite forces on each other, as Newton's third law alleges, why doesn't the earth "fall up" to meet the ball?

The key to understanding this and many similar puzzles is that **the forces are equal but the accelerations are not.** Equal causes can produce very unequal effects. **FIGURE 7.11** shows equal-magnitude forces on the ball and the earth. The force on ball B is simply the gravitational force of Chapter 6:

$$\vec{F}_{\text{earth on ball}} = (\vec{F}_{\text{G}})_{\text{B}} = -m_{\text{B}}g\hat{j} \qquad (7.1)$$

where m_{B} is the mass of the ball. According to Newton's second law, this force gives the ball an acceleration

$$\vec{a}_{\text{B}} = \frac{(\vec{F}_{\text{G}})_{\text{B}}}{m_{\text{B}}} = -g\hat{j} \qquad (7.2)$$

This is just the familiar free-fall acceleration.

According to Newton's third law, the ball pulls up on the earth with force $\vec{F}_{\text{ball on earth}}$. Because $\vec{F}_{\text{earth on ball}}$ and $\vec{F}_{\text{ball on earth}}$ are an action/reaction pair, $\vec{F}_{\text{ball on earth}}$ must be equal in magnitude and opposite in direction to $\vec{F}_{\text{earth on ball}}$. That is,

$$\vec{F}_{\text{ball on earth}} = -\vec{F}_{\text{earth on ball}} = -(\vec{F}_{\text{G}})_{\text{B}} = +m_{\text{B}}g\hat{j} \qquad (7.3)$$

Using this result in Newton's second law, we find the upward acceleration of the earth as a whole is

$$\vec{a}_{\text{E}} = \frac{\vec{F}_{\text{ball on earth}}}{m_{\text{E}}} = \frac{m_{\text{B}}g\hat{j}}{m_{\text{E}}} = \left(\frac{m_{\text{B}}}{m_{\text{E}}}\right)g\hat{j} \qquad (7.4)$$

The upward acceleration of the earth is less than the downward acceleration of the ball by the factor $m_{\text{B}}/m_{\text{E}}$. If we assume a 1 kg ball, we can estimate the magnitude of $\vec{a}_{\text{E}}$:

$$\vec{a}_{\text{E}} = \frac{\vec{F}_{\text{ball on earth}}}{m_{\text{E}}} = \frac{m_{\text{B}}g\hat{j}}{m_{\text{E}}} = \left(\frac{m_{\text{B}}}{m_{\text{E}}}\right)g\hat{j}$$

With this incredibly small acceleration, it would take the earth 8×10^{15} years, approximately 500,000 times the age of the universe, to reach a speed of 1 mph! So we certainly would not expect to see or feel the earth "fall up" after we drop a ball.

NOTE Newton's third law equates the size of two forces, not two accelerations. The acceleration continues to depend on the mass, as Newton's second law states. **In an interaction between two objects of different mass, the lighter mass will do essentially all of the accelerating even though the forces exerted on the two objects are equal.**

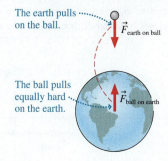

FIGURE 7.11 The action/reaction forces of a ball and the earth are equal in magnitude.

The earth pulls on the ball. $\vec{F}_{\text{earth on ball}}$

The ball pulls equally hard on the earth. $\vec{F}_{\text{ball on earth}}$

EXAMPLE 7.3 | **The forces on accelerating boxes**

The hand shown in **FIGURE 7.12** pushes boxes A and B to the right across a frictionless table. The mass of B is larger than the mass of A.

a. Draw free-body diagrams of A, B, and the hand H, showing only the *horizontal* forces. Connect action/reaction pairs with dashed lines.

b. Rank in order, from largest to smallest, the horizontal forces shown on your free-body diagrams.

FIGURE 7.12 Hand H pushes boxes A and B.

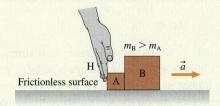

$m_{\text{B}} > m_{\text{A}}$

H

$\vec{a}$

Frictionless surface   A   B

VISUALIZE a. The hand H pushes on box A, and A pushes back on H. Thus $\vec{F}_{\text{H on A}}$ and $\vec{F}_{\text{A on H}}$ are an action/reaction pair. Similarly, A pushes on B and B pushes back on A. **The hand H does not touch box B, so there is no interaction between them.** There is no friction. **FIGURE 7.13** on the next page shows five horizontal forces and identifies two action/reaction pairs. Notice that each force is shown on the free-body diagram of the object that it acts *on*.

b. According to Newton's third law, $F_{\text{A on H}} = F_{\text{H on A}}$ and $F_{\text{A on B}} = F_{\text{B on A}}$. But the third law is not our only tool. The boxes are *accelerating* to the right, because there's no friction, so Newton's *second* law tells us that box A must have a net force to the right. Consequently, $F_{\text{H on A}} > F_{\text{B on A}}$. Similarly, $F_{\text{arm on H}} > F_{\text{A on H}}$ is needed to accelerate the hand. Thus

$$F_{\text{arm on H}} > F_{\text{A on H}} = F_{\text{H on A}} > F_{\text{B on A}} = F_{\text{A on B}}$$

Continued

ASSESS You might have expected $F_{\text{A on B}}$ to be larger than $F_{\text{H on A}}$ because $m_B > m_A$. It's true that the *net* force on B is larger than the *net* force on A, but we have to reason more closely to judge the individual forces. Notice how we used both the second and the third laws to answer this question.

FIGURE 7.13 The free-body diagrams, showing only the horizontal forces.

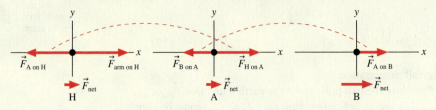

STOP TO THINK 7.2 A small car is pushing a larger truck that has a dead battery. The mass of the truck is larger than the mass of the car. Which of the following statements is true?

a. The car exerts a force on the truck, but the truck doesn't exert a force on the car.
b. The car exerts a larger force on the truck than the truck exerts on the car.
c. The car exerts the same amount of force on the truck as the truck exerts on the car.
d. The truck exerts a larger force on the car than the car exerts on the truck.
e. The truck exerts a force on the car, but the car doesn't exert a force on the truck.

Acceleration Constraints

Newton's third law is one quantitative relationship you can use to solve problems of interacting objects. In addition, we frequently have other information about the motion in a problem. For example, if two objects A and B move together, their accelerations are *constrained* to be equal: $\vec{a}_A = \vec{a}_B$. A well-defined relationship between the accelerations of two or more objects is called an **acceleration constraint.** It is an independent piece of information that can help solve a problem.

In practice, we'll express acceleration constraints in terms of the x- and y-components of $\vec{a}$. Consider the car being towed in FIGURE 7.14. This is one-dimensional motion, so we can write the acceleration constraint as

$$a_{Cx} = a_{Tx} = a_x$$

Because the accelerations of both objects are equal, we can drop the subscripts C and T and call both of them a_x.

Don't assume the accelerations of A and B will always have the same sign. Consider blocks A and B in FIGURE 7.15. The blocks are connected by a string, so they are constrained to move together and their accelerations have equal magnitudes. But A has a positive acceleration (to the right) in the x-direction while B has a negative acceleration (downward) in the y-direction. Thus the acceleration constraint is

$$a_{Ax} = -a_{By}$$

This relationship does *not* say that a_{Ax} is a negative number. It is simply a relational statement, saying that a_{Ax} is (-1) times whatever a_{By} happens to be. The acceleration a_{By} in Figure 7.15 is a negative number, so a_{Ax} is positive. In some problems, the signs of a_{Ax} and a_{By} may not be known until the problem is solved, but the *relationship* is known from the beginning.

FIGURE 7.14 The car and the truck have the same acceleration.

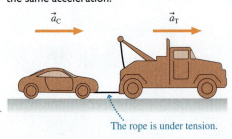

The rope is under tension.

FIGURE 7.15 The string constrains the two objects to accelerate together.

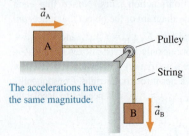

The accelerations have the same magnitude.

A Revised Strategy for Interacting-Objects Problems

Problems of interacting objects can be solved with a few modifications to the problem-solving strategy we developed in « Section 6.2.

PROBLEM-SOLVING STRATEGY 7.1

Interacting-objects problems

MODEL Identify which objects are part of the system and which are part of the environment. Make simplifying assumptions.

VISUALIZE Draw a pictorial representation.

- Show important points in the motion with a sketch. You may want to give each object a separate coordinate system. Define symbols, list acceleration constraints, and identify what the problem is trying to find.
- Draw an interaction diagram to identify the forces on each object and all action/reaction pairs.
- Draw a *separate* free-body diagram for each object showing only the forces acting *on* that object, not forces exerted by the object. Connect the force vectors of action/reaction pairs with dashed lines.

SOLVE Use Newton's second and third laws.

- Write the equations of Newton's second law for *each* object, using the force information from the free-body diagrams.
- Equate the magnitudes of action/reaction pairs.
- Include the acceleration constraints, the friction model, and other quantitative information relevant to the problem.
- Solve for the acceleration, then use kinematics to find velocities and positions.

ASSESS Check that your result has the correct units and significant figures, is reasonable, and answers the question.

You might be puzzled that the Solve step calls for the use of the third law to equate just the *magnitudes* of action/reaction forces. What about the "opposite in direction" part of the third law? You have already used it! Your free-body diagrams should show the two members of an action/reaction pair to be opposite in direction, and that information will have been utilized in writing the second-law equations. Because the directional information has already been used, all that is left is the magnitude information.

EXAMPLE 7.4 | **Keep the crate from sliding**

You and a friend have just loaded a 200 kg crate filled with priceless art objects into the back of a 2000 kg truck. As you press down on the accelerator, force $\vec{F}_{\text{surface on truck}}$ propels the truck forward. To keep things simple, call this just $\vec{F}_T$. What is the maximum magnitude $\vec{F}_T$ can have without the crate sliding? The static and kinetic coefficients of friction between the crate and the bed of the truck are 0.80 and 0.30. Rolling friction of the truck is negligible.

MODEL The crate and the truck are separate objects that form the system. We'll model them as particles. The earth and the road surface are part of the environment.

VISUALIZE The sketch in **FIGURE 7.16** on the next page establishes a coordinate system, lists the known information, and—new

to problems of interacting objects—identifies the acceleration constraint. As long as the crate doesn't slip, it must accelerate *with* the truck. Both accelerations are in the positive x-direction, so the acceleration constraint in this problem is $a_{Cx} = a_{Tx} = a_x$.

The interaction diagram of Figure 7.16 shows the crate interacting twice with the truck—a friction force parallel to the surface of the truck bed and a normal force perpendicular to this surface. The truck interacts similarly with the road surface, but notice that the crate does not interact with the ground; there's no contact between them. The two interactions within the system are each an action/reaction pair, so this is a total of four forces. You can also see four external forces crossing the system boundary, so the free-body diagrams should show a total of eight forces.

Continued

FIGURE 7.16 Pictorial representation of the crate and truck in Example 7.4.

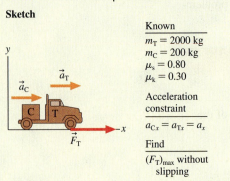

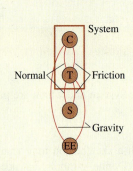

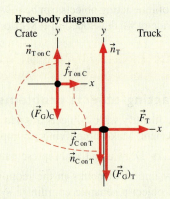

Sketch

Known
$m_T = 2000$ kg
$m_C = 200$ kg
$\mu_s = 0.80$
$\mu_k = 0.30$

Acceleration constraint

$a_{Cx} = a_{Tx} = a_x$

Find

$(F_T)_{max}$ without slipping

Interaction diagram

Free-body diagrams

Finally, the interaction information is transferred to the free-body diagrams, where we see friction between the crate and truck as an action/reaction pair and the normal forces (the truck pushes up on the crate, the crate pushes down on the truck) as another action/reaction pair. It's easy to overlook forces such as $\vec{f}_{C \text{ on } T}$, but you won't make this mistake if you first identify action/reaction pairs on an interaction diagram. Note that $\vec{f}_{C \text{ on } T}$ and $\vec{f}_{T \text{ on } C}$ are *static* friction forces because they are forces that prevent slipping; force $\vec{f}_{T \text{ on } C}$ must point forward to prevent the crate from sliding out the back of the truck.

SOLVE Now we're ready to write Newton's second law. For the crate:

$$\sum (F_{\text{on crate}})_x = f_{T \text{ on } C} = m_C a_{Cx} = m_C a_x$$
$$\sum (F_{\text{on crate}})_y = n_{T \text{ on } C} - (F_G)_C = n_{T \text{ on } C} - m_C g = 0$$

For the truck:

$$\sum (F_{\text{on truck}})_x = F_T - f_{C \text{ on } T} = m_T a_{Tx} = m_T a_x$$
$$\sum (F_{\text{on truck}})_y = n_T - (F_G)_T - n_{C \text{ on } T}$$
$$= n_T - m_T g - n_{C \text{ on } T} = 0$$

Be sure you agree with all the signs, which are based on the free-body diagrams. The net force in the y-direction is zero because there's no motion in the y-direction. It may seem like a lot of effort to write all the subscripts, but it is very important in problems with more than one object.

Notice that we've already used the acceleration constraint $a_{Cx} = a_{Tx} = a_x$. Another important piece of information is Newton's third law, which tells us that $f_{C \text{ on } T} = f_{T \text{ on } C}$ and $n_{C \text{ on } T} = n_{T \text{ on } C}$. Finally, we know that the maximum value of F_T will occur when the static friction on the crate reaches its maximum value:

$$f_{T \text{ on } C} = f_{s \text{ max}} = \mu_s n_{T \text{ on } C}$$

The friction depends on the normal force on the crate, not the normal force on the truck.

Now we can assemble all the pieces. From the y-equation of the crate, $n_{T \text{ on } C} = m_C g$. Thus

$$f_{T \text{ on } C} = \mu_s n_{T \text{ on } C} = \mu_s m_C g$$

Using this in the x-equation of the crate, we find that the acceleration is

$$a_x = \frac{f_{T \text{ on } C}}{m_C} = \mu_s g$$

This is the crate's maximum acceleration without slipping. Now use this acceleration *and* the fact that $f_{C \text{ on } T} = f_{T \text{ on } C} = \mu_s m_C g$ in the x-equation of the truck to find

$$F_T - f_{C \text{ on } T} = F_T - \mu_s m_C g = m_T a_x = m_T \mu_s g$$

Solving for F_T, we find the maximum propulsion without the crate sliding is

$$(F_T)_{max} = \mu_s (m_T + m_C)g$$
$$= (0.80)(2200 \text{ kg})(9.80 \text{ m/s}^2) = 17,000 \text{ N}$$

ASSESS This is a hard result to assess. Few of us have any intuition about the size of forces that propel cars and trucks. Even so, the fact that the forward force on the truck is a significant fraction (80%) of the combined weight of the truck and the crate seems plausible. We might have been suspicious if F_T had been only a tiny fraction of the weight or much greater than the weight.

As you can see, there are many equations and many pieces of information to keep track of when solving a problem of interacting objects. These problems are not inherently harder than the problems you learned to solve in Chapter 6, but they do require a high level of organization. Using the systematic approach of the problem-solving strategy will help you solve similar problems successfully.

STOP TO THINK 7.3 Boxes A and B are sliding to the right across a frictionless table. The hand H is slowing them down. The mass of A is larger than the mass of B. Rank in order, from largest to smallest, the *horizontal* forces on A, B, and H.

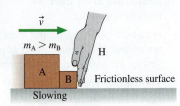

a. $F_{B \text{ on } H} = F_{H \text{ on } B} = F_{A \text{ on } B} = F_{B \text{ on } A}$
b. $F_{B \text{ on } H} = F_{H \text{ on } B} > F_{A \text{ on } B} = F_{B \text{ on } A}$
c. $F_{B \text{ on } H} = F_{H \text{ on } B} < F_{A \text{ on } B} = F_{B \text{ on } A}$
d. $F_{H \text{ on } B} = F_{H \text{ on } A} > F_{A \text{ on } B}$

7.4 Ropes and Pulleys

Many objects are connected by strings, ropes, cables, and so on. In single-particle dynamics, we defined *tension* as the force exerted on an object by a rope or string. Now we need to think more carefully about the string itself. Just what do we mean when we talk about the tension "in" a string?

Tension Revisited

FIGURE 7.17 shows a heavy safe hanging from a rope, placing the rope under tension. If you cut the rope, the safe and the lower portion of the rope will fall. Thus there must be a force *within* the rope by which the upper portion of the rope pulls upward on the lower portion to prevent it from falling.

Chapter 5 introduced an atomic-level model in which tension is due to the stretching of spring-like molecular bonds within the rope. Stretched springs exert pulling forces, and the combined pulling force of billions of stretched molecular springs in a string or rope is what we call *tension*.

An important aspect of tension is that it pulls equally *in both directions*. To gain a mental picture, imagine holding your arms outstretched and having two friends pull on them. You'll remain at rest—but "in tension"—as long as they pull with equal strength in opposite directions. But if one lets go, analogous to the breaking of molecular bonds if a rope breaks or is cut, you'll fly off in the other direction!

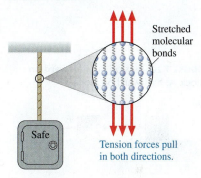

FIGURE 7.17 Tension forces within the rope are due to stretching the spring-like molecular bonds.

Stretched molecular bonds

Safe

Tension forces pull in both directions.

EXAMPLE 7.5 | **Pulling a rope**

FIGURE 7.18a shows a student pulling horizontally with a 100 N force on a rope that is attached to a wall. In **FIGURE 7.18b**, two students in a tug-of-war pull on opposite ends of a rope with 100 N each. Is the tension in the second rope larger than, smaller than, or the same as that in the first rope?

FIGURE 7.18 Which rope has a larger tension?

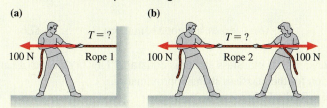

(a)

$T = ?$
100 N   Rope 1

(b)

$T = ?$
100 N   Rope 2   100 N

SOLVE Surely pulling on a rope from both ends causes more tension than pulling on one end. Right? Before jumping to conclusions, let's analyze the situation carefully.

FIGURE 7.19a shows the first student, the rope, and the wall as separate, interacting objects. Force $\vec{F}_{\text{S on R}}$ is the student pulling on the rope, so it has magnitude 100 N. Forces $\vec{F}_{\text{S on R}}$ and $\vec{F}_{\text{R on S}}$ are an action/reaction pair and must have equal magnitudes. Similarly for forces $\vec{F}_{\text{W on R}}$ and $\vec{F}_{\text{R on W}}$. Finally, because the rope is in static equilibrium, *force $\vec{F}_{\text{W on R}}$ has to balance force $\vec{F}_{\text{S on R}}$*. Thus

$$F_{\text{R on W}} = F_{\text{W on R}} = F_{\text{S on R}} = F_{\text{R on S}} = 100 \text{ N}$$

The first and third equalities are Newton's third law; the second equality is Newton's first law for the rope.

Forces $\vec{F}_{\text{R on S}}$ and $\vec{F}_{\text{R on W}}$ are the pulling forces exerted by the rope and are what we *mean* by "the tension in the rope." Thus the tension in the first rope is 100 N.

FIGURE 7.19b repeats the analysis for the rope pulled by two students. Each student pulls with 100 N, so $F_{\text{S1 on R}} = 100$ N and $F_{\text{S2 on R}} = 100$ N. Just as before, there are two action/reaction pairs and the rope is in static equilibrium. Thus

$$F_{\text{R on S2}} = F_{\text{S2 on R}} = F_{\text{S1 on R}} = F_{\text{R on S1}} = 100 \text{ N}$$

The tension in the rope—the pulling forces $\vec{F}_{\text{R on S1}}$ and $\vec{F}_{\text{R on S2}}$—is still 100 N!

You may have *assumed* that the student on the right in Figure 7.18b is doing something to the rope that the wall in Figure 7.18a does not do. But our analysis finds that the wall, just like the student, pulls to the right with 100 N. The rope doesn't care whether it's pulled by a wall or a hand. It experiences the same forces in both cases, so the rope's tension is the same in both.

ASSESS Ropes and strings exert forces at *both* ends. The force with which they pull—and thus the force pulling on them at each end—*is* the tension in the rope. Tension is not the sum of the pulling forces.

FIGURE 7.19 Analysis of tension forces.

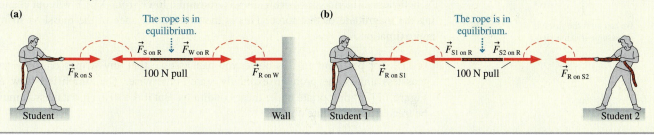

(a)

The rope is in equilibrium.

$\vec{F}_{\text{S on R}}$ $\vec{F}_{\text{W on R}}$

$\vec{F}_{\text{R on S}}$ 100 N pull $\vec{F}_{\text{R on W}}$

Student   Wall

(b)

The rope is in equilibrium.

$\vec{F}_{\text{S1 on R}}$ $\vec{F}_{\text{S2 on R}}$

$\vec{F}_{\text{R on S1}}$ 100 N pull $\vec{F}_{\text{R on S2}}$

Student 1   Student 2

STOP TO THINK 7.4 All three 50 kg blocks are at rest. Is the tension in rope 2 greater than, less than, or equal to the tension in rope 1?

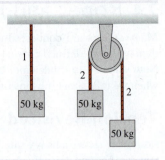

The Massless String Approximation

The tension is constant throughout a rope that is in equilibrium, but what happens if the rope is accelerating? For example, **FIGURE 7.20a** shows two connected blocks being pulled by force $\vec{F}$. Is the string's tension at the right end, where it pulls back on B, the same as the tension at the left end, where it pulls on A?

FIGURE 7.20b shows the horizontal forces acting on the blocks and the string. The only horizontal forces acting on the string are $\vec{T}_{\text{A on S}}$ and $\vec{T}_{\text{B on S}}$, so Newton's second law *for the string* is

$$(F_{\text{net}})_x = T_{\text{B on S}} - T_{\text{A on S}} = m_S a_x \qquad (7.5)$$

where m_S is the mass of the string. If the string is accelerating, then the tensions at the two ends can *not* be the same. The tension at the "front" of the string must be greater than the tension at the "back" in order to accelerate the string!

Often in physics and engineering problems the mass of the string or rope is much less than the masses of the objects that it connects. In such cases, we can adopt the **massless string approximation.** In the limit $m_S \rightarrow 0$, Equation 7.5 becomes

$$T_{\text{B on S}} = T_{\text{A on S}} \qquad \text{(massless string approximation)} \qquad (7.6)$$

In other words, **the tension in a massless string is constant.** This is nice, but it isn't the primary justification for the massless string approximation.

Look again at Figure 7.20b. If $T_{\text{B on S}} = T_{\text{A on S}}$, then

$$\vec{T}_{\text{S on A}} = -\vec{T}_{\text{S on B}} \qquad (7.7)$$

That is, the force on block A is equal and opposite to the force on block B. Forces $\vec{T}_{\text{S on A}}$ and $\vec{T}_{\text{S on B}}$ act *as if* they are an action/reaction pair of forces. Thus we can draw the simplified diagram of **FIGURE 7.21** in which the string is missing and blocks A and B interact directly with each other through forces that we can call $\vec{T}_{\text{A on B}}$ and $\vec{T}_{\text{B on A}}$.

In other words, **if objects A and B interact with each other through a massless string, we can omit the string and treat forces $\vec{F}_{\text{A on B}}$ and $\vec{F}_{\text{B on A}}$ as if they are an action/reaction pair.** This is not literally true because A and B are not in contact. Nonetheless, all a massless string does is transmit a force from A to B without changing the magnitude of that force. This is the real significance of the massless string approximation.

NOTE For problems in this book, you can assume that any strings or ropes are massless unless the problem explicitly states otherwise. The simplified view of Figure 7.21 is appropriate under these conditions. But if the string has a mass, it must be treated as a separate object.

FIGURE 7.20 Tension pulls forward on block A, backward on block B.

(a)

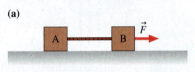

(b)

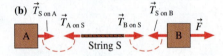

FIGURE 7.21 The massless string approximation allows objects A and B to act *as if* they are directly interacting.

This pair of forces acts *as if* it were an action/reaction pair.

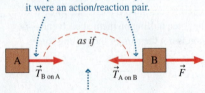

We can omit the string if we assume it is massless.

EXAMPLE 7.6 | Comparing two tensions

Blocks A and B in **FIGURE 7.22** are connected by massless string 2 and pulled across a frictionless table by massless string 1. B has a larger mass than A. Is the tension in string 2 larger than, smaller than, or equal to the tension in string 1?

FIGURE 7.22 Blocks A and B are pulled across a frictionless table by massless strings.

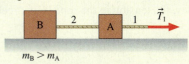

$$m_B > m_A$$

MODEL The massless string approximation allows us to treat A and B *as if* they interact directly with each other. The blocks are accelerating because there's a force to the right and no friction.

SOLVE B has a larger mass, so it may be tempting to conclude that string 2, which pulls B, has a greater tension than string 1, which pulls A. The flaw in this reasoning is that Newton's second law tells us only about the *net* force. The net force on B *is* larger than the net force on A, but the net force on A is *not* just the tension $\vec{T}_1$ in the forward direction. The tension in string 2 also pulls *backward* on A!

FIGURE 7.23 shows the horizontal forces in this frictionless situation. Because the string is massless, forces $\vec{T}_{A\,on\,B}$ and $\vec{T}_{B\,on\,A}$ act *as if* they are an action/reaction pair.

FIGURE 7.23 The horizontal forces on blocks A and B.

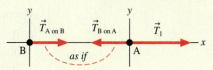

From Newton's third law,

$$T_{A\,on\,B} = T_{B\,on\,A} = T_2$$

where T_2 is the tension in string 2. From Newton's second law, the net force on A is

$$(F_{A\,net})_x = T_1 - T_{B\,on\,A} = T_1 - T_2 = m_A a_{Ax}$$

The net force on A is the *difference* in tensions. The blocks are accelerating to the right, making $a_{Ax} > 0$, so

$$T_1 > T_2$$

The tension in string 2 is *smaller* than the tension in string 1.

ASSESS This is not an intuitively obvious result. A careful study of the reasoning in this example is worthwhile. An alternative analysis would note that $\vec{T}_1$ accelerates *both* blocks, of combined mass $(m_A + m_B)$, whereas $\vec{T}_2$ accelerates only block B. Thus string 1 must have the larger tension.

Pulleys

Strings and ropes often pass over pulleys. The application might be as simple as lifting a heavy weight or as complex as the internal cable-and-pulley arrangement that precisely moves a robot arm.

FIGURE 7.24a shows a simple situation in which block B, as it falls, drags block A across a table. As the string moves, static friction between the string and pulley causes the pulley to turn. If we assume that

- The string *and* the pulley are both massless, and
- There is no friction where the pulley turns on its axle,

then no net force is needed to accelerate the string or turn the pulley. Thus **the tension in a massless string remains constant as it passes over a massless, frictionless pulley.**

Because of this, we can draw the simplified free-body diagram of **FIGURE 7.24b**, in which the string and pulley are omitted. Forces $\vec{T}_{A\,on\,B}$ and $\vec{T}_{B\,on\,A}$ act *as if* they are an action/reaction pair, even though they are not opposite in direction because the tension force gets "turned" by the pulley.

(a)   **(b)**

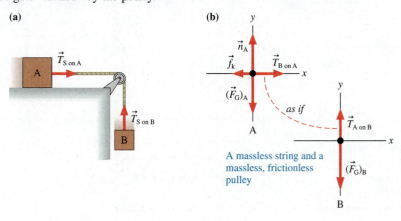

A massless string and a massless, frictionless pulley

◀ **FIGURE 7.24** Blocks A and B are connected by a string that passes over a pulley.

> ### TACTICS BOX 7.2
>
> #### Working with ropes and pulleys
>
> For massless ropes or strings and massless, frictionless pulleys:
>
> - If a force pulls on one end of a rope, the tension in the rope equals the magnitude of the pulling force.
> - If two objects are connected by a rope, the tension is the same at both ends.
> - If the rope passes over a pulley, the tension in the rope is unaffected.
>
> Exercises 17–22

STOP TO THINK 7.5 In Figure 7.24, on the previous page, is the tension in the string greater than, less than, or equal to the gravitational force acting on block B?

7.5 Examples of Interacting-Objects Problems

We will conclude this chapter with three extended examples. Although the mathematics will be more involved than in any of our work up to this point, we will continue to emphasize the *reasoning* one uses in approaching problems such as these. The solutions will be based on Problem-Solving Strategy 7.1. In fact, these problems are now reaching such a level of complexity that, for all practical purposes, it becomes impossible to work them unless you are following a well-planned strategy. Our earlier emphasis on forces and free-body diagrams will now really begin to pay off!

EXAMPLE 7.7 | **Placing a leg in traction**

Serious fractures of the leg often need a stretching force to keep contracting leg muscles from forcing the broken bones together too hard. This is done using *traction*, an arrangement of a rope, a weight, and pulleys as shown in **FIGURE 7.25**. The rope must make the same angle on both sides of the pulley so that the net force on the leg is horizontal, but the angle can be adjusted to control the amount of traction. The doctor has specified 50 N of traction for this patient with a 4.2 kg hanging mass. What is the proper angle?

FIGURE 7.25 A leg in traction.

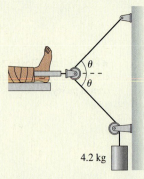

4.2 kg

MODEL Model the leg and the weight as particles. The other point where forces are applied is the pulley attached to the patient's foot, which we'll treat as a separate object. We'll assume massless ropes and a massless, frictionless pulley.

VISUALIZE **FIGURE 7.26** shows three free-body diagrams. Forces $\vec{T}_{P1}$ and $\vec{T}_{P2}$ are the tension forces of the rope as it pulls on the pulley. The pulley is in equilibrium, so these forces are balanced by $\vec{F}_{L\,on\,P}$, which forms an action/reaction pair with the 50 N traction force $\vec{F}_{P\,on\,L}$. Our model of the rope and pulley makes the tension force constant, $T_{P1} = T_{P2} = T_W$, so we'll call it simply T.

FIGURE 7.26 The free-body diagrams.

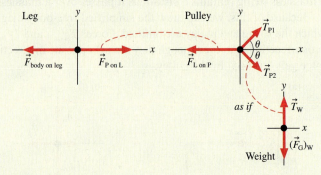

SOLVE The *x*-component equation of Newton's second law for the pulley is

$$\sum (F_{on\,P})_x = T_{P1}\cos\theta + T_{P2}\cos\theta - F_{L\,on\,P}$$
$$= 2T\cos\theta - F_{L\,on\,P} = 0$$

Thus the correct angle for the ropes is

$$\theta = \cos^{-1}\left(\frac{F_{\text{L on P}}}{2T}\right)$$

We know, from Newton's third law, that $F_{\text{L on P}} = F_{\text{P on L}} = 50$ N. We can determine the tension force by analyzing the weight. It also is in equilibrium, so the upward tension force exactly balances the downward gravitational force:

$$T = (F_G)_W = m_W g = (4.2\ \text{kg})(9.80\ \text{m/s}^2) = 41\ \text{N}$$

Thus the proper angle is

$$\theta = \cos^{-1}\left(\frac{50\ \text{N}}{2(41\ \text{N})}\right) = 52°$$

ASSESS The traction force would approach 82 N if angle θ approached zero because the two ropes would pull in parallel. Conversely, the traction would approach 0 N if θ approached 90°. The desired traction is roughly midway between these two extremes, so an angle near 45° seems reasonable.

EXAMPLE 7.8 | The show must go on!

A 200 kg set used in a play is stored in the loft above the stage. The rope holding the set passes up and over a pulley, then is tied backstage. The director tells a 100 kg stagehand to lower the set. When he unties the rope, the set falls and the unfortunate man is hoisted into the loft. What is the stagehand's acceleration?

MODEL The system is the stagehand M and the set S, which we will model as particles. Assume a massless rope and a massless, frictionless pulley.

VISUALIZE FIGURE 7.27 shows the pictorial representation. The man's acceleration a_{My} is positive, while the set's acceleration a_{Sy} is negative. These two accelerations have the same magnitude because the two objects are connected by a rope, but they have opposite signs. Thus the acceleration constraint is $a_{Sy} = -a_{My}$. Forces $\vec{T}_{\text{M on S}}$ and $\vec{T}_{\text{S on M}}$ are not literally an action/reaction pair, but they act *as if* they are because the rope is massless and the pulley is massless and frictionless. Notice that the pulley has "turned" the tension force so that $\vec{T}_{\text{M on S}}$ and $\vec{T}_{\text{S on M}}$ are *parallel* to each other rather than opposite, as members of a true action/reaction pair would have to be.

SOLVE Newton's second law for the man and the set is

$$\sum (F_{\text{on M}})_y = T_{\text{S on M}} - m_M g = m_M a_{My}$$

$$\sum (F_{\text{on S}})_y = T_{\text{M on S}} - m_S g = m_S a_{Sy} = -m_S a_{My}$$

Only the y-equations are needed. Notice that we used the acceleration constraint in the last step. Newton's third law is

$$T_{\text{M on S}} = T_{\text{S on M}} = T$$

where we can drop the subscripts and call the tension simply T. With this substitution, the two second-law equations can be written

$$T - m_M g = m_M a_{My}$$
$$T - m_S g = -m_S a_{My}$$

These are simultaneous equations in the two unknowns T and a_{My}. We can eliminate T by subtracting the second equation from the first to give

$$(m_S - m_M)g = (m_S + m_M)a_{My}$$

Finally, we can solve for the hapless stagehand's acceleration:

$$a_{My} = \frac{m_S - m_M}{m_S + m_M}g = \frac{100\ \text{kg}}{300\ \text{kg}}\,9.80\ \text{m/s}^2 = 3.27\ \text{m/s}^2$$

This is also the acceleration with which the set falls. If the rope's tension was needed, we could now find it from $T = m_M a_{My} + m_M g$.

ASSESS If the stagehand weren't holding on, the set would fall with free-fall acceleration g. The stagehand acts as a *counterweight* to reduce the acceleration.

FIGURE 7.27 Pictorial representation for Example 7.8.

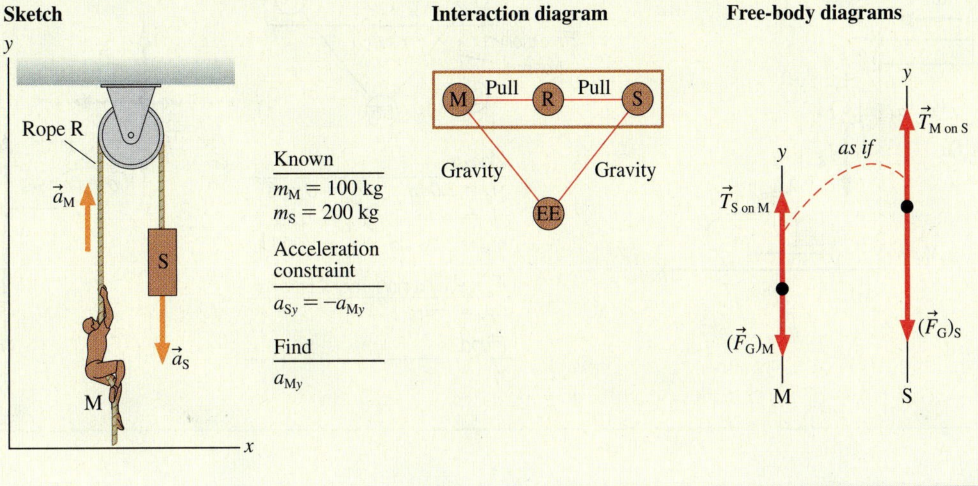

Sketch

Interaction diagram

Free-body diagrams

Rope R

$\vec{a}_M$

$\vec{a}_S$

S

M

M Pull R Pull S

Gravity   Gravity

EE

Known
$m_M = 100$ kg
$m_S = 200$ kg

Acceleration constraint
$a_{Sy} = -a_{My}$

Find
a_{My}

$\vec{T}_{\text{M on S}}$

as if

$\vec{T}_{\text{S on M}}$

$(\vec{F}_G)_M$

$(\vec{F}_G)_S$

M   S

CHALLENGE EXAMPLE 7.9 | A not-so-clever bank robbery

Bank robbers have pushed a 1000 kg safe to a second-story floor-to-ceiling window. They plan to break the window, then lower the safe 3.0 m to their truck. Not being too clever, they stack up 500 kg of furniture, tie a rope between the safe and the furniture, and place the rope over a pulley. Then they push the safe out the window. What is the safe's speed when it hits the truck? The coefficient of kinetic friction between the furniture and the floor is 0.50.

MODEL This is a continuation of the situation that we analyzed in Figures 7.15 and 7.24, which are worth reviewing. The system is the safe S and the furniture F, which we will model as particles. We will assume a massless rope and a massless, frictionless pulley.

VISUALIZE The safe and the furniture are tied together, so their accelerations have the same magnitude. The safe has a y-component of acceleration a_{Sy} that is negative because the safe accelerates in the negative y-direction. The furniture has an x-component a_{Fx} that is positive. Thus the acceleration constraint is

$$a_{Fx} = -a_{Sy}$$

The free-body diagrams shown in **FIGURE 7.28** are modeled after Figure 7.24 but now include a kinetic friction force on the furniture. Forces $\vec{T}_{\text{F on S}}$ and $\vec{T}_{\text{S on F}}$ act *as if* they are an action/reaction pair, so they have been connected with a dashed line.

SOLVE We can write Newton's second law directly from the free-body diagrams. For the furniture,

$$\sum (F_{\text{on F}})_x = T_{\text{S on F}} - f_k = T - f_k = m_F a_{Fx} = -m_F a_{Sy}$$

$$\sum (F_{\text{on F}})_y = n - m_F g = 0$$

And for the safe,

$$\sum (F_{\text{on S}})_y = T - m_S g = m_S a_{Sy}$$

Notice how we used the acceleration constraint in the first equation. We also went ahead and made use of Newton's third law:

$T_{\text{F on S}} = T_{\text{S on F}} = T$. We have one additional piece of information, the model of kinetic friction:

$$f_k = \mu_k n = \mu_k m_F g$$

where we used the y-equation of the furniture to deduce that $n = m_F g$. Substitute this result for f_k into the x-equation of the furniture, then rewrite the furniture's x-equation and the safe's y-equation:

$$T - \mu_k m_F g = -m_F a_{Sy}$$
$$T - m_S g = m_S a_{Sy}$$

We have succeeded in reducing our knowledge to two simultaneous equations in the two unknowns a_{Sy} and T. Subtract the second equation from the first to eliminate T:

$$(m_S - \mu_k m_F) g = -(m_S + m_F) a_{Sy}$$

Finally, solve for the safe's acceleration:

$$a_{Sy} = -\left(\frac{m_S - \mu_k m_F}{m_S + m_F}\right) g$$

$$= -\left(\frac{1000 \text{ kg} - (0.50)(500 \text{ kg})}{1000 \text{ kg} + 500 \text{ kg}}\right) 9.80 \text{ m/s}^2 = -4.9 \text{ m/s}^2$$

Now we need to calculate the kinematics of the falling safe. Because the time of the fall is not known or needed, we can use

$$v_{1y}^2 = v_{0y}^2 + 2a_{Sy} \Delta y = 0 + 2a_{Sy}(y_1 - y_0) = -2a_{Sy} y_0$$

$$v_1 = \sqrt{-2a_{Sy} y_0} = \sqrt{-2(-4.9 \text{ m/s}^2)(3.0 \text{ m})} = 5.4 \text{ m/s}$$

ASSESS The value of v_{1y} is negative, but we only needed to find the speed so we took the absolute value. This is about 12 mph, so it seems unlikely that the truck will survive the impact of the 1000 kg safe!

FIGURE 7.28 Pictorial representation for Challenge Example 7.9.

Sketch

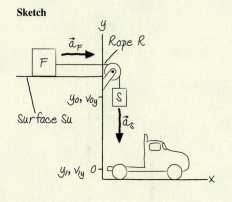

Interaction diagram

Free-body diagrams

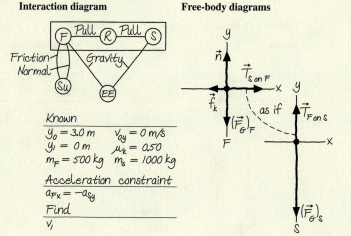

SUMMARY

The goal of Chapter 7 has been to use Newton's third law to understand how objects interact.

GENERAL PRINCIPLES

Newton's Third Law

Every force occurs as one member of an **action/reaction pair** of forces. The two members of an action/reaction pair:

- Act on two *different* objects.
- Are equal in magnitude but opposite in direction:

$$\vec{F}_{\text{A on B}} = -\vec{F}_{\text{B on A}}$$

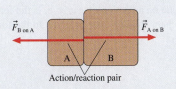

Action/reaction pair

Solving Interacting-Objects Problems

MODEL Identify which objects form the system.

VISUALIZE Draw a pictorial representation.

- Define symbols and coordinates.
- Identify acceleration constraints.
- Draw an interaction diagram.
- Draw a separate free-body diagram for each object.
- Connect action/reaction pairs with dashed lines.

SOLVE Write Newton's second law for each object.

- Use the free-body diagrams.
- Equate the magnitudes of action/reaction pairs.
- Include acceleration constraints and friction.

ASSESS Is the result reasonable?

IMPORTANT CONCEPTS

Objects, systems, and the environment

Objects whose motion is of interest are the system.

Objects whose motion is not of interest form the environment.

The objects of interest interact with the environment, but those interactions can be considered external forces.

Interaction diagram

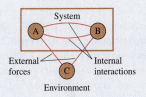

APPLICATIONS

Acceleration constraints

Objects that are constrained to move together must have accelerations of equal magnitude: $a_A = a_B$. This must be expressed in terms of components, such as $a_{Ax} = -a_{By}$.

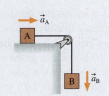

Strings and pulleys

The tension in a string or rope pulls in both directions. The tension is constant in a string if the string is:

- Massless, or
- In equilibrium

Objects connected by massless strings passing over massless, frictionless pulleys act *as if* they interact via an action/reaction pair of forces.

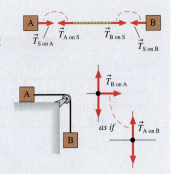

TERMS AND NOTATION

CONCEPTUAL QUESTIONS

1. You find yourself in the middle of a frozen lake with a surface so slippery ($\mu_s = \mu_k = 0$) you cannot walk. However, you happen to have several rocks in your pocket. The ice is extremely hard. It cannot be chipped, and the rocks slip on it just as much as your feet do. Can you think of a way to get to shore? Use pictures, forces, and Newton's laws to explain your reasoning.

2. How does a sprinter sprint? What is the forward force on a sprinter as she accelerates? Where does that force come from? Your explanation should include an interaction diagram and a free-body diagram.

3. How does a rocket take off? What is the upward force on it? Your explanation should include an interaction diagram and free-body diagrams of the rocket and of the parcel of gas being exhausted.

4. How do basketball players jump straight up into the air? Your explanation should include an interaction diagram and a free-body diagram.

5. A mosquito collides head-on with a car traveling 60 mph. Is the force of the mosquito on the car larger than, smaller than, or equal to the force of the car on the mosquito? Explain.

6. A mosquito collides head-on with a car traveling 60 mph. Is the magnitude of the mosquito's acceleration larger than, smaller than, or equal to the magnitude of the car's acceleration? Explain.

7. A small car is pushing a large truck. They are speeding up. Is the force of the truck on the car larger than, smaller than, or equal to the force of the car on the truck?

8. A very smart 3-year-old child is given a wagon for her birthday. She refuses to use it. "After all," she says, "Newton's third law says that no matter how hard I pull, the wagon will exert an equal but opposite force on me. So I will never be able to get it to move forward." What would you say to her in reply?

9. Teams red and blue are having a tug-of-war. According to Newton's third law, the force with which the red team pulls on the blue team exactly equals the force with which the blue team pulls on the red team. How can one team ever win? Explain.

10. Will hanging a magnet in front of the iron cart in **FIGURE Q7.10** make it go? Explain.

FIGURE Q7.10

11. **FIGURE Q7.11** shows two masses at rest. The string is massless and the pulley is frictionless. The spring scale reads in kg. What is the reading of the scale?

FIGURE Q7.11

12. **FIGURE Q7.12** shows two masses at rest. The string is massless and the pullies are frictionless. The spring scale reads in kg. What is the reading of the scale?

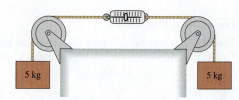

FIGURE Q7.12

13. The hand in **FIGURE Q7.13** is pushing on the back of block A. Blocks A and B, with $m_B > m_A$, are connected by a massless string and slide on a frictionless surface. Is the force of the string on B larger than, smaller than, or equal to the force of the hand on A? Explain.

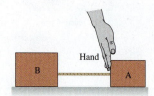

FIGURE Q7.13

14. Blocks A and B in **FIGURE Q7.14** are connected by a massless string over a massless, frictionless pulley. The blocks have just been released from rest. Will the pulley rotate clockwise, counterclockwise, or not at all? Explain.

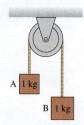

FIGURE Q7.14

15. In case a in **FIGURE Q7.15**, block A is accelerated across a frictionless table by a hanging 10 N weight (1.02 kg). In case b, block A is accelerated across a frictionless table by a steady 10 N tension in the string. The string is massless, and the pulley is massless and frictionless. Is A's acceleration in case b greater than, less than, or equal to its acceleration in case a? Explain.

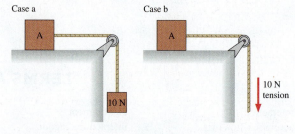

FIGURE Q7.15

EXERCISES AND PROBLEMS

Exercises

Section 7.2 Analyzing Interacting Objects

For Exercises 1 through 5:
  a. Draw an interaction diagram.
  b. Identify the "system" on your interaction diagram.
  c. Draw a free-body diagram for each object in the system. Use dashed lines to connect members of an action/reaction pair.

1. | A soccer ball and a bowling ball have a head-on collision at this instant. Rolling friction is negligible.

2. | A weightlifter stands up at constant speed from a squatting position while holding a heavy barbell across his shoulders.

3. | A steel cable with mass is lifting a girder. The girder is speeding up.

4. || Block A in **FIGURE EX7.4** is heavier than block B and is sliding down the incline. All surfaces have friction. The rope is massless, and the massless pulley turns on frictionless bearings. The rope and the pulley are among the interacting objects, but you'll have to decide if they're part of the system.

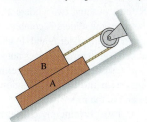

FIGURE EX7.4

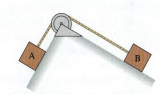

FIGURE EX7.5

5. || Block A in **FIGURE EX7.5** is sliding down the incline. The rope is massless, and the massless pulley turns on frictionless bearings, but the surface is not frictionless. The rope and the pulley are among the interacting objects, but you'll have to decide if they're part of the system.

Section 7.3 Newton's Third Law

6. | a. How much force does an 80 kg astronaut exert on his chair while sitting at rest on the launch pad?
    b. How much force does the astronaut exert on his chair while accelerating straight up at 10 m/s²?

7. | Block B in **FIGURE EX7.7** rests on a surface for which the static and kinetic coefficients of friction are 0.60 and 0.40, respectively. The ropes are massless. What is the maximum mass of block A for which the system remains in equilibrium?

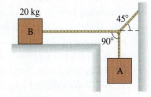

FIGURE EX7.7

8. || A 1000 kg car pushes a 2000 kg truck that has a dead battery. When the driver steps on the accelerator, the drive wheels of the car push against the ground with a force of 4500 N. Rolling friction can be neglected.
    a. What is the magnitude of the force of the car on the truck?
    b. What is the magnitude of the force of the truck on the car?

9. || Blocks with masses of 1 kg, 2 kg, and 3 kg are lined up in a row on a frictionless table. All three are pushed forward by a 12 N force applied to the 1 kg block.
    a. How much force does the 2 kg block exert on the 3 kg block?
    b. How much force does the 2 kg block exert on the 1 kg block?

10. | A 3000 kg meteorite falls toward the earth. What is the magnitude of the *earth's* acceleration just before impact? The earth's mass is 5.98×10^{24} kg.

11. | The foot of a 55 kg sprinter is on the ground for 0.25 s while her body accelerates from rest to 2.0 m/s.
    a. Is the friction between her foot and the ground static friction or kinetic friction?
    b. What is the magnitude of the friction force?

12. || A steel cable lying flat on the floor drags a 20 kg block across a horizontal, frictionless floor. A 100 N force applied to the cable causes the block to reach a speed of 4.0 m/s in a distance of 2.0 m. What is the mass of the cable?

13. || An 80 kg spacewalking astronaut pushes off a 640 kg satellite, exerting a 100 N force for the 0.50 s it takes him to straighten his arms. How far apart are the astronaut and the satellite after 1.0 min?

14. || The sled dog in **FIGURE EX7.14** drags sleds A and B across the snow. The coefficient of friction between the sleds and the snow is 0.10. If the tension in rope 1 is 150 N, what is the tension in rope 2?

FIGURE EX7.14

15. || Two-thirds of the weight of a 1500 kg car rests on the drive wheels. What is the maximum acceleration of this car on a concrete surface?

Section 7.4 Ropes and Pulleys

16. || **FIGURE EX7.16** shows two 1.0 kg blocks connected by a rope. A second rope hangs beneath the lower block. Both ropes have a mass of 250 g. The entire assembly is accelerated upward at 3.0 m/s² by force $\vec{F}$.
    a. What is F?
    b. What is the tension at the top end of rope 1?
    c. What is the tension at the bottom end of rope 1?
    d. What is the tension at the top end of rope 2?

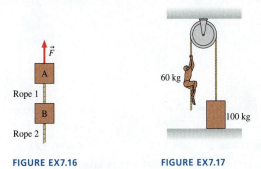

FIGURE EX7.16   **FIGURE EX7.17**

17. || What is the tension in the rope of **FIGURE EX7.17**?

18. ‖ A 2.0-m-long, 500 g rope pulls a 10 kg block of ice across a horizontal, frictionless surface. The block accelerates at $2.0 \, \text{m/s}^2$. How much force pulls forward on (a) the ice, (b) the rope? Assume that the rope is perfectly horizontal.

19. | A woman living in a third-story apartment is moving out. Rather than carrying everything down the stairs, she decides to pack her belongings into crates, attach a frictionless pulley to her balcony railing, and lower the crates by rope. How hard must she pull on the horizontal end of the rope to lower a 25 kg crate at steady speed?

20. ‖ Two blocks are attached to opposite ends of a massless rope that goes over a massless, frictionless, stationary pulley. One of the blocks, with a mass of 6.0 kg, accelerates downward at $\frac{3}{4}g$. What is the mass of the other block?

21. ‖ The cable cars in San Francisco are pulled along their tracks by an underground steel cable that moves along at 9.5 mph. The cable is driven by large motors at a central power station and extends, via an intricate pulley arrangement, for several miles beneath the city streets. The length of a cable stretches by up to 100 ft during its lifetime. To keep the tension constant, the cable passes around a 1.5-m-diameter "tensioning pulley" that rolls back and forth on rails, as shown in **FIGURE EX7.21**. A 2000 kg block is attached to the tensioning pulley's cart, via a rope and pulley, and is suspended in a deep hole. What is the tension in the cable car's cable?

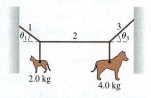

FIGURE EX7.21

22. ‖ A 2.0 kg rope hangs from the ceiling. What is the tension at the midpoint of the rope?

23. ‖ A mobile at the art museum has a 2.0 kg steel cat and a 4.0 kg steel dog suspended from a lightweight cable, as shown in **FIGURE EX7.23**. It is found that $\theta_1 = 20°$ when the center rope is adjusted to be perfectly horizontal. What are the tension and the angle of rope 3?

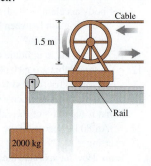

FIGURE EX7.23

24. ‖ The 1.0 kg block in **FIGURE EX7.24** is tied to the wall with a rope. It sits on top of the 2.0 kg block. The lower block is pulled to the right with a tension force of 20 N. The coefficient of kinetic friction at both the lower and upper surfaces of the 2.0 kg block is $\mu_k = 0.40$.
a. What is the tension in the rope attached to the wall?
b. What is the acceleration of the 2.0 kg block?

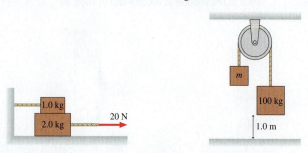

FIGURE EX7.24   **FIGURE EX7.25**

25. ‖ The 100 kg block in **FIGURE EX7.25** takes 6.0 s to reach the floor after being released from rest. What is the mass of the block on the left? The pulley is massless and frictionless.

Problems

26. ‖ **FIGURE P7.26** shows two strong magnets on opposite sides of a small table. The long-range attractive force between the magnets keeps the lower magnet in place.
a. Draw an interaction diagram and draw free-body diagrams for both magnets and the table. Use dashed lines to connect the members of an action/reaction pair.
b. The lower magnet is being pulled upward against the bottom of the table. Suppose that each magnet's weight is 2.0 N and that the magnetic force of the lower magnet on the upper magnet is 6.0 N. How hard does the lower magnet push against the table?

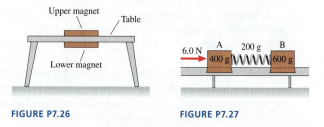

FIGURE P7.26   **FIGURE P7.27**

27. ‖ **FIGURE P7.27** shows a 6.0 N force pushing two gliders along an air track. The 200 g spring between the gliders is compressed. How much force does the spring exert on (a) glider A and (b) glider B? The spring is firmly attached to the gliders, and it does not sag.

28. ‖ A rope of length L and mass m is suspended from the ceiling. Find an expression for the tension in the rope at position y, measured upward from the free end of the rope.

29. ‖ While driving to work last year, I was holding my coffee mug in my left hand while changing the CD with my right hand. Then the cell phone rang, so I placed the mug on the flat part of my dashboard. Then, believe it or not, a deer ran out of the woods and on to the road right in front of me. Fortunately, my reaction time was zero, and I was able to stop from a speed of 20 m/s in a mere 50 m, just barely avoiding the deer. Later tests revealed that the static and kinetic coefficients of friction of the coffee mug on the dash are 0.50 and 0.30, respectively; the coffee and mug had a mass of 0.50 kg; and the mass of the deer was 120 kg. Did my coffee mug slide?

30. ‖ A Federation starship $(2.0 \times 10^6 \, \text{kg})$ uses its tractor beam to pull a shuttlecraft $(2.0 \times 10^4 \, \text{kg})$ aboard from a distance of 10 km away. The tractor beam exerts a constant force of $4.0 \times 10^4 \, \text{N}$ on the shuttlecraft. Both spacecraft are initially at rest. How far does the starship move as it pulls the shuttlecraft aboard?

31. ‖ Your forehead can withstand a force of about 6.0 kN before
BIO fracturing, while your cheekbone can withstand only about 1.3 kN. Suppose a 140 g baseball traveling at 30 m/s strikes your head and stops in 1.5 ms.
a. What is the magnitude of the force that stops the baseball?
b. What force does the baseball exert on your head? Explain.
c. Are you in danger of a fracture if the ball hits you in the forehead? On the cheek?

32. ‖ Bob, who has a mass of 75 kg, can throw a 500 g rock with a speed of 30 m/s. The distance through which his hand moves as he accelerates the rock from rest until he releases it is 1.0 m.
a. What constant force must Bob exert on the rock to throw it with this speed?
b. If Bob is standing on frictionless ice, what is his recoil speed after releasing the rock?

33. ||| Two packages at UPS start sliding down the 20° ramp shown in **FIGURE P7.33**. Package A has a mass of 5.0 kg and a coefficient of friction of 0.20. Package B has a mass of 10 kg and a coefficient of friction of 0.15. How long does it take package A to reach the bottom?

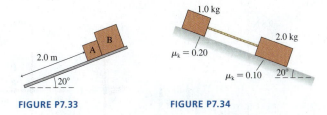

FIGURE P7.33   FIGURE P7.34

34. ||| The two blocks in **FIGURE P7.34** are sliding down the incline. What is the tension in the massless string?

35. || The coefficient of static friction is 0.60 between the two blocks in **FIGURE P7.35**. The coefficient of kinetic friction between the lower block and the floor is 0.20. Force $\vec{F}$ causes both blocks to cross a distance of 5.0 m, starting from rest. What is the least amount of time in which this motion can be completed without the top block sliding on the lower block?

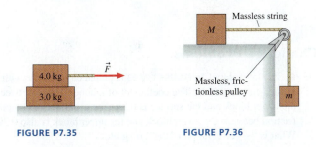

FIGURE P7.35   FIGURE P7.36

36. ||| The block of mass M in **FIGURE P7.36** slides on a frictionless surface. Find an expression for the tension in the string.

37. || The 10.2 kg block in **FIGURE P7.37** is held in place by a force applied to a rope passing over two massless, frictionless pulleys. Find the tensions T_1 to T_5 and the magnitude of force $\vec{F}$.

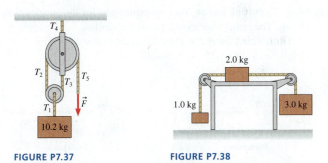

FIGURE P7.37   FIGURE P7.38

38. || The coefficient of kinetic friction between the 2.0 kg block in **FIGURE P7.38** and the table is 0.30. What is the acceleration of the 2.0 kg block?

39. ||| **FIGURE P7.39** shows a block of mass m resting on a 20° slope. The block has coefficients of friction $\mu_s = 0.80$ and $\mu_k = 0.50$ with the surface. It is connected via a massless string over a massless, frictionless pulley to a hanging block of mass 2.0 kg.
    a. What is the minimum mass m that will stick and not slip?
    b. If this minimum mass is nudged ever so slightly, it will start being pulled up the incline. What acceleration will it have?

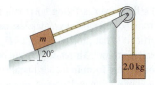

FIGURE P7.39

40. || A 4.0 kg box is on a frictionless 35° slope and is connected via a massless string over a massless, frictionless pulley to a hanging 2.0 kg weight. The picture for this situation is similar to **FIGURE P7.39**.
    a. What is the tension in the string if the 4.0 kg box is *held* in place, so that it cannot move?
    b. If the box is then released, which way will it move on the slope?
    c. What is the tension in the string once the box begins to move?

41. || The 1.0 kg physics book in **FIGURE P7.41** is connected by a string to a 500 g coffee cup. The book is given a push up the slope and released with a speed of 3.0 m/s. The coefficients of friction are $\mu_s = 0.50$ and $\mu_k = 0.20$.
    a. How far does the book slide?
    b. At the highest point, does the book stick to the slope, or does it slide back down?

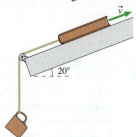

FIGURE P7.41

42. || The 2000 kg cable car shown in **FIGURE P7.42** descends a 200-m-high hill. In addition to its brakes, the cable car controls its speed by pulling an 1800 kg counterweight up the other side of the hill. The rolling friction of both the cable car and the counterweight are negligible.
    a. How much braking force does the cable car need to descend at constant speed?
    b. One day the brakes fail just as the cable car leaves the top on its downward journey. What is the runaway car's speed at the bottom of the hill?

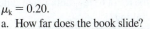

FIGURE P7.42

43. || The century-old *ascensores* in Valparaiso, Chile, are picturesque cable cars built on stilts to keep the passenger compartments level as they go up and down the steep hillsides. As **FIGURE P7.43** shows, one car ascends as the other descends. The cars use a two-cable arrangement to compensate for friction; one cable passing around a large pulley connects the cars, the second is pulled by a small motor. Suppose the mass of both cars (with passengers) is 1500 kg, the coefficient of rolling friction is 0.020, and the cars move at constant speed. What is the tension in (a) the connecting cable and (b) the cable to the motor?

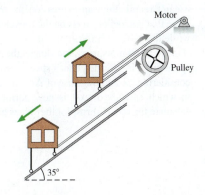

FIGURE P7.43

44. ‖ A 3200 kg helicopter is flying horizontally. A 250 kg crate is suspended from the helicopter by a massless cable that is constantly 20° from vertical. What propulsion force $\vec{F}_{\text{prop}}$ is being provided by the helicopter's rotor? Air resistance can be ignored. Give your answer in component form in a coordinate system where $\hat{\imath}$ points in the direction of motion and $\hat{\jmath}$ points upward.

45. ‖ A house painter uses the chair-and-pulley arrangement of **FIGURE P7.45** to lift himself up the side of a house. The painter's mass is 70 kg and the chair's mass is 10 kg. With what force must he pull down on the rope in order to accelerate upward at 0.20 m/s²?

FIGURE P7.45

46. ‖ A long, 1.0 kg rope hangs from a support that breaks, causing the rope to fall, if the pull exceeds 40 N. A student team has built a 2.0 kg robot "mouse" that runs up and down the rope. What maximum acceleration can the robot have—both magnitude and direction—without the rope falling?

47. ‖ A 50-cm-diameter, 400 g beach ball is dropped with a 4.0 mg ant riding on the top. The ball experiences air resistance, but the ant does not. What is the magnitude of the normal force exerted on the ant when the ball's speed is 2.0 m/s?

48. ‖ A 70 kg tightrope walker stands at the center of a rope. The rope supports are 10 m apart and the rope sags 10° at each end. The tightrope walker crouches down, then leaps straight up with an acceleration of 8.0 m/s² to catch a passing trapeze. What is the tension in the rope as he jumps?

49. ‖ Find an expression for the magnitude of the horizontal force F in **FIGURE P7.49** for which m_1 does not slip either up or down along the wedge. All surfaces are frictionless.

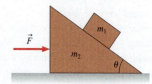

FIGURE P7.49

50. ‖ A rocket burns fuel at a rate of 5.0 kg/s, expelling the exhaust gases at a speed of 4.0 km/s *relative to the rocket*. We would like to find the thrust of the rocket engine.
   a. Model the fuel burning as a steady ejection of small pellets, each with the small mass Δm. Suppose it takes a short time Δt to accelerate a pellet (at constant acceleration) to the exhaust speed v_{ex}. Further, suppose the rocket is clamped down so that it can't recoil. Find an expression for the magnitude of the force that one pellet exerts on the rocket during the short time while the pellet is being expelled.
   b. If the rocket is moving, v_{ex} is no longer the pellet's speed through space but it is still the pellet's speed *relative to the rocket*. By considering the limiting case of Δm and Δt approaching zero, in which case the rocket is now burning fuel continuously, calculate the rocket thrust for the values given above.

Problems 51 and 52 show the free-body diagrams of two interacting systems. For each of these, you are to
   a. Write a realistic problem for which these are the correct free-body diagrams. Be sure that the answer your problem requests is consistent with the diagrams shown.
   b. Finish the solution of the problem.

51.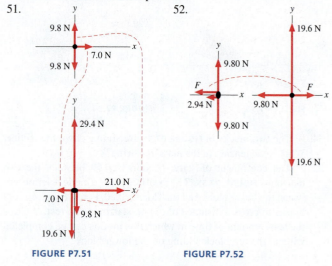

FIGURE P7.51   **FIGURE P7.52**

Challenge Problems

53. ‖‖ The lower block in **FIGURE CP7.53** is pulled on by a rope with a tension force of 20 N. The coefficient of kinetic friction between the lower block and the surface is 0.30. The coefficient of kinetic friction between the lower block and the upper block is also 0.30. What is the acceleration of the 2.0 kg block?

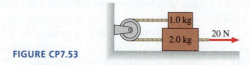

FIGURE CP7.53

54. ‖‖ In **FIGURE CP7.54**, find an expression for the acceleration of m_1. The pulleys are massless and frictionless.
   Hint: Think carefully about the acceleration constraint.

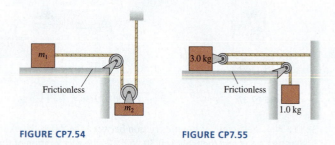

FIGURE CP7.54   **FIGURE CP7.55**

55. ‖‖ What is the acceleration of the 3.0 kg block in **FIGURE CP7.55** across the frictionless table?
   Hint: Think carefully about the acceleration constraint.

56. ||| A 40-cm-diameter, 50-cm-tall, 15 kg hollow cylinder is placed on top of a 40-cm-diameter, 30-cm-tall, 100 kg cylinder of solid aluminum, then the two are sent sliding across frictionless ice. The static and kinetic coefficients of friction between the cylinders are 0.45 and 0.25, respectively. Air resistance cannot be neglected. What is the maximum speed the cylinders can have without the top cylinder sliding off?

57. ||| FIGURE CP7.57 shows a 200 g hamster sitting on an 800 g wedge-shaped block. The block, in turn, rests on a spring scale. An extra-fine lubricating oil having $\mu_s = \mu_k = 0$ is sprayed on the top surface of the block, causing the hamster to slide down. Friction between the block and the scale is large enough that the block does *not* slip on the scale. What does the scale read, in grams, as the hamster slides down?

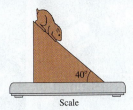

FIGURE CP7.57

58. ||| FIGURE CP7.58 shows three hanging masses connected by massless strings over two massless, frictionless pulleys.

a. Find the acceleration constraint for this system. It is a single equation relating a_{1y}, a_{2y}, and a_{3y}.
   Hint: y_A isn't constant.

b. Find an expression for the tension in string A.
   Hint: You should be able to write four second-law equations. These, plus the acceleration constraint, are five equations in five unknowns.

c. Suppose: $m_1 = 2.5$ kg, $m_2 = 1.5$ kg, and $m_3 = 4.0$ kg. Find the acceleration of each.

d. The 4.0 kg mass would appear to be in equilibrium. Explain why it accelerates.

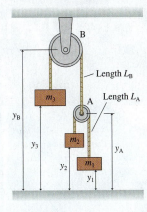

FIGURE CP7.58

8 Dynamics II: Motion in a Plane

Why doesn't the roller coaster fall off the track at the top of the loop?

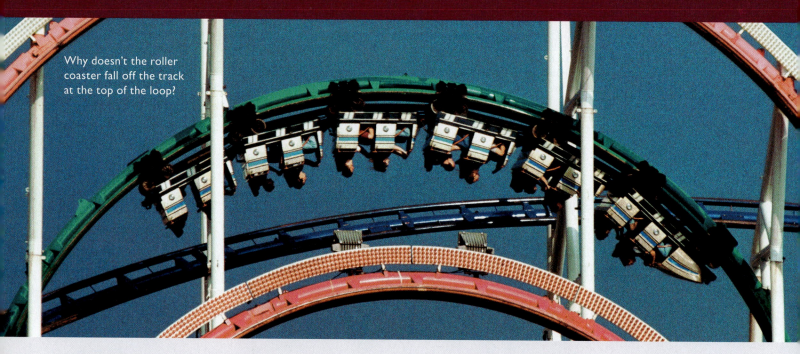

IN THIS CHAPTER, you will learn to solve problems about motion in two dimensions.

Are Newton's laws different in two dimensions?

No. Newton's laws are vector equations, and they work equally well in two and three dimensions. For motion in a plane, we'll focus on how a force *tangent* to a particle's trajectory changes its speed, while a force *perpendicular* to the trajectory changes the particle's direction.

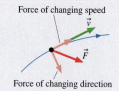

Force of changing speed $\vec{v}$

$\vec{F}$

Force of changing direction

« LOOKING BACK Chapter 4 Kinematics of projectile and circular motion

How do we analyze projectile-like motion?

For linear motion, one component of the acceleration was always zero. Motion in a plane generally has acceleration along two axes. If the accelerations are independent, we can use *x*- and *y*-coordinates and we will find motions analogous to the projectile motion we studied in Chapter 4.

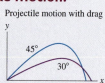

Projectile motion with drag
y
45°
30°
x

How do we analyze circular motion?

Circular motion must have a force component toward the center of the circle to create the centripetal acceleration. In this case the acceleration components are radial and, perhaps, tangential. We'll use a different coordinate system, *rtz* coordinates, to study the dynamics of circular motion

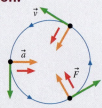

$\vec{v}$

$\vec{a}$

$\vec{F}$

Does this analysis apply to orbits?

Yes, it does. The circular orbit of a satellite or planet is motion in which the force of gravity is creating the inward centripetal acceleration. You'll see that an orbiting projectile is in free fall.

$\vec{F}_G$
Planet
$\vec{F}_G$ $\vec{F}_G$

« LOOKING BACK Section 6.3 Gravity and weight

Why doesn't the water fall out of the bucket?

How can you swing a bucket of water over your head without the water falling out? Why doesn't a car going around a loop-the-loop fall off at the top? Circular motion is not always intuitive, but you'll strengthen your ability to use Newtonian reasoning by thinking about some of these problems.

Why is planar motion important?

By starting with linear motion, we were able to develop the ideas and tools of Newtonian mechanics with minimal distractions. But planes and rockets move in a plane. Satellites and electrons orbit in a plane. The points on a rotating hard drive move in a plane. In fact, much of this chapter is a prelude to Chapter 12, where we will study rotational motion. This chapter gives you the tools you need to analyze more complex—and more realistic—forms of motion.

8.1 Dynamics in Two Dimensions

Newton's second law, $\vec{a} = \vec{F}_{net}/m$, determines an object's acceleration; it makes no distinction between linear motion and two-dimensional motion in a plane. We began with motion along a line, in order to focus on the essential physics, but now we turn our attention to the motion of projectiles, satellites, and other objects that move in two dimensions. We'll continue to follow « Problem-Solving Strategy 6.1, which is well worth a review, but we'll find that we need to think carefully about the appropriate coordinate system for each problem.

EXAMPLE 8.1 | Rocketing in the wind

A small rocket for gathering weather data has a mass of 30 kg and generates 1500 N of thrust. On a windy day, the wind exerts a 20 N horizontal force on the rocket. If the rocket is launched straight up, what is the shape of its trajectory, and by how much has it been deflected sideways when it reaches a height of 1.0 km? Because the rocket goes much higher than this, assume there's no significant mass loss during the first 1.0 km of flight.

MODEL Model the rocket as a particle. We need to find the *function* $y(x)$ describing the curve the rocket follows. Because rockets have aerodynamic shapes, we'll assume no vertical air resistance.

VISUALIZE FIGURE 8.1 shows a pictorial representation. We've chosen a coordinate system with a vertical y-axis. Three forces act on the rocket: two vertical and one horizontal.

FIGURE 8.1 Pictorial representation of the rocket launch.

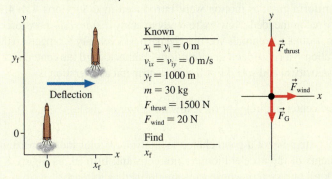

Known
$x_i = y_i = 0$ m
$v_{ix} = v_{iy} = 0$ m/s
$y_f = 1000$ m
$m = 30$ kg
$F_{thrust} = 1500$ N
$F_{wind} = 20$ N

Find
x_f

SOLVE In this problem, the vertical and horizontal forces are independent of each other. Newton's second law is

$$a_x = \frac{(F_{net})_x}{m} = \frac{F_{wind}}{m}$$

$$a_y = \frac{(F_{net})_y}{m} = \frac{F_{thrust} - mg}{m}$$

The primary difference from the linear-motion problems you've been solving is that the rocket accelerates along both axes. However, both accelerations are constant, so we can use kinematics to find

$$x = \tfrac{1}{2}a_x(\Delta t)^2 = \frac{F_{wind}}{2m}(\Delta t)^2$$

$$y = \tfrac{1}{2}a_y(\Delta t)^2 = \frac{F_{thrust} - mg}{2m}(\Delta t)^2$$

where we used the fact that all initial positions and velocities are zero. From the x-equation, $(\Delta t)^2 = 2mx/F_{wind}$. Substituting this into the y-equation, we find

$$y(x) = \left(\frac{F_{thrust} - mg}{F_{wind}}\right)x$$

This is the equation of the rocket's trajectory. It is a linear equation. Somewhat surprisingly, given that the rocket has both vertical and horizontal accelerations, its trajectory is a *straight line*. We can rearrange this result to find the deflection at height y:

$$x = \left(\frac{F_{wind}}{F_{thrust} - mg}\right)y$$

From the data provided, we can calculate a deflection of 17 m at a height of 1000 m.

ASSESS The solution depended on the fact that the time parameter Δt is the *same* for both components of the motion.

Projectile Motion

We found in Chapter 6 that the gravitational force on an object near the surface of a planet is $\vec{F}_G = (mg,\text{ down})$. For a coordinate system with a vertical y-axis,

$$\vec{F}_G = -mg\hat{\jmath} \tag{8.1}$$

Consequently, from Newton's second law, the acceleration is

$$a_x = \frac{(F_G)_x}{m} = 0$$
$$a_y = \frac{(F_G)_y}{m} = -g \tag{8.2}$$

Equations 8.2 justify the analysis of projectile motion in « Section 4.2—a downward acceleration $a_y = -g$ with no horizontal acceleration—where we found that a drag-free

FIGURE 8.2 A projectile is affected by drag. These are trajectories of a plastic ball launched at different angles.

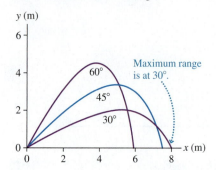

projectile follows a parabolic trajectory. The fact that the two components of acceleration are independent of each other allows us to solve for the vertical and horizontal motions.

However, the situation is quite different for a low-mass projectile, where the effects of drag are too large to ignore. We'll leave it as a homework problem for you to show that the acceleration of a projectile subject to drag is

$$a_x = -\frac{\rho CA}{2m} v_x \sqrt{v_x^2 + v_y^2}$$

$$a_y = -g - \frac{\rho CA}{2m} v_y \sqrt{v_x^2 + v_y^2}$$

(8.3)

Here the components of acceleration are neither constant nor independent of each other because a_x depends on v_y and vice versa. It turns out that these two equations cannot be solved exactly for the trajectory, but they can be solved numerically. **FIGURE 8.2** shows the numerical solution for the motion of a 5 g plastic ball that's been hit with an initial speed of 25 m/s. It doesn't travel very far (the maximum distance without drag would be more than 60 m), and the maximum range is no longer reached for a launch angle of 45°. Notice that the trajectories are not at all parabolic.

STOP TO THINK 8.1 This force will cause the particle to

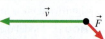

a. Speed up and curve upward.
c. Slow down and curve upward.
e. Move to the right and down.

b. Speed up and curve downward.
d. Slow down and curve downward.
f. Reverse direction.

8.2 Uniform Circular Motion

FIGURE 8.3 Uniform circular motion and the *rtz*-coordinate system.

Velocity has only a tangential component.

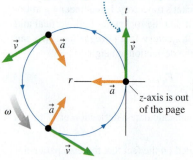

Acceleration has only a radial component.

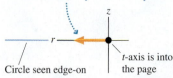

The kinematics of uniform circular motion were introduced in **«** Sections 4.4–4.5, and a review is *highly* recommended. Now we're ready to study *dynamics*—how forces *cause* circular motion. **FIGURE 8.3** reminds you that the particle's velocity is tangent to the circle, and its acceleration—a *centripetal acceleration*—points toward the center. If the particle has angular velocity ω and speed $v = \omega r$, its centripetal acceleration is

$$\vec{a} = \left(\frac{v^2}{r}, \text{ toward center of circle}\right) = (\omega^2 r, \text{ toward center of circle}) \quad (8.4)$$

An *xy*-coordinate system is not a good choice to analyze circular motion because the *x*- and *y*-components of the acceleration are not constant. Instead, as Figure 8.3 shows, we'll use a coordinate system whose axes are defined as follows:

- The origin is at the point where the particle is located.
- The *r*-axis (radial axis) points *from* the particle *toward* the center of the circle.
- The *t*-axis (tangential axis) is tangent to the circle, pointing in the counterclockwise direction.
- The *z*-axis is perpendicular to the plane of motion.

These three mutually perpendicular axes form the **rtz-coordinate system.**

You can see that the *rtz*-components of $\vec{v}$ and $\vec{a}$ are

$$
\begin{aligned}
v_r &= 0 & a_r &= \frac{v^2}{r} = \omega^2 r \\
v_t &= \omega r & a_t &= 0 \\
v_z &= 0 & a_z &= 0
\end{aligned}
\quad (8.5)
$$

where the angular velocity $\omega = d\theta/dt$ must be in rad/s. For uniform circular motion, **the velocity vector has only a tangential component and the acceleration vector has only a radial component.** Now you can begin to see the advantages of the *rtz*-coordinate system.

NOTE Recall that ω and v_t are positive for a counterclockwise (ccw) rotation, negative for a clockwise (cw) rotation. The particle's speed is $v = |v_t|$. ▶

Dynamics of Uniform Circular Motion

A particle in uniform circular motion is clearly not traveling at constant velocity in a straight line. Consequently, according to Newton's first law, the particle *must* have a net force acting on it. We've already determined the acceleration of a particle in uniform circular motion—the centripetal acceleration of Equation 8.4. Newton's second law tells us exactly how much net force is needed to cause this acceleration:

$$\vec{F}_{\text{net}} = m\vec{a} = \left(\frac{mv^2}{r}, \text{ toward center of circle}\right) \qquad (8.6)$$

In other words, a particle of mass m moving at constant speed v around a circle of radius r *must* have a net force of magnitude mv^2/r pointing toward the center of the circle. Without such a force, the particle would move off in a straight line tangent to the circle.

FIGURE 8.4 shows the net force $\vec{F}_{\text{net}}$ acting on a particle as it undergoes uniform circular motion. You can see that $\vec{F}_{\text{net}}$, like $\vec{a}$, **points along the radial axis of the *rtz*-coordinate system, toward the center of the circle.** The tangential and perpendicular components of $\vec{F}_{\text{net}}$ are zero.

NOTE The force described by Equation 8.6 is not a *new* force. The force itself must have an identifiable agent and will be one of our familiar forces, such as tension, friction, or the normal force. Equation 8.6 simply tells us how the force needs to act—how strongly and in which direction—to cause the particle to move with speed v in a circle of radius r. ▶

The usefulness of the *rtz*-coordinate system becomes apparent when we write Newton's second law, Equation 8.6, in terms of the r-, t-, and z-components:

$$(F_{\text{net}})_r = \sum F_r = ma_r = \frac{mv^2}{r} = m\omega^2 r$$
$$(F_{\text{net}})_t = \sum F_t = ma_t = 0 \qquad (8.7)$$
$$(F_{\text{net}})_z = \sum F_z = ma_z = 0$$

For uniform circular motion, the sum of the forces along the *t*-axis and along the *z*-axis must equal zero, and the sum of the forces along the *r*-axis *must* equal ma_r, where a_r is the centripetal acceleration.

On banked curves, the normal force of the road assists in providing the centripetal acceleration of the turn.

FIGURE 8.4 The net force points in the radial direction, toward the center of the circle.

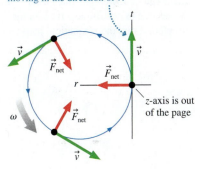

With no force, the particle would continue moving in the direction of $\vec{v}$.

z-axis is out of the page

EXAMPLE 8.2 | **Spinning in a circle**

An energetic father places his 20 kg child on a 5.0 kg cart to which a 2.0-m-long rope is attached. He then holds the end of the rope and spins the cart and child around in a circle, keeping the rope parallel to the ground. If the tension in the rope is 100 N, how many revolutions per minute (rpm) does the cart make? Rolling friction between the cart's wheels and the ground is negligible.

MODEL Model the child in the cart as a particle in uniform circular motion.

VISUALIZE **FIGURE 8.5** on the next page shows the pictorial representation. A circular-motion problem usually does not have starting and ending points like a projectile problem, so numerical subscripts such as x_1 or y_2 are usually not needed. Here we need to define the cart's speed v and the radius r of the circle. Further, a

motion diagram is not needed for uniform circular motion because we already know the acceleration $\vec{a}$ points to the center of the circle.

The essential part of the pictorial representation is the free-body diagram. **For uniform circular motion we'll draw the free-body diagram in the *rz*-plane, looking at the edge of the circle, because this is the plane of the forces.** The contact forces acting on the cart are the normal force of the ground and the tension force of the rope. The normal force is perpendicular to the plane of the motion and thus in the z-direction. The direction of $\vec{T}$ is determined by the statement that the rope is parallel to the ground. In addition, there is the long-range gravitational force $\vec{F}_G$.

SOLVE We defined the r-axis to point toward the center of the circle, so $\vec{T}$ points in the positive r-direction and has r-component $T_r = T$.

Continued

FIGURE 8.5 Pictorial representation of a cart spinning in a circle.

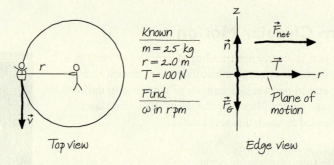

Newton's second law, using the *rtz*-components of Equations 8.7, is

$$\sum F_r = T = \frac{mv^2}{r}$$

$$\sum F_z = n - mg = 0$$

We've taken the *r*- and *z*-components of the forces directly from the free-body diagram, as you learned to do in Chapter 6. Then we've *explicitly* equated the sums to $a_r = v^2/r$ and $a_z = 0$. This is the basic strategy for all uniform circular-motion problems. From the *z*-equation we can find that $n = mg$. This would be useful if we needed to determine a friction force, but it's not needed in this problem. From the *r*-equation, the speed of the cart is

$$v = \sqrt{\frac{rT}{m}} = \sqrt{\frac{(2.0 \text{ m})(100 \text{ N})}{25 \text{ kg}}} = 2.83 \text{ m/s}$$

The cart's angular velocity is then found from Equations 8.5:

$$\omega = \frac{v_t}{r} = \frac{v}{r} = \frac{2.83 \text{ m/s}}{2.0 \text{ m}} = 1.41 \text{ rad/s}$$

This is another case where we inserted the radian unit because ω is specifically an *angular* velocity. Finally, we need to convert ω to rpm:

$$\omega = \frac{1.41 \text{ rad}}{1 \text{ s}} \times \frac{1 \text{ rev}}{2\pi \text{ rad}} \times \frac{60 \text{ s}}{1 \text{ min}} = 14 \text{ rpm}$$

ASSESS 14 rpm corresponds to a period $T \approx 4$ s. This result is reasonable.

The Central-Force Model

A force that is always directed toward the same point is called a **central force.** The tension in the rope of the last example is a central force, as is the gravitational force acting on an orbiting satellite. An object acted on by an attractive central force can undergo uniform circular motion around the central point. More complicated trajectories can occur in some situations—such as satellites following elliptical orbits—but for now we'll focus on circular motion, or motion with constant *r*. This **central-force model** is another important model of motion.

MODEL 8.1

Central force with constant *r*

For objects on which a constant net force points toward a central point.

- Model the object as a particle.
- The force causes a centripetal acceleration.
  - The motion is uniform circular motion.
- Mathematically:
  - Newton's second law is

$$\vec{F}_{\text{net}} = \left(\frac{mv^2}{r} \text{ or } m\omega^2 r, \text{ toward center} \right)$$

  - Use the kinematics of uniform circular motion.
- Limitations: Model fails if the force has a tangential component or if *r* changes.

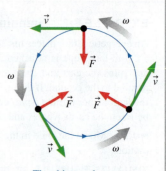

The object undergoes uniform circular motion.

Exercise 10

Let's look at more examples of the central-force model in action.

EXAMPLE 8.3 | Turning the corner I

What is the maximum speed with which a 1500 kg car can make a left turn around a curve of radius 50 m on a level (unbanked) road without sliding?

MODEL The car doesn't complete a full circle, but it is in uniform circular motion for a quarter of a circle while turning. We can model the car as a particle subject to a central force. Assume that rolling friction is negligible.

VISUALIZE FIGURE 8.6 shows the pictorial representation. The issue we must address is *how* a car turns a corner. What force or forces cause the direction of the velocity vector to change? Imagine driving on a completely frictionless road, such as a very icy road. You would not be able to turn a corner. Turning the steering wheel would be of no use; the car would slide straight ahead, in accordance with both Newton's first law and the experience of anyone who has ever driven on ice! So it must be *friction* that somehow allows the car to turn.

Figure 8.6 shows the top view of a tire as it turns a corner. If the road surface were frictionless, the tire would slide straight ahead. The force that prevents an object from sliding across a surface is *static friction*. Static friction $\vec{f}_s$ pushes *sideways* on the tire, toward the center of the circle. How do we know the direction is sideways? If $\vec{f}_s$ had a component either parallel to $\vec{v}$ or opposite to $\vec{v}$, it would cause the car to speed up or slow down. Because the car changes direction but not speed, static friction must be perpendicular to $\vec{v}$. $\vec{f}_s$ causes the centripetal acceleration of circular motion around the curve, and thus the free-body diagram, drawn from behind the car, shows the static friction force pointing toward the center of the circle.

SOLVE The maximum turning speed is reached when the static friction force reaches its maximum $f_{s\,max} = \mu_s n$. If the car enters the curve at a speed higher than the maximum, static friction will not

be large enough to provide the necessary centripetal acceleration and the car will slide.

The static friction force points in the positive r-direction, so its radial component is simply the magnitude of the vector: $(f_s)_r = f_s$. Newton's second law in the rtz-coordinate system is

$$\sum F_r = f_s = \frac{mv^2}{r}$$
$$\sum F_z = n - mg = 0$$

The only difference from Example 8.2 is that the tension force toward the center has been replaced by a static friction force toward the center. From the radial equation, the speed is

$$v = \sqrt{\frac{rf_s}{m}}$$

The speed will be a maximum when f_s reaches its maximum value:

$$f_s = f_{s\,max} = \mu_s n = \mu_s mg$$

where we used $n = mg$ from the z-equation. At that point,

$$v_{max} = \sqrt{\frac{rf_{s\,max}}{m}} = \sqrt{\mu_s rg}$$
$$= \sqrt{(1.0)(50\text{ m})(9.80\text{ m/s}^2)} = 22\text{ m/s}$$

where the coefficient of static friction was taken from Table 6.1.

ASSESS 22 m/s ≈ 45 mph, a reasonable answer for how fast a car can take an unbanked curve. Notice that the car's mass canceled out and that the final equation for v_{max} is quite simple. This is another example of why it pays to work algebraically until the very end.

FIGURE 8.6 Pictorial representation of a car turning a corner.

Because μ_s depends on road conditions, the maximum safe speed through turns can vary dramatically. Wet roads, in particular, lower the value of μ_s and thus lower the speed of turns. Icy conditions are even worse. The corner you turn every day at 45 mph will require a speed of no more than 15 mph if the coefficient of static friction drops to 0.1.

EXAMPLE 8.4 | Turning the corner II

A highway curve of radius 70 m is banked at a 15° angle. At what speed v_0 can a car take this curve without assistance from friction?

MODEL Model the car as a particle subject to a central force.

VISUALIZE Having just discussed the role of friction in turning corners, it is perhaps surprising to suggest that the same turn can also be accomplished without friction. Example 8.3 considered a level roadway, but real highway curves are *banked* by being tilted

Continued

up at the outside edge of the curve. The angle is modest on ordinary highways, but it can be quite large on high-speed racetracks. The purpose of banking becomes clear if you look at the free-body diagram in **FIGURE 8.7**. The normal force $\vec{n}$ is perpendicular to the road, so tilting the road causes $\vec{n}$ to have a component toward the center of the circle. **The radial component n_r is the central force that causes the centripetal acceleration needed to turn the car.** Notice that we are *not* using a tilted coordinate system, although this looks rather like an inclined-plane problem. The center of the circle is in the same horizontal plane as the car, and for circular-motion problems we need the r-axis to pass through the center. Tilted axes are for *linear* motion along an incline.

FIGURE 8.7 Pictorial representation of a car on a banked curve.

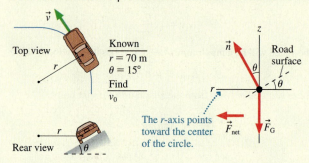

Top view

Known
$r = 70$ m
$\theta = 15°$

Find
v_0

Rear view

The r-axis points toward the center of the circle.

Road surface

SOLVE Without friction, $n_r = n \sin \theta$ is the only component of force in the radial direction. It is this inward component of the normal force on the car that causes it to turn the corner. Newton's second law is

$$\sum F_r = n \sin \theta = \frac{mv_0^2}{r}$$

$$\sum F_z = n \cos \theta - mg = 0$$

where θ is the angle at which the road is banked and we've assumed that the car is traveling at the correct speed v_0. From the z-equation,

$$n = \frac{mg}{\cos \theta}$$

Substituting this into the r-equation and solving for v_0 give

$$\frac{mg}{\cos \theta} \sin \theta = mg \tan \theta = \frac{mv_0^2}{r}$$

$$v_0 = \sqrt{rg \tan \theta} = 14 \text{ m/s}$$

ASSESS This is ≈ 28 mph, a reasonable speed. Only at this very specific speed can the turn be negotiated without reliance on friction forces.

It's interesting to explore what happens at other speeds on a banked curve. **FIGURE 8.8** shows that the car will need to rely on both the banking *and* friction if it takes the curve at a speed faster or slower than v_0.

FIGURE 8.8 Free-body diagrams for a car going around a banked curve at speeds slower and faster than the friction-free speed v_0.

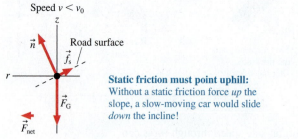

Speed $v < v_0$

Road surface

Static friction must point uphill:
Without a static friction force *up* the slope, a slow-moving car would slide *down* the incline!

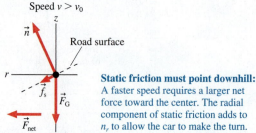

Speed $v > v_0$

Road surface

Static friction must point downhill:
A faster speed requires a larger net force toward the center. The radial component of static friction adds to n_r to allow the car to make the turn.

EXAMPLE 8.5 | A rock in a sling

A Stone Age hunter places a 1.0 kg rock in a sling and swings it in a horizontal circle around his head on a 1.0-m-long vine. If the vine breaks at a tension of 200 N, what is the maximum angular speed, in rpm, with which he can swing the rock?

MODEL Model the rock as a particle in uniform circular motion.

VISUALIZE This problem appears, at first, to be essentially the same as Example 8.2, where the father spun his child around on

a rope. However, the lack of a normal force from a supporting surface makes a *big* difference. In this case, the *only* contact force on the rock is the tension in the vine. Because the rock moves in a horizontal circle, you may be tempted to draw a free-body diagram like **FIGURE 8.9a**, where $\vec{T}$ is directed along the r-axis. You will quickly run into trouble, however, because this diagram has a net force in the z-direction and it is impossible to satisfy $\sum F_z = 0$. The

FIGURE 8.9 Pictorial representation of a rock in a sling.

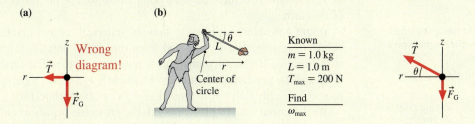

FIGURE 8.9 Pictorial representation of a rock in a sling.

gravitational force $\vec{F}_G$ certainly points vertically downward, so the difficulty must be with $\vec{T}$.

As an experiment, tie a small weight to a string, swing it over your head, and check the *angle* of the string. You will quickly discover that the string is *not* horizontal but, instead, is angled downward. The sketch of **FIGURE 8.9b** labels the angle θ. Notice that the rock moves in a *horizontal* circle, so the center of the circle is *not* at his hand. The r-axis points to the center of the circle, but the tension force is directed along the vine. Thus the correct free-body diagram is the one in Figure 8.9b.

SOLVE The free-body diagram shows that the downward gravitational force is balanced by an upward component of the tension, leaving the radial component of the tension to cause the centripetal acceleration. Newton's second law is

$$\sum F_r = T\cos\theta = \frac{mv^2}{r}$$

$$\sum F_z = T\sin\theta - mg = 0$$

where θ is the angle of the vine below horizontal. From the z-equation we find

$$\sin\theta = \frac{mg}{T}$$

$$\theta_{max} = \sin^{-1}\left(\frac{(1.0\ \text{kg})(9.8\ \text{m/s}^2)}{200\ \text{N}}\right) = 2.81°$$

where we've evaluated the angle at the maximum tension of 200 N. The vine's angle of inclination is small but not zero.

Turning now to the r-equation, we find the rock's speed is

$$v_{max} = \sqrt{\frac{rT\cos\theta_{max}}{m}}$$

Careful! The radius r of the circle is *not* the length L of the vine. You can see in Figure 8.9b that $r = L\cos\theta$. Thus

$$v_{max} = \sqrt{\frac{LT\cos^2\theta_{max}}{m}} = \sqrt{\frac{(1.0\ \text{m})(200\ \text{N})(\cos 2.81°)^2}{1.0\ \text{kg}}} = 14.1\ \text{m/s}$$

We can now find the maximum angular speed, the value of ω that brings the tension to the breaking point:

$$\omega_{max} = \frac{v_{max}}{r} = \frac{v_{max}}{L\cos\theta_{max}} = \frac{14.1\ \text{rad}}{1\ \text{s}} \times \frac{1\ \text{rev}}{2\pi\ \text{rad}} \times \frac{60\ \text{s}}{1\ \text{min}} = 135\ \text{rpm}$$

STOP TO THINK 8.2 A block on a string spins in a horizontal circle on a frictionless table. Rank in order, from largest to smallest, the tensions T_a to T_e acting on blocks a to e.

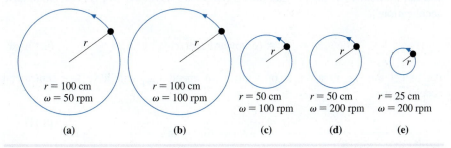

8.3 Circular Orbits

Satellites orbit the earth, the earth orbits the sun, and our entire solar system orbits the center of the Milky Way galaxy. Not all orbits are circular, but in this section we'll limit our analysis to circular orbits.

How does a satellite orbit the earth? What forces act on it? To answer these important questions, let's return, for a moment, to projectile motion. Projectile motion occurs when the only force on an object is gravity. Our analysis of projectiles assumed that

FIGURE 8.10 Projectiles being launched at increasing speeds from height h on a smooth, airless planet.

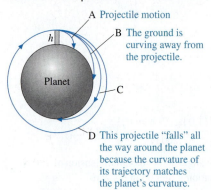

FIGURE 8.11 The "real" gravitational force is always directed toward the center of the planet.

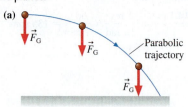

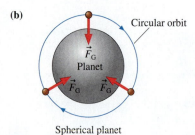

the earth is flat and that the acceleration due to gravity is everywhere straight down. This is an acceptable approximation for projectiles of limited range, such as baseballs or cannon balls, but there comes a point where we can no longer ignore the curvature of the earth.

FIGURE 8.10 shows a perfectly smooth, spherical, airless planet with one tower of height h. A projectile is launched from this tower parallel to the ground ($\theta = 0°$) with speed v_0. If v_0 is very small, as in trajectory A, the "flat-earth approximation" is valid and the problem is identical to Example 8.4 in which a car drove off a cliff. The projectile simply falls to the ground along a parabolic trajectory.

As the initial speed v_0 is increased, the projectile begins to notice that the ground is curving out from beneath it. It is falling the entire time, always getting closer to the ground, but the distance that the projectile travels before finally reaching the ground—that is, its range—increases because the projectile must "catch up" with the ground that is curving away from it. Trajectories B and C are of this type. The actual calculation of these trajectories is beyond the scope of this textbook, but you should be able to understand the factors that influence the trajectory.

If the launch speed v_0 is sufficiently large, there comes a point where the curve of the trajectory and the curve of the earth are parallel. In this case, the projectile "falls" but it never gets any closer to the ground! This is the situation for trajectory D. A closed trajectory around a planet or star, such as trajectory D, is called an **orbit**.

The most important point of this qualitative analysis is that **an orbiting projectile is in free fall**. This is, admittedly, a strange idea, but one worth careful thought. An orbiting projectile is really no different from a thrown baseball or a car driving off a cliff. The only force acting on it is gravity, but its tangential velocity is so large that the curvature of its trajectory matches the curvature of the earth. When this happens, the projectile "falls" under the influence of gravity but never gets any closer to the surface.

In the flat-earth approximation, shown in FIGURE 8.11a, the gravitational force acting on an object of mass m is

$$\vec{F}_G = (mg, \text{vertically downward}) \quad \text{(flat-earth approximation)} \quad (8.8)$$

But since stars and planets are actually spherical (or very close to it), the "real" force of gravity acting on an object is directed toward the *center* of the planet, as shown in FIGURE 8.11b. In this case the gravitational force is

$$\vec{F}_G = (mg, \text{toward center}) \quad \text{(spherical planet)} \quad (8.9)$$

That is, gravity is a central force causing the centripetal acceleration of uniform circular motion. Thus the gravitational force causes the object in Figure 8.11b to have acceleration

$$\vec{a} = \frac{\vec{F}_{net}}{m} = (g, \text{toward center}) \quad (8.10)$$

An object moving in a circle of radius r at speed v_{orbit} will have this centripetal acceleration if

$$a_r = \frac{(v_{orbit})^2}{r} = g \quad (8.11)$$

That is, if an object moves parallel to the surface with the speed

$$v_{orbit} = \sqrt{rg} \quad (8.12)$$

then the free-fall acceleration is exactly the centripetal acceleration needed for a circular orbit of radius r. An object with any other speed will not follow a circular orbit.

The earth's radius is $r = R_e = 6.37 \times 10^6$ m. (A table of useful astronomical data is inside the back cover of this book.) The orbital speed of a projectile just skimming the surface of an airless, bald earth is

$$v_{orbit} = \sqrt{rg} = \sqrt{(6.37 \times 10^6 \text{ m})(9.80 \text{ m/s}^2)} = 7900 \text{ m/s} \approx 16{,}000 \text{ mph}$$

Even if there were no trees and mountains, a real projectile moving at this speed would burn up from the friction of air resistance.

Satellites

Suppose, however, that we launched the projectile from a tower of height $h = 230$ mi $\approx 3.8 \times 10^5$ m, just above the earth's atmosphere. This is approximately the height of the International Space Station and other low-earth-orbit satellites. Note that $h \ll R_e$, so the radius of the orbit $r = R_e + h = 6.75 \times 10^6$ m is only 5% greater than the earth's radius. Many people have a mental image that satellites orbit far above the earth, but in fact many satellites come pretty close to skimming the surface. Our calculation of v_{orbit} thus turns out to be quite a good estimate of the speed of a satellite in low earth orbit.

We can use v_{orbit} to calculate the period of a satellite orbit:

$$T = \frac{2\pi r}{v_{orbit}} = 2\pi \sqrt{\frac{r}{g}} \qquad (8.13)$$

The International Space Station is in free fall.

For a low earth orbit, with $r = R_e + 230$ miles, we find $T = 5210$ s $= 87$ min. The period of the International Space Station at an altitude of 230 mi is, indeed, close to 87 minutes. (The actual period is 93 min. The difference, you'll learn in Chapter 13, arises because g is slightly less at a satellite's altitude.)

When we discussed *weightlessness* in Chapter 6, we discovered that it occurs during free fall. We asked the question, at the end of « Section 6.3, whether astronauts and their spacecraft were in free fall. We can now give an affirmative answer: They are, indeed, in free fall. They are falling continuously around the earth, under the influence of only the gravitational force, but never getting any closer to the ground because the earth's surface curves beneath them. Weightlessness in space is no different from the weightlessness in a free-falling elevator. It does *not* occur from an absence of gravity. Instead, the astronaut, the spacecraft, and everything in it are weightless because they are all falling together.

8.4 Reasoning About Circular Motion

Some aspects of circular motion are puzzling and counterintuitive. Examining a few of these will give us a chance to practice Newtonian reasoning.

Centrifugal Force?

If the car turns a corner quickly, you feel "thrown" against the door. But there's really no such force because there is no agent exerting it. FIGURE 8.12 shows a bird's-eye view of you riding in a car as it makes a left turn. You try to continue moving in a straight line, obeying Newton's first law, when—without having been provoked—the door suddenly turns in front of you and runs into you! You do, indeed, then feel the force of the door because it is now the normal force of the door, pointing *inward* toward the center of the curve, causing you to turn the corner. But you were not "thrown" into the door; the door ran into you.

The "force" that seems to push an object to the outside of a circle is commonly known as the *centrifugal force*. Despite having a name, the centrifugal force is fictitious. It describes your experience *relative to a noninertial reference frame,* but there really is no such force. **You must always use Newton's laws in an inertial reference frame.** There are no centrifugal forces in an inertial reference frame.

> **NOTE** You might wonder if the *rtz*-coordinate system is an inertial reference frame. It is. We're using the *rtz*-coordinates to establish directions for decomposing vectors, but we're not making measurements in the *rtz*-system. That is, velocities and accelerations are measured in the laboratory reference frame. The particle would always be at rest ($\vec{v} = \vec{0}$) if we measured velocities in a reference frame attached to the particle.

FIGURE 8.12 Bird's-eye view of a passenger as a car turns a corner.

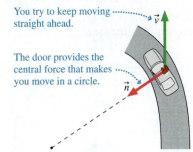

You try to keep moving straight ahead. $\vec{v}$

The door provides the central force that makes you move in a circle. $\vec{n}$

Gravity on a Rotating Earth

There is one small problem with the admonition that you must use Newton's laws in an inertial reference frame: A reference frame attached to the ground isn't truly inertial because of the earth's rotation. Fortunately, we can make a simple correction that allows us to continue using Newton's laws on the earth's surface.

FIGURE 8.13 shows an object being weighed by a spring scale on the earth's equator. An observer hovering above the north pole sees two forces on the object: the gravitational force $\vec{F}_{M\,\text{on}\,m}$, given by Newton's law of gravity, and the outward spring force $\vec{F}_{\text{Sp}}$. The object moves in a circle as the earth rotates, so Newton's second law is

$$\sum F_r = F_{M\,\text{on}\,m} - F_{\text{Sp}} = m\omega^2 R$$

where ω is the angular speed of the rotating earth. The spring-scale reading $F_{\text{Sp}} = F_{M\,\text{on}\,m} - m\omega^2 R$ is *less* than it would be on a nonrotating earth.

The blow-up in Figure 8.13 shows how we see things in a noninertial, flat-earth reference frame. For us the object is at rest, in equilibrium, hence the upward spring force must be exactly balanced by a downward gravitational force $\vec{F}_{\text{G}}$. Thus $F_{\text{Sp}} = F_{\text{G}}$.

Now both we and the hovering, inertial observer see the same reading on the scale. If F_{Sp} is the same for both of us, then

$$F_{\text{G}} = F_{M\,\text{on}\,m} - m\omega^2 R \qquad (8.14)$$

In other words, force $\vec{F}_{\text{G}}$—what we called the *effective* gravitational force in Chapter 6—is slightly less than the true gravitational force $\vec{F}_{M\,\text{on}\,m}$ because of the earth's rotation. In essence, $m\omega^2 R$ is the centrifugal force, a fictitious force trying—from our perspective in a noninertial reference frame—to "throw" us off the rotating platform. There really is no such force, but—this is the important point—**we can continue to use Newton's laws in our rotating reference frame if we pretend there is.**

Because $F_{\text{G}} = mg$ for an object at rest, the effect of the centrifugal term in Equation 8.14 is to make g a little smaller than it would be on a nonrotating earth:

$$g = \frac{F_{\text{G}}}{m} = \frac{F_{M\,\text{on}\,m} - m\omega^2 R}{m} = \frac{GM}{R^2} - \omega^2 R = g_{\text{earth}} - \omega^2 R \qquad (8.15)$$

We calculated $g_{\text{earth}} = 9.83 \text{ m/s}^2$ in Chapter 6. Using $\omega = 1$ rev/day (which must be converted to SI units) and $R = 6370$ km, we find $\omega^2 R = 0.033 \text{ m/s}^2$ at the equator. Thus the free-fall acceleration—what we actually measure in our rotating reference frame—is about 9.80 m/s^2, exactly what we measure in the laboratory.

Things are a little more complicated at other latitudes, but the bottom line is that we can safely use Newton's laws in our rotating, noninertial reference frame on the earth's surface if we calculate the gravitational force—as we've been doing—as $F_{\text{G}} = mg$ with g the measured free-fall value, a value that compensates for our rotation, rather than the purely gravitational g_{earth}.

Why Does the Water Stay in the Bucket?

If you swing a bucket of water over your head quickly, the water stays in, but you'll get a shower if you swing too slowly. Why? We'll answer this question by starting with an equivalent situation, a roller coaster doing a loop-the-loop.

FIGURE 8.14 shows a roller-coaster car going around a vertical loop-the-loop of radius r. Why doesn't the car fall off at the top of the circle? Now, motion in a vertical circle is *not* uniform circular motion; the car slows down as it goes up one side and speeds up as it comes back down the other. But at the very top and very bottom points, only the car's direction is changing, not its speed, so at those points the acceleration is purely centripetal. Thus **there must be a net force toward the center of the circle.**

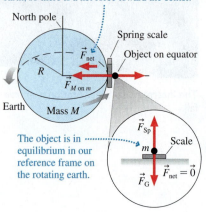

FIGURE 8.13 The earth's rotation affects the measured value of g.

The object is in circular motion on a rotating earth, so there is a net force toward the center.

North pole

Spring scale

Object on equator

$\vec{F}_{\text{net}}$

R

$\vec{F}_{M\,\text{on}\,m}$

Earth   Mass M

The object is in equilibrium in our reference frame on the rotating earth.

$\vec{F}_{\text{Sp}}$

m Scale

$\vec{F}_{\text{G}}$ $\vec{F}_{\text{net}} = \vec{0}$

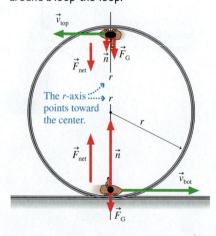

FIGURE 8.14 A roller-coaster car going around a loop-the-loop.

$\vec{v}_{\text{top}}$

$\vec{n}$ $\vec{F}_{\text{G}}$

$\vec{F}_{\text{net}}$

r

The r-axis points toward the center.

r

r

$\vec{F}_{\text{net}}$ $\vec{n}$

$\vec{v}_{\text{bot}}$

$\vec{F}_{\text{G}}$

First consider the very bottom of the loop. To have a net force toward the center—upward at this point—requires $n > F_G$. The normal force has to *exceed* the gravitational force to provide the net force needed to "turn the corner" at the bottom of the circle. This is why you "feel heavy" at the bottom of the circle or at the bottom of a valley on a roller coaster.

We can analyze the situation quantitatively by writing the *r*-component of Newton's second law. At the bottom of the circle, with the *r*-axis pointing upward, we have

$$\sum F_r = n_r + (F_G)_r = n - mg = ma_r = \frac{m(v_{bot})^2}{r} \tag{8.16}$$

From Equation 8.16 we find

$$n = mg + \frac{m(v_{bot})^2}{r} \tag{8.17}$$

The normal force at the bottom is *larger* than *mg*.

Things are a little trickier as the roller-coaster car crosses the top of the loop. Whereas the normal force of the track pushes up when the car is at the bottom of the circle, it *presses down* when the car is at the top and the track is above the car. Think about the free-body diagram to make sure you agree.

The car is still moving in a circle, so there *must* be a net force toward the center of the circle. The *r*-axis, which points toward the center of the circle, now points *downward*. Consequently, both forces have *positive* components. Newton's second law at the top of the circle is

$$\sum F_r = n_r + (F_G)_r = n + mg = \frac{m(v_{top})^2}{r} \tag{8.18}$$

Thus at the top the normal force of the track on the car is

$$n = \frac{m(v_{top})^2}{r} - mg \tag{8.19}$$

The normal force at the top can exceed *mg* if v_{top} is large enough. Our interest, however, is in what happens as the car goes slower and slower. As v_{top} decreases, there comes a point when *n* reaches zero. "No normal force" means "no contact," so at that speed, the track is *not* pushing against the car. Instead, the car is able to complete the circle because gravity alone provides sufficient centripetal acceleration.

The speed at which $n = 0$ is called the *critical speed* v_c:

$$v_c = \sqrt{\frac{rmg}{m}} = \sqrt{rg} \tag{8.20}$$

The critical speed is the slowest speed at which the car can complete the circle. Equation 8.19 would give a negative value for *n* if $v < v_c$, but that is physically impossible. The track can push against the wheels of the car ($n > 0$), but it can't pull on them. If $v < v_c$, the car cannot turn the full loop but, instead, comes off the track and becomes a projectile! **FIGURE 8.15** summarizes our reasoning.

Water stays in a bucket swung over your head for the same reason: Circular motion requires a net force toward the center of the circle. At the top of the circle—if you swing the bucket fast enough—the bucket adds to the force of gravity by pushing *down* on the water, just like the downward normal force of the track on the roller-coaster car. As long as the bucket is pushing against the water, the bucket and the water are in contact and thus the water is "in" the bucket. As you swing slower and slower, requiring the water to have less and less centripetal acceleration, the bucket-on-water normal force decreases until it becomes zero at the critical speed. At the critical speed, gravity alone provides sufficient centripetal acceleration. Below the critical speed, gravity provides *too much* downward force for circular motion, so the water leaves the bucket and becomes a projectile following a parabolic trajectory toward your head!

FIGURE 8.15 A roller-coaster car at the top of the loop.

The normal force adds to gravity to make a large enough force for the car to turn the circle.

$v > v_c$

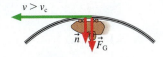

At v_c, gravity alone is enough force for the car to turn the circle. $\vec{n} = \vec{0}$ at the top point.

v_c

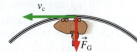

The gravitational force is too large for the car to stay in the circle!

Normal force became zero here.

$v < v_c$

Parabolic trajectory

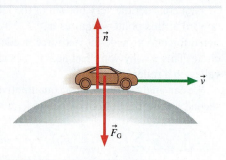

8.5 Nonuniform Circular Motion

Many interesting examples of circular motion involve objects whose speed changes. As we've already noted, a roller-coaster car doing a loop-the-loop slows down as it goes up one side, speeds up as it comes back down the other side. Circular motion with a changing speed is called *nonuniform circular motion*.

FIGURE 8.16 shows a particle moving in a circle of radius r. In addition to a radial force component—required for all circular motion—this particle experiences a *tangential* force component $(F_{net})_t$ and hence a tangential acceleration

$$a_t = \frac{dv_t}{dt} \tag{8.21}$$

Now v_t is the particle's velocity *around* the circle, with speed $v = |v_t|$, so a tangential force component causes the particle to change speed. That is, the particle is undergoing nonuniform circular motion. Note that $(F_{net})_t$, like v_t, is positive when ccw, negative when cw.

Force and acceleration are still related to each other through Newton's second law:

$$(F_{net})_r = \sum F_r = ma_r = \frac{mv_t^2}{r} = m\omega^2 r$$
$$(F_{net})_t = \sum F_t = ma_t \tag{8.22}$$
$$(F_{net})_z = \sum F_z = 0$$

If the tangential force is constant, you can apply what you know about constant-acceleration kinematics to solve for v_t at a later time.

NOTE Equations 8.22 differ from Equations 8.7 for uniform circular motion only in the fact that a_t is no longer zero.

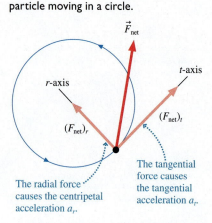

FIGURE 8.16 Net force $\vec{F}_{net}$ is applied to a particle moving in a circle.

The radial force causes the centripetal acceleration a_r.

The tangential force causes the tangential acceleration a_t.

EXAMPLE 8.6 | Sliding out of the curve

A 1500 kg car drives around a flat, 50-m-diameter track, starting from rest. The drive wheels supply a small but steady 525 N force in the forward direction. The coefficient of static friction between the car tires and the road is 0.90. How many revolutions of the track have been made when the car slides out of the curve?

MODEL Model the car as a particle in nonuniform circular motion. Assume that rolling friction and air resistance can be neglected.

VISUALIZE FIGURE 8.17 shows a pictorial representation. As in earlier examples, it's static friction, perpendicular to the tires, that causes the centripetal acceleration of circular motion. The propulsion force is a tangential force. For the first time, we need a free-body diagram showing forces in three dimensions.

FIGURE 8.17 Pictorial representation of a car speeding up around a circle.

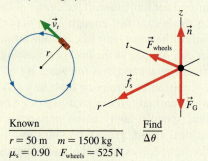

Known	Find
$r = 50$ m $m = 1500$ kg	$\Delta\theta$
$\mu_s = 0.90$ $F_{wheels} = 525$ N	

SOLVE At slow speeds, static friction in the radial direction keeps the car moving in a circle. But there's an upper limit to the size of the static friction force, and the car will begin to slide out of the curve when that limit is reached. The r-component of Newton's second law is

$$\sum F_r = f_s = \frac{mv_t^2}{r}$$

That is, static friction increases proportional to v_t^2 until the car reaches a velocity v_{max} at which the static friction is $f_{s\,max}$.

Recall that the maximum possible static friction is $f_{s\,max} = \mu_s n$. We can find the normal force from the z-component of Newton's second law:

$$\sum F_z = n - F_G = 0$$

Thus $n = F_G = mg$ and $f_{s\,max} = \mu_s mg$. Combining these two equations, we see that the mass cancels and we have

$$v_{max}^2 = \mu_s rg$$

How far does the car have to travel to reach this speed? We can find the car's tangential acceleration from the t-component of the second law: $a_t = F_{wheels}/m$. This is a constant acceleration, so we can use constant-acceleration kinematics. Let s measure the distance around the circle—the arc length. Thus, because the initial velocity is $v_0 = 0$, we have

$$v_t^2 = v_0^2 + 2a_s s = 2a_s s = \frac{2sF_{wheels}}{m}$$

You'll recall that the angular displacement, measured in radians, is $\Delta\theta = s/r$. So when the car reaches velocity v_t, it has revolved through an angle

$$\Delta\theta = \frac{s}{r} = \frac{mv_t^2}{2rF_{wheels}}$$

Using the maximum speed before sliding, we find that the car slides out of the curve after revolving through an angle

$$\Delta\theta_{max} = \frac{mv_{max}^2}{2rF_{wheels}} = \frac{m}{2rF_{wheels}} \times \mu_s rg = \frac{\mu_s mg}{2F_{wheels}}$$

For the car in this problem,

$$\Delta\theta_{max} = \frac{(0.90)(1500\text{ kg})(9.80\text{ m/s}^2)}{2(525\text{ N})}$$

$$= 12.6\text{ rad} \times \frac{1\text{ rev}}{2\pi\text{ rad}} = 2.0\text{ rev}$$

It completes 2.0 revolutions before it starts to slide.

ASSESS A 525 N force on a 1500 kg car causes a tangential acceleration of $a_t \approx 0.3$ m/s². That's a quite modest acceleration, so it seems reasonable that the car would complete 2 rev before gaining enough speed to start sliding.

We've come a long way since our first dynamics problems in Chapter 6, but our basic strategy has not changed.

STOP TO THINK 8.4 A ball on a string is swung in a vertical circle. The string happens to break when it is parallel to the ground and the ball is moving up. Which trajectory does the ball follow?

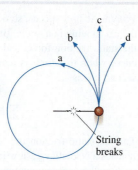

String breaks

CHALLENGE **EXAMPLE 8.7** | Swinging on two strings

The 250 g ball shown in **FIGURE 8.18** revolves in a horizontal plane as the vertical shaft spins. What is the critical angular speed, in rpm, that the shaft must exceed to keep both strings taut?

FIGURE 8.18 A ball revolving on two strings.

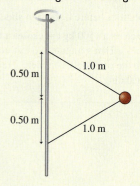

0.50 m

1.0 m

0.50 m

1.0 m

MODEL Model the ball as a particle in uniform circular motion. For both strings to be straight, as shown, both must be under tension. If the angular speed is slowly decreased, eventually the lower string will go slack and the ball will sag. The critical angular speed ω_c is the angular speed at which the tension in the lower string reaches zero. We need to find an expression for the tension in the lower string, then determine when that tension becomes zero.

VISUALIZE **FIGURE 8.19** is the ball's free-body diagram with the r-axis pointing toward the center of the circle. The ball is acted on by two tension forces, at equal angles above and below horizontal, and by gravity. The free-body diagram is similar to Example 8.5, the rock in the sling, but with an additional tension force.

FIGURE 8.19 Free-body diagram of the ball.

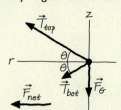

SOLVE This is uniform circular motion, so we need to consider only the r- and z-components of Newton's second law. All the information is on the free-body diagram, where we see that gravity has only a z-component but the tensions have both r- and z-components. The two equations are

$$\sum F_r = T_{top} \cos\theta + T_{bot} \cos\theta = m\omega^2 r$$
$$\sum F_z = T_{top} \sin\theta - T_{bot} \sin\theta - mg = 0$$

Factoring out the $\cos\theta$ and $\sin\theta$ terms, we have two simultaneous equations:

$$T_{top} + T_{bot} = \frac{m\omega^2 r}{\cos\theta}$$

$$T_{top} - T_{bot} = \frac{mg}{\sin\theta}$$

Subtracting the second equation from the first will eliminate T_{top}:

$$2T_{bot} = \frac{m\omega^2 r}{\cos\theta} - \frac{mg}{\sin\theta}$$

and thus

$$T_{bot} = \frac{m}{2}\left(\frac{\omega^2 r}{\cos\theta} - \frac{g}{\sin\theta}\right)$$

You can see that this expression becomes negative—a physically impossible situation—if ω is too small. The angular speed at which the tension reaches zero—the critical angular speed—is found by setting the expression in parentheses equal to zero. This gives

$$\omega_c = \sqrt{\frac{g}{r\tan\theta}}$$

For our situation,

$$r = \sqrt{(1.0\text{ m})^2 - (0.50\text{ m})^2} = 0.866\text{ m}$$

$$\theta = \sin^{-1}[(0.50\text{ m})/(1.0\text{ m})] = 30°$$

Thus the critical angular speed is

$$\omega_c = \sqrt{\frac{9.80\text{ m/s}^2}{(0.866\text{ m})\tan 30°}} = 4.40\text{ rad/s}$$

Converting to rpm:

$$\omega_c = 4.40\text{ rad/s} \times \frac{1\text{ rev}}{2\pi\text{ rad}} \times \frac{60\text{ s}}{1\text{ min}} = 42\text{ rpm}$$

ASSESS ω_c is the *minimum* angular speed needed to keep both strings taut. For a ball attached to meter-long strings, 42 rpm—a bit less than 1 revolution per second—seems plausible. Your intuition probably suggests that the bottom string wouldn't be taut if the shaft spun at only a few rpm, and hundreds of rpm seems much too high. Remember that the goal of Assess is not to prove that an answer is correct but to rule out answers that, with a little thought, are clearly wrong.

SUMMARY

The goal of Chapter 8 has been to learn to solve problems about motion in two dimensions.

GENERAL PRINCIPLES

Newton's Second Law

Expressed in x- and y-component form:

$$(F_{net})_x = \sum F_x = ma_x$$
$$(F_{net})_y = \sum F_y = ma_y$$

Expressed in rtz-component form:

$$(F_{net})_r = \sum F_r = ma_r = \frac{mv_t^2}{r} = m\omega^2 r$$

$$(F_{net})_t = \sum F_t = \begin{cases} 0 & \text{uniform circular motion} \\ ma_t & \text{nonuniform circular motion} \end{cases}$$

$$(F_{net})_z = \sum F_z = 0$$

Uniform Circular Motion

- Speed is constant.
- $\vec{F}_{net}$ points toward the center of the circle.
- The **centripetal acceleration** $\vec{a}$ points toward the center of the circle. It changes the particle's direction but not its speed.

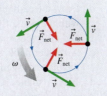

Nonuniform Circular Motion

- Speed changes.
- $\vec{F}_{net}$ and $\vec{a}$ have both radial and tangential components.
- The radial component changes the particle's direction.
- The tangential component changes the particle's speed.

IMPORTANT CONCEPTS

rtz-coordinates

- The r-axis points toward the center of the circle.
- The t-axis is tangent, pointing counterclockwise.

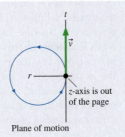

Plane of motion

Projectile motion

- With no drag, the x- and y-components of acceleration are independent. The trajectory is a parabola.
- With drag, the trajectory is not a parabola. Maximum range is achieved for an angle less than 45°.

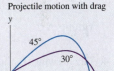

Projectile motion with drag

APPLICATIONS

Orbits

An object acted on only by gravity has a circular orbit of radius r if its speed is

$$v = \sqrt{rg}$$

The object is in free fall.

Circular motion on surfaces

Circular motion requires a net force pointing to the center. n must be > 0 for the object to be in contact with a surface.

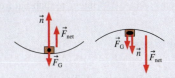

TERMS AND NOTATION

rtz-coordinate system	central force	central-force model	orbit

CONCEPTUAL QUESTIONS

1. In uniform circular motion, which of the following are constant: speed, velocity, angular velocity, centripetal acceleration, magnitude of the net force?

2. A car runs out of gas while driving down a hill. It rolls through the valley and starts up the other side. At the very bottom of the valley, which of the free-body diagrams in **FIGURE Q8.2** is correct? The car is moving to the right, and drag and rolling friction are negligible.

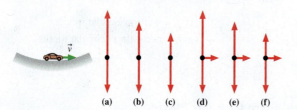

(a) **(b)** **(c)** **(d)** **(e)** **(f)**

FIGURE Q8.2

3. **FIGURE Q8.3** is a bird's-eye view of particles on strings moving in horizontal circles on a tabletop. All are moving at the same speed. Rank in order, from largest to smallest, the tensions T_a to T_d. Give your answer in the form a > b = c > d and explain your ranking.

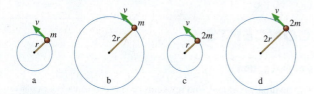

a   b   c   d

FIGURE Q8.3

4. Tarzan swings through the jungle on a massless vine. At the lowest point of his swing, is the tension in the vine greater than, less than, or equal to the gravitational force on Tarzan? Explain.

5. **FIGURE Q8.5** shows two balls of equal mass moving in vertical circles. Is the tension in string A greater than, less than, or equal to the tension in string B if the balls travel over the top of the circle (a) with equal speed and (b) with equal angular velocity?

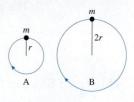

FIGURE Q8.5

6. Ramon and Sally are observing a toy car speed up as it goes around a circular track. Ramon says, "The car's speeding up, so there must be a net force parallel to the track." "I don't think so," replies Sally. "It's moving in a circle, and that requires centripetal acceleration. The net force has to point to the center of the circle." Do you agree with Ramon, Sally, or neither? Explain.

7. A jet plane is flying on a level course at constant speed. The engines are at full throttle.
   a. What is the net force on the plane? Explain.
   b. Draw a free-body diagram of the plane as seen from the side with the plane flying to the right. Name (don't just label) any and all forces shown on your diagram.
   c. Airplanes bank when they turn. Draw a free-body diagram of the plane as seen from behind as it makes a right turn.
   d. *Why* do planes bank as they turn? Explain.

8. A small projectile is launched parallel to the ground at height $h = 1$ m with sufficient speed to orbit a completely smooth, airless planet. A bug rides inside a small hole inside the projectile. Is the bug weightless? Explain.

9. You can swing a ball on a string in a vertical circle if you swing it fast enough. But if you swing too slowly, the string goes slack as the ball nears the top. Explain *why* there's a minimum speed to keep the ball moving in a circle.

10. A golfer starts with the club over her head and swings it to reach maximum speed as it contacts the ball. Halfway through her swing, when the golf club is parallel to the ground, does the acceleration vector of the club head point (a) straight down, (b) parallel to the ground, approximately toward the golfer's shoulders, (c) approximately toward the golfer's feet, or (d) toward a point above the golfer's head? Explain.

EXERCISES AND PROBLEMS

Problems labeled ▓ integrate material from earlier chapters.

Exercises

Section 8.1 Dynamics in Two Dimensions

1. ‖ As a science fair project, you want to launch an 800 g model rocket straight up and hit a horizontally moving target as it passes 30 m above the launch point. The rocket engine provides a constant thrust of 15.0 N. The target is approaching at a speed of 15 m/s. At what horizontal distance between the target and the rocket should you launch?

2. ‖ A 500 g model rocket is on a cart that is rolling to the right at a speed of 3.0 m/s. The rocket engine, when it is fired, exerts an 8.0 N vertical thrust on the rocket. Your goal is to have the rocket pass through a small horizontal hoop that is 20 m above the ground. At what horizontal distance left of the hoop should you launch?

3. ‖ A 4.0×10^{10} kg asteroid is heading directly toward the center of the earth at a steady 20 km/s. To save the planet, astronauts strap a giant rocket to the asteroid perpendicular to its direction of travel. The rocket generates 5.0×10^9 N of thrust. The rocket is fired when the asteroid is 4.0×10^6 km away from earth. You can ignore the earth's gravitational force on the asteroid and their rotation about the sun.

a. If the mission fails, how many hours is it until the asteroid impacts the earth?

b. The radius of the earth is 6400 km. By what minimum angle must the asteroid be deflected to just miss the earth?

c. What is the actual angle of deflection if the rocket fires at full thrust for 300 s before running out of fuel?

4. ‖ A 55 kg astronaut who weighs 180 N on a distant planet is pondering whether she can leap over a 3.5-m-wide chasm without falling in. If she leaps at a 15° angle, what initial speed does she need to clear the chasm?

Section 8.2 Uniform Circular Motion

5. | A 1500 kg car drives around a flat 200-m-diameter circular track at 25 m/s. What are the magnitude and direction of the net force on the car? What causes this force?

6. | A 1500 kg car takes a 50-m-radius unbanked curve at 15 m/s. What is the size of the friction force on the car?

7. ‖ A 200 g block on a 50-cm-long string swings in a circle on a horizontal, frictionless table at 75 rpm.
   a. What is the speed of the block?
   b. What is the tension in the string?

8. ‖ In the Bohr model of the hydrogen atom, an electron (mass $m = 9.1 \times 10^{-31}$ kg) orbits a proton at a distance of 5.3×10^{-11} m. The proton pulls on the electron with an electric force of 8.2×10^{-8} N. How many revolutions per second does the electron make?

9. | Suppose the moon were held in its orbit not by gravity but by a massless cable attached to the center of the earth. What would be the tension in the cable? Use the table of astronomical data inside the back cover of the book.

10. ‖ A highway curve of radius 500 m is designed for traffic moving at a speed of 90 km/h. What is the correct banking angle of the road?

11. ‖ It is proposed that future space stations create an artificial gravity by rotating. Suppose a space station is constructed as a 1000-m-diameter cylinder that rotates about its axis. The inside surface is the deck of the space station. What rotation period will provide "normal" gravity?

12. ‖ A 5.0 g coin is placed 15 cm from the center of a turntable. The coin has static and kinetic coefficients of friction with the turntable surface of $\mu_s = 0.80$ and $\mu_k = 0.50$. The turntable very slowly speeds up to 60 rpm. Does the coin slide off?

13. ‖ Mass m_1 on the frictionless table of **FIGURE EX8.13** is connected by a string through a hole in the table to a hanging mass m_2. With what speed must m_1 rotate in a circle of radius r if m_2 is to remain hanging at rest?

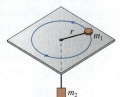

FIGURE EX8.13

Section 8.3 Circular Orbits

14. | A satellite orbiting the moon very near the surface has a period of 110 min. What is free-fall acceleration on the surface of the moon?

15. ‖ What is free-fall acceleration toward the sun at the distance of the earth's orbit? Astronomical data are inside the back cover of the book.

16. ‖ A 9.4×10^{21} kg moon orbits a distant planet in a circular orbit of radius 1.5×10^8 m. It experiences a 1.1×10^{19} N gravitational pull from the planet. What is the moon's orbital period in earth days?

17. ‖ Communications satellites are placed in circular orbits where they stay directly over a fixed point on the equator as the earth rotates. These are called *geosynchronous orbits*. The altitude of a geosynchronous orbit is 3.58×10^7 m ($\approx$22,000 miles).
   a. What is the period of a satellite in a geosynchronous orbit?
   b. Find the value of g at this altitude.
   c. What is the weight of a 2000 kg satellite in a geosynchronous orbit?

Section 8.4 Reasoning About Circular Motion

18. | A car drives over the top of a hill that has a radius of 50 m. What maximum speed can the car have at the top without flying off the road?

19. ‖ The weight of passengers on a roller coaster increases by 50% as the car goes through a dip with a 30 m radius of curvature. What is the car's speed at the bottom of the dip?

20. ‖ A roller coaster car crosses the top of a circular loop-the-loop at twice the critical speed. What is the ratio of the normal force to the gravitational force?

21. ‖ The normal force equals the magnitude of the gravitational force as a roller coaster car crosses the top of a 40-m-diameter loop-the-loop. What is the car's speed at the top?

22. ‖ A student has 65-cm-long arms. What is the minimum angular velocity (in rpm) for swinging a bucket of water in a vertical circle without spilling any? The distance from the handle to the bottom of the bucket is 35 cm.

23. | While at the county fair, you decide to ride the Ferris wheel. Having eaten too many candy apples and elephant ears, you find the motion somewhat unpleasant. To take your mind off your stomach, you wonder about the motion of the ride. You estimate the radius of the big wheel to be 15 m, and you use your watch to find that each loop around takes 25 s.
   a. What are your speed and the magnitude of your acceleration?
   b. What is the ratio of your weight at the top of the ride to your weight while standing on the ground?
   c. What is the ratio of your weight at the bottom of the ride to your weight while standing on the ground?

24. ‖ A 500 g ball swings in a vertical circle at the end of a 1.5-m-long string. When the ball is at the bottom of the circle, the tension in the string is 15 N. What is the speed of the ball at that point?

25. ‖ A 500 g ball moves in a vertical circle on a 102-cm-long string. If the speed at the top is 4.0 m/s, then the speed at the bottom will be 7.5 m/s. (You'll learn how to show this in Chapter 10.)
   a. What is the gravitational force acting on the ball?
   b. What is the tension in the string when the ball is at the top?
   c. What is the tension in the string when the ball is at the bottom?

26. ‖ A heavy ball with a weight of 100 N ($m = 10.2$ kg) is hung from the ceiling of a lecture hall on a 4.5-m-long rope. The ball is pulled to one side and released to swing as a pendulum, reaching a speed of 5.5 m/s as it passes through the lowest point. What is the tension in the rope at that point?

Section 8.5 Nonuniform Circular Motion

27. ‖ A toy train rolls around a horizontal 1.0-m-diameter track. The coefficient of rolling friction is 0.10. How long does it take the train to stop if it's released with an angular speed of 30 rpm?

28. ‖ A new car is tested on a 200-m-diameter track. If the car speeds up at a steady 1.5 m/s², how long after starting is the magnitude of its centripetal acceleration equal to the tangential acceleration?

29. ‖ An 85,000 kg stunt plane performs a loop-the-loop, flying in a 260-m-diameter vertical circle. At the point where the plane is flying straight down, its speed is 55 m/s and it is speeding up at a rate of 12 m/s per second.
    a. What is the magnitude of the net force on the plane? You can neglect air resistance.
    b. What angle does the net force make with the horizontal? Let an angle above horizontal be positive and an angle below horizontal be negative.

30. ‖ Three cars are driving at 25 m/s along the road shown in FIGURE EX8.30. Car B is at the bottom of a hill and car C is at the top. Both hills have a 200 m radius of curvature. Suppose each car suddenly brakes hard and starts to skid. What is the tangential acceleration (i.e., the acceleration parallel to the road) of each car? Assume $\mu_k = 1.0$.

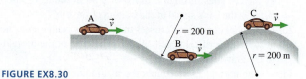

FIGURE EX8.30

Problems

31. ‖ Derive Equations 8.3 for the acceleration of a projectile subject to drag.

32. ‖ A 100 g bead slides along a frictionless wire with the parabolic
CALC shape $y = (2 \text{ m}^{-1})x^2$.
    a. Find an expression for a_y, the vertical component of acceleration, in terms of x, v_x, and a_x.
    Hint: Use the basic definitions of velocity and acceleration.
    b. Suppose the bead is released at some negative value of x and has a speed of 2.3 m/s as it passes through the lowest point of the parabola. What is the net force on the bead at this instant? Write your answer in component form.

33. ‖ Space scientists have a large test chamber from which all the air can be evacuated and in which they can create a horizontal uniform electric field. The electric field exerts a constant horizontal force on a charged object. A 15 g charged projectile is launched with a speed of 6.0 m/s at an angle 35° above the horizontal. It lands 2.9 m in front of the launcher. What is the magnitude of the electric force on the projectile?

34. ‖ A 5000 kg interceptor rocket is launched at an angle of 44.7°. The thrust of the rocket motor is 140,700 N.
    a. Find an equation $y(x)$ that describes the rocket's trajectory.
    b. What is the shape of the trajectory?
    c. At what elevation does the rocket reach the speed of sound, 330 m/s?

35. ‖ A motorcycle daredevil plans to ride up a 2.0-m-high, 20° ramp, sail across a 10-m-wide pool filled with hungry crocodiles, and land at ground level on the other side. He has done this stunt many times and approaches it with confidence. Unfortunately, the motorcycle engine dies just as he starts up the ramp. He is going 11 m/s at that instant, and the rolling friction of his rubber tires (coefficient 0.02) is not negligible. Does he survive, or does he become crocodile food? Justify your answer by calculating the distance he travels through the air after leaving the end of the ramp.

36. ‖ A rocket-powered hockey puck has a thrust of 2.0 N and a total mass of 1.0 kg. It is released from rest on a frictionless table, 4.0 m from the edge of a 2.0 m drop. The front of the rocket is pointed directly toward the edge. How far does the puck land from the base of the table?

37. ‖ A 500 g model rocket is resting horizontally at the top edge of a 40-m-high wall when it is accidentally bumped. The bump pushes it off the edge with a horizontal speed of 0.5 m/s and at the same time causes the engine to ignite. When the engine fires, it exerts a constant 20 N horizontal thrust away from the wall.
    a. How far from the base of the wall does the rocket land?
    b. Describe the rocket's trajectory as it travels to the ground.

38. ‖ A 2.0 kg projectile with initial velocity $\vec{v} = 8.0\,\hat{\imath}$ m/s experi-
CALC ences the variable force $\vec{F} = -2.0t\,\hat{\imath} + 4.0t^2\,\hat{\jmath}$ N, where t is in s.
    a. What is the projectile's speed at $t = 2.0$ s?
    b. At what instant of time is the projectile moving parallel to the y-axis?

39. ‖ A 75 kg man weighs himself at the north pole and at the equator. Which scale reading is higher? By how much? Assume the earth is spherical.

40. ‖ A concrete highway curve of radius 70 m is banked at a 15° angle. What is the maximum speed with which a 1500 kg rubber-tired car can take this curve without sliding?

41. ‖ a. An object of mass m swings in a horizontal circle on a string of length L that tilts downward at angle θ. Find an expression for the angular velocity ω.
    b. A student ties a 500 g rock to a 1.0-m-long string and swings it around her head in a horizontal circle. At what angular speed, in rpm, does the string tilt down at a 10° angle?

42. ‖‖ You've taken your neighbor's young child to the carnival to ride the rides. She wants to ride The Rocket. Eight rocket-shaped cars hang by chains from the outside edge of a large steel disk. A vertical axle through the center of the ride turns the disk, causing the cars to revolve in a circle. You've just finished taking physics, so you decide to figure out the speed of the cars while you wait. You estimate that the disk is 5 m in diameter and the chains are 6 m long. The ride takes 10 s to reach full speed, then the cars swing out until the chains are 20° from vertical. What is the cars' speed?

43. ‖ A 4.4-cm-diameter, 24 g plastic ball is attached to a 1.2-m-long string and swung in a vertical circle. The ball's speed is 6.1 m/s at the point where it is moving straight up. What is the magnitude of the net force on the ball? Air resistance is not negligible.

44. ‖ A charged particle of mass m moving with speed v in a plane perpendicular to a magnetic field experiences a force $\vec{F} = (qvB, \text{perpendicular to } \vec{v})$, where q is the amount of charge and B is the magnetic field strength. Because the force is always perpendicular to the particle's velocity, the particle undergoes uniform circular motion. Find an expression for the period of the motion. Gravity can be neglected.

45. ‖‖ Two wires are tied to the 2.0 kg sphere shown in FIGURE P8.45. The sphere revolves in a horizontal circle at constant speed.
    a. For what speed is the tension the same in both wires?
    b. What is the tension?

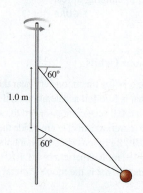

FIGURE P8.45

46. ‖ Two wires are tied to the 300 g sphere shown in **FIGURE P8.46**. The sphere revolves in a horizontal circle at a constant speed of 7.5 m/s. What is the tension in each of the wires?

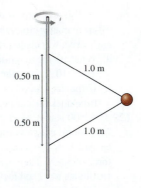

FIGURE P8.46

47. ‖ A conical pendulum is formed by attaching a ball of mass m to a string of length L, then allowing the ball to move in a horizontal circle of radius r. **FIGURE P8.47** shows that the string traces out the surface of a cone, hence the name.
    a. Find an expression for the tension T in the string.
    b. Find an expression for the ball's angular speed ω.
    c. What are the tension and angular speed (in rpm) for a 500 g ball swinging in a 20-cm-radius circle at the end of a 1.0-m-long string?

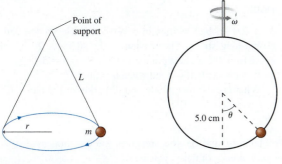

FIGURE P8.47 **FIGURE P8.48**

48. ‖ The 10 mg bead in **FIGURE P8.48** is free to slide on a frictionless wire loop. The loop rotates about a vertical axis with angular velocity ω. If ω is less than some critical value ω_c, the bead sits at the bottom of the spinning loop. When $\omega > \omega_c$, the bead moves out to some angle θ.
    a. What is ω_c in rpm for the loop shown in the figure?
    b. At what value of ω, in rpm, is $\theta = 30°$?

49. ‖‖ In an old-fashioned amusement park ride, passengers stand inside a 5.0-m-diameter hollow steel cylinder with their backs against the wall. The cylinder begins to rotate about a vertical axis. Then the floor on which the passengers are standing suddenly drops away! If all goes well, the passengers will "stick" to the wall and not slide. Clothing has a static coefficient of friction against steel in the range 0.60 to 1.0 and a kinetic coefficient in the range 0.40 to 0.70. A sign next to the entrance says "No children under 30 kg allowed." What is the minimum angular speed, in rpm, for which the ride is safe?

50. ‖ The ultracentrifuge is an important tool for separating and an-
BIO alyzing proteins. Because of the enormous centripetal accelerations, the centrifuge must be carefully balanced, with each sample matched by a sample of identical mass on the opposite side. Any difference in the masses of opposing samples creates a net force on the shaft of the rotor, potentially leading to a catastrophic failure of the apparatus. Suppose a scientist makes a slight error in sample preparation and one sample has a mass 10 mg larger than the opposing sample. If the samples are 12 cm from the axis of the rotor and the ultracentrifuge spins at 70,000 rpm, what is the magnitude of the net force on the rotor due to the unbalanced samples?

51. ‖ In an amusement park ride called The Roundup, passengers stand inside a 16-m-diameter rotating ring. After the ring has acquired sufficient speed, it tilts into a vertical plane, as shown in **FIGURE P8.51**.
    a. Suppose the ring rotates once every 4.5 s. If a rider's mass is 55 kg, with how much force does the ring push on her at the top of the ride? At the bottom?
    b. What is the longest rotation period of the wheel that will prevent the riders from falling off at the top?

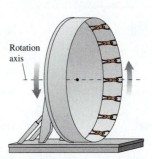

FIGURE P8.51

52. ‖ Suppose you swing a ball of mass m in a vertical circle on a string of length L. As you probably know from experience, there is a minimum angular velocity ω_{min} you must maintain if you want the ball to complete the full circle without the string going slack at the top.
    a. Find an expression for ω_{min}.
    b. Evaluate ω_{min} in rpm for a 65 g ball tied to a 1.0-m-long string.

53. ‖ A 30 g ball rolls around a 40-cm-diameter L-shaped track, shown in **FIGURE P8.53**, at 60 rpm. What is the magnitude of the *net* force that the track exerts on the ball? Rolling friction can be neglected.
    Hint: The track exerts more than one force on the ball.

FIGURE P8.53 **FIGURE P8.54**

54. ‖‖ **FIGURE P8.54** shows a small block of mass m sliding around the inside of an L-shaped track of radius r. The bottom of the track is frictionless; the coefficient of kinetic friction between the block and the wall of the track is μ_k. The block's speed is v_0 at $t_0 = 0$. Find an expression for the block's speed at a later time t.

55. ‖ The physics of circular motion sets an upper limit to the
BIO speed of human walking. (If you need to go faster, your gait changes from a walk to a run.) If you take a few steps and watch what's happening, you'll see that your body pivots in circular motion over your forward foot as you bring your rear foot forward for the next step. As you do so, the normal force of the ground on your foot decreases and your body tries to "lift off" from the ground.
    a. A person's center of mass is very near the hips, at the top of the legs. Model a person as a particle of mass m at the top of a leg of length L. Find an expression for the person's maximum walking speed v_{max}.
    b. Evaluate your expression for the maximum walking speed of a 70 kg person with a typical leg length of 70 cm. Give your answer in both m/s and mph, then comment, based on your experience, as to whether this is a reasonable result. A "normal" walking speed is about 3 mph.

56. ‖ A 100 g ball on a 60-cm-long string is swung in a vertical circle about a point 200 cm above the floor. The tension in the string when the ball is at the very bottom of the circle is 5.0 N. A very sharp knife is suddenly inserted, as shown in **FIGURE P8.56**, to cut the string directly below the point of support. How far to the right of where the string was cut does the ball hit the floor?

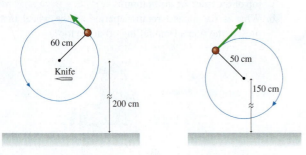

FIGURE P8.56   **FIGURE P8.57**

57. ‖ A 60 g ball is tied to the end of a 50-cm-long string and swung in a vertical circle. The center of the circle, as shown in **FIGURE P8.57**, is 150 cm above the floor. The ball is swung at the minimum speed necessary to make it over the top without the string going slack. If the string is released at the instant the ball is at the top of the loop, how far to the right does the ball hit the ground?

58. ‖ Elm Street has a pronounced dip at the bottom of a steep hill before going back uphill on the other side. Your science teacher has asked everyone in the class to measure the radius of curvature of the dip. Some of your classmates are using surveying equipment, but you decide to base your measurement on what you've learned in physics. To do so, you sit on a spring scale, drive through the dip at different speeds, and for each speed record the scale's reading as you pass through the bottom of the dip. Your data are as follows:

Speed (m/s)	Scale reading (N)
5	599
10	625
15	674
20	756
25	834

Sitting on the scale while the car is parked gives a reading of 588 N. Analyze your data, using a graph, to determine the dip's radius of curvature.

59. ‖ A 100 g ball on a 60-cm-long string is swung in a vertical circle about a point 200 cm above the floor. The string suddenly breaks when it is parallel to the ground and the ball is moving upward. The ball reaches a height 600 cm above the floor. What was the tension in the string an instant before it broke?

60. ‖ Scientists design a new particle accelerator in which protons (mass 1.7×10^{-27} kg) follow a circular trajectory given by $\vec{r} = c\cos(kt^2)\hat{\imath} + c\sin(kt^2)\hat{\jmath}$, where $c = 5.0$ m and $k = 8.0 \times 10^4$ rad/s^2 are constants and t is the elapsed time.
    CALC
    a. What is the radius of the circle?
    b. What is the proton's speed at $t = 3.0$ s?
    c. What is the force on the proton at $t = 3.0$ s? Give your answer in component form.

61. ‖‖ A 1500 kg car starts from rest and drives around a flat 50-m-diameter circular track. The forward force provided by the car's drive wheels is a constant 1000 N.
    a. What are the magnitude and direction of the car's acceleration at $t = 10$ s? Give the direction as an angle from the r-axis.
    b. If the car has rubber tires and the track is concrete, at what time does the car begin to slide out of the circle?

62. ‖‖ A 500 g steel block rotates on a steel table while attached to a 2.0-m-long massless rod. Compressed air fed through the rod is ejected from a nozzle on the back of the block, exerting a thrust force of 3.5 N. The nozzle is 70° from the radial line, as shown in **FIGURE P8.62**. The block starts from rest.
    a. What is the block's angular velocity after 10 rev?
    b. What is the tension in the rod after 10 rev?

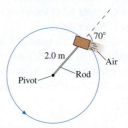

FIGURE P8.62

63. ‖‖ A 2.0 kg ball swings in a vertical circle on the end of an 80-cm-long string. The tension in the string is 20 N when its angle from the highest point on the circle is $\theta = 30°$.
    a. What is the ball's speed when $\theta = 30°$?
    b. What are the magnitude and direction of the ball's acceleration when $\theta = 30°$?

In Problems 64 and 65 you are given the equation used to solve a problem. For each of these, you are to
    a. Write a realistic problem for which this is the correct equation. Be sure that the answer your problem requests is consistent with the equation given.
    b. Finish the solution of the problem.

64. $60 \text{ N} = (0.30 \text{ kg})\omega^2(0.50 \text{ m})$

65. $(1500 \text{ kg})(9.8 \text{ m/s}^2) - 11,760 \text{ N} = (1500 \text{ kg}) v^2/(200 \text{ m})$

Challenge Problems

66. ‖‖ Sam (75 kg) takes off up a 50-m-high, 10° frictionless slope on his jet-powered skis. The skis have a thrust of 200 N. He keeps his skis tilted at 10° after becoming airborne, as shown in **FIGURE CP8.66**. How far does Sam land from the base of the cliff?

FIGURE CP8.66

67. ‖‖ In the absence of air resistance, a projectile that lands at the
    CALC elevation from which it was launched achieves maximum range when launched at a 45° angle. Suppose a projectile of mass m is launched with speed v_0 into a headwind that exerts a constant, horizontal retarding force $\vec{F}_{\text{wind}} = -F_{\text{wind}}\hat{\imath}$.
    a. Find an expression for the angle at which the range is maximum.
    b. By what percentage is the maximum range of a 0.50 kg ball reduced if $F_{\text{wind}} = 0.60$ N?

68. ||| The father of Example 8.2 stands at the summit of a conical hill as he spins his 20 kg child around on a 5.0 kg cart with a 2.0-m-long rope. The sides of the hill are inclined at 20°. He again keeps the rope parallel to the ground, and friction is negligible. What rope tension will allow the cart to spin with the same 14 rpm it had in the example?

69. ||| A small bead slides around a horizontal circle at height y inside the cone shown in FIGURE CP8.69. Find an expression for the bead's speed in terms of a, h, y, and g.

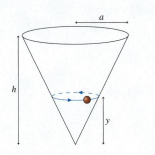

FIGURE CP8.69

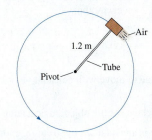

FIGURE CP8.70

70. ||| A 500 g steel block rotates on a steel table while attached to a 1.2-m-long hollow tube as shown in FIGURE CP8.70. Compressed air fed through the tube and ejected from a nozzle on the back of the block exerts a thrust force of 4.0 N perpendicular to the tube.

The maximum tension the tube can withstand without breaking is 50 N. If the block starts from rest, how many revolutions does it make before the tube breaks?

71. ||| If a vertical cylinder of water (or any other liquid) rotates about its axis, as shown in FIGURE CP8.71, the surface forms a smooth curve. Assuming that the water rotates as a unit (i.e., all the water rotates with the same angular velocity), show that the shape of the surface is a parabola described by the equation $z = (\omega^2/2g)r^2$.

Hint: Each particle of water on the surface is subject to only two forces: gravity and the normal force due to the water underneath it. The normal force, as always, acts perpendicular to the surface.

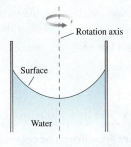

FIGURE CP8.71

Newton's Laws

KEY FINDINGS   What are the overarching findings of Part I?

- **Kinematics** is the description of motion. Motion can be described
  - Visually
  - Graphically
  - Mathematically

- **Forces** cause objects to *change* their motion—that is, to accelerate.
- Objects **interact** by exerting equal but opposite forces on each other.

LAWS   What laws of physics govern motion?

Newton's first law   An object will remain at rest or will continue to move with constant velocity if and only if $\vec{F}_{net} = \vec{0}$. The object is in **mechanical equilibrium**.

Newton's second law $\vec{F}_{net} = m\vec{a}$ A net force on an object causes the object to accelerate.

Newton's third law $\vec{F}_{A\,on\,B} = -\vec{F}_{B\,on\,A}$ For every action, there is an equal but opposite reaction.

MODELS   What are the most common models for applying the laws of physics to moving objects?

Constant force/Uniform acceleration

- Model the object as a particle.
  - Acceleration is in the direction of the net force and is constant.

- Mathematically:
  - Newton's second law is
$$\vec{F}_{net} = \sum_i \vec{F}_i = m\vec{a}$$
  - Use *xy*-coordinates.
  - Constant-acceleration kinematics:
$$v_{fs} = v_{is} + a_s\,\Delta t$$
$$s_f = s_i + v_{is}\,\Delta t + \tfrac{1}{2}a_s(\Delta t)^2$$
$$v_{fs}^2 = v_{is}^2 + 2a_s\,\Delta s$$

a_s Horizontal line
0 ─────────── t
The acceleration is constant.

v_s Straight line
v_{is} ─────── t
The slope is a_s.

s Parabola
s_i ─────── t
The slope is v_s.

- Special cases:
  - Uniform motion: $a_s = 0$. The displacement graph is a straight line with slope v_s.
  - Projectile motion: The only force is gravity. Horizontal motion is uniform; vertical motion has constant $a_y = -g$.

Central force/Uniform circular motion

- Model the object as a particle.
  - The force causes a constant centripetal acceleration. The particle moves around a circle at constant speed and with constant angular velocity.

- Mathematically: Newton's second law is
  - $\vec{F}_{net} = (mv^2/r$ or $m\omega^2 r$, toward the center$)$
  - Use *rtz*-coordinates.
  - Uniform-circular-motion kinematics:
    - The tangential velocity is $v_t = \omega r$.
    - The centripetal acceleration is v^2/r or $\omega^2 r$.
    - ω and v_t are positive for a ccw rotation, negative for a cw rotation.

- General case: Accelerated circular/rotational motion. Angular acceleration is contant.
$$\omega_f = \omega_i + \alpha\,\Delta t$$
$$\theta_f = \theta_i + \omega_i\,\Delta t + \tfrac{1}{2}\alpha(\Delta t)^2$$
$$\omega_f^2 = \omega_i^2 + 2\alpha\,\Delta\theta$$
These equations are analogous to constant-acceleration kinematics.

TOOLS   What are the most important tools for analyzing the physics of motion?

- The particle model and motion diagrams

- Vectors

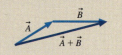

- Free-body diagrams

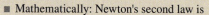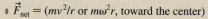

- Interaction diagrams

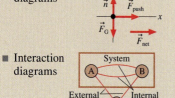

External forces / Internal interactions   Environment

- Calculus and graphical analysis

$$v_s = ds/dt = \text{ slope of the position graph}$$
$$a_s = dv_s/dt = \text{ slope of the velocity graph}$$
$$v_{fs} = v_{is} + \int_{t_i}^{t_f} a_s\,dt = v_{is} + \text{ area under acceleration curve}$$
$$s_f = s_i + \int_{t_i}^{t_f} v_s\,dt = s_i + \text{ area under the velocity curve}$$

PART
II Conservation Laws

OVERVIEW

Why Some Things Don't Change

Part I of this textbook was about *change*. One particular type of change—motion—is governed by Newton's second law. Although Newton's second law is a very powerful statement, it isn't the whole story. Part II will now focus on things that *stay the same* as other things around them change.

Consider, for example, an explosive chemical reaction taking place inside a closed, sealed box. No matter how violent the explosion, the total mass of the products—the final mass M_f —is the same as the initial mass M_i of the reactants. In other words, matter cannot be created or destroyed, only rearranged. This is an important and powerful statement about nature.

A quantity that stays the same throughout an interaction is said to be *conserved*. The most important such quantity is *energy*. If a system of interacting objects is *isolated*—an important qualification—then the energy of the system never changes no matter how complex the interactions. This description of how nature behaves, called the *law of conservation of energy*, is perhaps the most important physical law ever discovered.

But what is energy? How do you determine the energy of a system? These are not easy questions. Energy is an abstract idea, not as tangible or easy to picture as mass or force. Our modern concept of energy wasn't fully formulated until the middle of the 19th century, two hundred years after Newton, when the relationship between *energy* and *heat* was finally understood. That is a topic we will take up in Part V, where the concept of energy will be found to be the basis of thermodynamics. But all that in due time. In Part II we will be content to introduce the concept of energy and show how energy can be a useful problem-solving tool. We'll also meet another quantity—*momentum*—that is conserved under the proper circumstances.

Conservation laws give us a new and different perspective on motion. This is not insignificant. You've seen optical illusions where a figure appears first one way, then another, even though the information has not changed. Likewise with motion. Some situations are most easily analyzed from the perspective of Newton's laws; others make more sense from a conservation-law perspective. An important goal of Part II is to learn which is better for a given problem.

Energy is the lifeblood of modern society. These photovoltaic panels transform solar energy into electrical energy and, unavoidably, increased thermal energy.

9 Work and Kinetic Energy

The bow may be very contemporary, but it's still the bow string doing work on the arrow that makes the arrow fly.

IN THIS CHAPTER, you will begin your study of how energy is transferred and transformed.

How should we think about energy?

Chapters 9 and 10 will develop the basic energy model, a powerful set of ideas for using energy. A key distinction is between the system, which has energy, and the environment. Energy can be transferred between the system and the environment or transformed within the system.

« LOOKING BACK Section 7.1 Interacting objects

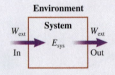

What are some important forms of energy?

Three important forms of energy:

■ Potential energy is energy associated with an object's *position*.
■ Kinetic energy is energy associated with an object's *motion*.
■ Thermal energy is the energy of the random motion of *atoms* within an object.

Energy is measured in joules.

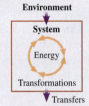

Potential energy

Kinetic energy

What is work?

A process that *changes the energy of a system by mechanical means*—pushing or pulling on it—is called work.

Work W is done when a force pushes or pulls a particle through a displacement, thus changing the particle's kinetic energy.

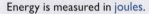

Force does work on the particle

$\vec{F}$

$\vec{F}$

$\Delta \vec{r}$

What laws govern energy?

Working with energy is very much like accounting: A system's energy E changes by the amount of work done on the system. The mathematical statement of this idea is called the energy principle:

$$\Delta E_{\text{sys}} = W_{\text{ext}}$$

Environment

W_{ext} In

System E_{sys}

W_{ext} Out

What is power?

Power is the rate at which energy is transferred or transformed. For machines, power is the rate at which they do work. For electricity, power is the rate at which electric energy is transformed into heat, sound, or light. Power is measured in watts, where 1 watt is a rate of 1 joule per second.

100 W

Why is energy important?

Energy is one of the most important concepts in science, engineering, and society. Some would say it is *the* most important. All life depends on energy, transformed from solar energy to chemical energy to us. Society depends on energy, from industry and transportation to heating and cooling our buildings. Using energy wisely and efficiently is a key concern of the 21st century.

9.1 Energy Overview

Energy. It's a word you hear all the time, and everyone has some sense of what *energy* means. Moving objects have energy; energy is the ability to make things happen; energy is associated with heat and with electricity; we're constantly told to conserve energy; living organisms need energy; and engineers harness energy to do useful things. Some scientists consider the *law of conservation of energy* to be the most important of all the laws of nature. But all that in due time—first we have to start with the basic ideas.

Just what is energy? The concept of energy has grown and changed with time, and it is not easy to define in a general way just what energy is. Rather than starting with a formal definition, we're going to let the concept of energy expand slowly over the course of several chapters. Our goal is to understand the characteristics of energy, how energy is used, and how energy is transformed from one form into another. It's a complex story, so we'll take it step by step until all the pieces are in place.

Some important forms of energy

Kinetic energy K	Potential energy U	Thermal energy E_{th}

Kinetic energy is the energy of motion. All moving objects have kinetic energy. The more massive an object or the faster it moves, the larger its kinetic energy.	Potential energy is stored energy associated with an object's position. The roller coaster's gravitational potential energy depends on its height above the ground.	Thermal energy is the sum of the microscopic kinetic and potential energies of all the atoms and bonds that make up the object. An object has more thermal energy when hot than when cold.

The Energy Principle

« Section 7.1 introduced *interaction diagrams* and the very important distinction between the **system,** those objects whose motion and interactions we wish to analyze, and the **environment,** objects external to the system but exerting forces on the system. The most important step in an energy analysis is to clearly define the system. Why? Because energy is not some disembodied, ethereal substance; it's the energy *of something.* Specifically, it's *the energy of a system.*

FIGURE 9.1 illustrates the idea pictorially. The system *has energy,* the **system energy,** which we'll designate E_{sys}. There are many kinds or forms of energy: kinetic energy K, potential energy U, thermal energy E_{th}, chemical energy, and so on. We'll introduce these one by one as we go along. Within the system, energy can be *transformed without loss.* Chemical energy can be transformed into kinetic energy, which is then transformed into thermal energy. As long as the system is not interacting with the environment, the total energy of the system is unchanged. You'll recognize this idea as an initial statement of the *law of conservation of energy.*

But systems often do interact with their environment. Those interactions *change* the energy of the system, either increasing it (energy added) or decreasing it (energy removed). We say that interactions with the environment *transfer* energy into or out of the system. Interestingly, there are only two ways to transfer energy. One is by mechanical means, using forces to push and pull on the system. A process that

FIGURE 9.1 A system-environment perspective on energy.

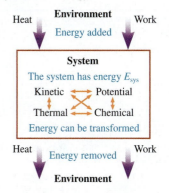

transfers energy to or from a system by mechanical means is called **work,** with the symbol *W*. We'll have a lot to say about work in this chapter. The second is by thermal means when the environment is hotter or colder than the system. A process that transfers energy to or from a system by thermal means is called **heat.** We'll defer a discussion of heat until Chapter 18, but we wanted to mention it now in order to gain an overview of what the energy story is all about.

Some energy transfers ...   **... and transformations**

Putting a shot

System: The shot

Transfer: $W \rightarrow K$

The athlete (the environment) does work pushing the shot to give it kinetic energy.

Pulling a slingshot

System: The slingshot

Transfer: $W \rightarrow U$

The boy (the environment) does work by stretching the rubber band to give it potential energy.

A falling diver

System: The diver and the earth

Transformation: $U \rightarrow K$

The diver is speeding up as gravitational potential energy is transformed into kinetic energy.

A speeding meteor

System: The meteor and the air

Transformation: $K \rightarrow E_{\text{th}}$

The meteor and the air get hot enough to glow as the meteor's kinetic energy is transformed into thermal energy.

The key ideas are **energy transfer** between the environment and the system and **energy transformation** within the system. This is much like what happens with money. You may have several accounts at the bank—perhaps a checking account and a couple of savings accounts. You can move money back and forth between the accounts, thus transforming it without changing the total amount of money. Of course, you can also transfer money into or out of your accounts by making deposits or withdrawals. If we treat a withdrawal as a negative deposit—which is exactly what accountants do—simple accounting tells you that

$$\Delta(\text{balance}) = \text{net deposit}$$

That is, the change in your bank balance is simply the sum of all your deposits.

Energy accounting works the same way. Transformations of energy within the system move the energy around but don't change the total energy of the system. *Change* occurs only when there's a transfer of energy between the system and the environment. If we treat incoming energy as a positive transfer and outgoing energy as a negative transfer, and with work being the only energy-transfer process that we consider for now, we can write

$$\Delta E_{\text{sys}} = W_{\text{ext}} \tag{9.1}$$

where the subscript on *W* refers to external work done by the environment. This very simple looking statement, which is just a statement of energy accounting, is called the **energy principle.** But don't let the simplicity fool you; this will turn out to be an incredibly powerful tool for analyzing physical situations and solving problems.

The Basic Energy Model

We'll complete our energy overview—a roadmap of the next two chapters—with the **basic energy model.**

MODEL 9.1

Basic energy model

Energy is a property of the system.

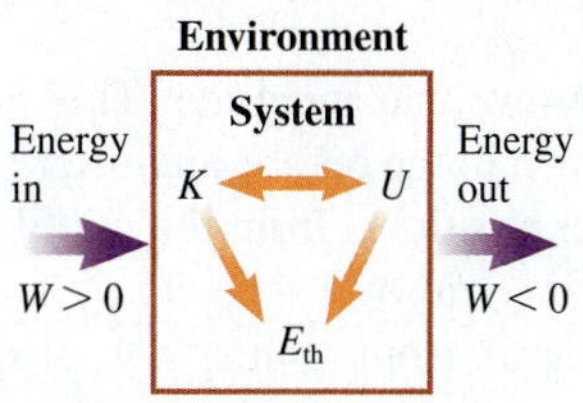

- Energy is *transformed* within the system without loss.
- Energy is *transferred* to and from the system by forces from the environment.
  - The forces do *work* on the system.
  - $W > 0$ for energy added.
  - $W < 0$ for energy removed.
- The energy of an *isolated system*—one that doesn't interact with its environment—does not change. We say it is *conserved*.
- The energy principle is $\Delta E_{sys} = W_{ext}$.
- Limitations: Model fails if there is energy transfer via thermal processes (heat).

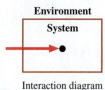

Exercise 1

We call this model *basic* because, for now, the only forms of energy we'll consider are kinetic energy, potential energy, and thermal energy, and the only energy-transfer process we'll consider is work. This is an excellent model for a mechanical process, but it's not complete. We'll expand the model when we get to thermodynamics by adding chemical energy, another form of energy, and heat, another energy-transfer process. And this model, although basic, still has many complexities, so we'll be developing it piece by piece in this chapter and the next.

9.2 Work and Kinetic Energy for a Single Particle

Let's start our investigation of energy with the simplest possible situation: One particle of mass m is acted on by one constant force $\vec{F}$ that acts parallel to the direction of motion, pushing or pulling on the particle as it undergoes a displacement Δs. We define the particle to be the system—a one-particle system—while the agent of the force is in the environment. **FIGURE 9.2** shows both an interaction diagram and a new kind of pictorial representation, a **before-and-after representation,** in which we show an object *before* and *after* an interaction and, as usual, establish a coordinate system and define appropriate symbols.

You know what's going to happen. If the force is in the direction of motion—the situation shown in the figure—the particle will speed up and its "energy of motion" will increase. Conversely, if the force opposes the motion, the particle will slow down and lose energy. Our goal is to make this idea precise by discovering exactly how the changing energy is related to the applied force.

We'll start by writing Newton's second law for the particle:

$$F_s = ma_s = m\frac{dv_s}{dt} \tag{9.2}$$

Newton's second law tells us how the particle's velocity changes with time. But suppose we want to know how the velocity changes with position. To answer that question, we can use the chain rule that you've learned in calculus:

$$\frac{dv_s}{dt} = \frac{dv_s}{ds}\frac{ds}{dt} = v_s\frac{dv_s}{ds} \tag{9.3}$$

where in the last step we used $ds/dt = v_s$. With this, we can write Newton's second law as

$$F_s = mv_s\frac{dv_s}{ds} \tag{9.4}$$

FIGURE 9.2 The interaction diagram and before-and-after representation for a one-particle system.

Interaction diagram

A before-and-after representation shows the object's position and velocity before and after an interaction.

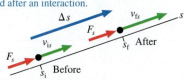

Before-and-after representation

Now we have an alternative version of Newton's second law in terms of the rate of change of velocity with position.

To use Equation 9.4, we first rewrite it as

$$mv_s\, dv_s = F_s\, ds \tag{9.5}$$

Now we can integrate. This is going to be a definite integral over just the motion shown in the before-and-after representation. That is, the right side will be an integral over position s from the initial position s_i to the final position s_f. The left side will be an integral over velocity v_s, and its limits have to match the limits of the right-hand integral: from v_{is} at s_i to v_{fs} at s_f. Thus we have

$$\int_{v_{is}}^{v_{fs}} mv_s\, dv_s = \int_{s_i}^{s_f} F_s\, ds \tag{9.6}$$

We have two integrals to examine, and we'll do them one by one. We can start with the integral on the left, which is of the form $\int x\, dx$. Factoring out m, which is a constant, we find

$$m \int_{v_{is}}^{v_{fs}} v_s\, dv_s = m \left[\tfrac{1}{2} v_s^2 \right]_{v_{is}}^{v_{fs}} = \tfrac{1}{2} m v_{fs}^2 - \tfrac{1}{2} m v_{is}^2 = \Delta \left(\tfrac{1}{2} m v^2 \right) = \Delta K \tag{9.7}$$

You'll notice that we dropped the subscript s in the next-to-last step. v_s is a vector component, with a sign to indicate direction, but the sign makes no difference after v_s is squared. All that matters is the particle's *speed* v.

The last step in Equation 9.7 introduces a new quantity

$$K = \tfrac{1}{2} m v^2 \qquad \text{(kinetic energy)} \tag{9.8}$$

which is called the **kinetic energy** of the particle. **Kinetic energy is energy of motion.** It depends on the particle's mass and speed but not on its position. Furthermore, kinetic energy is a property or characteristic of the system. So what we've calculated with the left-hand integral is $\Delta K = K_f - K_i$, the *change* in the system's kinetic energy as the force pushes the particle through the displacement Δs. ΔK is positive if the particle speeds up (gain of kinetic energy), negative if it slows down (loss of kinetic energy).

NOTE By its definition, kinetic energy can *never* be negative. Finding a negative value for K while solving a problem is an indication that you've made a mistake somewhere.

The unit of kinetic energy is mass multiplied by velocity squared. In SI units, this is $\text{kg m}^2/\text{s}^2$. Because energy is so important, the unit of energy is given its own name, the **joule.** We define

$$1 \text{ joule} = 1 \text{ J} = 1 \text{ kg m}^2/\text{s}^2$$

All other forms of energy are also measured in joules.

To give you an idea about the size of a joule, consider a 0.5 kg mass (≈ 1 lb on earth) moving at 4 m/s (≈ 10 mph). Its kinetic energy is

$$K = \tfrac{1}{2} m v^2 = \tfrac{1}{2} (0.5 \text{ kg})(4 \text{ m/s})^2 = 4 \text{ J}$$

This suggests that everyday objects moving at ordinary speeds will have energies from a fraction of a joule up to, perhaps, a few thousand joules. A running person has $K \approx 1000$ J, while a high-speed truck might have $K \approx 10^6$ J.

NOTE You *must* have masses in kilograms and velocities in m/s before doing energy calculations.

STOP TO THINK 9.1 A 1000 kg car has a speed of 20 m/s. A 2000 kg truck has a speed of 10 m/s. Which has more kinetic energy?

a. The car.   b. The truck.   c. Their kinetic energies are the same.

Work

Now let's turn to the integral on the right-hand side of Equation 9.6. This integral is telling us *by how much* the kinetic energy changes due to the force. That is, it is the energy transferred to or from the system by the force. Earlier we said that a process that transfers energy to or from a system by mechanical means—by forces—is called work. So the integral on the right-hand side of Equation 9.6 must be the *work W done by force* $\vec{F}$.

Having identified the left side of Equation 9.6 with the changing kinetic energy of the system and the right side with the work done on the system, we can rewrite Equation 9.6 as

Change in the system's kinetic energy   Amount of work done by an external force

$$\Delta K = K_f - K_i = W \qquad (9.9)$$

Final kinetic energy   Initial kinetic energy

(Energy principle for a one-particle system)

This is our first version of the energy principle. Notice that it's a cause-and-effect statement: **The work done on a one-particle system causes the system's kinetic energy to change.**

We'll study work thoroughly in the next section, but for now we're considering only the simplest case of a constant force parallel to the direction of motion (the *s*-axis). A constant force can be factored out of the integral, giving

$$W = \int_{s_i}^{s_f} F_s \, ds = F_s \int_{s_i}^{s_f} ds = F_s s \Big|_{s_i}^{s_f} = F_s(s_f - s_i)$$
$$= F_s \Delta s \qquad (9.10)$$

The unit of work, that of force multiplied by distance, is the N m. Recall that $1\,\text{N} = 1\,\text{kg m/s}^2$. Thus

$$1\,\text{N m} = 1\,(\text{kg m/s}^2)\,\text{m} = 1\,\text{kg m}^2/\text{s}^2 = 1\,\text{J}$$

Thus the unit of work is really the unit of energy. This is consistent with the idea that work is a transfer of energy. Rather than use N m, we will measure work in joules.

EXAMPLE 9.1 | **Firing a cannonball**

A 5.0 kg cannonball is fired straight up at 35 m/s. What is its speed after rising 45 m?

MODEL Let the system consist of only the cannonball, which we model as a particle. Assume that air resistance is negligible.

VISUALIZE FIGURE 9.3 is a before-and-after pictorial representation. Because before-and-after representations are usually simpler than

FIGURE 9.3 Before-and-after representation of the cannonball.

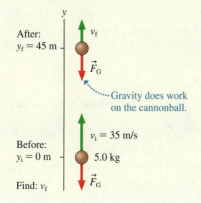

the pictorial representation used in dynamics problems, you can include known information right on the diagram instead of making a Known table.

SOLVE It isn't necessary to use work and energy to solve this problem. You could solve it as a free-fall problem. Or you might have previously learned to solve problems like this using potential energy, a topic we'll take up in the next chapter. But using work and energy emphasizes how these two key ideas are related, and it gives us a simple example of the *problem-solving process* before we get to more complex problems. The energy principle is $\Delta K = W$, where work is done by the force of gravity. The cannonball is rising, so its displacement Δy is positive. But the force vector points down, with component $F_y = -mg$. Thus gravity does work

$$W = F_y \Delta y = -mg\,\Delta y = -(5.0\,\text{kg})(9.80\,\text{m/s}^2)(45\,\text{m}) = -2210\,\text{J}$$

as the cannonball rises 45 m. A negative work means that the system is losing energy, which is what we expect as the cannonball slows.

Continued

The cannonball's change of kinetic energy is $\Delta K = K_f - K_i$. The initial kinetic energy is

$$K_i = \tfrac{1}{2}mv_i^2 = \tfrac{1}{2}(5.0 \text{ kg})(35 \text{ m/s})^2 = 3060 \text{ J}$$

Using the energy principle, we find the final kinetic energy to be $K_f = K_i + W = 3060 \text{ J} - 2210 \text{ J} = 850 \text{ J}$. Then

$$v_f = \sqrt{\frac{2K_f}{m}} = \sqrt{\frac{2(850 \text{ J})}{5.0 \text{ kg}}} = 18 \text{ m/s}$$

ASSESS 35 m/s ≈ 70 mph. A cannonball fired upward at that speed is going to go fairly high. To have lost half its speed at a height of 45 m ≈ 150 ft seems reasonable.

Signs of Work

FIGURE 9.4 How to determine the sign of W.

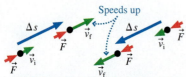

Work is *positive* when the force acts in the *same direction* as the displacement.

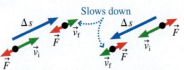

Work is *negative* when the force and displacement are in *opposite* directions.

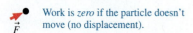

Work is *zero* if the particle doesn't move (no displacement).

Work can be either positive or negative, but some care is needed to get the sign right when calculating work. The key is to remember that work is an energy transfer. If the force causes the particle to speed up, then the work done by that force is positive. Similarly, negative work means that the force is causing the object to slow and lose energy.

The sign of W is *not* determined by the direction the force vector points. That's only half the issue. The displacement Δs also has a sign, so you have to consider both the force direction *and* the displacement direction. As **FIGURE 9.4** shows, **work is positive when the force acts in the direction of the displacement** (causing the particle to speed up). Similarly, **work is negative when force and displacement are in opposite directions** (causing the particle to slow). And there's no work at all ($W = 0$) if the particle doesn't move!

STOP TO THINK 9.2   A rock falls to the bottom of a deep canyon. Is the work done on the rock by gravity positive, negative, or zero?

Extending the Model

Our initial model has been of a single particle acted on by a constant force parallel to the displacement. We can easily make some straightforward extensions of this model to slightly more complex—and interesting—situations:

- **Force perpendicular to the displacement:** A force parallel to a particle's displacement causes the particle to speed up or slow down, changing its energy. But a force *perpendicular* to the displacement does *not* change the particle's speed; it is neither speeding up nor slowing down. Its energy is not changing, so no work is being done on it. **A force perpendicular to the displacement does no work.**

- **Multiple forces:** If multiple forces act on a system, their works add. That is, $\Delta K = W_{tot}$, where the total work done is

$$W_{tot} = W_1 + W_2 + W_3 + \cdots \tag{9.11}$$

- **Multiparticle systems:** If a system has more than one particle, the system's energy is the total kinetic energy of all the particles:

$$E_{sys} = K_{tot} = K_1 + K_2 + K_3 + \cdots \tag{9.12}$$

K_{tot} is truly a *system* energy, not the energy of any one particle. How does K_{tot} change when work is done? You can see from its definition that ΔK_{tot} is the sum of all the individual kinetic-energy changes, and each of those changes is the work done on that particular particle. Thus

$$\Delta K_{tot} = W_{tot} \tag{9.13}$$

where now W_{tot} is the total work done on *all* the particles in the system.

NOTE You might expect $W_{tot} = (F_{net})_s \Delta s$, where $\vec{F}_{net}$ is the net work on the system. This is true for a one-particle system (if all the forces are constant), but in general it is *not* true for a multiparticle system because each particle undergoes a different displacement. You must find the work done on each particle, then sum those to find the total work done on the system.

STOP TO THINK 9.3 Two equal-mass pucks on frictionless ice are pushed toward each other by two equal but opposite forces. Is the total work positive, negative, or zero?

9.3 Calculating the Work Done

Section 9.2 introduced two key ideas: (1) a system has energy and (2) work is a mechanical process that changes the system's energy. Now we're ready to look more closely at how to calculate the work done in different situations. Although we'll be focusing on the mathematical techniques of calculating work, it's important to keep in mind that our real goal is to learn how the energy of a system changes when forces are applied to it.

"Work" is a common word in the English language, with many meanings. Work might refer to physical exertion, to your job or occupation, or even to a work of art. But set aside those ideas about work because they are *not* what work means in physics. Work, as we'll use the word, is a *process*. Specifically, it is a process that changes a system's energy by mechanical means—pushing or pulling on it with forces. We say that work *transfers* energy between the environment and the system.

Equation 9.6 defined work as

$$W = \int_{s_i}^{s_f} F_s \, ds \qquad (9.14)$$

(work done by force $\vec{F}$ as a particle is displaced from s_i to s_f)

where, to remind you, F_s is the component of $\vec{F}$ in the direction of motion (the s-direction). We began by looking at a force that was parallel to the displacement, but such a restriction is not required because any force component perpendicular to the motion does no work. Equation 9.14 is, in fact, a general definition of work.

We'll start by learning how to calculate work for constant forces, and we'll introduce a new mathematical idea, the *dot product* of two vectors, that will allow us to write the work in a compact notation. Then we'll consider the work done by a variable force that changes as the particle moves.

Constant Force

FIGURE 9.5 shows a particle moving in a straight line. A constant force $\vec{F}$, which makes an angle θ with respect to the particle's displacement $\Delta \vec{r}$, acts on the particle throughout its motion. We've established an s-axis in the direction of motion, and you can see that the force component along the direction of motion is $F_s = F \cos \theta$. According to Equation 9.14, the work done on the particle by this force is

$$W = \int_{s_i}^{s_f} F_s \, ds = \int_{s_i}^{s_f} F \cos \theta \, ds \qquad (9.15)$$

Both F and $\cos \theta$ are constant, so they can be taken outside the integral. Thus

$$W = F \cos \theta \int_{s_i}^{s_f} ds = F \cos \theta (s_f - s_i) \qquad (9.16)$$

Now s_i and s_f are specific to this coordinate system, but their *difference* $s_f - s_i$ is Δr, the magnitude of the particle's displacement vector. Thus we can write more generally, independent of any specific coordinate system, that the work done by the constant force $\vec{F}$ is

$$W = F(\Delta r) \cos \theta \qquad \text{(work done by a constant force)} \qquad (9.17)$$

where θ is the angle between the force and the particle's displacement $\Delta \vec{r}$.

NOTE You may have learned in an earlier physics course that work is "force times distance." This is *not* the definition of work, merely a special case. Work is "force times distance" only if the force is constant *and* parallel to the displacement ($\theta = 0°$).

FIGURE 9.5 Work being done by a constant force.

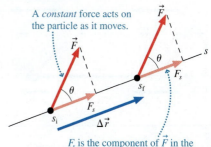

A *constant* force acts on the particle as it moves.

F_s is the component of $\vec{F}$ in the direction of motion. It causes the particle to speed up or slow down.

EXAMPLE 9.2 | **Pulling a suitcase**

A strap inclined upward at a 45° angle pulls a suitcase 100 m through the airport. The tension in the strap is 20 N. How much work does the tension force do on the suitcase?

MODEL Let the system consist of only the suitcase, which we model as a particle.

VISUALIZE FIGURE 9.6 is a before-and-after pictorial representation.

SOLVE The motion is along the x-axis, so in this case $\Delta r = \Delta x$. We can use Equation 9.17 to find that the tension does work:

$$W = T(\Delta x)\cos\theta = (20\text{ N})(100\text{ m})\cos 45° = 1400\text{ J}$$

FIGURE 9.6 Pictorial representation of the suitcase.

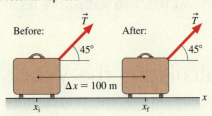

ASSESS Because a person pulls the strap, we would say informally that the person does 1400 J of work on the suitcase.

According to the basic energy model, work can be either positive or negative to indicate energy transfer into or out of the system. The quantities F and Δr are always positive, so **the sign of W is determined entirely by the angle θ between the force $\vec{F}$ and the displacement $\Delta\vec{r}$.**

TACTICS BOX 9.1 ⓂⓅ

Calculating the work done by a constant force

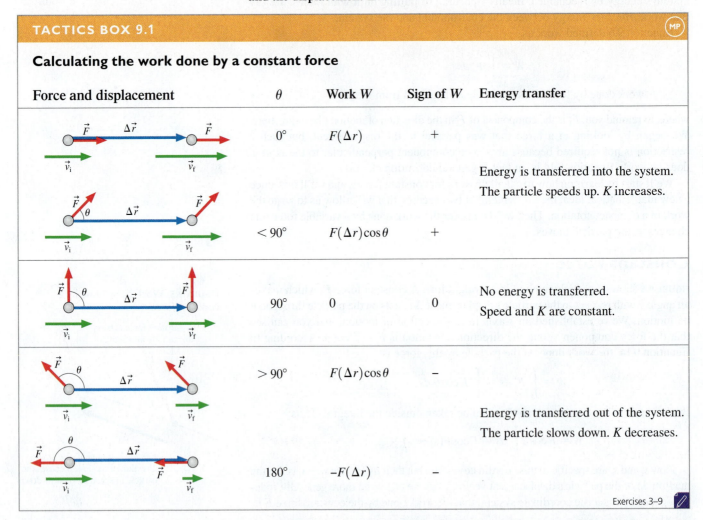

Force and displacement	θ	Work W	Sign of W	Energy transfer
	0°	$F(\Delta r)$	+	
				Energy is transferred into the system. The particle speeds up. K increases.
	< 90°	$F(\Delta r)\cos\theta$	+	
	90°	0	0	No energy is transferred. Speed and K are constant.
	> 90°	$F(\Delta r)\cos\theta$	–	
				Energy is transferred out of the system. The particle slows down. K decreases.
	180°	$-F(\Delta r)$	–	

Exercises 3–9 ✐

NOTE The sign of W depends on the angle between the force vector and the displacement vector, *not* on the coordinate axes. A force to the left does *positive* work if it pushes a particle to the left (the force and the displacement are in the same direction) even though the force component F_x is negative. Think about whether the force is trying to increase the particle's speed ($W > 0$) or decrease the particle's speed ($W < 0$).

EXAMPLE 9.3 | Launching a rocket

A 150,000 kg rocket is launched straight up. The rocket motor generates a thrust of 4.0×10^6 N. What is the rocket's speed at a height of 500 m?

MODEL Let the system consist of only the rocket, which we model as a particle. Thrust and gravity are constant forces that do work on the rocket. We'll ignore air resistance and any slight mass loss.

VISUALIZE FIGURE 9.7 shows a before-and-after representation and a free-body diagram.

FIGURE 9.7 Before-and-after representation and free-body diagram of a rocket launch.

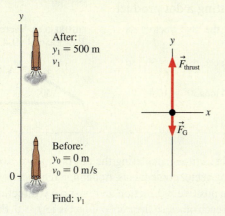

SOLVE We can solve this problem with the energy principle, $\Delta K = W_{tot}$. Both forces do work on the rocket. The thrust is in the direction of motion, with $\theta = 0°$, and thus

$$W_{thrust} = F_{thrust}(\Delta r) = (4.0 \times 10^6 \text{ N})(500 \text{ m}) = 2.00 \times 10^9 \text{ J}$$

The gravitational force points downward, opposite the displacement $\Delta \vec{r}$, so $\theta = 180°$. Thus the work done by gravity is

$$W_{grav} = -F_G(\Delta r) = -mg(\Delta r)$$
$$= -(1.5 \times 10^5 \text{ kg})(9.8 \text{ m/s}^2)(500 \text{ m}) = -0.74 \times 10^9 \text{ J}$$

The work done by the thrust is positive. By itself, the thrust would cause the rocket to speed up. The work done by gravity is negative, not because $\vec{F}_G$ points down but because $\vec{F}_G$ is opposite the displacement. By itself, gravity would cause the rocket to slow down. The energy principle, using $K_i = 0$, is

$$\Delta K = \tfrac{1}{2}mv_1^2 - 0 = W_{tot} = W_{thrust} + W_{grav} = 1.26 \times 10^9 \text{ J}$$

Solving for the speed, we find

$$v_1 = \sqrt{\frac{2W_{tot}}{m}} = 130 \text{ m/s}$$

ASSESS The total work is positive, meaning that energy is transferred *to* the rocket. In response, the rocket speeds up.

STOP TO THINK 9.4 A crane uses a single cable to lower a steel girder into place. The girder moves with constant speed. The cable tension does work W_T and gravity does work W_G. Which statement is true?

a. W_T is positive and W_G is positive.
b. W_T is positive and W_G is negative.
c. W_T is negative and W_G is positive.
d. W_T is negative and W_G is negative.
e. W_T and W_G are both zero.

Work as a Dot Product of Two Vectors

There's something different about the quantity $F(\Delta r)\cos\theta$ in Equation 9.17. We've spent many chapters adding vectors, but this is the first time we've *multiplied* two vectors. Multiplying vectors is not like multiplying scalars. In fact, there is more than one way to multiply vectors. We will introduce one way now, the *dot product*.

FIGURE 9.8 shows two vectors, $\vec{A}$ and $\vec{B}$, with angle α between them. We define the **dot product** of $\vec{A}$ and $\vec{B}$ as

$$\vec{A} \cdot \vec{B} = AB\cos\alpha \tag{9.18}$$

A dot product *must have* the dot symbol · between the vectors. The notation $\vec{A}\vec{B}$, without the dot, is *not* the same thing as $\vec{A} \cdot \vec{B}$. The dot product is also called the **scalar product** because the value is a scalar. Later, when we need it, we'll introduce a different way to multiply vectors called the *cross product*.

The dot product of two vectors depends on the orientation of the vectors. FIGURE 9.9 shows five different situations, including the three "special cases" where $\alpha = 0°, 90°$, and $180°$.

FIGURE 9.8 Vectors $\vec{A}$ and $\vec{B}$, with angle α between them.

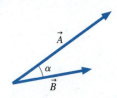

NOTE The dot product of a vector with itself is well defined. If $\vec{B} = \vec{A}$ (i.e., $\vec{B}$ is a copy of $\vec{A}$), then $\alpha = 0°$. Thus $\vec{A} \cdot \vec{A} = A^2$.

FIGURE 9.9 The dot product $\vec{A} \cdot \vec{B}$ as α ranges from $0°$ to $180°$.

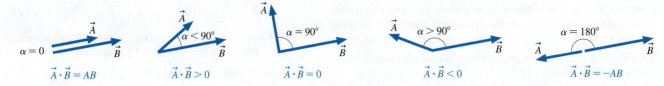

$\alpha = 0$ $\vec{A} \cdot \vec{B} = AB$

$\alpha < 90°$ $\vec{A} \cdot \vec{B} > 0$

$\alpha = 90°$ $\vec{A} \cdot \vec{B} = 0$

$\alpha > 90°$ $\vec{A} \cdot \vec{B} < 0$

$\alpha = 180°$ $\vec{A} \cdot \vec{B} = -AB$

EXAMPLE 9.4 | **Calculating a dot product**

Compute the dot product of the two vectors in **FIGURE 9.10**.

SOLVE The angle between the vectors is $\alpha = 30°$, so

$$\vec{A} \cdot \vec{B} = AB \cos \alpha = (3)(4)\cos 30° = 10.4$$

FIGURE 9.10 Vectors $\vec{A}$ and $\vec{B}$ of Example 9.4.

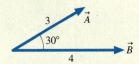

FIGURE 9.11 The unit vectors $\hat{\imath}$ and $\hat{\jmath}$.

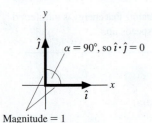

$\alpha = 90°$, so $\hat{\imath} \cdot \hat{\jmath} = 0$

Magnitude = 1

Like vector addition and subtraction, calculating the dot product of two vectors is often performed most easily using vector components. **FIGURE 9.11** reminds you of the unit vectors $\hat{\imath}$ and $\hat{\jmath}$ that point in the positive x-direction and positive y-direction. The two unit vectors are perpendicular to each other, so their dot product is $\hat{\imath} \cdot \hat{\jmath} = 0$. Furthermore, because the magnitudes of $\hat{\imath}$ and $\hat{\jmath}$ are 1, $\hat{\imath} \cdot \hat{\imath} = 1$ and $\hat{\jmath} \cdot \hat{\jmath} = 1$.

In terms of components, we can write the dot product of vectors $\vec{A}$ and $\vec{B}$ as

$$\vec{A} \cdot \vec{B} = (A_x \hat{\imath} + A_y \hat{\jmath}) \cdot (B_x \hat{\imath} + B_y \hat{\jmath})$$

Multiplying this out, and using the results for the dot products of the unit vectors:

$$\vec{A} \cdot \vec{B} = A_x B_x \hat{\imath} \cdot \hat{\imath} + (A_x B_y + A_y B_x)\hat{\imath} \cdot \hat{\jmath} + A_y B_y \hat{\jmath} \cdot \hat{\jmath}$$
$$= A_x B_x + A_y B_y \tag{9.19}$$

That is, the **dot product is the sum of the products of the components.**

EXAMPLE 9.5 | **Calculating a dot product using components**

Compute the dot product of $\vec{A} = 3\hat{\imath} + 3\hat{\jmath}$ and $\vec{B} = 4\hat{\imath} - \hat{\jmath}$.

SOLVE **FIGURE 9.12** shows vectors $\vec{A}$ and $\vec{B}$. We could calculate the dot product by first doing the geometry needed to find the angle between the vectors and then using Equation 9.18. But calculating the dot product from the vector components is much easier. It is

$$\vec{A} \cdot \vec{B} = A_x B_x + A_y B_y = (3)(4) + (3)(-1) = 9$$

FIGURE 9.12 Vectors $\vec{A}$ and $\vec{B}$.

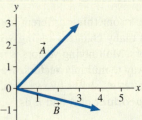

Looking at Equation 9.17, the work done by a constant force, you should recognize that it is the dot product of the force vector and the displacement vector:

$$W = \vec{F} \cdot \Delta\vec{r} \quad \text{(work done by a constant force)} \tag{9.20}$$

This definition of work is valid for a constant force.

EXAMPLE 9.6 | Calculating work using the dot product

A 70 kg skier is gliding at 2.0 m/s when he starts down a very slippery 50-m-long, 10° slope. What is his speed at the bottom?

MODEL Model the skier as a particle and interpret "very slippery" to mean frictionless. Use the energy principle to find his final speed.

VISUALIZE **FIGURE 9.13** shows a pictorial representation.

FIGURE 9.13 Pictorial representation of the skier.

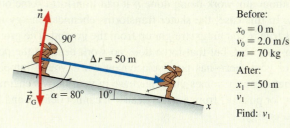

Before:
$x_0 = 0$ m
$v_0 = 2.0$ m/s
$m = 70$ kg

After:
$x_1 = 50$ m
v_1
Find: v_1

SOLVE The only forces on the skier are $\vec{F}_G$ and $\vec{n}$. The normal force is perpendicular to the motion and thus does no work. The work done by gravity is easily calculated as a dot product:

$$W = \vec{F}_G \cdot \Delta \vec{r} = mg(\Delta r)\cos\alpha$$

$$= (70\text{ kg})(9.8\text{ m/s}^2)(50\text{ m})\cos 80° = 5960\text{ J}$$

Notice that the angle *between* the vectors is 80°, not 10°. Then, from the energy principle, we find

$$\Delta K = \tfrac{1}{2}mv_1^2 - \tfrac{1}{2}mv_0^2 = W$$

$$v_1 = \sqrt{v_0^2 + \frac{2W}{m}} = \sqrt{(2.0\text{ m/s})^2 + \frac{2(5960\text{ J})}{70\text{ kg}}} = 13\text{ m/s}$$

NOTE While in the midst of the mathematics of calculating work, do not lose sight of what the energy principle is all about. It is a statement about *energy transfer:* Work causes a particle's kinetic energy to either increase or decrease.

STOP TO THINK 9.5 Which force does the most work as a particle undergoes displacement $\Delta\vec{r}$?

a. The 10 N force.
b. The 8 N force.
c. The 6 N force.
d. They all do the same amount of work.

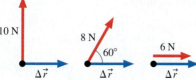

Zero-Work Situations

There are three common situations where *no* work is done. The most obvious is when the object doesn't move ($\Delta s = 0$). If you were to hold a 200 lb weight over your head, you might break out in a sweat and your arms would tire. You might feel that you had done a lot of work, but you would have done *zero* work in the physics sense because the weight was not displaced and thus you transferred no energy to it. **A force acting on a particle does no work unless the particle is displaced.**

FIGURE 9.14 shows a particle moving in uniform circular motion. As you learned in Chapter 8, uniform circular motion requires a force pointing toward the center of the circle. How much work does this force do?

Zero! You can see that the force is everywhere perpendicular to the small displacement $d\vec{s}$, so the dot product is zero and the force does *no* work on the particle. This shouldn't be surprising. The particle's speed, and hence its kinetic energy, doesn't change in uniform circular motion, so no energy is transferred to or from the system. **A force everywhere perpendicular to the motion does no work.** The friction force on a car turning a corner does no work. Neither does the tension force when a ball on a string is in circular motion.

Last, consider the roller skater in **FIGURE 9.15** who straightens her arms and pushes off from a wall. She applies a force to the wall and thus, by Newton's third law, the wall applies a force $\vec{F}_{W\text{ on }S}$ to her. How much work does this force do?

Surprisingly, zero. The reason is subtle but worth discussing because it gives us insight into how energy is transferred and transformed. The skater differs from suitcases and rockets in two important ways. First, the skater, as she extends her arms,

FIGURE 9.14 A perpendicular force does no work.

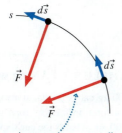

The force is everywhere perpendicular to the displacement, so it does no work.

FIGURE 9.15 Does the wall do work on the skater?

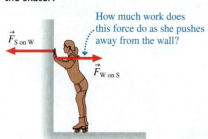

How much work does this force do as she pushes away from the wall?

is a *deformable object*. We cannot use the particle model for a deformable object. Second, the skater has an *internal source of energy*. Because she's a living object, she has an internal store of chemical energy that is available through metabolic processes.

Although the skater's center of mass is displaced, *the palms of her hands—where the force is exerted—are not*. The particles on which force $F_{W\ on\ S}$ acts have no displacement, and we've just seen that there's no work without displacement. The force acts, but the force doesn't push any physical thing through a displacement. Hence no work is done.

But the skater indisputably gains kinetic energy. How? Recall, from the energy overview that started this chapter, that the full energy principle is $\Delta E_{sys} = W_{ext}$. A system can gain kinetic energy without any work being done *if* it can transform some other energy into kinetic energy. In this case, the skater transforms chemical energy into kinetic energy. The same is true if you jump straight up from the ground. The ground applies an upward force to your feet, but that force does no work because the point of application—the soles of your feet—has no displacement while you're jumping. Instead, your increased kinetic energy comes via a decrease in your body's chemical energy. A brick cannot jump or push off from a wall because it cannot deform and has no usable source of internal energy.

STOP TO THINK 9.6 A car accelerates smoothly away from a stop sign. Is the work done on the car positive, negative, or zero?

Variable Force

We've learned how to calculate the work done on an object by a constant force, but what about a force that changes as the object moves? Equation 9.14, the definition of work, is all we need:

$$W = \int_{s_i}^{s_f} F_s\, ds = \text{area under the force-versus-position graph} \qquad (9.21)$$

(work done by a variable force)

The integral sums up the small amounts of work $F_s\, ds$ done in each step along the trajectory. The only new feature, because F_s now varies with position, is that we cannot take F_s outside the integral. We must evaluate the integral either geometrically by finding the area under the curve (which we'll do in the next example) or by actually doing the integration (which we'll do in the next section).

EXAMPLE 9.7 | **Using work to find the speed of a car**

A 1500 kg car is towed, starting from rest. **FIGURE 9.16** shows the tension force in the tow rope as the car travels from $x = 0$ m to $x = 200$ m. What is the car's speed after being pulled 200 m?

MODEL Let the system consist of only the car, which we model as a particle. We'll neglect rolling friction. Two vertical forces, the normal force and gravity, are perpendicular to the motion and thus do no work.

FIGURE 9.16 Force-versus-position graph for a car.

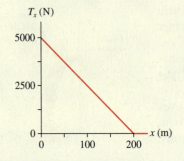

SOLVE We can solve this problem with the energy principle, $\Delta K = K_f - K_i = W$, where W is the work done by the tension force, but the force is not constant so we have to use the full definition of work as an integral. In this case, we can do the integral graphically:

$$W = \int_{0\,m}^{200\,m} T_x\, dx$$

= area under the force curve from 0 m to 200 m

$= \frac{1}{2}(5000\text{ N})(200\text{ m}) = 500{,}000$ J

The initial kinetic energy is zero, so the final kinetic energy is simply the energy transferred to the system by the work of the tension: $K_f = W = 500{,}000$ J. Then, from the definition of kinetic energy,

$$v_f = \sqrt{\frac{2K_f}{m}} = \sqrt{\frac{2(500{,}000\text{ J})}{1500\text{ kg}}} = 26\text{ m/s}$$

ASSESS 26 m/s ≈ 55 mph is a reasonable final speed after being towed 200 m.

9.4 Restoring Forces and the Work Done by a Spring

If you stretch a rubber band, a force tries to pull the rubber band back to its equilibrium, or unstretched, length. A force that restores a system to an equilibrium position is called a **restoring force.** Objects that exert restoring forces are called **elastic.** The most basic examples of elasticity are things like springs and rubber bands, but other examples of elasticity and restoring forces abound. For example, the steel beams flex slightly as you drive your car over a bridge, but they are restored to equilibrium after your car passes by. Nearly everything that stretches, compresses, flexes, bends, or twists exhibits a restoring force and can be called elastic.

We didn't introduce restoring forces in Part I of this textbook because we didn't have the mathematical tools to deal with them. But now—using work and energy—we do. We're going to use a simple spring as our model of elasticity. Suppose you have a spring whose **equilibrium length** is L_0. This is the length of the spring when it is neither pushing nor pulling. If you stretch (or compress) the spring, how hard does it pull (or push) back? Measurements show that

- The force is *opposite the displacement*. This is what we *mean* by a restoring force.
- If you don't stretch or compress the spring too much, the force is *proportional to the displacement from equilibrium*. The farther you push or pull, the larger the force.

FIGURE 9.17 shows a spring along a generic s-axis exerting force $\vec{F}_{\text{Sp}}$. Notice that s_{eq} is the position, or coordinate, of the free end of the spring, *not* the spring's equilibrium length L_0. When the spring is stretched, the **spring displacement** $\Delta s = s - s_{\text{eq}}$ is positive while $(F_{\text{Sp}})_s$, the s-component of the restoring force, is negative. Similarly, compressing the spring makes $\Delta s < 0$ and $(F_{\text{Sp}})_s > 0$. The graph of force versus displacement is a straight line with negative slope, showing that the spring force is proportional to but *opposite* the displacement.

The equation of the straight-line graph passing through the origin is

$$(F_{\text{Sp}})_s = -k\,\Delta s \qquad \text{(Hooke's law)} \qquad (9.22)$$

The minus sign is the mathematical indication of a *restoring* force, and the constant k—the absolute value of the slope of the line—is called the **spring constant** of the spring. The units of the spring constant are N/m. This relationship between the force and displacement of a spring was discovered by Robert Hooke, a contemporary (and sometimes bitter rival) of Newton. **Hooke's law** is not a true "law of nature," in the sense that Newton's laws are, but is actually just a *model* of a restoring force. It works well for *small* displacements from equilibrium, but Hooke's law will fail for any real spring that is compressed or stretched too far. A hypothetical massless spring for which Hooke's law is true at all displacements is called an **ideal spring.**

NOTE The force does not depend on the spring's physical length L but, instead, on the *displacement* Δs of the end of the spring.

The spring constant k is a property that characterizes a spring, just as mass m characterizes a particle. For a given spring, k is a constant—it does not change as the spring is stretched or compressed. If k is large, it takes a large pull to cause a significant stretch, and we call the spring a "stiff" spring. A spring with small k can be stretched with very little force, and we call it a "soft" spring.

NOTE In an earlier physics course, you may have learned Hooke's law as $F_{\text{Sp}} = -kx$ rather than as $-k\,\Delta s$. This can be misleading, and it is a common source of errors. The restoring force is $-kx$ *only* if the coordinate system in the problem is chosen such that the origin is at the equilibrium position of the free end of the spring. That is, $x = \Delta s$ only if $x_{\text{eq}} = 0$. This choice of origin is often made, but in some problems it will be more convenient to locate the origin of the coordinate system elsewhere.

FIGURE 9.17 Properties of a spring.

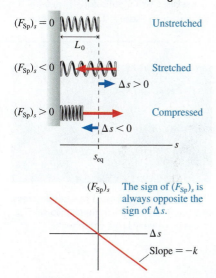

$(F_{\text{Sp}})_s = 0$ Unstretched L_0

$(F_{\text{Sp}})_s < 0$ Stretched $\Delta s > 0$

$(F_{\text{Sp}})_s > 0$ Compressed $\Delta s < 0$

s_{eq} s

$(F_{\text{Sp}})_s$ The sign of $(F_{\text{Sp}})_s$ is always opposite the sign of Δs.

Δs Slope $= -k$

EXAMPLE 9.8 | Pull until it slips

FIGURE 9.18 shows a spring attached to a 2.0 kg block. The other end of the spring is pulled by a motorized toy train that moves forward at 5.0 cm/s. The spring constant is 50 N/m, and the coefficient of static friction between the block and the surface is 0.60. The spring is at its equilibrium length at $t = 0$ s when the train starts to move. When does the block slip?

FIGURE 9.18 A toy train stretches the spring until the block slips.

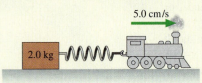

5.0 cm/s

2.0 kg

MODEL Model the block as a particle and the spring as an ideal spring obeying Hooke's law.

FIGURE 9.19 The free-body diagram.

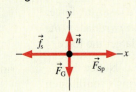

VISUALIZE FIGURE 9.19 is a free-body diagram for the block.

SOLVE Recall that the tension in a massless string pulls equally at *both* ends of the string. The same is true

for the spring force: It pulls (or pushes) equally at *both* ends. This is the key to solving the problem. As the right end of the spring moves, stretching the spring, the spring pulls backward on the train *and* forward on the block with equal strength. As the spring stretches, the static friction force on the block increases in magnitude to keep the block at rest. The block is in static equilibrium, so

$$\sum (F_{net})_x = (F_{Sp})_x + (f_s)_x = F_{Sp} - f_s = 0$$

where F_{Sp} is the *magnitude* of the spring force. The magnitude is $F_{Sp} = k\,\Delta x$, where $\Delta x = v_x t$ is the distance the train has moved. Thus

$$f_s = F_{Sp} = k\,\Delta x$$

The block slips when the static friction force reaches its maximum value $f_{s\,max} = \mu_s n = \mu_s mg$. This occurs when the train has moved

$$\Delta x = \frac{f_{s\,max}}{k} = \frac{\mu_s mg}{k} = \frac{(0.60)(2.0\,\text{kg})(9.80\,\text{m/s}^2)}{50\,\text{N/m}}$$

$$= 0.235\,\text{m} = 23.5\,\text{cm}$$

The time at which the block slips is

$$t = \frac{\Delta x}{v_x} = \frac{23.5\,\text{cm}}{5.0\,\text{cm/s}} = 4.7\,\text{s}$$

The slip can range from a few centimeters in a relatively small earthquake to several meters in a very large earthquake.

This example illustrates a class of motion called *stick-slip motion*. Once the block slips, it will shoot forward some distance, then stop and stick again. As the train continues, there will be a recurring sequence of stick, slip, stick, slip, stick. . . .

Earthquakes are an important example of stick-slip motion. The large tectonic plates making up the earth's crust are attempting to slide past each other, but friction causes the edges of the plates to stick together. You may think of rocks as rigid and brittle, but large masses of rock are somewhat elastic and can be "stretched." Eventually the elastic force of the deformed rocks exceeds the friction force between the plates. An earthquake occurs as the plates slip and lurch forward. Once the tension is released, the plates stick together again and the process starts all over.

STOP TO THINK 9.7 The graph shows the force magnitude versus displacement for three springs. Rank in order, from largest to smallest, the spring constants k_a, k_b, and k_c.

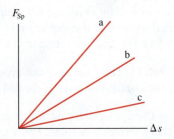

Work Done by Springs

The primary goal of this section is to calculate the work done by a spring. FIGURE 9.20 shows a spring acting on an object as it moves from s_i to s_f. The spring force on the object varies as the object moves, but we can calculate the spring's work by using Equation 9.21 for a variable force. Hooke's law for the spring is $(F_{Sp})_s = -k\,\Delta s = -k(s - s_{eq})$. Thus

$$W = \int_{s_i}^{s_f} (F_{Sp})_s\, ds = -k \int_{s_i}^{s_f} (s - s_{eq})\, ds \qquad (9.23)$$

This is an integration best carried out with a change of variables. Define $u = s - s_{eq}$, in which case $ds = du$. This changes the integrand from $(s - s_{eq})\,ds$ to $u\,du$. When we change variables, we also have to change the integration limits. At the lower limit, where $s = s_i$, the new variable u is $s_i - s_{eq} = \Delta s_i$. The lower limit becomes the initial *displacement*. Similarly, $s = s_f$ makes $u = s_f - s_{eq} = \Delta s_f$ at the upper limit. With these changes, the integral is

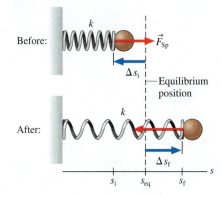

FIGURE 9.20 The spring does work on the object.

$$W = -k \int_{\Delta s_i}^{\Delta s_f} u\,ds = -\tfrac{1}{2}ku^2 \Big|_{\Delta s_i}^{\Delta s_f} = -\tfrac{1}{2}k(\Delta s_f)^2 + \tfrac{1}{2}k(\Delta s_i)^2 \qquad (9.24)$$

With a small rearrangement of the right side, we see that the work done by a spring is

$$W = -\left(\tfrac{1}{2}k(\Delta s_f)^2 - \tfrac{1}{2}k(\Delta s_i)^2\right) \qquad \text{(work done by a spring)} \qquad (9.25)$$

Because the displacements are squared, it makes no difference whether the initial and final displacements are stretches or compressions.

The work done by a spring is energy transferred to the object by the force of the spring. We can use this—and the energy principle—to solve problems that we were unable to solve with a direct application of Newton's laws.

EXAMPLE 9.9 | **Using the energy principle for a spring**

The "pincube machine" was an ill-fated predecessor of the pinball machine. A 100 g cube is launched by pulling a spring back 12 cm and releasing it. The spring's spring constant is 65 N/m. What is the cube's launch speed as it leaves the spring? Assume that the surface is frictionless.

MODEL Let the system consist of only the cube, which we model as a particle. Two vertical forces, the normal force and gravity, are perpendicular to the cube's displacement, and we've seen that perpendicular forces do no work. Only the spring force does work.

VISUALIZE FIGURE 9.21 is a before-and-after pictorial representation in which, for horizontal motion, we've replaced the generic s-axis with an x-axis. Notice that for problem solving we use numerical subscripts in place of the generic i and f.

SOLVE We can solve this problem with the energy principle, $\Delta K = K_1 - K_0 = W$, where W is the work done by the spring. The initial displacement is $\Delta x_0 = -0.12$ m. The cube will separate from the spring when the spring has expanded back to its equilibrium length, so the final displacement is $\Delta x_1 = 0$ m. From Equation 9.25, the spring does work

$$W = -\left(\tfrac{1}{2}k(\Delta x_1)^2 - \tfrac{1}{2}k(\Delta x_0)^2\right) = \tfrac{1}{2}(65 \text{ N/m})(-0.12 \text{ m})^2 - 0$$
$$= 0.468 \text{ J}$$

FIGURE 9.21 Pictorial representation of the pincube machine.

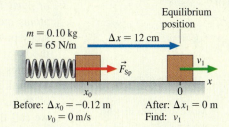

$m = 0.10$ kg
$k = 65$ N/m
$\Delta x = 12$ cm
Equilibrium position

Before: $\Delta x_0 = -0.12$ m
$v_0 = 0$ m/s

After: $\Delta x_1 = 0$ m
Find: v_1

The initial kinetic energy is zero, so the final kinetic energy is simply the energy transferred to the system by the work of the spring: $K_1 = W = 0.468$ J. Then, from the definition of kinetic energy,

$$v_1 = \sqrt{\frac{2K_1}{m}} = \sqrt{\frac{2(0.468 \text{ J})}{0.10 \text{ kg}}} = 3.1 \text{ m/s}$$

ASSESS 3.1 m/s $\approx$ 6 mph seems a reasonable final speed for a small, spring-launched cube.

STOP TO THINK 9.8 A block is attached to a spring, the spring is stretched, and the block is released at the position shown. As the block moves to the right, is the work done by the wall positive, negative, or zero?

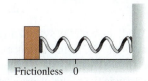

Frictionless 0

9.5 Dissipative Forces and Thermal Energy

Suppose you drag a heavy sofa across the floor at a steady speed. You are doing work, but the sofa is not gaining kinetic energy. And when you stop pulling, the sofa almost instantly stops moving. Where is the energy going that you're adding to the system? And what happens to the sofa's kinetic energy when you stop pulling?

You know that rubbing things together raises their temperature, in extreme cases making them hot enough to start a fire. As the sofa slides across the floor, friction causes the bottom of the sofa and the floor to get hotter. An increasing temperature is associated with increasing *thermal energy*, so in this situation the work done by pulling is increasing the system's thermal energy instead of its kinetic energy. Our goal in this section is to understand what thermal energy is and how it is related to *dissipative forces*.

Energy at the Microscopic Level

FIGURE 9.22 shows two different perspectives of an object. In the macrophysics perspective of Figure 9.22a you see an object of mass m moving as a whole with velocity v_{obj}. As a consequence of its motion, the object has macroscopic kinetic energy $K_{macro} = \frac{1}{2}mv_{obj}^2$.

FIGURE 9.22 Two perspectives of motion and energy.

(a) The macroscopic motion of the system as a whole

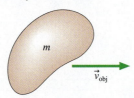

(b) The microscopic motion of the atoms inside

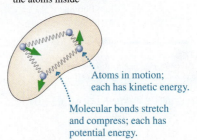

Atoms in motion; each has kinetic energy.

Molecular bonds stretch and compress; each has potential energy.

NOTE You recognize the prefix *micro,* meaning "small." You may not be familiar with *macro,* which means "large." Everyday objects, which consist of vast numbers of particle-like atoms, are *macroscopic objects*. We will use the term **macrophysics** to refer to the motion and dynamics of an object as a whole and **microphysics** to refer to the motions of atoms within an object.

Figure 9.22b is a microphysics view of the same object, where now we see a *system of particles*. Each of these atoms is jiggling about and has kinetic energy. As the atoms move, they stretch and compress the spring-like bonds between them. We'll study potential energy in Chapter 10, but you'll recall from the energy overview at the beginning of this chapter that potential energy is *stored* energy. Stretched and compressed springs store energy, so the bonds have potential energy.

The kinetic energy of one atom is exceedingly small, but there are enormous numbers of atoms in a macroscopic object. The total kinetic energy of all the atoms is what we call the *microscopic kinetic energy*, K_{micro}. The total potential energy of all the bonds is the *microscopic potential energy*, U_{micro}. These energies are distinct from the macroscopic energy of the object as a whole.

The combined microscopic kinetic and potential energy of the atoms—the energy of the jiggling atoms and stretching bonds—is called the **thermal energy** of the system:

$$E_{th} = K_{micro} + U_{micro} \qquad (9.26)$$

This energy is hidden from view in our macrophysics perspective, but it is quite real. We will discover later, when we reach thermodynamics, that the thermal energy is related to the *temperature* of the system. Raising the temperature causes the atoms to move faster and the bonds to stretch more, giving the system more thermal energy.

With the inclusion of thermal energy, the system has both macroscopic kinetic energy *and* thermal energy: $E_{sys} = K + E_{th}$. K is understood to be the total macroscopic kinetic energy; we'll use the subscript "macro" only if there's a need to distinguish macroscopic energy from microscopic energy. With this, the energy principle becomes

$$\Delta E_{sys} = \Delta K + \Delta E_{th} = W_{ext} \qquad (9.27)$$

Work done on the system might increase the system's kinetic energy, its thermal energy, or both. Or, in the absence of work, kinetic energy can be transformed into thermal energy as long as the total energy change is zero. Recognizing thermal energy greatly expands the range of problems we can analyze with the energy principle.

NOTE The microscopic energy of atoms is *not* called "heat." As we've already seen, heat is a *process*, similar to work, for transferring energy between the system and the environment. We'll have a lot more to say about heat in future chapters. For the time being we want to use the correct term "thermal energy" to describe the random, thermal motions of the atoms in a system. If the temperature of a system goes up (i.e., it gets hotter), it is because the system's thermal energy has increased.

Dissipative Forces

Forces such as friction and drag cause the macroscopic kinetic energy of a system to be *dissipated* as thermal energy. Hence these are called **dissipative forces**. FIGURE 9.23 shows how microscopic interactions are responsible for transforming macroscopic kinetic energy into thermal energy when two objects slide against each other. Because friction causes *both* objects to get warmer, with increased thermal energy, **we must define the system to include both objects whose temperature changes**—both the sofa *and* the floor.

For example, FIGURE 9.24 shows a box being pulled at constant speed across a horizontal surface with friction. As you can imagine, both the surface and the box are getting warmer—increasing thermal energy—but the kinetic energy is not changing. If we define the system to be box + surface, then the increasing thermal energy of the system is entirely due to the work being done on the system by tension in the rope: $\Delta E_{th} = W_{tension}$.

The work done by tension in pulling the box through distance Δs is simply $W_{tension} = T\Delta s$; thus $\Delta E_{th} = T\Delta s$. Because the box is moving with constant velocity, and thus no net force, the tension force has to exactly balance the friction force: $T = f_k$. Consequently, the increase in thermal energy due to the dissipative force of friction is

$$\Delta E_{th} = f_k \Delta s \qquad (9.28)$$

Notice that the increase in thermal energy is directly proportional to the total distance of sliding. **Dissipative forces always increase the thermal energy; they never decrease it.**

You might wonder why we didn't simply calculate the work done by friction. The rather subtle reason is that work is defined only for forces acting on a *particle*. There is work being done on individual atoms at the boundary as they are pulled this way and that, but we would need a detailed knowledge of atomic-level friction forces to calculate this work. The friction force $\vec{f}_k$ is an average force on the object as a whole; it is not a force on any particular particle, so we cannot use it to calculate work. Furthermore, increasing thermal energy is not an energy transfer—the definition of work—from the box to the surface or from the surface to the box; both the box *and* the surface are gaining thermal energy. **The techniques used to calculate the work done on a particle cannot be used to calculate the work done by dissipative forces.**

NOTE The considerations that led to Equation 9.28 allow us to calculate the total increase in thermal energy of the entire system, but we cannot determine what fraction of ΔE_{th} goes to the box and what fraction goes to the surface.

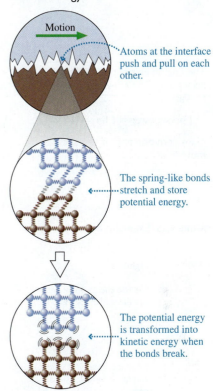

FIGURE 9.23 Motion with friction leads to thermal energy.

Motion

Atoms at the interface push and pull on each other.

The spring-like bonds stretch and store potential energy.

The potential energy is transformed into kinetic energy when the bonds break.

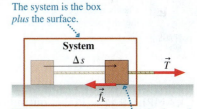

FIGURE 9.24 Work done by tension is dissipated as thermal energy.

The system is the box *plus* the surface.

System

Δs

$\vec{T}$

$\vec{f}_k$

Work done by tension increases the thermal energy of the box *and* the surface.

EXAMPLE 9.10 | **Increasing kinetic and thermal energy**

A rope with 30 N of tension pulls a 10 kg crate 3.0 m across a horizontal floor, starting from rest. The coefficient of friction between the crate and the floor is 0.20. What is the increase in thermal energy? What is the crate's final speed?

MODEL Let the system consist of both the crate and the floor. The tension in the rope does work on the system, but the vertical normal force and gravitational force do not.

SOLVE The energy principle, Equation 9.27, is $\Delta K + \Delta E_{th} = W_{ext}$. The friction force on an object moving on a horizontal surface is $f_k = \mu_k n = \mu_k mg$, so the increase in thermal energy, given by Equation 9.28, is

$$\Delta E_{th} = f_k \Delta s = \mu_k mg \Delta s$$
$$= (0.20)(10\,kg)(9.80\,m/s^2)(3.0\,m) = 59\,J$$

The tension force does work $W_{ext} = T\Delta s = (30\,N)(3.0\,m) = 90\,J$. 59 J of this goes to increasing the thermal energy, so $\Delta K = 31$ J is the crate's change of kinetic energy. Because $K_i = 0$ J, $K_f = \Delta K = 31$ J. Using the definition of kinetic energy, we find that the crate's final speed is

$$v_f = \sqrt{\frac{2K_f}{m}} = \sqrt{\frac{2(31\,J)}{10\,kg}} = 2.5\,m/s$$

ASSESS The thermal energy of the crate *and* floor increases by 59 J. We cannot determine ΔE_{th} for the crate (or floor) alone.

9.6 Power

Work is a transfer of energy between the environment and a system. In many situations we would like to know *how quickly* the energy is transferred. Does the force act quickly and transfer the energy very rapidly, or is it a slow and lazy transfer of energy? If you need to buy a motor to lift 1000 kg of bricks up 20 m, it makes a *big* difference whether the motor has to do this in 30 s or 30 min!

The question How quickly? implies that we are talking about a *rate*. For example, the velocity of an object—how quickly it is moving—is the *rate of change* of position. So when we raise the issue of how quickly the energy is transferred, we are talking about the *rate of transfer* of energy. The rate at which energy is transferred or transformed is called the **power** P, and it is defined as

$$P = \frac{dE_{sys}}{dt} \tag{9.29}$$

The unit of power is the **watt**, which is defined as $1 \text{ watt} = 1 \text{ W} = 1 \text{ J/s}$. Common prefixes used with power are mW (milliwatts), kW (kilowatts), and MW (megawatts).

For example, the rope in Example 9.10 pulled with a tension of 30 N and, by doing work, transferred 90 J of energy to the system. If it took 10 s to drag the crate 3.0 m, then energy was being transferred at the rate of 9 J/s. We would say that whatever was supplying this energy—whether a human or a motor—has a "power output" of 9 W.

The idea of power as a *rate* of energy transfer applies no matter what the form of energy. **FIGURE 9.25** shows three examples of the idea of power. For now, we want to focus primarily on *work* as the source of energy transfer. Within this more limited scope, power is simply the **rate of doing work**: $P = dW/dt$. If a particle moves through a small displacement $d\vec{r}$ while acted on by force $\vec{F}$, the force does a small amount of work dW given by

$$dW = \vec{F} \cdot d\vec{r}$$

Dividing both sides by dt, to give a rate of change, yields

$$\frac{dW}{dt} = \vec{F} \cdot \frac{d\vec{r}}{dt}$$

But $d\vec{r}/dt$ is the velocity $\vec{v}$, so we can write the power as

$$P = \vec{F} \cdot \vec{v} = Fv \cos\theta \tag{9.30}$$

In other words, the power delivered to a particle by a force acting on it is the dot product of the force and the particle's velocity. These ideas will become clearer with some examples.

The English unit of power is the *horsepower*. The conversion factor to watts is

$$1 \text{ horsepower} = 1 \text{ hp} = 746 \text{ W}$$

Many common appliances, such as motors, are rated in hp.

FIGURE 9.25 Examples of power.

100 W

Lightbulb

Electric energy $\longrightarrow$ light and thermal energy at 100 J/s

$\frac{1}{2}$ hp

Athlete

Chemical energy of glucose and fat $\longrightarrow$ mechanical energy at ≈ 350 J/s $\approx \frac{1}{2}$ hp

20 kW

Gas furnace

Chemical energy of gas $\longrightarrow$ thermal energy at 20,000 J/s

---

EXAMPLE 9.11 | Power output of a motor

A factory uses a motor and a cable to drag a 300 kg machine to the proper place on the factory floor. What power must the motor supply to drag the machine at a speed of 0.50 m/s? The coefficient of friction between the machine and the floor is 0.60.

SOLVE The force applied by the motor, through the cable, is the tension force $\vec{T}$. This force does work on the machine with power $P = Tv$. The machine is in equilibrium, because the motion is at

constant velocity, hence the tension in the rope balances the friction and is

$$T = f_k = \mu_k mg$$

The motor's power output is

$$P = Tv = \mu_k mgv = 882 \text{ W}$$

EXAMPLE 9.12 | Power output of a car engine

A 1500 kg car has a front profile that is 1.6 m wide by 1.4 m high and a drag coefficient of 0.50. The coefficient of rolling friction is 0.02. What power must the engine provide to drive at a steady 30 m/s (≈ 65 mph) if 25% of the power is "lost" before reaching the drive wheels?

SOLVE The net force on a car moving at a steady speed is zero. The motion is opposed both by rolling friction and by air resistance. The forward force on the car $\vec{F}_{car}$ (recall that this is really $\vec{F}_{ground \, on \, car}$, a

reaction to the drive wheels pushing backward on the ground with $\vec{F}_{car \, on \, ground}$) exactly balances the two opposing forces:

$$F_{car} = f_r + F_{drag}$$

where $\vec{F}_{drag}$ is the drag due to the air. Using the results of Chapter 6, where both rolling friction and drag were introduced, this becomes

$$F_{\text{car}} = \mu_r mg + \tfrac{1}{2} C\rho A v^2 = 294\ \text{N} + 655\ \text{N} = 949\ \text{N}$$

Here $A = (1.6\ \text{m}) \times (1.4\ \text{m})$ is the front cross-section area of the car, and we used 1.3 kg/m³ as the density of air. The power required to push the car forward at 30 m/s is

$$P_{\text{car}} = F_{\text{car}}\, v = (949\ \text{N})(30\ \text{m/s}) = 28{,}500\ \text{W} = 38\ \text{hp}$$

This is the power *needed* at the drive wheels to push the car against the dissipative forces of friction and air resistance. The power output

of the engine is larger because some energy is used to run the water pump, the power steering, and other accessories. In addition, energy is lost to friction in the drive train. If 25% of the power is lost (a typical value), leading to $P_{\text{car}} = 0.75 P_{\text{engine}}$, the engine's power output is

$$P_{\text{engine}} = \frac{P_{\text{car}}}{0.75} = 38{,}000\ \text{W} = 51\ \text{hp}$$

ASSESS Automobile engines are typically rated at ≈ 200 hp. Most of that power is reserved for fast acceleration and climbing hills.

STOP TO THINK 9.9   Four students run up the stairs in the time shown. Rank in order, from largest to smallest, their power outputs P_a to P_d.

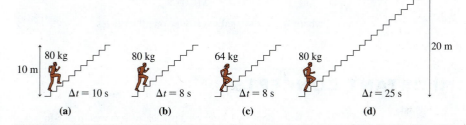

80 kg $\Delta t = 10$ s   (a)
80 kg $\Delta t = 8$ s   (b)
64 kg $\Delta t = 8$ s   (c)
80 kg $\Delta t = 25$ s   (d)
10 m   20 m

CHALLENGE EXAMPLE 9.13 | Stopping a brick

A 25.0-cm-long spring stands vertically on the ground, with its lower end secured in a base. A 1.5 kg brick is held 40 cm directly above the spring and dropped onto the spring. The spring compresses to a length of 17.0 cm before starting to launch the brick back upward. What is the spring's spring constant?

MODEL Let the system consist of only the brick, which is modeled as a particle. Assume that the spring is ideal. Both gravity and, after contact, the spring do work on the brick.

VISUALIZE FIGURE 9.26 is a before-and-after pictorial representation. We've chosen to place the origin of the y-axis at the equilibrium position of the spring's upper end. The length of the spring is not relevant, but the *difference* between its before and after lengths—8.0 cm—is the spring's maximum compression. The point of maximum compression is a turning point for the brick, as it reverses direction and starts moving back up, so the brick's instantaneous velocity is zero.

SOLVE At first, this seems like a two-part problem: free fall until hitting the spring, then deceleration as the spring compresses. But by using the energy principle, $\Delta E_{\text{sys}} = \Delta K = W_{\text{tot}} = W_G + W_{\text{Sp}}$, we can do it in one step. Interestingly, $\Delta K = 0$ because the brick is instantaneously at rest at the beginning and again at the point of maximum spring compression. Consequently, $W_{\text{tot}} = 0$. That's not a difficulty because gravity does positive work (the downward gravitational force is in the direction of the brick's displacement) while

FIGURE 9.26 Pictorial representation of the brick and spring.

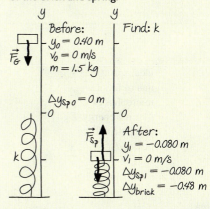

Before:
$y_0 = 0.40$ m
$v_0 = 0$ m/s
$m = 1.5$ kg
$\vec{F}_G$
$\Delta y_{Sp\,0} = 0$ m
0

After:
$y_1 = -0.080$ m
$v_1 = 0$ m/s
$\Delta y_{Sp\,1} = -0.080$ m
$\Delta y_{\text{brick}} = -0.48$ m
$\vec{F}_{Sp}$
k

Find: k

the spring does negative work (the upward spring force is opposite the displacement).

The work done by gravity is

$$W_G = (F_G)_y \Delta y_{\text{brick}} = -mg\,\Delta y_{\text{brick}}$$

where Δy_{brick} is the brick's total displacement. The negative sign comes from $(F_G)_y = -mg$ but W_G is positive because $\Delta y_{\text{brick}} = y_1 - y_0 = -0.48$ m is also negative. It may seem strange that calculating the work done by gravity is so simple when the brick first accelerates, then slows quickly after hitting the spring. But work depends on only the displacement, not how fast or slow the object is moving.

We have to be careful with the spring because its displacement is not the same as the brick's displacement. The spring begins compressing only when contact is made, so $\Delta y_{Sp\,0} = 0$ m at the instant the brick is released. The work done by the spring then continues until maximum compression, when the spring's displacement is $\Delta y_{Sp\,1} = -0.080$ m. The work done by the spring, Equation 9.25, is thus

$$W_{\text{Sp}} = -\left(\tfrac{1}{2}k(\Delta y_{Sp\,1})^2 - \tfrac{1}{2}k(\Delta y_{Sp\,0})^2\right) = -\tfrac{1}{2}k(\Delta y_{Sp\,1})^2$$

With this information about the two works, the energy principle is

$$\Delta K = 0 = W_G + W_{\text{Sp}} = -mg\,\Delta y_{\text{brick}} - \tfrac{1}{2}k(\Delta y_{Sp\,1})^2$$

Solving for the spring constant gives

$$k = -\frac{2mg\,\Delta y_{\text{brick}}}{(\Delta y_{Sp\,1})^2} = -\frac{2(1.5\ \text{kg})(9.80\ \text{m/s}^2)(-0.48\ \text{m})}{(-0.080\ \text{m})^2}$$

$$= 2200\ \text{N/m}$$

ASSESS 2200 N/m is a fairly large spring constant, but that's to be expected for a spring that's going to stop a falling ≈ 3 pound brick. The complexity of this problem was not the math, which was fairly simple, but the reasoning. It's a good illustration of how to apply energy reasoning to other problems.

SUMMARY

The goal of Chapter 9 has been to begin your study of how energy is transferred and transformed.

GENERAL PRINCIPLES

Basic Energy Model

- Energy is a property of the system.
- Energy is *transformed* within the system without loss.
- Energy is *transferred* to and from the system by forces that do work *W*.
- $W > 0$ for energy added.
- $W < 0$ for energy removed.

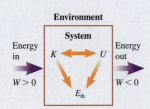

The Energy Principle

Doing work on a system changes the **system energy:**

$$\Delta E_{sys} = W_{ext}$$

For systems containing only particles, no interactions, $E_{sys} = K + E_{th}$. All forces are external forces, so

$$\Delta K + \Delta E_{th} = W_{tot}$$

where W_{tot} is the total work done on all particles.

IMPORTANT CONCEPTS

Kinetic energy is an energy of motion: $K = \frac{1}{2}mv^2$

Potential energy is stored energy.

Thermal energy is the **microscopic** energy of moving atoms and stretched bonds.

Dissipative forces, such as friction and drag, transform **macroscopic** energy into thermal energy. For friction:

$$\Delta E_{th} = f_k \Delta s$$

The **work** done by a force on a particle as it moves from s_i to s_f is

$$W = \int_{s_i}^{s_f} F_s \, ds = \text{area under the force curve}$$

The work done by a constant force is

$$W = \vec{F} \cdot \Delta \vec{r}$$

The work done by a spring is

$$W = -\left(\frac{1}{2}k(\Delta s_f)^2 - \frac{1}{2}k(\Delta s_i)^2\right)$$

where Δs is the displacement of the end of the spring.

APPLICATIONS

Hooke's law

The **restoring force** of an ideal spring is

$$(F_{Sp})_s = -k\,\Delta s$$

where k is the **spring constant** and Δs is the displacement of the end of the spring from equilibrium.

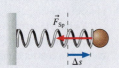

Power is the rate at which energy is transferred or transformed:

$$P = dE_{sys}/dt$$

For a particle with velocity $\vec{v}$, the power delivered to the particle by force $\vec{F}$ is $P = \vec{F} \cdot \vec{v} = Fv\cos\theta$.

Dot product

$$\vec{A} \cdot \vec{B} = AB\cos\theta = A_x B_x + A_y B_y$$

TERMS AND NOTATION

energy	energy transformation	scalar product	ideal spring
system	energy principle	restoring force	macrophysics
environment	basic energy model	elastic	microphysics
system energy, E_{sys}	before-and-after representation	equilibrium length, L_0	thermal energy, E_{th}
work, W	kinetic energy, K	spring displacement, Δs	dissipative force
heat	joule, J	spring constant, k	power, P
energy transfer	dot product	Hooke's law	watt, W

CONCEPTUAL QUESTIONS

1. If a particle's speed increases by a factor of 3, by what factor does its kinetic energy change?

2. Particle A has half the mass and eight times the kinetic energy of particle B. What is the speed ratio v_A/v_B?

3. An elevator held by a single cable is ascending but slowing down. Is the work done by tension positive, negative, or zero? What about the work done by gravity? Explain.

4. The rope in **FIGURE Q9.4** pulls the box to the left across a rough surface. Is the work done by tension positive, negative, or zero? Explain.

FIGURE Q9.4

5. A 0.2 kg plastic cart and a 20 kg lead cart both roll without friction on a horizontal surface. Equal forces are used to push both carts forward a distance of 1 m, starting from rest. After traveling 1 m, is the kinetic energy of the plastic cart greater than, less than, or equal to the kinetic energy of the lead cart? Explain.

6. A particle moving to the left is slowed by a force pushing to the right. Is the work done on the particle positive or negative? Or is there not enough information to tell? Explain.

7. A particle moves in a vertical plane along the *closed* path seen in **FIGURE Q9.7**, starting at A and eventually returning to its starting point. Is the work done by gravity positive, negative, or zero? Explain.

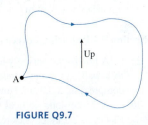

FIGURE Q9.7

8. You need to raise a heavy block by pulling it with a massless rope. You can either (a) pull the block straight up height h, or (b) pull it up a long, frictionless plane inclined at a 15° angle until its height has increased by h. Assume you will move the block at constant speed either way. Will you do more work in case a or case b? Or is the work the same in both cases? Explain.

9. A ball on a string travels once around a circle with a circumference of 2.0 m. The tension in the string is 5.0 N. How much work is done by tension?

10. A sprinter accelerates from rest. Is the work done on the sprinter positive, negative, or zero? Explain.

11. A spring has an unstretched length of 10 cm. It exerts a restoring force F when stretched to a length of 11 cm.
   a. For what length of the spring is its restoring force $3F$?
   b. At what compressed length is the restoring force $2F$?

12. The left end of a spring is attached to a wall. When Bob pulls on the right end with a 200 N force, he stretches the spring by 20 cm. The same spring is then used for a tug-of-war between Bob and Carlos. Each pulls on his end of the spring with a 200 N force. How far does the spring stretch? Explain.

13. The driver of a car traveling at 60 mph slams on the brakes, and the car skids to a halt. What happened to the kinetic energy the car had just before stopping?

14. The motor of a crane uses power P to lift a steel beam. By what factor must the motor's power increase to lift the beam twice as high in half the time?

EXERCISES AND PROBLEMS

Problems labeled   integrate material from earlier chapters.

Exercises

Section 9.2 Work and Kinetic Energy for a Single Particle

1. | Which has the larger kinetic energy, a 10 g bullet fired at 500 m/s or a 75 kg student running at 5.5 m/s?

2. ‖ At what speed does a 1000 kg compact car have the same kinetic energy as a 20,000 kg truck going 25 km/h?

3. ‖ A mother has four times the mass of her young son. Both are running with the same kinetic energy. What is the ratio v_{son}/v_{mother} of their speeds?

4. | A horizontal rope with 15 N tension drags a 25 kg box 2.0 m to the left across a horizontal surface. How much work is done by (a) tension and (b) gravity?

5. | A 25 kg box sliding to the left across a horizontal surface is brought to a halt in a distance of 35 cm by a horizontal rope pulling to the right with 15 N tension. How much work is done by (a) tension and (b) gravity?

6. | A 2.0 kg book is lying on a 0.75-m-high table. You pick it up and place it on a bookshelf 2.25 m above the floor.
   a. How much work does gravity do on the book?
   b. How much work does your hand do on the book?

7. ‖ A 20 g particle is moving to the left at 30 m/s. A force on the particle causes it to move the the right at 30 m/s. How much work is done by the force?

8. ‖ **FIGURE EX9.8** is the kinetic-energy graph for a 2.0 kg object moving along the x-axis. Determine the work done on the object during each of the four intervals AB, BC, CD, and DE.

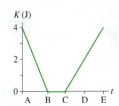

FIGURE EX9.8

9. | You throw a 5.5 g coin straight down at 4.0 m/s from a 35-m-high bridge.
   a. How much work does gravity do as the coin falls to the water below?
   b. What is the speed of the coin just as it hits the water?

10. ‖ The cable of a crane is lifting a 750 kg girder. The girder increases its speed from 0.25 m/s to 0.75 m/s in a distance of 3.5 m.
   a. How much work is done by gravity?
   b. How much work is done by tension?

Section 9.3 Calculating the Work Done

11. | Evaluate the dot product $\vec{A} \cdot \vec{B}$ if
   a. $\vec{A} = 4\hat{\imath} - 2\hat{\jmath}$ and $\vec{B} = -2\hat{\imath} - 3\hat{\jmath}$.
   b. $\vec{A} = -4\hat{\imath} + 2\hat{\jmath}$ and $\vec{B} = 2\hat{\imath} + 4\hat{\jmath}$.

12. | Evaluate the dot product $\vec{A} \cdot \vec{B}$ if
   a. $\vec{A} = 3\hat{\imath} + 4\hat{\jmath}$ and $\vec{B} = 2\hat{\imath} - 6\hat{\jmath}$.
   b. $\vec{A} = 3\hat{\imath} - 2\hat{\jmath}$ and $\vec{B} = 6\hat{\imath} + 4\hat{\jmath}$.

13. | What is the angle θ between vectors $\vec{A}$ and $\vec{B}$ in each part of Exercise 12?

14. | Evaluate the dot product of the three pairs of vectors in **FIGURE EX9.14**.

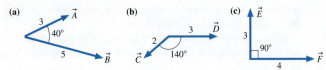

FIGURE EX9.14

15. | Evaluate the dot product of the three pairs of vectors in **FIGURE EX9.15**.

FIGURE EX9.15

16. ‖ A 25 kg air compressor is dragged up a rough incline from $\vec{r}_1 = (1.3\hat{\imath} + 1.3\hat{\jmath})$ m to $\vec{r}_2 = (8.3\hat{\imath} + 2.9\hat{\jmath})$ m, where the y-axis is vertical. How much work does gravity do on the compressor during this displacement?

17. ‖ A 45 g bug is hovering in the air. A gust of wind exerts a force $\vec{F} = (4.0\hat{\imath} - 6.0\hat{\jmath}) \times 10^{-2}$ N on the bug.
   a. How much work is done by the wind as the bug undergoes displacement $\Delta\vec{r} = (2.0\hat{\imath} - 2.0\hat{\jmath})$ m?
   b. What is the bug's speed at the end of this displacement? Assume that the speed is due entirely to the wind.

18. ‖ The two ropes seen in **FIGURE EX9.18** are used to lower a 255 kg piano 5.00 m from a second-story window to the ground. How much work is done by each of the three forces?

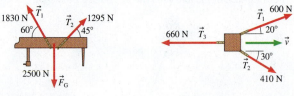

FIGURE EX9.18   **FIGURE EX9.19**

19. ‖ The three ropes shown in the bird's-eye view of **FIGURE EX9.19** are used to drag a crate 3.0 m across the floor. How much work is done by each of the three forces?

20. | **FIGURE EX9.20** is the force-versus-position graph for a particle moving along the x-axis. Determine the work done on the particle during each of the three intervals 0–1 m, 1–2 m, and 2–3 m.

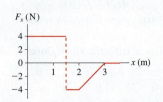

FIGURE EX9.20

21. ‖ A 500 g particle moving along the x-axis experiences the force shown in **FIGURE EX9.21**. The particle's velocity is 2.0 m/s at $x = 0$ m. What is its velocity at $x = 3$ m?

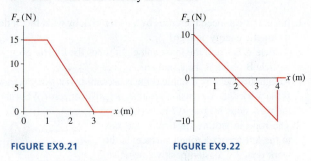

FIGURE EX9.21   **FIGURE EX9.22**

22. ‖ A 2.0 kg particle moving along the x-axis experiences the force shown in **FIGURE EX9.22**. The particle's velocity is 4.0 m/s at $x = 0$ m. What is its velocity at $x = 2$ m and $x = 4$ m?

23. ‖ A particle moving on the x-axis experiences a force given by **CALC** $F_x = qx^2$, where q is a constant. How much work is done on the particle as it moves from $x = 0$ to $x = d$?

24. ‖ A 150 g particle at $x = 0$ is moving at 2.00 m/s in the **CALC** $+x$-direction. As it moves, it experiences a force given by $F_x = (0.250 \text{ N}) \sin(x/2.00 \text{ m})$. What is the particle's speed when it reaches $x = 3.14$ m?

Section 9.4 Restoring Forces and the Work Done by a Spring

25. | A horizontal spring with spring constant 750 N/m is attached to a wall. An athlete presses against the free end of the spring, compressing it 5.0 cm. How hard is the athlete pushing?

26. | A 35-cm-long vertical spring has one end fixed on the floor. Placing a 2.2 kg physics textbook on the spring compresses it to a length of 29 cm. What is the spring constant?

27. ‖ A 10-cm-long spring is attached to the ceiling. When a 2.0 kg mass is hung from it, the spring stretches to a length of 15 cm.
   a. What is the spring constant?
   b. How long is the spring when a 3.0 kg mass is suspended from it?

28. ‖ A 60 kg student is standing atop a spring in an elevator as it accelerates upward at 3.0 m/s². The spring constant is 2500 N/m. By how much is the spring compressed?

29. ‖ A 5.0 kg mass hanging from a spring scale is slowly lowered onto a vertical spring, as shown in **FIGURE EX9.29**. The scale reads in newtons.
   a. What does the spring scale read just before the mass touches the lower spring?
   b. The scale reads 20 N when the lower spring has been compressed by 2.0 cm. What is the value of the spring constant for the lower spring?
   c. At what compression will the scale read zero?

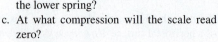

FIGURE EX9.29

30. ‖ A horizontal spring with spring constant 85 N/m extends outward from a wall just above floor level. A 1.5 kg box sliding across a frictionless floor hits the end of the spring and compresses it 6.5 cm before the spring expands and shoots the box back out. How fast was the box going when it hit the spring?

Section 9.5 Dissipative Forces and Thermal Energy

31. | One mole $(6.02 \times 10^{23}$ atoms) of helium atoms in the gas phase has 3700 J of microscopic kinetic energy at room temperature. If we assume that all atoms move with the same speed, what is that speed? The mass of a helium atom is 6.68×10^{-27} kg.

32. | A 55 kg softball player slides into second base, generating 950 J of thermal energy in her legs and the ground. How fast was she running?

33. ‖ A baggage handler throws a 15 kg suitcase along the floor of an airplane luggage compartment with a speed of 1.2 m/s. The suitcase slides 2.0 m before stopping. Use work and energy to find the suitcase's coefficient of kinetic friction on the floor.

34. ‖‖ An 8.0 kg crate is pulled 5.0 m up a 30° incline by a rope angled 18° above the incline. The tension in the rope is 120 N, and the crate's coefficient of kinetic friction on the incline is 0.25.
    a. How much work is done by tension, by gravity, and by the normal force?
    b. What is the increase in thermal energy of the crate and incline?

35. ‖ Justin, with a mass of 30 kg, is going down an 8.0-m-high water slide. He starts at rest, and his speed at the bottom is 11 m/s. How much thermal energy is created by friction during his descent?

Section 9.6 Power

36. | a. How much work does an elevator motor do to lift a 1000 kg elevator a height of 100 m?
    b. How much power must the motor supply to do this in 50 s at constant speed?

37. | a. How much work must you do to push a 10 kg block of steel across a steel table at a steady speed of 1.0 m/s for 3.0 s?
    b. What is your power output while doing so?

38. ‖ How much energy is consumed by (a) a 1.2 kW hair dryer used for 10 min and (b) a 10 W night light left on for 24 h?

39. | At midday, solar energy strikes the earth with an intensity of about 1 kW/m^2. What is the area of a solar collector that could collect 150 MJ of energy in 1 h? This is roughly the energy content of 1 gallon of gasoline.

40. ‖‖ A 50 kg sprinter, starting from rest, runs 50 m in 7.0 s at constant acceleration.
    a. What is the magnitude of the horizontal force acting on the sprinter?
    b. What is the sprinter's power output at 2.0 s, 4.0 s, and 6.0 s?

41. ‖ A 70 kg human sprinter can accelerate from rest to 10 m/s in
BIO 3.0 s. During the same time interval, a 30 kg greyhound can go from rest to 20 m/s. What is the average power output of each? *Average* power over a time interval Δt is $\Delta E/\Delta t$.

42. ‖ The human heart pumps the average adult's 6.0 L (6000 cm^3)
BIO of blood through the body every minute. The heart must do work to overcome frictional forces that resist blood flow. The average adult blood pressure is 1.3×10^4 N/m^2.
    a. How much work does the heart do to move the 6.0 L of blood completely through the body?
    b. What power output must the heart have to do this task once a minute?
    Hint: When the heart contracts, it applies force to the blood. Pressure is force/area. Model the circulatory system as a single closed tube, with cross-section area A and volume $V = 6.0$ L, filled with blood to which the heart applies a force.

Problems

43. ‖ A 1000 kg elevator accelerates upward at 1.0 m/s^2 for 10 m, starting from rest.
    a. How much work does gravity do on the elevator?
    b. How much work does the tension in the elevator cable do on the elevator?
    c. What is the elevator's kinetic energy after traveling 10 m?

44. ‖ a. Starting from rest, a crate of mass m is pushed up a frictionless slope of angle θ by a *horizontal* force of magnitude F. Use work and energy to find an expression for the crate's speed v when it is at height h above the bottom of the slope.
    b. Doug uses a 25 N horizontal force to push a 5.0 kg crate up a 2.0-m-high, 20° frictionless slope. What is the speed of the crate at the top of the slope?

45. ‖‖ Susan's 10 kg baby brother Paul sits on a mat. Susan pulls the mat across the floor using a rope that is angled 30° above the floor. The tension is a constant 30 N and the coefficient of friction is 0.20. Use work and energy to find Paul's speed after being pulled 3.0 m.

46. ‖ A particle of mass m moving along the x-axis has velocity $v_x = v_0 \sin(\pi x/2L)$. How much work is done on the particle as it moves (a) from $x = 0$ to $x = L$ and (b) from $x = 0$ to $x = 2L$?

47. | A ball shot straight up with kinetic energy K_0 reaches height h. What height will it reach if the initial kinetic energy is doubled?

48. ‖ A pile driver lifts a 250 kg weight and then lets it fall onto the end of a steel pipe that needs to be driven into the ground. A fall of 1.5 m drives the pipe in 35 cm. What is the average force exerted on the pipe?

49. ‖ A 50 kg ice skater is gliding along the ice, heading due north at 4.0 m/s. The ice has a small coefficient of static friction, to prevent the skater from slipping sideways, but $\mu_k = 0$. Suddenly, a wind from the northeast exerts a force of 4.0 N on the skater.
    a. Use work and energy to find the skater's speed after gliding 100 m in this wind.
    b. What is the minimum value of μ_s that allows her to continue moving straight north?

50. | You're fishing from a tall pier and have just caught a 1.5 kg fish. As it breaks the surface, essentially at rest, the tension in the vertical fishing line is 16 N. Use work and energy to find the fish's upward speed after you've lifted it 2.0 m.

51. ‖ Hooke's law describes an ideal spring. Many real springs are
CALC better described by the restoring force $(F_{Sp})_s = -k\Delta s - q(\Delta s)^3$, where q is a constant. Consider a spring with $k = 250$ N/m and $q = 800$ N/m^3.
    a. How much work must you do to compress this spring 15 cm? Note that, by Newton's third law, the work you do *on* the spring is the negative of the work done *by* the spring.
    b. By what percent has the cubic term increased the work over what would be needed to compress an ideal spring?
    Hint: Let the spring lie along the s-axis with the equilibrium position of the end of the spring at $s = 0$. Then $\Delta s = s$.

52. ‖ The force acting on a particle is $F_x = F_0 e^{-x/L}$. How much work
CALC does this force do as the particle moves along the x-axis from $x = 0$ to $x = L$?

53. ‖ The gravitational attraction between two objects with masses
CALC m_1 and m_2, separated by distance x, is $F = Gm_1m_2/x^2$, where G is the *gravitational constant*.
    a. How much work is done by gravity when the separation changes from x_1 to x_2? Assume $x_2 < x_1$.
    b. If one mass is much greater than the other, the larger mass stays essentially at rest while the smaller mass moves toward it. Suppose a 1.5×10^{13} kg comet is passing the orbit of Mars, heading straight for the sun at a speed of 3.5×10^4 m/s. What will its speed be when it crosses the orbit of Mercury? Astronomical data are given in the tables at the back of the book, and $G = 6.67 \times 10^{-11}$ N m^2/kg^2.

54. ‖ An *electric dipole* consists of two equal but opposite electric
CALC charges, $+q$ and $-q$, separated by a small distance d. If another charge Q is distance x from the center of the dipole, in the plane perpendicular to the line between $+q$ and $-q$, and if $x \gg d$, the dipole exerts an electric force $F = KqQd/x^3$ on charge Q, where K is a constant. How much work does the electric force do if charge Q moves from distance x_1 to distance x_2?

55. ‖ A 50 g rock is placed in a slingshot and the rubber band is stretched. The magnitude of the force of the rubber band on the rock is shown by the graph in **FIGURE P9.55**. The rubber band is stretched 30 cm and then released. What is the speed of the rock?

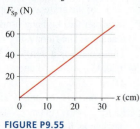

FIGURE P9.55

56. ‖ When a 65 kg cheerleader stands on a vertical spring, the spring compresses by 5.5 cm. When a second cheerleader stands on the shoulders of the first, the spring compresses an additional 4.5 cm. What is the mass of the second cheerleader?

57. ‖ Two identical horizontal springs are attached to opposite sides of a box that sits on a frictionless table. The outer ends of the springs are clamped while the springs are at their equilibrium lengths. Then a 2.0 N force applied to the box, parallel to the springs, compresses one spring by 3.0 cm while stretching the other by the same amount. What is the spring constant of the springs?

58. ‖ A spring of equilibrium length L_1 and spring constant k_1 hangs from the ceiling. Mass m_1 is suspended from its lower end. Then a second spring, with equilibrium length L_2 and spring constant k_2, is hung from the bottom of m_1. Mass m_2 is suspended from this second spring. How far is m_2 below the ceiling?

59. ‖ A horizontal spring with spring constant 250 N/m is compressed by 12 cm and then used to launch a 250 g box across the floor. The coefficient of kinetic friction between the box and the floor is 0.23. What is the box's launch speed?

60. ‖ A 90 kg firefighter needs to climb the stairs of a 20-m-tall building while carrying a 40 kg backpack filled with gear. How much power does he need to reach the top in 55 s?

61. ‖ A hydroelectric power plant uses spinning turbines to transform the kinetic energy of moving water into electric energy with 80% efficiency. That is, 80% of the kinetic energy becomes electric energy. A small hydroelectric plant at the base of a dam generates 50 MW of electric power when the falling water has a speed of 18 m/s. What is the water flow rate—kilograms of water per second—through the turbines?

62. ‖ When you ride a bicycle at constant speed, nearly all the energy
BIO you expend goes into the work you do against the drag force of the air. Model a cyclist as having cross-section area 0.45 m² and, because the human body is not aerodynamically shaped, a drag coefficient of 0.90.
   a. What is the cyclist's power output while riding at a steady 7.3 m/s (16 mph)?
   b. *Metabolic power* is the rate at which your body "burns" fuel to power your activities. For many activities, your body is roughly 25% efficient at converting the chemical energy of food into mechanical energy. What is the cyclist's metabolic power while cycling at 7.3 m/s?
   c. The food calorie is equivalent to 4190 J. How many calories does the cyclist burn if he rides over level ground at 7.3 m/s for 1 h?

63. ‖ A farmer uses a tractor to pull a 150 kg bale of hay up a 15° incline to the barn at a steady 5.0 km/h. The coefficient of kinetic friction between the bale and the ramp is 0.45. What is the tractor's power output?

64. ‖ A Porsche 944 Turbo has a rated engine power of 217 hp. 30% of the power is lost in the drive train, and 70% reaches the wheels. The total mass of the car and driver is 1480 kg, and two-thirds of the weight is over the drive wheels.
   a. What is the maximum acceleration of the Porsche on a concrete surface where $\mu_s = 1.00$?
      Hint: What force pushes the car forward?
   b. If the Porsche accelerates at a_{max}, what is its speed when it reaches maximum power output?
   c. How long does it take the Porsche to reach the maximum power output?

65. ‖ Astronomers using a 2.0-m-diameter telescope observe a distant supernova—an exploding star. The telescope's detector records 9.1×10^{-11} J of light energy during the first 10 s. It's known that this type of supernova has a visible-light power output of 5.0×10^{37} W for the first 10 s of the explosion. How distant is the supernova? Give your answer in *light years*, where one light year is the distance light travels in one year. The speed of light is 3.0×10^8 m/s.

66. ‖ Six dogs pull a two-person sled with a total mass of 220 kg. The coefficient of kinetic friction between the sled and the snow is 0.080. The sled accelerates at 0.75 m/s² until it reaches a cruising speed of 12 km/h. What is the team's (a) maximum power output during the acceleration phase and (b) power output during the cruising phase?

In Problems 67 through 69 you are given the equation(s) used to solve a problem. For each of these, you are to
   a. Write a realistic problem for which this is the correct equation(s).
   b. Draw a pictorial representation.
   c. Finish the solution of the problem.

67. $\frac{1}{2}(2.0 \text{ kg})(4.0 \text{ m/s})^2 + 0$
   $\quad + (0.15)(2.0 \text{ kg})(9.8 \text{ m/s}^2)(2.0 \text{ m}) = 0 + 0 + T(2.0 \text{ m})$

68. $F_{push} - (0.20)(30 \text{ kg})(9.8 \text{ m/s}^2) = 0$
   $75 \text{ W} = F_{push}v$

69. $T - (1500 \text{ kg})(9.8 \text{ m/s}^2) = (1500 \text{ kg})(1.0 \text{ m/s}^2)$
   $P = T(2.0 \text{ m/s})$

Challenge Problems

70. ‖‖ A 12 kg weather rocket generates a thrust of 200 N. The rocket, pointing upward, is clamped to the top of a vertical spring. The bottom of the spring, whose spring constant is 550 N/m, is anchored to the ground.
   a. Initially, before the engine is ignited, the rocket sits at rest on top of the spring. How much is the spring compressed?
   b. After the engine is ignited, what is the rocket's speed when the spring has stretched 40 cm?

71. ‖‖ A gardener pushes a 12 kg lawnmower whose handle is tilted up 37° above horizontal. The lawnmower's coefficient of rolling friction is 0.15. How much power does the gardener have to supply to push the lawnmower at a constant speed of 1.2 m/s? Assume his push is parallel to the handle.

72. ‖‖ A uniform solid bar with mass m and length L rotates with
CALC angular velocity ω about an axle at one end of the bar. What is the bar's kinetic energy?

10 Interactions and Potential Energy

These windmills are transforming the kinetic energy of the wind into electric energy.

IN THIS CHAPTER, you will develop a better understanding of energy and its conservation.

How do interactions affect energy?

We continue our investigation of energy by allowing interactions to be part of the system, rather than external forces. You will learn that interactions can store energy within the system. Further, this interaction energy can be transformed—via the interaction forces—into kinetic energy.

Environment

System

The interaction is part of the system.

What is potential energy?

Interaction energy is usually called potential energy. There are many kinds of potential energy, each associated with *position*.

- Gravitational potential energy changes with height.
- Elastic potential energy changes with stretching.

« LOOKING BACK Section 9.1 Energy overview

When is energy conserved?

- If a system is isolated, its total energy is conserved.
- If a system both is isolated and has *no dissipative forces*, its mechanical energy, $K + U$, is conserved.

Energy bar charts are a tool for visualizing energy conservation.

$$K_i + U_i = K_f + U_f$$

What is an energy diagram?

An energy diagram is a graphical representation of how the energy of a particle changes as it moves. Turning points occur where the total energy line crosses the potential-energy curve. And potential-energy minima are points of stable equilibrium.

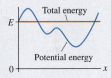

Total energy

E

Potential energy

How is force related to potential energy?

Only certain types of forces, called conservative forces, are associated with a potential energy. For these forces,

- The work done changes the potential energy by $\Delta U = -W$.
- Force is the negative of the slope of the potential-energy curve.

U

$F_s = -\text{slope}$

s

Where are we now in our study of energy?

Energy is a big topic, not one that can be presented in a single chapter. Chapters 9 and 10 are primarily about mechanical energy and the mechanical transfer of energy via work. And we've touched on thermal energy because it's unavoidable in realistic mechanical systems with friction. These are related by the energy principle:

$$\Delta E_{sys} = \Delta K + \Delta U + \Delta E_{th} = W_{ext}$$

Part V of this book, Thermodynamics, will expand our energy ideas to include heat and a deeper understanding of thermal energy. Then we'll add another form of energy—electric energy—in Part VI.

10.1 Potential Energy

If you press a ball against a spring and release it, the ball shoots forward. It certainly seems like the spring had a supply of stored energy that was transferred to the ball. Or imagine tossing the ball straight up. Where does its kinetic energy go as it slows? And from where does it acquire kinetic energy as it falls? There's again a sense that the energy is stored somewhere as the ball rises, then released as the ball falls. But is energy really stored? And if so, where? And how? Answering these questions is key to expanding our understanding of the basic energy model.

Chapter 9 emphasized the importance of the *system* and the *environment*. The system has energy E_{sys}, and forces from the environment—external forces—change the system's energy by doing work on the system. In Chapter 9, we considered only systems of particles, and all forces originated in the environment. But that's not the only way to define the system. What happens if we bring some of the interactions inside the system?

FIGURE 10.1 shows two particle-like objects A and B that interact with each other and nothing else. For example, these might be two objects connected by a spring, two masses exerting gravitational forces on each other, or two charged particles exerting electric forces on each other. Regardless of what the interaction is, this is an action/reaction pair of forces that obeys Newton's third law. There are two ways to define a system.

System 1 has been chosen to consist of only the two particles; the forces are external forces. This is exactly the analysis we did in Chapter 9, so we know that the energy principle for system 1 is

$$\Delta E_{sys\,1} = \Delta K_{tot} = W_{ext} = W_A + W_B \qquad (10.1)$$

where K_{tot} is the combined kinetic energies of A and B. Work W_A is the work done on A by force $\vec{F}_{B\,on\,A}$, and similarly W_B is the work done on B by force $\vec{F}_{A\,on\,B}$. The work of these two forces changes the system's kinetic energy.

Now consider the same two particles but with a different choice of system, system 2, where we've included the interaction within the system. It's important to recognize that a system is not a physical thing. It's an analysis tool that we can define however we wish, and our choice doesn't change the behavior of physical objects. Objects A and B are oblivious to our choice of system, so ΔK_{tot} for system 2 is exactly the same as for system 1. But W_{ext} *has* changed. System 2 has no external forces to transfer energy to or from the system, so $W_{ext} = 0$. Consequently, the energy principle for system 2 is

$$\Delta E_{sys\,2} = W_{ext} = 0 \qquad (10.2)$$

Now the fact that $\Delta E_{sys\,1} \neq \Delta E_{sys\,2}$ is not an issue; after all, they are different systems. But we know that system 2 has a changing kinetic energy, so how can $\Delta E_{sys\,2} = 0$?

Because system 2 has an interaction inside the system that system 1 lacks, let's postulate that system 2 has an additional form of energy associated with the interaction. That is, system 1 has $E_{sys\,1} = K_{tot}$, because particles have only kinetic energy, but system 2 has $E_{sys\,2} = K_{tot} + U$, where U, called **potential energy,** is the energy of the interaction.

If this is true, we can combine $\Delta E_{sys\,2} = 0$, from Equation 10.2, with our knowledge of ΔK_{tot} from Equation 10.1 to write

$$\Delta E_{sys\,2} = \Delta K_{tot} + \Delta U = (W_A + W_B) + \Delta U = 0 \qquad (10.3)$$

That is, system 2 can have $\Delta E_{sys} = 0$ if it has a potential energy that changes by

$$\Delta U = -(W_A + W_B) = -W_{int} \qquad (10.4)$$

where W_{int} is the total work done *inside the system* by the interaction forces.

Equation 10.3 tells us that the system's kinetic energy can increase ($\Delta K > 0$) if its potential energy decreases ($\Delta U < 0$) by the same amount. In effect, **the interaction stores energy inside the system** with the *potential* to be converted to kinetic energy (or, in other situations, thermal energy)—hence the name *potential energy*. This

FIGURE 10.1 Two choices of the system and the environment.

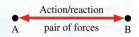

Action/reaction pair of forces

A   B

External forces do work on the system.

System 1:

$\vec{F}_{B\,on\,A}$ $\vec{F}_{A\,on\,B}$

The interaction is part of the system.

System 2:

A   A–B interaction   B

idea will become more concrete as we start looking at specific examples. And, since we *postulated* the existence of an energy associated with interactions, we'll need to investigate the types of interactions for which this is true.

> **NOTE** Kinetic energy is the energy of an object. In contrast, potential energy is the energy of an interaction. You can say "The ball has kinetic energy" but not "The ball has potential energy." We'll look at the best way to describe potential energy when we get to specific examples.

Systems Matter

When solving a problem, *you* get to define the system. But your choice has consequences! E_{sys} is the energy *of* the system, so a different system will have a different energy. Similarly, W_{ext} is the work done on the system by forces originating in the environment, and that will depend on the boundary between the system and the environment.

In Figure 10.1, system 1 is a restricted system of just the particles, so system 1 has only kinetic energy. All the interaction forces are external forces that do work. Thus system 1 obeys

$$\Delta E_{sys} = \Delta K_{tot} = W_A + W_B$$

System 2 includes the interaction, so system 2 has both kinetic and potential energy. But the choice of the system boundary means that no work is done by external forces. So for system 2,

$$\Delta E_{sys} = \Delta K_{tot} + \Delta U = 0$$

Both mathematical statements are true because they refer to different systems. Notice that, for system 2, kinetic energy can be transformed into potential energy, or vice versa, but **the total energy of the system does not change.** This is our first glimpse of the idea of *conservation of energy*.

The point to remember is that **any choice of system is acceptable, but you can't mix and match.** You can define the system so that you have to calculate work, or you can define the system where you use potential energy, but using both work *and* potential energy is incorrect because it double counts the contribution of the interaction. Thus the most critical step in an energy analysis is to clearly define the system you're working with.

10.2 Gravitational Potential Energy

We'll start our exploration of potential energy with **gravitational potential energy,** the interaction energy associated with the gravitational interaction between two masses. The symbol for gravitational potential energy is U_G. We'll restrict ourselves to the "flat-earth approximation" $\vec{F}_G = -mg\hat{\jmath}$. The gravitational potential energy of two astronomical bodies will be taken up in Chapter 13.

FIGURE 10.2 shows a ball of mass m moving upward from an initial vertical position y_i to a final vertical position y_f. The earth exerts force $\vec{F}_{E \text{ on } B}$ on the ball and, by Newton's third law, the ball exerts an equal-but-opposite force $\vec{F}_{B \text{ on } E}$ on the earth.

We could define the system to consist of only the ball, in which case the force of gravity is an external force that does work on the ball, changing its kinetic energy. We did exactly this in Chapter 9. Now let's define the system to be ball + earth. This brings the interaction inside the system, so (ignoring any gravitational forces from distant astronomical bodies) there's no external work. Instead, we have an energy of interaction—the gravitational potential energy—described by Equation 10.4:

$$\Delta U_G = -(W_B + W_E) \tag{10.5}$$

where W_B is the work gravity does on the ball and W_E is the work gravity does on the earth. The latter, practically speaking, is zero. $\vec{F}_{E \text{ on } B}$ and $\vec{F}_{B \text{ on } E}$ have equal magnitudes, by Newton's third law, but the earth's displacement is completely insignificant

FIGURE 10.2 The ball + earth system has a gravitational potential energy.

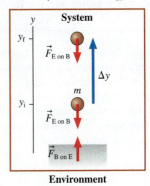

Environment

compared to the ball's displacement. Because work is a product of force and displacement, the work done on the earth is essentially zero and we can write

$$\Delta U_G = -W_B \tag{10.6}$$

You learned in Chapter 9 to compute the work of gravity on the ball: $W_B = (F_G)_y \Delta y = -mg\,\Delta y$. So if the ball changes its vertical position by Δy, the gravitational potential energy changes by

$$\Delta U_G = -W_B = mg\,\Delta y \tag{10.7}$$

Notice that increasing the ball's height ($\Delta y > 0$) increases the gravitational potential energy ($\Delta U_G > 0$), as we would expect.

Our energy analysis has given us an expression for ΔU_G, the *change* in potential energy, but not an expression for U_G itself. If we write out what the Δ in Equation 10.7 means—final value minus initial value—we have

$$U_{Gf} - U_{Gi} = mgy_f - mgy_i \tag{10.8}$$

Consequently, we define the gravitational potential energy to be

$$U_G = mgy \quad \text{(gravitational potential energy)} \tag{10.9}$$

Notice that **gravitational potential energy is an energy of position.** It depends on the object's position but not on its speed. You should convince yourself that the units of mass times acceleration times position are joules, the unit of energy.

STOP TO THINK 10.1 Rank in order, from largest to smallest, the gravitational potential energies of balls a to d.

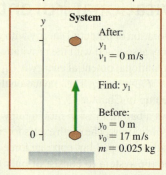

EXAMPLE 10.1 | **Launching a pebble**

Rafael uses a slingshot to shoot a 25 g pebble straight up at 17 m/s. How high does the pebble go?

MODEL Let the system consist of both the earth and the pebble, which we model as a particle. Assume that air resistance is negligible. There are no external forces to do work, but the system does have gravitational potential energy.

VISUALIZE FIGURE 10.3 is a before-and-after pictorial representation. The before-and-after representation will continue to be our primary visualization tool.

SOLVE The energy principle for the pebble + earth system is

$$\Delta E_{sys} = \Delta K + \Delta U_G = W_{ext} = 0$$

That is, the system energy does not change at all. Instead, kinetic energy is transformed into potential energy without loss inside the system. In principle, the kinetic energy is that of the ball plus the kinetic energy of the earth. But as we just noted, the enormous mass difference means that the earth is effectively at rest while the pebble does all the moving, so the only kinetic energy we need to consider is that of the pebble. Thus we have

$$0 = \Delta K + \Delta U_G = (\tfrac{1}{2}mv_1^2 - \tfrac{1}{2}mv_0^2) + (mgy_1 - mgy_0)$$

FIGURE 10.3 Pictorial representation of the pebble + earth system.

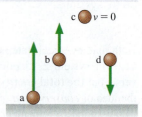

We know that $v_1 = 0$ m/s, and we chose a coordinate system in which $y_0 = 0$ m, so we're left with

$$y_1 = \frac{v_0^2}{2g} = \frac{(17 \text{ m/s})^2}{2(9.80 \text{ m/s}^2)} = 15 \text{ m}$$

The answer did not depend on the pebble's mass, which is not surprising after our earlier practice with free-fall problems.

ASSESS A height of 15 m ≈ 45 ft seems reasonable for a slingshot.

The Zero of Potential Energy

Our expression for the gravitational potential energy $U_G = mgy$ seems straightforward. But you might notice, on further reflection, that the value of U_G depends on where you choose to put the origin of your coordinate system. Consider **FIGURE 10.4**, where Amber and Carlos are attempting to determine the potential energy when a 1 kg rock is 1 m above the ground. Amber chooses to put the origin of her coordinate system on the ground, measures $y_{rock} = 1$ m, and quickly computes $U_G = mgy = 9.8$ J. Carlos, on the other hand, read Chapter 1 very carefully and recalls that it is entirely up to him where to locate the origin of his coordinate system. So he places his origin next to the rock, measures $y_{rock} = 0$ m, and declares that $U_G = mgy = 0$ J!

How can the potential energy have two different values? The source of this apparent difficulty comes from our interpretation of Equation 10.7. Our energy analysis found that the potential energy *changes* by $\Delta U_G = mg(y_f - y_i)$. Our claim that $U_G = mgy$ is consistent with this finding, but so also would be a claim that $U_G = mgy + C$, where C is any constant.

In other words, potential energy does not have a uniquely defined value. Adding or subtracting the same constant from all potential energies in a problem has no physical consequences because our analysis uses only ΔU_G, the *change* in the potential energy, never the actual value of U_G. In practice, we work with potential energies by setting a *reference point* or *reference level* where $U_G = 0$. This is the **zero of potential energy.** Where you place the reference point is entirely up to you; it makes no difference as long as every potential energy in the problem uses the same reference point. For gravitational potential energy, we choose the reference level by placing the origin of the y-axis at that point. Where $y = 0$, $U_G = 0$. In Figure 10.4, Amber has placed her zero of potential energy at the ground, whereas Carlos has set a reference level 1 m above the ground. Either is perfectly acceptable as long as Amber and Carlos use their reference levels consistently.

But what happens when the rock falls? When it gets to the ground, Amber measures $y = 0$ m and computes $U_G = 0$ J. No problem. But Carlos measures $y = -1$ m and thus computes $U_G = -9.8$ J. A negative potential energy may seem surprising, but it's not wrong; it simply means that the potential energy is less than at the reference point. The potential energy with the rock on the ground is certainly less than when the rock was 1 m above the ground, so for Carlos—with an elevated reference level—the potential energy is negative. The important point is that both Amber and Carlos agree that the gravitational potential energy *changes* by $\Delta U_G = -9.8$ J as the rock falls.

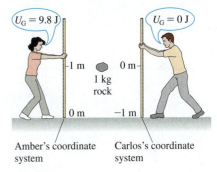

FIGURE 10.4 Amber and Carlos use different coordinate systems to determine the gravitational potential energy.

Amber's coordinate system

Carlos's coordinate system

Energy Bar Charts

If an object of mass m interacts with the earth (or other astronomical body) and there are no other forces, the energy principle for the object + earth system, Equation 10.3, is

$$\Delta K + \Delta U_G = (K_f - K_i) + (U_{Gf} - U_{Gi}) = 0 \qquad (10.10)$$

We can rewrite this as

$$K_i + U_{Gi} = K_f + U_{Gf} \qquad (10.11)$$

The quantity $E_{mech} = K_{tot} + U_{int}$, the total macroscopic kinetic and potential energy, is called the **mechanical energy** of the system. Equation 10.11 is telling us that—in this situation—**the mechanical energy does not change** as the object undergoes vertical motion. Whatever initial mechanical energy the system had before the vertical motion, it has exactly the same mechanical energy after the motion. Kinetic energy may be transformed into potential energy during the motion, or vice versa, but their sum remains unchanged.

A quantity that is unchanged during an interaction is said to be *conserved*, and Equation 10.11 is our first statement of the *law of conservation of energy*. Now this is a highly restricted situation: only the gravitational force, no other forces, and only vertical motion. We'll explore energy conservation thoroughly later in this chapter and see to what extent these restrictions can be lifted, but we're already beginning to see the power of thinking about mechanical systems in terms of energy.

Equation 10.11, which is really just energy accounting, can be represented graphically with an **energy bar chart.** For example, **FIGURE 10.5** is a bar chart showing how energy is transformed when a ball is tossed straight up. Kinetic energy is gradually transformed into potential energy as the ball rises, then potential energy is transformed into kinetic energy as it falls, but **the combined height of the bars does not change.** That is, the mechanical energy of the ball + earth system is conserved.

FIGURE 10.5 Energy bar charts for a ball tossed into the air.

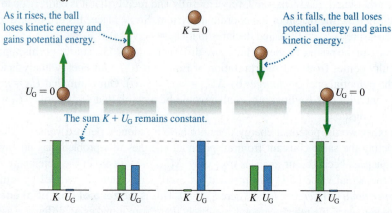

As it rises, the ball loses kinetic energy and gains potential energy.

$K = 0$

As it falls, the ball loses potential energy and gains kinetic energy.

$U_G = 0$

$U_G = 0$

The sum $K + U_G$ remains constant.

K U_G K U_G K U_G K U_G K U_G

NOTE Most bar charts have no numbers. Their purpose is to think about the relative changes—what's increasing, what's decreasing, and what remains constant; there's no significance to how tall a bar is.

EXAMPLE 10.2 | Dropping a watermelon

A 5.0 kg watermelon is dropped from a third-story balcony, 11 m above the street. Unfortunately, the water department forgot to replace the cover on a manhole, and the watermelon falls into a 3.0-m-deep hole. How fast is the watermelon going when it hits bottom?

MODEL Let the system consist of both the earth and the watermelon, which we model as a particle. Assume that air resistance is negligible. There are no external forces, and the motion is vertical, so the system's mechanical energy is conserved.

VISUALIZE FIGURE 10.6 shows both a before-and-after pictorial representation and an energy bar chart. Initially the system has gravitational potential energy but no kinetic energy. Potential

FIGURE 10.6 Pictorial representation and energy bar chart of the watermelon + earth system.

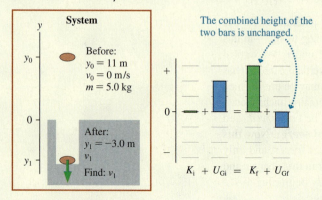

System

Before:
$y_0 = 11$ m
$v_0 = 0$ m/s
$m = 5.0$ kg

After:
$y_1 = -3.0$ m
v_1
Find: v_1

The combined height of the two bars is unchanged.

$K_i + U_{Gi} = K_f + U_{Gf}$

energy is transformed into kinetic energy as the watermelon falls. Our choice of the y-axis origin has placed the zero of potential energy at ground level, so the potential energy is negative when the watermelon reaches the bottom of the hole. Even so, the combined height of the two bars has not changed.

SOLVE The energy principle for the watermelon + earth system, written as a conservation statement, is

$$K_i + U_{Gi} = 0 + mgy_0 = K_f + U_{Gf} = \tfrac{1}{2}mv_1^2 + mgy_1$$

Solving for the impact speed, we find

$$v_1 = \sqrt{2g(y_0 - y_1)}$$
$$= \sqrt{2(9.80 \text{ m/s}^2)(11.0 \text{ m} - (-3.0 \text{ m}))}$$
$$= 17 \text{ m/s}$$

ASSESS A speed of 17 m/s ≈ 35 mph seems reasonable for the watermelon after falling ≈ 4 stories. In thinking about this problem, you might be concerned that, once below ground level, potential energy continues being transformed into kinetic energy even though the potential energy is "less than none." Keep in mind that the actual value of U is not relevant because we can place the zero of potential energy anywhere we wish, so a negative potential energy is just a number with no implication that it's "less than none." There's no "storehouse" of potential energy that might run dry. As long as the interaction acts, potential energy can continue being transformed into kinetic energy.

Digging Deeper into Gravitational Potential Energy

The concept of gravitational potential energy would be of little interest or use if it applied only to vertical free fall. Let's begin to expand the idea. **FIGURE 10.7** shows a particle of mass m moving at an angle while acted on by gravity. How much work does gravity do?

Gravity is a constant force. In Chapter 9 you learned that, in general, the work done by a constant force is $W = \vec{F} \cdot \Delta \vec{r}$. If we write both $\vec{F}_G$ and $\Delta \vec{r}$ in terms of components, and use the Chapter 9 result for calculating the dot product with components, we find that the work done by gravity is

$$W_{\text{by grav}} = \vec{F}_G \cdot \Delta \vec{r} = (F_G)_x(\Delta r_x) + (F_G)_y(\Delta r_y) = 0 + (-mg)(\Delta y)$$

$$= -mg\,\Delta y \qquad (10.12)$$

Because $\vec{F}_G$ has no x-component, the work depends only on the vertical displacement Δy.

Consequently, **the change in gravitational potential energy depends only on an object's vertical displacement.** This is true not only for motion along a straight line, as in Figure 10.7, but also for motion along a *curved* trajectory because a curve can be represented as the limit of a very large number of very short straight-line segments.

For example, **FIGURE 10.8** shows an object sliding down a curved, frictionless surface. The change in gravitational potential energy of the object + earth system depends only on Δy, the distance the object descends, *not* on the shape of the curve. But now there's an additional force—the normal force of the surface. Does this force affect the system's energy? No! The normal force is always perpendicular to the box's instantaneous displacement, and you learned in Chapter 9 that forces perpendicular to the displacement do no work. **Forces always perpendicular to the motion do not affect the system's energy.** They can be ignored during an energy analysis.

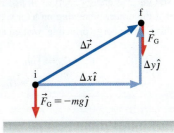

FIGURE 10.7 Gravity does work on a particle moving at an angle.

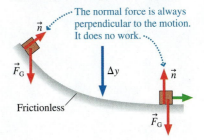

FIGURE 10.8 For motion on any frictionless surface, only the vertical displacement Δy affects the energy.

The normal force is always perpendicular to the motion. It does no work.

Frictionless

STOP TO THINK 10.2 Two identical projectiles are fired with the same speed but at different angles. Neglect air resistance. At the elevation shown as a dashed line,

a. The speed of A is greater than the speed of B.
b. The speed of A is the same as the speed of B.
c. The speed of A is less than the speed of B.

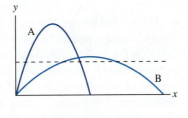

EXAMPLE 10.3 | The speed of a sled

Christine runs forward with her sled at 2.0 m/s. She hops onto the sled at the top of a 5.0-m-high, very slippery slope. What is her speed at the bottom?

MODEL Let the system consist of the earth and the sled, which we model as a particle. Because the slope is "very slippery," we'll assume that friction is negligible. The slope exerts a normal force on the sled, but it is always perpendicular to the motion and does not affect the energy.

VISUALIZE FIGURE 10.9a shows a before-and-after pictorial representation. We are not told the angle of the slope, or even if it is a straight slope, but the change in potential energy depends only on the vertical distance Christine descends and *not* on the shape of the hill. FIGURE 10.9b is an energy bar chart in which we see an initial

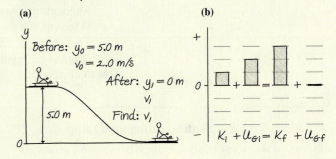

FIGURE 10.9 Pictorial representation and energy bar chart of the sled + earth system.

Continued

kinetic *and* potential energy being transformed into entirely kinetic energy as Christine goes down the slope.

SOLVE The energy analysis is just like in Example 10.2; the fact that the object is moving on a curved surface hasn't changed anything. The energy principle, written as a conservation statement, is

$$K_i + U_{Gi} = \tfrac{1}{2}mv_0^2 + mgy_0$$
$$= K_f + U_{Gf} = \tfrac{1}{2}mv_1^2 + 0$$

Her speed at the bottom is

$$v_1 = \sqrt{v_0^2 + 2gy_0}$$
$$= \sqrt{(2.0 \text{ m/s})^2 + 2(9.80 \text{ m/s}^2)(5.0 \text{ m})}$$
$$= 10 \text{ m/s}$$

ASSESS 10 m/s ≈ 20 mph is fast but believable for a 5 m ≈ 15 ft descent.

STOP TO THINK 10.3 A small child slides down the four frictionless slides a–d. Each has the same height. Rank in order, from largest to smallest, her speeds v_a to v_d at the bottom.

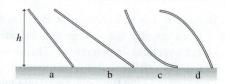

Motion with Gravity and Friction

What if there's friction? You learned in « Section 9.5 that friction increases the thermal energy of the system—defined to include *both* objects—by $\Delta E_{th} = f_k \Delta s$. For a system with both gravitational potential energy and friction, the energy principle becomes

$$\Delta K + \Delta U_G + \Delta E_{th} = 0 \tag{10.13}$$

or, equivalently,

$$K_i + U_{Gi} = K_f + U_{Gf} + \Delta E_{th} \tag{10.14}$$

Mechanical energy $K + U_G$ is *not* conserved if there is friction. Because $\Delta E_{th} > 0$ (friction always makes surfaces hotter, never cooler), the final mechanical energy is less than the initial mechanical energy. That is, some fraction of the initial kinetic and potential energy is transformed into thermal energy during the motion. Friction causes objects to slow down, and motion ceases when all the mechanical energy has been transformed into thermal energy. Mechanical energy is conserved only when there are no dissipative forces and thus $\Delta E_{th} = 0$.

> **NOTE** We can write the energy principle in terms of initial and final values of the kinetic energy and the potential energy, but *not* the thermal energy. Objects always have thermal energy—the atoms are constantly in motion—but we have no way to know how much. All we can calculate is the *change* in thermal energy.

Although mechanical energy is not conserved, *the system's energy is.* Equation 10.13 tells us that the sum of kinetic, potential, and thermal energy—the energy of the system—does not change as the object moves on a surface with friction. The initial mechanical energy does not disappear; it's merely transformed into an equal amount of thermal energy.

EXAMPLE 10.4 | **Skateboarding up a ramp**

During the skateboard finals, Isabella encounters a 6.0-m-long, 15° upward ramp. Isabella's mass, including the skateboard, is 55 kg, and the coefficient of rolling friction between her wheels and the ramp is 0.025. With what speed must she start up the ramp to reach the top at 2.5 m/s? What percentage of her mechanical energy is "lost" to friction?

MODEL Let the system consist of the earth (including the ramp) and Isabella on the skateboard.

VISUALIZE FIGURE 10.10 shows a before-and-after pictorial representation.

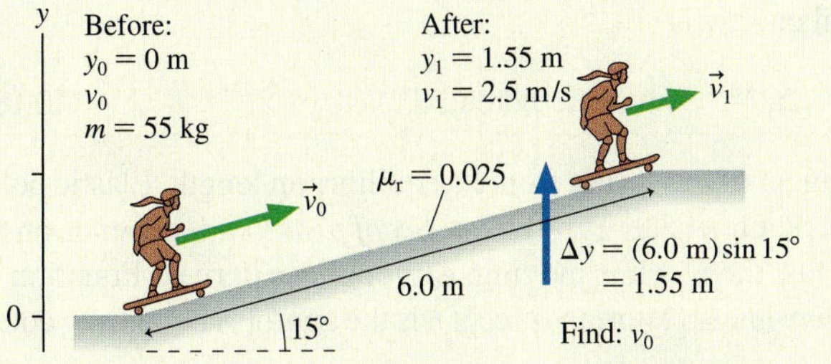

FIGURE 10.10 Pictorial representation of Isabella on the ramp.

Before:
$y_0 = 0$ m
v_0
$m = 55$ kg

After:
$y_1 = 1.55$ m
$v_1 = 2.5$ m/s
$\vec{v}_1$

$\mu_r = 0.025$

6.0 m

$\Delta y = (6.0 \text{ m}) \sin 15°$
$= 1.55$ m

$15°$

Find: v_0

SOLVE Isabella's kinetic energy is transformed into potential energy as she gains height, but some of her kinetic energy is also transformed into increased thermal energy of her wheels and the ramp because of rolling friction. The energy principle including friction is

$$K_i + U_{Gi} = \tfrac{1}{2}mv_0^2 + 0$$
$$= K_f + U_{Gf} + \Delta E_{th} = \tfrac{1}{2}mv_1^2 + mgy_1 + f_r \Delta s$$

where we've used rolling friction f_r rather than the kinetic friction of sliding. Rolling friction is $f_r = \mu_r n$, and recall—from Chapter 6 —that the normal force of an object on a slope is $n = mg \cos\theta$. (Draw a free-body diagram if you're not sure.) Thus

$$\tfrac{1}{2}mv_0^2 = \tfrac{1}{2}mv_1^2 + mgy_1 + \mu_r mg \, \Delta s \cos\theta$$

The mass cancels. Solving for Isabella's speed at the bottom of the ramp, we find

$$v_0 = \sqrt{v_1^2 + 2gy_1 + 2\mu_r g \, \Delta s \cos\theta} = 6.3 \text{ m/s}$$

Isabella's initial mechanical energy is entirely kinetic energy: $K_0 = \tfrac{1}{2}mv_0^2 = 1090$ J. The thermal energy of the ramp and her wheels increases by $\Delta E_{th} = \mu_r mg \, \Delta s \cos\theta = 78$ J. Thus the percentage of mechanical energy transformed into thermal energy as Isabella ascends the ramp is

$$\frac{78 \text{ J}}{1090 \text{ J}} \times 100 = 7.2\%$$

This energy is not truly lost—it's still in the system—but it's no longer available for motion.

ASSESS The ramp is $1.55 \text{ m} \approx 5$ ft high. Starting up the ramp at $6.3 \text{ m/s} \approx 12$ mph in order to reach the top at $2.5 \text{ m/s} \approx 5$ mph seems reasonable.

STOP TO THINK 10.4 A skier glides down a gentle slope at constant speed. What energy transformation is taking place?

a. $U_G \rightarrow K$
b. $U_G \rightarrow K + E_{th}$
c. $U_G \rightarrow E_{th}$
d. $K \rightarrow E_{th}$
e. No energy transformation is occurring.

10.3 Elastic Potential Energy

Much of what you've just learned about gravitational potential energy carries over to the *elastic potential energy* of a spring. **FIGURE 10.11** shows a spring exerting a force on a block while the block moves on a frictionless, horizontal surface. In Chapter 9, we analyzed this problem by defining the system to consist of only the block, and we calculated the work of the spring on the block. Now let's define the system to be block + spring + wall. That is, the system is the spring and the objects connected by the spring. The surface and the earth exert forces on the block—the normal force and gravity—but those forces are always perpendicular to the displacement and do not transfer any energy to the system.

We'll assume the spring to be massless, so it has no kinetic energy. Instead, the spring is the *interaction* between the block and the wall. Because the interaction is inside the system, it has an interaction energy, the **elastic potential energy,** given by

$$\Delta U_{Sp} = -(W_B + W_W) \tag{10.15}$$

where W_B is the work the spring does on the block and W_W is the work done on the wall. But the wall is rigid and has no displacement, so $W_W = 0$ and thus $\Delta U_{Sp} = -W_B$.

We calculated the work done by an ideal spring—one that produces a linear restoring force for all displacement—in « Section 9.4. If the block moves from an initial position

FIGURE 10.11 The block + spring + wall system has an elastic potential energy.

The system is the spring and the objects the spring is attached to.

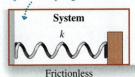

System

k

Frictionless

s_i, where the spring's displacement is $\Delta s_i = s_i - s_{eq}$, to a final position s_f with displacement $\Delta s_f = s_f - s_{eq}$, the spring does work

$$W_B = -\left(\tfrac{1}{2}k(\Delta s_f)^2 - \tfrac{1}{2}k(\Delta s_i)^2\right) \tag{10.16}$$

With the minus sign of Equation 10.15, we have

$$\Delta U_{Sp} = U_f - U_i = -W_B = \tfrac{1}{2}k(\Delta s_f)^2 - \tfrac{1}{2}k(\Delta s_i)^2 \tag{10.17}$$

Thus the elastic potential energy is

$$U_{Sp} = \tfrac{1}{2}k(\Delta s)^2 \quad \text{(elastic potential energy)} \tag{10.18}$$

where Δs is the displacement of the spring from its equilibrium length. Elastic potential energy, like gravitational potential energy, is an *energy of position*. It depends on where the block is, not on how fast the block is moving. Although we derived Equation 10.18 for a spring, it applies to *any* linear restoring force if k is the appropriate "spring constant" for that force.

The energy principle for a system with elastic potential energy and no external interactions is either $\Delta E_{sys} = \Delta K + \Delta U_{Sp} = 0$ or, recognizing that mechanical energy is again conserved,

$$K_i + U_{Sp\,i} = K_f + U_{Sp\,f} \tag{10.19}$$

EXAMPLE 10.5 | An air-track glider compresses a spring

In a laboratory experiment, your instructor challenges you to figure out how fast a 500 g air-track glider is traveling when it collides with a horizontal spring attached to the end of the track. He pushes the glider, and you notice that the spring compresses 2.7 cm before the glider rebounds. After discussing the situation with your lab partners, you decide to hang the spring on a hook and suspend the glider from the bottom end of the spring. This stretches the spring by 3.5 cm. Based on your measurements, how fast was the glider moving?

MODEL Let the system consist of the track, the spring, and the glider. The spring is inside the system, so the elastic interaction will be treated as a potential energy. Gravity and the normal force of the track on the glider are perpendicular to the glider's displacement, so they do no work and do not enter into an energy analysis. An air track is essentially frictionless, and there are no other external forces.

VISUALIZE FIGURE 10.12 shows a before-and-after pictorial representation of the collision, an energy bar chart, and a free-body diagram of the suspended glider.

SOLVE The glider's kinetic energy is gradually transformed into elastic potential energy as it compresses the spring. The point of maximum compression—After in Figure 10.12—is a turning point in the motion. The velocity is instantaneously zero, the glider's kinetic energy is zero, and thus—as the bar chart shows—all the energy has been transformed into potential energy. The spring will expand and cause the glider to rebound, but that's not part of this problem. The energy principle with elastic potential energy, in conservation form, is

$$K_i + U_{Sp\,i} = \tfrac{1}{2}mv_0^2 + 0 = K_f + U_{Sp\,f} = 0 + \tfrac{1}{2}k(\Delta x_1)^2$$

where we utilized our knowledge that the initial elastic potential energy and the final kinetic energy, at the turning point, are zero. Solving for the glider's initial speed, we find

$$v_0 = \sqrt{\frac{k}{m}}\,\Delta x_1$$

FIGURE 10.12 Pictorial representation of the experiment.

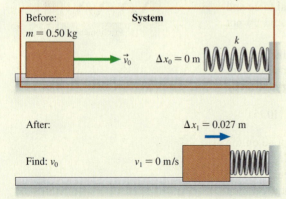

Before:
$m = 0.50$ kg
$\Delta x_0 = 0$ m
k
$\vec{v}_0$

After:
$\Delta x_1 = 0.027$ m
Find: v_0
$v_1 = 0$ m/s

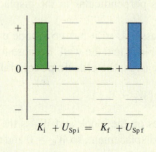

$K_i + U_{Sp\,i} = K_f + U_{Sp\,f}$

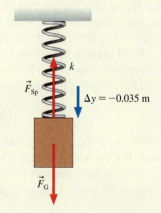

k
$\vec{F}_{Sp}$
$\Delta y = -0.035$ m
$\vec{F}_G$

It was at this point that you and your lab partners realized you needed to determine the spring constant k. One way to do so is to measure the stretch caused by a suspended mass. The hanging glider is in equilibrium with no net force, and the free-body diagram shows that the upward spring force exactly balances the downward gravitational force. From Hooke's law, the *magnitude* of the spring force is $F_{Sp} = k|\Delta y|$. Thus Newton's first law for the suspended glider is

$$F_{Sp} = k|\Delta y| = F_G = mg$$

from which the spring's spring constant is

$$k = \frac{mg}{|\Delta y|} = \frac{(0.50 \text{ kg})(9.80 \text{ m/s}^2)}{0.035 \text{ m}} = 140 \text{ N/m}$$

Knowing k, you can now find that the glider's speed was

$$v_0 = \sqrt{\frac{k}{m}}\, \Delta x_1 = \sqrt{\frac{140 \text{ N/m}}{0.50 \text{ kg}}}\, (0.027 \text{ m}) = 0.45 \text{ m/s}$$

ASSESS A speed of ≈ 0.5 m/s is typical for gliders on an air track.

Including Gravity

Now that we see how the basic energy model works, it's easy to extend it to new situations. If a problem has both a spring *and* a vertical displacement, we define the system so that both the gravitational interaction and the elastic interaction are inside the system. Then we have both elastic *and* gravitational potential energy. That is,

$$U = U_G + U_{Sp} \tag{10.20}$$

You have to be careful with the energy accounting because there are more ways that energy can be transformed, but nothing fundamental has changed by having two potential energies rather than one.

And we know how to include the increased thermal energy if there's friction. Thus for a system that has gravitational interactions, elastic interactions, and friction, but no external forces that do work, the energy principle is

$$\Delta E_{sys} = \Delta K + \Delta U_G + \Delta U_{Sp} + \Delta E_{th} = 0 \tag{10.21}$$

or, in conservation form,

$$K_i + U_{Gi} + U_{Sp\,i} = K_f + U_{Gf} + U_{Sp\,f} + \Delta E_{th} \tag{10.22}$$

This is looking a bit more complex as we have more and more energies to keep track of, but the message of Equations 10.21 and 10.22 is both simple and profound: **For a system that has no other interactions with its environment, the total energy of the system does not change.** It can be transformed in many ways by the interactions, but the total does not change.

EXAMPLE 10.6 | A spring-launched block

Your lab assignment for the week is to devise an innovative method to determine the spring constant of a spring. You see several small blocks of different mass lying around, so you decide to measure how high the compressed spring will launch each of the blocks. You and your lab partners realize that you need to compress the spring the same amount each time, so that only the mass is varying, and you choose to use a compression of 4.0 cm. You decide to measure height from the point on the compressed spring at which the block is released. Four launches generate the data in the table:

Mass (g)	Height (m)
50	2.07
100	1.11
150	0.65
200	0.51

What value will you report for the spring constant?

MODEL Let the system consist of the earth, the block, the spring, and the floor, so there will be two potential energies. There's no friction and we'll assume no drag; hence the mechanical energy of this system is conserved. Model the spring as ideal.

VISUALIZE FIGURE 10.13 shows a pictorial representation, including an energy bar chart. We've chosen to place the origin of the

FIGURE 10.13 Pictorial representation of the experiment.

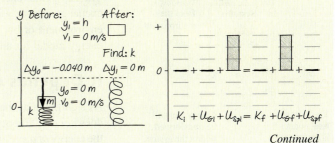

Continued

coordinate system at the point of launch. The projectile reaches height $y_1 = h$, at which point $v_1 = 0$ m/s.

SOLVE You might think we would need to find the block's speed as it leaves the spring. That would be true if we were solving this problem with Newton's laws of motion. But with an energy analysis, we can compare the system's pre-launch energy to its energy when the block reaches its highest point, completely bypassing the launch speed. The block certainly has kinetic energy *during* the motion, but the net energy transfer, shown on the energy bar chart, is from elastic potential energy to gravitational potential energy.

With both elastic and gravitational potential energy included, the energy principle is

$$K_i + U_{Gi} + U_{Sp\,i} = 0 + 0 + \tfrac{1}{2}k(\Delta y_0)^2$$
$$= K_f + U_{Gf} + U_{Sp\,f} = 0 + mgy_1 + 0$$

The block travels to position y_1, but the end of the spring does not! Be careful in spring problems not to mistake the position of an object for the position of the end of the spring; sometimes they are the same, but not always. Here the final elastic potential energy is that of an empty, unstretched spring: zero. Solving for the height, we find

$$y_1 = h = \frac{k(\Delta y_0)^2}{2mg} = \frac{k(\Delta y_0)^2}{2g} \times \frac{1}{m}$$

The first expression is correct as an algebraic expression, but here we want to use the result to analyze an experiment in which we measure h as m is varied. By isolating the mass term, we see that plotting h versus $1/m$ (that is, using $1/m$ as the x-variable) should yield a straight line with slope $k(\Delta y_0)^2/2g$.

FIGURE 10.14 is a graph of h versus $1/m$, with masses first converted to kg. The graph is linear and the best-fit line has a y-intercept very near zero, confirming our analysis of the situation. The experimentally determined slope is 0.105 m kg, with the units determined by rise over run. Thus the experimental value of the spring constant is

$$k = \frac{2g}{(\Delta y_0)^2} \times slope = 1290 \text{ N/m}$$

FIGURE 10.14 Graph of the block height versus the inverse of its mass.

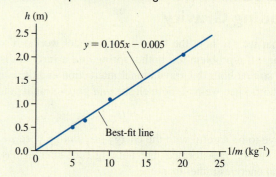

ASSESS 1290 N/m is a reasonably stiff spring, but that's to be expected if you're launching blocks a meter or more into the air.

STOP TO THINK 10.5 A spring-loaded pop gun shoots a plastic ball with a speed of 4 m/s. If the spring is compressed twice as far, the ball's speed will be

a. 2 m/s
b. 4 m/s
c. 8 m/s
d. 16 m/s

10.4 Conservation of Energy

One of the most powerful statements in physics is the **law of conservation of energy:**

> **Law of conservation of energy** The total energy $E_{sys} = E_{mech} + E_{th}$ of an isolated system is a constant. The kinetic, potential, and thermal energy within the system can be transformed into each other, but their sum cannot change. Further, the mechanical energy $E_{mech} = K + U$ is conserved if the system is both isolated and nondissipative.

FIGURE 10.15 The basic energy model for an isolated system.

The system is isolated from the environment.

System
$K \longleftrightarrow U$
E_{th}
$\rightarrow \Delta E_{sys} = 0$

The system's total energy E_{sys} is conserved.

Energy can still be transformed within the system.

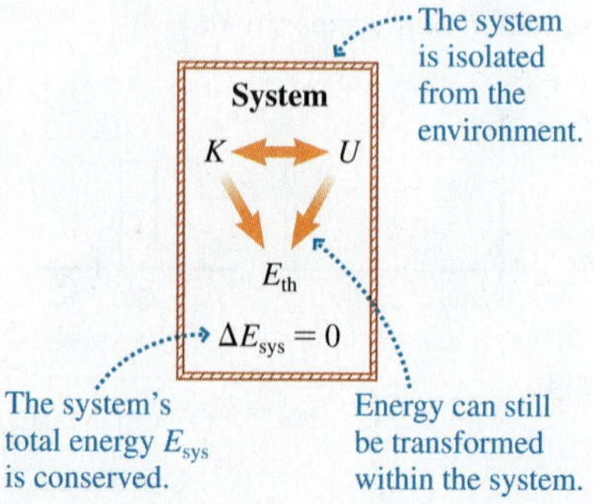

The key is that energy is conserved for an **isolated system,** a system that does not exchange energy with its environment either because it has no interactions with the environment or because those interactions do no work. **FIGURE 10.15** shows our basic energy model for an isolated system.

It Depends on the System

As significant as the law of conservation of energy is, it's critical to notice that the law does *not* say "Energy is always conserved." The law of conservation of energy refers to the energy *of a system*—hence our emphasis on systems in Chapters 9 and 10. Energy is conserved for some choices of system, but not others. For example, energy is conserved for a projectile moving near the earth if you define the system to be projectile + earth, but not if you define the system to be only the projectile.

In addition, the law of conservation of energy comes with an important qualification: Is the system isolated? Energy is certainly not conserved if an external force does work on the system. Thus the answer to the question "Is energy conserved?" is "It depends on the system."

A Strategy for Energy Problems

This is a good place to summarize the problem-solving strategy we've been developing for using the law of conservation of energy.

PROBLEM-SOLVING STRATEGY 10.1   (MP)

Energy-conservation problems

MODEL Define the system so that there are no external forces or so that any external forces do no work on the system. If there's friction, bring both surfaces into the system. Model objects as particles and springs as ideal.

VISUALIZE Draw a before-and-after pictorial representation and an energy bar chart. A free-body diagram may be needed to visualize forces.

SOLVE If the system is both isolated and nondissipative, then the mechanical energy is conserved:

$$K_i + U_i = K_f + U_f$$

where K is the total kinetic energy of all moving objects and U is the total potential energy of all interactions within the system. If there's friction, then

$$K_i + U_i = K_f + U_f + \Delta E_{th}$$

where the thermal energy increase due to friction is $\Delta E_{th} = f_k \Delta s$.

ASSESS Check that your result has correct units and significant figures, is reasonable, and answers the question.

Exercise 14

EXAMPLE 10.7 | The speed of a pendulum

A pendulum is created by attaching one end of a 78-cm-long string to the ceiling and tying a 150 g steel ball to the other end. The ball is pulled back until the string is 60° from vertical, then released. What is the speed of the ball at its lowest point?

MODEL Let the system consist of the earth and the ball. The tension force, like a normal force, is always perpendicular to the motion and does no work, so this is an isolated system with no friction. Its mechanical energy is conserved.

VISUALIZE FIGURE 10.16 shows a before-and-after pictorial representation, where we've placed the zero of potential energy at the lowest point of the ball's swing. Trigonometry is needed to determine the ball's initial height.

SOLVE Conservation of mechanical energy is

$$K_i + U_{Gi} = 0 + mgy_0 = K_f + U_{Gf} = \tfrac{1}{2}mv_1^2 + 0$$

Thus the ball's speed at the bottom is

$$v_1 = \sqrt{2gy_0} = \sqrt{2(9.80 \text{ m/s}^2)(0.39 \text{ m})} = 2.8 \text{ m/s}$$

The speed is exactly the same as if the ball had simply fallen 0.39 m.

FIGURE 10.16 Pictorial representation of a pendulum.

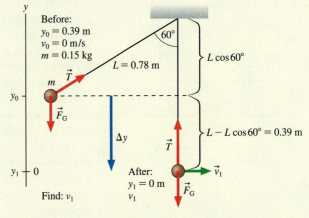

ASSESS To solve this problem directly from Newton's laws of motion requires advanced mathematics because of the complex way the net force changes with angle. But we can solve it in one line with an energy analysis!

Where Is Potential Energy?

Kinetic energy is the energy of a moving object. The basic energy model says that kinetic energy can be transformed into potential energy without loss, but where *is* the potential energy? If energy is real, not just an accounting fiction, what is it that has potential energy?

Potential energy is stored in *fields*. We've not yet introduced fields in this textbook, although we'll have a lot to say about electric and magnetic fields in later chapters. Even so, you've no doubt heard of magnetic fields and gravitational fields. Our modern understanding of the fundamental forces of nature, the long-range forces such as gravitational and electric forces, is that they are mediated by fields. How do two masses exert forces on each other through empty space? Or two electric charges? Through their fields!

When two masses move apart, the gravitational field changes to a new configuration that can store more energy. Thus the phrase "kinetic energy is transformed into gravitational potential energy" really means that the energy of a moving object is transformed into the energy of the gravitational field. At a later time, the field's energy can be transformed back into kinetic energy. The same holds true for the energy of charges and electric fields, a topic we'll take up in Part VII.

What about elastic potential energy? Remember that all solids, including springs, are held together by molecular bonds. Although quantum physics is needed for a complete understanding of bonds, they are essentially electric forces between neighboring atoms. When a solid is placed under tension, a vast number of molecular bonds stretch just a little and more energy is stored in their electric fields. What we call elastic potential energy at the macroscopic level is really energy stored in the electric fields of molecular bonds.

The theory of field energy is an advanced topic in physics. Nonetheless, this brief discussion helps complete our picture of what energy is and how it's associated with physical objects.

10.5 Energy Diagrams

Potential energy is an energy of position. The gravitational potential energy depends on the height of an object, and the elastic potential energy depends on a spring's displacement. Other potential energies you will meet in the future will depend in some way on position. Functions of position are easy to represent as graphs. A graph showing a system's potential energy and total energy as a function of position is called an **energy diagram.** Energy diagrams allow you to visualize motion based on energy considerations.

FIGURE 10.17 is the energy diagram of a particle in free fall. The gravitational potential energy $U_G = mgy$ is graphed as a line through the origin with slope mg. The *potential-energy curve* is labeled PE. The line labeled TE is the *total energy line*, $E = K + U_G$. It is horizontal because mechanical energy is conserved, meaning that the object's mechanical energy E has the same value at every position.

Suppose the particle is at position y_1. By definition, the distance from the axis to the potential-energy curve is the system's potential energy U_{G1} at that position. Because $K_1 = E - U_{G1}$, the distance between the potential-energy curve and the total energy line is the particle's kinetic energy.

The four-frame "movie" of **FIGURE 10.18** illustrates how an energy diagram is used to visualize motion. The first frame shows a particle projected upward from $y_a = 0$ with kinetic energy K_a. Initially the energy is entirely kinetic, with $U_{Ga} = 0$. A pictorial representation and an energy bar chart help to illustrate what the energy diagram is showing.

In the second frame, the particle has gained height but lost speed. The potential energy U_{Gb} is larger, and the distance K_b between the potential-energy curve and the total energy line is less. The particle continues rising and slowing until, in the third frame, it reaches the y-value where the total energy line crosses the potential-energy

FIGURE 10.17 The energy diagram of a particle in free fall.

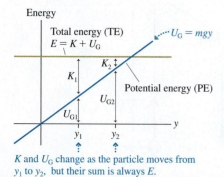

K and U_G change as the particle moves from y_1 to y_2, but their sum is always E.

FIGURE 10.18 A four-frame movie of a particle in free fall.

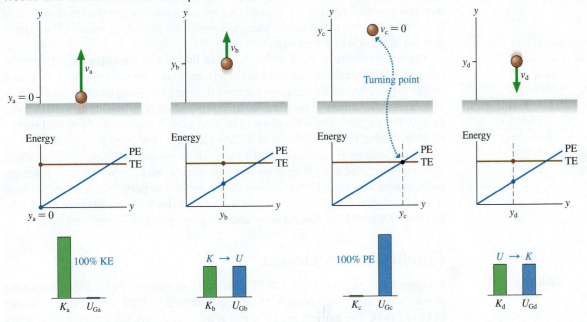

curve. This point, where $K = 0$ and the energy is entirely potential, is a *turning point* where the particle reverses direction. Finally, we see the particle speeding up as it falls.

A particle with this amount of total energy would need negative kinetic energy to be to the right of the point, at y_c, where the total energy line crosses the potential-energy curve. Negative K is not physically possible, so **the particle cannot be at positions with $U > E$.** Now, it's certainly true that you could make the particle reach a larger value of y simply by throwing it harder. But that would increase E and move the total energy line higher.

NOTE It's important to realize that the TE line is under your control. If you project an object with a different speed, or drop it from a different height, you're giving it a different total energy. You can give the object different *initial conditions* and use the energy diagram to explore how it will move with that amount of total energy.

FIGURE 10.19 shows the energy diagram of a mass on a horizontal spring, where x has been measured from the wall where the spring is attached. The equilibrium length of the spring is L_0 and the displacement of the end of the spring is $\Delta x = x - L_0$, so the spring's potential energy is $U_{Sp} = \frac{1}{2}k(\Delta x)^2 = \frac{1}{2}k(x - L_0)^2$. The potential-energy curve, a graph of U_{Sp} versus x, is a parabola centered at the equilibrium position. You can't change the PE curve—it's determined by the spring constant—but you can set the TE to any height you wish simply by stretching the spring to the proper length. The figure shows one possible TE line.

If you pull the mass out to position x_R and release it, the initial mechanical energy is entirely potential. As the restoring force of the spring pulls the mass to the left, the kinetic energy increases as the potential energy decreases. The mass has maximum speed at $x = L_0$, where $U_{Sp} = 0$, and then it slows down as the spring starts to compress. You should be able to visualize that x_L, where the PE curve crosses the TE line, is a turning point. It's the point of maximum compression where the mass instantaneously has $K = 0$. The mass will reverse direction, speed up until $x = L_0$, then slow down until reaching x_R, where it started. This is another turning point, so it will reverse direction again and start the process over. In other words, the mass will *oscillate* back and forth between the left and right turning points at x_L and x_R where the TE line crosses the PE curve. We'll study oscillations in Chapter 15, but we can already see from the energy diagram that a mass on a spring undergoes oscillatory motion.

FIGURE 10.19 The energy diagram of a mass on a horizontal spring.

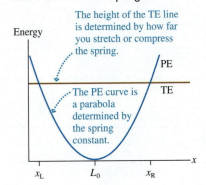

FIGURE 10.20 A more general energy diagram.

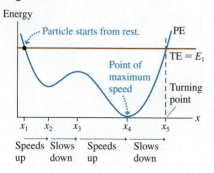

Speeds up | Slows down | Speeds up | Slows down

FIGURE 10.20 applies these ideas to a more general energy diagram. We don't know how this potential energy was created, but we can visualize the motion of a particle in a system that has this potential energy. Suppose the particle is released from rest at position x_1. How will it then move?

The initial conditions are $K = 0$ at x_1, hence the TE line must cross the PE curve at this point. The particle cannot move to the left, because that would require $K < 0$, so it begins to move toward the right. We see from the energy diagram that U decreases from x_1 to x_2, so the particle is speeding up as potential energy is transformed into kinetic energy. The particle then slows down from x_2 to x_3 as it goes up the "potential-energy hill," increasing U at the expense of K. The particle doesn't stop at x_3 because it still has kinetic energy. It speeds up from x_3 to x_4 (K increasing as U decreases), reaching its maximum speed at x_4, then slows down between x_4 and x_5. Position x_5 is a turning point, a point where the TE line crosses the PE curve. The particle is instantaneously at rest, then reverses direction. The particle will oscillate back and forth between x_1 and x_5, following the pattern of slowing down and speeding up that we've outlined.

Equilibrium Positions

FIGURE 10.21 Points of stable and unstable equilibrium.

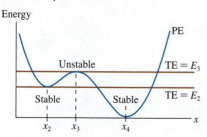

Positions x_2, x_3, and x_4 in Figure 10.20, where the potential energy has a local minimum or maximum, are special positions. Consider a particle with the total energy E_2 shown in **FIGURE 10.21**. The particle can be at rest at x_2, with $K = 0$, but it cannot move away from x_2. In other words, a particle with energy E_2 is in *equilibrium* at x_2. If you disturb the particle, giving it a small kinetic energy and a total energy just *slightly* larger than E_2, the particle will undergo a very small oscillation centered on x_2, like a marble in the bottom of a bowl. An equilibrium for which small disturbances cause small oscillations is called a point of **stable equilibrium.** You should recognize that *any* minimum in the PE curve is a point of stable equilibrium. Position x_4 is also a point of stable equilibrium, in this case for a particle with $E = 0$.

Figure 10.21 also shows a particle with energy E_3 that is tangent to the curve at x_3. If a particle is placed *exactly* at x_3, it will stay there at rest ($K = 0$). But if you disturb the particle at x_3, giving it an energy only slightly more than E_3, it will speed up as it moves away from x_3. This is like trying to balance a marble on top of a hill. The slightest displacement will cause the marble to roll down the hill. A point of equilibrium for which a small disturbance causes the particle to move away is called a point of **unstable equilibrium.** Any maximum in the PE curve, such as x_3, is a point of unstable equilibrium.

We can summarize these lessons as follows:

TACTICS BOX 10.1   (MP)

Interpreting an energy diagram

❶ The distance from the axis to the PE curve is the particle's potential energy. The distance from the PE curve to the TE line is its kinetic energy. These are transformed as the position changes, causing the particle to speed up or slow down, but the sum $K + U$ doesn't change.

❷ A point where the TE line crosses the PE curve is a turning point. The particle reverses direction.

❸ The particle cannot be at a point where the PE curve is above the TE line.

❹ The PE curve is determined by the properties of the system—mass, spring constant, and the like. You cannot change the PE curve. However, you can raise or lower the TE line simply by changing the initial conditions to give the particle more or less total energy.

❺ A minimum in the PE curve is a point of stable equilibrium. A maximum in the PE curve is a point of unstable equilibrium.

Exercises 15–17

EXAMPLE 10.8 | **Balancing a mass on a spring**

A spring of length L_0 and spring constant k is standing on one end. A block of mass m is placed on the spring, compressing it. What is the length of the compressed spring?

MODEL Assume an ideal spring obeying Hooke's law. The block + earth + spring system has both gravitational potential energy U_G *and* elastic potential energy U_{Sp}. The block sitting on top of the spring is at a point of stable equilibrium (small disturbances cause the block to oscillate slightly around the equilibrium position), so we can solve this problem by looking at the energy diagram.

VISUALIZE **FIGURE 10.22a** is a pictorial representation. We've used a coordinate system with the origin at ground level, so the displacement of the spring is $y - L_0$.

SOLVE **FIGURE 10.22b** shows the two potential energies separately and also shows the total potential energy:

$$U_{tot} = U_G + U_{Sp}$$

$$= mgy + \tfrac{1}{2}k(y - L_0)^2$$

The equilibrium position (the minimum of U_{tot}) has shifted from L_0 to a smaller value of y, closer to the ground. We can find the equilibrium by locating the position of the minimum in the PE curve. You know from calculus that the minimum of a function is at the point where the derivative (or slope) is zero. The derivative of U_{tot} is

$$\frac{dU_{tot}}{dy} = mg + k(y - L_0)$$

The derivative is zero at the point y_{eq}, so we can easily find

$$mg + k(y_{eq} - L_0) = 0$$

$$y_{eq} = L_0 - \frac{mg}{k}$$

The block compresses the spring by the length mg/k from its original length L_0, giving it a new equilibrium length $L_0 - mg/k$.

FIGURE 10.22 The block + earth + spring system has both gravitational and elastic potential energy.

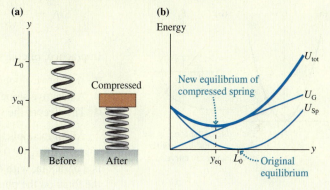

STOP TO THINK 10.6 A particle with the potential energy shown in the graph is moving to the right. It has 1 J of kinetic energy at $x = 1$ m. Where is the particle's turning point?

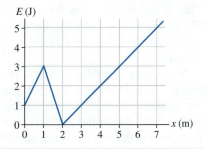

10.6 Force and Potential Energy

As you've seen, we can find the energy of an interaction—potential energy—by calculating the work the interaction force does inside the system. Can we reverse this procedure? That is, if we know a system's potential energy, can we find the interaction force?

We defined the change in potential energy to be $\Delta U = -W_{int}$. Suppose that an object undergoes a *very small* displacement Δs, so small that the interaction force $\vec{F}$ is essentially constant. The work done by a constant force is $W = F_s \Delta s$, where F_s is the force component parallel to the displacement. During this small displacement, the system's potential energy changes by

$$\Delta U = -W_{int} = -F_s \Delta s \qquad\qquad (10.23)$$

FIGURE 10.23 Relating force to the PE curve.

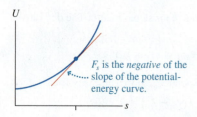

F_s is the *negative* of the slope of the potential-energy curve.

FIGURE 10.24 Elastic potential energy and force graphs.

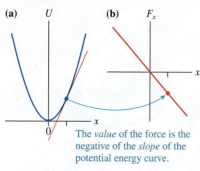

(a) U

(b) F_x

The *value* of the force is the negative of the *slope* of the potential energy curve.

FIGURE 10.25 A potential-energy curve and the associated force curve.

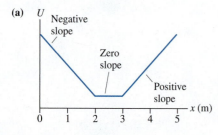

(a) U — Negative slope, Zero slope, Positive slope — x (m)

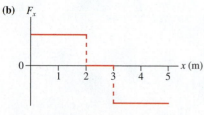

(b) F_x — x (m)

which we can rewrite as

$$F_s = -\frac{\Delta U}{\Delta s} \tag{10.24}$$

In the limit $\Delta s \to 0$, the force on the object is

$$F_s = \lim_{\Delta s \to 0}\left(-\frac{\Delta U}{\Delta s}\right) = -\frac{dU}{ds} \tag{10.25}$$

That is, the interaction force on an object is the *negative* of the derivative of the potential energy with respect to position.

Graphically, as **FIGURE 10.23** shows, force is the negative of the slope, at position s, of the potential-energy curve in an energy diagram:

$$F_s = -\frac{dU}{ds} = \text{the negative of the slope of the PE curve at } s \tag{10.26}$$

In practice, of course, we'll usually use either $F_x = -dU/dx$ or $F_y = -dU/dy$. Thus

- A positive slope corresponds to a negative force: to the left or downward.
- A negative slope corresponds to a positive force: to the right or upward.
- The steeper the slope, the larger the force.

As an example, consider the elastic potential energy $U_{Sp} = \frac{1}{2}kx^2$ for a horizontal spring with $x_{eq} = 0$ so that $\Delta x = x$. **FIGURE 10.24a** shows that the potential-energy curve is a parabola, with changing slope. If an object attached to the spring is at position x, the force on the object is

$$F_x = -\frac{dU_{Sp}}{dx} = -\frac{d}{dx}\left(\tfrac{1}{2}kx^2\right) = -kx$$

This is just Hooke's law for an ideal spring, with the minus sign indicating that Hooke's law is a restoring force. **FIGURE 10.24b** is a graph of force versus x. At each position x, the *value* of the force is equal to the negative of the *slope* of the PE curve.

We already knew Hooke's law, of course, so the point of this particular exercise was to illustrate the meaning of Equation 10.26. But if we had *not* known the force, we see that it's possible to find the force from the PE curve. For example, you'll learn in Part VIII that **FIGURE 10.25a** is a possible potential-energy function for a charged particle, one that we could create with suitably shaped electrodes. What force does the particle experience in this region of space? We find out by measuring the slope of the PE curve. The result is shown in **FIGURE 10.25b**. On the left side of this region of space $(x < 2 \text{ m})$, a negative slope, and thus a positive value of F_x, means that the force pushes the particle to the right. A negative force on the right side $(x > 3 \text{ m})$ tells us that F_x pushes the particle to the left. And there's no force at all in the center. This is a *restoring force* because a particle trying to leave this region is pushed back toward the center, but it's not a linear restoring force like that of a spring.

EXAMPLE 10.9 | **Finding equilibrium positions**

A system's potential energy is given by $U(x) = (2x^3 - 3x^2)$ J, where x is a particle's position in m. Where are the equilibrium positions for this system, and are they stable or unstable equilibria?

SOLVE You learned in Chapter 6 that a particle in equilibrium has $\vec{F}_{net} = \vec{0}$. Then, in the previous section, you learned that the maxima and minima of the PE curve are points of equilibrium.

These may seem to be two different criteria for equilibrium, but actually they are identical. The interaction force on the particle is $F_x = -dU/dx$. The force is zero—equilibrium—at positions where the derivative is zero. But you've learned in calculus that positions where the derivative of a function is zero are the maxima and minima of the function. At either a maximum or minimum of the PE curve, the slope is zero and hence the force is zero.

For this potential-energy function,

$$F_x = -\frac{dU}{dx} = (-6x^2 + 6x) \text{ N}$$

when x is in m. The force is zero—minima or maxima of U—when $6x_{eq}^2 = 6x_{eq}$. This has two solutions:

$$x_{eq} = 0 \text{ m} \quad \text{and} \quad x_{eq} = 1 \text{ m}$$

These are positions of equilibrium, where a particle at rest will remain at rest. But how do we know if these are positions of stable equilibrium or unstable equilibrium?

A *minimum* in the PE curve is a stable equilibrium. Recall, from calculus, that a minimum of a function has a first derivative equal to zero *and* a second derivative that's positive. Similarly, a maximum of a function has a first derivative equal to zero and a second derivative that's negative. The second derivative of U is

$$\frac{d^2U}{dx^2} = \frac{d}{dx}(6x^2 - 6x) = (12x - 6) \text{ N/m}$$

The second derivative evaluated at $x = 0$ m is -6 N/m < 0, so $x = 0$ m is a maximum of the PE curve. At $x = 1$ m, the second derivative is $+6$ N/m > 0, hence a minimum in the PE curve. Thus this system has an unstable equilibrium if the particle is at $x = 0$ m and a stable equilibrium if it is at $x = 1$ m. There's a force on the particle at all other positions.

STOP TO THINK 10.7 A particle moves along the x-axis with the potential energy shown. The x-component of the force on the particle when it is at $x = 4$ m is

a. 4 N
b. 2 N
c. 1 N
d. -4 N
e. -2 N
f. -1 N

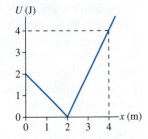

10.7 Conservative and Nonconservative Forces

A system in which particles interact gravitationally or elastically or, as we'll discover later, electrically has a potential energy. But do all forces have potential energies? Is there a "tension potential energy" or a "friction potential energy"? If not, what's special about the gravitational and elastic forces? What conditions must an interaction satisfy to have an associated potential energy?

FIGURE 10.26 shows a particle that can move from point A to point B along two possible paths while a force $\vec{F}$ acts on it. In general, the force experienced along path 1 is not the same as the force experienced along path 2. The force changes the particle's speed, so the particle's kinetic energy when it arrives at B differs from the kinetic energy it had when it left A.

Let's assume that there is a potential energy U associated with force $\vec{F}$ just as the gravitational potential energy $U_G = mgy$ is associated with the gravitational force $\vec{F}_G = -mg\hat{j}$. What restrictions does this assumption place on $\vec{F}$? There are three steps in the logic:

1. Potential energy is an energy of position. U depends only on where the particle is, not on how it got there. The system has one value of potential energy when the particle is at A, a different value when the particle is at B. Thus the change in potential energy, $\Delta U = U_B - U_A$, is the same whether the particle moves along path 1 or path 2.

FIGURE 10.26 A particle can move from A to B along either of two paths.

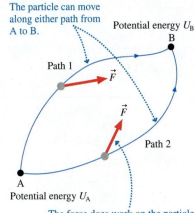

The particle can move along either path from A to B.

Potential energy U_B

Path 1

Path 2

Potential energy U_A

The force does work on the particle as it moves from A to B, changing the particle's kinetic energy.

2. Potential energy is transformed into kinetic energy, with $\Delta K = -\Delta U$. If ΔU is independent of the path followed, then ΔK is also independent of the path. The particle has the same kinetic energy at B no matter which path it follows.

3. According to the energy principle, the change in a particle's kinetic energy is equal to the work done on the particle by force $\vec{F}$. That is, $\Delta K = W$. Because ΔK is independent of the path followed, it *must* be the case that **the work done by force $\vec{F}$ as the particle moves from A to B is independent of the path followed.**

A force for which the work done on a particle as it moves from an initial to a final position is independent of the path followed is called a **conservative force.** The importance of conservative forces is that **a potential energy can be associated with any conservative force.** Specifically, the potential-energy difference between an initial position i and a final position f is

$$\Delta U = -W_c(i \rightarrow f) \tag{10.27}$$

where the notation $W_c(i \rightarrow f)$ is the work done by a conservative force as the particle moves along *any* path from i to f. Equation 10.27 is a general definition of the potential energy associated with a conservative force.

A force for which we can define a potential energy is called *conservative* because the mechanical energy $K + U$ is conserved for a system in which this is the only interaction. We've already shown that the gravitational force is a conservative force by showing that ΔU_G depends only on the vertical displacement, not on the path followed; hence mechanical energy is conserved when two masses interact gravitationally. Similarly, mechanical energy is conserved for a mass on a spring—an elastic interaction—if there are no other forces. Conservative forces do not contribute to any loss of mechanical energy.

Nonconservative Forces

A characteristic of a conservative force is that **an object returning to its starting point will return with no loss of kinetic energy** because $\Delta U = 0$ if the initial and final points are the same. If a ball is tossed into the air, energy is transformed from kinetic into potential and back such that the ball's kinetic energy is unchanged when it returns to its initial height. The same is true for a puck sliding up and back down a frictionless slope.

But not all forces are conservative forces. If the slope has friction, then the puck returns with *less* kinetic energy. Part of its kinetic energy is transformed into gravitational potential energy as it slides up, but part is transformed into some other form of energy—thermal energy—that lacks the "potential" to be transformed back into kinetic energy. A force for which we cannot define a potential energy is called a **nonconservative force.** Friction and drag, which transform mechanical energy into thermal energy, are nonconservative forces, so there is no "friction potential energy."

Similarly, forces like tension and thrust are nonconservative. If you pull an object with a rope, the work done by tension is proportional to the distance traveled. More work is done along a longer path between two points than along a shorter path, so tension fails the "Work is independent of the path followed" test and does not have a potential energy.

All in all, most forces are *not* conservative forces. Gravitational forces, linear restoring forces, and, later, electric forces turn out to be fairly special because they are among the few forces for which we *can* define a potential energy. Fortunately, these are some of the most important forces in nature, so the energy principle is powerful and useful despite there being only a small number of conservative forces.

10.8 The Energy Principle Revisited

We opened Chapter 9 by introducing the energy principle—basically a statement of energy accounting—but noted that we would need to develop many new ideas to make sense of energy. We've now explored kinetic energy, potential energy, work, conservative and nonconservative forces, and much more. It's time to return to the basic energy model and start pulling together the many ideas introduced in Chapters 9 and 10.

FIGURE 10.27 shows a system of three objects that interact with each other and are acted on by external forces from the environment. These forces cause the system's kinetic energy K to change. By how much? Kinetic energy is energy of motion, and the kinetic energy would be the same if we had defined the system—as we did in Chapter 9— to consist of only the objects, not the interactions. Thus $\Delta K = W_{tot} = W_c + W_{nc}$, where in the second step we've divided the total work done by all forces into the work W_c done by conservative forces and the work W_{nc} done by nonconservative forces.

Now let's make a further distinction by dividing the nonconservative forces into *dissipative* forces and *external* forces. Dissipative forces, like friction and drag, transform mechanical energy into thermal energy.

To illustrate what we mean by an external force, suppose you pick up a box at rest on the floor and place it at rest on a table. The box + earth system gains gravitational potential energy, but $\Delta K = 0$ and $\Delta E_{th} = 0$. So where did the energy come from? Or consider pulling the box across the table with a string. The box gains kinetic energy and possibly thermal energy, but not by transforming potential energy. The force of your hand and the tension of the string are forces that "reach in" from the environment to change the system. Thus they are *external forces*. They are nonconservative forces, with no potential energy, but they change the system's mechanical energy rather than its thermal energy.

With this distinction, the system's change in kinetic energy is

$$\Delta K = W_{tot} = W_c + W_{nc} = W_c + W_{diss} + W_{ext} \qquad (10.28)$$

When we bring the conservative interactions inside the system, the work done by conservative forces becomes potential energy: $W_c = -\Delta U$. And, as we learned in Chapter 9, the work done by dissipative forces becomes thermal energy: $W_{diss} = -\Delta E_{th}$. With these substitutions, Equation 10.28 becomes

$$\Delta K = -\Delta U - \Delta E_{th} + W_{ext} \qquad (10.29)$$

Separating energy terms from the work, we can write Equation 10.29 as

$$\Delta K + \Delta U + \Delta E_{th} = \Delta E_{mech} + \Delta E_{th} = \Delta E_{sys} = W_{ext} \qquad (10.30)$$

where $E_{sys} = K + U + E_{th} = E_{mech} + E_{th}$ is the energy of the system. Equation 10.30, the energy principle but with all the terms now defined, is our most general statement about how the energy of a mechanical system changes.

In Section 10.4 we defined an *isolated system* as a system that does not exchange energy with its environment. That is, an isolated system is one on which no work is done by external forces: $W_{ext} = 0$. Thus an immediate conclusion from Equation 10.30 is that **the total energy E_{sys} of an isolated system is conserved.** If, in addition, the system is nondissipative (i.e., no friction forces), then $\Delta E_{th} = 0$. In that case, the mechanical energy E_{mech} is conserved. You'll recognize this as the *law of conservation of energy* from Section 10.4. The law of conservation of energy is one of the most powerful statements in physics.

FIGURE 10.28 reproduces the basic energy model of Chapter 9. Now you can see that this is a pictorial representation of Equation 10.30. E_{sys}, the total energy of the system, changes only if external forces transfer energy into or out of the system by doing work on the system. The kinetic, potential, and thermal energies within the system can be

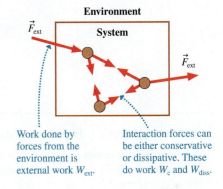

FIGURE 10.27 A system with both internal interactions and external forces.

Work done by forces from the environment is external work W_{ext}.

Interaction forces can be either conservative or dissipative. These do work W_c and W_{diss}.

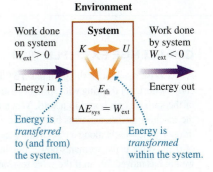

FIGURE 10.28 The basic energy model.

Work done on system $W_{ext} > 0$

Energy in

Work done by system $W_{ext} < 0$

Energy out

Energy is *transferred* to (and from) the system.

$\Delta E_{sys} = W_{ext}$

Energy is *transformed* within the system.

transformed into each other by forces within the system. And E_{sys} is conserved in the absence of interactions with the environment.

Energy Bar Charts Expanded

Energy bar charts can now be expanded to include the thermal energy and the work done by external forces. The energy principle, Equation 10.30, can be rewritten as

$$K_i + U_i + W_{ext} = K_f + U_f + \Delta E_{th} \tag{10.31}$$

The initial mechanical energy $(K_i + U_i)$ plus any energy added to or removed from the system (W_{ext}) becomes, without loss, the final mechanical energy $(K_f + U_f)$ plus any increase in the system's thermal energy (ΔE_{th}). Remember that we have no way to determine $E_{th\,i}$ or $E_{th\,f}$, only the *change* in thermal energy. ΔE_{th} is always positive when the system contains dissipative forces.

EXAMPLE 10.10 | **Hauling up supplies**

A mountain climber uses a rope to drag a bag of supplies up a slope at constant speed. Show the energy transfers and transformations on an energy bar chart.

MODEL Let the system consist of the earth, the bag of supplies, and the slope.

SOLVE The tension in the rope is an external force that does work on the bag of supplies. This is an energy transfer into the system. The bag has kinetic energy, but it moves at a steady speed and so K is not *changing*. Instead, the energy transfer into the system increases both gravitational potential energy (the bag is gaining height) and thermal energy (the bag and the slope are getting warmer due to friction). The overall process is $W_{ext} \rightarrow U + E_{th}$. This is shown in **FIGURE 10.29**.

FIGURE 10.29 The energy bar chart for Example 10.10.

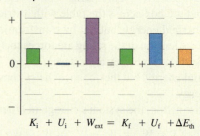

$$K_i \; + \; U_i \; + \; W_{ext} \; = \; K_f \; + \; U_f \; + \Delta E_{th}$$

STOP TO THINK 10.8 A weight attached to a rope is released from rest. As the weight falls, picking up speed, the rope spins a generator that causes a lightbulb to glow. Define the system to be the weight and the earth. In this situation,

    a. $U \rightarrow K + W_{ext}$. E_{mech} is not conserved but E_{sys} is.
    b. $U + W_{ext} \rightarrow K$. Both E_{mech} and E_{sys} are conserved.
    c. $U \rightarrow K + E_{th}$. E_{mech} is not conserved but E_{sys} is.
    d. $U \rightarrow K + W_{ext}$. Neither E_{mech} nor E_{sys} is conserved.
    e. $W_{ext} \rightarrow K + U$. E_{mech} is not conserved but E_{sys} is.

CHALLENGE EXAMPLE 10.11 | **A spring workout**

An exercise machine at the gym has a 5.0 kg weight attached to one end of a horizontal spring with spring constant 80 N/m. The other end of the spring is anchored to a wall. When a woman working out on the machine pushes her arms forward, a cable stretches the spring by dragging the weight along a track with a coefficient of kinetic friction of 0.30. What is the woman's power output at the moment when the weight has moved 50 cm if the cable tension is a constant 100 N?

MODEL This is a complex situation, but one that we can analyze. First, identify the weight, the spring, the wall, and the track as the system. We need to have the track inside the system because friction increases the temperature of both the weight *and* the track. The tension in the cable is an external force. The work W_{ext} done by the cable's tension transfers energy into the system, causing K, U_{Sp}, and E_{th} all to increase.

FIGURE 10.30 Pictorial representation and energy bar chart for Challenge Example 10.11.

(a)

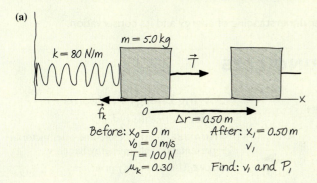

Before: $x_0 = 0$ m
$v_0 = 0$ m/s
$T = 100$ N
$\mu_k = 0.30$

After: $x_1 = 0.50$ m
v_1

Find: v_1 and P_1

(b)

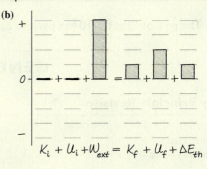

$$K_i + U_i + W_{ext} = K_f + U_f + \Delta E_{th}$$

VISUALIZE **FIGURE 10.30a** is a before-and-after pictorial representation. The energy transfers and transformations are shown in the energy bar chart of **FIGURE 10.30b**.

SOLVE You learned in « Section 9.6 that power is the *rate* at which work is done and that the power delivered by force $\vec{F}$ to an object moving with velocity $\vec{v}$ is $P = \vec{F} \cdot \vec{v}$. Here the tension $\vec{T}$ pulls parallel to the weight's velocity, so the power being supplied when the weight has velocity $\vec{v}$ is $P = Tv$. We know the cable's tension, so we need to use energy considerations to find the weight's speed v_1 after the spring has been stretched to $\Delta x_1 = 50$ cm.

The energy principle $K_i + U_i + W_{ext} = K_f + U_f + \Delta E_{th}$ is

$$\tfrac{1}{2}mv_0^2 + \tfrac{1}{2}k(\Delta x_0)^2 + W_{ext} = \tfrac{1}{2}mv_1^2 + \tfrac{1}{2}k(\Delta x_1)^2 + \Delta E_{th}$$

The initial displacement is $\Delta x_0 = 0$ m and we know that $v_0 = 0$ m/s, so the energy principle simplifies to

$$\tfrac{1}{2}mv_1^2 = W_{ext} - \tfrac{1}{2}k(\Delta x_1)^2 - \Delta E_{th}$$

The external work done by the cable's tension is

$$W_{ext} = T\Delta r = (100 \text{ N})(0.50 \text{ m}) = 50.0 \text{ J}$$

From Chapter 9, the increase in thermal energy due to friction is

$$\Delta E_{th} = f_k \Delta r = \mu_k mg \Delta r$$
$$= (0.30)(5.0 \text{ kg})(9.80 \text{ m/s}^2)(0.50 \text{ m}) = 7.4 \text{ J}$$

Solving for the speed v_1, when the spring's displacement is $\Delta x_1 = 50$ cm $= 0.50$ m, we have

$$v_1 = \sqrt{\frac{2\left(W_{ext} - \tfrac{1}{2}k(\Delta x_1)^2 - \Delta E_{th}\right)}{m}} = 3.6 \text{ m/s}$$

The power being supplied at this instant to keep stretching the spring is

$$P = Tv_1 = (100 \text{ N})(3.6 \text{ m/s}) = 360 \text{ W}$$

ASSESS The work done by the cable's tension is energy transferred to the system. Part of the energy increases the speed of the weight, part increases the potential energy stored in the spring, and part is transformed into increased thermal energy, thus increasing the temperature. We had to bring all these energy ideas together to solve this problem.

SUMMARY

The goal of Chapter 10 has been to develop a better understanding of energy and its conservation.

GENERAL PRINCIPLES

The Energy Principle Revisited

- Energy is *transformed* within the system.
- Energy is *transferred* to and from the system by work W.

Two variations of the energy principle are

$$\Delta E_{sys} = \Delta K + \Delta U + \Delta E_{th} = W_{ext}$$
$$K_i + U_i + W_{ext} = K_f + U_f + \Delta E_{th}$$

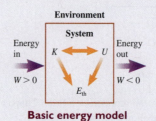

Environment

System

Energy in
$W > 0$

$K \longleftrightarrow U$

E_{th}

Energy out
$W < 0$

Basic energy model

Solving Energy Problems

MODEL Define the system.

VISUALIZE Draw a before-and-after pictorial representation and an energy bar chart.

SOLVE Use the energy principle:

$$K_i + U_i + W_{ext} = K_f + U_f + \Delta E_{th}$$

ASSESS Is the result reasonable?

Law of Conservation of Energy

- **Isolated system:** $W_{ext} = 0$. The total system energy $E_{sys} = K + U + E_{th}$ is conserved. $\Delta E_{sys} = 0$.
- **Isolated, nondissipative system:** $W_{ext} = 0$ and $W_{diss} = 0$. The **mechanical energy** $E_{mech} = K + U$ is conserved: $K_i + U_i = K_f + U_f$.

IMPORTANT CONCEPTS

Potential energy, or *interaction energy,* is energy stored inside a system via interaction forces. The energy is stored in *fields*.

- Potential energy is associated only with **conservative forces** for which the work done is independent of the path.
- Work W_{int} by the interaction forces causes $\Delta U = -W_{int}$.
- Force $F_s = -dU/ds = -$(slope of the potential energy curve).
- Potential energy is an energy of the system, not an energy of a specific object.

Energy diagrams show the potential-energy curve PE and the total mechanical energy line TE.

- From the axis to the curve is U. From the curve to the TE line is K.
- **Turning points** occur where the TE line crosses the PE curve.
- Minima and maxima in the PE curve are, respectively, positions of **stable** and **unstable equilibrium.**

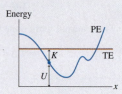

Energy

PE

K

TE

U

x

APPLICATIONS

Gravitational potential energy is an energy of the earth + object system:

$$U_G = mgy$$

Elastic potential energy is an energy of the spring + attached objects system:

$$U_{Sp} = \tfrac{1}{2}k(\Delta s)^2$$

y

x

Energy bar charts show the energy principle in graphical form.

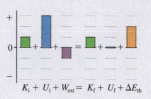

$+$

0

$-$

$K_i + U_i + W_{ext} = K_f + U_f + \Delta E_{th}$

TERMS AND NOTATION

potential energy, U
gravitational potential energy, U_G
zero of potential energy
mechanical energy, E_{mech}

energy bar chart
elastic potential energy, U_{Sp}
law of conservation of energy

isolated system
energy diagram
stable equilibrium

unstable equilibrium
conservative force
nonconservative force

CONCEPTUAL QUESTIONS

1. Upon what basic quantity does kinetic energy depend? Upon what basic quantity does potential energy depend?
2. Can kinetic energy ever be negative? Can gravitational potential energy ever be negative? For each, give a plausible *reason* for your answer without making use of any equations.
3. A roller-coaster car rolls down a frictionless track, reaching speed v_0 at the bottom. If you want the car to go twice as fast at the bottom, by what factor must you increase the height of the track? Explain.
4. The three balls in **FIGURE Q10.4**, which have equal masses, are fired with equal speeds from the same height above the ground. Rank in order, from largest to smallest, their speeds v_a, v_b, and v_c as they hit the ground. Explain.

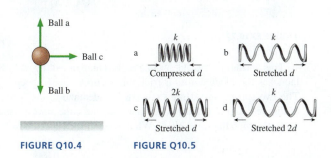

FIGURE Q10.4   **FIGURE Q10.5**

5. Rank in order, from most to least, the elastic potential energy $(U_{Sp})_a$ to $(U_{Sp})_d$ stored in the springs of **FIGURE Q10.5**. Explain.
6. A spring is compressed 1.0 cm. How far must you compress a spring with twice the spring constant to store the same amount of energy?
7. A spring gun shoots out a plastic ball at speed v_0. The spring is then compressed twice the distance it was on the first shot. By what factor is the ball's speed increased? Explain.
8. A particle with the potential energy shown in **FIGURE Q10.8** is moving to the right at $x = 5$ m with total energy E.
   a. At what value or values of x is this particle's speed a maximum?
   b. Does this particle have a turning point or points in the range of x covered by the graph? If so, where?
   c. If E is changed appropriately, could the particle remain at rest at any point or points in the range of x covered by the graph? If so, where?

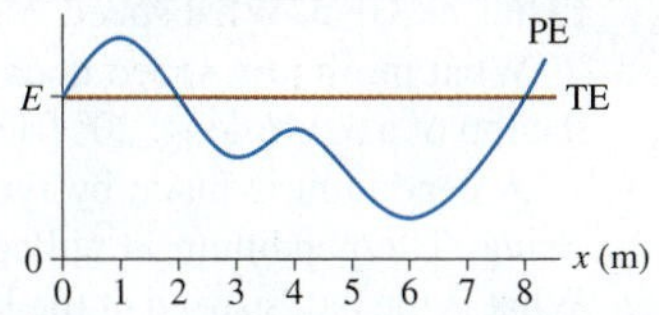

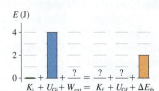

FIGURE Q10.8

9. A compressed spring launches a block up an incline. Which objects should be included within the system in order to make an energy analysis as easy as possible?
10. A process occurs in which a system's potential energy decreases while the system does work on the environment. Does the system's kinetic energy increase, decrease, or stay the same? Or is there not enough information to tell? Explain.
11. A process occurs in which a system's potential energy increases while the environment does work on the system. Does the system's kinetic energy increase, decrease, or stay the same? Or is there not enough information to tell? Explain.
12. **FIGURE Q10.12** is the energy bar chart for a firefighter sliding down a fire pole from the second floor to the ground. Let the system consist of the firefighter, the pole, and the earth. What are the bar heights of W_{ext}, K_f, and U_{Gf}?

FIGURE Q10.12

$K_i \;+\; U_{Gi} + W_{ext} = K_f \;+\; U_{Gf} + \Delta E_{th}$

13. a. If the force on a particle at some point in space is zero, must its potential energy also be zero at that point? Explain.
    b. If the potential energy of a particle at some point in space is zero, must the force on it also be zero at that point? Explain.

EXERCISES AND PROBLEMS

Problems labeled [ ] integrate material from earlier chapters.

Exercises

Section 10.1 Potential Energy

1. | Object A is stationary while objects B and C are in motion. Forces from object A do 10 J of work on object B and –5 J of work on object C. Forces from the environment do 4 J of work on object B and 8 J of work on object C. Objects B and C do not interact. What are ΔK_{tot} and ΔU_{int} if (a) objects A, B, and C are defined as separate systems and (b) one system is defined to include objects A, B, and C and their interactions?
2. | A system of two objects has $\Delta K_{tot} = 7$ J and $\Delta U_{int} = -5$ J.
   a. How much work is done by interaction forces?
   b. How much work is done by external forces?

Section 10.2 Gravitational Potential Energy

3. | The lowest point in Death Valley is 85 m below sea level. The summit of nearby Mt. Whitney has an elevation of 4420 m. What is the change in potential energy when an energetic 65 kg hiker makes it from the floor of Death Valley to the top of Mt. Whitney?
4. | a. What is the kinetic energy of a 1500 kg car traveling at a speed of 30 m/s (≈ 65 mph)?
   b. From what height would the car have to be dropped to have this same amount of kinetic energy just before impact?
5. | a. With what minimum speed must you toss a 100 g ball straight up to just touch the 10-m-high roof of the gymnasium if you release the ball 1.5 m above the ground? Solve this problem using energy.
   b. With what speed does the ball hit the ground?

6. | What height does a frictionless playground slide need so that a 35 kg child reaches the bottom at a speed of 4.5 m/s?

7. | A 55 kg skateboarder wants to just make it to the upper edge of a "quarter pipe," a track that is one-quarter of a circle with a radius of 3.0 m. What speed does he need at the bottom?

8. | What minimum speed does a 100 g puck need to make it to the top of a 3.0-m-long, 20° frictionless ramp?

9. || A pendulum is made by tying a 500 g ball to a 75-cm-long string. The pendulum is pulled 30° to one side, then released. What is the ball's speed at the lowest point of its trajectory?

10. || A 20 kg child is on a swing that hangs from 3.0-m-long chains. What is her maximum speed if she swings out to a 45° angle?

11. || A 1500 kg car traveling at 10 m/s suddenly runs out of gas while approaching the valley shown in **FIGURE EX10.11**. The alert driver immediately puts the car in neutral so that it will roll. What will be the car's speed as it coasts into the gas station on the other side of the valley?

FIGURE EX10.11

12. | The maximum energy a bone can absorb without breaking is
BIO surprisingly small. Experimental data show that the leg bones of a healthy, 60 kg human can absorb about 200 J. From what maximum height could a 60 kg person jump and land rigidly upright on both feet without breaking his legs? Assume that all energy is absorbed by the leg bones in a rigid landing.

13. || A cannon tilted up at a 30° angle fires a cannon ball at 80 m/s from atop a 10-m-high fortress wall. What is the ball's impact speed on the ground below?

14. || In a hydroelectric dam, water falls 25 m and then spins a turbine to generate electricity.
a. What is ΔU_G of 1.0 kg of water?
b. Suppose the dam is 80% efficient at converting the water's potential energy to electrical energy. How many kilograms of water must pass through the turbines each second to generate 50 MW of electricity? This is a typical value for a small hydroelectric dam.

Section 10.3 Elastic Potential Energy

15. | How far must you stretch a spring with $k = 1000$ N/m to store 200 J of energy?

16. || A stretched spring stores 2.0 J of energy. How much energy will be stored if the spring is stretched three times as far?

17. | A student places her 500 g physics book on a frictionless table. She pushes the book against a spring, compressing the spring by 4.0 cm, then releases the book. What is the book's speed as it slides away? The spring constant is 1250 N/m.

18. | A block sliding along a horizontal frictionless surface with speed v collides with a spring and compresses it by 2.0 cm. What will be the compression if the same block collides with the spring at a speed of $2v$?

19. | A 10 kg runaway grocery cart runs into a spring with spring constant 250 N/m and compresses it by 60 cm. What was the speed of the cart just before it hit the spring?

20. || As a 15,000 kg jet plane lands on an aircraft carrier, its tail hook snags a cable to slow it down. The cable is attached to a spring with spring constant 60,000 N/m. If the spring stretches 30 m to stop the plane, what was the plane's landing speed?

21. || The elastic energy stored in your tendons can contribute up to
BIO 35% of your energy needs when running. Sports scientists find that (on average) the knee extensor tendons in sprinters stretch 41 mm while those of nonathletes stretch only 33 mm. The spring constant of the tendon is the same for both groups, 33 N/mm. What is the difference in maximum stored energy between the sprinters and the nonathletes?

22. || The spring in **FIGURE EX10.22a** is compressed by Δx. It launches the block across a frictionless surface with speed v_0. The two springs in **FIGURE EX10.22b** are identical to the spring of Figure EX10.22a. They are compressed by the same Δx and used to launch the same block. What is the block's speed now?

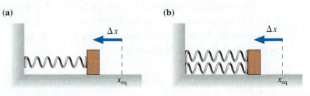

FIGURE EX10.22

23. ||| The spring in **FIGURE EX10.23a** is compressed by Δx. It launches the block across a frictionless surface with speed v_0. The two springs in **FIGURE EX10.23b** are identical to the spring of Figure EX10.23a. They are compressed the same *total* Δx and used to launch the same block. What is the block's speed now?

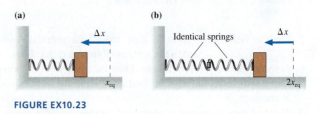

FIGURE EX10.23

Section 10.4 Conservation of Energy

Section 10.5 Energy Diagrams

24. || **FIGURE EX10.24** is the potential-energy diagram for a 20 g particle that is released from rest at $x = 1.0$ m.
a. Will the particle move to the right or to the left?
b. What is the particle's maximum speed? At what position does it have this speed?
c. Where are the turning points of the motion?

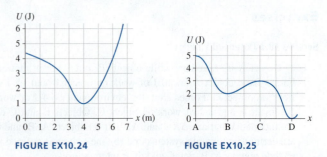

FIGURE EX10.24   **FIGURE EX10.25**

25. || **FIGURE EX10.25** is the potential-energy diagram for a 500 g particle that is released from rest at A. What are the particle's speeds at B, C, and D?

26. ‖ In **FIGURE EX10.26**, what is the maximum speed of a 2.0 g particle that oscillates between $x = 2.0$ mm and $x = 8.0$ mm?

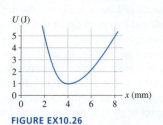

FIGURE EX10.26

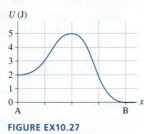

FIGURE EX10.27

27. | a. In **FIGURE EX10.27**, what minimum speed does a 100 g particle need at point A to reach point B?
    b. What minimum speed does a 100 g particle need at point B to reach point A?

28. ‖ **FIGURE EX10.28** shows the potential energy of a 500 g particle as it moves along the x-axis. Suppose the particle's mechanical energy is 12 J.
    a. Where are the particle's turning points?
    b. What is the particle's speed when it is at $x = 4.0$ m?
    c. What is the particle's maximum speed? At what position or positions does this occur?
    d. Suppose the particle's energy is lowered to 4.0 J. Can the particle ever be at $x = 2.0$ m? At $x = 4.0$ m?

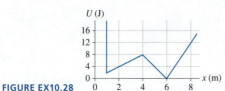

FIGURE EX10.28

29. ‖ In **FIGURE EX10.28**, what is the maximum speed a 200 g particle could have at $x = 2.0$ m and never reach $x = 6.0$ m?

Section 10.6 Force and Potential Energy

30. ‖ A system in which only one particle can move has the potential energy shown in **FIGURE EX10.30**. What is the x-component of the force on the particle at $x = 5$, 15, and 25 cm?

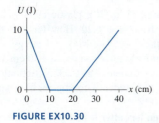

FIGURE EX10.30

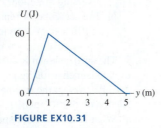

FIGURE EX10.31

31. ‖ A system in which only one particle can move has the potential energy shown in **FIGURE EX10.31**. What is the y-component of the force on the particle at $y = 0.5$ m and 4 m?

32. ‖ CALC A particle moving along the y-axis is in a system with potential energy $U = 4y^3$ J, where y is in m. What is the y-component of the force on the particle at $y = 0$ m, 1 m, and 2 m?

33. ‖ CALC A particle moving along the x-axis is in a system with potential energy $U = 10/x$ J, where x is in m. What is the x-component of the force on the particle at $x = 2$ m, 5 m, and 8 m?

34. ‖ **FIGURE EX10.34** shows the potential energy of a system in which a particle moves along the x-axis. Draw a graph of the force F_x as a function of position x.

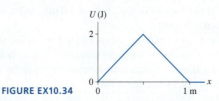

FIGURE EX10.34

Section 10.7 Conservative and Nonconservative Forces

35. | A particle moves from A to D in **FIGURE EX10.35** while experiencing force $\vec{F} = (6\hat{\imath} + 8\hat{\jmath})$ N. How much work does the force do if the particle follows path (a) ABD, (b) ACD, and (c) AD? Is this a conservative force? Explain.

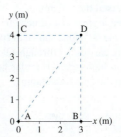

FIGURE EX10.35

36. | A force does work on a 50 g particle as the particle moves along the following straight paths in the xy-plane: 25 J from $(0$ m, 0 m$)$ to $(5$ m, 0 m$)$; 35 J from $(0$ m, 0 m$)$ to $(0$ m, 5 m$)$; -5 J from $(5$ m, 0 m$)$ to $(5$ m, 5 m$)$; -15 J from $(0$ m, 5 m$)$ to $(5$ m, 5 m$)$; and 20 J from $(0$ m, 0 m$)$ to $(5$ m, 5 m$)$.
    a. Is this a conservative force?
    b. If the zero of potential energy is at the origin, what is the potential energy at $(5$ m, 5 m$)$?

Section 10.8 The Energy Principle Revisited

37. | A system loses 400 J of potential energy. In the process, it does 400 J of work on the environment and the thermal energy increases by 100 J. Show this process on an energy bar chart.

38. | What is the final kinetic energy of the system for the process shown in **FIGURE EX10.38**?

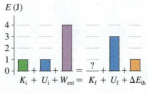

FIGURE EX10.38

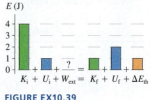

FIGURE EX10.39

39. | How much work is done by the environment in the process shown in **FIGURE EX10.39**? Is energy transferred from the environment to the system or from the system to the environment?

40. ‖ A cable with 20.0 N of tension pulls straight up on a 1.50 kg block that is initially at rest. What is the block's speed after being lifted 2.00 m? Solve this problem using work and energy.

Problems

41. ‖ A very slippery ice cube slides in a *vertical* plane around the inside of a smooth, 20-cm-diameter horizontal pipe. The ice cube's speed at the bottom of the circle is 3.0 m/s. What is the ice cube's speed at the top?

42. ‖ A 50 g ice cube can slide up and down a frictionless 30° slope. At the bottom, a spring with spring constant 25 N/m is compressed 10 cm and used to launch the ice cube up the slope. How high does it go above its starting point?

43. ‖ You have been hired to design a spring-launched roller coaster that will carry two passengers per car. The car goes up a 10-m-high hill, then descends 15 m to the track's lowest point. You've determined that the spring can be compressed a maximum of 2.0 m and that a loaded car will have a maximum mass of 400 kg. For safety reasons, the spring constant should be 10% larger than the minimum needed for the car to just make it over the top.
    a. What spring constant should you specify?
    b. What is the maximum speed of a 350 kg car if the spring is compressed the full amount?

44. ‖ It's been a great day of new, frictionless snow. Julie starts at the top of the 60° slope shown in **FIGURE P10.44**. At the bottom, a circular arc carries her through a 90° turn, and she then launches off a 3.0-m-high ramp. How far horizontally is her touchdown point from the end of the ramp?

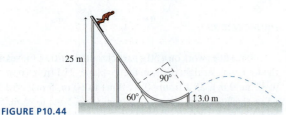

FIGURE P10.44

45. ‖ A block of mass m slides down a frictionless track, then around the inside of a circular loop-the-loop of radius R. From what minimum height h must the block start to make it around without falling off? Give your answer as a multiple of R.

46. ‖‖ A 1000 kg safe is 2.0 m above a heavy-duty spring when the rope holding the safe breaks. The safe hits the spring and compresses it 50 cm. What is the spring constant of the spring?

47. ‖‖‖ You have a ball of unknown mass, a spring with spring constant 950 N/m, and a meter stick. You use various compressions of the spring to launch the ball vertically, then use the meter stick to measure the ball's maximum height above the launch point. Your data are as follows:

Compression (cm)	Height (cm)
2.0	32
3.0	65
4.0	115
5.0	189

Use an appropriate graph of the data to determine the ball's mass.

48. ‖ Sam, whose mass is 75 kg, straps on his skis and starts down a 50-m-high, 20° frictionless slope. A strong headwind exerts a *horizontal* force of 200 N on him as he skies. Use work and energy to find Sam's speed at the bottom.

49. ‖ A horizontal spring with spring constant 100 N/m is compressed 20 cm and used to launch a 2.5 kg box across a frictionless, horizontal surface. After the box travels some distance, the surface becomes rough. The coefficient of kinetic friction of the box on the surface is 0.15. Use work and energy to find how far the box slides across the rough surface before stopping.

50. ‖ Truck brakes can fail if they get too hot. In some mountainous areas, ramps of loose gravel are constructed to stop runaway trucks that have lost their brakes. The combination of a slight upward slope and a large coefficient of rolling resistance as the truck tires sink into the gravel brings the truck safely to a halt. Suppose a gravel ramp slopes upward at 6.0° and the coefficient of rolling friction is 0.40. Use work and energy to find the length of a ramp that will stop a 15,000 kg truck that enters the ramp at 35 m/s ($\approx$ 75 mph).

51. ‖ A freight company uses a compressed spring to shoot 2.0 kg packages up a 1.0-m-high frictionless ramp into a truck, as **FIGURE P10.51** shows. The spring constant is 500 N/m and the spring is compressed 30 cm.
    a. What is the speed of the package when it reaches the truck?
    b. A careless worker spills his soda on the ramp. This creates a 50-cm-long sticky spot with a coefficient of kinetic friction 0.30. Will the next package make it into the truck?

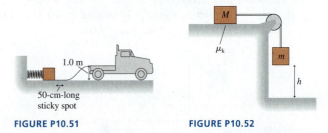

FIGURE P10.51   **FIGURE P10.52**

52. ‖ Use work and energy to find an expression for the speed of the block in **FIGURE P10.52** just before it hits the floor if (a) the coefficient of kinetic friction for the block on the table is μ_k and (b) the table is frictionless.

53. ‖ a. A 50 g ice cube can slide without friction up and down a 30° slope. The ice cube is pressed against a spring at the bottom of the slope, compressing the spring 10 cm. The spring constant is 25 N/m. When the ice cube is released, what total distance will it travel up the slope before reversing direction?
    b. The ice cube is replaced by a 50 g plastic cube whose coefficient of kinetic friction is 0.20. How far will the plastic cube travel up the slope? Use work and energy.

54. ‖ The spring shown in **FIGURE P10.54** is compressed 50 cm and used to launch a 100 kg physics student. The track is frictionless until it starts up the incline. The student's coefficient of kinetic friction on the 30° incline is 0.15.
    a. What is the student's speed just after losing contact with the spring?
    b. How far up the incline does the student go?

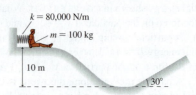

FIGURE P10.54

55. ‖ Protons and neutrons (together called *nucleons*) are held
CALC together in the nucleus of an atom by a force called the *strong force*. At very small separations, the strong force between two nucleons is larger than the repulsive electrical force between two protons—hence its name. But the strong force quickly weakens as the distance between the protons increases. A well-established model for the potential energy of two nucleons interacting via the strong force is

$$U = U_0\left[1 - e^{-x/x_0}\right]$$

where x is the distance between the centers of the two nucleons, x_0 is a constant having the value $x_0 = 2.0 \times 10^{-15}$ m, and $U_0 = 6.0 \times 10^{-11}$ J.

Quantum effects are essential for a proper understanding of nucleons, but let us innocently consider two neutrons as if they were small, hard, electrically neutral spheres of mass 1.67×10^{-27} kg and diameter 1.0×10^{-15} m. Suppose you hold two neutrons 5.0×10^{-15} m apart, measured between their centers, then release them. What is the speed of each neutron as they crash together? Keep in mind that *both* neutrons are moving.

56. ‖ A 2.6 kg block is attached to a horizontal rope that exerts a
CALC variable force $F_x = (20 - 5x)$ N, where x is in m. The coefficient of kinetic friction between the block and the floor is 0.25. Initially the block is at rest at $x = 0$ m. What is the block's speed when it has been pulled to $x = 4.0$ m?

57. ‖ A system has potential energy
CALC

$$U(x) = x + \sin\big((2 \text{ rad/m})x\big)$$

as a particle moves over the range $0 \text{ m} \le x \le \pi$ m.
a. Where are the equilibrium positions in this range?
b. For each, is it a point of stable or unstable equilibrium?

58. ‖ A particle that can move along the x-axis is part of a system
CALC with potential energy

$$U(x) = \frac{A}{x^2} - \frac{B}{x}$$

where A and B are positive constants.
a. Where are the particle's equilibrium positions?
b. For each, is it a point of stable or unstable equilibrium?

59. ‖ A 100 g particle experiences the one-dimensional, conservative force F_x shown in FIGURE P10.59.

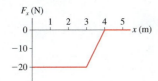

FIGURE P10.59

a. Let the zero of potential energy be at $x = 0$ m. What is the potential energy at $x = 1.0, 2.0, 3.0,$ and 4.0 m?
   Hint: Use the definition of potential energy and the geometric interpretation of work.
b. Suppose the particle is shot to the right from $x = 1.0$ m with a speed of 25 m/s. Where is its turning point?

60. ‖ A clever engineer designs a "sprong" that obeys the force law
CALC $F_x = -q(x - x_{eq})^3$, where x_{eq} is the equilibrium position of the end of the sprong and q is the sprong constant. For simplicity, we'll let $x_{eq} = 0$ m. Then $F_x = -qx^3$.

a. What are the units of q?
b. Find an expression for the potential energy of a stretched or compressed sprong.
c. A sprong-loaded toy gun shoots a 20 g plastic ball. What is the launch speed if the sprong constant is 40,000, with the units you found in part a, and the sprong is compressed 10 cm? Assume the barrel is frictionless.

61. ‖ The potential energy for a particle that can move along the
CALC x-axis is $U = Ax^2 + B\sin(\pi x/L)$, where A, B, and L are constants. What is the force on the particle at (a) $x = 0$, (b) $x = L/2$, and (c) $x = L$?

62. ‖ A particle that can move along the x-axis experiences an
CALC interaction force $F_x = (3x^2 - 5x)$ N, where x is in m. Find an expression for the system's potential energy.

63. ‖ An object moving in the xy-plane is subjected to the force
CALC $\vec{F} = (2xy\,\hat{\imath} + x^2\,\hat{\jmath})$ N, where x and y are in m.
a. The particle moves from the origin to the point with coordinates (a, b) by moving first along the x-axis to $(a, 0)$, then parallel to the y-axis. How much work does the force do?
b. The particle moves from the origin to the point with coordinates (a, b) by moving first along the y-axis to $(0, b)$, then parallel to the x-axis. How much work does the force do?
c. Is this a conservative force?

64. ‖ An object moving in the xy-plane is subjected to the force
CALC $\vec{F} = (2xy\,\hat{\imath} + 3y\,\hat{\jmath})$ N, where x and y are in m.
a. The particle moves from the origin to the point with coordinates (a, b) by moving first along the x-axis to $(a, 0)$, then parallel to the y-axis. How much work does the force do?
b. The particle moves from the origin to the point with coordinates (a, b) by moving first along the y-axis to $(0, b)$, then parallel to the x-axis. How much work does the force do?
c. Is this a conservative force?

65. Write a realistic problem for which the energy bar chart shown in FIGURE P10.65 correctly shows the energy at the beginning and end of the problem.

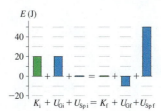

FIGURE P10.65

$$K_i + U_{Gi} + U_{Spi} = K_f + U_{Gf} + U_{Spf}$$

In Problems 66 through 68 you are given the equation used to solve a problem. For each of these, you are to
a. Write a realistic problem for which this is the correct equation.
b. Draw the before-and-after pictorial representation.
c. Finish the solution of the problem.

66. $\frac{1}{2}(1500 \text{ kg})(5.0 \text{ m/s})^2 + (1500 \text{ kg})(9.80 \text{ m/s}^2)(10 \text{ m})$
   $= \frac{1}{2}(1500 \text{ kg})v_i^2 + (1500 \text{ kg})(9.80 \text{ m/s}^2)(0 \text{ m})$

67. $\frac{1}{2}(0.20 \text{ kg})(2.0 \text{ m/s})^2 + \frac{1}{2}k(0 \text{ m})^2$
   $= \frac{1}{2}(0.20 \text{ kg})(0 \text{ m/s})^2 + \frac{1}{2}k(-0.15 \text{ m})^2$

68. $\frac{1}{2}(0.50 \text{ kg})v_f^2 + (0.50 \text{ kg})(9.80 \text{ m/s}^2)(0 \text{ m})$
   $+ \frac{1}{2}(400 \text{ N/m})(0 \text{ m})^2 = \frac{1}{2}(0.50 \text{ kg})(0 \text{ m/s})^2$
   $+ (0.50 \text{ kg})(9.80 \text{ m/s}^2)((-0.10 \text{ m}) \sin 30°)$
   $+ \frac{1}{2}(400 \text{ N/m})(-0.10 \text{ m})^2$

Challenge Problems

69. ‖ A pendulum is formed from a small ball of mass m on a string of length L. As **FIGURE CP10.69** shows, a peg is height $h = L/3$ above the pendulum's lowest point. From what minimum angle θ must the pendulum be released in order for the ball to go over the top of the peg without the string going slack?

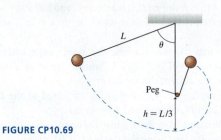

FIGURE CP10.69

70. ‖ In a physics lab experiment, a compressed spring launches a 20 g metal ball at a 30° angle. Compressing the spring 20 cm causes the ball to hit the floor 1.5 m below the point at which it leaves the spring after traveling 5.0 m horizontally. What is the spring constant?

71. ‖ It's your birthday, and to celebrate you're going to make your first bungee jump. You stand on a bridge 100 m above a raging river and attach a 30-m-long bungee cord to your harness. A bungee cord, for practical purposes, is just a long spring, and this cord has a spring constant of 40 N/m. Assume that your mass is 80 kg. After a long hesitation, you dive off the bridge. How far are you above the water when the cord reaches its maximum elongation?

72. ‖ A 10 kg box slides 4.0 m down the frictionless ramp shown in **FIGURE CP10.72**, then collides with a spring whose spring constant is 250 N/m.
    a. What is the maximum compression of the spring?
    b. At what compression of the spring does the box have its maximum speed?

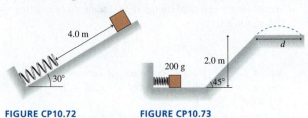

FIGURE CP10.72　　　　　**FIGURE CP10.73**

73. ‖ The spring in **FIGURE CP10.73** has a spring constant of 1000 N/m. It is compressed 15 cm, then launches a 200 g block. The horizontal surface is frictionless, but the block's coefficient of kinetic friction on the incline is 0.20. What distance d does the block sail through the air?

74. ‖ A sled starts from rest at the top of the frictionless, hemispherical, snow-covered hill shown in **FIGURE CP10.74**.
    a. Find an expression for the sled's speed when it is at angle ϕ.
    b. Use Newton's laws to find the maximum speed the sled can have at angle ϕ without leaving the surface.
    c. At what angle ϕ_{max} does the sled "fly off" the hill?

FIGURE CP10.74

11 Impulse and Momentum

An exploding firework is a dramatic event. Nonetheless, the explosion obeys some simple laws of physics.

IN THIS CHAPTER, you will learn to use the concepts of impulse and momentum.

What is momentum?

An object's momentum is the product of its mass and velocity. An object can have a large momentum by having a large mass or a large velocity. Momentum is a vector, and it is especially important to pay attention to the *signs* of the components of momentum.

What is impulse?

A force of short duration is an impulsive force. The impulse J_x that this force delivers to an object is the area under the force-versus-time graph. For time-dependent forces, impulse and momentum are often more useful than Newton's laws.

How are impulse and momentum related?

Working with momentum is similar to working with energy. It's important to clearly define the system. The momentum principle says that a system's momentum changes when an impulse is delivered:

$$\Delta p_x = J_x$$

A momentum bar chart, similar to an energy bar chart, shows this principle graphically.

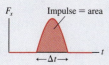

≪ LOOKING BACK Section 9.1 Energy overview

Is momentum conserved?

The total momentum of an isolated system is conserved. The particles of an isolated system interact with each other but not with the environment. Regardless of how intense the interactions are, the final momentum equals the initial momentum.

≪ LOOKING BACK Section 10.4 Energy conservation

How does momentum apply to collisions?

One important application of momentum conservation is the study of collisions.

- In a totally inelastic collision, the objects stick together. Momentum is conserved.
- In a perfectly elastic collision, the objects bounce apart. Both momentum and energy are conserved.

Where else is momentum used?

This chapter looks at two other important applications of momentum conservation:

- An explosion is a short interaction that drives two or more objects apart.

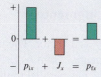

- In rocket propulsion the object's mass is changing continuously.

A tennis ball collides with a racket. Notice that the left side of the ball is flattened.

FIGURE 11.1 A collision.

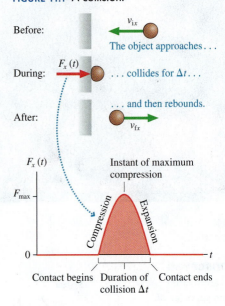

11.1 Momentum and Impulse

A **collision** is a short-duration interaction between two objects. The collision between a tennis ball and a racket, or a baseball and a bat, may seem instantaneous to your eye, but that is a limitation of your perception. A high-speed photograph reveals that the side of the ball is significantly flattened during the collision. It takes time to compress the ball, and more time for the ball to re-expand as it leaves the racket or bat.

The duration of a collision depends on the materials from which the objects are made, but 1 to 10 ms (0.001 to 0.010 s) is fairly typical. This is the time during which the two objects are in contact with each other. The harder the objects, the shorter the contact time. A collision between two steel balls lasts less than 200 microseconds.

FIGURE 11.1 shows an object colliding with a wall. The object approaches with an initial horizontal velocity v_{ix}, experiences a force of duration Δt, and leaves with final velocity v_{fx}. Notice that the object, as in the photo above, *deforms* during the collision. A particle cannot be deformed, so we cannot model colliding objects as particles. Instead, we model a colliding object as an *elastic object* that compresses and then expands, much like a spring. Indeed, that's exactly what happens during a collision at the microscopic level: Molecular bonds compress, store elastic potential energy, then transform some or all of that potential energy back into the kinetic energy of the rebounding object. We'll examine the energy issues of collisions later in this chapter.

The force of a collision is usually very large in comparison to other forces exerted on the object. A large force exerted for a small interval of time is called an **impulsive force**. The graph of Figure 11.1 shows how a typical impulsive force behaves, rapidly growing to a maximum at the instant of maximum compression, then decreasing back to zero. The force is zero before contact begins and after contact ends. Because an impulsive force is a function of time, we will write it as $F_x(t)$.

> **NOTE** Both v_x and F_x are components of vectors and thus have *signs* indicating which way the vectors point.

We can use Newton's second law to find how the object's velocity changes as a result of the collision. Acceleration in one dimension is $a_x = dv_x/dt$, so the second law is

$$ma_x = m\frac{dv_x}{dt} = F_x(t)$$

After multiplying both sides by dt, we can write the second law as

$$m\,dv_x = F_x(t)\,dt \tag{11.1}$$

The force is nonzero only during an interval of time from t_i to $t_f = t_i + \Delta t$, so let's integrate Equation 11.1 over this interval. The velocity changes from v_{ix} to v_{fx} during the collision; thus

$$m\int_{v_i}^{v_f} dv_x = mv_{fx} - mv_{ix} = \int_{t_i}^{t_f} F_x(t)\,dt \tag{11.2}$$

We need some new tools to help us make sense of Equation 11.2.

Momentum

The product of a particle's mass and velocity is called the *momentum* of the particle:

$$\boxed{\text{momentum} = \vec{p} \equiv m\vec{v}} \tag{11.3}$$

Momentum, like velocity, is a vector. The units of momentum are kg m/s. The plural of "momentum" is "momenta," from its Latin origin.

The momentum vector $\vec{p}$ is parallel to the velocity vector $\vec{v}$. **FIGURE 11.2** shows that $\vec{p}$, like any vector, can be decomposed into x- and y-components. Equation 11.3, which is a vector equation, is a shorthand way to write the simultaneous equations

$$p_x = mv_x$$
$$p_y = mv_y$$

FIGURE 11.2 The momentum $\vec{p}$ can be decomposed into x- and y-components.

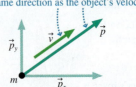

Momentum is a vector pointing in the same direction as the object's velocity.

NOTE One of the most common errors in momentum problems is a failure to use the appropriate signs. The momentum component p_x has the same sign as v_x. Momentum is *negative* for a particle moving to the left (on the x-axis) or down (on the y-axis). ◄

An object can have a large momentum by having either a small mass but a large velocity or a small velocity but a large mass. For example, a 5.5 kg (12 lb) bowling ball rolling at a modest 2 m/s has momentum of magnitude $p = (5.5 \text{ kg})(2 \text{ m/s}) = 11 \text{ kg m/s}$. This is almost exactly the same momentum as a 9 g bullet fired from a high-speed rifle at 1200 m/s.

Newton actually formulated his second law in terms of momentum rather than acceleration:

$$\vec{F} = m\vec{a} = m\frac{d\vec{v}}{dt} = \frac{d(m\vec{v})}{dt} = \frac{d\vec{p}}{dt} \tag{11.4}$$

This statement of the second law, saying that **force is the rate of change of momentum,** is more general than our earlier version $\vec{F} = m\vec{a}$. It allows for the possibility that the mass of the object might change, such as a rocket that is losing mass as it burns fuel.

Returning to Equation 11.2, you can see that mv_{ix} and mv_{fx} are p_{ix} and p_{fx}, the x-component of the particle's momentum before and after the collision. Further, $p_{\text{fx}} - p_{\text{ix}}$ is Δp_x, the *change* in the particle's momentum. In terms of momentum, Equation 11.2 is

$$\Delta p_x = p_{\text{fx}} - p_{\text{ix}} = \int_{t_i}^{t_f} F_x(t) \, dt \tag{11.5}$$

Now we need to examine the right-hand side of Equation 11.5.

Impulse

Equation 11.5 tells us that the particle's change in momentum is related to the time integral of the force. Let's define a quantity J_x called the *impulse* to be

$$\textbf{impulse } = J_x \equiv \int_{t_i}^{t_f} F_x(t) \, dt \tag{11.6}$$
$$= \text{area under the } F_x(t) \text{ curve between } t_i \text{ and } t_f$$

Strictly speaking, impulse has units of N s, but you should be able to show that N s are equivalent to kg m/s, the units of momentum.

The interpretation of the integral in Equation 11.6 as an area under a curve is especially important. **FIGURE 11.3a** portrays the impulse graphically. Because the force changes in a complicated way during a collision, it is often useful to describe the collision in terms of an *average* force F_{avg}. As **FIGURE 11.3b** shows, F_{avg} is the height of a rectangle that has the same area, and thus the same impulse, as the real force curve. The impulse exerted during the collision is

$$J_x = F_{\text{avg}} \, \Delta t \tag{11.7}$$

Equation 11.2, which we found by integrating Newton's second law, can now be rewritten in terms of impulse and momentum as

$$\Delta p_x = J_x \quad \text{(momentum principle)} \tag{11.8}$$

This result, called the **momentum principle,** says that **an impulse delivered to an object causes the object's momentum to change.** The momentum p_{fx} "after" an interaction, such as a collision or an explosion, is equal to the momentum p_{ix} "before" the interaction *plus* the impulse that arises from the interaction:

$$p_{\text{fx}} = p_{\text{ix}} + J_x \tag{11.9}$$

FIGURE 11.3 Looking at the impulse graphically.

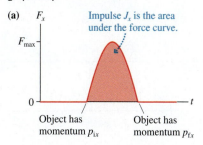

(a)

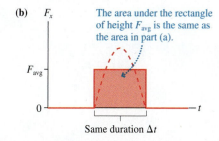

(b)

FIGURE 11.4 The momentum principle helps us understand a rubber ball bouncing off a wall.

The wall delivers an impulse to the ball.

Before: After:

$\vec{F}$

$v_{ix} > 0$ $v_{fx} < 0$

F_x

0 ──────────────── t

J_x = area under curve

Maximum compression

p_x

p_{ix}

Contact begins

0 ──────────────── t

$\Delta p_x = J_x$

p_{fx}

Contact ends

The impulse changes the ball's momentum.

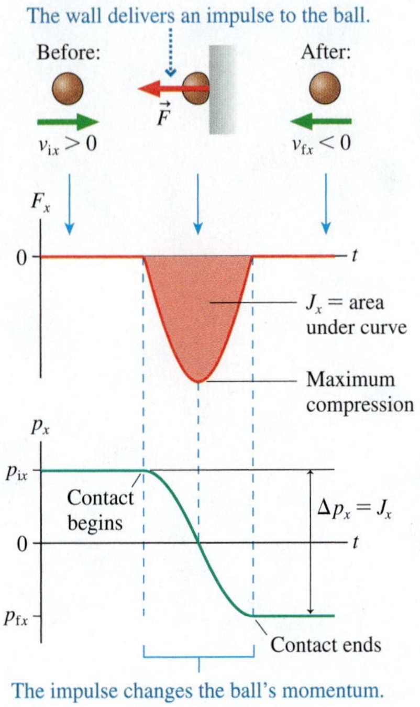

FIGURE 11.4 illustrates the momentum principle for a rubber ball bouncing off a wall. Notice the signs; they are very important. The ball is initially traveling toward the right, so v_{ix} and p_{ix} are positive. After the bounce, v_{fx} and p_{fx} are negative. The force *on the ball* is toward the left, so F_x is also negative. The graphs show how the force and the momentum change with time.

Although the interaction is very complex, the impulse—the area under the force graph—is all we need to know to find the ball's velocity as it rebounds from the wall. The final momentum is

$$p_{fx} = p_{ix} + J_x = p_{ix} + \text{area under the force curve} \qquad (11.10)$$

and the final velocity is $v_{fx} = p_{fx}/m$. In this example, the area has a negative value.

An Analogy with the Energy Principle

You've probably noticed that there is a similarity between the momentum principle and the energy principle of Chapters 9 and 10. For a system of one object acted on by a force:

$$\text{energy principle:} \qquad \Delta K = W = \int_{x_i}^{x_f} F_x \, dx$$

$$(11.11)$$

$$\text{momentum principle:} \qquad \Delta p_x = J_x = \int_{t_i}^{t_f} F_x \, dt$$

In both cases, a force acting on an object changes the state of the system. If the force acts over the spatial interval from x_i to x_f, it does *work* that changes the object's kinetic energy. If the force acts over a time interval from t_i to t_f, it delivers an *impulse* that changes the object's momentum. **FIGURE 11.5** shows that the geometric interpretation of work as the area under the *F*-versus-*x* graph parallels an interpretation of impulse as the area under the *F*-versus-*t* graph.

FIGURE 11.5 Impulse and work are both the area under a force curve, but it's very important to know what the horizontal axis is.

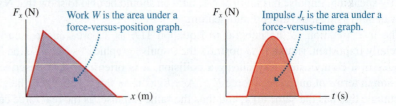

F_x (N)   Work W is the area under a force-versus-position graph.

x (m)

F_x (N)   Impulse J_x is the area under a force-versus-time graph.

t (s)

This does not mean that a force *either* creates an impulse *or* does work but does not do both. Quite the contrary. A force acting on a particle *both* creates an impulse *and* does work, changing both the momentum and the kinetic energy of the particle. Whether you use the energy principle or the momentum principle depends on the question you are trying to answer.

In fact, we can express the kinetic energy in terms of momentum as

$$K = \tfrac{1}{2}mv^2 = \frac{(mv)^2}{2m} = \frac{p^2}{2m} \qquad (11.12)$$

You cannot change a particle's kinetic energy without also changing its momentum.

Momentum Bar Charts

The momentum principle tells us that **impulse transfers momentum to an object.** If an object has 2 kg m/s of momentum, a 1 kg m/s impulse delivered to the object increases its momentum to 3 kg m/s. That is, $p_{fx} = p_{ix} + J_x$.

Just as we did with energy, we can represent this "momentum accounting" with a **momentum bar chart.** For example, the bar chart of **FIGURE 11.6** represents the ball colliding with a wall in Figure 11.4. Momentum bar charts are a tool for visualizing an interaction.

> **NOTE** The vertical scale of a momentum bar chart has no numbers; it can be adjusted to match any problem. However, be sure that all bars in a given problem use a consistent scale.

FIGURE 11.6 A momentum bar chart.

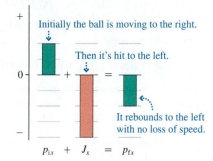

Initially the ball is moving to the right.

Then it's hit to the left.

It rebounds to the left with no loss of speed.

$$p_{ix} + J_x = p_{fx}$$

STOP TO THINK 11.1 The cart's change of momentum is

a. -30 kg m/s
b. -20 kg m/s
c. 0 kg m/s
d. 10 kg m/s
e. 20 kg m/s
f. 30 kg m/s

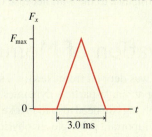

Before: 10 kg   2 m/s

After: 1 m/s

Solving Impulse and Momentum Problems

Impulse and momentum problems, like energy problems, relate the situation before an interaction to the situation afterward. Consequently, the *before-and-after pictorial representation* remains our primary visualization tool. Let's look at an example.

EXAMPLE 11.1 | Hitting a baseball

A 150 g baseball is thrown with a speed of 20 m/s. It is hit straight back toward the pitcher at a speed of 40 m/s. The interaction force between the ball and the bat is shown in **FIGURE 11.7**. What *maximum* force F_{max} does the bat exert on the ball? What is the *average* force of the bat on the ball?

FIGURE 11.7 The interaction force between the baseball and the bat.

F_x

F_{max}

0

t

3.0 ms

MODEL Model the baseball as an elastic object and the interaction as a collision.

VISUALIZE FIGURE 11.8 is a before-and-after pictorial representation. Because F_x is positive (a force to the right), we know the ball was initially moving toward the left and is hit back toward the right. Thus we converted the statements about *speeds* into information about *velocities,* with v_{ix} negative.

SOLVE The momentum principle is

$$\Delta p_x = J_x = \text{area under the force curve}$$

We know the velocities before and after the collision, so we can calculate the ball's momenta:

$$p_{ix} = mv_{ix} = (0.15 \text{ kg})(-20 \text{ m/s}) = -3.0 \text{ kg m/s}$$
$$p_{fx} = mv_{fx} = (0.15 \text{ kg})(40 \text{ m/s}) = 6.0 \text{ kg m/s}$$

FIGURE 11.8 A before-and-after pictorial representation.

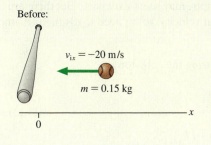

Before: $v_{ix} = -20$ m/s $m = 0.15$ kg x 0

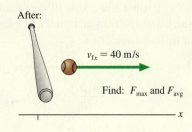

After: $v_{fx} = 40$ m/s   Find: F_{max} and F_{avg} x

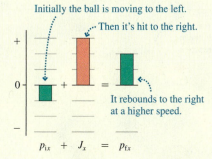

Initially the ball is moving to the left.

Then it's hit to the right.

It rebounds to the right at a higher speed.

$$p_{ix} + J_x = p_{fx}$$

Continued

Thus the *change* in momentum is

$$\Delta p_x = p_{fx} - p_{ix} = 9.0 \text{ kg m/s}$$

The force curve is a triangle with height F_{max} and width 3.0 ms. The area under the curve is

$$J_x = \text{area} = \tfrac{1}{2}(F_{max})(0.0030 \text{ s}) = (F_{max})(0.0015 \text{ s})$$

Using this information in the momentum principle, we have

$$9.0 \text{ kg m/s} = (F_{max})(0.0015 \text{ s})$$

Thus the *maximum* force is

$$F_{max} = \frac{9.0 \text{ kg m/s}}{0.0015 \text{ s}} = 6000 \text{ N}$$

The *average* force, which depends on the collision duration $\Delta t = 0.0030$ s, has the smaller value:

$$F_{avg} = \frac{J_x}{\Delta t} = \frac{\Delta p_x}{\Delta t} = \frac{9.0 \text{ kg m/s}}{0.0030 \text{ s}} = 3000 \text{ N}$$

ASSESS F_{max} is a large force, but quite typical of the impulsive forces during collisions. The main thing to focus on is our new perspective: An impulse changes the momentum of an object.

Other forces often act on an object during a collision or other brief interaction. In Example 11.1, for instance, the baseball is also acted on by gravity. Usually these other forces are *much* smaller than the interaction forces. The 1.5 N weight of the ball is vastly less than the 6000 N force of the bat on the ball. We can reasonably neglect these small forces *during* the brief time of the impulsive force by using what is called the **impulse approximation.**

When we use the impulse approximation, p_{ix} and p_{fx} (and v_{ix} and v_{fx}) are the momenta (and velocities) *immediately* before and *immediately* after the collision. For example, the velocities in Example 11.1 are those of the ball just before and after it collides with the bat. We could then do a follow-up problem, including gravity and drag, to find the ball's speed a second later as the second baseman catches it. We'll look at some two-part examples later in the chapter.

STOP TO THINK 11.2 A 10 g rubber ball and a 10 g clay ball are thrown at a wall with equal speeds. The rubber ball bounces, the clay ball sticks. Which ball delivers a larger impulse to the wall?

    a. The clay ball delivers a larger impulse because it sticks.
    b. The rubber ball delivers a larger impulse because it bounces.
    c. They deliver equal impulses because they have equal momenta.
    d. Neither delivers an impulse to the wall because the wall doesn't move.

11.2 Conservation of Momentum

The momentum principle was derived from Newton's second law and is really just an alternative way of looking at single-particle dynamics. To discover the real power of momentum for problem solving, we need also to invoke Newton's third law, which will lead us to one of the most important principles in physics: conservation of momentum.

FIGURE 11.9 shows two objects with initial velocities $(v_{ix})_1$ and $(v_{ix})_2$. The objects collide, then bounce apart with final velocities $(v_{fx})_1$ and $(v_{fx})_2$. The forces during the collision, as the objects are interacting, are the action/reaction pair $\vec{F}_{1 \text{ on } 2}$ and $\vec{F}_{2 \text{ on } 1}$. For now, we'll continue to assume that the motion is one dimensional along the x-axis.

NOTE The notation, with all the subscripts, may seem excessive. But there are two objects, and each has an initial and a final velocity, so we need to distinguish among four different velocities.

Newton's second law for each object *during* the collision is

$$\frac{d(p_x)_1}{dt} = (F_x)_{2 \text{ on } 1}$$

$$\frac{d(p_x)_2}{dt} = (F_x)_{1 \text{ on } 2} = -(F_x)_{2 \text{ on } 1}$$

(11.13)

We made explicit use of Newton's third law in the second equation.

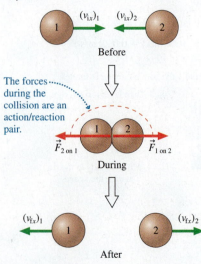

FIGURE 11.9 A collision between two objects.

$(v_{ix})_1$ $(v_{ix})_2$

Before

The forces during the collision are an action/reaction pair.

$\vec{F}_{2 \text{ on } 1}$ $\vec{F}_{1 \text{ on } 2}$

During

$(v_{fx})_1$ $(v_{fx})_2$

After

Although Equations 11.13 are for two different objects, suppose—just to see what happens—we were to *add* these two equations. If we do, we find that

$$\frac{d(p_x)_1}{dt} + \frac{d(p_x)_2}{dt} = \frac{d}{dt}\left[(p_x)_1 + (p_x)_2\right] = (F_x)_{2 \text{ on } 1} + (-(F_x)_{2 \text{ on } 1}) = 0 \quad (11.14)$$

If the time derivative of the quantity $(p_x)_1 + (p_x)_2$ is zero, it must be the case that

$$(p_x)_1 + (p_x)_2 = \text{constant} \quad (11.15)$$

Equation 11.15 is a conservation law! If $(p_x)_1 + (p_x)_2$ is a constant, then the sum of the momenta *after* the collision equals the sum of the momenta *before* the collision. That is,

$$(p_{fx})_1 + (p_{fx})_2 = (p_{ix})_1 + (p_{ix})_2 \quad (11.16)$$

Furthermore, this equality is independent of the interaction force. We don't need to know *anything* about $\vec{F}_{1 \text{ on } 2}$ and $\vec{F}_{2 \text{ on } 1}$ to make use of Equation 11.16.

As an example, **FIGURE 11.10** is a before-and-after pictorial representation of two equal-mass train cars colliding and coupling. Equation 11.16 relates the momenta of the cars after the collision to their momenta before the collision:

$$m_1(v_{fx})_1 + m_2(v_{fx})_2 = m_1(v_{ix})_1 + m_2(v_{ix})_2$$

Initially, car 1 is moving with velocity $(v_{ix})_1 = v_i$ while car 2 is at rest. Afterward, they roll together with the common final velocity v_f. Furthermore, $m_1 = m_2 = m$. With this information, the momentum equation is

$$mv_f + mv_f = mv_i + 0$$

The mass cancels, and we find that the train cars' final velocity is $v_f = \frac{1}{2}v_i$. That is, we can predict that the final speed is exactly half the initial speed of car 1 without knowing anything about the complex interaction between the two cars as they collide.

FIGURE 11.10 Two colliding train cars.

Before:

$(v_{ix})_1 = v_i$ $(v_{ix})_2 = 0$

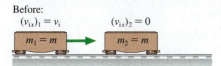

After:

$(v_{fx})_1 = (v_{fx})_2 = v_f$

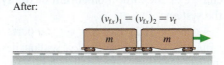

Systems of Particles

Equation 11.16 illustrates the idea of a conservation law for momentum, but it was derived for the specific case of two particles colliding in one dimension. Our goal is to develop a more general law of conservation of momentum, a law that will be valid in three dimensions and that will work for any type of interaction. The next few paragraphs are fairly mathematical, so you might want to begin by looking ahead to Equations 11.24 and the statement of the law of conservation of momentum to see where we're heading.

Our study of energy in the last two chapters has emphasized the importance of having a clearly defined system. The same is true for momentum. Consider a system consisting of N particles. **FIGURE 11.11** shows a simple case where $N = 3$, but N could be anything. The particles might be large entities (cars, baseballs, etc.), or they might be the microscopic atoms in a gas. We can identify each particle by an identification number k. Every particle in the system *interacts* with every other particle via action/reaction pairs of forces $\vec{F}_{j \text{ on } k}$ and $\vec{F}_{k \text{ on } j}$. In addition, every particle is subjected to possible *external forces* $\vec{F}_{\text{ext on } k}$ from agents outside the system.

If particle k has velocity $\vec{v}_k$, its momentum is $\vec{p}_k = m_k\vec{v}_k$. We define the **total momentum** $\vec{P}$ of the system as the vector sum

$$\vec{P} = \text{total momentum} = \vec{p}_1 + \vec{p}_2 + \vec{p}_3 + \cdots + \vec{p}_N = \sum_{k=1}^{N} \vec{p}_k \quad (11.17)$$

That is the total momentum *of the system* is the vector sum of the individual momenta.

FIGURE 11.11 A system of particles.

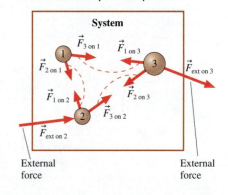

External force   External force

The total momentum of the rocket + gases system is conserved, so the rocket accelerates forward as the gases are expelled backward.

The time derivative of $\vec{P}$ tells us how the total momentum of the system changes with time:

$$\frac{d\vec{P}}{dt} = \sum_k \frac{d\vec{p}_k}{dt} = \sum_k \vec{F}_k \qquad (11.18)$$

where we used Newton's second law for each particle in the form $\vec{F}_k = d\vec{p}_k/dt$, which was Equation 11.4.

The net force acting on particle k can be divided into *external forces,* from outside the system, and *interaction forces* due to all the other particles in the system:

$$\vec{F}_k = \sum_{j \neq k} \vec{F}_{j \text{ on } k} + \vec{F}_{\text{ext on } k} \qquad (11.19)$$

The restriction $j \neq k$ expresses the fact that particle k does not exert a force on itself. Using this in Equation 11.18 gives the rate of change of the total momentum $\vec{P}$ of the system:

$$\frac{d\vec{P}}{dt} = \sum_k \sum_{j \neq k} \vec{F}_{j \text{ on } k} + \sum_k \vec{F}_{\text{ext on } k} \qquad (11.20)$$

The double sum on $\vec{F}_{j \text{ on } k}$ adds *every* interaction force within the system. But the interaction forces come in action/reaction pairs, with $\vec{F}_{k \text{ on } j} = -\vec{F}_{j \text{ on } k}$, so $\vec{F}_{k \text{ on } j} + \vec{F}_{j \text{ on } k} = \vec{0}$. Consequently, **the sum of all the interaction forces is zero.** As a result, Equation 11.20 becomes

$$\frac{d\vec{P}}{dt} = \sum_k \vec{F}_{\text{ext on } k} = \vec{F}_{\text{net}} \qquad (11.21)$$

where $\vec{F}_{\text{net}}$ is the net force exerted on the system by agents outside the system. But this is just Newton's second law written for the system as a whole! That is, **the rate of change of the total momentum of the system is equal to the net force applied to the system.**

Equation 11.21 has two very important implications. First, we can analyze the motion of the system as a whole without needing to consider interaction forces between the particles that make up the system. In fact, we have been using this idea all along as an *assumption* of the particle model. When we treat cars and rocks and baseballs as particles, we assume that the internal forces between the atoms—the forces that hold the object together—do not affect the motion of the object as a whole. Now we have *justified* that assumption.

Isolated Systems

The second implication of Equation 11.21, and the more important one from the perspective of this chapter, applies to an isolated system. In Chapter 10, we defined an *isolated system* as one that is not influenced or altered by external forces from the environment. For momentum, that means a system on which the *net* external force is zero: $\vec{F}_{\text{net}} = \vec{0}$. That is, an isolated system is one on which there are *no* external forces or for which the external forces are balanced and add to zero.

For an isolated system, Equation 11.21 is simply

$$\frac{d\vec{P}}{dt} = \vec{0} \qquad \text{(isolated system)} \qquad (11.22)$$

In other words, **the *total* momentum of an isolated system does not change.** The total momentum $\vec{P}$ remains constant, *regardless* of whatever interactions are going on *inside* the system. The importance of this result is sufficient to elevate it to a law of nature, alongside Newton's laws.

Law of conservation of momentum The total momentum $\vec{P}$ of an isolated system is a constant. Interactions within the system do not change the system's total momentum. Mathematically, the law of conservation of momentum is

$$\vec{P}_f = \vec{P}_i \qquad (11.23)$$

The total momentum *after* an interaction is equal to the total momentum *before* the interaction. Because Equation 11.23 is a vector equation, the equality is true for each of the components of the momentum vector. That is,

$$(p_{fx})_1 + (p_{fx})_2 + (p_{fx})_3 + \cdots = (p_{ix})_1 + (p_{ix})_2 + (p_{ix})_3 + \cdots$$
$$(p_{fy})_1 + (p_{fy})_2 + (p_{fy})_3 + \cdots = (p_{iy})_1 + (p_{iy})_2 + (p_{iy})_3 + \cdots \qquad (11.24)$$

The *x*-equation is an extension of Equation 11.16 to *N* interacting particles.

NOTE It is worth emphasizing the critical role of Newton's third law. The law of conservation of momentum is a direct consequence of the fact that interactions within an isolated system are action/reaction pairs.

EXAMPLE 11.2 | A glider collision

A 250 g air-track glider is pushed across a level track toward a 500 g glider that is at rest. **FIGURE 11.12** shows a position-versus-time graph of the 250 g glider as recorded by a motion detector. Best-fit lines have been found. What is the speed of the 500 g glider after the collision?

FIGURE 11.12 Position graph of the 250 g glider.

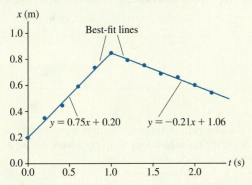

MODEL Let the system be the two gliders. The gliders interact with each other, but the external forces (normal force and gravity) balance to make $\vec{F}_{net} = \vec{0}$. Thus the gliders form an isolated system and their total momentum is conserved.

VISUALIZE **FIGURE 11.13** is a before-and-after pictorial representation. The graph of Figure 11.12 tells us that the 250 g glider initially moves to the right, collides at $t = 1.0$ s, then rebounds to the left (decreasing *x*). Note that the best-fit lines are written as a generic $y = \ldots$, which is what you would see in data-analysis software.

SOLVE Conservation of momentum for this one-dimensional problem requires that the final momentum equal the initial momentum: $P_{fx} = P_{ix}$. In terms of the individual components, conservation of momentum is

$$(p_{fx})_1 + (p_{fx})_2 = (p_{ix})_1 + (p_{ix})_2$$

Each momentum is mv_x, so conservation of momentum in terms of velocities is

FIGURE 11.13 Before-and-after representation of a collision.

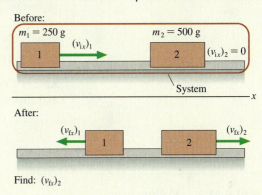

$$m_1(v_{fx})_1 + m_2(v_{fx})_2 = m_1(v_{ix})_1 + m_2(v_{ix})_2 = m_1(v_{ix})_1$$

where, in the last step, we used $(v_{ix})_2 = 0$ for the 500 g glider. Solving for the heavier glider's final velocity gives

$$(v_{fx})_2 = \frac{m_1}{m_2}\left[(v_{ix})_1 - (v_{fx})_1\right]$$

From Chapter 2 kinematics, the velocities of the 250 g glider before and after the collision are the slopes of the position-versus-time graph. Referring to Figure 11.12, we see that $(v_{ix})_1 = 0.75$ m/s and $(v_{fx})_1 = -0.21$ m/s. The latter is negative because the rebound motion is to the left. Thus

$$(v_{fx})_2 = \frac{250 \text{ g}}{500 \text{ g}}\left[0.75 \text{ m/s} - (-0.21 \text{ m/s})\right] = 0.48 \text{ m/s}$$

The 500 g glider moves away from the collision at 0.48 m/s.

ASSESS The 500 g glider has twice the mass of the glider that was pushed, so a somewhat smaller speed seems reasonable. Paying attention to the *signs*—which are positive and which negative—was very important for reaching a correct answer. We didn't convert the masses to kilograms because only the mass *ratio* of 0.50 was needed.

A Strategy for Conservation of Momentum Problems

PROBLEM-SOLVING STRATEGY 11.1

Conservation of momentum

MODEL Clearly define the *system*.

■ If possible, choose a system that is isolated ($\vec{F}_{net} = \vec{0}$) or within which the interactions are sufficiently short and intense that you can ignore external forces for the duration of the interaction (the impulse approximation). Momentum is conserved.

■ If it's not possible to choose an isolated system, try to divide the problem into parts such that momentum is conserved during one segment of the motion. Other segments of the motion can be analyzed using Newton's laws or conservation of energy.

VISUALIZE Draw a before-and-after pictorial representation. Define symbols that will be used in the problem, list known values, and identify what you're trying to find.

SOLVE The mathematical representation is based on the law of conservation of momentum: $\vec{P}_f = \vec{P}_i$. In component form, this is

$$(p_{fx})_1 + (p_{fx})_2 + (p_{fx})_3 + \cdots = (p_{ix})_1 + (p_{ix})_2 + (p_{ix})_3 + \cdots$$
$$(p_{fy})_1 + (p_{fy})_2 + (p_{fy})_3 + \cdots = (p_{iy})_1 + (p_{iy})_2 + (p_{iy})_3 + \cdots$$

ASSESS Check that your result has correct units and significant figures, is reasonable, and answers the question.

Exercise 17

EXAMPLE 11.3 | Rolling away

Bob sees a stationary cart 8.0 m in front of him. He decides to run to the cart as fast as he can, jump on, and roll down the street. Bob has a mass of 75 kg and the cart's mass is 25 kg. If Bob accelerates at a steady 1.0 m/s², what is the cart's speed just after Bob jumps on?

MODEL This is a two-part problem. First Bob accelerates across the ground. Then Bob lands on and sticks to the cart, a "collision" between Bob and the cart. The interaction forces between Bob and the cart (i.e., friction) act only over the fraction of a second it takes Bob's feet to become stuck to the cart. Using the impulse approximation allows the system Bob + cart to be treated as an isolated system during the brief interval of the "collision," and thus the total momentum of Bob + cart is conserved during this interaction. But the system Bob + cart is *not* an isolated system for the entire problem because Bob's initial acceleration has nothing to do with the cart.

VISUALIZE Our strategy is to divide the problem into an *acceleration* part, which we can analyze using kinematics, and a *collision* part, which we can analyze with momentum conservation. The pictorial representation of **FIGURE 11.14** includes information about both parts. Notice that Bob's velocity $(v_{1x})_B$ at the end of his run is his "before" velocity for the collision.

SOLVE The first part of the mathematical representation is kinematics. We don't know how long Bob accelerates, but we do know his acceleration and the distance. Thus

$$(v_{1x})_B^2 = (v_{0x})_B^2 + 2a_x \Delta x = 2a_x x_1$$

His velocity after accelerating for 8.0 m is

$$(v_{1x})_B = \sqrt{2a_x x_1} = 4.0 \text{ m/s}$$

The second part of the problem, the collision, uses conservation of momentum: $P_{2x} = P_{1x}$. Equation 11.24 is

$$m_B(v_{2x})_B + m_C(v_{2x})_C = m_B(v_{1x})_B + m_C(v_{1x})_C = m_B(v_{1x})_B$$

where we've used $(v_{1x})_C = 0$ m/s because the cart starts at rest. In this problem, Bob and the cart move together at the end with a common velocity, so we can replace both $(v_{2x})_B$ and $(v_{2x})_C$ with simply v_{2x}. Solving for v_{2x}, we find

$$v_{2x} = \frac{m_B}{m_B + m_C}(v_{1x})_B = \frac{75 \text{ kg}}{100 \text{ kg}} \times 4.0 \text{ m/s} = 3.0 \text{ m/s}$$

The cart's speed is 3.0 m/s immediately after Bob jumps on.

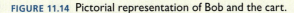

FIGURE 11.14 Pictorial representation of Bob and the cart.

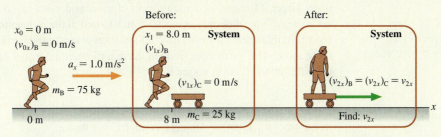

Notice how easy this was! No forces, no acceleration constraints, no simultaneous equations. Why didn't we think of this before? Conservation laws are indeed powerful, but they can answer only certain questions. Had we wanted to know how far Bob slid across the cart before sticking to it, how long the slide took, or what the cart's acceleration was during the collision, we would not have been able to answer such questions on the basis of the conservation law. There is a price to pay for finding a simple connection between before and after, and that price is the loss of information about the details of the interaction. If we are satisfied with knowing only about before and after, then conservation laws are a simple and straightforward way to proceed. But many problems *do* require us to understand the interaction, and for these there is no avoiding Newton's laws.

It Depends on the System

The first step in the problem-solving strategy asks you to clearly define the system. **The goal is to choose a system whose momentum will be conserved.** Even then, it is the *total* momentum of the system that is conserved, not the momenta of the individual objects within the system.

As an example, consider what happens if you drop a rubber ball and let it bounce off a hard floor. Is momentum conserved? You might be tempted to answer yes because the ball's rebound speed is very nearly equal to its impact speed. But there are two errors in this reasoning.

First, momentum depends on *velocity*, not speed. The ball's velocity and momentum change sign during the collision. Even if their magnitudes are equal, the ball's momentum after the collision is *not* equal to its momentum before the collision.

But more important, we haven't defined the system. The momentum of what? Whether or not momentum is conserved depends on the system. **FIGURE 11.15** shows two different choices of systems. In Figure 11.15a, where the ball itself is chosen as the system, the gravitational force of the earth on the ball is an external force. This force causes the ball to accelerate toward the earth, changing the ball's momentum. When the ball hits, the force of the floor on the ball is also an external force. The impulse of $\vec{F}_{\text{floor on ball}}$ changes the ball's momentum from "down" to "up" as the ball bounces. The momentum of this system is most definitely *not* conserved.

Figure 11.15b shows a different choice. Here the system is ball + earth. Now the gravitational forces and the impulsive forces of the collision are interactions *within* the system. This is an isolated system, so the *total* momentum $\vec{P} = \vec{p}_{\text{ball}} + \vec{p}_{\text{earth}}$ is conserved.

In fact, the total momentum (in this reference frame) is $\vec{P} = \vec{0}$ because both the ball and the earth are initially at rest. The ball accelerates toward the earth after you release it, while the earth—due to Newton's third law—accelerates toward the ball in such a way that their individual momenta are always equal but opposite.

FIGURE 11.15 Whether or not momentum is conserved as a ball falls to earth depends on your choice of the system.

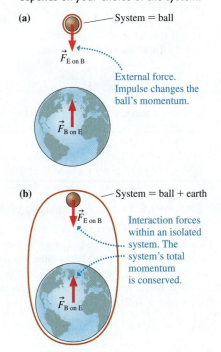

Why don't we notice the earth "leaping up" toward us each time we drop something? Because of the earth's enormous mass relative to everyday objects—roughly 10^{25} times larger. Momentum is the product of mass and velocity, so the earth would need an "upward" speed of only about 10^{-25} m/s to match the momentum of a typical falling object. At that speed, it would take 300 million years for the earth to move the diameter of an atom! The earth does, indeed, have a momentum equal and opposite to that of the ball, but we'll never notice it.

STOP TO THINK 11.3 Objects A and C are made of different materials, with different "springiness," but they have the same mass and are initially at rest. When ball B collides with object A, the ball ends up at rest. When ball B is thrown with the same speed and collides with object C, the ball rebounds to the left. Compare the velocities of A and C after the collisions. Is v_A greater than, equal to, or less than v_C?

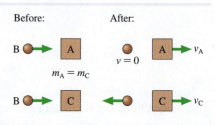

11.3 Collisions

Collisions can have different possible outcomes. A rubber ball dropped on the floor bounces, but a ball of clay sticks to the floor without bouncing. A golf club hitting a golf ball causes the ball to rebound away from the club, but a bullet striking a block of wood embeds itself in the block.

FIGURE 11.16 An inelastic collision.

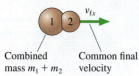

Inelastic Collisions

A collision in which the two objects stick together and move with a common final velocity is called a **perfectly inelastic collision.** The clay hitting the floor and the bullet embedding itself in the wood are examples of perfectly inelastic collisions. Other examples include railroad cars coupling together upon impact and darts hitting a dart board. As **FIGURE 11.16** shows, the key to analyzing a perfectly inelastic collision is the fact that **the two objects have a common final velocity.**

A system consisting of the two colliding objects is isolated, so its total momentum is conserved. However, mechanical energy is *not* conserved because some of the initial kinetic energy is transformed into thermal energy during the collision.

EXAMPLE 11.4 | **An inelastic glider collision**

In a laboratory experiment, a 200 g air-track glider and a 400 g air-track glider are pushed toward each other from opposite ends of the track. The gliders have Velcro tabs on the front and will stick together when they collide. The 200 g glider is pushed with an initial speed of 3.0 m/s. The collision causes it to reverse direction at 0.40 m/s. What was the initial speed of the 400 g glider?

MODEL Define the system to be the two gliders. This is an isolated system, so its total momentum is conserved in the collision. The gliders stick together, so this is a perfectly inelastic collision.

VISUALIZE FIGURE 11.17 shows a pictorial representation. We've chosen to let the 200 g glider (glider 1) start out moving to the right, so $(v_{ix})_1$ is a positive 3.0 m/s. The gliders move to the left after the collision, so their common final velocity is $v_{fx} = -0.40$ m/s.

SOLVE The law of conservation of momentum, $P_{fx} = P_{ix}$, is

$$(m_1 + m_2)v_{fx} = m_1(v_{ix})_1 + m_2(v_{ix})_2$$

where we made use of the fact that the combined mass $m_1 + m_2$ moves together after the collision. We can easily solve for the initial velocity of the 400 g glider:

$$(v_{ix})_2 = \frac{(m_1 + m_2)v_{fx} - m_1(v_{ix})_1}{m_2}$$

$$= \frac{(0.60\ \text{kg})(-0.40\ \text{m/s}) - (0.20\ \text{kg})(3.0\ \text{m/s})}{0.40\ \text{kg}} = -2.1\ \text{m/s}$$

The negative sign indicates that the 400 g glider started out moving to the left. The initial *speed* of the glider, which we were asked to find, is 2.1 m/s.

FIGURE 11.17 The before-and-after pictorial representation of an inelastic collision.

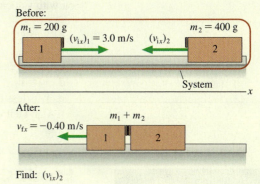

Find: $(v_{ix})_2$

EXAMPLE 11.5 | A ballistic pendulum

A 10 g bullet is fired into a 1200 g wood block hanging from a 150-cm-long string. The bullet embeds itself into the block, and the block then swings out to an angle of 40°. What was the speed of the bullet? (This is called a *ballistic pendulum.*)

MODEL This is a two-part problem. Part one, the impact of the bullet on the block, is an inelastic collision. For this part, we define the system to be bullet + block. Momentum is conserved, but mechanical energy is not because some of the energy is transformed into thermal energy. For part two, the subsequent swing, mechanical energy is conserved for the system bullet + block + earth (there's no friction). The *total* momentum is conserved, including the momentum of the earth, but that's not helpful. The momentum of the block with the bullet—which is all that we can calculate—is not conserved because the block is acted on by the external forces of tension and gravity.

VISUALIZE FIGURE 11.18 is a pictorial representation in which we've identified before-and-after quantities for both the collision and the swing.

FIGURE 11.18 A ballistic pendulum is used to measure the speed of a bullet.

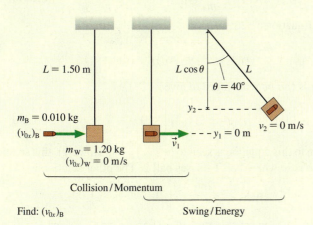

Find: $(v_{0x})_B$

SOLVE The momentum conservation equation $P_f = P_i$ applied to the inelastic collision gives

$$(m_W + m_B)v_{1x} = m_W(v_{0x})_W + m_B(v_{0x})_B$$

The wood block is initially at rest, with $(v_{0x})_W = 0$, so the bullet's velocity is

$$(v_{0x})_B = \frac{m_W + m_B}{m_B} v_{1x}$$

where v_{1x} is the velocity of the block + bullet *immediately* after the collision, as the pendulum begins to swing. If we can determine v_{1x} from an analysis of the swing, then we will be able to calculate the speed of the bullet. Turning our attention to the swing, the energy conservation equation $K_f + U_{Gf} = K_i + U_{Gi}$ is

$$\tfrac{1}{2}(m_W + m_B)v_2^2 + (m_W + m_B)gy_2 = \tfrac{1}{2}(m_W + m_B)v_1^2 + (m_W + m_B)gy_1$$

We used the *total* mass $(m_W + m_B)$ of the block and embedded bullet, but notice that it cancels out. We also dropped the x-subscript on v_1 because for energy calculations we need only speed, not velocity. The speed is zero at the top of the swing $(v_2 = 0)$, and we've defined the y-axis such that $y_1 = 0$ m. Thus

$$v_1 = \sqrt{2gy_2}$$

The initial speed is found simply from the maximum height of the swing. You can see from the geometry of Figure 11.18 that

$$y_2 = L - L\cos\theta = L(1 - \cos\theta) = 0.351\ \text{m}$$

With this, the initial velocity of the pendulum, immediately after the collision, is

$$v_{1x} = v_1 = \sqrt{2gy_2} = \sqrt{2(9.80\ \text{m/s}^2)(0.351\ \text{m})} = 2.62\ \text{m/s}$$

Having found v_{1x} from an energy analysis of the swing, we can now calculate that the speed of the bullet was

$$(v_{0x})_B = \frac{m_W + m_B}{m_B} v_{1x} = \frac{1.210\ \text{kg}}{0.010\ \text{kg}} \times 2.62\ \text{m/s} = 320\ \text{m/s}$$

ASSESS It would have been very difficult to solve this problem using Newton's laws, but it yielded to a straightforward analysis based on the concepts of momentum and energy.

STOP TO THINK 11.4 The two particles are both moving to the right. Particle 1 catches up with particle 2 and collides with it. The particles stick together and continue on with velocity v_f. Which of these statements is true?

a. v_f is greater than v_1.
b. $v_f = v_1$
c. v_f is greater than v_2 but less than v_1.
d. $v_f = v_2$
e. v_f is less than v_2.
f. Can't tell without knowing the masses.

A perfectly elastic collision conserves both momentum and mechanical energy.

FIGURE 11.19 A perfectly elastic collision.

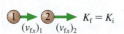

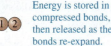

Elastic Collisions

In an inelastic collision, some of the mechanical energy is dissipated inside the objects as thermal energy and not all of the kinetic energy is recovered. We're now interested in "perfect bounce" collisions in which kinetic energy is stored as elastic potential energy in compressed molecular bonds, and then *all* of the stored energy is transformed back into the post-collision kinetic energy of the objects. A collision in which mechanical energy is conserved is called a **perfectly elastic collision.** A perfectly elastic collision is an idealization, like a frictionless surface, but collisions between two very hard objects, such as two billiard balls or two steel balls, come close to being perfectly elastic.

FIGURE 11.19 shows a head-on, perfectly elastic collision of a ball of mass m_1, having initial velocity $(v_{ix})_1$, with a ball of mass m_2 that is initially at rest. The balls' velocities after the collision are $(v_{fx})_1$ and $(v_{fx})_2$. These are velocities, not speeds, and have signs. Ball 1, in particular, might bounce backward and have a negative value for $(v_{fx})_1$.

The collision must obey two conservation laws: conservation of momentum (obeyed in any collision) and conservation of mechanical energy (because the collision is perfectly elastic). Although the energy is transformed into potential energy during the collision, the mechanical energy before and after the collision is purely kinetic energy. Thus

momentum conservation: $\quad m_1(v_{fx})_1 + m_2(v_{fx})_2 = m_1(v_{ix})_1 \quad$ (11.25)

energy conservation: $\quad \frac{1}{2}m_1(v_{fx})_1^2 + \frac{1}{2}m_2(v_{fx})_2^2 = \frac{1}{2}m_1(v_{ix})_1^2 \quad$ (11.26)

Momentum conservation alone is not sufficient to analyze the collision because there are two unknowns: the two final velocities. Energy conservation provides the additional information that we need. Isolating $(v_{fx})_1$ in Equation 11.25 gives

$$(v_{fx})_1 = (v_{ix})_1 - \frac{m_2}{m_1}(v_{fx})_2 \quad (11.27)$$

We substitute this into Equation 11.26:

$$\frac{1}{2}m_1\left[(v_{ix})_1 - \frac{m_2}{m_1}(v_{fx})_2\right]^2 + \frac{1}{2}m_2(v_{fx})_2^2 = \frac{1}{2}m_1(v_{ix})_1^2$$

With a bit of algebra, this can be rearranged to give

$$(v_{fx})_2\left[\left(1 + \frac{m_2}{m_1}\right)(v_{fx})_2 - 2(v_{ix})_1\right] = 0 \quad (11.28)$$

One possible solution to this equation is seen to be $(v_{fx})_2 = 0$. However, this solution is of no interest; it is the case where ball 1 misses ball 2. The other solution is

$$(v_{fx})_2 = \frac{2m_1}{m_1 + m_2}(v_{ix})_1$$

which, finally, can be substituted back into Equation 11.27 to yield $(v_{fx})_1$. The complete solution is

$$(v_{fx})_1 = \frac{m_1 - m_2}{m_1 + m_2}(v_{ix})_1 \qquad (v_{fx})_2 = \frac{2m_1}{m_1 + m_2}(v_{ix})_1 \qquad (11.29)$$

(perfectly elastic collision with ball 2 initially at rest)

Equations 11.29 allow us to compute the final velocity of each ball. These equations are a little difficult to interpret, so let us look at the three special cases shown in **FIGURE 11.20**.

Case a: $m_1 = m_2$. This is the case of one billiard ball striking another of equal mass. For this case, Equations 11.29 give

$$v_{f1} = 0 \qquad v_{f2} = v_{i1}$$

Case b: $m_1 \gg m_2$. This is the case of a bowling ball running into a Ping-Pong ball. We do not want an exact solution here, but an approximate solution for the limiting case that $m_1 \to \infty$. Equations 11.29 in this limit give

$$v_{f1} \approx v_{i1} \qquad v_{f2} \approx 2v_{i1}$$

Case c: $m_1 \ll m_2$. Now we have the reverse case of a Ping-Pong ball colliding with a bowling ball. Here we are interested in the limit $m_1 \to 0$, in which case Equations 11.29 become

$$v_{f1} \approx -v_{i1} \qquad v_{f2} \approx 0$$

These cases agree well with our expectations and give us confidence that Equations 11.29 accurately describe a perfectly elastic collision.

Using Reference Frames

Equations 11.29 assumed that ball 2 was at rest prior to the collision. Suppose, however, you need to analyze the perfectly elastic collision that is just about to take place in **FIGURE 11.21**. What are the direction and speed of each ball after the collision? You could solve the simultaneous momentum and energy equations, but the mathematics becomes quite messy when both balls have an initial velocity. Fortunately, there's an easier way.

You already know the answer—Equations 11.29—when ball 2 is initially at rest. And in Chapter 4 you learned the Galilean transformation of velocity. This transformation relates an object's velocity as measured in one reference frame to its velocity in a different reference frame that moves with respect to the first. The Galilean transformation provides an elegant and straightforward way to analyze the collision of Figure 11.21.

FIGURE 11.20 Three special elastic collisions.

Case a: $m_1 = m_2$

Ball 1 stops. Ball 2 goes forward with $v_{f2} = v_{i1}$.

Case b: $m_1 \gg m_2$

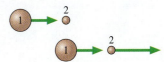

Ball 1 hardly slows down. Ball 2 is knocked forward at $v_{f2} \approx 2v_{i1}$.

Case c: $m_1 \ll m_2$

Ball 1 bounces off ball 2 with almost no loss of speed. Ball 2 hardly moves.

FIGURE 11.21 A perfectly elastic collision in which both balls have an initial velocity.

2.0 m/s   3.0 m/s

$m_1 = 200$ g $m_2 = 100$ g

TACTICS BOX 11.1	MP

Analyzing elastic collisions

❶ Use the Galilean transformation to transform the initial velocities of balls 1 and 2 from the "lab frame" to a reference frame in which ball 2 is at rest.

❷ Use Equations 11.29 to determine the outcome of the collision in the frame where ball 2 is initially at rest.

❸ Transform the final velocities back to the "lab frame."

FIGURE 11.22a on the next page shows the situation, just before the collision, in the lab frame L. Ball 1 has initial velocity $(v_{ix})_{1L} = 2.0$ m/s. Recall from Chapter 4 that the subscript notation means "velocity of ball 1 relative to the lab frame L." Because ball 2 is moving to the left, it has $(v_{ix})_{2L} = -3.0$ m/s. We would like to observe the collision from a reference frame in which ball 2 is at rest. That will be true if we choose a moving reference frame M that travels alongside ball 2 with the same velocity: $(v_x)_{ML} = -3.0$ m/s.

FIGURE 11.22 The collision seen in two reference frames: the lab frame L and a moving frame M in which ball 2 is initially at rest.

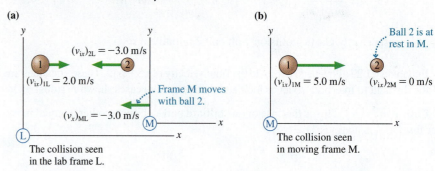

(a) The collision seen in the lab frame L.

(b) The collision seen in moving frame M.

We first need to transform the balls' velocities from the lab frame to the moving reference frame. From Chapter 4, the Galilean transformation of velocity for an object O is

$$(v_x)_{OM} = (v_x)_{OL} + (v_x)_{LM} \qquad (11.30)$$

That is, O's velocity in reference frame M is its velocity in reference frame L plus the velocity of frame L relative to frame M. Because reference frame M is moving to the left relative to L with $(v_x)_{ML} = -3.0$ m/s, reference frame L is moving to the right relative to M with $(v_x)_{LM} = +3.0$ m/s. Applying the transformation to the two initial velocities gives

$$
\begin{aligned}
(v_{ix})_{1M} &= (v_{ix})_{1L} + (v_x)_{LM} = 2.0 \text{ m/s} + 3.0 \text{ m/s} = 5.0 \text{ m/s} \\
(v_{ix})_{2M} &= (v_{ix})_{2L} + (v_x)_{LM} = -3.0 \text{ m/s} + 3.0 \text{ m/s} = 0 \text{ m/s}
\end{aligned} \qquad (11.31)
$$

$(v_{ix})_{2M} = 0$ m/s, as expected, because we chose a moving reference frame in which ball 2 would be at rest.

FIGURE 11.22b now shows a situation—with ball 2 initially at rest—in which we can use Equations 11.29 to find the post-collision velocities in frame M:

$$
\begin{aligned}
(v_{fx})_{1M} &= \frac{m_1 - m_2}{m_1 + m_2}(v_{ix})_{1M} = 1.7 \text{ m/s} \\
(v_{fx})_{2M} &= \frac{2m_1}{m_1 + m_2}(v_{ix})_{1M} = 6.7 \text{ m/s}
\end{aligned} \qquad (11.32)
$$

Reference frame M hasn't changed—it's still moving to the left in the lab frame at 3.0 m/s—but the collision has changed both balls' velocities in frame M.

To finish, we need to transform the post-collision velocities in frame M back to the lab frame L. We can do so with another application of the Galilean transformation:

$$
\begin{aligned}
(v_{fx})_{1L} &= (v_{fx})_{1M} + (v_x)_{ML} = 1.7 \text{ m/s} + (-3.0 \text{ m/s}) = -1.3 \text{ m/s} \\
(v_{fx})_{2L} &= (v_{fx})_{2M} + (v_x)_{ML} = 6.7 \text{ m/s} + (-3.0 \text{ m/s}) = 3.7 \text{ m/s}
\end{aligned} \qquad (11.33)
$$

FIGURE 11.23 shows the outcome of the collision in the lab frame. It's not hard to confirm that these final velocities do, indeed, conserve both momentum and energy.

FIGURE 11.23 The post-collision velocities in the lab frame.

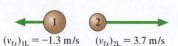

$(v_{fx})_{1L} = -1.3$ m/s $(v_{fx})_{2L} = 3.7$ m/s

Two Collision Models

No collision is perfectly elastic, although collisions between two very hard objects (metal spheres) or between two springs (such as a collision on an air track) come close. Collisions can be perfectly inelastic, although many real-world inelastic collisions exhibit a small residual bounce. Thus perfectly elastic and perfectly inelastic collisions are *models* of collisions in which we simplify reality in order to gain understanding without getting bogged down in the messy details of real collisions.

MODEL 11.1

Collisions

For two colliding objects.

- Represent the objects as elastic objects moving in a straight line.

- In a **perfectly inelastic collision,** the objects stick and move together. Kinetic energy is transformed into thermal energy. Mathematically:

Before:
After:

$$(m_1 + m_2)v_{fx} = m_1(v_{ix})_1 + m_2(v_{ix})_2$$

- In a **perfectly elastic collision,** the objects bounce apart with no loss of energy. Mathematically:

Before:
After:

- If object 2 is initially at rest, then

$$(v_{fx})_1 = \frac{m_1 - m_2}{m_1 + m_2}(v_{ix})_1 \qquad (v_{ix})_2 = \frac{2m_1}{m_1 + m_2}(v_{ix})_1$$

- If both objects are moving, use the Galilean transformation to transform the velocities to a reference frame in which object 2 is at rest.

- Limitations: Model fails if the collision is not head-on or cannot reasonably be approximated as a "thud" or as a "perfect bounce."

Exercise 22

11.4 Explosions

An **explosion,** where the particles of the system move apart from each other after a brief, intense interaction, is the opposite of a collision. The explosive forces, which could be from an expanding spring or from expanding hot gases, are *internal* forces. If the system is isolated, its total momentum during the explosion will be conserved.

EXAMPLE 11.6 | **Recoil**

A 10 g bullet is fired from a 3.0 kg rifle with a speed of 500 m/s. What is the recoil speed of the rifle?

MODEL The rifle causes a small mass of gunpowder to explode, and the expanding gas then exerts forces on *both* the bullet and the rifle. Let's define the system to be bullet + gas + rifle. The forces due to the expanding gas during the explosion are internal forces, within the system. Any friction forces between the bullet and the rifle as the bullet travels down the barrel are also internal forces. Gravity is balanced by the upward force of the person holding the rifle, so $\vec{F}_{net} = \vec{0}$. This is an isolated system and the law of conservation of momentum applies.

VISUALIZE **FIGURE 11.24** shows a pictorial representation before and after the bullet is fired.

FIGURE 11.24 Before-and-after pictorial representation of a rifle firing a bullet.

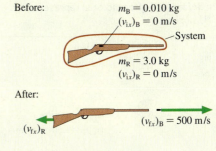

Before:
$m_B = 0.010$ kg
$(v_{ix})_B = 0$ m/s
System
$m_R = 3.0$ kg
$(v_{ix})_R = 0$ m/s

After:
$(v_{fx})_R$
$(v_{fx})_B = 500$ m/s

Find: $(v_{fx})_R$

Continued

SOLVE The x-component of the total momentum is $P_x = (p_x)_B + (p_x)_R + (p_x)_{gas}$. Everything is at rest before the trigger is pulled, so the initial momentum is zero. After the trigger is pulled, the momentum of the expanding gas is the sum of the momenta of all the molecules in the gas. For every molecule moving in the forward direction with velocity v and momentum mv there is, on average, another molecule moving in the opposite direction with velocity $-v$ and thus momentum $-mv$. When the values are summed over the enormous number of molecules in the gas, we will be left with $p_{gas} \approx 0$. In addition, the mass of the gas is much less than that of the rifle or bullet. For both reasons, we can reasonably neglect the momentum of the gas. The law of conservation of momentum is thus

$$P_{fx} = m_B(v_{fx})_B + m_R(v_{fx})_R = P_{ix} = 0$$

Solving for the rifle's velocity, we find

$$(v_{fx})_R = -\frac{m_B}{m_R}(v_{fx})_B = -\frac{0.010 \text{ kg}}{3.0 \text{ kg}} \times 500 \text{ m/s} = -1.7 \text{ m/s}$$

The minus sign indicates that the rifle's recoil is to the left. The recoil *speed* is 1.7 m/s.

We would not know where to begin to solve a problem such as this using Newton's laws. But Example 11.6 is a simple problem when approached from the before-and-after perspective of a conservation law. The selection of bullet + gas + rifle as "the system" was the critical step. For momentum conservation to be a useful principle, we had to select a system in which the complicated forces due to expanding gas and friction were all internal forces. The rifle by itself is *not* an isolated system, so its momentum is *not* conserved.

EXAMPLE 11.7 | Radioactivity

A ^{238}U uranium nucleus is radioactive. It spontaneously disintegrates into a small fragment that is ejected with a measured speed of 1.50×10^7 m/s and a "daughter nucleus" that recoils with a measured speed of 2.56×10^5 m/s. What are the atomic masses of the ejected fragment and the daughter nucleus?

MODEL The notation ^{238}U indicates the isotope of uranium with an atomic mass of 238 u, where u is the abbreviation for the *atomic mass unit*. The nucleus contains 92 protons (uranium is atomic number 92) and 146 neutrons. The disintegration of a nucleus is, in essence, an explosion. Only *internal* nuclear forces are involved, so the total momentum is conserved in the decay.

VISUALIZE FIGURE 11.25 shows the pictorial representation. The mass of the daughter nucleus is m_1 and that of the ejected fragment is m_2. Notice that we converted the speed information to velocity information, giving $(v_{fx})_1$ and $(v_{fx})_2$ opposite signs.

FIGURE 11.25 Before-and-after pictorial representation of the decay of a ^{238}U nucleus.

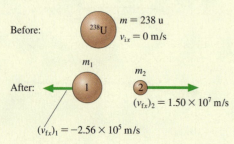

Before: ^{238}U $m = 238$ u $v_{ix} = 0$ m/s

After: m_1 1   2 m_2 $(v_{fx})_2 = 1.50 \times 10^7$ m/s

$(v_{fx})_1 = -2.56 \times 10^5$ m/s

Find: m_1 and m_2

SOLVE The nucleus was initially at rest, hence the total momentum is zero. The momentum after the decay is still zero if the two pieces fly apart in opposite directions with momenta equal in magnitude but opposite in sign. That is,

$$P_{fx} = m_1(v_{fx})_1 + m_2(v_{fx})_2 = P_{ix} = 0$$

Although we know both final velocities, this is not enough information to find the two unknown masses. However, we also have another conservation law, conservation of mass, that requires

$$m_1 + m_2 = 238 \text{ u}$$

Combining these two conservation laws gives

$$m_1(v_{fx})_1 + (238 \text{ u} - m_1)(v_{fx})_2 = 0$$

The mass of the daughter nucleus is

$$m_1 = \frac{(v_{fx})_2}{(v_{fx})_2 - (v_{fx})_1} \times 238 \text{ u}$$

$$= \frac{1.50 \times 10^7 \text{ m/s}}{(1.50 \times 10^7 - (-2.56 \times 10^5)) \text{ m/s}} \times 238 \text{ u} = 234 \text{ u}$$

With m_1 known, the mass of the ejected fragment is $m_2 = 238 - m_1 = 4$ u.

ASSESS All we learn from a momentum analysis is the masses. Chemical analysis shows that the daughter nucleus is the element thorium, atomic number 90, with two fewer protons than uranium. The ejected fragment carried away two protons as part of its mass of 4 u, so it must be a particle with two protons and two neutrons. This is the nucleus of a helium atom, ^{4}He, which in nuclear physics is called an *alpha particle* α. Thus the radioactive decay of ^{238}U can be written as ^{238}U $\rightarrow$ ^{234}Th $+ \alpha$.

Much the same reasoning explains how a rocket or jet aircraft accelerates. **FIGURE 11.26** shows a rocket with a parcel of fuel on board. Burning converts the fuel to hot gases that are expelled from the rocket motor. If we choose rocket + gases to be the system, the burning and expulsion are both internal forces. There are no other forces, so the total momentum of the rocket + gases system must be conserved. The rocket gains forward velocity and momentum as the exhaust gases are shot out the back, but the *total* momentum of the system remains zero.

Section 11.6 looks at rocket propulsion in more detail, but even without the details you should be able to understand that jet and rocket propulsion is a consequence of momentum conservation.

FIGURE 11.26 Rocket propulsion is an example of conservation of momentum.

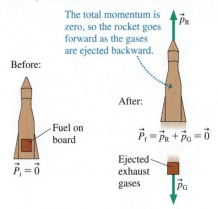

STOP TO THINK 11.5 An explosion in a rigid pipe shoots out three pieces. A 6 g piece comes out the right end. A 4 g piece comes out the left end with twice the speed of the 6 g piece. From which end, left or right, does the third piece emerge?

11.5 Momentum in Two Dimensions

The law of conservation of momentum $\vec{P}_f = \vec{P}_i$ is not restricted to motion along a line. Many interesting examples of collisions and explosions involve motion in a plane, and for these both the magnitude *and the direction* of the total momentum vector are unchanged. The total momentum is the vector sum of the individual momenta, so the total momentum is conserved only if each component is conserved:

$$(p_{fx})_1 + (p_{fx})_2 + (p_{fx})_3 + \cdots = (p_{ix})_1 + (p_{ix})_2 + (p_{ix})_3 + \cdots$$
$$(11.34)$$
$$(p_{fy})_1 + (p_{fy})_2 + (p_{fy})_3 + \cdots = (p_{iy})_1 + (p_{iy})_2 + (p_{iy})_3 + \cdots$$

Let's look at some examples of momentum conservation in two dimensions.

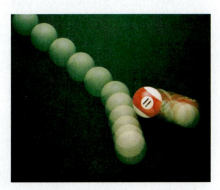

Collisions and explosions often involve motion in two dimensions.

EXAMPLE 11.8 | **A peregrine falcon strike**

Peregrine falcons often grab their prey from above while both falcon and prey are in flight. A 0.80 kg falcon, flying at 18 m/s, swoops down at a 45° angle from behind a 0.36 kg pigeon flying horizontally at 9.0 m/s. What are the speed and direction of the falcon (now holding the pigeon) immediately after impact?

MODEL The two birds, modeled as particles, are the system. This is a perfectly inelastic collision because after the collision the falcon and pigeon move at a common final velocity. The birds are not a perfectly isolated system because of external forces of the air, but during the brief collision the external impulse delivered by the air resistance will be negligible. Within this approximation, the total momentum of the falcon + pigeon system is conserved during the collision.

VISUALIZE **FIGURE 11.27** is a before-and-after pictorial representation. We've used angle ϕ to label the post-collision direction.

SOLVE The initial velocity components of the falcon are $(v_{ix})_F = v_F \cos\theta$ and $(v_{iy})_F = -v_F \sin\theta$. The pigeon's initial velocity is entirely along the x-axis. After the collision, when the falcon and pigeon have the common velocity $\vec{v}_f$, the components are $v_{fx} = v_f \cos\phi$ and $v_{fy} = -v_f \sin\phi$. Conservation of momentum in two

FIGURE 11.27 Pictorial representation of a falcon catching a pigeon.

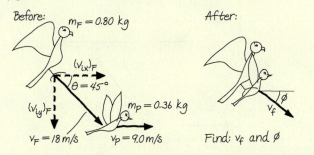

dimensions requires conservation of both the x- and y-components of momentum. This gives two conservation equations:

$$(m_F + m_P)v_{fx} = (m_F + m_P)v_f \cos\phi$$
$$= m_F(v_{ix})_F + m_P(v_{ix})_P = m_F v_F \cos\theta + m_P v_P$$
$$(m_F + m_P)v_{fy} = -(m_F + m_P)v_f \sin\phi$$
$$= m_F(v_{iy})_F + m_P(v_{iy})_P = -m_F v_F \sin\theta$$

Continued

The unknowns are v_f and ϕ. Dividing both equations by the total mass gives

$$v_f\cos\phi = \frac{m_F v_F \cos\theta + m_P v_P}{m_F + m_P} = 11.6 \text{ m/s}$$

$$v_f\sin\phi = \frac{m_F v_F \sin\theta}{m_F + m_P} = 8.78 \text{ m/s}$$

We can eliminate v_f by dividing the second equation by the first to give

$$\frac{v_f\sin\phi}{v_f\cos\phi} = \tan\phi = \frac{8.78 \text{ m/s}}{11.6 \text{ m/s}} = 0.757$$

$$\phi = \tan^{-1}(0.757) = 37°$$

Then $v_f = (11.6 \text{ m/s})/\cos(37°) = 15$ m/s. Immediately after impact, the falcon, with its meal, is traveling at 15 m/s at an angle 37° below the horizontal.

ASSESS It makes sense that the falcon would slow down after grabbing the slower-moving pigeon. And Figure 11.27 tells us that the total momentum is at an angle between 0° (the pigeon's momentum) and 45° (the falcon's momentum). Thus our answer seems reasonable.

EXAMPLE 11.9 | A three-piece explosion

A 10.0 g projectile is traveling east at 2.0 m/s when it suddenly explodes into three pieces. A 3.0 g fragment is shot due west at 10 m/s while another 3.0 g fragment travels 40° north of east at 12 m/s. What are the speed and direction of the third fragment?

MODEL Although many complex forces are involved in the explosion, they are all internal to the system. There are no external forces, so this is an isolated system and its total momentum is conserved.

VISUALIZE FIGURE 11.28 shows a before-and-after pictorial representation. We'll use uppercase M and V to distinguish the initial object from the three pieces into which it explodes.

FIGURE 11.28 Before-and-after pictorial representation of the three-piece explosion.

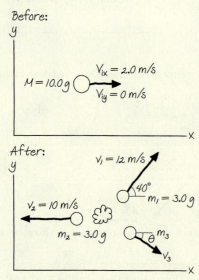

Before:

After:

Find: v_3 and θ

SOLVE The system is the initial object and the subsequent three pieces. Conservation of momentum requires

$$m_1(v_{fx})_1 + m_2(v_{fx})_2 + m_3(v_{fx})_3 = MV_{ix}$$

$$m_1(v_{fy})_1 + m_2(v_{fy})_2 + m_3(v_{fy})_3 = MV_{iy}$$

Conservation of mass implies that

$$m_3 = M - m_1 - m_2 = 4.0 \text{ g}$$

Neither the original object nor m_2 has any momentum along the y-axis. We can use Figure 11.28 to write out the x- and y-components of $\vec{v}_1$ and $\vec{v}_3$, leading to

$$m_1 v_1 \cos 40° - m_2 v_2 + m_3 v_3 \cos\theta = MV$$

$$m_1 v_1 \sin 40° - m_3 v_3 \sin\theta = 0$$

where we used $(v_{fx})_2 = -v_2$ because m_2 is moving in the negative x-direction. Inserting known values in these equations gives us

$$-2.42 + 4v_3 \cos\theta = 20$$

$$23.14 - 4v_3 \sin\theta = 0$$

We can leave the masses in grams in this situation because the conversion factor to kilograms appears on both sides of the equation and thus cancels out. To solve, first use the second equation to write $v_3 = 5.79/\sin\theta$. Substitute this result into the first equation, noting that $\cos\theta/\sin\theta = 1/\tan\theta$, to get

$$-2.42 + 4\left(\frac{5.79}{\sin\theta}\right)\cos\theta = -2.42 + \frac{23.14}{\tan\theta} = 20$$

Now solve for θ:

$$\tan\theta = \frac{23.14}{20 + 2.42} = 1.03$$

$$\theta = \tan^{-1}(1.03) = 45.8°$$

Finally, use this result in the earlier expression for v_3 to find

$$v_3 = \frac{5.79}{\sin 45.8°} = 8.1 \text{ m/s}$$

The third fragment, with a mass of 4.0 g, is shot 46° south of east at a speed of 8.1 m/s.

STOP TO THINK 11.6 An object traveling to the right with $\vec{p} = 2\,\hat{\imath}$ kg m/s suddenly explodes into two pieces. Piece 1 has the momentum $\vec{p}_1$ shown in the figure. What is the momentum $\vec{p}_2$ of the second piece?

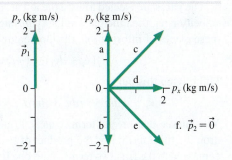

11.6 ADVANCED TOPIC Rocket Propulsion

Newton's second law $\vec{F} = m\vec{a}$ applies to objects whose mass does not change. That's an excellent assumption for balls and bicycles, but what about something like a rocket that loses a significant amount of mass as its fuel is burned? Problems of varying mass are solved with momentum rather than acceleration. We'll look at one important example.

FIGURE 11.29 shows a rocket being propelled by the thrust of burning fuel but *not* influenced by gravity or drag. Perhaps it is a rocket in deep space where gravity is very weak in comparison to the rocket's thrust. This may not be highly realistic, but ignoring gravity allows us to understand the essentials of rocket propulsion without making the mathematics too complicated. Rocket propulsion with gravity is a Challenge Problem in the end-of-chapter problems.

The system rocket + exhaust gases is an isolated system, so its total momentum is conserved. The basic idea is simple: As exhaust gases are shot out the back, the rocket "recoils" in the opposite direction. Putting this idea on a mathematical footing is fairly straightforward—it's basically the same as analyzing an explosion—but we have to be extremely careful with signs.

We'll use a before-and-after approach, as we do with all momentum problems. The Before state is a rocket of mass m (including all onboard fuel) moving with velocity v_x and having initial momentum $P_{ix} = mv_x$. During a small interval of time dt, the rocket burns a small mass of fuel m_{fuel} and expels the resulting gases from the back of the rocket at an exhaust speed v_{ex} *relative to the rocket*. That is, a space cadet on the rocket sees the gases leaving the rocket at speed v_{ex} regardless of how fast the rocket is traveling through space.

After this little packet of burned fuel has been ejected, the rocket has new velocity $v_x + dv_x$ and new mass $m + dm$. Now you're probably thinking that this can't be right; the rocket *loses* mass rather than gaining mass. But that's *our* understanding of the physical situation. The mathematical analysis knows only that the mass changes, not whether it increases or decreases. Saying that the mass is $m + dm$ at time $t + dt$ is a formal statement that the mass has changed, and that's how analysis of change is done in calculus. The fact that the rocket's mass is decreasing means that dm has a negative value. That is, the minus goes with the value of dm, not with the statement that the mass has changed.

After the gas has been ejected, both the rocket and the gas have momentum. Conservation of momentum tells us that

$$P_{fx} = m_{\text{rocket}}(v_x)_{\text{rocket}} + m_{\text{fuel}}(v_x)_{\text{fuel}} = P_{ix} = mv_x \qquad (11.35)$$

The mass of this little packet of burned fuel is the mass *lost* by the rocket: $m_{\text{fuel}} = -dm$. Mathematically, the minus sign tells us that the mass of the burned fuel (the gases) and the rocket mass are changing in opposite directions. Physically, we know that $dm < 0$, so the exhaust gases have a positive mass.

FIGURE 11.29 A before-and-after pictorial representation of a rocket burning a small amount of fuel.

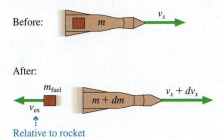

The gases are ejected toward the left at speed v_{ex} relative to the rocket. If the rocket's velocity is v_x, then the gas's velocity *through space* is $v_x - v_{ex}$. When we assemble all these pieces of information, the momentum conservation equation is

$$(m + dm)(v_x + dv_x) + (-dm)(v_x - v_{ex}) = mv_x \tag{11.36}$$

Multiplying this out gives

$$mv_x + v_x\,dm + m\,dv_x + dm\,dv_x - v_x\,dm + v_{ex}\,dm = mv_x \tag{11.37}$$

You can see that several terms cancel, leading to $m\,dv_x + v_{ex}\,dm + dm\,dv_x = 0$. We can drop the third term; it is the product of two infinitesimal terms and thus is negligible compared to the first two terms. With one final algebraic rearrangement, we're left with

$$dv_x = -v_{ex}\frac{dm}{m} \tag{11.38}$$

Remember that dm is negative—it's the mass *lost* by the rocket when a small amount of fuel is burned—and so dv_x is positive. Physically, Equation 11.38 is telling us the amount by which the rocket's velocity increases when it burns a small amount of fuel. Not surprisingly, a lighter rocket (smaller m) gains more velocity than a heavier rocket (larger m).

There are a couple of ways to use Equation 11.38. First, divide both sides by the small interval of time dt in which the fuel is burned, which will make this a rate equation:

$$\frac{dv_x}{dt} = \frac{-v_{ex}\,dm/dt}{m} = \frac{v_{ex}R}{m} \tag{11.39}$$

where $R = |dm/dt|$ is the rate—in kg/s—at which fuel is burned. The fuel burn rate is reasonably constant for most rocket engines.

The left side of Equation 11.39 is the rocket's acceleration: $a_x = dv_x/dt$. Thus from Newton's second law, $a_x = F_x/m$, the numerator on the right side of Equation 11.39 must be a force. This is the *thrust* of the rocket engine:

$$F_{thrust} = v_{ex}R \tag{11.40}$$

So Equation 11.39 is just Newton's second law, $a = F_{thrust}/m$, for the instantaneous acceleration, which will change as the rocket's mass m changes. But now we know how the thrust force is related to physical properties of the rocket engine.

Returning to Equation 11.38, we can find out how the rocket's velocity changes as fuel is burned by integrating. Suppose the rocket starts from rest ($v_x = 0$) with mass $m_0 = m_R + m_{F0}$, where m_R is the mass of the empty rocket and m_{F0} is the initial mass of the fuel. At a later time, when the mass has been reduced to m, the velocity is v.

Integrating between this Before and After, we find

$$\int_0^v dv_x = v = -v_{ex}\int_{m_0}^m \frac{dm}{m} = -v_{ex}\ln m \Big|_{m_0}^m \tag{11.41}$$

where $\ln m$ is the *natural logarithm* (logarithm with base e) of m. Evaluating this between the limits, and using the properties of logarithms, gives

$$-v_{ex}\ln m \Big|_{m_0}^m = -v_{ex}(\ln m - \ln m_0) = -v_{ex}\ln\left(\frac{m}{m_0}\right) = v_{ex}\ln\left(\frac{m_0}{m}\right) \tag{11.42}$$

Thus the rocket's velocity when its mass has decreased to m is

$$v = v_{ex}\ln\left(\frac{m_0}{m}\right) \tag{11.43}$$

Initially, when $m = m_0$, $v = 0$ because $\ln 1 = 0$. The maximum speed occurs when the fuel is completely gone and $m = m_R$. This is

$$v_{max} = v_{ex}\ln\left(\frac{m_R + m_{F0}}{m_R}\right) \tag{11.44}$$

Notice that the rocket speed can exceed v_{ex} if the fuel-mass-to-rocket-mass ratio is large enough. Also, because v_{max} is greatly improved by reducing m_R, we can see why rockets that need enough speed to go into orbit are usually *multistage rockets*, dropping off the mass of the lower stages when their fuel is depleted.

EXAMPLE 11.10 | Firing a rocket

Sounding rockets are small rockets used to gather weather data and do atmospheric research. One of the most popular sounding rockets has been the fairly small (10-in-diameter, 16-ft-long) Black Brant III. It is loaded with 210 kg of fuel, has a launch mass of 290 kg, and generates 49 kN of thrust for 9.0 s. What would be the maximum speed of a Black Brant III if launched from rest in deep space?

MODEL We define the system to be the rocket and its exhaust gases. This is an isolated system, its total momentum is conserved, and the rocket's maximum speed is given by Equation 11.44.

SOLVE We're given $m_{F0} = 210$ kg. Knowing that the launch mass is 290 kg, we can deduce that the mass of the empty rocket is $m_R = 80$ kg. Because the rocket burns 210 kg of fuel in 9.0 s, the fuel burn rate is

$$R = \frac{210 \text{ kg}}{9.0 \text{ s}} = 23.3 \text{ kg/s}$$

Knowing the burn rate and the thrust, we can use Equation 11.40 to calculate the exhaust velocity:

$$v_{ex} = \frac{F_{thrust}}{R} = \frac{49,000 \text{ N}}{23.3 \text{ kg/s}} = 2100 \text{ m/s}$$

Thus the rocket's maximum speed in deep space would be

$$v_{max} = v_{ex} \ln\left(\frac{m_R + m_{F0}}{m_R}\right) = (2100 \text{ m/s})\ln\left(\frac{290 \text{ kg}}{80 \text{ kg}}\right) = 2700 \text{ m/s}$$

ASSESS An actual sounding rocket doesn't reach this speed because it's affected both by gravity and by drag. Even so, the rocket's acceleration is so large that gravity plays a fairly minor role. A Black Brant III launched into the earth's atmosphere achieves a maximum speed of 2100 m/s and, because it continues to coast upward long after the fuel is exhausted, reaches a maximum altitude of 175 km (105 mi).

CHALLENGE EXAMPLE 11.11 | A rebounding pendulum

A 200 g steel ball hangs on a 1.0-m-long string. The ball is pulled sideways so that the string is at a 45° angle, then released. At the very bottom of its swing the ball strikes a 500 g steel paperweight that is resting on a frictionless table. To what angle does the ball rebound?

MODEL We can divide this problem into three parts. First the ball swings down as a pendulum. Second, the ball and paperweight have a collision. Steel balls bounce off each other very well, so we will model the collision as perfectly elastic. Third, the ball, after it bounces off the paperweight, swings back up as a pendulum.

For Parts 1 and 3, we define the system as the ball and the earth. This is an isolated, nondissipative system, so its mechanical energy is conserved. In Part 2, let the system consist of the ball and the paperweight, which have a perfectly elastic collision.

VISUALIZE FIGURE 11.30 shows four distinct moments of time: as the ball is released, an instant before the collision, an instant after the collision but before the ball and paperweight have had time to move, and as the ball reaches its highest point on the rebound. Call the ball A and the paperweight B, so $m_A = 0.20$ kg and $m_B = 0.50$ kg.

FIGURE 11.30 Four moments in the collision of a pendulum with a paperweight.

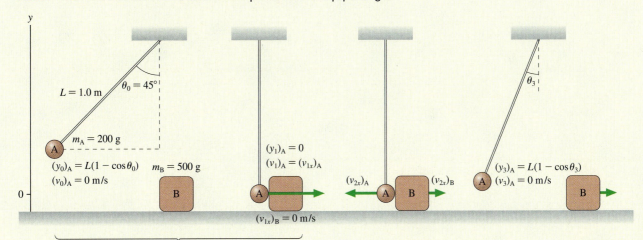

Part 1: Conservation of energy

Find: θ_3

Part 2: Perfectly elastic collision

Part 3: Conservation of energy

Continued

SOLVE Part 1: The first part involves the ball only. Its initial height is

$$(y_0)_A = L - L\cos\theta_0 = L(1 - \cos\theta_0) = 0.293 \text{ m}$$

We can use conservation of mechanical energy to find the ball's velocity at the bottom, just before impact on the paperweight:

$$\tfrac{1}{2}m_A(v_1)_A^2 + m_Ag(y_1)_A = \tfrac{1}{2}m_A(v_0)_A^2 + m_Ag(y_0)_A$$

We know $(v_0)_A = 0$. Solving for the velocity at the bottom, where $(y_1)_A = 0$, gives

$$(v_1)_A = \sqrt{2g(y_0)_A} = 2.40 \text{ m/s}$$

Part 2: The ball and paperweight undergo a perfectly elastic collision in which the paperweight is initially at rest. These are the conditions for which Equations 11.29 were derived. The velocities *immediately* after the collision, prior to any further motion, are

$$(v_{2x})_A = \frac{m_A - m_B}{m_A + m_B}(v_{1x})_A = -1.03 \text{ m/s}$$

$$(v_{2x})_B = \frac{2m_A}{m_A + m_B}(v_{1x})_A = +1.37 \text{ m/s}$$

The ball rebounds toward the left with a speed of 1.03 m/s while the paperweight moves to the right at 1.37 m/s. Kinetic energy has been conserved (you might want to check this), but it is now shared between the ball and the paperweight.

Part 3: Now the ball is a pendulum with an initial speed of 1.03 m/s. Mechanical energy is again conserved, so we can find its maximum height at the point where $(v_3)_A = 0$:

$$\tfrac{1}{2}m_A(v_3)_A^2 + m_Ag(y_3)_A = \tfrac{1}{2}m_A(v_2)_A^2 + m_Ag(y_2)_A$$

Solving for the maximum height gives

$$(y_3)_A = \frac{(v_2)_A^2}{2g} = 0.0541 \text{ m}$$

The height $(y_3)_A$ is related to angle θ_3 by $(y_3)_A = L(1 - \cos\theta_3)$. This can be solved to find the angle of rebound:

$$\theta_3 = \cos^{-1}\left(1 - \frac{(y_3)_A}{L}\right) = 19°$$

The paperweight speeds away at 1.37 m/s and the ball rebounds to an angle of 19°.

ASSESS The ball and the paperweight aren't hugely different in mass, so we expect the ball to transfer a significant fraction of its energy to the paperweight when they collide. Thus a rebound to roughly half the initial angle seems reasonable.

SUMMARY

The goals of Chapter 11 have been to learn to use the concepts of impulse and momentum.

GENERAL PRINCIPLES

Law of Conservation of Momentum

The total momentum $\vec{P} = \vec{p}_1 + \vec{p}_2 + \cdots$ of an isolated system is a constant. Thus

$$\vec{P}_f = \vec{P}_i$$

Newton's Second Law

In terms of momentum, Newton's second law is

$$\vec{F} = \frac{d\vec{p}}{dt}$$

Solving Momentum Conservation Problems

MODEL Choose an isolated system or a system that is isolated during at least part of the problem.

VISUALIZE Draw a pictorial representation of the system before and after the interaction.

SOLVE Write the law of conservation of momentum in terms of vector components:

$$(p_{fx})_1 + (p_{fx})_2 + \cdots = (p_{ix})_1 + (p_{ix})_2 + \cdots$$
$$(p_{fy})_1 + (p_{fy})_2 + \cdots = (p_{iy})_1 + (p_{iy})_2 + \cdots$$

ASSESS Is the result reasonable?

IMPORTANT CONCEPTS

Momentum $\vec{p} = m\vec{v}$

Impulse $J_x = \int_{t_i}^{t_f} F_x(t)\, dt$ = area under force curve

Impulse and momentum are related by the **momentum principle**

$$\Delta p_x = J_x$$

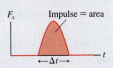

The impulse delivered to an object causes the object's momentum to change. This is an alternative statement of Newton's second law.

Momentum bar charts display the momentum principle $p_{fx} = p_{ix} + J_x$ in graphical form.

System A group of interacting particles.

Isolated system A system on which there are no external forces or the net external force is zero.

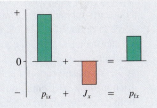

APPLICATIONS

Collisions In a **perfectly inelastic collision,** two objects stick together and move with a common final velocity. In a **perfectly elastic collision,** they bounce apart and conserve mechanical energy as well as momentum.

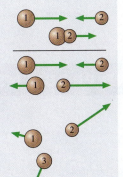

Explosions Two or more objects fly apart from each other. Their total momentum is conserved.

Two dimensions The same ideas apply in two dimensions. Both the x- and y-components of $\vec{P}$ must be conserved. This gives two simultaneous equations to solve.

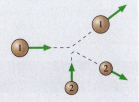

Rockets The momentum of the exhaust-gas + rocket system is conserved. **Thrust** is the product of the exhaust speed and the rate at which fuel is burned.

TERMS AND NOTATION

CONCEPTUAL QUESTIONS

1. Rank in order, from largest to smallest, the momenta $(p_x)_a$ to $(p_x)_e$ of the objects in **FIGURE Q11.1**.

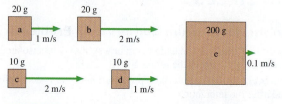

FIGURE Q11.1

2. A 2 kg object is moving to the right with a speed of 1 m/s when it experiences an impulse of 4 N s. What are the object's speed and direction after the impulse?

3. A 2 kg object is moving to the right with a speed of 1 m/s when it experiences an impulse of −4 N s. What are the object's speed and direction after the impulse?

4. A 0.2 kg plastic cart and a 20 kg lead cart can both roll without friction on a horizontal surface. Equal forces are used to push both carts forward for a time of 1 s, starting from rest. After the force is removed at $t = 1$ s, is the momentum of the plastic cart greater than, less than, or equal to the momentum of the lead cart? Explain.

5. A 0.2 kg plastic cart and a 20 kg lead cart can both roll without friction on a horizontal surface. Equal forces are used to push both carts forward for a distance of 1 m, starting from rest. After traveling 1 m, is the momentum of the plastic cart greater than, less than, or equal to the momentum of the lead cart? Explain.

6. Angie, Brad, and Carlos are discussing a physics problem in which two identical bullets are fired with equal speeds at equal-mass wood and steel blocks resting on a frictionless table. One bullet bounces off the steel block while the second becomes embedded in the wood block. "All the masses and speeds are the same," says Angie, "so I think the blocks will have equal speeds after the collisions." "But what about momentum?" asks Brad. "The bullet hitting the wood block transfers all its momentum and energy to the block, so the wood block should end up going faster

than the steel block." "I think the bounce is an important factor," replies Carlos. "The steel block will be faster because the bullet bounces off it and goes back the other direction." Which of these three do you agree with, and why?

7. It feels better to catch a hard ball while wearing a padded glove than to catch it bare handed. Use the ideas of this chapter to explain why.

8. Automobiles are designed with "crumple zones" intended to collapse in a collision. Use the ideas of this chapter to explain why.

9. A golf club continues forward after hitting the golf ball. Is momentum conserved in the collision? Explain, making sure you are careful to identify "the system."

10. Suppose a rubber ball collides head-on with a more massive steel ball traveling in the opposite direction with equal speed. Which ball, if either, receives the larger impulse? Explain.

11. Two particles collide, one of which was initially moving and the other initially at rest.
    a. Is it possible for *both* particles to be at rest after the collision? Give an example in which this happens, or explain why it can't happen.
    b. Is it possible for *one* particle to be at rest after the collision? Give an example in which this happens, or explain why it can't happen.

12. Two ice skaters, Paula and Ricardo, push off from each other. Ricardo weighs more than Paula.
    a. Which skater, if either, has the greater momentum after the push-off? Explain.
    b. Which skater, if either, has the greater speed after the push-off? Explain.

13. Two balls of clay of known masses hang from the ceiling on massless strings of equal length. They barely touch when both hang at rest. One ball is pulled back until its string is at 45°, then released. It swings down, collides with the second ball, and they stick together. To determine the angle to which the balls swing on the opposite side, would you invoke (a) conservation of momentum, (b) conservation of mechanical energy, (c) both, (d) either but not both, or (e) these laws alone are not sufficient to find the angle? Explain.

EXERCISES AND PROBLEMS

Problems labeled ▇ integrate material from earlier chapters.

Exercises

Section 11.1 Momentum and Impulse

1. | At what speed do a bicycle and its rider, with a combined mass of 100 kg, have the same momentum as a 1500 kg car traveling at 5.0 m/s?

2. | What is the magnitude of the momentum of
   a. A 3000 kg truck traveling at 15 m/s?
   b. A 200 g baseball thrown at 40 m/s?

3. ‖ What impulse does the force shown in **FIGURE EX11.3** exert on a 250 g particle?

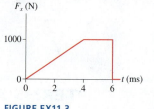

FIGURE EX11.3

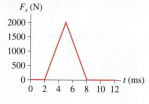

FIGURE EX11.4

4. ‖ What is the impulse on a 3.0 kg particle that experiences the force shown in **FIGURE EX11.4**?

5. ‖ In **FIGURE EX11.5**, what value of F_{max} gives an impulse of 6.0 N s?

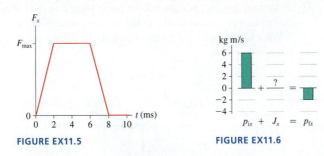

FIGURE EX11.5

FIGURE EX11.6

6. ‖ **FIGURE EX11.6** is an incomplete momentum bar chart for a collision that lasts 10 ms. What are the magnitude and direction of the average collision force exerted on the object?

7. ‖ **FIGURE EX11.7** is an incomplete momentum bar chart for a 50 g particle that experiences an impulse lasting 10 ms. What were the speed and direction of the particle before the impulse?

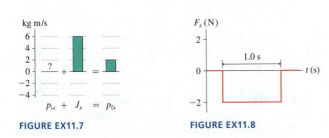

FIGURE EX11.7

FIGURE EX11.8

8. ∣ A 2.0 kg object is moving to the right with a speed of 1.0 m/s when it experiences the force shown in **FIGURE EX11.8**. What are the object's speed and direction after the force ends?

9. ∣ A 2.0 kg object is moving to the right with a speed of 1.0 m/s when it experiences the force shown in **FIGURE EX11.9**. What are the object's speed and direction after the force ends?

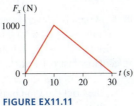

FIGURE EX11.9

10. ∣ A sled slides along a horizontal surface on which the coefficient of kinetic friction is 0.25. Its velocity at point A is 8.0 m/s and at point B is 5.0 m/s. Use the momentum principle to find how long the sled takes to travel from A to B.

11. ∣ Far in space, where gravity is negligible, a 425 kg rocket traveling at 75 m/s fires its engines. **FIGURE EX11.11** shows the thrust force as a function of time. The mass lost by the rocket during these 30 s is negligible.

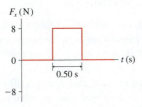

FIGURE EX11.11

   a. What impulse does the engine impart to the rocket?

   b. At what time does the rocket reach its maximum speed? What is the maximum speed?

12. ‖ A 600 g air-track glider collides with a spring at one end of the track. **FIGURE EX11.12** shows the glider's velocity and the force exerted on the glider by the spring. How long is the glider in contact with the spring?

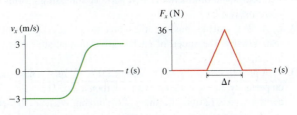

FIGURE EX11.12

13. ‖ A 250 g ball collides with a wall. **FIGURE EX11.13** shows the ball's velocity and the force exerted on the ball by the wall. What is v_{fx}, the ball's rebound velocity?

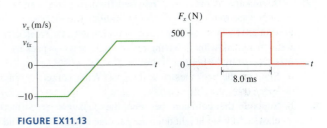

FIGURE EX11.13

Section 11.2 Conservation of Momentum

14. ∣ A 5000 kg open train car is rolling on frictionless rails at 22 m/s when it starts pouring rain. A few minutes later, the car's speed is 20 m/s. What mass of water has collected in the car?

15. ∣ A 10,000 kg railroad car is rolling at 2.0 m/s when a 4000 kg load of gravel is suddenly dropped in. What is the car's speed just after the gravel is loaded?

16. ‖ A 10-m-long glider with a mass of 680 kg (including the passengers) is gliding horizontally through the air at 30 m/s when a 60 kg skydiver drops out by releasing his grip on the glider. What is the glider's velocity just after the skydiver lets go?

17. ∣ Three identical train cars, coupled together, are rolling east at speed v_0. A fourth car traveling east at $2v_0$ catches up with the three and couples to make a four-car train. A moment later, the train cars hit a fifth car that was at rest on the tracks, and it couples to make a five-car train. What is the speed of the five-car train?

Section 11.3 Collisions

18. ∣ A 300 g bird flying along at 6.0 m/s sees a 10 g insect heading straight toward it at a speed of 30 m/s. The bird opens its mouth wide and enjoys a nice lunch. What is the bird's speed immediately after swallowing?

19. ∣ The parking brake on a 2000 kg Cadillac has failed, and it is rolling slowly, at 1.0 mph, toward a group of small children. Seeing the situation, you realize you have just enough time to drive your 1000 kg Volkswagen head-on into the Cadillac and save the children. With what speed should you impact the Cadillac to bring it to a halt?

20. ∣ A 1500 kg car is rolling at 2.0 m/s. You would like to stop the car by firing a 10 kg blob of sticky clay at it. How fast should you fire the clay?

21. ‖ Fred (mass 60 kg) is running with the football at a speed of 6.0 m/s when he is met head-on by Brutus (mass 120 kg), who is moving at 4.0 m/s. Brutus grabs Fred in a tight grip, and they fall to the ground. Which way do they slide, and how far? The coefficient of kinetic friction between football uniforms and Astroturf is 0.30.

22. ‖ A 50 g marble moving at 2.0 m/s strikes a 20 g marble at rest. What is the speed of each marble immediately after the collision?

23. ‖ A proton is traveling to the right at 2.0×10^7 m/s. It has a head-on perfectly elastic collision with a carbon atom. The mass of the carbon atom is 12 times the mass of the proton. What are the speed and direction of each after the collision?

24. ‖ A 50 g ball of clay traveling at speed v_0 hits and sticks to a 1.0 kg brick sitting at rest on a frictionless surface.
   a. What is the speed of the brick after the collision?
   b. What percentage of the mechanical energy is lost in this collision?

25. ‖ A package of mass m is released from rest at a warehouse loading dock and slides down the 3.0-m-high, frictionless chute of **FIGURE EX11.25** to a waiting truck. Unfortunately, the truck driver went on a break without having removed the previous package, of mass $2m$, from the bottom of the chute.
   a. Suppose the packages stick together. What is their common speed after the collision?
   b. Suppose the collision between the packages is perfectly elastic. To what height does the package of mass m rebound?

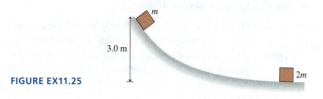

FIGURE EX11.25

Section 11.4 Explosions

26. ‖ A 50 kg archer, standing on frictionless ice, shoots a 100 g arrow at a speed of 100 m/s. What is the recoil speed of the archer?

27. ‖ A 70.0 kg football player is gliding across very smooth ice at 2.00 m/s. He throws a 0.450 kg football straight forward. What is the player's speed afterward if the ball is thrown at
   a. 15.0 m/s relative to the ground?
   b. 15.0 m/s relative to the player?

28. ‖ Dan is gliding on his skateboard at 4.0 m/s. He suddenly jumps backward off the skateboard, kicking the skateboard forward at 8.0 m/s. How fast is Dan going as his feet hit the ground? Dan's mass is 50 kg and the skateboard's mass is 5.0 kg.

29. ‖ Two ice skaters, with masses of 50 kg and 75 kg, are at the center of a 60-m-diameter circular rink. The skaters push off against each other and glide to opposite edges of the rink. If the heavier skater reaches the edge in 20 s, how long does the lighter skater take to reach the edge?

30. ‖ A ball of mass m and another ball of mass $3m$ are placed inside a smooth metal tube with a massless spring compressed between them. When the spring is released, the heavier ball flies out of one end of the tube with speed v_0. With what speed does the lighter ball emerge from the other end?

Section 11.5 Momentum in Two Dimensions

31. ‖ Two particles collide and bounce apart. **FIGURE EX11.31** shows the initial momenta of both and the final momentum of particle 2. What is the final momentum of particle 1? Write your answer using unit vectors.

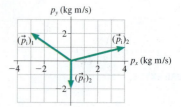

FIGURE EX11.31

32. ‖ An object at rest explodes into three fragments. **FIGURE EX11.32** shows the momentum vectors of two of the fragments. What is the momentum of the third fragment? Write your answer using unit vectors.

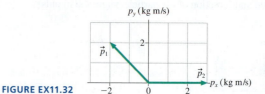

FIGURE EX11.32

33. ‖ A 20 g ball of clay traveling east at 3.0 m/s collides with a 30 g ball of clay traveling north at 2.0 m/s. What are the speed and the direction of the resulting 50 g ball of clay? Give your answer as an angle north of east.

34. ‖ At the center of a 50-m-diameter circular ice rink, a 75 kg skater traveling north at 2.5 m/s collides with and holds on to a 60 kg skater who had been heading west at 3.5 m/s.
   a. How long will it take them to glide to the edge of the rink?
   b. Where will they reach it? Give your answer as an angle north of west.

Section 11.6 Rocket Propulsion

35. ‖ A small rocket with 15 kN thrust burns 250 kg of fuel in 30 s. What is the exhaust speed of the hot gases?

36. ‖ A rocket in deep space has an empty mass of 150 kg and exhausts the hot gases of burned fuel at 2500 m/s. What mass of fuel is needed to reach a top speed of 4000 m/s?

37. ‖ A rocket in deep space has an exhaust-gas speed of 2000 m/s. When the rocket is fully loaded, the mass of the fuel is five times the mass of the empty rocket. What is the rocket's speed when half the fuel has been burned?

Problems

38. ‖ A tennis player swings her 1000 g racket with a speed of 10 m/s. She hits a 60 g tennis ball that was approaching her at a speed of 20 m/s. The ball rebounds at 40 m/s.
   a. How fast is her racket moving immediately after the impact? You can ignore the interaction of the racket with her hand for the brief duration of the collision.
   b. If the tennis ball and racket are in contact for 10 ms, what is the average force that the racket exerts on the ball? How does this compare to the gravitational force on the ball?

39. ‖ A 60 g tennis ball with an initial speed of 32 m/s hits a wall and rebounds with the same speed. **FIGURE P11.39** shows the force of the wall on the ball during the collision. What is the value of F_{max}, the maximum value of the contact force during the collision?

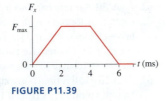

FIGURE P11.39

40. ‖ A 500 g cart is released from rest 1.00 m from the bottom of a frictionless, 30.0° ramp. The cart rolls down the ramp and bounces off a rubber block at the bottom. **FIGURE P11.40** shows the force during the collision. After the cart bounces, how far does it roll back up the ramp?

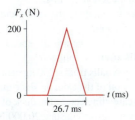

FIGURE P11.40

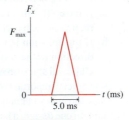

FIGURE P11.41

41. ‖‖ A 200 g ball is dropped from a height of 2.0 m, bounces on a hard floor, and rebounds to a height of 1.5 m. **FIGURE P11.41** shows the impulse received from the floor. What maximum force does the floor exert on the ball?

42. ‖ The flowers of the bunchberry plant open with astonishing force and speed, causing the pollen grains to be ejected out of the flower in a mere 0.30 ms at an acceleration of 2.5×10^4 m/s². If the acceleration is constant, what impulse is delivered to a pollen grain with a mass of 1.0×10^{-7} g?
BIO

43. ‖ A particle of mass m is at rest at $t = 0$. Its momentum for $t > 0$ is given by $p_x = 6t^2$ kg m/s, where t is in s. Find an expression for $F_x(t)$, the force exerted on the particle as a function of time.
CALC

44. ‖ Air-track gliders with masses 300 g, 400 g, and 200 g are lined up and held in place with lightweight springs compressed between them. All three are released at once. The 200 g glider flies off to the right while the 300 g glider goes left. Their position-versus-time graphs, as measured by motion detectors, are shown in **FIGURE P11.44**. What are the direction (right or left) and speed of the 400 g glider that was in the middle?

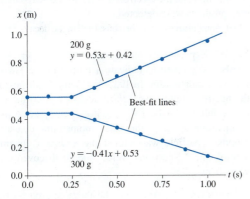

FIGURE P11.44

45. ‖ Most geologists believe that the dinosaurs became extinct 65 million years ago when a large comet or asteroid struck the earth, throwing up so much dust that the sun was blocked out for a period of many months. Suppose an asteroid with a diameter of 2.0 km and a mass of 1.0×10^{13} kg hits the earth (6.0×10^{24} kg) with an impact speed of 4.0×10^4 m/s.
   a. What is the earth's recoil speed after such a collision? (Use a reference frame in which the earth was initially at rest.)
   b. What percentage is this of the earth's speed around the sun? The earth orbits the sun at a distance of 1.5×10^{11} m.

46. ‖ Squids rely on jet propulsion to move around. A 1.50 kg squid (including the mass of water inside the squid) drifting at 0.40 m/s suddenly ejects 0.10 kg of water to get itself moving at 2.50 m/s. If drag is ignored over the small interval of time needed to expel the water (the impulse approximation), what is the water's ejection speed relative to the squid?
BIO

47. ‖ A firecracker in a coconut blows the coconut into three pieces. Two pieces of equal mass fly off south and west, perpendicular to each other, at speed v_0. The third piece has twice the mass as the other two. What are the speed and direction of the third piece? Give the direction as an angle east of north.

48. ‖ One billiard ball is shot east at 2.0 m/s. A second, identical billiard ball is shot west at 1.0 m/s. The balls have a glancing collision, not a head-on collision, deflecting the second ball by 90° and sending it north at 1.41 m/s. What are the speed and direction of the first ball after the collision? Give the direction as an angle south of east.

49. ‖ a. A bullet of mass m is fired into a block of mass M that is at rest. The block, with the bullet embedded, slides distance d across a horizontal surface. The coefficient of kinetic friction is μ_k. Find an expression for the bullet's speed v_{bullet}.
   b. What is the speed of a 10 g bullet that, when fired into a 10 kg stationary wood block, causes the block to slide 5.0 cm across a wood table?

50. ‖ You are part of a search-and-rescue mission that has been called out to look for a lost explorer. You've found the missing explorer, but, as **FIGURE P11.50** shows, you're separated from him by a 200-m-high cliff and a 30-m-wide raging river. To save his life, you need to get a 5.0 kg package of emergency supplies across the river. Unfortunately, you can't throw the package hard enough to make it across. Fortunately, you happen to have a 1.0 kg rocket intended for launching flares. Improvising quickly, you attach a sharpened stick to the front of the rocket, so that it will impale itself into the package of supplies, then fire the rocket at ground level toward the supplies. What minimum speed must the rocket have just before impact in order to save the explorer's life?

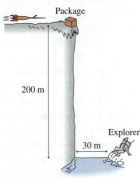

FIGURE P11.50

51. ‖ An object at rest on a flat, horizontal surface explodes into two fragments, one seven times as massive as the other. The heavier fragment slides 8.2 m before stopping. How far does the lighter fragment slide? Assume that both fragments have the same coefficient of kinetic friction.

52. ‖ A 1500 kg weather rocket accelerates upward at 10 m/s². It explodes 2.0 s after liftoff and breaks into two fragments, one twice as massive as the other. Photos reveal that the lighter fragment traveled straight up and reached a maximum height of 530 m. What were the speed and direction of the heavier fragment just after the explosion?

53. ‖ In a ballistics test, a 25 g bullet traveling horizontally at 1200 m/s goes through a 30-cm-thick 350 kg stationary target and emerges with a speed of 900 m/s. The target is free to slide on a smooth horizontal surface. What is the target's speed just after the bullet emerges?

54. ‖ Two 500 g blocks of wood are 2.0 m apart on a frictionless table. A 10 g bullet is fired at 400 m/s toward the blocks. It passes all the way through the first block, then embeds itself in the second block. The speed of the first block immediately afterward is 6.0 m/s. What is the speed of the second block after the bullet stops in it?

55. ‖‖‖ A 100 g granite cube slides down a 40° frictionless ramp. At the bottom, just as it exits onto a horizontal table, it collides with a 200 g steel cube at rest. How high above the table should the granite cube be released to give the steel cube a speed of 150 cm/s?

56. ‖‖‖ You have been asked to design a "ballistic spring system" to measure the speed of bullets. A spring whose spring constant is k is suspended from the ceiling. A block of mass M hangs from the spring. A bullet of mass m is fired vertically upward into the bottom of the block and stops in the block. The spring's maximum compression d is measured.
   a. Find an expression for the bullet's speed v_B in terms of m, M, k, and d.
   b. What was the speed of a 10 g bullet if the block's mass is 2.0 kg and if the spring, with $k = 50$ N/m, was compressed by 45 cm?

57. ‖ In FIGURE P11.57, a block of mass m slides along a frictionless track with speed v_m. It collides with a stationary block of mass M. Find an expression for the minimum value of v_m that will allow the second block to circle the loop-the-loop without falling off if the collision is (a) perfectly inelastic or (b) perfectly elastic.

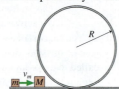

FIGURE P11.57

58. ‖ The stoplight had just changed and a 2000 kg Cadillac had entered the intersection, heading north at 3.0 m/s, when it was struck by a 1000 kg eastbound Volkswagen. The cars stuck together and slid to a halt, leaving skid marks angled 35° north of east. How fast was the Volkswagen going just before the impact?

59. ‖ Ann (mass 50 kg) is standing at the left end of a 15-m-long, 500 kg cart that has frictionless wheels and rolls on a frictionless track. Initially both Ann and the cart are at rest. Suddenly, Ann starts running along the cart at a speed of 5.0 m/s relative to the cart. How far will Ann have run *relative to the ground* when she reaches the right end of the cart?

60. ‖‖‖ Force $F_x = (10 \text{ N}) \sin(2\pi t/4.0 \text{ s})$ is exerted on a 250 g parti-
CALC cle during the interval $0 \text{ s} \leq t \leq 2.0 \text{ s}$. If the particle starts from rest, what is its speed at $t = 2.0$ s?

61. ‖‖‖ A 500 g particle has velocity $v_x = -5.0$ m/s at $t = -2$ s. Force
CALC $F_x = (4 - t^2)$ N, where t is in s, is exerted on the particle between $t = -2$ s and $t = 2$ s. This force increases from 0 N at $t = -2$ s to 4 N at $t = 0$ s and then back to 0 N at $t = 2$ s. What is the particle's velocity at $t = 2$ s?

62. ‖ A 30 ton rail car and a 90 ton rail car, initially at rest, are connected together with a giant but massless compressed spring between them. When released, the 30 ton car is pushed away at a speed of 4.0 m/s relative to the 90 ton car. What is the speed of the 30 ton car relative to the ground?

63. ‖‖‖ A 20 g ball is fired horizontally with speed v_0 toward a 100 g ball hanging motionless from a 1.0-m-long string. The balls undergo a head-on, perfectly elastic collision, after which the 100 g ball swings out to a maximum angle $\theta_{max} = 50°$. What was v_0?

64. ‖ A 100 g ball moving to the right at 4.0 m/s catches up and collides with a 400 g ball that is moving to the right at 1.0 m/s. If the collision is perfectly elastic, what are the speed and direction of each ball after the collision?

65. ‖ A 100 g ball moving to the right at 4.0 m/s collides head-on with a 200 g ball that is moving to the left at 3.0 m/s.
   a. If the collision is perfectly elastic, what are the speed and direction of each ball after the collision?
   b. If the collision is perfectly inelastic, what are the speed and direction of the combined balls after the collision?

66. ‖ Old naval ships fired 10 kg cannon balls from a 200 kg cannon. It was very important to stop the recoil of the cannon, since otherwise the heavy cannon would go careening across the deck of the ship. In one design, a large spring with spring constant 20,000 N/m was placed behind the cannon. The other end of the spring braced against a post that was firmly anchored to the ship's frame. What was the speed of the cannon ball if the spring compressed 50 cm when the cannon was fired?

67. ‖ A proton (mass 1 u) is shot toward an unknown target nucleus at a speed of 2.50×10^6 m/s. The proton rebounds with its speed reduced by 25% while the target nucleus acquires a speed of 3.12×10^5 m/s. What is the mass, in atomic mass units, of the target nucleus?

68. ‖ The nucleus of the polonium isotope ^{214}Po (mass 214 u) is radioactive and decays by emitting an alpha particle (a helium nucleus with mass 4 u). Laboratory experiments measure the speed of the alpha particle to be 1.92×10^7 m/s. Assuming the polonium nucleus was initially at rest, what is the recoil speed of the nucleus that remains after the decay?

69. ‖ A neutron is an electrically neutral subatomic particle with a mass just slightly greater than that of a proton. A free neutron is radioactive and decays after a few minutes into other subatomic particles. In one experiment, a neutron at rest was observed to decay into a proton (mass 1.67×10^{-27} kg) and an electron (mass 9.11×10^{-31} kg). The proton and electron were shot out back-to-back. The proton speed was measured to be 1.0×10^5 m/s and the electron speed was 3.0×10^7 m/s. No other decay products were detected.
   a. Did momentum seem to be conserved in the decay of this neutron?

   NOTE Experiments such as this were first performed in the 1930s and seemed to indicate a failure of the law of conservation of momentum. In 1933, Wolfgang Pauli postulated that the neutron might have a *third* decay product that is virtually impossible to detect. Even so, it can carry away just enough momentum to keep the total momentum conserved. This proposed particle was named the *neutrino,* meaning "little neutral one." Neutrinos were, indeed, discovered nearly 20 years later.

   b. If a neutrino was emitted in the above neutron decay, in which direction did it travel? Explain your reasoning.
   c. How much momentum did this neutrino "carry away" with it?

70. ‖ A 20 g ball of clay traveling east at 2.0 m/s collides with a 30 g ball of clay traveling 30° south of west at 1.0 m/s. What are the speed and direction of the resulting 50 g blob of clay?

71. ‖ **FIGURE P11.71** shows a collision between three balls of clay. The three hit simultaneously and stick together. What are the speed and direction of the resulting blob of clay?

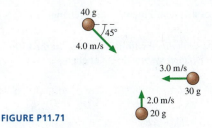

40 g

45°

4.0 m/s

3.0 m/s

30 g

2.0 m/s

20 g

FIGURE P11.71

72. ‖ A 2100 kg truck is traveling east through an intersection at 2.0 m/s when it is hit simultaneously from the side and the rear. (Some people have all the luck!) One car is a 1200 kg compact traveling north at 5.0 m/s. The other is a 1500 kg midsize traveling east at 10 m/s. The three vehicles become entangled and slide as one body. What are their speed and direction just after the collision?

73. ‖ A rocket in deep space has an empty mass of 150 kg and exhausts the hot gases of burned fuel at 2500 m/s. It is loaded with 600 kg of fuel, which it burns in 30 s. What is the rocket's speed 10 s, 20 s, and 30 s after launch?

74. ‖ a. To understand why rockets often have multiple stages, first consider a single-stage rocket with an empty mass of 200 kg, 800 kg of fuel, and a 2000 m/s exhaust speed. If fired in deep space, what is the rocket's maximum speed?

   b. Now divide the rocket into two stages, each with an empty mass of 100 kg, 400 kg of fuel, and a 2000 m/s exhaust speed. The first stage is released after it runs out of fuel. What is the top speed of the second stage? You'll need to consider how the equation for v_{max} should be altered when a rocket is not starting from rest.

In Problems 75 through 78 you are given the equation(s) used to solve a problem. For each of these, you are to

   a. Write a realistic problem for which this is the correct equation(s).
   b. Finish the solution of the problem, including a pictorial representation.

75. $(0.10 \text{ kg})(40 \text{ m/s}) - (0.10 \text{ kg})(-30 \text{ m/s}) = \frac{1}{2}(1400 \text{ N})\Delta t$

76. $(600 \text{ g})(4.0 \text{ m/s}) = (400 \text{ g})(3.0 \text{ m/s}) + (200 \text{ g})(v_{ix})_2$

77. $(3000 \text{ kg})v_{fx} = (2000 \text{ kg})(5.0 \text{ m/s}) + (1000 \text{ kg})(-4.0 \text{ m/s})$

78. $(0.10 \text{ kg} + 0.20 \text{ kg})v_{1x} = (0.10 \text{ kg})(3.0 \text{ m/s})$
$$\frac{1}{2}(0.30 \text{ kg})(0 \text{ m/s})^2 + \frac{1}{2}(3.0 \text{ N/m})(\Delta x_2)^2$$
$$= \frac{1}{2}(0.30 \text{ kg})(v_{1x})^2 + \frac{1}{2}(3.0 \text{ N/m})(0 \text{ m})^2$$

Challenge Problems

79. ‖‖‖ A 1000 kg cart is rolling to the right at 5.0 m/s. A 70 kg man is standing on the right end of the cart. What is the speed of the cart if the man suddenly starts running to the left with a speed of 10 m/s relative to the cart?

80. ‖‖‖ A spaceship of mass 2.0×10^6 kg is cruising at a speed of 5.0×10^6 m/s when the antimatter reactor fails, blowing the ship into three pieces. One section, having a mass of 5.0×10^5 kg, is blown straight backward with a speed of 2.0×10^6 m/s. A second piece, with mass 8.0×10^5 kg, continues forward at 1.0×10^6 m/s. What are the direction and speed of the third piece?

81. ‖‖‖ A 20 kg wood ball hangs from a 2.0-m-long wire. The maximum tension the wire can withstand without breaking is 400 N. A 1.0 kg projectile traveling horizontally hits and embeds itself in the wood ball. What is the greatest speed this projectile can have without causing the wire to break?

82. ‖‖‖ A two-stage rocket is traveling at 1200 m/s with respect to the earth when the first stage runs out of fuel. Explosive bolts release the first stage and push it backward with a speed of 35 m/s relative to the second stage. The first stage is three times as massive as the second stage. What is the speed of the second stage after the separation?

83. ‖‖‖ The air-track carts in **FIGURE CP11.83** are sliding to the right at 1.0 m/s. The spring between them has a spring constant of 120 N/m and is compressed 4.0 cm. The carts slide past a flame that burns through the string holding them together. Afterward, what are the speed and direction of each cart?

String holding carts together

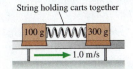

100 g   300 g

1.0 m/s

FIGURE P11.83

84. ‖‖‖ Section 11.6 found an equation for v_{max} of a rocket fired in deep
CALC space. What is v_{max} for a rocket fired vertically from the surface of an airless planet with free-fall acceleration g? Referring to Section 11.6, you can write an equation for ΔP_y, the change of momentum in the vertical direction, in terms of dm and dv_y. ΔP_y is no longer zero because now gravity delivers an impulse. Rewrite the momentum equation by including the impulse due to gravity during the time dt during which the mass changes by dm. Pay attention to signs! Your equation will have three differentials, but two are related through the fuel burn rate R. Use this relationship—again pay attention to signs; m is decreasing—to write your equation in terms of dm and dv_y. Then integrate to find a modified expression for v_{max} at the instant all the fuel has been burned.

   a. What is v_{max} for a vertical launch from an airless planet? Your answer will be in terms of m_R, the empty rocket mass; m_{F0}, the initial fuel mass; v_{ex}, the exhaust speed; R, the fuel burn rate; and g.

   b. A rocket with a total mass of 330,000 kg when fully loaded burns all 280,000 kg of fuel in 250 s. The engines generate 4.1 MN of thrust. What is this rocket's speed at the instant all the fuel has been burned if it is launched in deep space? If it is launched vertically from the earth?

Conservation Laws

KEY FINDINGS   What are the overarching findings of Part II?

- **Work** causes a system's energy to change. It is a *transfer* of energy to or from the environment.

- **Energy** can be *transformed* within a system, but the **total energy** of an isolated system does not change.

- An **impulse** causes a system's momentum to change.

- Momentum can be *exchanged* within a system, but the **total momentum** of an isolated system does not change.

LAWS   What laws of physics govern energy and momentum?

Energy principle

$$\Delta E_{sys} = \Delta K + \Delta U + \Delta E_{th} = W_{ext} \quad \text{or} \quad K_i + U_i + W_{ext} = K_f + U_f + \Delta E_{th}$$

Conservation of energy

For an isolated system ($W_{ext} = 0$), the total energy $E_{sys} = K + U + E_{th}$ is conserved. $\Delta E_{sys} = 0$.
For an isolated, nondissipative system, the **mechanical energy** $E_{mech} = K + U$ is conserved.

Momentum principle

$$\Delta \vec{p} = \vec{J}$$

Conservation of momentum

For an isolated system ($\vec{J} = \vec{0}$), the total momentum is conserved. $\Delta \vec{P} = \vec{0}$.

MODELS   What are the most common models for using conservation laws?

Basic energy model

- Energy is a property of the system.
- Energy is *transformed* within the system without loss.
- Energy is *transferred* to and from the system by forces that do *work*.
  - $W > 0$ for energy added.
  - $W < 0$ for energy removed.
  - $\Delta E_{sys} = \Delta K + \Delta U + \Delta E_{th} = W_{ext}$

Collision model

- Perfectly inelastic collision: Objects stick together and move with a common final velocity.
  - Momentum is conserved.
  - $(m_1 + m_2)v_{fx} = m_1(v_{ix})_1 + m_2(v_{ix})_2$

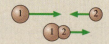

- Perfectly elastic collision: Objects bounce apart with no loss of energy.
  - Momentum and energy are both conserved.

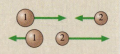

Other models and approximations

- An isolated system does not interact with its environment.
  - For energy, an isolated system has no work done on it.
  - For momentum, an isolated system experiences no impulse.
- Thermal energy is the microscopic energy of moving atoms and stretched bonds.

- An ideal spring obeys **Hooke's law** for all displacements: $(F_{Sp})_s = -k\,\Delta s$.

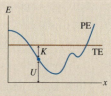

- The impulse approximation ignores forces that are small compared to impulsive forces during the brief time of a collision or explosion.

TOOLS   What are the most important tools for using energy and momentum?

- Kinetic energy:
  $K = \frac{1}{2}mv^2$
  - Kinetic energy is an energy of *motion*.

- Gravitational potential energy:
  $U_G = mgy$

- Elastic potential energy:
  $U_{Sp} = \frac{1}{2}k(\Delta s)^2$

  - Potential energy is an energy of *position*.

- Momentum: $\vec{p} = m\vec{v}$.

- Force acting through a displacement does work.
  - Constant force:
    $W = \vec{F} \cdot \Delta \vec{r} = F(\Delta r)\cos\theta$
  - Variable force:
    $$W = \int_{s_i}^{s_f} F_s\,ds$$
    = area under the F_s-versus-s curve

- Force acting over time delivers an impulse:
  $$J_x = \int_{t_i}^{t_f} F_x\,dt$$
  = area under the F_x-versus-t curve

- Energy diagrams
  - Visualize speed changes and turning points.
    F_x = negative of the slope of the curve.

- Bar charts
  - Momentum and energy conservation.

- Before-and-after pictorial representation.
  - Important problem-solving tool.

$p_{ix} + J_x = p_{fx}$

OVERVIEW

Power Over Our Environment

Early humans had to endure whatever nature provided. Only within the last few thousand years have agriculture and technology provided some level of control over the environment. And it has been a mere couple of centuries since machines, and later electronics, began to do much of our work and provide us with "creature comforts."

It's no coincidence that machines began to appear about a century after Galileo, Newton, and others ignited what we now call the *scientific revolution*. The machines and other devices we take for granted today are direct consequences of scientific knowledge and the scientific method.

Parts I and II have established Newton's theory of motion, the foundation of modern science. Most of the applications will be developed in other science and engineering courses, but we're now in a good position to examine a few of the more practical aspects of Newtonian mechanics.

Our goal for Part III is to apply our newfound theory to three important topics:

- **Rotation.** Rotation is a very important form of motion, but to understand rotational motion we'll need to introduce a new model—the *rigid-body model*. We'll then be able to study rolling wheels and spinning space stations. Rotation will also lead to the law of conservation of angular momentum.
- **Gravity.** By adding one more law, Newton's law of gravity, we'll be able to understand much about the physics of the space station, communication satellites, the solar system, and interplanetary travel.
- **Fluids.** Liquids and gases *flow*. Surprisingly, it takes no new physics to understand the basic mechanical properties of fluids. By applying our understanding of force, we'll be able to understand what pressure is, how a steel ship can float, and how fluids flow through pipes.

Newton's laws of motion and the conservation laws, especially conservation of energy, will be the tools that allow us to analyze and understand a variety of interesting and practical applications.

A hurricane is a fluid—the air—moving on a rotating sphere—the earth—under the influence of gravity. Understanding hurricanes is very much an application of Newtonian mechanics.

12 Rotation of a Rigid Body

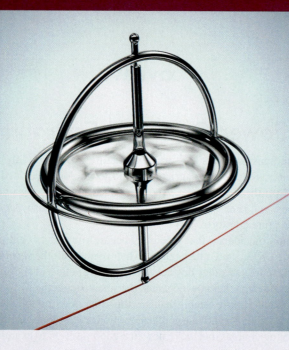

Not all motion can be described as that of a particle. Rotation requires the idea of an extended object.

IN THIS CHAPTER, you will learn to understand and apply the physics of rotation.

What is a rigid body?

An object whose size and shape don't change as it moves is called a rigid body. A rigid body is characterized by its moment of inertia I, which is the rotational equivalent of mass. We'll consider

- Rotation about an axle.
- Rolling without slipping.

« LOOKING BACK Section 6.1 Equilibrium

What is torque?

Torque is the tendency or ability of a force to rotate an object about a pivot point. You'll learn that torque depends on both the force *and* where the force is applied. A longer wrench provides a larger torque.

What does torque do?

Torque is to rotation what force is to linear motion. Torque τ causes an object to have angular acceleration. Newton's second law for rotation is $\alpha = \tau/I$. Much of rotational dynamics will look familiar because it is analogous to linear dynamics.

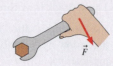

« LOOKING BACK Section 6.2 Newton's second law

What is angular momentum?

Angular momentum is to rotation what momentum is to linear motion. Angular momentum is an object's tendency to "keep rotating." Angular momentum $\vec{L}$ is a *vector* pointing along the rotation axis. You'll use angular momentum to understand the precession of a spinning top or gyroscope.

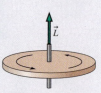

What is conserved in rotational motion?

The mechanical energy of a rotating object includes its rotational kinetic energy $\frac{1}{2}I\omega^2$. This is analogous to linear kinetic energy.

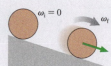

- Mechanical energy is conserved for frictionless, rotating systems.
- Angular momentum is conserved for isolated systems.

« LOOKING BACK Section 10.4 Conservation of energy; Section 11.2 Conservation of momentum

Why is rigid-body motion important?

The world is full of rotating objects, from windmill turbines to the gyroscopes used in navigation. The wheels on your bicycle or car roll without slipping. Scientists investigate rotating molecules and rotating galaxies. No understanding of motion is complete without understanding rotational motion, and this chapter will develop the tools you need. We'll also expand our understanding of equilibrium by exploring the conditions under which objects *don't* rotate.

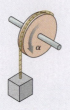

12.1 Rotational Motion

Thus far, our study of motion has focused almost exclusively on the *particle model,* in which an object is represented as a mass at a single point in space. As we expand our study of motion to rotation, we need to consider *extended objects* whose size and shape *do* matter. Thus this chapter will be based on the **rigid-body model:**

MODEL 12.1

Rigid-body model

A **rigid body** is an extended object whose size and shape do not change as it moves.

- Particle-like atoms are held together by rigid massless rods.
- A rigid body cannot be stretched, compressed, or deformed. All points on the body have the same angular velocity and angular acceleration.
- Limitations: Model fails if an object changes shape or is deformed.

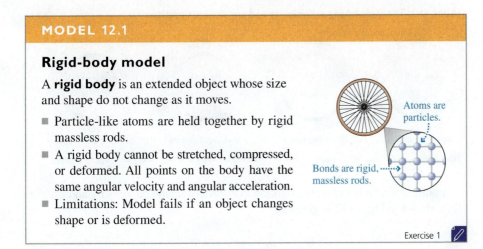

Atoms are particles.

Bonds are rigid, massless rods.

Exercise 1

FIGURE 12.1 illustrates the three basic types of motion of a rigid body: **translational motion, rotational motion,** and **combination motion.**

FIGURE 12.1 Three basic types of motion of a rigid body.

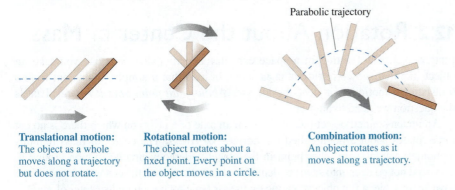

Parabolic trajectory

Translational motion:
The object as a whole moves along a trajectory but does not rotate.

Rotational motion:
The object rotates about a fixed point. Every point on the object moves in a circle.

Combination motion:
An object rotates as it moves along a trajectory.

Brief Review of Rotational Kinematics

Rotation is an extension of circular motion, so we begin with a brief summary of Chapter 4. **A review of « Sections 4.4–4.6 is highly recommended.** **FIGURE 12.2** shows a wheel rotating on an axle. Its angular velocity

$$\omega = \frac{d\theta}{dt} \tag{12.1}$$

is the rate at which the wheel rotates. The SI units of ω are radians per second (rad/s), but revolutions per second (rev/s) and revolutions per minute (rpm) are frequently used. Notice that all points have equal angular velocities, so we can refer to the angular velocity ω *of the wheel.*

If the wheel is speeding up or slowing down, its angular acceleration is

$$\alpha = \frac{d\omega}{dt} \tag{12.2}$$

The units of angular acceleration are rad/s². Angular acceleration is the *rate* at which the angular velocity ω changes, just as the linear acceleration is the rate at which the

FIGURE 12.2 Rotational motion.

Every point on the wheel rotates with the *same* angular velocity ω.

All points have a tangential velocity and a radial (centripetal) acceleration. They have a tangential acceleration if the wheel has angular acceleration.

ω v_t a_t a_r Axle

TABLE 12.1 Rotational kinematics for constant angular acceleration

$$\omega_f = \omega_i + \alpha\,\Delta t$$

$$\theta_f = \theta_i + \omega_i\,\Delta t + \tfrac{1}{2}\alpha(\Delta t)^2$$

$$\omega_f^2 = \omega_i^2 + 2\alpha\,\Delta\theta$$

linear velocity v changes. **TABLE 12.1** summarizes the kinematic equations for rotation with constant angular acceleration.

FIGURE 12.3 reminds you of the sign conventions for angular velocity and acceleration. They will be especially important in the present chapter. Be careful with the sign of α. Just as with linear acceleration, positive and negative values of α can't be interpreted as simply "speeding up" and "slowing down."

FIGURE 12.3 The signs of angular velocity and angular acceleration.

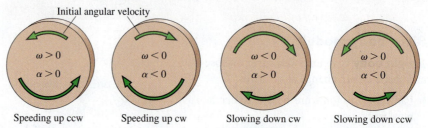

Initial angular velocity

| $\omega > 0$ $\alpha > 0$ | $\omega < 0$ $\alpha < 0$ | $\omega < 0$ $\alpha > 0$ | $\omega > 0$ $\alpha < 0$ |
| Speeding up ccw | Speeding up cw | Slowing down cw | Slowing down ccw |

The rotation is speeding up if ω and α have the same sign, slowing down if they have opposite signs.

A point at distance r from the rotation axis has instantaneous velocity and acceleration, shown in Figure 12.2, given by

$$v_r = 0 \qquad a_r = \frac{v_t^2}{r} = \omega^2 r$$

$$v_t = r\omega \qquad a_t = r\alpha \tag{12.3}$$

The sign convention for ω implies that v_t and a_t are positive if they point in the counterclockwise (ccw) direction, negative if they point in the clockwise (cw) direction.

12.2 Rotation About the Center of Mass

Imagine yourself floating in a space capsule deep in space. Suppose you take an object like that shown in **FIGURE 12.4a** and spin it so that it simply rotates but has no translational motion as it floats beside you. *About what point does it rotate?* That is the question we need to answer.

An unconstrained object (i.e., one not on an axle or a pivot) on which there is no net force rotates about a point called the **center of mass.** The center of mass remains motionless while every other point in the object undergoes circular motion around it. You need not go deep into space to demonstrate rotation about the center of mass. If you have an air table, a flat object rotating on the air table rotates about its center of mass.

To locate the center of mass, **FIGURE 12.4b** models the object as a set of particles numbered $i = 1, 2, 3, \ldots$. Particle i has mass m_i and is located at position (x_i, y_i). We'll prove later in this section that the center of mass is located at position

FIGURE 12.4 Rotation about the center of mass.

(a)

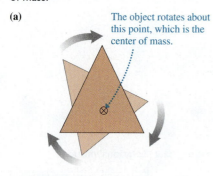

The object rotates about this point, which is the center of mass.

(b)

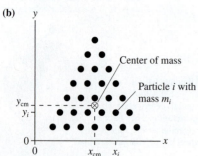

Center of mass

Particle i with mass m_i

$$x_{cm} = \frac{1}{M}\sum_i m_i x_i = \frac{m_1 x_1 + m_2 x_2 + m_3 x_3 + \cdots}{m_1 + m_2 + m_3 + \cdots}$$

$$y_{cm} = \frac{1}{M}\sum_i m_i y_i = \frac{m_1 y_1 + m_2 y_2 + m_3 y_3 + \cdots}{m_1 + m_2 + m_3 + \cdots} \tag{12.4}$$

where $M = m_1 + m_2 + m_3 + \cdots$ is the object's total mass.

Let's see if Equations 12.4 make sense. Suppose you have an object consisting of N particles, all with the same mass m. That is, $m_1 = m_2 = \cdots = m_N = m$. We can factor the m out of the numerator, and the denominator becomes simply Nm. The m cancels, and the x-coordinate of the center of mass is

$$x_{cm} = \frac{x_1 + x_2 + \cdots + x_N}{N} = x_{average}$$

In this case, x_{cm} is simply the *average* x-coordinate of all the particles. Likewise, y_{cm} will be the average of all the y-coordinates.

This *does* make sense! If the particle masses are all the same, the center of mass should be at the center of the object. And the "center of the object" is the average position of all the particles. To allow for *unequal* masses, Equations 12.4 are called *weighted averages.* Particles of higher mass count more than particles of lower mass, but the basic idea remains the same. **The center of mass is the mass-weighted center of the object.**

EXAMPLE 12.1 | **The center of mass of a barbell**

A barbell consists of a 500 g ball and a 2.0 kg ball connected by a massless 50-cm-long rod.

a. Where is the center of mass?

b. What is the speed of each ball if they rotate about the center of mass at 40 rpm?

MODEL Model the barbell as a rigid body.

VISUALIZE FIGURE 12.5 shows the two masses. We've chosen a coordinate system in which the masses are on the x-axis with the 2.0 kg mass at the origin.

SOLVE a. We can use Equations 12.4 to calculate that the center of mass is

$$x_{cm} = \frac{m_1 x_1 + m_2 x_2}{m_1 + m_2}$$

$$= \frac{(2.0 \text{ kg})(0.0 \text{ m}) + (0.50 \text{ kg})(0.50 \text{ m})}{2.0 \text{ kg} + 0.50 \text{ kg}} = 0.10 \text{ m}$$

$y_{cm} = 0$ because all the masses are on the x-axis. The center of mass is 20% of the way from the 2.0 kg ball to the 0.50 kg ball.

FIGURE 12.5 Finding the center of mass.

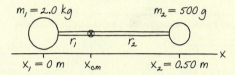

b. Each ball rotates about the center of mass. The radii of the circles are $r_1 = 0.10$ m and $r_2 = 0.40$ m. The tangential velocities are $(v_i)_t = r_i\omega$, but this equation requires ω to be in rad/s. The conversion is

$$\omega = 40 \frac{\text{rev}}{\text{min}} \times \frac{1 \text{ min}}{60 \text{ s}} \times \frac{2\pi \text{ rad}}{1 \text{ rev}} = 4.19 \text{ rad/s}$$

Consequently,

$$(v_1)_t = r_1\omega = (0.10 \text{ m})(4.19 \text{ rad/s}) = 0.42 \text{ m/s}$$
$$(v_2)_t = r_2\omega = (0.40 \text{ m})(4.19 \text{ rad/s}) = 1.68 \text{ m/s}$$

ASSESS The center of mass is closer to the heavier ball than to the lighter ball. We expected this because x_{cm} is a mass-weighted average of the positions. But the lighter mass moves faster because it is farther from the rotation axis.

Finding the Center of Mass by Integration

For any realistic object, carrying out the summations of Equations 12.4 over all the atoms in the object is not practical. Instead, as **FIGURE 12.6** shows, we can divide an extended object into many small cells or boxes, each with the same very small mass Δm. We will number the cells 1, 2, 3, ... , just as we did the particles. Cell i has coordinates (x_i, y_i) and mass $m_i = \Delta m$. The center-of-mass coordinates are then

$$x_{cm} = \frac{1}{M} \sum_i x_i \, \Delta m \quad \text{and} \quad y_{cm} = \frac{1}{M} \sum_i y_i \, \Delta m$$

Now, as you might expect, we'll let the cells become smaller and smaller, with the total number increasing. As each cell becomes infinitesimally small, we can replace Δm with dm and the sum by an integral. Then

$$x_{cm} = \frac{1}{M} \int x \, dm \quad \text{and} \quad y_{cm} = \frac{1}{M} \int y \, dm \qquad (12.5)$$

Equations 12.5 are a formal definition of the center of mass, but they are *not* ready to integrate in this form. First, integrals are carried out over *coordinates,* not over masses. Before we can integrate, we must replace dm by an equivalent expression involving a coordinate differential such as dx or dy. Second, no limits of integration have been specified. The procedure for using Equations 12.5 is best shown with an example.

FIGURE 12.6 Calculating the center of mass of an extended object.

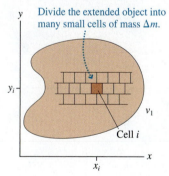

EXAMPLE 12.2 | The center of mass of a rod

Find the center of mass of a thin, uniform rod of length L and mass M. Use this result to find the tangential acceleration of one tip of a 1.60-m-long rod that rotates about its center of mass with an angular acceleration of 6.0 rad/s^2.

VISUALIZE **FIGURE 12.7** shows the rod. We've chosen a coordinate system such that the rod lies along the x-axis from 0 to L. Because the rod is "thin," we'll assume that $y_{cm} = 0$.

FIGURE 12.7 Finding the center of mass of a long, thin rod.

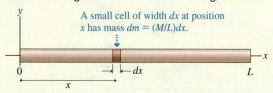

A small cell of width dx at position x has mass $dm = (M/L)dx$.

SOLVE Our first task is to find x_{cm}, which lies somewhere on the x-axis. To do this, we divide the rod into many small cells of mass dm. One such cell, at position x, is shown. The cell's width is dx. Because the rod is *uniform,* the mass of this little cell is the *same fraction* of the total mass M that dx is of the total length L. That is,

$$\frac{dm}{M} = \frac{dx}{L}$$

Consequently, we can express dm in terms of the coordinate differential dx as

$$dm = \frac{M}{L}\,dx$$

NOTE The change of variables from dm to the differential of a coordinate is *the* key step in calculating the center of mass.

With this expression for dm, Equation 12.5 for x_{cm} becomes

$$x_{cm} = \frac{1}{M}\left(\frac{M}{L}\int x\,dx\right) = \frac{1}{L}\int_0^L x\,dx$$

where in the last step we've noted that summing "all the mass in the rod" means integrating from $x = 0$ to $x = L$. This is a straightforward integral to carry out, giving

$$x_{cm} = \frac{1}{L}\left[\frac{x^2}{2}\right]_0^L = \frac{1}{L}\left[\frac{L^2}{2} - 0\right] = \tfrac{1}{2}L$$

The center of mass is at the center of the rod, as you probably guessed. For a 1.60-m-long rod, each tip of the rod rotates in a circle with $r = \tfrac{1}{2}L = 0.80$ m. The tangential acceleration, the rate at which the tip is speeding up, is

$$a_t = r\alpha = (0.80\text{ m})(6.0\text{ rad/s}^2) = 4.8\text{ m/s}^2$$

NOTE For any symmetrical object of uniform density, the center of mass is at the physical center of the object.

FIGURE 12.8 Finding the center of mass.

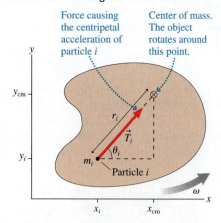

Force causing the centripetal acceleration of particle i

Center of mass. The object rotates around this point.

To see where the center-of-mass equations come from, **FIGURE 12.8** shows an object rotating about its center of mass. Particle i is moving in a circle, so it *must* have a centripetal acceleration. Acceleration requires a force, and this force is due to tension in the molecular bonds that hold the object together. Force $\vec{T}_i$ on particle i has magnitude

$$T_i = m_i(a_i)_r = m_i r_i \omega^2 \tag{12.6}$$

where we used Equation 12.3 for a_r. All points in a rigid rotating object have the *same* angular velocity, so ω doesn't need a subscript.

At every instant of time, the internal tension forces are all paired as action/reaction forces, equal in magnitude but opposite in direction, so the sum of all the tension forces must be zero. That is, $\sum \vec{T}_i = \vec{0}$. The x-component of this sum is

$$\sum_i (T_i)_x = \sum_i T_i \cos\theta_i = \sum_i (m_i r_i \omega^2)\cos\theta_i = 0 \tag{12.7}$$

You can see from Figure 12.8 that $\cos\theta_i = (x_{cm} - x_i)/r_i$. Thus

$$\sum_i (T_i)_x = \sum_i (m_i r_i \omega^2)\frac{x_{cm} - x_i}{r_i} = \left(\sum_i m_i x_{cm} - \sum_i m_i x_i\right)\omega^2 = 0 \tag{12.8}$$

This equation will be true if the term in parentheses is zero. x_{cm} is a constant, so we can bring it outside the summation to write

$$\sum_i m_i x_{cm} - \sum_i m_i x_i = \left(\sum_i m_i\right) x_{cm} - \sum_i m_i x_i = M x_{cm} - \sum_i m_i x_i = 0 \tag{12.9}$$

where we used the fact that $\sum m_i$ is simply the object's total mass M. Solving for x_{cm}, we find the x-coordinate of the object's center of mass to be

$$x_{cm} = \frac{1}{M}\sum_i m_i x_i = \frac{m_1 x_1 + m_2 x_2 + m_3 x_3 + \cdots}{m_1 + m_2 + m_3 + \cdots} \tag{12.10}$$

This was Equation 12.4. The y-equation is found similarly.

STOP TO THINK 12.1 A baseball bat is cut into two pieces at its center of mass. Which end is heavier?

a. The handle end (left end).
b. The hitting end (right end).
c. The two ends weigh the same.

12.3 Rotational Energy

A rotating rigid body—whether it's rotating freely about its center of mass or constrained to rotate on an axle—has kinetic energy because all atoms in the object are in motion. The kinetic energy due to rotation is called **rotational kinetic energy.**

FIGURE 12.9 shows a few of the particles making up a solid object that rotates with angular velocity ω. Particle i, which rotates in a circle of radius r_i, moves with speed $v_i = r_i\omega$. The object's rotational kinetic energy is the sum of the kinetic energies of each of the particles:

$$K_{rot} = \tfrac{1}{2}m_1v_1^2 + \tfrac{1}{2}m_2v_2^2 + \cdots \tag{12.11}$$

$$= \tfrac{1}{2}m_1r_1^2\omega^2 + \tfrac{1}{2}m_2r_2^2\omega^2 + \cdots = \tfrac{1}{2}\left(\sum_i m_ir_i^2\right)\omega^2$$

The quantity $\sum m_ir_i^2$ is called the object's **moment of inertia** I:

$$I = m_1r_1^2 + m_2r_2^2 + m_3r_3^2 + \cdots = \sum_i m_ir_i^2 \tag{12.12}$$

The units of moment of inertia are kg m^2. **An object's moment of inertia depends on the axis of rotation.** Once the axis is specified, allowing the values of r_i to be determined, the moment of inertia *about that axis* can be calculated from Equation 12.12.

Written using the moment of inertia I, the rotational kinetic energy is

$$K_{rot} = \tfrac{1}{2}I\omega^2 \tag{12.13}$$

NOTE Rotational kinetic energy is *not* a new form of energy. It is the familiar kinetic energy of motion, simply expressed in a form that is convenient for rotational motion. Notice the analogy with the familiar $\tfrac{1}{2}mv^2$.

FIGURE 12.9 Rotational kinetic energy is due to the motion of the particles.

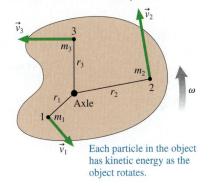

Each particle in the object has kinetic energy as the object rotates.

EXAMPLE 12.3 | A rotating widget

Students participating in an engineering project design the triangular widget seen in FIGURE 12.10. The three masses, held together by lightweight plastic rods, rotate in the plane of the page about an axle passing through the right-angle corner. At what angular velocity does the widget have 100 mJ of rotational energy?

FIGURE 12.10 The rotating widget.

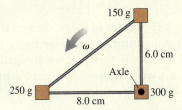

MODEL Model the widget as a rigid body consisting of three particles connected by massless rods.

SOLVE Rotational energy is $K = \tfrac{1}{2}I\omega^2$. The moment of inertia is measured about the rotation axis, thus

$$I = \sum_i m_ir_i^2 = (0.25\text{ kg})(0.080\text{ m})^2 + (0.15\text{ kg})(0.060\text{ m})^2$$
$$+ (0.30\text{ kg})(0\text{ m})^2$$
$$= 2.14 \times 10^{-3}\text{ kg m}^2$$

The largest mass makes no contribution to I because it is on the rotation axis with $r = 0$. With I known, the angular velocity is

$$\omega = \sqrt{\frac{2K}{I}} = \sqrt{\frac{2(0.10\text{ J})}{2.14 \times 10^{-3}\text{ kg m}^2}}$$

$$= 9.67\text{ rad/s} \times \frac{1\text{ rev}}{2\pi\text{ rad}} = 1.54\text{ rev/s} = 92\text{ rpm}$$

ASSESS The moment of inertia depends on the distance of each mass from the rotation axis. The moment of inertia would be different for an axle passing through either of the other two masses, and thus the required angular velocity would be different.

FIGURE 12.11 Moment of inertia depends on both the mass and how the mass is distributed.

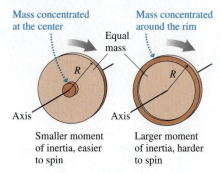

Before rushing to calculate moments of inertia, let's get a better understanding of the meaning. First, notice that **moment of inertia is the rotational equivalent of mass**. It plays the same role in Equation 12.13 as mass m in the now-familiar $K = \frac{1}{2}mv^2$. Recall that the quantity we call *mass* was actually defined as the *inertial mass*. Objects with larger mass have a larger *inertia,* meaning that they're harder to accelerate. Similarly, an object with a larger moment of inertia is harder to rotate. The fact that *moment of inertia* retains the word "inertia" reminds us of this.

Consider the two wheels shown in **FIGURE 12.11**. They have the same total mass M and the same radius R. As you probably know from experience, it's much easier to spin the wheel whose mass is concentrated at the center than to spin the one whose mass is concentrated around the rim. This is because having the mass near the center (smaller values of r_i) lowers the moment of inertia.

Moments of inertia for many solid objects are tabulated and found online. You would need to compute I yourself only for an object of unusual shape. **TABLE 12.2** is a short list of common moments of inertia. We'll see in the next section where these come from, but do notice how I depends on the rotation axis.

TABLE 12.2 Moments of inertia of objects with uniform density

Object and axis	Picture	I	Object and axis	Picture	I
Thin rod, about center		$\frac{1}{12}ML^2$	Cylinder or disk, about center		$\frac{1}{2}MR^2$
Thin rod, about end		$\frac{1}{3}ML^2$	Cylindrical hoop, about center		MR^2
Plane or slab, about center		$\frac{1}{12}Ma^2$	Solid sphere, about diameter		$\frac{2}{5}MR^2$
Plane or slab, about edge		$\frac{1}{3}Ma^2$	Spherical shell, about diameter		$\frac{2}{3}MR^2$

If the rotation axis is not through the center of mass, then rotation may cause the center of mass to move up or down. In that case, the object's gravitational potential energy $U_G = Mgy_{cm}$ will change. If there are no dissipative forces (i.e., if the axle is frictionless) and if no work is done by external forces, then the mechanical energy

$$E_{mech} = K_{rot} + U_G = \tfrac{1}{2}I\omega^2 + Mgy_{cm} \qquad (12.14)$$

is a conserved quantity.

EXAMPLE 12.4 | **The speed of a rotating rod**

A 1.0-m-long, 200 g rod is hinged at one end and connected to a wall. It is held out horizontally, then released. What is the speed of the tip of the rod as it hits the wall?

MODEL The mechanical energy is conserved if we assume the hinge is frictionless. The rod's gravitational potential energy is transformed into rotational kinetic energy as it "falls."

FIGURE 12.12 A before-and-after pictorial representation of the rod.

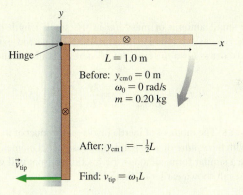

VISUALIZE FIGURE 12.12 is a familiar before-and-after pictorial representation of the rod.

SOLVE Mechanical energy is conserved, so we can equate the rod's final mechanical energy to its initial mechanical energy:

$$\tfrac{1}{2}I\omega_1^2 + Mgy_{cm1} = \tfrac{1}{2}I\omega_0^2 + Mgy_{cm0}$$

The initial conditions are $\omega_0 = 0$ and $y_{cm0} = 0$. The center of mass moves to $y_{cm1} = -\tfrac{1}{2}L$ as the rod hits the wall. From Table 12.2 we find $I = \tfrac{1}{3}ML^2$ for a rod rotating about one end. Thus

$$\tfrac{1}{2}I\omega_1^2 + Mgy_{cm1} = \tfrac{1}{6}ML^2\omega_1^2 - \tfrac{1}{2}MgL = 0$$

We can solve this for the rod's angular velocity as it hits the wall:

$$\omega_1 = \sqrt{\frac{3g}{L}}$$

The tip of the rod is moving in a circle with radius $r = L$. Its final speed is

$$v_{tip} = \omega_1 L = \sqrt{3gL} = 5.4 \text{ m/s}$$

ASSESS 5.4 m/s = 12 mph, which seems plausible for a meter-long rod swinging through 90°.

STOP TO THINK 12.2 A solid cylinder and a cylindrical shell, each with radius R and mass M, rotate about their axes with the same angular velocity ω. Which has more kinetic energy?

a. The solid cylinder.
b. The cylindrical shell.
c. They have the same kinetic energy.
d. Neither has kinetic energy because they are only rotating, not moving.

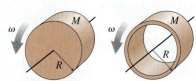

12.4 Calculating Moment of Inertia

The equation for rotational energy is easy to write, but we can't make use of it without knowing an object's moment of inertia. Unlike mass, we can't measure moment of inertia by putting an object on a scale. And while we can guess that the center of mass of a symmetrical object is at the physical center of the object, we can *not* guess the moment of inertia of even a simple object. To find I, we really must carry through the calculation.

Equation 12.12 defines the moment of inertia as a sum over all the particles in the system. As we did for the center of mass, we can replace the individual particles with cells 1, 2, 3, ... of mass Δm. Then the moment of inertia summation can be converted to an integration:

$$I = \sum_i r_i^2\, \Delta m \xrightarrow[\Delta m \to 0]{} I = \int r^2\, dm \qquad (12.15)$$

where r is the distance from the rotation axis. If we let the rotation axis be the z-axis, then we can write the moment of inertia as

$$I = \int (x^2 + y^2)\, dm \qquad \text{(rotation about the } z\text{-axis)} \qquad (12.16)$$

NOTE You *must* replace dm by an equivalent expression involving a coordinate differential such as dx or dy before you can carry out the integration.

You can use any coordinate system to calculate the coordinates x_{cm} and y_{cm} of the center of mass. But the moment of inertia is defined for rotation about a particular axis, and r is measured from that axis. Thus the coordinate system used for moment-of-inertia calculations *must* have its origin at the pivot point.

EXAMPLE 12.5 | Moment of inertia of a rod about a pivot at one end

Find the moment of inertia of a thin, uniform rod of length L and mass M that rotates about a pivot at one end.

MODEL An object's moment of inertia depends on the axis of rotation. In this case, the rotation axis is at the end of the rod.

VISUALIZE FIGURE 12.13 defines an x-axis with the origin at the pivot point.

FIGURE 12.13 Setting up the integral to find the moment of inertia of a rod.

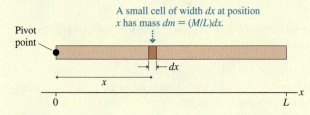

A small cell of width dx at position x has mass $dm = (M/L)dx$.

Pivot point

x

0 L

SOLVE Because the rod is thin, we can assume that $y \approx 0$ for all points on the rod. Thus

$$I = \int x^2 \, dm$$

The small amount of mass dm in the small length dx is $dm = (M/L) \, dx$, as we found in Example 12.2. The rod extends from $x = 0$ to $x = L$, so the moment of inertia about one end is

$$I_{end} = \frac{M}{L} \int_0^L x^2 \, dx = \frac{M}{L} \left[\frac{x^3}{3} \right]_0^L = \tfrac{1}{3} ML^2$$

ASSESS The moment of inertia involves a product of the total mass M with the *square* of a length, in this case L. All moments of inertia have a similar form, although the fraction in front will vary. This is the result shown earlier in Table 12.2.

EXAMPLE 12.6 | Moment of inertia of a circular disk about an axis through the center

Find the moment of inertia of a circular disk of radius R and mass M that rotates on an axis passing through its center.

VISUALIZE FIGURE 12.14 shows the disk and defines distance r from the axis.

FIGURE 12.14 Setting up the integral to find the moment of inertia of a disk.

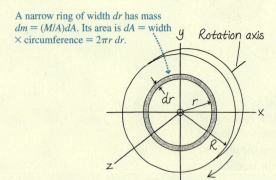

A narrow ring of width dr has mass $dm = (M/A)dA$. Its area is $dA =$ width $\times$ circumference $= 2\pi r \, dr$.

y Rotation axis

dr r

x

R

z

SOLVE This is a situation of great practical importance. To solve this problem, we need to use a two-dimensional integration scheme that you learned in calculus. Rather than dividing the disk into little boxes, let's divide it into narrow *rings* of mass dm. Figure 12.14 shows one such ring, of radius r and width dr. Let dA represent the area of this ring. The mass dm in this ring is the same fraction of the total mass M as dA is of the total area A. That is,

$$\frac{dm}{M} = \frac{dA}{A}$$

Thus the mass in the small area dA is

$$dm = \frac{M}{A} \, dA$$

This is the reasoning we used to find the center of mass of the rod in Example 12.2, only now we're using it in two dimensions.

The total area of the disk is $A = \pi R^2$, but what is dA? If we imagine unrolling the little ring, it would form a long, thin rectangle of length $2\pi r$ and height dr. Thus the *area* of this little ring is $dA = 2\pi r \, dr$. With this information we can write

$$dm = \frac{M}{\pi R^2} (2\pi r \, dr) = \frac{2M}{R^2} r \, dr$$

Now we have an expression for dm in terms of a coordinate differential dr, so we can proceed to carry out the integration for I. Using Equation 12.15, we find

$$I_{disk} = \int r^2 \, dm = \int r^2 \left(\frac{2M}{R^2} r \, dr \right) = \frac{2M}{R^2} \int_0^R r^3 \, dr$$

where in the last step we have used the fact that the disk extends from $r = 0$ to $r = R$. Performing the integration gives

$$I_{disk} = \frac{2M}{R^2} \left[\frac{r^4}{4} \right]_0^R = \tfrac{1}{2} MR^2$$

ASSESS Once again, the moment of inertia involves a product of the total mass M with the *square* of a length, in this case R.

If a complex object can be divided into simpler pieces 1, 2, 3, ... whose moments of inertia $I_1, I_2, I_3 \ldots$ are already known, the moment of inertia of the entire object is

$$I_{object} = I_1 + I_2 + I_3 + \cdots \qquad (12.17)$$

The Parallel-Axis Theorem

The moment of inertia depends on the rotation axis. Suppose you need to know the moment of inertia for rotation about the off-center axis in **FIGURE 12.15**. You can find this quite easily if you know the moment of inertia for rotation around a *parallel axis* through the center of mass.

If the axis of interest is distance d from a parallel axis through the center of mass, the moment of inertia is

$$I = I_{cm} + Md^2 \qquad (12.18)$$

Equation 12.18 is called the **parallel-axis theorem.** We'll give a proof for the one-dimensional object shown in **FIGURE 12.16**.

The x-axis has its origin at the rotation axis, and the x'-axis has its origin at the center of mass. You can see that the coordinates of dm along these two axes are related by $x = x' + d$. By definition, the moment of inertia about the rotation axis is

$$I = \int x^2 \, dm = \int (x' + d)^2 \, dm = \int (x')^2 \, dm + 2d \int x' \, dm + d^2 \int dm \qquad (12.19)$$

The first of the three integrals on the right, by definition, is the moment of inertia I_{cm} about the center of mass. The third is simply Md^2 because adding up (integrating) all the dm gives the total mass M.

If you refer back to Equations 12.5, the definition of the center of mass, you'll see that the middle integral on the right is equal to Mx'_{cm}. But $x'_{cm} = 0$ because we specifically chose the x'-axis to have its origin at the center of mass. Thus the second integral is zero and we end up with Equation 12.18. The proof in two dimensions is similar.

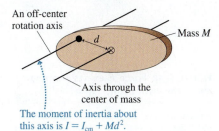

FIGURE 12.15 An off-center axis.

An off-center rotation axis

Mass M

Axis through the center of mass

The moment of inertia about this axis is $I = I_{cm} + Md^2$.

FIGURE 12.16 Proving the parallel-axis theorem.

Use this axis for calculating I about the pivot.

Pivot $x = x' + d$

$\otimes$ cm

d x'

Use this axis for calculating I_{cm}.

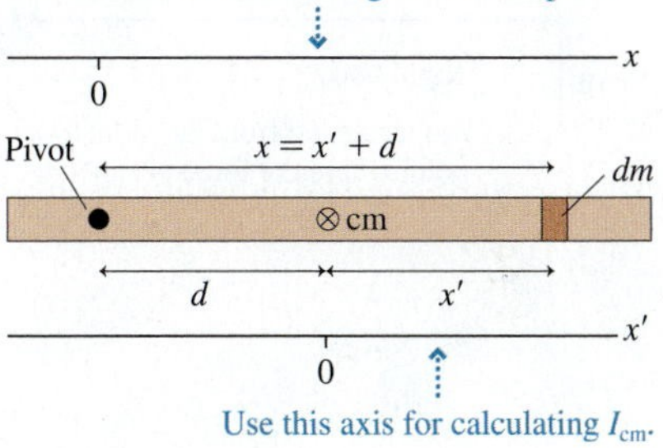

EXAMPLE 12.7 | **The moment of inertia of a thin rod**

Find the moment of inertia of a thin rod with mass M and length L about an axis one-third of the length from one end.

SOLVE From Table 12.2 we know the moment of inertia about the center of mass is $\frac{1}{12}ML^2$. The center of mass is at the center of the

rod. An axis $\frac{1}{3}L$ from one end is $d = \frac{1}{6}L$ from the center of mass. Using the parallel-axis theorem, we have

$$I = I_{cm} + Md^2 = \frac{1}{12}ML^2 + M\left(\frac{1}{6}L\right)^2 = \frac{1}{9}ML^2$$

STOP TO THINK 12.3 Four Ts are made from two identical rods of equal mass and length. Rank in order, from largest to smallest, the moments of inertia I_a to I_d for rotation about the dashed line.

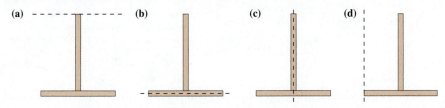

(a)   (b)   (c)   (d)

12.5 Torque

Consider the common experience of pushing open a door. **FIGURE 12.17** is a top view of a door hinged on the left. Four pushing forces are shown, all of equal strength. Which of these will be most effective at opening the door?

Force $\vec{F}_1$ will open the door, but force $\vec{F}_2$, which pushes straight at the hinge, will not. Force $\vec{F}_3$ will open the door, but not as easily as $\vec{F}_1$. What about $\vec{F}_4$? It is perpendicular to the door, it has the same magnitude as $\vec{F}_1$, but you know from

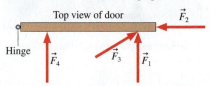

FIGURE 12.17 The four forces have different effects on the swinging door.

Top view of door

$\vec{F}_2$

Hinge $\vec{F}_4$ $\vec{F}_3$ $\vec{F}_1$

experience that pushing close to the hinge is not as effective as pushing at the outer edge of the door.

The ability of a force to cause a rotation depends on three factors:

1. The magnitude F of the force.
2. The distance r from the point of application to the pivot.
3. The angle at which the force is applied.

We can incorporate these three factors into a single quantity called the *torque*. **FIGURE 12.18** shows a force $\vec{F}$ trying to rotate the wrench and nut about a *pivot point*—the axis about which the nut will rotate. We say that this force exerts a **torque** τ (Greek tau), which we define as

$$\tau \equiv rF \sin\phi \qquad (12.20)$$

Torque depends on the three properties we just listed: the magnitude of the force, its distance from the pivot, and its angle. Loosely speaking, τ measures the "effectiveness" of the force at causing an object to rotate about a pivot. **Torque is the rotational equivalent of force.**

NOTE Angle ϕ is measured *counterclockwise* from the dashed line that extends outward along the radial line. This is consistent with our sign convention for the angular position θ.

The SI units of torque are newton-meters, abbreviated N m. Although we defined $1 \text{ N m} = 1 \text{ J}$ during our study of energy, torque is not an energy-related quantity and so we do *not* use joules as a measure of torque.

Torque, like force, has a sign. A torque that tries to rotate the object in a ccw direction is positive while a negative torque gives a cw rotation. **FIGURE 12.19** summarizes the signs. Notice that a force pushing straight toward the pivot or pulling straight out from the pivot exerts *no* torque.

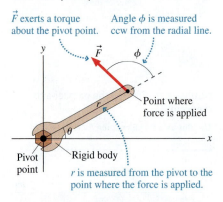

FIGURE 12.18 Force $\vec{F}$ exerts a torque about the pivot point.

$\vec{F}$ exerts a torque about the pivot point.

Angle ϕ is measured ccw from the radial line.

Point where force is applied

Rigid body

Pivot point

r is measured from the pivot to the point where the force is applied.

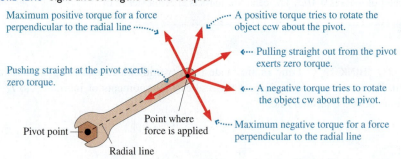

FIGURE 12.19 Signs and strengths of the torque.

Maximum positive torque for a force perpendicular to the radial line

A positive torque tries to rotate the object ccw about the pivot.

Pushing straight at the pivot exerts zero torque.

Pulling straight out from the pivot exerts zero torque.

A negative torque tries to rotate the object cw about the pivot.

Point where force is applied

Maximum negative torque for a force perpendicular to the radial line

Pivot point

Radial line

NOTE Torque differs from force in a very important way. Torque is calculated or measured *about a pivot point*. To say that a torque is 20 N m is meaningless. You need to say that the torque is 20 N m about a particular point. Torque can be calculated about any pivot point, but its value depends on the point chosen because this choice determines r and ϕ.

Returning to the door of Figure 12.17, you can see that $\vec{F}_1$ is most effective at opening the door because $\vec{F}_1$ exerts the largest torque *about the pivot point*. $\vec{F}_3$ has equal magnitude, but it is applied at an angle less than 90° and thus exerts less torque. $\vec{F}_2$, pushing straight at the hinge with $\phi = 0°$, exerts no torque at all. And $\vec{F}_4$, with a smaller value for r, exerts less torque than $\vec{F}_1$.

Your foot exerts a torque that rotates the crank.

Interpreting Torque

Torque can be interpreted from two perspectives, as shown in FIGURE 12.20. First, the quantity $F \sin \phi$ is the tangential force component F_t. Consequently, the torque is

$$\tau = rF_t \tag{12.21}$$

In other words, torque is the product of r with the force component F_t that is tangent to the circular path followed by this point on the wrench. This interpretation makes sense because the radial component of $\vec{F}$ points straight at the pivot point and cannot exert a torque.

A second perspective, widely used in applications, is based on the idea of a *moment arm*. Figure 12.20 shows the **line of action**, the line along which the force acts. The minimum distance between the pivot point and the line of action—the length of a line drawn *perpendicular to the line of action*—is called the **moment arm** (or the *lever arm*) d. Because $\sin(180° - \phi) = \sin \phi$, it is easy to see that $d = r \sin \phi$. Thus the torque $rF \sin \phi$ can also be written

$$|\tau| = dF \tag{12.22}$$

NOTE Equation 12.22 gives only $|\tau|$, the magnitude of the torque; the sign has to be supplied by observing the direction in which the torque acts.

FIGURE 12.20 Two useful interpretations of the torque.

1. Torque is due to the tangential component of force: $\tau = rF_t$.

$F_t = F \sin \phi$

Moment arm d

Line of action

2. Torque is the force multiplied by the moment arm: $|\tau| = dF$.

EXAMPLE 12.8 | Applying a torque

Luis uses a 20-cm-long wrench to turn a nut. The wrench handle is tilted 30° above the horizontal, and Luis pulls straight down on the end with a force of 100 N. How much torque does Luis exert on the nut?

VISUALIZE FIGURE 12.21 shows the situation. The angle is a negative $\phi = -120°$ because it is *clockwise* from the radial line.

FIGURE 12.21 A wrench being used to turn a nut.

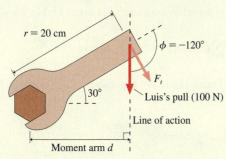

$r = 20$ cm

$\phi = -120°$

F_t

Luis's pull (100 N)

Line of action

Moment arm d

30°

SOLVE The tangential component of the force is

$$F_t = F \sin \phi = -86.6 \text{ N}$$

According to our sign convention, F_t is negative because it points in a cw direction. The torque, from Equation 12.21, is

$$\tau = rF_t = (0.20 \text{ m})(-86.6 \text{ N}) = -17 \text{ N m}$$

Alternatively, Figure 12.21 has drawn the *line of action* by extending the force vector forward and backward. The *moment arm*, the distance between the pivot point and the line of action, is

$$d = r \sin(60°) = 0.17 \text{ m}$$

Inserting the moment arm in Equation 12.22 gives

$$|\tau| = dF = (0.17 \text{ m})(100 \text{ N}) = 17 \text{ N m}$$

The torque acts to give a cw rotation, so we insert a minus sign to end up with

$$\tau = -17 \text{ N m}$$

ASSESS The largest possible torque, if Luis pulled perpendicular to the 20-cm-long wrench, would have a magnitude of 20 N m. Pulling at an angle will reduce this, so 17 N m is a reasonable answer.

STOP TO THINK 12.4 Rank in order, from largest to smallest, the five torques τ_a to τ_e. The rods all have the same length and are pivoted at the dot.

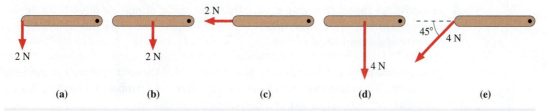

2 N

2 N

2 N

4 N

45°

4 N

(a)   (b)   (c)   (d)   (e)

FIGURE 12.22 The forces exert a net torque about the pivot point.

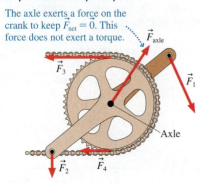

The axle exerts a force on the crank to keep $\vec{F}_{net} = \vec{0}$. This force does not exert a torque.

Net Torque

FIGURE 12.22 shows the forces acting on the crankset of a bicycle. The crankset is free to rotate about the axle, but the axle prevents it from having any translational motion relative to the bike frame. It does so by exerting force $\vec{F}_{axle}$ on the crankset to balance the other forces and keep $\vec{F}_{net} = \vec{0}$.

Forces $\vec{F}_1, \vec{F}_2, \vec{F}_3, \ldots$ exert torques $\tau_1, \tau_2, \tau_3, \ldots$ on the crankset, but $\vec{F}_{axle}$ does *not* exert a torque because it is applied at the pivot point and has zero moment arm. Thus the *net* torque about the axle is the sum of the torques due to the applied forces:

$$\tau_{net} = \tau_1 + \tau_2 + \tau_3 + \cdots = \sum_i \tau_i \qquad (12.23)$$

Gravitational Torque

Gravity exerts a torque on many objects. If the object in **FIGURE 12.23** is released, a torque due to gravity will cause it to rotate around the axle. To calculate the torque about the axle, we start with the fact that gravity acts on *every* particle in the object, exerting a downward force of magnitude $F_i = m_i g$ on particle i. The *magnitude* of the gravitational torque on particle i is $|\tau_i| = d_i m_i g$, where d_i is the moment arm. But we need to be careful with signs.

FIGURE 12.23 Gravitational torque.

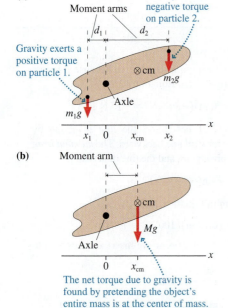

(a)

Moment arms

Gravity exerts a negative torque on particle 2.

Gravity exerts a positive torque on particle 1.

(b)

Moment arm

The net torque due to gravity is found by pretending the object's entire mass is at the center of mass.

A moment arm must be a positive number because it's a distance. If we establish a coordinate system with the origin at the axle, then you can see from Figure 12.23a that the moment arm d_i of particle i is $|x_i|$. A particle to the right of the axle (positive x_i) experiences a *negative* torque because gravity tries to rotate this particle in a clockwise direction. Similarly, a particle to the left of the axle (negative x_i) has a positive torque. The torque is opposite in sign to x_i, so we can get the sign right by writing

$$\tau_i = -x_i m_i g = -(m_i x_i)g \qquad (12.24)$$

The net torque due to gravity is found by summing Equation 12.24 over all particles:

$$\tau_{grav} = \sum_i \tau_i = \sum_i (-m_i x_i g) = -\left(\sum_i m_i x_i\right)g \qquad (12.25)$$

But according to the definition of center of mass, Equations 12.4, $\sum m_i x_i = M x_{cm}$. Thus the torque due to gravity is

$$\tau_{grav} = -Mg x_{cm} \qquad (12.26)$$

where x_{cm} is the position of the center of mass *relative to the axis of rotation*.

Equation 12.26 has the simple interpretation shown in Figure 12.23b. Mg is the net gravitational force on the entire object, and x_{cm} is the moment arm between the rotation axis and the center of mass. The gravitational torque on an extended object of mass M is equivalent to the torque of a *single* force vector $\vec{F}_G = -Mg\hat{\jmath}$ acting at the object's center of mass.

In other words, **the gravitational torque is found by treating the object as if all its mass were concentrated at the center of mass.** This is the basis for the well-known technique of finding an object's center of mass by balancing it. An object will balance on a pivot, as shown in **FIGURE 12.24**, only if the center of mass is directly above the pivot point. If the pivot is *not* under the center of mass, the gravitational torque will cause it to rotate.

FIGURE 12.24 An object balances on a pivot that is directly under the center of mass.

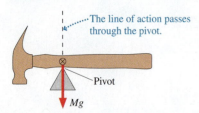

The line of action passes through the pivot.

Pivot

Mg

NOTE The point at which gravity acts is also called the *center of gravity*. As long as gravity is uniform over the object—always true for earthbound objects—there's no difference between center of mass and center of gravity.

EXAMPLE 12.9 | The gravitational torque on a beam

The 4.00-m-long, 500 kg steel beam shown in **FIGURE 12.25** is supported 1.20 m from the right end. What is the gravitational torque about the support?

FIGURE 12.25 A steel beam supported at one point.

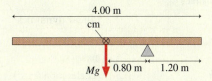

MODEL The center of mass of the beam is at the midpoint. $x_{cm} = -0.80$ m is measured from the pivot point.

SOLVE This is a straightforward application of Equation 12.26. The gravitational torque is

$$\tau_{grav} = -Mgx_{cm} = -(500 \text{ kg})(9.80 \text{ m/s}^2)(-0.80 \text{ m}) = 3920 \text{ N m}$$

ASSESS The torque is positive because gravity tries to rotate the beam ccw. Notice that the beam in Figure 12.25 is *not* in equilibrium. It will fall over unless other forces, not shown, are supporting it.

12.6 Rotational Dynamics

What does a torque do? **A torque causes an angular acceleration.** To see why, **FIGURE 12.26** shows a rigid body undergoing *pure rotational motion* about a fixed and unmoving axis. This might be a rotation about the object's center of mass, such as we considered in Section 12.2. Or it might be an object, such as a turbine, rotating on an axle.

The forces $\vec{F}_1$, $\vec{F}_2$, $\vec{F}_3$, . . . in Figure 12.26 are external forces acting on particles of masses $m_1, m_2, m_3, \ldots$ that are part of the rigid body. These forces exert torques $\tau_1, \tau_2, \tau_3, \ldots$ about the rotation axis. The *net* torque on the object is the sum of the torques:

$$\tau_{net} = \sum_i \tau_i \tag{12.27}$$

Focus on particle i, which is acted on by force $\vec{F}_i$ and undergoes circular motion with radius r_i. In Chapter 8, we found that the radial component of $\vec{F}_i$ is responsible for the centripetal acceleration of circular motion, while the tangential component $(F_i)_t$ causes the particle to speed up or slow down with a tangential acceleration $(a_i)_t$. Newton's second law is

$$(F_i)_t = m_i(a_i)_t = m_i r_i \alpha \tag{12.28}$$

where in the last step we used the relationship between tangential and angular acceleration: $a_t = r\alpha$. The angular acceleration α does not have a subscript because *all particles in the object have the same angular acceleration.* That is, α is the angular acceleration of the entire object.

Multiplying both sides by r_i gives

$$r_i(F_i)_t = m_i r_i^2 \alpha \tag{12.29}$$

But $r_i(F_i)_t$ is the torque τ_i about the axis on particle i; hence Newton's second law for a single particle in the object is

$$\tau_i = m_i r_i^2 \alpha \tag{12.30}$$

Returning now to Equation 12.27, we see that the net torque on the object is

$$\tau_{net} = \sum_i \tau_i = \sum_i m_i r_i^2 \alpha = \left(\sum_i m_i r_i^2 \right) \alpha \tag{12.31}$$

In the last step, we factored out α by using the key idea that every particle in a rotating rigid body has the *same* angular acceleration.

You'll recognize the quantity in parentheses as the moment of inertia. Substituting I into Equation 12.31 puts the final piece of the puzzle into place. An object that experiences a net torque τ_{net} about the axis of rotation undergoes an angular acceleration

$$\alpha = \frac{\tau_{net}}{I} \quad \text{(Newton's second law for rotational motion)} \tag{12.32}$$

where I is the object's moment of inertia *about the rotation axis.* This result, Newton's second law for rotation, is the fundamental equation of rigid-body dynamics.

FIGURE 12.26 The external forces exert a torque about the rotation axis.

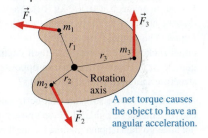

A net torque causes the object to have an angular acceleration.

In practice we often write $\tau_{net} = I\alpha$, but Equation 12.32 better conveys the idea that **torque is the cause of angular acceleration.** In the absence of a net torque ($\tau_{net} = 0$), the object either does not rotate ($\omega = 0$) or rotates with *constant* angular velocity ($\omega = $ constant).

TABLE 12.3 summarizes the analogies between linear and rotational dynamics.

TABLE 12.3 Rotational and linear dynamics

Rotational dynamics		Linear dynamics	
torque (N m)	τ_{net}	force (N)	$\vec{F}_{net}$
moment of inertia (kg m^2)	I	mass (kg)	m
angular acceleration (rad/s^2)	α	acceleration (m/s^2)	$\vec{a}$
second law	$\alpha = \tau_{net}/I$	second law	$\vec{a} = \vec{F}_{net}/m$

EXAMPLE 12.10 | Rotating rockets

Far out in space, a 100,000 kg rocket and a 200,000 kg rocket are docked at opposite ends of a motionless 90-m-long connecting tunnel. The tunnel is rigid and its mass is much less than that of either rocket. The rockets start their engines simultaneously, each generating 50,000 N of thrust in opposite directions. What is the structure's angular velocity after 30 s?

MODEL The entire structure can be modeled as two masses at the ends of a massless, rigid rod. There's no net force, so the structure does not undergo translational motion, but the thrusts do create torques that will give the structure angular acceleration and cause it to rotate. We'll assume the thrust forces are perpendicular to the connecting tunnel. This is an unconstrained rotation, so the structure will rotate about its center of mass.

VISUALIZE FIGURE 12.27 shows the rockets and defines distances r_1 and r_2 from the center of mass.

FIGURE 12.27 The thrusts exert a torque on the structure.

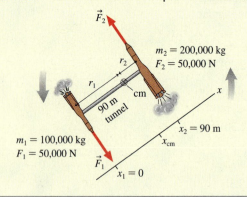

$\vec{F}_2$
$m_2 = 200,000$ kg
$F_2 = 50,000$ N
r_2
r_1
cm
90 m tunnel
x
$x_2 = 90$ m
$m_1 = 100,000$ kg
$F_1 = 50,000$ N
$\vec{F}_1$
x_{cm}
$x_1 = 0$

SOLVE Our strategy will be to use Newton's second law to find the angular acceleration, followed by rotational kinematics to find ω. We'll need to determine the moment of inertia, and that requires knowing the distances of the two rockets from the rotation axis. As we did in Example 12.1, we choose a coordinate system in which the masses are on the x-axis and in which m_1 is at the origin. Then

$$x_{cm} = \frac{m_1 x_1 + m_2 x_2}{m_1 + m_2}$$

$$= \frac{(100,000 \text{ kg})(0 \text{ m}) + (200,000 \text{ kg})(90 \text{ m})}{100,000 \text{ kg} + 200,000 \text{ kg}} = 60 \text{ m}$$

The structure's center of mass is $r_1 = 60$ m from the 100,000 kg rocket and $r_2 = 30$ m from the 200,000 kg rocket. The moment of inertia about the center of mass is

$$I = m_1 r_1^2 + m_2 r_2^2 = 540,000,000 \text{ kg m}^2$$

The two rocket thrusts exert net torque

$$\tau_{net} = r_1 F_1 + r_2 F_2 = (60 \text{ m})(50,000 \text{ N}) + (30 \text{ m})(50,000 \text{ N})$$
$$= 4,500,000 \text{ N m}$$

With I and τ_{net} now known, we can use Newton's second law to find the angular acceleration:

$$\alpha = \frac{\tau}{I} = \frac{4,500,000 \text{ N m}}{540,000,000 \text{ kg m}^2} = 0.00833 \text{ rad/s}^2$$

After 30 seconds, the structure's angular velocity is

$$\omega = \alpha \, \Delta t = 0.25 \text{ rad/s}$$

STOP TO THINK 12.5 Rank in order, from largest to smallest, the angular accelerations α_a to α_d.

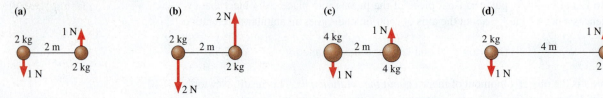

(a)
2 kg — 2 m — 2 kg
1 N
1 N

(b)
2 N
2 kg — 2 m — 2 kg
2 N

(c)
4 kg — 2 m — 4 kg
1 N
1 N

(d)
2 kg — 4 m — 2 kg
1 N
1 N

12.7 Rotation About a Fixed Axis

In this section we'll look at rigid bodies that rotate about a fixed axis. The problem-solving strategy for rotational dynamics is very similar to that for linear dynamics.

PROBLEM-SOLVING STRATEGY 12.1

Rotational dynamics problems

MODEL Model the object as a rigid body.

VISUALIZE Draw a pictorial representation to clarify the situation, define coordinates and symbols, and list known information.

- Identify the axis about which the object rotates.
- Identify forces and determine their distances from the axis. For most problems it will be useful to draw a free-body diagram.
- Identify any torques caused by the forces and the signs of the torques.

SOLVE The mathematical representation is based on Newton's second law for rotational motion:

$$\tau_{net} = I\alpha \quad \text{or} \quad \alpha = \frac{\tau_{net}}{I}$$

- Find the moment of inertia in Table 12.2 or, if needed, calculate it as an integral or by using the parallel-axis theorem.
- Use rotational kinematics to find angles and angular velocities.

ASSESS Check that your result has correct units and significant figures, is reasonable, and answers the question.

Exercise 28

EXAMPLE 12.11 | **Starting an airplane engine**

The engine in a small airplane is specified to have a torque of 60 N m. This engine drives a 2.0-m-long, 40 kg propeller. On start-up, how long does it take the propeller to reach 200 rpm?

MODEL The propeller can be modeled as a rigid rod that rotates about its center. The engine exerts a torque on the propeller.

VISUALIZE FIGURE 12.28 shows the propeller and the rotation axis.

FIGURE 12.28 A rotating airplane propeller.

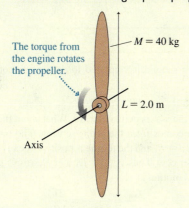

The torque from the engine rotates the propeller.

$M = 40$ kg

$L = 2.0$ m

Axis

SOLVE The moment of inertia of a rod rotating about its center is found from Table 12.2:

$$I = \tfrac{1}{12}ML^2 = \tfrac{1}{12}(40 \text{ kg})(2.0 \text{ m})^2 = 13.33 \text{ kg m}^2$$

The 60 N m torque of the engine causes an angular acceleration

$$\alpha = \frac{\tau}{I} = \frac{60 \text{ N m}}{13.33 \text{ kg m}^2} = 4.50 \text{ rad/s}^2$$

The time needed to reach $\omega_f = 200$ rpm $= 3.33$ rev/s $= 20.9$ rad/s is

$$\Delta t = \frac{\Delta\omega}{\alpha} = \frac{\omega_f - \omega_i}{\alpha} = \frac{20.9 \text{ rad/s} - 0 \text{ rad/s}}{4.5 \text{ rad/s}^2} = 4.6 \text{ s}$$

ASSESS We've assumed a constant angular acceleration, which is reasonable for the first few seconds while the propeller is still turning slowly. Eventually, air resistance and friction will cause opposing torques and the angular acceleration will decrease. At full speed, the negative torque due to air resistance and friction cancels the torque of the engine. Then $\tau_{net} = 0$ and the propeller turns at *constant* angular velocity with no angular acceleration.

Constraints Due to Ropes and Pulleys

FIGURE 12.29 The rope's motion must match the motion of the rim of the pulley.

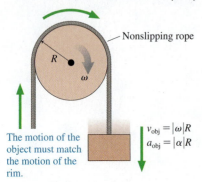

Nonslipping rope

R

ω

The motion of the object must match the motion of the rim.

$v_{obj} = |\omega|R$
$a_{obj} = |\alpha|R$

Many important applications of rotational dynamics involve objects, such as pulleys, that are connected via ropes or belts to other objects. **FIGURE 12.29** shows a rope passing over a pulley and connected to an object in linear motion. If the rope does not slip as the pulley rotates, then the rope's speed v_{rope} must exactly match the speed of the rim of the pulley, which is $v_{rim} = |\omega|R$. If the pulley has an angular acceleration, the rope's acceleration a_{rope} must match the *tangential* acceleration of the rim of the pulley, $a_t = |\alpha|R$.

The object attached to the other end of the rope has the same speed and acceleration as the rope. Consequently, an object connected to a pulley of radius R by a rope that does not slip must obey the constraints

$$v_{obj} = |\omega|R$$

(motion constraints for a nonslipping rope)   (12.33)

$$a_{obj} = |\alpha|R$$

These constraints are very similar to the acceleration constraints introduced in Chapter 7 for two objects connected by a string or rope.

> **NOTE** The constraints are given as magnitudes. Specific problems will need to introduce signs that depend on the direction of motion and on the choice of coordinate system.

The Constant-Torque Model

If all the torques exerted on an object are constant, the object rotates with constant angular acceleration. Even if the torques aren't perfectly constant, there are many situations where it's reasonable to model them as if they were. The **constant-torque model,** analogous to the constant-force model of Section 6.2, is the most important model of rotational dynamics.

MODEL 12.2

Constant torque

For objects on which the net torque is constant.

- Model the object as a rigid body with constant angular acceleration.
- Take into account constraints due to ropes and pulleys.
- Mathematically:
  - **Newton's second law** is $\tau_{net} = I\alpha$.
  - Use the kinematics of constant angular acceleration.
- Limitations: Model fails if the torque isn't constant.

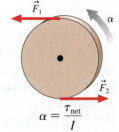

$\vec{F}_1$

α

$\vec{F}_2$

$$\alpha = \frac{\tau_{net}}{I}$$

The object has constant angular acceleration.

EXAMPLE 12.12 | **Lowering a bucket**

A 2.0 kg bucket is attached to a massless string that is wrapped around a 1.0 kg, 4.0-cm-diameter cylinder, as shown in **FIGURE 12.30a**. The cylinder rotates on an axle through the center. The bucket is released from rest 1.0 m above the floor. How long does it take to reach the floor?

MODEL Assume the string does not slip.

VISUALIZE **FIGURE 12.30b** shows the free-body diagram for the cylinder and the bucket. The string tension exerts an upward force on the bucket and a downward force on the outer edge of the cylinder. The string is massless, so these two tension forces act as if they are an action/reaction pair: $T_b = T_c = T$.

SOLVE Newton's second law applied to the linear motion of the bucket is

$$ma_y = T - mg$$

where, as usual, the y-axis points upward. What about the cylinder? The only torque comes from the string tension. The moment arm for the tension is $d = R$, and the torque is positive because the string turns the cylinder ccw. Thus $\tau_{string} = TR$ and Newton's second law for the rotational motion is

$$\alpha = \frac{\tau_{net}}{I} = \frac{TR}{\frac{1}{2}MR^2} = \frac{2T}{MR}$$

FIGURE 12.30 The falling bucket turns the cylinder.

(a)

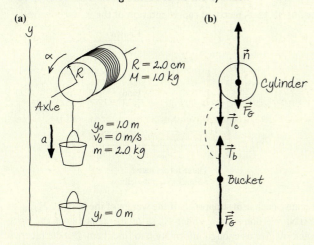

(b)

The moment of inertia of a cylinder rotating about a center axis was taken from Table 12.2.

The last piece of information we need is the constraint due to the fact that the string doesn't slip. Equation 12.33 relates only the absolute values, but in this problem α is positive (ccw acceleration) while a_y is negative (downward acceleration). Hence

$$a_y = -\alpha R$$

Using α from the cylinder's equation in the constraint, we find

$$a_y = -\alpha R = -\frac{2T}{MR}R = -\frac{2T}{M}$$

Thus the tension is $T = -\frac{1}{2}Ma_y$. If we use this value of the tension in the bucket's equation, we can solve for the acceleration:

$$ma_y = -\frac{1}{2}Ma_y - mg$$
$$a_y = -\frac{g}{(1 + M/2m)} = -7.84 \text{ m/s}^2$$

The time to fall through $\Delta y = -1.0$ m is found from kinematics:

$$\Delta y = \frac{1}{2}a_y(\Delta t)^2$$
$$\Delta t = \sqrt{\frac{2\,\Delta y}{a_y}} = \sqrt{\frac{2(-1.0 \text{ m})}{-7.84 \text{ m/s}^2}} = 0.50 \text{ s}$$

ASSESS The expression for the acceleration gives $a_y = -g$ if $M = 0$. This makes sense because the bucket would be in free fall if there were no cylinder. When the cylinder has mass, the downward force of gravity on the bucket has to accelerate the bucket *and* spin the cylinder. Consequently, the acceleration is reduced and the bucket takes longer to fall.

12.8 Static Equilibrium

An extended object that is completely stationary is in *static equilibrium*. It has no linear acceleration ($\vec{a} = \vec{0}$) and no angular acceleration ($\alpha = 0$). Thus, from Newton's laws, the conditions for static equilibrium are no net force *and* no net torque. These two rules are the basis for a branch of engineering, called *statics*, that analyzes buildings, dams, bridges, and other structures in static equilibrium.

Section 6.1 introduced the model of mechanical equilibrium for objects that can be represented as particles. For extended objects, we have the **static equilibrium model.**

Structures such as bridges are analyzed in engineering statics.

MODEL 12.3

Static equilibrium

For extended objects at rest.

- Model the object as a rigid body with no acceleration.
- Mathematically:
  - No net force: $\vec{F}_{\text{net}} = \sum \vec{F}_i = \vec{0}$, and
  - No net torque: $\tau_{\text{net}} = \sum \tau_i = 0$
- The torque is zero about *every* point, so use any point that is convenient for the pivot point.
- Limitations: Model fails if either the forces or the torques aren't balanced.

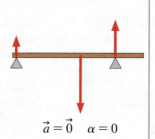

$\vec{a} = \vec{0} \quad \alpha = 0$

The object is at rest.

For any point you choose, an object that is not rotating is not rotating about that point. This seems to be a trivial statement, but it has an important implication: For a rigid body in static equilibrium, the net torque is zero about *any* point. You can use any point you wish as the pivot point for calculating torque. Even so, some choices are better than others for problem solving. As the examples will show, it's often best to choose a point at which several forces act because the torques exerted by those forces will be zero.

EXAMPLE 12.13 | Lifting weights

Weightlifting can exert extremely large forces on the body's joints and tendons. In the *strict curl* event, a standing athlete uses both arms to lift a barbell by moving only his forearms, which pivot at the elbows. The record weight lifted in the strict curl is over 200 pounds (about 900 N). **FIGURE 12.31** shows the arm bones and the biceps, the main lifting muscle when the forearm is horizontal. What is the tension in the tendon connecting the biceps muscle to the bone while a 900 N barbell is held stationary in this position?

FIGURE 12.31 An arm holding a barbell.

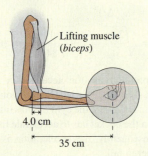

Lifting muscle (biceps)

4.0 cm

35 cm

MODEL Model the arm as two rigid rods connected by a hinge. We'll ignore the arm's weight because it is so much less than that of the barbell. Although the tendon pulls at a slight angle, it is close enough to vertical that we'll treat it as such.

VISUALIZE **FIGURE 12.32** shows the forces acting on our simplified model of the forearm. The biceps pulls the forearm up against the upper arm at the elbow, so the force $\vec{F}_{elbow}$ *on* the forearm at the elbow—a force due to the upper arm—is a downward force.

SOLVE Static equilibrium requires both the net force *and* the net torque on the forearm to be zero. Only the y-component of force is relevant, and setting it to zero gives a first equation:

$$\sum F_y = F_{tendon} - F_{elbow} - F_{barbell} = 0$$

FIGURE 12.32 A pictorial representation of the forces involved.

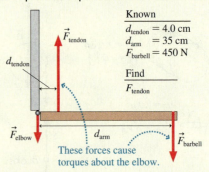

Known
$d_{tendon} = 4.0$ cm
$d_{arm} = 35$ cm
$F_{barbell} = 450$ N

Find
F_{tendon}

These forces cause torques about the elbow.

Because each arm supports half the weight of the barbell, $F_{barbell} = 450$ N. We don't know either F_{tendon} or F_{elbow}, nor does the force equation give us enough information to find them. But the fact that the net torque also must be zero gives us that extra information. The torque is zero about *every* point, so we can choose any point we wish to calculate the torque. The elbow joint is a convenient point because force $\vec{F}_{elbow}$ exerts no torque about this point; its moment arm is zero. Thus the torque equation is

$$\tau_{net} = d_{tendon}F_{tendon} - d_{arm}F_{barbell} = 0$$

The tension in the tendon tries to rotate the arm ccw, so it produces a positive torque. Similarly, the torque due to the barbell is negative. We can solve the torque equation for F_{tendon} to find

$$F_{tendon} = F_{barbell}\frac{d_{arm}}{d_{tendon}} = (450 \text{ N})\frac{35 \text{ cm}}{4.0 \text{ cm}} = 3900 \text{ N}$$

ASSESS The short distance d_{tendon} from the tendon to the elbow joint means that the force supplied by the biceps has to be very large to counter the torque generated by a force applied at the opposite end of the forearm. Although we ended up not needing the force equation in this problem, we could now use it to calculate that the force exerted at the elbow is $F_{elbow} = 3450$ N. These large forces can easily damage the tendon or the elbow.

EXAMPLE 12.14 | Walking the plank

Adrienne (50 kg) and Bo (90 kg) are playing on a 100 kg rigid plank resting on the supports seen in **FIGURE 12.33**. If Adrienne stands on the left end, can Bo walk all the way to the right end without the plank tipping over? If not, how far can he get past the support on the right?

FIGURE 12.33 Adrienne and Bo on the plank.

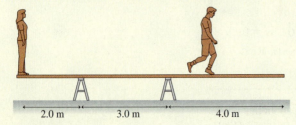

2.0 m   3.0 m   4.0 m

MODEL Model the plank as a uniform rigid body with its center of mass at the center.

VISUALIZE **FIGURE 12.34** shows the forces acting on the plank. Both supports exert upward forces. $\vec{n}_A$ and $\vec{n}_B$ are the normal forces of Adrienne's and Bo's feet pushing down on the board.

SOLVE Because the plank is resting on the supports, not held down, forces $\vec{n}_1$ and $\vec{n}_2$ must point upward. (The supports could pull down if the plank were nailed to them, but that's not the case here.) Force $\vec{n}_1$ will decrease as Bo moves to the right, and the tipping point occurs when $n_1 = 0$. The plank remains in static equilibrium right up to the tipping point, so both the net force and the net torque on it are zero. The force equation is

$$\sum F_y = n_1 + n_2 - n_A - n_B - Mg$$
$$= n_1 + n_2 - m_Ag - m_Bg - Mg = 0$$

Adrienne is at rest, with zero net force, so her downward force on the board, an action/reaction pair with the upward normal force of the board on her, equals her weight: $n_A = m_Ag$. Bo's center of

FIGURE 12.34 A pictorial representation of the forces on the plank.

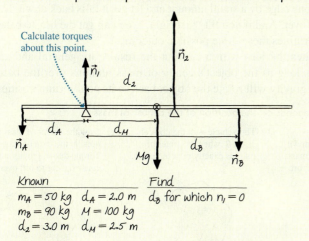

Known
$m_A = 50$ kg $d_A = 2.0$ m
$m_B = 90$ kg $M = 100$ kg
$d_2 = 3.0$ m $d_M = 2.5$ m

Find
d_B for which $n_l = 0$

mass oscillates up and down as he walks, so he's *not* in equilibrium and, strictly speaking, $n_B \neq m_B g$. But we'll assume that he edges out onto the board slowly, with minimal bouncing, in which case $n_B = m_B g$ is a reasonable approximation.

We can again choose any point we wish for calculating torque. Let's use the support on the left. Adrienne and the support on the right exert positive torques about this point; the other forces exert negative torques. Force $\vec{n}_1$ exerts no torque because it acts at the pivot point. Thus the torque equation is

$$\tau_{net} = d_A m_A g - d_B m_B g - d_M M g + d_2 n_2 = 0$$

At the tipping point, where $n_1 = 0$, the force equation gives $n_2 = (m_A + m_B + M)g$. Substituting this into the torque equation and then solving for Bo's position give

$$d_B = \frac{d_A m_A - d_M M + d_2(m_A + m_B + M)}{m_B} = 6.3 \text{ m}$$

Bo doesn't quite make it to the end. The plank tips when he's 6.3 m past the left support, our pivot point, and thus 3.3 m past the support on the right.

ASSESS We could have solved this problem somewhat more simply had we chosen the support on the right for calculating the torques. However, you might not recognize the "best" point for calculating the torques in a problem. The point of this example is that it doesn't matter which point you choose.

EXAMPLE 12.15 | Will the ladder slip?

A 3.0-m-long ladder leans against a frictionless wall at an angle of 60°. What is the minimum value of μ_s, the coefficient of static friction with the ground, that prevents the ladder from slipping?

MODEL The ladder is a rigid rod of length L. To not slip, it must be in both translational equilibrium $(\vec{F}_{net} = \vec{0})$ and rotational equilibrium $(\tau_{net} = 0)$.

VISUALIZE FIGURE 12.35 shows the ladder and the forces acting on it.

FIGURE 12.35 A ladder in total equilibrium.

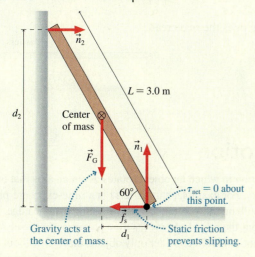

Gravity acts at the center of mass.

Static friction prevents slipping.

SOLVE The x- and y-components of $\vec{F}_{net} = \vec{0}$ are

$$\sum F_x = n_2 - f_s = 0$$
$$\sum F_y = n_1 - Mg = 0$$

The net torque is zero about *any* point, so which should we choose? The bottom corner of the ladder is a good choice because two forces pass through this point and have no torque about it. The torque about the bottom corner is

$$\tau_{net} = d_1 F_G - d_2 n_2 = \tfrac{1}{2}(L\cos 60°)Mg - (L\sin 60°)n_2 = 0$$

The signs are based on the observation that $\vec{F}_G$ would cause the ladder to rotate ccw while $\vec{n}_2$ would cause it to rotate cw. All together, we have three equations in the three unknowns n_1, n_2, and f_s. If we solve the third for n_2,

$$n_2 = \frac{\tfrac{1}{2}(L\cos 60°)Mg}{L\sin 60°} = \frac{Mg}{2\tan 60°}$$

we can then substitute this into the first to find

$$f_s = \frac{Mg}{2\tan 60°}$$

Our model of friction is $f_s \leq f_{s\,max} = \mu_s n_1$. We can find n_1 from the second equation: $n_1 = Mg$. Using this, the model of static friction tells us that

$$f_s \leq \mu_s Mg$$

Comparing these two expressions for f_s, we see that μ_s must obey

$$\mu_s \geq \frac{1}{2\tan 60°} = 0.29$$

Thus the minimum value of the coefficient of static friction is 0.29.

ASSESS You know from experience that you can lean a ladder or other object against a wall if the ground is "rough," but it slips if the surface is too smooth. 0.29 is a "medium" value for the coefficient of static friction, which is reasonable.

Balance and Stability

If you tilt a box up on one edge by a small amount and let go, it falls back down. If you tilt it too much, it falls over. And if you tilt "just right," you can get the box to balance on its edge. What determines these three possible outcomes?

FIGURE 12.36 illustrates the idea with a car, but the results are general and apply in many situations. As long as the object's center of mass remains over the base of support, torque due to gravity will rotate the object back to its equilibrium position.

FIGURE 12.36 Stability depends on the position of the center of mass.

(a) The torque due to gravity will bring the car back down as long as the center of mass is above the base of support.

(b) The vehicle is at the critical angle θ_c when its center of mass is exactly over the pivot.

(c) Now the center of mass is outside the base of support. Torque due to gravity will cause the car to roll over.

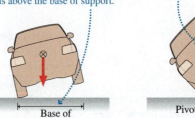
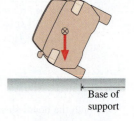

A *critical angle* θ_c is reached when the center of mass is directly over the pivot point. This is the point of balance, with no net torque. For vehicles, the distance between the tires is called the track width t. If the height of the center of mass is h, you can see from Figure 12.36b that the critical angle is

$$\theta_c = \tan^{-1}\left(\frac{t}{2h}\right)$$

For passenger cars with $h \approx 0.33t$, the critical angle is $\theta_c \approx 57°$. But for a sport utility vehicle (SUV) with $h \approx 0.47t$, a higher center of mass, the critical angle is only $\theta_c \approx 47°$. Various automobile safety groups have determined that a vehicle with $\theta_c > 50°$ is unlikely to roll over in an accident. A rollover becomes increasingly likely when θ_c is reduced below 50°. The general rule is that **a wider base of support and/or a lower center of mass improve stability.**

This dancer balances *en pointe* by having her center of mass directly over her toes, her base of support.

STOP TO THINK 12.6 What does the scale read?

a. 500 N
b. 1000 N
c. 2000 N
d. 4000 N

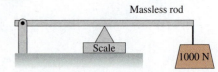

12.9 Rolling Motion

Rolling is a *combination motion* in which an object rotates about an axis that is moving along a straight-line trajectory. For example, **FIGURE 12.37** is a time-exposure photo of a rolling wheel with one lightbulb on the axis and a second lightbulb at the edge. The axis light moves straight ahead, but the edge light moves along a curve. Let's see if we can understand this interesting motion. We'll consider only objects that roll without slipping.

FIGURE 12.38 shows a round object—a wheel or a sphere—that rolls forward exactly one revolution. The point that had been on the bottom follows the curve you saw in Figure 12.37 to the top and back to the bottom. *Because the object doesn't slip,* the center of mass moves forward exactly one circumference: $\Delta x_{cm} = 2\pi R$.

We can also write the distance traveled in terms of the velocity of the center of mass: $\Delta x_{cm} = v_{cm}\Delta t$. But Δt, the time it takes the object to make one complete revolution, is nothing other than the rotation period T. In other words, $\Delta x_{cm} = v_{cm}T$.

FIGURE 12.37 The trajectories of the center of a wheel and of a point on the rim are seen in a time-exposure photograph.

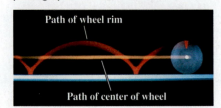

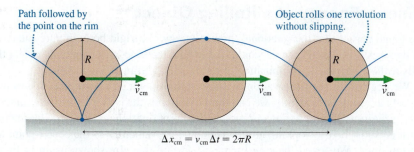

Path followed by the point on the rim

Object rolls one revolution without slipping.

$\Delta x_{cm} = v_{cm} \Delta t = 2\pi R$

◄ FIGURE 12.38 An object rolling through one revolution.

These two expressions for Δx_{cm} come from two perspectives on the motion: one looking at the rotation and the other looking at the translation of the center of mass. But it's the same distance no matter how you look at it, so these two expressions must be equal. Consequently,

$$\Delta x_{cm} = 2\pi R = v_{cm} T \qquad (12.34)$$

If we divide by T, we can write the center-of-mass velocity as

$$v_{cm} = \frac{2\pi}{T} R \qquad (12.35)$$

But $2\pi/T$ is the angular velocity ω, as you learned in Chapter 4, leading to

$$v_{cm} = R\omega \qquad (12.36)$$

Equation 12.36 is the **rolling constraint,** the basic link between translation and rotation for objects that roll without slipping.

Let's look carefully at a particle in the rolling object. As **FIGURE 12.39a** shows, the position vector $\vec{r}_i$ for particle i is the vector sum $\vec{r}_i = \vec{r}_{cm} + \vec{r}_{i,\,rel}$. Taking the time derivative of this equation, we can write the velocity of particle i as

$$\vec{v}_i = \vec{v}_{cm} + \vec{v}_{i,\,rel} \qquad (12.37)$$

In other words, the velocity of particle i can be divided into two parts: the velocity $\vec{v}_{cm}$ of the object as a whole plus the velocity $\vec{v}_{i,\,rel}$ of particle i relative to the center of mass (i.e., the velocity that particle i would have if the object were only rotating and had no translational motion).

FIGURE 12.39b applies this idea to point P at the very bottom of the rolling object, the point of contact between the object and the surface. This point is moving around the center of the object at angular velocity ω, so $v_{i,\,rel} = -R\omega$. The negative sign indicates that the motion is cw. At the same time, the center-of-mass velocity, Equation 12.36, is $v_{cm} = R\omega$. Adding these, we find that the velocity of point P, the lowest point, is $v_i = 0$. In other words, **the point on the bottom of a rolling object is instantaneously at rest.**

Although this seems surprising, it is really what we mean by "rolling without slipping." If the bottom point had a velocity, it would be moving horizontally relative to the surface. In other words, it would be slipping or sliding across the surface. To roll without slipping, the bottom point, the point touching the surface, must be at rest.

FIGURE 12.40 shows how the velocity vectors at the top, center, and bottom of a rotating wheel are found by adding the rotational velocity vectors to the center-of-mass velocity. You can see that $v_{bottom} = 0$ and that $v_{top} = 2R\omega = 2v_{cm}$.

FIGURE 12.39 The motion of a particle in the rolling object.

(a)

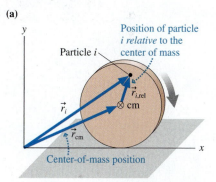

Particle i

Position of particle i *relative* to the center of mass

$\vec{r}_{i,rel}$

$\otimes$ cm

$\vec{r}_i$

$\vec{r}_{cm}$

Center-of-mass position

(b)

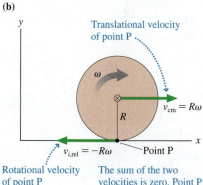

Translational velocity of point P

ω

$v_{cm} = R\omega$

R

$v_{i,rel} = -R\omega$

Point P

Rotational velocity of point P

The sum of the two velocities is zero. Point P is instantaneously at rest.

FIGURE 12.40 Rolling without slipping is a combination of translation and rotation.

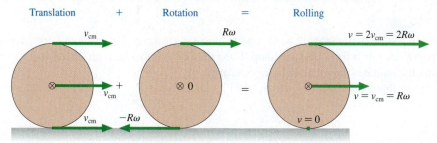

Translation   +   Rotation   =   Rolling

v_{cm}

$R\omega$

$v = 2v_{cm} = 2R\omega$

$\otimes$

$+$

v_{cm}

$\otimes 0$

$=$

$\otimes$

$v = v_{cm} = R\omega$

v_{cm}

$-R\omega$

$v = 0$

Kinetic Energy of a Rolling Object

We found earlier that the rotational kinetic energy of a rigid body in pure rotational motion is $K_{rot} = \frac{1}{2}I\omega^2$. Now we would like to find the kinetic energy of an object that rolls without slipping, a combination of rotational and translation motion.

We begin with the observation that the bottom point in **FIGURE 12.41** is instantaneously at rest. Consequently, we can think of an axis through P as an *instantaneous axis of rotation*. The idea of an instantaneous axis of rotation seems a little far-fetched, but it is confirmed by looking at the instantaneous velocities of the center point and the top point. We found these in Figure 12.40 and they are shown again in Figure 12.41. They are exactly what you would expect as the tangential velocity $v_t = r\omega$ for rotation about P at distances R and $2R$.

From this perspective, the object's motion is pure rotation about point P. Thus the kinetic energy is that of pure rotation:

$$K = K_{\text{rotation about P}} = \tfrac{1}{2}I_P \omega^2 \tag{12.38}$$

I_P is the moment of inertia for rotation about point P. We can use the parallel-axis theorem to write I_P in terms of the moment of inertia I_{cm} about the center of mass. Point P is displaced by distance $d = R$; thus

$$I_P = I_{cm} + MR^2$$

Using this expression in Equation 12.38 gives us the kinetic energy:

$$K = \tfrac{1}{2}I_{cm}\omega^2 + \tfrac{1}{2}M(R\omega)^2 \tag{12.39}$$

We know from the rolling constraint that $R\omega$ is the center-of-mass velocity v_{cm}. Thus the kinetic energy of a rolling object is

$$K_{rolling} = \tfrac{1}{2}I_{cm}\omega^2 + \tfrac{1}{2}Mv_{cm}^2 = K_{rot} + K_{cm} \tag{12.40}$$

In other words, **the rolling motion of a rigid body can be described as a translation of the center of mass (with kinetic energy K_{cm}) plus a rotation about the center of mass (with kinetic energy K_{rot}).**

The Great Downhill Race

FIGURE 12.42 shows a contest in which a sphere, a cylinder, and a circular hoop, all of mass M and radius R, are placed at height h on a slope of angle θ. All three are released from rest at the same instant of time and roll down the ramp without slipping. To make things more interesting, they are joined by a particle of mass M that slides down the ramp without friction. Which one will win the race to the bottom of the hill? Does rotation affect the outcome?

An object's initial gravitational potential energy is transformed into kinetic energy as it rolls (or slides, in the case of the particle). The kinetic energy, as we just discovered, is a combination of translational and rotational kinetic energy. If we choose the bottom of the ramp as the zero point of potential energy, the statement of energy conservation $K_f = U_i$ can be written

$$\tfrac{1}{2}I_{cm}\omega^2 + \tfrac{1}{2}Mv_{cm}^2 = Mgh \tag{12.41}$$

The translational and rotational velocities are related by $\omega = v_{cm}/R$. In addition, notice from Table 12.2 that the moments of inertia of all the objects can be written in the form

$$I_{cm} = cMR^2 \tag{12.42}$$

where c is a constant that depends on the object's geometry. For example, $c = \frac{2}{5}$ for a sphere but $c = 1$ for a circular hoop. Even the particle can be represented by $c = 0$, which eliminates the rotational kinetic energy.

With this information, Equation 12.41 becomes

$$\tfrac{1}{2}(cMR^2)\left(\frac{v_{cm}}{R}\right)^2 + \tfrac{1}{2}Mv_{cm}^2 = \tfrac{1}{2}M(1+c)v_{cm}^2 = Mgh$$

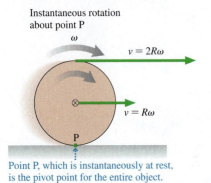

FIGURE 12.41 Rolling motion is an instantaneous rotation about point P.

Instantaneous rotation about point P

ω

$v = 2R\omega$

$v = R\omega$

P

Point P, which is instantaneously at rest, is the pivot point for the entire object.

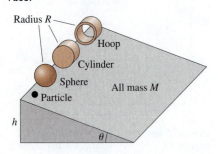

FIGURE 12.42 Which will win the downhill race?

Radius R

Hoop

Cylinder

Sphere

All mass M

Particle

h

θ

Thus the finishing speed of an object with $I = cMR^2$ is

$$v_{cm} = \sqrt{\frac{2gh}{1+c}} \qquad (12.43)$$

The final speed is independent of both M and R, but it does depend on the *shape* of the rolling object. The particle, with the smallest value of c, will finish with the highest speed, while the circular hoop, with the largest c, will be the slowest. In other words, the rolling aspect of the motion *does* matter!

We can use Equation 12.43 to find the acceleration a_{cm} of the center of mass. The objects move through distance $\Delta x = h/\sin\theta$, so we can use constant-acceleration kinematics to find

$$v_{cm}^2 = 2a_{cm}\,\Delta x$$

$$a_{cm} = \frac{v_{cm}^2}{2\Delta x} = \frac{2gh/(1+c)}{2h/\sin\theta} = \frac{g\sin\theta}{1+c} \qquad (12.44)$$

Recall, from Chapter 2, that $a_{particle} = g\sin\theta$ is the acceleration of a particle sliding down a frictionless incline. We can use this fact to write Equation 12.44 in an interesting form:

$$a_{cm} = \frac{a_{particle}}{1+c} \qquad (12.45)$$

This analysis leads us to the conclusion that **the acceleration of a rolling object is less—in some cases significantly less—than the acceleration of a particle.** The reason is that the energy has to be shared between translational kinetic energy and rotational kinetic energy. A particle, by contrast, can put all its energy into translational kinetic energy.

FIGURE 12.43 shows the results of the race. The simple particle wins by a fairly wide margin. Of the solid objects, the sphere has the largest acceleration. Even so, its acceleration is only 71% the acceleration of a particle. The acceleration of the circular hoop, which comes in last, is a mere 50% that of a particle.

> **NOTE** The objects having the largest acceleration are those whose mass is most concentrated near the center. Placing the mass far from the center, as in the hoop, increases the moment of inertia. Thus it requires a larger effort to get a hoop rolling than to get a sphere of equal mass rolling.

FIGURE 12.43 And the winner is …

$c = 0$
$a = a_{particle}$
Particle

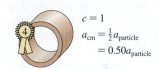

$c = \frac{2}{5}$
$a_{cm} = \frac{5}{7}a_{particle}$
$= 0.71a_{particle}$
Solid sphere

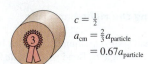

$c = \frac{1}{2}$
$a_{cm} = \frac{2}{3}a_{particle}$
$= 0.67a_{particle}$
Solid cylinder

$c = 1$
$a_{cm} = \frac{1}{2}a_{particle}$
$= 0.50a_{particle}$
Circular hoop

12.10 The Vector Description of Rotational Motion

Rotation about a fixed axis, such as an axle, can be described in terms of a scalar angular velocity ω and a scalar torque τ, using a plus or minus sign to indicate the direction of rotation. This is very much analogous to the one-dimensional kinematics of Chapter 2. For more general rotational motion, angular velocity, torque, and other quantities must be treated as *vectors*. We won't go into much detail because the subject rapidly gets very complicated, but we will sketch some important basic ideas.

The Angular Velocity Vector

FIGURE 12.44 shows a rotating rigid body. We can define an angular velocity vector $\vec{\omega}$ as follows:

- The magnitude of $\vec{\omega}$ is the object's angular velocity ω.
- $\vec{\omega}$ points along the axis of rotation in the direction given by the *right-hand rule* illustrated in Figure 12.44.

FIGURE 12.44 The angular velocity vector $\vec{\omega}$ is found using the right-hand rule.

1. Using your right hand, curl your fingers in the direction of rotation with your thumb along the rotation axis.

2. Your thumb is then pointing in the direction of $\vec{\omega}$.

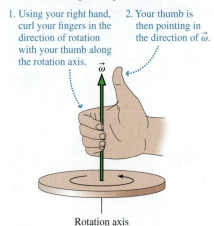

Rotation axis

If the object rotates in the xy-plane, the vector $\vec{\omega}$ points along the z-axis. The scalar angular velocity $\omega = v_t/r$ that we've been using is now seen to be ω_z, the z-component of the vector $\vec{\omega}$. You should convince yourself that the sign convention for ω (positive for ccw rotation, negative for cw rotation) is equivalent to having the vector $\vec{\omega}$ pointing in the positive z-direction or the negative z-direction.

The Cross Product of Two Vectors

We defined the torque exerted by force $\vec{F}$ to be $\tau = rF \sin \phi$. The quantity F is the magnitude of the force vector $\vec{F}$, and the distance r is really the magnitude of the position vector $\vec{r}$. Hence torque looks very much like a product of the two vectors $\vec{r}$ and $\vec{F}$. Previously, in conjunction with the definition of work, we introduced the dot product of two vectors: $\vec{A} \cdot \vec{B} = AB \cos \alpha$, where α is the angle between the vectors. $\tau = rF \sin \phi$ is a different way of multiplying vectors that depends on the *sine* of the angle between them.

FIGURE 12.45 shows two vectors, $\vec{A}$ and $\vec{B}$, with angle α between them. We define the **cross product** of $\vec{A}$ and $\vec{B}$ as the vector

$$\vec{A} \times \vec{B} \equiv (AB \sin \alpha, \text{ in the direction given by the right-hand rule}) \quad (12.46)$$

The symbol $\times$ between the vectors is *required* to indicate a cross product. The cross product is also called the **vector product** because the result is a vector.

The **right-hand rule,** which specifies the direction of $\vec{A} \times \vec{B}$, can be stated in three different but equivalent ways:

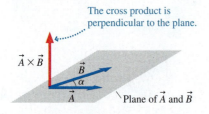

FIGURE 12.45 The cross product $\vec{A} \times \vec{B}$, is a vector perpendicular to the plane of vectors $\vec{A}$ and $\vec{B}$.

The cross product is perpendicular to the plane.

$\vec{A} \times \vec{B}$

$\vec{B}$

α

$\vec{A}$

Plane of $\vec{A}$ and $\vec{B}$

Using the right-hand rule

Spread your *right* thumb and index finger apart by angle α. Bend your middle finger so that it is *perpendicular* to your thumb and index finger. Orient your hand so that your thumb points in the direction of $\vec{A}$ and your index finger in the direction of $\vec{B}$. Your middle finger now points in the direction of $\vec{A} \times \vec{B}$.

Make a loose fist with your *right* hand with your thumb extended outward. Orient your hand so that your thumb is perpendicular to the plane of $\vec{A}$ and $\vec{B}$ and your fingers are curling *from* the line of vector $\vec{A}$ *toward* the line of vector $\vec{B}$. Your thumb now points in the direction of $\vec{A} \times \vec{B}$.

Imagine using a screwdriver to turn the slot in the head of a screw from the direction of $\vec{A}$ to the direction of $\vec{B}$. The screw will move either "in" or "out." The direction in which the screw moves is the direction of $\vec{A} \times \vec{B}$.

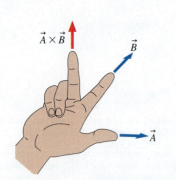

$\vec{A} \times \vec{B}$

$\vec{B}$

$\vec{A}$

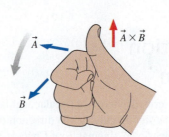

$\vec{A}$

$\vec{A} \times \vec{B}$

$\vec{B}$

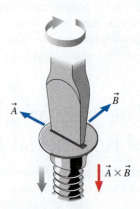

$\vec{A}$

$\vec{B}$

$\vec{A} \times \vec{B}$

These methods are easier to demonstrate than to describe in words! Your instructor will show you how they work. Some individuals find one method of thinking about the direction of the cross product easier than the others, but they all work, and you'll soon find the method that works best for you.

Referring back to Figure 12.45, you should use the right-hand rule to convince yourself that the cross product $\vec{A} \times \vec{B}$ is a vector that points *upward,* perpendicular to the plane of $\vec{A}$ and $\vec{B}$. FIGURE 12.46 shows that the cross product, like the dot product, depends on the angle between the two vectors. Notice the two special cases: $\vec{A} \times \vec{B} = \vec{0}$ when $\alpha = 0°$ (parallel vectors) and $\vec{A} \times \vec{B}$ has its maximum magnitude AB when $\alpha = 90°$ (perpendicular vectors).

FIGURE 12.46 The magnitude of the cross-product vector increases from 0 to AB as α increases from 0° to 90°.

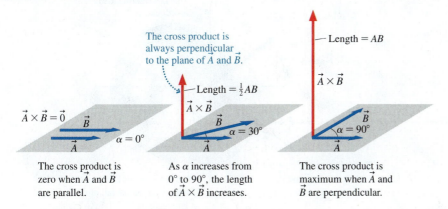

The cross product is zero when $\vec{A}$ and $\vec{B}$ are parallel.

As α increases from 0° to 90°, the length of $\vec{A} \times \vec{B}$ increases.

The cross product is maximum when $\vec{A}$ and $\vec{B}$ are perpendicular.

EXAMPLE 12.16 | **Calculating a cross product**

FIGURE 12.47 shows vectors $\vec{C}$ and $\vec{D}$ in the plane of the page. What is the cross product $\vec{E} = \vec{C} \times \vec{D}$?

FIGURE 12.47 Vectors $\vec{C}$ and $\vec{D}$.

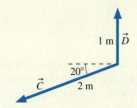

SOLVE The angle between the two vectors is $\alpha = 110°$. Consequently, the magnitude of the cross product is

$$E = CD \sin \alpha = (2\text{ m})(1\text{ m}) \sin(110°) = 1.88\text{ m}^2$$

The direction of $\vec{E}$ is given by the right-hand rule. To curl your right fingers from $\vec{C}$ to $\vec{D}$, you have to point your thumb *into* the page. Alternatively, if you turned a screwdriver from $\vec{C}$ to $\vec{D}$ you would be driving a screw *into* the page. Thus

$$\vec{E} = (1.88\text{ m}^2, \text{ into page})$$

ASSESS Notice that $\vec{E}$ has units of m².

The cross product has three important properties:

1. The product $\vec{A} \times \vec{B}$ is *not* equal to the product $\vec{B} \times \vec{A}$. That is, the cross product does not obey the commutative rule $ab = ba$ that you know from arithmetic. In fact, you can see from the right-hand rule that the product $\vec{B} \times \vec{A}$ points in exactly the opposite direction from $\vec{A} \times \vec{B}$. Thus, as **FIGURE 12.48a** shows,

$$\vec{B} \times \vec{A} = -\vec{A} \times \vec{B}$$

2. In a *right-handed coordinate system,* which is the standard coordinate system of science and engineering, the z-axis is oriented relative to the xy-plane such that the unit vectors obey $\hat{\imath} \times \hat{\jmath} = \hat{k}$. This is shown in **FIGURE 12.48b**. You can also see from this figure that $\hat{\jmath} \times \hat{k} = \hat{\imath}$ and $\hat{k} \times \hat{\imath} = \hat{\jmath}$.

3. The derivative of a cross product is

$$\frac{d}{dt}(\vec{A} \times \vec{B}) = \frac{d\vec{A}}{dt} \times \vec{B} + \vec{A} \times \frac{d\vec{B}}{dt} \qquad (12.47)$$

Torque

Now let's return to torque. As a concrete example, **FIGURE 12.49** on the next page shows a long wrench being used to loosen the nuts holding a car wheel on. We've established a right-handed coordinate system with its origin at the nut, so force $\vec{F}$ exerts a torque about the origin. Let's define a *torque vector*

$$\vec{\tau} \equiv \vec{r} \times \vec{F} \qquad (12.48)$$

If we place the vector tails together in order to use the right-hand rule, we see that the torque vector is perpendicular to the plane of $\vec{r}$ and $\vec{F}$. The angle between the vectors is ϕ, so the magnitude of the torque is $\tau = rF|\sin\phi|$.

FIGURE 12.48 Properties of the cross product.

(a)

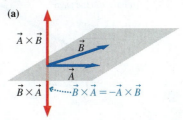

(b)

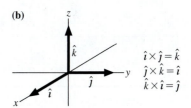

$$\hat{\imath} \times \hat{\jmath} = \hat{k}$$
$$\hat{\jmath} \times \hat{k} = \hat{\imath}$$
$$\hat{k} \times \hat{\imath} = \hat{\jmath}$$

FIGURE 12.49 The torque vector.

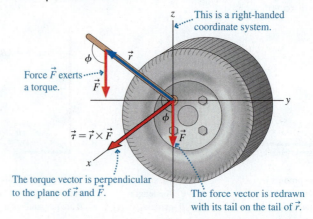

FIGURE 12.49 The torque vector.

You can see that the scalar torque $\tau = rF\sin\phi$ we've been using is really the component along the rotation axis—in this case τ_x—of the vector $\vec{\tau}$. This is the basis for our earlier sign convention for τ. In Figure 12.49, where the force causes a ccw rotation, the torque vector points in the positive x-direction, and thus τ_x is positive.

EXAMPLE 12.17 | **Wrench torque revisited**

Example 12.8 found the torque that Luis exerts on a nut by pulling on the end of a wrench. What is the torque vector?

VISUALIZE **FIGURE 12.50** shows the position vector $\vec{r}$, drawn from the pivot point to the point where the force is applied. The figure

FIGURE 12.50 Calculating the torque vector.

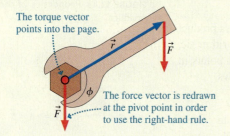

also redraws the force vector $\vec{F}$ at the pivot point, not because force is applied there but because it's easiest to use the right-hand rule if the vectors are drawn with their tails together.

SOLVE We already know the magnitude of the torque, 17 N m, from Example 12.8. Now we need to apply the right-hand rule. If you place your right thumb along $\vec{r}$ and your index finger along $\vec{F}$, which is somewhat awkward, you'll see that your middle finger points into the page. Alternatively, make a loose fist of your right hand, then orient your fist so that your fingers curl *from* $\vec{r}$ *toward* $\vec{F}$. Doing so requires your thumb to point into the page. Using either method, we conclude that

$$\vec{\tau} = (17\text{ N m, into page})$$

12.11 Angular Momentum

FIGURE 12.51 shows a particle that, at this instant, is located at position $\vec{r}$ and is moving with momentum $\vec{p} = m\vec{v}$. Together, $\vec{r}$ and $\vec{p}$ define the *plane of motion*. We define the particle's **angular momentum** $\vec{L}$ relative to the origin to be the vector

$$\vec{L} \equiv \vec{r} \times \vec{p} = (mrv\sin\beta, \text{ direction of right-hand rule}) \qquad (12.49)$$

Because of the cross product, **the angular momentum vector is perpendicular to the plane of motion.** The units of angular momentum are kg m²/s.

NOTE Angular momentum is the rotational equivalent of linear momentum in much the same way that torque is the rotational equivalent of force. Notice that the vector definitions are parallel: $\vec{\tau} \equiv \vec{r} \times \vec{F}$ and $\vec{L} \equiv \vec{r} \times \vec{p}$.

FIGURE 12.51 The angular momentum vector $\vec{L}$.

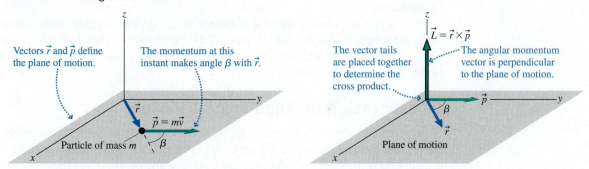

FIGURE 12.51 The angular momentum vector $\vec{L}$.

Angular momentum, like torque, is *about* the point from which $\vec{r}$ is measured. A different origin would yield a different angular momentum. Angular momentum is especially simple for a particle in circular motion. As **FIGURE 12.52** shows, the angle β between $\vec{p}$ (or $\vec{v}$) and $\vec{r}$ is always 90° if we make the obvious choice of measuring $\vec{r}$ from the center of the circle. For motion in the *xy*-plane, the angular momentum vector $\vec{L}$—which must be perpendicular to the plane of motion—is entirely along the *z*-axis:

$$L_z = mrv_t \qquad \text{(particle in circular motion)} \qquad (12.50)$$

where v_t is the tangential component of velocity. Our sign convention for v_t makes L_z, like ω, positive for a ccw rotation, negative for a cw rotation.

In Chapter 11, we found that Newton's second law for a particle can be written $\vec{F}_{net} = d\vec{p}/dt$. There's a similar connection between torque and angular momentum. To show this, we take the time derivative of $\vec{L}$:

$$\frac{d\vec{L}}{dt} = \frac{d}{dt}(\vec{r} \times \vec{p}) = \frac{d\vec{r}}{dt} \times \vec{p} + \vec{r} \times \frac{d\vec{p}}{dt} \qquad (12.51)$$
$$= \vec{v} \times \vec{p} + \vec{r} \times \vec{F}_{net}$$

where we used Equation 12.47 for the derivative of a cross product. We also used the definitions $\vec{v} = d\vec{r}/dt$ and $\vec{F}_{net} = d\vec{p}/dt$.

Vectors $\vec{v}$ and $\vec{p}$ are parallel, and the cross product of two parallel vectors is $\vec{0}$. Thus the first term in Equation 12.51 vanishes. The second term $\vec{r} \times \vec{F}_{net}$ is the net torque, $\vec{\tau}_{net} = \vec{\tau}_1 + \vec{\tau}_2 + \cdots$, so we arrive at

$$\frac{d\vec{L}}{dt} = \vec{\tau}_{net} \qquad (12.52)$$

Equation 12.52, which says **a net torque causes the particle's angular momentum to change**, is the rotational equivalent of $d\vec{p}/dt = \vec{F}_{net}$.

Angular Momentum of a Rigid Body

Equation 12.52 is the angular momentum of a single particle. The angular momentum of a rigid body composed of particles with individual angular momenta $\vec{L}_1, \vec{L}_2, \vec{L}_3, \ldots$ is the vector sum

$$\vec{L} = \vec{L}_1 + \vec{L}_2 + \vec{L}_3 + \cdots = \sum_i \vec{L}_i \qquad (12.53)$$

We can combine Equations 12.52 and 12.53 to find the rate of change of the system's angular momentum:

$$\frac{d\vec{L}}{dt} = \sum_i \frac{d\vec{L}_i}{dt} = \sum_i \vec{\tau}_i = \vec{\tau}_{net} \qquad (12.54)$$

Because any internal forces are action/reaction pairs of forces, acting with the same strength in opposite directions, the net torque due to internal forces is zero. Thus the

FIGURE 12.52 Angular momentum of circular motion.

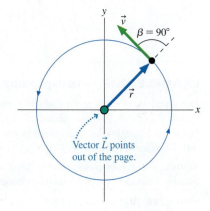

only forces that contribute to the net torque are external forces exerted on the system by the environment.

For a system of particles, **the rate of change of the system's angular momentum is the net torque on the system.** Equation 12.54 is analogous to the Chapter 11 result $d\vec{P}/dt = \vec{F}_{net}$, which says that the rate of change of a system's total linear momentum is the net force on the system.

Conservation of Angular Momentum

A net torque on a rigid body causes its angular momentum to change. Conversely, the angular momentum does *not* change—it is *conserved*—for a system with no net torque. This is the basis of the law of conservation of angular momentum.

> **Law of conservation of angular momentum** The angular momentum $\vec{L}$ of an isolated system ($\vec{\tau}_{net} = \vec{0}$) is conserved. The final angular momentum $\vec{L}_f$ is equal to the initial angular momentum $\vec{L}_i$. Both the magnitude *and* the direction of $\vec{L}$ are unchanged.

EXAMPLE 12.18 | An expanding rod

Two equal masses are at the ends of a massless 50-cm-long rod. The rod spins at 2.0 rev/s about an axis through its midpoint. Suddenly, a compressed gas expands the rod out to a length of 160 cm. What is the rotation frequency after the expansion?

MODEL The forces push outward from the pivot and exert no torques. Thus the system's angular momentum is conserved.

VISUALIZE FIGURE 12.53 is a before-and-after pictorial representation. The angular momentum vectors $\vec{L}_i$ and $\vec{L}_f$ are perpendicular to the plane of motion.

SOLVE The particles are moving in circles, so each has angular momentum $L = mrv_t = mr^2\omega = \frac{1}{4}ml^2\omega$, where we used $r = \frac{1}{2}l$. Thus the initial angular momentum of the system is

$$L_i = \tfrac{1}{4}ml_i^2\omega_i + \tfrac{1}{4}ml_i^2\omega_i = \tfrac{1}{2}ml_i^2\omega_i$$

Similarly, the angular momentum after the expansion is $L_f = \frac{1}{2}ml_f^2\omega_f$. Angular momentum is conserved as the rod expands, thus

$$\tfrac{1}{2}ml_f^2\omega_f = \tfrac{1}{2}ml_i^2\omega_i$$

Solving for ω_f, we find

$$\omega_f = \left(\frac{l_i}{l_f}\right)^2 \omega_i = \left(\frac{50 \text{ cm}}{160 \text{ cm}}\right)^2 (2.0 \text{ rev/s}) = 0.20 \text{ rev/s}$$

ASSESS The values of the masses weren't needed. All that matters is the ratio of the lengths.

FIGURE 12.53 The system before and after the rod expands.

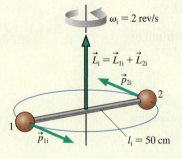
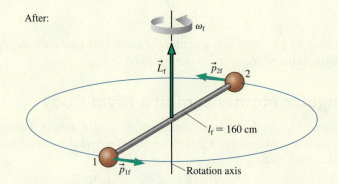

Before:

After:

Angular Momentum and Angular Velocity

The analogy between linear and rotational motion has been so consistent that you might expect one more. The Chapter 9 result $\vec{P} = M\vec{v}$, which we can now write as $M\vec{v}_{cm}$ because it is translational motion of the object as a whole, might give us reason to anticipate that angular momentum and angular velocity are related by $\vec{L} = I\vec{\omega}$. Unfortunately, the analogy breaks down here. For an arbitrarily shaped object, the angular momentum vector and the angular velocity vector don't necessarily point in the same direction. The general relationship between $\vec{L}$ and $\vec{\omega}$ is beyond the scope of this text.

The good news is that the analogy *does* continue to hold in two important situations: the rotation of a *symmetrical* object about the symmetry axis and the rotation of any object about a fixed axle. For example, the axis of a cylinder or disk is a symmetry axis, as is any diameter through a sphere. In these two situations, the angular momentum and angular velocity are related by

$$\vec{L} = I\vec{\omega} \qquad \text{(rotation about a fixed axle or axis of symmetry)} \qquad (12.55)$$

This relationship is shown for a spinning disk in **FIGURE 12.54**. Equation 12.55 is particularly important for applying the law of conservation of angular momentum.

If an object's angular momentum is conserved, its angular speed is inversely proportional to its moment of inertia. The rotation of the rod in Example 12.18 slowed dramatically as it expanded because its moment of inertia increased. Similarly, the ice skater in **FIGURE 12.55** uses her moment of inertia to control her spin. She spins faster if she pulls in her arms, decreasing her moment of inertia. Similarly, extending her arms increases her moment of inertia, and her angular velocity drops until she can skate out of the spin. It's all a matter of conserving angular momentum.

TABLE 12.4 summarizes the analogies between linear and angular quantities.

FIGURE 12.54 The angular momentum vector about an axis of symmetry.

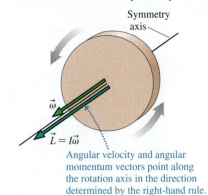

Angular velocity and angular momentum vectors point along the rotation axis in the direction determined by the right-hand rule.

TABLE 12.4 Angular and linear momentum and energy

Angular motion	Linear motion
$K_{rot} = \frac{1}{2}I\omega^2$	$K_{cm} = \frac{1}{2}Mv_{cm}^2$
$\vec{L} = I\vec{\omega}$ *	$\vec{P} = M\vec{v}_{cm}$
$d\vec{L}/dt = \vec{\tau}_{net}$	$d\vec{P}/dt = \vec{F}_{net}$
The angular momentum of a system is conserved if there is no net torque.	The linear momentum of a system is conserved if there is no net force.

*Rotation about an axis of symmetry.

FIGURE 12.55 An ice skater's rotation speed depends on her moment of inertia.

Large moment of inertia; slow spin

Small moment of inertia; fast spin

EXAMPLE 12.19 | Two interacting disks

A 20-cm-diameter, 2.0 kg solid disk is rotating at 200 rpm. A 20-cm-diameter, 1.0 kg circular loop is dropped straight down onto the rotating disk. Friction causes the loop to accelerate until it is "riding" on the disk. What is the final angular velocity of the combined system?

MODEL The friction between the two objects creates torques that speed up the loop and slow down the disk. But these torques are internal to the combined disk + loop system, so $\tau_{net} = 0$ and the *total* angular momentum of the disk + loop system is conserved.

VISUALIZE FIGURE 12.56 is a before-and-after pictorial representation. Initially only the disk is rotating, at angular velocity $\vec{\omega}_i$. The rotation is about an axis of symmetry, so the angular momentum $\vec{L} = I\vec{\omega}$ is parallel to $\vec{\omega}$. At the end of the problem, $\vec{\omega}_{disk} = \vec{\omega}_{loop} = \vec{\omega}_f$.

SOLVE Both angular momentum vectors point along the rotation axis. Conservation of angular momentum tells us that the magnitude of $\vec{L}$ is unchanged. Thus

$$L_f = I_{disk}\omega_f + I_{loop}\omega_f = L_i = I_{disk}\omega_i$$

Solving for ω_f gives

$$\omega_f = \frac{I_{disk}}{I_{disk} + I_{loop}}\omega_i$$

FIGURE 12.56 The circular loop drops onto the rotating disk.

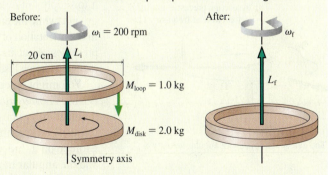

The moments of inertia for a disk and a loop can be found in Table 12.2, leading to

$$\omega_f = \frac{\frac{1}{2}M_{disk}R^2}{\frac{1}{2}M_{disk}R^2 + M_{loop}R^2}\omega_i = 100 \text{ rpm}$$

ASSESS The angular velocity has been reduced to half its initial value, which seems reasonable.

STOP TO THINK 12.7 Two buckets spin around in a horizontal circle on frictionless bearings. Suddenly, it starts to rain. As a result,

a. The buckets continue to rotate at constant angular velocity because the rain is falling vertically while the buckets move in a horizontal plane.
b. The buckets continue to rotate at constant angular velocity because the total mechanical energy of the bucket + rain system is conserved.
c. The buckets speed up because the potential energy of the rain is transformed into kinetic energy.
d. The buckets slow down because the angular momentum of the bucket + rain system is conserved.
e. Both a and b.   f. None of the above.

12.12 ADVANCED TOPIC Precession of a Gyroscope

Rotating objects can exhibit surprising and unexpected behaviors. For example, a common lecture demonstration makes use of a bicycle wheel with two handles along the axis. The wheel is spun, then handed to an unsuspecting student who is asked to turn the spinning wheel 90°. Surprisingly, this is *very hard to do.* The reason is that the angular momentum is a *vector*, so the wheel's rotation axis—the direction of $\vec{L}$—is highly resistant to change. If the wheel is spinning fast, a *large* torque is required to turn the wheel's axis.

We'll look at a related example: the precession of a gyroscope. A **gyroscope**—whether it's a toy or a precision instrument used for navigation—is a rapidly spinning wheel or disk whose axis of rotation can assume any orientation. As it spins, it has angular momentum $\vec{L} = I\vec{\omega}$ along the rotation axis. A navigation gyroscope is mounted in gimbals that allow it to spin with virtually no torque from the environment. Once its axis is pointed north, conservation of angular momentum will ensure that the axis continues to point north no matter how the ship or plane moves.

We want to consider a horizontal gyroscope, with the disk spinning in a vertical plane, that is supported at only one end of its axle, as shown in **FIGURE 12.57**. You would expect it to simply fall over—but it doesn't. Instead, the axle remains horizontal, parallel to the ground, while the entire gyroscope slowly rotates in a horizontal plane. This steady change in the orientation of the rotation axis is called **precession,** and we say that the gyroscope precesses about its point of support. The **precession frequency** Ω (capital Greek omega) is much less than the disk's rotation frequency ω. Note that Ω, like ω, is in rad/s.

You might object that angular momentum is not conserved during precession. This is true. The *magnitude* of $\vec{L}$ is constant, but its *direction* is changing. However, angular momentum is conserved only for an isolated system, one on which there is no net torque. The spinning gyroscope is *not* an isolated system because gravity is exerting a torque on it. Indeed, understanding the relationship between the gravitational torque and the angular momentum is the key to understanding why the gyroscope precesses.

FIGURE 12.58a shows a gyroscope that is *not* spinning. When released, it most definitely falls over by rotating about the point of support until the disk hits the table. Because the motion is *rotation*, rather than the translational motion of a gyroscope that is simply dropped, we can analyze it using the concepts of torque and angular momentum.

There are two forces acting on the gyroscope: gravity pulling downward at the disk's center of mass (we'll assume that the axle is massless) and the normal force of the support pushing upward. The normal force exerts no torque about the pivot point because it acts at the pivot point, so **the net torque on the gyroscope is entirely a gravitational torque:**

$$\vec{\tau} = \vec{r} \times \vec{F}_G = Mgd\,\hat{\imath} \qquad (12.56)$$

FIGURE 12.57 A spinning gyroscope precesses in a horizontal plane.

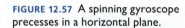
The gyroscope precesses around a horizontal circle at frequency Ω.

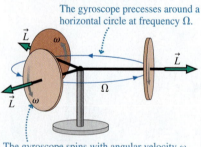
The gyroscope spins with angular velocity ω.

where $d = |\vec{r}|$ is the distance from the pivot to the center of the disk. To evaluate the cross product, we redrew $\vec{F}_G$ at the pivot and established a coordinate system with the z-axis along the axle. Vectors $\vec{r}$ and $\vec{F}_G$ are perpendicular ($\sin\alpha = 1$), and by using the right-hand rule we see that $\vec{\tau}$ points along the x-axis.

We found in the previous section that a torque causes the angular momentum to change. In particular,

$$\frac{d\vec{L}}{dt} = \vec{\tau} \tag{12.57}$$

So in a small interval of time dt, the torque causes the gyroscope's angular momentum about the point of support to change by $d\vec{L} = \vec{\tau}\, dt$.

FIGURE 12.58b shows graphically what happens. Initially, when the gyroscope is first released, $\vec{L} = \vec{0}$. After a small interval of time, the gyroscope acquires a small amount of angular momentum $d\vec{L}$ in the direction of $\vec{\tau}$, the $\hat{\imath}$ direction. An angular momentum along the x-axis means that the gyroscope is rotating in the yz-plane—which is exactly what it does as it starts to fall. During the next interval of time, $\vec{L}$ increases a bit more in the $\hat{\imath}$ direction, and then a bit more. This is what we expect as the falling gyroscope picks up speed, increasing its angular momentum.

Now the magnitude of $\vec{\tau}$ does not remain constant—the angle between $\vec{r}$ and $\vec{F}_G$ changes as the gyroscope falls, changing the cross product—so integrating Equation 12.57 symbolically is very difficult. Nonetheless, the *direction* of $\vec{\tau}$ is always the $\hat{\imath}$ direction, so we can see that the angular momentum keeps increasing in the $\hat{\imath}$ direction as the gyroscope falls.

What's different about a spinning gyroscope that causes it to precess rather than fall? In **FIGURE 12.59a** we've again just released the gyroscope, its axle is again along the z-axis, but now it's spinning with angular velocity $\vec{\omega} = \omega\hat{k}$. Consequently, the gyroscope has initial angular momentum $\vec{L} = I\vec{\omega} = I\omega\hat{k}$ along the z-axis. The torque is exactly as we calculated above, and that torque again causes the angular momentum to change by $d\vec{L} = \vec{\tau}\,dt$. **The only difference is that the gyroscope starts with initial angular momentum—but that makes all the difference.**

FIGURE 12.59b, looking down from above, shows the initial angular momentum $\vec{L}$. A very small time interval dt after we release the gyroscope, its angular momentum will have changed to $\vec{L} + d\vec{L}$. The small *change* in angular momentum, $d\vec{L}$, is parallel to the torque and thus *perpendicular* to the spinning gyroscope's angular momentum $\vec{L}$. Because we're adding vectors, not scalars, the "new" angular momentum has rotated to a new position but *not* increased in magnitude. So during dt, the angular momentum—and thus the entire gyroscope—rotates through a small angle $d\phi$ in the horizontal plane.

The torque vector is always perpendicular to the axle, and thus $d\vec{L}$ is always perpendicular to $\vec{L}$. With each subsequent time interval dt, the gyroscope rotates through another small angle $d\phi$ while the magnitude of the angular momentum (and hence the disk's angular velocity ω) is unchanged. The gyroscope is precessing in the horizontal plane!

You've encountered a similar situation previously. If a ball is initially at rest, pulling on it with a string causes the ball to accelerate (increasing $\vec{v}$) in the direction of the pull. That is, $\vec{v}$ increases in magnitude but doesn't change direction. But if a ball on a string is in uniform circular motion, a force directed to the center—the string tension—has a very different effect. $d\vec{v}$ points toward the center, because that's the direction of the centripetal acceleration, but now $d\vec{v}$ is perpendicular to $\vec{v}$. Adding them as vectors to get $\vec{v} + d\vec{v}$ changes the *direction* of the velocity vector but not its magnitude. Then, as now, having an initial vector ($\vec{v}$ or $\vec{L}$) leads to very different behavior than not having an initial vector.

The small horizontal rotation $d\phi$ is a small piece of the precession. Because it occurs during the small time interval dt, the *rate* of horizontal rotation—the precession frequency—is

$$\Omega = \frac{d\phi}{dt} \tag{12.58}$$

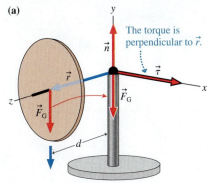

FIGURE 12.58 The gravitational torque on a nonspinning gyroscope causes it to fall over.

(a)

The torque is perpendicular to $\vec{r}$.

The gyroscope falls.

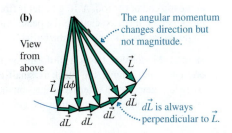

(b)

The initial angular momentum is zero.

The angular momentum increases in the direction of the torque.

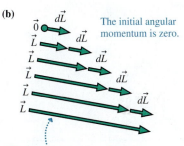

FIGURE 12.59 For a spinning gyroscope, the gravitational torque changes the direction but not the magnitude of the angular momentum.

(a)

The torque hasn't changed.

The gyroscope precesses.

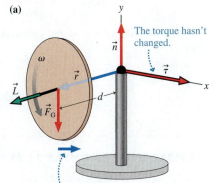

(b)

View from above

The angular momentum changes direction but not magnitude.

$d\vec{L}$ is always perpendicular to $\vec{L}$.

In Figure 12.59b the small length dL is the arc length spanned by $d\phi$, hence $d\phi = dL/L$ and the precession frequency is

$$\Omega = \frac{dL/L}{dt} = \frac{dL/dt}{L} \qquad (12.59)$$

From Equations 12.56 and 12.57, $dL/dt = \tau = Mgd$. Further, the gyroscope's angular momentum has magnitude $L = I\omega$, where I is the moment of inertia of the disk rotating about the axle. Thus the precession frequency of the gyroscope—in rad/s—is

$$\Omega = \frac{Mgd}{I\omega} \qquad (12.60)$$

Because the spin angular velocity ω is in the denominator, a very rapidly spinning gyroscope precesses very slowly. As the gyroscope runs down, due to any little bit of friction, it begins to precess faster and faster.

We have made one tacit assumption. As the gyroscope precesses, the precessional motion has its own angular momentum along the vertical axis. The gyroscope's angular momentum $\vec{L}$ is not simply the angular momentum of the spinning disk, as we assumed, but the vector sum $\vec{L}_{\text{spin}} + \vec{L}_{\text{precess}}$. As long as the gyroscope precesses slowly, with $\Omega \ll \omega$, the precessional angular momentum is very small compared to the spin angular momentum and our assumption is well justified. But toward the end of the gyroscope's motion, as ω decreases and Ω increases, our model of precession breaks down and the gyroscope's motion becomes more complex.

EXAMPLE 12.20 | A precessing gyroscope

A gyroscope used in a lecture demonstration consists of a 120 g, 7.0-cm-diameter solid disk that rotates on a lightweight axle. From the center of the disk to the end of the axle is 5.0 cm. When spun, placed on a stand, and released, the gyroscope is observed to precess with a period of 1.0 s. How fast, in rpm, is it spinning?

SOLVE The precession frequency is given by Equation 12.60. The moment of inertia of a disk of mass M and radius R about an axis through its center is $I = \frac{1}{2}MR^2$. Inserting this into Equation 12.60, we see that the precession frequency

$$\Omega = \frac{Mgd}{I\omega} = \frac{Mgd}{\frac{1}{2}MR^2\omega} = \frac{2gd}{\omega R^2}$$

is actually independent of the gyroscope's mass. Solving for ω gives

$$\omega = \frac{2gd}{\Omega R^2}$$

A precession period of 1.0 s corresponds to the precession frequency

$$\Omega = \frac{2\pi \text{ rad}}{1.0 \text{ s}} = 6.28 \text{ rad/s}$$

Thus the gyroscope's spin angular velocity is

$$\omega = \frac{2(9.80 \text{ m/s}^2)(0.050 \text{ m})}{(6.28 \text{ rad/s})(0.035 \text{ m})^2} = 127 \text{ rad/s}$$

Converting to rpm gives

$$\omega = 127 \text{ rad/s} \times \frac{1 \text{ rev}}{2\pi \text{ rad}} \times \frac{60 \text{ s}}{1 \text{ min}} = 1200 \text{ rpm}$$

ASSESS 1200 rpm is 20 rev/s. That seems reasonable for a spinning top or gyroscope. And $\Omega \ll \omega$, so our precession model of the gyroscope is valid.

CHALLENGE EXAMPLE 12.21 | The ballistic pendulum revisited

A 2.0 kg block hangs from the end of a 1.5 kg, 1.0-m-long rod, together forming a pendulum that swings from a frictionless pivot at the top end of the rod. A 10 g bullet is fired horizontally into the block, where it sticks, causing the pendulum to swing out to a 30° angle. What was the speed of the bullet?

MODEL Model the rod as a uniform rod that can rotate around one end, and assume the block is small enough to model as a particle. There are no external torques on the bullet + block + rod system, so angular momentum is conserved in the inelastic

collision. Further, the mechanical energy of the system is conserved after (but not during) the collision as the pendulum swings outward.

VISUALIZE FIGURE 12.60 is a pictorial representation. This is a two-part problem, so we've separated the collision's before-and-after from the pendulum swing's before-and-after. The end of the collision is the beginning of the swing.

SOLVE This is a *ballistic pendulum*. Example 11.5 considered a simpler ballistic pendulum with a mass on a string, rather than on

FIGURE 12.60 Pictorial representation of the bullet hitting the pendulum.

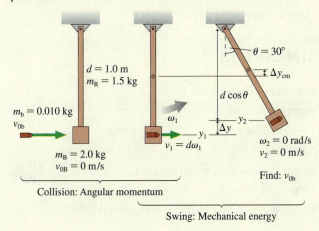

Collision: Angular momentum

Swing: Mechanical energy

a rod, and a review of that example is highly recommended. The key to both is that a different conservation law applies to each part of the problem.

Angular momentum is conserved in the collision, thus $L_1 = L_0$. Before the collision, the angular momentum—which we'll measure about the pendulum's pivot point—is entirely that of the bullet. The angular momentum of a particle is $L = mrv \sin \beta$. An instant before the collision, just as the bullet reaches the block, $r = d$ and, because $\vec{v}$ is perpendicular to $\vec{r}$ at that instant, $\beta = 90°$. Thus $L_0 = m_b d v_{0b}$. (This is the magnitude of the angular momentum; from the right-hand rule, the angular momentum vector points out of the page.)

An instant after the collision, but before the pendulum has had time to move, the rod has angular velocity ω_1 and the block, with the embedded bullet, is moving in a circle with speed $v_1 = \omega_1 r = \omega_1 d$. The angular momentum of the block + bullet system is that of a particle, still with $\beta = 90°$, while that of the rod—an object rotating on a fixed axle—is $I_{rod}\omega_1$. Thus the post-collision angular momentum is

$$L_1 = (m_B + m_b)v_1 r + I_{rod}\omega_1 = (m_B + m_b)d^2\omega_1 + \tfrac{1}{3}m_R d^2\omega_1$$

The moment of inertia of the rod was taken from Table 12.2.

Equating the before-and-after angular momenta, then solving for v_{0b}, gives

$$m_b d v_{0b} = (m_B + m_b)d^2\omega_1 + \tfrac{1}{3}m_R d^2\omega_1$$

$$v_{0b} = \frac{m_B + m_b + \tfrac{1}{3}m_R}{m_b} d\omega_1 = 251 d\omega_1$$

Once we know ω_1, which we'll find from energy conservation in the swing, we'll be able to compute the bullet's speed.

Mechanical energy is conserved during the swing, but you must be careful to include all the energies. The kinetic energy has two components: the translational kinetic energy of the block + bullet system and the rotational kinetic energy of the rod. The gravitational potential energy also has two components: the potential energy of the block + bullet system and the potential energy of the rod. The latter changes because the center of mass moves upward as the rod swings. Thus the energy conservation statement is

$$\tfrac{1}{2}(m_B + m_b)v_2^2 + \tfrac{1}{2}I_{rod}\omega_2^2 + (m_B + m_b)gy_2 + m_R g y_{cm2} =$$
$$\tfrac{1}{2}(m_B + m_b)v_1^2 + \tfrac{1}{2}I_{rod}\omega_1^2 + (m_B + m_b)gy_1 + m_R g y_{cm1}$$

Although this looks very complicated, you should convince yourself that we've done nothing more than add up two kinetic energies and two potential energies before and after the swing.

We know that $v_2 = 0$ and $\omega_2 = 0$ at the end of the swing, and that $v_1 = d\omega_1$ at the beginning. We also know the moment of inertia of a rod pivoted at one end. Combining the potential energy terms and using $\Delta y = y_f - y_i$, we thus have

$$\tfrac{1}{2}(m_B + m_b + \tfrac{1}{3}m_R)d^2\omega_1^2 = (m_B + m_b)g\,\Delta y + m_R g\,\Delta y_{cm}$$

We see from Figure 12.60 that the block, at its highest point, is distance $d \cos\theta$ *below* the pivot. It started distance d below the pivot, so the bullet + block system *gained* height $\Delta y = d - d\cos\theta = d(1 - \cos\theta)$. The rod's center of mass started distance $d/2$ below the pivot and rises only half as much as the block, so $\Delta y_{cm} = \tfrac{1}{2}d(1 - \cos\theta)$. With these, the energy equation becomes

$$\tfrac{1}{2}(m_B + m_b + \tfrac{1}{3}m_R)d^2\omega_1^2 = (m_B + m_b + \tfrac{1}{2}m_R)gd(1 - \cos\theta)$$

We can now solve for ω_1:

$$\omega_1 = \sqrt{\frac{m_B + m_b + \tfrac{1}{2}m_R}{m_B + m_b + \tfrac{1}{3}m_R} \frac{2\,g(1 - \cos\theta)}{d}} = 1.70 \text{ rad/s}$$

and with that

$$v_{0b} = 251 d\omega_1 = 430 \text{ m/s}$$

ASSESS 430 m/s seems a reasonable speed for a bullet. This was a challenging problem, but one that you can solve if you focus on the problem-solving strategies—drawing a careful pictorial representation, defining the system, and thinking about which conservation laws apply—rather than hunting for the "right" equation.

SUMMARY

The goal of Chapter 12 has been to understand and apply the physics of rotation.

GENERAL PRINCIPLES

Solving Rotational Dynamics Problems

MODEL Model the object as a **rigid body**.

VISUALIZE Draw a pictorial representation.

SOLVE Use **Newton's second law** for rotational motion:

$$\alpha = \frac{\tau_{net}}{I}$$

Use rotational kinematics to find angles and angular velocities.

ASSESS Is the result reasonable?

Conservation Laws

Energy is conserved for an isolated system.

- Pure rotation $E = K_{rot} + U_G = \frac{1}{2}I\omega^2 + Mgy_{cm}$
- Rolling $E = K_{rot} + K_{cm} + U_G = \frac{1}{2}I\omega^2 + \frac{1}{2}Mv_{cm}^2 + Mgy_{cm}$

Angular momentum is conserved if $\vec{\tau}_{net} = \vec{0}$.

- Particle $\vec{L} = \vec{r} \times \vec{p}$
- Rotation about a symmetry axis or fixed axle $\vec{L} = I\vec{\omega}$

IMPORTANT CONCEPTS

Torque is the rotational equivalent of force:

$$\tau = rF \sin\phi = rF_t = dF$$

The vector description of torque is

$$\vec{\tau} = \vec{r} \times \vec{F}$$

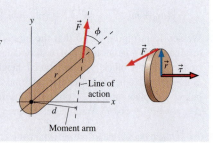

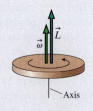

Vector description of rotation

Angular velocity $\vec{\omega}$ points along the rotation axis in the direction of the right-hand rule.

For a rigid body rotating about a fixed axle or an axis of symmetry, the angular momentum is $\vec{L} = I\vec{\omega}$.

Newton's second law is $\dfrac{d\vec{L}}{dt} = \vec{\tau}_{net}$.

A system of particles on which there is no net force undergoes unconstrained rotation about the **center of mass**:

$$x_{cm} = \frac{1}{M}\int x \, dm \qquad y_{cm} = \frac{1}{M}\int y \, dm$$

The gravitational torque on a body can be found by treating the body as a particle with all the mass M concentrated at the center of mass.

The **moment of inertia**

$$I = \sum_i m_i r_i^2 = \int r^2 \, dm$$

is the rotational equivalent of mass. The moment of inertia depends on how the mass is distributed around the axis. If I_{cm} is known, I about a parallel axis distance d away is given by the **parallel-axis theorem:** $I = I_{cm} + Md^2$.

APPLICATIONS

Rigid-body model

- Size and shape do not change as the object moves.
- The object is modeled as particle-like atoms connected by massless, rigid rods.

Rigid-body equilibrium

An object is in total equilibrium only if both $\vec{F}_{net} = \vec{0}$ and $\vec{\tau}_{net} = \vec{0}$.

No rotational or translational motion

Rolling motion

For an object that rolls without slipping

$$v_{cm} = R\omega$$
$$K = K_{rot} + K_{cm}$$

TERMS AND NOTATION

rigid-body model	rotational kinetic energy, K_{rot}	constant-torque model	angular momentum, $\vec{L}$
rigid body	moment of inertia, I	static equilibrium model	law of conservation of angular
translational motion	parallel-axis theorem	rolling constraint	momentum
rotational motion	torque, τ	cross product	gyroscope
combination motion	line of action	vector product	precession
center of mass	moment arm, d	right-hand rule	precession frequency

CONCEPTUAL QUESTIONS

1. Is the center of mass of the dumbbell in **FIGURE Q12.1** at point a, b, or c? Explain.

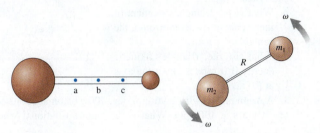

FIGURE Q12.1

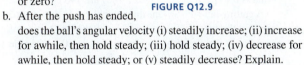

FIGURE Q12.2

2. If the angular velocity ω is held constant, by what *factor* must R change to double the rotational kinetic energy of the dumbbell in **FIGURE Q12.2**?

3. **FIGURE Q12.3** shows three rotating disks, all of equal mass. Rank in order, from largest to smallest, their rotational kinetic energies K_a to K_c.

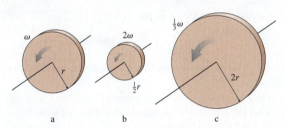

FIGURE Q12.3

4. Must an object be rotating to have a moment of inertia? Explain.

5. The moment of inertia of a uniform rod about an axis through its center is $\frac{1}{12} mL^2$. The moment of inertia about an axis at one end is $\frac{1}{3} mL^2$. Explain *why* the moment of inertia is larger about the end than about the center.

6. You have two solid steel spheres. Sphere 2 has twice the radius of sphere 1. By what *factor* does the moment of inertia I_2 of sphere 2 exceed the moment of inertia I_1 of sphere 1?

7. The professor hands you two spheres. They have the same mass, the same radius, and the same exterior surface. The professor claims that one is a solid sphere and the other is hollow. Can you determine which is which without cutting them open? If so, how? If not, why not?

8. Six forces are applied to the door in **FIGURE Q12.8**. Rank in order, from largest to smallest, the six torques τ_a to τ_f about the hinge on the left. Explain.

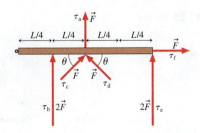

FIGURE Q12.8

9. A student gives a quick push to a ball at the end of a massless, rigid rod, as shown in **FIGURE Q12.9**, causing the ball to rotate clockwise in a *horizontal* circle. The rod's pivot is frictionless.

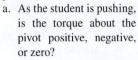

FIGURE Q12.9

   a. As the student is pushing, is the torque about the pivot positive, negative, or zero?

   b. After the push has ended, does the ball's angular velocity (i) steadily increase; (ii) increase for awhile, then hold steady; (iii) hold steady; (iv) decrease for awhile, then hold steady; or (v) steadily decrease? Explain.

   c. Right after the push has ended, is the torque positive, negative, or zero?

10. Rank in order, from largest to smallest, the angular accelerations α_a to α_d in **FIGURE Q12.10**. Explain.

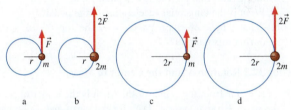

FIGURE Q12.10

11. The solid cylinder and cylindrical shell in **FIGURE Q12.11** have the same mass, same radius, and turn on frictionless, horizontal axles. (The cylindrical shell has lightweight spokes connecting the shell to the axle.) A rope is wrapped around each cylinder and tied to a block. The blocks have the same mass and are held the same height above the ground. Both blocks are released simultaneously. Which hits the ground first? Or is it a tie? Explain.

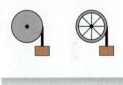

FIGURE Q12.11

12. A diver in the pike position (legs straight, hands on ankles) usually makes only one or one-and-a-half rotations. To make two or three rotations, the diver goes into a tuck position (knees bent, body curled up tight). Why?

13. Is the angular momentum of disk a in **FIGURE Q12.13** larger than, smaller than, or equal to the angular momentum of disk b? Explain.

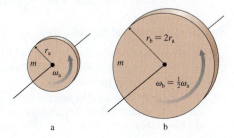

FIGURE Q12.13

EXERCISES AND PROBLEMS

Problems labeled ▨ integrate material from earlier chapters.

Exercises

Section 12.1 Rotational Motion

1. ‖ A high-speed drill reaches 2000 rpm in 0.50 s.
   a. What is the drill's angular acceleration?
   b. Through how many revolutions does it turn during this first 0.50 s?

2. ‖ A skater holds her arms outstretched as she spins at 180 rpm. What is the speed of her hands if they are 140 cm apart?

3. ‖ A ceiling fan with 80-cm-diameter blades is turning at 60 rpm. Suppose the fan coasts to a stop 25 s after being turned off.
   a. What is the speed of the tip of a blade 10 s after the fan is turned off?
   b. Through how many revolutions does the fan turn while stopping?

4. ‖‖ An 18-cm-long bicycle crank arm, with a pedal at one end, is attached to a 20-cm-diameter sprocket, the toothed disk around which the chain moves. A cyclist riding this bike increases her pedaling rate from 60 rpm to 90 rpm in 10 s.
   a. What is the tangential acceleration of the pedal?
   b. What length of chain passes over the top of the sprocket during this interval?

Section 12.2 Rotation About the Center of Mass

5. ‖ How far from the center of the earth is the center of mass of the earth + moon system? Data for the earth and moon can be found inside the back cover of the book.

6. ‖ The three masses shown in FIGURE EX12.6 are connected by massless, rigid rods. What are the coordinates of the center of mass?

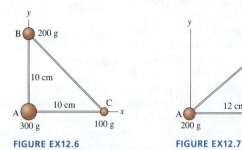

FIGURE EX12.6   FIGURE EX12.7

7. ‖ The three masses shown in FIGURE EX12.7 are connected by massless, rigid rods. What are the coordinates of the center of mass?

8. ‖ A 100 g ball and a 200 g ball are connected by a 30-cm-long, massless, rigid rod. The balls rotate about their center of mass at 120 rpm. What is the speed of the 100 g ball?

Section 12.3 Rotational Energy

9. ‖ A thin, 100 g disk with a diameter of 8.0 cm rotates about an axis through its center with 0.15 J of kinetic energy. What is the speed of a point on the rim?

10. ‖ What is the rotational kinetic energy of the earth? Assume the earth is a uniform sphere. Data for the earth can be found inside the back cover of the book.

11. ‖ The three 200 g masses in FIGURE EX12.11 are connected by massless, rigid rods.
   a. What is the triangle's moment of inertia about the axis through the center?
   b. What is the triangle's kinetic energy if it rotates about the axis at 5.0 rev/s?

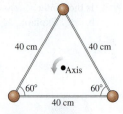

FIGURE EX12.11

12. ‖ A drum major twirls a 96-cm-long, 400 g baton about its center of mass at 100 rpm. What is the baton's rotational kinetic energy?

Section 12.4 Calculating Moment of Inertia

13. ‖ The four masses shown in FIGURE EX12.13 are connected by massless, rigid rods.
   a. Find the coordinates of the center of mass.
   b. Find the moment of inertia about an axis that passes through mass A and is perpendicular to the page.

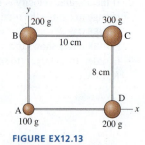

FIGURE EX12.13

14. ‖ The four masses shown in FIGURE EX12.13 are connected by massless, rigid rods.
   a. Find the coordinates of the center of mass.
   b. Find the moment of inertia about a diagonal axis that passes through masses B and D.

15. ‖ The three masses shown in FIGURE EX12.15 are connected by massless, rigid rods.
   a. Find the coordinates of the center of mass.
   b. Find the moment of inertia about an axis that passes through mass A and is perpendicular to the page.
   c. Find the moment of inertia about an axis that passes through masses B and C.

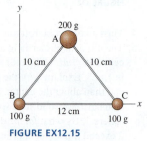

FIGURE EX12.15

16. ‖ A 12-cm-diameter CD has a mass of 21 g. What is the CD's moment of inertia for rotation about a perpendicular axis (a) through its center and (b) through the edge of the disk?

17. ‖ A 25 kg solid door is 220 cm tall, 91 cm wide. What is the door's moment of inertia for (a) rotation on its hinges and (b) rotation about a vertical axis inside the door, 15 cm from one edge?

Section 12.5 Torque

18. ‖ In FIGURE EX12.18, what is the net torque about the axle?

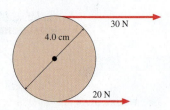

FIGURE EX12.18

19. ‖ In **FIGURE EX12.19**, what magnitude force provides 5.0 N m net torque about the axle?

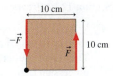

FIGURE EX12.19

20. ‖ The 20-cm-diameter disk in **FIGURE EX12.20** can rotate on an axle through its center. What is the net torque about the axle?

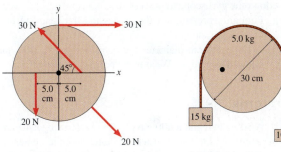

FIGURE EX12.20

FIGURE EX12.21

21. ‖ The axle in **FIGURE EX12.21** is half the distance from the center to the rim. What is the net torque about the axle?

22. ‖ A 4.0-m-long, 500 kg steel beam extends horizontally from the point where it has been bolted to the framework of a new building under construction. A 70 kg construction worker stands at the far end of the beam. What is the magnitude of the torque about the bolt due to the worker and the weight of the beam?

23. ‖ An athlete at the gym holds a 3.0 kg steel ball in his hand. His
BIO arm is 70 cm long and has a mass of 3.8 kg, with the center of mass at 40% of the arm length. What is the magnitude of the torque about his shoulder due to the ball and the weight of his arm if he holds his arm
 a. Straight out to his side, parallel to the floor?
 b. Straight, but 45° below horizontal?

Section 12.6 Rotational Dynamics

Section 12.7 Rotation About a Fixed Axis

24. ‖ An object's moment of inertia is 2.0 kg m². Its angular velocity is increasing at the rate of 4.0 rad/s per second. What is the net torque on the object?

25. ‖ An object whose moment of inertia is 4.0 kg m² experiences the torque shown in **FIGURE EX12.25**. What is the object's angular velocity at $t = 3.0$ s? Assume it starts from rest.

FIGURE EX12.25

26. ‖‖ A 1.0 kg ball and a 2.0 kg ball are connected by a 1.0-m-long rigid, massless rod. The rod is rotating cw about its center of mass at 20 rpm. What net torque will bring the balls to a halt in 5.0 s?

27. ‖ Starting from rest, a 12-cm-diameter compact disk takes 3.0 s to reach its operating angular velocity of 2000 rpm. Assume that the angular acceleration is constant. The disk's moment of inertia is 2.5×10^{-5} kg m².
 a. How much net torque is applied to the disk?
 b. How many revolutions does it make before reaching full speed?

28. ‖ A 4.0 kg, 36-cm-diameter metal disk, initially at rest, can rotate on an axle along its axis. A steady 5.0 N tangential force is applied to the edge of the disk. What is the disk's angular velocity, in rpm, 4.0 s later?

Section 12.8 Static Equilibrium

29. ‖ The two objects in **FIGURE EX12.29** are balanced on the pivot. What is distance d?

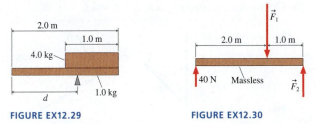

FIGURE EX12.29

FIGURE EX12.30

30. ‖ The object shown in **FIGURE EX12.30** is in equilibrium. What are the magnitudes of $\vec{F}_1$ and $\vec{F}_2$?

31. ‖ The 3.0-m-long, 100 kg rigid beam of **FIGURE EX12.31** is supported at each end. An 80 kg student stands 2.0 m from support 1. How much upward force does each support exert on the beam?

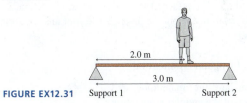

FIGURE EX12.31   Support 1   Support 2

32. ‖ A 5.0 kg cat and a 2.0 kg bowl of tuna fish are at opposite ends of the 4.0-m-long seesaw of **FIGURE EX12.32**. How far to the left of the pivot must a 4.0 kg cat stand to keep the seesaw balanced?

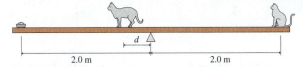

FIGURE EX12.32

Section 12.9 Rolling Motion

33. ‖ A car tire is 60 cm in diameter. The car is traveling at a speed of 20 m/s.
 a. What is the tire's angular velocity, in rpm?
 b. What is the speed of a point at the top edge of the tire?
 c. What is the speed of a point at the bottom edge of the tire?

34. ‖ A 500 g, 8.0-cm-diameter can is filled with uniform, dense food. It rolls across the floor at 1.0 m/s. What is the can's kinetic energy?

35. ‖ An 8.0-cm-diameter, 400 g solid sphere is released from rest at the top of a 2.1-m-long, 25° incline. It rolls, without slipping, to the bottom.
 a. What is the sphere's angular velocity at the bottom of the incline?
 b. What fraction of its kinetic energy is rotational?

36. ‖ A solid sphere of radius R is placed at a height of 30 cm on a 15° slope. It is released and rolls, without slipping, to the bottom. From what height should a circular hoop of radius R be released on the same slope in order to equal the sphere's speed at the bottom?

Section 12.10 The Vector Description of Rotational Motion

37. | Evaluate the cross products $\vec{A} \times \vec{B}$ and $\vec{C} \times \vec{D}$.

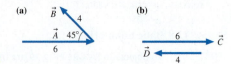

FIGURE EX12.37

38. | Evaluate the cross products $\vec{A} \times \vec{B}$ and $\vec{C} \times \vec{D}$.

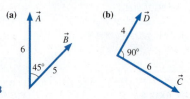

FIGURE EX12.38

39. | Vector $\vec{A} = 3\hat{\imath} + \hat{\jmath}$ and vector $\vec{B} = 3\hat{\imath} - 2\hat{\jmath} + 2\hat{k}$. What is the cross product $\vec{A} \times \vec{B}$?

40. || Force $\vec{F} = -10\hat{\jmath}$ N is exerted on a particle at $\vec{r} = (5\hat{\imath} + 5\hat{\jmath})$ m. What is the torque on the particle about the origin?

41. || A 1.3 kg ball on the end of a lightweight rod is located at $(x, y) = (3.0 \text{ m}, 2.0 \text{ m})$, where the y-axis is vertical. The other end of the rod is attached to a pivot at $(x, y) = (0 \text{ m}, 3.0 \text{ m})$. What is the torque about the pivot? Write your answer using unit vectors.

Section 12.11 Angular Momentum

42. || What are the magnitude and direction of the angular momentum relative to the origin of the 200 g particle in **FIGURE EX12.42**?

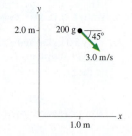

FIGURE EX12.42

43. || What is the angular momentum vector of the 2.0 kg, 4.0-cm-diameter rotating disk in **FIGURE EX12.43**?

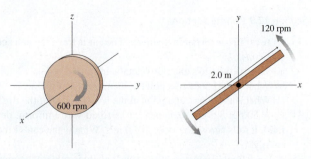

FIGURE EX12.43   **FIGURE EX12.44**

44. || What is the angular momentum vector of the 500 g rotating bar in **FIGURE EX12.44**?

45. || How fast, in rpm, would a 5.0 kg, 22-cm-diameter bowling ball have to spin to have an angular momentum of 0.23 kg m²/s?

46. || A 2.0 kg, 20-cm-diameter turntable rotates at 100 rpm on frictionless bearings. Two 500 g blocks fall from above, hit the turntable simultaneously at opposite ends of a diameter, and stick. What is the turntable's angular velocity, in rpm, just after this event?

Section 12.12 Precession of a Gyroscope

47. || A 75 g, 6.0-cm-diameter solid spherical top is spun at 1200 rpm on an axle that extends 1.0 cm past the edge of the sphere. The tip of the axle is placed on a support. What is the top's precession frequency in rpm?

48. || A toy gyroscope has a ring of mass M and radius R attached to the axle by lightweight spokes. The end of the axle is distance R from the center of the ring. The gyroscope is spun at angular velocity ω, then the end of the axle is placed on a support that allows the gyroscope to precess.
    a. Find an expression for the precession frequency Ω in terms of M, R, ω, and g.
    b. A 120 g, 8.0-cm-diameter gyroscope is spun at 1000 rpm and allowed to precess. What is the precession period?

Problems

49. || A 300 g ball and a 600 g ball are connected by a 40-cm-long massless, rigid rod. The structure rotates about its center of mass at 100 rpm. What is its rotational kinetic energy?

50. || An 800 g steel plate has the shape of the isosceles triangle
CALC shown in **FIGURE P12.50**. What are the x- and y-coordinates of the center of mass?
    Hint: Divide the triangle into vertical strips of width dx, then relate the mass dm of a strip at position x to the values of x and dx.

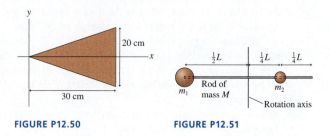

FIGURE P12.50   **FIGURE P12.51**

51. || Determine the moment of inertia about the axis of the object shown in **FIGURE P12.51**.

52. || What is the moment of inertia of a 2.0 kg, 20-cm-diameter disk for rotation about an axis (a) through the center, and (b) through the edge of the disk?

53. || Calculate by direct integration the moment of inertia for a thin
CALC rod of mass M and length L about an axis located distance d from one end. Confirm that your answer agrees with Table 12.2 when $d = 0$ and when $d = L/2$.

54. || Calculate the moment
CALC of inertia of the rectangular plate in **FIGURE P12.54** for rotation about a perpendicular axis through the center.

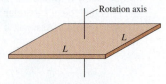

FIGURE P12.54

55. || a. A disk of mass M and radius R has a hole of radius r centered
CALC   on the axis. Calculate the moment of inertia of the disk.
    b. Confirm that your answer agrees with Table 12.2 when $r = 0$ and when $r = R$.
    c. A 4.0-cm-diameter disk with a 3.0-cm-diameter hole rolls down a 50-cm-long, 20° ramp. What is its speed at the bottom? What percent is this of the speed of a particle sliding down a frictionless ramp?

56. ⫼ Consider a solid cone of radius R, height H, and mass M. The
CALC volume of a cone is $\frac{1}{3}\pi HR^2$.
   a. What is the distance from the apex (the point) to the center of mass?
   b. What is the moment of inertia for rotation about the axis of the cone?
   Hint: The moment of inertia can be calculated as the sum of the moments of inertia of lots of small pieces.

57. ⫼ A person's center of mass is easily found by having the per-
BIO son lie on a *reaction board*. A horizontal, 2.5-m-long, 6.1 kg reaction board is supported only at the ends, with one end resting on a scale and the other on a pivot. A 60 kg woman lies on the reaction board with her feet over the pivot. The scale reads 25 kg. What is the distance from the woman's feet to her center of mass?

58. ⫼ A 3.0-m-long ladder, as shown in Figure 12.35, leans against a frictionless wall. The coefficient of static friction between the ladder and the floor is 0.40. What is the minimum angle the ladder can make with the floor without slipping?

59. ⫼ In **FIGURE P12.59**, an 80 kg construction worker sits down 2.0 m from the end of a 1450 kg steel beam to eat his lunch. What is the tension in the cable?

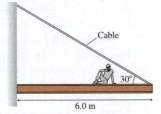

FIGURE P12.59

60. ⫼ A 40 kg, 5.0-m-long beam is supported by, but not attached to, the two posts in **FIGURE P12.60**. A 20 kg boy starts walking along the beam. How close can he get to the right end of the beam without it falling over?

FIGURE P12.60   **FIGURE P12.61**

61. ⫼ Your task in a science contest is to stack four identical uniform bricks, each of length L, so that the top brick is as far to the right as possible without the stack falling over. Is it possible, as **FIGURE P12.61** shows, to stack the bricks such that no part of the top brick is over the table? Answer this question by determining the maximum possible value of d.

62. ⫼ A 120-cm-wide sign hangs from a 5.0 kg, 200-cm-long pole. A cable of negligible mass supports the end of the rod as shown in **FIGURE P12.62**. What is the maximum mass of the sign if the maximum tension in the cable without breaking is 300 N?

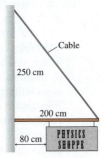

FIGURE P12.62

63. ⫼ A piece of modern sculpture consists of an 8.0-m-long, 150 kg stainless steel bar passing diametrically through a 50 kg copper sphere. The center of the sphere is 2.0 m from one end of the bar. To be mounted for display, the bar is oriented vertically, with the copper sphere at the lower end, then tilted 35° from vertical and held in place by one horizontal steel cable attached to the bar 2.0 m from the top end. What is the tension in the cable?

64. ⫼ Flywheels are large, massive wheels used to store energy. They can be spun up slowly, then the wheel's energy can be released quickly to accomplish a task that demands high power. An industrial flywheel has a 1.5 m diameter and a mass of 250 kg. Its maximum angular velocity is 1200 rpm.
   a. A motor spins up the flywheel with a constant torque of 50 N m. How long does it take the flywheel to reach top speed?
   b. How much energy is stored in the flywheel?
   c. The flywheel is disconnected from the motor and connected to a machine to which it will deliver energy. Half the energy stored in the flywheel is delivered in 2.0 s. What is the average power delivered to the machine?
   d. How much torque does the flywheel exert on the machine?

65. ⫼ Blocks of mass m_1 and m_2 are connected by a massless string that passes over the pulley in **FIGURE P12.65**. The pulley turns on frictionless bearings. Mass m_1 slides on a horizontal, frictionless surface. Mass m_2 is released while the blocks are at rest.
   a. Assume the pulley is massless. Find the acceleration of m_1 and the tension in the string. This is a Chapter 7 review problem.
   b. Suppose the pulley has mass m_p and radius R. Find the acceleration of m_1 and the tensions in the upper and lower portions of the string. Verify that your answers agree with part a if you set $m_p = 0$.

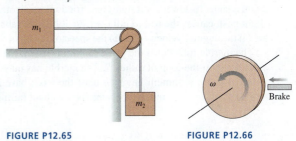

FIGURE P12.65   **FIGURE P12.66**

66. ⫼ The 2.0 kg, 30-cm-diameter disk in **FIGURE P12.66** is spinning at 300 rpm. How much friction force must the brake apply to the rim to bring the disk to a halt in 3.0 s?

67. ⫼ A 30-cm-diameter, 1.2 kg solid turntable rotates on a 1.2-cm-diameter, 450 g shaft at a constant 33 rpm. When you hit the stop switch, a brake pad presses against the shaft and brings the turntable to a halt in 15 seconds. How much friction force does the brake pad apply to the shaft?

68. ⫼ Your engineering team has been assigned the task of measuring the properties of a new jet-engine turbine. You've previously determined that the turbine's moment of inertia is 2.6 kg m². The next job is to measure the frictional torque of the bearings. Your plan is to run the turbine up to a predetermined rotation speed, cut the power, and time how long it takes the turbine to reduce its rotation speed by 50%. Your data are given in the table. Draw an appropriate graph of the data and, from the slope of the best-fit line, determine the frictional torque.

Rotation (rpm)	Time (s)
1500	19
1800	22
2100	25
2400	30
2700	34

69. ‖ A hollow sphere is rolling along a horizontal floor at 5.0 m/s when it comes to a 30° incline. How far up the incline does it roll before reversing direction?

70. ‖ A 750 g disk and a 760 g ring, both 15 cm in diameter, are rolling along a horizontal surface at 1.5 m/s when they encounter a 15° slope. How far up the slope does each travel before rolling back down?

71. ‖‖ A cylinder of radius R, length L, and mass M is released from rest on a slope inclined at angle θ. It is oriented to roll straight down the slope. If the slope were frictionless, the cylinder would *slide* down the slope without rotating. What minimum coefficient of static friction is needed for the cylinder to roll down without slipping?

72. ‖ The 5.0 kg, 60-cm-diameter disk in **FIGURE P12.72** rotates on an axle passing through one edge. The axle is parallel to the floor. The cylinder is held with the center of mass at the same height as the axle, then released.

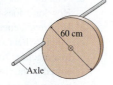

FIGURE P12.72

   a. What is the cylinder's initial angular acceleration?
   b. What is the cylinder's angular velocity when it is directly below the axle?

73. ‖ A thin, uniform rod of length L and mass M is placed vertically on a horizontal table. If tilted ever so slightly, the rod will fall over.
   a. What is the speed of the center of mass just as the rod hits the table if there's so much friction that the bottom tip of the rod does not slide?
   b. What is the speed of the center of mass just as the rod hits the table if the table is frictionless?

74. ‖ A long, thin rod of mass M and length L is standing straight up on a table. Its lower end rotates on a frictionless pivot. A very slight push causes the rod to fall over. As it hits the table, what are (a) the angular velocity and (b) the speed of the tip of the rod?

75. ‖‖ The marble rolls down the track shown in **FIGURE P12.75** and around a loop-the-loop of radius R. The marble has mass m and radius r. What minimum height h must the track have for the marble to make it around the loop-the-loop without falling off?

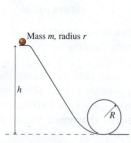

FIGURE P12.75

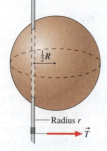

FIGURE P12.76

76. ‖ The sphere of mass M and radius R in **FIGURE P12.76** is rigidly attached to a thin rod of radius r that passes through the sphere at distance $\frac{1}{2}R$ from the center. A string wrapped around the rod pulls with tension T. Find an expression for the sphere's angular acceleration. The rod's moment of inertia is negligible.

77. ‖ A satellite follows the elliptical orbit shown in **FIGURE P12.77**. The only force on the satellite is the gravitational attraction of the planet. The satellite's speed at point a is 8000 m/s.
   a. Does the satellite experience any torque about the center of the planet? Explain.
   b. What is the satellite's speed at point b?
   c. What is the satellite's speed at point c?

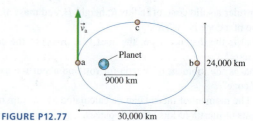

FIGURE P12.77

78. ‖‖ A 10 g bullet traveling at 400 m/s strikes a 10 kg, 1.0-m-wide door at the edge opposite the hinge. The bullet embeds itself in the door, causing the door to swing open. What is the angular velocity of the door just after impact?

79. ‖ A 200 g, 40-cm-diameter turntable rotates on frictionless bearings at 60 rpm. A 20 g block sits at the center of the turntable. A compressed spring shoots the block radially outward along a frictionless groove in the surface of the turntable. What is the turntable's rotation angular velocity when the block reaches the outer edge?

80. ‖ Luc, who is 1.80 m tall and weighs 950 N, is standing at the center of a playground merry-go-round with his arms extended, holding a 4.0 kg dumbbell in each hand. The merry-go-round can be modeled as a 4.0-m-diameter disk with a weight of 1500 N. Luc's body can be modeled as a uniform 40-cm-diameter cylinder with massless arms extending to hands that are 85 cm from his center. The merry-go-round is coasting at a steady 35 rpm when Luc brings his hands in to his chest. Afterward, what is the angular velocity, in rpm, of the merry-go-round?

81. ‖‖ A merry-go-round is a common piece of playground equipment. A 3.0-m-diameter merry-go-round with a mass of 250 kg is spinning at 20 rpm. John runs tangent to the merry-go-round at 5.0 m/s, in the same direction that it is turning, and jumps onto the outer edge. John's mass is 30 kg. What is the merry-go-round's angular velocity, in rpm, after John jumps on?

82. ‖‖ A 45 kg figure skater is spinning on the toes of her skates at 1.0 rev/s. Her arms are outstretched as far as they will go. In this orientation, the skater can be modeled as a cylindrical torso (40 kg, 20 cm average diameter, 160 cm tall) plus two rod-like arms (2.5 kg each, 66 cm long) attached to the outside of the torso. The skater then raises her arms straight above her head, where she appears to be a 45 kg, 20-cm-diameter, 200-cm-tall cylinder. What is her new angular velocity, in rev/s?

83. ‖‖ During most of its lifetime, a star maintains an equilibrium size in which the inward force of gravity on each atom is balanced by an outward pressure force due to the heat of the nuclear reactions in the core. But after all the hydrogen "fuel" is consumed by nuclear fusion, the pressure force drops and the star undergoes a *gravitational collapse* until it becomes a *neutron star*. In a neutron star, the electrons and protons of the atoms are squeezed together by gravity until they fuse into neutrons. Neutron stars spin very rapidly and emit intense pulses of radio and light waves, one pulse per rotation. These "pulsing stars" were discovered in the 1960s and are called *pulsars*.
   a. A star with the mass ($M = 2.0 \times 10^{30}$ kg) and size ($R = 7.0 \times 10^8$ m) of our sun rotates once every 30·days. After undergoing gravitational collapse, the star forms a pulsar that is observed by astronomers to emit radio pulses every 0.10 s. By treating the neutron star as a solid sphere, deduce its radius.
   b. What is the speed of a point on the equator of the neutron star? Your answers will be somewhat too large because a star cannot be accurately modeled as a solid sphere. Even so, you will be able to show that a star, whose mass is 10^6 larger than the earth's, can be compressed by gravitational forces to a size smaller than a typical state in the United States!

84. ‖ The earth's rotation axis, which is tilted 23.5° from the plane of the earth's orbit, today points to Polaris, the north star. But Polaris has not always been the north star because the earth, like a spinning gyroscope, precesses. That is, a line extending along the earth's rotation axis traces out a 23.5° cone as the earth precesses with a period of 26,000 years. This occurs because the earth is not a perfect sphere. It has an *equatorial bulge*, which allows both the moon and the sun to exert a gravitational torque on the earth. Our expression for the precession frequency of a gyroscope can be written $\Omega = \tau/I\omega$. Although we derived this equation for a specific situation, it's a valid result, differing by at most a constant close to 1, for the precession of any rotating object. What is the average gravitational torque on the earth due to the moon and the sun?

Challenge Problems

85. ‖ The bunchberry flower has the fastest-moving parts ever
BIO observed in a plant. Initially, the stamens are held by the petals in a bent position, storing elastic energy like a coiled spring. When the petals release, the tips of the stamen act like medieval catapults, flipping through a 60° angle in just 0.30 ms to launch pollen from anther sacs at their ends. The human eye just sees a burst of pollen; only high-speed photography reveals the details. As **FIGURE CP12.85** shows, we can model the stamen tip as a 1.0-mm-long, 10 μg rigid rod with a 10 μg anther sac at the end. Although oversimplifying, we'll assume a constant angular acceleration.
    a. How large is the "straightening torque"?
    b. What is the speed of the anther sac as it releases its pollen?

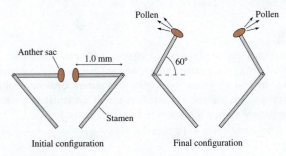

FIGURE CP12.85

86. ‖ The two blocks in **FIGURE CP12.86** are connected by a massless rope that passes over a pulley. The pulley is 12 cm in diameter and has a mass of 2.0 kg. As the pulley turns, friction at the axle exerts a torque of magnitude 0.50 N m. If the blocks are released from rest, how long does it take the 4.0 kg block to reach the floor?

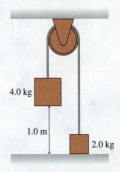

FIGURE CP12.86

87. ‖ A rod of length L and mass M has a nonuniform mass distribu-
CALC tion. The linear mass density (mass per length) is $\lambda = cx^2$, where x is measured from the *center* of the rod and c is a constant.
    a. What are the units of c?
    b. Find an expression for c in terms of L and M.
    c. Find an expression in terms of L and M for the moment of inertia of the rod for rotation about an axis through the center.

88. ‖ In **FIGURE CP12.88**, a 200 g toy car is placed on a narrow 60-cm-diameter track with wheel grooves that keep the car going in a circle. The 1.0 kg track is free to turn on a frictionless, vertical axis. The spokes have negligible mass. After the car's switch is turned on, it soon reaches a steady speed of 0.75 m/s relative to the track. What then is the track's angular velocity, in rpm?

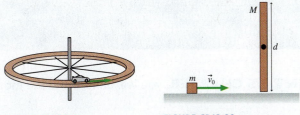

FIGURE CP12.88   **FIGURE CP12.89**

89. ‖ **FIGURE CP12.89** shows a cube of mass m sliding without friction at speed v_0. It undergoes a perfectly elastic collision with the bottom tip of a rod of length d and mass $M = 2m$. The rod is pivoted about a frictionless axle through its center, and initially it hangs straight down and is at rest. What is the cube's velocity—both speed and direction—after the collision?

90. ‖ A 75 g, 30-cm-long rod hangs vertically on a frictionless, horizontal axle passing through its center. A 10 g ball of clay traveling horizontally at 2.5 m/s hits and sticks to the very bottom tip of the rod. To what maximum angle, measured from vertical, does the rod (with the attached ball of clay) rotate?

13 Newton's Theory of Gravity

The Andromeda Galaxy—seen here in a false-color image showing ultraviolet (blue) and infrared (red) light—consists of billions of stars orbiting the galactic center.

IN THIS CHAPTER, you will understand the motion of satellites and planets.

What are Kepler's laws?

Before Galileo and the telescope, Kepler used naked-eye measurements to make three major discoveries:

- The planets move in elliptical orbits.
- Planets "sweep out" equal areas in equal times.
- For circular orbits, the square of a planet's period is proportional to the cube of the orbit's radius.

Kepler's discoveries were the first solid proof of Copernicus's assertion that earth and the other planets orbit the sun.

What is Newton's theory?

Newton proposed that *any* two masses M and m are attracted toward each other by a gravitational force of magnitude

$$F_{M \text{ on } m} = F_{m \text{ on } M} = \frac{GMm}{r^2}$$

where r is the distance between the masses and G is the gravitational constant.

- Newton's law is an inverse-square law.
- Newton's law predicts the value of g.

In addition to a specific force law for gravity, Newton's three laws of motion apply to all objects in the universe.

What is gravitational energy?

The gravitational potential energy of two masses is

$$U_{\text{G}} = -\frac{GMm}{r}$$

Gravitational potential energy is negative, with a zero point at infinity. The $U_{\text{G}} = mgy$ that you learned in Chapter 10 is a special case for objects very near the surface of a planet. Gravitationally interacting stars, planets, and satellites are always isolated systems, so mechanical energy is conserved.

« LOOKING BACK Chapter 10 Potential energy and energy conservation

What does the theory say about orbits?

Kepler's laws can be *derived* from Newton's theory of gravity:

- Orbits can be circular or elliptical.
- Orbits conserve energy and angular momentum.
- Geosynchronous orbits have the same period as the rotating planet.

« LOOKING BACK Sections 8.2–8.3 Circular motion

Why is the theory of gravity important?

Satellites, space stations, the GPS system, and future missions to planets all depend on Newton's theory of gravity. Our modern understanding of the cosmos—from stars and galaxies to the Big Bang—is based on understanding gravity. Newton's theories of motion and gravity were the beginnings of modern science.

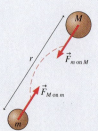
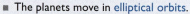

13.1 A Little History

The study of the structure of the universe is called **cosmology.** The ancient Greeks developed a cosmological model, with the earth at the center of the universe while the moon, the sun, the planets, and the stars were points of light turning about the earth on large "celestial spheres." This viewpoint was further expanded by the second-century Egyptian astronomer Ptolemy (the P is silent). He developed an elaborate mathematical model of the solar system that quite accurately predicted the complex planetary motions.

Then, in 1543, the medieval world was turned on its head with the publication of Nicholas Copernicus's *De Revolutionibus.* Copernicus argued that it is not the earth at rest in the center of the universe—it is the sun! Furthermore, Copernicus asserted that all of the planets revolve about the sun (hence his title) in circular orbits.

Tycho and Kepler

The greatest medieval astronomer was Tycho Brahe. For 30 years, from 1570 to 1600, Tycho compiled the most accurate astronomical observations the world had known. The invention of the telescope was still to come, but Tycho developed ingenious mechanical sighting devices that allowed him to determine the positions of stars and planets in the sky with unprecedented accuracy.

Tycho had a young mathematical assistant named Johannes Kepler. Kepler had become one of the first outspoken defenders of Copernicus, and his goal was to find evidence for circular planetary orbits in Tycho's records. To appreciate the difficulty of this task, keep in mind that Kepler was working before the development of graphs or of calculus—and certainly before calculators! His mathematical tools were algebra, geometry, and trigonometry, and he was faced with thousands upon thousands of individual observations of planetary positions measured as angles above the horizon.

Many years of work led Kepler to discover that the orbits are not circles, as Copernicus claimed, but *ellipses.* Furthermore, the speed of a planet is not constant but varies as it moves around the ellipse.

Kepler's laws, as we call them today, state that

1. Planets move in elliptical orbits, with the sun at one focus of the ellipse.
2. A line drawn between the sun and a planet sweeps out equal areas during equal intervals of time.
3. The square of a planet's orbital period is proportional to the cube of the semimajor-axis length.

FIGURE 13.1a shows that an ellipse has two *foci* (plural of *focus*), and the sun occupies one of these. The long axis of the ellipse is the *major axis,* and half the length of this axis is called the *semimajor-axis length.* As the planet moves, a line drawn from the sun to the planet "sweeps out" an area. **FIGURE 13.1b** shows two such areas. Kepler's discovery that the areas are equal for equal Δt implies that the planet moves faster when near the sun, slower when farther away.

All the planets except Mercury have elliptical orbits that are only very slightly distorted circles. As **FIGURE 13.2** shows, a circle is an ellipse in which the two foci move to the center, effectively making one focus, and the semimajor-axis length becomes the radius. Because the mathematics of ellipses is difficult, this chapter will focus on circular orbits.

Kepler published the first two of his laws in 1609, the same year in which Galileo first turned a telescope to the heavens. Through his telescope Galileo could *see* moons orbiting Jupiter, just as Copernicus had suggested the planets orbit the sun. He could *see* that Venus has phases, like the moon, which implied its orbital motion about the sun. By the time of Galileo's death in 1642, the Copernican revolution was complete.

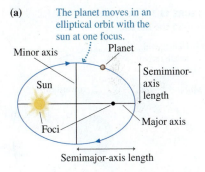

FIGURE 13.1 The elliptical orbit of a planet about the sun.

(a) The planet moves in an elliptical orbit with the sun at one focus.

Minor axis · Planet · Sun · Semiminor-axis length · Foci · Major axis · Semimajor-axis length

(b) The line between the sun and the planet sweeps out equal areas during equal intervals of time.

Slower · Faster

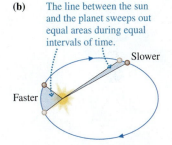

FIGURE 13.2 A circular orbit is a special case of an elliptical orbit.

Kepler's second law: Equal areas in equal times imply the speed is constant.

Kepler's first law: The sun is at the center.

The motion is uniform circular motion.

Sun · r

Circular orbit

Kepler's third law: The square of the period is proportional to r^3.

Isaac Newton, 1642–1727.

13.2 Isaac Newton

A popular image has Newton thinking of the idea of gravity after an apple fell on his head. This amusing story is at least close to the truth. Newton himself said that the "notion of gravitation" came to him as he "sat in a contemplative mood" and "was occasioned by the fall of an apple." It occurred to him that, perhaps, the apple was attracted to the *center* of the earth but was prevented from getting there by the earth's surface. And if the apple was so attracted, why not the moon? In other words, perhaps gravitation is a *universal* force between all objects in the universe! This is not shocking today, but no one before Newton had ever thought that the mundane motion of objects on earth had any connection at all with the stately motion of the planets through the heavens.

Suppose the moon's circular motion around the earth is due to the pull of the earth's gravity. Then, as you learned in Chapter 8 and is shown in **FIGURE 13.3**, the moon must be in *free fall* with the free-fall acceleration $g_{\text{at moon}}$.

NOTE The symbol g_{moon} is the free-fall acceleration caused by the *moon's* gravity—that is, the acceleration of a falling object on the moon. Here we're interested in the acceleration *of* the moon by the earth's gravity, which we'll call $g_{\text{at moon}}$.

The centripetal acceleration of an object in uniform circular motion is

$$a_r = g_{\text{at moon}} = \frac{v_m^2}{r_m} \tag{13.1}$$

The moon's speed is related to the radius r_m and period T_m of its orbit by $v_m = $ circumference/period $= 2\pi r_m/T_m$. Combining these, Newton found

$$g_{\text{at moon}} = \frac{4\pi^2 r_m}{T_m^2} = \frac{4\pi^2(3.84 \times 10^8 \text{ m})}{(2.36 \times 10^6 \text{ s})^2} = 0.00272 \text{ m/s}^2$$

Astronomical measurements had established a reasonably good value for r_m by the time of Newton, and the period $T_m = 27.3$ days (2.36×10^6 s) was quite well known.

The moon's centripetal acceleration is significantly less than the free-fall acceleration on the earth's surface. In fact,

$$\frac{g_{\text{at moon}}}{g_{\text{on earth}}} = \frac{0.00272 \text{ m/s}^2}{9.80 \text{ m/s}^2} = \frac{1}{3600}$$

This is an interesting result, but it was Newton's next step that was critical. He compared the radius of the moon's orbit to the radius of the earth:

$$\frac{r_m}{R_e} = \frac{3.84 \times 10^8 \text{ m}}{6.37 \times 10^6 \text{ m}} = 60.2$$

NOTE We'll use a lowercase r, as in r_m, to indicate the radius of an orbit. We'll use an uppercase R, as in R_e, to indicate the radius of a star or planet.

Newton recognized that $(60.2)^2$ is almost exactly 3600. Thus, he reasoned:

- If g has the value 9.80 at the earth's surface, and
- If the force of gravity and g decrease in size depending inversely on the square of the distance from the center of the earth,
- Then g will have exactly the value it needs at the distance of the moon to cause the moon to orbit the earth with a period of 27.3 days.

His two ratios were not identical, but he found them to "answer pretty nearly" and knew that he had to be on the right track.

FIGURE 13.3 The moon is in free fall around the earth.

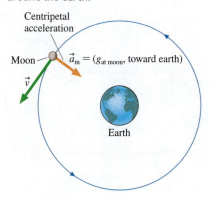

I deduced that the forces which keep the planets in their orbs must be reciprocally as the squares of their distances from the centers about which they revolve; and thereby compared the force requisite to keep the Moon in her orb with the force of gravity at the surface of the Earth; and found them answer pretty nearly.

Isaac Newton

STOP TO THINK 13.1 A satellite orbits the earth with constant speed at a height above the surface equal to the earth's radius. The magnitude of the satellite's acceleration is

a. $4g_{\text{on earth}}$ b. $2g_{\text{on earth}}$ c. $g_{\text{on earth}}$

d. $\frac{1}{2}g_{\text{on earth}}$ e. $\frac{1}{4}g_{\text{on earth}}$ f. 0

13.3 Newton's Law of Gravity

Newton proposed that *every* object in the universe attracts *every other* object with a force that is

1. Inversely proportional to the square of the distance between the objects.
2. Directly proportional to the product of the masses of the two objects.

To make these ideas more specific, **FIGURE 13.4** shows masses m_1 and m_2 separated by distance r. Each mass exerts an attractive force on the other, a force that we call the **gravitational force.** These two forces form an action/reaction pair, so $\vec{F}_{1 \text{ on } 2}$ is equal and opposite to $\vec{F}_{2 \text{ on } 1}$. The magnitude of the forces is given by Newton's law of gravity.

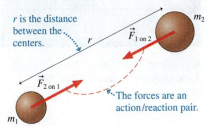

FIGURE 13.4 The gravitational forces on masses m_1 and m_2.

r is the distance between the centers.

$\vec{F}_{1 \text{ on } 2}$

$\vec{F}_{2 \text{ on } 1}$

The forces are an action/reaction pair.

> **Newton's law of gravity** If two objects with masses m_1 and m_2 are a distance r apart, the objects exert attractive forces on each other of magnitude
>
> $$F_{1 \text{ on } 2} = F_{2 \text{ on } 1} = \frac{Gm_1m_2}{r^2} \qquad (13.2)$$
>
> The forces are directed along the straight line joining the two objects.

The constant G, called the **gravitational constant,** is a proportionality constant necessary to relate the masses, measured in kilograms, to the force, measured in newtons. In the SI system of units, G has the value

$$G = 6.67 \times 10^{-11} \text{ N m}^2/\text{kg}^2$$

FIGURE 13.5 is a graph of the gravitational force as a function of the distance between the two masses. As you can see, an inverse-square force decreases rapidly.

Strictly speaking, Equation 13.2 is valid only for particles. However, Newton was able to show that this equation also applies to spherical objects, such as planets, if r is the distance between their centers. Our intuition and common sense suggest this to us, as they did to Newton. The rather difficult proof is not essential, so we will omit it.

FIGURE 13.5 The gravitational force is an inverse-square force.

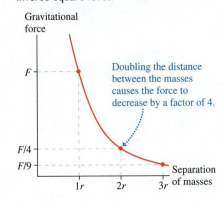

Gravitational force

Doubling the distance between the masses causes the force to decrease by a factor of 4.

Separation of masses

Gravitational Force and Weight

Knowing G, we can calculate the size of the gravitational force. Consider two 1.0 kg masses that are 1.0 m apart. According to Newton's law of gravity, these two masses exert an attractive gravitational force on each other of magnitude

$$F_{1 \text{ on } 2} = F_{2 \text{ on } 1} = \frac{Gm_1m_2}{r^2}$$

$$= \frac{(6.67 \times 10^{-11} \text{ N m}^2/\text{kg}^2)(1.0 \text{ kg})(1.0 \text{ kg})}{(1.0 \text{ m})^2} = 6.67 \times 10^{-11} \text{ N}$$

This is an exceptionally tiny force, especially when compared to the gravitational force of the entire earth on each mass: $F_G = mg = 9.8$ N.

The fact that the gravitational force between two ordinary-size objects is so small is the reason we are not aware of it. You are being attracted to every object around you, but the forces are so tiny in comparison to the normal forces and friction forces acting on you that they are completely undetectable. Only when one (or both) of the masses is exceptionally large—planet-size—does the force of gravity become important.

We find a more respectable result if we calculate the force *of the earth* on a 1.0 kg mass at the earth's surface:

$$F_{\text{earth on 1 kg}} = \frac{GM_e m_{1 \text{ kg}}}{R_e^2}$$

$$= \frac{(6.67 \times 10^{-11} \text{ N m}^2/\text{kg}^2)(5.98 \times 10^{24} \text{ kg})(1.0 \text{ kg})}{(6.37 \times 10^6 \text{ m})^2} = 9.8 \text{ N}$$

where the distance between the mass and the center of the earth is the earth's radius.

The dynamics of stellar motions, spanning many thousands of light years, are governed by Newton's law of gravity.

A galaxy of $\approx 10^{11}$ stars spanning a distance greater than 100,000 light years.

The earth's mass M_e and radius R_e were taken from Table 13.2 in Section 13.6. This table, which is also printed inside the back cover of the book, contains astronomical data that will be used for examples and homework.

The force $F_{\text{earth on 1 kg}} = 9.8$ N is exactly the weight of a stationary 1.0 kg mass: $F_G = mg = 9.8$ N. Is this a coincidence? Of course not. Weight—the upward force of a spring scale—exactly balances the downward gravitational force, so numerically they must be equal.

Although weak, gravity is a *long-range* force. No matter how far apart two objects may be, there is a gravitational attraction between them given by Equation 13.2. Consequently, gravity is the most ubiquitous force in the universe. It not only keeps your feet on the ground, it also keeps the earth orbiting the sun, the solar system orbiting the center of the Milky Way galaxy, and the entire Milky Way galaxy performing an intricate orbital dance with other galaxies making up what is called the "local cluster" of galaxies.

The Principle of Equivalence

Newton's law of gravity depends on a rather curious assumption. The concept of *mass* was introduced in Chapter 5 by considering the relationship between force and acceleration. The *inertial mass* of an object, which is the mass that appears in Newton's second law, is found by measuring the object's acceleration a in response to force F:

$$m_{\text{inert}} = \text{inertial mass} = \frac{F}{a} \tag{13.3}$$

Gravity plays no role in this definition of mass.

The quantities m_1 and m_2 in Newton's law of gravity are being used in a very different way. Masses m_1 and m_2 govern the strength of the gravitational attraction between two objects. The mass used in Newton's law of gravity is called the **gravitational mass.** The gravitational mass of an object can be determined by measuring the attractive force exerted on it by another mass M a distance r away:

$$m_{\text{grav}} = \text{gravitational mass} = \frac{r^2 F_{M \text{ on } m}}{GM} \tag{13.4}$$

Acceleration does not enter into the definition of the gravitational mass.

These are two very different concepts of mass. Yet Newton, in his theory of gravity, asserts that the inertial mass in his second law is the very same mass that governs the strength of the gravitational attraction between two objects. The assertion that $m_{\text{grav}} = m_{\text{inert}}$ is called the **principle of equivalence.** It says that inertial mass is *equivalent to* gravitational mass.

As a hypothesis about nature, the principle of equivalence is subject to experimental verification or disproof. Many exceptionally clever experiments have looked for any difference between the gravitational mass and the inertial mass, and they have shown that any difference, if it exists at all, is less than 10 parts in a trillion! As far as we know today, the gravitational mass and the inertial mass are exactly the same thing.

But why should a quantity associated with the dynamics of motion, relating force to acceleration, have anything at all to do with the gravitational attraction? This is a question that intrigued Einstein and eventually led to his general theory of relativity, the theory about curved spacetime and black holes. General relativity is beyond the scope of this textbook, but it explains the principle of equivalence as a property of space itself.

Newton's Theory of Gravity

Newton's theory of gravity is more than just Equation 13.2. The *theory* of gravity consists of:

1. A specific force law for gravity, given by Equation 13.2, *and*
2. The principle of equivalence, *and*
3. An assertion that Newton's three laws of motion are universally applicable. These laws are as valid for heavenly bodies, the planets and stars, as for earthly objects.

Consequently, everything we have learned about forces, motion, and energy is relevant to the dynamics of satellites, planets, and galaxies.

STOP TO THINK 13.2 The figure shows a binary star system. The mass of star 2 is twice the mass of star 1. Compared to $\vec{F}_{1\text{ on }2}$, the magnitude of the force $\vec{F}_{2\text{ on }1}$ is

    a. Four times as big.
    b. Twice as big.
    c. The same size.
    d. Half as big.
    e. One-quarter as big.

13.4 Little g and Big G

The familiar equation $F_G = mg$ works well when an object is on the surface of a planet, but *mg* will not help us find the force exerted on the same object if it is in orbit around the planet. Neither can we use *mg* to find the force of attraction between the earth and the moon. Newton's law of gravity provides a more fundamental starting point because it describes a *universal* force that exists between all objects.

To illustrate the connection between Newton's law of gravity and the familiar $F_G = mg$, **FIGURE 13.6** shows an object of mass *m* on the surface of Planet X. Planet X inhabitant Mr. Xhzt, standing on the surface, finds that the downward gravitational force is $F_G = mg_X$, where g_X is the free-fall acceleration on Planet X.

We, taking a more cosmic perspective, reply, "Yes, that is the force *because* of a universal force of attraction between your planet and the object. The size of the force is determined by Newton's law of gravity."

We and Mr. Xhzt are both correct. Whether you think locally or globally, we and Mr. Xhzt must arrive at the *same numerical value* for the magnitude of the force. Suppose an object of mass *m* is on the surface of a planet of mass *M* and radius *R*. The local gravitational force is

$$F_G = mg_{\text{surface}} \qquad (13.5)$$

where g_{surface} is the free-fall acceleration at the planet's surface. The force of gravitational attraction for an object on the surface ($r = R$), as given by Newton's law of gravity, is

$$F_{M\text{ on }m} = \frac{GMm}{R^2} \qquad (13.6)$$

Because these are two names and two expressions for the same force, we can equate the right-hand sides to find that

$$g_{\text{surface}} = \frac{GM}{R^2} \qquad (13.7)$$

We have used Newton's law of gravity to *predict* the value of *g* at the surface of a planet. The value depends on the mass and radius of the planet as well as on the value of *G*, which establishes the overall strength of the gravitational force.

The expression for g_{surface} in Equation 13.7 is valid for any planet or star. Using the mass and radius of Mars, we can predict the Martian value of *g*:

$$g_{\text{Mars}} = \frac{GM_{\text{Mars}}}{R_{\text{Mars}}^2} = \frac{(6.67 \times 10^{-11}\ \text{N}\,\text{m}^2/\text{kg}^2)(6.42 \times 10^{23}\ \text{kg})}{(3.37 \times 10^6\ \text{m})^2} = 3.8\ \text{m/s}^2$$

NOTE We noted in Chapter 6 that measured values of *g* are very slightly smaller on a rotating planet. We'll ignore rotation in this chapter.

FIGURE 13.6 Weighing an object of mass *m* on Planet X.

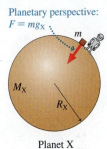

Planetary perspective:
$F = mg_X$

Planet X

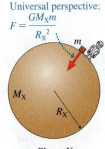

Universal perspective:
$F = \dfrac{GM_X m}{R_X^2}$

Planet X

Decrease of g with Distance

Equation 13.7 gives $g_{surface}$ at the surface of a planet. More generally, imagine an object of mass m at distance $r > R$ from the center of a planet. Further, suppose that gravity from the planet is the only force acting on the object. Then its acceleration, the free-fall acceleration, is given by Newton's second law:

$$g = \frac{F_{M \text{ on } m}}{m} = \frac{GM}{r^2} \tag{13.8}$$

This more general result agrees with Equation 13.7 if $r = R$, but it allows us to determine the "local" free-fall acceleration at distances $r > R$. Equation 13.8 expresses Newton's discovery, with regard to the moon, that g decreases inversely with the square of the distance.

FIGURE 13.7 shows a satellite orbiting at height h above the earth's surface. Its distance from the center of the earth is $r = R_e + h$. Most people have a mental image that satellites orbit "far" from the earth, but in reality h is typically 200 miles $\approx 3 \times 10^5$ m, while $R_e = 6.37 \times 10^6$ m. Thus the satellite is barely "skimming" the earth at a height only about 5% of the earth's radius!

The value of g at height h above the earth (i.e., above sea level) is

$$g = \frac{GM_e}{(R_e + h)^2} = \frac{GM_e}{R_e^2(1 + h/R_e)^2} = \frac{g_{earth}}{(1 + h/R_e)^2} \tag{13.9}$$

where $g_{earth} = 9.83 \text{ m/s}^2$ is the value calculated from Equation 13.7 for $h = 0$ on a nonrotating earth. **TABLE 13.1** shows the value of g evaluated at several values of h.

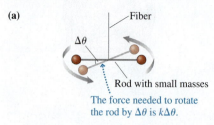

FIGURE 13.7 A satellite orbits the earth at height h.

Satellite

h is typically very small compared with R_e.

r
h
R_e
Earth
M_e

TABLE 13.1 Variation of g with height above the ground

Height h	Example	g (m/s^2)
0 m	ground	9.83
4500 m	Mt. Whitney	9.82
10,000 m	jet airplane	9.80
300,000 m	space station	8.90
35,900,000 m	communications satellite	0.22

NOTE The free-fall acceleration of a satellite such as the space station is only slightly less than the ground-level value. An object in orbit is not "weightless" because there is no gravity in space but because it is in free fall, as you learned in Chapter 8.

Weighing the Earth

We can predict g if we know the earth's mass. But how do we know the value of M_e? We cannot place the earth on a giant pan balance, so how is its mass known? Furthermore, how do we know the value of G?

Newton did not know the value of G. He could say that the gravitational force is proportional to the product $m_1 m_2$ and inversely proportional to r^2, but he had no means of knowing the value of the proportionality constant.

Determining G requires a *direct* measurement of the gravitational force between two known masses at a known separation. The small size of the gravitational force between ordinary-size objects makes this quite a feat. Yet the English scientist Henry Cavendish came up with an ingenious way of doing so with a device called a *torsion balance*. Two fairly small masses m, typically about 10 g, are placed on the ends of a lightweight rod. The rod is hung from a thin fiber, as shown in **FIGURE 13.8a**, and allowed to reach equilibrium.

If the rod is then rotated slightly and released, a *restoring force* will return it to equilibrium. This is analogous to displacing a spring from equilibrium, and in fact the restoring force and the angle of displacement obey a version of Hooke's law:

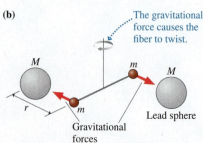

FIGURE 13.8 Cavendish's experiment to measure G.

(a)

Fiber

$\Delta\theta$

Rod with small masses

The force needed to rotate the rod by $\Delta\theta$ is $k\Delta\theta$.

(b)

The gravitational force causes the fiber to twist.

M
m
M
r
m

Lead sphere

Gravitational forces

$F_{\text{restore}} = k\,\Delta\theta$. The "torsion constant" k can be determined by timing the period of oscillations. Once k is known, a force that twists the rod slightly away from equilibrium can be measured by the product $k\,\Delta\theta$. It is possible to measure very small angular deflections, so this device can be used to determine very small forces.

Two larger masses M (typically lead spheres with $M \approx 10$ kg) are then brought close to the torsion balance, as shown in **FIGURE 13.8b**. The gravitational attraction that they exert on the smaller hanging masses causes a very small but measurable twisting of the balance, enough to measure $F_{M\text{ on }m}$. Because m, M, and r are all known, Cavendish was able to determine G from

$$G = \frac{F_{M\text{ on }m}\, r^2}{Mm} \tag{13.10}$$

His first results were not highly accurate, but improvements over the years in this and similar experiments have produced the value of G accepted today.

With an independently determined value of G, we can return to Equation 13.7 to find

$$M_e = \frac{g_{\text{earth}} R_e^{\,2}}{G} \tag{13.11}$$

We have weighed the earth! The value of g_{earth} at the earth's surface is known with great accuracy from kinematics experiments. The earth's radius R_e is determined by surveying techniques. Combining our knowledge from these very different measurements has given us a way to determine the mass of the earth.

The gravitational constant G is what we call a *universal constant*. Its value establishes the strength of one of the fundamental forces of nature. As far as we know, the gravitational force between two masses would be the same anywhere in the universe. Universal constants tell us something about the most basic and fundamental properties of nature. You will soon meet other universal constants.

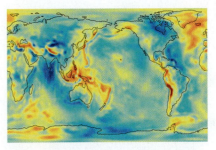

The free-fall acceleration varies slightly due to mountains and to variation in the density of the earth's crust. This map shows the *gravitational anomaly*, with red regions of slightly stronger gravity and blue regions of slightly weaker gravity. The variation is tiny, less than 0.001 m/s^2.

STOP TO THINK 13.3 A planet has four times the mass of the earth, but the acceleration due to gravity on the planet's surface is the same as on the earth's surface. The planet's radius is

a. $4R_e$ b. $2R_e$ c. R_e d. $\frac{1}{2}R_e$ e. $\frac{1}{4}R_e$

13.5 Gravitational Potential Energy

Gravitational problems are ideal for the conservation-law tools we developed in Chapters 9 through 11. Because gravity is the only force, and it is a conservative force, both the momentum and the mechanical energy of the system $m_1 + m_2$ are conserved. To employ conservation of energy, however, we need to determine an appropriate form for the gravitational potential energy for two interacting masses.

The definition of potential energy that we developed in Chapter 11 is

$$\Delta U = U_f - U_i = -W_c(i \rightarrow f) \tag{13.12}$$

where $W_c(i \rightarrow f)$ is the work done by a conservative force as a particle moves from position i to position f. For a flat earth, we used $F = -mg$ and the choice that $U = 0$ at the surface $(y = 0)$ to arrive at the now-familiar $U_G = mgy$. This result for U_G is valid only for $y \ll R_e$, when the earth's curvature and size are not apparent. We now need to find an expression for the gravitational potential energy of masses that interact over *large* distances.

FIGURE 13.9 shows two particles of mass m_1 and m_2. Let's calculate the work done on mass m_2 by the conservative force $\vec{F}_{1\text{ on }2}$ as m_2 moves from an initial position at distance r to a final position very far away. The force, which points to the left, is opposite the displacement; hence this force does *negative* work. Consequently, due to the minus sign in Equation 13.12, ΔU is *positive*. A pair of masses *gains* potential energy as the masses move farther apart, just as a particle near the earth's surface gains potential energy as it moves to a higher altitude.

FIGURE 13.9 Calculating the work done by the gravitational force as mass m_2 moves from r to ∞.

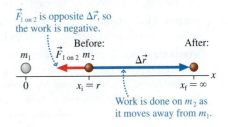

$\vec{F}_{1\text{ on }2}$ is opposite $\Delta \vec{r}$, so the work is negative.

Work is done on m_2 as it moves away from m_1.

We can establish a coordinate system with m_1 at the origin and m_2 moving along the x-axis. The gravitational force is a variable force, so we need the full definition of work:

$$W(i \rightarrow f) = \int_{x_i}^{x_f} F_x \, dx \tag{13.13}$$

$\vec{F}_{1 \text{ on } 2}$ points toward the left, so its x-component is $(F_{1 \text{ on } 2})_x = -Gm_1m_2/x^2$. As mass m_2 moves from $x_i = r$ to $x_f = \infty$, the potential energy changes by

$$\Delta U = U_{\text{at} \infty} - U_{\text{at} r} = -\int_r^\infty (F_{1 \text{ on } 2})_x \, dx = -\int_r^\infty \left(\frac{-Gm_1m_2}{x^2} \right) dx$$

$$= +Gm_1m_2 \int_r^\infty \frac{dx}{x^2} = -\frac{Gm_1m_2}{x} \Big|_r^\infty = -\frac{Gm_1m_2}{r} \tag{13.14}$$

NOTE We chose to integrate along the x-axis, but the fact that gravity is a conservative force means that ΔU will have this value if m_2 moves from r to ∞ along *any* path.

To proceed further, we need to choose the point where $U = 0$. We would like our choice to be valid for any star or planet, regardless of its mass and radius. This will be the case if we set $U = 0$ at the point where the interaction between the masses vanishes. According to Newton's law of gravity, the strength of the interaction is zero only when $r = \infty$. Two masses infinitely far apart will have no tendency, or potential, to move together, so we will *choose* to place the zero point of potential energy at $r = \infty$. That is, $U_{\text{at} \infty} = 0$.

This choice gives us the gravitational potential energy of masses m_1 and m_2:

$$U_G = -\frac{Gm_1m_2}{r} \tag{13.15}$$

This is the potential energy of masses m_1 and m_2 when their *centers* are separated by distance r. **FIGURE 13.10** is a graph of U_G as a function of the distance r between the masses. Notice that it asymptotically approaches 0 as $r \rightarrow \infty$.

NOTE Although Equation 13.15 looks rather similar to Newton's law of gravity, it depends only on $1/r$, *not* on $1/r^2$.

It may seem disturbing that the potential energy is negative, but we encountered similar situations in Chapter 10. All a negative potential energy means is that the potential energy of the two masses at separation r is *less* than their potential energy at infinite separation. Only the *change* in U has physical significance, and the change will be the same no matter where we place the zero of potential energy.

To illustrate, suppose two masses a distance r_1 apart are released from rest. How will they move? From a force perspective, you would note that each mass experiences an attractive force and accelerates toward the other. The energy perspective of **FIGURE 13.11** tells the same thing. By moving toward smaller r (that is, $r_1 \rightarrow r_2$), the system *loses* potential energy and *gains* kinetic energy while conserving E_{mech}. The system is "falling downhill," although in a more general sense than we think about on a flat earth.

FIGURE 13.10 The gravitational potential-energy curve.

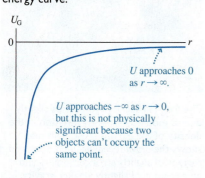

FIGURE 13.11 Two masses gain kinetic energy as their separation decreases.

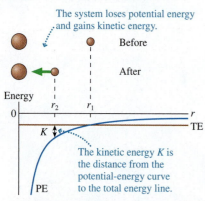

EXAMPLE 13.1 | **Crashing into the sun**

Suppose the earth suddenly came to a halt and ceased revolving around the sun. The gravitational force would then pull it directly into the sun. What would be the earth's speed as it crashed?

MODEL Model the earth and the sun as spherical masses. This is an isolated system, so its mechanical energy is conserved.

VISUALIZE **FIGURE 13.12** is a before-and-after pictorial representation for this gruesome cosmic event. The "crash" occurs as the earth touches the sun, at which point the distance between their centers is $r_2 = R_s + R_e$. The initial separation r_1 is the radius of the earth's *orbit* about the sun, not the radius of the earth.

FIGURE 13.12 Before-and-after pictorial representation of the earth crashing into the sun (not to scale).

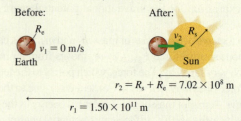

SOLVE Strictly speaking, the kinetic energy is the sum $K = K_{earth} + K_{sun}$. However, the sun is so much more massive than the earth that the lightweight earth does almost all of the moving. It is a reasonable approximation to consider the sun as remaining at rest. In that case, the energy conservation equation $K_2 + U_2 = K_1 + U_1$ is

$$\tfrac{1}{2}M_e v_2^{\,2} - \frac{GM_s M_e}{R_s + R_e} = 0 - \frac{GM_s M_e}{r_1}$$

This is easily solved for the earth's speed at impact. Using data from Table 13.2, we find

$$v_2 = \sqrt{2GM_s\left(\frac{1}{R_s + R_e} - \frac{1}{r_1}\right)} = 6.13 \times 10^5 \text{ m/s}$$

ASSESS The earth would be really flying along at over 1 million miles per hour as it crashed into the sun! It is worth noting that we do not have the mathematical tools to solve this problem using Newton's second law because the acceleration is not constant. But the solution is straightforward when we use energy conservation.

EXAMPLE 13.2 | **Escape speed**

A 1000 kg rocket is fired straight away from the surface of the earth. What speed does the rocket need to "escape" from the gravitational pull of the earth and never return? Assume a nonrotating earth.

MODEL In a simple universe, consisting of only the earth and the rocket, an insufficient launch speed will cause the rocket eventually to fall back to earth. Once the rocket finally slows to a halt, gravity will ever so slowly pull it back. The only way the rocket can escape is to never stop $(v = 0)$ and thus never have a turning point! That is, the rocket must continue moving away from the earth forever. The *minimum* launch speed for escape, which is called the **escape speed,** will cause the rocket to stop $(v = 0)$ only as it reaches $r = \infty$. Now ∞, of course, is not a "place," so a statement like this means that we want the rocket's speed to approach $v = 0$ asymptotically as $r \to \infty$.

VISUALIZE **FIGURE 13.13** is a before-and-after pictorial representation.

SOLVE Energy conservation $K_2 + U_2 = K_1 + U_1$ is

$$0 + 0 = \tfrac{1}{2}mv_1^{\,2} - \frac{GM_e m}{R_e}$$

FIGURE 13.13 Pictorial representation of a rocket launched with sufficient speed to escape the earth's gravity.

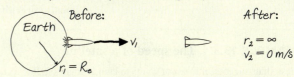

Before: After:

where we used the fact that both the kinetic and potential energy are zero at $r = \infty$. Thus the escape speed is

$$v_{escape} = v_1 = \sqrt{\frac{2GM_e}{R_e}} = 11{,}200 \text{ m/s} \approx 25{,}000 \text{ mph}$$

ASSESS It is difficult to assess the answer to a problem with which we have no direct experience, although we certainly expected the escape speed to be very large. Notice that the problem was mathematically easy; the difficulty was deciding how to interpret it. That is why—as you have now seen many times—the "physics" of a problem consists of thinking, interpreting, and modeling. We will see variations on this problem in the future, with both gravity and electricity, so you might want to review the *reasoning* involved.

The Flat-Earth Approximation

How does Equation 13.15 for the gravitational potential energy relate to our previous use of $U_G = mgy$ on a flat earth? **FIGURE 13.14** shows an object of mass m located at height y above the surface of the earth. The object's distance from the earth's center is $r = R_e + y$ and its gravitational potential energy is

$$U_G = -\frac{GM_e m}{r} = -\frac{GM_e m}{R_e + y} = -\frac{GM_e m}{R_e(1 + y/R_e)} \tag{13.16}$$

where, in the last step, we factored R_e out of the denominator.

Suppose the object is very close to the earth's surface $(y \ll R_e)$. In that case, the ratio $y/R_e \ll 1$. There is an approximation you will learn about in calculus, called the *binomial approximation,* that says

$$(1 + x)^n \approx 1 + nx \qquad \text{if } x \ll 1 \tag{13.17}$$

As an illustration, you can easily use your calculator to find that $1/1.01 = 0.9901$, to four significant figures. But suppose you wrote $1.01 = 1 + 0.01$. You could then use the binomial approximation to calculate

$$\frac{1}{1.01} = \frac{1}{1 + 0.01} = (1 + 0.01)^{-1} \approx 1 + (-1)(0.01) = 0.9900$$

You can see that the approximate answer is off by only 0.01%.

FIGURE 13.14 Gravity on a flat earth.

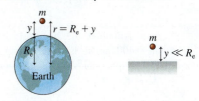

For a spherical earth:

$$U_G = -\frac{GM_e m}{R_e + y}$$

We can treat the earth as flat if $y \ll R_e$:

$$U_G = mgy$$

If we call $y/R_e = x$ in Equation 13.16 and use the binomial approximation, with $n = -1$, we find

$$U_G(\text{if } y \ll R_e) \approx -\frac{GM_e m}{R_e}\left(1 - \frac{y}{R_e}\right) = -\frac{GM_e m}{R_e} + m\left(\frac{GM_e}{R_e^2}\right)y \qquad (13.18)$$

Now the first term is just the gravitational potential energy U_0 when the object is at ground level $(y = 0)$. In the second term, you can recognize $GM_e/R_e^2 = g_{earth}$ from the definition of g in Equation 13.7. Thus we can write Equation 13.18 as

$$U_G(\text{if } y \ll R_e) = U_0 + mg_{earth}y \qquad (13.19)$$

Although we chose U_G to be zero when $r = \infty$, we are always free to change our minds. If we change the zero point of potential energy to be $U_0 = 0$ at the surface, which is the choice we made in Chapter 10, then Equation 13.19 becomes

$$U_G(\text{if } y \ll R_e) = mg_{earth}y \qquad (13.20)$$

We can sleep easier knowing that Equation 13.15 for the gravitational potential energy is consistent with our earlier "flat-earth" expression for the potential energy.

EXAMPLE 13.3 | **The speed of a satellite**

A less-than-successful inventor wants to launch small satellites into orbit by launching them straight up from the surface of the earth at very high speed.

a. With what speed should he launch the satellite if it is to have a speed of 500 m/s at a height of 400 km? Ignore air resistance.

b. By what percentage would your answer be in error if you used a flat-earth approximation?

MODEL Mechanical energy is conserved if we ignore drag.

VISUALIZE FIGURE 13.15 shows a pictorial representation.

FIGURE 13.15 Pictorial representation of a satellite launched straight up.

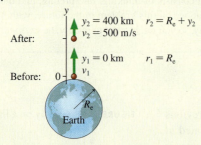

SOLVE a. Although the height is exaggerated in the figure, 400 km = 400,000 m is high enough that we cannot ignore the earth's spherical shape. The energy conservation equation $K_2 + U_2 = K_1 + U_1$ is

$$\frac{1}{2}mv_2^2 - \frac{GM_e m}{R_e + y_2} = \frac{1}{2}mv_1^2 - \frac{GM_e m}{R_e + y_1}$$

where we've written the distance between the satellite and the earth's center as $r = R_e + y$. The initial height is $y_1 = 0$. Notice that the satellite mass m cancels and is not needed. Solving for the launch speed, we have

$$v_1 = \sqrt{v_2^2 + 2GM_e\left(\frac{1}{R_e} - \frac{1}{R_e + y_2}\right)} = 2770 \text{ m/s}$$

This is about 6000 mph, much less than the escape speed.

b. The calculation is the same in the flat-earth approximation except that we use $U_G = mgy$. Thus

$$\frac{1}{2}mv_2^2 + mgy_2 = \frac{1}{2}mv_1^2 + mgy_1$$

$$v_1 = \sqrt{v_2^2 + 2gy_2} = 2840 \text{ m/s}$$

The flat-earth value of 2840 m/s is 70 m/s too big. The error, as a percentage of the correct 2770 m/s, is

$$\text{error} = \frac{70}{2770} \times 100 = 2.5\%$$

ASSESS The true speed is less than the flat-earth approximation because the force of gravity decreases with height. Launching a rocket against a decreasing force takes less effort than it would with the flat-earth force of mg at all heights.

STOP TO THINK 13.4 Rank in order, from largest to smallest, the absolute values of the gravitational potential energies of these pairs of masses. The numbers give the relative masses and distances.

(a) $m_1 = 2$ ---- $r = 4$ ---- $m_2 = 2$

(b) $m_1 = 1$ -- $r = 1$ -- $m_2 = 1$

(c) $m_1 = 1$ --- $r = 2$ --- $m_2 = 1$

(d) $m_1 = 1$ ---- $r = 4$ ---- $m_2 = 4$

(e) $m_1 = 4$ ------ $r = 8$ ------ $m_2 = 4$

13.6 Satellite Orbits and Energies

Solving Newton's second law to find the trajectory of a mass moving under the influence of gravity is mathematically beyond this textbook. It turns out that the solution is a set of elliptical orbits, which is Kepler's first law. Kepler had no *reason* why orbits should be ellipses rather than some other shape. Newton was able to show that ellipses are a *consequence* of his theory of gravity.

The mathematics of ellipses is rather difficult, so we will restrict most of our analysis to the limiting case in which an ellipse becomes a circle. Most planetary orbits differ only very slightly from being circular. The earth's orbit, for example has a (semiminor axis/semimajor axis) ratio of 0.99986—very close to a true circle!

FIGURE 13.16 shows a massive body M, such as the earth or the sun, with a lighter body m orbiting it. The lighter body is called a **satellite**, even though it may be a planet orbiting the sun. For circular motion, where the gravitational force provides the centripetal acceleration v^2/r, Newton's second law for the satellite is

$$F_{M \text{ on } m} = \frac{GMm}{r^2} = ma_r = \frac{mv^2}{r} \qquad (13.21)$$

Thus the speed of a satellite in a circular orbit is

$$v = \sqrt{\frac{GM}{r}} \qquad (13.22)$$

A satellite must have this specific speed in order to have a circular orbit of radius r about the larger mass M. If the velocity differs from this value, the orbit will become elliptical rather than circular. Notice that the orbital speed does *not* depend on the satellite's mass m. This is consistent with our previous discovery, for motion on a flat earth, that motion due to gravity is independent of the mass.

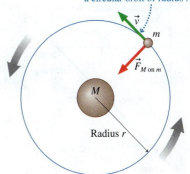

FIGURE 13.16 The orbital motion of a satellite due to the force of gravity.

The satellite must have speed $\sqrt{GM/r}$ to maintain a circular orbit of radius r.

$\vec{v}$
m
$\vec{F}_{M \text{ on } m}$
M
Radius r

EXAMPLE 13.4 | **The speed of the space station**

A small supply satellite needs to dock with the International Space Station. The ISS is in a near-circular orbit at a height of 420 km. What are the speeds of the ISS and the supply satellite in this orbit?

SOLVE Despite their different masses, the satellite and the ISS travel side by side with the same speed. They are simply in free fall together. Using $r = R_e + h$ with $h = 420$ km $= 4.20 \times 10^5$ m, we find the speed

$$v = \sqrt{\frac{(6.67 \times 10^{-11} \text{ N m}^2/\text{kg}^2)(5.98 \times 10^{24} \text{ kg})}{6.79 \times 10^6 \text{ m}}}$$

$$= 7660 \text{ m/s} \approx 17,000 \text{ mph}$$

ASSESS The answer depends on the mass of the earth but *not* on the mass of the satellite.

Kepler's Third Law

An important parameter of circular motion is the *period*. Recall that the period T is the time to complete one full orbit. The relationship among speed, radius, and period is

$$v = \frac{\text{circumference}}{\text{period}} = \frac{2\pi r}{T} \qquad (13.23)$$

We can find a relationship between a satellite's period and the radius of its orbit by using Equation 13.22 for v:

$$v = \frac{2\pi r}{T} = \sqrt{\frac{GM}{r}} \qquad (13.24)$$

Squaring both sides and solving for T give

$$T^2 = \left(\frac{4\pi^2}{GM}\right)r^3 \qquad (13.25)$$

The International Space Station appears to be floating, but it's actually traveling at nearly 8000 m/s as it orbits the earth.

In other words, the *square* of the period is proportional to the *cube* of the radius. This is Kepler's third law. You can see that Kepler's third law is a direct consequence of Newton's law of gravity.

TABLE 13.2 contains astronomical information about the solar system. We can use these data to check the validity of Equation 13.25. **FIGURE 13.17** is a graph of log T versus log r for all the planets in Table 13.2 except Mercury. Notice that the scales on each axis are increasing logarithmically—by *factors* of 10—rather than linearly. (Also, the vertical axis has converted T to the SI units of s.) As you can see, the graph is a straight line with a best-fit equation

$$\log T = 1.500 \log r - 9.264$$

Taking the logarithm of both sides of Equation 13.25, and using the logarithm properties $\log a^n = n \log a$ and $\log(ab) = \log a + \log b$, we have

$$\log T = \tfrac{3}{2} \log r + \tfrac{1}{2} \log\left(\frac{4\pi^2}{GM}\right)$$

In other words, theory predicts that the slope of a log T-versus-log r graph should be exactly $\frac{3}{2}$. As Figure 13.17 shows, the solar-system data agree to an impressive four significant figures. A homework problem will let you use the y-intercept of the graph to determine the mass of the sun.

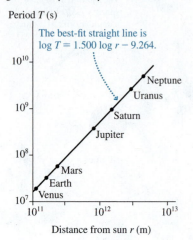

FIGURE 13.17 The graph of log T versus log r for the planetary data of Table 13.2.

TABLE 13.2 Useful astronomical data

Planetary body	Mean distance from sun (m)	Period (years)	Mass (kg)	Mean radius (m)
Sun	—	—	1.99×10^{30}	6.96×10^8
Moon	3.84×10^8*	27.3 days	7.36×10^{22}	1.74×10^6
Mercury	5.79×10^{10}	0.241	3.18×10^{23}	2.43×10^6
Venus	1.08×10^{11}	0.615	4.88×10^{24}	6.06×10^6
Earth	1.50×10^{11}	1.00	5.98×10^{24}	6.37×10^6
Mars	2.28×10^{11}	1.88	6.42×10^{23}	3.37×10^6
Jupiter	7.78×10^{11}	11.9	1.90×10^{27}	6.99×10^7
Saturn	1.43×10^{12}	29.5	5.68×10^{26}	5.85×10^7
Uranus	2.87×10^{12}	84.0	8.68×10^{25}	2.33×10^7
Neptune	4.50×10^{12}	165	1.03×10^{26}	2.21×10^7

*Distance from earth.

A particularly interesting application of Equation 13.25 is to communications satellites that are in **geosynchronous orbits** above the earth. These satellites have a period of 24 h = 86,400 s, making their orbital motion synchronous with the earth's rotation. As a result, a satellite in such an orbit appears to remain stationary over one point on the earth's equator. Equation 13.25 allows us to compute the radius of an orbit with this period:

$$r_{\text{geo}} = R_\text{e} + h_{\text{geo}} = \left[\left(\frac{GM}{4\pi^2}\right) T^2 \right]^{1/3}$$

$$= \left[\left(\frac{(6.67 \times 10^{-11}\,\text{N m}^2/\text{kg}^2)(5.98 \times 10^{24}\,\text{kg})}{4\pi^2} \right) (86,400\,\text{s})^2 \right]^{1/3}$$

$$= 4.225 \times 10^7\,\text{m}$$

The height of the orbit is

$$h_{\text{geo}} = r_{\text{geo}} - R_\text{e} = 3.59 \times 10^7\,\text{m} = 35,900\,\text{km} \approx 22,300\,\text{mi}$$

NOTE When you use Equation 13.25, the period *must* be in SI units of s.

Geosynchronous orbits are much higher than the low-earth orbits used by the space station and remote-sensing satellites, where $h \approx 300$ km. Communications satellites in geosynchronous orbits were first proposed in 1948 by science fiction writer Arthur C. Clarke, 10 years before the first artificial satellite of any type!

EXAMPLE 13.5 | Extrasolar planets

In recent years, astronomers have discovered thousands of planets orbiting nearby stars. These are called *extrasolar planets*. Suppose a planet is observed to have a 1200 day period as it orbits a star at the same distance that Jupiter is from the sun. What is the mass of the star in solar masses? (1 *solar mass* is defined to be the mass of the sun.)

SOLVE Here "day" means earth days, as used by astronomers to measure the period. Thus the planet's period in SI units is $T = 1200$ days $= 1.037 \times 10^8$ s. The orbital radius is that of Jupiter,

which we can find in Table 13.2 to be $r = 7.78 \times 10^{11}$ m. Solving Equation 13.25 for the mass of the star gives

$$M = \frac{4\pi^2 r^3}{GT^2} = 2.59 \times 10^{31} \text{ kg} \times \frac{1 \text{ solar mass}}{1.99 \times 10^{30} \text{ kg}}$$

$$= 13 \text{ solar masses}$$

ASSESS This is a large, but not extraordinary, star.

STOP TO THINK 13.5 Two planets orbit a star. Planet 1 has orbital radius r_1 and planet 2 has $r_2 = 4r_1$. Planet 1 orbits with period T_1. Planet 2 orbits with period

a. $T_2 = 8T_1$ b. $T_2 = 4T_1$ c. $T_2 = 2T_1$

d. $T_2 = \frac{1}{2}T_1$ e. $T_2 = \frac{1}{4}T_1$ f. $T_2 = \frac{1}{8}T_1$

Kepler's Second Law

FIGURE 13.18a shows a planet moving in an elliptical orbit. In Chapter 12 we defined a particle's *angular momentum* to be

$$L = mrv \sin \beta \qquad (13.26)$$

where β is the angle between $\vec{r}$ and $\vec{v}$. For a circular orbit, where β is always 90°, this reduces to simply $L = mrv$.

The only force on the satellite, the gravitational force, points directly toward the star or planet that the satellite is orbiting and exerts no torque; thus **the satellite's angular momentum is conserved as it orbits.**

The satellite moves forward a small distance $\Delta s = v\,\Delta t$ during the small interval of time Δt. This motion defines the triangle of area ΔA shown in **FIGURE 13.18b**. ΔA is the area "swept out" by the satellite during Δt. You can see that the height of the triangle is $h = \Delta s \sin \beta$, so the triangle's area is

$$\Delta A = \frac{1}{2} \times \text{base} \times \text{height} = \frac{1}{2} \times r \times \Delta s \sin \beta = \frac{1}{2} rv \sin \beta \, \Delta t \qquad (13.27)$$

The *rate* at which the area is swept out by the satellite as it moves is

$$\frac{\Delta A}{\Delta t} = \frac{1}{2} rv \sin \beta = \frac{mrv \sin \beta}{2m} = \frac{L}{2m} \qquad (13.28)$$

The angular momentum L is conserved, so it has the same value at every point in the orbit. Consequently, the rate at which the area is swept out by the satellite is constant. This is Kepler's second law, which says that a line drawn between the sun and a planet sweeps out equal areas during equal intervals of time. We see that Kepler's second law is a consequence of the conservation of angular momentum.

Another consequence of angular momentum is that the orbital speed is constant only for a circular orbit. Consider the "ends" of an elliptical orbit, where r is a minimum or maximum. At these points, $\beta = 90°$ and thus $L = mrv$. Because L is constant, the satellite's speed at the farthest point must be less than its speed at the nearest point. In general, a satellite slows as r increases, then speeds up as r decreases, to keep its angular momentum (and its energy) constant.

FIGURE 13.18 Angular momentum is conserved for a planet in an elliptical orbit.

(a)

The gravitational force points straight at the sun and exerts no torque.

Sun

(b)

Area ΔA is swept out during Δt.

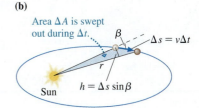

$\Delta s = v\Delta t$

$h = \Delta s \sin\beta$

Sun

Kepler's laws summarize observational data about the motions of the planets. They were an outstanding achievement, but they did not form a theory. Newton put forward a *theory,* a specific set of relationships between force and motion that allows *any* motion to be understood and calculated. Newton's theory of gravity has allowed us to *deduce* Kepler's laws and, thus, to understand them at a more fundamental level.

Orbital Energetics

Let us conclude this chapter by thinking about the energetics of orbital motion. We found, with Equation 13.24, that a satellite in a circular orbit must have $v^2 = GM/r$. A satellite's speed is determined entirely by the size of its orbit. The satellite's kinetic energy is thus

$$K = \tfrac{1}{2}mv^2 = \frac{GMm}{2r} \tag{13.29}$$

But $-GMm/r$ is the potential energy, U_G, so

$$K = -\tfrac{1}{2}U_G \tag{13.30}$$

This is an interesting result. In all our earlier examples, the kinetic and potential energy were two independent parameters. In contrast, a satellite can move in a circular orbit *only* if there is a very specific relationship between K and U. It is not that K and U *have* to have this relationship, but if they do not, the trajectory will be elliptical rather than circular.

Equation 13.30 gives us the mechanical energy of a satellite in a circular orbit:

$$E_{mech} = K + U_G = \tfrac{1}{2}U_G \tag{13.31}$$

The gravitational potential energy is negative, hence the *total* mechanical energy is also negative. Negative total energy is characteristic of a **bound system,** a system in which the satellite is bound to the central mass by the gravitational force and cannot get away. The total energy of an unbound system must be ≥ 0 because the satellite can reach infinity, where $U = 0$, while still having kinetic energy. A negative value of E_{mech} tells us that the satellite is unable to escape the central mass.

FIGURE 13.19 shows the energies of a satellite in a circular orbit as a function of the orbit's radius. Notice how $E_{mech} = \tfrac{1}{2}U_G$. This figure can help us understand the energetics of transferring a satellite from one orbit to another. Suppose a satellite is in an orbit of radius r_1 and we'd like it to be in a larger orbit of radius r_2. The kinetic energy at r_2 is less than at r_1 (the satellite moves more slowly in the larger orbit), but you can see that the total energy *increases* as r increases. Consequently, transferring a satellite to a larger orbit requires a net energy increase $\Delta E > 0$. Where does this increase of energy come from?

Artificial satellites are raised to higher orbits by firing their rocket motors to create a forward thrust. This force does work on the satellite, and the energy principle of Chapter 10 tells us that this work increases the satellite's energy by $\Delta E_{mech} = W_{ext}$. Thus the energy to "lift" a satellite into a higher orbit comes from the chemical energy stored in the rocket fuel.

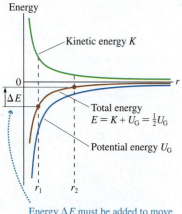

FIGURE 13.19 The kinetic, potential, and total energy of a satellite in a circular orbit.

Energy ΔE must be added to move a satellite from an orbit with radius r_1 to radius r_2.

EXAMPLE 13.6 | **Raising a satellite**

How much work must be done to boost a 1000 kg communications satellite from a low earth orbit with $h = 300$ km to a geosynchronous orbit?

SOLVE The required work is $W_{ext} = \Delta E_{mech}$, and from Equation 13.31 we see that $\Delta E_{mech} = \tfrac{1}{2}\Delta U_G$. The initial orbit has radius $r_{low} = R_e + h = 6.67 \times 10^6$ m. We earlier found the radius of a geosynchronous orbit to be 4.22×10^7 m. Thus

$$W_{ext} = \Delta E_{mech} = \tfrac{1}{2}\Delta U_G = \tfrac{1}{2}(-GM_em)\left(\frac{1}{r_{geo}} - \frac{1}{r_{low}}\right) = 2.52 \times 10^{10} \text{ J}$$

ASSESS It takes a lot of energy to boost satellites to high orbits!

You might think that the way to get a satellite into a larger orbit would be to point the thrusters toward the earth and blast outward. That would work fine *if* the satellite were initially at rest and moved straight out along a linear trajectory. But an orbiting satellite is already moving and has significant inertia. A force directed straight outward would *change* the satellite's velocity vector in that direction but would not cause it to *move* along that line. (Remember all those earlier motion diagrams for motion along curved trajectories.) In addition, a force directed outward would be almost at right angles to the motion and would do essentially zero work. Navigating in space is not as easy as it appears in *Star Wars!*

To move the satellite in **FIGURE 13.20** from the orbit with radius r_1 to the larger circular orbit of radius r_2, the thrusters are turned on at point 1 to apply a brief *forward* thrust force in the direction of motion, *tangent* to the circle. This force does a significant amount of work because the force is parallel to the displacement, so the satellite quickly gains kinetic energy ($\Delta K > 0$). But $\Delta U_G = 0$ because the satellite does not have time to change its distance from the earth during a thrust of short duration. With the kinetic energy increased, but not the potential energy, the satellite no longer meets the requirement $K = -\frac{1}{2}U_G$ for a circular orbit. Instead, it goes into an elliptical orbit.

In the elliptical orbit, the satellite moves "uphill" toward point 2 by transforming kinetic energy into potential energy. At point 2, the satellite has arrived at the desired distance from earth and has the "right" value of the potential energy, but its kinetic energy is now *less* than needed for a circular orbit. (The analysis is more complex than we want to pursue here. It will be left for a homework Challenge Problem.) If no action is taken, the satellite will continue on its elliptical orbit and "fall" back to point 1. But another *forward* thrust at point 2 increases its kinetic energy, without changing U_G, until the kinetic energy reaches the value $K = -\frac{1}{2}U_G$ required for a circular orbit. Presto! The second burn kicks the satellite into the desired circular orbit of radius r_2. The work $W_{ext} = \Delta E_{mech}$ is the *total* work done in both burns. It takes a more extended analysis to see how the work has to be divided between the two burns, but even without those details you now have enough knowledge about orbits and energy to understand the ideas that are involved.

FIGURE 13.20 Transferring a satellite to a larger circular orbit.

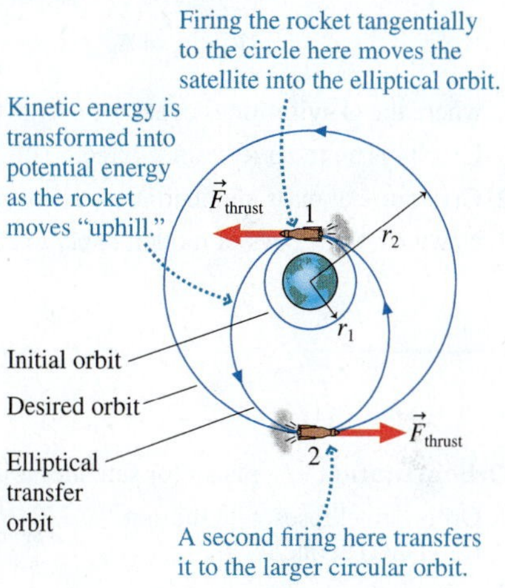

Firing the rocket tangentially to the circle here moves the satellite into the elliptical orbit.

Kinetic energy is transformed into potential energy as the rocket moves "uphill."

$\vec{F}_{thrust}$ 1 r_2

r_1

Initial orbit

Desired orbit

Elliptical transfer orbit

$\vec{F}_{thrust}$ 2

A second firing here transfers it to the larger circular orbit.

CHALLENGE EXAMPLE 13.7 | A binary star system

Astronomers discover a binary star system with a period of 90 days. Both stars have a mass twice that of the sun. How far apart are the two stars?

MODEL Model the stars as spherical masses exerting gravitational forces on each other.

VISUALIZE An isolated system rotates around its center of mass. **FIGURE 13.21** shows the orbits and the forces. If r is the distance of each star to the center of mass—the radius of that star's orbit— then the distance between the stars is $d = 2r$.

FIGURE 13.21 The binary star system.

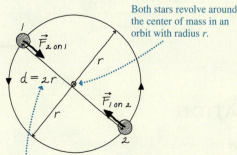

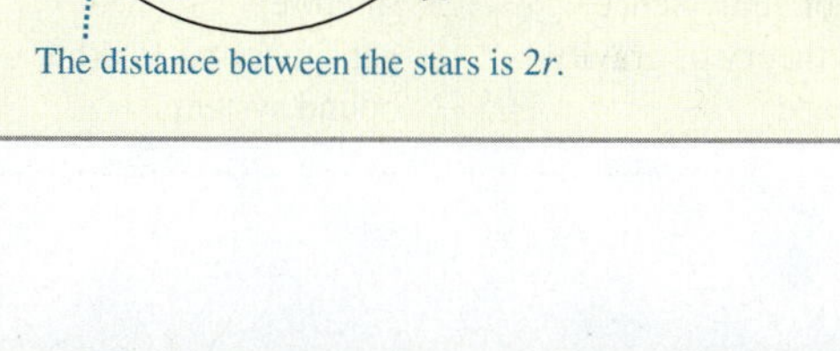

Both stars revolve around the center of mass in an orbit with radius r.

$\vec{F}_{2\,on\,1}$

r

$d = 2r$

$\vec{F}_{1\,on\,2}$

r

The distance between the stars is $2r$.

SOLVE Star 2 has only one force acting on it, $\vec{F}_{1\,on\,2}$, and that force has to provide the centripetal acceleration v^2/r of circular motion. Newton's second law for star 2 is

$$F_{1\,on\,2} = \frac{GM_1 M_2}{d^2} = \frac{GM^2}{4r^2} = Ma_r = \frac{Mv^2}{r}$$

where we used $M_1 = M_2 = M$. The equation for star 1 is identical. The star's speed is related to the period and the circumference of its orbit by $v = 2\pi r/T$. With this, the force equation becomes

$$\frac{GM^2}{4r^2} = \frac{4\pi^2 Mr}{T^2}$$

Solving for r gives

$$r = \left[\frac{GMT^2}{16\pi^2}\right]^{1/3}$$

$$= \left[\frac{(6.67 \times 10^{-11}\,\text{N}\,\text{m}^2/\text{kg}^2)(2 \times 1.99 \times 10^{30}\,\text{kg})(7.78 \times 10^6\,\text{s})^2}{16\pi^2}\right]^{1/3}$$

$$= 4.67 \times 10^{10}\,\text{m}$$

The distance between the stars is $d = 2r = 9.3 \times 10^{10}$ m.

ASSESS The result is in the range of solar-system distances and thus is reasonable.

SUMMARY

The goal of Chapter 13 has been to understand the motion of satellites and planets.

GENERAL PRINCIPLES

Newton's Theory of Gravity

1. Two objects with masses M and m a distance r apart exert attractive **gravitational forces** on each other of magnitude

$$F_{M \text{ on } m} = F_{m \text{ on } M} = \frac{GMm}{r^2}$$

where the **gravitational constant** is $G = 6.67 \times 10^{-11} \text{ N m}^2/\text{kg}^2$.
Gravity is an **inverse-square** force.

2. Gravitational mass and inertial mass are equivalent.

3. Newton's three laws of motion apply to all objects in the universe.

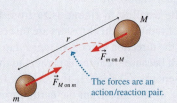

The forces are an action/reaction pair.

IMPORTANT CONCEPTS

Orbital motion of a planet (or satellite) is described by **Kepler's laws:**

1. Orbits are ellipses with the sun (or planet) at one focus.

2. A line between the sun and the planet sweeps out equal areas during equal intervals of time.

3. The square of the planet's period T is proportional to the cube of the orbit's semimajor axis.

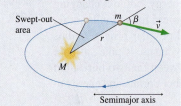

Swept-out area

Semimajor axis

Circular orbits are a special case of an ellipse. For a circular orbit around a mass M,

$$v = \sqrt{\frac{GM}{r}} \quad \text{and} \quad T^2 = \left(\frac{4\pi^2}{GM}\right)r^3$$

Conservation of angular momentum

The angular momentum $L = mrv \sin \beta$ remains constant throughout the orbit. Kepler's second law is a consequence of this law.

Orbital energetics

A satellite's mechanical energy $E_{\text{mech}} = K + U_G$ is conserved, where the gravitational potential energy is

$$U_G = -\frac{GMm}{r}$$

For circular orbits, $K = -\frac{1}{2}U_G$ and $E_{\text{mech}} = \frac{1}{2}U_G$. Negative total energy is characteristic of a **bound system.**

APPLICATIONS

For a planet of mass M and radius R:

- Free-fall acceleration on the surface is $g_{\text{surface}} = \dfrac{GM}{R^2}$

- **Escape speed** is $v_{\text{escape}} = \sqrt{\dfrac{2GM}{R}}$

- Radius of a **geosynchronous orbit** is $r_{\text{geo}} = \left(\dfrac{GM}{4\pi^2}T^2\right)^{1/3}$

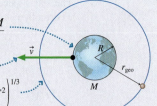

To move a satellite to a larger orbit:

- Forward thrust to move to an elliptical transfer orbit

- A second forward thrust to move to the new circular orbit

TERMS AND NOTATION

cosmology	Newton's law of gravity	principle of equivalence	satellite
Kepler's laws	gravitational constant, G	Newton's theory of gravity	geosynchronous orbit
gravitational force	gravitational mass	escape speed	bound system

CONCEPTUAL QUESTIONS

1. Is the earth's gravitational force on the sun larger than, smaller than, or equal to the sun's gravitational force on the earth? Explain.

2. The gravitational force of a star on orbiting planet 1 is F_1. Planet 2, which is twice as massive as planet 1 and orbits at twice the distance from the star, experiences gravitational force F_2. What is the ratio F_1/F_2?

3. A 1000 kg satellite and a 2000 kg satellite follow exactly the same orbit around the earth.
   a. What is the ratio F_1/F_2 of the gravitational force on the first satellite to that on the second satellite?
   b. What is the ratio a_1/a_2 of the acceleration of the first satellite to that of the second satellite?

4. How far away from the earth must an orbiting spacecraft be for the astronauts inside to be weightless? Explain.

5. A space station astronaut is working outside the station as it orbits the earth. If he drops a hammer, will it fall to earth? Explain why or why not.

6. The free-fall acceleration at the surface of planet 1 is 20 m/s². The radius and the mass of planet 2 are twice those of planet 1. What is g on planet 2?

7. *Why* is the gravitational potential energy of two masses negative? Note that saying "because that's what the equation gives" is *not* an explanation.

8. The escape speed from Planet X is 10,000 m/s. Planet Y has the same radius as Planet X but is twice as dense. What is the escape speed from Planet Y?

9. The mass of Jupiter is 300 times the mass of the earth. Jupiter orbits the sun with $T_{Jupiter} = 11.9$ yr in an orbit with $r_{Jupiter} = 5.2r_{earth}$. Suppose the earth could be moved to the distance of Jupiter and placed in a circular orbit around the sun. Which of the following describes the earth's new period? Explain.
   a. 1 yr
   b. Between 1 yr and 11.9 yr
   c. 11.9 yr
   d. More than 11.9 yr
   e. It would depend on the earth's speed.
   f. It's impossible for a planet of earth's mass to orbit at the distance of Jupiter.

10. Satellites in near-earth orbit experience a very slight drag due to the extremely thin upper atmosphere. These satellites slowly but surely spiral inward, where they finally burn up as they reach the thicker lower levels of the atmosphere. The radius decreases so slowly that you can consider the satellite to have a circular orbit at all times. As a satellite spirals inward, does it speed up, slow down, or maintain the same speed? Explain.

EXERCISES AND PROBLEMS

Problems labeled ▊ integrate material from earlier chapters.

Exercises

Section 13.3 Newton's Law of Gravity

1. | What is the ratio of the sun's gravitational force on you to the earth's gravitational force on you?

2. ‖ What is the ratio of the sun's gravitational force on the moon to the earth's gravitational force on the moon?

3. ‖ The centers of a 10 kg lead ball and a 100 g lead ball are separated by 10 cm.
   a. What gravitational force does each exert on the other?
   b. What is the ratio of this gravitational force to the gravitational force of the earth on the 100 g ball?

4. | What is the force of attraction between a 50 kg woman and a 70 kg man sitting 1.0 m apart?

5. ‖ The International Space Station orbits 300 km above the surface of the earth. What is the gravitational force on a 1.0 kg sphere inside the International Space Station?

6. ‖ Two 65 kg astronauts leave earth in a spacecraft, sitting 1.0 m apart. How far are they from the center of the earth when the gravitational force between them is as strong as the gravitational force of the earth on one of the astronauts?

7. ‖ A 20 kg sphere is at the origin and a 10 kg sphere is at $x = 20$ cm. At what position on the x-axis could you place a small mass such that the net gravitational force on it due to the spheres is zero?

Section 13.4 Little g and Big G

8. | a. What is the free-fall acceleration at the surface of the sun?
   b. What is the free-fall acceleration toward the sun at the distance of the earth?

9. ‖ What is the free-fall acceleration at the surface of (a) the moon and (b) Jupiter?

10. ‖ A sensitive gravimeter at a mountain observatory finds that the free-fall acceleration is 0.0075 m/s² less than that at sea level. What is the observatory's altitude?

11. | Saturn's moon Titan has a mass of 1.35×10^{23} kg and a radius of 2580 km. What is the free-fall acceleration on Titan?

12. ‖ A newly discovered planet has a radius twice as large as earth's and a mass five times as large. What is the free-fall acceleration on its surface?

13. | Suppose we could shrink the earth without changing its mass. At what fraction of its current radius would the free-fall acceleration at the surface be three times its present value?

14. ‖ Planet Z is 10,000 km in diameter. The free-fall acceleration on Planet Z is 8.0 m/s².
   a. What is the mass of Planet Z?
   b. What is the free-fall acceleration 10,000 km above Planet Z's north pole?

Section 13.5 Gravitational Potential Energy

15. | An astronaut on earth can throw a ball straight up to a height of 15 m. How high can he throw the ball on Mars?

16. ‖ What is the escape speed from Jupiter?

17. ‖ A rocket is launched straight up from the earth's surface at a speed of 15,000 m/s. What is its speed when it is very far away from the earth?

18. | A space station orbits the sun at the same distance as the earth but on the opposite side of the sun. A small probe is fired away from the station. What minimum speed does the probe need to escape the solar system?

19. ‖ A proposed *space elevator* would consist of a cable stretching from the earth's surface to a satellite, orbiting far in space, that would keep the cable taut. A motorized climber could slowly carry rockets to the top, where they could be launched away from the earth using much less energy. What would be the escape speed for a craft launched from a space elevator at a height of 36,000 km? Ignore the earth's rotation.

20. ‖ Nothing can escape the *event horizon* of a black hole, not even light. You can think of the event horizon as being the distance from a black hole at which the escape speed is the speed of light, 3.00×10^8 m/s, making all escape impossible. What is the radius of the event horizon for a black hole with a mass 5.0 times the mass of the sun? This distance is called the *Schwarzschild radius*.

21. ‖ You have been visiting a distant planet. Your measurements have determined that the planet's mass is twice that of earth but the free-fall acceleration at the surface is only one-fourth as large.
   a. What is the planet's radius?
   b. To get back to earth, you need to escape the planet. What minimum speed does your rocket need?

22. ‖ Two meteoroids are heading for earth. Their speeds as they cross the moon's orbit are 2.0 km/s.
   a. The first meteoroid is heading straight for earth. What is its speed of impact?
   b. The second misses the earth by 5000 km. What is its speed at its closest point?

23. ‖ A binary star system has two stars, each with the same mass as our sun, separated by 1.0×10^{12} m. A comet is very far away and essentially at rest. Slowly but surely, gravity pulls the comet toward the stars. Suppose the comet travels along a trajectory that passes through the midpoint between the two stars. What is the comet's speed at the midpoint?

Section 13.6 Satellite Orbits and Energies

24. ‖ The *asteroid belt* circles the sun between the orbits of Mars and Jupiter. One asteroid has a period of 5.0 earth years. What are the asteroid's orbital radius and speed?

25. ‖ You are the science officer on a visit to a distant solar system. Prior to landing on a planet you measure its diameter to be 1.8×10^7 m and its rotation period to be 22.3 hours. You have previously determined that the planet orbits 2.2×10^{11} m from its star with a period of 402 earth days. Once on the surface you find that the free-fall acceleration is 12.2 m/s². What is the mass of (a) the planet and (b) the star?

26. ‖ Three satellites orbit a planet of radius R, as shown in FIGURE EX13.26. Satellites S_1 and S_3 have mass m. Satellite S_2 has mass $2m$. Satellite S_1 orbits in 250 minutes and the force on S_1 is 10,000 N.
   a. What are the periods of S_2 and S_3?
   b. What are the forces on S_2 and S_3?
   c. What is the kinetic-energy ratio K_1/K_3 for S_1 and S_3?

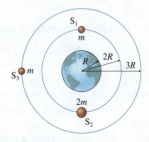

FIGURE EX13.26

27. ‖ A satellite orbits the sun with a period of 1.0 day. What is the radius of its orbit?

28. ‖ A new planet is discovered orbiting the star Vega in a circular orbit. The planet takes 55 earth years to complete one orbit around the star. Vega's mass is 2.1 times the sun's mass. What is the radius of the planet's orbit? Give your answer as a multiple of the radius of the earth's orbit.

29. ‖ A small moon orbits its planet in a circular orbit at a speed of 7.5 km/s. It takes 28 hours to complete one full orbit. What is the mass of the planet?

30. ‖ An earth satellite moves in a circular orbit at a speed of 5500 m/s. What is its orbital period?

31. ‖ What are the speed and altitude of a geosynchronous satellite orbiting Mars? Mars rotates on its axis once every 24.8 hours.

32. ‖ a. At what height above the earth is the free-fall acceleration 10% of its value at the surface?
   b. What is the speed of a satellite orbiting at that height?

33. ‖ In 2000, NASA placed a satellite in orbit around an asteroid. Consider a spherical asteroid with a mass of 1.0×10^{16} kg and a radius of 8.8 km.
   a. What is the speed of a satellite orbiting 5.0 km above the surface?
   b. What is the escape speed from the asteroid?

34. ‖ Pluto moves in a fairly elliptical orbit around the sun. Pluto's speed at its closest approach of 4.43×10^9 km is 6.12 km/s. What is Pluto's speed at the most distant point in its orbit, where it is 7.30×10^9 km from the sun?

Problems

35. ‖ FIGURE P13.35 shows three masses. What are the magnitude and the direction of the net gravitational force on (a) the 20.0 kg mass and (b) the 5.0 kg mass? Give the direction as an angle cw or ccw from the y-axis.

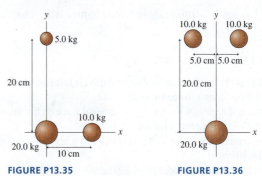

FIGURE P13.35   FIGURE P13.36

36. ‖ What are the magnitude and direction of the net gravitational force on the 20.0 kg mass in FIGURE P13.36?

37. ‖ What is the total gravitational potential energy of the three masses in FIGURE P13.35?

38. ‖ What is the total gravitational potential energy of the three masses in FIGURE P13.36?

39. ‖ Two spherical objects have a combined mass of 150 kg. The gravitational attraction between them is 8.00×10^{-6} N when their centers are 20 cm apart. What is the mass of each?

40. ‖‖ Two 100 kg lead spheres are suspended from 100-m-long massless cables. The tops of the cables have been carefully anchored *exactly* 1 m apart. By how much is the distance between the centers of the spheres less than 1 m?

41. ‖ A rogue black hole with a mass 12 times the mass of the sun drifts into the solar system on a collision course with earth. How far is the black hole from the center of the earth when objects on the earth's surface begin to lift into the air and "fall" up into the black hole? Give your answer as a multiple of the earth's radius.

42. ▮▮▮ An object of mass m is dropped from height h above a planet of mass M and radius R. Find an expression for the object's speed as it hits the ground.

43. ▮▮▮ A projectile is shot straight up from the earth's surface at a speed of 10,000 km/h. How high does it go?

44. ▮▮ The unexplored planet Alpha Centauri III has a radius of 7.0×10^6 m. A visiting astronaut drops a rock, from rest, into a 100-m-deep crevasse. She records that it takes 6.0 s for the rock to reach the bottom. What is the mass of Alpha Centauri III?

45. ▮▮ An astronaut circling the earth at an altitude of 400 km is horrified to discover that a cloud of space debris is moving in the exact same orbit as his spacecraft, but in the opposite direction. The astronaut detects the debris when it is 25 km away. How much time does he have to fire his rockets and change orbits?

46. ▮▮ Suppose that on earth you can jump straight up a distance of 45 cm. Asteroids are made of material with mass density 2800 kg/m³. What is the maximum diameter of a spherical asteroid from which you could escape by jumping?

47. ▮▮▮ A rogue band of colonists on the moon declares war and
CALC prepares to use a catapult to launch large boulders at the earth. Assume that the boulders are launched from the point on the moon nearest the earth. For this problem you can ignore the rotation of the two bodies and the orbiting of the moon.
   a. What is the minimum speed with which a boulder must be launched to reach the earth?
   Hint: The minimum speed is not the escape speed. You need to analyze a three-body system.
   b. Ignoring air resistance, what is the impact speed on earth of a boulder launched at this minimum speed?

48. ▮▮▮ Two spherical asteroids have the same radius R. Asteroid 1 has mass M and asteroid 2 has mass $2M$. The two asteroids are released from rest with distance $10R$ between their centers. What is the speed of each asteroid just before they collide?
   Hint: You will need to use two conservation laws.

49. ▮▮ A starship is circling a distant planet of radius R. The astronauts find that the free-fall acceleration at their altitude is half the value at the planet's surface. How far above the surface are they orbiting? Your answer will be a multiple of R.

50. ▮▮▮ The two stars in a binary star system have masses 2.0×10^{30} kg and 6.0×10^{30} kg. They are separated by 2.0×10^{12} m. What are
   a. The system's rotation period, in years?
   b. The speed of each star?

51. ▮▮▮ A 4000 kg lunar lander is in orbit 50 km above the surface of the moon. It needs to move out to a 300-km-high orbit in order to link up with the mother ship that will take the astronauts home. How much work must the thrusters do?

52. ▮▮▮ The 75,000 kg space shuttle used to fly in a 250-km-high circular orbit. It needed to reach a 610-km-high circular orbit to service the Hubble Space Telescope. How much energy was required to boost it to the new orbit?

53. ▮▮▮ How much energy would be required to move the earth into a circular orbit with a radius 1.0 km larger than its current radius?

54. ▮▮ NASA would like to place a satellite in orbit around the moon such that the satellite always remains in the same position over the lunar surface. What is the satellite's altitude?

55. ▮▮ In 2014, the European Space Agency placed a satellite in orbit around comet 67P/Churyumov-Gerasimenko and then landed a probe on the surface. The actual orbit was elliptical, but we'll approximate it as a 50-km-diameter circular orbit with a period of 11 days.
   a. What was the satellite's orbital speed around the comet? (Both the comet and the satellite are orbiting the sun at a much higher speed.)

   b. What is the mass of the comet?
   c. The lander was pushed from the satellite, toward the comet, at a speed of 70 cm/s, and it then fell—taking about 7 hours—to the surface. What was its landing speed? The comet's shape is irregular, but on average it has a diameter of 3.6 km.

56. ▮▮ A satellite orbiting the earth is directly over a point on the equator at 12:00 midnight every two days. It is not over that point at any time in between. What is the radius of the satellite's orbit?

57. ▮▮▮ **FIGURE P13.57** shows two planets of mass m orbiting a star of mass M. The planets are in the same orbit, with radius r, but are always at opposite ends of a diameter. Find an exact expression for the orbital period T.
   Hint: Each planet feels two forces.

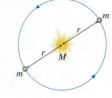

FIGURE P13.57

58. ▮▮ Figure 13.17 showed a graph of log T versus log r for the planetary data given in Table 13.2. Such a graph is called a *log-log graph*. The scales in Figure 13.17 are logarithmic, not linear, meaning that each division along the axis corresponds to a *factor* of 10 increase in the value. Strictly speaking, the "correct" labels on the y-axis should be 7, 8, 9, and 10 because these are the logarithms of $10^7, \ldots, 10^{10}$.
   a. Consider two quantities u and v that are related by the expression $v^p = Cu^q$, where C is a constant. The exponents p and q are not necessarily integers. Define $x = \log u$ and $y = \log v$. Find an expression for y in terms of x.
   b. What *shape* will a graph of y versus x have? Explain.
   c. What *slope* will a graph of y versus x have? Explain.
   d. Use the experimentally determined "best-fit" line in Figure 13.17 to find the mass of the sun.

59. ▮▮ Large stars can explode as they finish burning their nuclear fuel, causing a *supernova*. The explosion blows away the outer layers of the star. According to Newton's third law, the forces that push the outer layers away have *reaction forces* that are inwardly directed on the core of the star. These forces compress the core and can cause the core to undergo a *gravitational collapse*. The gravitational forces keep pulling all the matter together tighter and tighter, crushing atoms out of existence. Under these extreme conditions, a proton and an electron can be squeezed together to form a neutron. If the collapse is halted when the neutrons all come into contact with each other, the result is an object called a *neutron star*, an entire star consisting of solid nuclear matter. Many neutron stars rotate about their axis with a period of ≈ 1 s and, as they do so, send out a pulse of electromagnetic waves once a second. These stars were discovered in the 1960s and are called *pulsars*.
   a. Consider a neutron star with a mass equal to the sun, a radius of 10 km, and a rotation period of 1.0 s. What is the speed of a point on the equator of the star?
   b. What is g at the surface of this neutron star?
   c. A stationary 1.0 kg mass has a weight on earth of 9.8 N. What would be its weight on the star?
   d. How many revolutions per minute are made by a satellite orbiting 1.0 km above the surface?
   e. What is the radius of a geosynchronous orbit?

60. ▮▮ The solar system is 25,000 light years from the center of our Milky Way galaxy. One *light year* is the distance light travels in one year at a speed of 3.0×10^8 m/s. Astronomers have determined that the solar system is orbiting the center of the galaxy at a speed of 230 km/s.
   a. Assuming the orbit is circular, what is the period of the solar system's orbit? Give your answer in years?

b. Our solar system was formed roughly 5 billion years ago. How many orbits has it completed?

c. The gravitational force on the solar system is the net force due to all the matter inside our orbit. Most of that matter is concentrated near the center of the galaxy. Assume that the matter has a spherical distribution, like a giant star. What is the approximate mass of the galactic center?

d. Assume that the sun is a typical star with a typical mass. If galactic matter is made up of stars, approximately how many stars are in the center of the galaxy?

Astronomers have spent many years trying to determine how many stars there are in the Milky Way. The number of stars seems to be only about 10% of what you found in part d. In other words, about 90% of the mass of the galaxy appears to be in some form other than stars. This is called the *dark matter* of the universe. No one knows what the dark matter is. This is one of the outstanding scientific questions of our day.

61. ‖ Three stars, each with the mass of our sun, form an equilateral triangle with sides 1.0×10^{12} m long. (This triangle would just about fit within the orbit of Jupiter.) The triangle has to rotate, because otherwise the stars would crash together in the center. What is the period of rotation?

62. ‖ Comets move around the sun in very elliptical orbits. At its closet approach, in 1986, Comet Halley was 8.79×10^7 km from the sun and moving with a speed of 54.6 km/s. What was the comet's speed when it crossed Neptune's orbit in 2006?

63. ‖‖ A 55,000 kg space capsule is in a 28,000-km-diameter circular orbit around the moon. A brief but intense firing of its engine in the forward direction suddenly decreases its speed by 50%. This causes the space capsule to go into an elliptical orbit. What are the space capsule's (a) maximum and (b) minimum distances from the center of the moon in its new orbit?

Hint: You will need to use two conservation laws.

In Problems 64 through 66 you are given the equation(s) used to solve a problem. For each of these, you are to

a. Write a realistic problem for which this is the correct equation(s).

b. Draw a pictorial representation.

c. Finish the solution of the problem.

64. $\dfrac{(6.67 \times 10^{-11}\,\mathrm{N\,m^2/kg^2})(5.68 \times 10^{26}\,\mathrm{kg})}{r^2}$

$= \dfrac{(6.67 \times 10^{-11}\,\mathrm{N\,m^2/kg^2})(5.98 \times 10^{24}\,\mathrm{kg})}{(6.37 \times 10^{6}\,\mathrm{m})^2}$

65. $\dfrac{(6.67 \times 10^{-11}\,\mathrm{N\,m^2/kg^2})(5.98 \times 10^{24}\,\mathrm{kg})(1000\,\mathrm{kg})}{r^2}$

$= \dfrac{(1000\,\mathrm{kg})(1997\,\mathrm{m/s})^2}{r}$

66. $\frac{1}{2}(100\,\mathrm{kg})v_2^2$

$\quad - \dfrac{(6.67 \times 10^{-11}\,\mathrm{N\,m^2/kg^2})(7.36 \times 10^{22}\,\mathrm{kg})(100\,\mathrm{kg})}{1.74 \times 10^{6}\,\mathrm{m}}$

$= 0 - \dfrac{(6.67 \times 10^{-11}\,\mathrm{N\,m^2/kg^2})(7.36 \times 10^{22}\,\mathrm{kg})(100\,\mathrm{kg})}{3.48 \times 10^{6}\,\mathrm{m}}$

Challenge Problems

67. ‖‖ Two Jupiter-size planets are released from rest 1.0×10^{11} m apart. What are their speeds as they crash together?

68. ‖ A satellite in a circular orbit of radius r has period T. A satellite in a nearby orbit with radius $r + \Delta r$, where $\Delta r \ll r$, has the very slightly different period $T + \Delta T$.

a. Show that

$$\frac{\Delta T}{T} = \frac{3}{2}\frac{\Delta r}{r}$$

b. Two earth satellites are in parallel orbits with radii 6700 km and 6701 km. One day they pass each other, 1 km apart, along a line radially outward from the earth. How long will it be until they are again 1 km apart?

69. ‖‖ While visiting Planet Physics, you toss a rock straight up at 11 m/s and catch it 2.5 s later. While you visit the surface, your cruise ship orbits at an altitude equal to the planet's radius every 230 min. What are the (a) mass and (b) radius of Planet Physics?

70. ‖‖ A moon lander is orbiting the moon at an altitude of 1000 km. By what percentage must it decrease its speed so as to just graze the moon's surface one-half period later?

71. ‖‖ Let's look in more detail at how a satellite is moved from one circular orbit to another. **FIGURE CP13.71** shows two circular orbits, of radii r_1 and r_2, and an elliptical orbit that connects them. Points 1 and 2 are at the ends of the semimajor axis of the ellipse.

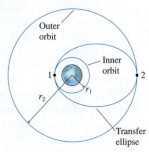

FIGURE CP13.71

a. A satellite moving along the elliptical orbit has to satisfy two conservation laws. Use these two laws to prove that the velocities at points 1 and 2 are

$$v_1' = \sqrt{\frac{2GM(r_2/r_1)}{r_1 + r_2}} \quad \text{and} \quad v_2' = \sqrt{\frac{2GM(r_1/r_2)}{r_1 + r_2}}$$

The prime indicates that these are the velocities on the elliptical orbit. Both reduce to Equation 13.22 if $r_1 = r_2 = r$.

b. Consider a 1000 kg communications satellite that needs to be boosted from an orbit 300 km above the earth to a geosynchronous orbit 35,900 km above the earth. Find the velocity v_1 on the inner circular orbit and the velocity v_1' at the low point on the elliptical orbit that spans the two circular orbits.

c. How much work must the rocket motor do to transfer the satellite from the circular orbit to the elliptical orbit?

d. Now find the velocity v_2' at the high point of the elliptical orbit and the velocity v_2 of the outer circular orbit.

e. How much work must the rocket motor do to transfer the satellite from the elliptical orbit to the outer circular orbit?

f. Compute the total work done and compare your answer to the result of Example 13.6.

72. ‖‖ **FIGURE CP13.72** shows a particle of mass m at distance x from
CALC the center of a very thin cylinder of mass M and length L. The particle is outside the cylinder, so $x > L/2$.

a. Calculate the gravitational potential energy of these two masses.

b. Use what you know about the relationship between force and potential energy to find the magnitude of the gravitational force on m when it is at position x.

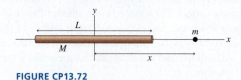

FIGURE CP13.72

14 Fluids and Elasticity

A defining characteristic of any fluid is that—like this waterfall—it flows.

IN THIS CHAPTER, you will learn about systems that flow or deform.

What is a fluid?

A fluid is a substance that flows. Both gases and liquids are fluids.

- A gas is compressible. The molecules move freely with few interactions.
- A liquid is incompressible. The molecules are weakly bound to one another.

What is pressure?

Fluids exert forces on the walls of their containers. Pressure is the force-to-area ratio F/A.

- Pressure in liquids, called hydrostatic pressure, is due to gravity. Pressure increases with depth.
- Pressure in gases is primarily thermal. Pressure is constant in a container.

What is buoyancy?

Buoyancy is the upward force a fluid exerts on an object.

- Archimedes' principle says that the buoyant force equals the weight of the displaced fluid.
- An object floats if the buoyant force is sufficient to balance the object's weight.

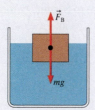

« LOOKING BACK Section 6.1 Equilibrium

How does a fluid flow?

An ideal fluid—an incompressible, nonviscous fluid flowing smoothly—flows along streamlines. Bernoulli's equation, a statement of energy conservation, relates the pressures, speeds, and heights at two points on a streamline.

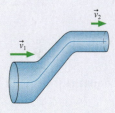

What is elasticity?

Elasticity describes how objects deform under stress. An object's

- Young's modulus characterizes how it stretches when pulled.
- Bulk modulus tells us how much it is compressed by pressure.

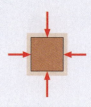

« LOOKING BACK Section 9.4 Restoring forces

Why are fluids important?

Gases and liquids are two of the three common states of matter. Scientists study atmospheres and oceans, while engineers use the controlled flow of fluids in a vast number of applications. This chapter will let you see how Newton's laws can be applied to systems that can flow or deform.

14.1 Fluids

Quite simply, a **fluid** is a substance that flows. Because they flow, fluids take the shape of their container rather than retaining a shape of their own. You may think that gases and liquids are quite different, but both are fluids, and their similarities are often more important than their differences.

The detailed behavior of gases and, especially, liquids can be complex. Fortunately, the essential characteristics of gases and liquids are captured in simple molecular models that are related to the ball-and-spring model of solids that we presented in « Section 5.2.

MODEL 14.1

Molecular model of gases and liquids

Gases and liquids are fluids—they flow and exert pressure.

■ **Gases**
- Molecules move freely through space.
- Molecules do not interact except for occasional collisions with each other or the walls.
- Molecules are far apart, so a gas is *compressible*.

A gas fills the container.

■ **Liquids**
- Molecules are weakly bound and stay close together.
- A liquid is *incompressible* because the molecules can't get any closer.
- Weak bonds allow the molecules to move around.

A liquid has a surface.

Volume and Density

One important parameter that characterizes a macroscopic system is its volume V, the amount of space the system occupies. The SI unit of volume is m^3. Nonetheless, both cm^3 and, to some extent, liters (L) are widely used metric units of volume. In most cases, you *must* convert these to m^3 before doing calculations.

While it is true that $1\ m = 100\ cm$, it is *not* true that $1\ m^3 = 100\ cm^3$. **FIGURE 14.1** shows that the volume conversion factor is $1\ m^3 = 10^6\ cm^3$. A liter is $1000\ cm^3$, so $1\ m^3 = 10^3\ L$. A milliliter (1 mL) is the same as $1\ cm^3$.

A system is also characterized by its *density*. Suppose you have several blocks of copper, each of different size. Each block has a different mass m and a different volume V. Nonetheless, all the blocks are copper, so there should be some quantity that has the *same* value for all the blocks, telling us, "This is copper, not some other material." The most important such parameter is the *ratio* of mass to volume, which we call the **mass density** ρ (lowercase Greek rho):

$$\rho = \frac{m}{V} \quad \text{(mass density)} \qquad (14.1)$$

Conversely, an object of density ρ has mass $m = \rho V$.

The SI units of mass density are kg/m^3. Nonetheless, units of g/cm^3 are widely used. You need to convert these to SI units before doing most calculations. You must convert both the grams to kilograms and the cubic centimeters to cubic meters. The net result is the conversion factor

$$1\ g/cm^3 = 1000\ kg/m^3$$

The mass density is usually called simply "the density" if there is no danger of confusion. However, we will meet other types of density as we go along, and sometimes it is important to be explicit about which density you are using. **TABLE 14.1** provides a

FIGURE 14.1 There are $10^6\ cm^3$ in $1\ m^3$.

Subdivide the $1\ m \times 1\ m \times 1\ m$ cube into little cubes 1 cm on a side. You will get 100 subdivisions along each edge.

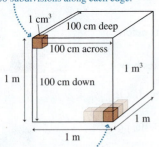

There are $100 \times 100 \times 100 = 10^6$ little $1\ cm^3$ cubes in the big $1\ m^3$ cube.

short list of mass densities of various fluids. Notice the enormous difference between the densities of gases and liquids. Gases have lower densities because the molecules in gases are farther apart than in liquids.

What does it *mean* to say that the density of gasoline is 680 kg/m³ or, equivalently, 0.68 g/cm³? Density is a mass-to-volume ratio. It is often described as the "mass per unit volume," but for this to make sense you have to know what is meant by "unit volume." Regardless of which system of length units you use, a **unit volume** is one of those units cubed. For example, if you measure lengths in meters, a unit volume is 1 m³. But 1 cm³ is a unit volume if you measure lengths in centimeters, and 1 mi³ is a unit volume if you measure lengths in miles.

Density is the mass of one unit of volume, whatever the units happen to be. To say that the density of gasoline is 680 kg/m³ is to say that the mass of 1 m³ of gasoline is 680 kg. The mass of 1 cm³ of gasoline is 0.68 g, so the density of gasoline in those units is 0.68 g/cm³.

The mass density is independent of the object's size. Mass and volume are parameters that characterize a *specific piece* of some substance—say copper—whereas the mass density characterizes the substance itself. All pieces of copper have the same mass density, which differs from the mass density of any other substance.

TABLE 14.1 Densities of fluids at standard temperature (0°C) and pressure (1 atm)

Substance	ρ (kg/m³)
Helium gas	0.18
Air	1.29
Gasoline	680
Ethyl alcohol	790
Benzene	880
Oil (typical)	900
Water	1000
Seawater	1030
Glycerin	1260
Mercury	13,600

EXAMPLE 14.1 | **Weighing the air**

What is the mass of air in a living room with dimensions 4.0 m × 6.0 m × 2.5 m?

MODEL Table 14.1 gives air density at a temperature of 0°C. The air density doesn't vary significantly over a small range of temperatures (we'll study this issue in a later chapter), so we'll use this value even though most people keep their living room warmer than 0°C.

SOLVE The room's volume is

$$V = (4.0 \text{ m}) \times (6.0 \text{ m}) \times (2.5 \text{ m}) = 60 \text{ m}^3$$

The mass of the air is

$$m = \rho V = (1.29 \text{ kg/m}^3)(60 \text{ m}^3) = 77 \text{ kg}$$

ASSESS This is perhaps more mass than you might have expected from a substance that hardly seems to be there. For comparison, a swimming pool this size would contain 60,000 kg of water.

STOP TO THINK 14.1 A piece of glass is broken into two pieces of different size. Rank in order, from largest to smallest, the mass densities of pieces a, b, and c.

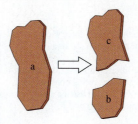

14.2 Pressure

"Pressure" is a word we all know and use. You probably have a commonsense idea of what pressure is. For example, you feel the effects of varying pressure against your eardrums when you swim underwater or take off in an airplane. Cans of whipped cream are "pressurized" to make the contents squirt out when you press the nozzle. It's hard to open a "vacuum sealed" jar of jelly the first time, but easy after the seal is broken.

You've probably seen water squirting out of a hole in the side of a container, as in **FIGURE 14.2**. Notice that the water emerges at greater speed from a hole at greater depth. And you've probably felt the air squirting out of a hole in a bicycle tire or inflatable air mattress. These observations suggest that

- "Something" pushes the water or air *sideways,* out of the hole.
- In a liquid, the "something" is larger at greater depths. In a gas, the "something" appears to be the same everywhere.

Our goal is to turn these everyday observations into a precise definition of pressure.

FIGURE 14.2 Water pressure pushes the water *sideways,* out of the holes.

FIGURE 14.3 The fluid presses against area A with force $\vec{F}$.

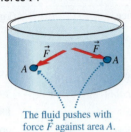

The fluid pushes with force $\vec{F}$ against area A.

FIGURE 14.3 shows a fluid—either a liquid or a gas—pressing against a small area A with force $\vec{F}$. This is the force that pushes the fluid out of a hole. In the absence of a hole, $\vec{F}$ pushes against the wall of the container. Let's define the **pressure** at this point in the fluid to be the ratio of the force to the area on which the force is exerted:

$$p = \frac{F}{A} \tag{14.2}$$

Notice that pressure is a scalar, not a vector. You can see, from Equation 14.2, that a fluid exerts a force of magnitude

$$F = pA \tag{14.3}$$

on a surface of area A. The force is *perpendicular* to the surface.

> **NOTE** Pressure itself is *not* a force, even though we sometimes talk informally about "the force exerted by the pressure." The correct statement is that the *fluid* exerts a force on a surface.

From its definition, pressure has units of N/m². The SI unit of pressure is the **pascal,** defined as

$$1 \text{ pascal} = 1 \text{ Pa} \equiv 1 \text{ N/m}^2$$

This unit is named for the 17th-century French scientist Blaise Pascal, who was one of the first to study fluids. Large pressures are often given in kilopascals, where 1 kPa = 1000 Pa.

Equation 14.2 is the basis for the simple pressure-measuring device shown in **FIGURE 14.4a**. Because the spring constant k and the area A are known, we can determine the pressure by measuring the compression of the spring. Once we've built such a device, we can place it in various liquids and gases to learn about pressure. **FIGURE 14.4b** shows what we can learn from a series of simple experiments.

FIGURE 14.4 Learning about pressure.

(a) Piston attached to spring

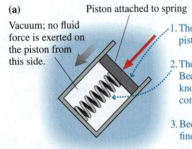

Vacuum; no fluid force is exerted on the piston from this side.

1. The fluid exerts force $\vec{F}$ on a piston with surface area A.

2. The force compresses the spring. Because the spring constant k is known, we can use the spring's compression to find F.

3. Because A is known, we can find the pressure from $p = F/A$.

(b) Pressure-measuring device in fluid

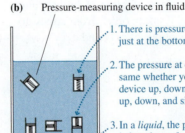

1. There is pressure *everywhere* in a fluid, not just at the bottom or at the walls of the container.

2. The pressure at one point in the fluid is the same whether you point the pressure-measuring device up, down, or sideways. The fluid pushes up, down, and sideways with equal strength.

3. In a *liquid*, the pressure increases with depth below the surface. In a *gas*, the pressure is nearly the same at all points (at least in laboratory-size containers).

The first statement in Figure 14.4b is especially important. Pressure exists at *all* points within a fluid, not just at the walls of the container. You may recall that tension exists at *all* points in a string, not only at its ends where it is tied to an object. We understood tension as the different parts of the string *pulling* against each other. Pressure is an analogous idea, except that the different parts of a fluid are *pushing* against each other.

Causes of Pressure

Gases and liquids are both fluids, but they have some important differences. Liquids are nearly incompressible; gases are highly compressible. The molecules in a liquid attract each other via molecular bonds; the molecules in a gas do not interact other than through occasional collisions. These differences affect how we think about pressure in gases and liquids.

Imagine that you have two sealed jars, each containing a small amount of mercury and nothing else. All the air has been removed from the jars. Suppose you take the two jars into orbit on the space station, where they are weightless. One jar you keep cool, so that the mercury is a liquid. The other you heat until the mercury becomes a gas. What can we say about the pressure in these two jars?

As **FIGURE 14.5** shows, molecular bonds hold the liquid mercury together. It might quiver like Jello, but it remains a cohesive drop floating in the center of the jar. The liquid drop exerts no forces on the walls, so there's *no* pressure in the jar containing the liquid. (If we actually did this experiment, a very small fraction of the mercury would be in the vapor phase and create what is called *vapor pressure*.)

The gas is different. The gas molecules collide with the wall of the container, and each bounce exerts a force on the wall. The force from any one collision is incredibly small, but there are an extraordinarily large number of collisions every second. These collisions cause the gas to have a pressure. We will do the calculation in Chapter 20.

FIGURE 14.6 shows the jars back on earth. Because of gravity, the liquid now fills the bottom of the jar and exerts a force on the bottom and the sides. Liquid mercury is incompressible, so the volume of liquid in Figure 14.6 is the same as in Figure 14.5. There is still no pressure on the top of the jar (other than the very small vapor pressure).

At first glance, the situation in the gas-filled jar seems unchanged from Figure 14.5. However, the earth's gravitational pull causes the gas density to be *slightly* more at the bottom of the jar than at the top. Because the pressure due to collisions is proportional to the density, the pressure is *slightly* larger at the bottom of the jar than at the top.

Thus there appear to be two contributions to the pressure in a container of fluid:

1. A *gravitational contribution* that arises from gravity pulling down on the fluid. Because a fluid can flow, forces are exerted on both the bottom and sides of the container. The gravitational contribution depends on the strength of the gravitational force.
2. A *thermal contribution* due to the collisions of freely moving gas molecules with the walls. The thermal contribution depends on the absolute temperature of the gas.

A detailed analysis finds that these two contributions are not entirely independent of each other, but the distinction is useful for a basic understanding of pressure. Let's see how these two contributions apply to different situations.

Pressure in Gases

The pressure in a laboratory-size container of gas is due almost entirely to the thermal contribution. A container would have to be ≈ 100 m tall for gravity to cause the pressure at the top to be even 1% less than the pressure at the bottom. Laboratory-size containers are much less than 100 m tall, so we can quite reasonably assume that p has the *same* value at all points in a laboratory-size container of gas.

Decreasing the number of molecules in a container decreases the gas pressure simply because there are fewer collisions with the walls. If a container is completely empty, with no atoms or molecules, then the pressure is $p = 0$ Pa. This is a *perfect vacuum*. No perfect vacuum exists in nature, not even in the most remote depths of outer space, because it is impossible to completely remove every atom from a region of space. In practice, a **vacuum** is an enclosed space in which $p \ll 1$ atm. Using $p = 0$ Pa is then a very good approximation.

Atmospheric Pressure

The earth's atmosphere is *not* a laboratory-size container. The height of the atmosphere is such that the gravitational contribution to pressure *is* important. As **FIGURE 14.7** shows, the density of air slowly decreases with increasing height until approaching

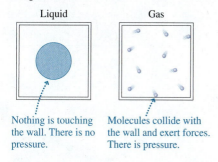

FIGURE 14.5 A liquid and a gas in a weightless environment.

Liquid | Gas

Nothing is touching the wall. There is no pressure. | Molecules collide with the wall and exert forces. There is pressure.

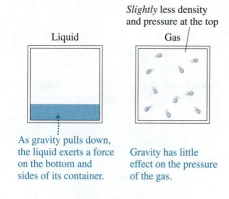

FIGURE 14.6 Gravity affects the pressure of the fluids.

Slightly less density and pressure at the top

Liquid | Gas

As gravity pulls down, the liquid exerts a force on the bottom and sides of its container. | Gravity has little effect on the pressure of the gas.

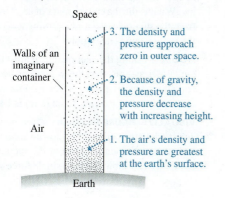

FIGURE 14.7 The pressure and density decrease with increasing height in the atmosphere.

Space

3. The density and pressure approach zero in outer space.

Walls of an imaginary container

2. Because of gravity, the density and pressure decrease with increasing height.

Air

1. The air's density and pressure are greatest at the earth's surface.

Earth

zero in the vacuum of space. Consequently, the pressure of the air, what we call the *atmospheric pressure* p_{atmos}, decreases with height. The air pressure is less in Denver than in Miami.

The atmospheric pressure *at sea level* varies slightly with the weather, but the global average sea-level pressure is 101,300 Pa. Consequently, we define the **standard atmosphere** as

$$1 \text{ standard atmosphere} = 1 \text{ atm} \equiv 101,300 \text{ Pa} = 101.3 \text{ kPa}$$

The standard atmosphere, usually referred to simply as "atmospheres," is a commonly used unit of pressure. But it is not an SI unit, so you must convert atmospheres to pascals before doing most calculations with pressure.

> **NOTE** Unless you happen to live right at sea level, the atmospheric pressure around you is somewhat less than 1 atm. Pressure experiments use a barometer to determine the actual atmospheric pressure. For simplicity, this textbook will always assume that the pressure of the air is $p_{atmos} = 1$ atm unless stated otherwise.

Given that the pressure of the air at sea level is 101.3 kPa, you might wonder why the weight of the air doesn't crush your forearm when you rest it on a table. Your forearm has a surface area of $\approx 200 \text{ cm}^2 = 0.02 \text{ m}^2$, so the force of the air pressing against it is ≈ 2000 N (≈ 450 pounds). How can you even lift your arm?

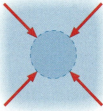

FIGURE 14.8 Pressure forces in a fluid push with equal strength in all directions.

The forces of a fluid push in *all* directions.

The reason, as **FIGURE 14.8** shows, is that a fluid exerts pressure forces in *all* directions. There *is* a downward force of ≈ 2000 N on your forearm, but the air underneath your arm exerts an upward force of the same magnitude. The *net* force is very close to zero. (To be accurate, there is a net *upward* force called the buoyant force. We'll study buoyancy in Section 14.4. The buoyant force of the air is usually too small to notice.)

But, you say, there isn't any air under my arm if I rest it on a table. Actually, there is. There would be a *vacuum* under your arm if there were no air. Imagine placing your arm on the top of a large vacuum cleaner suction tube. What happens? You feel a downward force as the vacuum cleaner "tries to suck your arm in." However, the downward force you feel is not a *pulling* force from the vacuum cleaner. It is the *pushing* force of the air above your arm *when the air beneath your arm is removed and cannot push back*. Air molecules do not have hooks! They have no ability to "pull" on your arm. The air can only push.

Vacuum cleaners, suction cups, and other similar devices are powerful examples of how strong atmospheric pressure forces can be *if* the air is removed from one side of an object so as to produce an unbalanced force. The fact that we are *surrounded* by the fluid allows us to move around in the air, just as we swim underwater, oblivious of these strong forces.

EXAMPLE 14.2 | A suction cup

A 10.0-cm-diameter suction cup is pushed against a smooth ceiling. What is the maximum mass of an object that can be suspended from the suction cup without pulling it off the ceiling? The mass of the suction cup is negligible.

MODEL Pushing the suction cup against the ceiling pushes the air out. We'll assume that the volume enclosed between the suction cup and the ceiling is a perfect vacuum with $p = 0$ Pa. We'll also assume that the pressure in the room is 1 atm.

VISUALIZE FIGURE 14.9 shows a free-body diagram of the suction cup stuck to the ceiling. The downward normal force of the ceiling is distributed around the rim of the suction cup, but in the particle model we can show this as a single force vector.

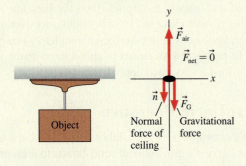

FIGURE 14.9 A suction cup is held to the ceiling by air pressure pushing upward on the bottom.

SOLVE The suction cup remains stuck to the ceiling, in static equilibrium, as long as $F_{air} = n + F_G$. The magnitude of the upward force exerted by the air is

$$F_{air} = pA = p\pi r^2 = (101{,}300 \text{ Pa})\pi(0.050 \text{ m})^2 = 796 \text{ N}$$

There is no downward force from the air in this case because there is no air inside the cup. Increasing the hanging mass decreases the normal force n by an equal amount. The maximum weight has been reached when n is reduced to zero. Thus

$$(F_G)_{max} = mg = F_{air} = 796 \text{ N}$$

$$m = \frac{796 \text{ N}}{g} = 81 \text{ kg}$$

ASSESS The suction cup can support a mass of up to 81 kg if all the air is pushed out, leaving a perfect vacuum inside. A real suction cup won't achieve a perfect vacuum, but suction cups can hold substantial weight.

Pressure in Liquids

Gravity causes a liquid to fill the bottom of a container. Thus it's not surprising that the pressure in a liquid is due almost entirely to the gravitational contribution. We'd like to determine the pressure at depth d below the surface of the liquid. We will assume that the liquid is at rest; flowing liquids will be considered later in this chapter.

The shaded cylinder of liquid in **FIGURE 14.10** extends from the surface to depth d. This cylinder, like the rest of the liquid, is in static equilibrium with $\vec{F}_{net} = \vec{0}$. Three forces act on this cylinder: the gravitational force mg on the liquid in the cylinder, a downward force $p_0 A$ due to the pressure p_0 at the surface of the liquid, and an upward force pA due to the liquid beneath the cylinder pushing up on the bottom of the cylinder. This third force is a consequence of our earlier observation that different parts of a fluid push against each other. Pressure p, which is what we're trying to find, is the pressure at the bottom of the cylinder.

The upward force balances the two downward forces, so

$$pA = p_0 A + mg \tag{14.4}$$

The liquid is a cylinder of cross-section area A and height d. Its volume is $V = Ad$ and its mass is $m = \rho V = \rho Ad$. Substituting this expression for the mass of the liquid into Equation 14.4, we find that the area A cancels from all terms. The pressure at depth d in a liquid is

$$p = p_0 + \rho gd \qquad \text{(hydrostatic pressure at depth } d) \tag{14.5}$$

where ρ is the liquid's density. Because the fluid is at rest, the pressure given by Equation 14.5 is called the **hydrostatic pressure**. The fact that g appears in Equation 14.5 reminds us that this is a gravitational contribution to the pressure.

As expected, $p = p_0$ at the surface, where $d = 0$. Pressure p_0 is often due to the air or other gas above the liquid. $p_0 = 1 \text{ atm} = 101.3 \text{ kPa}$ for a liquid that is open to the air. However, p_0 can also be the pressure due to a piston or a closed surface pushing down on the top of the liquid.

NOTE Equation 14.5 assumes that the liquid is *incompressible;* that is, its density ρ doesn't increase with depth. This is an excellent assumption for liquids, but not a good one for a gas, which *is* compressible.

FIGURE 14.10 Measuring the pressure at depth d in a liquid.

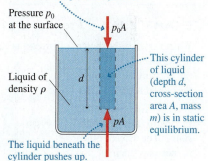

Whatever is above the liquid pushes down on the top of the cylinder.

Pressure p_0 at the surface

Liquid of density ρ

This cylinder of liquid (depth d, cross-section area A, mass m) is in static equilibrium.

The liquid beneath the cylinder pushes up. Pressure at depth d is p.

Free-body diagram of the column of liquid

$\vec{F}_{net} = \vec{0}$

EXAMPLE 14.3 | **The pressure on a submarine**

A submarine cruises at a depth of 300 m. What is the pressure at this depth? Give the answer in both pascals and atmospheres.

SOLVE The density of seawater, from Table 14.1, is $\rho = 1030 \text{ kg/m}^3$. The pressure at depth $d = 300$ m is found from Equation 14.5 to be

$$p = p_0 + \rho gd = 1.013 \times 10^5 \text{ Pa}$$
$$+ (1030 \text{ kg/m}^3)(9.80 \text{ m/s}^2)(300 \text{ m}) = 3.13 \times 10^6 \text{ Pa}$$

Converting the answer to atmospheres gives

$$p = 3.13 \times 10^6 \text{ Pa} \times \frac{1 \text{ atm}}{1.013 \times 10^5 \text{ Pa}} = 30.9 \text{ atm}$$

ASSESS The pressure deep in the ocean is very large. Windows on submersibles must be very thick to withstand the large forces.

FIGURE 14.11 Some properties of a liquid in hydrostatic equilibrium are not what you might expect.

(a)

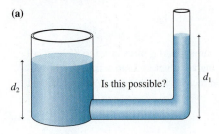

(b)

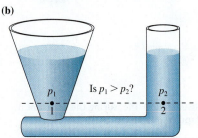

The hydrostatic pressure in a liquid depends only on the depth and the pressure at the surface. This observation has some important implications. **FIGURE 14.11a** shows two connected tubes. It's certainly true that the larger volume of liquid in the wide tube weighs more than the liquid in the narrow tube. You might think that this extra weight would push the liquid in the narrow tube higher than in the wide tube. But it doesn't. If d_1 were larger than d_2, then, according to the hydrostatic pressure equation, the pressure at the bottom of the narrow tube would be higher than the pressure at the bottom of the wide tube. This *pressure difference* would cause the liquid to *flow* from right to left until the heights were equal.

Thus a first conclusion: **A connected liquid in hydrostatic equilibrium rises to the same height in all open regions of the container.**

FIGURE 14.11b shows two connected tubes of different shape. The conical tube holds more liquid above the dotted line, so you might think that $p_1 > p_2$. But it isn't. Both points are at the same depth, thus $p_1 = p_2$. If p_1 were larger than p_2, the pressure at the bottom of the left tube would be larger than the pressure at the bottom of the right tube. This would cause the liquid to flow until the pressures were equal.

Thus a second conclusion: **The pressure is the same at all points on a horizontal line through a connected liquid in hydrostatic equilibrium.**

> **NOTE** Both of these conclusions are restricted to liquids in hydrostatic equilibrium. The situation is different for flowing fluids, as we'll see later in the chapter.

EXAMPLE 14.4 | **Pressure in a closed tube**

Water fills the tube shown in **FIGURE 14.12**. What is the pressure at the top of the closed tube?

MODEL This is a liquid in hydrostatic equilibrium. The closed tube is not an open region of the container, so the water cannot rise to an equal height. Nevertheless, the pressure is still the same at all points on a horizontal line. In particular, the pressure at the top of the closed tube equals the

FIGURE 14.12 A water-filled tube.

pressure in the open tube at the height of the dashed line. Assume $p_0 = 1.00$ atm.

SOLVE A point 40 cm above the bottom of the open tube is at a depth of 60 cm. The pressure at this depth is

$$p = p_0 + \rho g d$$
$$= 1.013 \times 10^5 \text{ Pa} + (1000 \text{ kg/m}^3)(9.80 \text{ m/s}^2)(0.60 \text{ m})$$
$$= 1.072 \times 10^5 \text{ Pa} = 1.06 \text{ atm}$$

This is the pressure at the top of the closed tube.

ASSESS The water in the open tube *pushes* the water in the closed tube up against the top of the tube, which is why the pressure is greater than 1 atm.

We can draw one more conclusion from the hydrostatic pressure equation $p = p_0 + \rho g d$. If we change the pressure p_0 at the surface to p_1, the pressure at depth d becomes $p' = p_1 + \rho g d$. The *change* in pressure $\Delta p = p_1 - p_0$ is the same at all points in the fluid, independent of the size or shape of the container. This idea, that **a change in the pressure at one point in an incompressible fluid appears undiminished at all points in the fluid,** was first recognized by Blaise Pascal and is called **Pascal's principle.**

For example, if we compressed the air above the open tube in Example 14.4 to a pressure of 1.5 atm, an increase of 0.5 atm, the pressure at the top of the closed tube would increase to 1.56 atm. Pascal's principle is the basis for hydraulic systems, as we'll see in the next section.

STOP TO THINK 14.2 Water is slowly poured into the container until the water level has risen into tubes A, B, and C. The water doesn't overflow from any of the tubes. How do the water depths in the three columns compare to each other?

a. $d_A > d_B > d_C$
b. $d_A < d_B < d_C$
c. $d_A = d_B = d_C$
d. $d_A = d_C > d_B$
e. $d_A = d_C < d_B$

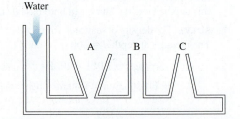

14.3 Measuring and Using Pressure

The pressure in a fluid is measured with a *pressure gauge*. The fluid pushes against some sort of spring, and the spring's displacement is registered by a pointer on a dial.

Many pressure gauges, such as tire gauges and the gauges on air tanks, measure not the actual or absolute pressure p but what is called **gauge pressure.** The gauge pressure, denoted p_g, is the pressure *in excess* of 1 atm. That is,

$$p_g = p - 1 \text{ atm} \qquad (14.6)$$

You must add 1 atm = 101.3 kPa to the reading of a pressure gauge to find the absolute pressure p that you need for doing most science or engineering calculations: $p = p_g + 1$ atm.

A tire-pressure gauge reads the gauge pressure p_g, not the absolute pressure p.

EXAMPLE 14.5 | **An underwater pressure gauge**

An underwater pressure gauge reads 60 kPa. What is its depth?

MODEL The gauge reads gauge pressure, not absolute pressure.

SOLVE The hydrostatic pressure at depth d, with $p_0 = 1$ atm, is $p = 1$ atm $+ \rho g d$. Thus the gauge pressure is

$$p_g = p - 1 \text{ atm} = (1 \text{ atm} + \rho g d) - 1 \text{ atm} = \rho g d$$

The term $\rho g d$ is the pressure *in excess* of atmospheric pressure and thus *is* the gauge pressure. Solving for d, we find

$$d = \frac{60,000 \text{ Pa}}{(1000 \text{ kg/m}^3)(9.80 \text{ m/s}^2)} = 6.1 \text{ m}$$

Solving Hydrostatic Problems

We now have enough information to formulate a set of rules for thinking about hydrostatic problems.

TACTICS BOX 14.1

Hydrostatics

❶ **Draw a picture.** Show open surfaces, pistons, boundaries, and other features that affect pressure. Include height and area measurements and fluid densities. Identify the points at which you need to find the pressure.

❷ **Determine the pressure at surfaces.**
- Surface open to the air: $p_0 = p_{atmos}$, usually 1 atm.
- Surface covered by a gas: $p_0 = p_{gas}$.
- Closed surface: $p = F/A$, where F is the force the surface, such as a piston, exerts on the fluid.

❸ **Use horizontal lines.** Pressure in a connected fluid is the same at any point along a horizontal line.

❹ **Allow for gauge pressure.** Pressure gauges read $p_g = p - 1$ atm.

❺ **Use the hydrostatic pressure equation.** $p = p_0 + \rho g d$.

Exercises 4–13

Manometers and Barometers

Gas pressure is sometimes measured with a device called a *manometer*. A manometer, shown in **FIGURE 14.13**, is a U-shaped tube connected to the gas at one end and open to the air at the other end. The tube is filled with a liquid—usually mercury—of density ρ. The liquid is in static equilibrium. A scale allows the user to measure the height h of the right side above the left side.

Steps 1–3 from Tactics Box 14.1 lead to the conclusion that the pressures p_1 and p_2 must be equal. Pressure p_1, at the surface on the left, is simply the gas pressure: $p_1 = p_{gas}$. Pressure p_2 is the hydrostatic pressure at depth $d = h$ in the liquid on the right: $p_2 = 1$ atm $+ \rho g h$. Equating these two pressures gives

$$p_{gas} = 1 \text{ atm} + \rho g h \qquad (14.7)$$

FIGURE 14.13 A manometer is used to measure gas pressure.

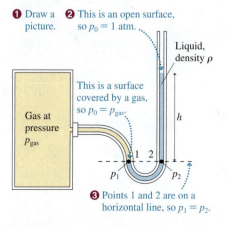

❶ Draw a picture. ❷ This is an open surface, so $p_0 = 1$ atm.

Liquid, density ρ

This is a surface covered by a gas, so $p_0 = p_{gas}$.

Gas at pressure p_{gas}

h

p_1 p_2

❸ Points 1 and 2 are on a horizontal line, so $p_1 = p_2$.

Figure 14.13 assumed $p_{gas} > 1$ atm, so the right side of the liquid is higher than the left. Equation 14.7 is also valid for $p_{gas} < 1$ atm if the distance of the right side *below* the left side is considered to be a negative value of h.

EXAMPLE 14.6 | **Using a manometer**

The pressure of a gas cell is measured with a mercury manometer. The mercury is 36.2 cm higher in the outside arm than in the arm connected to the gas cell.

a. What is the gas pressure?

b. What is the reading of a pressure gauge attached to the gas cell?

SOLVE a. From Table 14.1, the density of mercury is $\rho = 13,600$ kg/m³. Equation 14.7 with $h = 0.362$ m gives

$$p_{gas} = 1 \text{ atm} + \rho gh = 149.5 \text{ kPa}$$

We had to change 1 atm to 101,300 Pa before adding. Converting the result to atmospheres, we have $p_{gas} = 1.476$ atm.

b. The pressure gauge reads gauge pressure: $p_g = p - 1$ atm $= 0.476$ atm or 48.2 kPa.

FIGURE 14.14 A barometer.

(a) Seal and invert tube.

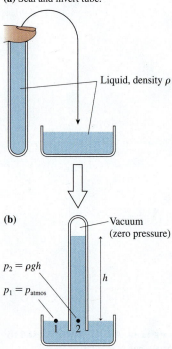

Liquid, density ρ

(b)

Vacuum (zero pressure)

$p_2 = \rho gh$

$p_1 = p_{atmos}$

h

1   2

Another important pressure-measuring instrument is the *barometer*, which is used to measure the atmospheric pressure p_{atmos}. **FIGURE 14.14a** shows a glass tube, sealed at the bottom, that has been completely filled with a liquid. If we temporarily seal the top end, we can invert the tube and place it in a beaker of the same liquid. When the temporary seal is removed, some, but not all, of the liquid runs out, leaving a liquid column in the tube that is a height h above the surface of the liquid in the beaker. This device, shown in **FIGURE 14.14b**, is a barometer. What does it measure? And why doesn't *all* the liquid in the tube run out?

We can analyze the barometer much as we did the manometer. Points 1 and 2 in Figure 14.14b are on a horizontal line drawn even with the surface of the liquid. The liquid is in hydrostatic equilibrium, so the pressure at these two points must be equal. Liquid runs out of the tube only until a balance is reached between the pressure at the base of the tube and the pressure of the air.

You can think of a barometer as rather like a seesaw. If the pressure of the atmosphere increases, it presses down on the liquid in the beaker. This forces liquid up the tube until the pressures at points 1 and 2 are equal. If the atmospheric pressure falls, liquid has to flow out of the tube to keep the pressures equal at these two points.

The pressure at point 2 is the pressure due to the weight of the liquid in the tube plus the pressure of the gas above the liquid. But in this case there is no gas above the liquid! Because the tube had been completely full of liquid when it was inverted, the space left behind when the liquid ran out is a vacuum (ignoring a very slight *vapor pressure* of the liquid, negligible except in extremely precise measurements). Thus pressure p_2 is simply $p_2 = \rho gh$.

Equating p_1 and p_2 gives

$$p_{atmos} = \rho gh \qquad (14.8)$$

Thus we can measure the atmosphere's pressure by measuring the height of the liquid column in a barometer.

The average air pressure at sea level causes a column of mercury in a mercury barometer to stand 760 mm above the surface. Knowing that the density of mercury is 13,600 kg/m³ (at 0°C), we can use Equation 14.8 to find that the average atmospheric pressure is

$$p_{atmos} = \rho_{Hg}gh = (13,600 \text{ kg/m}^3)(9.80 \text{ m/s}^2)(0.760 \text{ m})$$
$$= 1.013 \times 10^5 \text{ Pa} = 101.3 \text{ kPa}$$

This is the value given earlier as "one standard atmosphere."

The barometric pressure varies slightly from day to day as the weather changes. Weather systems are called *high-pressure systems* or *low-pressure systems*, depending on whether the local sea-level pressure is higher or lower than one standard atmosphere. Higher pressure is usually associated with fair weather, while lower pressure portends rain.

Pressure Units

In practice, pressure is measured in several different units. This plethora of units and abbreviations has arisen historically as scientists and engineers working on different subjects (liquids, high-pressure gases, low-pressure gases, weather, etc.) developed what seemed to them the most convenient units. These units continue in use through tradition, so it is necessary to become familiar with converting back and forth between them. TABLE 14.2 gives the basic conversions.

TABLE 14.2 Pressure units

Unit	Abbreviation	Conversion to 1 atm	Uses
pascal	Pa	101.3 kPa	SI unit: $1 \text{ Pa} = 1 \text{ N/m}^2$
atmosphere	atm	1 atm	general
millimeters of mercury	mm of Hg	760 mm of Hg	gases and barometric pressure
inches of mercury	in	29.92 in	barometric pressure in U.S. weather forecasting
pounds per square inch	psi	14.7 psi	engineering and industry

Blood Pressure

The last time you had a medical checkup, the doctor may have told you something like "Your blood pressure is 120 over 80." What does that mean?

About every 0.8 s, assuming a pulse rate of 75 beats per minute, your heart "beats." The heart muscles contract and push blood out into your aorta. This contraction, like squeezing a balloon, raises the pressure in your heart. The pressure increase, in accordance with Pascal's principle, is transmitted through all your arteries.

FIGURE 14.15 is a pressure graph showing how blood pressure changes during one cycle of the heartbeat. The medical condition of *high blood pressure* usually means that your resting systolic pressure is higher than necessary for blood circulation. The high pressure causes undue stress and strain on your entire circulatory system, often leading to serious medical problems. Low blood pressure can cause you to get dizzy if you stand up quickly because the pressure isn't adequate to pump the blood up to your brain.

Blood pressure is measured with a cuff that goes around your arm. The doctor or nurse pressurizes the cuff, places a stethoscope over the artery in your arm, then slowly releases the pressure while watching a pressure gauge. Initially, the cuff squeezes the artery shut and cuts off the blood flow. When the cuff pressure drops below the systolic pressure, the pressure pulse during each beat of your heart forces the artery open briefly and a squirt of blood goes through. You can feel this, and the doctor or nurse records the pressure when she hears the blood start to flow. This is your systolic pressure.

This pulsing of the blood through your artery lasts until the cuff pressure reaches the diastolic pressure. Then the artery remains open continuously and the blood flows smoothly. This transition is easily heard in the stethoscope, and the doctor or nurse records your diastolic pressure.

Blood pressure is measured in millimeters of mercury. And it is a gauge pressure, the pressure in excess of 1 atm. A fairly typical blood pressure of a healthy young adult is 120/80, meaning that the systolic pressure is $p_g = 120$ mm of Hg (absolute pressure $p = 880$ mm of Hg) and the diastolic pressure is 80 mm of Hg.

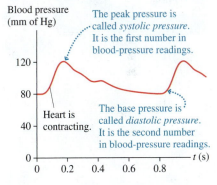

FIGURE 14.15 Blood pressure during one cycle of a heartbeat.

The Hydraulic Lift

The use of pressurized liquids to do useful work is a technology known as **hydraulics.** Pascal's principle is the fundamental idea underlying hydraulic devices. If you increase the pressure at one point in a liquid by pushing a piston in, that pressure increase is transmitted to all points in the liquid. A second piston at some other point in the fluid can then push outward and do useful work.

The brake system in your car is a hydraulic system. Stepping on the brake pushes a piston into the *master brake cylinder* and increases the pressure in the *brake fluid*. The fluid itself hardly moves, but the pressure increase is transmitted to the four wheels where it pushes the brake pads against the spinning brake disk. You've used a pressurized liquid to achieve the useful goal of stopping your car.

One advantage of hydraulic systems over simple mechanical linkages is the possibility of *force multiplication*. To see how this works, we'll analyze a *hydraulic lift,* such as the one that lifts your car at the repair shop. **FIGURE 14.16a** shows force $\vec{F}_2$, due to the weight of the car, pressing down on a liquid via a piston of area A_2. A much smaller force $\vec{F}_1$ presses down on a piston of area A_1. Can this system possibly be in equilibrium?

As you now know, the hydrostatic pressure is the same at all points along a horizontal line through a fluid. Consider the line passing through the liquid/piston interface on the left in Figure 14.16a. Pressures p_1 and p_2 must be equal, thus

$$p_0 + \frac{F_1}{A_1} = p_0 + \frac{F_2}{A_2} + \rho g h \qquad (14.9)$$

The atmosphere presses equally on both sides, so p_0 cancels. The system is in static equilibrium if

$$F_2 = \frac{A_2}{A_1} F_1 - \rho g h A_2 \qquad (14.10)$$

NOTE Force $\vec{F}_2$ is the force of the heavy object pushing *down* on the liquid. According to Newton's third law, the liquid pushes *up* on the object with a force of equal magnitude. Thus F_2 in Equation 14.10 is the "lifting force."

Suppose we need to lift the car higher. If piston 1 is pushed down distance d_1, as in **FIGURE 14.16b**, it displaces volume $V_1 = A_1 d_1$ of liquid. Because the liquid is incompressible, V_1 must equal the volume $V_2 = A_2 d_2$ added beneath piston 2 as it rises distance d_2. That is,

$$d_2 = \frac{d_1}{A_2/A_1} \qquad (14.11)$$

The distance is *divided* by the same factor as that by which force is multiplied. A small force may be able to support a heavy weight, but you have to push the small piston a large distance to raise the heavy weight by a small amount.

This conclusion is really just a statement of energy conservation. Work is done *on* the liquid by a small force pushing the liquid through a large displacement. Work is done *by* the liquid when it lifts the heavy weight through a small distance. A full analysis must consider the fact that the gravitational potential energy of the liquid is also changing, so we can't simply equate the output work to the input work, but you can see that energy considerations require piston 1 to move farther than piston 2.

FIGURE 14.16 A hydraulic lift.

(a)

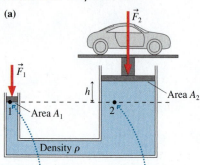

Pressure p_1 is due to atmospheric pressure p_0 plus pressure F_1/A_1, due to $\vec{F}_1$.

Pressure p_2 is p_0 plus F_2/A_2 plus $\rho g h$ from the liquid column of height h.

(b)

Because the fluid is incompressible, $A_1 d_1 = A_2 d_2$.

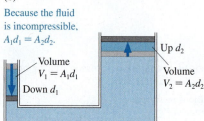

Volume $V_1 = A_1 d_1$

Down d_1

Up d_2

Volume $V_2 = A_2 d_2$

EXAMPLE 14.7 | Lifting a car

The hydraulic lift at a car repair shop is filled with oil. The car rests on a 25-cm-diameter piston. To lift the car, compressed air is used to push down on a 6.0-cm-diameter piston. What does the pressure gauge read when a 1300 kg car is 2.0 m above the compressed-air piston?

MODEL Assume that the oil is incompressible. Its density, from Table 14.1, is 900 kg/m³.

SOLVE F_2 is the weight of the car pressing down on the piston: $F_2 = mg = 12,700$ N. The piston areas are $A_1 = \pi(0.030 \text{ m})^2 = 0.00283 \text{ m}^2$ and $A_2 = \pi(0.125 \text{ m})^2 = 0.0491 \text{ m}^2$. The force required to hold the car at height $h = 2.0$ m is found by solving Equation 14.10 for F_1:

$$F_1 = \frac{A_1}{A_2} F_2 + \rho g h A_1$$

$$= \frac{0.00283 \text{ m}^2}{0.0491 \text{ m}^2} \times 12{,}700 \text{ N} + (900 \text{ kg/m}^3)(9.8 \text{ m/s}^2)(2.0 \text{ m})(0.00283 \text{ m}^2)$$

$$= 782 \text{ N}$$

The pressure applied to the fluid by the compressed-air piston is

$$\frac{F_1}{A_1} = \frac{782 \text{ N}}{0.00283 \text{ m}^2} = 2.76 \times 10^5 \text{ Pa} = 2.7 \text{ atm}$$

This is the pressure *in excess* of atmospheric pressure, which is what a pressure gauge measures, so the gauge reads, depending on its units, 276 kPa or 2.7 atm.

ASSESS 782 N is roughly the weight of an average adult man. The multiplication factor $A_2/A_1 = 17$ makes it quite easy for this much force to lift the car.

STOP TO THINK 14.3 Rank in order, from largest to smallest, the magnitudes of the forces $\vec{F}_a$, $\vec{F}_b$, and $\vec{F}_c$ required to balance the masses. The masses are in kilograms, the three smaller cylinders have the same diameter, and the four larger cylinders have the same diameter.

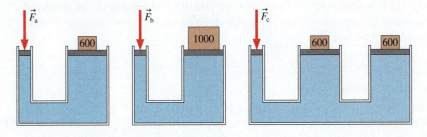

14.4 Buoyancy

A rock, as you know, sinks like a rock. Wood floats on the surface of a lake. A penny with a mass of a few grams sinks, but a massive steel aircraft carrier floats. How can we understand these diverse phenomena?

An air mattress floats effortlessly on the surface of a swimming pool. But if you've ever tried to push an air mattress underwater, you know it is nearly impossible. As you push down, the water pushes up. This net upward force of a fluid is called the **buoyant force.**

The basic reason for the buoyant force is easy to understand. **FIGURE 14.17** shows a cylinder submerged in a liquid. The pressure in the liquid increases with depth, so the pressure at the bottom of the cylinder is larger than at the top. Both cylinder ends have equal area, so force $\vec{F}_{up}$ is larger than force $\vec{F}_{down}$. (Remember that pressure forces push in *all* directions.) Consequently, the pressure in the liquid exerts a *net upward force* on the cylinder of magnitude $F_{net} = F_{up} - F_{down}$. This is the buoyant force.

The submerged cylinder illustrates the idea in a simple way, but the result is not limited to cylinders or to liquids. Suppose we isolate a parcel of fluid of arbitrary shape and volume by drawing an imaginary boundary around it, as shown in **FIGURE 14.18a** on the next page. This parcel is in static equilibrium. Consequently, the gravitational force pulling down on the parcel must be balanced by an upward force. The upward force, which is exerted on this parcel of fluid by the surrounding fluid, is the buoyant force $\vec{F}_B$. The buoyant force matches the weight of the fluid: $F_B = mg$.

Imagine that we could somehow remove this parcel of fluid and instantaneously replace it with an object of exactly the same shape and size, as shown in **FIGURE 14.18b**. Because the buoyant force is exerted by the *surrounding* fluid, and the surrounding

FIGURE 14.17 The buoyant force arises because the fluid pressure at the bottom of the cylinder is larger than at the top.

The net force of the fluid on the cylinder is the buoyant force $\vec{F}_B$.

Increasing pressure $\quad \vec{F}_{down}$

$\vec{F}_{net} = \vec{F}_B$

$\vec{F}_{up}$

$F_{up} > F_{down}$ because the pressure increases with depth. Hence the fluid exerts a net upward force.

FIGURE 14.18 The buoyant force.

(a) Imaginary boundary around a parcel of fluid

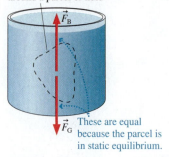

$\vec{F}_B$

$\vec{F}_G$ These are equal because the parcel is in static equilibrium.

(b) Real object with same size and shape as the parcel of fluid

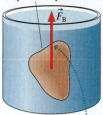

$\vec{F}_B$

The buoyant force on the object is the same as on the parcel of fluid because the *surrounding* fluid has not changed.

fluid hasn't changed, the buoyant force on this new object is *exactly the same* as the buoyant force on the parcel of fluid that we removed.

When an object (or a portion of an object) is immersed in a fluid, it *displaces* fluid that would otherwise fill that region of space. This fluid is called the **displaced fluid.** The displaced fluid's volume is exactly the volume of the portion of the object that is immersed in the fluid. Figure 14.18 leads us to conclude that the magnitude of the upward buoyant force matches the weight of this displaced fluid.

This idea was first recognized by the ancient Greek mathematician and scientist Archimedes, perhaps the greatest scientist of antiquity, and today we know it as *Archimedes' principle.*

> **Archimedes' principle** A fluid exerts an upward buoyant force $\vec{F}_B$ on an object immersed in or floating on the fluid. The magnitude of the buoyant force equals the weight of the fluid displaced by the object.

Suppose the fluid has density ρ_f and the object displaces volume V_f of fluid. The mass of the displaced fluid is $m_f = \rho_f V_f$ and so its weight is $m_f g = \rho_f V_f g$. Thus Archimedes' principle in equation form is

$$F_B = \rho_f V_f g \qquad (14.12)$$

NOTE It is important to distinguish the density and volume of the displaced fluid from the density and volume of the object. To do so, we'll use subscript f for the fluid and o for the object.

EXAMPLE 14.8 | Holding a block of wood underwater

A 10 cm × 10 cm × 10 cm block of wood with density 700 kg/m³ is held underwater by a string tied to the bottom of the container. What is the tension in the string?

MODEL The buoyant force is given by Archimedes' principle.

VISUALIZE FIGURE 14.19 shows the forces acting on the wood.

FIGURE 14.19 The forces acting on the submerged wood.

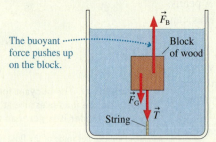

The buoyant force pushes up on the block.

Block of wood

$\vec{F}_B$

$\vec{F}_G$

String $\vec{T}$

SOLVE The block is in equilibrium, so

$$\sum F_y = F_B - T - m_o g = 0$$

Thus the tension is $T = F_B - m_o g$. The mass of the block is $m_o = \rho_o V_o$, and the buoyant force, given by Equation 14.12, is $F_B = \rho_f V_f g$. Thus

$$T = \rho_f V_f g - \rho_o V_o g = (\rho_f - \rho_o) V_o g$$

where we've used the fact that $V_f = V_o$ for a completely submerged object. The volume is $V_o = 1000 \text{ cm}^3 = 1.0 \times 10^{-3} \text{ m}^3$, and hence the tension in the string is

$$T = ((1000 \text{ kg/m}^3) - (700 \text{ kg/m}^3))$$
$$\times (1.0 \times 10^{-3} \text{ m}^3)(9.8 \text{ m/s}^2) = 2.9 \text{ N}$$

ASSESS The tension depends on the *difference* in densities. The tension would vanish if the wood density matched the water density.

Float or Sink?

If you *hold* an object underwater and then release it, it floats to the surface, sinks, or remains "hanging" in the water. How can we predict which it will do? The net force on the object an instant after you release it is $\vec{F}_{net} = (F_B - m_o g)\hat{k}$. Whether it heads for the surface or the bottom depends on whether the buoyant force F_B is larger or smaller than the object's weight $m_o g$.

The magnitude of the buoyant force is $\rho_f V_f g$. The weight of a uniform object, such as a block of steel, is simply $\rho_o V_o g$. But a compound object, such as a scuba diver, may have pieces of varying density. If we define the **average density** to be $\rho_{avg} = m_o/V_o$, the weight of a compound object is $\rho_{avg} V_o g$.

Comparing $\rho_f V_f g$ to $\rho_{avg} V_o g$, and noting that $V_f = V_o$ for an object that is fully submerged, we see that an object floats or sinks depending on whether the fluid density ρ_f is larger or smaller than the object's average density ρ_{avg}. If the densities are equal, the object is in static equilibrium and hangs motionless. This is called **neutral buoyancy.** These conditions are summarized in Tactics Box 14.2.

TACTICS BOX 14.2

Finding whether an object floats or sinks

❶ Object sinks   **❷** Object floats   **❸** Neutral buoyancy

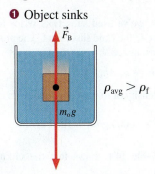 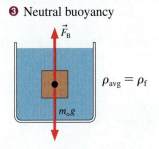

$\rho_{avg} > \rho_f$ $\rho_{avg} < \rho_f$ $\rho_{avg} = \rho_f$

An object sinks if it weighs more than the fluid it displaces—that is, if its average density is greater than the density of the fluid.	An object floats on the surface if it weighs less than the fluid it displaces—that is, if its average density is less than the density of the fluid.	An object hangs motionless if it weighs exactly the same as the fluid it displaces—that is, if its average density equals the density of the fluid.

Exercises 14–18

As an example, steel is denser than water, so a chunk of steel sinks. Oil is less dense than water, so oil floats on water. Fish use *swim bladders* filled with air and scuba divers use weighted belts to adjust their average density to match the density of water. Both are examples of neutral buoyancy.

If you release a block of wood underwater, the net upward force causes the block to shoot to the surface. Then what? Let's begin with a *uniform* object such as the block shown in **FIGURE 14.20.** This object contains nothing tricky, like indentations or voids. Because it's floating, it must be the case that $\rho_o < \rho_f$.

Now that the object is floating, it's in static equilibrium. The upward buoyant force, given by Archimedes' principle, exactly balances the downward weight of the object. That is,

$$F_B = \rho_f V_f g = m_o g = \rho_o V_o g \qquad (14.13)$$

In this case, the volume of the displaced fluid is *not* the same as the volume of the object. In fact, we can see from Equation 14.13 that the volume of fluid displaced by a floating object of uniform density is

$$V_f = \frac{\rho_o}{\rho_f} V_o < V_o \qquad (14.14)$$

You've often heard it said that "90% of an iceberg is underwater." Equation 14.14 is the basis for that statement. Most icebergs break off glaciers and are fresh-water ice with a density of 917 kg/m^3. The density of seawater is 1030 kg/m^3. Thus

$$V_f = \frac{917 \text{ kg/m}^3}{1030 \text{ kg/m}^3} V_o = 0.89 V_o$$

V_f, the displaced water, is the volume of the iceberg that is underwater. You can see that, indeed, 89% of the volume of an iceberg is underwater.

NOTE ▶ Equation 14.14 applies only to *uniform* objects. It does not apply to boats, hollow spheres, or other objects of nonuniform composition. ◀

FIGURE 14.20 A floating object is in static equilibrium.

An object of density ρ_o and volume V_o is floating on a fluid of density ρ_f.

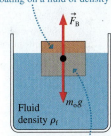

Fluid density ρ_f

The submerged volume of the object is equal to the volume V_f of displaced fluid.

About 90% of an iceberg is underwater.

EXAMPLE 14.9 | **Measuring the density of an unknown liquid**

You need to determine the density of an unknown liquid. You notice that a block floats in this liquid with 4.6 cm of the side of the block submerged. When the block is placed in water, it also floats but with 5.8 cm submerged. What is the density of the unknown liquid?

MODEL The block is an object of uniform composition.

VISUALIZE FIGURE 14.21 shows the block and defines the cross-section area A and submerged lengths h_u in the unknown liquid and h_w in water.

FIGURE 14.21 More of the block is submerged in water than in an unknown liquid.

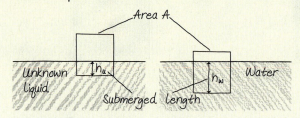

SOLVE The block is floating, so Equation 14.14 applies. The block displaces volume $V_u = Ah_u$ of the unknown liquid. Thus

$$V_u = Ah_u = \frac{\rho_o}{\rho_u} V_o$$

Similarly, the block displaces volume $V_w = Ah_w$ of the water, leading to

$$V_w = Ah_w = \frac{\rho_o}{\rho_w} V_o$$

Because there are two fluids, we've used subscripts w for water and u for the unknown in place of the fluid subscript f. The product $\rho_o V_o$ appears in both equations; hence

$$\rho_u Ah_u = \rho_w Ah_w$$

The area A cancels, and the density of the unknown liquid is

$$\rho_u = \frac{h_w}{h_u}\rho_w = \frac{5.8 \text{ cm}}{4.6 \text{ cm}} \times 1000 \text{ kg/m}^3 = 1260 \text{ kg/m}^3$$

ASSESS Comparison with Table 14.1 shows that the unknown liquid is likely to be glycerin.

Boats

FIGURE 14.22 A physicist's boat.

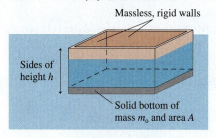

We'll conclude by designing a boat. FIGURE 14.22 is a physicist's idea of a boat. Four massless but rigid walls are attached to a solid steel plate of mass m_o and area A. As the steel plate settles down into the water, the sides allow the boat to displace a volume of water much larger than that displaced by the steel alone. The boat will float if the weight of the displaced water equals the weight of the boat.

In terms of density, the boat will float if $\rho_{avg} < \rho_f$. If the sides of the boat are height h, the boat's volume is $V_o = Ah$ and its average density is $\rho_{avg} = m_o/V_o = m_o/Ah$. The boat will float if

$$\rho_{avg} = \frac{m_o}{Ah} < \rho_f \tag{14.15}$$

Thus the minimum height of the sides, a height that would allow the boat to float (in perfectly still water!) with water right up to the rails, is

$$h_{min} = \frac{m_o}{\rho_f A} \tag{14.16}$$

As a quick example, a 5 m × 10 m steel "barge" with a 2-cm-thick floor has an area of 50 m² and a mass of 7900 kg. The minimum height of the massless walls, as given by Equation 14.16, is 16 cm.

Real ships and boats are more complicated, but the same idea holds true. Whether it's made of concrete, steel, or lead, **a boat will float if its geometry allows it to displace enough water to equal the weight of the boat.**

STOP TO THINK 14.4 An ice cube is floating in a glass of water that is filled entirely to the brim. When the ice cube melts, the water level will

a. Fall.
b. Stay the same, right at the brim.
c. Rise, causing the water to spill.

14.5 Fluid Dynamics

The wind blowing through your hair, a white-water river, and oil gushing from an oil well are examples of fluids in motion. We've focused thus far on fluid statics, but it's time to turn our attention to fluid dynamics.

Fluid flow is a complex subject. Many aspects, especially turbulence and the formation of eddies, are still not well understood and are areas of current science and engineering research. We will avoid these difficulties by using a simplified model. The **ideal-fluid model** provides a good, though not perfect, description of fluid flow in many situations. It captures the essence of fluid flow while eliminating unnecessary details.

MODEL 14.2

Ideal fluid

Applies to liquids and gases. A fluid can be considered *ideal* if

- The fluid is *incompressible*.
- The fluid is *nonviscous*.
  - **Viscosity,** a fluid's resistance to flow, is analogous to kinetic friction.
  - Nonviscous flow is analogous to friction-free motion.
- The flow is *laminar*.
  - **Laminar flow** occurs when the fluid velocity at each point in the fluid is constant. The flow is smooth; it doesn't change or fluctuate.
- Limitations: The model fails if
  - the fluid has significant viscosity.
  - the flow is *turbulent* rather than laminar.

Exercise 20

The rising smoke in the photograph of **FIGURE 14.23** begins as laminar flow, recognizable by the smooth contours, but at some point undergoes a transition to turbulent flow. A laminar-to-turbulent transition is not uncommon in fluid flow. The ideal-fluid model can be applied to the laminar flow, but not to the turbulent flow.

The Equation of Continuity

FIGURE 14.24a shows smoke being used to help engineers visualize the airflow around a car in a wind tunnel. The smoothness of the flow tells us this is laminar flow. But notice also how the individual smoke trails retain their identity. They don't cross or get mixed together. Each smoke trail represents a **streamline** in the fluid. **FIGURE 14.24b** illustrates three important properties of streamlines.

FIGURE 14.24 Particles in an ideal fluid move along streamlines.

(a)   Streamline

(b) 1. Streamlines never cross.

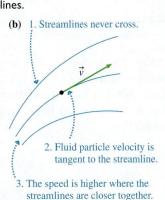

$\vec{v}$

2. Fluid particle velocity is tangent to the streamline.

3. The speed is higher where the streamlines are closer together.

FIGURE 14.23 Rising smoke changes from laminar flow to turbulent flow.

Turbulent flow

Laminar flow

FIGURE 14.25 The flow speed changes as a tube's cross-section area changes.

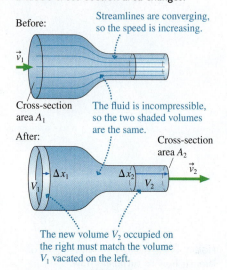

Before:

Streamlines are converging, so the speed is increasing.

$\vec{v}_1$

Cross-section area A_1

The fluid is incompressible, so the two shaded volumes are the same.

After:

Cross-section area A_2

Δx_1 Δx_2 $\vec{v}_2$

V_1 V_2

The new volume V_2 occupied on the right must match the volume V_1 vacated on the left.

FIGURE 14.26 Narrowing the cross section increases the water's speed.

A key observation is that the flow speed changes as the streamlines get closer together or farther apart. In **FIGURE 14.25**, where an ideal fluid flows through a tube, the streamlines get closer together as the tube gets narrower, so the flow speed must increase. To understand quantitatively how the speed changes, focus on the shaded volume of fluid as it moves through the tube.

An ideal fluid is incompressible, so this shaded portion of the fluid moves forward *without changing volume*. If the flow speed on the left side of the tube is v_1, then during a time interval Δt the shaded fluid moves forward distance $\Delta x_1 = v_1 \Delta t$ and "vacates" volume $V_1 = A_1 \Delta x_1 = A_1 v_1 \Delta t$. (New fluid moves in to fill this volume, but we're focusing on the shaded portion of the fluid.) On the right side of the tube, where the flow speed is v_2, the shaded fluid moves forward distance $\Delta x_2 = v_2 \Delta t$ and occupies a new volume $V_2 = A_2 \Delta x_2 = A_2 v_2 \Delta t$. These two volumes must be equal, leading to the conclusion that

$$v_1 A_1 = v_2 A_2 \qquad (14.17)$$

Equation 14.17 is called the **equation of continuity,** and it is one of two important equations for the flow of an ideal fluid. The equation of continuity says that **flow is faster in narrower parts of a tube, slower in wider parts**. You're familiar with this conclusion from many everyday observations. For example, water is shot from the narrow nozzle in **FIGURE 14.26** much faster than it's flowing through the wider hose.

The quantity

$$Q = vA \qquad (14.18)$$

is called the **volume flow rate.** The SI units of Q are m^3/s, although in practice Q may be measured in cm^3/s, liters per minute, or, in the United States, gallons per minute. Another way to express the meaning of the equation of continuity is to say that **the volume flow rate is constant at all points in a tube.**

EXAMPLE 14.10 | **River rafting**

A 20-m-wide, 4.0-m-deep river with a triangular cross section flows at a lazy 1.0 m/s. As it makes its way to the sea, the river enters a gorge where rocky walls confine it to a width of 4.0 m and a depth of 6.0 m, again with a triangular cross section. What is the river's volume flow rate? And what is its speed through the gorge?

MODEL Model the flowing river as an ideal fluid.

SOLVE A triangular river cross section of width w and depth d has cross-section area $A = \frac{1}{2}wd$. In the first part of the river,

$$A_1 = \frac{1}{2}(20 \text{ m})(4.0 \text{ m}) = 40 \text{ m}^2$$

and thus the river's flow rate is

$$Q = v_1 A_1 = (1.0 \text{ m/s})(40 \text{ m}^2) = 40 \text{ m}^3/s$$

The volume flow rate is the same at all points along the river, including the gorge. The gorge has cross-section area

$$A_2 = \frac{1}{2}(4.0 \text{ m})(6.0 \text{ m}) = 12 \text{ m}^2$$

Thus the river's speed through the gorge is

$$v_2 = \frac{Q}{A_2} = \frac{40 \text{ m}^3/s}{12 \text{ m}^2} = 3.3 \text{ m/s}$$

ASSESS A river speed of 3.3 m/s, or about 7 mph, is typical of a Class III whitewater-rafting river. For a 6-m-deep river, some whitewater and waves on the surface don't greatly affect our assumption of smooth flow through a perfectly triangular cross section.

STOP TO THINK 14.5 The figure shows volume flow rates (in cm^3/s) for all but one tube. What is the volume flow rate through the unmarked tube? Is the flow direction in or out?

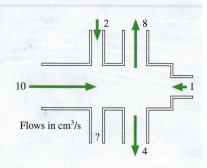

Flows in cm^3/s

Bernoulli's Equation

The equation of continuity is one of two important relationships for ideal fluids. The other is a statement of energy conservation. The general statement of the energy principle that you learned in Chapter 10 is

$$\Delta K + \Delta U = W_{ext} \tag{14.19}$$

where W_{ext} is the work done by any external forces.

Let's see how this applies to the fluid flowing through the tube of **FIGURE 14.27**. This is the situation we considered in Figure 14.25, but now the tube changes height. The more darkly shaded volume of fluid is the system to which we will apply Equation 14.19. Work is done on this system by pressure forces from the *surrounding* fluid in the tube. The unseen fluid to the left of our system exerts force $\vec{F}_1 = (p_1 A_1, \text{to the right})$, where p_1 is the fluid pressure at this point in the tube and A_1 is the cross-section area. During a small time interval Δt, this force pushes the fluid through displacement $\Delta \vec{r}_1 = (\Delta x_1, \text{to the right})$ and does work

$$W_1 = \vec{F}_1 \cdot \Delta \vec{r}_1 = F_1 \Delta x_1 = (p_1 A_1)\Delta x_1 = p_1(A_1 \Delta x_1) = p_1 V \tag{14.20}$$

The A_1 and Δx_1 enter the equation from different terms, but they conveniently combine to give $V = A_1 \Delta x_1$, the volume "vacated" by the shaded fluid as it's pushed forward.

The situation is much the same on the right edge of the system, where the surrounding fluid exerts a pressure force $\vec{F}_2 = (p_2 A_2, \text{to the left})$. However, force $\vec{F}_2$ is *opposite* the displacement $\Delta \vec{r}_2$, which introduces a minus sign into the dot product for the work, giving

$$W_2 = \vec{F}_2 \cdot \Delta \vec{r}_2 = -F_2 \Delta x_2 = -(p_2 A_2)\Delta x_2 = -p_2(A_2 \Delta x_2) = -p_2 V \tag{14.21}$$

Because the fluid is incompressible, the volume $V = A_2 \Delta x_2$ "gained" on the right side is exactly the same as that lost on the left. Altogether, the work done on the system by the surrounding fluid is

$$W_{ext} = W_1 + W_2 = (p_1 - p_2)V \tag{14.22}$$

The work depends on the *pressure difference* $p_1 - p_2$.

Now let's see what happens to the system's potential and kinetic energy. Most of the system does not change during time interval Δt; it's fluid at the same height moving at the same speed. All we need to consider are the volumes V at the two ends. On the right, the system *gains* kinetic and gravitational potential energy as the fluid moves into volume V. Simultaneously, the system *loses* kinetic and gravitational potential energy on the left as fluid vacates volume V.

The mass of fluid in volume V is $m = \rho V$, where ρ is the fluid density. Thus the net change in the system's gravitational potential energy during Δt is

$$\Delta U_G = mgy_2 - mgy_1 = \rho V g y_2 - \rho V g y_1 \tag{14.23}$$

Similarly, the system's change in kinetic energy is

$$\Delta K = \tfrac{1}{2}mv_2^2 - \tfrac{1}{2}mv_1^2 = \tfrac{1}{2}\rho V v_2^2 - \tfrac{1}{2}\rho V v_1^2 \tag{14.24}$$

Combining Equations 14.22, 14.23, and 14.24 gives us the energy equation for the fluid in the flow tube:

$$\tfrac{1}{2}\rho V v_2^2 - \tfrac{1}{2}\rho V v_1^2 + \rho V g y_2 - \rho V g y_1 = p_1 V - p_2 V \tag{14.25}$$

The volume V cancels out of all the terms. Regrouping terms, we have

$$p_1 + \tfrac{1}{2}\rho v_1^2 + \rho g y_1 = p_2 + \tfrac{1}{2}\rho v_2^2 + \rho g y_2 \tag{14.26}$$

Equation 14.26 is called **Bernoulli's equation.** It is named for the 18th-century Swiss scientist Daniel Bernoulli, who made some of the earliest studies of fluid dynamics.

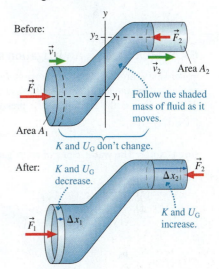

FIGURE 14.27 Energy analysis of fluid flow through a tube.

Before:

K and U_G don't change.

Follow the shaded mass of fluid as it moves.

After: K and U_G decrease.

K and U_G increase.

Bernoulli's equation is really nothing more than a statement about work and energy. It is sometimes useful to express Bernoulli's equation in the alternative form

$$p + \tfrac{1}{2}\rho v^2 + \rho g y = \text{constant} \tag{14.27}$$

This version of Bernoulli's equation tells us that the quantity $p + \tfrac{1}{2}\rho v^2 + \rho g y$ remains constant along a streamline.

NOTE Using Bernoulli's equation is very much like using the law of conservation of energy. Rather than identifying a "before" and "after," you want to identify two points on a streamline. As the following examples show, Bernoulli's equation is often used in conjunction with the equation of continuity.

EXAMPLE 14.11 | An irrigation system

Water flows through the pipes shown in **FIGURE 14.28**. The water's speed through the lower pipe is 5.0 m/s and a pressure gauge reads 75 kPa. What is the reading of the pressure gauge on the upper pipe?

FIGURE 14.28 The water pipes of an irrigation system.

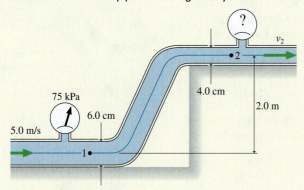

MODEL Treat the water as an ideal fluid obeying Bernoulli's equation. Consider a streamline connecting point 1 in the lower pipe with point 2 in the upper pipe.

SOLVE Bernoulli's equation, Equation 14.26, relates the pressure, fluid speed, and heights at points 1 and 2. It is easily solved for the pressure p_2 at point 2:

$$p_2 = p_1 + \tfrac{1}{2}\rho v_1^2 - \tfrac{1}{2}\rho v_2^2 + \rho g y_1 - \rho g y_2$$
$$= p_1 + \tfrac{1}{2}\rho(v_1^2 - v_2^2) + \rho g(y_1 - y_2)$$

All quantities on the right are known except v_2, and that is where the equation of continuity will be useful. The cross-section areas and water speeds at points 1 and 2 are related by

$$v_1 A_1 = v_2 A_2$$

from which we find

$$v_2 = \frac{A_1}{A_2} v_1 = \frac{r_1^2}{r_2^2} v_1 = \frac{(0.030 \text{ m})^2}{(0.020 \text{ m})^2}(5.0 \text{ m/s}) = 11.25 \text{ m/s}$$

The pressure at point 1 is $p_1 = 75$ kPa + 1 atm = 176,300 Pa. We can now use the above expression for p_2 to calculate $p_2 = 105,900$ Pa. This is the absolute pressure; the pressure gauge on the upper pipe will read

$$p_2 = 105,900 \text{ Pa} - 1 \text{ atm} = 4.6 \text{ kPa}$$

ASSESS Reducing the pipe size decreases the pressure because it makes $v_2 > v_1$. Gaining elevation also reduces the pressure.

EXAMPLE 14.12 | Hydroelectric power

Small hydroelectric plants in the mountains sometimes bring the water from a reservoir down to the power plant through enclosed tubes. In one such plant, the 100-cm-diameter intake tube in the base of the dam is 50 m below the reservoir surface. The water drops 200 m through the tube before flowing into the turbine through a 50-cm-diameter nozzle.

a. What is the water speed into the turbine?

b. By how much does the inlet pressure differ from the hydrostatic pressure at that depth?

MODEL Treat the water as an ideal fluid obeying Bernoulli's equation. Consider a streamline that begins at the surface of the reservoir and ends at the exit of the nozzle. The pressure at the surface is $p_1 = p_\text{atmos}$ and $v_1 \approx 0$ m/s. The water discharges into air, so $p_3 = p_\text{atmos}$ at the exit.

VISUALIZE **FIGURE 14.29** is a pictorial representation of the situation.

FIGURE 14.29 Pictorial representation of the water flow to a hydroelectric plant.

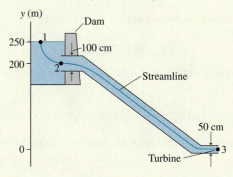

SOLVE a. Bernoulli's equation, with $v_1 = 0$ m/s and $y_3 = 0$ m, is

$$p_{atmos} + \rho g y_1 = p_{atmos} + \tfrac{1}{2}\rho v_3^2$$

The power plant is in the mountains, where $p_{atmos} < 1$ atm, but p_{atmos} occurs on both sides of Bernoulli's equation and cancels. Solving for v_3 gives

$$v_3 = \sqrt{2 g y_1} = \sqrt{2(9.80 \text{ m/s}^2)(250 \text{ m})} = 70 \text{ m/s}$$

b. You might expect the pressure p_2 at the intake to be the hydrostatic pressure $p_{atmos} + \rho g d$ at depth d. But the water is *flowing* into the intake tube, so it's not in static equilibrium. We can find the intake speed v_2 from the equation of continuity:

$$v_2 = \frac{A_3}{A_2} v_3 = \frac{r_3^2}{r_2^2} \sqrt{2 g y_1}$$

The intake is along the streamline between points 1 and 3, so we can apply Bernoulli's equation to points 1 and 2:

$$p_{atmos} + \rho g y_1 = p_2 + \tfrac{1}{2}\rho v_2^2 + \rho g y_2$$

Solving this equation for p_2, and noting that $y_1 - y_2 = d$, we find

$$p_2 = p_{atmos} + \rho g (y_1 - y_2) - \tfrac{1}{2}\rho v_2^2$$

$$= p_{atmos} + \rho g d - \tfrac{1}{2}\rho \left(\frac{r_3}{r_2}\right)^4 (2 g y_1)$$

$$= p_{static} - \rho g y_1 \left(\frac{r_3}{r_2}\right)^4$$

The intake pressure is less than hydrostatic pressure by the amount

$$\rho g y_1 \left(\frac{r_3}{r_2}\right)^4 = 153{,}000 \text{ Pa} = 1.5 \text{ atm}$$

ASSESS The water's exit speed from the nozzle is the same as if it fell 250 m from the surface of the reservoir. This isn't surprising because we've assumed a nonviscous (i.e., frictionless) liquid. "Real" water would have less speed but still flow very fast.

Two Applications

The speed of a flowing gas is often measured with a device called a **Venturi tube.** Venturi tubes measure gas speeds in environments as different as chemistry laboratories, wind tunnels, and jet engines.

FIGURE 14.30 shows gas flowing through a tube that changes from cross-section area A_1 to area A_2. A U-shaped glass tube containing liquid of density ρ_{liq} connects the two segments of the flow tube. When gas flows through the horizontal tube, the liquid stands height h higher in the side of the U tube connected to the narrow segment of the flow tube.

Figure 14.30 shows how a Venturi tube works. We can make this analysis quantitative and determine the gas-flow speed from the liquid height h. Two pieces of information we have to work with are Bernoulli's equation

$$p_1 + \tfrac{1}{2}\rho v_1^2 + \rho g y_1 = p_2 + \tfrac{1}{2}\rho v_2^2 + \rho g y_2 \tag{14.28}$$

and the equation of continuity

$$v_2 A_2 = v_1 A_1 \tag{14.29}$$

In addition, the hydrostatic equation for the liquid tells us that the pressure p_2 above the right tube differs from the pressure p_1 above the left tube by $\rho_{liq} g h$. That is,

$$p_2 = p_1 - \rho_{liq} g h \tag{14.30}$$

First we use Equations 14.29 and 14.30 to eliminate v_2 and p_2 in Bernoulli's equation:

$$p_1 + \tfrac{1}{2}\rho v_1^2 = (p_1 - \rho_{liq} g h) + \tfrac{1}{2}\rho \left(\frac{A_1}{A_2}\right)^2 v_1^2 \tag{14.31}$$

The potential energy terms have disappeared because $y_1 = y_2$ for a horizontal tube. Equation 14.31 can now be solved for v_1, then v_2 is obtained from Equation 14.29. We'll skip a few algebraic steps and go right to the result:

$$v_1 = A_2 \sqrt{\frac{2 \rho_{liq} g h}{\rho (A_1^2 - A_2^2)}}$$

$$\tag{14.32}$$

$$v_2 = A_1 \sqrt{\frac{2 \rho_{liq} g h}{\rho (A_1^2 - A_2^2)}}$$

Equations 14.32 are reasonably accurate as long as the flow speeds are much less than the speed of sound, about 340 m/s. The Venturi tube is an example of the power of Bernoulli's equation.

FIGURE 14.30 A Venturi tube measures gas-flow speeds.

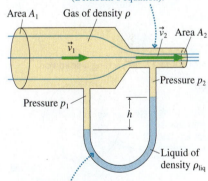

1. As the gas flows into a smaller cross section, it speeds up (equation of continuity). As it speeds up, the pressure decreases (Bernoulli's equation).

Area A_1 Gas of density ρ

$\vec{v}_1$ $\vec{v}_2$ Area A_2

Pressure p_2

Pressure p_1

h

Liquid of density ρ_{liq}

2. The U tube acts like a manometer. The liquid level is higher on the side where the pressure is lower.

FIGURE 14.31 Airflow over a wing generates lift by creating unequal pressures above and below.

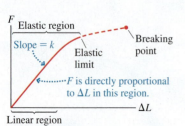

1. The streamlines in the flow tube are compressed, indicating that the air speeds up as it flows over the top of the wing. This lowers the pressure to $p < p_{atmos}$.

2. The pressure difference exerts an upward force on the wing.

$\vec{F}_{lift}$

$p \approx p_{atmos}$ beneath wing

As a final example, we can use Bernoulli's equation to understand, at least qualitatively, how airplane wings generate *lift*. **FIGURE 14.31** shows the cross section of an airplane wing. This shape is called an *airfoil*.

Although you usually think of an airplane moving through the air, in the airplane's reference frame it is the air that flows across a stationary wing. As it does, the streamlines must separate. The bottom of the wing does not significantly alter the streamlines going under the wing. But the streamlines going over the top of the wing get bunched together. As we've seen, with the equation of continuity, the flow speed has to increase when streamlines get closer together. Consequently, the air speed increases as it flows across the top of the wing.

If the air speed increases, then, from Bernoulli's equation, the air pressure must decrease. And if the air pressure above the wing is less than the air pressure below, the air will exert a net *upward* force on the wing. The upward force of the air due to the pressure difference across the wing is called **lift.** A full understanding of lift in aerodynamics involves other, more complicated factors, such as the creation of vortices on the trailing edge of the wing, but our introduction to fluid dynamics has given you enough tools to at least begin to understand how airplanes stay aloft.

STOP TO THINK 14.6 Rank in order, from highest to lowest, the liquid heights h_a to h_d. The airflow is from left to right.

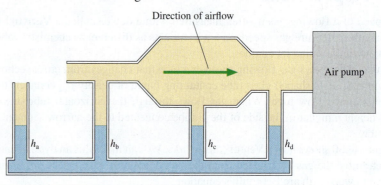

Direction of airflow

Air pump

h_a h_b h_c h_d

14.6 Elasticity

The final subject to explore in this chapter is elasticity. Although elasticity applies primarily to solids rather than fluids, you will see that similar ideas come into play.

Tensile Stress and Young's Modulus

Suppose you clamp one end of a solid rod while using a strong machine to pull on the other with force $\vec{F}$. **FIGURE 14.32a** shows the experimental arrangement. We usually think of solids as being, well, solid. But any material, be it plastic, concrete, or steel, will stretch as the spring-like molecular bonds expand.

FIGURE 14.32b shows graphically the amount of force needed to stretch the rod by the amount ΔL. This graph contains several regions of interest. First is the *elastic region,* ending at the *elastic limit*. As long as ΔL is less than the elastic limit, the rod will return to its initial length L when the force is removed. Just such a reversible stretch is what we mean when we say a material is *elastic*. A stretch beyond the elastic limit will permanently deform the object; it will not return to its initial length when the force is removed. And, not surprisingly, there comes a point when the rod breaks.

For most materials, the graph begins with a *linear region,* which is where we will focus our attention. If ΔL is within the linear region, the force needed to stretch the rod is

$$F = k \Delta L \qquad (14.33)$$

FIGURE 14.32 Stretching a solid rod.

(a)

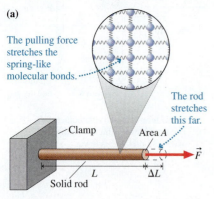

The pulling force stretches the spring-like molecular bonds.

Clamp

Area A

$\vec{F}$

L ΔL

Solid rod

The rod stretches this far.

(b)

F

Elastic region

Slope $= k$

Elastic limit

Breaking point

F is directly proportional to ΔL in this region.

ΔL

Linear region

where k is the slope of the graph. You'll recognize Equation 14.33 as none other than Hooke's law.

The difficulty with Equation 14.33 is that the proportionality constant k depends both on the composition of the rod—whether it is, say, steel or aluminum—and on the rod's length and cross-section area. It would be useful to characterize the elastic properties of steel in general, or aluminum in general, without needing to know the dimensions of a specific rod.

We can meet this goal by thinking about Hooke's law at the atomic scale. The elasticity of a material is directly related to the spring constant of the molecular bonds between neighboring atoms. As **FIGURE 14.33** shows, the force pulling each bond is proportional to the quantity F/A. This force causes each bond to stretch by an amount proportional to $\Delta L/L$. We don't know what the proportionality constants are, but we don't need to. Hooke's law applied to a molecular bond tells us that the force pulling on a bond is proportional to the amount that the bond stretches. Thus F/A must be proportional to $\Delta L/L$. We can write their proportionality as

$$\frac{F}{A} = Y \frac{\Delta L}{L} \qquad (14.34)$$

The proportionality constant Y is called **Young's modulus.** It is directly related to the spring constant of the molecular bonds, so it depends on the material from which the object is made but *not* on the object's geometry.

A comparison of Equations 14.33 and 14.34 shows that Young's modulus can be written as $Y = kL/A$. This is not a definition of Young's modulus but simply an expression for making an experimental determination of the value of Young's modulus. This k is the spring constant of the rod seen in Figure 14.32. It is a quantity easily measured in the laboratory.

The quantity F/A, where A is the cross-section area, is called **tensile stress.** Notice that it is essentially the same definition as pressure. Even so, tensile stress differs in that the stress is applied in a particular direction whereas pressure forces are exerted in all directions. Another difference is that stress is measured in N/m^2 rather than pascals. The quantity $\Delta L/L$, the fractional increase in the length, is called **strain.** Strain is dimensionless. The numerical values of strain are always very small because solids cannot be stretched very much before reaching the breaking point.

With these definitions, Equation 14.34 can be written

$$\text{stress} = Y \times \text{strain} \qquad (14.35)$$

Because strain is dimensionless, Young's modulus Y has the same dimensions as stress, namely N/m^2. **TABLE 14.3** gives values of Young's modulus for several common materials. Large values of Y characterize materials that are stiff and rigid. "Softer" materials, at least relatively speaking, have smaller values of Y. You can see that steel has a larger Young's modulus than aluminum.

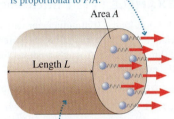

FIGURE 14.33 A material's elasticity is directly related to the spring constant of the molecular bonds.

The number of bonds is proportional to area A. If the rod is pulled with force F, the force pulling on each bond is proportional to F/A.

Area A

Length L

The number of bonds along the rod is proportional to length L. If the rod stretches by ΔL, the stretch of each bond is proportional to $\Delta L/L$.

TABLE 14.3 Elastic properties of various materials

Substance	Young's modulus (N/m^2)	Bulk modulus (N/m^2)
Steel	20×10^{10}	16×10^{10}
Copper	11×10^{10}	14×10^{10}
Aluminum	7×10^{10}	7×10^{10}
Concrete	3×10^{10}	–
Wood (Douglas fir)	1×10^{10}	–
Plastic (polystyrene)	0.3×10^{10}	–
Mercury	–	3×10^{10}
Water	–	0.2×10^{10}

Concrete is a widely used building material because it is relatively inexpensive and, with its large Young's modulus, it has tremendous compressional strength.

We introduced Young's modulus by considering how materials stretch. But Equation 14.35 and Young's modulus also apply to the compression of materials. Compression is particularly important in engineering applications, where beams, columns, and support foundations are compressed by the load they bear. Concrete is often compressed, as in columns that support highway overpasses, but rarely stretched.

> **NOTE** Whether the rod is stretched or compressed, Equation 14.35 is valid only in the linear region of the graph in Figure 14.32b. The breaking point is usually well outside the linear region, so you can't use Young's modulus to compute the maximum possible stretch or compression.

EXAMPLE 14.13 | **Stretching a wire**

A 2.0-m-long, 1.0-mm-diameter wire is suspended from the ceiling. Hanging a 4.5 kg mass from the wire stretches the wire's length by 1.0 mm. What is Young's modulus for this wire? Can you identify the material?

MODEL The hanging mass creates tensile stress in the wire.

SOLVE The force pulling on the wire, which is simply the weight of the hanging mass, produces tensile stress

$$\frac{F}{A} = \frac{mg}{\pi r^2} = \frac{(4.5\ \text{kg})(9.80\ \text{m/s}^2)}{\pi(0.0005\ \text{m})^2} = 5.6 \times 10^7\ \text{N/m}^2$$

The resulting stretch of 1.0 mm is a strain of $\Delta L/L = (1.0\ \text{mm})/(2000\ \text{mm}) = 5.0 \times 10^{-4}$. Thus Young's modulus for the wire is

$$Y = \frac{F/A}{\Delta L/L} = 11 \times 10^{10}\ \text{N/m}^2$$

Referring to Table 14.3, we see that the wire is made of copper.

Volume Stress and the Bulk Modulus

FIGURE 14.34 An object is compressed by pressure forces pushing equally on all sides.

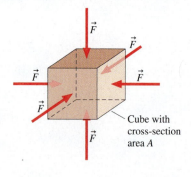

FIGURE 14.34 An object is compressed by pressure forces pushing equally on all sides.

Cube with cross-section area A

Young's modulus characterizes the response of an object to being pulled in one direction. **FIGURE 14.34** shows an object being squeezed in all directions. For example, objects under water are squeezed from all sides by the water pressure. The force per unit area F/A applied to *all* surfaces of an object is called the **volume stress.** Because the force pushes equally on all sides, the volume stress (unlike the tensile stress) really is the same as pressure p.

No material is perfectly rigid. A volume stress applied to an object compresses its volume slightly. The **volume strain** is defined as $\Delta V/V$. The volume strain is a *negative* number because the volume stress *decreases* the volume.

Volume stress, or pressure, is linearly proportional to the volume strain, much as the tensile stress is linearly proportional to the strain in a rod. That is,

$$\frac{F}{A} = p = -B\frac{\Delta V}{V} \tag{14.36}$$

where B is called the **bulk modulus.** The negative sign in Equation 14.36 ensures that the pressure is a positive number. Table 14.3 gives values of the bulk modulus for several materials. Smaller values of B correspond to materials that are more easily compressed. Both solids and liquids can be compressed and thus have a bulk modulus, whereas Young's modulus applies only to solids.

EXAMPLE 14.14 | **Compressing a sphere**

A 1.00-m-diameter solid steel sphere is lowered to a depth of 10,000 m in a deep ocean trench. By how much does its diameter shrink?

MODEL The water pressure applies a volume stress to the sphere.

SOLVE The water pressure at $d = 10,000$ m is

$$p = p_0 + \rho g d = 1.01 \times 10^8\ \text{Pa}$$

where we used the density of seawater. The bulk modulus of steel, taken from Table 14.3, is $16 \times 10^{10}\ \text{N/m}^2$. Thus the volume strain is

$$\frac{\Delta V}{V} = -\frac{p}{B} = -\frac{1.01 \times 10^8\ \text{Pa}}{16 \times 10^{10}\ \text{Pa}} = -6.3 \times 10^{-4}$$

The volume of a sphere is $V = \frac{4}{3}\pi r^3$. If the radius changes by the infinitesimal amount dr, we can use calculus to find that the volume changes by

$$dV = \frac{4}{3}\pi d(r^3) = \frac{4}{3}\pi \times 3r^2 dr = 4\pi r^2 dr$$

Thus $\Delta V = 4\pi r^2 \Delta r$ is a quite good approximation for very small changes in the sphere's radius and volume. Using this, the volume strain is

$$\frac{\Delta V}{V} = \frac{4\pi r^2 \Delta r}{\frac{4}{3}\pi r^3} = \frac{3\,\Delta r}{r} = -6.3 \times 10^{-4}$$

Solving for Δr gives $\Delta r = -1.05 \times 10^{-4}$ m $= -0.105$ mm. The diameter changes by twice this, decreasing 0.21 mm.

ASSESS The immense pressure of the deep ocean causes only a tiny change in the sphere's diameter. You can see that treating solids and liquids as incompressible is an excellent approximation under nearly all circumstances.

CHALLENGE EXAMPLE 14.15 | Draining a cone

A conical tank of radius R and height H, pointed end down, is full of water. A small hole of radius r is opened at the bottom of the tank, with $r \ll R$ so that the tank drains slowly. Find an expression for the time T it takes to drain the tank completely.

MODEL Model the water as an ideal fluid. We can use Bernoulli's equation to relate the flow speed from the hole to the height of the water in the cone.

VISUALIZE FIGURE 14.35 is a pictorial representation. Because the tank drains slowly, we've assumed that the water velocity at the top surface is always very close to zero: $v_1 = 0$. The pressure at the surface is $p_1 = p_{atmos}$. The water discharges into air, so we also have $p_2 = p_{atmos}$ at the exit.

FIGURE 14.35 Pictorial representation of water draining from a tank.

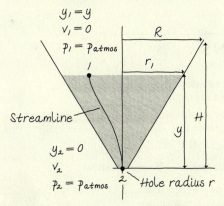

SOLVE As the tank drains, the water height y decreases from H to 0. If we can find an expression for dy/dt, the *rate* at which the water height changes, we'll be able to find T by integrating from "full tank" at $t = 0$ to "empty tank" at $t = T$. Our starting point is the rate at which water flows out of the hole at the bottom—the volume flow rate $Q = v_2 A_2 = \pi r^2 v_2$, where v_2 is the exit speed. The volume of water inside the tank is changing at the rate

$$\frac{dV_{water}}{dt} = -Q = -\pi r^2 v_2$$

where the minus sign shows that the volume is *decreasing* with time.

We need to relate both V_{water} and v_2 to the height y of the water surface. The volume of a cone is $V = \frac{1}{3} \times$ base $\times$ height, so the cone of water has volume $V_{water} = \frac{1}{3}\pi r_1^2 y$. Based on the similar triangles in Figure 14.35, $r_1/R = y/H$. Thus $r_1 = (R/H)y$ and

$$V_{water} = \frac{\pi R^2}{3H^2}y^3$$

Taking the time derivative, we find

$$\frac{dV_{water}}{dt} = \frac{d}{dt}\left[\frac{\pi R^2}{3H^2}y^3\right] = \frac{\pi R^2}{H^2}y^2\frac{dy}{dt}$$

This relates the rate at which the volume changes to the rate at which the height changes.

We can next relate v_2 to the water height y by using Bernoulli's equation to connect the conditions at the surface (point 1) to conditions at the exit (point 2):

$$p_1 + \frac{1}{2}\rho v_1^2 + \rho g y_1 = p_2 + \frac{1}{2}\rho v_2^2 + \rho g y_2$$

With $p_1 = p_2$, $v_1 = 0$, $y_1 = y$, and $y_2 = 0$ at the bottom, Bernoulli's equation simplifies to $\rho g y = \frac{1}{2}\rho v_2^2$. Thus the exit speed of the water is

$$v_2 = \sqrt{2gy}$$

The exit speed decreases as the water height drops because the pressure at the bottom is less.

With this information, our equation for the rate at which the volume is changing becomes

$$\frac{dV_{water}}{dt} = \frac{\pi R^2}{H^2}y^2\frac{dy}{dt} = -\pi r^2 v_2 = -\pi r^2 \sqrt{2gy}$$

In preparation for integration, we need to get all the y's on one side of the equation and dt on the other. Rearranging gives

$$dt = -\frac{R^2}{r^2 H^2 \sqrt{2g}}y^{3/2}\,dy$$

We need to integrate this from the beginning, with $y = H$ at $t = 0$, to the moment the tank is empty, with $y = 0$ at $t = T$:

$$\int_0^T dt = T = -\frac{R^2}{r^2 H^2 \sqrt{2g}}\int_H^0 y^{3/2}\,dy = \frac{R^2}{r^2 H^2 \sqrt{2g}}\int_0^H y^{3/2}\,dy$$

The minus sign was eliminated by reversing the integration limits. Performing the integration gives us the desired result for the time to drain the tank:

$$T = \frac{R^2}{r^2 H^2 \sqrt{2g}}\int_0^H y^{3/2}\,dy = \frac{R^2}{r^2 H^2 \sqrt{2g}}\left[\frac{2}{5}y^{5/2}\right]_0^H$$

$$= \frac{2}{5}\frac{R^2}{r^2}\sqrt{\frac{H}{2g}}$$

ASSESS Making the tank larger by increasing R or H increases the time needed to drain. Making the hole at the bottom larger—a larger value of r—decreases the time. These are as we would have expected, giving us confidence in our result.

SUMMARY

The goal of Chapter 14 has been to learn about systems that flow or deform.

GENERAL PRINCIPLES

Fluid Statics

Gases

- Freely moving particles
- Compressible
- Pressure primarily thermal
- Pressure is constant in a laboratory-size container

Liquids

- Loosely bound particles
- Incompressible
- Pressure primarily gravitational
- Hydrostatic pressure at depth d is $p = p_0 + \rho g d$

Fluid Dynamics

Ideal-fluid model

- Incompressible
- Nonviscous
- Smooth, laminar flow

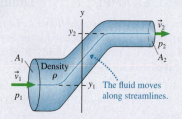

The fluid moves along streamlines.

Equation of continuity

$$v_1 A_1 = v_2 A_2$$

Bernoulli's equation

$$p_1 + \tfrac{1}{2}\rho v_1^2 + \rho g y_1 = p_2 + \tfrac{1}{2}\rho v_2^2 + \rho g y_2$$

Bernoulli's equation is a statement of energy conservation.

IMPORTANT CONCEPTS

Density $\rho = m/V$, where m is mass and V is volume.

Pressure $p = F/A$, where F is the magnitude of the fluid force and A is the area on which the force acts.

- Pressure exists at all points in a fluid.
- Pressure pushes equally in all directions.
- Pressure is constant along a horizontal line.
- Gauge pressure is $p_g = p - 1$ atm.

APPLICATIONS

Buoyancy is the upward force of a fluid on an object.

Archimedes' principle

The magnitude of the buoyant force equals the weight of the fluid displaced by the object.

Sink	$\rho_{avg} > \rho_f$	$F_B < m_o g$
Rise to surface	$\rho_{avg} < \rho_f$	$F_B > m_o g$
Neutrally buoyant	$\rho_{avg} = \rho_f$	$F_B = m_o g$

Elasticity describes the deformation of solids and liquids under stress.

Linear stretch and compression

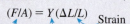

$$(F/A) = Y(\Delta L/L)$$

Tensile stress   Young's modulus   Strain

Volume compression

$$p = -B(\Delta V/V)$$

Bulk modulus   Volume strain

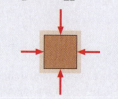

TERMS AND NOTATION

fluid	hydrostatic pressure	neutral buoyancy	Venturi tube
gas	Pascal's principle	ideal-fluid model	lift
liquid	gauge pressure, p_g	viscosity	Young's modulus, Y
mass density, ρ	hydraulics	laminar flow	tensile stress
unit volume	buoyant force	streamline	strain
pressure, p	displaced fluid	equation of continuity	volume stress
pascal, Pa	Archimedes' principle	volume flow rate, Q	volume strain
vacuum	average density, ρ_{avg}	Bernoulli's equation	bulk modulus, B
standard atmosphere, atm			

CONCEPTUAL QUESTIONS

1. An object has density ρ.
   a. Suppose each of the object's three dimensions is increased by a factor of 2 without changing the material of which the object is made. Will the density change? If so, by what factor? Explain.
   b. Suppose each of the object's three dimensions is increased by a factor of 2 without changing the object's mass. Will the density change? If so, by what factor? Explain.

2. Rank in order, from largest to smallest, the pressures at a, b, and c in **FIGURE Q14.2**. Explain.

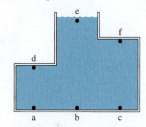

FIGURE Q14.2

3. Rank in order, from largest to smallest, the pressures at d, e, and f in **FIGURE Q14.2**. Explain.

4. **FIGURE Q14.4** shows two rectangular tanks, A and B, full of water. They have equal depths and equal thicknesses (the dimension into the page) but different widths.

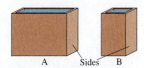

FIGURE Q14.4

   a. Compare the forces the water exerts on the bottoms of the tanks. Is F_A larger than, smaller than, or equal to F_B? Explain.
   b. Compare the forces the water exerts on the sides of the tanks. Is F_A larger than, smaller than, or equal to F_B? Explain.

5. In **FIGURE Q14.5**, is p_A larger than, smaller than, or equal to p_B? Explain.

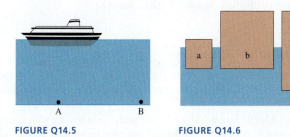

FIGURE Q14.5   **FIGURE Q14.6**

6. Rank in order, from largest to smallest, the densities of blocks a, b, and c in **FIGURE Q14.6**. Explain.

7. Blocks a, b, and c in **FIGURE Q14.7** have the same volume. Rank in order, from largest to smallest, the sizes of the buoyant forces F_a, F_b, and F_c on a, b, and c. Explain.

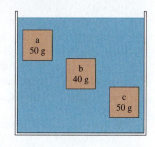

FIGURE Q14.7

8. Blocks a, b, and c in **FIGURE Q14.7** have the same density. Rank in order, from largest to smallest, the sizes of the buoyant forces F_a, F_b, and F_c on a, b, and c. Explain.

9. The two identical beakers in **FIGURE Q14.9** are filled to the same height with water. Beaker B has a plastic sphere floating in it. Which beaker, with all its contents, weighs more? Or are they equal? Explain.

   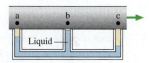

FIGURE Q14.9   **FIGURE Q14.10**

10. Gas flows through the pipe of **FIGURE Q14.10**. You can't see into the pipe to know how the inner diameter changes. Rank in order, from largest to smallest, the gas speeds v_a, v_b, and v_c at points a, b, and c. Explain.

11. Wind blows over the house in **FIGURE Q14.11**. A window on the ground floor is open. Is there an airflow through the house? If so, does the air flow in the window and out the chimney, or in the chimney and out the window? Explain.

FIGURE Q14.11

12. A 2000 N force stretches a wire by 1 mm. A second wire of the same material is twice as long and has twice the diameter. How much force is needed to stretch it by 1 mm? Explain.

13. A wire is stretched right to the breaking point by a 5000 N force. A longer wire made of the same material has the same diameter. Is the force that will stretch it right to the breaking point larger than, smaller than, or equal to 5000 N? Explain.

EXERCISES AND PROBLEMS

Problems labeled ▓ integrate material from earlier chapters.

Exercises

Section 14.1 Fluids

1. ‖ What is the volume in mL of 55 g of a liquid with density 1100 kg/m³?

2. | Cylinders A and B have equal heights. Cylinder A is filled with helium gas at 1.0 atm pressure and 0°C. The diameter of cylinder B is half that of cylinder A, and cylinder B is filled with glycerin. What is the ratio of the fluid mass in cylinder B to that in cylinder A?

3. ‖ a. 50 g of gasoline are mixed with 50 g of water. What is the average density of the mixture?
   b. 50 cm³ of gasoline are mixed with 50 cm³ of water. What is the average density of the mixture?

4. ‖ A 6.0 m × 12.0 m swimming pool slopes linearly from a 1.0 m depth at one end to a 3.0 m depth at the other. What is the mass of water in the pool?

Section 14.2 Pressure

5. ‖ A 1.0-m-diameter vat of liquid is 2.0 m deep. The pressure at the bottom of the vat is 1.3 atm. What is the mass of the liquid in the vat?

6. ‖ The deepest point in the ocean is 11 km below sea level, deeper than Mt. Everest is tall. What is the pressure in atmospheres at this depth?

7. ‖ A 3.0-cm-diameter tube is held upright and filled to the top with mercury. The mercury pressure at the bottom of the tube— the pressure in excess of atmospheric pressure—is 50 kPa. How tall is the tube?

8. ‖ a. What volume of water has the same mass as 8.0 m³ of ethyl alcohol?
   b. If this volume of water is in a cubic tank, what is the pressure at the bottom?

9. ‖ A 50-cm-thick layer of oil floats on a 120-cm-thick layer of water. What is the pressure at the bottom of the water layer?

10. ‖ A research submarine has a 20-cm-diameter window 8.0 cm thick. The manufacturer says the window can withstand forces up to 1.0×10^6 N. What is the submarine's maximum safe depth? The pressure inside the submarine is maintained at 1.0 atm.

11. ‖ A 20-cm-diameter circular cover is placed over a 10-cm-diameter hole that leads into an evacuated chamber. The pressure in the chamber is 20 kPa. How much force is required to pull the cover off?

12. ‖ The container shown in **FIGURE EX14.12** is filled with oil. It is open to the atmosphere on the left.
   a. What is the pressure at point A?
   b. What is the pressure difference between points A and B? Between points A and C?

FIGURE EX14.12

Section 14.3 Measuring and Using Pressure

13. ‖ What is the height of a water barometer at atmospheric pressure?

14. ‖ What is the minimum hose diameter of an ideal vacuum cleaner that could lift a 10 kg (22 lb) dog off the floor?

15. ‖ How far must a 2.0-cm-diameter piston be pushed down into one cylinder of a hydraulic lift to raise an 8.0-cm-diameter piston by 20 cm?

Section 14.4 Buoyancy

16. ‖ A 6.00-cm-diameter sphere with a mass of 89.3 g is neutrally buoyant in a liquid. Identify the liquid.

17. ‖ A 2.0 cm × 2.0 cm × 6.0 cm block floats in water with its long axis vertical. The length of the block above water is 2.0 cm. What is the block's mass density?

18. ‖ Astronauts visiting a new planet find a lake filled with an unknown liquid. They have with them a plastic cube, 6.0 cm on each side, with a density of 840 kg/m³. First they weigh the cube with a spring scale, measuring a weight of 21 N. Then they float the cube in the lake and find that two-thirds of the cube is submerged. At what depth in the lake will the pressure be twice the atmospheric pressure of 85 kPa?

19. ‖ A sphere completely submerged in water is tethered to the bottom with a string. The tension in the string is one-third the weight of the sphere. What is the density of the sphere?

20. ‖ A 5.0 kg rock whose density is 4800 kg/m³ is suspended by a string such that half of the rock's volume is under water. What is the tension in the string?

21. ‖ What is the tension of the string in **FIGURE EX14.21**?

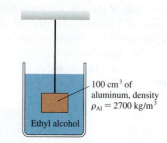

100 cm³ of aluminum, density $\rho_{Al} = 2700$ kg/m³

Ethyl alcohol

FIGURE EX14.21

22. ‖ A 10-cm-diameter, 20-cm-tall steel cylinder ($\rho_{steel} = 7900$ kg/m³) floats in mercury. The axis of the cylinder is perpendicular to the surface. What length of steel is above the surface?

23. ‖ You need to determine the density of a ceramic statue. If you suspend it from a spring scale, the scale reads 28.4 N. If you then lower the statue into a tub of water, so that it is completely submerged, the scale reads 17.0 N. What is the statue's density?

24. ‖ Styrofoam has a density of 150 kg/m³. What is the maximum mass that can hang without sinking from a 50-cm-diameter Styrofoam sphere in water? Assume the volume of the mass is negligible compared to that of the sphere.

25. ‖ You and your friends are playing in the swimming pool with a 60-cm-diameter beach ball. How much force would be needed to push the ball completely under water?

Section 14.5 Fluid Dynamics

26. ‖ Water flows through a 2.5-cm-diameter hose at 3.0 m/s. How long, in minutes, will it take to fill a 600 L child's wading pool?

27. ‖ A long horizontal tube has a square cross section with sides of width L. A fluid moves through the tube with speed v_0. The tube then changes to a circular cross section with diameter L. What is the fluid's speed in the circular part of the tube?

28. ‖ A 1.0-cm-diameter pipe widens to 2.0 cm, then narrows to 5.0 mm. Liquid flows through the first segment at a speed of 4.0 m/s.
   a. What is the speed in the second and third segments?
   b. What is the volume flow rate through the pipe?

29. ‖ A bucket is filled with water to a height of 23 cm, then a plug is removed from a 4.0-mm-diameter hole in the bottom of the bucket. As the water begins to pour out of the hole, how fast is it moving?

30. ‖ What does the top pressure gauge read in **FIGURE EX14.30**?

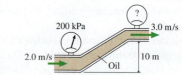

200 kPa   3.0 m/s
2.0 m/s   10 m
Oil

FIGURE EX14.30

31. ‖ A 2.0 mL syringe has an inner diameter of 6.0 mm, a needle
BIO inner diameter of 0.25 mm, and a plunger pad diameter (where
you place your finger) of 1.2 cm. A nurse uses the syringe
to inject medicine into a patient whose blood pressure is 140/100.
   a. What is the minimum force the nurse needs to apply to the
syringe?
   b. The nurse empties the syringe in 2.0 s. What is the flow speed
of the medicine through the needle?

Section 14.6 Elasticity

32. ‖ A 70 kg mountain climber dangling in a crevasse stretches
a 50-m-long, 1.0-cm-diameter rope by 8.0 cm. What is Young's
modulus for the rope?
33. ‖ An 80-cm-long, 1.0-mm-diameter steel guitar string must be
tightened to a tension of 2000 N by turning the tuning screws. By
how much is the string stretched?
34. ‖ A 3.0-m-tall, 50-cm-diameter concrete column supports a
200,000 kg load. By how much is the column compressed?
35. ‖ a. What is the pressure at a depth of 5000 m in the ocean?
   b. What is the fractional volume change $\Delta V/V$ of seawater at
this pressure?
   c. What is the density of seawater at this pressure?
36. ‖ A large 10,000 L aquarium is supported by four wood
posts (Douglas fir) at the corners. Each post has a square
4.0 cm × 4.0 cm cross section and is 80 cm tall. By how much is
each post compressed by the weight of the aquarium?
37. ‖ A 5.0-m-diameter solid aluminum sphere is launched into space.
By how much does its diameter increase? Give your answer in μm.

Problems

38. ‖ a. In **FIGURE P14.38**, how much force does the fluid exert on
the end of the cylinder at A?
   b. How much force does the fluid exert on the end of the
cylinder at B?

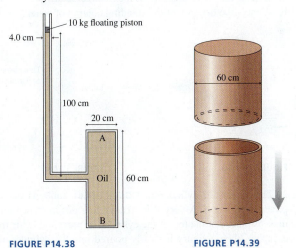

FIGURE P14.38   **FIGURE P14.39**

39. ‖ The two 60-cm-diameter cylinders in **FIGURE P14.39**, closed at
one end, open at the other, are joined to form a single cylinder,
then the air inside is removed.
   a. How much force does the atmosphere exert on the flat end of
each cylinder?
   b. Suppose one cylinder is bolted to a sturdy ceiling. How many
100 kg football players would need to hang from the lower
cylinder to pull the two cylinders apart?

40. ‖ *Postural hypotension* is the occurrence of low systolic blood
BIO pressure when a person stands up too quickly from a reclining
position. A brain blood pressure lower than 90 mm of Hg can
cause fainting or lightheadedness. In a healthy adult, the
automatic constriction and expansion of blood vessels keep the
brain blood pressure constant while posture is changing, but
disease or aging can weaken this response. If the blood pressure
in your brain is 118 mm of Hg while lying down, what would it
be when you stand up if this automatic response failed? Assume
your brain is 40 cm from your heart and the density of blood is
1060 kg/m^3.
41. ‖ A friend asks you how much pressure is in your car tires.
You know that the tire manufacturer recommends 30 psi,
but it's been a while since you've checked. You can't find
a tire gauge in the car, but you do find the owner's manual
and a ruler. Fortunately, you've just finished taking physics, so
you tell your friend, "I don't know, but I can figure it out." From
the owner's manual you find that the car's mass is 1500 kg. It
seems reasonable to assume that each tire supports one-fourth of
the weight. With the ruler you find that the tires are 15 cm wide
and the flattened segment of the tire in contact with the road is
13 cm long. What answer—in psi—will you give your friend?
42. ‖ a. The 70 kg student in **FIGURE P14.42** balances a 1200 kg
elephant on a hydraulic lift. What is the diameter of the
piston the student is standing on?
   b. When a second student joins the first, the piston sinks
35 cm. What is the second student's mass?

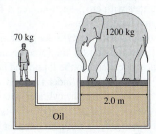

FIGURE P14.42

43. ‖ A 55 kg cheerleader uses an oil-filled hydraulic lift to hold four
110 kg football players at a height of 1.0 m. If her piston is 16 cm in
diameter, what is the diameter of the football players' piston?
44. ‖ A U-shaped tube, open to the air on both ends, contains
mercury. Water is poured into the left arm until the water column
is 10.0 cm deep. How far upward from its initial position does
the mercury in the right arm rise?
45. ‖ Geologists place *tiltmeters* on the sides of volcanoes to
measure the displacement of the surface as magma moves inside
the volcano. Although most tiltmeters today are electronic, the
traditional tiltmeter, used for decades, consisted of two or more
water-filled metal cans placed some distance apart and connected
by a hose. **FIGURE P14.45** shows two such cans, each having a window
to measure the water height. Suppose the cans are placed so that
the water level in both is initially at the 5.0 cm mark. A week later,
the water level in can 2 is at the 6.5 cm mark.
   a. Did can 2 move up or down relative to can 1? By what distance?
   b. Where is the water level now in can 1?

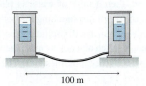

FIGURE P14.45

46. ‖ Glycerin is poured into an open U-shaped tube until the height in both sides is 20 cm. Ethyl alcohol is then poured into one arm until the height of the alcohol column is 20 cm. The two liquids do not mix. What is the difference in height between the top surface of the glycerin and the top surface of the alcohol?

47. ‖ An aquarium of length L, width (front to back) W, and depth
CALC D is filled to the top with liquid of density ρ.
   a. Find an expression for the force of the liquid on the bottom of the aquarium.
   b. Find an expression for the force of the liquid on the front window of the aquarium.
   c. Evaluate the forces for a 100-cm-long, 35-cm-wide, 40-cm-deep aquarium filled with water.

48. ‖ It's possible to use the ideal-gas law to show that the density of
CALC the earth's atmosphere decreases exponentially with height. That is, $\rho = \rho_0 \exp(-z/z_0)$, where z is the height above sea level, ρ_0 is the density at sea level (you can use the Table 14.1 value), and z_0 is called the *scale height* of the atmosphere.
   a. Determine the value of z_0.
   Hint: What is the weight of a column of air?
   b. What is the density of the air in Denver, at an elevation of 1600 m? What percent of sea-level density is this?

49. ‖ The average density of the body of a fish is 1080 kg/m³. To
BIO keep from sinking, a fish increases its volume by inflating an internal air bladder, known as a swim bladder, with air. By what percent must the fish increase its volume to be neutrally buoyant in fresh water? The density of air at 20°C is 1.19 kg/m³.

50. ‖ A cylinder with cross-section area A floats with its long axis
CALC vertical in a liquid of density ρ.
   a. Pressing down on the cylinder pushes it deeper into the liquid. Find an expression for the force needed to push the cylinder distance x deeper into the liquid and hold it there.
   b. A 4.0-cm-diameter cylinder floats in water. How much work must be done to push the cylinder 10 cm deeper into the water?

51. ‖ A less-dense liquid of density ρ_1 floats on top of a more-dense liquid of density ρ_2. A uniform cylinder of length l and density ρ, with $\rho_1 < \rho < \rho_2$, floats at the interface with its long axis vertical. What fraction of the length is in the more-dense liquid?

52. ‖ A 30-cm-tall, 4.0-cm-diameter plastic tube has a sealed bottom. 250 g of lead pellets are poured into the bottom of the tube, whose mass is 30 g, then the tube is lowered into a liquid. The tube floats with 5.0 cm extending above the surface. What is the density of the liquid?

53. ‖ One day when you come into physics lab you find several plastic hemispheres floating like boats in a tank of fresh water. Each lab group is challenged to determine the heaviest rock that can be placed in the bottom of a plastic boat without sinking it. You get one try. Sinking the boat gets you no points, and the maximum number of points goes to the group that can place the heaviest rock without sinking. You begin by measuring one of the hemispheres, finding that it has a mass of 21 g and a diameter of 8.0 cm. What is the mass of the heaviest rock that, in perfectly still water, won't sink the plastic boat?

54. ‖‖‖ A spring with spring constant 35 N/m is attached to the ceiling, and a 5.0-cm-diameter, 1.0 kg metal cylinder is attached to its lower end. The cylinder is held so that the spring is neither stretched nor compressed, then a tank of water is placed underneath with the surface of the water just touching the bottom of the cylinder. When released, the cylinder will oscillate a few times but, damped by the water, quickly reach an equilibrium position. When in equilibrium, what length of the cylinder is submerged?

55. ‖‖‖ A plastic "boat" with a 25 cm² square cross section floats in a liquid. One by one, you place 50 g masses inside the boat and measure how far the boat extends below the surface. Your data are as follows:

Mass added (g)	Depth (cm)
50	2.9
100	5.0
150	6.6
200	8.6

Draw an appropriate graph of the data and, from the slope and intercept of the best-fit line, determine the mass of the boat and the density of the liquid.

56. ‖‖‖ A 355 mL soda can is 6.2 cm in diameter and has a mass of 20 g. Such a soda can half full of water is floating upright in water. What length of the can is above the water level?

57. ‖ A nuclear power plant draws 3.0×10^6 L/min of cooling water from the ocean. If the water is drawn in through two parallel, 3.0-m-diameter pipes, what is the water speed in each pipe?

58. ‖ a. A liquid of density ρ flows at speed v_0 through a horizontal pipe that expands smoothly from diameter d_0 to a larger diameter d_1. The pressure in the narrower section is p_0. Find an expression for the pressure p_1 in the wider section.
   b. A pressure gauge reads 50 kPa as water flows at 10.0 m/s through a 16.8-cm-diameter horizontal pipe. What is the reading of a pressure gauge after the pipe has expanded to 20.0 cm in diameter?

59. ‖‖ A tree loses water to the air by the process of *transpiration* at
BIO the rate of 110 g/h. This water is replaced by the upward flow of sap through vessels in the trunk. If the trunk contains 2000 vessels, each 100 μm in diameter, what is the upward speed of the sap in each vessel? The density of tree sap is 1040 kg/m³.

60. ‖ Water flows from the pipe shown in **FIGURE P14.60** with a speed of 4.0 m/s.
   a. What is the water pressure as it exits into the air?
   b. What is the height h of the standing column of water?

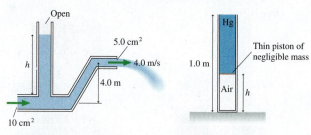

FIGURE P14.60   **FIGURE P14.61**

61. ‖ The 1.0-m-tall cylinder in **FIGURE P14.61** contains air at a pressure of 1 atm. A very thin, frictionless piston of negligible mass is placed at the top of the cylinder, to prevent any air from escaping, then mercury is slowly poured into the cylinder until no more can be added without the cylinder overflowing. What is the height h of the column of compressed air?
   Hint: Boyle's law, which you learned in chemistry, says $p_1V_1 = p_2V_2$ for a gas compressed at constant temperature, which we will assume to be the case.

62. ‖‖‖ Water flowing out of a 16-mm-diameter faucet fills a 2.0 L bottle in 10 s. At what distance below the faucet has the water stream narrowed to 10 mm diameter?

63. ‖ Water from a vertical pipe emerges as a 10-cm-diameter cylinder and falls straight down 7.5 m into a bucket. The water exits the pipe with a speed of 2.0 m/s. What is the diameter of the column of water as it hits the bucket?

64. ‖ A 2.0-m-diameter vertical cylinder, open at the top, is filled with ethyl alcohol to the 2.5 m mark. A worker pulls a plug from a hole 55 cm from the bottom, causing the fluid level to begin dropping at a rate of 3.4 mm/min. What is the diameter of the hole?

65. | A hurricane wind blows across a 6.0 m × 15.0 m flat roof at a speed of 130 km/h.
   a. Is the air pressure above the roof higher or lower than the pressure inside the house? Explain.
   b. What is the pressure difference?
   c. How much force is exerted on the roof? If the roof cannot withstand this much force, will it "blow in" or "blow out"?

66. ‖ Air flows through the tube shown in **FIGURE P14.66** at a rate of 1200 cm³/s. Assume that air is an ideal fluid. What is the height h of mercury in the right side of the U-tube?

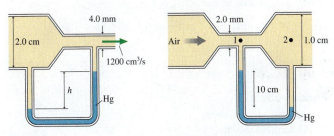

FIGURE P14.66　　　　　**FIGURE P14.67**

67. ‖ Air flows through the tube shown in **FIGURE P14.67**. Assume that air is an ideal fluid.
   a. What are the air speeds v_1 and v_2 at points 1 and 2?
   b. What is the volume flow rate?

68. ‖ A water tank of height h has a small hole at height y. The water is replenished to keep h from changing. The water squirting from the hole has range x. The range approaches zero as $y \to 0$ because the water squirts right onto the ground. The range also approaches zero as $y \to h$ because the horizontal velocity becomes zero. Thus there must be some height y between 0 and h for which the range is a maximum.
   a. Find an algebraic expression for the flow speed v with which the water exits the hole at height y.
   b. Find an algebraic expression for the range of a particle shot horizontally from height y with speed v.
   c. Combine your expressions from parts a and b. Then find the maximum range x_{max} and the height y of the hole. "Real" water won't achieve quite this range because of viscosity, but it will be close.

69. ‖ a. A cylindrical tank of radius R, filled to the top with a liquid, has a small hole in the side, of radius r, at distance d below the surface. Find an expression for the volume flow rate through the hole.
   b. A 4.0-mm-diameter hole is 1.0 m below the surface of a 2.0-m-diameter tank of water. What is the rate, in mm/min, at which the water level will initially drop if the water is not replenished?

70. ‖ There is a disk of cartilage between each pair of vertebrae in
BIO your spine. Young's modulus for cartilage is $1.0 \times 10^6 \text{ N/m}^2$.

Suppose a relaxed disk is 4.0 cm in diameter and 5.0 mm thick. If a disk in the lower spine supports half the weight of a 66 kg person, by how many mm does the disk compress?

Challenge Problems

71. ‖ The bottom of a steel "boat" is a 5.0 m × 10 m × 2.0 cm piece of steel ($\rho_{steel} = 7900 \text{ kg/m}^3$). The sides are made of 0.50-cm-thick steel. What minimum height must the sides have for this boat to float in perfectly calm water?

72. ‖ The tank shown in **FIGURE CP14.72** is completely filled with a
CALC liquid of density ρ. The right face is not permanently attached to the tank but, instead, is held against a rubber seal by the tension in a spring. To prevent leakage, the spring must both pull with sufficient strength *and* prevent a torque from pushing the bottom of the right face out.
   a. What minimum spring tension is needed?
   b. If the spring has the minimum tension, at what height d from the bottom must it be attached?

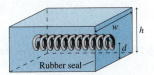

FIGURE CP14.72　　　　　**FIGURE CP14.73**

73. ‖ In **FIGURE CP14.73**, a cone of density ρ_0 and total height l floats in a liquid of density ρ_f. The height of the cone above the liquid is h. What is the ratio h/l of the exposed height to the total height?

74. ‖ Disk brakes, such as those in your car, operate by using pressurized oil to push outward on a piston. The piston, in turn, presses brake pads against a spinning rotor or wheel, as seen in **FIGURE CP14.74**. Consider a 15 kg industrial grinding wheel, 26 cm in diameter, spinning at 900 rpm. The brake pads are actuated by 2.0-cm-diameter pistons, and they contact the wheel an average distance 12 cm from the axis. If the coefficient of kinetic friction between the brake pad and the wheel is 0.60, what oil pressure is needed to stop the wheel in 5.0 s?

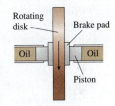

FIGURE CP14.74

75. ‖ In addition to the buoyant force, an object moving in a liquid
CALC experiences a *linear* drag force $\vec{F}_{drag} = (bv, \text{direction opposite the motion})$, where b is a constant. For a sphere of radius R, the drag constant can be shown to be $b = 6\pi\eta R$, where η is the *viscosity* of the liquid. Consider a sphere of radius R and density ρ that is released from rest at the surface of a liquid with density ρ_f.
   a. Find an expression in terms of R, η, g, and the densities for the sphere's *terminal speed* v_{term} as it falls through the liquid.
   b. Solve Newton's second law to find an expression for $v_y(t)$, the sphere's vertical velocity as a function of time as it falls. Pay careful attention to signs!
   c. Water at 20°C has viscosity $\eta = 1.0 \times 10^{-3} \text{ Pa s}$. Aluminum has density 2700 kg/m³. If a 3.0-mm-diameter aluminum pellet is dropped into water, what is its terminal speed, and how long does it take to reach 90% of its terminal speed?

Applications of Newtonian Mechanics

KEY FINDINGS   What are the overarching findings of Part III?

- Newton's laws of motion together with the conservation laws for energy and momentum form what is called **Newtonian mechanics.** Part III has introduced relatively little new physics and instead has focused on applying Newtonian mechanics to new situations.

- Rotational motion is analogous to linear motion.
- With Newton's law of gravity, we can use Newtonian mechanics to understand the motions of satellites and planets.
- Liquids and gases cannot be modeled as particles. Even so, fluids still obey Newtonian laws of statics and motion.

LAWS   What laws of physics govern these applications?

Newton's second law for rotation   A net torque causes an extended object to have angular acceleration: $\tau_{\text{net}} = I\alpha$

Conservation of angular momentum   For an isolated system ($\tau_{\text{net}} = 0$), the total angular momentum is conserved: $\Delta \vec{L} = \vec{0}$.

Newton's law of gravity   The attractive force between two objects separated by r is $F_{m\text{ on }M} = F_{M\text{ on }m} = \dfrac{GMm}{r^2}$.

Bernoulli's equation   For two points along a streamline of a flowing fluid, $p_1 + \frac{1}{2}\rho v_1^2 + \rho g y_1 = p_2 + \frac{1}{2}\rho v_2^2 + \rho g y_2$.

MODELS   What are the most important models of Part III?

Rotation models

- A **rigid body** is an extended object whose size and shape do not change.
  - Particle-like atoms held together by bonds that are massless, rigid rods.

- A rigid body that experiences a **constant torque** undergoes constant angular acceleration.
  - Obeys Newton's second law for rotation.

- An extended object is in **static equilibrium** if and only if there are no net force *and* no net torque.

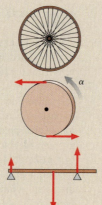

Fluid models

- A **gas** fills its container.
  - Molecules move freely.
  - Molecules are far apart.
  - A gas is compressible.

- A **liquid** has a well-defined surface.
  - Molecules are loosely bound.
  - Molecules are close together.
  - A liquid is incompressible.

- An **ideal fluid** obeys Bernoulli's equation.
  - Incompressible
  - Nonviscous
  - Smooth, laminar flow

TOOLS   What are the most important tools introduced in Part III?

Rotation

- Rotational/linear analogs:

Moment of inertia I	Mass m
Torque τ	Force F
Angular acceleration α	Acceleration a
Angular momentum $\vec{L}$	Momentum $\vec{p}$

- Rotational kinetic energy: $K = \frac{1}{2}I\omega^2$
- **Torque:** $\tau = rF\sin\phi = rF_{\parallel} = dF$
  - d is the moment arm or lever arm.
  - Vector torque is $\vec{\tau} = \vec{r} \times \vec{F}$.
- **Angular momentum:** $\vec{L} = I\vec{\omega}$
- Rolling motion:
  - Rolling without slipping: $v_{\text{cm}} = R\omega$
  - $K = K_{\text{cm}} + K_{\text{rot}}$

Gravity

- Newton's theory of gravity predicts Kepler's three observational laws. We can summarize them as:
  - Planets and satellites move in elliptical orbits.
  - Angular momentum is conserved.
  - For circular orbits, the square of the period is proportional to the cube of the orbit's radius.
- Gravitational potential energy:

  $$U_{\text{G}} = -\frac{GMm}{r}$$

  - The potential energy is negative with a zero at infinity
  - **Escape speed** is the minimum speed needed to reach infinity.

Fluids

- **Pressure** $p = F/A$ exists at all points in a fluid.
  - Gas pressure is primarily thermal.
  - Liquid pressure is primarily gravitational.
  - Hydrostatic pressure at depth d:
    $$p = p_0 + \rho g d$$
- **Archimedes' principle** says that the upward buoyant force equals the weight of the displaced liquid.
  - An object floats if its average density is less than the fluid density.
- **Equation of continuity:** $v_1 A_1 = v_2 A_2$
  - Relates two points on a streamline.
  - v is the flow speed.
  - A is the cross-section area.

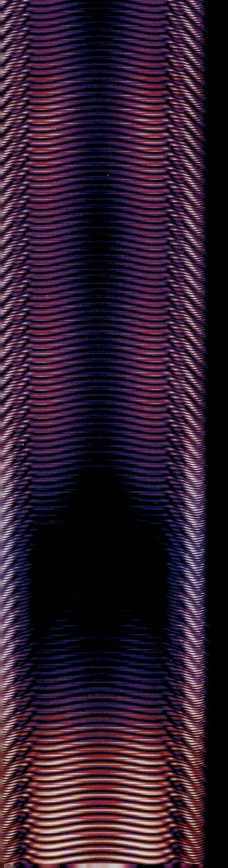

OVERVIEW

The Wave Model

Parts I–III of this text have been primarily about the physics of particles. You've seen that systems ranging from balls and rockets to planets can be thought of as particles or as systems of particles. A *particle* is one of the two fundamental models of classical physics. The other, to which we now turn our attention, is a *wave*.

Waves are ubiquitous in nature. Familiar examples include

- Undulating ripples on a pond.
- The swaying ground of an earthquake.
- A vibrating guitar string.
- The sweet sound of a flute.
- The colors of the rainbow.

The physics of oscillations and waves is the subject of Part IV. The most basic oscillation, called *simple harmonic motion,* is still the motion of a particle. But oscillators are the sources of waves, and the mathematical description of oscillation will carry over to the mathematical description of waves.

A wave, in contrast with a particle, is diffuse, spread out, not to be found at a single point in space. We will start with waves traveling outward through some medium, like the spreading ripples after a pebble hits a pool of water. These are called *traveling waves.* An investigation of what happens when waves travel through each other will lead us to *standing waves,* which are essential for understanding phenomena ranging from those as common as musical instruments to as complex as lasers and the electrons in atoms. We'll also study one of the most important defining characteristics of waves—their ability to exhibit *interference.*

Our exploration of wave phenomena will call upon sound waves, light waves, and vibrating strings for examples, but our goal will be to emphasize the unity and coherence that are common to *all* types of waves. Later, in Part VII, we will devote three chapters to light and optics, perhaps the most important application of waves. Although light is an electromagnetic wave, your understanding of those chapters will depend on nothing more than the "waviness" of light. If you wish, you can proceed to those chapters immediately after finishing Part IV. The electromagnetic aspects of light waves will be taken up in Chapter 31.

The song of a humpback whale can travel hundreds of kilometers underwater. This graph uses a procedure called wavelet analysis to study the frequency structure of a humpback whale song.

15 Oscillations

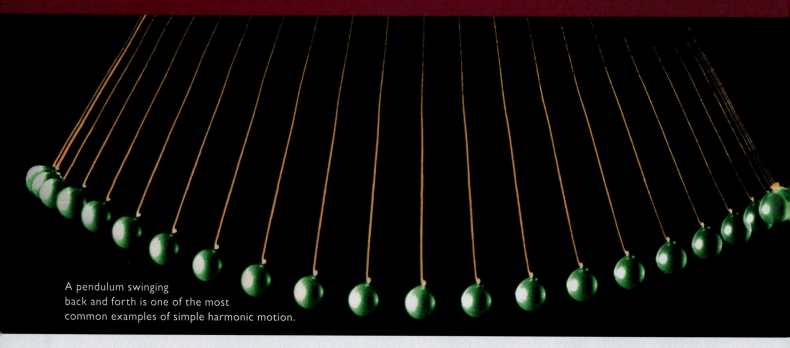

A pendulum swinging back and forth is one of the most common examples of simple harmonic motion.

IN THIS CHAPTER, you will learn about systems that oscillate in simple harmonic motion.

What are oscillations?

Oscillatory motion is a repetitive motion back and forth around an equilibrium position. We'll describe oscillations in terms of their amplitude, period, and frequency. The most important oscillation is simple harmonic motion (SHM), where the position and velocity graphs are sinusoidal.

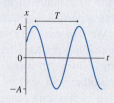

What things undergo SHM?

The prototype of SHM is a mass oscillating on a spring. Lessons learned from this system apply to all SHM.

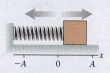

- A pendulum is a classic example of SHM.
- Any system with a linear restoring force undergoes SHM.

« LOOKING BACK Section 9.4 Restoring forces

How is SHM related to circular motion?

The projection of uniform circular motion onto a line oscillates back and forth in SHM.

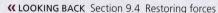

- This link to circular motion will help us develop the mathematics of SHM.
- A phase constant, based on the angle on a circle, will describe the initial conditions.

« LOOKING BACK Section 4.4 Circular motion

Is energy conserved in SHM?

If there is no friction or other dissipative force, the mechanical energy of an oscillating system is conserved. Energy is transformed back and forth between kinetic and potential energy. Energy conservation is an important problem-solving strategy.

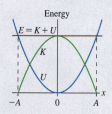

« LOOKING BACK Sections 10.3–10.5 Elastic potential energy and energy diagrams

What if there's friction?

If there's dissipation, the system "runs down." This is called a damped oscillation. The oscillation amplitude undergoes exponential decay. But the amplitude can grow very large when an oscillatory system is *driven* at its natural frequency. This is called resonance.

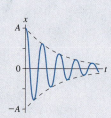

Why is SHM important?

Simple harmonic motion is one of the most common and important motions in science and engineering.

- Oscillations and vibrations occur in mechanical, electrical, chemical, and atomic systems. Understanding how a system might oscillate is an important part of engineering design.
- More complex oscillations can be understood in terms of SHM.
- Oscillations are the sources of waves, which we'll study in the next two chapters.

15 Oscillations

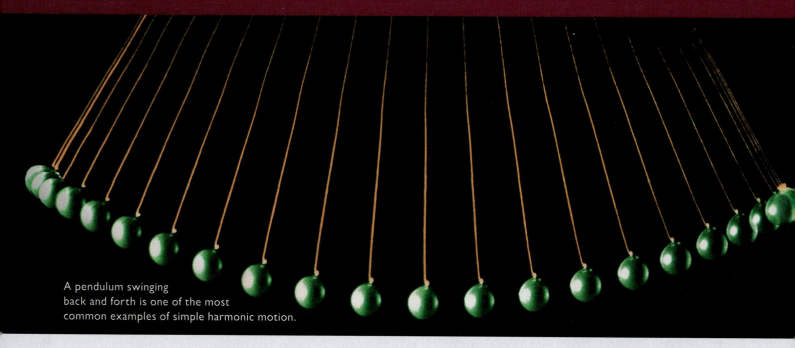

A pendulum swinging back and forth is one of the most common examples of simple harmonic motion.

IN THIS CHAPTER, you will learn about systems that oscillate in simple harmonic motion.

What are oscillations?

Oscillatory motion is a repetitive motion back and forth around an equilibrium position. We'll describe oscillations in terms of their amplitude, period, and frequency. The most important oscillation is simple harmonic motion (SHM), where the position and velocity graphs are sinusoidal.

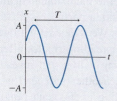

What things undergo SHM?

The prototype of SHM is a mass oscillating on a spring. Lessons learned from this system apply to all SHM.

- A pendulum is a classic example of SHM.
- Any system with a linear restoring force undergoes SHM.

« LOOKING BACK Section 9.4 Restoring forces

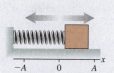

How is SHM related to circular motion?

The projection of uniform circular motion onto a line oscillates back and forth in SHM.

- This link to circular motion will help us develop the mathematics of SHM.
- A phase constant, based on the angle on a circle, will describe the initial conditions.

« LOOKING BACK Section 4.4 Circular motion

Is energy conserved in SHM?

If there is no friction or other dissipative force, the mechanical energy of an oscillating system is conserved. Energy is transformed back and forth between kinetic and potential energy. Energy conservation is an important problem-solving strategy.

« LOOKING BACK Sections 10.3–10.5 Elastic potential energy and energy diagrams

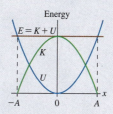

What if there's friction?

If there's dissipation, the system "runs down." This is called a damped oscillation. The oscillation amplitude undergoes exponential decay. But the amplitude can grow very large when an oscillatory system is *driven* at its natural frequency. This is called resonance.

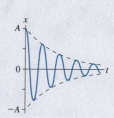

Why is SHM important?

Simple harmonic motion is one of the most common and important motions in science and engineering.

- Oscillations and vibrations occur in mechanical, electrical, chemical, and atomic systems. Understanding how a system might oscillate is an important part of engineering design.
- More complex oscillations can be understood in terms of SHM.
- Oscillations are the sources of waves, which we'll study in the next two chapters.

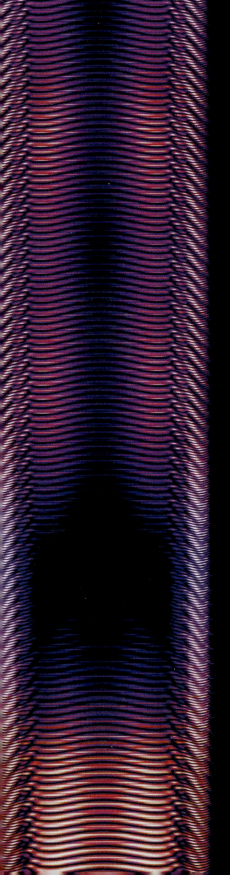

The Wave Model

Parts I–III of this text have been primarily about the physics of particles. You've seen that systems ranging from balls and rockets to planets can be thought of as particles or as systems of particles. A *particle* is one of the two fundamental models of classical physics. The other, to which we now turn our attention, is a *wave*.

Waves are ubiquitous in nature. Familiar examples include

- Undulating ripples on a pond.
- The swaying ground of an earthquake.
- A vibrating guitar string.
- The sweet sound of a flute.
- The colors of the rainbow.

The physics of oscillations and waves is the subject of Part IV. The most basic oscillation, called *simple harmonic motion,* is still the motion of a particle. But oscillators are the sources of waves, and the mathematical description of oscillation will carry over to the mathematical description of waves.

A wave, in contrast with a particle, is diffuse, spread out, not to be found at a single point in space. We will start with waves traveling outward through some medium, like the spreading ripples after a pebble hits a pool of water. These are called *traveling waves.* An investigation of what happens when waves travel through each other will lead us to *standing waves,* which are essential for understanding phenomena ranging from those as common as musical instruments to as complex as lasers and the electrons in atoms. We'll also study one of the most important defining characteristics of waves—their ability to exhibit *interference.*

Our exploration of wave phenomena will call upon sound waves, light waves, and vibrating strings for examples, but our goal will be to emphasize the unity and coherence that are common to *all* types of waves. Later, in Part VII, we will devote three chapters to light and optics, perhaps the most important application of waves. Although light is an electromagnetic wave, your understanding of those chapters will depend on nothing more than the "waviness" of light. If you wish, you can proceed to those chapters immediately after finishing Part IV. The electromagnetic aspects of light waves will be taken up in Chapter 31.

The song of a humpback whale can travel hundreds of kilometers underwater. This graph uses a procedure called wavelet analysis to study the frequency structure of a humpback whale song.

15.1 Simple Harmonic Motion

One of the most common and important types of motion is **oscillatory motion**—a repetitive motion back and forth around an equilibrium position. Swinging chandeliers, vibrating guitar strings, the electrons in cell phone circuits, and even atoms in solids are all undergoing oscillatory motion. In addition, oscillations are the sources of *waves*, our subject in the following two chapters. Oscillations play a major role in all fields of science and engineering.

FIGURE 15.1 shows position-versus-time graphs for two different oscillating systems. The shape of the graph depends on the details of the oscillator, but all oscillators have two things in common:

1. The oscillations take place around an equilibrium position.
2. The motion is *periodic,* repeating at regular intervals of time.

The time to complete one full cycle, or one oscillation, is called the **period** of the oscillation. Period is represented by the symbol *T*.

A system can oscillate in many ways, but the most fundamental oscillation is the smooth *sinusoidal* oscillation (i.e., like a sine or cosine) of Figure 15.1. This sinusoidal oscillation is called **simple harmonic motion,** abbreviated SHM. You'll learn in more advanced courses that *any* oscillation can be represented as a sum of sinusoidal oscillations, so SHM is the basis for understanding all oscillatory motion.

The prototype of simple harmonic motion is a mass oscillating on a spring. **FIGURE 15.2** shows an air-track glider attached to a spring. If the glider is pulled out a few centimeters and released, it oscillates back and forth. The graph shows an actual air-track measurement in which the glider's position was recorded 20 times per second. This is a position-versus-time graph that has been rotated 90° from its usual orientation to match the motion of the glider. You can see that it's a sinusoidal oscillation—simple harmonic motion.

As Figures 15.1 and 15.2 show, an oscillator moves back and forth between $x = -A$ and $x = +A$, where A, the **amplitude** of the motion, is the maximum displacement from equilibrium. Notice that the amplitude is the distance from the *axis* to a maximum or minimum, *not* the distance from the minimum to the maximum.

Period and amplitude are two important characteristics of oscillatory motion. A third is the **frequency,** *f*, which is the number of cycles or oscillations completed per second. If one cycle takes *T* seconds, the oscillator can complete $1/T$ cycles each second. That is, period and frequency are inverses of each other:

$$f = \frac{1}{T} \quad \text{or} \quad T = \frac{1}{f} \tag{15.1}$$

The units of frequency are **hertz,** abbreviated Hz, named in honor of the German physicist Heinrich Hertz, who produced the first artificially generated radio waves in 1887. By definition,

$$1 \text{ Hz} \equiv 1 \text{ cycle per second} = 1 \text{ s}^{-1}$$

We will often deal with very rapid oscillations and make use of the units shown in **TABLE 15.1**. For example, electrons oscillating back and forth at 101 MHz in an FM radio circuit have an oscillation period $T = 1/(101 \times 10^6 \text{ Hz}) = 9.9 \times 10^{-9} \text{ s} = 9.9 \text{ ns}$.

NOTE Uppercase and lowercase letters *are* important. 1 MHz is 1 megahertz = 10^6 Hz, but 1 mHz is 1 millihertz = 10^{-3} Hz!

Kinematics of Simple Harmonic Motion

We'll start by *describing* simple harmonic motion mathematically—that is, with kinematics. Then in Section 15.4 we'll take up the dynamics of how forces *cause* simple harmonic motion.

FIGURE 15.1 Examples of oscillatory motion.

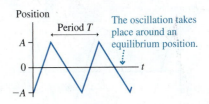
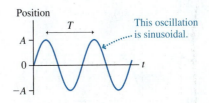

FIGURE 15.2 A prototype simple-harmonic-motion experiment.

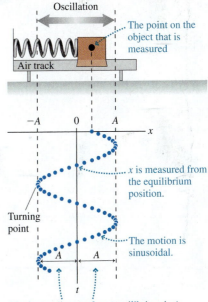

TABLE 15.1 Units of frequency

Frequency	Period
10^3 Hz = 1 kilohertz = 1 kHz	1 ms
10^6 Hz = 1 megahertz = 1 MHz	1 μs
10^9 Hz = 1 gigahertz = 1 GHz	1 ns

FIGURE 15.3 Position and velocity graphs for simple harmonic motion.

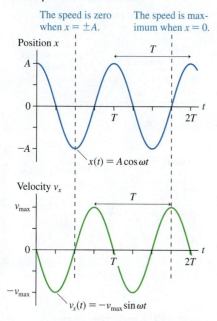

The speed is zero when $x = \pm A$.

The speed is maximum when $x = 0$.

$x(t) = A\cos\omega t$

$v_x(t) = -v_{max}\sin\omega t$

FIGURE 15.3 shows a SHM position graph—such as the one generated by the air-track glider—in its "normal" position. For the moment we'll assume that the oscillator starts at maximum displacement ($x = +A$) at $t = 0$. Also shown is the oscillator's velocity-versus-time graph, which we can deduce from the slope of the position graph.

■ The instantaneous velocity is zero at the instants when $x = \pm A$ because the slope of the position graph is zero. These are the *turning points* in the motion.

■ The position graph has maximum slope when $x = 0$, so these are points of maximum speed. When $x = 0$ with a positive slope—maximum speed to the right—the instantaneous velocity is $v_x = +v_{max}$, where v_{max} is the amplitude of the velocity curve. Similarly, $v_x = -v_{max}$ when $x = 0$ with a negative slope—maximum speed to the left.

Although these are empirical observations (we don't yet have any "theory" of oscillation), we can see that the position graph, with a maximum at $t = 0$, is a cosine function with amplitude A and period T. We can write this as

$$x(t) = A\cos\left(\frac{2\pi t}{T}\right) \qquad (15.2)$$

where the notation $x(t)$ indicates that x is a *function* of time t.

NOTE The arguments of sine and cosine functions are in *radians*. This will be true throughout our study of oscillations and waves. Be sure to set your calculator to radian mode before doing calculations.

Because $\cos(0 \text{ rad}) = \cos(2\pi \text{ rad}) = 1$, we can see that $x = A$ at $t = 0$ and again at $t = T$. In other words, this is a cosine function with amplitude A and period T. Notice that x passes through zero at $t = \frac{1}{4}T$ and $t = \frac{3}{4}T$ because $\cos\left(\frac{1}{2}\pi\right) = \cos\left(\frac{3}{2}\pi\right) = 0$.

We can write Equation 15.2 in two alternative forms. First, because the oscillation frequency is $f = 1/T$, we can write

$$x(t) = A\cos(2\pi ft) \qquad (15.3)$$

Second, recall from Chapter 4 that a particle in circular motion has *angular velocity* ω that is related to the period by $\omega = 2\pi/T$, where ω is in rad/s. For oscillations and waves, ω is called the **angular frequency.** We can write the position in terms of ω as

$$x(t) = A\cos(\omega t) \qquad (15.4)$$

Most of our work with oscillations and waves will be in terms of the angular frequency.

Now that we've defined $f = 1/T$, we see that ω, f, and T are related by

$$\omega \text{ (in rad/s)} = \frac{2\pi}{T} = 2\pi f \text{ (in Hz)} \qquad (15.5)$$

Be careful! Both f and ω are frequencies, but they're not the same and they're not interchangeable. Frequency f is the true frequency, in cycles per second, and it's always measured in Hz. Angular frequency ω is useful because it's related to the angle of the cosine function and, as you'll learn in the next section, to a circular-motion analog of SHM, but it's always in rad/s.

Just as the position graph is a cosine function, you can see that the velocity graph in Figure 15.3 is an "upside-down" sine function with the same period. We can write the velocity function as

$$v_x(t) = -v_{max}\sin\left(\frac{2\pi t}{T}\right) \qquad (15.6)$$

where the minus sign inverts the graph. This function is zero at $t = 0$ and again at $t = T$.

NOTE v_{max} is the maximum *speed* and thus is a *positive* number.

We deduced Equation 15.6 from the experimental results, but we could equally well find it from the position function of Equation 15.2. After all, velocity is the time derivative of position. TABLE 15.2 reminds you of the derivatives of the sine and cosine functions. Using the derivative of the position function, we find

$$v_x(t) = \frac{dx}{dt} = -\frac{2\pi A}{T}\sin\left(\frac{2\pi t}{T}\right) = -2\pi fA\sin(2\pi ft) = -\omega A\sin\omega t \quad (15.7)$$

Comparing Equation 15.7, the mathematical definition of velocity, to Equation 15.6, the empirical description, we see that the maximum speed of an oscillation is

$$v_{max} = \frac{2\pi A}{T} = 2\pi fA = \omega A \quad (15.8)$$

Not surprisingly, the object has a greater maximum speed if you stretch the spring farther and give the oscillation a larger amplitude.

TABLE 15.2 Derivatives of sine and cosine functions

$$\frac{d}{dt}(a\sin(bt+c)) = +ab\cos(bt+c)$$

$$\frac{d}{dt}(a\cos(bt+c)) = -ab\sin(bt+c)$$

EXAMPLE 15.1 | **A system in simple harmonic motion**

An air-track glider is attached to a spring, pulled 20.0 cm to the right, and released at $t = 0$ s. It makes 15 oscillations in 10.0 s.

a. What is the period of oscillation?
b. What is the object's maximum speed?
c. What are the position and velocity at $t = 0.800$ s?

MODEL An object oscillating on a spring is in SHM.

SOLVE a. The oscillation frequency is

$$f = \frac{15 \text{ oscillations}}{10.0 \text{ s}} = 1.50 \text{ oscillations/s} = 1.50 \text{ Hz}$$

Thus the period is $T = 1/f = 0.667$ s.
b. The oscillation amplitude is $A = 0.200$ m. Thus

$$v_{max} = \frac{2\pi A}{T} = \frac{2\pi(0.200 \text{ m})}{0.667 \text{ s}} = 1.88 \text{ m/s}$$

c. The object starts at $x = +A$ at $t = 0$ s. This is exactly the oscillation described by Equations 15.2 and 15.6. The position at $t = 0.800$ s is

$$x = A\cos\left(\frac{2\pi t}{T}\right) = (0.200 \text{ m})\cos\left(\frac{2\pi(0.800 \text{ s})}{0.667 \text{ s}}\right)$$

$$= (0.200 \text{ m})\cos(7.54 \text{ rad}) = 0.0625 \text{ m} = 6.25 \text{ cm}$$

The velocity at this instant of time is

$$v_x = -v_{max}\sin\left(\frac{2\pi t}{T}\right) = -(1.88 \text{ m/s})\sin\left(\frac{2\pi(0.800 \text{ s})}{0.667 \text{ s}}\right)$$

$$= -(1.88 \text{ m/s})\sin(7.54 \text{ rad}) = -1.79 \text{ m/s} = -179 \text{ cm/s}$$

At $t = 0.800$ s, which is slightly more than one period, the object is 6.25 cm to the right of equilibrium and moving to the *left* at 179 cm/s. Notice the use of radians in the calculations.

EXAMPLE 15.2 | **Finding the time**

A mass oscillating in simple harmonic motion starts at $x = A$ and has period T. At what time, as a fraction of T, does the object first pass through $x = \frac{1}{2}A$?

SOLVE Figure 15.3 showed that the object passes through the equilibrium position $x = 0$ at $t = \frac{1}{4}T$. This is one-quarter of the total distance in one-quarter of a period. You might expect it to take $\frac{1}{8}T$ to reach $\frac{1}{2}A$, but this is not the case because the SHM graph is not linear between $x = A$ and $x = 0$. We need to use $x(t) = A\cos(2\pi t/T)$. First, we write the equation with $x = \frac{1}{2}A$:

$$x = \frac{A}{2} = A\cos\left(\frac{2\pi t}{T}\right)$$

Then we solve for the time at which this position is reached:

$$t = \frac{T}{2\pi}\cos^{-1}\left(\frac{1}{2}\right) = \frac{T}{2\pi}\frac{\pi}{3} = \frac{1}{6}T$$

ASSESS The motion is slow at the beginning and then speeds up, so it takes longer to move from $x = A$ to $x = \frac{1}{2}A$ than it does to move from $x = \frac{1}{2}A$ to $x = 0$. Notice that the answer is independent of the amplitude A.

STOP TO THINK 15.1 An object moves with simple harmonic motion. If the amplitude and the period are both doubled, the object's maximum speed is

a. Quadrupled.   b. Doubled.   c. Unchanged.
d. Halved.   e. Quartered.

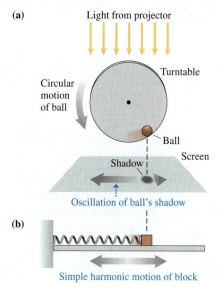

FIGURE 15.4 A projection of the circular motion of a rotating ball matches the simple harmonic motion of an object on a spring.

(a)

Light from projector

Turntable

Circular motion of ball

Ball

Shadow

Screen

Oscillation of ball's shadow

(b)

Simple harmonic motion of block

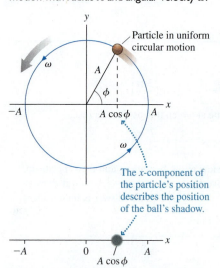

FIGURE 15.5 A particle in uniform circular motion with radius A and angular velocity ω.

Particle in uniform circular motion

ω

A

ϕ

$-A$ $\quad A \cos \phi$ $\quad A$ $\quad x$

ω

The x-component of the particle's position describes the position of the ball's shadow.

$-A$ $\quad 0$ $\quad A$ $\quad x$

$A \cos \phi$

15.2 SHM and Circular Motion

The graphs of Figure 15.3 and the position function $x(t) = A \cos \omega t$ are for an oscillation in which the object just happened to be at $x_0 = A$ at $t = 0$. But you will recall that $t = 0$ is an arbitrary choice, the instant of time when you or someone else starts a stopwatch. What if you had started the stopwatch when the object was at $x_0 = -A$, or when the object was somewhere in the middle of an oscillation? In other words, what if the oscillator had different *initial conditions?* The position graph would still show an oscillation, but neither Figure 15.3 nor $x(t) = A \cos \omega t$ would describe the motion correctly.

To learn how to describe the oscillation for other initial conditions it will help to turn to a topic you studied in Chapter 4—circular motion. There's a very close connection between simple harmonic motion and circular motion.

Imagine you have a turntable with a small ball glued to the edge. **FIGURE 15.4a** shows how to make a "shadow movie" of the ball by projecting a light past the ball and onto a screen. The ball's shadow oscillates back and forth as the turntable rotates. This is certainly periodic motion, with the same period as the turntable, but is it simple harmonic motion?

To find out, you could place a real object on a real spring directly below the shadow, as shown in **FIGURE 15.4b**. If you did so, and if you adjusted the turntable to have the same period as the spring, you would find that the shadow's motion exactly matches the simple harmonic motion of the object on the spring. **Uniform circular motion projected onto one dimension is simple harmonic motion.**

To understand this, consider the particle in **FIGURE 15.5**. It is in uniform circular motion, moving *counterclockwise* in a circle with radius A. As in Chapter 4, we can locate the particle by the angle ϕ measured counterclockwise (ccw) from the x-axis. Projecting the ball's shadow onto a screen in Figure 15.4 is equivalent to observing just the x-component of the particle's motion. Figure 15.5 shows that the x-component, when the particle is at angle ϕ, is

$$x = A \cos \phi \tag{15.9}$$

Recall that the particle's *angular velocity,* in rad/s, is

$$\omega = \frac{d\phi}{dt} \tag{15.10}$$

This is the rate at which the angle ϕ is increasing. If the particle starts from $\phi_0 = 0$ at $t = 0$, its angle at a later time t is simply

$$\phi = \omega t \tag{15.11}$$

As ϕ increases, the particle's x-component is

$$x(t) = A \cos \omega t \tag{15.12}$$

This is identical to Equation 15.4 for the position of a mass on a spring! Thus the x-component of a particle in uniform circular motion is simple harmonic motion.

NOTE When used to describe oscillatory motion, ω is called the *angular frequency* rather than the angular velocity. The angular frequency of an oscillator has the same numerical value, in rad/s, as the angular velocity of the corresponding particle in circular motion.

The names and units can be a bit confusing until you get used to them. It may help to notice that *cycle* and *oscillation* are not true units. Unlike the "standard meter" or the "standard kilogram," to which you could compare a length or a mass, there is no "standard cycle" to which you can compare an oscillation. Cycles and oscillations are simply counted events. Thus the frequency f has units of hertz, where $1 \text{ Hz} = 1 \text{ s}^{-1}$. We may *say* "cycles per second" to be clear, but the actual units are only "per second."

The radian is the SI unit of angle. However, the radian is a *defined* unit. Further, its definition as a ratio of two lengths ($\theta = s/r$) makes it a pure number without dimensions. As we noted in Chapter 4, the unit of angle, be it radians or degrees, is really just a *name* to remind us that we're dealing with an angle. The 2π in the equation $\omega = 2\pi f$ (and in similar situations), which is stated without units, *means* 2π rad/cycle. When multiplied by the frequency f in cycles/s, it gives the frequency in rad/s. That is why, in this context, ω is called the angular *frequency*.

NOTE *Hertz* is specifically "cycles per second" or "oscillations per second." It is used for f but *not* for ω. We'll always be careful to use rad/s for ω, but you should be aware that many books give the units of ω as simply s^{-1}.

Initial Conditions: The Phase Constant

Now we're ready to consider the issue of other initial conditions. The particle in Figure 15.5 started at $\phi_0 = 0$. This was equivalent to an oscillator starting at the far right edge, $x_0 = A$. **FIGURE 15.6** shows a more general situation in which the initial angle ϕ_0 can have any value. The angle at a later time t is then

$$\phi = \omega t + \phi_0 \tag{15.13}$$

In this case, the particle's projection onto the x-axis at time t is

$$x(t) = A\cos(\omega t + \phi_0) \tag{15.14}$$

If Equation 15.14 describes the particle's projection, then it must also be the position of an oscillator in simple harmonic motion. The oscillator's velocity v_x is found by taking the derivative dx/dt. The resulting equations,

$$x(t) = A\cos(\omega t + \phi_0)$$
$$v_x(t) = -\omega A\sin(\omega t + \phi_0) = -v_{max}\sin(\omega t + \phi_0) \tag{15.15}$$

are the two primary kinematic equations of simple harmonic motion.

The quantity $\phi = \omega t + \phi_0$, which steadily increases with time, is called the **phase** of the oscillation. The phase is simply the *angle* of the circular-motion particle whose shadow matches the oscillator. The constant ϕ_0 is called the **phase constant.** It is determined by the *initial conditions* of the oscillator.

To see what the phase constant means, set $t = 0$ in Equations 15.15:

$$x_0 = A\cos\phi_0$$
$$v_{0x} = -\omega A\sin\phi_0 \tag{15.16}$$

The position x_0 and velocity v_{0x} at $t = 0$ are the initial conditions. **Different values of the phase constant correspond to different starting points on the circle and thus to different initial conditions.**

The cosine function of Figure 15.3 and the equation $x(t) = A\cos\omega t$ are for an oscillation with $\phi_0 = 0$ rad. You can see from Equations 15.16 that $\phi_0 = 0$ rad implies $x_0 = A$ and $v_0 = 0$. That is, the particle starts from rest at the point of maximum displacement.

FIGURE 15.7, on the next page, illustrates these ideas by looking at three values of the phase constant: $\phi_0 = \pi/3$ rad $(60°)$, $-\pi/3$ rad $(-60°)$, and π rad $(180°)$. Notice that $\phi_0 = \pi/3$ rad and $\phi_0 = -\pi/3$ rad have the same starting position, $x_0 = \frac{1}{2}A$. This is a property of the cosine function in Equation 15.16. But these are *not* the same initial conditions. In one case the oscillator starts at $\frac{1}{2}A$ while moving to the left, in the other case it starts at $\frac{1}{2}A$ while moving to the right. You can distinguish between the two by visualizing the motion.

All values of the phase constant ϕ_0 between 0 and π rad correspond to a particle in the upper half of the circle and *moving to the left.* Thus v_{0x} is negative. All values of the phase constant ϕ_0 between π and 2π rad (or, as they are usually stated, between $-\pi$ and 0 rad) have the particle in the lower half of the circle and *moving to the right.* Thus v_{0x} is positive. If you're told that the oscillator is at $x = \frac{1}{2}A$ and moving to the right at $t = 0$, then the phase constant must be $\phi_0 = -\pi/3$ rad, not $+\pi/3$ rad.

FIGURE 15.6 A particle in uniform circular motion with initial angle ϕ_0.

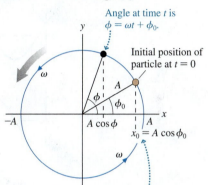

FIGURE 15.6 A particle in uniform circular motion with initial angle ϕ_0.

Angle at time t is $\phi = \omega t + \phi_0$.

Initial position of particle at $t = 0$

The initial x-component of the particle's position can be anywhere between $-A$ and A, depending on ϕ_0.

FIGURE 15.7 Different initial conditions are described by different values of the phase constant.

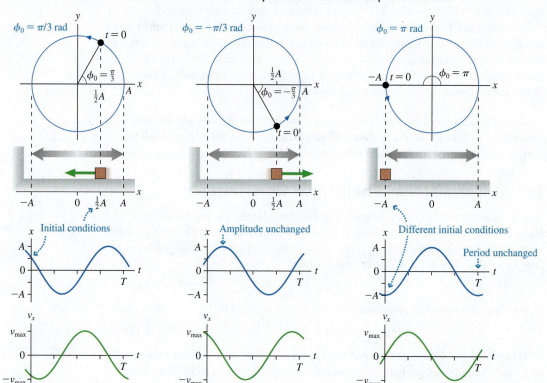

EXAMPLE 15.3 | Using the initial conditions

An object on a spring oscillates with a period of 0.80 s and an amplitude of 10 cm. At $t = 0$ s, it is 5.0 cm to the left of equilibrium and moving to the left. What are its position and direction of motion at $t = 2.0$ s?

MODEL An object oscillating on a spring is in simple harmonic motion.

SOLVE We can find the phase constant ϕ_0 from the initial condition $x_0 = -5.0$ cm $= A \cos \phi_0$. This condition gives

$$\phi_0 = \cos^{-1}\left(\frac{x_0}{A}\right) = \cos^{-1}\left(-\frac{1}{2}\right) = \pm \frac{2}{3}\pi \text{ rad} = \pm 120°$$

Because the oscillator is moving to the *left* at $t = 0$, it is in the upper half of the circular-motion diagram and must have a phase constant between 0 and π rad. Thus ϕ_0 is $\frac{2}{3}\pi$ rad. The angular frequency is

$$\omega = \frac{2\pi}{T} = \frac{2\pi}{0.80 \text{ s}} = 7.85 \text{ rad/s}$$

Thus the object's position at time $t = 2.0$ s is

$$\begin{aligned}
x(t) &= A \cos(\omega t + \phi_0) \\
&= (10 \text{ cm}) \cos\left((7.85 \text{ rad/s})(2.0 \text{ s}) + \frac{2}{3}\pi\right) \\
&= (10 \text{ cm}) \cos(17.8 \text{ rad}) = 5.0 \text{ cm}
\end{aligned}$$

The object is now 5.0 cm to the right of equilibrium. But which way is it moving? There are two ways to find out. The direct way is to calculate the velocity at $t = 2.0$ s:

$$v_x = -\omega A \sin(\omega t + \phi_0) = +68 \text{ cm/s}$$

The velocity is positive, so the motion is to the right. Alternatively, we could note that the phase at $t = 2.0$ s is $\phi = 17.8$ rad. Dividing by π, you can see that

$$\phi = 17.8 \text{ rad} = 5.67\pi \text{ rad} = (4\pi + 1.67\pi) \text{ rad}$$

The 4π rad represents two complete revolutions. The "extra" phase of 1.67π rad falls between π and 2π rad, so the particle in the circular-motion diagram is in the lower half of the circle and moving to the right.

NOTE The inverse-cosine function $\cos^{-1}$ is a *two-valued* function. Your calculator returns a single value, an angle between 0 rad and π rad. But the negative of this angle is also a solution. As Example 15.3 demonstrates, you must use additional information to choose between them.

STOP TO THINK 15.2 The figure shows four oscillators at $t = 0$. Which one has the phase constant $\phi_0 = \pi/4$ rad?

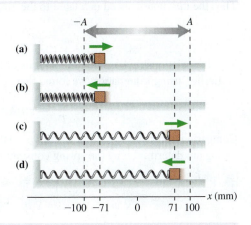

15.3 Energy in SHM

We've begun to develop the mathematical language of simple harmonic motion, but thus far we haven't included any physics. We've made no mention of the mass of the object or the spring constant of the spring. An energy analysis, using the tools of Chapter 10, is a good starting place.

FIGURE 15.8 shows an object oscillating on a spring, our prototype of simple harmonic motion. Now we'll specify that the object has mass m, the spring has spring constant k, and the motion takes place on a frictionless surface. You learned in Chapter 10 that the elastic potential energy when the object is at position x is $U_{Sp} = \frac{1}{2}k(\Delta x)^2$, where $\Delta x = x - x_{eq}$ is the displacement from the equilibrium position x_{eq}. In this chapter we'll always use a coordinate system in which $x_{eq} = 0$, making $\Delta x = x$. There's no chance for confusion with gravitational potential energy, so we can omit the subscript Sp and write the elastic potential energy as

$$U = \tfrac{1}{2}kx^2 \tag{15.17}$$

Thus the mechanical energy of an object oscillating on a spring is

$$E = K + U = \tfrac{1}{2}mv^2 + \tfrac{1}{2}kx^2 \tag{15.18}$$

The lower portion of Figure 15.8 is an energy diagram, showing the parabolic potential-energy curve $U = \frac{1}{2}kx^2$ and the kinetic energy $K = E - U$. Recall that a particle oscillates between the *turning points* where the total energy line E crosses the potential-energy curve. The left turning point is at $x = -A$, and the right turning point is at $x = +A$. To go beyond these points would require a negative kinetic energy, which is physically impossible.

You can see that **the particle has purely potential energy at $x = \pm A$ and purely kinetic energy as it passes through the equilibrium point at $x = 0$.** At maximum displacement, with $x = \pm A$ and $v = 0$, the energy is

$$E(\text{at } x = \pm A) = U = \tfrac{1}{2}kA^2 \tag{15.19}$$

At $x = 0$, where $v = \pm v_{max}$, the energy is

$$E(\text{at } x = 0) = K = \tfrac{1}{2}m(v_{max})^2 \tag{15.20}$$

The system's mechanical energy is conserved because the surface is frictionless and there are no external forces, so the energy at maximum displacement and the energy at maximum speed, Equations 15.19 and 15.20, must be equal. That is

$$\tfrac{1}{2}m(v_{max})^2 = \tfrac{1}{2}kA^2 \tag{15.21}$$

FIGURE 15.8 Energy transformations during SHM.

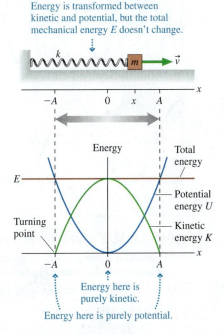

Energy is transformed between kinetic and potential, but the total mechanical energy E doesn't change.

Thus the maximum speed is related to the amplitude by

$$v_{max} = \sqrt{\frac{k}{m}}\,A \qquad (15.22)$$

This is a relationship based on the physics of the situation.

Earlier, using kinematics, we found that

$$v_{max} = \frac{2\pi A}{T} = 2\pi f A = \omega A \qquad (15.23)$$

Comparing Equations 15.22 and 15.23, we see that frequency and period of an oscillating spring are determined by the spring constant k and the object's mass m:

$$\omega = \sqrt{\frac{k}{m}} \qquad f = \frac{1}{2\pi}\sqrt{\frac{k}{m}} \qquad T = 2\pi\sqrt{\frac{m}{k}} \qquad (15.24)$$

These three expressions are really only one equation. They say the same thing, but each expresses it in slightly different terms.

Equations 15.24 tell us that the period and frequency are related to the object's mass m and the spring constant k. It is perhaps surprising, but **the period and frequency do not depend on the amplitude A**. A small oscillation and a large oscillation have the same period.

Conservation of Energy

Because energy is conserved, we can combine Equations 15.18, 15.19, and 15.20 to write

$$E = \tfrac{1}{2}mv^2 + \tfrac{1}{2}kx^2 = \tfrac{1}{2}kA^2 = \tfrac{1}{2}m(v_{max})^2 \quad \text{(conservation of energy)} \qquad (15.25)$$

Any pair of these expressions may be useful, depending on the known information. For example, you can use the amplitude A to find the speed at any point x by combining the first and second expressions for E. The speed v at position x is

$$v = \sqrt{\frac{k}{m}(A^2 - x^2)} = \omega\sqrt{A^2 - x^2} \qquad (15.26)$$

FIGURE 15.9 shows graphically how the kinetic and potential energy change with time. They both oscillate but remain *positive* because x and v are squared. Energy is continuously being transformed back and forth between the kinetic energy of the moving block and the stored potential energy of the spring, but their sum remains constant. Notice that K and U both oscillate *twice* each period; make sure you understand why.

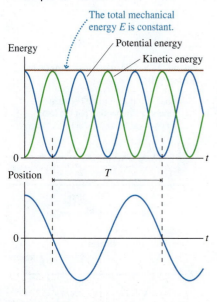

FIGURE 15.9 Kinetic energy, potential energy, and the total mechanical energy for simple harmonic motion.

The total mechanical energy E is constant.

Energy
Potential energy
Kinetic energy

0   t

Position T

0   t

EXAMPLE 15.4 | **Using conservation of energy**

A 500 g block on a spring is pulled a distance of 20 cm and released. The subsequent oscillations are measured to have a period of 0.80 s.

a. At what position or positions is the block's speed 1.0 m/s?

b. What is the spring constant?

MODEL The motion is SHM. Energy is conserved.

SOLVE a. The block starts from the point of maximum displacement, where $E = U = \tfrac{1}{2}kA^2$. At a later time, when the position is x and the speed is v, energy conservation requires

$$\tfrac{1}{2}mv^2 + \tfrac{1}{2}kx^2 = \tfrac{1}{2}kA^2$$

Solving for x, we find

$$x = \sqrt{A^2 - \frac{mv^2}{k}} = \sqrt{A^2 - \left(\frac{v}{\omega}\right)^2}$$

where we used $k/m = \omega^2$ from Equation 15.24. The angular frequency is easily found from the period: $\omega = 2\pi/T = 7.85$ rad/s. Thus

$$x = \sqrt{(0.20\text{ m})^2 - \left(\frac{1.0\text{ m/s}}{7.85\text{ rad/s}}\right)^2} = \pm 0.15\text{ m} = \pm 15\text{ cm}$$

There are two positions because the block has this speed on either side of equilibrium.

b. Although part a did not require that we know the spring constant, it is straightforward to find from Equation 15.24:

$$T = 2\pi\sqrt{\frac{m}{k}}$$

$$k = \frac{4\pi^2 m}{T^2} = \frac{4\pi^2(0.50\text{ kg})}{(0.80\text{ s})^2} = 31\text{ N/m}$$

STOP TO THINK 15.3 The four springs shown here have been compressed from their equilibrium position at $x = 0$ cm. When released, the attached mass will start to oscillate. Rank in order, from highest to lowest, the maximum speeds of the masses.

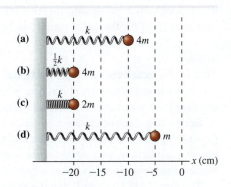

15.4 The Dynamics of SHM

Our analysis thus far has been based on the experimental observation that the oscillation of a spring "looks" sinusoidal. It's time to look at force and acceleration and to see that Newton's second law *predicts* sinusoidal motion.

A motion diagram will help us visualize the object's acceleration. **FIGURE 15.10** shows one cycle of the motion, separating motion to the left and motion to the right to make the diagram clear. As you can see, the object's velocity is large as it passes through the equilibrium point at $x = 0$, but $\vec{v}$ is *not changing* at that point. Acceleration measures the *change* of the velocity; hence $\vec{a} = \vec{0}$ at $x = 0$.

FIGURE 15.10 Motion diagram of simple harmonic motion. The left and right motions are separated vertically for clarity but really occur along the same line.

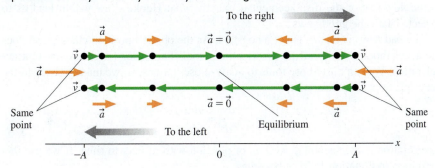

In contrast, the velocity is changing rapidly at the turning points. At the right turning point, $\vec{v}$ changes from a right-pointing vector to a left-pointing vector. Thus the acceleration $\vec{a}$ at the right turning point is large and *to the left*. In one-dimensional motion, the acceleration component a_x has a large *negative* value at the right turning point. Similarly, the acceleration $\vec{a}$ at the left turning point is large and *to the right*. Consequently, a_x has a large positive value at the left turning point.

Our motion-diagram analysis suggests that the acceleration a_x is most positive when the displacement is most negative, most negative when the displacement is a maximum, and zero when $x = 0$. This is confirmed by taking the derivative of the velocity:

$$a_x = \frac{dv_x}{dt} = \frac{d}{dt}(-\omega A \sin \omega t) = -\omega^2 A \cos \omega t \qquad (15.27)$$

then graphing it.

FIGURE 15.11 shows the position graph that we started with in Figure 15.3 and the corresponding acceleration graph. Comparing the two, you can see that the acceleration

FIGURE 15.11 Position and acceleration graphs for an oscillating spring. We've chosen $\phi_0 = 0$.

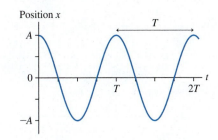

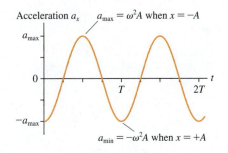

graph looks like an upside-down position graph. In fact, because $x = A\cos\omega t$, Equation 15.27 for the acceleration can be written

$$a_x = -\omega^2 x \qquad (15.28)$$

That is, **the acceleration is proportional to the negative of the displacement.** The acceleration is, indeed, most positive when the displacement is most negative and is most negative when the displacement is most positive.

Recall that the acceleration is related to the net force by Newton's second law. Consider again our prototype mass on a spring, shown in **FIGURE 15.12**. This is the simplest possible oscillation, with no distractions due to friction or gravitational forces. We will assume the spring itself to be massless.

You learned in Chapter 9 that the spring force is given by Hooke's law:

$$(F_{Sp})_x = -k\,\Delta x \qquad (15.29)$$

The minus sign indicates that the spring force is a **restoring force,** a force that always points back toward the equilibrium position. If we place the origin of the coordinate system at the equilibrium position, as we've done throughout this chapter, then $\Delta x = x$ and Hooke's law is simply $(F_{Sp})_x = -kx$.

The x-component of Newton's second law for the object attached to the spring is

$$(F_{net})_x = (F_{Sp})_x = -kx = ma_x \qquad (15.30)$$

Equation 15.30 is easily rearranged to read

$$a_x = -\frac{k}{m}x \qquad (15.31)$$

You can see that Equation 15.31 is identical to Equation 15.28 if the system oscillates with angular frequency $\omega = \sqrt{k/m}$. We previously found this expression for ω from an energy analysis. Our experimental observation that the acceleration is proportional to the *negative* of the displacement is exactly what Hooke's law would lead us to expect. That's the good news.

The bad news is that a_x is not a constant. As the object's position changes, so does the acceleration. Nearly all of our kinematic tools have been based on constant acceleration. We can't use those tools to analyze oscillations, so we must go back to the very definition of acceleration:

$$a_x = \frac{dv_x}{dt} = \frac{d^2x}{dt^2}$$

Acceleration is the second derivative of position with respect to time. If we use this definition in Equation 15.31, it becomes

$$\frac{d^2x}{dt^2} = -\frac{k}{m}x \quad \text{(equation of motion for a mass on a spring)} \qquad (15.32)$$

Equation 15.32, which is called the **equation of motion,** is a second-order differential equation. Unlike other equations we've dealt with, Equation 15.32 cannot be solved by direct integration. We'll need to take a different approach.

Solving the Equation of Motion

The solution to an algebraic equation such as $x^2 = 4$ is a number. The solution to a differential equation is a *function.* The x in Equation 15.32 is really $x(t)$, the position as a function of time. The solution to this equation is a function $x(t)$ whose second derivative is the function itself multiplied by $(-k/m)$.

One important property of differential equations that you will learn about in math is that the solutions are *unique.* That is, there is only *one* solution to Equation 15.32 that satisfies the initial conditions. If we were able to *guess* a solution, the uniqueness property would tell us that we had found the *only* solution. That might seem a rather

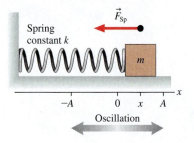

FIGURE 15.12 The prototype of simple harmonic motion: a mass oscillating on a horizontal spring without friction.

strange way to solve equations, but in fact differential equations are frequently solved by using your knowledge of what the solution needs to look like to guess an appropriate function. Let us give it a try!

We know from experimental evidence that the oscillatory motion of a spring appears to be sinusoidal. Let us *guess* that the solution to Equation 15.32 should have the functional form

$$x(t) = A\cos(\omega t + \phi_0) \tag{15.33}$$

where A, ω, and ϕ_0 are unspecified constants that we can adjust to any values that might be necessary to satisfy the differential equation.

If you were to guess that a solution to the algebraic equation $x^2 = 4$ is $x = 2$, you would verify your guess by substituting it into the original equation to see if it works. We need to do the same thing here: Substitute our guess for $x(t)$ into Equation 15.32 to see if, for an appropriate choice of the three constants, it works. To do so, we need the second derivative of $x(t)$. That is straightforward:

$$x(t) = A\cos(\omega t + \phi_0)$$

$$\frac{dx}{dt} = -\omega A \sin(\omega t + \phi_0) \tag{15.34}$$

$$\frac{d^2x}{dt^2} = -\omega^2 A \cos(\omega t + \phi_0)$$

If we now substitute the first and third of Equations 15.34 into Equation 15.32, we find

$$-\omega^2 A \cos(\omega t + \phi_0) = -\frac{k}{m}A\cos(\omega t + \phi_0) \tag{15.35}$$

Equation 15.35 will be true at all instants of time if and only if $\omega^2 = k/m$. There do not seem to be any restrictions on the two constants A and ϕ_0—they are determined by the initial conditions.

So we have found—by guessing!—that the solution to the equation of motion for a mass oscillating on a spring is

$$x(t) = A\cos(\omega t + \phi_0) \tag{15.36}$$

where the angular frequency

$$\omega = 2\pi f = \sqrt{\frac{k}{m}} \tag{15.37}$$

is determined by the mass and the spring constant.

NOTE Once again we see that the oscillation frequency is independent of the amplitude A.

Equations 15.36 and 15.37 seem somewhat anticlimactic because we've been using these results for the last several pages. But keep in mind that we had been *assuming* $x = A\cos\omega t$ simply because the experimental observations "looked" like a cosine function. We've now justified that assumption by showing that Equation 15.36 really is the solution to Newton's second law for a mass on a spring. **The *theory* of oscillation, based on Hooke's law for a spring and Newton's second law, is in good agreement with the experimental observations.**

An optical technique called *interferometry* reveals the bell-like vibrations of a wine glass.

EXAMPLE 15.5 | **Analyzing an oscillator**

At $t = 0$ s, a 500 g block oscillating on a spring is observed moving to the right at $x = 15$ cm. It reaches a maximum displacement of 25 cm at $t = 0.30$ s.

a. Draw a position-versus-time graph for one cycle of the motion.

b. What is the maximum force on the block, and what is the first time at which this occurs?

Continued

MODEL The motion is simple harmonic motion.

SOLVE a. The position equation of the block is $x(t) = A\cos(\omega t + \phi_0)$. We know that the amplitude is $A = 0.25$ m and that $x_0 = 0.15$ m. From these two pieces of information we obtain the phase constant:

$$\phi_0 = \cos^{-1}\left(\frac{x_0}{A}\right) = \cos^{-1}(0.60) = \pm 0.927 \text{ rad}$$

The object is initially moving to the right, which tells us that the phase constant must be between $-\pi$ and 0 rad. Thus $\phi_0 = -0.927$ rad. The block reaches its maximum displacement $x_{max} = A$ at time $t = 0.30$ s. At that instant of time

$$x_{max} = A = A\cos(\omega t + \phi_0)$$

This equation can be true only if $\cos(\omega t + \phi_0) = 1$, which requires $\omega t + \phi_0 = 0$. Thus

$$\omega = \frac{-\phi_0}{t} = \frac{-(-0.927 \text{ rad})}{0.30 \text{ s}} = 3.09 \text{ rad/s}$$

Now that we know ω, it is straightforward to compute the period:

$$T = \frac{2\pi}{\omega} = 2.0 \text{ s}$$

FIGURE 15.13 graphs $x(t) = (25 \text{ cm})\cos(3.09t - 0.927)$, where t is in s, from $t = 0$ s to $t = 2.0$ s.

FIGURE 15.13 Position-versus-time graph for the oscillator of Example 15.5.

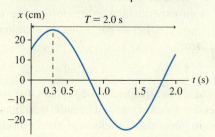

b. We found the acceleration of SHM to be $a_x = -\omega^2 x$, so the force on the block at position x is $F_x = -m\omega^2 x$. The force will be a maximum, $F_{max} = m\omega^2 A$, when x reaches its minimum displacement $x = -A$. For the block,

$$F_{max} = (0.50 \text{ kg})(3.09 \text{ rad/s})^2(0.25 \text{ m}) = 1.2 \text{ N}$$

This occurs exactly half a period (1.0 s) after the block reaches its maximum displacement, thus at $t = 1.3$ s.

ASSESS A 2 s period is a modest oscillation, so we don't expect the block's acceleration to be extreme. A maximum force of 1.2 N on a 0.5 kg block causes a maximum acceleration of 2.4 m/s², which seems reasonable.

STOP TO THINK 15.4 This is the position graph of a mass on a spring. What can you say about the velocity and the force at the instant indicated by the dashed line?

a. Velocity positive; force to the right.
b. Velocity negative; force to the right.
c. Velocity zero; force to the right.
d. Velocity positive; force to the left.
e. Velocity negative; force to the left.
f. Velocity zero; force to the left.
g. Velocity and force both zero.

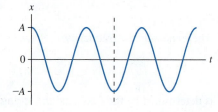

15.5 Vertical Oscillations

We have focused our analysis on a horizontally oscillating spring. But the typical demonstration you'll see in class is a mass bobbing up and down on a spring hung vertically from a support. Is it safe to assume that a vertical oscillation has the same mathematical description as a horizontal oscillation? Or does the additional force of gravity change the motion? Let us look at this more carefully.

FIGURE 15.14 shows a block of mass m hanging from a spring of spring constant k. An important fact to notice is that the equilibrium position of the block is *not* where the spring is at its unstretched length. At the equilibrium position of the block, where it hangs motionless, the spring has stretched by ΔL.

Finding ΔL is an equilibrium problem in which the upward spring force balances the downward gravitational force on the block. The y-component of the spring force is given by Hooke's law:

$$(F_{Sp})_y = -k\,\Delta y = +k\,\Delta L \tag{15.38}$$

FIGURE 15.14 Gravity stretches the spring.

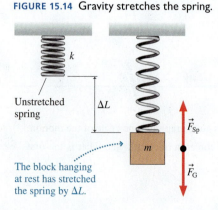

Unstretched spring

The block hanging at rest has stretched the spring by ΔL.

Equation 15.38 makes a distinction between ΔL, which is simply a *distance* and is a positive number, and the displacement Δy. The block is displaced downward, so $\Delta y = -\Delta L$. Newton's first law for the block in equilibrium is

$$(F_{net})_y = (F_{Sp})_y + (F_G)_y = k\,\Delta L - mg = 0 \qquad (15.39)$$

from which we can find

$$\Delta L = \frac{mg}{k} \qquad (15.40)$$

This is the distance the spring stretches when the block is attached to it.

Let the block oscillate around this equilibrium position, as shown in **FIGURE 15.15**. We've now placed the origin of the y-axis at the block's equilibrium position in order to be consistent with our analyses of oscillations throughout this chapter. If the block moves upward, as the figure shows, the spring gets shorter compared to its equilibrium length, but the spring is still *stretched* compared to its unstretched length in Figure 15.14. When the block is at position y, the spring is stretched by an amount $\Delta L - y$ and hence exerts an *upward* spring force $F_{Sp} = k(\Delta L - y)$. The net force on the block at this point is

$$(F_{net})_y = (F_{Sp})_y + (F_G)_y = k(\Delta L - y) - mg = (k\,\Delta L - mg) - ky \qquad (15.41)$$

But $k\,\Delta L - mg$ is zero, from Equation 15.40, so the net force on the block is simply

$$(F_{net})_y = -ky \qquad (15.42)$$

Equation 15.42 for vertical oscillations is *exactly* the same as Equation 15.30 for horizontal oscillations, where we found $(F_{net})_x = -kx$. That is, the restoring force for vertical oscillations is identical to the restoring force for horizontal oscillations. The role of gravity is to determine where the equilibrium position is, but it doesn't affect the oscillatory motion around the equilibrium position.

Because the net force is the same, Newton's second law has exactly the same oscillatory solution:

$$y(t) = A\cos(\omega t + \phi_0) \qquad (15.43)$$

with, again, $\omega = \sqrt{k/m}$. **The vertical oscillations of a mass on a spring are the same simple harmonic motion as those of a block on a horizontal spring.** This is an important finding because it was not obvious that the motion would still be simple harmonic motion when gravity was included.

FIGURE 15.15 The block oscillates around the equilibrium position.

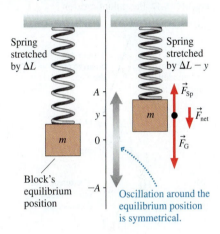

EXAMPLE 15.6 | Bungee oscillations

An 83 kg student hangs from a bungee cord with spring constant 270 N/m. The student is pulled down to a point where the cord is 5.0 m longer than its unstretched length, then released. Where is the student, and what is his velocity 2.0 s later?

MODEL A bungee cord can be modeled as a spring. Vertical oscillations on the bungee cord are SHM.

VISUALIZE FIGURE 15.16 shows the situation.

SOLVE Although the cord is stretched by 5.0 m when the student is released, this is *not* the amplitude of the oscillation. Oscillations occur around the equilibrium position, so we have to begin by finding the equilibrium point where the student hangs motionless. The cord stretch at equilibrium is given by Equation 15.40:

$$\Delta L = \frac{mg}{k} = 3.0 \text{ m}$$

Stretching the cord 5.0 m pulls the student 2.0 m below the equilibrium point, so $A = 2.0$ m. That is, the student oscillates with amplitude $A = 2.0$ m about a point 3.0 m beneath the bungee cord's

FIGURE 15.16 A student on a bungee cord oscillates about the equilibrium position.

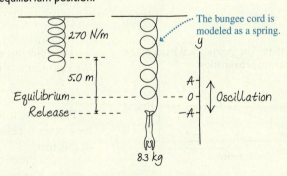

original end point. The student's position as a function of time, as measured from the equilibrium position, is

$$y(t) = (2.0 \text{ m})\cos(\omega t + \phi_0)$$

where $\omega = \sqrt{k/m} = 1.80$ rad/s.

Continued

The initial condition

$$y_0 = A \cos\phi_0 = -A$$

requires the phase constant to be $\phi_0 = \pi$ rad. At $t = 2.0$ s the student's position and velocity are

$$y = (2.0 \text{ m}) \cos\big((1.80 \text{ rad/s})(2.0 \text{ s}) + \pi \text{ rad}\big) = 1.8 \text{ m}$$

$$v_y = -\omega A \sin(\omega t + \phi_0) = -1.6 \text{ m/s}$$

The student is 1.8 m *above* the equilibrium position, or 1.2 m *below* the original end of the cord. Because his velocity is negative, he's passed through the highest point and is heading down.

15.6 The Pendulum

FIGURE 15.17 Pendulum motion

(a)

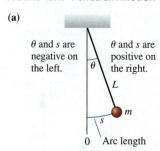

θ and s are negative on the left.

θ and s are positive on the right.

L

m

s

0   Arc length

(b)

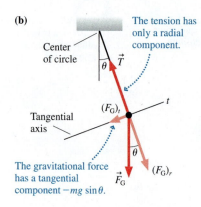

The tension has only a radial component.

Center of circle

θ $\vec{T}$

Tangential axis

$(F_G)_t$

t

θ

$(F_G)_r$

The gravitational force has a tangential component $-mg\sin\theta$.

$\vec{F}_G$

Now let's look at another very common oscillator: a pendulum. **FIGURE 15.17a** shows a mass m attached to a string of length L and free to swing back and forth. The pendulum's position can be described by the arc of length s, which is zero when the pendulum hangs straight down. Because angles are measured ccw, s and θ are positive when the pendulum is to the right of center, negative when it is to the left.

Two forces are acting on the mass: the string tension $\vec{T}$ and gravity $\vec{F}_G$. As we did with circular motion, it will be useful to divide the forces into tangential components, parallel to the motion, and radial components parallel to the string. These are shown on the free-body diagram of **FIGURE 15.17b**.

Newton's second law for the tangential component, parallel to the motion, is

$$(F_{\text{net}})_t = \sum F_t = (F_G)_t = -mg\sin\theta = ma_t \qquad (15.44)$$

Using $a_t = d^2s/dt^2$ for acceleration "around" the circle, and noting that the mass cancels, we can write Equation 15.44 as

$$\frac{d^2s}{dt^2} = -g\sin\theta \qquad (15.45)$$

This is the equation of motion for an oscillating pendulum. The sine function makes this equation more complicated than the equation of motion for an oscillating spring.

The Small-Angle Approximation

Suppose we restrict the pendulum's oscillations to *small angles* of less than about $10°$. This restriction allows us to make use of an interesting and important piece of geometry.

FIGURE 15.18 shows an angle θ and a circular arc of length $s = r\theta$. A right triangle has been constructed by dropping a perpendicular from the top of the arc to the axis. The height of the triangle is $h = r\sin\theta$. Suppose that the angle θ is "small," which, in practice, means $\theta \ll 1$ rad. In that case there is very little difference between h and s. If $h \approx s$, then $r\sin\theta \approx r\theta$. It follows that

$$\sin\theta \approx \theta \quad \text{if} \quad \theta \ll 1 \text{ rad}$$

The result that $\sin\theta \approx \theta$ for small angles is called the **small-angle approximation.** We can similarly note that $l \approx r$ for small angles. Because $l = r\cos\theta$, it follows that

$$\cos\theta \approx 1 \quad \text{if} \quad \theta \ll 1 \text{ rad}$$

Finally, we can take the ratio of sine and cosine to find $\tan\theta \approx \sin\theta \approx \theta$. We will have other occasions to use the small-angle approximation throughout the remainder of this text.

FIGURE 15.18 The geometrical basis of the small-angle approximation.

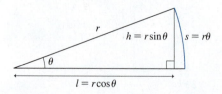

r

$h = r\sin\theta$ $s = r\theta$

θ

$l = r\cos\theta$

| **NOTE** The small-angle approximation is valid *only* if angle θ is in radians!

How small does θ have to be to justify using the small-angle approximation? It's easy to use your calculator to find that the small-angle approximation is good to three significant figures, an error of $\leq 0.1\%$, up to angles of ≈ 0.10 rad ($\approx 5°$). In practice, we will use the approximation up to about $10°$, but for angles any larger it rapidly loses validity and produces unacceptable results.

If we restrict the pendulum to $\theta < 10°$, we can use $\sin \theta \approx \theta$. In that case, Equation 15.44 for the net force on the mass is

$$(F_{net})_t = -mg \sin \theta \approx -mg\theta = -\frac{mg}{L}s$$

where, in the last step, we used the fact that angle θ is related to the arc length by $\theta = s/L$. Then the equation of motion becomes

$$\frac{d^2s}{dt^2} = \frac{(F_{net})_t}{m} = -\frac{g}{L}s \qquad (15.46)$$

This is *exactly* the same as Equation 15.32 for a mass oscillating on a spring. The names are different, with x replaced by s and k/m by g/L, but that does not make it a different equation.

Because we know the solution to the spring problem, we can immediately write the solution to the pendulum problem just by changing variables and constants:

$$s(t) = A\cos(\omega t + \phi_0) \quad \text{or} \quad \theta(t) = \theta_{max}\cos(\omega t + \phi_0) \qquad (15.47)$$

The angular frequency

$$\omega = 2\pi f = \sqrt{\frac{g}{L}} \qquad (15.48)$$

is determined by the length of the string. The pendulum is interesting in that **the frequency, and hence the period, is independent of the mass.** It depends only on the length of the pendulum. The amplitude A and the phase constant ϕ_0 are determined by the initial conditions, just as they were for an oscillating spring.

EXAMPLE 15.7 | The maximum angle of a pendulum

A 300 g mass on a 30-cm-long string oscillates as a pendulum. It has a speed of 0.25 m/s as it passes through the lowest point. What maximum angle does the pendulum reach?

MODEL Assume that the angle remains small, in which case the motion is simple harmonic motion.

SOLVE The angular frequency of the pendulum is

$$\omega = \sqrt{\frac{g}{L}} = \sqrt{\frac{9.8 \text{ m/s}^2}{0.30 \text{ m}}} = 5.72 \text{ rad/s}$$

The speed at the lowest point is $v_{max} = \omega A$, so the amplitude is

$$A = s_{max} = \frac{v_{max}}{\omega} = \frac{0.25 \text{ m/s}}{5.72 \text{ rad/s}} = 0.0437 \text{ m}$$

The maximum angle, at the maximum arc length s_{max}, is

$$\theta_{max} = \frac{s_{max}}{L} = \frac{0.0437 \text{ m}}{0.30 \text{ m}} = 0.146 \text{ rad} = 8.3°$$

ASSESS Because the maximum angle is less than 10°, our analysis based on the small-angle approximation is reasonable.

EXAMPLE 15.8 | The gravimeter

Deposits of minerals and ore can alter the local value of the free-fall acceleration because they tend to be denser than surrounding rocks. Geologists use a *gravimeter*—an instrument that accurately measures the local free-fall acceleration—to search for ore deposits. One of the simplest gravimeters is a pendulum. To achieve the highest accuracy, a stopwatch is used to time 100 oscillations of a pendulum of different lengths. At one location in the field, a geologist makes the following measurements:

Length (m)	Time (s)
0.500	141.7
1.000	200.6
1.500	245.8
2.000	283.5

What is the local value of g?

MODEL Assume the oscillation angle is small, in which case the motion is simple harmonic motion with a period independent of the mass of the pendulum. Because the data are known to four significant figures (± 1 mm on the length and ± 0.1 s on the timing, both of which are easily achievable), we expect to determine g to four significant figures.

SOLVE From Equation 15.48, using $f = 1/T$, we find

$$T^2 = \left(2\pi\sqrt{\frac{L}{g}}\right)^2 = \frac{4\pi^2}{g}L$$

That is, the square of a pendulum's period is proportional to its length. Consequently, a graph of T^2 versus L should be a straight line passing through the origin with slope $4\pi^2/g$. We can use the

Continued

experimentally measured slope to determine g. **FIGURE 15.19** is a graph of the data, with the period found by dividing the measured time by 100.

As expected, the graph is a straight line passing through the origin. The slope of the best-fit line is 4.021 s²/m. Consequently,

$$g = \frac{4\pi^2}{\text{slope}} = \frac{4\pi^2}{4.021 \text{ s}^2/\text{m}} = 9.818 \text{ m/s}^2$$

ASSESS The fact that the graph is linear and passes through the origin confirms our model of the situation. Had this *not* been the case, we would have had to conclude either that our model of the pendulum as a simple, small-angle pendulum was not valid or that our measurements were bad. This is an important reason for having multiple data points rather than using only one length.

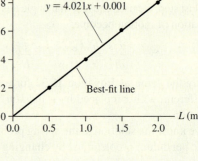

FIGURE 15.19 Graph of the square of the pendulum's period versus its length.

The Simple-Harmonic-Motion Model

You can begin to see how, in a sense, we have solved *all* simple-harmonic-motion problems once we have solved the problem of the horizontal spring. The restoring force of a spring, $F_{Sp} = -kx$, is directly proportional to the displacement x from equilibrium. The pendulum's restoring force, in the small-angle approximation, is directly proportional to the displacement s. A restoring force that is directly proportional to the displacement from equilibrium is called a **linear restoring force.** For *any* linear restoring force, the equation of motion is identical to the spring equation (other than perhaps using different symbols). Consequently, **any system with a linear restoring force will undergo simple harmonic motion around the equilibrium position**.

This is why an oscillating spring is the prototype of SHM. Everything that we learn about an oscillating spring can be applied to the oscillations of any other linear restoring force, ranging from the vibration of airplane wings to the motion of electrons in electric circuits.

MODEL 15.1

Simple harmonic motion

For any system with a restoring force that's linear or can be well approximated as linear.

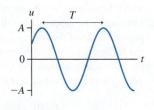

- Motion is SHM around the equilibrium position.
- Frequency and period are independent of the amplitude.
- Mathematically:
  - For an appropriate position variable u, the equation of motion can be written
  $$d^2u/dt^2 = -Cu$$
  where C is a collection of constants.
  - The angular frequency is $\omega = \sqrt{C}$.
  - The position and velocity are
  $$u = A\cos(\omega t + \phi_0) \quad v_u = -v_{max}\sin(\omega t + \phi_0)$$
  where A and ϕ_0 are determined by the initial conditions.
  - Mechanical energy is conserved.
- Limitations: Model fails if the restoring force deviates significantly from linear.

Exercise 22

The Physical Pendulum

A mass on a string is often called a *simple pendulum*. But you can also make a pendulum from any solid object that swings back and forth on a pivot under the influence of gravity. This is called a *physical pendulum*.

FIGURE 15.20 shows a physical pendulum of mass M for which the distance between the pivot and the center of mass is l. The moment arm of the gravitational force acting at the center of mass is $d = l\sin\theta$, so the gravitational torque is

$$\tau = -Mgd = -Mgl\sin\theta$$

The torque is negative because, for positive θ, it's causing a clockwise rotation. If we restrict the angle to being small ($\theta < 10°$), as we did for the simple pendulum, we can use the small-angle approximation to write

$$\tau = -Mgl\theta \qquad (15.49)$$

Gravity exerts a linear restoring torque on the pendulum—that is, the torque is directly proportional to the angular displacement θ—so we expect the physical pendulum to undergo SHM.

From Chapter 12, Newton's second law for rotational motion is

$$\alpha = \frac{d^2\theta}{dt^2} = \frac{\tau}{I}$$

where I is the object's moment of inertia about the pivot point. Using Equation 15.49 for the torque, we find

$$\frac{d^2\theta}{dt^2} = \frac{-Mgl}{I}\theta \qquad (15.50)$$

The equation of motion is of the form $d^2\theta/dt^2 = -C\theta$, so the model for simple harmonic motion tells us that the motion is SHM with angular frequency

$$\omega = 2\pi f = \sqrt{\frac{Mgl}{I}} \qquad (15.51)$$

It appears that the frequency depends on the mass of the pendulum, but recall that the moment of inertia is directly proportional to M. Thus M cancels and the frequency of a physical pendulum, like that of a simple pendulum, is independent of mass.

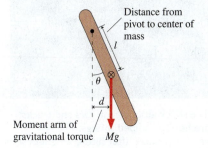

FIGURE 15.20 A physical pendulum.

Distance from pivot to center of mass

Moment arm of gravitational torque

EXAMPLE 15.9 | A swinging leg as a pendulum

A student in a biomechanics lab measures the length of his leg, from hip to heel, to be 0.90 m. What is the frequency of the pendulum motion of the student's leg? What is the period?

MODEL We can model a human leg reasonably well as a rod of uniform cross section, pivoted at one end (the hip) to form a physical pendulum. For small-angle oscillations it will undergo SHM. The center of mass of a uniform leg is at the midpoint, so $l = L/2$.

SOLVE The moment of inertia of a rod pivoted about one end is $I = \frac{1}{3}ML^2$, so the pendulum frequency is

$$f = \frac{1}{2\pi}\sqrt{\frac{Mgl}{I}} = \frac{1}{2\pi}\sqrt{\frac{Mg(L/2)}{ML^2/3}} = \frac{1}{2\pi}\sqrt{\frac{3g}{2L}} = 0.64 \text{ Hz}$$

The corresponding period is $T = 1/f = 1.6$ s. Notice that we didn't need to know the mass.

ASSESS As you walk, your legs do swing as physical pendulums as you bring them forward. The frequency is fixed by the length of your legs and their distribution of mass; it doesn't depend on amplitude. Consequently, you don't increase your walking speed by taking more rapid steps—changing the frequency is difficult. You simply take longer strides, changing the amplitude but not the frequency.

STOP TO THINK 15.5 One person swings on a swing and finds that the period is 3.0 s. A second person of equal mass joins him. With two people swinging, the period is

a. 6.0 s

b. >3.0 s but not necessarily 6.0 s

c. 3.0 s

d. <3.0 s but not necessarily 1.5 s

e. 1.5 s

f. Can't tell without knowing the length

The shock absorbers in cars and trucks are heavily damped springs. The vehicle's vertical motion, after hitting a rock or a pothole, is a damped oscillation.

15.7 Damped Oscillations

A pendulum left to itself gradually slows down and stops. The sound of a ringing bell gradually dies away. All real oscillators do run down—some very slowly but others quite quickly—as friction or other dissipative forces transform their mechanical energy into the thermal energy of the oscillator and its environment. An oscillation that runs down and stops is called a **damped oscillation.**

There are many possible reasons for the dissipation of energy, such as air resistance, friction, and internal forces within a metal spring as it flexes. The forces involved in dissipation are complex, but a simple *linear drag* model gives a quite accurate description of most damped oscillations. That is, we'll assume a drag force that depends linearly on the velocity as

$$\vec{F}_{drag} = -b\vec{v} \quad \text{(model of the drag force)} \tag{15.52}$$

where the minus sign is the mathematical statement that the force is always opposite in direction to the velocity in order to slow the object.

The **damping constant** b depends in a complicated way on the shape of the object *and* on the viscosity of the air or other medium in which the particle moves. The damping constant plays the same role in our model of drag that the coefficient of friction does in our model of friction.

The units of b need to be such that they will give units of force when multiplied by units of velocity. As you can confirm, these units are kg/s. A value $b = 0$ kg/s corresponds to the limiting case of no resistance, in which case the mechanical energy is conserved. A typical value of b for a spring or a pendulum in air is ≤ 0.10 kg/s. Objects moving in a liquid can have significantly larger values of b.

FIGURE 15.21 shows a mass oscillating on a spring in the presence of a drag force. With the drag included, Newton's second law is

$$(F_{net})_x = (F_{Sp})_x + (F_{drag})_x = -kx - bv_x = ma_x \tag{15.53}$$

Using $v_x = dx/dt$ and $a_x = d^2x/dt^2$, we can write Equation 15.53 as

$$\frac{d^2x}{dt^2} + \frac{b}{m}\frac{dx}{dt} + \frac{k}{m}x = 0 \tag{15.54}$$

Equation 15.54 is the equation of motion of a damped oscillator. If you compare it to Equation 15.32, the equation of motion for a block on a frictionless surface, you'll see that it differs by the inclusion of the term involving dx/dt.

Equation 15.54 is another second-order differential equation. We will simply assert (and, as a homework problem, you can confirm) that the solution is

$$x(t) = Ae^{-bt/2m}\cos(\omega t + \phi_0) \quad \text{(damped oscillator)} \tag{15.55}$$

where the angular frequency is given by

$$\omega = \sqrt{\frac{k}{m} - \frac{b^2}{4m^2}} = \sqrt{\omega_0^2 - \frac{b^2}{4m^2}} \tag{15.56}$$

Here $\omega_0 = \sqrt{k/m}$ is the angular frequency of an undamped oscillator ($b = 0$). The constant e is the base of natural logarithms, so $e^{-bt/2m}$ is an *exponential function.* Because $e^0 = 1$, Equation 15.55 reduces to our previous $x(t) = A\cos(\omega t + \phi_0)$ when $b = 0$. This makes sense and gives us confidence in Equation 15.55.

Lightly Damped Oscillators

A *lightly damped* oscillator, which oscillates many times before stopping, is one for which $b/2m \ll \omega_0$. In that case, $\omega \approx \omega_0$ is a good approximation. That is, light damping does not affect the oscillation frequency.

FIGURE 15.22 is a graph of the position $x(t)$ for a lightly damped oscillator, as given by Equation 15.55. To keep things simple, we've assumed that the phase constant is

FIGURE 15.21 An oscillating mass in the presence of a drag force.

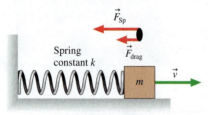

Spring constant k

$\vec{F}_{Sp}$

$\vec{F}_{drag}$

m

$\vec{v}$

FIGURE 15.22 Position-versus-time graph for a lightly damped oscillator.

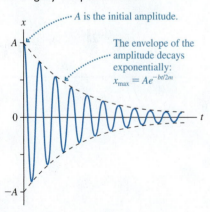

A is the initial amplitude.

The envelope of the amplitude decays exponentially:

$x_{max} = Ae^{-bt/2m}$

zero. Notice that the term $Ae^{-bt/2m}$, which is shown by the dashed line, acts as a slowly varying amplitude:

$$x_{\text{max}}(t) = Ae^{-bt/2m} \tag{15.57}$$

where A is the *initial* amplitude, at $t = 0$. The oscillation keeps bumping up against this line, slowly dying out with time.

A slowly changing line that provides a border to a rapid oscillation is called the **envelope** of the oscillations. In this case, the oscillations have an *exponentially decaying envelope*. Make sure you study Figure 15.22 long enough to see how both the oscillations and the decaying amplitude are related to Equation 15.55.

Changing the amount of damping, by changing the value of b, affects how quickly the oscillations decay. **FIGURE 15.23** shows just the envelope $x_{\text{max}}(t)$ for several oscillators that are identical except for the value of the damping constant b. (You need to imagine a rapid oscillation within each envelope, as in Figure 15.22.) Increasing b causes the oscillations to damp more quickly, while decreasing b makes them last longer.

FIGURE 15.23 Oscillation envelopes for several values of b, mass 1.0 kg.

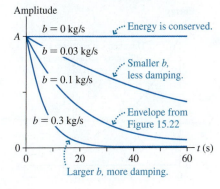

MATHEMATICAL ASIDE

Exponential decay

Exponential decay occurs in a vast number of physical systems of importance in science and engineering. Mechanical vibrations, electric circuits, and nuclear radioactivity all exhibit exponential decay.

The mathematical analysis of physical systems frequently leads to solutions of the form

$$u = Ae^{-v/v_0} = A\exp(-v/v_0)$$

where exp is the *exponential function*. The number $e = 2.71828\ldots$ is the base of natural logarithms in the same way that 10 is the base of ordinary logarithms.

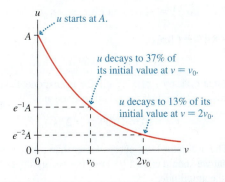

A graph of u illustrates what we mean by exponential decay. It starts with $u = A$ at $v = 0$ (because $e^0 = 1$) and then steadily decays, asymptotically approaching zero. The quantity v_0 is called the *decay constant*. When $v = v_0$, $u = e^{-1}A = 0.37A$. When $v = 2v_0$, $u = e^{-2}A = 0.13A$.

The decay constant v_0 must have the same units as v. If v represents position, then v_0 is a length; if v represents time, then v_0 is a time interval. In a specific situation, v_0 is often called the *decay length* or the *decay time*. It is the length or time in which the quantity decays to 37% of its initial value.

No matter what the process is or what u represents, **a quantity that decays exponentially decays to 37% of its initial value when one decay constant has passed.** Thus exponential decay is a universal behavior. The decay curve always looks exactly like the figure shown here. Once you've learned the properties of exponential decay, you'll immediately know how to apply this knowledge to a new situation.

The mechanical energy of a damped oscillator is *not* conserved because of the drag force. We previously found the energy of an undamped oscillator to be $E = \frac{1}{2}kA^2$. This is still valid for a lightly damped oscillator if we replace A with the slowly decaying amplitude x_{max}. Thus

$$E(t) = \frac{1}{2}k(x_{\text{max}})^2 = \frac{1}{2}k(Ae^{-bt/2m})^2 = \frac{1}{2}kA^2e^{-bt/m} \tag{15.58}$$

Here A is the initial amplitude, so $\frac{1}{2}kA^2$ is the initial energy, which we call E_0. Let's define the **time constant** τ (also called the *decay constant* or the *decay time*) to be

$$\tau = \frac{m}{b}. \tag{15.59}$$

FIGURE 15.24 Energy decay of a lightly damped oscillator.

Energy

E_0

The energy has decreased to 37% of its initial value at $t = \tau$.

$0.37E_0$

$0.13E_0$

0

$0 \qquad t = \tau \qquad t = 2\tau \qquad t$

Because b has units of kg/s, τ has units of seconds. With this, we can write the energy decay as

$$E(t) = E_0 e^{-t/\tau} \qquad (15.60)$$

In other words, **a lightly damped oscillator's mechanical energy decays exponentially with time constant τ.**

As **FIGURE 15.24** shows, the time constant is the amount of time needed for the energy to decay to e^{-1}, or 37%, of its initial value. We say that the time constant τ measures the "characteristic time" during which the energy of the oscillation is dissipated. Roughly two-thirds of the initial energy is gone after one time constant has elapsed, and nearly 90% has dissipated after two time constants have gone by.

For practical purposes, we can speak of the time constant as the *lifetime* of the oscillation—about how long it lasts. Mathematically, there is never a time when the oscillation is "over." The decay approaches zero asymptotically, but it never gets there in any finite time. The best we can do is define a characteristic time when the motion is "almost over," and that is what the time constant τ does.

EXAMPLE 15.10 | A damped pendulum

A 500 g mass swings on a 60-cm-string as a pendulum. The amplitude is observed to decay to half its initial value after 35 oscillations.

a. What is the time constant for this oscillator?

b. At what time will the *energy* have decayed to half its initial value?

MODEL The motion is a damped oscillation.

SOLVE a. The initial amplitude at $t = 0$ is $x_{max} = A$. After 35 oscillations the amplitude is $x_{max} = \frac{1}{2}A$. The period of the pendulum is

$$T = 2\pi \sqrt{\frac{L}{g}} = 2\pi \sqrt{\frac{0.60 \text{ m}}{9.8 \text{ m/s}^2}} = 1.55 \text{ s}$$

so 35 oscillations have occurred at $t = 54.2$ s.

The amplitude of oscillation at time t is given by Equation 15.57: $x_{max}(t) = Ae^{-bt/2m} = Ae^{-t/2\tau}$. In this case,

$$\tfrac{1}{2}A = Ae^{-(54.2 \text{ s})/2\tau}$$

Notice that we do not need to know A itself because it cancels out. To solve for τ, we take the natural logarithm of both sides of the equation:

$$\ln\left(\tfrac{1}{2}\right) = -\ln 2 = \ln e^{-(54.2 \text{ s})/2\tau} = -\frac{54.2 \text{ s}}{2\tau}$$

This is easily rearranged to give

$$\tau = \frac{54.2 \text{ s}}{2 \ln 2} = 39 \text{ s}$$

If desired, we could now determine the damping constant to be $b = m/\tau = 0.013$ kg/s.

b. The energy at time t is given by

$$E(t) = E_0 e^{-t/\tau}$$

The time at which an exponential decay is reduced to $\frac{1}{2}E_0$, half its initial value, has a special name. It is called the **half-life** and given the symbol $t_{1/2}$. The concept of the half-life is widely used in applications such as radioactive decay. To relate $t_{1/2}$ to τ, we first write

$$E(\text{at } t = t_{1/2}) = \tfrac{1}{2}E_0 = E_0 e^{-t_{1/2}/\tau}$$

The E_0 cancels, giving

$$\tfrac{1}{2} = e^{-t_{1/2}/\tau}$$

Again, we take the natural logarithm of both sides:

$$\ln\left(\tfrac{1}{2}\right) = -\ln 2 = \ln e^{-t_{1/2}/\tau} = -t_{1/2}/\tau$$

Finally, we solve for $t_{1/2}$:

$$t_{1/2} = \tau \ln 2 = 0.693\tau$$

This result that $t_{1/2}$ is 69% of τ is valid for any exponential decay. In this particular problem, half the energy is gone at

$$t_{1/2} = (0.693)(39 \text{ s}) = 27 \text{ s}$$

ASSESS The oscillator loses energy faster than it loses amplitude. This is what we should expect because the energy depends on the *square* of the amplitude.

STOP TO THINK 15.6 Rank in order, from largest to smallest, the time constants τ_a to τ_d of the decays shown in the figure. All the graphs have the same scale.

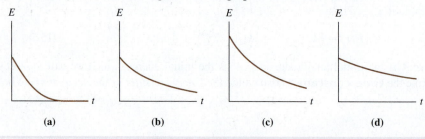

E $\qquad$ E $\qquad$ E $\qquad$ E

(a) $\qquad$ (b) $\qquad$ (c) $\qquad$ (d)

15.8 Driven Oscillations and Resonance

Thus far we have focused on the free oscillations of an isolated system. Some initial disturbance displaces the system from equilibrium, and it then oscillates freely until its energy is dissipated. These are very important situations, but they do not exhaust the possibilities. Another important situation is an oscillator that is subjected to a periodic external force. Its motion is called a **driven oscillation.**

A simple example of a driven oscillation is pushing a child on a swing, where your push is a periodic external force applied to the swing. A more complex example is a car driving over a series of equally spaced bumps. Each bump causes a periodic upward force on the car's shock absorbers, which are big, heavily damped springs. The electromagnetic coil on the back of a loudspeaker cone provides a periodic magnetic force to drive the cone back and forth, causing it to send out sound waves. Air turbulence moving across the wings of an aircraft can exert periodic forces on the wings and other aerodynamic surfaces, causing them to vibrate if they are not properly designed.

As these examples suggest, driven oscillations have many important applications. However, driven oscillations are a mathematically complex subject. We will simply hint at some of the results, saving the details for more advanced classes.

Consider an oscillating system that, when left to itself, oscillates at a frequency f_0. We will call this the **natural frequency** of the oscillator. The natural frequency for a mass on a spring is $\sqrt{k/m}/2\pi$, but it might be given by some other expression for another type of oscillator. Regardless of the expression, f_0 is simply the frequency of the system if it is displaced from equilibrium and released.

Suppose that this system is subjected to a *periodic* external force of frequency f_{ext}. This frequency, which is called the **driving frequency,** is completely independent of the oscillator's natural frequency f_0. Somebody or something in the environment selects the frequency f_{ext} of the external force, causing the force to push on the system f_{ext} times every second.

Although it is possible to solve Newton's second law with an external driving force, we will be content to look at a graphical representation of the solution. The most important result is that the oscillation amplitude depends very sensitively on the frequency f_{ext} of the driving force. The response to the driving frequency is shown in **FIGURE 15.25** for a system with $m = 1.0$ kg, a natural frequency $f_0 = 2.0$ Hz, and a damping constant $b = 0.20$ kg/s. This graph of amplitude versus driving frequency, called the **response curve,** occurs in many different applications.

When the driving frequency is substantially different from the oscillator's natural frequency, at the right and left edges of Figure 15.25, the system oscillates but the amplitude is very small. The system simply does not respond well to a driving frequency that differs much from f_0. As the driving frequency gets closer and closer to the natural frequency, the amplitude of the oscillation rises dramatically. After all, f_0 is the frequency at which the system "wants" to oscillate, so it is quite happy to respond to a driving frequency near f_0. Hence the amplitude reaches a maximum when the driving frequency exactly matches the system's natural frequency: $f_{ext} = f_0$.

The amplitude can become exceedingly large when the frequencies match, especially if the damping constant is very small. **FIGURE 15.26** shows the same oscillator with three different values of the damping constant. There's very little response if the damping constant is increased to 0.80 kg/s, but the amplitude for $f_{ext} = f_0$ becomes very large when the damping constant is reduced to 0.08 kg/s. This large-amplitude response to a driving force whose frequency matches the natural frequency of the system is a phenomenon called **resonance.** The condition for resonance is

$$f_{ext} = f_0 \quad \text{(resonance condition)} \qquad (15.61)$$

Within the context of driven oscillations, the natural frequency f_0 is often called the **resonance frequency.**

An important feature of Figure 15.26 is how the amplitude and width of the resonance depend on the damping constant. A heavily damped system responds fairly

FIGURE 15.25 The response curve of a driven oscillator at frequencies near its natural frequency of 2.0 Hz.

Amplitude

Maximum amplitude when $f_{ext} = f_0$.

Amplitude is small when f_{ext} differs substantially from f_0.

f_{ext} (Hz)

Natural frequency f_0

FIGURE 15.26 The resonance amplitude becomes higher and narrower as the damping constant decreases.

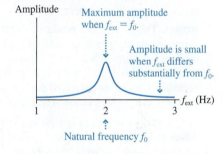

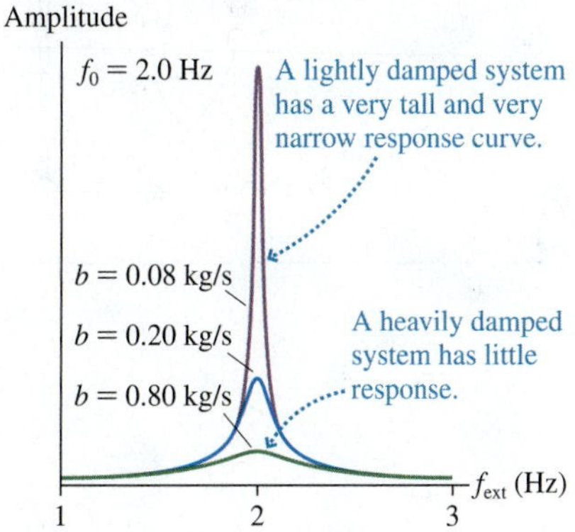

Amplitude

$f_0 = 2.0$ Hz

A lightly damped system has a very tall and very narrow response curve.

$b = 0.08$ kg/s

$b = 0.20$ kg/s

$b = 0.80$ kg/s

A heavily damped system has little response.

f_{ext} (Hz)

A singer or musical instrument can shatter a crystal goblet by matching the goblet's natural oscillation frequency.

little, even at resonance, but it responds to a wide range of driving frequencies. Very lightly damped systems can reach exceptionally high amplitudes, but notice that the range of frequencies to which the system responds becomes narrower and narrower as b decreases.

This allows us to understand why a few singers can break crystal goblets but not inexpensive, everyday glasses. An inexpensive glass gives a "thud" when tapped, but a fine crystal goblet "rings" for several seconds. In physics terms, the goblet has a much longer time constant than the glass. That, in turn, implies that the goblet is very lightly damped while the ordinary glass is heavily damped (because the internal forces within the glass are not those of a high-quality crystal structure).

The singer causes a sound wave to impinge on the goblet, exerting a small driving force at the frequency of the note she is singing. If the singer's frequency matches the natural frequency of the goblet—resonance! Only the lightly damped goblet, like the top curve in Figure 15.26, can reach amplitudes large enough to shatter. The restriction, though, is that its natural frequency has to be matched very precisely. The sound also has to be very loud.

CHALLENGE EXAMPLE 15.11 | A swinging pendulum

A pendulum consists of a massless, rigid rod with a mass at one end. The other end is pivoted on a frictionless pivot so that the rod can rotate in a complete circle. The pendulum is inverted, with the mass directly above the pivot point, then released. The speed of the mass as it passes through the lowest point is 5.0 m/s. If the pendulum later undergoes small-amplitude oscillations at the bottom of the arc, what will its frequency be?

MODEL This is a simple pendulum because the rod is massless. However, our analysis of a pendulum used the small-angle approximation. It applies only to the small-amplitude oscillations at the end, *not* to the pendulum swinging down from the inverted position. Fortunately, energy is conserved throughout, so we can analyze the big swing using conservation of mechanical energy.

VISUALIZE FIGURE 15.27 is a pictorial representation of the pendulum swinging down from the inverted position. The pendulum length is L, so the initial height is $2L$.

FIGURE 15.27 Before-and-after pictorial representation of the pendulum swinging down from an inverted position.

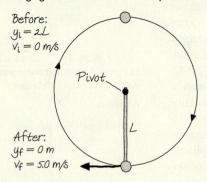

Before:
$y_i = 2L$
$v_i = 0$ m/s

Pivot

After:
$y_f = 0$ m
$v_f = 5.0$ m/s

L

SOLVE The frequency of a simple pendulum is $f = \sqrt{g/L}/2\pi$. We're not given L, but we can find it by analyzing the pendulum's swing down from an inverted position. Mechanical energy is conserved, and the only potential energy is gravitational potential energy. Conservation of mechanical energy $K_f + U_{Gf} = K_i + U_{Gi}$, with $U_G = mgy$, is

$$\tfrac{1}{2}mv_f^2 + mgy_f = \tfrac{1}{2}mv_i^2 + mgy_i$$

The mass cancels, which is good since we don't know it, and two terms are zero. Thus

$$\tfrac{1}{2}v_f^2 = g(2L) = 2gL$$

Solving for L, we find

$$L = \frac{v_f^2}{4g} = \frac{(5.0 \text{ m/s})^2}{4(9.80 \text{ m/s}^2)} = 0.638 \text{ m}$$

Now we can calculate the frequency:

$$f = \frac{1}{2\pi}\sqrt{\frac{g}{L}} = \frac{1}{2\pi}\sqrt{\frac{9.80 \text{ m/s}^2}{0.638 \text{ m}}} = 0.62 \text{ Hz}$$

ASSESS The frequency corresponds to a period of about 1.5 s, which seems reasonable.

SUMMARY

The goal of Chapter 15 has been to learn about systems that oscillate in simple harmonic motion.

GENERAL PRINCIPLES

Dynamics

SHM occurs when a **linear restoring force** acts to return a system to an equilibrium position.

Horizontal spring

$(F_{net})_x = -kx$

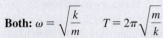

Vertical spring

The origin is at the equilibrium position $\Delta L = mg/k$.

$(F_{net})_y = -ky$

Both: $\omega = \sqrt{\dfrac{k}{m}}$ $T = 2\pi\sqrt{\dfrac{m}{k}}$

Simple pendulum

$\omega = \sqrt{\dfrac{g}{L}}$

$T = 2\pi\sqrt{\dfrac{L}{g}}$

Physical pendulum

$\omega = \sqrt{\dfrac{Mgl}{I}}$

$T = 2\pi\sqrt{\dfrac{I}{Mgl}}$

Energy

If there is no friction or dissipation, kinetic and potential energy are alternately transformed into each other, but the total mechanical energy $E = K + U$ is conserved.

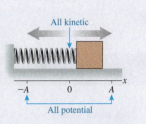

$$E = \tfrac{1}{2}mv^2 + \tfrac{1}{2}kx^2$$
$$= \tfrac{1}{2}m(v_{max})^2$$
$$= \tfrac{1}{2}kA^2$$

The energy of a lightly damped oscillator decays exponentially

$$E = E_0 e^{-t/\tau}$$

where τ is the **time constant**.

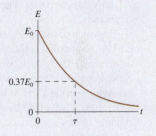

IMPORTANT CONCEPTS

Simple harmonic motion (SHM) is a sinusoidal oscillation with period T and amplitude A.

Frequency $f = \dfrac{1}{T}$

Angular frequency

$\omega = 2\pi f = \dfrac{2\pi}{T}$

Position $x(t) = A\cos(\omega t + \phi_0)$

$$= A\cos\left(\dfrac{2\pi t}{T} + \phi_0\right)$$

Velocity $v_x(t) = -v_{max}\sin(\omega t + \phi_0)$ with maximum speed $v_{max} = \omega A$

Acceleration $a_x(t) = -\omega^2 x(t) = -\omega^2 A\cos(\omega t + \phi_0)$

SHM is the projection onto the x-axis of **uniform circular motion.**

$\phi = \omega t + \phi_0$ is the **phase**

The position at time t is

$$x(t) = A\cos\phi$$
$$= A\cos(\omega t + \phi_0)$$

The **phase constant** ϕ_0 is determined by the initial conditions:

$$x_0 = A\cos\phi_0 \qquad v_{0x} = -\omega A\sin\phi_0$$

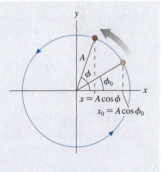

APPLICATIONS

Resonance

When a system is driven by a periodic external force, it responds with a large-amplitude oscillation if $f_{ext} \approx f_0$, where f_0 is the system's natural oscillation frequency, or **resonant frequency.**

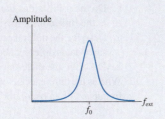

Damping

If there is a drag force $\vec{F}_{drag} = -b\vec{v}$, where b is the damping constant, then (for lightly damped systems)

$$x(t) = Ae^{-bt/2m}\cos(\omega t + \phi_0)$$

The time constant for energy loss is $\tau = m/b$.

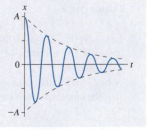

TERMS AND NOTATION

oscillatory motion	angular frequency, ω	damped oscillation	natural frequency, f_0
period, T	phase, ϕ	damping constant, b	driving frequency, f_{ext}
simple harmonic motion, SHM	phase constant, ϕ_0	envelope	response curve
	restoring force	exponential decay	resonance
amplitude, A	equation of motion	time constant, τ	resonance frequency, f_0
frequency, f	small-angle approximation	half-life, $t_{1/2}$	
hertz, Hz	linear restoring force	driven oscillation	

CONCEPTUAL QUESTIONS

1. A block oscillating on a spring has period $T = 2$ s. What is the period if:
   a. The block's mass is doubled? Explain. Note that you do not know the value of either m or k, so do *not* assume any particular values for them. The required analysis involves thinking about ratios.
   b. The value of the spring constant is quadrupled?
   c. The oscillation amplitude is doubled while m and k are unchanged?

2. A pendulum on Planet X, where the value of g is unknown, oscillates with a period $T = 2$ s. What is the period of this pendulum if:
   a. Its mass is doubled? Explain. Note that you do not know the value of m, L, or g, so do not assume any specific values. The required analysis involves thinking about ratios.
   b. Its length is doubled?
   c. Its oscillation amplitude is doubled?

3. **FIGURE Q15.3** shows a position-versus-time graph for a particle in SHM. What are (a) the amplitude A, (b) the angular frequency ω, and (c) the phase constant ϕ_0?

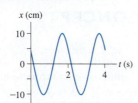

FIGURE Q15.3

4. **FIGURE Q15.4** shows a position-versus-time graph for a particle in SHM.
   a. What is the phase constant ϕ_0? Explain.
   b. What is the phase of the particle at each of the three numbered points on the graph?

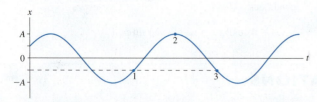

FIGURE Q15.4

5. Equation 15.25 states that $\frac{1}{2}kA^2 = \frac{1}{2}m(v_{max})^2$. What does this mean? Write a couple of sentences explaining how to interpret this equation.

6. A block oscillating on a spring has an amplitude of 20 cm. What will the amplitude be if the total energy is doubled? Explain.

7. A block oscillating on a spring has a maximum speed of 20 cm/s. What will the block's maximum speed be if the total energy is doubled? Explain.

8. The solid disk and circular hoop in **FIGURE Q15.8** have the same radius and the same mass. Each can swing back and forth as a pendulum from a pivot at the top edge. Which, if either, has the larger period of oscillation?

FIGURE Q15.8

9. **FIGURE Q15.9** shows the potential-energy diagram and the total energy line of a particle oscillating on a spring.
   a. What is the spring's equilibrium length?
   b. Where are the turning points of the motion? Explain.
   c. What is the particle's maximum kinetic energy?
   d. What will be the turning points if the particle's total energy is doubled?

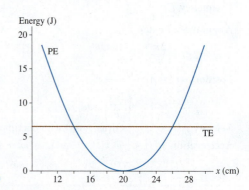

FIGURE Q15.9

10. Suppose the damping constant b of an oscillator increases.
    a. Is the medium more resistive or less resistive?
    b. Do the oscillations damp out more quickly or less quickly?
    c. Is the time constant τ increased or decreased?

11. a. Describe the difference between τ and T. Don't just *name* them; say what is different about the physical concepts they represent.
    b. Describe the difference between τ and $t_{1/2}$.

12. What is the difference between the driving frequency and the natural frequency of an oscillator?

EXERCISES AND PROBLEMS

Problems labeled   integrate material from earlier chapters.

Exercises

Section 15.1 Simple Harmonic Motion

1. | An air-track glider attached to a spring oscillates between the 10 cm mark and the 60 cm mark on the track. The glider completes 10 oscillations in 33 s. What are the (a) period, (b) frequency, (c) angular frequency, (d) amplitude, and (e) maximum speed of the glider?

2. || An air-track glider is attached to a spring. The glider is pulled to the right and released from rest at $t = 0$ s. It then oscillates with a period of 2.0 s and a maximum speed of 40 cm/s.
   a. What is the amplitude of the oscillation?
   b. What is the glider's position at $t = 0.25$ s?

3. | When a guitar string plays the note "A," the string vibrates at 440 Hz. What is the period of the vibration?

4. || An object in SHM oscillates with a period of 4.0 s and an amplitude of 10 cm. How long does the object take to move from $x = 0.0$ cm to $x = 6.0$ cm?

Section 15.2 SHM and Circular Motion

5. | What are the (a) amplitude, (b) frequency, and (c) phase constant of the oscillation shown in **FIGURE EX15.5**?

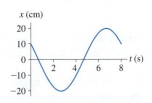

FIGURE EX15.5   **FIGURE EX15.6**

6. || What are the (a) amplitude, (b) frequency, and (c) phase constant of the oscillation shown in **FIGURE EX15.6**?

7. || **FIGURE EX15.7** is the position-versus-time graph of a particle in simple harmonic motion.
   a. What is the phase constant?
   b. What is the velocity at $t = 0$ s?
   c. What is v_{max}?

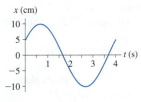   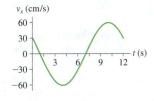

FIGURE EX15.7   **FIGURE EX15.8**

8. || **FIGURE EX15.8** is the velocity-versus-time graph of a particle in simple harmonic motion.
   a. What is the amplitude of the oscillation?
   b. What is the phase constant?
   c. What is the position at $t = 0$ s?

9. || An object in simple harmonic motion has an amplitude of 4.0 cm, a frequency of 2.0 Hz, and a phase constant of $2\pi/3$ rad. Draw a position graph showing two cycles of the motion.

10. || An object in simple harmonic motion has an amplitude of 8.0 cm, a frequency of 0.25 Hz, and a phase constant of $-\pi/2$ rad. Draw a position graph showing two cycles of the motion.

11. | An object in simple harmonic motion has amplitude 4.0 cm and frequency 4.0 Hz, and at $t = 0$ s it passes through the equilibrium point moving to the right. Write the function $x(t)$ that describes the object's position.

12. | An object in simple harmonic motion has amplitude 8.0 cm and frequency 0.50 Hz. At $t = 0$ s it has its most negative position. Write the function $x(t)$ that describes the object's position.

13. || An air-track glider attached to a spring oscillates with a period of 1.5 s. At $t = 0$ s the glider is 5.00 cm left of the equilibrium position and moving to the right at 36.3 cm/s.
   a. What is the phase constant?
   b. What is the phase at $t = 0$ s, 0.5 s, 1.0 s, and 1.5 s?

Section 15.3 Energy in SHM

Section 15.4 The Dynamics of SHM

14. | A block attached to a spring with unknown spring constant oscillates with a period of 2.0 s. What is the period if
   a. The mass is doubled?
   b. The mass is halved?
   c. The amplitude is doubled?
   d. The spring constant is doubled?
   Parts a to d are independent questions, each referring to the initial situation.

15. || A 200 g air-track glider is attached to a spring. The glider is pushed in 10 cm and released. A student with a stopwatch finds that 10 oscillations take 12.0 s. What is the spring constant?

16. || A 200 g mass attached to a horizontal spring oscillates at a frequency of 2.0 Hz. At $t = 0$ s, the mass is at $x = 5.0$ cm and has $v_x = -30$ cm/s. Determine:
   a. The period.   b. The angular frequency.
   c. The amplitude.   d. The phase constant.
   e. The maximum speed.   f. The maximum acceleration.
   g. The total energy.   h. The position at $t = 0.40$ s.

17. || The position of a 50 g oscillating mass is given by $x(t) = (2.0 \text{ cm})\cos(10t - \pi/4)$, where t is in s. Determine:
   a. The amplitude.   b. The period.
   c. The spring constant.   d. The phase constant.
   e. The initial conditions.   f. The maximum speed.
   g. The total energy.   h. The velocity at $t = 0.40$ s.

18. || A 1.0 kg block is attached to a spring with spring constant 16 N/m. While the block is sitting at rest, a student hits it with a hammer and almost instantaneously gives it a speed of 40 cm/s. What are
   a. The amplitude of the subsequent oscillations?
   b. The block's speed at the point where $x = \frac{1}{2}A$?

19. || A student is bouncing on a trampoline. At her highest point, her feet are 55 cm above the trampoline. When she lands, the trampoline sags 15 cm before propelling her back up. For how long is she in contact with the trampoline?

20. ‖ **FIGURE EX15.20** is a kinetic-energy graph of a mass oscillating on a *very* long horizontal spring. What is the spring constant?

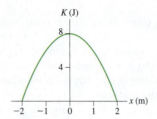

FIGURE EX15.20

Section 15.5 Vertical Oscillations

21. | A spring is hanging from the ceiling. Attaching a 500 g physics book to the spring causes it to stretch 20 cm in order to come to equilibrium.
    a. What is the spring constant?
    b. From equilibrium, the book is pulled down 10 cm and released. What is the period of oscillation?
    c. What is the book's maximum speed?

22. ‖ A spring with spring constant 15 N/m hangs from the ceiling. A ball is attached to the spring and allowed to come to rest. It is then pulled down 6.0 cm and released. If the ball makes 30 oscillations in 20 s, what are its (a) mass and (b) maximum speed?

23. ‖ A spring is hung from the ceiling. When a block is attached to its end, it stretches 2.0 cm before reaching its new equilibrium length. The block is then pulled down slightly and released. What is the frequency of oscillation?

Section 15.6 The Pendulum

24. | A grandfather clock ticks each time the pendulum passes through the lowest point. If the pendulum is modeled as a simple pendulum, how long must it be for the ticks to occur once a second?

25. ‖ A 200 g ball is tied to a string. It is pulled to an angle of 8.0° and released to swing as a pendulum. A student with a stopwatch finds that 10 oscillations take 12 s. How long is the string?

26. | A mass on a string of unknown length oscillates as a pendulum with a period of 4.0 s. What is the period if
    a. The mass is doubled?
    b. The string length is doubled?
    c. The string length is halved?
    d. The amplitude is doubled?
    Parts a to d are independent questions, each referring to the initial situation.

27. | What is the length of a pendulum whose period on the moon matches the period of a 2.0-m-long pendulum on the earth?

28. | What is the period of a 1.0-m-long pendulum on (a) the earth and (b) Venus?

29. | Astronauts on the first trip to Mars take along a pendulum that has a period on earth of 1.50 s. The period on Mars turns out to be 2.45 s. What is the free-fall acceleration on Mars?

30. ‖‖ A 100 g mass on a 1.0-m-long string is pulled 8.0° to one side and released. How long does it take for the pendulum to reach 4.0° on the opposite side?

31. ‖ A uniform steel bar swings from a pivot at one end with a period of 1.2 s. How long is the bar?

Section 15.7 Damped Oscillations

Section 15.8 Driven Oscillations and Resonance

32. | A 2.0 g spider is dangling at the end of a silk thread. You can make the spider bounce up and down on the thread by tapping lightly on his feet with a pencil. You soon discover that you can give the spider the largest amplitude on his little bungee cord if you tap exactly once every second. What is the spring constant of the silk thread?

33. ‖ The amplitude of an oscillator decreases to 36.8% of its initial value in 10.0 s. What is the value of the time constant?

34. ‖ In a science museum, a 110 kg brass pendulum bob swings at the end of a 15.0-m-long wire. The pendulum is started at exactly 8:00 A.M. every morning by pulling it 1.5 m to the side and releasing it. Because of its compact shape and smooth surface, the pendulum's damping constant is only 0.010 kg/s. At exactly 12:00 noon, how many oscillations will the pendulum have completed and what is its amplitude?

35. | Vision is blurred if the head is vibrated at 29 Hz because the
BIO vibrations are resonant with the natural frequency of the eyeball in its socket. If the mass of the eyeball is 7.5 g, a typical value, what is the effective spring constant of the musculature that holds the eyeball in the socket?

36. ‖ A 350 g mass on a 45-cm-long string is released at an angle of 4.5° from vertical. It has a damping constant of 0.010 kg/s. After 25 s, (a) how many oscillations has it completed and (b) how much energy has been lost?

Problems

37. ‖ The motion of a particle is given by $x(t) = (25 \text{ cm})\cos(10t)$, where *t* is in s. What is the first time at which the kinetic energy is twice the potential energy?

38. ‖ a. When the displacement of a mass on a spring is $\frac{1}{2}A$, what fraction of the energy is kinetic energy and what fraction is potential energy?
    b. At what displacement, as a fraction of *A*, is the energy half kinetic and half potential?

39. ‖ For a particle in simple harmonic motion, show that $v_{\text{max}} = (\pi/2)v_{\text{avg}}$, where v_{avg} is the average speed during one cycle of the motion.

40. ‖ A 100 g block attached to a spring with spring constant 2.5 N/m oscillates horizontally on a frictionless table. Its velocity is 20 cm/s when $x = -5.0$ cm.
    a. What is the amplitude of oscillation?
    b. What is the block's maximum acceleration?
    c. What is the block's position when the acceleration is maximum?
    d. What is the speed of the block when $x = 3.0$ cm?

41. ‖ A 0.300 kg oscillator has a speed of 95.4 cm/s when its displacement is 3.00 cm and 71.4 cm/s when its displacement is 6.00 cm. What is the oscillator's maximum speed?

42. ‖ An ultrasonic transducer, of the type used in medical ultra-
BIO sound imaging, is a very thin disk ($m = 0.10$ g) driven back and forth in SHM at 1.0 MHz by an electromagnetic coil.
    a. The maximum restoring force that can be applied to the disk without breaking it is 40,000 N. What is the maximum oscillation amplitude that won't rupture the disk?
    b. What is the disk's maximum speed at this amplitude?

43. || Astronauts in space cannot weigh themselves by standing on a
BIO bathroom scale. Instead, they determine their mass by oscillating
on a large spring. Suppose an astronaut attaches one end of a large
spring to her belt and the other end to a hook on the wall of the
space capsule. A fellow astronaut then pulls her away from the
wall and releases her. The spring's length as a function of time is
shown in **FIGURE P15.43**.
a. What is her mass if the spring constant is 240 N/m?
b. What is her speed when the spring's length is 1.2 m?

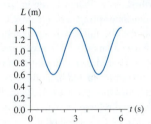

FIGURE P15.43

44. || Your lab instructor has asked you to measure a spring constant
using a dynamic method—letting it oscillate—rather than a static
method of stretching it. You and your lab partner suspend the
spring from a hook, hang different masses on the lower end, and
start them oscillating. One of you uses a meter stick to measure
the amplitude, the other uses a stopwatch to time 10 oscillations.
Your data are as follows:

Mass (g)	Amplitude (cm)	Time (s)
100	6.5	7.8
150	5.5	9.8
200	6.0	10.9
250	3.5	12.4

Use the best-fit line of an appropriate graph to determine the spring
constant.

45. || A 5.0 kg block hangs from a spring with spring constant
2000 N/m. The block is pulled down 5.0 cm from the equilibrium
position and given an initial velocity of 1.0 m/s back toward equilibrium. What are the (a) frequency, (b) amplitude, and (c) total
mechanical energy of the motion?

46. ||| A 200 g block hangs from a spring with spring constant 10 N/m.
At $t = 0$ s the block is 20 cm below the equilibrium point and
moving upward with a speed of 100 cm/s. What are the block's
a. Oscillation frequency?
b. Distance from equilibrium when the speed is 50 cm/s?
c. Distance from equilibrium at $t = 1.0$ s?

47. || A block hangs in equilibrium from a vertical spring. When a
second identical block is added, the original block sags by 5.0 cm.
What is the oscillation frequency of the two-block system?

48. || A 75 kg student jumps off a bridge with a 12-m-long bungee
cord tied to his feet. The massless bungee cord has a spring
constant of 430 N/m.
a. How far below the bridge is the student's lowest point?
b. How long does it take the student to reach his lowest point?
You can assume that the bungee cord exerts no force until it
begins to stretch.

49. || Scientists are measuring the properties of a newly discovered
elastic material. They create a 1.5-m-long, 1.6-mm-diameter cord,
attach an 850 g mass to the lower end, then pull the mass down
2.5 mm and release it. Their high-speed video camera records 36
oscillations in 2.0 s. What is Young's modulus of the material?

50. ||| A mass hanging from a spring oscillates with a period of 0.35 s.
Suppose the mass and spring are swung in a horizontal circle, with
the free end of the spring at the pivot. What rotation frequency, in
rpm, will cause the spring's length to stretch by 15%?

51. || A compact car has a mass of 1200 kg. Assume that the car has
one spring on each wheel, that the springs are identical, and that
the mass is equally distributed over the four springs.
a. What is the spring constant of each spring if the empty car
bounces up and down 2.0 times each second?
b. What will be the car's oscillation frequency while carrying
four 70 kg passengers?

52. || The two blocks in **FIGURE P15.52** oscillate on a frictionless sur-
face with a period of 1.5 s. The upper block just begins to slip
when the amplitude is increased to 40 cm. What is the coefficient
of static friction between the two blocks?

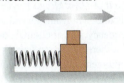

FIGURE P15.52

53. || A 1.00 kg block is attached to a horizontal spring with spring
constant 2500 N/m. The block is at rest on a frictionless surface. A
10 g bullet is fired into the block, in the face opposite the spring,
and sticks. What was the bullet's speed if the subsequent oscillations have an amplitude of 10.0 cm?

54. ||| It has recently become possible to "weigh" DNA molecules
BIO by measuring the influence of their mass on a nano-oscillator.
FIGURE P15.54 shows a thin rectangular cantilever etched out of
silicon (density 2300 kg/m³) with a small gold dot (not visible)
at the end. If pulled down and released, the end of the cantilever
vibrates with SHM, moving up and down like a diving board after
a jump. When bathed with DNA molecules whose ends have been
modified to bind with gold, one or more molecules may attach to
the gold dot. The addition of their mass causes a very slight—but
measurable—decrease in the oscillation frequency.

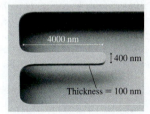

FIGURE P15.54

A vibrating cantilever of mass M can be modeled as a block of
mass $\frac{1}{3}M$ attached to a spring. (The factor of $\frac{1}{3}$ arises from the moment of inertia of a bar pivoted at one end.) Neither the mass nor
the spring constant can be determined very accurately— perhaps
to only two significant figures—but the oscillation frequency can
be measured with very high precision simply by counting the oscillations. In one experiment, the cantilever was initially vibrating
at exactly 12 MHz. Attachment of a DNA molecule caused the
frequency to decrease by 50 Hz. What was the mass of the DNA?

55. || It is said that Galileo discovered a basic principle of the pendu-
lum—that the period is independent of the amplitude—by using
his pulse to time the period of swinging lamps in the cathedral
as they swayed in the breeze. Suppose that one oscillation of a
swinging lamp takes 5.5 s.
a. How long is the lamp chain?
b. What maximum speed does the lamp have if its maximum
angle from vertical is 3.0°?

56. ‖ Orangutans can move by *brachiation,* swinging like a pendu-
BIO lum beneath successive handholds. If an orangutan has arms that
are 0.90 m long and repeatedly swings to a 20° angle, taking one
swing after another, estimate its speed of forward motion in m/s.
While this is somewhat beyond the range of validity of the
small-angle approximation, the standard results for a pendulum
are adequate for making an estimate.

57. ‖ The pendulum shown in **FIGURE P15.57**
is pulled to a 10° angle on the left side and
released.
a. What is the period of this pendulum?
b. What is the pendulum's maximum angle
on the right side?

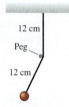

12 cm

Peg

12 cm

FIGURE P15.57

58. ‖ A uniform rod of mass M and length L swings as a pendulum
on a pivot at distance $L/4$ from one end of the rod. Find an expression for the frequency f of small-angle oscillations.

59. ‖ Interestingly, there have been several studies using cadavers to
BIO determine the moments of inertia of human body parts, information that is important in biomechanics. In one study, the center of
mass of a 5.0 kg lower leg was found to be 18 cm from the knee.
When the leg was allowed to pivot at the knee and swing freely as
a pendulum, the oscillation frequency was 1.6 Hz. What was the
moment of inertia of the lower leg about the knee joint?

60. ‖ A 500 g air-track glider attached to a spring with spring constant 10 N/m is sitting at rest on a frictionless air track. A 250 g
glider is pushed toward it from the far end of the track at a speed of
120 cm/s. It collides with and sticks to the 500 g glider. What are
the amplitude and period of the subsequent oscillations?

61. ‖ A 200 g block attached to a horizontal spring is oscillating with
an amplitude of 2.0 cm and a frequency of 2.0 Hz. Just as it passes
through the equilibrium point, moving to the right, a sharp blow
directed to the left exerts a 20 N force for 1.0 ms. What are the new
(a) frequency and (b) amplitude?

62. ‖ **FIGURE P15.62** is a top view of an object of mass m connected
between two stretched rubber bands of length L. The object rests
on a frictionless surface. At equilibrium, the tension in each rubber band is T. Find an expression for the frequency of oscillations *perpendicular* to the rubber bands. Assume the amplitude is
sufficiently small that the magnitude of the tension in the rubber
bands is essentially unchanged as the mass oscillates.

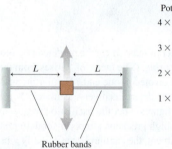

Rubber bands

FIGURE P15.62

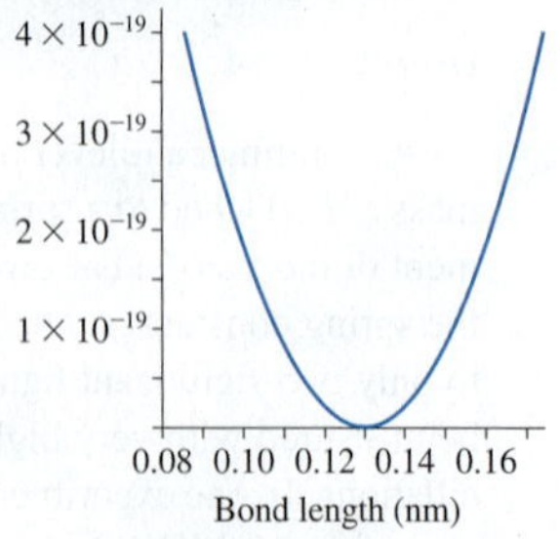

Potential energy (J)

4×10^{-19}

3×10^{-19}

2×10^{-19}

1×10^{-19}

0.08 0.10 0.12 0.14 0.16
Bond length (nm)

FIGURE P15.63

63. ‖ A molecular bond can be modeled as a spring between two
atoms that vibrate with simple harmonic motion. **FIGURE P15.63**
shows an SHM approximation for the potential energy of an HCl
molecule. Because the chlorine atom is so much more massive
than the hydrogen atom, it is reasonable to assume that the hydrogen atom ($m = 1.67 \times 10^{-27}$ kg) vibrates back and forth while
the chlorine atom remains at rest. Use the graph to estimate the
vibrational frequency of the HCl molecule.

64. ‖ A penny rides on top of a piston as it undergoes vertical simple harmonic motion with an amplitude of 4.0 cm. If the frequency is low, the penny rides up and down without difficulty. If the
frequency is steadily increased, there comes a point at which the
penny leaves the surface.
a. At what point in the cycle does the penny first lose contact
with the piston?
b. What is the maximum frequency for which the penny just
barely remains in place for the full cycle?

65. ‖ On your first trip to Planet X you happen to take along a 200 g
mass, a 40-cm-long spring, a meter stick, and a stopwatch. You're
curious about the free-fall acceleration on Planet X, where ordinary
tasks seem easier than on earth, but you can't find this information
in your Visitor's Guide. One night you suspend the spring from
the ceiling in your room and hang the mass from it. You find that
the mass stretches the spring by 31.2 cm. You then pull the mass
down 10.0 cm and release it. With the stopwatch you find that 10
oscillations take 14.5 s. Based on this information, what is g?

66. ‖ Suppose a large spherical object, such as a planet, with radius R
and mass M has a narrow tunnel passing diametrically through it. A
particle of mass m is inside the tunnel at a distance $x \leq R$ from the
center. It can be shown that the net gravitational force on the particle is
due entirely to the sphere of mass with radius $r \leq x$; there is no net
gravitational force from the mass in the spherical shell with $r > x$.
a. Find an expression for the gravitational force on the particle,
assuming the object has uniform density. Your expression will
be in terms of x, R, m, M, and any necessary constants.
b. You should have found that the gravitational force is a linear
restoring force. Consequently, in the absence of air resistance,
objects in the tunnel will oscillate with SHM. Suppose an in-
trepid astronaut exploring a 150-km-diameter, 3.5×10^{18} kg
asteroid discovers a tunnel through the center. If she jumps
into the hole, how long will it take her to fall all the way
through the asteroid and emerge on the other side?

67. ‖ The 15 g head of a bobble-head doll oscillates in SHM at a
frequency of 4.0 Hz.
a. What is the spring constant of the spring on which the head
is mounted?
b. The amplitude of the head's oscillations decreases to 0.5 cm
in 4.0 s. What is the head's damping constant?

68. ‖ An oscillator with a mass of 500 g and a period of 0.50 s has
an amplitude that decreases by 2.0% during each complete oscillation. If the initial amplitude is 10 cm, what will be the amplitude
after 25 oscillations?

69. ‖‖ A spring with spring constant 15.0 N/m hangs from the ceiling.
A 500 g ball is attached to the spring and allowed to come to
rest. It is then pulled down 6.0 cm and released. What is the time
constant if the ball's amplitude has decreased to 3.0 cm after
30 oscillations?

70. ‖‖ A captive James Bond is strapped to a table beneath a huge
pendulum made of a 2.0-m-diameter uniform circular metal blade
rigidly attached, at its top edge, to a 6.0-m-long, massless rod.
The pendulum is set swinging with a 10° amplitude when its lower
edge is 3.0 m above the prisoner, then the table slowly starts ascending at 1.0 mm/s. After 25 minutes, the pendulum's amplitude
has decreased to 7.0°. Fortunately, the prisoner is freed with a
mere 30 s to spare. What was the speed of the lower edge of the
blade as it passed over him for the last time?

71. ‖ A 250 g air-track glider is attached to a spring with spring
constant 4.0 N/m. The damping constant due to air resistance is
0.015 kg/s. The glider is pulled out 20 cm from equilibrium and
released. How many oscillations will it make during the time in
which the amplitude decays to e^{-1} of its initial value?

72. ‖ A 200 g oscillator in a vacuum chamber has a frequency of 2.0 Hz. When air is admitted, the oscillation decreases to 60% of its initial amplitude in 50 s. How many oscillations will have been completed when the amplitude is 30% of its initial value?

73. ‖ Prove that the expression for $x(t)$ in Equation 15.55 is a solution to the equation of motion for a damped oscillator, Equation 15.54, if and only if the angular frequency ω is given by the expression in Equation 15.56.

74. ‖ A block on a frictionless table is connected as shown in **FIGURE P15.74** to two springs having spring constants k_1 and k_2. Show that the block's oscillation frequency is given by

$$f = \sqrt{f_1^2 + f_2^2}$$

where f_1 and f_2 are the frequencies at which it would oscillate if attached to spring 1 or spring 2 alone.

FIGURE P15.74

75. ‖ A block on a frictionless table is connected as shown in **FIGURE P15.75** to two springs having spring constants k_1 and k_2. Find an expression for the block's oscillation frequency f in terms of the frequencies f_1 and f_2 at which it would oscillate if attached to spring 1 or spring 2 alone.

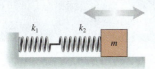

FIGURE P15.75

Challenge Problems

76. ‖ A 15-cm-long, 200 g rod is pivoted at one end. A 20 g ball of clay is stuck on the other end. What is the period if the rod and clay swing as a pendulum?

77. ‖ A solid sphere of mass M and radius R is suspended from a thin rod, as shown in **FIGURE CP15.77**. The sphere can swing back and forth at the bottom of the rod. Find an expression for the frequency f of small-angle oscillations.

FIGURE CP15.77

78. ‖ A uniform rod of length L oscillates as a pendulum about a
CALC pivot that is a distance x from the center.
   a. For what value of x, in terms of L, is the oscillation period a minimum?
   b. What is the minimum oscillation period of a 15 kg, 1.0-m-long steel bar?

79. ‖ A spring is standing upright on a table with its bottom end fastened to the table. A block is dropped from a height 3.0 cm above the top of the spring. The block sticks to the top end of the spring and then oscillates with an amplitude of 10 cm. What is the oscillation frequency?

80. ‖ The analysis of a simple pendulum assumed that the mass was a particle, with no size. A realistic pendulum is a small, uniform sphere of mass M and radius R at the end of a massless string, with L being the distance from the pivot to the center of the sphere.
   a. Find an expression for the period of this pendulum.
   b. Suppose $M = 25$ g, $R = 1.0$ cm, and $L = 1.0$ m, typical values for a real pendulum. What is the ratio T_{real}/T_{simple}, where T_{real} is your expression from part a and T_{simple} is the expression derived in this chapter?

81. ‖ **FIGURE CP15.81** shows a 200 g uniform rod pivoted at one end. The other end is attached to a horizontal spring. The spring is neither stretched nor compressed when the rod hangs straight down. What is the rod's oscillation period? You can assume that the rod's angle from vertical is always small.

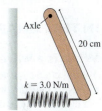

FIGURE CP15.81

16 Traveling Waves

This surfer is "catching a wave." At the same time, he's seeing light waves and hearing sound waves.

IN THIS CHAPTER, you will learn the basic properties of traveling waves.

What is a wave?

A wave is a disturbance traveling through a medium. In a transverse wave, the displacement is perpendicular to the direction of travel. In a longitudinal wave, the displacement is parallel to the direction of travel.

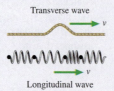

Transverse wave

Longitudinal wave

What are some wave properties?

A wave is characterized by:

- Wave speed: How fast it travels through the medium.
- Wavelength: The distance between two neighboring crests.
- Frequency: The number of oscillations per second.
- Amplitude: The maximum displacement.

《 LOOKING BACK Sections 15.1–15.2 Properties of simple harmonic motion

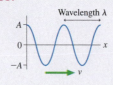

Are sound and light waves?

Yes! Very important waves.

- Sound waves are longitudinal waves.
- Light waves are transverse waves.

The colors of visible light correspond to different wavelengths.

Visible light

700   600   500   400
Wavelength (nm)

Do waves carry energy?

They do. The rate at which a wave delivers energy to a surface is the intensity of the wave. For sound waves, we'll use a logarithmic decibel scale to characterize the loudness of a sound.

Intensity is decreasing

What is the Doppler effect?

The frequency and wavelength of a wave are shifted if there is relative motion between the source and the observer of the waves. This is called the Doppler effect. It explains why the pitch of an ambulance siren drops as it races past you.

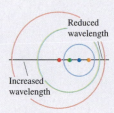

Reduced wavelength

Increased wavelength

How will I use waves?

Waves are literally everywhere. Communications systems from radios to cell phones to fiber optics use waves. Sonar and radar and medical ultrasound use waves. Music and musical instruments are all about waves. Waves are present in the oceans, the atmosphere, and the earth. This chapter and the next will allow you to understand and work with a wide variety of waves that you may meet in your career.

16.1 An Introduction to Waves

From sound and light to ocean waves and seismic waves, we're surrounded by waves. Understanding musical instruments, cell phones, or lasers requires a knowledge of waves. With this chapter we shift our focus from the particle model to a new **wave model** that emphasizes those aspects of wave behavior common to all waves.

The wave model is built around the idea of a **traveling wave,** which is an organized disturbance traveling with a well-defined wave speed. We'll begin our study of traveling waves by looking at two distinct wave motions.

Two types of traveling waves

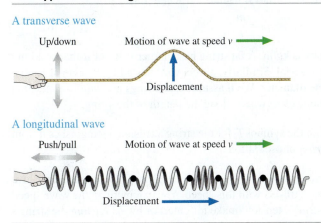

A transverse wave

Up/down   Motion of wave at speed v →

Displacement

A longitudinal wave

Push/pull   Motion of wave at speed v →

Displacement →

A **transverse wave** is a wave in which the displacement is *perpendicular* to the direction in which the wave travels. For example, a wave travels along a string in a horizontal direction while the particles that make up the string oscillate vertically. Electromagnetic waves are also transverse waves because the electromagnetic fields oscillate perpendicular to the direction in which the wave travels.

In a **longitudinal wave,** the particles in the medium are displaced *parallel* to the direction in which the wave travels. Here we see a chain of masses connected by springs. If you give the first mass in the chain a sharp push, a disturbance travels down the chain by compressing and expanding the springs. Sound waves in gases and liquids are the most well known examples of longitudinal waves.

We can also classify waves on the basis of what is "waving":

1. **Mechanical waves** travel only within a material *medium,* such as air or water. Two familiar mechanical waves are sound waves and water waves.
2. **Electromagnetic waves,** from radio waves to visible light to x rays, are a self-sustaining oscillation of the *electromagnetic field.* Electromagnetic waves require no material medium and can travel through a vacuum.

The **medium** of a mechanical wave is the substance through or along which the wave moves. For example, the medium of a water wave is the water, the medium of a sound wave is the air, and the medium of a wave on a stretched string is the string. A medium must be *elastic.* That is, a restoring force of some sort brings the medium back to equilibrium after it has been displaced or disturbed. The tension in a stretched string pulls the string back straight after you pluck it. Gravity restores the level surface of a lake after the wave generated by a boat has passed by.

As a wave passes through a medium, the atoms of the medium—we'll simply call them the particles of the medium—are displaced from equilibrium. This is a **disturbance** of the medium. The water ripples of **FIGURE 16.1** are a disturbance of the water's surface. A pulse traveling down a string is a disturbance, as are the wake of a boat and the sonic boom created by a jet traveling faster than the speed of sound. **The disturbance of a wave is an *organized* motion of the particles in the medium,** in contrast to the *random* molecular motions of thermal energy.

FIGURE 16.1 Ripples on a pond are a traveling wave.

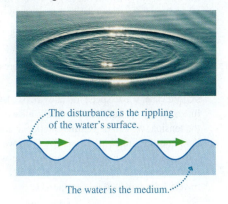

The disturbance is the rippling of the water's surface.

The water is the medium.

Wave Speed

A wave disturbance is created by a *source.* The source of a wave might be a rock thrown into water, your hand plucking a stretched string, or an oscillating loudspeaker cone pushing on the air. Once created, the disturbance travels outward through the medium at the **wave speed** v. This is the speed with which a ripple moves across the water or a pulse travels down a string.

NOTE The disturbance propagates through the medium, but the medium as a whole does not move! The ripples on the pond (the disturbance) move outward from the splash of the rock, but there is no outward flow of water from the splash. **A wave transfers energy, but it does not transfer any material or substance outward from the source.**

As an example, we'll prove in Section 16.4 that the wave speed on a string stretched with tension T_s is

$$v_{\text{string}} = \sqrt{\frac{T_s}{\mu}} \qquad \text{(wave speed on a stretched string)} \qquad (16.1)$$

where μ is the string's **linear density,** its mass-to-length ratio:

$$\mu = \frac{m}{L} \qquad (16.2)$$

The SI unit of linear density is kg/m. A fat string has a larger value of μ than a skinny string made of the same material. Similarly, a steel wire has a larger value of μ than a plastic string of the same diameter. We'll assume that strings are *uniform,* meaning the linear density is the same everywhere along the length of the string.

NOTE The subscript s on the symbol T_s for the string's tension distinguishes it from the symbol T for the *period* of oscillation.

Equation 16.1 is the wave *speed,* not the wave velocity, so v_{string} always has a positive value. Every point on a wave travels with this speed. You can increase the wave speed either by *increasing* the string's tension (make it tighter) or by *decreasing* the string's linear density (make it skinnier).

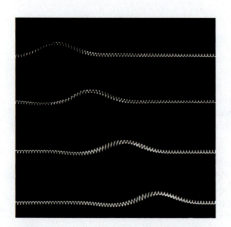

This sequence of photographs shows a wave pulse traveling along a spring.

EXAMPLE 16.1 | **Measuring the linear density**

A 2.00-m-long metal wire is attached to a motion sensor, stretched horizontally to a pulley 1.50 m away, then connected to a 2.00 kg hanging mass that provides tension. A mechanical pick plucks a horizontal segment of wire right at the pulley, creating a small wave pulse that travels along the wire. The plucking motion starts a timer that is stopped by the motion sensor when the pulse reaches the end of the wire. What is the wire's linear density if the pulse takes 18.0 ms to travel the length of the wire?

MODEL Model the pulse as a traveling wave and the pulley as frictionless.

VISUALIZE FIGURE 16.2 is a pictorial representation. The free-body diagram is for the hanging mass.

SOLVE The wave speed on a wire is determined by the wire's linear density μ and tension T_s. By measuring the wave speed and the tension, we can determine the linear density. The hanging mass is in equilibrium, with no net force, so we see from the free-body diagram that the tension throughout the wire (because the pulley is frictionless) is $T_s = F_G = Mg = 19.6$ N. Because the wave pulse travels 1.50 m in 18.0 ms, its speed is

$$v = \frac{1.50 \text{ m}}{0.0180 \text{ s}} = 83.3 \text{ m/s}$$

FIGURE 16.2 A wave pulse on a wire.

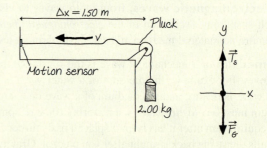

Thus, from Equation 16.1, the wire's linear density is

$$\mu = \frac{T_s}{v^2} = \frac{19.6 \text{ N}}{(83.3 \text{ m/s})^2} = 2.82 \times 10^{-3} \text{ kg/m} = 2.82 \text{ g/m}$$

Linear densities of strings are often stated in g/m, although these are not SI units. You must use kg/m in any calculations.

ASSESS A meter of thin wire is likely to have a mass of a few grams, so a linear density of a few g/m is reasonable. Note that the total length of the wire was not relevant.

The wave speed on a string is a property of the string—its tension and linear density. In general, **the wave speed is a property** *of the medium.* The wave speed depends on the restoring forces within the medium but not at all on the shape or size of the pulse, how the pulse was generated, or how far it has traveled.

STOP TO THINK 16.1 Which of the following actions would make a pulse travel faster along a stretched string? More than one answer may be correct. If so, give all that are correct.

a. Move your hand up and down more quickly as you generate the pulse.
b. Move your hand up and down a larger distance as you generate the pulse.
c. Use a heavier string of the same length, under the same tension.
d. Use a lighter string of the same length, under the same tension.
e. Stretch the string tighter to increase the tension.
f. Loosen the string to decrease the tension.
g. Put more force into the wave.

16.2 One-Dimensional Waves

To understand waves we must deal with functions of *two* variables. Until now, we have been concerned with quantities that depend only on time, such as $x(t)$ or $v(t)$. Functions of the one variable t are appropriate for a particle because a particle is only in one place at a time, but a wave is not localized. It is spread out through space at each instant of time. To describe a wave mathematically requires a function that specifies not only an instant of time (when) but also a point in space (where).

Rather than leaping into mathematics, we will start by thinking about waves graphically. Consider the wave pulse shown moving along a stretched string in **FIGURE 16.3**. (We will consider somewhat artificial triangular and square-shaped pulses in this section to make clear where the edges of the pulse are.) The graph shows the string's displacement Δy at a particular instant of time t_1 as a function of position x along the string. This is a "snapshot" of the wave, much like what you might make with a camera whose shutter is opened briefly at t_1. A graph that shows the wave's displacement as a function of position at a single instant of time is called a **snapshot graph.** For a wave on a string, a snapshot graph is literally a picture of the wave at this instant.

FIGURE 16.4 shows a sequence of snapshot graphs as the wave of Figure 16.3 continues to move. These are like successive frames from a video. Notice that the wave pulse moves forward distance $\Delta x = v \Delta t$ during the time interval Δt. That is, the wave moves with constant speed.

A snapshot graph tells only half the story. It tells us *where* the wave is and how it varies with position, but only at one instant of time. It gives us no information about how the wave *changes* with time. As a different way of portraying the wave, suppose we follow the dot marked on the string in Figure 16.4 and produce a graph showing how the displacement of this dot changes with time. The result, shown in **FIGURE 16.5**, is a displacement-versus-time graph at a single position in space. A graph that shows the wave's displacement as a function of time at a single position in space is called a **history graph.** It tells the history of that particular point in the medium.

FIGURE 16.3 A snapshot graph of a wave pulse on a string.

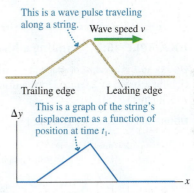

This is a wave pulse traveling along a string. Wave speed v

Trailing edge   Leading edge

Δy This is a graph of the string's displacement as a function of position at time t_1.

FIGURE 16.4 A sequence of snapshot graphs shows the wave in motion.

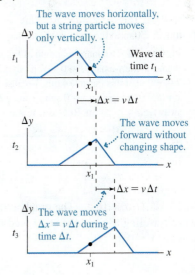

The wave moves horizontally, but a string particle moves only vertically.

Δy

t_1 Wave at time t_1

x_1

$\Delta x = v \Delta t$

Δy

t_2 The wave moves forward without changing shape.

x_1

$\Delta x = v \Delta t$

Δy The wave moves $\Delta x = v \Delta t$ during time Δt.

t_3

x_1

FIGURE 16.5 A history graph for the dot on the string in Figure 16.4.

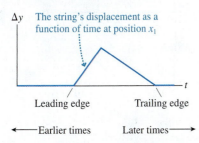

Δy The string's displacement as a function of time at position x_1

Leading edge   Trailing edge

Earlier times   Later times

FIGURE 16.6 An alternative look at a traveling wave.

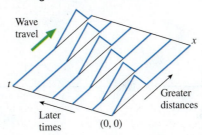

Wave travel

x

t

Later times   (0, 0)   Greater distances

You might think we have made a mistake; the graph of Figure 16.5 is reversed compared to Figure 16.4. It is not a mistake, but it requires careful thought to see why. As the wave moves toward the dot, the steep **leading edge** causes the dot to rise quickly. On the displacement-versus-time graph, *earlier* times (smaller values of t) are to the *left* and later times (larger t) to the right. Thus the leading edge of the wave is on the *left* side of the Figure 16.5 history graph. As you move to the right on Figure 16.5 you see the slowly falling **trailing edge** of the wave as it moves past the dot at later times.

The snapshot graph of Figure 16.3 and the history graph of Figure 16.5 portray complementary information. The snapshot graph tells us how things look throughout all of space, but at only one instant of time. The history graph tells us how things look at all times, but at only one position in space. We need them both to have the full story of the wave. An alternative representation of the wave is the series of graphs in **FIGURE 16.6**, where we can get a clearer sense of the wave moving forward. But graphs like these are essentially impossible to draw by hand, so it is necessary to move back and forth between snapshot graphs and history graphs.

EXAMPLE 16.2 | **Finding a history graph from a snapshot graph**

FIGURE 16.7 is a snapshot graph at $t = 0$ s of a wave moving to the right at a speed of 2.0 m/s. Draw a history graph for the position $x = 8.0$ m.

FIGURE 16.7 A snapshot graph at $t = 0$ s.

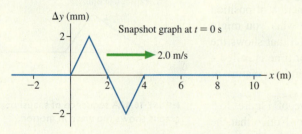

MODEL This is a wave traveling at constant speed. The pulse moves 2.0 m to the right every second.

VISUALIZE The snapshot graph of Figure 16.7 shows the wave at all points on the x-axis at $t = 0$ s. You can see that nothing is happening at $x = 8.0$ m at this instant of time because the wave has not yet reached $x = 8.0$ m. In fact, at $t = 0$ s the leading edge of the wave is still 4.0 m away from $x = 8.0$ m. Because the wave is traveling at 2.0 m/s, it will

take 2.0 s for the leading edge to reach $x = 8.0$ m. Thus the history graph for $x = 8.0$ m will be zero until $t = 2.0$ s. The first part of the wave causes a *downward* displacement of the medium, so immediately after $t = 2.0$ s the displacement at $x = 8.0$ m will be negative. The negative portion of the wave pulse is 2.0 m wide and takes 1.0 s to pass $x = 8.0$ m, so the midpoint of the pulse reaches $x = 8.0$ m at $t = 3.0$ s. The positive portion takes another 1.0 s to go past, so the trailing edge of the pulse arrives at $t = 4.0$ s. You could also note that the trailing edge was initially 8.0 m away from $x = 8.0$ m and needed 4.0 s to travel that distance at 2.0 m/s. The displacement at $x = 8.0$ m returns to zero at $t = 4.0$ s and remains zero for all later times. This information is all portrayed on the history graph of **FIGURE 16.8**.

FIGURE 16.8 The corresponding history graph at $x = 8.0$ m.

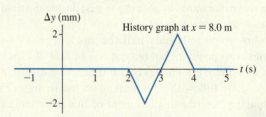

STOP TO THINK 16.2 The graph at the right is the history graph at $x = 4.0$ m of a wave traveling to the right at a speed of 2.0 m/s. Which is the history graph of this wave at $x = 0$ m?

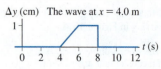

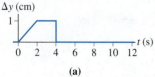

(a)

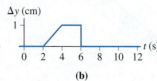

(b)

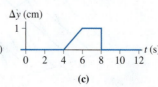

(c)

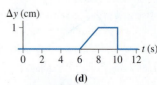

(d)

which is a *partial differential equation*. We have reason to think that sinusoidal waves can travel on stretched strings, so let's guess that a solution to Equation 16.28 is

$$D(x, t) = A\sin(kx - \omega t + \phi_0) \tag{16.29}$$

where the minus sign gives a wave traveling in the +x-direction and—from Equation 16.13—the wave speed is $v = \omega/k$.

To evaluate this possible solution we need its second partial derivatives. With respect to position we have

$$\frac{\partial D}{\partial x} = kA\cos(kx - \omega t + \phi_0)$$

$$\frac{\partial^2 D}{\partial x^2} = \frac{\partial}{\partial x}\left(\frac{\partial D}{\partial x}\right) = -k^2 A\sin(kx - \omega t + \phi_0) \tag{16.30}$$

and with respect to time we have

$$\frac{\partial D}{\partial t} = -\omega A\cos(kx - \omega t + \phi_0)$$

$$\frac{\partial^2 D}{\partial t^2} = \frac{\partial}{\partial t}\left(\frac{\partial D}{\partial t}\right) = -\omega^2 A\sin(kx - \omega t + \phi_0) \tag{16.31}$$

Substituting the second partial derivatives into the wave equation, Equation 16.28, gives

$$-\omega^2 A\sin(kx - \omega t + \phi_0) = \frac{T_\text{s}}{\mu}(-k^2 A\sin(kx - \omega t + \phi_0)) \tag{16.32}$$

This will be true only if

$$\omega^2 = \frac{T_\text{s}}{\mu}k^2 \tag{16.33}$$

But ω/k is the wave speed v, so what we've found is that the sinusoidal wave of Equation 16.1 *is* a solution to the wave equation, but only if the wave travels with speed

$$v = \frac{\omega}{k} = \sqrt{\frac{T_\text{s}}{\mu}} \tag{16.34}$$

You should be able to convince yourself that we would have arrived at the same result if we had started with $D(x, t) = A\sin(kx + \omega t + \phi_0)$ for a wave traveling in the –x-direction.

Let's summarize. We used Newton's second law for a small piece of the string to come up with an equation for the dynamics of motion on a string. We then showed that a solution to this equation is a sinusoidal traveling wave, and we made a specific prediction for the wave speed in terms of two properties or characteristics of the string: its tension and its mass density. Thus the answer to the question with which we started this section—*why* do waves propagate along a string—is that wave motion is simply a consequence of Newton's second law, the relationship between force and acceleration, when applied to a continuous object.

With Equation 16.9 in hand, we can write Equation 16.28 as

$$\frac{\partial^2 D}{\partial t^2} = v^2\frac{\partial^2 D}{\partial x^2} \quad \text{(the general wave equation)} \tag{16.35}$$

We derived Equation 16.28 specifically for a string, but *any* physical system that obeys Equation 16.35 for some type of displacement D will have sinusoidal waves traveling with speed v. Equation 16.35 is called the **wave equation,** and it occurs over and over in science and engineering. We will see it again in this chapter in our analysis of sound waves. And much later, in Chapter 31, we'll discover that electromagnetic fields also obey this equation. Thus electromagnetic waves exist, and we'll be able to predict that all electromagnetic waves, regardless of wavelength, travel through vacuum with the same speed—the speed of light.

EXAMPLE 16.4 | **Generating a sinusoidal wave**

A very long string with $\mu = 2.0$ g/m is stretched along the x-axis with a tension of 5.0 N. At $x = 0$ m it is tied to a 100 Hz simple harmonic oscillator that vibrates perpendicular to the string with an amplitude of 2.0 mm. The oscillator is at its maximum positive displacement at $t = 0$ s.

a. Write the displacement equation for the traveling wave on the string.

b. At $t = 5.0$ ms, what is the string's displacement at a point 2.7 m from the oscillator?

MODEL The oscillator generates a sinusoidal traveling wave on a string. The displacement of the wave has to match the displacement of the oscillator at $x = 0$ m.

SOLVE a. The equation for the displacement is

$$D(x, t) = A \sin(kx - \omega t + \phi_0)$$

with A, k, ω, and ϕ_0 to be determined. The wave amplitude is the same as the amplitude of the oscillator that generates the wave, so $A = 2.0$ mm. The oscillator has its maximum displacement $y_{osc} = A = 2.0$ mm at $t = 0$ s, thus

$$D(0 \text{ m}, 0 \text{ s}) = A \sin(\phi_0) = A$$

This requires the phase constant to be $\phi_0 = \pi/2$ rad. The wave's frequency is $f = 100$ Hz, the frequency of the source; therefore

the angular frequency is $\omega = 2\pi f = 200\pi$ rad/s. We still need $k = 2\pi/\lambda$, but we do not know the wavelength. However, we have enough information to determine the wave speed, and we can then use either $\lambda = v/f$ or $k = \omega/v$. The speed is

$$v = \sqrt{\frac{T_s}{\mu}} = \sqrt{\frac{5.0 \text{ N}}{0.0020 \text{ kg/m}}} = 50 \text{ m/s}$$

Using v, we find $\lambda = 0.50$ m and $k = 2\pi/\lambda = 4\pi$ rad/m. Thus the wave's displacement equation is

$$D(x, t) = (2.0 \text{ mm}) \times$$
$$\sin[2\pi((2.0 \text{ m}^{-1})x - (100 \text{ s}^{-1})t) + \pi/2 \text{ rad}]$$

Notice that we have separated out the 2π. This step is not essential, but for some problems it makes subsequent steps easier.

b. The wave's displacement at $t = 5.0$ ms $= 0.0050$ s is

$$D(x, t = 5.0 \text{ ms}) = (2.0 \text{ mm}) \sin(4\pi x - \pi \text{ rad} + \pi/2 \text{ rad})$$
$$= (2.0 \text{ mm}) \sin(4\pi x - \pi/2 \text{ rad})$$

At $x = 2.7$ m (calculator set to radians!), the displacement is

$$D(2.7 \text{ m}, 5.0 \text{ ms}) = 1.6 \text{ mm}$$

16.5 Sound and Light

Although there are many kinds of waves in nature, two are especially significant for us as humans. These are sound waves and light waves, the basis of hearing and seeing.

Sound Waves

FIGURE 16.19 A sound wave is a sequence of compressions and rarefactions.

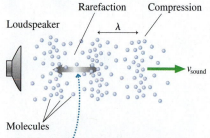

Individual molecules oscillate back and forth. As they do so, the compressions propagate forward at speed v_{sound}. Compressions are regions of higher pressure, so a sound wave is a pressure wave.

We usually think of sound waves traveling in air, but sound can travel through any gas, through liquids, and even through solids. **FIGURE 16.19** shows a loudspeaker cone vibrating back and forth in a fluid such as air or water. Each time the cone moves forward, it collides with the molecules and pushes them closer together. A half cycle later, as the cone moves backward, the fluid has room to expand and the density decreases a little. These regions of higher and lower density (and thus higher and lower pressure) are called **compressions** and **rarefactions.**

This periodic sequence of compressions and rarefactions travels outward from the loudspeaker as a longitudinal sound wave. When the wave reaches your ear, the oscillating pressure causes your eardrum to vibrate. These vibrations are transferred into your inner ear and perceived as sound.

The speed of sound waves depends on the compressibility of the medium. As **TABLE 16.1** shows, the speed is faster in liquids and solids (relatively incompressible) than in gases (highly compressible). For sound waves in air, the speed at temperature T (in °C) is

$$v_{\text{sound in air}} = 331 \text{ m/s} \times \sqrt{\frac{T(°C) + 273}{273}} \tag{16.36}$$

We'll derive this result in Section 16.6, but recall from chemistry that adding 273 to a Celsius temperature converts it to an absolute temperature in kelvins. The speed of sound increases with increasing temperature but, interestingly, does *not* depend on the air pressure. For air at room temperature (20°C),

$$v_{\text{sound in air}} = 343 \text{ m/s} \quad \text{(sound speed in air at 20°C)}$$

TABLE 16.1 The speed of sound

Medium	Speed (m/s)
Air (0°C)	331
Air (20°C)	343
Helium (0°C)	970
Ethyl alcohol	1170
Water (20°C)	1480
Granite	6000
Aluminum	6420

This is the value you should use when solving problems unless you're given temperature information.

A speed of 343 m/s is high, but not extraordinarily so. A distance as small as 100 m is enough to notice a slight delay between when you see something, such as a person hammering a nail, and when you hear it. The time required for sound to travel 1 km is $t = (1000 \text{ m})/(343 \text{ m/s}) \approx 3$ s. You may have learned to estimate the distance to a bolt of lightning by timing the number of seconds between when you see the flash and when you hear the thunder. Because sound takes 3 s to travel 1 km, the time divided by 3 gives the distance in kilometers. Or, in English units, the time divided by 5 gives the distance in miles.

Your ears are able to detect sound waves with frequencies between roughly 20 Hz and 20,000 Hz, or 20 kHz. You can use the fundamental relationship $v_{sound} = \lambda f$ to calculate that a 20 Hz sound wave has a 17-m-long wavelength, while the wavelength of a 20 kHz note is a mere 17 mm. Low frequencies are perceived as "low-pitch" bass notes, while high frequencies are heard as "high-pitch" treble notes. Your high-frequency range of hearing can deteriorate with age (10 kHz is the average upper limit at age 65) or as a result of exposure to very loud sounds.

Sound waves exist at frequencies well above 20 kHz, even though humans can't hear them. These are called *ultrasonic* frequencies. Oscillators vibrating at frequencies of many MHz generate the ultrasonic waves used in ultrasound medical imaging. A 3 MHz wave traveling through water (which is basically what your body is) at a sound speed of 1480 m/s has a wavelength of about 0.5 mm. It is this very small wavelength that allows ultrasound to image very small objects. We'll see why when we study *diffraction* in Chapter 33.

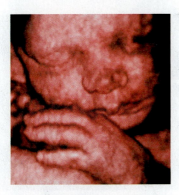

This ultrasound image is an example of using high-frequency sound waves to "see" within the human body.

Electromagnetic Waves

A light wave is an *electromagnetic wave,* a self-sustaining oscillation of the electromagnetic field. Other electromagnetic waves, such as radio waves, microwaves, and ultraviolet light, have the same physical characteristics as light waves even though we cannot sense them with our eyes. It is easy to demonstrate that light will pass unaffected through a container from which all the air has been removed, and light reaches us from distant stars through the vacuum of interstellar space. Such observations raise interesting but difficult questions. If light can travel through a region in which there is no matter, then what is the *medium* of a light wave? What is it that is waving?

It took scientists over 50 years, most of the 19th century, to answer this question. We will examine the answers in more detail in Chapter 31 after we introduce the ideas of electric and magnetic fields. For now we can say that light waves are a "self-sustaining oscillation of the electromagnetic field." That is, the displacement D is an electric or magnetic field. Being self-sustaining means that electromagnetic waves require *no material medium* in order to travel; hence electromagnetic waves are not mechanical waves. Fortunately, we can learn about the wave properties of light without having to understand electromagnetic fields.

It was predicted theoretically in the late 19th century, and has been subsequently confirmed, that all electromagnetic waves travel through vacuum with the same speed, called the *speed of light*. The value of the speed of light is

$$v_{light} = c = 299,792,458 \text{ m/s} \qquad \text{(electromagnetic wave speed in vacuum)}$$

where the special symbol c is used to designate the speed of light. (This is the c in Einstein's famous formula $E = mc^2$.) Now *this* is really moving—about one million times faster than the speed of sound in air!

NOTE ▶ $c = 3.00 \times 10^8$ m/s is the appropriate value to use in calculations.

The wavelengths of light are extremely small. You will learn in Chapter 33 how these wavelengths are determined, but for now we will note that visible light

White light passing through a prism is spread out into a band of colors called the *visible spectrum*.

is an electromagnetic wave with a wavelength (in air) in the range of roughly 400 nm (400×10^{-9} m) to 700 nm (700×10^{-9} m). Each wavelength is perceived as a different color, with the longer wavelengths seen as orange or red light and the shorter wavelengths seen as blue or violet light. A prism is able to spread the different wavelengths apart, from which we learn that "white light" is all the colors, or wavelengths, combined. The spread of colors seen with a prism, or seen in a rainbow, is called the *visible spectrum*.

If the wavelengths of light are unbelievably small, the oscillation frequencies are unbelievably large. The frequency for a 600 nm wavelength of light (orange) is

$$f = \frac{v}{\lambda} = \frac{3.00 \times 10^8 \text{ m/s}}{600 \times 10^{-9} \text{ m}} = 5.00 \times 10^{14} \text{ Hz}$$

The frequencies of light waves are roughly a factor of a trillion (10^{12}) higher than sound frequencies.

Electromagnetic waves exist at many frequencies other than the rather limited range that our eyes detect. One of the major technological advances of the 20th century was learning to generate and detect electromagnetic waves at many frequencies, ranging from low-frequency radio waves to the extraordinarily high frequencies of x rays. FIGURE 16.20 shows that the visible spectrum is a small slice of the much broader **electromagnetic spectrum.**

FIGURE 16.20 The electromagnetic spectrum from 10^6 Hz to 10^{18} Hz.

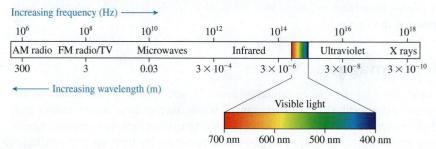

EXAMPLE 16.5 | Traveling at the speed of light

A satellite exploring Jupiter transmits data to the earth as a radio wave with a frequency of 200 MHz. What is the wavelength of the electromagnetic wave, and how long does it take the signal to travel 800 million kilometers from Jupiter to the earth?

SOLVE Radio waves are sinusoidal electromagnetic waves traveling with speed c. Thus

$$\lambda = \frac{c}{f} = \frac{3.00 \times 10^8 \text{ m/s}}{2.00 \times 10^8 \text{ Hz}} = 1.5 \text{ m}$$

The time needed to travel 800×10^6 km $= 8.0 \times 10^{11}$ m is

$$\Delta t = \frac{\Delta x}{c} = \frac{8.0 \times 10^{11} \text{ m}}{3.00 \times 10^8 \text{ m/s}} = 2700 \text{ s} = 45 \text{ min}$$

The Index of Refraction

Light waves travel with speed c in a vacuum, but they slow down as they pass through transparent materials such as water or glass or even, to a very slight extent, air. The slowdown is a consequence of interactions between the electromagnetic field of the wave and the electrons in the material. The speed of light in a material is characterized by the material's **index of refraction** n, defined as

$$n = \frac{\text{speed of light in a vacuum}}{\text{speed of light in the material}} = \frac{c}{v} \tag{16.37}$$

TABLE 16.2 Typical indices of refraction

Material	Index of refraction
Vacuum	1 exactly
Air	1.0003
Water	1.33
Glass	1.50
Diamond	2.42

The index of refraction of a material is always greater than 1 because $v < c$. A vacuum has $n = 1$ exactly. TABLE 16.2 shows the index of refraction for several materials. You can see that liquids and solids have larger indices of refraction than gases.

NOTE An accurate value for the index of refraction of air is relevant only in very precise measurements. We will assume $n_{air} = 1.00$ in this text.

If the speed of a light wave changes as it enters into a transparent material, such as glass, what happens to the light's frequency and wavelength? Because $v = \lambda f$, either λ or f or both have to change when v changes.

As an analogy, think of a sound wave in the air as it impinges on the surface of a pool of water. As the air oscillates back and forth, it periodically pushes on the surface of the water. These pushes generate the compressions of the sound wave that continues on into the water. Because each push of the air causes one compression of the water, the frequency of the sound wave in the water must be *exactly the same* as the frequency of the sound wave in the air. In other words, **the frequency of a wave is the frequency of the source. It does not change as the wave moves from one medium to another.**

The same is true for electromagnetic waves; the frequency does not change as the wave moves from one material to another.

FIGURE 16.21 shows a light wave passing through a transparent material with index of refraction n. As the wave travels through vacuum it has wavelength λ_{vac} and frequency f_{vac} such that $\lambda_{vac} f_{vac} = c$. In the material, $\lambda_{mat} f_{mat} = v = c/n$. The frequency does not change as the wave enters $(f_{mat} = f_{vac})$, so the wavelength must. The wavelength in the material is

$$\lambda_{mat} = \frac{v}{f_{mat}} = \frac{c}{nf_{mat}} = \frac{c}{nf_{vac}} = \frac{\lambda_{vac}}{n} \qquad (16.38)$$

The wavelength in the transparent material is less than the wavelength in vacuum. This makes sense. Suppose a marching band is marching at one step per second at a speed of 1 m/s. Suddenly they slow their speed to $\frac{1}{2}$ m/s but maintain their march at one step per second. The only way to go slower while marching at the same pace is to take *smaller steps*. When a light wave enters a material, the only way it can go slower while oscillating at the same frequency is to have a *smaller wavelength*.

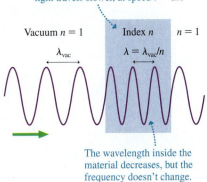

FIGURE 16.21 Light passing through a transparent material with index of refraction n.

A transparent material in which light travels slower, at speed $v = c/n$

Vacuum $n = 1$ Index n $n = 1$

λ_{vac} $\lambda = \lambda_{vac}/n$

The wavelength inside the material decreases, but the frequency doesn't change.

EXAMPLE 16.6 | **Light traveling through glass**

Orange light with a wavelength of 600 nm is incident upon a 1.00-mm-thick glass microscope slide.

a. What is the light speed in the glass?

b. How many wavelengths of the light are inside the slide?

SOLVE a. From Table 16.2 we see that the index of refraction of glass is $n_{glass} = 1.50$. Thus the speed of light in glass is

$$v_{glass} = \frac{c}{n_{glass}} = \frac{3.00 \times 10^8 \text{ m/s}}{1.50} = 2.00 \times 10^8 \text{ m/s}$$

b. The wavelength inside the glass is

$$\lambda_{glass} = \frac{\lambda_{vac}}{n_{glass}} = \frac{600 \text{ nm}}{1.50} = 400 \text{ nm} = 4.00 \times 10^{-7} \text{ m}$$

N wavelengths span a distance $d = N\lambda$, so the number of wavelengths in $d = 1.00$ mm is

$$N = \frac{d}{\lambda} = \frac{1.00 \times 10^{-3} \text{ m}}{4.00 \times 10^{-7} \text{ m}} = 2500$$

ASSESS The fact that 2500 wavelengths fit within 1 mm shows how small the wavelengths of light are.

STOP TO THINK 16.4 A light wave travels from left to right through three transparent materials of equal thickness. Rank in order, from largest to smallest, the indices of refraction n_a, n_b, and n_c.

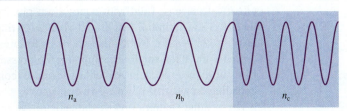

n_a n_b n_c

The Wave Model

We introduced the concept of a *wave model* at the beginning of this chapter. Now we're in a position to articulate what this means.

MODEL 16.1

The wave model

A wave is an organized disturbance that travels.

- Two classes of waves:
  - **Mechanical waves** travel through a medium.
  - **Electromagnetic waves** travel through vacuum.
- Two types of waves:
  - **Transverse waves** are displaced perpendicular to the direction in which the wave travels.
  - **Longitudinal waves** are displaced parallel to the direction in which the wave travels.
- The **wave speed** is a property of the medium.
- Sinusoidal waves are periodic in both time (period) and space (wavelength).
  - The wave frequency is the oscillation frequency of the source.
  - The fundamental relationship for periodic waves is $v = \lambda f$. This says that wave moves forward one wavelength during one period.

Transverse wave

Longitudinal wave

History graph

Snapshot graph

16.6 ADVANCED TOPIC The Wave Equation in a Fluid

In Section 16.4 we used Newton's second law to show that traveling waves can propagate on a stretched string *and* to predict the wave speed in terms of properties of the string. Now we wish to do the same for sound waves—longitudinal waves propagating through a fluid.

A sound wave is a sequence of compressions and rarefactions in which the fluid is alternately compressed and expanded. A substance's compressibility is characterized by its *bulk modulus B*, which you met in **« Section 14.6** when we looked at the elastic properties of materials. If excess pressure p is applied to an object of volume V, then the fractional change in volume—the fraction by which it's compressed—is

$$\frac{\Delta V}{V} = -\frac{p}{B} \tag{16.39}$$

The minus sign indicates that the volume *decreases* when pressure is applied. Gases are much more compressible than liquids, so gases have much smaller values of B than liquids.

Let's apply this to a fluid—either a liquid or a gas. **FIGURE 16.22** shows a small cylindrical piece of fluid with equilibrium pressure p_0 located between positions x and $x + \Delta x$. The initial length and volume of this little piece of fluid are $L_i = \Delta x$ and $V_i = aL_i = a\,\Delta x$. Notice that we use a for area in this chapter so that there is no conflict with A for amplitude. Suppose the pressure changes to $p_0 + p$. The volume of this little piece of fluid will either decrease (compression) or increase (expansion), depending on whether p is positive or negative.

The volume changes only if the ends of the cylinder undergo *different* displacements. (Equal displacements would shift the cylinder but not change its volume.) In

FIGURE 16.22 An element of fluid changes volume as the pressure changes.

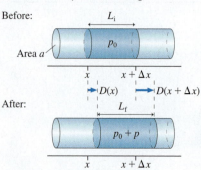

the bottom half of Figure 16.22 we see that the left end of the cylinder has undergone displacement $D(x, t)$ while the displacement at the right end is $D(x + \Delta x, t)$. Now the cylinder has length

$$L_f = L_i + (D(x + \Delta x, t) - D(x, t)) \tag{16.40}$$

and consequently its volume has *changed* by

$$\Delta V = a(L_f - L_i) = a(D(x + \Delta x, t) - D(x, t)) \tag{16.41}$$

Substituting both the initial volume and the volume change into Equation 16.39, we have

$$\frac{\Delta V}{V} = \frac{a(D(x + \Delta x, t) - D(x, t))}{a\,\Delta x} = -\frac{p}{B} \tag{16.42}$$

After canceling the a, you can see that, just as in Section 16.4, we're left—in the limit $\Delta x \to 0$—with the definition of the derivative of D with respect to x. It is again a *partial* derivative because we're holding the variable t constant. Thus we find that the fluid pressure (or, to be exact, the pressure deviation from p_0) at position x is related to the displacement of the medium by

$$p(x, t) = -B\frac{\partial D}{\partial x} \tag{16.43}$$

The pressure depends on how rapidly the fluid's displacement changes with position.

We anticipate that we'll discover sinusoidal sound waves later in this section, so a displacement wave of amplitude A,

$$D(x, t) = A\sin(kx - \omega t + \phi_0) \tag{16.44}$$

is associated with a pressure wave

$$p(x, t) = -B\frac{\partial D}{\partial x} = -kBA\cos(kx - \omega t + \phi_0)$$
$$= -p_{max}\cos(kx - \omega t + \phi_0) \tag{16.45}$$

The *pressure amplitude,* or maximum pressure, is

$$p_{max} = kBA = \frac{2\pi fBA}{v_{sound}} \tag{16.46}$$

where we used $\omega = 2\pi f = vk$ (Equation 16.13) in the last step to write the result in terms of the wave's speed and frequency. In other words, a sound wave is not just a traveling wave of molecular displacement. **A sound wave is also a traveling pressure wave.**

As an example, a quite loud 100 decibel, 500 Hz sound wave in air has a pressure amplitude of 2 Pa. That is, the pressure varies around atmospheric pressure by ± 2 Pa. You can use Equation 16.46 and $B_{air} = 1.42 \times 10^5$ Pa to find that the amplitude of the oscillating air molecules is a microscopic 1.5 μm.

FIGURE 16.23 uses Equations 16.44 and 16.45 to draw snapshot graphs of displacement and pressure for a sound wave propagating to the right. Positive displacement pushes molecules to the right while negative displacement pushes them to the left, so molecules pile up (a compression) at points where the displacement is changing from positive to negative. These are the points where the displacement has the *most negative slope* and thus, from Equation 16.43, the greatest pressure.

In general, the pressure wave has a maximum or minimum at points where the displacement wave is zero, and vice versa. This observation will help us understand standing sound waves in Chapter 17.

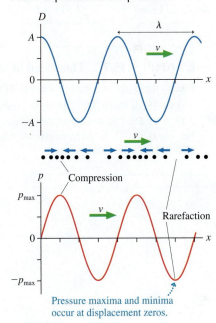

FIGURE 16.23 Snapshot graphs of the sound displacement and pressure.

Pressure maxima and minima occur at displacement zeros.

Predicting the Speed of Sound

Let's return to our small, cylindrical piece of fluid and this time, in **FIGURE 16.24** on the next page, apply Newton's second law to it. The fluid pressure at position x is $p(x, t)$,

FIGURE 16.24 Fluid pressure exerts a net force on the cylinder.

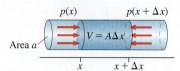

and this pressure pushes the cylinder to the right with force $ap(x, t)$. At the same time, the fluid at position $x + \Delta x$ has pressure $p(x + \Delta x, t)$, and it pushes the cylinder to the left with force $ap(x + \Delta x, t)$. The net force is

$$F_{\text{net }x} = ap(x, t) - ap(x + \Delta x, t) = -a(p(x + \Delta x, t) - p(x, t)) \quad (16.47)$$

The minus sign arises because the pressure forces push in opposite directions.

Newton's second law is $F_{\text{net }x} = ma_x$. The cylinder's mass is $m = \rho V = \rho a \Delta x$, where ρ is the fluid density. (Don't confuse area a with acceleration a_x!) The acceleration, just as in our analysis of a string, is the second partial derivative of displacement with respect to time. Thus the second law for our little cylinder of fluid is

$$F_{\text{net }x} = -a(p(x + \Delta x, t) - p(x, t)) = ma_x = \rho a \Delta x \frac{\partial^2 D}{\partial t^2} \quad (16.48)$$

The area a cancels, and a slight rearrangement gives

$$\frac{\partial^2 D}{\partial t^2} = -\frac{1}{\rho} \frac{p(x + \Delta x, t) - p(x, t)}{\Delta x} \rightarrow -\frac{1}{\rho} \frac{\partial p}{\partial x} \quad (16.49)$$

where, once again, we have a derivative in the limit $\Delta x \rightarrow 0$.

Fortunately, we already found, in Equation 16.43, that $p = -B \, \partial D/\partial x$. Substituting this for p in Equation 16.49 gives

$$\frac{\partial^2 D}{\partial t^2} = \frac{B}{\rho} \frac{\partial^2 D}{\partial x^2} \quad (16.50)$$

Equation 16.50 is a wave equation! Just as in our analysis of a string, applying Newton's second law to a small piece of the medium has led to a wave equation. We've already shown that a sinusoidal traveling wave, Equation 16.44, is a solution, so we don't need to prove it again. Further, by comparing Equation 16.50 to the general wave equation, Equation 16.35, we can predict the speed of sound in a fluid:

$$v_{\text{sound}} = \sqrt{\frac{B}{\rho}} \quad (16.51)$$

TABLE 16.3 Bulk moduli of common fluids

Medium	B (Pa)
Mercury (20°C)	2.85×10^{10}
Water (20°C)	2.18×10^{9}
Ethyl alcohol (20°C)	1.06×10^{9}
Helium (0°C, 1 atm)	1.688×10^{5}
Air (0°C, 1 atm)	1.418×10^{5}

TABLE 16.3 gives values of the bulk modulus for several common fluids.

EXAMPLE 16.7 | **The speed of sound in water**

Predict the speed of sound in water at 20°C.

SOLVE From Table 16.3, the bulk modulus of water at 20°C is 2.18×10^9 Pa. The density of water is usually given as 1000 kg/m³, but this is at 4°C. To three significant figures, the density at 20°C is 998 kg/m³. Thus we predict

$$v_{\text{sound}} = \sqrt{\frac{2.18 \times 10^9 \text{ Pa}}{998 \text{ kg/m}^3}} = 1480 \text{ m/s}$$

This is exactly the value given earlier in Table 16.1.

For gases, both B and ρ are proportional to the pressure, so their ratio is independent of pressure. At 0°C and 1 atm, the density of air is $\rho_0 = 1.292$ kg/m³. Thus the speed of sound in air at 0°C is

$$v_{\text{sound in air}} = \sqrt{\frac{B_0}{\rho_0}} = \sqrt{\frac{1.418 \times 10^5 \text{ Pa}}{1.292 \text{ kg/m}^3}} = 331 \text{ m/s} \quad (\text{at } 0°C) \quad (16.52)$$

exactly as shown in Table 16.1.

You can use the ideal-gas law to show that the density (at constant pressure) of a gas is inversely proportional to its absolute temperature T in kelvins. If the density at 0°C and 1 atm is ρ_0, then the density at temperature T is

$$\rho_T = \rho_0 \frac{273}{T(\text{K})} = \rho_0 \frac{273}{T(°C) + 273} \quad (16.53)$$

where we used 0°C = 273 K to convert kelvin to °C. Thus a general expression for the speed of sound in air is

$$v_{\text{sound in air}} = \sqrt{\frac{B}{\rho}} = \sqrt{\frac{B_0}{\rho_0}\frac{T(°C) + 273}{273}} = 331 \text{ m/s} \times \sqrt{\frac{T(°C) + 273}{273}} \quad (16.54)$$

This was the expression given without proof in Section 16.5. Now we see that it comes from the wave equation for sound with a little help from the ideal-gas law.

16.7 Waves in Two and Three Dimensions

Suppose you were to take a photograph of ripples spreading on a pond. If you mark the location of the *crests* on the photo, your picture would look like **FIGURE 16.25a.** The lines that locate the crests are called **wave fronts,** and they are spaced precisely one wavelength apart. The diagram shows only a single instant of time, but you can imagine a movie in which you would see the wave fronts moving outward from the source at speed *v*. A wave like this is called a **circular wave.** It is a two-dimensional wave that spreads across a surface.

Although the wave fronts are circles, you would hardly notice the curvature if you observed a small section of the wave front very, very far away from the source. The wave fronts would appear to be parallel lines, still spaced one wavelength apart and traveling at speed *v*. A good example is an ocean wave reaching a beach. Ocean waves are generated by storms and wind far out at sea, hundreds or thousands of miles away. By the time they reach the beach where you are working on your tan, the crests appear to be straight lines. An aerial view of the ocean would show a wave diagram like **FIGURE 16.25b.**

Many waves of interest, such as sound waves or light waves, move in three dimensions. For example, loudspeakers and lightbulbs emit **spherical waves.** That is, the crests of the wave form a series of concentric spherical shells separated by the wavelength λ. In essence, the waves are three-dimensional ripples. It will still be useful to draw wave-front diagrams such as Figure 16.25, but now the circles are slices through the spherical shells locating the wave crests.

If you observe a spherical wave very, very far from its source, the small piece of the wave front that you can see is a little patch on the surface of a very large sphere. If the radius of the sphere is sufficiently large, you will not notice the curvature and this little patch of the wave front appears to be a plane. **FIGURE 16.26** illustrates the idea of a **plane wave.**

To visualize a plane wave, imagine standing on the *x*-axis facing a sound wave as it comes toward you from a very distant loudspeaker. Sound is a longitudinal wave, so the particles of medium oscillate toward you and away from you. If you were to locate all of the particles that, at one instant of time, were at their maximum displacement toward you, they would all be located in a plane perpendicular to the travel direction. This is one of the wave fronts in Figure 16.26, and all the particles in this plane are doing exactly the same thing at that instant of time. This plane is moving toward you at speed *v*. There is another plane one wavelength behind it where the molecules are also at maximum displacement, yet another two wavelengths behind the first, and so on.

Because a plane wave's displacement depends on *x* but not on *y* or *z*, the displacement function $D(x, t)$ describes a plane wave just as readily as it does a one-dimensional wave. Once you specify a value for *x*, the displacement is the same at every point in the *yz*-plane that slices the *x*-axis at that value (i.e., one of the planes shown in Figure 16.26).

NOTE There are no perfect plane waves in nature, but many waves of practical interest can be modeled as plane waves.

We can describe a circular wave or a spherical wave by changing the mathematical description from $D(x, t)$ to $D(r, t)$, where *r* is the radial distance measured outward from the source. Then the displacement of the medium will be the same at every point

FIGURE 16.25 The wave fronts of a circular or spherical wave.

(a) Wave fronts are the crests of the wave. They are spaced one wavelength apart.

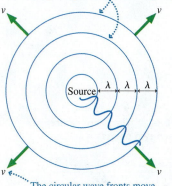

The circular wave fronts move outward from the source at speed *v*.

(b)

Very far away from the source, small sections of the wave fronts appear to be straight lines.

FIGURE 16.26 A plane wave.

Very far from the source, small segments of spherical wave fronts appear to be planes. The wave is cresting at every point in these planes.

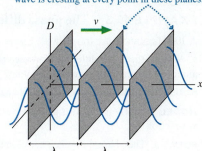

on a spherical surface. In particular, a sinusoidal spherical wave with wave number k and angular frequency ω is written

$$D(r, t) = A(r) \sin(kr - \omega t + \phi_0) \qquad (16.55)$$

Other than the change of x to r, the only difference is that the amplitude is now a function of r. A one-dimensional wave propagates with no change in the wave amplitude. But circular and spherical waves spread out to fill larger and larger volumes of space. To conserve energy, an issue we'll look at later in the chapter, the wave's amplitude has to decrease with increasing distance r. This is why sound and light decrease in intensity as you get farther from the source. We don't need to specify exactly how the amplitude decreases with distance, but you should be aware that it does.

Phase and Phase Difference

« Section 15.2 introduced the concept of *phase* for an oscillator in simple harmonic motion. Phase is also important for waves. The **phase** of a sinusoidal wave, denoted ϕ, is the quantity $(kx - \omega t + \phi_0)$. Phase will be an important concept in Chapter 17, where we will explore the consequences of adding various waves together. For now, we can note that the wave fronts seen in Figures 16.25 and 16.26 are "surfaces of constant phase." To see this, write the displacement as simply $D(x, t) = A \sin \phi$. Because each point on a wave front has the same displacement, the phase must be the same at every point.

It will be useful to know the *phase difference* $\Delta \phi$ between two different points on a sinusoidal wave. **FIGURE 16.27** shows two points on a sinusoidal wave at time t. The phase difference between these points is

$$\Delta \phi = \phi_2 - \phi_1 = (kx_2 - \omega t + \phi_0) - (kx_1 - \omega t + \phi_0)$$
$$= k(x_2 - x_1) = k \Delta x = 2\pi \frac{\Delta x}{\lambda} \qquad (16.56)$$

That is, **the phase difference between two points on a wave depends on only the ratio of their separation Δx to the wavelength λ.** For example, two points on a wave separated by $\Delta x = \frac{1}{2} \lambda$ have a phase difference $\Delta \phi = \pi$ rad.

An important consequence of Equation 16.56 is that **the phase difference between two adjacent wave fronts is $\Delta \phi = 2\pi$ rad.** This follows from the fact that two adjacent wave fronts are separated by $\Delta x = \lambda$. This is an important idea. Moving from one crest of the wave to the next corresponds to changing the *distance* by λ and changing the *phase* by 2π rad.

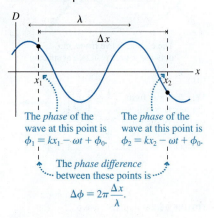

FIGURE 16.27 The phase difference between two points on a wave.

The *phase* of the wave at this point is $\phi_1 = kx_1 - \omega t + \phi_0$.

The *phase* of the wave at this point is $\phi_2 = kx_2 - \omega t + \phi_0$.

The *phase difference* between these points is

$$\Delta \phi = 2\pi \frac{\Delta x}{\lambda}.$$

EXAMPLE 16.8 | **The phase difference between two points on a sound wave**

A 100 Hz sound wave travels with a wave speed of 343 m/s.

a. What is the phase difference between two points 60.0 cm apart along the direction the wave is traveling?

b. How far apart are two points whose phase differs by 90°?

MODEL Treat the wave as a plane wave traveling in the positive x-direction.

SOLVE a. The phase difference between two points is

$$\Delta \phi = 2\pi \frac{\Delta x}{\lambda}$$

In this case, $\Delta x = 60.0$ cm $= 0.600$ m. The wavelength is

$$\lambda = \frac{v}{f} = \frac{343 \text{ m/s}}{100 \text{ Hz}} = 3.43 \text{ m}$$

and thus

$$\Delta \phi = 2\pi \frac{0.600 \text{ m}}{3.43 \text{ m}} = 0.350\pi \text{ rad} = 63.0°$$

b. A phase difference $\Delta \phi = 90°$ is $\pi/2$ rad. This will be the phase difference between two points when $\Delta x/\lambda = \frac{1}{4}$, or when $\Delta x = \lambda/4$. Here, with $\lambda = 3.43$ m, $\Delta x = 85.8$ cm.

ASSESS The phase difference increases as Δx increases, so we expect the answer to part b to be larger than 60 cm.

STOP TO THINK 16.5   What is the phase difference between the crest of a wave and the adjacent trough?

a. -2π rad
d. $\pi/2$ rad

b. 0 rad
e. π rad

c. $\pi/4$ rad
f. 3π rad

16.8 Power, Intensity, and Decibels

A traveling wave transfers energy from one point to another. The sound wave from a loudspeaker sets your eardrum into motion. Light waves from the sun warm the earth. The *power* of a wave is the rate, in joules per second, at which the wave transfers energy. As you learned in Chapter 9, power is measured in watts. A loudspeaker might emit 2 W of power, meaning that energy in the form of sound waves is radiated at the rate of 2 joules per second.

A focused light, like that of a projector, is more *intense* than the diffuse light that goes in all directions. Similarly, a loudspeaker that beams its sound forward into a small area produces a louder sound in that area than a speaker of equal power that radiates the sound in all directions. Quantities such as brightness and loudness depend not only on the rate of energy transfer, or power, but also on the *area* that receives that power.

FIGURE 16.28 shows a wave impinging on a surface of area a. The surface is perpendicular to the direction in which the wave is traveling. This might be a real, physical surface, such as your eardrum or a photovoltaic cell, but it could equally well be a mathematical surface in space that the wave passes right through. If the wave has power P, we define the **intensity** I of the wave to be

$$I = \frac{P}{a} = \text{power-to-area ratio} \qquad (16.57)$$

The SI units of intensity are W/m². Because intensity is a power-to-area ratio, a wave focused into a small area will have a larger intensity than a wave of equal power that is spread out over a large area.

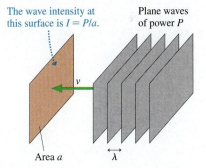

FIGURE 16.28 Plane waves of power P impinge on area a with intensity $I = P/a$.

The wave intensity at this surface is $I = P/a$.

Plane waves of power P

v

Area a

λ

EXAMPLE 16.9 | **The intensity of a laser beam**

A typical red laser pointer emits 1.0 mW of light power into a 1.0-mm-diameter laser beam. What is the intensity of the laser beam?

MODEL The laser beam is a light wave.

SOLVE The light waves of the laser beam pass through a mathematical surface that is a circle of diameter 1.0 mm. The intensity of the laser beam is

$$I = \frac{P}{a} = \frac{P}{\pi r^2} = \frac{0.0010 \text{ W}}{\pi (0.00050 \text{ m})^2} = 1300 \text{ W/m}^2$$

ASSESS This is roughly the intensity of sunlight at noon on a summer day. The difference between the sun and a small laser is not their intensities, which are about the same, but their powers. The laser has a small power of 1 mW. It can produce a very intense wave only because the area through which the wave passes is very small. The sun, by contrast, radiates a total power $P_{\text{sun}} \approx 4 \times 10^{26}$ W. This immense power is spread through *all* of space, producing an intensity of 1400 W/m² at a distance of 1.5×10^{11} m, the radius of the earth's orbit.

If a source of spherical waves radiates uniformly in all directions, then, as **FIGURE 16.29** on the next page shows, the power at distance r is spread uniformly over the surface of a sphere of radius r. The surface area of a sphere is $a = 4\pi r^2$, so the intensity of a uniform spherical wave is

$$I = \frac{P_{\text{source}}}{4\pi r^2} \qquad \text{(intensity of a uniform spherical wave)} \qquad (16.58)$$

The inverse-square dependence of r is really just a statement of energy conservation. The source emits energy at the rate P joules per second. The energy is spread over a

FIGURE 16.29 A source emitting uniform spherical waves.

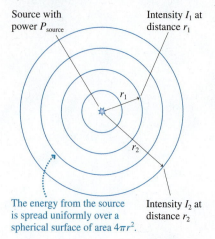

Source with power P_{source}

Intensity I_1 at distance r_1

r_1

r_2

The energy from the source is spread uniformly over a spherical surface of area $4\pi r^2$.

Intensity I_2 at distance r_2

larger and larger area as the wave moves outward. Consequently, the energy *per unit area* must decrease in proportion to the surface area of a sphere.

If the intensity at distance r_1 is $I_1 = P_{source}/4\pi r_1^2$ and the intensity at r_2 is $I_2 = P_{source}/4\pi r_2^2$, then you can see that the intensity *ratio* is

$$\frac{I_1}{I_2} = \frac{r_2^2}{r_1^2} \tag{16.59}$$

You can use Equation 16.59 to compare the intensities at two distances from a source without needing to know the power of the source.

> **NOTE** Wave intensities are strongly affected by reflections and absorption. Equations 16.58 and 16.59 apply to situations such as the light from a star or the sound from a firework exploding high in the air. Indoor sound does *not* obey a simple inverse-square law because of the many reflecting surfaces.

For a sinusoidal wave, each particle in the medium oscillates back and forth in simple harmonic motion. You learned in Chapter 15 that a particle in SHM with amplitude A has energy $E = \frac{1}{2}kA^2$, where k is the spring constant of the medium, not the wave number. It is this oscillatory energy of the medium that is transferred, particle to particle, as the wave moves through the medium.

Because a wave's intensity is proportional to the rate at which energy is transferred through the medium, and because the oscillatory energy in the medium is proportional to the *square* of the amplitude, we can infer that

$$I \propto A^2 \tag{16.60}$$

That is, **the intensity of a wave is proportional to the square of its amplitude.** If you double the amplitude of a wave, you increase its intensity by a factor of 4.

Sound Intensity Level

Human hearing spans an extremely wide range of intensities, from the *threshold of hearing* at $\approx 1 \times 10^{-12}$ W/m^2 (at midrange frequencies) to the *threshold of pain* at ≈ 10 W/m^2. If we want to make a scale of loudness, it's convenient and logical to place the zero of our scale at the threshold of hearing. To do so, we define the **sound intensity level,** expressed in **decibels** (dB), as

$$\beta = (10 \text{ dB}) \log_{10}\left(\frac{I}{I_0}\right) \tag{16.61}$$

where $I_0 = 1.0 \times 10^{-12}$ W/m^2. The symbol β is the Greek letter beta. Notice that β is computed as a base-10 logarithm, not a natural logarithm.

The decibel is named after Alexander Graham Bell, inventor of the telephone. Sound intensity level is actually dimensionless because it's formed from the ratio of two intensities, so decibels are just a *name* to remind us that we're dealing with an intensity *level* rather than a true intensity.

Right at the threshold of hearing, where $I = I_0$, the sound intensity level is

$$\beta = (10 \text{ dB}) \log_{10}\left(\frac{I_0}{I_0}\right) = (10 \text{ dB}) \log_{10}(1) = 0 \text{ dB}$$

Note that 0 dB doesn't mean no sound; it means that, for most people, no sound is heard. Dogs have more sensitive hearing than humans, and most dogs can easily perceive a 0 dB sound. The sound intensity level at the pain threshold is

$$\beta = (10 \text{ dB}) \log_{10}\left(\frac{10 \text{ W/m}^2}{10^{-12} \text{ W/m}^2}\right) = (10 \text{ dB}) \log_{10}(10^{13}) = 130 \text{ dB}$$

The major point to notice is that the sound intensity level increases by 10 dB each time the actual intensity increases by a *factor* of 10. For example, the sound intensity level increases from 70 dB to 80 dB when the sound intensity increases from 10^{-5} W/m^2 to 10^{-4} W/m^2. Perception experiments find that sound is perceived as "twice as loud" when the intensity increases by a factor of 10. In terms of decibels, we can say that the perceived loudness of a sound doubles with each increase in the sound intensity level by 10 dB.

TABLE 16.4 gives the sound intensity levels for a number of sounds. Although 130 dB is the threshold of pain, quieter sounds can damage your hearing. A fairly short exposure to 120 dB can cause damage to the hair cells in the ear, but lengthy exposure to sound intensity levels of over 85 dB can produce damage as well.

TABLE 16.4 Sound intensity levels of common sounds

Sound	β (dB)
Threshold of hearing	0
Person breathing, at 3 m	10
A whisper, at 1 m	20
Quiet room	30
Outdoors, no traffic	40
Quiet restaurant	50
Normal conversation, at 1 m	60
Busy traffic	70
Vacuum cleaner, for user	80
Niagara Falls, at viewpoint	90
Snowblower, at 2 m	100
Stereo, at maximum volume	110
Rock concert	120
Threshold of pain	130
Loudest football stadium	140

EXAMPLE 16.10 | **Blender noise**

The blender making a smoothie produces a sound intensity level of 83 dB. What is the intensity of the sound? What will the sound intensity level be if a second blender is turned on?

SOLVE We can solve Equation 16.61 for the sound intensity, finding $I = I_0 \times 10^{\beta/10\ dB}$. Here we used the fact that 10 raised to a power is an "antilogarithm." In this case,

$$I = (1.0 \times 10^{-12}\ \text{W/m}^2) \times 10^{8.3} = 2.0 \times 10^{-4}\ \text{W/m}^2$$

A second blender doubles the sound power and thus raises the intensity to $I = 4.0 \times 10^{-4}$ W/m^2. The new sound intensity level is

$$\beta = (10\ \text{dB}) \log_{10}\left(\frac{4.0 \times 10^{-4}\ \text{W/m}^2}{1.0 \times 10^{-12}\ \text{W/m}^2}\right) = 86\ \text{dB}$$

ASSESS In general, doubling the actual sound intensity increases the decibel level by 3 dB.

STOP TO THINK 16.6 Four trumpet players are playing the same note. If three of them suddenly stop, the sound intensity level decreases by

a. 40 dB   b. 12 dB   c. 6 dB   d. 4 dB

16.9 The Doppler Effect

Our final topic for this chapter is an interesting effect that occurs when you are in motion relative to a wave source. It is called the *Doppler effect*. You've likely noticed that the pitch of an ambulance's siren drops as it goes past you. Why?

FIGURE 16.30a on the next page shows a source of sound waves moving away from Pablo and toward Nancy at a steady speed v_s. The subscript s indicates that this is the speed of the source, not the speed of the waves. The source is emitting sound waves of frequency f_0 as it travels. The figure is a motion diagram showing the position of the source at times $t = 0$, T, $2T$, and $3T$, where $T = 1/f_0$ is the period of the waves.

Nancy measures the frequency of the wave emitted by the *approaching source* to be f_+. At the same time, Pablo measures the frequency of the wave emitted by the *receding source* to be f_-. Our task is to relate f_+ and f_- to the source frequency f_0 and speed v_s.

After a wave crest leaves the source, its motion is governed by the properties of the medium. That is, the motion of the source cannot affect a wave that has already been emitted. Thus each circular wave front in FIGURE 16.30b is centered on the point from which it was emitted. The wave crest from point 3 was emitted just as this figure was made, but it hasn't yet had time to travel any distance.

FIGURE 16.30 A motion diagram showing the wave fronts emitted by a source as it moves to the right at speed v_s.

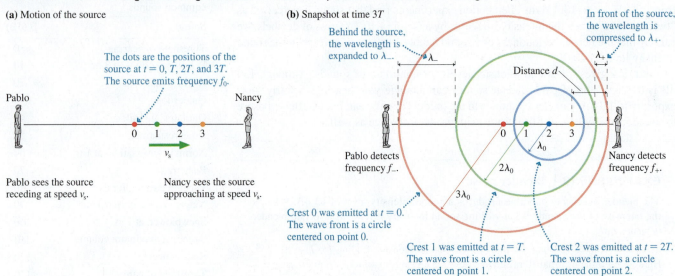

(a) Motion of the source

The dots are the positions of the source at $t = 0$, T, $2T$, and $3T$. The source emits frequency f_0.

Pablo

Nancy

v_s

Pablo sees the source receding at speed v_s.

Nancy sees the source approaching at speed v_s.

(b) Snapshot at time $3T$

Behind the source, the wavelength is expanded to λ_-.

λ_-

In front of the source, the wavelength is compressed to λ_+.

λ_+

Distance d

λ_0

$2\lambda_0$

$3\lambda_0$

Pablo detects frequency f_-.

Nancy detects frequency f_+.

Crest 0 was emitted at $t = 0$. The wave front is a circle centered on point 0.

Crest 1 was emitted at $t = T$. The wave front is a circle centered on point 1.

Crest 2 was emitted at $t = 2T$. The wave front is a circle centered on point 2.

The wave crests are bunched up in the direction the source is moving, stretched out behind it. The distance between one crest and the next is one wavelength, so the wavelength λ_+ Nancy measures is *less* than the wavelength $\lambda_0 = v/f_0$ that would be emitted if the source were at rest. Similarly, λ_- behind the source is larger than λ_0.

These crests move through the medium at the wave speed v. Consequently, the frequency $f_+ = v/\lambda_+$ detected by the observer whom the source is approaching is *higher* than the frequency f_0 emitted by the source. Similarly, $f_- = v/\lambda_-$ detected behind the source is *lower* than frequency f_0. This change of frequency when a source moves relative to an observer is called the **Doppler effect.**

The distance labeled d in Figure 16.30b is the difference between how far the wave has moved and how far the source has moved at time $t = 3T$. These distances are

$$\Delta x_{\text{wave}} = vt = 3vT$$
$$\Delta x_{\text{source}} = v_s t = 3v_s T \tag{16.62}$$

The distance d spans three wavelengths; thus the wavelength of the wave emitted by an approaching source is

$$\lambda_+ = \frac{d}{3} = \frac{\Delta x_{\text{wave}} - \Delta x_{\text{source}}}{3} = \frac{3vT - 3v_s T}{3} = (v - v_s)T \tag{16.63}$$

You can see that our arbitrary choice of three periods was not relevant because the 3 cancels. The frequency detected in Nancy's direction is

$$f_+ = \frac{v}{\lambda_+} = \frac{v}{(v - v_s)T} = \frac{v}{(v - v_s)}f_0 \tag{16.64}$$

where $f_0 = 1/T$ is the frequency of the source and is the frequency you would detect if the source were at rest. We'll find it convenient to write the detected frequency as

$$f_+ = \frac{f_0}{1 - v_s/v} \qquad \text{(Doppler effect for an approaching source)}$$
$$\tag{16.65}$$
$$f_- = \frac{f_0}{1 + v_s/v} \qquad \text{(Doppler effect for a receding source)}$$

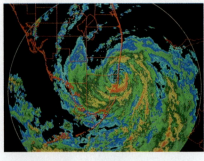

Doppler weather radar uses the Doppler shift of reflected radar signals to measure wind speeds and thus better gauge the severity of a storm.

Proof of the second version, for the frequency f_- of a receding source, is similar. You can see that $f_+ > f_0$ in front of the source, because the denominator is less than 1, and $f_- < f_0$ behind the source.

EXAMPLE 16.11 | **How fast are the police traveling?**

A police siren has a frequency of 550 Hz as the police car approaches you, 450 Hz after it has passed you and is receding. How fast are the police traveling? The temperature is 20°C.

MODEL The siren's frequency is altered by the Doppler effect. The frequency is f_+ as the car approaches and f_- as it moves away.

SOLVE To find v_s, we rewrite Equations 16.65 as

$$f_0 = (1 + v_s/v)f_-$$

$$f_0 = (1 - v_s/v)f_+$$

We subtract the second equation from the first, giving

$$0 = f_- - f_+ + \frac{v_s}{v}(f_- + f_+)$$

This is easily solved to give

$$v_s = \frac{f_+ - f_-}{f_+ + f_-}v = \frac{100 \text{ Hz}}{1000 \text{ Hz}} \times 343 \text{ m/s} = 34.3 \text{ m/s}$$

ASSESS If you now solve for the siren frequency when at rest, you will find $f_0 = 495$ Hz. Surprisingly, the at-rest frequency is not halfway between f_- and f_+.

A Stationary Source and a Moving Observer

Suppose the police car in Example 16.11 is at rest while you drive toward it at 34.3 m/s. You might think that this is equivalent to having the police car move toward you at 34.3 m/s, but it isn't. Mechanical waves move through a medium, and the Doppler effect depends not just on how the source and the observer move with respect to each other but also on how they move with respect to the medium. We'll omit the proof, but it's not hard to show that the frequencies heard by an observer moving at speed v_o relative to a stationary source emitting frequency f_0 are

$$f_+ = (1 + v_o/v)f_0 \quad \text{(observer approaching a source)}$$
$$f_- = (1 - v_o/v)f_0 \quad \text{(observer receding from a source)} \tag{16.66}$$

A quick calculation shows that the frequency of the police siren as you approach it at 34.3 m/s is 545 Hz, not the 550 Hz you heard as it approached you at 34.3 m/s.

The Doppler Effect for Light Waves

The Doppler effect is observed for all types of waves, not just sound waves. If a source of light waves is receding from you, the wavelength λ_- that you detect is longer than the wavelength λ_0 emitted by the source.

Although the reason for the Doppler shift for light is the same as for sound waves, there is one fundamental difference. We derived Equations 16.65 for the Doppler-shifted frequencies by measuring the wave speed v relative to the medium. For electromagnetic waves in empty space, there is no medium. Consequently, we need to turn to Einstein's theory of relativity to determine the frequency of light waves from a moving source. The result, which we state without proof, is

$$\lambda_- = \sqrt{\frac{1 + v_s/c}{1 - v_s/c}}\,\lambda_0 \quad \text{(receding source)}$$

$$\lambda_+ = \sqrt{\frac{1 - v_s/c}{1 + v_s/c}}\,\lambda_0 \quad \text{(approaching source)} \tag{16.67}$$

Here v_s is the speed of the source *relative to* the observer.

The light waves from a receding source are shifted to longer wavelengths ($\lambda_- > \lambda_0$). Because the longest visible wavelengths are perceived as the color red, the light from a receding source is **red shifted.** That is *not* to say that the light is red, simply that its wavelength is shifted toward the red end of the spectrum. If $\lambda_0 = 470$ nm (blue) light emitted by a rapidly receding source is detected at $\lambda_- = 520$ nm (green), we would say that the light has been red shifted. Similarly, light from an approaching source is **blue shifted,** meaning that the detected wavelengths are shorter than the emitted wavelengths ($\lambda_+ < \lambda_0$) and thus are shifted toward the blue end of the spectrum.

EXAMPLE 16.12 | Measuring the velocity of a galaxy

Hydrogen atoms in the laboratory emit red light with wavelength 656 nm. In the light from a distant galaxy, this "spectral line" is observed at 691 nm. What is the speed of this galaxy relative to the earth?

MODEL The observed wavelength is longer than the wavelength emitted by atoms at rest with respect to the observer (i.e., red shifted), so we are looking at light emitted from a galaxy that is receding from us.

SOLVE Squaring the expression for λ_- in Equations 16.67 and solving for v_s give

$$v_s = \frac{(\lambda_-/\lambda_0)^2 - 1}{(\lambda_-/\lambda_0)^2 + 1} c$$

$$= \frac{(691 \text{ nm}/656 \text{ nm})^2 - 1}{(691 \text{ nm}/656 \text{ nm})^2 + 1} c$$

$$= 0.052c = 1.56 \times 10^7 \text{ m/s}$$

ASSESS The galaxy is moving away from the earth at about 5% of the speed of light!

FIGURE 16.31 A Hubble Space Telescope picture of a quasar.

In the 1920s, an analysis of the red shifts of many galaxies led the astronomer Edwin Hubble to the conclusion that the galaxies of the universe are *all* moving apart from each other. Extrapolating backward in time must bring us to a point when all the matter of the universe—and even space itself, according to the theory of relativity—began rushing out of a primordial fireball. Many observations and measurements since have given support to the idea that the universe began in a *Big Bang* about 14 billion years ago.

As an example, **FIGURE 16.31** is a Hubble Space Telescope picture of a *quasar,* short for *quasistellar object.* Quasars are extraordinarily powerful sources of light and radio waves. The light reaching us from quasars is highly red shifted, corresponding in some cases to objects that are moving away from us at greater than 90% of the speed of light. Astronomers have determined that some quasars are 10 to 12 *billion* light years away from the earth, hence the light we see was emitted when the universe was only about 25% of its present age. Today, the red shifts of distant quasars and supernovae (exploding stars) are being used to refine our understanding of the structure and evolution of the universe.

STOP TO THINK 16.7 Amy and Zack are both listening to the source of sound waves that is moving to the right. Compare the frequencies each hears.

a. $f_{Amy} > f_{Zack}$
b. $f_{Amy} = f_{Zack}$
c. $f_{Amy} < f_{Zack}$

Amy   10 m/s f_0 10 m/s   10 m/s   Zack

CHALLENGE EXAMPLE 16.13 | Decreasing the sound

The loudspeaker on a homecoming float—mounted on a pole—is stuck playing an annoying 210 Hz tone. When the speaker is 10 m away, you measure the sound to be a loud 95 dB at 208 Hz. How long will it take for the sound intensity level to drop to a tolerable 55 dB?

MODEL The source is on a pole, so model the sound waves as uniform spherical waves. Assume a temperature of 20°C.

SOLVE The 208 Hz frequency you measure is less than the 210 Hz frequency that was emitted, so the float must be moving away from you. The Doppler effect for a receding source is

$$f_- = \frac{f_0}{1 + v_s/v}$$

We can solve this to find the speed of the float:

$$v_s = \left(\frac{f_0}{f_-} - 1\right)v = \left(\frac{210 \text{ Hz}}{208 \text{ Hz}} - 1\right) \times 343 \text{ m/s} = 3.3 \text{ m/s}$$

The sound intensity of a spherical wave decreases with the inverse square of the distance from the source. A sound intensity level β corresponds to an intensity $I = I_0 \times 10^{\beta/10 \text{ dB}}$, where $I_0 = 1.0 \times 10^{-12}$ W/m². At the initial 95 dB, the intensity is

$$I_1 = I_0 \times 10^{9.5} = 3.2 \times 10^{-3} \text{ W/m}^2$$

At the desired 55 dB, the intensity will have dropped to

$$I_2 = I_0 \times 10^{5.5} = 3.2 \times 10^{-7} \text{ W/m}^2$$

The intensity ratio is related to the distances by

$$\frac{I_1}{I_2} = \frac{r_2^2}{r_1^2}$$

Thus the sound will have dropped to 55 dB when the distance to the speaker is

$$r_2 = \sqrt{\frac{I_1}{I_2}} r_1 = \sqrt{10^4} \times 10 \text{ m} = 1000 \text{ m}$$

The float has to travel $\Delta x = 990$ m, which will take

$$\Delta t = \frac{\Delta x}{v_s} = \frac{990 \text{ m}}{3.3 \text{ m/s}} = 300 \text{ s} = 5.0 \text{ min}$$

ASSESS To drop the sound intensity level by 40 dB requires decreasing the intensity by a factor of 10^4. And with the intensity depending on the inverse square of the distance, that requires increasing the distance by a factor of 100. Floats don't move very fast—3.3 m/s is about 7 mph—so needing several minutes to travel the ≈ 1000 m seems reasonable.

SUMMARY

The goal of Chapter 16 has been to learn the basic properties of traveling waves.

GENERAL PRINCIPLES

The Wave Model

This model is based on the idea of a traveling wave, which is an organized disturbance traveling at a well-defined **wave speed** v.

- In transverse waves the displacement is perpendicular to the direction in which the wave travels.
- In longitudinal waves the particles of the medium are displaced parallel to the direction in which the wave travels.

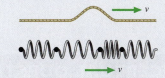

A wave transfers **energy,** but no material or substance is transferred outward from the source.

Two basic classes of waves:

- Mechanical waves travel through a material medium such as water or air.
- Electromagnetic waves require no material medium and can travel through a vacuum.

For mechanical waves, such as sound waves and waves on strings, the speed of the wave is a property of the medium. Speed does not depend on the size or shape of the wave.

IMPORTANT CONCEPTS

The **displacement** D of a wave is a function of both position (where) and time (when).

- A snapshot graph shows the wave's displacement as a function of position at a single instant of time.

- A history graph shows the wave's displacement as a function of time at a single point in space.

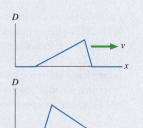

For a transverse wave on a string, the snapshot graph is a picture of the wave. The displacement of a longitudinal wave is parallel to the motion; thus the snapshot graph of a longitudinal sound wave is *not* a picture of the wave.

Sinusoidal waves are periodic in both time (period T) and space (wavelength λ):

$$D(x, t) = A \sin\left[2\pi(x/\lambda - t/T) + \phi_0 \right]$$
$$= A \sin(kx - \omega t + \phi_0)$$

where A is the **amplitude,** $k = 2\pi/\lambda$ is the **wave number,** $\omega = 2\pi f = 2\pi/T$ is the **angular frequency,** and ϕ_0 is the **phase constant** that describes initial conditions.

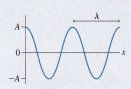

One-dimensional waves   Two- and three-dimensional waves

The fundamental relationship for any sinusoidal wave is $v = \lambda f$.

APPLICATIONS

- **String** (transverse): $v = \sqrt{T_s/\mu}$
- **Sound** (longitudinal): $v = \sqrt{B/\rho} = 343$ m/s in 20°C air
- **Light** (transverse): $v = c/n$, where $c = 3.00 \times 10^8$ m/s is the speed of light in a vacuum and n is the material's **index of refraction**

The wave intensity is the power-to-area ratio: $I = P/a$

For a circular or spherical wave: $I = P_{\text{source}}/4\pi r^2$

The sound intensity level is

$$\beta = (10 \text{ dB}) \log_{10}(I/1.0 \times 10^{-12} \text{ W/m}^2)$$

The Doppler effect occurs when a wave source and detector are moving with respect to each other: the frequency detected differs from the frequency f_0 emitted.

Approaching source

$$f_+ = \frac{f_0}{1 - v_s/v}$$

Receding source

$$f_- = \frac{f_0}{1 + v_s/v}$$

Observer approaching a source

$$f_+ = (1 + v_o/v)f_0$$

Observer receding from a source

$$f_- = (1 - v_o/v)f_0$$

The Doppler effect for light uses a result derived from the theory of relativity.

TERMS AND NOTATION

wave model	linear density, μ	partial derivative	plane wave
traveling wave	snapshot graph	wave equation	phase, ϕ
transverse wave	history graph	compression	intensity, I
longitudinal wave	leading edge	rarefaction	sound intensity level, β
mechanical wave	trailing edge	electromagnetic spectrum	decibels
electromagnetic wave	sinusoidal wave	index of refraction, n	Doppler effect
medium	amplitude, A	wave front	red shifted
disturbance	wavelength, λ	circular wave	blue shifted
wave speed, v	wave number, k	spherical wave	

CONCEPTUAL QUESTIONS

1. The three wave pulses in **FIGURE Q16.1** travel along the same stretched string. Rank in order, from largest to smallest, their wave speeds v_a, v_b, and v_c. Explain.

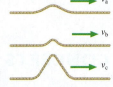

FIGURE Q16.1

2. A wave pulse travels along a stretched string at a speed of 200 cm/s. What will be the speed if:
   a. The string's tension is doubled?
   b. The string's mass is quadrupled (but its length is unchanged)?
   c. The string's length is quadrupled (but its mass is unchanged)?
   Note: Each part is independent and refers to changes made to the original string.

3. **FIGURE Q16.3** is a history graph showing the displacement as a function of time at one point on a string. Did the displacement at this point reach its maximum of 2 mm *before* or *after* the interval of time when the displacement was a constant 1 mm?

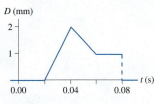

FIGURE Q16.3

4. **FIGURE Q16.4** shows a snapshot graph *and* a history graph for a wave pulse on a stretched string. They describe the same wave from two perspectives.
   a. In which direction is the wave traveling? Explain.
   b. What is the speed of this wave?

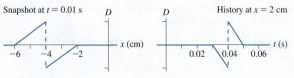

FIGURE Q16.4

5. Rank in order, from largest to smallest, the wavelengths λ_a, λ_b, and λ_c for sound waves having frequencies $f_a = 100$ Hz, $f_b = 1000$ Hz, and $f_c = 10,000$ Hz. Explain.

6. A sound wave with wavelength λ_0 and frequency f_0 moves into a new medium in which the speed of sound is $v_1 = 2v_0$. What are the new wavelength λ_1 and frequency f_1? Explain.

7. What are the amplitude, wavelength, frequency, and phase constant of the traveling wave in **FIGURE Q16.7**?

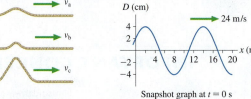

Snapshot graph at $t = 0$ s

FIGURE Q16.7

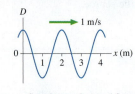

Snapshot graph at $t = 1.0$ s

FIGURE Q16.8

8. **FIGURE Q16.8** is a snapshot graph of a sinusoidal wave at $t = 1.0$ s. What is the phase constant of this wave?

9. **FIGURE Q16.9** shows the wave fronts of a circular wave. What is the phase difference between (a) points A and B, (b) points C and D, and (c) points E and F?

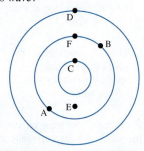

FIGURE Q16.9

10. Sound wave A delivers 2 J of energy in 2 s. Sound wave B delivers 10 J of energy in 5 s. Sound wave C delivers 2 mJ of energy in 1 ms. Rank in order, from largest to smallest, the sound powers P_A, P_B, and P_C of these three sound waves. Explain.

11. One physics professor talking produces a sound intensity level of 52 dB. It's a frightening idea, but what would be the sound intensity level of 100 physics professors talking simultaneously?

12. You are standing at $x = 0$ m, listening to a sound that is emitted at frequency f_0. The graph of **FIGURE Q16.12** shows the frequency you hear during a 4-second interval. Which of the following describes the sound source? Explain your choice.
   A. It moves from left to right and passes you at $t = 2$ s.
   B. It moves from right to left and passes you at $t = 2$ s.
   C. It moves toward you but doesn't reach you. It then reverses direction at $t = 2$ s.
   D. It moves away from you until $t = 2$ s. It then reverses direction and moves toward you but doesn't reach you.

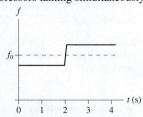

FIGURE Q16.12

EXERCISES AND PROBLEMS

Problems labeled integrate material from earlier chapters.

Exercises

Section 16.1 An Introduction to Waves

1. | The wave speed on a string under tension is 200 m/s. What is the speed if the tension is halved?
2. | The wave speed on a string is 150 m/s when the tension is 75 N. What tension will give a speed of 180 m/s?
3. || A 25 g string is under 20 N of tension. A pulse travels the length of the string in 50 ms. How long is the string?

Section 16.2 One-Dimensional Waves

4. || Draw the history graph $D(x = 4.0 \text{ m}, t)$ at $x = 4.0$ m for the wave shown in **FIGURE EX16.4**.

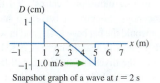

FIGURE EX16.4

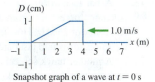

FIGURE EX16.5

5. || Draw the history graph $D(x = 0 \text{ m}, t)$ at $x = 0$ m for the wave shown in **FIGURE EX16.5**.
6. || Draw the snapshot graph $D(x, t = 0 \text{ s})$ at $t = 0$ s for the wave shown in **FIGURE EX16.6**.

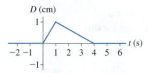

FIGURE EX16.6

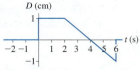

FIGURE EX16.7

7. || Draw the snapshot graph $D(x, t = 1.0 \text{ s})$ at $t = 1.0$ s for the wave shown in **FIGURE EX16.7**.
8. || **FIGURE EX16.8** is a picture at $t = 0$ s of the particles in a medium as a longitudinal wave is passing through. The equilibrium spacing between the particles is 1.0 cm. Draw the snapshot graph $D(x, t = 0 \text{ s})$ of this wave at $t = 0$ s.

FIGURE EX16.8

9. || **FIGURE EX16.9** is the snapshot graph at $t = 0$ of a *longitudinal* wave. Draw the corresponding picture of the particle positions, as was done in Figure 16.9b. Let the equilibrium spacing between the particles be 1.0 cm.

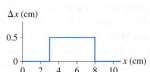

FIGURE EX16.9

Section 16.3 Sinusoidal Waves

10. | A wave has angular frequency 30 rad/s and wavelength 2.0 m. What are its (a) wave number and (b) wave speed?

11. | A wave travels with speed 200 m/s. Its wave number is 1.5 rad/m. What are its (a) wavelength and (b) frequency?
12. | The displacement of a wave traveling in the negative y-direction is $D(y, t) = (5.2 \text{ cm}) \sin(5.5y + 72t)$, where y is in m and t is in s. What are the (a) frequency, (b) wavelength, and (c) speed of this wave?
13. | The displacement of a wave traveling in the positive x-direction is $D(x, t) = (3.5 \text{ cm}) \sin(2.7x - 124t)$, where x is in m and t is in s. What are the (a) frequency, (b) wavelength, and (c) speed of this wave?
14. || What are the amplitude, frequency, and wavelength of the wave in **FIGURE EX16.14**?

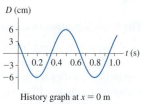

FIGURE EX16.14

History graph at $x = 0$ m
Wave traveling left at 2.0 m/s

Section 16.4 The Wave Equation on a String

15. || **CALC** Show that the displacement $D(x, t) = cx^2 + dt^2$, where c and d are constants, is a solution to the wave equation. Then find an expression in terms of c and d for the wave speed.
16. || **CALC** Show that the displacement $D(x, t) = \ln(ax + bt)$, where a and b are constants, is a solution to the wave equation. Then find an expression in terms of a and b for the wave speed.

Section 16.5 Sound and Light

17. || a. What is the wavelength of a 2.0 MHz ultrasound wave traveling through aluminum?
    b. What frequency of electromagnetic wave would have the same wavelength as the ultrasound wave of part a?
18. | a. What is the frequency of an electromagnetic wave with a wavelength of 20 cm?
    b. What would be the wavelength of a sound wave in water with the same frequency as the electromagnetic wave of part a?
19. | a. What is the frequency of blue light that has a wavelength of 450 nm?
    b. What is the frequency of red light that has a wavelength of 650 nm?
    c. What is the index of refraction of a material in which the red-light wavelength is 450 nm?
20. | a. An FM radio station broadcasts at a frequency of 101.3 MHz. What is the wavelength?
    b. What is the frequency of a sound source that produces the same wavelength in 20°C air?
21. | a. Telephone signals are often transmitted over long distances by microwaves. What is the frequency of microwave radiation with a wavelength of 3.0 cm?
    b. Microwave signals are beamed between two mountaintops 50 km apart. How long does it take a signal to travel from one mountaintop to the other?
22. || A hammer taps on the end of a 4.00-m-long metal bar at room temperature. A microphone at the other end of the bar picks up two pulses of sound, one that travels through the metal and one that travels through the air. The pulses are separated in time by 9.00 ms. What is the speed of sound in this metal?

23. | Cell phone conversations are transmitted by high-frequency radio waves. Suppose the signal has wavelength 35 cm while traveling through air. What are the (a) frequency and (b) wavelength as the signal travels through 3-mm-thick window glass into your room?

24. || a. How long does it take light to travel through a 3.0-mm-thick piece of window glass?
    b. Through what thickness of water could light travel in the same amount of time?

25. | A light wave has a 670 nm wavelength in air. Its wavelength in a transparent solid is 420 nm.
    a. What is the speed of light in this solid?
    b. What is the light's frequency in the solid?

26. | A 440 Hz sound wave in 20°C air propagates into the water of a swimming pool. What are the wave's (a) frequency and (b) wavelength in the water?

Section 16.6 The Wave Equation in a Fluid

27. | What is the speed of sound in air (a) on a cold winter day in Minnesota when the temperature is –25°F, and (b) on a hot summer day in Death Valley when the temperature is 125°F?

28. | The density of mercury is 13,600 kg/m³. What is the speed of sound in mercury at 20°C?

Section 16.7 Waves in Two and Three Dimensions

29. | A circular wave travels outward from the origin. At one instant of time, the phase at $r_1 = 20$ cm is 0 rad and the phase at $r_2 = 80$ cm is 3π rad. What is the wavelength of the wave?

30. || A spherical wave with a wavelength of 2.0 m is emitted from the origin. At one instant of time, the phase at $r = 4.0$ m is π rad. At that instant, what is the phase at $r = 3.5$ m and at $r = 4.5$ m?

31. || A loudspeaker at the origin emits a 120 Hz tone on a day when the speed of sound is 340 m/s. The phase difference between two points on the x-axis is 5.5 rad. What is the distance between these two points?

32. || A sound source is located somewhere along the x-axis. Experiments show that the same wave front simultaneously reaches listeners at $x = -7.0$ m and $x = +3.0$ m.
    a. What is the x-coordinate of the source?
    b. A third listener is positioned along the positive y-axis. What is her y-coordinate if the same wave front reaches her at the same instant it does the first two listeners?

Section 16.8 Power, Intensity, and Decibels

33. || A sound wave with intensity 2.0×10^{-3} W/m² is perceived to
BIO be modestly loud. Your eardrum is 6.0 mm in diameter. How much energy will be transferred to your eardrum while listening to this sound for 1.0 min?

34. || The intensity of electromagnetic waves from the sun is 1.4 kW/m² just above the earth's atmosphere. Eighty percent of this reaches the surface at noon on a clear summer day. Suppose you think of your back as a 30 cm × 50 cm rectangle. How many joules of solar energy fall on your back as you work on your tan for 1.0 h?

35. || A concert loudspeaker suspended high above the ground emits 35 W of sound power. A small microphone with a 1.0 cm² area is 50 m from the speaker.
    a. What is the sound intensity at the position of the microphone?
    b. How much sound energy impinges on the microphone each second?

36. || During takeoff, the sound intensity level of a jet engine is 140 dB at a distance of 30 m. What is the sound intensity level at a distance of 1.0 km?

37. | The sun emits electromagnetic waves with a power of 4.0×10^{26} W. Determine the intensity of electromagnetic waves from the sun just outside the atmospheres of Venus, the earth, and Mars.

38. | What are the sound intensity levels for sound waves of intensity (a) 3.0×10^{-6} W/m² and (b) 3.0×10^{-2} W/m²?

39. || A loudspeaker on a tall pole broadcasts sound waves equally in all directions. What is the speaker's power output if the sound intensity level is 90 dB at a distance of 20 m?

40. || The sound intensity level 5.0 m from a large power saw is 100 dB. At what distance will the sound be a more tolerable 80 dB?

Section 16.9 The Doppler Effect

41. | A friend of yours is loudly singing a single note at 400 Hz while racing toward you at 25.0 m/s on a day when the speed of sound is 340 m/s.
    a. What frequency do you hear?
    b. What frequency does your friend hear if you suddenly start singing at 400 Hz?

42. | An opera singer in a convertible sings a note at 600 Hz while cruising down the highway at 90 km/h. What is the frequency heard by
    a. A person standing beside the road in front of the car?
    b. A person on the ground behind the car?

43. || A bat locates insects by emitting ultrasonic "chirps" and then
BIO listening for echoes from the bugs. Suppose a bat chirp has a frequency of 25 kHz. How fast would the bat have to fly, and in what direction, for you to just barely be able to hear the chirp at 20 kHz?

44. || A mother hawk screeches as she dives at you. You recall from biology that female hawks screech at 800 Hz, but you hear the screech at 900 Hz. How fast is the hawk approaching?

Problems

45. || **FIGURE P16.45** is a history graph at $x = 0$ m of a wave traveling in the positive x-direction at 4.0 m/s.
    a. What is the wavelength?
    b. What is the phase constant of the wave?
    c. Write the displacement equation for this wave.

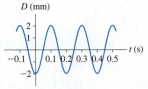

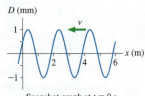

History graph at $x = 0$ m
Wave traveling right at 4.0 m/s

Snapshot graph at $t = 0$ s

FIGURE P16.45   **FIGURE P16.46**

46. || **FIGURE P16.46** is a snapshot graph at $t = 0$ s of a 5.0 Hz wave traveling to the left.
    a. What is the wave speed?
    b. What is the phase constant of the wave?
    c. Write the displacement equation for this wave.

47. | String 1 in **FIGURE P16.47** has linear density 2.0 g/m and string 2 has linear density 4.0 g/m. A student sends pulses in both directions by quickly pulling up on the knot, then releasing it.

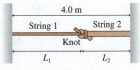

FIGURE P16.47

What should the string lengths L_1 and L_2 be if the pulses are to reach the ends of the strings simultaneously?

48. || Oil explorers set off explosives to make loud sounds, then listen for the echoes from underground oil deposits. Geologists suspect that there is oil under 500-m-deep Lake Physics. It's known that Lake Physics is carved out of a granite basin. Explorers detect a weak echo 0.94 s after exploding dynamite at the lake surface. If it's really oil, how deep will they have to drill into the granite to reach it?

49. ∥ One cue your hearing system uses to localize a sound (i.e., to
BIO tell where a sound is coming from) is the slight difference in the
arrival times of the sound at your ears. Your ears are spaced
approximately 20 cm apart. Consider a sound source 5.0 m from
the center of your head along a line 45° to your right. What is the
difference in arrival times? Give your answer in microseconds.
Hint: You are looking for the difference between two numbers that
are nearly the same. What does this near equality imply about the
necessary precision during intermediate stages of the calculation?

50. ∥ A helium-neon laser beam has a wavelength in air of 633 nm. It
takes 1.38 ns for the light to travel through 30 cm of an unknown
liquid. What is the wavelength of the laser beam in the liquid?

51. ∥ Earthquakes are essentially sound waves—called seismic
waves—traveling through the earth. Because the earth is solid,
it can support both longitudinal and transverse seismic waves.
The speed of longitudinal waves, called P waves, is 8000 m/s.
Transverse waves, called S waves, travel at a slower 4500 m/s.
A seismograph records the two waves from a distant earthquake.
If the S wave arrives 2.0 min after the P wave, how far away was
the earthquake? You can assume that the waves travel in straight
lines, although actual seismic waves follow more complex routes.

52. ∥ Helium (density 0.18 kg/m^3 at 0°C and 1 atm pressure)
remains a gas until the extraordinarily low temperature of 4.2 K.
What is the speed of sound in helium at 5 K?

53. ∥ A 20.0-cm-long, 10.0-cm-diameter cylinder with a piston at
one end contains 1.34 kg of an unknown liquid. Using the piston
to compress the length of the liquid by 1.00 mm increases the
pressure by 41.0 atm. What is the speed of sound in the liquid?

54. ∥ A sound wave is described by $D(y, t) = (0.0200 \text{ mm}) \times$
$\sin[(8.96 \text{ rad/m})y + (3140 \text{ rad/s})t + \pi/4 \text{ rad}]$, where y is in m
and t is in s.
   a. In what direction is this wave traveling?
   b. Along which axis is the air oscillating?
   c. What are the wavelength, the wave speed, and the period of
   oscillation?

55. ∥ A wave on a string is described by $D(x, t) = (3.0 \text{ cm}) \times$
$\sin[2\pi(x/(2.4 \text{ m}) + t/(0.20 \text{ s}) + 1)]$, where x is in m and t is in s.
   a. In what direction is this wave traveling?
   b. What are the wave speed, the frequency, and the wave number?
   c. At $t = 0.50$ s, what is the displacement of the string at
   $x = 0.20$ m?

56. ∥ A wave on a string is described by $D(x, t) = (2.00 \text{ cm}) \times$
$\sin[(12.57 \text{ rad/m})x - (638 \text{ rad/s})t]$, where x is in m and t in s.
The linear density of the string is 5.00 g/m. What are
   a. The string tension?
   b. The maximum displacement of a point on the string?
   c. The maximum speed of a point on the string?

57. ∥ **FIGURE P16.57** shows a snapshot graph of a wave traveling
to the right along a string at 45 m/s. At this instant, what is the
velocity of points 1, 2, and 3 on the string?

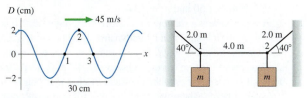

FIGURE P16.57   FIGURE P16.58

58. ∥∥ **FIGURE P16.58** shows two masses hanging from a steel wire.
The mass of the wire is 60.0 g. A wave pulse travels along the
wire from point 1 to point 2 in 24.0 ms. What is mass m?

59. ∥ A wire is made by welding
together two metals having different densities. **FIGURE P16.59**
shows a 2.00-m-long section of
wire centered on the junction,

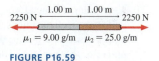

FIGURE P16.59

but the wire extends much farther in both directions. The wire is
placed under 2250 N tension, then a 1500 Hz wave with an amplitude
of 3.00 mm is sent down the wire. How many wavelengths (complete
cycles) of the wave are in this 2.00-m-long section of the wire?

60. ∥ The string in **FIGURE P16.60** has linear
density μ. Find an expression in terms of
M, μ, and θ for the speed of waves on the
string.

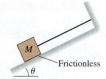

FIGURE P16.60

61. ∥ A string that is under 50.0 N of tension has linear density
5.0 g/m. A sinusoidal wave with amplitude 3.0 cm and wavelength
2.0 m travels along the string. What is the maximum speed of a
particle on the string?

62. ∥ The G string on a guitar is a 0.46-mm-diameter steel string
with a linear density of 1.3 g/m. When the string is properly
tuned to 196 Hz, the wave speed on the string is 250 m/s. Tuning
is done by turning the tuning screw, which slowly tightens—and
stretches—the string. By how many mm does a 75-cm-long G
string stretch when it's first tuned?

63. ∥ A sinusoidal wave travels along a stretched string. A particle on the
string has a maximum speed of 2.0 m/s and a maximum acceleration
of 200 m/s^2. What are the frequency and amplitude of the wave?

64. ∥∥ Is the displacement $D(x, t) = (0.10 - 0.10x^2 + xt - 2.5t^2)$ m,
CALC where x is in m and t is in s, a possible traveling wave? If so, what
is the wave speed?

65. ∥∥ Is the displacement $D(x, t) = (3.0 \text{ mm}) \, e^{i(2.0x + 8.0t + 5.0)}$, where
CALC x is in m, t is in s, and $i = \sqrt{-1}$, a possible traveling wave? If so,
what is the wave speed? *Complex exponentials* are often used to
represent waves in more advanced treatments.

66. ∥ An AM radio station broadcasts with a power of 25 kW at a
frequency of 920 kHz. Estimate the intensity of the radio wave at
a point 10 km from the broadcast antenna.

67. ∥ LASIK eye surgery uses pulses of laser light to shave off tissue
BIO from the cornea, reshaping it. A typical LASIK laser emits a
1.0-mm-diameter laser beam with a wavelength of 193 nm. Each
laser pulse lasts 15 ns and contains 1.0 mJ of light energy.
   a. What is the power of one laser pulse?
   b. During the very brief time of the pulse, what is the intensity
   of the light wave?

68. ∥ The sound intensity 50 m from a wailing tornado siren is
0.10 W/m^2.
   a. What is the intensity at 1000 m?
   b. The weakest intensity likely to be heard over background
   noise is $\approx 1 \, \mu\text{W/m}^2$. Estimate the maximum distance at
   which the siren can be heard.

69. ∥ A distant star system is discovered in which a planet with
twice the radius of the earth and rotating 3.0 times as fast as the
earth orbits a star with a total power output of 6.8×10^{29} W.
   a. If the star's radius is 6.0 times that of the sun, what is the
   electromagnetic wave intensity at the surface? Astronomers
   call this the *surface flux*. Astronomical data are provided
   inside the back cover of the book.
   b. Every planet-day (one rotation), the planet receives 9.4×10^{22} J
   of energy. What is the planet's distance from its star? Give
   your answer in *astronomical units* (AU), where 1 AU is the
   distance of the earth from the sun.

70. ‖ A compact sound source radiates 25 W of sound energy uniformly in all directions. What is the ratio of the sound intensity at a distance of 1.0 m to that at 5.0 m in (a) a two-dimensional universe, (b) our normal three-dimensional universe, and (c) a hypothetical four-dimensional universe?

71. ‖ A loudspeaker, mounted on a tall pole, is engineered to emit 75% of its sound energy into the forward hemisphere, 25% toward the back. You measure an 85 dB sound intensity level when standing 3.5 m in front of and 2.5 m below the speaker. What is the speaker's power output?

72. ‖ Your ears are sensitive to differences in pitch, but they are not
BIO very sensitive to differences in intensity. You are not capable of detecting a difference in sound intensity level of less than 1 dB. By what factor does the sound intensity increase if the sound intensity level increases from 60 dB to 61 dB?

73. ‖ The intensity of a sound source is described by an inverse-square law only if the source is very small (a point source) and only if the waves can travel unimpeded in all directions. For an extended source or in a situation where obstacles absorb or reflect the waves, the intensity at distance r can often be expressed as $I = cP_{source}/r^x$, where c is a constant and the exponent x—which would be 2 for an ideal spherical wave—depends on the situation. In one such situation, you use a sound meter to measure the sound intensity level at different distances from a source, acquiring the data in the table. Use the best-fit line of an appropriate graph to determine the exponent x that characterizes this sound source.

Distance (m)	Intensity level (dB)
1	100
3	93
10	85
30	78
100	70

74. ‖ A physics professor demonstrates the Doppler effect by tying a 600 Hz sound generator to a 1.0-m-long rope and whirling it around her head in a horizontal circle at 100 rpm. What are the highest and lowest frequencies heard by a student in the classroom?

75. ‖ An avant-garde composer wants to use the Doppler effect in his new opera. As the soprano sings, he wants a large bat to fly toward her from the back of the stage. The bat will be outfitted with a microphone to pick up the singer's voice and a loudspeaker to rebroadcast the sound toward the audience. The composer wants the sound the audience hears from the bat to be, in musical terms, one half-step higher in frequency than the note they are hearing from the singer. Two notes a half-step apart have a frequency ratio of $2^{1/12} = 1.059$. With what speed must the bat fly toward the singer?

76. ‖‖ A loudspeaker on a pole is radiating 100 W of sound energy in all
CALC directions. You are walking directly toward the speaker at 0.80 m/s. When you are 20 m away, what are (a) the sound intensity level and (b) the rate (dB/s) at which the sound intensity level is increasing? Hint: Use the chain rule and the relationship $\log_{10} x = \ln x / \ln 10$.

77. ‖ Show that the Doppler frequency f_- of a receding source is $f_- = f_0/(1 + v_s/v)$.

78. ‖ A starship approaches its home planet at a speed of $0.10c$. When it is 54×10^6 km away, it uses its green laser beam ($\lambda = 540$ nm) to signal its approach.
   a. How long does the signal take to travel to the home planet?
   b. At what wavelength is the signal detected on the home planet?

79. ‖ Wavelengths of light from a distant galaxy are found to be 0.50% longer than the corresponding wavelengths measured in a terrestrial laboratory. Is the galaxy approaching or receding from the earth? At what speed?

80. ‖ You have just been pulled over for running a red light, and the police officer has informed you that the fine will be $250. In desperation, you suddenly recall an idea that your physics professor recently discussed in class. In your calmest voice, you tell the officer that the laws of physics prevented you from knowing that the light was red. In fact, as you drove toward it, the light was Doppler shifted to where it appeared green to you. "OK," says the officer, "Then I'll ticket you for speeding. The fine is $1 for every 1 km/h over the posted speed limit of 50 km/h." How big is your fine? Use 650 nm as the wavelength of red light and 540 nm as the wavelength of green light.

Challenge Problems

81. ‖‖‖ One way to monitor global warming is to measure the average temperature of the ocean. Researchers are doing this by measuring the time it takes sound pulses to travel underwater over large distances. At a depth of 1000 m, where ocean temperatures hold steady near 4°C, the average sound speed is 1480 m/s. It's known from laboratory measurements that the sound speed increases 4.0 m/s for every 1.0°C increase in temperature. In one experiment, where sounds generated near California are detected in the South Pacific, the sound waves travel 8000 km. If the smallest time change that can be reliably detected is 1.0 s, what is the smallest change in average temperature that can be measured?

82. ‖‖‖ A rope of mass m and length L hangs from a ceiling.
CALC a. Show that the wave speed on the rope a distance y above the lower end is $v = \sqrt{gy}$.
   b. Show that the time for a pulse to travel the length of the string is $\Delta t = 2\sqrt{L/g}$.

83. ‖‖‖ A communications truck with a 44-cm-diameter dish receiver
CALC on the roof starts out 10 km from its base station. It drives directly away from the base station at 50 km/h for 1.0 h, keeping the receiver pointed at the base station. The base station antenna broadcasts continuously with 2.5 kW of power, radiated uniformly in all directions. How much electromagnetic energy does the truck's dish receive during that 1.0 h?

84. ‖‖‖ Some modern optical devic-
CALC es are made with glass whose index of refraction changes with distance from the front surface. FIGURE CP16.84 shows the index of refraction as a function of the distance into a slab of glass of thickness L. The index of refraction increases linearly from n_1 at the front surface to n_2 at the rear surface.
   a. Find an expression for the time light takes to travel through this piece of glass.
   b. Evaluate your expression for a 1.0-cm-thick piece of glass for which $n_1 = 1.50$ and $n_2 = 1.60$.

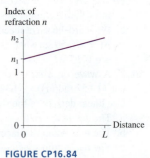

FIGURE CP16.84

85. ‖‖‖ A water wave is a *shallow-water wave* if the water depth d is less
CALC than $\approx \lambda/10$. It is shown in hydrodynamics that the speed of a shallow-water wave is $v = \sqrt{gd}$, so waves slow down as they move into shallower water. Ocean waves, with wavelengths of typically 100 m, are shallow-water waves when the water depth is less than ≈ 10 m. Consider a beach where the depth increases linearly with distance from the shore until reaching a depth of 5.0 m at a distance of 100 m. How long does it take a wave to move the last 100 m to the shore? Assume that the waves are so small that they don't break before reaching the shore.

17 Superposition

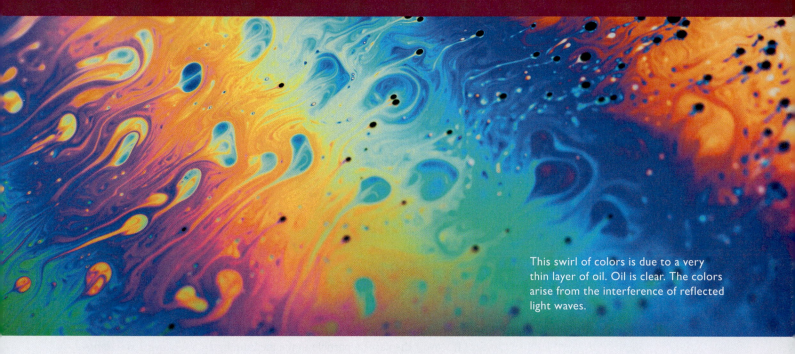

This swirl of colors is due to a very thin layer of oil. Oil is clear. The colors arise from the interference of reflected light waves.

IN THIS CHAPTER, you will understand and use the ideas of superposition.

What is superposition?

Waves can pass through each other. When they do, their displacements add together at each point. This is called the principle of superposition. It is a property of waves but not of particles.

《 LOOKING BACK Sections 16.1–16.4 Properties of traveling waves

What is a standing wave?

A standing wave is created when two waves travel in opposite directions between two boundaries.

- Standing waves have well-defined patterns called modes.
- Some points on the wave, called nodes, do not oscillate at all.

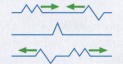

How are standing waves related to music?

The notes played by musical instruments are standing waves.

- Guitars have string standing waves.
- Flutes have pressure standing waves.

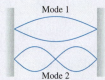

Changing the length of a standing wave changes its frequency and the note played.

《 LOOKING BACK Section 16.5 Sound waves

What is interference?

When two sources emit waves with the same wavelength, the overlapped waves create an interference pattern.

- Constructive interference (red) occurs where waves add to produce a wave with a larger amplitude.
- Destructive interference (black) occurs where waves cancel.

What are beats?

The superposition of two waves with slightly different frequencies produces a loud-soft-loud-soft modulation of the intensity called beats. Beats have important applications in music, ultrasonics, and telecommunications.

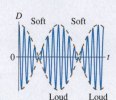

Why is superposition important?

Superposition and standing waves occur often in the world around us, especially when there are reflections. Musical instruments, microwave systems, and lasers all depend on standing waves. Standing waves are also important for large structures such as buildings and bridges. Superposition of light waves causes interference, which is used in electro-optic devices and precision measuring techniques.

17.1 The Principle of Superposition

FIGURE 17.1a shows two baseball players, Alan and Bill, at batting practice. Unfortunately, someone has turned the pitching machines so that pitching machine A throws baseballs toward Bill while machine B throws toward Alan. If two baseballs are launched at the same time, and with the same speed, they collide at the crossing point. Two particles cannot occupy the same point of space at the same time.

FIGURE 17.1 Unlike particles, two waves can pass directly through each other.

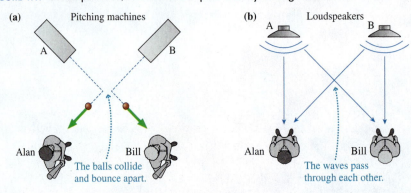

But waves, unlike particles, can pass directly through each other. In **FIGURE 17.1b** Alan and Bill are listening to the stereo system in the locker room after practice. Because both hear the music quite well, the sound wave that travels from loudspeaker A toward Bill must pass through the wave traveling from loudspeaker B toward Alan.

What happens to the medium at a point where two waves are present simultaneously? If wave 1 displaces a particle in the medium by D_1 and wave 2 *simultaneously* displaces it by D_2, the net displacement of the particle is simply $D_1 + D_2$. This is a very important idea because it tells us how to combine waves. It is known as the *principle of superposition*.

> **Principle of superposition** When two or more waves are *simultaneously* present at a single point in space, the displacement of the medium at that point is the sum of the displacements due to each individual wave.

Mathematically, the net displacement of a particle in the medium is

$$D_{net} = D_1 + D_2 + \cdots = \sum_i D_i \tag{17.1}$$

where D_i is the displacement that would be caused by wave i alone. We will make the simplifying assumption that the displacements of the individual waves are along the same line so that we can add displacements as scalars rather than vectors.

To use the principle of superposition you must know the displacement caused by each wave if traveling alone. Then you go through the medium *point by point* and add the displacements due to each wave *at that point* to find the net displacement at that point.

To illustrate, **FIGURE 17.2** shows snapshot graphs taken 1 s apart of two waves traveling at the same speed (1 m/s) in opposite directions. The principle of superposition comes into play wherever the waves overlap. The solid line is the sum *at each point* of the two displacements at that point. This is the displacement that you would actually observe as the two waves pass through each other.

Notice how two overlapping positive displacements add to give a displacement twice that of the individual waves. This is called *constructive interference*. Similarly, *destructive interference* is occurring at the points where positive and negative displacements add to give a superposition with zero displacement. We will defer the main discussion until later in this chapter, but you can already see that *interference is a consequence of superposition*.

FIGURE 17.2 The superposition of two waves as they pass through each other.

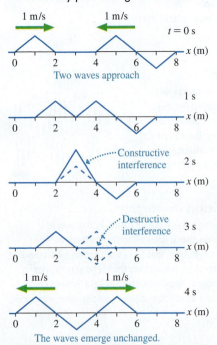

STOP TO THINK 17.1 Two pulses on a string approach each other at speeds of 1 m/s. What is the shape of the string at $t = 6$ s?

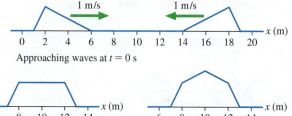

Approaching waves at $t = 0$ s

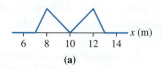
(a)

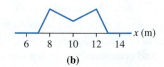
(b)

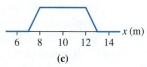
(c)

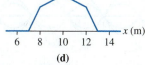
(d)

17.2 Standing Waves

FIGURE 17.3 is a time-lapse photograph of a *standing wave* on a vibrating string. It's not obvious from the photograph, but this is actually a superposition of two waves. To understand this, consider two sinusoidal waves **with the same frequency, wavelength, and amplitude** traveling in opposite directions. For example, **FIGURE 17.4a** shows two waves on a string, and **FIGURE 17.4b** shows nine snapshot graphs, at intervals of $\frac{1}{8}T$. The dots identify two of the crests to help you visualize the wave movement.

At *each point,* the net displacement—the superposition—is found by adding the red displacement and the green displacement. **FIGURE 17.4c** shows the result. It is the wave you would actually observe. The blue dot shows that the blue wave is moving neither right nor left. The wave of Figure 17.4c is called a **standing wave** because the crests and troughs "stand in place" as the wave oscillates.

FIGURE 17.3 A vibrating string is an example of a standing wave.

FIGURE 17.4 The superposition of two sinusoidal waves traveling in opposite directions.

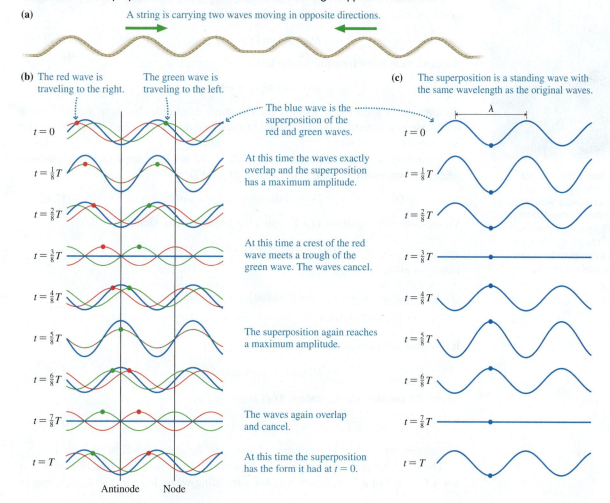

(a)   A string is carrying two waves moving in opposite directions.

(b) The red wave is traveling to the right. The green wave is traveling to the left.

The blue wave is the superposition of the red and green waves.

(c) The superposition is a standing wave with the same wavelength as the original waves.

$t = 0$

$t = \frac{1}{8}T$ At this time the waves exactly overlap and the superposition has a maximum amplitude.

$t = \frac{2}{8}T$

$t = \frac{3}{8}T$ At this time a crest of the red wave meets a trough of the green wave. The waves cancel.

$t = \frac{4}{8}T$

$t = \frac{5}{8}T$ The superposition again reaches a maximum amplitude.

$t = \frac{6}{8}T$

$t = \frac{7}{8}T$ The waves again overlap and cancel.

$t = T$ At this time the superposition has the form it had at $t = 0$.

Antinode   Node

FIGURE 17.5 The intensity of a standing wave is maximum at the antinodes, zero at the nodes.

(a) Nodes are spaced λ/2 apart.

(b) The intensity is maximum at the antinodes.

The intensity is zero at the nodes.

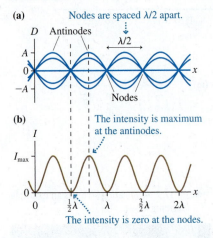

This photograph shows the Tacoma Narrows suspension bridge on the day in 1940 when it experienced a wind-induced standing-wave oscillation that led to its collapse. The red line shows the original line of the deck of the bridge. You can clearly see the large amplitude of the oscillation and the node at the center of the span.

Nodes and Antinodes

FIGURE 17.5a has collapsed the nine graphs of Figure 17.4b into a single graphical representation of a standing wave. Compare this to the Figure 17.3 photograph of a vibrating string. A striking feature of a standing-wave pattern is the existence of **nodes,** points that *never move!* **The nodes are spaced λ/2 apart.** Halfway between the nodes are the points where the particles in the medium oscillate with maximum displacement. These points of maximum amplitude are called **antinodes,** and you can see that they are also spaced λ/2 apart.

It seems surprising and counterintuitive that some particles in the medium have no motion at all. To understand this, look closely at the two traveling waves in Figure 17.4a. You will see that the nodes occur at points where at *every instant* of time the displacements of the two traveling waves have equal magnitudes but *opposite signs.* That is, nodes are points of destructive interference where the net displacement is always zero. In contrast, antinodes are points of constructive interference where two displacements of the same sign always add to give a net displacement larger than that of the individual waves.

In Chapter 16 you learned that the *intensity* of a wave is proportional to the square of the amplitude: $I \propto A^2$. You can see in **FIGURE 17.5b** that maximum intensity occurs at the antinodes and that the intensity is zero at the nodes. If this is a sound wave, the loudness is maximum at the antinodes and zero at the nodes. A standing light wave is bright at the antinodes, dark at the nodes. The key idea is that **the intensity is maximum at points of constructive interference and zero at points of destructive interference.**

The Mathematics of Standing Waves

A sinusoidal wave traveling to the right along the x-axis with angular frequency $\omega = 2\pi f$, wave number $k = 2\pi/\lambda$, and amplitude a is

$$D_R = a \sin(kx - \omega t) \tag{17.2}$$

An equivalent wave traveling to the left is

$$D_L = a \sin(kx + \omega t) \tag{17.3}$$

We previously used the symbol A for the wave amplitude, but here we will use a lowercase a to represent the amplitude of each individual wave and reserve A for the amplitude of the net wave.

According to the principle of superposition, the net displacement of the medium when both waves are present is the sum of D_R and D_L:

$$D(x, t) = D_R + D_L = a \sin(kx - \omega t) + a \sin(kx + \omega t) \tag{17.4}$$

We can simplify Equation 17.4 by using the trigonometric identity

$$\sin(\alpha \pm \beta) = \sin\alpha\cos\beta \pm \cos\alpha\sin\beta$$

Doing so gives

$$D(x, t) = a(\sin kx \cos \omega t - \cos kx \sin \omega t) + a(\sin kx \cos \omega t + \cos kx \sin \omega t)$$
$$= (2a \sin kx) \cos \omega t \tag{17.5}$$

It is useful to write Equation 17.5 as

$$D(x, t) = A(x) \cos \omega t \tag{17.6}$$

where the **amplitude function** $A(x)$ is defined as

$$A(x) = 2a \sin kx \tag{17.7}$$

The amplitude reaches a maximum value $A_{max} = 2a$ at points where $\sin kx = 1$.

The displacement $D(x, t)$ given by Equation 17.6 is neither a function of $x - vt$ nor a function of $x + vt$; hence it is *not* a traveling wave. Instead, the $\cos \omega t$ term in

Equation 17.6 describes a medium in which each point oscillates in simple harmonic motion with frequency $f = \omega/2\pi$. The function $A(x) = 2a \sin kx$ gives the amplitude of the oscillation for a particle at position x.

FIGURE 17.6 graphs Equation 17.6 at several different instants of time. Notice that the graphs are identical to those of Figure 17.5a, showing us that Equation 17.6 is the mathematical description of a standing wave.

The nodes of the standing wave are the points at which the amplitude is zero. They are located at positions x for which

$$A(x) = 2a \sin kx = 0 \tag{17.8}$$

The sine function is zero if the angle is an integer multiple of π rad, so Equation 17.8 is satisfied if

$$kx_m = \frac{2\pi x_m}{\lambda} = m\pi \qquad m = 0, 1, 2, 3, \ldots \tag{17.9}$$

Thus the position x_m of the mth node is

$$x_m = m\frac{\lambda}{2} \qquad m = 0, 1, 2, 3, \ldots \tag{17.10}$$

You can see that the spacing between two adjacent nodes is $\lambda/2$, in agreement with Figure 17.5b. The nodes are *not* spaced by λ, as you might have expected.

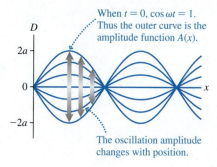

FIGURE 17.6 The net displacement resulting from two counter-propagating sinusoidal waves.

When $t = 0$, $\cos \omega t = 1$. Thus the outer curve is the amplitude function $A(x)$.

The oscillation amplitude changes with position.

EXAMPLE 17.1 | **Node spacing on a string**

A very long string has a linear density of 5.0 g/m and is stretched with a tension of 8.0 N. 100 Hz waves with amplitudes of 2.0 mm are generated at the ends of the string.

a. What is the node spacing along the resulting standing wave?

b. What is the maximum displacement of the string?

MODEL Two counter-propagating waves of equal frequency create a standing wave.

VISUALIZE The standing wave will look like Figure 17.5a.

SOLVE a. The speed of the waves on the string is

$$v = \sqrt{\frac{T_s}{\mu}} = \sqrt{\frac{8.0 \text{ N}}{0.0050 \text{ kg/m}}} = 40 \text{ m/s}$$

and the wavelength is

$$\lambda = \frac{v}{f} = \frac{40 \text{ m/s}}{100 \text{ Hz}} = 0.40 \text{ m} = 40 \text{ cm}$$

Thus the spacing between adjacent nodes is $\lambda/2 = 20$ cm.

b. The maximum displacement is $A_{\max} = 2a = 4.0$ mm.

17.3 Standing Waves on a String

Wiggling both ends of a very long string is not a practical way to generate standing waves. Instead, as in the photograph in Figure 17.3, standing waves are usually seen on a string that is fixed at both ends. To understand why this condition causes standing waves, we need to examine what happens when a traveling wave encounters a discontinuity.

FIGURE 17.7a shows a *discontinuity* between a string with a larger linear density and one with a smaller linear density. The tension is the same in both strings, so the wave speed is slower on the left, faster on the right. Whenever a wave encounters a discontinuity, some of the wave's energy is *transmitted* forward and some is *reflected*.

FIGURE 17.7 A wave reflects when it encounters a discontinuity or a boundary.

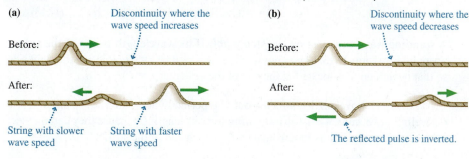

(a) Discontinuity where the wave speed increases

Before:

After:

String with slower wave speed   String with faster wave speed

(b) Discontinuity where the wave speed decreases

Before:

After:

The reflected pulse is inverted.

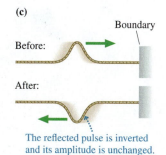

(c) Boundary

Before:

After:

The reflected pulse is inverted and its amplitude is unchanged.

Light waves exhibit an analogous behavior when they encounter a piece of glass. Most of the light wave's energy is transmitted through the glass, which is why glass is transparent, but a small amount of energy is reflected. That is how you see your reflection dimly in a storefront window.

In **FIGURE 17.7b**, an incident wave encounters a discontinuity at which the wave speed decreases. In this case, the reflected pulse is *inverted*. A positive displacement of the incident wave becomes a negative displacement of the reflected wave. Because $\sin(\phi + \pi) = -\sin\phi$, we say that the reflected wave has a *phase change of π upon reflection*. This aspect of reflection will be important later in the chapter when we look at the interference of light waves.

The wave in **FIGURE 17.7c** reflects from a *boundary*. This is like Figure 17.7b in the limit that the string on the right becomes infinitely massive. Thus the reflection in Figure 17.7c looks like that of Figure 17.7b with one exception: Because there is no transmitted wave, *all* the wave's energy is reflected. Hence **the amplitude of a wave reflected from a boundary is unchanged.**

Creating Standing Waves

FIGURE 17.8 Reflections at the two boundaries cause a standing wave on the string.

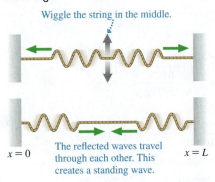

Wiggle the string in the middle.

$x = 0$ The reflected waves travel through each other. This creates a standing wave. $x = L$

FIGURE 17.8 shows a string of length L tied at $x = 0$ and $x = L$. If you wiggle the string in the middle, sinusoidal waves travel outward in both directions and soon reach the boundaries. Because the speed of a reflected wave does not change, **the wavelength and frequency of a reflected sinusoidal wave are unchanged.** Consequently, reflections at the ends of the string cause two waves of *equal amplitude and wavelength* to travel in opposite directions along the string. As we've just seen, these are the conditions that cause a standing wave!

To connect the mathematical analysis of standing waves in Section 17.2 with the physical reality of a string tied down at the ends, we need to impose *boundary conditions*. A **boundary condition** is a mathematical statement of any constraint that *must* be obeyed at the boundary or edge of a medium. Because the string is tied down at the ends, the displacements at $x = 0$ and $x = L$ must be zero at all times. Thus the standing-wave boundary conditions are $D(x = 0, t) = 0$ and $D(x = L, t) = 0$. Stated another way, we require nodes at both ends of the string.

We found that the displacement of a standing wave is $D(x, t) = (2a \sin kx) \cos \omega t$. This equation already satisfies the boundary condition $D(x = 0, t) = 0$. That is, the origin has already been located at a node. The second boundary condition, at $x = L$, requires $D(x = L, t) = 0$. This condition will be met at all times if

$$2a \sin kL = 0 \qquad \text{(boundary condition at } x = L) \qquad (17.11)$$

Equation 17.11 will be true if $\sin kL = 0$, which in turn requires

$$kL = \frac{2\pi L}{\lambda} = m\pi \qquad m = 1, 2, 3, 4, \ldots \qquad (17.12)$$

kL must be a multiple of $m\pi$, but $m = 0$ is excluded because L can't be zero.

For a string of fixed length L, the only quantity in Equation 17.12 that can vary is λ. That is, the boundary condition is satisfied only if the wavelength has one of the values

$$\lambda_m = \frac{2L}{m} \qquad m = 1, 2, 3, 4, \ldots \qquad (17.13)$$

A standing wave can exist on the string *only* if its wavelength is one of the values given by Equation 17.13. The mth possible wavelength $\lambda_m = 2L/m$ is just the right size so that its mth node is located at the end of the string (at $x = L$).

NOTE Other wavelengths, which would be perfectly acceptable wavelengths for a traveling wave, cannot exist as a *standing* wave of length L because they cannot meet the boundary conditions requiring a node at each end of the string.

If standing waves are possible only for certain wavelengths, then only a few specific oscillation frequencies are allowed. Because $\lambda f = v$ for a sinusoidal wave, the oscillation frequency corresponding to wavelength λ_m is

$$f_m = \frac{v}{\lambda_m} = \frac{v}{2L/m} = m\frac{v}{2L} \qquad m = 1, 2, 3, 4, \ldots \qquad (17.14)$$

The lowest allowed frequency

$$f_1 = \frac{v}{2L} \qquad \text{(fundamental frequency)} \qquad (17.15)$$

which corresponds to wavelength $\lambda_1 = 2L$, is called the **fundamental frequency** of the string. The allowed frequencies can be written in terms of the fundamental frequency as

$$f_m = mf_1 \qquad m = 1, 2, 3, 4, \ldots \qquad (17.16)$$

The allowed standing-wave frequencies are all integer multiples of the fundamental frequency. The higher-frequency standing waves are called **harmonics,** with the $m = 2$ wave at frequency f_2 called the *second harmonic,* the $m = 3$ wave called the *third harmonic,* and so on.

FIGURE 17.9 graphs the first four possible standing waves on a string of fixed length L. These possible standing waves are called the **modes** of the string, or sometimes the *normal modes*. Each mode, numbered by the integer m, has a unique wavelength and frequency. Keep in mind that these drawings simply show the *envelope,* or outer edge, of the oscillations. The string is continuously oscillating at all positions between these edges, as we showed in more detail in Figure 17.5a.

There are three things to note about the modes of a string.

1. m is the number of *antinodes* on the standing wave, not the number of nodes. You can tell a string's mode of oscillation by counting the number of antinodes.
2. The *fundamental mode,* with $m = 1$, has $\lambda_1 = 2L$, not $\lambda_1 = L$. Only half of a wavelength is contained between the boundaries, a direct consequence of the fact that the spacing between nodes is $\lambda/2$.
3. The frequencies of the normal modes form a series: $f_1, 2f_1, 3f_1, 4f_1, \ldots$. The fundamental frequency f_1 can be found as the *difference* between the frequencies of any two adjacent modes. That is, $f_1 = \Delta f = f_{m+1} - f_m$.

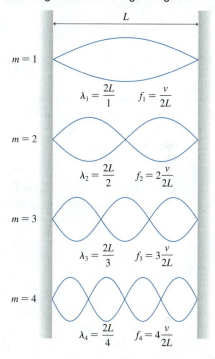

FIGURE 17.9 The first four modes for standing waves on a string of length L.

L

$m = 1$

$\lambda_1 = \dfrac{2L}{1} \qquad f_1 = \dfrac{v}{2L}$

$m = 2$

$\lambda_2 = \dfrac{2L}{2} \qquad f_2 = 2\dfrac{v}{2L}$

$m = 3$

$\lambda_3 = \dfrac{2L}{3} \qquad f_3 = 3\dfrac{v}{2L}$

$m = 4$

$\lambda_4 = \dfrac{2L}{4} \qquad f_4 = 4\dfrac{v}{2L}$

EXAMPLE 17.2 | **Measuring g**

Standing-wave frequencies can be measured very accurately. Consequently, standing waves are often used in experiments to make accurate measurements of other quantities. One such experiment, shown in FIGURE 17.10, uses standing waves to measure the free-fall acceleration g. A heavy mass is suspended from a 1.65-m-long, 5.85 g steel wire; then an oscillating magnetic field (because steel is magnetic) is used to excite the $m = 3$ standing wave on the wire. Measuring the frequency for different masses yields the data given in the table. Analyze these data to determine the local value of g.

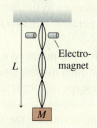

FIGURE 17.10 An experiment to measure g.

L

Electro-magnet

M

Mass (kg)	f_3 (Hz)
2.00	68
4.00	97
6.00	117
8.00	135
10.00	152

MODEL The hanging mass creates tension in the wire. This establishes the wave speed along the wire and thus the frequencies of standing waves. Masses of a few kg might stretch the wire a mm or so, but that doesn't change the length L until the third decimal place. The mass of the wire itself is insignificant in comparison to that of the hanging mass. We'll be justified in determining g to three significant figures.

SOLVE The frequency of the third harmonic is

$$f_3 = \frac{3}{2}\frac{v}{L}$$

The wave speed on the wire is

$$v = \sqrt{\frac{T_s}{\mu}} = \sqrt{\frac{Mg}{m/L}} = \sqrt{\frac{MgL}{m}}$$

where Mg is the weight of the hanging mass, and thus the tension in the wire, while m is the mass of the wire. Combining these two equations, we have

Continued

$$f_3 = \frac{3}{2}\sqrt{\frac{Mg}{mL}} = \frac{3}{2}\sqrt{\frac{g}{mL}}\sqrt{M}$$

Squaring both sides gives

$$f_3^2 = \frac{9g}{4mL}M$$

A graph of the square of the standing-wave frequency versus mass M should be a straight line passing through the origin with slope $9g/4mL$. We can use the experimental slope to determine g.

FIGURE 17.11 is a graph of f_3^2 versus M. The slope of the best-fit line is $2289 \text{ kg}^{-1}\text{s}^{-2}$, from which we find

$$g = \text{slope} \times \frac{4mL}{9}$$

$$= 2289 \text{ kg}^{-1}\text{s}^{-2} \times \frac{4(0.00585 \text{ kg})(1.65 \text{ m})}{9} = 9.82 \text{ m/s}^2$$

FIGURE 17.11 Graph of the data.

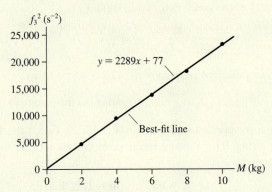

ASSESS The fact that the graph is linear and passes through the origin confirms our model. This is an important reason for having multiple data points rather than using only one mass.

STOP TO THINK 17.2 A standing wave on a string vibrates as shown at the right. Suppose the string tension is quadrupled while the frequency and the length of the string are held constant. Which standing-wave pattern is produced?

Original standing wave

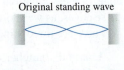

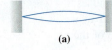 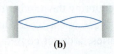 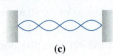 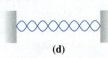

(a)   (b)   (c)   (d)

Standing Electromagnetic Waves

Because electromagnetic waves are transverse waves, a standing electromagnetic wave is very much like a standing wave on a string. Standing electromagnetic waves can be established between two parallel mirrors that reflect light back and forth. The mirrors are boundaries, analogous to the boundaries at the ends of a string. In fact, this is exactly how a laser operates. The two facing mirrors in **FIGURE 17.12** form what is called a *laser cavity*.

Because the mirrors act like the points to which a string is tied, the light wave must have a node at the surface of each mirror. One of the mirrors is only partially reflective, to allow some light to escape and form the laser beam, but this doesn't affect the boundary condition.

Because the boundary conditions are the same, Equations 17.13 and 17.14 for λ_m and f_m apply to a laser just as they do to a vibrating string. The primary difference is the size of the wavelength. A typical laser cavity has a length $L \approx 30$ cm, and visible light has a wavelength $\lambda \approx 600$ nm. The standing light wave in a laser cavity has a mode number m that is approximately

$$m = \frac{2L}{\lambda} \approx \frac{2 \times 0.30 \text{ m}}{6.00 \times 10^{-7} \text{ m}} = 1,000,000$$

In other words, the standing light wave inside a laser cavity has approximately one million antinodes! This is a consequence of the very short wavelength of light.

FIGURE 17.12 A laser contains a standing light wave between two parallel mirrors.

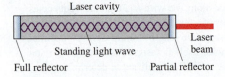

Laser cavity

Standing light wave

Laser beam

Full reflector   Partial reflector

EXAMPLE 17.3 | **The standing light wave inside a laser**

Helium-neon lasers emit the red laser light commonly used in classroom demonstrations and supermarket checkout scanners. A helium-neon laser operates at a wavelength of precisely 632.9924 nm when the spacing between the mirrors is 310.372 mm.

a. In which mode does this laser operate?

b. What is the next longest wavelength that could form a standing wave in this laser cavity?

MODEL The light wave forms a standing wave between the two mirrors.

VISUALIZE The standing wave looks like Figure 17.12.

SOLVE a. We can use $\lambda_m = 2L/m$ to find that m (the mode) is

$$m = \frac{2L}{\lambda_m} = \frac{2(0.310372 \text{ m})}{6.329924 \times 10^{-7} \text{ m}} = 980{,}650$$

There are 980,650 antinodes in the standing light wave.

b. The next longest wavelength that can fit in this laser cavity will have one fewer node. It will be the $m = 980{,}649$ mode and its wavelength will be

$$\lambda = \frac{2L}{m} = \frac{2(0.310372 \text{ m})}{980{,}649} = 632.9930 \text{ nm}$$

ASSESS The wavelength increases by a mere 0.0006 nm when the mode number is decreased by 1.

Microwaves, with a wavelength of a few centimeters, can also set up standing waves. This is not always good. If the microwaves in a microwave oven form a standing wave, there are nodes where the electromagnetic field intensity is always zero. These nodes cause cold spots where the food does not heat. Although designers of microwave ovens try to prevent standing waves, ovens usually do have cold spots spaced $\lambda/2$ apart at nodes in the microwave field. A turntable in a microwave oven keeps the food moving so that no part of your dinner remains at a node.

17.4 Standing Sound Waves and Musical Acoustics

A long, narrow column of air, such as the air in a tube or pipe, can support a *longitudinal* standing sound wave. Longitudinal waves are somewhat trickier than string waves because a graph—showing displacement *parallel* to the tube—is not a picture of the wave.

To illustrate the ideas, **FIGURE 17.13** is a series of three graphs and pictures that show the $m = 2$ standing wave inside a column of air closed at both ends. We call this a *closed-closed tube*. The air at the closed ends cannot oscillate because the air molecules are pressed up against the wall, unable to move; hence **a closed end of a column of air must be a displacement node.** Thus the boundary conditions—nodes at the ends—are the same as for a standing wave on a string.

FIGURE 17.13 The $m = 2$ standing sound wave in a closed-closed tube of air.

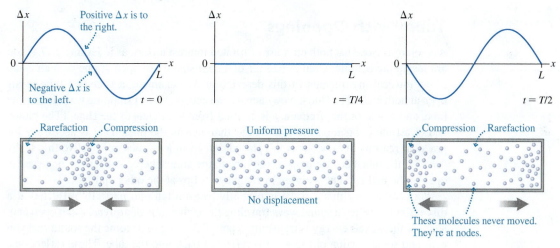

FIGURE 17.14 The $m = 2$ longitudinal standing wave can be represented as a displacement wave or as a pressure wave.

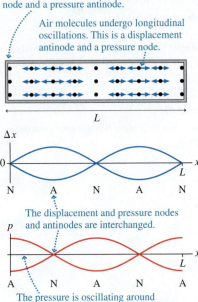

The closed end is a displacement node and a pressure antinode.

Air molecules undergo longitudinal oscillations. This is a displacement antinode and a pressure node.

The displacement and pressure nodes and antinodes are interchanged.

The pressure is oscillating around atmospheric pressure p_{atmos}.

Although the graph looks familiar, it is now a graph of *longitudinal* displacement. At $t = 0$, positive displacements in the left half and negative displacements in the right half cause all the air molecules to converge at the center of the tube. The density and pressure rise at the center and fall at the ends—a *compression* and *rarefaction* in the terminology of Chapter 16. A half cycle later, the molecules have rushed to the ends of the tube. Now the pressure is maximum at the ends, minimum in the center. Try to visualize the air molecules sloshing back and forth this way.

FIGURE 17.14 combines these illustrations into a single picture showing where the molecules are oscillating (antinodes) and where they're not (nodes). A graph of the displacement Δx looks just like the $m = 2$ graph of a standing wave on a string. Because the boundary conditions are the same, the possible wavelengths and frequencies of standing waves in a closed-closed tube are the same as for a string of the same length.

It is often useful to think of sound as a *pressure wave* rather than a displacement wave, and the bottom graph in Figure 17.14 shows the $m = 2$ pressure standing wave in a closed-closed tube. Notice that the pressure is oscillating around p_{atmos}, its equilibrium value.

We showed in « Section 16.6 that the pressure in a sound wave is minimum or maximum at points where the displacement is zero, and vice versa. Consequently, **the nodes and antinodes of the pressure wave are interchanged with those of the displacement wave.** You can see in Figure 17.13 that the gas molecules are alternately pushed up against the ends of the tube, then pulled away, causing the pressure at the closed ends to oscillate with maximum amplitude—an antinode—at a point where the displacement is a node.

EXAMPLE 17.4 | Singing in the shower

A shower stall is 2.45 m (8 ft) tall. For what frequencies less than 500 Hz are there standing sound waves in the shower stall?

MODEL The shower stall, to a first approximation, is a column of air 2.45 m long. It is closed at the ends by the ceiling and floor. Assume a 20°C speed of sound.

VISUALIZE A standing sound wave will have nodes at the ceiling and the floor. The $m = 2$ mode will look like Figure 17.14 rotated 90°.

SOLVE The fundamental frequency for a standing sound wave in this air column is

$$f_1 = \frac{v}{2L} = \frac{343 \text{ m/s}}{2(2.45 \text{ m})} = 70 \text{ Hz}$$

The possible standing-wave frequencies are integer multiples of the fundamental frequency. These are 70 Hz, 140 Hz, 210 Hz, 280 Hz, 350 Hz, 420 Hz, and 490 Hz.

ASSESS The many possible standing waves in a shower cause the sound to *resonate*, which helps explain why some people like to sing in the shower. Our approximation of the shower stall as a one-dimensional tube is actually a bit too simplistic. A three-dimensional analysis would find additional modes, making the "sound spectrum" even richer.

Tubes with Openings

Air columns closed at both ends are of limited interest unless, as in Example 17.4, you are inside the column. Columns of air that *emit* sound are open at one or both ends. Many musical instruments fit this description. For example, a flute is a tube of air open at both ends. The flutist blows across one end to create a standing wave inside the tube, and a note of that frequency is emitted from both ends of the flute. (The blown end of a flute is open on the side, rather than across the tube. That is necessary for practical reasons of how flutes are played, but from a physics perspective this is the "end" of the tube because it opens the tube to the atmosphere.) A trumpet, however, is open at the bell end but is *closed* by the player's lips at the other end.

You saw earlier that a wave is partially transmitted and partially reflected at a discontinuity. When a sound wave traveling through a tube of air reaches an open end, some of the wave's energy is transmitted out of the tube to become the sound that you hear and some portion of the wave is reflected back into the tube. These reflections, analogous to the reflection of a string wave from a boundary, allow standing sound waves to exist in a tube of air that is open at one or both ends.

Not surprisingly, the *boundary condition* at the open end of a column of air is not the same as the boundary condition at a closed end. The air pressure at the open end of a tube is constrained to match the atmospheric pressure of the surrounding air. Consequently, the open end of a tube must be a pressure node. Because pressure nodes and antinodes are interchanged with those of the displacement wave, **an open end of an air column is required to be a displacement antinode.** (A careful analysis shows that the antinode is actually just outside the open end, but for our purposes we'll assume the antinode is exactly at the open end.)

FIGURE 17.15 The first three standing sound wave modes in columns of air with different boundary conditions.

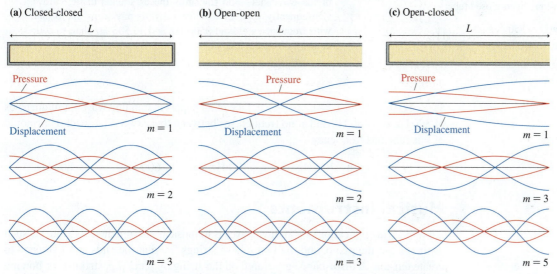

FIGURE 17.15 shows displacement and pressure graphs of the first three standing-wave modes of a tube closed at both ends (a *closed-closed tube*), a tube open at both ends (an *open-open tube*), and a tube open at one end but closed at the other (an *open-closed tube*), all with the same length L. Notice the pressure and displacement boundary conditions. The standing wave in the open-open tube looks like the closed-closed tube except that the positions of the nodes and antinodes are interchanged. In both cases there are m half-wavelength segments between the ends; thus the wavelengths and frequencies of an open-open tube and a closed-closed tube are the same as those of a string tied at both ends:

$$\lambda_m = \frac{2L}{m} \qquad m = 1, 2, 3, 4, \ldots$$
$$f_m = m \frac{v}{2L} = mf_1 \qquad \text{(open-open or closed-closed tube)} \qquad (17.17)$$

The open-closed tube is different. The fundamental mode has only one-quarter of a wavelength in a tube of length L; hence the $m = 1$ wavelength is $\lambda_1 = 4L$. This is twice the λ_1 wavelength of an open-open or a closed-closed tube. Consequently, **the fundamental frequency of an open-closed tube is half that of an open-open or a closed-closed tube of the same length.** It will be left as a homework problem for you to show that the possible wavelengths and frequencies of an open-closed tube of length L are

$$\lambda_m = \frac{4L}{m} \qquad m = 1, 3, 5, 7, \ldots$$
$$f_m = m \frac{v}{4L} = mf_1 \qquad \text{(open-closed tube)} \qquad (17.18)$$

Notice that m in Equation 17.18 takes on only *odd* values.

EXAMPLE 17.5 | Resonances of the ear canal

The eardrum, which transmits sound vibrations to the sensory organs of the inner ear, lies at the end of the ear canal. For adults, the ear canal is about 2.5 cm in length. What frequency standing waves can occur in the ear canal that are within the range of human hearing? The speed of sound in the warm air of the ear canal is 350 m/s.

MODEL The ear canal is open to the air at one end, closed by the eardrum at the other. We can model it as an open-closed tube. The standing waves will be those of Figure 17.15c.

SOLVE The lowest standing-wave frequency is the fundamental frequency for a 2.5-cm-long open-closed tube:

$$f_1 = \frac{v}{4L} = \frac{350 \text{ m/s}}{4(0.025 \text{ m})} = 3500 \text{ Hz}$$

Standing waves also occur at the harmonics, but an open-closed tube has only odd harmonics. These are

$$f_3 = 3f_1 = 10,500 \text{ Hz}$$
$$f_5 = 5f_1 = 17,500 \text{ Hz}$$

Higher harmonics are beyond the range of human hearing, as discussed in Section 16.5.

ASSESS The ear canal is short, so we expected the standing-wave frequencies to be relatively high. The air in your ear canal responds readily to sounds at these frequencies—what we call a *resonance* of the ear canal—and transmits theses sounds to the eardrum. Consequently, your ear actually is slightly more sensitive to sounds with frequencies around 3500 Hz and 10,500 Hz than to sounds at nearby frequencies.

STOP TO THINK 17.3 An open-open tube of air supports standing waves at frequencies of 300 Hz and 400 Hz and at no frequencies between these two. The second harmonic of this tube has frequency

a. 100 Hz   b. 200 Hz   c. 400 Hz   d. 600 Hz   e. 800 Hz

Musical Instruments

An important application of standing waves is to musical instruments. Instruments such as the guitar, the piano, and the violin have strings fixed at the ends and tightened to create tension. A disturbance generated on the string by plucking, striking, or bowing it creates a standing wave on the string.

The fundamental frequency of a vibrating string is

$$f_1 = \frac{v}{2L} = \frac{1}{2L}\sqrt{\frac{T_s}{\mu}}$$

where T_s is the tension in the string and μ is its linear density. The fundamental frequency is the musical note you hear when the string is sounded. Increasing the tension in the string raises the fundamental frequency, which is how stringed instruments are tuned.

NOTE v is the wave speed *on the string*, not the speed of sound in air.

For the guitar or the violin, the strings are all the same length and under approximately the same tension. Were that not the case, the neck of the instrument would tend to twist toward the side of higher tension. The strings have different frequencies because they differ in linear density: The lower-pitched strings are "fat" while the higher-pitched strings are "skinny." This difference changes the frequency by changing the wave speed. *Small* adjustments are then made in the tension to bring each string to the exact desired frequency. Once the instrument is tuned, you play it by using your fingertips to alter the effective length of the string. As you shorten the string's length, the frequency and pitch go up.

A piano covers a much wider range of frequencies than a guitar or violin. This range cannot be produced by changing only the linear densities of the strings. The high end would have strings too thin to use without breaking, and the low end would have solid rods rather than flexible wires! So a piano is tuned through a combination of changing the linear density *and* the length of the strings. The bass note strings are not only fatter, they are also longer.

The strings on a harp vibrate as standing waves. Their frequencies determine the notes that you hear.

With a wind instrument, blowing into the mouthpiece creates a standing sound wave inside a tube of air. The player changes the notes by using her fingers to cover holes or open valves, changing the length of the tube and thus its frequency. The fact that the holes are on the side makes very little difference; the first open hole becomes an antinode because the air is free to oscillate in and out of the opening.

A wind instrument's frequency depends on the speed of sound *inside* the instrument. But the speed of sound depends on the temperature of the air. When a wind player first blows into the instrument, the air inside starts to rise in temperature. This increases the sound speed, which in turn raises the instrument's frequency for each note until the air temperature reaches a steady state. Consequently, wind players must "warm up" before tuning their instrument.

Many wind instruments have a "buzzer" at one end of the tube, such as a vibrating reed on a saxophone or vibrating lips on a trombone. Buzzers generate a continuous range of frequencies rather than single notes, which is why they sound like a "squawk" if you play on just the mouthpiece without the rest of the instrument. When a buzzer is connected to the body of the instrument, most of those frequencies cause no response of the air molecules. But the frequency from the buzzer that matches the fundamental frequency of the instrument causes the buildup of a large-amplitude response at just that frequency—a standing-wave resonance. This is the energy input that generates and sustains the musical note.

EXAMPLE 17.6 | **Flutes and clarinets**

A clarinet is 66.0 cm long. A flute is nearly the same length, with 63.6 cm between the hole the player blows across and the end of the flute. What are the frequencies of the lowest note and the next higher harmonic on a flute and on a clarinet? The speed of sound in warm air is 350 m/s.

MODEL The flute is an open-open tube, open at the end as well as at the hole the player blows across. A clarinet is an open-closed tube because the player's lips and the reed seal the tube at the upper end.

SOLVE The lowest frequency is the fundamental frequency. For the flute, an open-open tube, this is

$$f_1 = \frac{v}{2L} = \frac{350 \text{ m/s}}{2(0.636 \text{ m})} = 275 \text{ Hz}$$

The clarinet, an open-closed tube, has

$$f_1 = \frac{v}{4L} = \frac{350 \text{ m/s}}{4(0.660 \text{ m})} = 133 \text{ Hz}$$

The next higher harmonic on the flute's open-open tube is $m = 2$ with frequency $f_2 = 2f_1 = 550$ Hz. An open-closed tube has only odd harmonics, so the next higher harmonic of the clarinet is $f_3 = 3f_1 = 399$ Hz.

ASSESS The clarinet plays a much lower note than the flute—musically, about an octave lower—because it is an open-closed tube. It's worth noting that neither of our fundamental frequencies is exactly correct because our open-open and open-closed tube models are a bit too simplified to adequately describe a real instrument. However, both calculated frequencies are close because our models do capture the essence of the physics.

A vibrating string plays the musical note corresponding to the fundamental frequency f_1, so stringed instruments must use several strings to obtain a reasonable range of notes. In contrast, wind instruments can sound at the second or third harmonic of the tube of air (f_2 or f_3). These higher frequencies are sounded by *overblowing* (flutes, brass instruments) or with keys that open small holes to encourage the formation of an antinode at that point (clarinets, saxophones). The controlled use of these higher harmonics gives wind instruments a wide range of notes.

17.5 Interference in One Dimension

One of the most basic characteristics of waves is the ability of two waves to combine into a single wave whose displacement is given by the principle of superposition. The pattern resulting from the superposition of two waves is often called **interference.** A standing wave is the interference pattern produced when two waves of equal frequency travel in opposite directions. In this section we will look at the interference of two waves traveling in the *same* direction.

FIGURE 17.16 Two overlapped waves travel along the x-axis.

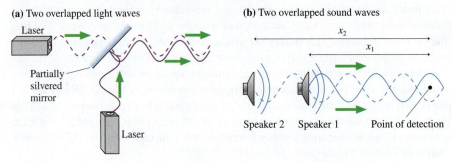

(a) Two overlapped light waves

Laser

Partially silvered mirror

Laser

(b) Two overlapped sound waves

x_2

x_1

Speaker 2   Speaker 1   Point of detection

FIGURE 17.16a shows two light waves impinging on a partially silvered mirror. Such a mirror partially transmits and partially reflects each wave, causing two *overlapped* light waves to travel along the x-axis to the right of the mirror. Or consider the two loudspeakers in FIGURE 17.16b. The sound wave from loudspeaker 2 passes just to the side of loudspeaker 1; hence two overlapped sound waves travel to the right along the x-axis. We want to find out what happens when two overlapped waves travel in the same direction along the same axis.

Figure 17.16b shows a point on the x-axis where the overlapped waves are detected, either by your ear or by a microphone. This point is distance x_1 from loudspeaker 1 and distance x_2 from loudspeaker 2. (We will use loudspeakers and sound waves for most of our examples, but our analysis is valid for any wave.) What is the amplitude of the combined waves at this point?

Throughout this section, **we will assume that the waves are sinusoidal, have the same frequency and amplitude, and travel to the right along the x-axis.** Thus we can write the displacements of the two waves as

$$D_1(x_1, t) = a\sin(kx_1 - \omega t + \phi_{10}) = a\sin\phi_1$$
$$D_2(x_2, t) = a\sin(kx_2 - \omega t + \phi_{20}) = a\sin\phi_2$$

(17.19)

where ϕ_1 and ϕ_2 are the *phases* of the waves. Both waves have the same wave number $k = 2\pi/\lambda$ and the same angular frequency $\omega = 2\pi f$.

The phase constants ϕ_{10} and ϕ_{20} are characteristics of *the sources,* not the medium. FIGURE 17.17 shows snapshot graphs at $t = 0$ of waves emitted by three sources with phase constants $\phi_0 = 0$ rad, $\phi_0 = \pi/2$ rad, and $\phi_0 = \pi$ rad. You can see that **the phase constant tells us what the source is doing at $t = 0$.** For example, a loudspeaker at its center position and moving backward at $t = 0$ has $\phi_0 = 0$ rad. Looking back at Figure 17.16b, you can see that loudspeaker 1 has phase constant $\phi_{10} = 0$ rad and loudspeaker 2 has $\phi_{20} = \pi$ rad.

> **NOTE** We will often consider *identical sources,* by which we mean that $\phi_{20} = \phi_{10}$. That is, the sources oscillate in phase. ◀

Let's examine overlapped waves graphically before diving into the mathematics. FIGURE 17.18 shows two important situations. In part a, the crests of the two waves are aligned as they travel along the x-axis. In part b, the crests of one wave align with the troughs of the other wave. The graphs and the wave fronts are slightly displaced from each other so that you can see what each wave is doing, but the *physical situation* is one in which the waves are traveling *on top of* each other.

The two waves of Figure 17.18a have the same displacement at every point: $D_1(x) = D_2(x)$. Two waves that are aligned crest to crest and trough to trough are said to be **in phase.** Waves that are in phase march along "in step" with each other.

When we combine two in-phase waves, using the principle of superposition, the net displacement at each point is twice the displacement of each individual wave. The superposition of two waves to create a traveling wave with an amplitude *larger* than either individual wave is called **constructive interference.** When the waves are exactly in phase, giving $A = 2a$, we have *maximum constructive interference.*

FIGURE 17.17 Waves from three sources having phase constants $\phi_0 = 0$ rad, $\phi_0 = \pi/2$ rad, and $\phi_0 = \pi$ rad.

(a) Snapshot graph at $t = 0$ for $\phi_0 = 0$ rad

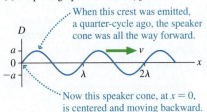

When this crest was emitted, a quarter-cycle ago, the speaker cone was all the way forward.

Now this speaker cone, at $x = 0$, is centered and moving backward.

(b) Snapshot graph at $t = 0$ for $\phi_0 = \pi/2$ rad

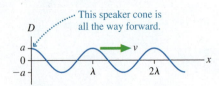

This speaker cone is all the way forward.

(c) Snapshot graph at $t = 0$ for $\phi_0 = \pi$ rad

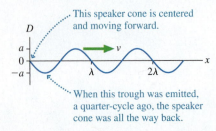

This speaker cone is centered and moving forward.

When this trough was emitted, a quarter-cycle ago, the speaker cone was all the way back.

In Figure 17.18b, where the crests of one wave align with the troughs of the other, the waves march along "out of step" with $D_1(x) = -D_2(x)$ at every point. Two waves that are aligned crest to trough are said to be *180° out of phase* or, more generally, just **out of phase.** A superposition of two waves to create a wave with an amplitude smaller than either individual wave is called **destructive interference.** In this case, because $D_1 = -D_2$, the net displacement is *zero* at *every point* along the axis. The combination of two waves that cancel each other to give no wave is called *perfect destructive interference.*

NOTE Perfect destructive interference occurs only if the two waves have exactly equal amplitudes, as we're assuming. A 180° phase difference always produces *maximum destructive interference*, but the cancellation won't be perfect if there is any difference in the amplitudes.

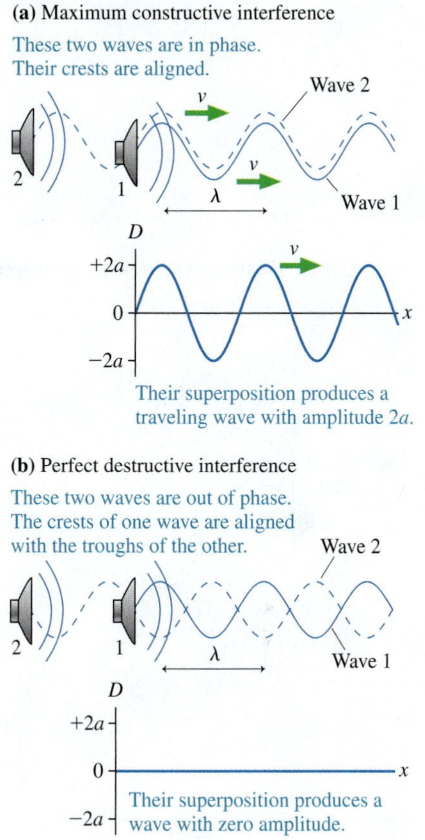

The Phase Difference

To understand interference, we need to focus on the *phases* of the two waves, which are

$$\phi_1 = kx_1 - \omega t + \phi_{10}$$
$$\phi_2 = kx_2 - \omega t + \phi_{20} \tag{17.20}$$

The difference between the two phases ϕ_1 and ϕ_2, called the **phase difference** $\Delta\phi$, is

$$\Delta\phi = \phi_2 - \phi_1 = (kx_2 - \omega t + \phi_{20}) - (kx_1 - \omega t + \phi_{10})$$
$$= k(x_2 - x_1) + (\phi_{20} - \phi_{10}) \tag{17.21}$$
$$= 2\pi \frac{\Delta x}{\lambda} + \Delta\phi_0$$

You can see that there are two contributions to the phase difference. $\Delta x = x_2 - x_1$, the distance between the two sources, is called **path-length difference.** It is the extra distance traveled by wave 2 on the way to the point where the two waves are combined. $\Delta\phi_0 = \phi_{20} - \phi_{10}$ is the *inherent phase difference* between the sources.

The condition of being in phase, where crests are aligned with crests and troughs with troughs, is $\Delta\phi = 0, 2\pi, 4\pi$, or any integer multiple of 2π rad. Thus the condition for maximum constructive interference is

> Maximum constructive interference:
> $$\Delta\phi = 2\pi \frac{\Delta x}{\lambda} + \Delta\phi_0 = m \cdot 2\pi \text{ rad} \qquad m = 0, 1, 2, 3, \ldots \tag{17.22}$$

For identical sources, which have $\Delta\phi_0 = 0$ rad, maximum constructive interference occurs when $\Delta x = m\lambda$. That is, **two identical sources produce maximum constructive interference when the path-length difference is an integer number of wavelengths.**

FIGURE 17.19 shows two identical sources (i.e., the two loudspeakers are doing the same thing at the same time), so $\Delta\phi_0 = 0$ rad. The path-length difference Δx is the extra distance traveled by the wave from loudspeaker 2 before it combines with loudspeaker 1. In this case, $\Delta x = \lambda$. Because a wave moves forward exactly one wavelength during one period, loudspeaker 1 emits a crest exactly as a crest of wave 2 passes by. The two waves are "in step," with $\Delta\phi = 2\pi$ rad, so the two waves interfere constructively to produce a wave of amplitude $2a$.

Maximum destructive interference, where the crests of one wave are aligned with the troughs of the other, occurs when two waves are *out of phase*, meaning that $\Delta\phi = \pi, 3\pi, 5\pi$, or any odd multiple of π rad. Thus the condition for maximum destructive interference is

> Maximum destructive interference:
> $$\Delta\phi = 2\pi \frac{\Delta x}{\lambda} + \Delta\phi_0 = \left(m + \tfrac{1}{2}\right) \cdot 2\pi \text{ rad} \qquad m = 0, 1, 2, 3, \ldots \tag{17.23}$$

FIGURE 17.18 Interference of two waves traveling along the x-axis.

(a) Maximum constructive interference

These two waves are in phase. Their crests are aligned.

Their superposition produces a traveling wave with amplitude $2a$.

(b) Perfect destructive interference

These two waves are out of phase. The crests of one wave are aligned with the troughs of the other.

Their superposition produces a wave with zero amplitude.

FIGURE 17.19 Two identical sources one wavelength apart.

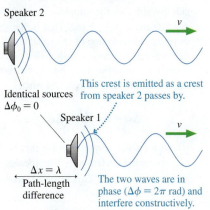

This crest is emitted as a crest from speaker 2 passes by.

Identical sources $\Delta\phi_0 = 0$

Speaker 1

$\Delta x = \lambda$
Path-length difference

The two waves are in phase ($\Delta\phi = 2\pi$ rad) and interfere constructively.

For identical sources, which have $\Delta\phi_0 = 0$ rad, maximum destructive interference occurs when $\Delta x = (m + \frac{1}{2})\lambda$. **That is, two identical sources produce maximum destructive interference when the path-length difference is a half-integer number of wavelengths.**

Two waves can be out of phase because the sources are located at different positions, because the sources themselves are out of phase, or because of a combination of these two. **FIGURE 17.20** illustrates these ideas by showing three different ways in which two waves interfere destructively. Each of these three arrangements creates waves with $\Delta\phi = \pi$ rad.

FIGURE 17.20 Destructive interference three ways.

(a) The sources are out of phase.

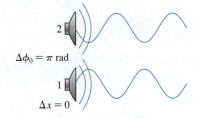

$\Delta\phi_0 = \pi$ rad

$\Delta x = 0$

(b) Identical sources are separated by half a wavelength.

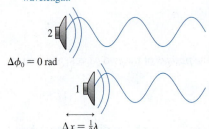

$\Delta\phi_0 = 0$ rad

$\Delta x = \frac{1}{2}\lambda$

(c) The sources are both separated and partially out of phase.

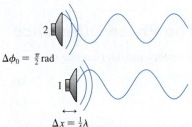

$\Delta\phi_0 = \frac{\pi}{2}$ rad

$\Delta x = \frac{1}{4}\lambda$

NOTE Don't confuse the phase difference of the waves ($\Delta\phi$) with the phase difference of the sources ($\Delta\phi_0$). It is $\Delta\phi$, the phase difference of the waves, that governs interference.

EXAMPLE 17.7 | **Interference between two sound waves**

You are standing in front of two side-by-side loudspeakers playing sounds of the same frequency. Initially there is almost no sound at all. Then one of the speakers is moved slowly away from you. The sound intensity increases as the separation between the speakers increases, reaching a maximum when the speakers are 0.75 m apart. Then, as the speaker continues to move, the intensity starts to decrease. What is the distance between the speakers when the sound intensity is again a minimum?

MODEL The changing sound intensity is due to the interference of two overlapped sound waves.

VISUALIZE Moving one speaker relative to the other changes the phase difference between the waves.

SOLVE A minimum sound intensity implies that the two sound waves are interfering destructively. Initially the loudspeakers are side by side, so the situation is as shown in Figure 17.20a with $\Delta x = 0$ and $\Delta\phi_0 = \pi$ rad. That is, the speakers themselves are out of phase. Moving one of the speakers does not change $\Delta\phi_0$, but it does change the path-length difference Δx and thus increases the overall phase difference $\Delta\phi$. Constructive interference, causing maximum intensity, is reached when

$$\Delta\phi = 2\pi \frac{\Delta x}{\lambda} + \Delta\phi_0 = 2\pi \frac{\Delta x}{\lambda} + \pi = 2\pi \text{ rad}$$

where we used $m = 1$ because this is the first separation giving constructive interference. The speaker separation at which this occurs is $\Delta x = \lambda/2$. This is the situation shown in **FIGURE 17.21**.

FIGURE 17.21 Two out-of-phase sources generate waves that are in phase if the sources are one half-wavelength apart.

The sources are out of phase, $\Delta\phi_0 = \pi$ rad.

$\Delta x = \frac{1}{2}\lambda$

The sources are separated by half a wavelength.

As a result, the waves are in phase.

Because $\Delta x = 0.75$ m is $\lambda/2$, the sound's wavelength is $\lambda = 1.50$ m. The next point of destructive interference, with $m = 1$, occurs when

$$\Delta\phi = 2\pi \frac{\Delta x}{\lambda} + \Delta\phi_0 = 2\pi \frac{\Delta x}{\lambda} + \pi = 3\pi \text{ rad}$$

Thus the distance between the speakers when the sound intensity is again a minimum is

$$\Delta x = \lambda = 1.50 \text{ m}$$

ASSESS A separation of λ gives constructive interference for two *identical* speakers ($\Delta\phi_0 = 0$). Here the phase difference of π rad between the speakers (one is pushing forward as the other pulls back) gives destructive interference at this separation.

STOP TO THINK 17.4 Two loudspeakers emit waves with $\lambda = 2.0$ m. Speaker 2 is 1.0 m in front of speaker 1. What, if anything, can be done to cause constructive interference between the two waves?

a. Move speaker 1 forward (to the right) 1.0 m.
b. Move speaker 1 forward (to the right) 0.5 m.
c. Move speaker 1 backward (to the left) 0.5 m.
d. Move speaker 1 backward (to the left) 1.0 m.
e. Nothing. The situation shown already causes constructive interference.
f. Constructive interference is not possible for any placement of the speakers.

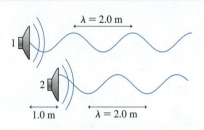

17.6 The Mathematics of Interference

Let's look more closely at the superposition of two waves. As two waves of equal amplitude and frequency travel together along the x-axis, the net displacement of the medium is

$$D = D_1 + D_2 = a\sin(kx_1 - \omega t + \phi_{10}) + a\sin(kx_2 - \omega t + \phi_{20})$$
$$= a\sin\phi_1 + a\sin\phi_2 \tag{17.24}$$

where the phases ϕ_1 and ϕ_2 were defined in Equations 17.20.

A useful trigonometric identity is

$$\sin\alpha + \sin\beta = 2\cos\left[\tfrac{1}{2}(\alpha - \beta)\right]\sin\left[\tfrac{1}{2}(\alpha + \beta)\right] \tag{17.25}$$

This identity is certainly not obvious, although it is easily proven by working backward from the right side. We can use this identity to write the net displacement of Equation 17.24 as

$$D = \left[2a\cos\left(\frac{\Delta\phi}{2}\right)\right]\sin\left(kx_{\text{avg}} - \omega t + (\phi_0)_{\text{avg}}\right) \tag{17.26}$$

where $\Delta\phi = \phi_2 - \phi_1$ is the phase difference between the two waves, exactly as in Equation 17.21. $x_{\text{avg}} = (x_1 + x_2)/2$ is the average distance to the two sources and $(\phi_0)_{\text{avg}} = (\phi_{10} + \phi_{20})/2$ is the average phase constant of the sources.

The sine term shows that the superposition of the two waves is still a traveling wave. An observer would see a sinusoidal wave moving along the x-axis with the *same* wavelength and frequency as the original waves.

But how *big* is this wave compared to the two original waves? They each had amplitude a, but the amplitude of their superposition is

$$A = \left|2a\cos\left(\frac{\Delta\phi}{2}\right)\right| \tag{17.27}$$

where we have used an absolute value sign because amplitudes must be positive. Depending upon the phase difference of the two waves, the amplitude of their superposition can be anywhere from zero (perfect destructive interference) to $2a$ (maximum constructive interference).

The amplitude has its maximum value $A = 2a$ if $\cos(\Delta\phi/2) = \pm 1$. This occurs when

$$\Delta\phi = m\cdot 2\pi \quad \text{(maximum amplitude } A = 2a) \tag{17.28}$$

where m is an integer. Similarly, the amplitude is zero if $\cos(\Delta\phi/2) = 0$, which occurs when

$$\Delta\phi = \left(m + \tfrac{1}{2}\right)\cdot 2\pi \quad \text{(minimum amplitude } A = 0) \tag{17.29}$$

Equations 17.28 and 17.29 are identical to the conditions of Equations 17.22 and 17.23 for constructive and destructive interference. We initially found these conditions by

FIGURE 17.22 The interference of two waves for three different values of the phase difference.

For $\Delta\phi = 40°$, the interference is constructive but not maximum constructive.

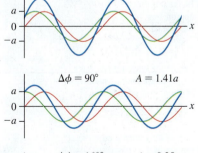

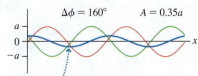

For $\Delta\phi = 160°$, the interference is destructive but not perfect destructive.

considering the alignment of the crests and troughs. Now we have confirmed them with an algebraic addition of the waves.

It is entirely possible, of course, that the two waves are neither exactly in phase nor exactly out of phase. Equation 17.27 allows us to calculate the amplitude of the superposition for any value of the phase difference. As an example, **FIGURE 17.22** on the previous page shows the calculated interference of two waves that differ in phase by 40°, by 90°, and by 160°.

EXAMPLE 17.8 | **More interference of sound waves**

Two loudspeakers emit 500 Hz sound waves with an amplitude of 0.10 mm. Speaker 2 is 1.00 m behind speaker 1, and the phase difference between the speakers is 90°. What is the amplitude of the sound wave at a point 2.00 m in front of speaker 1?

MODEL The amplitude is determined by the interference of the two waves. Assume that the speed of sound has a room-temperature (20°C) value of 343 m/s.

SOLVE The amplitude of the sound wave is

$$A = |2a \cos(\Delta\phi/2)|$$

where $a = 0.10$ mm and the phase difference between the waves is

$$\Delta\phi = \phi_2 - \phi_1 = 2\pi \frac{\Delta x}{\lambda} + \Delta\phi_0$$

The sound's wavelength is

$$\lambda = \frac{v}{f} = \frac{343 \text{ m/s}}{500 \text{ Hz}} = 0.686 \text{ m}$$

Distances $x_1 = 2.00$ m and $x_2 = 3.00$ m are measured from the speakers, so the path-length difference is $\Delta x = 1.00$ m. We're given that the inherent phase difference between the speakers is $\Delta\phi_0 = \pi/2$ rad. Thus the phase difference at the observation point is

$$\Delta\phi = 2\pi \frac{\Delta x}{\lambda} + \Delta\phi_0 = 2\pi \frac{1.00 \text{ m}}{0.686 \text{ m}} + \frac{\pi}{2} \text{ rad} = 10.73 \text{ rad}$$

and the amplitude of the wave at this point is

$$A = \left| 2a \cos\left(\frac{\Delta\phi}{2}\right) \right| = \left| (0.200 \text{ mm}) \cos\left(\frac{10.73}{2}\right) \right| = 0.121 \text{ mm}$$

ASSESS The interference is constructive because $A > a$, but less than maximum constructive interference.

Application: Thin-Film Optical Coatings

The shimmering colors of soap bubbles and oil slicks, as seen in the photo at the beginning of the chapter, are due to the interference of light waves. In fact, the idea of light-wave interference in one dimension has an important application in the optics industry, namely the use of **thin-film optical coatings**. These films, less than 1 μm (10^{-6} m) thick, are placed on glass surfaces, such as lenses, to control reflections from the glass. Antireflection coatings on the lenses in cameras, microscopes, and other optical equipment are examples of thin-film coatings.

FIGURE 17.23 shows a light wave of wavelength λ approaching a piece of glass that has been coated with a transparent film of thickness d whose index of refraction is n. The air-film boundary is a discontinuity at which the wave speed suddenly decreases, and you saw earlier, in Figure 17.7, that a discontinuity causes a reflection. Most of the light is transmitted into the film, but a little bit is reflected.

Furthermore, you saw in Figure 17.7 that the wave reflected from a discontinuity at which the speed decreases is *inverted* with respect to the incident wave. For a sinusoidal wave, which we're now assuming, the inversion is represented mathematically as a phase shift of π rad. The speed of a light wave decreases when it enters a material with a *larger* index of refraction. Thus **a light wave that reflects from a boundary at which the index of refraction increases has a phase shift of π rad.** There is no phase shift for the reflection from a boundary at which the index of refraction decreases. The reflection in Figure 17.23 is from a boundary between air ($n_{\text{air}} = 1.00$) and a transparent film with $n_{\text{film}} > n_{\text{air}}$, so the reflected wave is inverted due to the phase shift of π rad.

When the transmitted wave reaches the glass, most of it continues on into the glass but a portion is reflected back to the left. We'll assume that the index of refraction of the glass is larger than that of the film, $n_{\text{glass}} > n_{\text{film}}$, so this reflection also has a phase shift of π rad. This second reflection, after traveling back through the film, passes back into the air. There are now *two* equal-frequency reflected waves traveling to the left, and these waves will interfere. If the two reflected waves are *in phase*, they will interfere constructively to cause a *strong reflection*. If the two reflected waves are *out of*

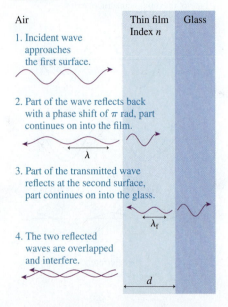

FIGURE 17.23 The two reflections, one from the coating and one from the glass, interfere.

Air | Thin film | Glass
Index n

1. Incident wave approaches the first surface.

2. Part of the wave reflects back with a phase shift of π rad, part continues on into the film.

λ

3. Part of the transmitted wave reflects at the second surface, part continues on into the glass.

λ_f

4. The two reflected waves are overlapped and interfere.

d

phase, they will interfere destructively to cause a *weak reflection* or, if their amplitudes are equal, *no reflection* at all.

This suggests practical uses for thin-film optical coatings. The reflections from glass surfaces, even if weak, are often undesirable. For example, reflections degrade the performance of optical equipment. These reflections can be eliminated by coating the glass with a film whose thickness is chosen to cause *destructive* interference of the two reflected waves. This is an *antireflection coating.*

The amplitude of the reflected light depends on the phase difference between the two reflected waves. This phase difference is

$$\Delta\phi = \phi_2 - \phi_1 = (kx_2 + \phi_{20} + \pi \text{ rad}) - (kx_1 + \phi_{10} + \pi \text{ rad})$$

$$= 2\pi\frac{\Delta x}{\lambda_f} + \Delta\phi_0 \tag{17.30}$$

where we explicitly included the reflection phase shift of each wave. In this case, because *both* waves had a phase shift of π rad, the reflection phase shifts cancel.

The wavelength λ_f is the wavelength *in the film* because that's where the path-length difference Δx occurs. You learned in Chapter 16 that the wavelength in a transparent material with index of refraction n is $\lambda_f = \lambda/n$, where the unsubscripted λ is the wavelength in vacuum or air. That is, λ is the wavelength that we measure on "our" side of the air-film boundary.

The path-length difference between the two waves is $\Delta x = 2d$ because wave 2 travels through the film *twice* before rejoining wave 1. The two waves have a common origin—the initial division of the incident wave at the front surface of the film—so the inherent phase difference is $\Delta\phi_0 = 0$. Thus the phase difference of the two reflected waves is

$$\Delta\phi = 2\pi\frac{2d}{\lambda/n} = 2\pi\frac{2nd}{\lambda} \tag{17.31}$$

The interference is constructive, causing a strong reflection, when $\Delta\phi = m \cdot 2\pi$ rad. So when both reflected waves have a phase shift of π rad, constructive interference occurs for wavelengths

$$\lambda_C = \frac{2nd}{m} \qquad m = 1, 2, 3, \ldots \qquad \text{(constructive interference)} \tag{17.32}$$

You will notice that m starts with 1, rather than 0, in order to give meaningful results. Destructive interference, with minimum reflection, requires $\Delta\phi = (m - \frac{1}{2}) \cdot 2\pi$ rad. This—again, when both waves have a phase shift of π rad—occurs for wavelengths

$$\lambda_D = \frac{2nd}{m - \frac{1}{2}} \qquad m = 1, 2, 3, \ldots \qquad \text{(destructive interference)} \tag{17.33}$$

We've used $m - \frac{1}{2}$, rather than $m + \frac{1}{2}$, so that m can start with 1 to match the condition for constructive interference.

NOTE The exact condition for constructive or destructive interference is satisfied for only a few discrete wavelengths λ. Nonetheless, reflections are strongly enhanced (nearly constructive interference) for a range of wavelengths near λ_C. Likewise, there is a range of wavelengths near λ_D for which the reflection is nearly canceled.

Antireflection coatings use the interference of light waves to nearly eliminate reflections from glass surfaces.

EXAMPLE 17.9 | Designing an antireflection coating

Magnesium fluoride (MgF_2) is used as an antireflection coating on lenses. The index of refraction of MgF_2 is 1.39. What is the thinnest film of MgF_2 that works as an antireflection coating at $\lambda = 510$ nm, near the center of the visible spectrum?

MODEL Reflection is minimized if the two reflected waves interfere destructively.

SOLVE The film thicknesses that cause destructive interference at wavelength λ are

$$d = \left(m - \frac{1}{2}\right)\frac{\lambda}{2n}$$

The thinnest film has $m = 1$. Its thickness is

$$d = \frac{\lambda}{4n} = \frac{510 \text{ nm}}{4(1.39)} = 92 \text{ nm}$$

The film thickness is significantly less than the wavelength of visible light!

Continued

ASSESS The reflected light is completely eliminated (perfect destructive interference) only if the two reflected waves have equal amplitudes. In practice, they don't. Nonetheless, the reflection is reduced from ≈ 4% of the incident intensity for "bare glass" to well under 1%. Furthermore, the intensity of reflected light is much reduced across most of the visible spectrum (400–700 nm), even though the phase difference deviates more and more from π rad

as the wavelength moves away from 510 nm. It is the increasing reflection at the ends of the visible spectrum ($\lambda \approx 400$ nm and $\lambda \approx 700$ nm), where $\Delta\phi$ deviates significantly from π rad, that gives a reddish-purple tinge to the lenses on cameras and binoculars. Homework problems will let you explore situations where only one of the two reflections has a reflection phase shift of π rad.

17.7 Interference in Two and Three Dimensions

FIGURE 17.24 A circular or spherical wave.

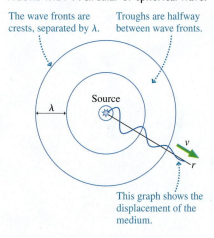

The wave fronts are crests, separated by λ.

Troughs are halfway between wave fronts.

Source

λ

v

r

This graph shows the displacement of the medium.

Ripples on a lake move in two dimensions. The glow from a lightbulb spreads outward as a spherical wave. A circular or spherical wave, illustrated in **FIGURE 17.24**, can be written

$$D(r, t) = a\sin(kr - \omega t + \phi_0) \tag{17.34}$$

where r is the distance measured outward from the source. Equation 17.34 is our familiar wave equation with the one-dimensional coordinate x replaced by a more general radial coordinate r. Recall that the wave fronts represent the *crests* of the wave and are spaced by the wavelength λ.

What happens when two circular or spherical waves overlap? For example, imagine two paddles oscillating up and down on the surface of a pond. We will assume that the two paddles oscillate with the same frequency and amplitude and that they are in phase. **FIGURE 17.25** shows the wave fronts of the two waves. The ripples overlap as they travel, and, as was the case in one dimension, this causes interference. An important difference, though, is that amplitude decreases with distance as waves spread out in two or three dimensions—a consequence of energy conservation—so the two overlapped waves generally do *not* have equal amplitudes. Consequently, destructive interference rarely produces perfect cancellation.

Maximum constructive interference occurs where two crests align or two troughs align. Several locations of constructive interference are marked in Figure 17.25. Intersecting wave fronts are points where two crests are aligned. It's a bit harder to visualize, but two troughs are aligned when a midpoint between two wave fronts is overlapped with another midpoint between two wave fronts. Maximum, but usually not perfect, destructive interference occurs where the crest of one wave aligns with a trough of the other wave. Several points of destructive interference are also indicated in Figure 17.25.

FIGURE 17.25 The overlapping ripple patterns of two sources. Several points of constructive and destructive interference are noted.

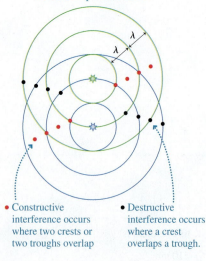

Two in-phase sources emit circular or spherical waves.

λ

λ

- Constructive interference occurs where two crests or two troughs overlap

- Destructive interference occurs where a crest overlaps a trough.

A picture on a page is static, but **the wave fronts are in motion**. Try to imagine the wave fronts of Figure 17.25 expanding outward as new circular rings are born at the sources. The waves will move forward half a wavelength during half a period, causing the crests in Figure 17.25 to be replaced by troughs while the troughs become crests.

The important point to recognize is that **the motion of the waves does not affect the points of constructive and destructive interference.** Points in the figure where two crests overlap will become points where two troughs overlap, but this overlap is still constructive interference. Similarly, points in the figure where a crest and a trough overlap will become a point where a trough and a crest overlap—still destructive interference.

The mathematical description of interference in two or three dimensions is very similar to that of one-dimensional interference. The net displacement of a particle in the medium is

$$D = D_1 + D_2 = a_1\sin(kr_1 - \omega t + \phi_{10}) + a_2\sin(kr_2 - \omega t + \phi_{20}) \tag{17.35}$$

The only differences between Equation 17.35 and the earlier one-dimensional Equation 17.24 are that the linear coordinates have been changed to radial coordinates and we've allowed the amplitudes to differ. These changes do not affect the phase difference, which, with x replaced by r, is now

$$\Delta\phi = 2\pi \frac{\Delta r}{\lambda} + \Delta\phi_0 \qquad (17.36)$$

The term $2\pi(\Delta r/\lambda)$ is the phase difference that arises when the waves travel different distances from the sources to the point at which they combine. Δr itself is the *path-length difference*. As before, $\Delta\phi_0$ is any inherent phase difference of the sources themselves.

Maximum constructive interference occurs, just as in one dimension, at those points where $\cos(\Delta\phi/2) = \pm 1$. Similarly, maximum destructive interference occurs at points where $\cos(\Delta\phi/2) = 0$. The conditions for constructive and destructive interference are

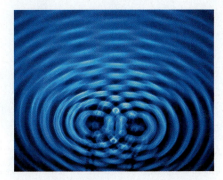

Two overlapping water waves create an interference pattern.

> **Maximum constructive interference:**
>
> $$\Delta\phi = 2\pi \frac{\Delta r}{\lambda} + \Delta\phi_0 = m \cdot 2\pi$$
>
> $$m = 0, 1, 2, \ldots \qquad (17.37)$$
>
> **Maximum destructive interference:**
>
> $$\Delta\phi = 2\pi \frac{\Delta r}{\lambda} + \Delta\phi_0 = \left(m + \tfrac{1}{2}\right) \cdot 2\pi$$

Identical Sources

For two identical sources (i.e., sources that oscillate in phase with $\Delta\phi_0 = 0$), the conditions for constructive and destructive interference are simple:

$$\text{Constructive:} \quad \Delta r = m\lambda$$
$$\text{(identical sources)} \qquad (17.38)$$
$$\text{Destructive:} \quad \Delta r = \left(m + \frac{1}{2}\right)\lambda$$

The waves from two identical sources interfere constructively at points where the path-length difference is an integer number of wavelengths because, for these values of Δr, crests are aligned with crests and troughs with troughs. **The waves interfere destructively at points where the path-length difference is a half-integer number of wavelengths** because, for these values of Δr, crests are aligned with troughs. These two statements are the essence of interference.

> **NOTE** Equation 17.38 applies only if the sources are in phase. If the sources are not in phase, you must use the more general Equation 17.37 to locate the points of constructive and destructive interference.

Wave fronts are spaced exactly one wavelength apart; hence we can measure the distances r_1 and r_2 simply by counting the rings in the wave-front pattern. In **FIGURE 17.26**, which is based on Figure 17.25, point A is distance $r_1 = 3\lambda$ from the first source and $r_2 = 2\lambda$ from the second. The path-length difference is $\Delta r_A = 1\lambda$, the condition for the maximum constructive interference of identical sources. Point B has $\Delta r_B = \frac{1}{2}\lambda$, so it is a point of maximum destructive interference.

> **NOTE** Interference is determined by Δr, the path-length *difference,* rather than by r_1 or r_2.

FIGURE 17.26 For identical sources, the path-length difference Δr determines whether the interference at a particular point is constructive or destructive.

- At A, $\Delta r_A = \lambda$, so this is a point of constructive interference.

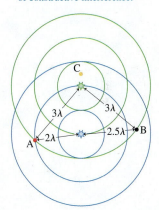

- At B, $\Delta r_B = \frac{1}{2}\lambda$, so this is a point of destructive interference.

STOP TO THINK 17.5 The interference at point C in Figure 17.26 is

a. Maximum constructive.
b. Constructive, but less than maximum.
c. Maximum destructive.
d. Destructive, but less than maximum.
e. There is no interference at point C.

We can now locate the points of maximum constructive interference by drawing a line through *all* the points at which $\Delta r = 0$, another line through all the points at which $\Delta r = \lambda$, and so on. These lines, shown in red in **FIGURE 17.27**, are called **antinodal lines.** They are analogous to the antinodes of a standing wave, hence the name. An antinode is a *point* of maximum constructive interference; for circular waves, oscillation at maximum amplitude occurs along a continuous *line*. Similarly, destructive interference occurs along lines called **nodal lines.** The amplitude is a minimum along a nodal line, usually close to zero, just as it is at a node in a standing-wave pattern.

FIGURE 17.27 The points of constructive and destructive interference fall along antinodal and nodal lines.

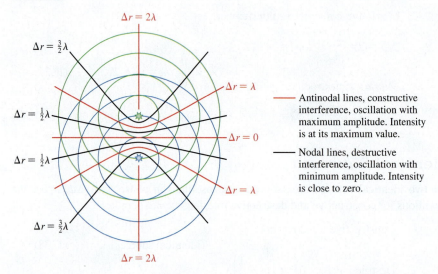

A Problem-Solving Strategy for Interference Problems

The information in this section is the basis of a strategy for solving interference problems. This strategy applies equally well to interference in one dimension if you use Δx instead of Δr.

PROBLEM-SOLVING STRATEGY 17.1   (MP)

Interference of two waves

MODEL Model the waves as linear, circular, or spherical.

VISUALIZE Draw a picture showing the sources of the waves and the point where the waves interfere. Give relevant dimensions. Identify the distances r_1 and r_2 from the sources to the point. Note any phase difference $\Delta\phi_0$ between the two sources.

SOLVE The interference depends on the path-length difference $\Delta r = r_2 - r_1$ and the source phase difference $\Delta\phi_0$.

Constructive: $\Delta\phi = 2\pi \dfrac{\Delta r}{\lambda} + \Delta\phi_0 = m \cdot 2\pi$

$m = 0, 1, 2, \ldots$

Destructive: $\Delta\phi = 2\pi \dfrac{\Delta r}{\lambda} + \Delta\phi_0 = \left(m + \tfrac{1}{2}\right) \cdot 2\pi$

For identical sources ($\Delta\phi_0 = 0$), the interference is maximum constructive if $\Delta r = m\lambda$, maximum destructive if $\Delta r = (m + \tfrac{1}{2})\lambda$.

ASSESS Check that your result has correct units and significant figures, is reasonable, and answers the question.

Exercise 18

EXAMPLE 17.10 | Two-dimensional interference between two loudspeakers

Two loudspeakers in a plane are 2.0 m apart and in phase with each other. Both emit 700 Hz sound waves into a room where the speed of sound is 341 m/s. A listener stands 5.0 m in front of the loudspeakers and 2.0 m to one side of the center. Is the interference at this point maximum constructive, maximum destructive, or in between? How will the situation differ if the loudspeakers are out of phase?

MODEL The two speakers are sources of in-phase, spherical waves. The overlap of these waves causes interference.

VISUALIZE FIGURE 17.28 shows the loudspeakers and defines the distances r_1 and r_2 to the point of observation. The figure includes dimensions and notes that $\Delta\phi_0 = 0$ rad.

FIGURE 17.28 Pictorial representation of the interference between two loudspeakers.

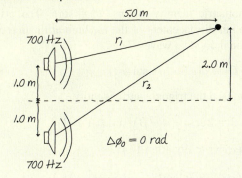

SOLVE It's not r_1 and r_2 that matter, but the *difference* Δr between them. From the geometry of the figure we can calculate that

$$r_1 = \sqrt{(5.0\text{ m})^2 + (1.0\text{ m})^2} = 5.10\text{ m}$$
$$r_2 = \sqrt{(5.0\text{ m})^2 + (3.0\text{ m})^2} = 5.83\text{ m}$$

Thus the path-length difference is $\Delta r = r_2 - r_1 = 0.73$ m. The wavelength of the sound waves is

$$\lambda = \frac{v}{f} = \frac{341\text{ m/s}}{700\text{ Hz}} = 0.487\text{ m}$$

In terms of wavelengths, the path-length difference is $\Delta r/\lambda = 1.50$, or

$$\Delta r = \tfrac{3}{2}\lambda$$

Because the sources are in phase $(\Delta\phi_0 = 0)$, this is the condition for *destructive* interference. If the sources were out of phase $(\Delta\phi_0 = \pi$ rad$)$, then the phase difference of the waves at the listener would be

$$\Delta\phi = 2\pi\frac{\Delta r}{\lambda} + \Delta\phi_0 = 2\pi\left(\frac{3}{2}\right) + \pi\text{ rad} = 4\pi\text{ rad}$$

This is an integer multiple of 2π rad, so in this case the interference would be *constructive*.

ASSESS Both the path-length difference and any inherent phase difference of the sources must be considered when evaluating interference.

STOP TO THINK 17.6 These two loudspeakers are in phase. They emit equal-amplitude sound waves with a wavelength of 1.0 m. At the point indicated, is the interference maximum constructive, maximum destructive, or something in between?

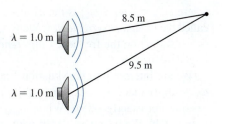

17.8 Beats

Thus far we have looked at the superposition of sources having the same wavelength and frequency. We can also use the principle of superposition to investigate a phenomenon that is easily demonstrated with two sources of slightly different frequency.

If you listen to two sounds with very different frequencies, such as a high note and a low note, you hear two distinct tones. But if the frequency difference is very small, just one or two hertz, then you hear a single tone whose intensity is *modulated* once or twice every second. That is, the sound goes up and down in volume, loud, soft, loud, soft, ... , making a distinctive sound pattern called **beats.**

Consider two sinusoidal waves traveling along the x-axis with angular frequencies $\omega_1 = 2\pi f_1$ and $\omega_2 = 2\pi f_2$. The two waves are

$$D_1 = a\sin(k_1 x - \omega_1 t + \phi_{10})$$
$$D_2 = a\sin(k_2 x - \omega_2 t + \phi_{20})$$

$$(17.39)$$

where the subscripts 1 and 2 indicate that the frequencies, wave numbers, and phase constants of the two waves may be different.

To simplify the analysis, let's make several assumptions:

1. The two waves have the same amplitude a,
2. A detector, such as your ear, is located at the origin $(x = 0)$,
3. The two sources are in phase $(\phi_{10} = \phi_{20})$, and
4. The source phases happen to be $\phi_{10} = \phi_{20} = \pi$ rad.

None of these assumptions is essential to the outcome. All could be otherwise and we would still come to basically the same conclusion, but the mathematics would be far messier. Making these assumptions allows us to emphasize the physics with the least amount of mathematics.

With these assumptions, the two waves as they reach the detector at $x = 0$ are

$$D_1 = a\sin(-\omega_1 t + \pi) = a\sin\omega_1 t$$
$$D_2 = a\sin(-\omega_2 t + \pi) = a\sin\omega_2 t \tag{17.40}$$

where we've used the trigonometric identity $\sin(\pi - \theta) = \sin\theta$. The principle of superposition tells us that the *net* displacement of the medium at the detector is the sum of the displacements of the individual waves. Thus

$$D = D_1 + D_2 = a(\sin\omega_1 t + \sin\omega_2 t) \tag{17.41}$$

Earlier, for interference, we used the trigonometric identity

$$\sin\alpha + \sin\beta = 2\cos\left[\tfrac{1}{2}(\alpha - \beta)\right]\sin\left[\tfrac{1}{2}(\alpha + \beta)\right]$$

We can use this identity again to write Equation 17.41 as

$$D = 2a\cos\left[\tfrac{1}{2}(\omega_1 - \omega_2)t\right]\sin\left[\tfrac{1}{2}(\omega_1 + \omega_2)t\right]$$
$$= \left[2a\cos(\omega_{mod}t)\right]\sin(\omega_{avg}t) \tag{17.42}$$

where $\omega_{avg} = \tfrac{1}{2}(\omega_1 + \omega_2)$ is the *average* angular frequency and $\omega_{mod} = \tfrac{1}{2}|\omega_1 - \omega_2|$ is called the *modulation frequency*. We've used the absolute value because the modulation depends only on the frequency *difference* between the sources, not on which has the larger frequency.

We are interested in the situation when the two frequencies are very nearly equal: $\omega_1 \approx \omega_2$. In that case, ω_{avg} hardly differs from either ω_1 or ω_2 while ω_{mod} is very near to—but not exactly—zero. When ω_{mod} is very small, the term $\cos(\omega_{mod}t)$ oscillates *very* slowly. We have grouped it with the $2a$ term because, together, they provide a slowly changing "amplitude" for the rapid oscillation at frequency ω_{avg}.

FIGURE 17.29 is a history graph of the wave at the detector $(x = 0)$. It shows the oscillation of the air against your eardrum at frequency $f_{avg} = \omega_{avg}/2\pi = \tfrac{1}{2}(f_1 + f_2)$. This oscillation determines the note you hear; it differs little from the two notes at frequencies f_1 and f_2. We are especially interested in the time-dependent amplitude, shown as a dashed line, that is given by the term $2a\cos(\omega_{mod}t)$. This periodically varying amplitude is called a **modulation** of the wave, which is where ω_{mod} gets its name.

As the amplitude rises and falls, the sound alternates as loud, soft, loud, soft, and so on. But that is exactly what you hear when you listen to beats! The alternating loud and soft sounds arise from the two waves being alternately in phase and out of phase, causing constructive and then destructive interference.

Imagine two people walking side by side at just slightly different paces. Initially both of their right feet hit the ground together, but after a while they get out of step. A little bit later they are back in step and the process alternates. The sound waves are doing the same. Initially the crests of each wave, of amplitude a, arrive together at your ear and the net displacement is doubled to $2a$. But after a while the two waves, being of slightly different frequency, get out of step and a crest of one arrives with a

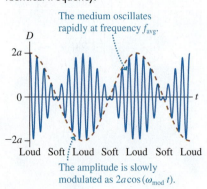

FIGURE 17.29 Beats are caused by the superposition of two waves of nearly identical frequency.

The medium oscillates rapidly at frequency f_{avg}.

Loud Soft Loud Soft Loud Soft Loud

The amplitude is slowly modulated as $2a\cos(\omega_{mod}t)$.

trough of the other. When this happens, the two waves cancel each other to give a net displacement of zero. This process alternates over and over, loud and soft.

Notice, in Figure 17.29, that the sound intensity rises and falls *twice* during one cycle of the modulation envelope. Each "loud-soft-loud" is one beat, so the **beat frequency** f_{beat}, which is the number of beats per second, is *twice* the modulation frequency $f_{mod} = \omega_{mod}/2\pi$. From the above definition of ω_{mod}, the beat frequency is

$$f_{beat} = 2f_{mod} = 2\frac{\omega_{mod}}{2\pi} = 2 \cdot \frac{1}{2}\left(\frac{\omega_1}{2\pi} - \frac{\omega_2}{2\pi}\right) = |f_1 - f_2| \qquad (17.43)$$

The beat frequency is simply the *difference* between the two individual frequencies.

EXAMPLE 17.11 | Detecting bats with beats

The little brown bat is a common species in North America. It emits echolocation pulses at a frequency of 40 kHz, well above the range of human hearing. To allow researchers to "hear" these bats, the bat detector shown in **FIGURE 17.30** combines the bat's sound wave at frequency f_1 with a wave of frequency f_2 from a tunable oscillator. The resulting beat frequency is then amplified and sent to a loudspeaker. To what frequency should the tunable oscillator be set to produce an audible beat frequency of 3 kHz?

SOLVE Combining two waves with different frequencies gives a beat frequency

$$f_{beat} = |f_1 - f_2|$$

A beat frequency will be generated at 3 kHz if the oscillator frequency and the bat frequency *differ* by 3 kHz. An oscillator frequency of either 37 kHz or 43 kHz will work nicely.

ASSESS The electronic circuitry of radios, televisions, and cell phones makes extensive use of *mixers* to generate difference frequencies.

FIGURE 17.30 The operation of a bat detector.

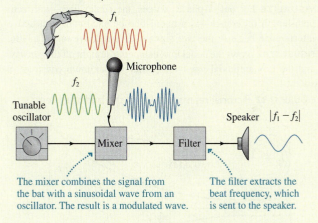

The mixer combines the signal from the bat with a sinusoidal wave from an oscillator. The result is a modulated wave.

The filter extracts the beat frequency, which is sent to the speaker.

Beats aren't limited to sound waves. **FIGURE 17.31** shows a graphical example of beats. Two "fences" of slightly different frequencies are superimposed on each other. The difference in the two frequencies is two lines per inch. You can confirm, with a ruler, that the figure has two "beats" per inch, in agreement with Equation 17.43.

Beats are important in many other situations. For example, you have probably seen movies where rotating wheels seem to turn slowly backward. Why is this? Suppose the movie camera is shooting at 30 frames per second but the wheel is rotating 32 times per second. The combination of the two produces a "beat" of 2 Hz, meaning that the wheel appears to rotate only twice per second. The same is true if the wheel is rotating 28 times per second, but in this case, where the wheel frequency slightly lags the camera frequency, it appears to rotate *backward* twice per second!

FIGURE 17.31 A graphical example of beats.

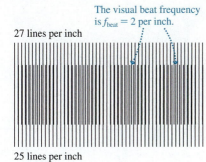

The visual beat frequency is $f_{beat} = 2$ per inch.

27 lines per inch

25 lines per inch

STOP TO THINK 17.7 You hear three beats per second when two sound tones are generated. The frequency of one tone is 610 Hz. The frequency of the other is

a. 604 Hz   b. 607 Hz   c. 613 Hz
d. 616 Hz   e. Either a or d.   f. Either b or c.

CHALLENGE **EXAMPLE** 17.12 | An airplane landing system

Your firm has been hired to design a system that allows airplane pilots to make instrument landings in rain or fog. You've decided to place two radio transmitters 50 m apart on either side of the runway. These two transmitters will broadcast the same frequency, but out of phase with each other. This will cause a nodal line to extend straight off the end of the runway. As long as the airplane's receiver is silent, the pilot knows she's directly in line with the runway. If she drifts to one side or the other, the radio will pick up a signal and sound a warning beep. To have sufficient accuracy, the first intensity maxima need to be 60 m on either side of the nodal line at a distance of 3.0 km. What frequency should you specify for the transmitters?

MODEL The two transmitters are sources of out-of-phase, circular waves. The overlap of these waves produces an interference pattern.

VISUALIZE For out-of-phase sources, the center line—with zero path-length difference—is a nodal line of maximum destructive interference because the two signals always arrive out of phase. **FIGURE 17.32** shows the nodal line, extending straight off the runway, and the first antinodal line—the points of maximum constructive

FIGURE 17.32 Pictorial representation of the landing system.

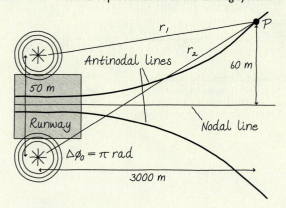

interference—on either side. Comparing this to Figure 17.27, where the two sources were in phase, you can see that the nodal and antinodal lines have been reversed.

SOLVE Point P, 60 m to the side at a distance of 3000 m, needs to be a point of maximum constructive interference. The distances are

$$r_1 = \sqrt{(3000 \text{ m})^2 + (60 \text{ m} - 25 \text{ m})^2} = 3000.204 \text{ m}$$
$$r_2 = \sqrt{(3000 \text{ m})^2 + (60 \text{ m} + 25 \text{ m})^2} = 3001.204 \text{ m}$$

We needed to keep several extra significant figures because we're looking for the difference between two numbers that are almost the same. The path-length difference at P is

$$\Delta r = r_2 - r_1 = 1.000 \text{ m}$$

We know, for out-of-phase transmitters, that the phase difference of the sources is $\Delta\phi_0 = \pi$ rad. The first maximum will occur where the phase difference between the waves is $\Delta\phi = 1 \cdot 2\pi$ rad. Thus the condition that we must satisfy at P is

$$\Delta\phi = 2\pi \text{ rad} = 2\pi \frac{\Delta r}{\lambda} + \pi \text{ rad}$$

Solving for λ, we find

$$\lambda = 2 \Delta r = 2.00 \text{ m}$$

Consequently, the required frequency is

$$f = \frac{c}{\lambda} = \frac{3.00 \times 10^8 \text{ m/s}}{2.00 \text{ m}} = 1.50 \times 10^8 \text{ Hz} = 150 \text{ MHz}$$

ASSESS 150 MHz is slightly higher than the frequencies of FM radio (≈ 100 MHz) but is well within the radio frequency range. Notice that the condition to be satisfied at P is that the path-length difference must be $\frac{1}{2}\lambda$. This makes sense. A path-length difference of $\frac{1}{2}\lambda$ contributes π rad to the phase difference. When combined with the π rad from the out-of-phase sources, the total phase difference of 2π rad creates constructive interference.

SUMMARY

The goal of Chapter 17 has been to understand and use the idea of superposition.

GENERAL PRINCIPLES

Principle of Superposition

The displacement of a medium when more than one wave is present is the sum at each point of the displacements due to each individual wave.

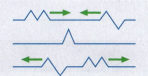

IMPORTANT CONCEPTS

Standing Waves

Standing waves are due to the superposition of two traveling waves moving in opposite directions.

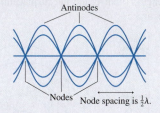

Antinodes

Nodes   Node spacing is $\frac{1}{2}\lambda$.

The amplitude at position x is

$$A(x) = 2a\sin kx$$

where a is the amplitude of each wave.

The boundary conditions determine which standing-wave frequencies and wavelengths are allowed. The allowed standing waves are **modes** of the system.

$m = 1$

$m = 2$

$m = 3$

Standing waves on a string

Solving Interference Problems

Maximum constructive interference occurs where crests are aligned with crests and troughs with troughs. The waves are in phase.

Maximum destructive interference occurs where crests are aligned with troughs. The waves are out of phase.

MODEL Model the wave as linear, circular, or spherical.

VISUALIZE Find distances to the sources.

SOLVE Interference depends on the **phase difference** $\Delta\phi$ between the waves:

$$\text{Constructive: } \Delta\phi = 2\pi\frac{\Delta r}{\lambda} + \Delta\phi_0 = m \cdot 2\pi$$

$$\text{Destructive: } \Delta\phi = 2\pi\frac{\Delta r}{\lambda} + \Delta\phi_0 = \left(m + \tfrac{1}{2}\right) \cdot 2\pi$$

Δr is the path-length difference of the two waves, and $\Delta\phi_0$ is any phase difference between the sources. For identical (in-phase) sources:

$$\text{Constructive: } \Delta r = m\lambda \qquad \text{Destructive: } \Delta r = \left(m + \tfrac{1}{2}\right)\lambda$$

ASSESS Is the result reasonable?

Antinodal lines, maximum constructive interference.

Nodal lines, maximum destructive interference.

APPLICATIONS

Boundary conditions

Strings, electromagnetic waves, and sound waves in closed-closed tubes must have nodes at both ends:

$$\lambda_m = \frac{2L}{m} \qquad f_m = m\frac{v}{2L} = mf_1 \qquad m = 1, 2, 3, \ldots.$$

The frequencies and wavelengths are the same for a sound wave in an open-open tube, which has antinodes at both ends.

A sound wave in an open-closed tube must have a node at the closed end but an antinode at the open end. This leads to

$$\lambda_m = \frac{4L}{m} \qquad f_m = m\frac{v}{4L} = mf_1 \qquad m = 1, 3, 5, 7, \ldots.$$

Beats (loud-soft-loud-soft modulations of intensity) occur when two waves of slightly different frequency are superimposed.

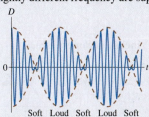

Soft Loud Soft Loud Soft

The beat frequency between waves of frequencies f_1 and f_2 is

$$f_{\text{beat}} = |f_1 - f_2|$$

TERMS AND NOTATION

principle of superposition	fundamental frequency, f_1	out of phase	antinodal line
standing wave	harmonic	destructive interference	nodal line
node	mode	phase difference, $\Delta\phi$	beats
antinode	interference	path-length difference, Δx	modulation
amplitude function, $A(x)$	in phase	or Δr	beat frequency, f_{beat}
boundary condition	constructive interference	thin-film optical coating	

CONCEPTUAL QUESTIONS

1. **FIGURE Q17.1** shows a standing wave oscillating on a string at frequency f_0.

   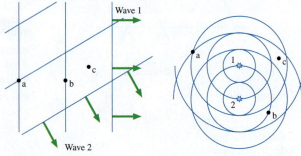

   FIGURE Q17.1

   a. What mode (m-value) is this?

   b. How many antinodes will there be if the frequency is doubled to $2f_0$?

2. If you take snapshots of a standing wave on a string, there are certain instants when the string is totally flat. What has happened to the energy of the wave at those instants?

3. **FIGURE Q17.3** shows the displacement of a standing sound wave in a 32-cm-long horizontal tube of air open at both ends.

   a. What mode (m-value) is this?

   b. Are the air molecules moving horizontally or vertically? Explain.

   c. At what distances from the left end of the tube do the molecules oscillate with maximum amplitude?

   d. At what distances from the left end of the tube does the air pressure oscillate with maximum amplitude?

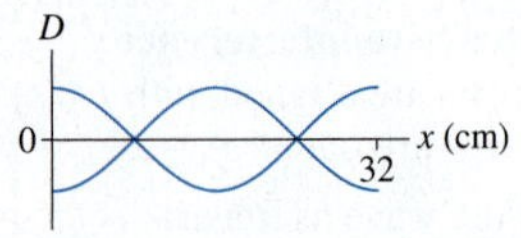

4. An organ pipe is tuned to exactly 384 Hz when the room temperature is 20°C. If the room temperature later increases to 22°C, does the pipe's frequency increase, decrease, or stay the same? Explain.

5. If you pour liquid into a tall, narrow glass, you may hear sound with a steadily rising pitch. What is the source of the sound? And why does the pitch rise as the glass fills?

6. A flute filled with helium will, until the helium escapes, play notes at a much higher pitch than normal. Why?

7. In music, two notes are said to be an *octave* apart when one note is exactly twice the frequency of the other. Suppose you have a guitar string playing frequency f_0. To increase the frequency by an octave, to $2f_0$, by what factor would you have to (a) increase the tension or (b) decrease the length?

8. **FIGURE Q17.8** is a snapshot graph of two plane waves passing through a region of space. Each wave has a 2.0 mm amplitude and the same wavelength. What is the net displacement of the medium at points a, b, and c?

FIGURE Q17.8   **FIGURE Q17.9**

9. **FIGURE Q17.9** shows the circular waves emitted by two in-phase sources. Are a, b, and c points of maximum constructive interference, maximum destructive interference, or in between?

10. A trumpet player hears 5 beats per second when she plays a note and simultaneously sounds a 440 Hz tuning fork. After pulling her tuning valve out to slightly increase the length of her trumpet, she hears 3 beats per second against the tuning fork. Was her initial frequency 435 Hz or 445 Hz? Explain.

EXERCISES AND PROBLEMS

Problems labeled ▨ integrate material from earlier chapters.

Exercises

Section 17.1 The Principle of Superposition

1. | **FIGURE EX17.1** is a snapshot graph at $t = 0$ s of two waves approaching each other at 1.0 m/s. Draw six snapshot graphs, stacked vertically, showing the string at 1 s intervals from $t = 1$ s to $t = 6$ s.

   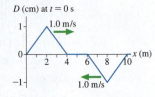

   FIGURE EX17.1

2. | **FIGURE EX17.2** is a snapshot graph at $t = 0$ s of two waves approaching each other at 1.0 m/s. Draw six snapshot graphs, stacked vertically, showing the string at 1 s intervals from $t = 1$ s to $t = 6$ s.

   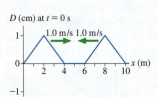

   FIGURE EX17.2

3. ‖ **FIGURE EX17.3a** is a snapshot graph at $t = 0$ s of two waves approaching each other at 1.0 m/s. At what time was the snapshot graph in **FIGURE EX17.3b** taken?

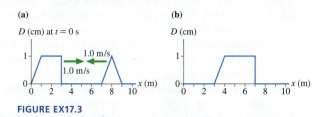

(a)

D (cm) at $t = 0$ s

(b)

D (cm)

FIGURE EX17.3

Section 17.2 Standing Waves

Section 17.3 Standing Waves on a String

4. ‖ **FIGURE EX17.4** is a snapshot graph at $t = 0$ s of two waves moving to the right at 1.0 m/s. The string is fixed at $x = 8.0$ m. Draw four snapshot graphs, stacked vertically, showing the string at $t = 2, 4, 6$, and 8 s.

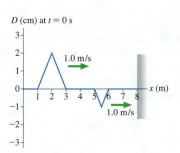

D (cm) at $t = 0$ s

FIGURE EX17.4

5. ‖ **FIGURE EX17.5** shows a standing wave oscillating at 100 Hz on a string. What is the wave speed?

50 cm

FIGURE EX17.5

FIGURE EX17.6

6. ‖ **FIGURE EX17.6** shows a standing wave on a 2.0-m-long string that has been fixed at both ends and tightened until the wave speed is 40 m/s. What is the frequency?

7. ‖ **FIGURE EX17.7** shows a standing wave on a string that is oscillating at 100 Hz.

   FIGURE EX17.7
   a. How many antinodes will there be if the frequency is increased to 200 Hz?
   b. If the tension is increased by a factor of 4, at what frequency will the string continue to oscillate as a standing wave that looks like the one in the figure?

8. ‖ a. What are the three longest wavelengths for standing waves on a 240-cm-long string that is fixed at both ends?
   b. If the frequency of the second-longest wavelength is 50 Hz, what is the frequency of the third-longest wavelength?

9. ‖ Standing waves on a 1.0-m-long string that is fixed at both ends are seen at successive frequencies of 36 Hz and 48 Hz.
   a. What are the fundamental frequency and the wave speed?
   b. Draw the standing-wave pattern when the string oscillates at 48 Hz.

10. ‖ The two highest-pitch strings on a violin are tuned to 440 Hz (the A string) and 659 Hz (the E string). What is the ratio of the mass of the A string to that of the E string? Violin strings are all the same length and under essentially the same tension.

11. ‖ A heavy piece of hanging sculpture is suspended by a 90-cm-long, 5.0 g steel wire. When the wind blows hard, the wire hums at its fundamental frequency of 80 Hz. What is the mass of the sculpture?

12. ‖ A carbon dioxide laser is an infrared laser. A CO_2 laser with a cavity length of 53.00 cm oscillates in the $m = 100,000$ mode. What are the wavelength and frequency of the laser beam?

13. ‖ Microwaves pass through a small hole into the "microwave cavity" of **FIGURE EX17.13**. What frequencies between 10 GHz and 20 GHz will create standing waves in the cavity?

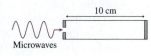

10 cm

Microwaves

FIGURE EX17.13

Section 17.4 Standing Sound Waves and Musical Acoustics

14. ‖ What are the three longest wavelengths for standing sound waves in a 121-cm-long tube that is (a) open at both ends and (b) open at one end, closed at the other?

15. ‖ **FIGURE EX17.15** shows a standing sound wave in an 80-cm-long tube. The tube is filled with an unknown gas. What is the speed of sound in this gas?

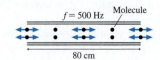

$f = 500$ Hz   Molecule

FIGURE EX17.15   80 cm

16. ‖ The fundamental frequency of an open-open tube is 1500 Hz when the tube is filled with 0°C helium. What is its frequency when filled with 0°C air?

17. ‖ We can make a simple model of the human vocal tract as an
BIO open-closed tube extending from the opening of the mouth to the diaphragm. What is the length of this tube if its fundamental frequency equals a typical speech frequency of 250 Hz? The speed of sound in the warm air is 350 m/s.

18. ‖ The lowest note on a grand piano has a frequency of 27.5 Hz. The entire string is 2.00 m long and has a mass of 400 g. The vibrating section of the string is 1.90 m long. What tension is needed to tune this string properly?

19. ‖ A bass clarinet can be modeled as a 120-cm-long open-closed tube. A bass clarinet player starts playing in a 20°C room, but soon the air inside the clarinet warms to where the speed of sound is 352 m/s. Does the fundamental frequency increase or decrease? By how much?

20. ‖ A violin string is 30 cm long. It sounds the musical note A (440 Hz) when played without fingering. How far from the end of the string should you place your finger to play the note C (523 Hz)?

21. ‖ A longitudinal standing wave can be created in a long, thin aluminum rod by stroking the rod with very dry fingers. This is often done as a physics demonstration, creating a high-pitched, very annoying whine. From a wave perspective, the standing wave is equivalent to a sound standing wave in an open-open tube. As **FIGURE EX17.21** shows, both ends of the rod are antinodes. What is the fundamental frequency of a 2.0-m-long aluminum rod?

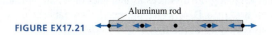

Aluminum rod

FIGURE EX17.21

Section 17.5 Interference in One Dimension

Section 17.6 The Mathematics of Interference

22. | Two loudspeakers emit sound waves along the *x*-axis. The sound has maximum intensity when the speakers are 20 cm apart. The sound intensity decreases as the distance between the speakers is increased, reaching zero at a separation of 60 cm.
    a. What is the wavelength of the sound?
    b. If the distance between the speakers continues to increase, at what separation will the sound intensity again be a maximum?

23. ‖ Two loudspeakers in a 20°C room emit 686 Hz sound waves along the *x*-axis.
    a. If the speakers are in phase, what is the smallest distance between the speakers for which the interference of the sound waves is maximum destructive?
    b. If the speakers are out of phase, what is the smallest distance between the speakers for which the interference of the sound waves is maximum constructive?

24. | Noise-canceling headphones are an application of destructive interference. Each side of the headphones uses a microphone to pick up noise, delays it slightly, then rebroadcasts the noise next to your ear where it can interfere with the incoming sound wave of the noise. Suppose you are sitting 1.8 m from an annoying, 110 Hz buzzing sound. What is the minimum headphone delay, in ms, that will cancel this noise?

25. | What is the thinnest film of MgF_2 ($n = 1.39$) on glass that produces a strong reflection for orange light with a wavelength of 600 nm?

26. ‖ A very thin oil film ($n = 1.25$) floats on water ($n = 1.33$). What is the thinnest film that produces a strong reflection for green light with a wavelength of 500 nm?

Section 17.7 Interference in Two and Three Dimensions

27. ‖ **FIGURE EX17.27** shows the circular wave fronts emitted by two wave sources.
    a. Are these sources in phase or out of phase? Explain.
    b. Make a table with rows labeled P, Q, and R and columns labeled r_1, r_2, Δr, and C/D. Fill in the table for points P, Q, and R, giving the distances as multiples of λ and indicating, with a C or a D, whether the interference at that point is constructive or destructive.

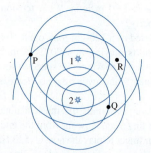

FIGURE EX17.27

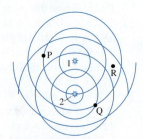

FIGURE EX17.28

28. ‖ **FIGURE EX17.28** shows the circular wave fronts emitted by two wave sources.
    a. Are these sources in phase or out of phase? Explain.
    b. Make a table with rows labeled P, Q, and R and columns labeled r_1, r_2, Δr, and C/D. Fill in the table for points P, Q, and R, giving the distances as multiples of λ and indicating, with a C or a D, whether the interference at that point is constructive or destructive.

29. ‖ Two in-phase loudspeakers, which emit sound in all directions, are sitting side by side. One of them is moved sideways by 3.0 m, then forward by 4.0 m. Afterward, constructive interference is observed $\frac{1}{4}$ and $\frac{3}{4}$ of the distance between the speakers along the line that joins them. What is the maximum possible wavelength of the sound waves?

30. ‖ Two in-phase speakers 2.0 m apart in a plane are emitting 1800 Hz sound waves into a room where the speed of sound is 340 m/s. Is the point 4.0 m in front of one of the speakers, perpendicular to the plane of the speakers, a point of maximum constructive interference, maximum destructive interference, or something in between?

31. ‖ Two out-of-phase radio antennas at $x = \pm 300$ m on the *x*-axis are emitting 3.0 MHz radio waves. Is the point $(x, y) = (300$ m, 800 m$)$ a point of maximum constructive interference, maximum destructive interference, or something in between?

Section 17.8 Beats

32. | Two strings are adjusted to vibrate at exactly 200 Hz. Then the tension in one string is increased slightly. Afterward, three beats per second are heard when the strings vibrate at the same time. What is the new frequency of the string that was tightened?

33. | A flute player hears four beats per second when she compares her note to a 523 Hz tuning fork (the note C). She can match the frequency of the tuning fork by pulling out the "tuning joint" to lengthen her flute slightly. What was her initial frequency?

34. | Traditional Indonesian music uses an ensemble called a *gamelan* that is based on tuned percussion instruments somewhat like gongs. In Bali, the gongs are often grouped in pairs that are slightly out of tune with each other. When both are played at once, the beat frequency lends a distinctive vibrating quality to the music. Suppose a pair of gongs are tuned to produce notes at 151 Hz and 155 Hz. How many beats are heard if the gongs are struck together and both ring for 2.5 s?

35. ‖ Two microwave signals of nearly equal wavelengths can generate a beat frequency if both are directed onto the same microwave detector. In an experiment, the beat frequency is 100 MHz. One microwave generator is set to emit microwaves with a wavelength of 1.250 cm. If the second generator emits the longer wavelength, what is that wavelength?

Problems

36. | A 2.0-m-long string vibrates at its second-harmonic frequency with a maximum amplitude of 2.0 cm. One end of the string is at $x = 0$ cm. Find the oscillation amplitude at $x = 10, 20, 30, 40,$ and 50 cm.

37. ‖ A string vibrates at its third-harmonic frequency. The amplitude at a point 30 cm from one end is half the maximum amplitude. How long is the string?

38. ‖ Tendons are, essentially, elastic cords stretched between two **BIO** fixed ends. As such, they can support standing waves. A woman has a 20-cm-long Achilles tendon—connecting the heel to a muscle in the calf—with a cross-section area of 90 mm². The density of tendon tissue is 1100 kg/m³. For a reasonable tension of 500 N, what will be the fundamental frequency of her Achilles tendon?

39. ‖ Biologists think that some spiders "tune" strands of their web **BIO** to give enhanced response at frequencies corresponding to those at which desirable prey might struggle. Orb spider web silk has a typical diameter of 20 μm, and spider silk has a density of

1300 kg/m³. To have a fundamental frequency at 100 Hz, to what tension must a spider adjust a 12-cm-long strand of silk?

40. ‖ A particularly beautiful note reaching your ear from a rare Stradivarius violin has a wavelength of 39.1 cm. The room is slightly warm, so the speed of sound is 344 m/s. If the string's linear density is 0.600 g/m and the tension is 150 N, how long is the vibrating section of the violin string?

41. ‖ A violinist places her finger so that the vibrating section of a 1.0 g/m string has a length of 30 cm, then she draws her bow across it. A listener nearby in a 20°C room hears a note with a wavelength of 40 cm. What is the tension in the string?

42. ‖ A steel wire is used to stretch the spring of **FIGURE P17.42**. An oscillating magnetic field drives the steel wire back and forth. A standing wave with three antinodes is created when the spring is stretched 8.0 cm. What stretch of the spring produces a standing wave with two antinodes?

FIGURE P17.42

43. ‖ Astronauts visiting Planet X have a 250-cm-long string whose mass is 5.00 g. They tie the string to a support, stretch it horizontally over a pulley 2.00 m away, and hang a 4.00 kg mass on the free end. Then the astronauts begin to excite standing waves on the horizontal portion of the string. Their data are as follows:

m	Frequency (Hz)
1	31
2	66
3	95
4	130
5	162

Use the best-fit line of an appropriate graph to determine the value of g, the free-fall acceleration on Planet X.

44. ‖ A 75 g bungee cord has an equilibrium length of 1.20 m. The cord is stretched to a length of 1.80 m, then vibrated at 20 Hz. This produces a standing wave with two antinodes. What is the spring constant of the bungee cord?

45. ‖ A metal wire under tension T_0 vibrates at its fundamental frequency. For what tension will the second-harmonic frequency be the same as the fundamental frequency at tension T_0?

46. ‖‖ In a laboratory experiment, one end of a horizontal string is tied to a support while the other end passes over a frictionless pulley and is tied to a 1.5 kg sphere. Students determine the frequencies of standing waves on the horizontal segment of the string, then they raise a beaker of water until the hanging 1.5 kg sphere is completely submerged. The frequency of the fifth harmonic with the sphere submerged exactly matches the frequency of the third harmonic before the sphere was submerged. What is the diameter of the sphere?

47. ‖‖ A vibrating standing wave on a string radiates a sound wave with intensity proportional to the square of the standing-wave amplitude. When a piano key is struck and held down, so that the string continues to vibrate, the sound level decreases by 8.0 dB in 1.0 s. What is the string's damping time constant τ?

48. ‖ What is the fundamental frequency of the steel wire in **FIGURE P17.48**?

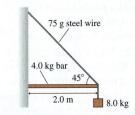

FIGURE P17.48

49. ‖ The two strings in **FIGURE P17.49** are of equal length and are being driven at equal frequencies. The linear density of the left string is 5.0 g/m. What is the linear density of the right string?

Stretched spring

FIGURE P17.49

50. ‖ Western music uses a musical scale with *equal temperament* tuning, which means that any two adjacent notes have the same frequency ratio r. That is, notes n and n + 1 are related by $f_{n+1} = rf_n$ for all n. In this system, the frequency doubles every 12 notes—an interval called an *octave*.
   a. What is the value of r?
   b. Orchestras tune to the note A, which has a frequency of 440 Hz. What is the frequency of the next note of the scale (called A-sharp)?

51. ‖ An open-open organ pipe is 78.0 cm long. An open-closed pipe has a fundamental frequency equal to the third harmonic of the open-open pipe. How long is the open-closed pipe?

52. ‖ Deep-sea divers often breathe a mixture of helium and oxygen
BIO to avoid getting the "bends" from breathing high-pressure nitrogen. The helium has the side effect of making the divers' voices sound odd. Although your vocal tract can be roughly described as an open-closed tube, the way you hold your mouth and position your lips greatly affects the standing-wave frequencies of the vocal tract. This is what allows different vowels to sound different. The "ee" sound is made by shaping your vocal tract to have standing-wave frequencies at, normally, 270 Hz and 2300 Hz. What will these frequencies be for a helium-oxygen mixture in which the speed of sound at body temperature is 750 m/s? The speed of sound in air at body temperature is 350 m/s.

53. ‖ In 1866, the German scientist Adolph Kundt developed a technique for accurately measuring the speed of sound in various gases. A long glass tube, known today as a Kundt's tube, has a vibrating piston at one end and is closed at the other. Very finely ground particles of cork are sprinkled in the bottom of the tube before the piston is inserted. As the vibrating piston is slowly moved forward, there are a few positions that cause the cork particles to collect in small, regularly spaced piles along the bottom. **FIGURE P17.53** shows an experiment in which the tube is filled with pure oxygen and the piston is driven at 400 Hz. What is the speed of sound in oxygen?

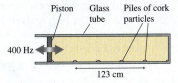

FIGURE P17.53

54. ‖ The 40-cm-long tube of FIG-
URE P17.54 has a 40-cm-long insert
that can be pulled in and out. A vibrating tuning fork is held next to the
tube. As the insert is slowly pulled
out, the sound from the tuning fork
creates standing waves in the tube

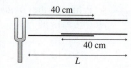

FIGURE P17.54

when the total length L is 42.5 cm, 56.7 cm, and 70.9 cm. What is
the frequency of the tuning fork? Assume $v_{sound} = 343$ m/s.

55. ‖ A 1.0-m-tall vertical tube is filled with 20°C water. A tuning
fork vibrating at 580 Hz is held just over the top of the tube as the
water is slowly drained from the bottom. At what water heights,
measured from the bottom of the tube, will there be a standing
wave in the tube above the water?

56. ‖‖ A 44-cm-diameter water tank is filled with 35 cm of water. A
CALC 3.0-mm-diameter spigot at the very bottom of the tank is opened
and water begins to flow out. The water falls into a 2.0-cm-diameter,
40-cm-tall glass cylinder. As the water falls and creates noise, resonance causes the column of air in the cylinder to produce a tone
at the column's fundamental frequency. What are (a) the frequency
and (b) the rate at which the frequency is changing (Hz/s) when the
cylinder has been filling for 4.0 s? You can assume that the height
of the water in the tank does not appreciably change in 4.0 s.

57. ‖ A 25-cm-long wire with a linear density of 20 g/m passes
across the open end of an 85-cm-long open-closed tube of air. If
the wire, which is fixed at both ends, vibrates at its fundamental
frequency, the sound wave it generates excites the second vibrational mode of the tube of air. What is the tension in the wire?
Assume $v_{sound} = 340$ m/s.

58. ‖ An old mining tunnel disappears into a hillside. You would like
to know how long the tunnel is, but it's too dangerous to go inside.
Recalling your recent physics class, you decide to try setting up
standing-wave resonances inside the tunnel. Using your subsonic
amplifier and loudspeaker, you find resonances at 4.5 Hz and 6.3 Hz,
and at no frequencies between these. It's rather chilly inside the
tunnel, so you estimate the sound speed to be 335 m/s. Based on
your measurements, how far is it to the end of the tunnel?

59. ‖ Two in-phase loudspeakers emit identical 1000 Hz sound
waves along the x-axis. What distance should one speaker be placed
behind the other for the sound to have an amplitude 1.5 times that
of each speaker alone?

60. ‖ Analyze the standing sound waves in an open-closed tube to
show that the possible wavelengths and frequencies are given by
Equation 17.18.

61. ‖ Two loudspeakers emit sound waves of the same frequency
along the x-axis. The amplitude of each wave is a. The sound
intensity is minimum when speaker 2 is 10 cm behind speaker 1.
The intensity increases as speaker 2 is moved forward and first
reaches maximum, with amplitude $2a$, when it is 30 cm in front
of speaker 1. What is
 a. The wavelength of the sound?
 b. The phase difference between the two loudspeakers?
 c. The amplitude of the sound (as a multiple of a) if the speakers
    are placed side by side?

62. ‖ Two loudspeakers emit sound waves along the x-axis. A listener in front of both speakers hears a maximum sound intensity
when speaker 2 is at the origin and speaker 1 is at $x = 0.50$ m. If
speaker 1 is slowly moved forward, the sound intensity decreases
and then increases, reaching another maximum when speaker 1 is
at $x = 0.90$ m.
 a. What is the frequency of the sound? Assume $v_{sound} = 340$ m/s.
 b. What is the phase difference between the speakers?

63. ‖ A sheet of glass is coated with a 500-nm-thick layer of oil
($n = 1.42$).
 a. For what *visible* wavelengths of light do the reflected waves
    interfere constructively?
 b. For what *visible* wavelengths of light do the reflected waves
    interfere destructively?
 c. What is the color of reflected light? What is the color of trans-
    mitted light?

64. ‖ A manufacturing firm has hired your company, Acoustical Con-
sulting, to help with a problem. Their employees are complaining
about the annoying hum from a piece of machinery. Using a frequency meter, you quickly determine that the machine emits a rather
loud sound at 1200 Hz. After investigating, you tell the owner that
you cannot solve the problem entirely, but you can at least improve
the situation by eliminating reflections of this sound from the walls.
You propose to do this by installing mesh screens in front of the
walls. A portion of the sound will reflect from the mesh; the rest will
pass through the mesh and reflect from the wall. How far should the
mesh be placed in front of the wall for this scheme to work?

65. ‖ A soap bubble is essentially a very thin film of water ($n =
1.33$) surrounded by air. The colors that you see in soap bubbles
are produced by interference.
 a. Derive an expression for the wavelengths λ_C for which con-
    structive interference causes a strong reflection from a soap
    bubble of thickness d.
 Hint: Think about the reflection phase shifts at both boundaries.
 b. What visible wavelengths of light are strongly reflected from
    a 390-nm-thick soap bubble? What color would such a soap
    bubble appear to be?

66. ‖ Engineers are testing a new thin-film coating whose index of refraction is less than that of glass. They deposit a 560-nm-thick layer on
glass, then shine lasers on it. A red laser with a wavelength of 640 nm
has no reflection at all, but a violet laser with a wavelength of 400 nm
has a maximum reflection. How the coating behaves at other wavelengths is unknown. What is the coating's index of refraction?

67. ‖ Scientists are testing a transparent material whose index of refraction for visible light varies with wavelength as
$n = 30.0 \text{ nm}^{1/2}/\lambda^{1/2}$, where λ is in nm. If a 295-nm-thick coating
is placed on glass ($n = 1.50$) for what visible wavelengths will
the reflected light have maximum constructive interference?

68. ‖ You are standing 2.5 m
directly in front of one of the
two loudspeakers shown in
FIGURE P17.68. They are 3.0
m apart and both are playing a
686 Hz tone in phase. As you
begin to walk directly away
from the speaker, at what distances from the speaker do you
hear a *minimum* sound intensity?
The room temperature is 20°C.

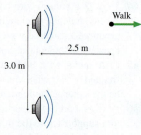

FIGURE P17.68

69. ‖ Two loudspeakers in a plane, 5.0 m apart, are playing the same
frequency. If you stand 12.0 m in front of the plane of the speakers, centered between them, you hear a sound of maximum intensity. As you walk parallel to the plane of the speakers, staying
12.0 m in front of them, you first hear a minimum of sound intensity when you are directly in front of one of the speakers. What is
the frequency of the sound? Assume a sound speed of 340 m/s.

70. ‖ Two identical loudspeakers separated by distance Δx each
emit sound waves of wavelength λ and amplitude a along the
x-axis. What is the minimum value of the ratio $\Delta x/\lambda$ for which
the amplitude of their superposition is also a?

71. ‖ The three identical loudspeakers in FIGURE P17.71 play a 170 Hz tone in a room where the speed of sound is 340 m/s. You are standing 4.0 m in front of the middle speaker. At this point, the amplitude of the wave from each speaker is a.

a. What is the amplitude at this point?

b. How far must speaker 2 be moved to the left to produce a maximum amplitude at the point where you are standing?

c. When the amplitude is maximum, by what factor is the sound intensity greater than the sound intensity from a single speaker?

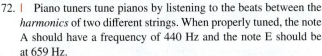

FIGURE P17.71

72. ‖ Piano tuners tune pianos by listening to the beats between the *harmonics* of two different strings. When properly tuned, the note A should have a frequency of 440 Hz and the note E should be at 659 Hz.

a. What is the frequency difference between the third harmonic of the A and the second harmonic of the E?

b. A tuner first tunes the A string very precisely by matching it to a 440 Hz tuning fork. She then strikes the A and E strings simultaneously and listens for beats between the harmonics. What beat frequency indicates that the E string is properly tuned?

c. The tuner starts with the tension in the E string a little low, then tightens it. What is the frequency of the E string when she hears four beats per second?

73. ‖ A flutist assembles her flute in a room where the speed of sound is 342 m/s. When she plays the note A, it is in perfect tune with a 440 Hz tuning fork. After a few minutes, the air inside her flute has warmed to where the speed of sound is 346 m/s.

a. How many beats per second will she hear if she now plays the note A as the tuning fork is sounded?

b. How far does she need to extend the "tuning joint" of her flute to be in tune with the tuning fork?

74. ‖ You have two small, identical boxes that generate 440 Hz notes. While holding one, you drop the other from a 20-m-high balcony. How many beats will you hear before the falling box hits the ground? You can ignore air resistance.

CALC

75. ‖ Two loudspeakers emit 400 Hz notes. One speaker sits on the ground. The other speaker is in the back of a pickup truck. You hear eight beats per second as the truck drives away from you. What is the truck's speed?

Challenge Problems

76. ‖‖ Two radio antennas are separated by 2.0 m. Both broadcast identical 750 MHz waves. If you walk around the antennas in a circle of radius 10 m, how many maxima will you detect?

77. ‖‖ A 280 Hz sound wave is directed into one end of the trombone slide seen in FIGURE CP17.77. A microphone is placed at the other end to record the intensity of sound waves that are transmitted through the tube. The straight sides of the slide are 80 cm in length and 10 cm apart with a semicircular bend at the end. For what slide extensions s will the microphone detect a maximum of sound intensity?

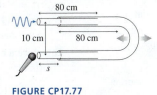

FIGURE CP17.77

78. ‖‖ As the captain of the scientific team sent to Planet Physics, one of your tasks is to measure g. You have a long, thin wire labeled 1.00 g/m and a 1.25 kg weight. You have your accurate space cadet chronometer but, unfortunately, you seem to have forgotten a meter stick. Undeterred, you first find the midpoint of the wire by folding it in half. You then attach one end of the wire to the wall of your laboratory, stretch it horizontally to pass over a pulley at the midpoint of the wire, then tie the 1.25 kg weight to the end hanging over the pulley. By vibrating the wire, and measuring time with your chronometer, you find that the wire's second-harmonic frequency is 100 Hz. Next, with the 1.25 kg weight still tied to one end of the wire, you attach the other end to the ceiling to make a pendulum. You find that the pendulum requires 314 s to complete 100 oscillations. Pulling out your trusty calculator, you get to work. What value of g will you report back to headquarters?

79. ‖‖ When mass M is tied to the bottom of a long, thin wire suspended from the ceiling, the wire's second-harmonic frequency is 200 Hz. Adding an additional 1.0 kg to the hanging mass increases the second-harmonic frequency to 245 Hz. What is M?

80. ‖‖ Ultrasound has many medical applications, one of which is to monitor fetal heartbeats by reflecting ultrasound off a fetus in the womb.

BIO

a. Consider an object moving at speed v_0 toward an at-rest source that is emitting sound waves of frequency f_0. Show that the reflected wave (i.e., the echo) that returns to the source has a Doppler-shifted frequency

$$f_{echo} = \left(\frac{v + v_0}{v - v_0} \right) f_0$$

where v is the speed of sound in the medium.

b. Suppose the object's speed is much less than the wave speed: $v_0 \ll v$. Then $f_{echo} \approx f_0$, and a microphone that is sensitive to these frequencies will detect a beat frequency if it listens to f_0 and f_{echo} simultaneously. Use the binomial approximation and other appropriate approximations to show that the beat frequency is $f_{beat} \approx (2v_0/v) f_0$.

c. The reflection of 2.40 MHz ultrasound waves from the surface of a fetus's beating heart is combined with the 2.40 MHz wave to produce a beat frequency that reaches a maximum of 65 Hz. What is the maximum speed of the surface of the heart? The speed of ultrasound waves within the body is 1540 m/s.

d. Suppose the surface of the heart moves in simple harmonic motion at 90 beats/min. What is the amplitude in mm of the heartbeat?

81. ‖‖ A water wave is called a *deep-water wave* if the water's depth is more than one-quarter of the wavelength. Unlike the waves we've considered in this chapter, the speed of a deep-water wave depends on its wavelength:

$$v = \sqrt{\frac{g\lambda}{2\pi}}$$

Longer wavelengths travel faster. Let's apply this to standing waves. Consider a diving pool that is 5.0 m deep and 10.0 m wide. Standing water waves can set up across the width of the pool. Because water sloshes up and down at the sides of the pool, the boundary conditions require antinodes at $x = 0$ and $x = L$. Thus a standing water wave resembles a standing sound wave in an open-open tube.

a. What are the wavelengths of the first three standing-wave modes for water in the pool? Do they satisfy the condition for being deep-water waves?

b. What are the wave speeds for each of these waves?

c. Derive a general expression for the frequencies f_m of the possible standing waves. Your expression should be in terms of m, g, and L.

d. What are the oscillation *periods* of the first three standing wave modes?

IV Oscillations and Waves

KEY FINDINGS   What are the overarching findings of Part IV?

- Particles are
  - Localized
  - Discrete
  - Two particles cannot occupy the same point in space.

- Waves are
  - Diffuse
  - Spread out
  - Two waves can pass through each other.

LAWS   What laws of physics govern oscillations and waves?

Newton's second law	SHM: $d^2x/dt^2 = -\omega^2 x$	Wave equation: $\partial^2 D/\partial t^2 = v^2 \partial^2 D/\partial x^2$
Conservation of energy	For SHM: $E = \frac{1}{2}mv^2 + \frac{1}{2}kx^2 = \frac{1}{2}m(v_{\max})^2 = \frac{1}{2}kA^2$	
Fundamental relationship for sinusoidal waves	$v = \lambda f = \omega/k$	
Principle of superposition	The net displacement is the sum of the displacements due to each wave.	

MODELS   What are the most important models of Part IV?

Simple harmonic motion

- Any object with a **linear restoring force** can undergo SHM. This is sinusoidal motion with

$$x = A\cos(\omega t + \phi_0)$$
$$v = -v_{\max}\sin(\omega t + \phi_0)$$

- Mechanical energy is conserved if there's no friction.

- With friction, the oscillations are damped. A simple model of **damping** predicts oscillations that decay exponentially with time.

- **Resonance** is a large-amplitude response when an oscillator is driven at its natural frequency.

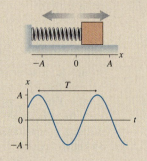

Waves

- A wave is a disturbance that travels.
  - **Mechanical waves** travel through a medium.
  - **Electromagnetic waves** travel through a vacuum.
  - Waves can be **transverse** or **longitudinal.**

- Wave speed is a property of the medium.

- Sinusoidal waves are periodic in both time (period) and space (wavelength). They obey $v = \lambda f$.

- Waves obey the principle of superposition.

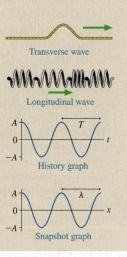

Transverse wave

Longitudinal wave

History graph

Snapshot graph

TOOLS   What are the most important tools introduced in Part IV?

Oscillation period
- Frequency $f = 1/T$
- Angular frequency $\omega = 2\pi f = 2\pi/T$
- Spring $T = 2\pi\sqrt{m/k}$
- Pendulum $T = 2\pi\sqrt{L/g}$
- Wave $f = v/\lambda$

Sinusoidal wave
- Displacement is a function of x and t:

$$D(x,t) = A\sin(kx \mp \omega t + \phi_0)$$

- $-\omega t$ for motion to the right
- $+\omega t$ for motion to the left
- The *wave number* is $k = 2\pi/\lambda$

Sound intensity level

$$\beta = (10\ \text{dB})\log_{10}(I/10^{-12}\ \text{W/m}^2)$$

Wave speed
- String $v = \sqrt{T_s/\mu}$
- Sound $v = \sqrt{B/\rho}$

Phase
- The quantity $\omega t + \phi_0$ is called the phase ϕ of SHM.
- The quantity $k - \omega t + \phi_0$ is the phase ϕ of a sinusoidal wave.
- The *phase constant ϕ_0* is given by the initial conditions.

Doppler effect
A frequency shift when the source moves relative to an observer:

$$f = f_0/(1 \mp v_s/v)$$

for an approaching/receding source.

Standing waves
Two waves traveling in opposite directions.

$m = 1$

$m = 2$

- Strings, open-open tubes, and closed-closed tubes:

$$f = mf_1$$

$f_1 = v/2L$ = fundamental frequency

- Standing waves have points that never move called *nodes*.

Interference
- *Constructive interference* if in phase
$$\Delta\phi = m \cdot 2\pi\ \text{rad}$$
- *Destructive interference* if out of phase
$$\Delta\phi = \left(m + \tfrac{1}{2}\right) \cdot 2\pi\ \text{rad}$$
- *Beats* $f_{\text{beat}} = |f_1 - f_2|$ if frequencies differ.

Thermodynamics

OVERVIEW

It's All About Energy

Thermodynamics—the science of energy in its broadest context—arose hand in hand with the industrial revolution as the systematic study of converting heat energy into mechanical motion and work. Hence the name *thermo + dynamics*. Indeed, the analysis of engines and generators of various kinds remains the focus of engineering thermodynamics. But thermodynamics, as a science, now extends to all forms of energy conversions, including those involving living organisms. For example:

- **Engines** convert the energy of a fuel into the mechanical energy of moving pistons, gears, and wheels.
- **Fuel cells** convert chemical energy into electrical energy.
- **Photovoltaic cells** convert the electromagnetic energy of light into electrical energy.
- **Lasers** convert electrical energy into the electromagnetic energy of light.
- **Organisms** convert the chemical energy of food into a variety of other forms of energy, including kinetic energy, sound energy, and thermal energy.

The major goals of Part V are to understand both *how* energy transformations such as these take place and *how efficient* they are. We'll discover that the laws of thermodynamics place limits on the efficiency of energy transformations, and understanding these limits is essential for analyzing the very real energy needs of society in the 21st century.

Our ultimate destination in Part V is an understanding of the thermodynamics of *heat engines*. A heat engine is a device, such as a power plant or an internal combustion engine, that transforms heat energy into useful work. These are the devices that power our modern society.

Understanding how to transform heat into work will be a significant achievement, but we first have many steps to take along the way. We need to understand the concepts of temperature and pressure. We need to learn about the properties of solids, liquids, and gases. Most important, we need to expand our view of energy to include *heat,* the energy that is transferred between two systems at different temperatures.

At a deeper level, we need to see how these concepts are connected to the underlying microphysics of randomly moving molecules. We will find that the familiar concepts of thermodynamics, such as temperature and pressure, have their roots in atomic-level motion and collisions. This *micro/macro connection* will lead to the second law of thermodynamics, one of the most subtle but also one of the most profound and far-reaching statements in physics.

Only after all these steps have been taken will we be able to analyze a real heat engine. It is an ambitious goal, but one we can achieve.

Smoke particles allow us to visualize *convection,* one of the ways in which heat is transferred from one place to another.

18 A Macroscopic Description of Matter

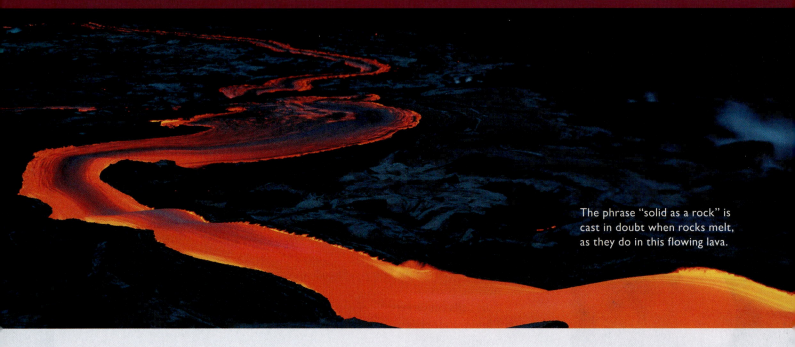

The phrase "solid as a rock" is cast in doubt when rocks melt, as they do in this flowing lava.

IN THIS CHAPTER, you will learn some of the characteristics of macroscopic systems.

What are the phases of matter?

Most materials can exist as a solid, a liquid, or a gas. These are the most common phases of matter.

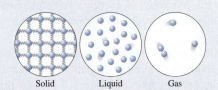

Solid   Liquid   Gas

Starting with this chapter and continuing through Part V you will come to understand that the macroscopic properties of matter, such as volume, density, pressure, and temperature, can often be understood in terms of the microscopic motions of their atoms and molecules. This micro/macro connection is an important part of our modern understanding of matter.

« LOOKING BACK Sections 14.1–14.3 Fluids and pressure

What is temperature?

You're familiar with temperature, but what does it actually measure? We'll start with the simple idea that temperature measures "hotness" and "coldness," but we'll come to recognize that temperature measures a system's thermal energy. We'll study the well-known fact that objects expand or contract when the temperature changes.

What is an ideal gas?

We'll model a gas as consisting of tiny, hard spheres that occasionally collide but otherwise do not interact with each other. This ideal gas obeys a law relating four state variables—the ideal-gas law:

$$pV = nRT$$

You will use the ideal-gas law to analyze what happens when a gas changes state.

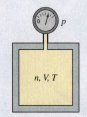

What is an ideal-gas process?

Heating or compressing a gas is a *process* that changes the state of the gas. An ideal-gas process can be shown as a trajectory through a pV diagram. We'll study three basic processes:

- Constant-volume process
- Constant-pressure process
- Constant-temperature process

Why are macroscopic properties important?

Physicists, chemists, biologists, and engineers all work with matter at the macroscopic level. Everything from basic science to engineering design depends on knowing how materials respond when they are heated, compressed, melted, or otherwise changed by factors in their environment. Changing states of matter underlie devices ranging from car engines to power plants to spacecraft.

18.1 Solids, Liquids, and Gases

In Part V we will study the properties of matter itself, as opposed to the motion of matter. We will focus on a *macroscopic description* of large quantities of matter. Even so, part of our modern understanding of matter is that macroscopic properties, such as pressure and temperature, have their basis in the microscopic motions of atoms and molecules, and we'll spend some time exploring this **micro/macro connection.**

As you know, each of the elements and most compounds can exist as a solid, liquid, or gas—the three most common **phases** of matter. The change between liquid and solid (freezing or melting) or between liquid and gas (boiling or condensing) is called a **phase change.** Water is the only substance for which all three phases—ice, liquid, and steam—are everyday occurrences.

NOTE This use of the word "phase" has no relationship at all to the *phase* or *phase constant* of simple harmonic motion and waves.

MODEL 18.1

Solids, liquids, and gases

Atoms vibrate around equilibrium positions.

Atoms are held close together by weak molecular bonds, but they can slide around each other.

Atoms are far apart and travel freely through space except for occasional collisions.

A **solid** is a rigid macroscopic system consisting of particle-like atoms connected by spring-like molecular bonds. Solids are nearly *incompressible,* which tells us that the atoms in a solid are just about as close together as they can get.

The solid shown here is a **crystal,** meaning that the atoms are arranged in a periodic array. Elements and many compounds have a crystal structure in their solid phase.

A **liquid** is a system in which the molecules are loosely held together by weak molecular bonds. The bonds are strong enough that the molecules never get far apart but not strong enough to prevent the molecules from sliding around each other.

A liquid is more complicated than either a solid or a gas. Like a solid, a liquid is nearly *incompressible.* Like a gas, a liquid flows and deforms to fit the shape of its container.

A **gas** is a system in which each molecule moves through space as a free, noninteracting particle until, on occasion, it collides with another molecule or with the wall of the container. A gas is a *fluid.* A gas is also highly *compressible,* which tells us that there is lots of space between the molecules.

Gases are fairly simple macroscopic systems; hence many of our examples in Part V will be based on gases.

State Variables

The parameters used to characterize or describe a macroscopic system are known as **state variables** because, taken all together, they describe the *state* of the macroscopic system. You met some state variables in earlier chapters: volume, pressure, mass, mass density, and thermal energy. We'll soon introduce several new state variables.

One important state variable, the mass density, is defined as the ratio of two other state variables:

$$\rho = \frac{M}{V} \quad \text{(mass density)} \quad (18.1)$$

In this chapter we'll use an uppercase *M* for the system mass and a lowercase *m* for the mass of an atom. **TABLE 18.1** is a short list of mass densities.

A system is said to be in **thermal equilibrium** if its state variables are constant and not changing. As an example, a gas is in thermal equilibrium if it has been left undisturbed long enough for *p, V,* and *T* to reach steady values.

TABLE 18.1 Densities of materials

Substance	ρ (kg/m³)
Air at STP*	1.29
Ethyl alcohol	790
Water (solid)	920
Water (liquid)	1000
Aluminum	2700
Copper	8920
Gold	19,300
Iron	7870
Lead	11,300
Mercury	13,600
Silicon	2330

*$T = 0°C, p = 1$ atm

EXAMPLE 18.1 | **The mass of a lead pipe**

A project on which you are working uses a cylindrical lead pipe with outer and inner diameters of 4.0 cm and 3.5 cm, respectively, and a length of 50 cm. What is its mass?

SOLVE The mass density of lead is $\rho_{lead} = 11,300$ kg/m^3. The volume of a circular cylinder of length l is $V = \pi r^2 l$. In this case we need to find the volume of the outer cylinder, of radius r_2, *minus*

the volume of air in the inner cylinder, of radius r_1. The volume of the pipe is

$$V = \pi r_2^2 l - \pi r_1^2 l = \pi (r_2^2 - r_1^2) l = 1.47 \times 10^{-4} \text{ m}^3$$

Hence the pipe's mass is

$$M = \rho_{lead} V = 1.7 \text{ kg}$$

STOP TO THINK 18.1 The pressure in a system is measured to be 60 kPa. At a later time the pressure is 40 kPa. The value of Δp is

   a. 60 kPa   b. 40 kPa   c. 20 kPa   d. −20 kPa

18.2 Atoms and Moles

The mass of a macroscopic system is directly related to the total number of atoms or molecules in the system, denoted N. Because N is determined simply by counting, it is a number with no units. A typical macroscopic system has $N \sim 10^{25}$ atoms, an incredibly large number.

The symbol $\sim$, if you are not familiar with it, stands for "has the order of magnitude." It means that the number is known only to within a factor of 10 or so. The statement $N \sim 10^{25}$, which is read "N is of order 10^{25}," implies that N is somewhere in the range 10^{24} to 10^{26}. It is far less precise than the "approximately equal" symbol $\approx$. Saying $N \sim 10^{25}$ gives us a rough idea of how large N is and allows us to know that it differs significantly from 10^5 or even 10^{15}.

It is often useful to know the number of atoms or molecules per cubic meter in a system. We call this quantity the **number density.** It characterizes how densely the atoms are packed together within the system. In an N-atom system that fills volume V, the number density is

$$\frac{N}{V} \quad \text{(number density)} \qquad (18.2)$$

The SI units of number density are m^{-3}. The number density of atoms in a solid is $(N/V)_{solid} \sim 10^{29}$ m^{-3}. The number density of a gas depends on the pressure, but is usually less than 10^{27} m^{-3}. As **FIGURE 18.1** shows, **the value of N/V in a *uniform* system is independent of the volume V.** That is, the number density is the same whether you look at the whole system or just a portion of it.

NOTE While we might say "There are 100 tennis balls per cubic meter," or "There are 10^{29} atoms per cubic meter," tennis balls and atoms are not units. The units of N/V are simply m^{-3}.

FIGURE 18.1 The number density of a uniform system is independent of the volume.

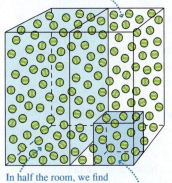

A 100 m^3 room contains 10,000 tennis balls. The number density of balls in the room is $N/V = 10,000/100 \text{ m}^3 = 100 \text{ m}^{-3}$

In half the room, we find 5000 balls in 50 m^3: $N/V = 5000/50 \text{ m}^3 = 100 \text{ m}^{-3}$

In one-tenth of the room, we find 1000 balls in 10 m^3: $N/V = 1000/10 \text{ m}^3 = 100 \text{ m}^{-3}$

Atomic Mass and Atomic Mass Number

You will recall from chemistry that atoms of different elements have different masses. The mass of an atom is determined primarily by its most massive constituents, the protons and neutrons in its nucleus. The *sum* of the number of protons and neutrons is called the **atomic mass number** A:

$$A = \text{number of protons} + \text{number of neutrons}$$

A, which by definition is an integer, is written as a leading superscript on the atomic symbol. For example, the common isotope of hydrogen, with one proton and no neutrons, is ^{1}H. The "heavy hydrogen" isotope called *deuterium,* which includes one neutron, is ^{2}H. The primary isotope of carbon, with six protons (which makes it carbon) and six neutrons, is ^{12}C. The radioactive isotope ^{14}C, used for carbon dating of archeological finds, contains six protons and eight neutrons.

The **atomic mass** scale is established by defining the mass of ^{12}C to be exactly 12 u, where u is the symbol for the **atomic mass unit.** That is, $m(^{12}C) = 12$ u. The atomic mass of any other atom is its mass relative to ^{12}C. For example, careful experiments with hydrogen find that the mass *ratio* $m(^1H)/m(^{12}C)$ is 1.0078/12. Thus the atomic mass of hydrogen is $m(^1H) = 1.0078$ u.

The numerical value of the atomic mass of ^{1}H is close to, but not exactly, its atomic mass number $A = 1$. For our purposes, it will be sufficient to overlook the slight difference and **use the integer atomic mass numbers as the values of the atomic mass.** That is, we'll use $m(^1H) = 1$ u, $m(^4He) = 4$ u, and $m(^{16}O) = 16$ u. For molecules, the **molecular mass** is the sum of the atomic masses of the atoms forming the molecule. Thus the molecular mass of O_2, the constituent of oxygen gas, is $m(O_2) = 32$ u.

NOTE An element's atomic mass number is *not* the same as its atomic number. The *atomic number,* the element's position in the periodic table, is the number of protons in the nucleus.

TABLE 18.2 shows the atomic mass numbers of some of the elements that we'll use for examples and homework problems. A complete periodic table of the elements, including atomic masses, is found in Appendix B.

Moles and Molar Mass

One way to specify the amount of substance in a macroscopic system is to give its mass. Another is to measure the amount of substance in *moles*. By definition, one **mole** of matter, be it solid, liquid, or gas, is the amount of substance containing as many basic particles as there are atoms in 0.012 kg (12 g) of ^{12}C. Many ingenious experiments have determined that there are 6.02×10^{23} atoms in 0.012 kg of ^{12}C, so we can say that 1 mole of substance, abbreviated 1 mol, is 6.02×10^{23} basic particles.

The basic particle depends on the substance. Helium is a **monatomic gas,** meaning that the basic particle is the helium atom. Thus 6.02×10^{23} helium atoms are 1 mol of helium. But oxygen gas is a **diatomic gas** because the basic particle is the two-atom diatomic molecule O_2. 1 mol of oxygen gas contains 6.02×10^{23} *molecules* of O_2 and thus $2 \times 6.02 \times 10^{23}$ oxygen atoms. **TABLE 18.3** lists the monatomic and diatomic gases that we will use for examples and homework problems.

The number of basic particles per mole of substance is called **Avogadro's number,** N_A. The value of Avogadro's number is

$$N_A = 6.02 \times 10^{23} \text{ mol}^{-1}$$

Despite its name, Avogadro's number is not simply "a number"; it has units. Because there are N_A particles per mole, the number of moles in a substance containing N basic particles is

$$n = \frac{N}{N_A} \quad \text{(moles of substance)} \quad (18.3)$$

Avogadro's number allows us to determine atomic masses in kilograms. Knowing that N_A ^{12}C atoms have a mass of 0.012 kg, the mass of one ^{12}C atom must be

$$m(^{12}C) = \frac{0.012 \text{ kg}}{6.02 \times 10^{23}} = 1.993 \times 10^{-26} \text{ kg}$$

We defined the atomic mass scale such that $m(^{12}C) = 12$ u. Thus the conversion factor between atomic mass units and kilograms is

$$1 \text{ u} = \frac{m(^{12}C)}{12} = 1.66 \times 10^{-27} \text{ kg}$$

One mole of helium, sulfur, copper, and mercury.

TABLE 18.2 Some atomic mass numbers

Element		A
^{1}H	Hydrogen	1
^{4}He	Helium	4
^{12}C	Carbon	12
^{14}N	Nitrogen	14
^{16}O	Oxygen	16
^{20}Ne	Neon	20
^{27}Al	Aluminum	27
^{40}Ar	Argon	40
^{207}Pb	Lead	207

TABLE 18.3 Monatomic and diatomic gases

Monatomic		Diatomic	
He	Helium	H_2	Hydrogen
Ne	Neon	N_2	Nitrogen
Ar	Argon	O_2	Oxygen

This conversion factor allows us to calculate the mass in kg of any atom. For example, a ^{20}Ne atom has atomic mass $m(^{20}\text{Ne}) = 20$ u. Multiplying by 1.66×10^{-27} kg/u gives $m(^{20}\text{Ne}) = 3.32 \times 10^{-26}$ kg. If the atomic mass is specified in kilograms, the number of atoms in a system of mass M can be found from

$$N = \frac{M}{m} \tag{18.4}$$

The **molar mass** of a substance is the mass of 1 mol of substance. The molar mass, which we'll designate M_{mol}, has units kg/mol. By definition, the molar mass of ^{12}C is 0.012 kg/mol. For other substances, whose atomic or molecular masses are given relative to ^{12}C, the numerical value of the molar mass is the numerical value of the atomic or molecular mass divided by 1000. For example, the molar mass of He, with $m = 4$ u, is $M_{mol}(\text{He}) = 0.004$ kg/mol and the molar mass of diatomic O_2 is $M_{mol}(O_2) = 0.032$ kg/mol.

Equation 18.4 uses the atomic mass to find the number of atoms in a system. Similarly, you can use the molar mass to determine the number of moles. For a system of mass M consisting of atoms or molecules with molar mass M_{mol},

$$n = \frac{M}{M_{mol}} \tag{18.5}$$

EXAMPLE 18.2 | **Moles of oxygen**

100 g of oxygen gas is how many moles of oxygen?

SOLVE We can do the calculation two ways. First, let's determine the number of molecules in 100 g of oxygen. The diatomic oxygen molecule O_2 has molecular mass $m = 32$ u. Converting this to kg, we get the mass of one molecule:

$$m = 32 \text{ u} \times \frac{1.66 \times 10^{-27} \text{ kg}}{1 \text{ u}} = 5.31 \times 10^{-26} \text{ kg}$$

Thus the number of molecules in 100 g = 0.100 kg is

$$N = \frac{M}{m} = \frac{0.100 \text{ kg}}{5.31 \times 10^{-26} \text{ kg}} = 1.88 \times 10^{24}$$

Knowing the number of molecules gives us the number of moles:

$$n = \frac{N}{N_A} = 3.13 \text{ mol}$$

Alternatively, we can use Equation 18.5 to find

$$n = \frac{M}{M_{mol}} = \frac{0.100 \text{ kg}}{0.032 \text{ kg/mol}} = 3.13 \text{ mol}$$

STOP TO THINK 18.2 Which system contains more atoms: 5 mol of helium $(A = 4)$ or 1 mol of neon $(A = 20)$?

a. Helium.   b. Neon.   c. They have the same number of atoms.

18.3 Temperature

We are all familiar with the idea of temperature. Mass is a measure of the amount of substance in a system. Velocity is a measure of how fast a system moves. What physical property of the system have you determined if you measure its temperature?

We will begin with the commonsense idea that temperature is a measure of how "hot" or "cold" a system is. As we develop these ideas, we'll find that **temperature** T is related to a system's *thermal energy*. We defined thermal energy in Chapter 9 as the kinetic and potential energy of the atoms and molecules in a system as they vibrate (a solid) or move around (a gas). A system has more thermal energy when it is "hot" than when it is "cold." In Chapter 20, we'll replace these vague notions of hot and cold with a precise relationship between temperature and thermal energy.

To start, we need a means to measure the temperature of a system. This is what a *thermometer* does. A thermometer can be any small macroscopic system that undergoes

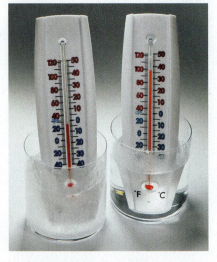

Thermal expansion of the liquid in the thermometer tube pushes it higher in the hot water than in the ice water.

a measurable change as it exchanges thermal energy with its surroundings. It is placed in contact with a larger system whose temperature it will measure. In a common glass-tube thermometer, for example, a small volume of mercury or alcohol expands or contracts when placed in contact with a "hot" or "cold" object. The object's temperature is determined by the length of the column of liquid.

A thermometer needs a *temperature scale* to be a useful measuring device. In 1742, the Swedish astronomer Anders Celsius sealed mercury into a small capillary tube and observed how it moved up and down the tube as the temperature changed. He selected two temperatures that anyone could reproduce, the freezing and boiling points of pure water, and labeled them 0 and 100. He then marked off the glass tube into one hundred equal intervals between these two reference points. By doing so, he invented the temperature scale that we today call the *Celsius scale*. The units of the Celsius temperature scale are "degrees Celsius," which we abbreviate °C. Note that the degree symbol ° is part of the unit, not part of the number.

The *Fahrenheit scale*, still widely used in the United States, is related to the Celsius scale by

$$T_F = \frac{9}{5} T_C + 32° \qquad (18.6)$$

FIGURE 18.2 shows several temperatures measured on the Celsius and Fahrenheit scales and also on the Kelvin scale.

Absolute Zero and Absolute Temperature

Any physical property that changes with temperature can be used as a thermometer. In practice, the most useful thermometers have a physical property that changes *linearly* with temperature. One of the most important scientific thermometers is the **constant-volume gas thermometer** shown in **FIGURE 18.3a**. This thermometer depends on the fact that the *absolute* pressure (not the gauge pressure) of a gas in a sealed container increases linearly as the temperature increases.

A gas thermometer is first calibrated by recording the pressure at two reference temperatures, such as the boiling and freezing points of water. These two points are plotted on a pressure-versus-temperature graph and a straight line is drawn through them. The gas bulb is then brought into contact with the system whose temperature is to be measured. The pressure is measured, then the corresponding temperature is read off the graph.

FIGURE 18.3b shows the pressure-temperature relationship for three different gases. Notice two important things about this graph.

1. There is a *linear* relationship between temperature and pressure.
2. All gases extrapolate to *zero pressure* at the same temperature: $T_0 = -273°C$. No gas actually gets that cold without condensing, although helium comes very close, but it is surprising that you get the same zero-pressure temperature for any gas and any starting pressure.

The pressure in a gas is due to collisions of the molecules with each other and the walls of the container. A pressure of zero would mean that all motion, and thus all collisions, had ceased. If there were no atomic motion, the system's thermal energy would be zero. The temperature at which all motion would cease, and at which $E_{th} = 0$, is called **absolute zero.** Because temperature is related to thermal energy, absolute zero is the lowest temperature that has physical meaning. We see from the gas-thermometer data that $T_0 = -273°C$.

It is useful to have a temperature scale with the zero point at absolute zero. Such a temperature scale is called an **absolute temperature scale.** Any system whose temperature is measured on an absolute scale will have $T > 0$. The absolute temperature scale having the same unit size as the Celsius scale is called the *Kelvin scale*. It is the SI scale of temperature. The units of the Kelvin scale are *kelvins,* abbreviated as K. The conversion between the Celsius scale and the Kelvin scale is

$$T_K = T_C + 273 \qquad (18.7)$$

FIGURE 18.2 Temperatures measured with different scales.

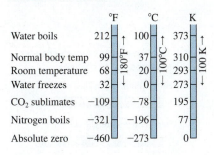

FIGURE 18.3 The pressure in a constant-volume gas thermometer extrapolates to zero at $T_0 = -273°C$.

(a)

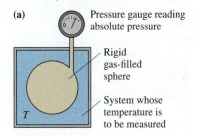

(b)

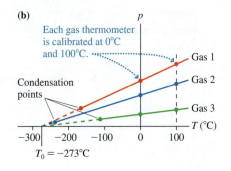

On the Kelvin scale, absolute zero is 0 K, the freezing point of water is 273 K, and the boiling point of water is 373 K.

NOTE ▶ The units are simply "kelvins," *not* "degrees Kelvin." ◀

STOP TO THINK 18.3 The temperature of a glass of water increases from 20°C to 30°C. What is ΔT?

a. 10 K   b. 283 K   c. 293 K   d. 303 K

18.4 Thermal Expansion

Objects expand when heated. This **thermal expansion** is why the liquid rises in a thermometer and why pipes, highways, and bridges have expansion joints. FIGURE 18.4 shows an object of length L that changes by ΔL when the temperature is changed from T to $T + \Delta T$. For most solids, the *fractional* change in length, $\Delta L/L$, is proportional to the temperature change ΔT with a proportionality coefficient that depends on the material. That is,

$$\frac{\Delta L}{L} = \alpha \, \Delta T \tag{18.8}$$

where α (Greek alpha) is the material's **coefficient of linear expansion.** Equation 18.8 characterizes both expansion ($\Delta L > 0$ if the temperature increases) and contraction ($\Delta L < 0$ if the temperature falls).

TABLE 18.4 gives the coefficients of linear expansion for a few common materials and also for a metal alloy called *invar* that is specifically designed to have extremely low thermal expansion. Because the fractional change in length is dimensionless, α has units of °C^{-1}, read as "per degree Celsius." The units may also be written K^{-1}, because a temperature *change* ΔT is the same in °C and K, but practical measurements are made in °C, not K.

FIGURE 18.4 An object's length changes when the temperature changes.

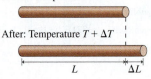

Before: Temperature T

After: Temperature $T + \Delta T$

TABLE 18.4 Coefficients of linear and volume expansion

Material	α (°C^{-1})
Aluminum	2.3×10^{-5}
Brass	1.9×10^{-5}
Concrete	1.2×10^{-5}
Steel	1.1×10^{-5}
Invar	0.09×10^{-5}

Material	β (°C^{-1})
Gasoline	9.6×10^{-4}
Mercury	1.8×10^{-4}
Ethyl alcohol	1.1×10^{-4}

EXAMPLE 18.3 │ An expanding pipe

A 55-m-long steel pipe runs from one side of a refinery to the other. By how much does the pipe expand on a 5°C winter day when 155°C oil is pumped through it?

SOLVE The expansion is given by Equation 18.8, with the coefficient of linear expansion for steel taken from Table 18.4:

$$\Delta L = \alpha L \, \Delta T = (1.1 \times 10^{-5} \, °C^{-1})(55 \text{ m})(150°C)$$
$$= 0.091 \text{ m} = 9.1 \text{ cm}$$

ASSESS 9.1 cm is a very small fraction of 55 m, so the pipe as a whole has expanded very little. Nevertheless, 9.1 cm ≈ 3.5 in is a huge expansion in terms of the engineering of the pipe. Pipes like this have to be designed with flexible expansion joints that allow for thermal expansion and contraction.

Volume expansion is treated the same way. If an object's volume changes by ΔV during a temperature change ΔT, the *fractional* change in volume is

$$\frac{\Delta V}{V} = \beta \, \Delta T \tag{18.9}$$

where β (Greek beta) is the material's **coefficient of volume expansion**.

Liquids are constrained by the shape of their container, so liquids are characterized by a volume-expansion coefficient but not by a linear-expansion coefficient. A few values are given in Table 18.4, where you can see that β, like α, has units of °C^{-1}.

Solids expand linearly in all three directions, as given by Equation 18.8, and in the process a solid changes its volume. Imagine a cube of edge length L and volume $V = L^3$. If the edge length changes by a very small amount dL, the volume changes by

$$dV = 3L^2\,dL \qquad (18.10)$$

If we divide both sides by $V = L^3$, we see that

$$\frac{dV}{V} = 3\frac{dL}{L} \qquad (18.11)$$

That is, the fractional change in volume is three times the fractional change in the length of the sides. Consequently, a solid's coefficient of volume expansion is

$$\beta_{\text{solid}} = 3\alpha \qquad (18.12)$$

You may have noticed that Table 18.4 does not include water. Water is a very common substance, but—due to its molecular structure—water has some unusual properties that set it apart from other liquids. If you lower the temperature of water, its volume contracts as expected. But only until the temperature reaches 4°C. If you continue cooling water, from 4°C down to the freezing point at 0°C, the volume expands! Water has maximum density at 4°C, with slightly less density—due to the expanded volume—at both higher and lower temperatures. Consequently, the thermal expansion of water cannot be characterized by a single coefficient of thermal expansion.

STOP TO THINK 18.4 A steel plate has a 2.000-cm-diameter hole through it. If the plate is heated, does the diameter of the hole increase or decrease?

18.5 Phase Changes

The temperature inside the freezer compartment of a refrigerator is typically about −20°C. Suppose you were to remove a few ice cubes from the freezer, place them in a sealed container with a thermometer, then heat them, as **FIGURE 18.5a** shows. We'll assume that the heating is done so slowly that the inside of the container always has a single, well-defined temperature.

FIGURE 18.5b shows the temperature as a function of time. After steadily rising from the initial −20°C, the temperature remains fixed at 0°C for an extended period of time. This is the interval of time during which the ice melts. As it's melting, the ice temperature is 0°C and the liquid water temperature is 0°C. Even though the system is being heated, the liquid water temperature doesn't begin to rise until all the ice has melted. If you were to turn off the flame at any point, the system would remain a mixture of ice and liquid water at 0°C.

NOTE In everyday language, the three phases of water are called *ice, water,* and *steam.* That is, the term *water* implies the liquid phase. Scientifically, these are the solid, liquid, and gas phases of the compound called *water.* To be clear, we'll use the term *water* in the scientific sense of a collection of H_2O molecules. We'll say either *liquid* or *liquid water* to denote the liquid phase.

The thermal energy of a solid is the kinetic energy of the vibrating atoms plus the potential energy of the stretched and compressed molecular bonds. Melting occurs when the thermal energy gets so large that molecular bonds begin to break, allowing the atoms to move around. The temperature at which a solid becomes a liquid or, if the thermal energy is reduced, a liquid becomes a solid is called the **melting point** or the **freezing point.** Melting and freezing are *phase changes.*

A system at the melting point is in **phase equilibrium,** meaning that any amount of solid can coexist with any amount of liquid. Raise the temperature ever so slightly

FIGURE 18.5 Water is transformed from solid to liquid to gas.

(a)

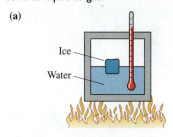

(b)

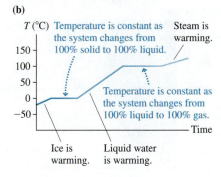

and the entire system becomes liquid. Lower it slightly and it all becomes solid. But exactly at the melting point the system has no tendency to move one way or the other. That is why the temperature remains constant at the melting point until the phase change is complete.

You can see the same thing happening in Figure 18.5b at 100°C, the boiling point. This is a phase equilibrium between the liquid phase and the gas phase, and any amount of liquid can coexist with any amount of gas at this temperature. Above this temperature, the thermal energy is too large for bonds to be established between molecules, so the system is a gas. If the thermal energy is reduced, the molecules begin to bond with each other and stick together. In other words, the gas condenses into a liquid. The temperature at which a gas becomes a liquid or, if the thermal energy is increased, a liquid becomes a gas is called the **condensation point** or the **boiling point.**

> **NOTE** Liquid water becomes solid ice at 0°C, but that doesn't mean the temperature of ice is always 0°C. Ice reaches the temperature of its surroundings. If the air temperature in a freezer is −20°C, then the ice temperature is −20°C. Likewise, steam can be heated to temperatures above 100°C. That doesn't happen when you boil water on the stove because the steam escapes, but steam can be heated far above 100°C in a sealed container.

Phase Diagrams

A **phase diagram** is used to show how the phases and phase changes of a substance vary with both temperature and pressure. **FIGURE 18.6** shows the phase diagrams for water and carbon dioxide. You can see that each diagram is divided into three regions corresponding to the solid, liquid, and gas phases. The boundary lines separating the regions indicate the phase transitions. The system is in phase equilibrium at a pressure-temperature point that falls on one of these lines.

Phase diagrams contain a great deal of information. Notice on the water phase diagram that the dashed line at $p = 1$ atm crosses the solid-liquid boundary at 0°C and the liquid-gas boundary at 100°C. These well-known melting and boiling point temperatures of water apply only at standard atmospheric pressure. You can see that in Denver, where $p_{atmos} < 1$ atm, water melts at slightly above 0°C and boils at a temperature below 100°C. A *pressure cooker* works by allowing the pressure inside to exceed 1 atm. This raises the boiling point, so foods that are in boiling water are at a temperature above 100°C and cook faster.

Crossing the solid-liquid boundary corresponds to melting or freezing while crossing the liquid-gas boundary corresponds to boiling or condensing. But there's another possibility—crossing the solid-gas boundary. The phase change in which a solid becomes a gas is called **sublimation.** It is not an everyday experience with water, but you probably are familiar with the sublimation of dry ice. Dry ice is solid carbon dioxide. You can see on the carbon dioxide phase diagram that the dashed line at $p = 1$ atm crosses the solid-*gas* boundary, rather than the solid-liquid boundary, at $T = -78$°C. This is the *sublimation temperature* of dry ice.

Liquid carbon dioxide does exist, but only at pressures greater than 5 atm and temperatures greater than −56°C. A CO_2 fire extinguisher contains *liquid* carbon dioxide under high pressure. (You can hear the liquid slosh if you shake a CO_2 fire extinguisher.)

One important difference between the water and carbon dioxide phase diagrams is the slope of the solid-liquid boundary. For most substances, the solid phase is denser than the liquid phase and the liquid is denser than the gas. Pressurizing the substance compresses it and increases the density. If you start compressing CO_2 gas at room temperature, thus moving upward through the phase diagram along a vertical line, you'll first condense it to a liquid and eventually, if you keep compressing, change it into a solid.

FIGURE 18.6 Phase diagrams (not to scale) for water and carbon dioxide.

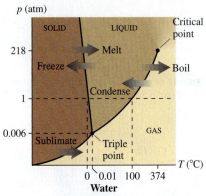

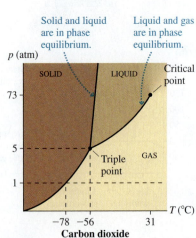

Water is a very unusual substance in that the density of ice is *less* than the density of liquid water. That is why ice floats. If you compress ice, making it denser, you eventually cause a phase transition in which the ice turns to liquid water! Consequently, the solid-liquid boundary for water slopes to the left.

The liquid-gas boundary ends at a point called the **critical point.** Below the critical point, liquid and gas are clearly distinct and there is a phase change if you go from one to the other. But there is no clear distinction between liquid and gas at pressures or temperatures above the critical point. The system is a *fluid,* but it can be varied continuously between high density and low density without a phase change.

The final point of interest on the phase diagram is the **triple point** where the phase boundaries meet. Two phases are in phase equilibrium along the boundaries. The triple point is the *one* value of temperature and pressure for which all three phases can coexist in phase equilibrium. That is, any amounts of solid, liquid, and gas can happily coexist at the triple point. For water, the triple point occurs at $T_3 = 0.01°C$ and $p_3 = 0.006$ atm.

The significance of the triple point of water is its connection to the Kelvin temperature scale. The Celsius scale required two *reference points,* the boiling and melting points of water. We can now see that these are not very satisfactory reference points because their values vary as the pressure changes. In contrast, there's only one temperature at which ice, liquid water, and water vapor will coexist in equilibrium. If you produce this equilibrium in the laboratory, then you *know* the system is at the triple-point temperature.

The triple-point temperature of water is an ideal reference point, hence the Kelvin temperature scale is *defined* to be a linear temperature scale starting from 0 K at absolute zero and passing through 273.16 K at the triple point of water. Because $T_3 = 0.01°C$, absolute zero on the Celsius scale is $T_0 = -273.15°C$.

NOTE To be consistent with our use of significant figures, $T_0 = -273$ K is the appropriate value to use in calculations *unless* you know other temperatures with an accuracy of better than 1°C.

Food takes longer to cook at high altitudes because the boiling point of water is less than 100°C.

STOP TO THINK 18.5 For which is there a sublimation temperature that is higher than a melting temperature?

a. Water   b. Carbon dioxide   c. Both   d. Neither

18.6 Ideal Gases

We noted earlier in the chapter that solids and liquids are nearly incompressible, an observation suggesting that atoms are fairly hard and cannot be pressed together once they come into contact with each other. Based on this observation, suppose we were to model atoms as "hard spheres" that do not interact except for occasional elastic collisions when two atoms come into contact and bounce apart.

This is a *model* of an atom—what we might call the *ideal atom*—because it ignores the weak attractive interactions that hold liquids and solids together. A gas of these noninteracting atoms is called an **ideal gas.** It is a gas of small, hard, randomly moving atoms that bounce off each other and the walls of their container but otherwise do not interact. The ideal gas is a somewhat simplified description of a real gas, but experiments show that the ideal-gas model is quite good for real gases if two conditions are met:

1. The density is low (i.e., the atoms occupy a volume much smaller than that of the container), and
2. The temperature is well above the condensation point.

If the density gets too high, or the temperature too low, then the attractive forces between the atoms begin to play an important role and our model, which ignores those attractive forces, fails. These are the forces that are responsible, under the right conditions, for the gas condensing into a liquid.

We've been using the term "atoms," but many gases, as you know, consist of molecules rather than atoms. Only helium, neon, argon, and the other inert elements in the far-right column of the periodic table of the elements form monatomic gases. Hydrogen (H_2), nitrogen (N_2), and oxygen (O_2) are diatomic gases. As far as translational motion is concerned, the ideal-gas model does not distinguish between a monatomic gas and a diatomic gas; both are considered as simply small, hard spheres. Hence the terms "atoms" and "molecules" can be used interchangeably to mean the basic constituents of the gas.

The Ideal-Gas Law

Section 18.1 introduced the idea of *state variables,* those parameters that describe the state of a macroscopic system. The state variables for an ideal gas are the volume V of its container, the number of moles n of the gas present in the container, the temperature T of the gas and its container, and the pressure p that the gas exerts on the walls of the container. These four state parameters are not independent of each other. If you change the value of one—by, say, raising the temperature—then one or more of the others will change as well. Each change of the parameters is a *change of state* of the system.

Experiments during the 17th and 18th centuries found a very specific relationship between the four state variables. Suppose you change the state of a gas, by heating it or compressing it or doing something else to it, and measure p, V, n, and T. Repeat this many times, changing the state of the gas each time, until you have a large table of p, V, n, and T values.

Then make a graph on which you plot pV, the product of the pressure and volume, on the vertical axis and nT, the product of the number of moles and temperature (in kelvins), on the horizontal axis. The very surprising result is that for *any* gas, whether it is hydrogen or helium or oxygen or methane, **you get exactly the same graph,** the linear graph shown in **FIGURE 18.7**. In other words, nothing about the graph indicates what gas was used because all gases give the same result.

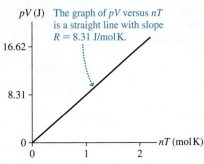

FIGURE 18.7 A graph of pV versus nT for an ideal gas.

pV (J) The graph of pV versus nT is a straight line with slope $R = 8.31$ J/mol K.

NOTE No real gas could extend to $nT = 0$ because it would condense. But an ideal gas never condenses because the only interactions among the molecules are hard-sphere collisions.

As you can see, there is a very clear proportionality between the quantity pV and the quantity nT. If we designate the slope of the line in this graph as R, then we can write the relationship as

$$pV = R \times (nT)$$

It is customary to write this relationship in a slightly different form, namely

$$pV = nRT \qquad \text{(ideal-gas law)} \qquad (18.13)$$

Equation 18.13 is the **ideal-gas law. The ideal-gas law is a relationship among the four state variables—p, V, n, and T—that characterize a gas in thermal equilibrium.**

The constant R, which is determined experimentally as the slope of the graph in Figure 18.7, is called the **universal gas constant.** Its value, in SI units, is

$$R = 8.31 \text{ J/mol K}$$

The units of R seem puzzling. The denominator mol K is clear because R multiplies nT. But what about the joules? The left side of the ideal-gas law, pV, has units

$$\text{Pa} \, \text{m}^3 = \frac{\text{N}}{\text{m}^2} \, \text{m}^3 = \text{N} \, \text{m} = \text{joules}$$

The product pV has units of joules, as shown on the vertical axis in Figure 18.7.

NOTE You perhaps learned in chemistry to work gas problems using units of atmospheres and liters. To do so, you had a different numerical value of R expressed in those units. In physics, however, we always work gas problems in SI units. Pressures *must* be in Pa, volumes in m^3, and temperatures in K.

The surprising fact, and one worth commenting upon, is that *all* gases have the *same* graph and the *same* value of R. There is no obvious reason a very simple atomic gas such as helium should have the same slope as a more complex gas such as methane (CH_4). Nonetheless, both turn out to have the same value for R. The ideal-gas law, within its limits of validity, describes *all* gases with a single value of the constant R.

EXAMPLE 18.4 | Calculating a gas pressure

100 g of oxygen gas is distilled into an evacuated $600\ cm^3$ container. What is the gas pressure at a temperature of 150°C?

MODEL The gas can be treated as an ideal gas. Oxygen is a diatomic gas of O_2 molecules.

SOLVE From the ideal-gas law, the pressure is $p = nRT/V$. In Example 18.2 we calculated the number of moles in 100 g of O_2 and found $n = 3.13$ mol. Gas problems typically involve several conversions to get quantities into the proper units, and this example is no exception. The SI units of V and T are m^3 and K, respectively, thus

$$V = (600\ cm^3)\left(\frac{1\ m}{100\ cm}\right)^3 = 6.00 \times 10^{-4}\ m^3$$

$$T = (150 + 273)\ K = 423\ K$$

With this information, the pressure is

$$p = \frac{nRT}{V} = \frac{(3.13\ mol)(8.31\ J/mol\,K)(423\ K)}{6.00 \times 10^{-4}\ m^3}$$

$$= 1.83 \times 10^7\ Pa = 181\ atm$$

In this text we will consider only gases in sealed containers. The number of moles (and number of molecules) will not change during a problem. In that case,

$$\frac{pV}{T} = nR = \text{constant} \tag{18.14}$$

If the gas is initially in state i, characterized by the state variables p_i, V_i, and T_i, and at some later time in a final state f, the state variables for these two states are related by

$$\frac{p_f V_f}{T_f} = \frac{p_i V_i}{T_i} \qquad \text{(ideal gas in a sealed container)} \tag{18.15}$$

This before-and-after relationship between the two states, reminiscent of a conservation law, will be valuable for many problems.

EXAMPLE 18.5 | Calculating a gas temperature

A cylinder of gas is at 0°C. A piston compresses the gas to half its original volume and three times its original pressure. What is the final gas temperature?

MODEL Treat the gas as an ideal gas in a sealed container.

SOLVE The before-and-after relationship of Equation 18.15 can be written

$$T_2 = T_1 \frac{p_2}{p_1} \frac{V_2}{V_1}$$

In this problem, the compression of the gas results in $V_2/V_1 = \frac{1}{2}$ and $p_2/p_1 = 3$. The initial temperature is $T_1 = 0°C = 273\ K$. With this information,

$$T_2 = 273\ K \times 3 \times \tfrac{1}{2} = 409\ K = 136°C$$

ASSESS We did not need to know actual values of the pressure and volume, just the *ratios* by which they change.

We will often want to refer to the number of molecules N in a gas rather than the number of moles n. This is an easy change to make. Because $n = N/N_A$, the ideal-gas law in terms of N is

$$pV = nRT = \frac{N}{N_A} RT = N \frac{R}{N_A} T \tag{18.16}$$

R/N_A, the ratio of two known constants, is known as **Boltzmann's constant k_B**:

$$k_B = \frac{R}{N_A} = 1.38 \times 10^{-23} \text{ J/K}$$

The subscript B distinguishes Boltzmann's constant from a spring constant or other uses of the symbol k.

Ludwig Boltzmann was an Austrian physicist who did some of the pioneering work in statistical physics during the mid-19th century. Boltzmann's constant k_B can be thought of as the "gas constant per molecule," whereas R is the "gas constant per mole." With this definition, the ideal-gas law in terms of N is

$$pV = Nk_B T \quad \text{(ideal-gas law)} \tag{18.17}$$

Equations 18.13 and 18.17 are both the ideal-gas law, just expressed in terms of different state variables.

Recall that the number density (molecules per m³) was defined as N/V. A rearrangement of Equation 18.17 gives the number density as

$$\frac{N}{V} = \frac{p}{k_B T} \tag{18.18}$$

This is a useful consequence of the ideal-gas law, but keep in mind that the pressure *must* be in SI units of pascals and the temperature *must* be in SI units of kelvins.

EXAMPLE 18.6 | **The distance between molecules**

"Standard temperature and pressure," abbreviated **STP,** are $T = 0°C$ and $p = 1$ atm. Estimate the average distance between gas molecules at STP.

MODEL Consider the gas to be an ideal gas.

SOLVE Suppose a container of volume V holds N molecules at STP. How do we estimate the distance between them? Imagine placing an imaginary sphere around each molecule, separating it from its neighbors. This divides the total volume V into N little spheres of volume v_i, where $i = 1$ to N. The spheres of two neighboring molecules touch each other, like a crate full of Ping-Pong balls of somewhat different sizes all touching their neighbors, so the distance between two molecules is the sum of the radii of their two spheres. Each of these spheres is somewhat different, but a reasonable *estimate* of the distance between molecules would be twice the *average* radius of a sphere.

The average volume of one of these little spheres is

$$v_{avg} = \frac{V}{N} = \frac{1}{N/V}$$

That is, the average volume per molecule (m³ per molecule) is the inverse of the number density, the number of molecules per m³. This is not the volume of the molecule itself, which is much smaller, but the average volume of space that each molecule can claim as its own. We can use Equation 18.18 to calculate the number density:

$$\frac{N}{V} = \frac{p}{k_B T} = \frac{1.01 \times 10^5 \text{ Pa}}{(1.38 \times 10^{-23} \text{ J/K})(273 \text{ K})}$$

$$= 2.69 \times 10^{25} \text{ molecules/m}^3$$

where we used the definition of STP in SI units. Thus the average volume per molecule is

$$v_{avg} = \frac{1}{N/V} = 3.72 \times 10^{-26} \text{ m}^3$$

The volume of a sphere is $\frac{4}{3}\pi r^3$, so the average radius of a sphere is

$$r_{avg} = \left(\frac{3}{4\pi} v_{avg}\right)^{1/3} = 2.1 \times 10^{-9} \text{ m} = 2.1 \text{ nm}$$

The average distance between two molecules, with their spheres touching, is twice r_{avg}. Thus

$$\text{average distance} = 2r_{avg} \approx 4 \text{ nm}$$

This is a simple estimate, so we've given the answer with only one significant figure.

ASSESS One of the assumptions of the ideal-gas model is that atoms or molecules are "far apart" in comparison to the sizes of atoms and molecules. Chemistry experiments find that small molecules, such as N_2 and O_2, are roughly 0.3 nm in diameter. For a gas at STP, we see that the average distance between molecules is more than 10 times the size of a molecule. Thus the ideal-gas model works very well for a gas at STP.

STOP TO THINK 18.6 You have two containers of equal volume. One is full of helium gas. The other holds an equal mass of nitrogen gas. Both gases have the same pressure. How does the temperature of the helium compare to the temperature of the nitrogen?

a. $T_{helium} > T_{nitrogen}$ b. $T_{helium} = T_{nitrogen}$ c. $T_{helium} < T_{nitrogen}$

18.7 Ideal-Gas Processes

The ideal-gas law is the connection between the state variables pressure, temperature, and volume. If the state variables change, as they would from heating or compressing the gas, the state of the gas changes. An **ideal-gas process** is the means by which the gas changes from one state to another.

> **NOTE** Even in a sealed container, the ideal-gas law is a relationship among *three* variables. In general, *all three change* during an ideal-gas process. As a result, thinking about cause and effect can be rather tricky. Don't make the mistake of thinking that one variable is constant unless you're sure, beyond a doubt, that it is.

The *pV* Diagram

It will be very useful to represent ideal-gas processes on a graph called a *pV* **diagram.** This is nothing more than a graph of pressure versus volume. The important idea behind the *pV* diagram is that *each point* on the graph represents a single, unique state of the gas. That seems surprising at first, because a point on the graph only directly specifies the values of *p* and *V*. But knowing *p* and *V*, and assuming that *n* is known for a sealed container, we can find the temperature by using the ideal-gas law. Thus each point actually represents a triplet of values (p, V, T) specifying the state of the gas.

For example, **FIGURE 18.8** is a *pV* diagram showing three states of a system consisting of 1 mol of gas. The values of *p* and *V* can be read from the axes for each point, then the temperature at that point determined from the ideal-gas law.

An ideal-gas *process* is a "trajectory" in the *pV* diagram showing all the intermediate states through which the gas passes. Figure 18.8 shows two different processes by which the gas can be changed from state 1 to state 3.

There are infinitely many ways to change the gas from state 1 to state 3. Although the initial and final states are the same for each of them, the particular process by which the gas changes—that is, the particular trajectory—will turn out to have very real consequences. The *pV* diagram is an important graphical representation of the process.

FIGURE 18.8 The state of the gas and ideal-gas processes can be shown on a *pV* diagram.

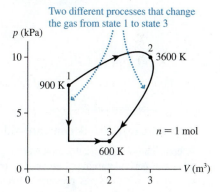

Quasi-Static Processes

Strictly speaking, the ideal-gas law applies only to gases in *thermal equilibrium,* meaning that the state variables are constant and not changing. But, by definition, an ideal-gas process causes some of the state variables to change. The gas is *not* in thermal equilibrium while the process of changing from state 1 to state 2 is under way.

To use the ideal-gas law throughout, we will assume that the process occurs *so slowly* that the system is never far from equilibrium. In other words, the values of *p*, *V*, and *T* at any point in the process are essentially the same as the equilibrium values they would assume if we stopped the process at that point. A process that is essentially in thermal equilibrium at all times is called a **quasi-static process.** It is an idealization, like a frictionless surface, but one that is a very good approximation in many real situations.

An important characteristic of a quasi-static process is that the trajectory through the *pV* diagram can be *reversed.* If you quasi-statically expand a gas by slowly pulling a piston out, as shown in **FIGURE 18.9a**, you can reverse the process by slowly pushing the piston in. The gas retraces its *pV* trajectory until it has returned to its initial state. Contrast this with what happens when the membrane bursts in **FIGURE 18.9b**. That is a sudden process, not at all quasi-static. The expanding gas is *not* in thermal equilibrium until some later time when it has completely filled the larger container, so the *irreversible* process of Figure 18.9b cannot be represented on a *pV* diagram.

The critical question is: How slow must a process be to qualify as quasi-static? That is a difficult question to answer. This textbook will always assume that processes are quasi-static. It turns out to be a reasonable assumption for the types of examples and homework problems we will look at. Irreversible processes will be left to more advanced courses.

FIGURE 18.9 The slow motion of the piston is a quasi-static process. The bursting of the membrane is not.

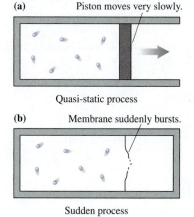

Constant-Volume Process

Many important gas processes take place in a container of constant, unchanging volume. A constant-volume process is called an **isochoric process,** where *iso* is a prefix meaning "constant" or "equal" while *choric* is from a Greek root meaning "volume." An isochoric process is one for which

$$V_f = V_i \qquad (18.19)$$

For example, suppose that you have a gas in the closed, rigid container shown in FIGURE 18.10a. Warming the gas with a Bunsen burner will raise its pressure without changing its volume. This process is shown as the vertical line $1 \rightarrow 2$ on the pV diagram of FIGURE 18.10b. A constant-volume cooling, by placing the container on a block of ice, would lower the pressure and be represented as the vertical line from 2 to 1. **Any isochoric process appears on a pV diagram as a vertical line.**

FIGURE 18.10 A constant-volume (isochoric) process.

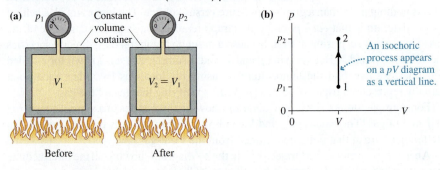

Before   After

EXAMPLE 18.7 | **A constant-volume gas thermometer**

A constant-volume gas thermometer is placed in contact with a reference cell containing water at the triple point. After reaching equilibrium, the gas pressure is recorded as 55.78 kPa. The thermometer is then placed in contact with a sample of unknown temperature. After the thermometer reaches a new equilibrium, the gas pressure is 65.12 kPa. What is the temperature of this sample?

MODEL The thermometer's volume doesn't change, so this is an isochoric process.

SOLVE The temperature at the triple point of water is $T_1 = 0.01°C = 273.16$ K. The ideal-gas law for a closed system

is $p_2 V_2 / T_2 = p_1 V_1 / T_1$. The volume doesn't change, so $V_2/V_1 = 1$. Thus

$$T_2 = T_1 \frac{V_2}{V_1} \frac{p_2}{p_1} = T_1 \frac{p_2}{p_1} = (273.16 \text{ K}) \frac{65.12 \text{ kPa}}{55.78 \text{ kPa}}$$

$$= 318.90 \text{ K} = 45.75°C$$

The temperature *must* be in kelvins to do this calculation, although it is common to convert the final answer to °C. The fact that the pressures were given to four significant figures justified using $T_K = T_C + 273.15$ rather than the usual $T_C + 273$.

ASSESS $T_2 > T_1$, which we expected from the increase in pressure.

Constant-Pressure Process

Other gas processes take place at a constant, unchanging pressure. A constant-pressure process is called an **isobaric process,** where *baric* is from the same root as "barometer" and means "pressure." An isobaric process is one for which

$$p_f = p_i \qquad (18.20)$$

FIGURE 18.11a shows one method of changing the state of a gas while keeping the pressure constant. A cylinder of gas has a tight-fitting piston of mass M that can slide up and down but seals the container so that no atoms enter or escape. As the free-body diagram of FIGURE 18.11b shows, the piston and the air press down with force $p_{atmos}A + Mg$ while the gas inside pushes up with force $p_{gas}A$. In equilibrium, the gas pressure inside the cylinder is

$$p_{gas} = p_{atmos} + \frac{Mg}{A} \qquad (18.21)$$

In other words, the gas pressure is determined by the requirement that the gas must support both the mass of the piston and the air pressing inward. **This pressure is independent of the temperature of the gas or the height of the piston, so it stays constant as long as M is unchanged.**

If the cylinder is warmed, the gas will expand and push the piston up. But the pressure, determined by mass M, will not change. This process is shown on the pV diagram of **FIGURE 18.11c** as the horizontal line $1 \rightarrow 2$. We call this an *isobaric expansion*. An *isobaric compression* occurs if the gas is cooled, lowering the piston. **Any isobaric process appears on a pV diagram as a horizontal line.**

FIGURE 18.11 A constant-pressure (isobaric) process.

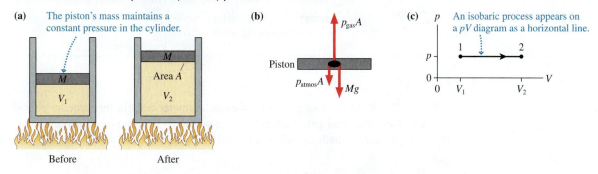

EXAMPLE 18.8 | **Comparing pressure**

The two cylinders in **FIGURE 18.12** contain ideal gases at 20°C. Each cylinder is sealed by a frictionless piston of mass M.

a. How does the pressure of gas 2 compare to that of gas 1? Is it larger, smaller, or the same?

b. Suppose gas 2 is warmed to 80°C. Describe what happens to the pressure and volume.

FIGURE 18.12 Compare the pressures of the two gases.

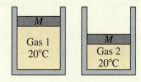

MODEL Treat the gases as ideal gases.

SOLVE a. The pressure in the gas is determined by the requirement that the piston be in mechanical equilibrium. The pressure of the gas inside pushes up on the piston; the air pressure and the weight of the piston press down. The gas pressure $p = p_{atmos} + Mg/A$ depends on the mass of the piston, but not at all on how high the piston is or what type of gas is inside the cylinder. Thus both pressures are the same.

b. Neither does the pressure depend on temperature. Warming the gas increases the temperature, but the pressure—determined by the mass and area of the piston—is unchanged. Because $pV/T = $ constant, and p is constant, it must be true that $V/T = $ constant. As T increases, the volume V also must increase to keep V/T unchanged. In other words, increasing the gas temperature causes the volume to expand—the piston goes up—but with no change in pressure. This is an isobaric process.

EXAMPLE 18.9 | **Identifying a gas**

Your lab assistant distilled 50 g of a gas into a cylinder, but he left without writing down what kind of gas it is. The cylinder has a pressure regulator that adjusts a piston to keep the pressure at a constant 2.00 atm. To identify the gas, you measure the cylinder volume at several different temperatures, acquiring the data shown at the right. What is the gas?

T (°C)	V (L)
−50	11.6
0	14.0
50	16.2
100	19.4
150	21.8

MODEL The pressure doesn't change, so heating the gas is an isobaric process.

SOLVE The ideal-gas law is $pV = nRT$. Writing this as

$$V = \frac{nR}{p} T$$

we see that a graph of V versus T should be a straight line passing through the origin. Further, we can use the slope of the graph, nR/p, to measure the number of moles of gas, and from that we can identify the gas by determining its molar mass.

FIGURE 18.13 on the next page is a graph of the data, with the volumes and temperatures converted to SI units of m^3 ($1 \, m^3 = 1000 \, L$) and kelvins. The y-intercept of the graph is

Continued

FIGURE 18.13 A graph of the gas volume versus its temperature.

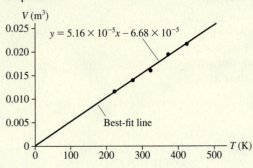

essentially zero, confirming the behavior of the gas as ideal, and the slope of the best-fit line is $5.16 \times 10^{-5} \, \text{m}^3/\text{K}$. The number of moles of gas is

$$n = \frac{p}{R} \times \text{slope} = \frac{2 \times 101{,}300 \, \text{Pa}}{8.31 \, \text{J/mol K}} \times 5.16 \times 10^{-5} \, \text{m}^3/\text{K} = 1.26 \, \text{mol}$$

From this, the molar mass is

$$M = \frac{0.050 \, \text{kg}}{1.26 \, \text{mol}} = 0.040 \, \text{kg/mol}$$

Thus the atomic mass is 40 u, identifying the gas as argon.

ASSESS The atomic mass is that of a well-known gas, which gives us confidence in the result.

STOP TO THINK 18.7 Two cylinders of equal diameter contain the same number of moles of the same ideal gas. Each cylinder is sealed by a frictionless piston. To have the same pressure in both cylinders, which piston would you use in cylinder 2?

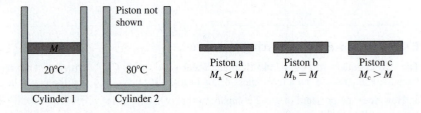

FIGURE 18.14 A constant-temperature (isothermal) process.

(a)

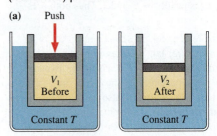

(b)

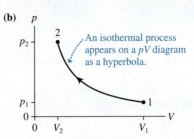

(c)

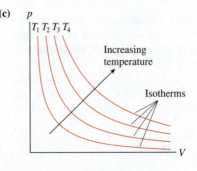

Constant-Temperature Process

The last process we wish to look at for now is one that takes place at a constant temperature. A constant-temperature process is called an **isothermal process.** An isothermal process is one for which $T_f = T_i$. Because $pV = nRT$, a constant-temperature process in a closed system (constant n) is one for which the product pV doesn't change. Thus

$$p_f V_f = p_i V_i \tag{18.22}$$

in an isothermal process.

One possible isothermal process is illustrated in **FIGURE 18.14a**, where a piston is being pushed down to compress a gas. If the piston is pushed *slowly,* then heat energy transfer through the walls of the cylinder keeps the gas at the same temperature as the surrounding liquid. This is an *isothermal compression.* The reverse process, with the piston slowly pulled out, would be an *isothermal expansion.*

Representing an isothermal process on the pV diagram is a little more complicated than the two previous processes because both p and V change. As long as T remains fixed, we have the relationship

$$p = \frac{nRT}{V} = \frac{\text{constant}}{V} \tag{18.23}$$

The inverse relationship between p and V causes the graph of an isothermal process to be a *hyperbola.* As one state variable goes up, the other goes down.

The process shown as $1 \rightarrow 2$ in **FIGURE 18.14b** represents the *isothermal compression* shown in Figure 18.14a. An *isothermal expansion* would move in the opposite direction along the hyperbola.

The location of the hyperbola depends on the value of T. A lower-temperature process is represented by a hyperbola closer to the origin than a higher-temperature process. **FIGURE 18.14c** shows four hyperbolas representing the temperatures T_1 to T_4, where $T_4 > T_3 > T_2 > T_1$. These are called **isotherms.** A gas undergoing an isothermal process moves along the isotherm of the appropriate temperature.

EXAMPLE 18.10 | **Compressing air in the lungs**

An ocean snorkeler takes a deep breath at the surface, filling his lungs with 4.0 L of air. He then descends to a depth of 5.0 m. At this depth, what is the volume of air in the snorkeler's lungs?

MODEL At the surface, the pressure in the lungs is 1.00 atm. Because the body cannot sustain large pressure differences between inside and outside, the air pressure in the lungs rises—and the volume decreases—to match the surrounding water pressure as he descends.

SOLVE The ideal-gas law for a sealed container is

$$V_2 = \frac{p_1}{p_2}\frac{T_2}{T_1}V_1$$

Air is quickly warmed to body temperature as it enters through the nose and mouth, and it remains at body temperature as the snorkeler dives, so $T_2/T_1 = 1$. We know $p_1 = 1.00$ atm $= 101,300$ Pa at the surface. We can find p_2 from the hydrostatic pressure equation, using the density of seawater:

$$p_2 = p_1 + \rho g d = 101,300 \text{ Pa} + (1030 \text{ kg/m}^3)(9.80 \text{ m/s}^2)(5.0 \text{ m})$$
$$= 151,800 \text{ Pa}$$

With this, the volume of the lungs at a depth of 5.0 m is

$$V_2 = \frac{101,300 \text{ Pa}}{151,800 \text{ Pa}} \times 1 \times 4.0 \text{ L} = 2.7 \text{ L}$$

ASSESS The air inside your lungs does compress—significantly—as you dive below the surface.

EXAMPLE 18.11 | **A multistep process**

A gas at 2.0 atm pressure and a temperature of 200°C is first expanded isothermally until its volume has doubled. It then undergoes an isobaric compression until it returns to its original volume. First show this process on a pV diagram. Then find the final temperature (in °C) and pressure.

MODEL The final state of the isothermal expansion is the initial state for an isobaric compression.

VISUALIZE **FIGURE 18.15** shows the process. As the gas expands isothermally, it moves downward along an isotherm until it reaches volume $V_2 = 2V_1$. The gas is then compressed at constant pressure p_2 until its final volume V_3 equals its original volume V_1. State 3 is on an isotherm closer to the origin, so we expect to find $T_3 < T_1$.

SOLVE $T_2/T_1 = 1$ during the isothermal expansion and $V_2 = 2V_1$, so the pressure at point 2 is

$$p_2 = p_1 \frac{T_2}{T_1}\frac{V_1}{V_2} = p_1 \frac{V_1}{2V_1} = \tfrac{1}{2}p_1 = 1.0 \text{ atm}$$

We have $p_3/p_2 = 1$ during the isobaric compression and $V_3 = V_1 = \tfrac{1}{2}V_2$, so

FIGURE 18.15 A pV diagram for the process of Example 18.11.

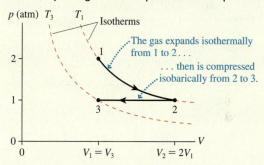

$$T_3 = T_2 \frac{p_3}{p_2}\frac{V_3}{V_2} = T_2 \frac{\tfrac{1}{2}V_2}{V_2} = \tfrac{1}{2}T_2 = 236.5 \text{ K} = -36.5°C$$

where we converted T_2 to 473 K before doing calculations and then converted T_3 back to °C. The final state, with $T_3 = -36.5°C$ and $p_3 = 1.0$ atm, is one in which both the pressure and the absolute temperature are half their original values.

STOP TO THINK 18.8 What is the ratio T_2/T_1 for this process?

a. $\frac{1}{4}$

b. $\frac{1}{2}$

c. 1 (no change)

d. 2

e. 4

f. There's not enough information to tell.

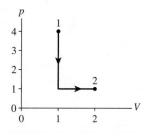

CHALLENGE EXAMPLE 18.12 | Depressing a piston

A large, 50.0-cm-diameter metal cylinder filled with air supports a 20.0 kg piston that can slide up and down without friction. The piston is 100.0 cm above the bottom when the temperature is 20°C. An 80.0 kg student then stands on the piston. After several minutes have elapsed, by how much has the piston been depressed?

MODEL The metal walls of the cylinder are a good thermal conductor, so after several minutes the gas temperature—even if it initially changed—will return to room temperature. The final temperature matches the initial temperature. Assume that the atmospheric pressure is 1 atm.

VISUALIZE FIGURE 18.16 shows the cylinder before and after the student stands on it. The volume of the cylinder is $V = Ah$, and only h changes.

SOLVE The ideal-gas law for a sealed container is

$$\frac{p_2 A h_2}{T_2} = \frac{p_1 A h_1}{T_1}$$

Because $T_2 = T_1$, the final height of the piston is

$$h_2 = \frac{p_1}{p_2} h_1$$

The pressure of the gas is determined by the mass of the piston (and anything on the piston) and the pressure of the air above. In equilibrium,

$$p = p_{atmos} + \frac{Mg}{A} = \begin{cases} 1.023 \times 10^5 \text{ Pa} & \text{piston only} \\ 1.063 \times 10^5 \text{ Pa} & \text{piston and student} \end{cases}$$

where we used $p_{atmos} = 1$ atm $= 1.013 \times 10^5$ Pa and $A = \pi r^2 = 0.196$ m². The final height of the piston is

FIGURE 18.16 The student compresses the gas.

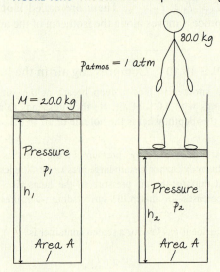

$$h_2 = \frac{1.023 \times 10^5 \text{ Pa}}{1.063 \times 10^5 \text{ Pa}} \times 100.0 \text{ cm} = 96.2 \text{ cm}$$

The question, however, was by how much the piston is depressed. This is $h_1 - h_2 = 3.8$ cm.

ASSESS Neither the piston nor the student increases the gas pressure to much above 1 atm, so it's not surprising that the added weight of the student doesn't push the piston down very far.

SUMMARY

The goal of Chapter 18 has been to learn some of the characteristics of macroscopic systems.

GENERAL PRINCIPLES

Three Common Phases of Matter

Solid   Rigid, definite shape. Nearly incompressible.

Liquid   Molecules loosely held together by molecular bonds, but able to move around. Nearly incompressible.

Gas   Molecules moving freely through space. Compressible.

The different phases exist for different conditions of temperature T and pressure p. The boundaries separating the regions of a **phase diagram** are lines of phase equilibrium. Any amounts of the two phases can coexist in equilibrium. The **triple point** is the one value of temperature and pressure at which all three phases can coexist in equilibrium.

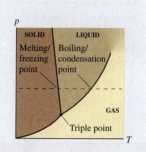

IMPORTANT CONCEPTS

Ideal-Gas Model

- Atoms and molecules are small, hard spheres that travel freely through space except for occasional collisions with each other or the walls.
- The model is valid when the density is low and the temperature well above the condensation point.

Ideal-Gas Law

The **state variables** of an ideal gas are related by the ideal-gas law

$$pV = nRT \quad \text{or} \quad pV = Nk_BT$$

where $R = 8.31$ J/mol K is the universal gas constant and $k_B = 1.38 \times 10^{-23}$ J/K is Boltzmann's constant. p, V, and T *must* be in SI units of Pa, m³, and K.

For a gas in a sealed container, with constant n:

$$\frac{p_2V_2}{T_2} = \frac{p_1V_1}{T_1}$$

Counting atoms and moles

A macroscopic sample of matter consists of N atoms (or molecules), each of mass m (the **atomic** or **molecular mass**):

$$N = \frac{M}{m}$$

Alternatively, we can state that the sample consists of n **moles**:

$$n = \frac{N}{N_A} \quad \text{or} \quad \frac{M}{M_{mol}}$$

where. $N_A = 6.02 \times 10^{23}$ mol^{-1} is **Avogadro's number.**

The molar mass M_{mol}, in kg/mol, is the numerical value of the atomic or molecular mass in u divided by 1000. The atomic or molecular mass, in atomic mass units u, is well approximated by the **atomic mass number** A. The atomic mass unit is

$$1 \text{ u} = 1.66 \times 10^{-27} \text{ kg}$$

The **number density** of the sample is $\frac{N}{V}$.

APPLICATIONS

Temperature scales

The Kelvin scale has absolute zero at $T_0 = 0$ K and the triple point of water at $T_3 = 273.16$ K.

$$T_K = T_C + 273$$

Thermal expansion

For a temperature change ΔT,

$$\Delta L/L = \alpha \Delta T \qquad \Delta V/V = \beta \Delta T$$

For a solid, $\beta = 3\alpha$.

Three basic gas processes

Ideal-gas processes are shown as trajectories through the pV diagram.

1. **Isochoric,** or constant volume
2. **Isobaric,** or constant pressure
3. **Isothermal,** or constant temperature

pV diagram

TERMS AND NOTATION

micro/macro connection	atomic mass unit, u	thermal expansion	ideal gas
phase	molecular mass	coefficient of linear expansion, α	ideal-gas law
phase change	mole, n	coefficient of volume expansion, β	universal gas constant, R
solid	monatomic gas	melting point	Boltzmann's constant, k_B
crystal	diatomic gas	freezing point	STP
liquid	Avogadro's number, N_A	phase equilibrium	ideal-gas process
gas	molar mass, M_{mol}	condensation point	pV diagram
state variable	temperature, T	boiling point	quasi-static process
thermal equilibrium	constant-volume gas	phase diagram	isochoric process
number density, N/V	thermometer	sublimation	isobaric process
atomic mass number, A	absolute zero, T_0	critical point	isothermal process
atomic mass	absolute temperature scale	triple point	isotherm

CONCEPTUAL QUESTIONS

1. Rank in order, from highest to lowest, the temperatures $T_1 = 0$ K, $T_2 = 0°C$, and $T_3 = 0°F$.

2. The sample in an experiment is initially at 10°C. If the sample's temperature is doubled, what is the new temperature in °C?

3. a. Is there a highest temperature at which ice can exist? If so, what is it? If not, why not?
   b. Is there a lowest temperature at which water vapor can exist? If so, what is it? If not, why not?

4. The cylinder in FIGURE Q18.4 is divided into two compartments by a frictionless piston that can slide back and forth. If the piston is in equilibrium, is the pressure on the left side greater than, less than, or equal to the pressure on the right? Explain.

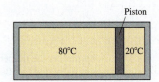

FIGURE Q18.4

5. A gas is in a sealed container. By what factor does the gas temperature change if:
   a. The volume is doubled and the pressure is tripled?
   b. The volume is halved and the pressure is tripled?

6. A gas is in a sealed container. The gas pressure is tripled and the temperature is doubled.
   a. Does the number of moles of gas in the container increase, decrease, or stay the same?
   b. By what factor does the number density of the gas increase?

7. An aquanaut lives in an underwater apartment 100 m beneath the surface of the ocean. Compare the freezing and boiling points of water in the aquanaut's apartment to their values at the surface. Are they higher, lower, or the same? Explain.

8. a. A sample of water vapor in an enclosed cylinder has an initial pressure of 500 Pa at an initial temperature of −0.01°C. A piston squeezes the sample smaller and smaller, without limit. Describe what happens to the water as the squeezing progresses.
   b. Repeat part a if the initial temperature is 0.03°C warmer.

9. A gas is in a sealed container. By what factor does the gas pressure change if:
   a. The volume is doubled and the temperature is tripled?
   b. The volume is halved and the temperature is tripled?

10. A gas undergoes the process shown in FIGURE Q18.10. By what factor does the temperature change?

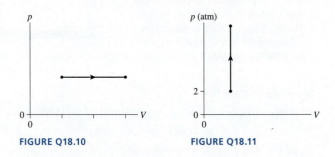

FIGURE Q18.10   **FIGURE Q18.11**

11. The temperature increases from 300 K to 1200 K as a gas undergoes the process shown in FIGURE Q18.11. What is the final pressure?

12. A student is asked to sketch a pV diagram for a gas that goes through a cycle consisting of (a) an isobaric expansion, (b) a constant-volume reduction in temperature, and (c) an isothermal process that returns the gas to its initial state. The student draws the diagram shown in FIGURE Q18.12. What, if anything, is wrong with the student's diagram?

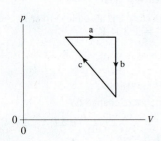

FIGURE Q18.12

EXERCISES AND PROBLEMS

Problems labeled [ ] integrate material from earlier chapters.

Exercises

Section 18.1 Solids, Liquids, and Gases

1. ‖ What volume of water has the same mass as 100 cm^3 of gold?
2. ‖ The nucleus of a uranium atom has a diameter of 1.5×10^{-14} m and a mass of 4.0×10^{-25} kg. What is the density of the nucleus?
3. ‖ What is the diameter of a copper sphere that has the same mass as a 10 cm $\times$ 10 cm $\times$ 10 cm cube of aluminum?
4. ‖ A hollow aluminum sphere with outer diameter 10.0 cm has a mass of 690 g. What is the sphere's inner diameter?

Section 18.2 Atoms and Moles

5. ‖ How many moles are in a 2.0 cm $\times$ 2.0 cm $\times$ 2.0 cm cube of copper?
6. ‖ How many atoms are in a 2.0 cm $\times$ 2.0 cm $\times$ 2.0 cm cube of aluminum?
7. ‖ What is the number density of (a) aluminum and (b) lead?
8. ‖ An element in its solid phase has mass density 1750 kg/m^3 and number density 4.39×10^{28} atoms/m^3. What is the element's atomic mass number?
9. ‖ 1.0 mol of gold is shaped into a sphere. What is the sphere's diameter?
10. ‖ What volume of aluminum has the same number of atoms as 10 cm^3 of mercury?

Section 18.3 Temperature

11. ‖ The lowest and highest natural temperatures ever recorded on earth are $-129°$F in Antarctica and $134°$F in Death Valley. What are these temperatures in °C and in K?
12. ‖ At what temperature does the numerical value in °F match the numerical value in °C?
13. ‖ A demented scientist creates a new temperature scale, the "Z scale." He decides to call the boiling point of nitrogen 0°Z and the melting point of iron 1000°Z.
    a. What is the boiling point of water on the Z scale?
    b. Convert 500°Z to degrees Celsius and to kelvins.

Section 18.4 Thermal Expansion

Section 18.5 Phase Changes

14. ‖ A concrete bridge is built of 325-cm-long concrete slabs with an expansion joint between them. The slabs just touch on a 115°F day, the hottest day for which the bridge is designed. What is the gap between the slabs when the temperature is 0°F?
15. ‖ A surveyor has a steel measuring tape that is calibrated to be 100.000 m long (i.e., accurate to ± 1 mm) at 20°C. If she measures the distance between two stakes to be 65.175 m on a 3°C day, does she need to add or subtract a correction factor to get the true distance? How large, in mm, is the correction factor?
16. ‖ Two students each build a piece of scientific equipment that uses a 655-mm-long metal rod. One student uses a brass rod, the other an invar rod. If the temperature increases by 5.0°C, how much more does the brass rod expand than the invar rod?

17. ‖ A 60.00 L fuel tank is filled with gasoline on a $-10°$C day, then rolled into a storage shed where the temperature is 20°C. If the tank is not vented, what minimum volume needs to be left empty at filling time so that the tank doesn't rupture as it warms?
18. ‖ What is the temperature in °F and the pressure in Pa at the triple point of (a) water and (b) carbon dioxide?

Section 18.6 Ideal Gases

19. ‖ A cylinder contains nitrogen gas. A piston compresses the gas to half its initial volume. Afterward,
    a. Has the mass density of the gas changed? If so, by what factor? If not, why not?
    b. Has the number of moles of gas changed? If so, by what factor? If not, why not?
20. ‖ 3.0 mol of gas at a temperature of $-120°$C fills a 2.0 L container. What is the gas pressure?
21. ‖ A rigid container holds 2.0 mol of gas at a pressure of 1.0 atm and a temperature of 30°C.
    a. What is the container's volume?
    b. What is the pressure if the temperature is raised to 130°C?
22. ‖ A gas at 100°C fills volume V_0. If the pressure is held constant, what is the volume if (a) the Celsius temperature is doubled and (b) the Kelvin temperature is doubled?
23. ‖ The total lung capacity of a typical adult is 5.0 L. Approximately
    BIO 20% of the air is oxygen. At sea level and at a body temperature of 37°C, how many oxygen molecules do the lungs contain at the end of a strong inhalation?
24. ‖ A 20-cm-diameter cylinder that is 40 cm long contains 50 g of oxygen gas at 20°C.
    a. How many moles of oxygen are in the cylinder?
    b. How many oxygen molecules are in the cylinder?
    c. What is the number density of the oxygen?
    d. What is the reading of a pressure gauge attached to the tank?
25. ‖ The solar corona is a very hot atmosphere surrounding the visible surface of the sun. X-ray emissions from the corona show that its temperature is about 2×10^6 K. The gas pressure in the corona is about 0.03 Pa. Estimate the number density of particles in the solar corona.
26. ‖ A gas at temperature T_0 and atmospheric pressure fills a cylinder. The gas is transferred to a new cylinder with three times the volume, after which the pressure is half the original pressure. What is the new temperature of the gas?

Section 18.7 Ideal-Gas Processes

27. ‖ A gas with initial state variables p_1, V_1, and T_1 expands isothermally until $V_2 = 2V_1$. What are (a) T_2 and (b) p_2?
28. ‖ A gas with initial state variables p_1, V_1, and T_1 is cooled in an isochoric process until $p_2 = \frac{1}{3}p_1$. What are (a) V_2 and (b) T_2?
29. ‖ A rigid, hollow sphere is submerged in boiling water in a room where the air pressure is 1.0 atm. The sphere has an open valve with its inlet just above the water level. After a long period of time has elapsed, the valve is closed. What will be the pressure inside the sphere if it is then placed in (a) a mixture of ice and water and (b) an insulated box filled with dry ice?
30. ‖ A rigid container holds hydrogen gas at a pressure of 3.0 atm and a temperature of 20°C. What will the pressure be if the temperature is lowered to $-20°$C?

31. ‖ A 24-cm-diameter vertical cylinder is sealed at the top by a frictionless 20 kg piston. The piston is 84 cm above the bottom when the gas temperature is 303°C. The air above the piston is at 1.00 atm pressure.
   a. What is the gas pressure inside the cylinder?
   b. What will the height of the piston be if the temperature is lowered to 15°C?

32. ‖ 0.10 mol of argon gas is admitted to an evacuated 50 cm³ container at 20°C. The gas then undergoes an isochoric heating to a temperature of 300°C.
   a. What is the final pressure of the gas?
   b. Show the process on a pV diagram. Include a proper scale on both axes.

33. ‖ 0.10 mol of argon gas is admitted to an evacuated 50 cm³ container at 20°C. The gas then undergoes an isobaric heating to a temperature of 300°C.
   a. What is the final volume of the gas?
   b. Show the process on a pV diagram. Include a proper scale on both axes.

34. ‖ 0.10 mol of argon gas is admitted to an evacuated 50 cm³ container at 20°C. The gas then undergoes an isothermal expansion to a volume of 200 cm³.
   a. What is the final pressure of the gas?
   b. Show the process on a pV diagram. Include a proper scale on both axes.

35. ‖ 0.0040 mol of gas undergoes the process shown in FIGURE EX18.35.
   a. What type of process is this?
   b. What are the initial and final temperatures in °C?

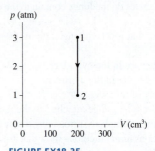

FIGURE EX18.35

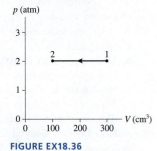

FIGURE EX18.36

36. ‖ A gas with an initial temperature of 900°C undergoes the process shown in FIGURE EX18.36.
   a. What type of process is this?
   b. What is the final temperature in °C?
   c. How many moles of gas are there?

37. ‖ 0.020 mol of gas undergoes the process shown in FIGURE EX18.37.
   a. What type of process is this?
   b. What is the final temperature in °C?
   c. What is the final volume V_2?

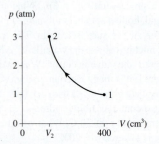

FIGURE EX18.37

38. ‖ 0.0050 mol of gas undergoes the process $1 \rightarrow 2 \rightarrow 3$ shown in FIGURE P16.38. What are (a) temperature T_1, (b) pressure p_2, and (c) volume V_3?

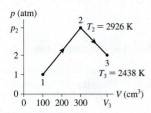

FIGURE P16.38

39. ‖ An ideal gas starts with pressure p_1 and volume V_1. Draw a pV diagram showing the process in which the gas undergoes an isochoric process that doubles the pressure, then an isobaric process that doubles the volume, followed by an isothermal process that doubles the volume again. Label each of the three processes.

40. ‖ An ideal gas starts with pressure p_1 and volume V_1. Draw a pV diagram showing the process in which the gas undergoes an isothermal process during which the volume is halved, then an isochoric process during which the pressure is halved, followed by an isobaric process during which the volume is doubled. Label each of the three processes.

Problems

41. ‖ The atomic mass number of copper is $A = 64$. Assume that atoms in solid copper form a cubic crystal lattice. To envision this, imagine that you place atoms at the centers of tiny sugar cubes, then stack the little sugar cubes to form a big cube. If you dissolve the sugar, the atoms left behind are in a cubic crystal lattice. What is the smallest distance between two copper atoms?

42. ‖ The molecular mass of water (H_2O) is $A = 18$. How many protons are there in 1.0 L of liquid water?

43. ‖ A brass ring with inner diameter 2.00 cm and outer diameter 3.00 cm needs to fit over a 2.00-cm-diameter steel rod, but at 20°C the hole through the brass ring is 50 μm too small in diameter. To what temperature, in °C, must the rod *and* ring be heated so that the ring just barely slips over the rod?

44. ‖ A 15°C, 2.0-cm-diameter aluminum bar just barely slips between two rigid steel walls 10.0 cm apart. If the bar is warmed to 25°C, how much force does it exert on each wall?

45. ‖ The semiconductor industry manufactures integrated circuits in large vacuum chambers where the pressure is 1.0×10^{-10} mm of Hg.
   a. What fraction is this of atmospheric pressure?
   b. At $T = 20$°C, how many molecules are in a cylindrical chamber 40 cm in diameter and 30 cm tall?

46. ‖ A 6.0-cm-diameter, 10-cm-long cylinder contains 100 mg of oxygen (O_2) at a pressure less than 1 atm. The cap on one end of the cylinder is held in place only by the pressure of the air. One day when the atmospheric pressure is 100 kPa, it takes a 184 N force to pull the cap off. What is the temperature of the gas?

47. ‖ A nebula—a region of the galaxy where new stars are forming—contains a very tenuous gas with 100 atoms/cm³. This gas is heated to 7500 K by ultraviolet radiation from nearby stars. What is the gas pressure in atm?

48. ‖ On average, each person in the industrialized world is responsible for the emission of 10,000 kg of carbon dioxide (CO_2) every year. This includes CO_2 that you generate directly, by burning fossil fuels to operate your car or your furnace, as well as CO_2 generated on your behalf by electric generating stations and manufacturing plants. CO_2 is a greenhouse gas that contributes to global warming. If you were to store your yearly CO_2 emissions in a cube at STP, how long would each edge of the cube be?

49. ‖ To determine the mass of neon contained in a rigid, 2.0 L cylinder, you vary the cylinder's temperature while recording the reading of a pressure gauge. Your data are as follows:

Temperature (°C)	Pressure gauge (atm)
100	6.52
150	7.80
200	8.83
250	9.59

Use the best-fit line of an appropriate graph to determine the mass of the neon.

50. ‖ The 3.0-m-long pipe in FIGURE P18.50 is closed at the top end. It is slowly pushed straight down into the water until the top end of the pipe is level with the water's surface. What is the length L of the trapped volume of air?

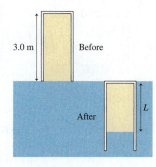

FIGURE P18.50

51. ‖ A diving bell is a 3.0-m-tall cylinder closed at the upper end but open at the lower end. The temperature of the air in the bell is 20°C. The bell is lowered into the ocean until its lower end is 100 m deep. The temperature at that depth is 10°C.
a. How high does the water rise in the bell after enough time has passed for the air inside to reach thermal equilibrium?
b. A compressed-air hose from the surface is used to expel all the water from the bell. What minimum air pressure is needed to do this?

52. ‖ An electric generating plant boils water to produce high-pressure steam. The steam spins a turbine that is connected to the generator.
a. How many liters of water must be boiled to fill a 5.0 m³ boiler with 50 atm of steam at 400°C?
b. The steam has dropped to 2.0 atm pressure at 150°C as it exits the turbine. How much volume does it now occupy?

53. ‖ On a cool morning, when the temperature is 15°C, you measure the pressure in your car tires to be 30 psi. After driving 20 mi on the freeway, the temperature of your tires is 45°C. What pressure will your tire gauge now show?

54. ‖ The air temperature and pressure in a laboratory are 20°C and 1.0 atm. A 1.0 L container is open to the air. The container is then sealed and placed in a bath of boiling water. After reaching thermal equilibrium, the container is opened. How many moles of air escape?

55. ‖ 10,000 cm³ of 200°C steam at a pressure of 20 atm is cooled until it condenses. What is the volume of the liquid water? Give your answer in cm³.

56. ‖ The mercury manometer shown in FIGURE P18.56 is attached to a gas cell. The mercury height h is 120 mm when the cell is placed in an ice-water mixture. The mercury height drops to 30 mm when the device is carried into an industrial freezer. What is the freezer temperature?
Hint: The right tube of the manometer is much narrower than the left tube. What reasonable assumption can you make about the gas volume?

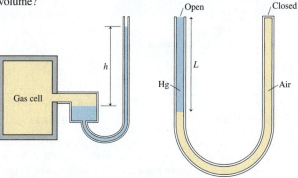

FIGURE P18.56   **FIGURE P18.57**

57. ‖ The U-shaped tube in FIGURE P18.57 has a total length of 1.0 m. It is open at one end, closed at the other, and is initially filled with air at 20°C and 1.0 atm pressure. Mercury is poured slowly into the open end without letting any air escape, thus compressing the air. This is continued until the open side of the tube is completely filled with mercury. What is the length L of the column of mercury?

58. ‖ The 50 kg circular piston shown in FIGURE P18.58 floats on 0.12 mol of compressed air.
a. What is the piston height h if the temperature is 30°C?
b. How far does the piston move if the temperature is increased by 100°C?

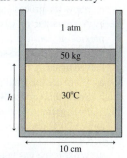

FIGURE P18.58

59. ‖ A diver 50 m deep in 10°C fresh water exhales a 1.0-cm-diameter bubble. What is the bubble's diameter just as it reaches the surface of the lake, where the water temperature is 20°C?
Hint: Assume that the air bubble is always in thermal equilibrium with the surrounding water.

60. | 8.0 g of helium gas follows the process $1 \rightarrow 2 \rightarrow 3$ shown in FIGURE P18.60. Find the values of V_1, V_3, p_2, and T_3.

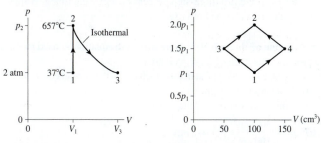

FIGURE P18.60   **FIGURE P18.61**

61. ‖ FIGURE P18.61 shows two different processes by which 1.0 g of nitrogen gas moves from state 1 to state 2. The temperature of state 1 is 25°C. What are (a) pressure p_1 and (b) temperatures (in °C) T_2, T_3, and T_4?

62. ‖ **FIGURE P18.62** shows two different processes by which 80 mol of gas move from state 1 to state 2. The dashed line is an isotherm.
    a. What is the temperature of the isothermal process?
    b. What maximum temperature is reached along the straight-line process?

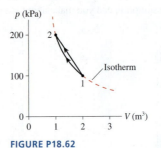

FIGURE P18.62

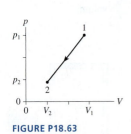

FIGURE P18.63

63. ‖ 4.0 g of oxygen gas, starting at 820°C, follow the process $1 \rightarrow 2$ shown in **FIGURE P18.63**. What is temperature T_2 (in °C)?
64. ‖ 10 g of dry ice (solid CO_2) is placed in a 10,000 cm^3 container, then all the air is quickly pumped out and the container sealed. The container is warmed to 0°C, a temperature at which CO_2 is a gas.
    a. What is the gas pressure? Give your answer in atm.
    The gas then undergoes an isothermal compression until the pressure is 3.0 atm, immediately followed by an isobaric compression until the volume is 1000 cm^3.
    b. What is the final temperature of the gas (in °C)?
    c. Show the process on a pV diagram.
65. ‖ A container of gas at 2.0 atm pressure and 127°C is compressed at constant temperature until the volume is halved. It is then further compressed at constant pressure until the volume is halved again.
    a. What are the final pressure and temperature of the gas?
    b. Show this process on a pV diagram.
66. ‖ Five grams of nitrogen gas at an initial pressure of 3.0 atm and at 20°C undergo an isobaric expansion until the volume has tripled.
    a. What is the gas volume after the expansion?
    b. What is the gas temperature after the expansion (in °C)?
    The gas pressure is then decreased at constant volume until the original temperature is reached.
    c. What is the gas pressure after the decrease?
    Finally, the gas is isothermally compressed until it returns to its initial volume.
    d. What is the final gas pressure?
    e. Show the full three-step process on a pV diagram. Use appropriate scales on both axes.

In Problems 67 through 70 you are given the equation(s) used to solve a problem. For each of these, you are to
    a. Write a realistic problem for which this is the correct equation(s).
    b. Draw a pV diagram.
    c. Finish the solution of the problem.

67. $p_2 = \dfrac{300 \text{ cm}^3}{100 \text{ cm}^3} \times 1 \times 2 \text{ atm}$

68. $(T_2 + 273) \text{ K} = \dfrac{200 \text{ kPa}}{500 \text{ kPa}} \times 1 \times (400 + 273) \text{ K}$

69. $V_2 = \dfrac{(400 + 273) \text{ K}}{(50 + 273) \text{ K}} \times 1 \times 200 \text{ cm}^3$

70. $(2.0 \times 101{,}300 \text{ Pa})(100 \times 10^{-6} \text{ m}^3) = n(8.31 \text{ J/mol K})T_1$

$n = \dfrac{0.12 \text{ g}}{20 \text{ g/mol}}$

$T_2 = \dfrac{200 \text{ cm}^3}{100 \text{ cm}^3} \times 1 \times T_1$

Challenge Problems

71. ⦀ An inflated bicycle inner tube is 2.2 cm in diameter and 200 cm in circumference. A small leak causes the gauge pressure to decrease from 110 psi to 80 psi on a day when the temperature is 20°C. What mass of air is lost? Assume the air is pure nitrogen.
72. ⦀ The cylinder in **FIGURE CP18.72** has a moveable piston attached to a spring. The cylinder's cross-section area is 10 cm^2, it contains 0.0040 mol of gas, and the spring constant is 1500 N/m. At 20°C the spring is neither compressed nor stretched. How far is the spring compressed if the gas temperature is raised to 100°C?

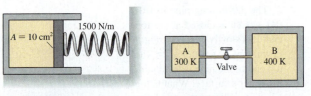

FIGURE CP18.72   **FIGURE CP18.73**

73. ⦀ Containers A and B in **FIGURE CP18.73** hold the same gas. The volume of B is four times the volume of A. The two containers are connected by a thin tube (negligible volume) and a valve that is closed. The gas in A is at 300 K and pressure of 1.0×10^5 Pa. The gas in B is at 400 K and pressure of 5.0×10^5 Pa. Heaters will maintain the temperatures of A and B even after the valve is opened. After the valve is opened, gas will flow one way or the other until A and B have equal pressure. What is this final pressure?
74. ⦀ The closed cylinder of **FIGURE CP18.74** has a tight-fitting but frictionless piston of mass M. The piston is in equilibrium when the left chamber has pressure p_0 and length L_0 while the spring on the right is compressed by ΔL.
    a. What is ΔL in terms of p_0, L_0, A, M, and k?
    b. Suppose the piston is moved a small distance x to the right. Find an expression for the net force $(F_x)_{net}$ on the piston. Assume all motions are slow enough for the gas to remain at the same temperature as its surroundings.
    c. If released, the piston will oscillate around the equilibrium position. Assuming $x \ll L_0$ find an expression for the oscillation period T.
    Hint: Use the binomial approximation.

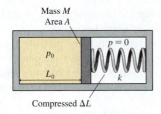

FIGURE CP18.74   Compressed ΔL

19 Work, Heat, and the First Law of Thermodynamics

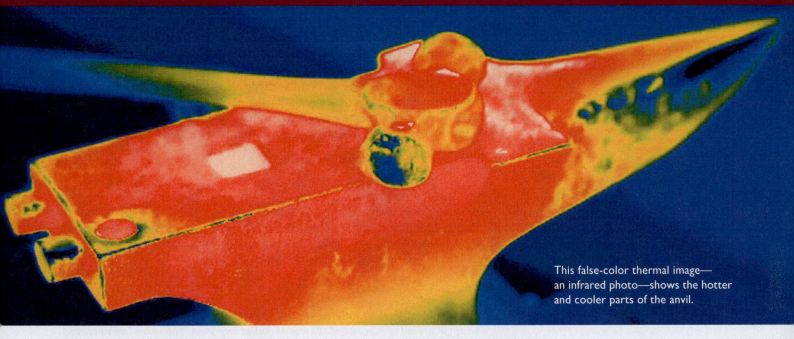

This false-color thermal image—an infrared photo—shows the hotter and cooler parts of the anvil.

IN THIS CHAPTER, you will learn and apply the first law of thermodynamics.

What is the first law of thermodynamics?

The first law of thermodynamics is a very general statement about energy and its conservation. A system's thermal energy changes if energy is transferred into or out of the system as *work* or as *heat*. That is,

$$\Delta E_{th} = W + Q$$

« LOOKING BACK Sections 10.4 and 10.8 Conservation of energy

How is work done on a gas?

Work W is the transfer of energy in a mechanical interaction when forces push or pull. Work is done on a gas by changing its volume.

■ $W > 0$ (energy added) in a compression.
■ $W < 0$ (energy extracted) in an expansion.

« LOOKING BACK Sections 9.2–9.3 Work

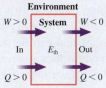

What is heat?

Heat Q is the transfer of energy in a thermal interaction when the system and its environment are at different temperatures.

■ $Q > 0$ (energy added) when the environment is hotter than the system.
■ $Q < 0$ (energy extracted) when the system is hotter than the environment.

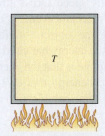

How is heat transferred?

Heat energy can be transferred between a system and its environment by

■ Conduction
■ Convection
■ Radiation
■ Evaporation

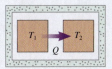

What are some thermal properties of matter?

Heat can cause either a temperature change or a phase change. A material's response to heat is governed by its specific heat, its heat of fusion, its heat of vaporization, and its thermal conductivity. These are thermal properties of matter. An important application is calorimetry—determining the final temperature when two or more systems interact thermally.

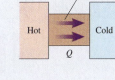

Why is the first law important?

The first law of thermodynamics is a very general statement that energy cannot be created or destroyed—merely moved from one place to another. Much of the machinery of modern society, from automobile engines to electric power plants to spacecraft, depends on and is an application of the first law. We'll look at some of these applications in more detail in Chapter 21.

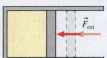

19.1 It's All About Energy

In « Section 10.8 we wrote the *energy principle* as

$$\Delta K = W_c + W_{diss} + W_{ext} \tag{19.1}$$

Recall that energy is about *systems* and their interactions. Equation 19.1 tells us that the kinetic energy of a system of particles is changed when forces do work on the particles by pushing or pulling them through a distance. Here

1. W_c is the work done by conservative forces. This work can be represented as a change in the system's potential energy: $\Delta U = -W_c$.
2. W_{diss} is the work done by friction-like dissipative forces within the system. This work increases the system's thermal energy: $\Delta E_{th} = -W_{diss}$.
3. W_{ext} is the work done by external forces that originate in the environment. The push of a piston rod would be an external force.

With these definitions, Equation 19.1 becomes

$$\Delta K + \Delta U + \Delta E_{th} = W_{ext} \tag{19.2}$$

The system's *mechanical energy* was defined as $E_{mech} = K + U$. **FIGURE 19.1** reminds you that the mechanical energy is the *macroscopic* energy of the system as a whole, while E_{th} is the *microscopic* energy of the atoms and molecules within the system. This led to our final energy statement of Chapter 10:

$$\Delta E_{sys} = \Delta E_{mech} + \Delta E_{th} = W_{ext} \tag{19.3}$$

Thus the total energy of an *isolated system,* for which $W_{ext} = 0$, is constant.

The emphasis in Chapters 9 and 10 was on isolated systems. There we were interested in learning how kinetic and potential energy were *transformed* into each other and, where there is friction, into thermal energy. Now we want to focus on how energy is *transferred* between the system and its environment when W_{ext} is *not* zero.

> **NOTE** Strictly speaking, Equation 19.3 should use the *internal energy* E_{int} rather than the thermal energy E_{th}, where $E_{int} = E_{th} + E_{chem} + E_{nuc} + \cdots$ includes all the various kinds of energies that can be stored inside a system. This textbook will focus on simple thermodynamics systems in which the internal energy is entirely thermal: $E_{int} = E_{th}$. We'll leave other forms of internal energy to more advanced courses.

Energy Transfer

Doing work on a system can have very different consequences. **FIGURE 19.2a** shows an object being lifted at steady speed by a rope. The rope's tension is an external force doing work W_{ext} on the system. In this case, the energy transferred into the system goes entirely to increasing the system's macroscopic potential energy U_{grav}, part of the mechanical energy. The energy-transfer process $W_{ext} \rightarrow E_{mech}$ is shown graphically in the energy bar chart of Figure 19.2a.

Contrast this with **FIGURE 19.2b**, where the same rope with the same tension now drags the object at steady speed across a rough surface. The tension does the same amount of work, but the mechanical energy does not change. Instead, friction increases the thermal energy of the object + surface system. The energy-transfer process $W_{ext} \rightarrow E_{th}$ is shown in the energy bar chart of Figure 19.2b.

The point of this example is that the energy transferred to a system can go entirely to the system's mechanical energy, entirely to its thermal energy, or (imagine dragging the object up an incline) some combination of the two. The energy isn't lost, but where it ends up depends on the circumstances.

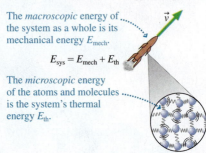

FIGURE 19.1 The total energy of a system.

The *macroscopic* energy of the system as a whole is its mechanical energy E_{mech}.

$$E_{sys} = E_{mech} + E_{th}$$

The *microscopic* energy of the atoms and molecules is the system's thermal energy E_{th}.

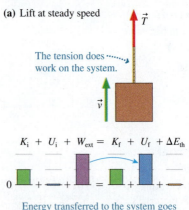

FIGURE 19.2 The work done by tension can have very different consequences.

(a) Lift at steady speed

The tension does work on the system.

$$K_i + U_i + W_{ext} = K_f + U_f + \Delta E_{th}$$

Energy transferred to the system goes entirely to mechanical energy.

(b) Drag at steady speed

The tension does work on the system.

$$K_i + U_i + W_{ext} = K_f + U_f + \Delta E_{th}$$

Energy transferred to the system goes entirely to thermal energy.

The Missing Piece: Heat

You can transfer energy into a system by the mechanical process of doing work on the system. But that can't be all there is to energy transfer. What happens when you place a pan of water on the stove and light the burner? The water temperature increases, so $\Delta E_{th} > 0$. But no work is done ($W_{ext} = 0$) and there is no change in the water's mechanical energy ($\Delta E_{mech} = 0$). This process clearly violates the energy equation $\Delta E_{mech} + \Delta E_{th} = W_{ext}$. What's wrong?

Nothing is wrong. The energy equation is correct as far as it goes, but it is incomplete. Work is energy transferred in a mechanical interaction, but that is not the only way a system can interact with its environment. Energy can also be transferred between the system and the environment if they have a *thermal interaction*. The energy transferred in a thermal interaction is called *heat*.

The symbol for heat is Q. When heat is included, the energy equation becomes

$$\Delta E_{sys} = \Delta E_{mech} + \Delta E_{th} = W + Q \qquad (19.4)$$

Heat and work are both energy transferred between the system and the environment.

> **NOTE** We've dropped the subscript "ext" from W. The work that we consider in thermodynamics is *always* the work done by the environment on the system. We won't need to distinguish this work from W_c or W_{diss}, so the subscript is superfluous.

We'll return to Equation 19.4 in Section 19.4 after we look at how work is calculated for ideal-gas processes and at what heat is.

STOP TO THINK 19.1 A gas cylinder and piston are covered with heavy insulation. The piston is pushed into the cylinder, compressing the gas. In this process the gas temperature

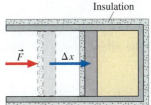

a. Increases.
b. Decreases.
c. Doesn't change.
d. There's not sufficient information to tell.

19.2 Work in Ideal-Gas Processes

We introduced the idea of *work* in Chapter 9. **Work** is the energy transferred between a system and the environment when a net force acts on the system over a distance. The process itself is a **mechanical interaction,** meaning that the system and the environment interact via macroscopic pushes and pulls. Loosely speaking, we say that the environment (or a particular force from the environment) "does work" on the system. A system is in **mechanical equilibrium** if there is no net force on the system.

FIGURE 19.3 on the next page reminds you that work can be either positive or negative. **The sign of the work is *not* just an arbitrary convention, nor does it have anything to do with the choice of coordinate system.** The sign of the work tells us which way energy is being transferred.

In contrast to the mechanical energy or the thermal energy, **work is not a state variable.** That is, work is not a number characterizing the system. Instead, work is the amount of energy that moves between the system and the environment during a mechanical interaction. We can measure the *change* in a state variable, such as a temperature change $\Delta T = T_f - T_i$, but it would make no sense to talk about a "change of work." Consequently, work always appears as W, never as ΔW.

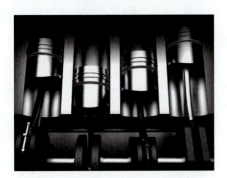

The pistons in a car engine do work on the air-fuel mixture by compressing it.

FIGURE 19.3 The sign of work.

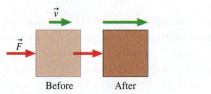
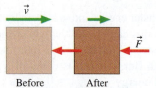

| Before | After | Before | After |

Work is *positive* when the force is in the direction of motion.
- The force causes the object to speed up.
- Energy is transferred from the environment to the system.

Work is *negative* when the force is opposite to the motion.
- The force causes the object to slow down.
- Energy is transferred from the system to the environment.

You learned in **« Section 9.3** how to calculate work. The small amount of work dW done by force $\vec{F}$ as a system moves through the small displacement $d\vec{s}$ is $dW = \vec{F} \cdot d\vec{s}$. If we restrict ourselves to situations where $\vec{F}$ is either parallel or opposite to $d\vec{s}$, then the total work done on the system as it moves from s_i to s_f is

$$W = \int_{s_i}^{s_f} F_s \, ds \tag{19.5}$$

Let's apply this definition to calculate the work done when a gas is expanded or compressed. **FIGURE 19.4a** shows a gas of pressure p whose volume can be changed by a moving piston. The *system* is the gas in the cylinder, which exerts a force $\vec{F}_{gas}$ of magnitude pA on the left side of the piston. To prevent the pressure from blowing the piston out of the cylinder, there must be an equal but opposite external force $\vec{F}_{ext}$ pushing on the right side of the piston. In practice, this would likely be the force exerted by a piston rod. Using the coordinate system of Figure 19.4a, we have

$$(F_{ext})_x = -(F_{gas})_x = -pA \tag{19.6}$$

Suppose the piston moves a small distance dx, as shown in **FIGURE 19.4b**. As it does so, the external force (i.e., the environment) does work

$$dW_{ext} = (F_{ext})_x \, dx = -pA \, dx \tag{19.7}$$

If dx is positive (expanding gas), then dW_{ext} is negative. This is to be expected because the force is opposite the displacement. dW_{ext} is positive if the gas is slightly compressed $(dx < 0)$.

If the piston moves dx, the gas volume changes by $dV = A \, dx$. Thus the work is

$$dW_{ext} = -p \, dV \tag{19.8}$$

If we allow the gas to change in a slow, quasi-static process from an initial volume V_i to a final volume V_f, the total work done by the environment is found by integrating Equation 19.8:

$$W_{ext} = -\int_{V_i}^{V_f} p \, dV \quad \text{(work done on a gas)} \tag{19.9}$$

Equation 19.9 is a key result of thermodynamics. Although we used a cylinder to derive it, it turns out to be true for a container of any shape.

NOTE The pressure of a gas usually changes as the gas expands or contracts. Consequently, p is *not* a constant that can be brought outside the integral. You need to know how the pressure changes with volume before you can carry out the integration.

What is this work done *on*? Not, as you might think, the piston. Work is an energy transfer, but the piston's energy is not changing. It is in equilibrium, with equal forces exerted on both faces. Further, the motion of the piston is very slow—it's a quasi-static process—so the piston has negligible kinetic energy. The piston is merely a movable boundary of the gas, so **the work is done by the environment on the gas.** When a gas expands or contracts, energy is transferred between the gas and the environment.

FIGURE 19.4 The external force does work on the gas as the piston moves.

(a) The gas pushes with force $\vec{F}_{gas}$. To keep the piston in place, an external force must be equal and opposite to $\vec{F}_{gas}$.

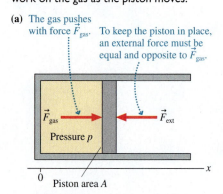

$\vec{F}_{gas}$ $\vec{F}_{ext}$

Pressure p

0
Piston area A

(b) As the piston moves dx, the external force does work $(F_{ext})_x \, dx$ on the gas.

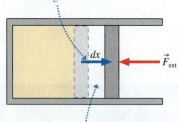

dx $\vec{F}_{ext}$

The volume changes by $dV = A \, dx$ as the piston moves dx.

We could also look at the moving piston from the perspective of the gas. The gas is exerting a force on the piston as it's displaced, so the gas does work W_{gas}. In fact, because $\vec{F}_{gas} = -\vec{F}_{ext}$, the work done by the gas is $W_{gas} = -W_{ext}$. This is really just another way of saying that there's no net work done on the piston. However, the work that appears in the energy principle, and now in the laws of thermodynamics, is W_{ext}, the work done by the environment—either positive or negative—on the system, and that is the work that will appear (without a subscript) in our equations.

But some care is needed. Both the environment *and* the gas do work, with opposite signs. It is *not* an either-or situation. When a gas is compressed, it is often said that "Work is done *on* the gas." This is a situation with $W_{ext} > 0$ and $W_{gas} < 0$. When a gas expands, it is often said that "Work is done *by* the gas." But this doesn't mean that $W_{ext} = 0$! It simply means that $W_{gas} > 0$ and so, in this case, you would use $W_{ext} < 0$.

We can give the work done on a gas a nice geometric interpretation. You learned in Chapter 18 how to represent an ideal-gas process as a curve in the pV diagram. FIGURE 19.5 shows that the work done on a gas is the negative of the area under the pV curve as the volume changes from V_i to V_f. That is,

> W = the negative of the area under the pV curve between V_i and V_f

Figure 19.5a shows a process in which a gas *expands* from V_i to a larger volume V_f. The area under the curve is positive, so the environment does a negative amount of work on an expanding gas. Figure 19.5b shows a process in which a gas is compressed to a smaller volume. This one is a little trickier because we have to integrate "backward" along the V-axis. You learned in calculus that integrating from a larger limit to a smaller limit gives a negative result, so the area in Figure 19.5b is a negative area. Consequently, as the minus sign in Equation 19.9 indicates, the environment does positive work on a gas to compress it.

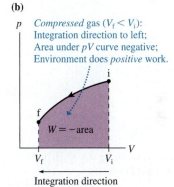

FIGURE 19.5 The work done on a gas is the negative of the area under the curve.

(a)

Expanding gas ($V_f > V_i$):
Integration direction to right;
Area under pV curve positive;
Environment does *negative* work.

$W = -\text{area}$

Integration direction

(b)

Compressed gas ($V_f < V_i$):
Integration direction to left;
Area under pV curve negative;
Environment does *positive* work.

$W = -\text{area}$

Integration direction

EXAMPLE 19.1 | **The work done on an expanding gas**

How much work is done in the ideal-gas process of FIGURE 19.6?

FIGURE 19.6 The ideal-gas process of Example 19.1.

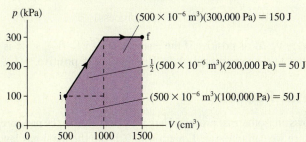

$(500 \times 10^{-6} \text{ m}^3)(300{,}000 \text{ Pa}) = 150 \text{ J}$

$\frac{1}{2}(500 \times 10^{-6} \text{ m}^3)(200{,}000 \text{ Pa}) = 50 \text{ J}$

$(500 \times 10^{-6} \text{ m}^3)(100{,}000 \text{ Pa}) = 50 \text{ J}$

MODEL The work done on a gas is the negative of the area under the pV curve. The gas is *expanding,* so we expect the work to be negative.

SOLVE As Figure 19.6 shows, the area under the curve can be divided into two rectangles and a triangle. Volumes *must* be converted to SI units of m^3. The total area under the curve is 250 J, so the work done on the gas as it expands is

$$W = -(\text{area under the } pV \text{ curve}) = -250 \text{ J}$$

ASSESS We noted previously that the product $\text{Pa} \, m^3$ is equivalent to joules. The work is negative, as expected, because the external force pushing on the piston is opposite the direction of the piston's displacement.

Equation 19.9 is the basis for a problem-solving strategy.

Work in ideal-gas processes

MODEL Model the gas as ideal and the process as quasi-static.

VISUALIZE Show the process on a pV diagram. Note whether it happens to be one of the basic gas processes: isochoric, isobaric, or isothermal.

SOLVE Calculate the work as the area under the pV curve either geometrically or by carrying out the integration:

$$\text{Work done on the gas } W = -\int_{V_i}^{V_f} p \, dV = -(\text{area under } pV \text{ curve})$$

ASSESS Check your signs.

- $W > 0$ when the gas is compressed. Energy is transferred from the environment to the gas.
- $W < 0$ when the gas expands. Energy is transferred from the gas to the environment.
- No work is done if the volume doesn't change. $W = 0$.

Exercise 4

Isochoric Process

The isochoric process in **FIGURE 19.7a** is one in which the volume does not change. Consequently,

$$W = 0 \qquad \text{(isochoric process)} \tag{19.10}$$

An isochoric process is the *only* ideal-gas process in which no work is done.

Isobaric Process

FIGURE 19.7b shows an isobaric process in which the volume changes from V_i to V_f. The rectangular area under the curve is $p \, \Delta V$, so the work done during this process is

$$W = -p \, \Delta V \qquad \text{(isobaric process)} \tag{19.11}$$

where $\Delta V = V_f - V_i$. ΔV is positive if the gas expands ($V_f > V_i$), so W is negative. ΔV is *negative* if the gas is compressed ($V_f < V_i$), making W positive.

Isothermal Process

FIGURE 19.8 shows an isothermal process. Here we need to know the pressure as a function of volume before we can integrate Equation 19.9. From the ideal-gas law, $p = nRT/V$. Thus the work on the gas as the volume changes from V_i to V_f is

$$W = -\int_{V_i}^{V_f} p \, dV = -\int_{V_i}^{V_f} \frac{nRT}{V} \, dV = -nRT \int_{V_i}^{V_f} \frac{dV}{V} \tag{19.12}$$

where we could take the T outside the integral because temperature is constant during an isothermal process. This is a straightforward integration, giving

$$W = -nRT \int_{V_i}^{V_f} \frac{dV}{V} = -nRT \ln V \Big|_{V_i}^{V_f}$$

$$= -nRT(\ln V_f - \ln V_i) = -nRT \ln\left(\frac{V_f}{V_i}\right) \tag{19.13}$$

FIGURE 19.7 Calculating the work done during ideal-gas processes.

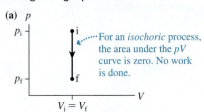

(a) For an *isochoric* process, the area under the pV curve is zero. No work is done.

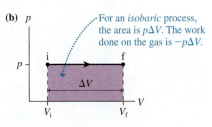

(b) For an *isobaric* process, the area is $p\Delta V$. The work done on the gas is $-p\Delta V$.

FIGURE 19.8 An isothermal process.

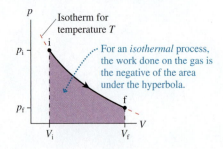

Isotherm for temperature T

For an *isothermal* process, the work done on the gas is the negative of the area under the hyperbola.

Because $nRT = p_iV_i = p_fV_f$ during an isothermal process, the work is:

$$W = -nRT \ln\left(\frac{V_f}{V_i}\right) = -p_iV_i \ln\left(\frac{V_f}{V_i}\right) = -p_fV_f \ln\left(\frac{V_f}{V_i}\right) \qquad (19.14)$$

(isothermal process)

Which version of Equation 19.14 is easiest to use will depend on the information you're given. The pressure, volume, and temperature *must* be in SI units.

EXAMPLE 19.2 | **The work of an isothermal compression**

A cylinder contains 7.0 g of nitrogen gas. How much work must be done to compress the gas at a constant temperature of 80°C until the volume is halved?

MODEL This is an isothermal ideal-gas process.

SOLVE Nitrogen gas is N_2, with molar mass $M_{mol} = 28$ g/mol, so 7.0 g is 0.25 mol of gas. The temperature is $T = 353$ K. Although we don't know the actual volume, we do know that $V_f = \frac{1}{2}V_i$. The volume ratio is all we need to calculate the work:

$$W = -nRT \ln\left(\frac{V_f}{V_i}\right)$$

$$= -(0.25 \text{ mol})(8.31 \text{ J/mol K})(353 \text{ K})\ln(1/2) = 508 \text{ J}$$

ASSESS The work is positive because a force from the environment pushes the piston inward to compress the gas.

Work Depends on the Path

FIGURE 19.9a shows two different processes that take a gas from an initial state i to a final state f. Although the initial and final states are the same, the work done during these two processes is *not* the same. **The work done during an ideal-gas process depends on the path followed through the pV diagram.**

You may recall that "work is independent of the path," but that referred to a different situation. In Chapter 10, we found that the work done by a conservative force is independent of the physical path of the object through space. For an ideal-gas process, the "path" is a sequence of thermodynamic states on a pV diagram. It is a figurative path because we can draw a picture of it on a pV diagram, but it is not a literal path.

The path dependence of work has an important implication for multistep processes. For a process $1 \to 2 \to 3$, the work has to be calculated separately for each step in the process, giving $W_{tot} = W_{1 \text{ to } 2} + W_{2 \text{ to } 3}$. In most cases, calculating the work for a process that goes directly from 1 to 3 will given an incorrect answer. The initial and final states are the same, but the work is *not* the same because work depends on the path followed through the pV diagram.

FIGURE 19.9 The work done during an ideal-gas process depends on the path.

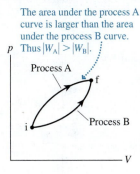

The area under the process A curve is larger than the area under the process B curve. Thus $|W_A| > |W_B|$.

STOP TO THINK 19.2 Two processes take an ideal gas from state 1 to state 3. Compare the work done by process A to the work done by process B.

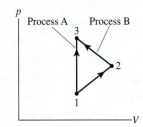

a. $W_A = W_B = 0$
b. $W_A = W_B$ but neither is zero
c. $W_A > W_B$
d. $W_A < W_B$

19.3 Heat

Heat is a more elusive concept than work. We use the word "heat" very loosely in the English language, often as synonymous with *hot*. We might say on a very hot day, "This heat is oppressive." If your apartment is cold, you may say, "Turn up the heat." These expressions date to a time long ago when it was thought that heat was a *substance* with fluid-like properties.

FIGURE 19.10 Joule's experiments to show the equivalence of heat and work.

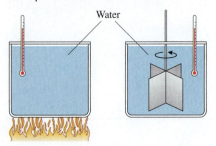

Water

The flame heats the water. The temperature increases.

The paddle does work on the water. The temperature increases.

Our concept of heat changed with the work of British physicist James Joule in the 1840s. Joule was the first to carry out careful experiments to learn how it is that systems change their temperature. Using experiments like those shown in **FIGURE 19.10**, Joule found that you can raise the temperature of a beaker of water by two entirely different means:

1. Heating it with a flame, or
2. Doing work on it with a rapidly spinning paddle wheel.

The final state of the water is *exactly* the same in both cases. This implies that heat and work are essentially equivalent. In other words, heat is not a substance. Instead, **heat is energy.** Heat and work, which previously had been regarded as two completely different phenomena, were now seen to be simply two different ways of transferring energy to or from a system.

Thermal Interactions

To be specific, **heat** is the energy transferred between a system and the environment as a consequence of a *temperature difference* between them. Unlike a mechanical interaction in which work is done, heat requires no macroscopic motion of the system. Instead (we'll look at the details in Chapter 20), energy is transferred when the *faster* molecules in the hotter object collide with the *slower* molecules in the cooler object. On average, these collisions cause the faster molecules to lose energy and the slower molecules to gain energy. The net result is that energy is transferred from the hotter object to the colder object. The process itself, whereby energy is transferred between the system and the environment via atomic-level collisions, is called a **thermal interaction.**

When you place a pan of water on the stove, heat is the energy transferred *from* the hotter flame *to* the cooler water. If you place the water in a freezer, heat is the energy transferred from the warmer water to the colder air in the freezer. A system is in **thermal equilibrium** with the environment, or two systems are in thermal equilibrium with each other, if there is no temperature difference.

Like work, **heat is not a state variable. That is, heat is not a property of the system.** Instead, heat is the amount of energy that moves between the system and the environment during a thermal interaction. It would not be meaningful to talk about a "change of heat." Thus heat appears in the energy equation simply as a value Q, never as ΔQ. **FIGURE 19.11** shows how to interpret the sign of Q.

FIGURE 19.11 The sign of heat.

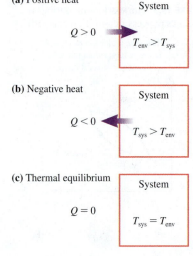

(a) Positive heat

System

$Q > 0$

$T_{env} > T_{sys}$

(b) Negative heat

System

$Q < 0$

$T_{sys} > T_{env}$

(c) Thermal equilibrium

System

$Q = 0$

$T_{sys} = T_{env}$

NOTE For both heat and work, a positive value indicates energy being transferred from the environment to the system. **TABLE 19.1** summarizes the similarities and differences between work and heat.

TABLE 19.1 Understanding work and heat

	Work	Heat
Interaction:	Mechanical	Thermal
Requires:	Force and displacement	Temperature difference
Process:	Macroscopic pushes and pulls	Microscopic collisions
Positive value:	$W > 0$ when a gas is compressed. Energy is transferred in.	$Q > 0$ when the environment is at a higher temperature than the system. Energy is transferred in.
Negative value:	$W < 0$ when a gas expands. Energy is transferred out.	$Q < 0$ when the system is at a higher temperature than the environment. Energy is transferred out.
Equilibrium:	A system is in mechanical equilibrium when there is no net force or torque on it.	A system is in thermal equilibrium when it is at the same temperature as the environment.

Units of Heat

Heat is energy transferred between the system and the environment. Consequently, the SI unit of heat is the joule. Historically, before the connection between heat and work had been recognized, a unit for measuring heat, the calorie, had been defined as

1 calorie = 1 cal = the quantity of heat needed to change the
temperature of 1 g of water by 1°C

Once Joule established that heat is energy, it was apparent that the calorie is really a unit of energy. In today's SI units, the conversion is

1 cal = 4.186 J

The calorie you know in relation to food is not the same as the heat calorie. The *food calorie,* abbreviated Cal with a capital C, is

1 food calorie = 1 Cal = 1000 cal = 1 kcal = 4186 J

We will not use calories in this textbook, but there are some fields of science and engineering where calories are still widely used. All the calculations you learn to do with joules can equally well be done with calories.

Heat, Temperature, and Thermal Energy

It is important to distinguish among *heat, temperature,* and *thermal energy.* These three ideas are related, but the distinctions among them are crucial. In brief,

- Thermal energy is an energy *of the system* due to the motion of its atoms and molecules and the stretching/compressing of spring-like molecular bonds. It is a *form* of energy. Thermal energy is a state variable, and it makes sense to talk about how E_{th} changes during a process. The system's thermal energy continues to exist even if the system is isolated and not interacting thermally with its environment.
- Heat is energy transferred *between the system* and the environment as they interact. Heat is *not* a particular form of energy, nor is it a state variable. It makes no sense to talk about how heat changes. $Q = 0$ if a system does not interact thermally with its environment. Heat may cause the system's thermal energy to change, but that doesn't make heat and thermal energy the same.
- Temperature is a state variable that quantifies the "hotness" or "coldness" of a system. We haven't yet given a precise definition of temperature, but it is related to the thermal energy *per molecule.* A temperature difference is a requirement for a thermal interaction in which heat energy is transferred between the system and the environment.

Heat is the energy transferred in a thermal interaction.

It is especially important not to associate an observed temperature increase with heat. Heating a system is one way to change its temperature, but, as Joule showed, not the only way. You can also change the system's temperature by doing work on the system or, as is the case with friction, transforming mechanical energy into thermal energy. **Observing the system tells us nothing about the process by which energy enters or leaves the system.**

STOP TO THINK 19.3   Which one or more of the following processes involves heat?

a. The brakes in your car get hot when you stop.
b. A steel block is held over a candle.
c. You push a rigid cylinder of gas across a frictionless surface.
d. You push a piston into a cylinder of gas, increasing the temperature of the gas.
e. You place a cylinder of gas in hot water. The gas expands, causing a piston to rise and lift a weight. The temperature of the gas does not change.

19.4 The First Law of Thermodynamics

Heat was the missing piece that we needed to arrive at a completely general statement of the law of conservation of energy. Restating Equation 19.4, we have

$$\Delta E_{sys} = \Delta E_{mech} + \Delta E_{th} = W + Q$$

Work and heat, two ways of transferring energy between a system and the environment, cause the system's energy to change.

At this point in the text we are not interested in systems that have a macroscopic motion of the system as a whole. Moving macroscopic systems were important to us for many chapters, but now, as we investigate the thermal properties of a system, we would like the system as a whole to rest peacefully on the laboratory bench while we study it. So we will assume, throughout the remainder of Part V, that $\Delta E_{mech} = 0$.

With this assumption clearly stated, the law of conservation of energy becomes

$$\Delta E_{th} = W + Q \qquad \text{(first law of thermodynamics)} \qquad (19.15)$$

The energy equation, in this form, is called the **first law of thermodynamics** or simply "the first law." The first law is a very general statement about energy.

Chapters 9 and 10 introduced the basic energy model—*basic* because it included work but not heat. The first law of thermodynamics is the basis for a more general energy model, the **thermodynamic energy model,** in which work and heat are on an equal footing.

MODEL 19.1

Thermodynamic energy model

Thermal energy is a property of the system.

- **Work** and **heat** are energies transferred between the system and the environment.
  - Work is energy transferred in a mechanical interaction.
  - Heat is energy transferred in a thermal interaction.
- The **first law of thermodynamics** says that transferring energy to the system changes the system's thermal energy.
  - $\Delta E_{th} = W + Q$
  - W and $Q > 0$ for energy added.
  - W and $Q < 0$ for energy removed.
- Limitations: Model fails if the system's mechanical energy also changes.

Keep in mind that thermal energy isn't the only thing that changes. Work or heat that changes the system's thermal energy also changes other state variables, such as pressure, volume, or temperature. The first law tells us only about ΔE_{th}. Other laws and relationships, such as the ideal-gas law, are needed to learn how the other state variables change.

Three Special Ideal-Gas Processes

The ideal-gas law relates the state variables p, V, and T, but it tells us nothing about how these variables change if we *do* something to the gas. The first law of thermodynamics is an additional law that gases must obey, one focused specifically on the process by which a gas is changed. In general, you need to consider both laws to solve thermodynamics problems about gases.

There are three ideal-gas processes in which one of the three terms in the first law—ΔE_{th}, W, or Q—is zero:

- **Isothermal process ($\Delta E_{th} = 0$):** If the temperature of a gas doesn't change, neither does its thermal energy. In this case, the first law is $W + Q = 0$. For example, a gas can be compressed ($W > 0$) without a temperature increase if all the energy is returned to the environment as heat ($Q < 0$ for heat energy leaving the system). An isothermal process exchanges work and heat (one entering the system, the other leaving) without changing the gas temperature. But pressure and volume *do* change. Equation 19.14 for the work done in an isothermal process, together with the ideal-gas law, will allow you to calculate what happens to p and V. **FIGURE 19.12a** shows an energy bar chart for an isothermal process.

- **Isochoric process ($W = 0$):** We saw in Section 19.2 that work is done on (or by) a gas when its volume changes. An isochoric process has $\Delta V = 0$; thus no work is done and the first law can be written $\Delta E_{th} = Q$. This is a process in which the system is *mechanically isolated* from its environment. Heating (or cooling) the gas increases (or decreases) its thermal energy, so the temperature goes up (or down). You'll learn later in this chapter how to calculate the temperature change ΔT. After using the first law to find ΔT, *then* you can use the ideal-gas law to calculate the pressure change. **FIGURE 19.12b** is an energy bar chart for a cooling process that lowers the temperature.

- **Adiabatic process ($Q = 0$):** A process in which no heat is transferred—perhaps the system is extremely well insulated—is called an **adiabatic process.** The system is *thermally isolated* from its environment. With $Q = 0$, the first law is $\Delta E_{th} = W$. But just because there's no heat does not mean that the temperature remains constant. Compressing an insulated gas—a mechanical interaction with $W > 0$—raises its temperature! Similarly, an adiabatic expansion (one with no heat exchanged) lowers the gas temperature. **$Q = 0$ does *not* mean $\Delta T = 0$.** We'll examine adiabatic processes and their pV curves later in the chapter, but notice that *all three* state variables p, V, and T change during an adiabatic process. **FIGURE 19.12c** is an energy bar chart for an adiabatic process with an increasing temperature.

FIGURE 19.12 First-law bar charts for three special ideal-gas processes.

(a) Isothermal process: $\Delta E_{th} = 0$

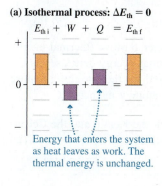

Energy that enters the system as heat leaves as work. The thermal energy is unchanged.

(b) Isochoric process: $W = 0$

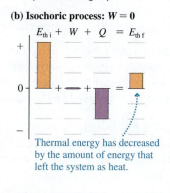

Thermal energy has decreased by the amount of energy that left the system as heat.

(c) Adiabatic process: $Q = 0$

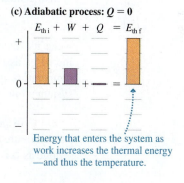

Energy that enters the system as work increases the thermal energy —and thus the temperature.

STOP TO THINK 19.4 Which first-law bar chart describes the process shown in the pV diagram?

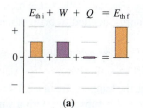

(a)

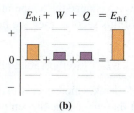

(b)

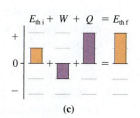

(c)

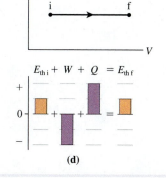

(d)

19.5 Thermal Properties of Matter

Heat and work are equivalent in the sense that the change of the system is *exactly the same* whether you transfer heat energy to it or do an equal amount of work on it. Adding energy to the system, or removing it, changes the system's thermal energy.

What happens to a system when you change its thermal energy? In this section we'll consider two distinct possibilities:

■ The temperature of the system changes.
■ The system undergoes a phase change, such as melting or freezing.

Temperature Change and Specific Heat

Suppose you do an experiment in which you add energy to water, either by doing work on it or by transferring heat to it. Either way, you will find that adding 4190 J of energy raises the temperature of 1 kg of water by 1 K. If you were fortunate enough to have 1 kg of gold, you would need to add only 129 J of energy to raise its temperature by 1 K.

The amount of energy that raises the temperature of 1 kg of a substance by 1 K is called the **specific heat** of that substance. The symbol for specific heat is c. Water has specific heat $c_{water} = 4190$ J/kg K. The specific heat of gold is $c_{gold} = 129$ J/kg K. Specific heat depends only on the material from which an object is made. **TABLE 19.2** provides some specific heats for common liquids and solids.

NOTE The term *specific heat* does not use the word "heat" in the way that we have defined it. Specific heat is an old idea, dating back to the days of the caloric theory when heat was thought to be a substance contained in the object. The term has continued in use even though our understanding of heat has changed.

If energy c is required to raise the temperature of 1 kg of a substance by 1 K, then energy Mc is needed to raise the temperature of mass M by 1 K and $(Mc)\Delta T$ is needed to raise the temperature of mass M by ΔT. In other words, the thermal energy of the system changes by

$$\Delta E_{th} = Mc\,\Delta T \quad \text{(temperature change)} \quad (19.16)$$

when its temperature changes by ΔT. ΔE_{th} can be either positive (thermal energy increases as the temperature goes up) or negative (thermal energy decreases as the temperature goes down). Recall that uppercase M is used for the mass of an entire system while lowercase m is reserved for the mass of an atom or molecule.

NOTE In practice, ΔT is usually measured in °C. But the Kelvin and the Celsius temperature scales have the same step size, so ΔT in K has exactly the same numerical value as ΔT in °C. Thus

■ You do not need to convert temperatures from °C to K if you need only a temperature change ΔT.
■ You do need to convert anytime you need the actual temperature T.

The first law of thermodynamics, $\Delta E_{th} = W + Q$, allows us to write Equation 19.16 as $Mc\,\Delta T = W + Q$. In other words, **we can change the system's temperature either by heating it or by doing an equivalent amount of work on it.** In working with solids and liquids, we almost always change the temperature by heating. If $W = 0$, which we will assume for the rest of this section, then the heat energy needed to bring about a temperature change ΔT is

$$Q = Mc\,\Delta T \quad \text{(temperature change)} \quad (19.17)$$

Because $\Delta T = \Delta E_{th}/Mc$, it takes more energy to change the temperature of a substance with a large specific heat than to change the temperature of a substance with a small

TABLE 19.2 Specific heats and molar specific heats of solids and liquids

Substance	c (J/kg K)	C (J/mol K)
Solids		
Aluminum	900	24.3
Copper	385	24.4
Iron	449	25.1
Gold	129	25.4
Lead	128	26.5
Ice	2090	37.6
Liquids		
Ethyl alcohol	2400	110.4
Mercury	140	28.1
Water	4190	75.4

specific heat. You can think of specific heat as measuring the *thermal inertia* of a substance. Metals, with small specific heats, warm up and cool down quickly. A piece of aluminum foil can be safely held within seconds of removing it from a hot oven. Water, with a very large specific heat, is slow to warm up and slow to cool down. This is fortunate for us. The large thermal inertia of water is essential for the biological processes of life. We wouldn't be here studying physics if water had a small specific heat!

EXAMPLE 19.3 | Running a fever

A 70 kg student catches the flu, and his body temperature increases from 37.0°C (98.6°F) to 39.0°C (102.2°F). How much energy is required to raise his body's temperature? The specific heat of a mammalian body is 3400 J/kg K, nearly that of water because mammals are mostly water.

MODEL Energy is supplied by the chemical reactions of the body's metabolism. These exothermic reactions transfer heat to the body. Normal metabolism provides enough heat energy to offset energy losses (radiation, evaporation, etc.) while maintaining a normal

body temperature of 37°C. We need to calculate the additional energy needed to raise the body's temperature by 2.0°C, or 2.0 K.

SOLVE The necessary heat energy is

$$Q = Mc\,\Delta T = (70\text{ kg})(3400\text{ J/kg K})(2.0\text{ K}) = 4.8 \times 10^5\text{ J}$$

ASSESS This appears to be a lot of energy, but a joule is actually a very small amount of energy. It is only 110 Cal, approximately the energy gained by eating an apple.

The **molar specific heat** is the amount of energy that raises the temperature of 1 mol of a substance by 1 K. We'll use an uppercase C for the molar specific heat. The heat energy needed to bring about a temperature change ΔT of n moles of substance is

$$Q = nC\,\Delta T \qquad (19.18)$$

Molar specific heats are listed in Table 19.2. Look at the five elemental solids (excluding ice). All have C very near 25 J/mol K. If we were to expand the table, we would find that most elemental solids have $C \approx 25$ J/mol K. This can't be a coincidence, but what is it telling us? This is a puzzle we will address in Chapter 20, where we will explore thermal energy at the atomic level.

Phase Change and Heat of Transformation

Suppose you start with a system in its solid phase and heat it at a steady rate. FIGURE 19.13, which you saw in Chapter 18, shows how the system's temperature changes. At first, the temperature increases linearly. This is not hard to understand because Equation 19.17 can be written

$$\text{slope of the }T\text{-versus-}Q\text{ graph} = \frac{\Delta T}{Q} = \frac{1}{Mc} \qquad (19.19)$$

The slope of the graph depends inversely on the system's specific heat. A constant specific heat implies a constant slope and thus a linear graph. In fact, you can measure c from such a graph.

NOTE The different slopes indicate that the solid, liquid, and gas phases of a substance have different specific heats.

But there are times, shown as horizontal line segments, during which heat is being transferred to the system but the temperature isn't changing. These are *phase changes*. The thermal energy continues to increase during a phase change, but the additional energy goes into breaking molecular bonds rather than speeding up the molecules. **A phase change is characterized by a change in thermal energy without a change in temperature.**

The amount of heat energy that causes 1 kg of a substance to undergo a phase change is called the **heat of transformation** of that substance. For example, laboratory experiments show that 333,000 J of heat are needed to melt 1 kg of ice at 0°C.

FIGURE 19.13 The temperature of a system that is heated at a steady rate.

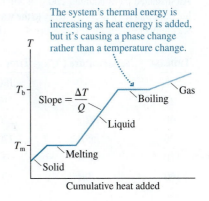

Lava—molten rock—undergoes a phase change when it contacts the much colder water. This is one way in which new islands are formed.

The symbol for heat of transformation is L. The heat required for the entire system of mass M to undergo a phase change is

$$Q = ML \quad \text{(phase change)} \quad (19.20)$$

Heat of transformation is a generic term that refers to any phase change. Two specific heats of transformation are the **heat of fusion** L_f, the heat of transformation between a solid and a liquid, and the **heat of vaporization** L_v, the heat of transformation between a liquid and a gas. The heat needed for these phase changes is

$$Q = \begin{cases} \pm ML_f & \text{melt/freeze} \\ \pm ML_v & \text{boil/condense} \end{cases} \quad (19.21)$$

where the $\pm$ indicates that heat must be *added* to the system during melting or boiling but *removed* from the system during freezing or condensing. **You must explicitly include the minus sign when it is needed.**

TABLE 19.3 gives the heats of transformation of a few substances. Notice that the heat of vaporization is always much larger than the heat of fusion. We can understand this. Melting breaks just enough molecular bonds to allow the system to lose rigidity and flow. Even so, the molecules in a liquid remain close together and loosely bonded. Vaporization breaks all bonds completely and sends the molecules flying apart. This process requires a larger increase in the thermal energy and thus a larger quantity of heat.

TABLE 19.3 Melting/boiling temperatures and heats of transformation

Substance	T_m (°C)	L_f (J/kg)	T_b (°C)	L_v (J/kg)
Nitrogen (N_2)	−210	0.26×10^5	−196	1.99×10^5
Ethyl alcohol	−114	1.09×10^5	78	8.79×10^5
Mercury	−39	0.11×10^5	357	2.96×10^5
Water	0	3.33×10^5	100	22.6×10^5
Lead	328	0.25×10^5	1750	8.58×10^5

EXAMPLE 19.4 | Melting wax

An insulated jar containing 200 g of solid candle wax is placed on a hot plate that supplies heat energy to the wax at the rate of 220 J/s. The wax temperature is measured every 30 s, yielding the following data:

Time (s)	Temperature (°C)	Time (s)	Temperature (°C)
0	20.0	180	70.5
30	31.7	210	70.5
60	42.2	240	70.6
90	55.0	270	70.5
120	64.7	300	70.4
150	70.4	330	74.5

What are the specific heat of the solid wax, the melting point, and the wax's heat of fusion?

MODEL The wax is in an insulated jar, so assume that heat loss to the environment is negligible.

VISUALIZE Heat energy is being supplied at the rate of 220 J/s, so the total heat energy that has been transferred into the wax at time t is $Q = 220t$ J. FIGURE 19.14 shows the temperature graphed

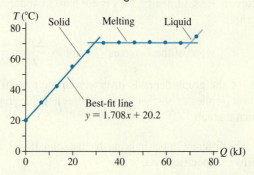

FIGURE 19.14 The heating curve of the wax.

against the cumulative heat Q, although notice that the horizontal axis is in kJ, not J. The initial linear slope corresponds to raising the wax's temperature to the melting point. Temperature remains constant during a phase change, even though the sample is still being heated, so the horizontal section of the graph is when the wax is melting. The temperature increase at the end shows that the temperature of the liquid wax is beginning to rise after melting is complete.

SOLVE From $Q = Mc\,\Delta T$, the slope of the T-versus-Q graph is $\Delta T/Q = 1/Mc$. The experimental slope of the best-fit line is $1.708°C/kJ = 0.001708$ K/J. Thus the specific heat of the solid wax is

$$c = \frac{1}{M \times \text{slope}} = \frac{1}{(0.200 \text{ kg})(0.001708 \text{ K/J})} = 2930 \text{ J/kg K}$$

From the table, we see that the melting temperature—which remains constant during the phase change—is 70.5°C. The heat required for the phase change is $Q = ML_f$, so the heat of fusion is $L_f = Q/M$. With data recorded only every 30 s, it's not exactly clear when the melting began and when it ended. The extension of the initial

slope shows that the temperature reached the melting point about halfway between 120 s and 150 s, so the melting started at about 135 s. We'll assume it was complete about halfway between 300 s and 330 s, or at about 315 s. Thus the melting took 180 s, during which, at 220 J/s, 39,600 J of heat energy was transferred from the hot plate to the wax. With this value of Q, the heat of fusion is

$$L_f = \frac{Q}{M} = \frac{39,600 \text{ J}}{0.200 \text{ kg}} = 2.0 \times 10^5 \text{ J/kg}$$

ASSESS Both the specific heat and the heat of fusion are similar to values in Tables 19.2 and 19.3, which gives us confidence in our results.

STOP TO THINK 19.5 Objects A and B are brought into close thermal contact with each other, but they are well isolated from their surroundings. Initially $T_A = 0°C$ and $T_B = 100°C$. The specific heat of A is less than the specific heat of B. The two objects will soon reach a common final temperature T_f. The final temperature is

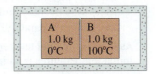

a. $T_f > 50°C$ b. $T_f = 50°C$ c. $T_f < 50°C$

19.6 Calorimetry

At one time or another you've probably put an ice cube into a hot drink to cool it quickly. You were engaged, in a somewhat trial-and-error way, in a practical aspect of heat transfer known as **calorimetry.**

FIGURE 19.15 shows two systems thermally interacting with each other but isolated from everything else. Suppose they start at different temperatures T_1 and T_2. As you know from experience, heat energy will be transferred from the hotter to the colder system until they reach a common final temperature T_f. The systems will then be in thermal equilibrium and the temperature will not change further.

The insulation prevents any heat energy from being transferred to or from the environment, so energy conservation tells us that any energy leaving the hotter system must enter the colder system. That is, the systems *exchange* energy with no net loss or gain. The concept is straightforward, but to state the idea mathematically we need to be careful with signs.

Let Q_1 be the energy transferred to system 1 as heat. Similarly, Q_2 is the energy transferred to system 2. The fact that the systems are merely exchanging energy can be written $|Q_1| = |Q_2|$. The energy *lost* by the hotter system is the energy *gained* by the colder system, so Q_1 and Q_2 have opposite signs: $Q_1 = -Q_2$. No energy is exchanged with the environment, hence it makes more sense to write this relationship as

$$Q_{net} = Q_1 + Q_2 = 0 \qquad (19.22)$$

This idea is not limited to the interaction of only two systems. If three or more systems are combined in isolation from the rest of their environment, each at a different initial temperature, they will all come to a common final temperature that can be found from the relationship

$$Q_{net} = Q_1 + Q_2 + Q_3 + \cdots = 0 \qquad (19.23)$$

NOTE The signs are very important in calorimetry problems. ΔT is always $T_f - T_i$, so ΔT and Q are negative for any system whose temperature decreases. The proper sign of Q for any phase change must be supplied *by you*, depending on the direction of the phase change.

FIGURE 19.15 Two systems interact thermally.

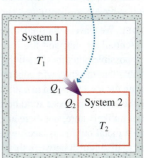

Heat energy is transferred from system 1 to system 2.

Energy conservation requires $|Q_1| = |Q_2|$

PROBLEM-SOLVING STRATEGY 19.2

Calorimetry problems

MODEL Model the systems as interacting with each other but isolated from the larger environment.

VISUALIZE List known information and identify what you need to find. Convert all quantities to SI units.

SOLVE The mathematical representation, a statement of energy conservation, is

$$Q_{net} = Q_1 + Q_2 + \cdots = 0$$

- For systems that undergo a temperature change, $Q = Mc(T_f - T_i)$. Be sure to have the temperatures T_i and T_f in the correct order.
- For systems that undergo a phase change, $Q = \pm ML$. Supply the correct sign by observing whether energy enters or leaves the system.
- Some systems may undergo a temperature change *and* a phase change. Treat the changes separately. The heat energy is $Q = Q_{\Delta T} + Q_{phase}$.

ASSESS Is the final temperature in the middle? T_f that is higher or lower than all initial temperatures is an indication that something is wrong, usually a sign error.

Exercise 15

NOTE You may have learned to solve calorimetry problems in other courses by writing $Q_{gained} = Q_{lost}$. That is, by balancing heat gained with heat lost. That approach works in simple problems, but it has two drawbacks. First, you often have to "fudge" the signs to make them work. Second, and more serious, you can't extend this approach to a problem with three or more interacting systems. Using $Q_{net} = 0$ is much preferred.

EXAMPLE 19.5 | Calorimetry with a phase change

Your 500 mL soda is at 20°C, room temperature, so you add 100 g of ice from the −20°C freezer. Does all the ice melt? If so, what is the final temperature? If not, what fraction of the ice melts? Assume that you have a well-insulated cup.

MODEL We have a thermal interaction between the soda, which is essentially water, and the ice. We need to distinguish between three possible outcomes. If all the ice melts, then $T_f > 0$°C. It's also possible that the soda will cool to 0°C before all the ice has melted, leaving the ice and liquid in equilibrium at 0°C. A third possibility is that the soda will freeze solid before the ice warms up to 0°C. That seems unlikely here, but there are situations, such as the pouring of molten metal out of furnaces, when all the liquid does solidify. We need to distinguish between these before knowing how to proceed.

VISUALIZE All the initial temperatures, masses, and specific heats are known. The final temperature of the combined soda + ice system is unknown.

SOLVE Let's first calculate the heat needed to melt all the ice and leave it as liquid water at 0°C. To do so, we must warm the ice to 0°C, then change it to water. The heat input for this two-stage process is

$$Q_{melt} = M_i c_i(20 \text{ K}) + M_i L_f = 37,500 \text{ J}$$

where L_f is the heat of fusion of water. It is used as a *positive* quantity because we must *add* heat to melt the ice. Next, let's calculate how much heat energy will leave the soda if it cools all

the way to 0°C. The volume is $V = 500 \text{ mL} = 5.00 \times 10^{-4} \text{ m}^3$ and thus the mass is $M_s = \rho V = 0.500 \text{ kg}$. The heat loss is

$$Q_{cool} = M_s c_w(-20 \text{ K}) = -41,900 \text{ J}$$

where $\Delta T = -20$ K because the temperature decreases. Because $|Q_{cool}| > Q_{melt}$, the soda has sufficient energy to melt all the ice. Hence the final state will be all liquid at $T_f > 0$. (Had we found $|Q_{cool}| < Q_{melt}$, then the final state would have been an ice-liquid mixture at 0°C.)

Energy conservation requires $Q_{ice} + Q_{soda} = 0$. The heat Q_{ice} consists of three terms: warming the ice to 0°C, melting the ice to water at 0°C, then warming the 0°C water to T_f. The mass will still be M_i in the last of these steps because it is the "ice system," but we need to use the specific heat of *liquid water*. Thus

$$Q_{ice} + Q_{soda} = [M_i c_i(20 \text{ K}) + M_i L_f + M_i c_w(T_f - 0°C)]$$
$$+ M_s c_w(T_f - 20°C) = 0$$

We've already done part of the calculation, allowing us to write

$$37,500 \text{ J} + M_i c_w(T_f - 0°C) + M_s c_w(T_f - 20°C) = 0$$

Solving for T_f gives

$$T_f = \frac{20 M_s c_w - 37,500}{M_i c_w + M_s c_w} = 1.7°C$$

ASSESS As expected, the soda has been cooled to nearly the freezing point.

EXAMPLE 19.6 | Three interacting systems

A 200 g piece of iron at 120°C and a 150 g piece of copper at −50°C are dropped into an insulated beaker containing 300 g of ethyl alcohol at 20°C. What is the final temperature?

MODEL Here you can't use a simple $Q_{\text{gained}} = Q_{\text{lost}}$ approach because you don't know whether the alcohol is going to warm up or cool down. In principle, the alcohol could freeze or boil, but the masses and temperatures of the metals suggest that the temperature will not change greatly. We will assume that the alcohol remains a liquid.

VISUALIZE All the initial temperatures, masses, and specific heats are known. We need to find the final temperature.

SOLVE Energy conservation requires

$$Q_i + Q_c + Q_e = M_i c_i (T_f - 120°C) + M_c c_c (T_f - (-50°C))$$
$$+ M_e c_e (T_f - 20°C) = 0$$

Solving for T_f gives

$$T_f = \frac{120 M_i c_i - 50 M_c c_c + 20 M_e c_e}{M_i c_i + M_c c_c + M_e c_e} = 26°C$$

ASSESS The temperature is between the initial iron and copper temperatures, as expected, and well below the boiling temperature of the alcohol, validating our assumption of no phase change. It turns out that the alcohol warms up ($Q_e > 0$), but we had no way to know this without doing the calculation.

19.7 The Specific Heats of Gases

Specific heats are given in Table 19.2 for solids and liquids. Gases are harder to characterize because the heat required to cause a specified temperature change depends on the *process* by which the gas changes state.

FIGURE 19.16 shows two isotherms on the pV diagram for a gas. Processes A and B, which start on the T_i isotherm and end on the T_f isotherm, have the *same* temperature change $\Delta T = T_f - T_i$. But process A, which takes place at constant volume, requires a *different* amount of heat than does process B, which occurs at constant pressure. The reason is that work is done in process B but not in process A. This is a situation that we are now equipped to analyze.

It is useful to define two different versions of the specific heat of gases, one for constant-volume (isochoric) processes and one for constant-pressure (isobaric) processes. We will define these as molar specific heats because we usually do gas calculations using moles instead of mass. The quantity of heat needed to change the temperature of n moles of gas by ΔT is

$Q = nC_V \Delta T$	(temperature change at constant volume)	
$Q = nC_P \Delta T$	(temperature change at constant pressure)	(19.24)

where C_V is the **molar specific heat at constant volume** and C_P is the **molar specific heat at constant pressure**. **TABLE 19.4** gives the values of C_V and C_P for a few common monatomic and diatomic gases. The units are J/mol K. Molar specific heats do vary somewhat with temperature—we'll look at an example in the next chapter—but the values in the table are adequate for temperatures from 200 K to 800 K.

NOTE Equations 19.24 apply to two specific ideal-gas processes. In a general gas process, for which neither p nor V is constant, we have no direct way to relate Q to ΔT. In that case, the heat must be found indirectly from the first law as $Q = \Delta E_{\text{th}} - W$.

FIGURE 19.16 Processes A and B have the same ΔT and the same ΔE_{th}, but they require different amounts of heat.

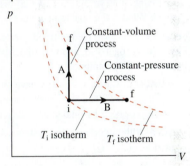

TABLE 19.4 Molar specific heats of gases (J/mol K) at $T = 0°C$

Gas	C_P	C_V	$C_P - C_V$
Monatomic Gases			
He	20.8	12.5	8.3
Ne	20.8	12.5	8.3
Ar	20.8	12.5	8.3
Diatomic Gases			
H_2	28.7	20.4	8.3
N_2	29.1	20.8	8.3
O_2	29.2	20.9	8.3

EXAMPLE 19.7 | Heating and cooling a gas

Three moles of O_2 gas are at 20.0°C. 600 J of heat energy are transferred to the gas at constant pressure, then 600 J are removed at constant volume. What is the final temperature? Show the process on a pV diagram.

MODEL O_2 is a diatomic ideal gas. The gas is heated as an isobaric process, then cooled as an isochoric process.

SOLVE The heat transferred during the constant-pressure process causes a temperature rise

$$\Delta T = T_2 - T_1 = \frac{Q}{nC_P} = \frac{600\,\text{J}}{(3.0\,\text{mol})(29.2\,\text{J/mol K})} = 6.8°C$$

Continued

where C_P for oxygen was taken from Table 19.4. Heating leaves the gas at temperature $T_2 = T_1 + \Delta T = 26.8°C$. The temperature then falls as heat is removed during the constant-volume process:

$$\Delta T = T_3 - T_2 = \frac{Q}{nC_V} = \frac{-600 \text{ J}}{(3.0 \text{ mol})(20.9 \text{ J/mol K})} = -9.5°C$$

We used a *negative* value for Q because heat energy is transferred from the gas to the environment. The final temperature of the gas is $T_3 = T_2 + \Delta T = 17.3°C$. **FIGURE 19.17** shows the process on a pV diagram. The gas expands (moves horizontally on the diagram) as heat is added, then cools at constant volume (moves vertically on the diagram) as heat is removed.

ASSESS The final temperature is lower than the initial temperature because $C_P > C_V$.

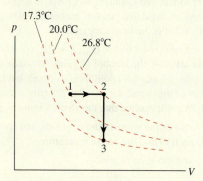

FIGURE 19.17 The pV diagram for Example 19.7.

EXAMPLE 19.8 | Calorimetry with a gas and a solid

The interior volume of a 200 g hollow aluminum box is 800 cm³. The box contains nitrogen gas at STP. A 20 cm³ block of copper at a temperature of 300°C is placed inside the box, then the box is sealed. What is the final temperature?

MODEL This example has three interacting systems: the aluminum box, the nitrogen gas, and the copper block. They must all come to a common final temperature T_f.

VISUALIZE The box and gas have the same initial temperature: $T_{Al} = T_{N2} = 0°C$. The box doesn't change size, so this is a constant-volume process. The final temperature is unknown.

SOLVE Although one of the systems is now a gas, the calorimetry equation $Q_{net} = Q_{Al} + Q_{N2} + Q_{Cu} = 0$ is still appropriate. In this case,

$$Q_{net} = m_{Al}c_{Al}(T_f - T_{Al}) + n_{N2}C_V(T_f - T_{N2})$$

$$+ m_{Cu}c_{Cu}(T_f - T_{Cu}) = 0$$

Notice that we used masses and specific heats for the solids but moles and the molar specific heat for the gas. We used C_V because this is a constant-volume process. Solving for T_f gives

$$T_f = \frac{m_{Al}c_{Al}T_{Al} + n_{N2}C_VT_{N2} + m_{Cu}c_{Cu}T_{Cu}}{m_{Al}c_{Al} + n_{N2}C_V + m_{Cu}c_{Cu}}$$

The specific heat values are found in Tables 19.2 and 19.4. The mass of the copper is

$$m_{Cu} = \rho_{Cu}V_{Cu} = (8920 \text{ kg/cm}^3)(20 \times 10^{-6} \text{ m}^3) = 0.178 \text{ kg}$$

The number of moles of the gas is found from the ideal-gas law, using the initial conditions. Notice that inserting the copper block *displaces* 20 cm³ of gas; hence the gas volume is only $V = 780 \text{ cm}^3 = 7.80 \times 10^{-4} \text{ m}^3$. Thus

$$n_{N2} = \frac{pV}{RT} = 0.0348 \text{ mol}$$

Computing the final temperature gives $T_f = 83°C$.

Cp and Cv

You may have noticed two curious features in Table 19.4. First, the molar specific heats of monatomic gases are *all alike*. And the molar specific heats of diatomic gases, while different from monatomic gases, are again *very nearly alike*. We saw a similar feature in Table 19.2 for the molar specific heats of solids. Second, the *difference* $C_P - C_V = 8.3 \text{ J/mol K}$ is the same in every case. And, most puzzling of all, the value of $C_P - C_V$ appears to be equal to the universal gas constant R! Why should this be?

The relationship between C_V and C_P hinges on one crucial idea: ΔE_{th}, **the change in the thermal energy of a gas, is the same for *any* two processes that have the same ΔT.** The thermal energy of a gas is associated with temperature, so any process that changes the gas temperature from T_i to T_f has the same ΔE_{th} as any other process that goes from T_i to T_f. Furthermore, the first law $\Delta E_{th} = Q + W$ tells us that a gas cannot distinguish between heat and work. The system's thermal energy changes in response to energy added to or removed from the system, but the response of the gas is the same whether you heat the system, do work on the system, or do some combination of both. Thus **any two processes that change the thermal energy of the gas by ΔE_{th} will cause the same temperature change ΔT.**

With that in mind, look back at Figure 19.16. Both gas processes have the same ΔT, so both have the same value of ΔE_{th}. Process A is an isochoric process in which no work is done (the piston doesn't move), so the first law for this process is

$$(\Delta E_{th})_A = W + Q = 0 + Q_{const\ vol} = nC_V\,\Delta T \qquad (19.25)$$

Process B is an isobaric process. You learned earlier that the work done on the gas during an isobaric process is $W = -p\,\Delta V$. Thus

$$(\Delta E_{th})_B = W + Q = -p\,\Delta V + Q_{const\ press} = -p\,\Delta V + nC_P\,\Delta T \qquad (19.26)$$

$(\Delta E_{th})_B = (\Delta E_{th})_A$ because both have the same ΔT, so we can equate the right sides of Equations 19.25 and 19.26:

$$-p\,\Delta V + nC_P\,\Delta T = nC_V\,\Delta T \qquad (19.27)$$

For the final step, we can use the ideal-gas law $pV = nRT$ to relate ΔV and ΔT during process B. For any gas process,

$$\Delta(pV) = \Delta(nRT) \qquad (19.28)$$

For a constant-pressure process, where p is constant, Equation 19.28 becomes

$$p\,\Delta V = nR\,\Delta T \qquad (19.29)$$

Substituting this expression for $p\,\Delta V$ into Equation 19.27 gives

$$-nR\,\Delta T + nC_P\,\Delta T = nC_V\,\Delta T \qquad (19.30)$$

The $n\,\Delta T$ cancels, and we are left with

$$C_P = C_V + R \qquad (19.31)$$

This result, which applies to all ideal gases, is exactly what we see in the data of Table 19.4.

But that's not the only conclusion we can draw. Equation 19.25 found that $\Delta E_{th} = nC_V\,\Delta T$ for a constant-volume process. But we had just noted that ΔE_{th} is the same for *all* gas processes that have the same ΔT. Consequently, this expression for ΔE_{th} is equally true for any other process. That is

$$\Delta E_{th} = nC_V\,\Delta T \qquad \text{(any ideal-gas process)} \qquad (19.32)$$

Compare this result to Equations 19.24. We first made a distinction between constant-volume and constant-pressure processes, but now we're saying that Equation 19.32 is true for any process. Are we contradicting ourselves? No, the difference lies in what you need to calculate.

- The change in thermal energy when the temperature changes by ΔT is the same for any process. That's Equation 19.32.
- The *heat* required to bring about the temperature change depends on what the process is. That's Equations 19.24. An isobaric process requires more heat than an isochoric process that produces the same ΔT.

The reason for the difference is seen by writing the first law as $Q = \Delta E_{th} - W$. In an isochoric process, where $W = 0$, *all* the heat input is used to increase the gas temperature. But in an isobaric process, some of the energy that enters the system as heat then leaves the system as work ($W < 0$) done by the expanding gas. Thus more heat is needed to produce the same ΔT.

Heat Depends on the Path

Consider the two ideal-gas processes shown in **FIGURE 19.18**. Even though the initial and final states are the same, the heat added during these two processes is *not* the same. We can use the first law $\Delta E_{th} = W + Q$ to see why.

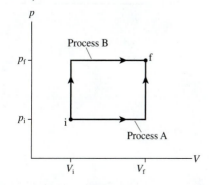

FIGURE 19.18 Is the heat input along these two paths the same or different?

The thermal energy is a state variable. That is, its value depends on the state of the gas, not the process by which the gas arrived at that state. Thus $\Delta E_{th} = E_{th\,f} - E_{th\,i}$ is the same for both processes. If ΔE_{th} is the same for processes A and B, then $W_A + Q_A = W_B + Q_B$.

You learned in Section 19.2 that the work done during an ideal-gas process depends on the path in the pV diagram. There's more area under the process B curve, so $|W_B| > |W_A|$. Both values of W are negative because the gas expands, so W_B is more negative than W_A. Consequently, $W_A + Q_A$ can equal $W_B + Q_B$ only if $Q_B > Q_A$. **The heat added or removed during an ideal-gas process depends on the path followed through the pV diagram.**

Adiabatic Processes

Section 19.4 introduced the idea of an *adiabatic process,* a process in which no heat energy is transferred ($Q = 0$). **FIGURE 19.19** compares an adiabatic process with isothermal and isochoric processes. We're now prepared to look at adiabatic processes in more detail.

In practice, there are two ways that an adiabatic process can come about. First, a gas cylinder can be completely surrounded by thermal insulation, such as thick pieces of Styrofoam. The environment can interact mechanically with the gas by pushing or pulling on the insulated piston, but there is no thermal interaction.

Second, the gas can be expanded or compressed very rapidly in what we call an *adiabatic expansion* or an *adiabatic compression.* In a rapid process there is essentially no time for heat to be transferred between the gas and the environment. We've already alluded to the idea that heat is transferred via atomic-level collisions. These collisions take time. If you stick one end of a copper rod into a flame, the other end will eventually get too hot to hold—but not instantly. Some amount of time is required for heat to be transferred from one end to the other. A process that takes place faster than the heat can be transferred is adiabatic.

> **NOTE** You may recall reading in Chapter 18 that we are going to study only quasi-static processes, processes that proceed slowly enough to remain essentially in equilibrium at all times. Now we're proposing to study processes that take place very rapidly. Isn't this a contradiction? Yes, to some extent it is. What we need to establish are the appropriate time scales. How slow must a process go to be quasi-static? How fast must it go to be adiabatic? These types of calculations must be deferred to a more advanced course. It turns out—fortunately!—that many practical applications, such as the compression strokes in gasoline and diesel engines, are fast enough to be adiabatic yet slow enough to still be considered quasi-static.

For an adiabatic process, with $Q = 0$, the first law of thermodynamics is $\Delta E_{th} = W$. Compressing a gas adiabatically ($W > 0$) increases the thermal energy. Thus an adiabatic compression raises the temperature of a gas. A gas that expands adiabatically ($W < 0$) gets colder as its thermal energy decreases. Thus an adiabatic expansion lowers the temperature of a gas. You can use an adiabatic process to change the gas temperature without using heat!

The work done in an adiabatic process goes entirely to changing the thermal energy of the gas. But we just found that $\Delta E_{th} = nC_V \Delta T$ for *any* process. Thus

$$W = nC_V \Delta T \qquad \text{(adiabatic process)} \qquad (19.33)$$

Equation 19.33 joins with the equations we derived earlier for the work done in isochoric, isobaric, and isothermal processes.

Gas processes can be represented as trajectories in the pV diagram. For example, a gas moves along a hyperbola during an isothermal process. How does an adiabatic process appear in a pV diagram? The result is more important than the derivation,

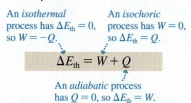

FIGURE 19.19 The relationship of three important processes to the first law of thermodynamics.

An *isothermal* process has $\Delta E_{th} = 0$, so $W = -Q$.

An *isochoric* process has $W = 0$, so $\Delta E_{th} = Q$.

$$\Delta E_{th} = W + Q$$

An *adiabatic* process has $Q = 0$, so $\Delta E_{th} = W$.

which is a bit tedious, so we'll begin with the answer and then, at the end of this section, show where it comes from.

First, we define the **specific heat ratio** γ (lowercase Greek gamma) to be

$$\gamma = \frac{C_P}{C_V} = \begin{cases} 1.67 & \text{monatomic gas} \\ 1.40 & \text{diatomic gas} \end{cases} \qquad (19.34)$$

The specific heat ratio has many uses in thermodynamics. Notice that γ is dimensionless.

An adiabatic process is one in which

$$pV^{\gamma} = \text{constant} \qquad \text{or} \qquad p_f V_f^{\gamma} = p_i V_i^{\gamma} \qquad (19.35)$$

This is similar to the isothermal $pV = \text{constant}$, but somewhat more complex due to the exponent γ.

The curves found by graphing $p = \text{constant}/V^{\gamma}$ are called **adiabats.** In **FIGURE 19.20** you see that the two adiabats are steeper than the hyperbolic isotherms. An adiabatic process moves along an adiabat in the same way that an isothermal process moves along an isotherm. You can see that the temperature falls during an adiabatic expansion and rises during an adiabatic compression.

If we use the ideal-gas-law expression $p = nRT/V$ in the adiabatic equation $pV^{\gamma} = \text{constant}$, we see that $TV^{\gamma-1}$ is also constant during an adiabatic process. Thus another useful equation for adiabatic processes is

$$T_f V_f^{\gamma-1} = T_i V_i^{\gamma-1} \qquad (19.36)$$

FIGURE 19.20 An adiabatic process moves along pV curves called *adiabats.*

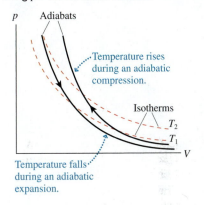

Temperature rises during an adiabatic compression.

Temperature falls during an adiabatic expansion.

EXAMPLE 19.9 | An adiabatic compression

Air containing gasoline vapor is admitted into the cylinder of an internal combustion engine at 1.00 atm pressure and 30°C. The piston rapidly compresses the gas from 500 cm³ to 50 cm³, a *compression ratio* of 10.

a. What are the final temperature and pressure of the gas?
b. Show the compression process on a pV diagram.
c. How much work is done to compress the gas?

MODEL The compression is rapid, with insufficient time for heat to be transferred from the gas to the environment, so we will model it as an adiabatic compression. We'll treat the gas as if it were 100% air.

SOLVE a. We know the initial pressure and volume, and we know the volume after the compression. For an adiabatic process, where pV^{γ} remains constant, the final pressure is

$$p_f = p_i\left(\frac{V_i}{V_f}\right)^{\gamma} = (1.00 \text{ atm})(10)^{1.40} = 25.1 \text{ atm}$$

Air is a mixture of N_2 and O_2, diatomic gases, so we used $\gamma = 1.40$. We can now find the temperature by using the ideal-gas law:

$$T_f = T_i \frac{p_f V_f}{p_i V_i} = (303 \text{ K})(25.1)\left(\frac{1}{10}\right) = 761 \text{ K} = 488°C$$

Temperature *must* be in kelvins for doing gas calculations such as these.

b. **FIGURE 19.21** shows the pV diagram. The 30°C and 488°C isotherms are included to show how the temperature changes during the process.

FIGURE 19.21 The adiabatic compression of the gas in an internal combustion engine.

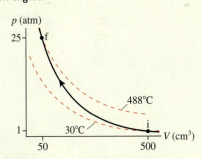

c. The work done is $W = nC_V \Delta T$, with $\Delta T = 458$ K. The number of moles is found from the ideal-gas law and the initial conditions:

$$n = \frac{p_i V_i}{RT_i} = 0.0201 \text{ mol}$$

Thus the work done to compress the gas is

$$W = nC_V \Delta T = (0.0201 \text{ mol})(20.8 \text{ J/mol K})(458 \text{ K}) = 192 \text{ J}$$

ASSESS The temperature rises dramatically during the compression stroke of an engine. But the higher temperature has nothing to do with heat! **The temperature and thermal energy of the gas are increased not by heating the gas but by doing work on it.** This is an important idea to understand.

Proof of Equation 19.35

Now let's see where Equation 19.35 comes from. Consider an adiabatic process in which an infinitesimal amount of work dW done on a gas causes an infinitesimal change in the thermal energy. For an adiabatic process, with $dQ = 0$, the first law of thermodynamics is

$$dE_{th} = dW \tag{19.37}$$

We can use Equation 19.32, which is valid for *any* gas process, to write $dE_{th} = nC_V dT$. Earlier in the chapter we found that the work done during a small volume change is $dW = -p\, dV$. With these substitutions, Equation 19.37 becomes

$$nC_V dT = -p\, dV \tag{19.38}$$

The ideal-gas law can now be used to write $p = nRT/V$. The n cancels, and the C_V can be moved to the other side of the equation to give

$$\frac{dT}{T} = -\frac{R}{C_V}\frac{dV}{V} \tag{19.39}$$

We're going to integrate Equation 19.39, but anticipating the need for $\gamma = C_P/C_V$ we can first use the fact that $C_P = C_V + R$ to write

$$\frac{R}{C_V} = \frac{C_P - C_V}{C_V} = \frac{C_P}{C_V} - 1 = \gamma - 1 \tag{19.40}$$

Now we integrate Equation 19.39 from the initial state i to the final state f:

$$\int_{T_i}^{T_f} \frac{dT}{T} = -(\gamma - 1)\int_{V_i}^{V_f} \frac{dV}{V} \tag{19.41}$$

Carrying out the integration gives

$$\ln\left(\frac{T_f}{T_i}\right) = \ln\left(\frac{V_i}{V_f}\right)^{\gamma-1} \tag{19.42}$$

where we used the logarithm properties $\log a - \log b = \log(a/b)$ and $c \log a = \log(a^c)$.

Taking the exponential of both sides now gives

$$\left(\frac{T_f}{T_i}\right) = \left(\frac{V_i}{V_f}\right)^{\gamma-1} \tag{19.43}$$

$$T_f V_f^{\gamma-1} = T_i V_i^{\gamma-1}$$

This was Equation 19.36. Writing $T = pV/nR$ and canceling $1/nR$ from both sides of the equation give Equation 19.35:

$$p_f V_f^{\gamma} = p_i V_i^{\gamma} \tag{19.44}$$

This was a lengthy derivation, but it is good practice at seeing how the ideal-gas law and the first law of thermodynamics can work together to yield results of great importance.

STOP TO THINK 19.6 For the two processes shown, which of the following is true:

a. $Q_A > Q_B$
b. $Q_A = Q_B$
c. $Q_A < Q_B$

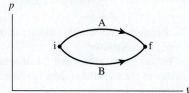

19.8 Heat-Transfer Mechanisms

You feel warmer when the sun is shining on you, colder when sitting on a metal bench or when the wind is blowing, especially if your skin is wet. This is due to the transfer of heat. Although we've talked about heat a lot in this chapter, we haven't said much about *how* heat is transferred from a hotter object to a colder object. There are four basic mechanisms by which objects exchange heat with their surroundings. Evaporation was treated in Section 19.5; in this section, we will consider the other mechanisms.

Heat-transfer mechanisms

When two objects are in direct contact, such as the soldering iron and the circuit board, heat is transferred by *conduction*.

Air currents near a lighted candle rise, taking thermal energy with them in a process known as *convection*.

The lamp at the top shines on the lambs huddled below, warming them. The energy is transferred by *radiation*.

Blowing on a hot cup of tea or coffee cools it by *evaporation*.

Conduction

FIGURE 19.22 shows an object sandwiched between a higher temperature T_H and a lower temperature T_C. The temperature *difference* causes heat energy to be transferred from the hot side to the cold side in a process known as **conduction.**

It is not surprising that more heat is transferred if the temperature difference ΔT is larger. A material with a larger cross section A (a fatter pipe) transfers more heat, while a thicker material, increasing the distance L between the hot and cold sources, decreases the rate of heat transfer.

These observations about heat conduction can be summarized in a single formula. If a small amount of heat dQ is transferred in a small time interval dt, the *rate* of heat transfer is dQ/dt. For a material of cross-section area A and length L, spanning a temperature difference $\Delta T = T_H - T_C$, the rate of heat transfer is

$$\frac{dQ}{dt} = k\frac{A}{L}\Delta T \qquad (19.45)$$

The quantity k, which characterizes whether the material is a good conductor of heat or a poor conductor, is called the **thermal conductivity** of the material. Because the heat-transfer rate J/s is a *power,* measured in watts, the units of k are W/m K. Values of k for common materials are given in **TABLE 19.5**; a material with a larger value of k is a better conductor of heat.

NOTE Heat conductivity is yet *another* use of the symbol k. Whenever you see a k, be very alert to the context in which it is used.

Most good heat conductors are metals, which are also good conductors of electricity. One exception is diamond, in which the strong bonds among atoms that make diamond such a hard material lead to a rapid transfer of thermal energy. Air and other gases are poor conductors of heat because there are no bonds between adjacent molecules.

FIGURE 19.22 Conduction of heat through a solid.

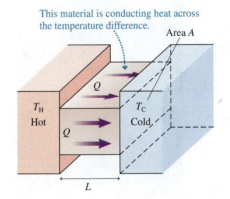

This material is conducting heat across the temperature difference.

TABLE 19.5 Thermal conductivities

Material	k (W/m K)
Diamond	2000
Silver	430
Copper	400
Aluminum	240
Iron	80
Stainless steel	14
Ice	1.7
Concrete	0.8
Glass	0.8
Styrofoam	0.035
Air (20°C, 1 atm)	0.023

Some of our perceptions of hot and cold have more to do with thermal conductivity than with temperature. For example, a metal chair feels colder to your bare skin than a wooden chair not because it has a lower temperature—both are at room temperature—but because it has a much larger thermal conductivity that conducts heat away from your body at a much higher rate.

EXAMPLE 19.10 | **Keeping a freezer cold**

A 1.8-m-wide by 1.0-m-tall by 0.65-m-deep home freezer is insulated with 5.0-cm-thick Styrofoam insulation. At what rate must the compressor remove heat from the freezer to keep the inside at $-20°C$ in a room where the air temperature is $25°C$?

MODEL Heat is transferred through each of the six sides by conduction. The compressor must remove heat at the same rate it enters to maintain a steady temperature inside. The heat conduction is determined primarily by the thick insulation, so we'll neglect the thin inner and outer panels.

SOLVE Each of the six sides is a slab of Styrofoam with cross-section area A_i and thickness $L = 5.0$ cm. The total rate of heat transfer is

$$\frac{dQ}{dt} = \sum_{i=1}^{6} k \frac{A_i}{L} \Delta T = \frac{k \Delta T}{L} \sum_{i=1}^{6} A_i = \frac{k \Delta T}{L} A_{total}$$

The total surface area is

$$A_{total} = 2 \times (1.8 \text{ m} \times 1.0 \text{ m} + 1.8 \text{ m} \times 0.65 \text{ m}$$
$$+ 1.0 \text{ m} \times 0.65 \text{ m}) = 7.24 \text{ m}^2$$

Using $k = 0.035$ W/m K from Table 19.5, we find

$$\frac{dQ}{dt} = \frac{k \Delta T}{L} A_{total} = \frac{(0.035 \text{ W/m K})(45 \text{ K})(7.24 \text{ m}^2)}{0.050 \text{ m}} = 230 \text{ W}$$

Heat enters the freezer through the walls at the rate 230 J/s; thus the compressor must remove 230 J of heat energy every second to keep the temperature at $-20°C$.

ASSESS We'll learn in Chapter 19 how the compressor does this and how much work it must do. A typical freezer uses electric energy at a rate of about 150 W, so our result seems reasonable.

Warm water (colored) moves by convection.

Convection

Air is a poor conductor of heat, but thermal energy is easily transferred through air, water, and other fluids because the air and water can flow. A pan of water on the stove is heated at the bottom. This heated water expands, becomes less dense than the water above it, and thus rises to the surface, while cooler, denser water sinks to take its place. The same thing happens to air. This transfer of thermal energy by the motion of a fluid—the well-known idea that "heat rises"—is called **convection.**

Convection is usually the main mechanism for heat transfer in fluid systems. On a small scale, convection mixes the pan of water that you heat on the stove; on a large scale, convection is responsible for making the wind blow and ocean currents circulate. Air is a very poor thermal conductor, but it is very effective at transferring energy by convection. To use air for thermal insulation, it is necessary to trap the air in small pockets to limit convection. And that's exactly what feathers, fur, double-paned windows, and fiberglass insulation do. Convection is much more rapid in water than in air, which is why people can die of hypothermia in 68°F (20°C) water but can live quite happily in 68°F air.

Because convection involves the often-turbulent motion of fluids, there is no simple equation for energy transfer by convection. Our description must remain qualitative.

Radiation

The sun *radiates* energy to earth through the vacuum of space. Similarly, you feel the warmth from the glowing red coals in a fireplace.

All objects emit energy in the form of **radiation,** electromagnetic waves generated by oscillating electric charges in the atoms that form the object. These waves transfer energy from the object that emits the radiation to the object that absorbs it. Electromagnetic waves carry energy from the sun; this energy is absorbed when sunlight falls on your skin, warming you by increasing your thermal energy. Your skin also emits electromagnetic radiation, helping to keep your body cool by decreasing your thermal energy. Radiation is a significant part of the *energy balance* that keeps your body at the proper temperature.

> **NOTE** The word "radiation" comes from "radiate," meaning "to beam." Radiation can refer to x rays or to the radioactive decay of nuclei, but it also can refer simply to light and other forms of electromagnetic waves that "beam" from an object. Here we are using this second meaning of the term.

You are familiar with radiation from objects hot enough to glow "red hot" or, at a high enough temperature, "white hot." The sun is simply a very hot ball of glowing gas, and the white light from an incandescent lightbulb is radiation emitted by a thin wire filament heated to a very high temperature by an electric current. Objects at lower temperatures also radiate, but at infrared wavelengths. You can't see this radiation (although you can sometimes feel it), but infrared-sensitive detectors can measure it and are used to make thermal images.

The energy radiated by an object depends strongly on temperature. If a small amount of heat energy dQ is radiated during a small time interval dt by an object with surface area A and absolute temperature T, the *rate* of heat transfer is found to be

$$\frac{dQ}{dt} = e\sigma A T^4 \qquad (19.46)$$

Because the rate of energy transfer is power ($1 \text{ J/s} = 1 \text{ W}$), dQ/dt is often called the *radiated power.* Notice the very strong fourth-power dependence on temperature. Doubling the absolute temperature of an object increases the radiated power by a factor of 16!

The parameter e in Equation 19.46 is the **emissivity** of the surface, a measure of how effectively it radiates. The value of e ranges from 0 to 1. σ is a constant, known as the Stefan-Boltzmann constant, with the value

$$\sigma = 5.67 \times 10^{-8} \text{ W/m}^2 \text{ K}^4$$

> **NOTE** Just as in the ideal-gas law, the temperature in Equation 19.46 *must* be in kelvins.

Objects not only emit radiation, they also *absorb* radiation emitted by their surroundings. Suppose an object at temperature T is surrounded by an environment at temperature T_0. The *net* rate at which the object radiates heat energy—that is, radiation emitted minus radiation absorbed—is

$$\frac{dQ_{\text{net}}}{dt} = e\sigma A(T^4 - T_0^4) \qquad (19.47)$$

This makes sense. An object should have no *net* radiation if it's in thermal equilibrium ($T = T_0$) with its surroundings.

Notice that the emissivity e appears for absorption as well as emission; good emitters are also good absorbers. A perfect absorber ($e = 1$), one absorbing all light and radiation impinging on it but reflecting none, would appear completely black. Thus a perfect absorber is sometimes called a **black body.** But a perfect absorber would also be a perfect emitter, so thermal radiation from an ideal emitter is called **black-body radiation.** It seems strange that black objects are perfect emitters, but think of black charcoal glowing bright red in a fire. At room temperature, it "glows" equally bright with infrared.

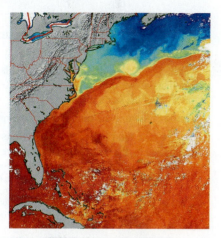

This satellite image shows radiation emitted by the ocean waters off the east coast of the United States. You can clearly see the warm waters of the Gulf Stream, a large-scale convection that transfers heat to northern latitudes.

EXAMPLE 19.11 | Taking the sun's temperature

The radius of the sun is 6.96×10^8 m. At the distance of the earth, 1.50×10^{11} m, the intensity of solar radiation (measured by satellites above the atmosphere) is 1370 W/m². What is the temperature of the sun's surface?

MODEL Assume the sun to be an ideal radiator with $e = 1$.

SOLVE The total power radiated by the sun is the power per m² multiplied by the surface area of a sphere extending to the earth:

$$P = \frac{1370 \text{ W}}{1 \text{ m}^2} \times 4\pi (1.50 \times 10^{11} \text{ m})^2 = 3.87 \times 10^{26} \text{ W}$$

That is, the sun radiates energy at the rate $dQ/dt = 3.87 \times 10^{26}$ J/s. That's a lot of power! This energy is radiated from the surface of a sphere of radius R_S. Using this information in Equation 19.46, we find that the sun's surface temperature is

$$T = \left[\frac{dQ/dt}{e\sigma(4\pi R_S^2)} \right]^{1/4}$$

$$= \left[\frac{3.87 \times 10^{26} \text{ W}}{(1)(5.67 \times 10^{-8} \text{ W/m}^2 \text{ K}^4)4\pi(6.96 \times 10^8 \text{ m})^2} \right]^{1/4}$$

$$= 5790 \text{ K}$$

ASSESS This temperature is confirmed by measurements of the solar spectrum, a topic we'll explore in Part VIII.

Thermal radiation plays a prominent role in climate and global warming. The earth as a whole is in thermal equilibrium. Consequently, it must radiate back into space exactly as much energy as it receives from the sun. The incoming radiation from the hot sun is mostly visible light. The earth's atmosphere is transparent to visible light, so this radiation reaches the surface and is absorbed. The cooler earth radiates infrared radiation, but the atmosphere is *not* completely transparent to infrared. Some components of the atmosphere, notably water vapor and carbon dioxide, are strong absorbers of infrared radiation. They hinder the emission of radiation and, rather like a blanket, keep the earth's surface warmer than it would be without these gases in the atmosphere.

The **greenhouse effect,** as it's called, is a natural part of the earth's climate. The earth would be much colder and mostly frozen were it not for naturally occurring carbon dioxide in the atmosphere. But carbon dioxide also results from the burning of fossil fuels, and human activities since the beginning of the industrial revolution have increased the atmospheric concentration of carbon dioxide by nearly 50%. This human contribution has amplified the greenhouse effect and is the primary cause of global warming.

STOP TO THINK 19.7 Suppose you are an astronaut in space, hard at work in your sealed spacesuit. The only way that you can transfer excess heat to the environment is by

a. Conduction.   b. Convection.   c. Radiation.   d. Evaporation.

CHALLENGE EXAMPLE 19.12 | Boiling water

400 mL of water is poured into a covered 8.0-cm-diameter, 150 g glass beaker with a 2.0-mm-thick bottom; then the beaker is placed on a 400°C hot plate. Once the water reaches the boiling point, how long will it take to boil away all the water?

MODEL The bottom of the beaker is a heat-conducting material transferring heat energy from the 400°C hot plate to the 100°C boiling water. The temperature of both the water and the beaker remains constant until the water has boiled away. We'll assume that heat losses due to convection and radiation are negligible, in which case the heat energy entering the system is used entirely for the phase change of the water. The beaker's mass isn't relevant because its temperature isn't changing.

SOLVE The heat energy required to boil mass M of water is

$$Q = ML_v$$

where $L_v = 2.26 \times 10^6$ J/kg is the heat of vaporization. The heat energy transferred through the bottom of the beaker during a time interval Δt is

$$Q = k\frac{A}{L}\Delta T \Delta t$$

where $k = 0.80$ W/m K is the thermal conductivity of glass. Because the heat transferred by conduction is used entirely for boiling the water, we can combine these two expressions:

$$k\frac{A}{L}\Delta T \Delta t = ML_v$$

and then solve for Δt:

$$\Delta t = \frac{MLL_v}{kA\,\Delta T} = \frac{(0.40 \text{ kg})(0.0020 \text{ m})(2.26 \times 10^6 \text{ J/kg})}{(0.80 \text{ W/m K})(0.0050 \text{ m}^2)(300 \text{ K})}$$

$$= 1500 \text{ s} = 25 \text{ min}$$

We used the density of water to find that $M = 400$ g $= 0.40$ kg and calculated $A = \pi r^2 = 0.0050$ m² as the area through which heat conduction occurs.

ASSESS 400 mL is roughly 2 cups, a small hot plate can bring 2 cups of water to a boil in 5 min or so, and boiling the water away takes quite a bit longer than bringing it to a boil. 25 min is a slight underestimate since we neglected energy losses due to convection and radiation, but it seems reasonable. A stove could boil the water away much faster because the burner temperature (gas flame or red-hot heating coil) is much higher.

SUMMARY

The goal of Chapter 19 has been to learn and apply the first law of thermodynamics.

GENERAL PRINCIPLES

First Law of Thermodynamics

$$\Delta E_{th} = W + Q$$

The first law is a general statement of energy conservation.

Work W and heat Q depend on the process by which the system is changed.

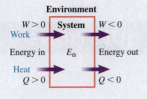

Energy

Thermal energy E_{th} Microscopic energy of moving molecules and stretched molecular bonds. ΔE_{th} depends on the initial/final states but is independent of the process.

Work W Energy transferred to the system by forces in a mechanical interaction.

Heat Q Energy transferred to the system via atomic-level collisions in a thermal interaction.

IMPORTANT CONCEPTS

Solving Problems of Work on an Ideal Gas

The work done on a gas is

$$W = -\int_{V_i}^{V_f} p \, dV$$

$$= -(\text{area under the } pV \text{ curve})$$

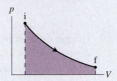

Solving Calorimetry Problems

When two or more systems interact thermally, they come to a common final temperature determined by

$$Q_{net} = Q_1 + Q_2 + \cdots = 0$$

An **adiabatic process** has $Q = 0$. Gases move along an **adiabat** for which $pV^\gamma =$ constant, where $\gamma = C_P/C_V$ is the **specific heat ratio**. An adiabatic process changes the temperature of the gas without heating it.

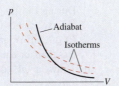

The **heat of transformation L** is the energy needed to cause 1 kg of substance to undergo a phase change

$$Q = \pm ML$$

The **specific heat c** of a substance is the energy needed to raise the temperature of 1 kg by 1 K:

$$Q = Mc \, \Delta T$$

The **molar specific heat C** is the energy needed to raise the temperature of 1 mol by 1 K:

$$Q = nC \, \Delta T$$

The molar specific heat of gases depends on the *process* by which the temperature is changed:

$C_V =$ molar specific heat at **constant volume**

$C_P = C_V + R =$ molar specific heat at **constant pressure**

Heat is transferred by **conduction, convection, radiation,** and **evaporation.**

Conduction: $dQ/dt = (kA/L)\Delta T$

Radiation: $dQ/dt = e\sigma AT^4$

SUMMARY OF BASIC GAS PROCESSES

Process	Definition	Stays constant	Work	Heat
Isochoric	$\Delta V = 0$	V and p/T	$W = 0$	$Q = nC_V \Delta T$
Isobaric	$\Delta p = 0$	p and V/T	$W = -p \, \Delta V$	$Q = nC_P \Delta T$
Isothermal	$\Delta T = 0$	T and pV	$W = -nRT \ln(V_f/V_i)$	$\Delta E_{th} = 0$
Adiabatic	$Q = 0$	pV^γ	$W = \Delta E_{th}$	$Q = 0$
All gas processes	First law $\Delta E_{th} = W + Q = nC_V \Delta T$		Ideal-gas law $pV = nRT$	

TERMS AND NOTATION

work, W	adiabatic process	molar specific heat at constant	thermal conductivity, k
mechanical interaction	specific heat, c	volume, C_V	convection
mechanical equilibrium	molar specific heat, C	molar specific heat at constant	radiation
heat, Q	heat of transformation, L	pressure, C_P	emissivity, e
thermal interaction	heat of fusion, L_f	specific heat ratio, γ	black body
thermal equilibrium	heat of vaporization, L_v	adiabat	black-body radiation
first law of thermodynamics	calorimetry	conduction	greenhouse effect
thermodynamic energy model			

CONCEPTUAL QUESTIONS

1. When a space capsule returns to earth, its surfaces get very hot as it passes through the atmosphere at high speed. Has the space capsule been heated? If so, what was the source of the heat? If not, why is it hot?

2. Do (a) temperature, (b) heat, and (c) thermal energy describe a property of a system, an interaction of the system with its environment, or both? Explain.

3. Two containers hold equal masses of nitrogen gas at equal temperatures. You supply 10 J of heat to container A while not allowing its volume to change, and you supply 10 J of heat to container B while not allowing its pressure to change. Afterward, is temperature T_A greater than, less than, or equal to T_B? Explain.

4. You need to raise the temperature of a gas by 10°C. To use the least amount of heat energy, should you heat the gas at constant pressure or at constant volume? Explain.

5. *Why* is the molar specific heat of a gas at constant pressure larger than the molar specific heat at constant volume?

6. FIGURE Q19.6 shows an adiabatic process.
   a. Is the final temperature higher than, lower than, or equal to the initial temperature?
   b. Is any heat energy added to or removed from the system in this process? Explain.

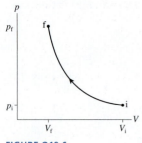

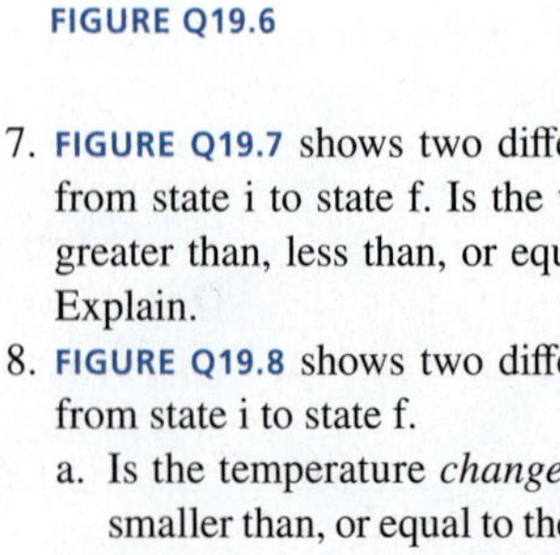

FIGURE Q19.6

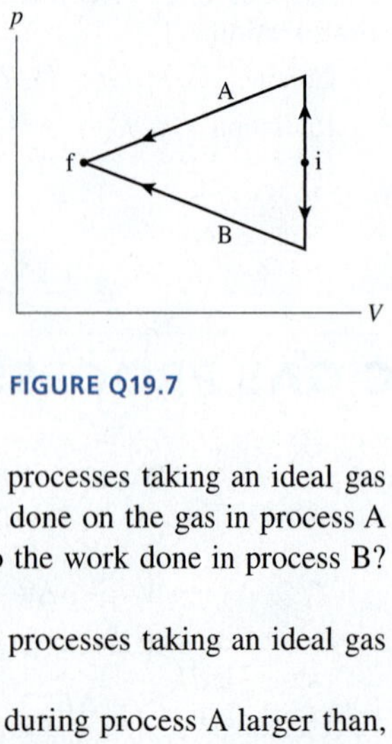

FIGURE Q19.7

7. FIGURE Q19.7 shows two different processes taking an ideal gas from state i to state f. Is the work done on the gas in process A greater than, less than, or equal to the work done in process B? Explain.

8. FIGURE Q19.8 shows two different processes taking an ideal gas from state i to state f.
   a. Is the temperature *change* ΔT during process A larger than, smaller than, or equal to the change during process B? Explain.
   b. Is the heat energy added during process A greater than, less than, or equal to the heat added during process B? Explain.

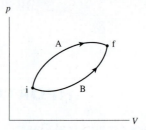

FIGURE Q19.8

FIGURE Q19.9

9. The gas cylinder in FIGURE Q19.9 is a rigid container that is well insulated except for the bottom surface, which is in contact with a block of ice. The initial gas temperature is $> 0°C$.
   a. During the process that occurs until the gas reaches a new equilibrium, are (i) ΔT, (ii) W, and (iii) Q greater than, less than, or equal to zero? Explain.
   b. Draw a pV diagram showing the process.

10. The gas cylinder in FIGURE Q19.10 is well insulated except for the bottom surface, which is in contact with a block of ice. The piston can slide without friction. The initial gas temperature is $> 0°C$.
    a. During the process that occurs until the gas reaches a new equilibrium, are (i) ΔT, (ii) W, and (iii) Q greater than, less than, or equal to zero? Explain.
    b. Draw a pV diagram showing the process.

FIGURE Q19.10

11. The gas cylinder in FIGURE Q19.11 is well insulated on all sides. The piston can slide without friction. Many small masses on top of the piston are removed one by one until the total mass is reduced by 50%.
    a. During this process, are (i) ΔT, (ii) W, and (iii) Q greater than, less than, or equal to zero? Explain.
    b. Draw a pV diagram showing the process.

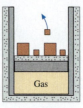

FIGURE Q19.11

EXERCISES AND PROBLEMS

Problems labeled ▦ integrate material from earlier chapters.

Exercises

Section 19.1 It's All About Energy

Section 19.2 Work in Ideal-Gas Processes

1. ‖ How much work is done on the gas in the process shown in **FIGURE EX19.1**?

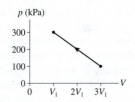

FIGURE EX19.1   FIGURE EX19.2

2. ‖ How much work is done on the gas in the process shown in **FIGURE EX19.2**?

3. ‖ 80 J of work are done on the gas in the process shown in **FIGURE EX19.3**. What is V_1 in cm³?

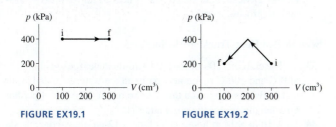

FIGURE EX19.3

4. ‖ A 2000 cm³ container holds 0.10 mol of helium gas at 300°C. How much work must be done to compress the gas to 1000 cm³ at (a) constant pressure and (b) constant temperature?

5. ‖ 500 J of work must be done to compress a gas to half its initial volume at constant temperature. How much work must be done to compress the gas by a factor of 10, starting from its initial volume?

Section 19.3 Heat

Section 19.4 The First Law of Thermodynamics

6. | Draw a first-law bar chart (see Figure 19.12) for the gas process in **FIGURE EX19.6**.

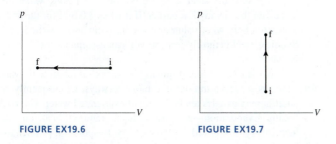

FIGURE EX19.6   FIGURE EX19.7

7. | Draw a first-law bar chart (see Figure 19.12) for the gas process in **FIGURE EX19.7**.

8. | Draw a first-law bar chart (see Figure 19.12) for the gas process in **FIGURE EX19.8**.

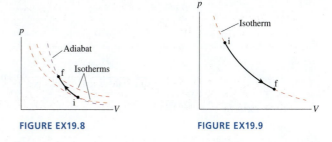

FIGURE EX19.8   FIGURE EX19.9

9. | Draw a first-law bar chart (see Figure 19.12) for the gas process in **FIGURE EX19.9**.

10. | A gas is compressed from 600 cm³ to 200 cm³ at a constant pressure of 400 kPa. At the same time, 100 J of heat energy is transferred out of the gas. What is the change in thermal energy of the gas during this process?

11. | 500 J of work are done on a system in a process that decreases the system's thermal energy by 200 J. How much heat energy is transferred to or from the system?

Section 19.5 Thermal Properties of Matter

12. ‖ How much heat energy must be added to a 6.0-cm-diameter copper sphere to raise its temperature from −50°C to 150°C?

13. ‖ A rapidly spinning paddle wheel raises the temperature of 200 mL of water from 21°C to 25°C. How much (a) heat is transferred and (b) work is done in this process?

14. | a. 100 J of heat energy are transferred to 20 g of mercury. By how much does the temperature increase?
    b. How much heat is needed to raise the temperature of 20 g of water by the same amount?

15. ‖ How much heat is needed to change 20 g of mercury at 20°C into mercury vapor at the boiling point?

16. ‖ What is the maximum mass of ethyl alcohol you could boil with 1000 J of heat, starting from 20°C?

17. | **BIO** One way you keep from overheating is by perspiring. Evaporation—a phase change—requires heat, and the heat energy is removed from your body. Evaporation is much like boiling, only water's heat of vaporization at 35°C is a somewhat larger 24×10^5 J/kg because at lower temperatures more energy is required to break the molecular bonds. Very strenuous activity can cause an adult human to produce 30 g of perspiration per minute. If all the perspiration evaporates, rather than dripping off, at what rate (in J/s) is it possible to exhaust heat by perspiring?

18. | A scientist whose scale is broken but who has a working 2.5 kW heating coil and a thermometer decides to improvise to determine the mass of a block of aluminum she has recently acquired. She heats the aluminum for 30 s and finds that its temperature increases from 20°C to 35°C. What is the mass of the aluminum?

19. ‖ Two cars collide head-on while each is traveling at 80 km/h. Suppose all their kinetic energy is transformed into the thermal energy of the wrecks. What is the temperature increase of each car? You can assume that each car's specific heat is that of iron.

20. ‖ An experiment measures the temperature of a 500 g substance while steadily supplying heat to it. **FIGURE EX19.20** shows the results of the experiment. What are the (a) specific heat of the solid phase, (b) specific heat of the liquid phase, (c) melting and boiling temperatures, and (d) heats of fusion and vaporization?

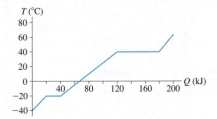

FIGURE EX19.20

Section 19.6 Calorimetry

21. ‖ 30 g of copper pellets are removed from a 300°C oven and immediately dropped into 100 mL of water at 20°C in an insulated cup. What will the new water temperature be?

22. ‖ A 750 g aluminum pan is removed from the stove and plunged into a sink filled with 10.0 L of water at 20.0°C. The water temperature quickly rises to 24.0°C. What was the initial temperature of the pan in °C and in °F?

23. ‖ A 50.0 g thermometer is used to measure the temperature of 200 mL of water. The specific heat of the thermometer, which is mostly glass, is 750 J/kg K, and it reads 20.0°C while lying on the table. After being completely immersed in the water, the thermometer's reading stabilizes at 71.2°C. What was the actual water temperature before it was measured?

24. ‖ A 500 g metal sphere is heated to 300°C, then dropped into a beaker containing 300 cm³ of mercury at 20.0°C. A short time later the mercury temperature stabilizes at 99.0°C. Identify the metal.

25. ‖‖ A 65 cm³ block of iron is removed from an 800°C furnace and immediately dropped into 200 mL of 20°C water. What fraction of the water boils away?

Section 19.7 The Specific Heats of Gases

26. ‖ A container holds 1.0 g of argon at a pressure of 8.0 atm.
   a. How much heat is required to increase the temperature by 100°C at constant volume?
   b. How much will the temperature increase if this amount of heat energy is transferred to the gas at constant pressure?

27. ‖ A container holds 1.0 g of oxygen at a pressure of 8.0 atm.
   a. How much heat is required to increase the temperature by 100°C at constant pressure?
   b. How much will the temperature increase if this amount of heat energy is transferred to the gas at constant volume?

28. ‖ The volume of a gas is halved during an adiabatic compression that increases the pressure by a factor of 2.5.
   a. What is the specific heat ratio γ?
   b. By what factor does the temperature increase?

29. ‖ A gas cylinder holds 0.10 mol of O_2 at 150°C and a pressure of 3.0 atm. The gas expands adiabatically until the pressure is halved. What are the final (a) volume and (b) temperature?

30. ‖ A gas cylinder holds 0.10 mol of O_2 at 150°C and a pressure of 3.0 atm. The gas expands adiabatically until the volume is doubled. What are the final (a) pressure and (b) temperature?

31. ‖ 0.10 mol of nitrogen gas follow the two processes shown in **FIGURE EX19.31**. How much heat is required for each?

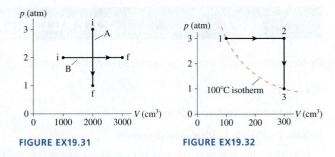

FIGURE EX19.31   **FIGURE EX19.32**

32. ‖ A monatomic gas follows the process $1 \rightarrow 2 \rightarrow 3$ shown in **FIGURE EX19.32**. How much heat is needed for (a) process $1 \rightarrow 2$ and (b) process $2 \rightarrow 3$?

Section 19.8 Heat-Transfer Mechanisms

33. ‖ The ends of a 20-cm-long, 2.0-cm-diameter rod are maintained at 0°C and 100°C by immersion in an ice-water bath and boiling water. Heat is conducted through the rod at 4.5×10^4 J per hour. Of what material is the rod made?

34. ‖ A 10 m × 14 m house is built on a 12-cm-thick concrete slab. What is the heat-loss rate through the slab if the ground temperature is 5°C while the interior of the house is 22°C?

35. ‖ You are boiling pasta and absentmindedly grab a copper stirring spoon rather than your wooden spoon. The copper spoon has a 20 mm × 1.5 mm rectangular cross section, and the distance from the boiling water to your 35°C hand is 18 cm. How long does it take the spoon to transfer 25 J of energy to your hand?

36. ‖ What maximum power can be radiated by a 10-cm-diameter solid lead sphere? Assume an emissivity of 1.

37. ‖‖ Radiation from the head is a major source of heat loss from the
BIO human body. Model a head as a 20-cm-diameter, 20-cm-tall cylinder with a flat top. If the body's surface temperature is 35°C, what is the net rate of heat loss on a chilly 5°C day? All skin, regardless of color, is effectively black in the infrared where the radiation occurs, so use an emissivity of 0.95.

Problems

38. ‖ A 5.0 g ice cube at −20°C is in a rigid, sealed container from which all the air has been evacuated. How much heat is required to change this ice cube into steam at 200°C? Steam has $C_V = 1500$ J/kg K and $C_P = 1960$ J/kg K.

39. ‖ A 5.0-m-diameter garden pond is 30 cm deep. Solar energy is incident on the pond at an average rate of 400 W/m². If the water absorbs all the solar energy and does not exchange energy with its surroundings, how many hours will it take to warm from 15°C to 25°C?

40. ‖ The burner on an electric stove has a power output of 2.0 kW. A 750 g stainless steel teakettle is filled with 20°C water and placed on the already hot burner. If it takes 3.0 min for the water to reach a boil, what volume of water, in cm³, was in the kettle? Stainless steel is mostly iron, so you can assume its specific heat is that of iron.

41. ‖ When air is inhaled, it quickly becomes saturated with water
BIO vapor as it passes through the moist airways. Consequently, an adult human exhales about 25 mg of evaporated water with each breath. Evaporation—a phase change—requires heat, and the heat energy is removed from your body. Evaporation is much like boiling, only water's heat of vaporization at 35°C is a somewhat larger 24×10^5 J/kg because at lower temperatures more energy is required to break the molecular bonds. At 12 breaths/min, on a

dry day when the inhaled air has almost no water content, what is the body's rate of energy loss (in J/s) due to exhaled water? (For comparison, the energy loss from radiation, usually the largest loss on a cool day, is about 100 J/s.)

42. ‖ 512 g of an unknown metal at a temperature of 15°C is dropped into a 100 g aluminum container holding 325 g of water at 98°C. A short time later, the container of water and metal stabilizes at a new temperature of 78°C. Identify the metal.

43. ‖ A 150 L ($\approx$ 40 gal) electric hot-water tank has a 5.0 kW heater. How many minutes will it take to raise the water temperature from 65°F to 140°F?

44. ‖ The specific heat of most solids is nearly constant over a wide
CALC temperature range. Not so for diamond. Between 200 K and 600 K, the specific heat of diamond is reasonably well described by $c = 2.8T - 350$ J/kg K, where T is in K. For gemstone diamonds, 1 carat = 200 mg. How much heat energy is needed to raise the temperature of a 3.5 carat diamond from -50°C to 250°C?

45. ‖ A lava flow is threatening to engulf a small town. A 400-m-wide, 35-cm-thick tongue of 1200°C lava is advancing at the rate of 1.0 m per minute. The mayor devises a plan to stop the lava in its tracks by flying in large quantities of 20°C water and dousing it. The lava has density 2500 kg/m³, specific heat 1100 J/kg K, melting temperature 800°C, and heat of fusion 4.0×10^5 J/kg. How many liters of water per minute, at a minimum, will be needed to save the town?

46. ‖ Suppose you take and hold a deep breath on a chilly day, inhaling
BIO 3.0 L of air at 0°C and 1 atm.
 a. How much heat must your body supply to warm the air to your internal body temperature of 37°C?
 b. By how much does the air's volume increase as it warms?

47. ‖ Your 300 mL cup of coffee is too hot to drink when served at 90°C. What is the mass of an ice cube, taken from a -20°C freezer, that will cool your coffee to a pleasant 60°C?

48. ‖ A typical nuclear reactor generates 1000 MW (1000 MJ/s) of electrical energy. In doing so, it produces 2000 MW of "waste heat" that must be removed from the reactor to keep it from melting down. Many reactors are sited next to large bodies of water so that they can use the water for cooling. Consider a reactor where the intake water is at 18°C. State regulations limit the temperature of the output water to 30°C so as not to harm aquatic organisms. How many liters of cooling water have to be pumped through the reactor each minute?

49. ‖‖ 2.0 mol of gas are at 30°C and a pressure of 1.5 atm. How much work must be done on the gas to compress it to one third of its initial volume at (a) constant temperature and (b) constant pressure? (c) Show both processes on a single pV diagram.

50. ‖‖ A 6.0-cm-diameter cylinder of nitrogen gas has a 4.0-cm-thick movable copper piston. The cylinder is oriented vertically, as shown in FIGURE P19.50, and the air above the piston is evacuated. When the gas temperature is 20°C, the piston floats 20 cm above the bottom of the cylinder.
 a. What is the gas pressure?
 b. How many gas molecules are in the cylinder?
 Then 2.0 J of heat energy are transferred to the gas.
 c. What is the new equilibrium temperature of the gas?
 d. What is the final height of the piston?
 e. How much work is done on the gas as the piston rises?

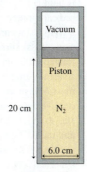

FIGURE P19.50

51. ‖ A 560 kg concrete table needs to be supported at the four corners by compressed-air cylinders. Each cylinder is 25 cm in diameter and has a 1.20 m initial length when the pressure inside is 1.0 atm. A hoist lowers the table very slowly, compressing the cylinders while allowing them to stay in thermal equilibrium with their surroundings. How much work has been done on the gas of the four cylinders when the table reaches its equilibrium position?

52. ‖ An ideal-gas process is described by $p = cV^{1/2}$, where c is a
CALC constant.
 a. Find an expression for the work done on the gas in this process as the volume changes from V_1 to V_2.
 b. 0.033 mol of gas at an initial temperature of 150°C is compressed, using this process, from 300 cm³ to 200 cm³. How much work is done on the gas?
 c. What is the final temperature of the gas in °C?

53. ‖ A 10-cm-diameter cylinder contains argon gas at 10 atm pressure and a temperature of 50°C. A piston can slide in and out of the cylinder. The cylinder's initial length is 20 cm. 2500 J of heat are transferred to the gas, causing the gas to expand at constant pressure. What are (a) the final temperature and (b) the final length of the cylinder?

54. ‖‖ A cube 20 cm on each side contains 3.0 g of helium at 20°C. 1000 J of heat energy are transferred to this gas. What are (a) the final pressure if the process is at constant volume and (b) the final volume if the process is at constant pressure? (c) Show and label both processes on a single pV diagram.

55. ‖ An 8.0-cm-diameter, well-insulated vertical cylinder containing nitrogen gas is sealed at the top by a 5.1 kg frictionless piston. The air pressure above the piston is 100 kPa.
 a. What is the gas pressure inside the cylinder?
 b. Initially, the piston height above the bottom of the cylinder is 26 cm. What will be the piston height if an additional 3.5 kg are placed on top of the piston?

56. ‖ n moles of an ideal gas at temperature T_1 and volume V_1 expand isothermally until the volume has doubled. In terms of n, T_1, and V_1, what are (a) the final temperature, (b) the work done on the gas, and (c) the heat energy transferred to the gas?

57. ‖‖ 5.0 g of nitrogen gas at 20°C and an initial pressure of 3.0 atm undergo an isobaric expansion until the volume has tripled.
 a. What are the gas volume and temperature after the expansion?
 b. How much heat energy is transferred to the gas to cause this expansion?
 The gas pressure is then decreased at constant volume until the original temperature is reached.
 c. What is the gas pressure after the decrease?
 d. What amount of heat energy is transferred from the gas as its pressure decreases?
 e. Show the total process on a pV diagram. Provide an appropriate scale on both axes.

58. ‖ 0.10 mol of nitrogen gas follow the two processes shown in FIGURE P19.58. How much heat is required for each?

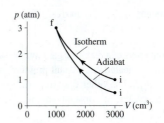

FIGURE P19.58

59. ||| You come into lab one day and find a well-insulated 2000 mL thermos bottle containing 500 mL of boiling liquid nitrogen. The remainder of the thermos has nitrogen gas at a pressure of 1.0 atm. The gas and liquid are in thermal equilibrium. While waiting for lab to start, you notice a piece of iron on the table with "197 g" written on it. Just for fun, you drop the iron into the thermos and seal the cap tightly so that no gas can escape. After a few seconds have passed, what is the pressure inside the thermos? The density of liquid nitrogen is 810 kg/m³.

60. ||| Your laboratory assignment for the week is to measure the specific heat ratio γ of carbon dioxide. The gas is contained in a cylinder with a movable piston and a thermometer. When the piston is withdrawn as far as possible, the cylinder's length is 20 cm. You decide to push the piston in very rapidly by various amounts and, for each push, to measure the temperature of the carbon dioxide. Before each push, you withdraw the piston all the way and wait several minutes for the gas to come to the room temperature of 21°C. Your data are as follows:

Push (cm)	Temperature (°C)
5	35
10	68
13	110
15	150

Use the best-fit line of an appropriate graph to determine γ for carbon dioxide.

61. || Two cylinders each contain 0.10 mol of a diatomic gas at 300 K and a pressure of 3.0 atm. Cylinder A expands isothermally and cylinder B expands adiabatically until the pressure of each is 1.0 atm.
   a. What are the final temperature and volume of each?
   b. Show both processes on a single pV diagram. Use an appropriate scale on both axes.

62. ||| FIGURE P19.62 shows a thermodynamic process followed by 120 mg of helium.
   a. Determine the pressure (in atm), temperature (in °C), and volume (in cm³) of the gas at points 1, 2, and 3. Put your results in a table for easy reading.
   b. How much work is done on the gas during each of the three segments?
   c. How much heat energy is transferred to or from the gas during each of the three segments?

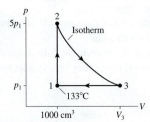

FIGURE P19.62

63. || Two containers of a diatomic gas have the same initial conditions. One container, heated at constant pressure, has a temperature increase of 20°C. The other container receives the same quantity of heat energy, but at constant volume. What is its temperature increase?

64. || 14 g of nitrogen gas at STP are adiabatically compressed to a pressure of 20 atm. What are (a) the final temperature, (b) the work done on the gas, (c) the heat transfer to the gas, and (d) the compression ratio V_{max}/V_{min} ? (e) Show the process on a pV diagram, using proper scales on both axes.

65. || 14 g of nitrogen gas at STP are pressurized in an isochoric process to a pressure of 20 atm. What are (a) the final temperature, (b) the work done on the gas, (c) the heat transfer to the gas, and (d) the pressure ratio p_{max}/p_{min} ? (e) Show the process on a pV diagram, using proper scales on both axes.

66. || A cylinder with a movable piston contains n moles of gas at
CALC a temperature higher than that of the surrounding environment. An external force on the piston keeps the pressure constant while the gas cools as $\Delta T = (\Delta T)_0 e^{-t/\tau}$, where ΔT is the temperature *difference* between the gas and the environment, $(\Delta T)_0$ is the initial temperature difference, and τ is the time constant.
   a. Find an expression for the *rate* at which the environment does work on the gas. Recall that the rate of doing work is *power*.
   b. What power is initially supplied by the environment if 0.15 mol of gas are initially 12°C warmer than the surroundings and cool with a time constant of 60 s?

67. || When strong winds rapidly carry air down from mountains to a lower elevation, the air has no time to exchange heat with its surroundings. The air is compressed as the pressure rises, and its temperature can increase dramatically. These warm winds are called Chinook winds in the Rocky Mountains and Santa Ana winds in California. Suppose the air temperature high in the mountains behind Los Angeles is 0°C at an elevation where the air pressure is 60 kPa. What will the air temperature be, in °C and °F, when the Santa Ana winds have carried this air down to an elevation near sea level where the air pressure is 100 kPa?

68. || You would like to put a solar hot water system on your roof, but you're not sure it's feasible. A reference book on solar energy shows that the ground-level solar intensity in your city is 800 W/m² for at least 5 hours a day throughout most of the year. Assuming that a completely black collector plate loses energy only by radiation, and that the air temperature is 20°C, what is the equilibrium temperature of a collector plate directly facing the sun? Note that while a plate has two sides, only the side facing the sun will radiate because the opposite side will be well insulated.

69. || A cubical box 20 cm on a side is constructed from 1.2-cm-thick concrete panels. A 100 W lightbulb is sealed inside the box. What is the air temperature inside the box when the light is on if the surrounding air temperature is 20°C?

70. || A cylindrical copper rod and an iron rod with exactly the same dimensions are welded together end to end. The outside end of the copper rod is held at 100°C, and the outside end of the iron rod is held at 0°C. What is the temperature at the midpoint where the rods are joined together?

71. ||| Most stars are *main-sequence* stars, a group of stars for which size, mass, surface temperature, and radiated power are closely related. The sun, for instance, is a yellow main-sequence star with a surface temperature of 5800 K. For a main-sequence star whose mass M is more than twice that of the sun, the total radiated power, relative to the sun, is approximately $P/P_{sun} = 1.5(M/M_{sun})^{3.5}$. The star Regulus A is a bluish main-sequence star with mass $3.8M_{sun}$ and radius $3.1R_{sun}$. What is the surface temperature of Regulus A?

72. ||| A satellite to reflect radar is a 2.0-m-diameter, 2.0-mm-thick
CALC spherical copper shell. While orbiting the earth, the satellite absorbs sunlight and is warmed to 50°C. When it passes into the

earth's shadow, the satellite radiates energy to deep space. The temperature of deep space is actually 3 K, as a result of the Big Bang 14 billion years ago, but it is so much colder than the satellite that you can assume a deep-space temperature of 0 K. If the satellite's emissivity is 0.75, to what temperature, in °C, will it drop during the 45 minutes it takes to move through the earth's shadow?

73. ‖ The sun's intensity at the distance of the earth is 1370 W/m². 30% of this energy is reflected by water and clouds; 70% is absorbed. What would be the earth's average temperature (in °C) if the earth had no atmosphere? The emissivity of the surface is very close to 1. (The actual average temperature of the earth, about 15°C, is higher than your calculation because of the greenhouse effect.)

In Problems 74 through 76 you are given the equation used to solve a problem. For each of these, you are to
   a. Write a realistic problem for which this is the correct equation.
   b. Finish the solution of the problem.

74. $50 \text{ J} = -n(8.31 \text{ J/mol K})(350 \text{ K})\ln\left(\frac{1}{3}\right)$

75. $(200 \times 10^{-6} \text{ m}^3)(13{,}600 \text{ kg/m}^3)$
   $\times (140 \text{ J/kg K})(90°\text{C} - 15°\text{C})$
   $+ (0.50 \text{ kg})(449 \text{ J/kg K})(90°\text{C} - T_i) = 0$

76. $(10 \text{ atm})V_2^{1.40} = (1.0 \text{ atm})V_1^{1.40}$

Challenge Problems

77. ‖ 10 g of aluminum at 200°C and 20 g of copper are dropped into 50 cm³ of ethyl alcohol at 15°C. The temperature quickly comes to 25°C. What was the initial temperature of the copper?

78. ‖ A beaker with a metal bottom is filled with 20 g of water at 20°C. It is brought into good thermal contact with a 4000 cm³ container holding 0.40 mol of a monatomic gas at 10 atm pressure. Both containers are well insulated from their surroundings.

What is the gas pressure after a long time has elapsed? You can assume that the containers themselves are nearly massless and do not affect the outcome.

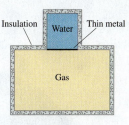

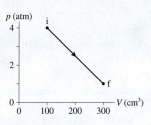

FIGURE CP19.78   **FIGURE CP19.79**

79. ‖ **FIGURE CP19.79** shows a thermodynamic process followed by 0.015 mol of hydrogen. How much heat energy is transferred to the gas?

80. ‖ One cylinder in the diesel engine of a truck has an initial volume of 600 cm³. Air is admitted to the cylinder at 30°C and a pressure of 1.0 atm. The piston rod then does 400 J of work to rapidly compress the air. What are its final temperature and volume?

81. ‖ 0.020 mol of a diatomic gas, with initial temperature 20°C,
CALC are compressed from 1500 cm³ to 500 cm³ in a process in which $pV^2 = $ constant. How much heat energy is added during this process?

82. ‖ A monatomic gas fills the left end of the cylinder in
CALC **FIGURE CP19.82**. At 300 K, the gas cylinder length is 10.0 cm and the spring is compressed by 2.0 cm. How much heat energy must be added to the gas to expand the cylinder length to 16.0 cm?

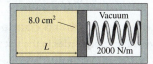

FIGURE CP19.82

20 The Micro/Macro Connection

Heating the air in a hot-air balloon increases the thermal energy of the molecules. This causes the gas to expand, lowering its density and allowing it to float in the cooler surrounding air.

IN THIS CHAPTER, you will see how macroscopic properties depend on the motion of atoms.

What is the micro/macro connection?

We've discovered several puzzles in the last two chapters:

- Why does the ideal-gas law work for every gas?
- Why is the molar specific heat the same for every monatomic gas? And for every diatomic gas? And for every elemental solid?
- Just what does temperature actually measure?

We can resolve these puzzles and understand many properties of macroscopic systems by studying the microscopic behavior of the system's atoms and molecules. This is the micro/macro connection.

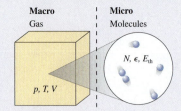

« LOOKING BACK Sections 19.3–19.5 Heat, the first law, and specific heats

Why do gases have pressure?

Gases have pressure due to the collisions of the molecules with the walls of the container. We'll find that we can calculate an average molecular speed, the root-mean-square speed, by relating the ideal-gas law to a microscopic calculation of the gas pressure.

What is temperature?

At a microscopic level, temperature measures the average translational kinetic energy of moving atoms and molecules. We will use this discovery to explain why all monatomic (and diatomic) gases have exactly the same molar specific heat.

« LOOKING BACK Section 19.7 The specific heats of gases

How do interacting systems reach equilibrium?

Two thermally interacting systems reach a common final temperature because they exchange energy via collisions. On average, more-energetic atoms transfer energy to less-energetic atoms until both systems have the same *average* translational kinetic energy.

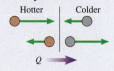

What is the second law of thermodynamics?

The second law of thermodynamics governs how systems evolve in time. One statement of the second law is that heat energy is transferred spontaneously from a hotter system to a colder system, but never from colder to hotter. Heat transfer is an irreversible process. The concept of entropy will help us see that irreversible processes occur because some macroscopic states are vastly more likely to occur than others.

20.1 Molecular Speeds and Collisions

If matter really consists of atoms and molecules, then the macroscopic properties of matter, such as temperature, pressure, and specific heat, ought to be related to the microscopic motion of those atoms and molecules. Our goal in this chapter is to explore this *micro/macro connection*. Let's begin by looking at a gas. Do all molecules in the gas move with the same speed, or is there a range of speeds?

FIGURE 20.1 shows an experiment to measure the speeds of molecules in a gas. The two rotating disks form a *velocity selector*. Once every revolution, the slot in the first disk allows a small pulse of molecules to pass through. By the time these molecules reach the second disk, the slots have rotated. The molecules can pass through the second slot and be detected *only* if they have exactly the right speed $v = L/\Delta t$ to travel between the two disks during time interval Δt it takes the axle to complete one revolution. Molecules having any other speed are blocked by the second disk. By changing the rotation speed of the axle, this apparatus can measure how many molecules have each of many possible speeds.

FIGURE 20.2 shows the results for nitrogen gas (N_2) at $T = 20°C$. The data are presented in the form of a **histogram,** a bar chart in which the height of each bar tells how many (or, in this case, what percentage) of the molecules have a speed in the *range* of speeds shown below the bar. For example, 16% of the molecules have speeds in the range from 600 m/s to 700 m/s. All the bars sum to 100%, showing that this histogram describes *all* of the molecules leaving the source.

It turns out that the molecules have what is called a *distribution* of speeds, ranging from as low as ≈ 100 m/s to as high as ≈ 1200 m/s. But not all speeds are equally likely; there is a *most likely speed* of ≈ 550 m/s. This is really fast, ≈ 1200 mph! Changing the temperature or changing to a different gas changes the most likely speed, as we'll learn later in the chapter, but it does not change the *shape* of the distribution.

If you were to repeat the experiment, you would again find the most likely speed to be ≈ 550 m/s and that 16% of the molecules have speeds between 600 m/s and 700 m/s. This is an important lesson. Although a gas consists of a vast number of molecules, each moving randomly, *averages,* such as the average number of molecules in the speed range 600 to 700 m/s, have precise, predictable values. **The micro/macro connection is built on the idea that the macroscopic properties of a system, such as temperature or pressure, are related to the *average* behavior of the atoms and molecules.**

Mean Free Path

Imagine someone opening a bottle of strong perfume a few feet away from you. If molecular speeds are hundreds of meters per second, you might expect to smell the perfume almost instantly. But that isn't what happens. As you know, it takes many seconds for the molecules to *diffuse* across the room. Let's see why this is.

FIGURE 20.3 shows a "movie" of one molecule. Instead of zipping along in a straight line, as it would in a vacuum, the molecule follows a convoluted zig-zag path in which it frequently collides with other molecules. A question we could ask is: What is the *average* distance between collisions? If a molecule has N_{coll} collisions as it travels distance L, the average distance between collisions, which is called the **mean free path** λ (lowercase Greek lambda), is

$$\lambda = \frac{L}{N_{coll}} \tag{20.1}$$

FIGURE 20.4a on the next page shows two molecules approaching each other. We will assume that the molecules are spherical and of radius r. We will also continue the ideal-gas assumption that the molecules undergo hard-sphere collisions, like billiard balls. In that case, the molecules will collide if the distance between their *centers* is less than $2r$. They will miss if the distance is greater than $2r$.

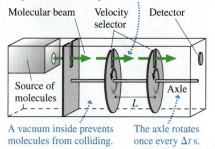

FIGURE 20.1 An experiment to measure the speeds of molecules in a gas.

Only molecules with speed $L/\Delta t$ reach the detector.

Molecular beam | Velocity selector | Detector

Source of molecules

Axle

L

A vacuum inside prevents molecules from colliding.

The axle rotates once every Δt s.

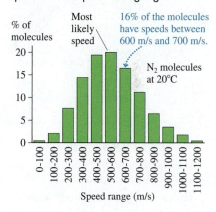

FIGURE 20.2 The distribution of molecular speeds in a sample of nitrogen gas.

% of molecules

Most likely speed

16% of the molecules have speeds between 600 m/s and 700 m/s.

N_2 molecules at 20°C

Speed range (m/s)

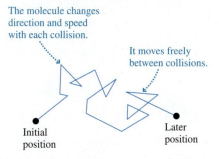

FIGURE 20.3 A single molecule follows a zig-zag path through a gas.

The molecule changes direction and speed with each collision.

It moves freely between collisions.

Initial position

Later position

FIGURE 20.4 A "sample" molecule collides with "target" molecules.

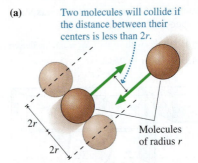

(a)

Two molecules will collide if the distance between their centers is less than $2r$.

$2r$

$2r$

Molecules of radius r

(b)

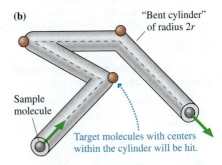

"Bent cylinder" of radius $2r$

Sample molecule

Target molecules with centers within the cylinder will be hit.

In **FIGURE 20.4b** we've drawn a cylinder of radius $2r$ centered on the trajectory of a "sample" molecule. The sample molecule collides with any "target" molecule whose center is located within the cylinder, causing the cylinder to bend at that point. Hence the number of collisions N_{coll} is equal to the number of molecules in a cylindrical volume of length L.

The volume of a cylinder is $V_{cyl} = AL = \pi(2r)^2 L$. If the number density of the gas is N/V particles per m³, then the number of collisions along a trajectory of length L is

$$N_{coll} = \frac{N}{V} V_{cyl} = \frac{N}{V} \pi(2r)^2 L = 4\pi \frac{N}{V} r^2 L \qquad (20.2)$$

Thus the mean free path between collisions is

$$\lambda = \frac{L}{N_{coll}} = \frac{1}{4\pi (N/V) r^2}$$

We made a tacit assumption in this derivation that the target molecules are at rest. While the general idea behind our analysis is correct, a more detailed calculation with all the molecules moving introduces an extra factor of $\sqrt{2}$, giving

$$\lambda = \frac{1}{4\sqrt{2}\,\pi (N/V) r^2} \qquad \text{(mean free path)} \qquad (20.3)$$

Laboratory measurements are necessary to determine atomic and molecular radii, but a reasonable rule of thumb is to assume that atoms in a monatomic gas have $r \approx 0.5 \times 10^{-10}$ m and diatomic molecules have $r \approx 1.0 \times 10^{-10}$ m.

EXAMPLE 20.1 | **The mean free path at room temperature**

What is the mean free path of a nitrogen molecule at 1.0 atm pressure and room temperature (20°C)?

SOLVE Nitrogen is a diatomic molecule, so $r \approx 1.0 \times 10^{-10}$ m. We can use the ideal-gas law in the form $pV = Nk_BT$ to determine the number density:

$$\frac{N}{V} = \frac{p}{k_BT} = \frac{101{,}300 \text{ Pa}}{(1.38 \times 10^{-23} \text{ J/K})(293 \text{ K})} = 2.5 \times 10^{25} \text{ m}^{-3}$$

Thus the mean free path is

$$\lambda = \frac{1}{4\sqrt{2}\,\pi (N/V) r^2}$$

$$= \frac{1}{4\sqrt{2}\,\pi (2.5 \times 10^{25} \text{ m}^{-3})(1.0 \times 10^{-10} \text{ m})^2}$$

$$= 2.3 \times 10^{-7} \text{ m} = 230 \text{ nm}$$

ASSESS You learned in Example 18.6 that the average separation between gas molecules at STP is ≈ 4 nm. It seems that any given molecule can slip between its neighbors, which are spread out in three dimensions, and travel—on average—about 60 times the average spacing before it collides with another molecule.

STOP TO THINK 20.1 The table shows the properties of four gases, each having the same number of molecules. Rank in order, from largest to smallest, the mean free paths λ_A to λ_D of molecules in these gases.

Gas	A	B	C	D
Volume	V	$2V$	V	V
Atomic mass	m	m	$2m$	m
Atomic radius	r	r	r	$2r$

20.2 Pressure in a Gas

Why does a gas have pressure? In Chapter 14, where pressure was introduced, we suggested that the pressure in a gas is due to collisions of the molecules with the walls of its container. The force due to one such collision may be unmeasurably tiny, but the steady rain of a vast number of molecules striking a wall each second exerts a measurable macroscopic force. The gas pressure is the force per unit area ($p = F/A$) resulting from these molecular collisions.

Our task in this section is to calculate the pressure by doing the appropriate averaging over molecular motions and collisions. This task can be divided into three main pieces:

1. Calculate the momentum change of a single molecule during a collision.
2. Find the force due to all collisions.
3. Introduce an appropriate average speed.

Force Due to Collisions

FIGURE 20.5 shows a molecule with an x-component of velocity v_x having a perfectly elastic collision with a wall and rebounding with its velocity changed to $-v_x$. This one molecule has a momentum *change*

$$\Delta p_x = m(-v_x) - mv_x = -2mv_x \tag{20.4}$$

Suppose there are N_{coll} collisions with the wall during a small interval of time Δt. Further, suppose that all the molecules have the same speed. (This latter assumption isn't really necessary, and we'll soon remove this constraint, but it helps us focus on the physics without getting lost in the math.) Then the total momentum change of the gas during Δt is

$$\Delta P_x = N_{coll}\,\Delta p_x = -2N_{coll}mv_x \tag{20.5}$$

You learned in Section 11.1 that the momentum principle can be written as

$$\Delta P_x = (F_{avg})_x\Delta t \tag{20.6}$$

Thus the average force of the wall *on the gas* is

$$(F_{on\ gas})_x = \frac{\Delta P_x}{\Delta t} = -\frac{2N_{coll}mv_x}{\Delta t} \tag{20.7}$$

Equation 20.7 has a negative sign because, as we've set it up, the collision force of the wall on the gas molecules is to the left. But Newton's third law is $(F_{on\ gas})_x = -(F_{on\ wall})_x$, so the force *on the wall* due to these collisions is

$$(F_{on\ wall})_x = \frac{2N_{coll}mv_x}{\Delta t} \tag{20.8}$$

We need to determine how many collisions occur during Δt. Assume that Δt is smaller than the mean time between collisions, so no collisions alter the molecular speeds during this interval. FIGURE 20.6 has shaded a volume of the gas of length $\Delta x = v_x\Delta t$. *Every one* of the molecules in this shaded region *that is moving to the right* will reach and collide with the wall during Δt. Molecules outside this region will not reach the wall and will not collide.

The shaded region has volume $A\,\Delta x$, where A is the area of the wall. If the gas has number density N/V, the number of molecules in the shaded region is $(N/V)A\,\Delta x = (N/V)Av_x\Delta t$. But only half these molecules are moving to the right, so the number of collisions during Δt is

$$N_{coll} = \frac{1}{2}\frac{N}{V}Av_x\Delta t \tag{20.9}$$

Substituting Equation 20.9 into Equation 20.8, we see that Δt cancels and that the force of the molecules on the wall is

$$F_{on\ wall} = \frac{N}{V}mv_x^2A \tag{20.10}$$

Notice that this expression for $F_{on\ wall}$ does not depend on any details of the collisions.

We can relax the assumption that all molecules have the same speed by replacing the squared velocity v_x^2 in Equation 20.10 with its average value. That is,

$$F_{on\ wall} = \frac{N}{V}m(v_x^2)_{avg}A \tag{20.11}$$

where $(v_x^2)_{avg}$ is the quantity v_x^2 averaged over all the molecules in the container.

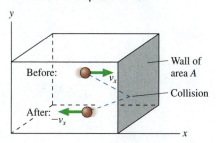

FIGURE 20.5 A molecule colliding with the wall exerts an impulse on it.

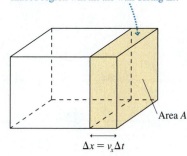

FIGURE 20.6 Determining the rate of collisions.

Only molecules moving to the right in the shaded region will hit the wall during Δt.

Area A

$\Delta x = v_x\Delta t$

The Root-Mean-Square Speed

We need to be somewhat careful when averaging velocities. The velocity component v_x has a sign. At any instant of time, half the molecules in a container move to the right and have positive v_x while the other half move to the left and have negative v_x. Thus the *average velocity* is $(v_x)_{avg} = 0$. If this weren't true, the entire container of gas would move away!

The speed of a molecule is $v = (v_x^2 + v_y^2 + v_z^2)^{1/2}$. Thus the average of the speed squared is

$$(v^2)_{avg} = (v_x^2 + v_y^2 + v_z^2)_{avg} = (v_x^2)_{avg} + (v_y^2)_{avg} + (v_z^2)_{avg} \qquad (20.12)$$

The square root of $(v^2)_{avg}$ is called the **root-mean-square speed** v_{rms}:

$$v_{rms} = \sqrt{(v^2)_{avg}} \qquad \text{(root-mean-square speed)} \qquad (20.13)$$

This is usually called the *rms speed*. You can remember its definition by noting that its name is the *opposite* of the sequence of operations: First you square all the speeds, then you average the squares (find the mean), then you take the square root. Because the square root "undoes" the square, v_{rms} must, in some sense, give an average speed.

> **NOTE** We could compute a true average speed v_{avg}, but that calculation is difficult. More important, the root-mean-square speed tends to arise naturally in many scientific and engineering calculations. It turns out that v_{rms} differs from v_{avg} by less than 10%, so for practical purposes we can interpret v_{rms} as being essentially the average speed of a molecule in a gas.

There's nothing special about the x-axis. The coordinate system is something that *we* impose on the problem, so *on average* it must be the case that

$$(v_x^2)_{avg} = (v_y^2)_{avg} = (v_z^2)_{avg} \qquad (20.14)$$

Hence we can use Equation 20.12 and the definition of v_{rms} to write

$$v_{rms}^2 = (v_x^2)_{avg} + (v_y^2)_{avg} + (v_z^2)_{avg} = 3(v_x^2)_{avg} \qquad (20.15)$$

Consequently, $(v_x^2)_{avg}$ is

$$(v_x^2)_{avg} = \tfrac{1}{3} v_{rms}^2 \qquad (20.16)$$

Using this result in Equation 20.11 gives us the net force on the wall of the container:

$$F_{on\ wall} = \frac{1}{3} \frac{N}{V} m v_{rms}^2 A \qquad (20.17)$$

Thus the pressure on the wall of the container due to all the molecular collisions is

$$p = \frac{F_{on\ wall}}{A} = \frac{1}{3} \frac{N}{V} m v_{rms}^2 \qquad (20.18)$$

We have met our goal. Equation 20.18 expresses the macroscopic pressure in terms of the microscopic physics. The pressure depends on the number density of molecules in the container and on how fast, on average, the molecules are moving.

EXAMPLE 20.2 | The rms speed of helium atoms

A container holds helium at a pressure of 200 kPa and a temperature of 60.0°C. What is the rms speed of the helium atoms?

SOLVE The rms speed can be found from the pressure and the number density. Using the ideal-gas law gives us the number density:

$$\frac{N}{V} = \frac{p}{k_B T} = \frac{200,000 \text{ Pa}}{(1.38 \times 10^{-23} \text{ J/K})(333 \text{ K})} = 4.35 \times 10^{25} \text{ m}^{-3}$$

The mass of a helium atom is $m = 4$ u $= 6.64 \times 10^{-27}$ kg. Thus

$$v_{rms} = \sqrt{\frac{3p}{(N/V)m}} = 1440 \text{ m/s}$$

ASSESS We found in Chapter 17 that the speed of sound in helium is roughly 1000 m/s. Individual atoms probably move somewhat faster than a wave front, so 1440 m/s seems quite reasonable.

STOP TO THINK 20.2 The speed of every molecule in a gas is suddenly increased by a factor of 4. As a result, v_{rms} increases by a factor of

a. 2.
b. <4 but not necessarily 2.
c. 4.
d. >4 but not necessarily 16.
e. 16.
f. v_{rms} doesn't change.

20.3 Temperature

A molecule of mass m and velocity v has translational kinetic energy

$$\epsilon = \tfrac{1}{2}mv^2 \qquad (20.19)$$

We'll use ϵ (lowercase Greek epsilon) to distinguish the energy of a molecule from the system energy E. Thus the average translational kinetic energy is

ϵ_{avg} = average translational kinetic energy of a molecule

$$= \tfrac{1}{2}m(v^2)_{avg} = \tfrac{1}{2}mv_{rms}^2 \qquad (20.20)$$

We've included the word "translational" to distinguish ϵ from rotational kinetic energy, which we will consider later in this chapter.

We can write the gas pressure, Equation 20.18, in terms of the average translational kinetic energy as

$$p = \frac{2}{3}\frac{N}{V}\left(\tfrac{1}{2}mv_{rms}^2\right) = \frac{2}{3}\frac{N}{V}\epsilon_{avg} \qquad (20.21)$$

The pressure is directly proportional to the average molecular translational kinetic energy. This makes sense. More-energetic molecules will hit the walls harder as they bounce and thus exert more force on the walls.

It's instructive to write Equation 20.21 as

$$pV = \tfrac{2}{3}N\epsilon_{avg} \qquad (20.22)$$

We know, from the ideal-gas law, that

$$pV = Nk_BT \qquad (20.23)$$

Comparing these two equations, we reach the significant conclusion that the average translational kinetic energy per molecule is

$$\epsilon_{avg} = \tfrac{3}{2}k_BT \qquad \text{(average translational kinetic energy)} \qquad (20.24)$$

where the temperature T is in kelvins. For example, the average translational kinetic energy of a molecule at room temperature (20°C) is

$$\epsilon_{avg} = \tfrac{3}{2}(1.38 \times 10^{-23} \text{ J/K})(293 \text{ K}) = 6.1 \times 10^{-21} \text{ J}$$

NOTE A molecule's average translational kinetic energy depends *only* on the temperature, not on the molecule's mass. If two gases have the same temperature, their molecules have the same average translational kinetic energy.

Equation 20.24 is especially satisfying because it finally gives real meaning to the concept of temperature. Writing it as

$$T = \frac{2}{3k_B}\epsilon_{avg} \qquad (20.25)$$

we can see that, for a gas, this **thing we call *temperature* measures the average translational kinetic energy.** A higher temperature corresponds to a larger value of ϵ_{avg} and thus to higher molecular speeds. This concept of temperature also gives meaning to *absolute zero* as the temperature at which $\epsilon_{avg} = 0$ and all molecular motion

FIGURE 20.7 The micro/macro connection for pressure and temperature.

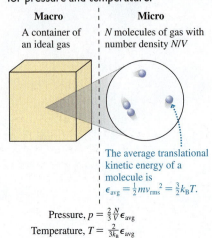

Macro	Micro
A container of an ideal gas	N molecules of gas with number density N/V

The average translational kinetic energy of a molecule is
$\epsilon_{avg} = \frac{1}{2}mv_{rms}^2 = \frac{3}{2}k_BT$.

Pressure, $p = \frac{2}{3}\frac{N}{V}\epsilon_{avg}$

Temperature, $T = \frac{2}{3k_B}\epsilon_{avg}$

ceases. (Quantum effects at very low temperatures prevent the motions from actually stopping, but our classical theory predicts that they would.) **FIGURE 20.7** summarizes what we've learned thus far about the micro/macro connection.

We can now justify our assumption that molecular collisions are perfectly elastic. Suppose they were not. If kinetic energy was lost in collisions, the average translational kinetic energy ϵ_{avg} of the gas would decrease and we would see a steadily decreasing temperature. But that doesn't happen. The temperature of an isolated system remains constant, indicating that ϵ_{avg} is not changing with time. Consequently, the collisions must be perfectly elastic.

EXAMPLE 20.3 | **Total microscopic kinetic energy**

What is the total translational kinetic energy of the molecules in 1.0 mol of gas at STP?

SOLVE The average translational kinetic energy of each molecule is

$$\epsilon_{avg} = \frac{3}{2}k_BT = \frac{3}{2}(1.38 \times 10^{-23} \text{ J/K})(273 \text{ K})$$
$$= 5.65 \times 10^{-21} \text{ J}$$

1.0 mol of gas contains N_A molecules; hence the total kinetic energy is

$$K_{micro} = N_A\epsilon_{avg} = 3400 \text{ J}$$

ASSESS The energy of any one molecule is incredibly small. Nonetheless, a macroscopic system has substantial thermal energy because it consists of an incredibly large number of molecules.

By definition, $\epsilon_{avg} = \frac{1}{2}mv_{rms}^2$. Using the ideal-gas law, we found $\epsilon_{avg} = \frac{3}{2}k_BT$. By equating these expressions we find that the rms speed of molecules in a gas is

$$v_{rms} = \sqrt{\frac{3k_BT}{m}} \qquad (20.26)$$

The rms speed depends on the square root of the temperature and inversely on the square root of the molecular mass. For example, room-temperature nitrogen (molecular mass 28 u) has rms speed

$$v_{rms} = \sqrt{\frac{3(1.38 \times 10^{-23} \text{ J/K})(293 \text{ K})}{28(1.66 \times 10^{-27} \text{ kg})}} = 509 \text{ m/s}$$

This is in excellent agreement with the experimental results of Figure 20.2.

EXAMPLE 20.4 | **Mean time between collisions**

Estimate the mean time between collisions for a nitrogen molecule at 1.0 atm pressure and room temperature (20°C).

MODEL Because v_{rms} is essentially the average molecular speed, the *mean time between collisions* is simply the time needed to travel distance λ, the mean free path, at speed v_{rms}.

SOLVE We found $\lambda = 2.3 \times 10^{-7}$ m in Example 20.1 and $v_{rms} = 509$ m/s above. Thus the mean time between collisions is

$$(\Delta t)_{avg} = \frac{\lambda}{v_{rms}} = \frac{2.3 \times 10^{-7} \text{ m}}{509 \text{ m/s}} = 4.5 \times 10^{-10} \text{ s}$$

ASSESS The air molecules around us move very fast, they collide with their neighbors about two billion times every second, and they manage to move, on average, only about 230 nm between collisions.

STOP TO THINK 20.3 The speed of every molecule in a gas is suddenly increased by a factor of 4. As a result, the temperature T increases by a factor of

a. 2.   b. <4 but not necessarily 2.

c. 4.   d. >4 but not necessarily 16.

e. 16.   f. T doesn't change.

20.4 Thermal Energy and Specific Heat

We defined the thermal energy of a system to be $E_{th} = K_{micro} + U_{micro}$, where K_{micro} is the microscopic kinetic energy of the moving molecules and U_{micro} is the potential energy of the stretched and compressed molecular bonds. We're now ready to take a microscopic look at thermal energy.

Monatomic Gases

FIGURE 20.8 shows a monatomic gas such as helium or neon. The atoms in an ideal gas have no molecular bonds with their neighbors; hence $U_{micro} = 0$. Furthermore, the kinetic energy of a monatomic gas particle is entirely translational kinetic energy ϵ. Thus the thermal energy of a monatomic gas of N atoms is

$$E_{th} = K_{micro} = \epsilon_1 + \epsilon_2 + \epsilon_3 + \cdots + \epsilon_N = N\epsilon_{avg} \qquad (20.27)$$

where ϵ_i is the translational kinetic energy of atom i. We found that $\epsilon_{avg} = \frac{3}{2}k_B T$; hence the thermal energy is

$$E_{th} = \tfrac{3}{2}Nk_B T = \tfrac{3}{2}nRT \qquad \text{(thermal energy of a monatomic gas)} \qquad (20.28)$$

where we used $N = nN_A$ and the definition of Boltzmann's constant, $k_B = R/N_A$.

We've noted for the last two chapters that thermal energy is associated with temperature. Now we have an explicit result for a monatomic gas: E_{th} is directly proportional to the temperature. Notice that E_{th} is independent of the atomic mass. Any two monatomic gases will have the same thermal energy if they have the same temperature and the same number of atoms (or moles).

If the temperature of a monatomic gas changes by ΔT, its thermal energy changes by

$$\Delta E_{th} = \tfrac{3}{2}nR\,\Delta T \qquad (20.29)$$

In Chapter 19 we found that the change in thermal energy for *any* ideal-gas process is related to the molar specific heat at constant volume by

$$\Delta E_{th} = nC_V\,\Delta T \qquad (20.30)$$

Equation 20.29 is a microscopic result that we obtained by relating the temperature to the average translational kinetic energy of the atoms. Equation 20.30 is a macroscopic result that we arrived at from the first law of thermodynamics. We can make a micro/macro connection by combining these two equations. Doing so gives us a *prediction* for the molar specific heat:

$$C_V = \tfrac{3}{2}R = 12.5 \text{ J/mol K} \qquad \text{(monatomic gas)} \qquad (20.31)$$

This was exactly the value of C_V for all three monatomic gases in Table 19.4. The perfect agreement of theory and experiment is strong evidence that gases really do consist of moving, colliding molecules.

The Equipartition Theorem

The particles of a monatomic gas are atoms. Their energy consists exclusively of their translational kinetic energy. A particle's translational kinetic energy can be written

$$\epsilon = \tfrac{1}{2}mv^2 = \tfrac{1}{2}mv_x^2 + \tfrac{1}{2}mv_y^2 + \tfrac{1}{2}mv_z^2 = \epsilon_x + \epsilon_y + \epsilon_z \qquad (20.32)$$

where we have written separately the energy associated with translational motion along the three axes. Because each axis in space is independent, we can think of ϵ_x, ϵ_y, and ϵ_z as independent *modes* of storing energy within the system.

FIGURE 20.8 The atoms in a monatomic gas have only translational kinetic energy.

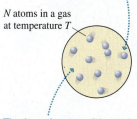

Atom i has translational kinetic energy ϵ_i but no potential energy or rotational kinetic energy.

N atoms in a gas at temperature T

The thermal energy of the gas is $E_{th} = \epsilon_1 + \epsilon_2 + \epsilon_3 + \cdots = N\epsilon_{avg}$.

Other systems have additional modes of energy storage. For example,

- Two atoms joined by a spring-like molecular bond can vibrate back and forth. Both kinetic and potential energy are associated with this vibration.
- A diatomic molecule, in addition to translational kinetic energy, has rotational kinetic energy if it rotates end-over-end like a dumbbell.

We define the number of **degrees of freedom** as the number of distinct and independent modes of energy storage. A monatomic gas has three degrees of freedom, the three modes of translational kinetic energy. Systems that can vibrate or rotate have more degrees of freedom.

An important result of statistical physics says that the energy in a system is distributed so that all modes of energy storage have equal amounts of energy. This conclusion is known as the *equipartition theorem,* meaning that the energy is equally divided. The proof is beyond what we can do in this textbook, so we will state the theorem without proof:

> **Equipartition theorem** The thermal energy of a system of particles is equally divided among all the possible degrees of freedom. For a system of N particles at temperature T, the energy stored in each mode (each degree of freedom) is $\frac{1}{2}Nk_BT$ or, in terms of moles, $\frac{1}{2}nRT$.

A monatomic gas has three degrees of freedom and thus, as we found above, $E_{th} = \frac{3}{2}Nk_BT$.

Solids

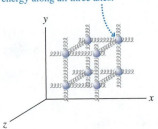

FIGURE 20.9 A simple model of a solid.

Each atom has microscopic translational kinetic energy *and* microscopic potential energy along all three axes.

FIGURE 20.9 reminds you of our "bedspring model" of a solid with particle-like atoms connected by a lattice of spring-like molecular bonds. How many degrees of freedom does a solid have? Three degrees of freedom are associated with the kinetic energy, just as in a monatomic gas. In addition, the molecular bonds can be compressed or stretched independently along the x-, y-, and z-axes. Three additional degrees of freedom are associated with these three modes of potential energy. Altogether, a solid has six degrees of freedom.

The energy stored in each of these six degrees of freedom is $\frac{1}{2}Nk_BT$. The thermal energy of a solid is the total energy stored in all six modes, or

$$E_{th} = 3Nk_BT = 3nRT \quad \text{(thermal energy of a solid)} \quad (20.33)$$

We can use this result to predict the molar specific heat of a solid. If the temperature changes by ΔT, then the thermal energy changes by

$$\Delta E_{th} = 3nR\,\Delta T \quad (20.34)$$

In Chapter 19 we defined the molar specific heat of a solid such that

$$\Delta E_{th} = nC\,\Delta T \quad (20.35)$$

By comparing Equations 20.34 and 20.35 we can predict that the molar specific heat of a solid is

$$C = 3R = 25.0 \text{ J/mol K} \quad \text{(solid)} \quad (20.36)$$

Not bad. The five elemental solids in Table 19.2 had molar specific heats clustered right around 25 J/mol K. They ranged from 24.3 J/mol K for aluminum to 26.5 J/mol K for lead. There are two reasons the agreement between theory and experiment isn't quite as perfect as it was for monatomic gases. First, our simple bedspring model of a solid isn't quite as accurate as our model of a monatomic gas. Second, quantum effects are beginning to make their appearance. More on this shortly. Nonetheless, our ability to predict C to within a few percent from a simple model of a solid is further evidence for the atomic structure of matter.

Diatomic Molecules

Diatomic molecules are a bigger challenge. How many degrees of freedom does a diatomic molecule have? **FIGURE 20.10** shows a diatomic molecule, such as molecular nitrogen N_2, oriented along the x-axis. Three degrees of freedom are associated with the molecule's translational kinetic energy. The molecule can have a dumbbell-like end-over-end rotation about either the y-axis or the z-axis. It can also rotate about its own axis. These are three rotational degrees of freedom. The two atoms can also vibrate back and forth, stretching and compressing the molecular bond. This vibrational motion has both kinetic and potential energy—thus two more degrees of freedom.

Altogether, then, a diatomic molecule has eight degrees of freedom, and we would expect the thermal energy of a gas of diatomic molecules to be $E_{th} = 4k_B T$. The analysis we followed for a monatomic gas would then lead to the prediction $C_V = 4R = 33.2$ J/mol K. As compelling as this reasoning seems to be, this is *not* the experimental value of C_V that was reported for diatomic gases in Table 19.4. Instead, we found $C_V = 20.8$ J/mol K.

Why should a theory that works so well for monatomic gases and solids fail so miserably for diatomic molecules? To see what's going on, notice that 20.8 J/mol K = $\frac{5}{2}R$. A monatomic gas, with three degrees of freedom, has $C_V = \frac{3}{2}R$. A solid, with six degrees of freedom, has $C = 3R$. A diatomic gas would have $C_V = \frac{5}{2}R$ if it had five degrees of freedom, not eight.

This discrepancy was a major conundrum as statistical physics developed in the late 19th century. Although it was not recognized as such at the time, we are here seeing our first evidence for the breakdown of classical Newtonian physics. Classically, a diatomic molecule has eight degrees of freedom. The equipartition theorem doesn't distinguish between them; all eight should have the same energy. But atoms and molecules are not classical particles. It took the development of quantum theory in the 1920s to accurately characterize the behavior of atoms and molecules. We don't yet have the tools needed to see why, but quantum effects prevent three of the modes—the two vibrational modes and the rotation of the molecule about its own axis—from being active at room temperature.

FIGURE 20.11 shows C_V as a function of temperature for hydrogen gas. C_V is right at $\frac{5}{2}R$ for temperatures from ≈ 200 K up to ≈ 800 K. But at very low temperatures C_V drops to the monatomic-gas value $\frac{3}{2}R$. The two rotational modes become "frozen out" and the nonrotating molecule has only translational kinetic energy. Quantum physics can explain this, but not Newtonian physics. You can also see that the two vibrational modes *do* become active at very high temperatures, where C_V rises to $\frac{7}{2}R$. Thus the real answer to What's wrong? is that Newtonian physics is not the right physics for describing atoms and molecules. We are somewhat fortunate that Newtonian physics is adequate to understand monatomic gases and solids, at least at room temperature.

Accepting the quantum result that a diatomic gas has only five degrees of freedom at commonly used temperatures (the translational degrees of freedom and the two end-over-end rotations), we find

$$E_{th} = \frac{5}{2}Nk_B T = \frac{5}{2}nRT$$
$$C_V = \frac{5}{2}R = 20.8 \text{ J/mol K}$$

(diatomic gases)   (20.37)

A diatomic gas has more thermal energy than a monatomic gas at the same temperature because the molecules have rotational as well as translational kinetic energy.

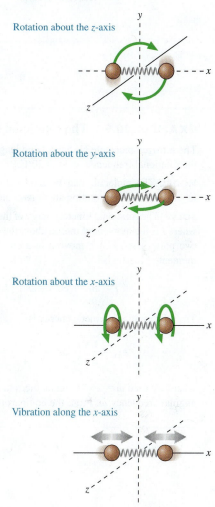

FIGURE 20.10 A diatomic molecule can rotate or vibrate.

Rotation about the z-axis

Rotation about the y-axis

Rotation about the x-axis

Vibration along the x-axis

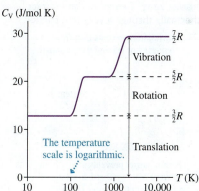

FIGURE 20.11 Hydrogen molar specific heat at constant volume as a function of temperature.

C_V (J/mol K)

While the micro/macro connection firmly establishes the atomic structure of matter, it also heralds the need for a new theory of matter at the atomic level. That is a task we will take up in Part VIII. For now, **TABLE 20.1** summarizes what we have learned from kinetic theory about thermal energy and molar specific heats.

TABLE 20.1 Kinetic theory predictions for the thermal energy and the molar specific heat

System	Degrees of freedom	E_{th}	C_V
Monatomic gas	3	$\frac{3}{2}Nk_BT = \frac{3}{2}nRT$	$\frac{3}{2}R = 12.5$ J/mol K
Diatomic gas	5	$\frac{5}{2}Nk_BT = \frac{5}{2}nRT$	$\frac{5}{2}R = 20.8$ J/mol K
Elemental solid	6	$3Nk_BT = 3nRT$	$3R = 25.0$ J/mol K

EXAMPLE 20.5 | **The rotational frequency of a molecule**

The nitrogen molecule N_2 has a bond length of 0.12 nm. Estimate the rotational frequency of N_2 at 20°C.

MODEL The molecule can be modeled as a rigid dumbbell of length $L = 0.12$ nm rotating about its center.

SOLVE The rotational kinetic energy of the molecule is $\epsilon_{rot} = \frac{1}{2}I\omega^2$, where I is the moment of inertia about the center. Because we have two point masses each moving in a circle of radius $r = L/2$, the moment of inertia is

$$I = mr^2 + mr^2 = 2m\left(\frac{L}{2}\right)^2 = \frac{mL^2}{2}$$

Thus the rotational kinetic energy is

$$\epsilon_{rot} = \frac{1}{2}\frac{mL^2}{2}\omega^2 = \frac{mL^2\omega^2}{4} = \pi^2 mL^2 f^2$$

where we used $\omega = 2\pi f$ to relate the rotational frequency f to the angular frequency ω. From the equipartition theorem, the energy

associated with this mode is $\frac{1}{2}Nk_BT$, so the *average* rotational kinetic energy per molecule is

$$(\epsilon_{rot})_{avg} = \frac{1}{2}k_BT$$

Equating these two expressions for ϵ_{rot} gives us

$$\pi^2 mL^2 f^2 = \frac{1}{2}k_BT$$

Thus the rotational frequency is

$$f = \sqrt{\frac{k_BT}{2\pi^2 mL^2}} = 7.8 \times 10^{11} \text{ rev/s}$$

We evaluated f at $T = 293$ K, using $m = 14$ u $= 2.34 \times 10^{-26}$ kg for each *atom*.

ASSESS This is a *very* high frequency, but these values are typical of molecular rotations.

STOP TO THINK 20.4 How many degrees of freedom does a bead on a rigid rod have?

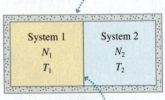

a. 1   b. 2   c. 3   d. 4   e. 5   f. 6

20.5 Thermal Interactions and Heat

We can now look in more detail at what happens when two systems at different temperatures interact with each other. **FIGURE 20.12** shows a rigid, insulated container divided into two sections by a very thin membrane. The left side, which we'll call system 1, has N_1 atoms at an initial temperature T_{1i}. System 2 on the right has N_2 atoms at an initial temperature T_{2i}. The membrane is so thin that atoms can collide at the boundary as if the membrane were not there, yet it is a barrier that prevents atoms from moving from one side to the other. The situation is analogous, on an atomic scale, to basketballs colliding through a shower curtain.

Suppose that system 1 is initially at a higher temperature: $T_{1i} > T_{2i}$. This is not an equilibrium situation. The temperatures will change with time until the systems eventually reach a common final temperature T_f. If you *watch* the gases as one warms and the other cools, you see nothing happening. This interaction is quite

FIGURE 20.12 Two gases can interact thermally through a very thin barrier.

Insulation prevents heat from entering or leaving the container.

System 1	System 2
N_1	N_2
T_1	T_2

A thin barrier prevents atoms from moving from system 1 to 2 but still allows them to collide. The barrier is clamped in place and cannot move.

different from a mechanical interaction in which, for example, you might see a piston move from one side toward the other. The only way in which the gases can interact is via molecular collisions at the boundary. This is a *thermal interaction,* and our goal is to understand how thermal interactions bring the systems to thermal equilibrium.

System 1 and system 2 begin with thermal energies

$$E_{1i} = \tfrac{3}{2}N_1 k_B T_{1i} = \tfrac{3}{2}n_1 R T_{1i}$$
$$E_{2i} = \tfrac{3}{2}N_2 k_B T_{2i} = \tfrac{3}{2}n_2 R T_{2i}$$

(20.38)

We've written the energies for monatomic gases; you could do the same calculation if one or both of the gases is diatomic by replacing the $\tfrac{3}{2}$ with $\tfrac{5}{2}$. Notice that we've omitted the subscript "th" to keep the notation manageable.

The total energy of the combined systems is $E_{tot} = E_{1i} + E_{2i}$. As systems 1 and 2 interact, their individual thermal energies E_1 and E_2 can change but their sum E_{tot} remains constant. The system will have reached thermal equilibrium when the individual thermal energies reach final values E_{1f} and E_{2f} that no longer change.

The Systems Exchange Energy

FIGURE 20.13 shows a fast atom and a slow atom approaching the barrier from opposite sides. They undergo a perfectly elastic collision at the barrier. Although no net energy is lost in a perfectly elastic collision, in most such collisions the more-energetic atom loses energy while the less-energetic atom gains energy. In other words, there's an energy *transfer* from the more-energetic atom's side to the less-energetic atom's side.

The average translational kinetic energy per atom is directly proportional to the temperature: $\epsilon_{avg} = \tfrac{3}{2}k_B T$. Because $T_{1i} > T_{2i}$, the atoms in system 1 are, on average, more energetic than the atoms in system 2. Thus *on average* the collisions transfer energy from system 1 to system 2. Not in every collision: sometimes a fast atom in system 2 collides with a slow atom in system 1, transferring energy from 2 to 1. But the net energy transfer, from all collisions, is from the warmer system 1 to the cooler system 2. In other words, **heat is the energy transferred *via collisions* between the more-energetic (warmer) atoms on one side and the less-energetic (cooler) atoms on the other.**

How do the systems "know" when they've reached thermal equilibrium? Energy transfer continues until the atoms on both sides of the barrier have the *same average translational kinetic energy.* Once the average translational kinetic energies are the same, there is no tendency for energy to flow in either direction. This is the state of thermal equilibrium, so the condition for thermal equilibrium is

$$\left(\epsilon_1\right)_{avg} = \left(\epsilon_2\right)_{avg} \qquad \text{(thermal equilibrium)} \qquad (20.39)$$

where, as before, ϵ is the translational kinetic energy of an atom.

Because the average energies are directly proportional to the final temperatures, $\epsilon_{avg} = \tfrac{3}{2}k_B T_f$, thermal equilibrium is characterized by the macroscopic condition

$$T_{1f} = T_{2f} = T_f \qquad \text{(thermal equilibrium)} \qquad (20.40)$$

In other words, **two thermally interacting systems reach a common final temperature *because* they exchange energy via collisions until the atoms on each side have, on average, equal translational kinetic energies.** This is a very important idea.

Equation 20.40 can be used to determine the equilibrium thermal energies. Because these are monatomic gases, $E_{th} = N\epsilon_{avg}$. Thus the equilibrium condition $\left(\epsilon_1\right)_{avg} = \left(\epsilon_2\right)_{avg} = \left(\epsilon_{tot}\right)_{avg}$ implies

$$\frac{E_{1f}}{N_1} = \frac{E_{2f}}{N_2} = \frac{E_{tot}}{N_1 + N_2}$$

(20.41)

FIGURE 20.13 On average, collisions transfer energy from more-energetic atoms to less-energetic atoms.

FIGURE 20.14 Equilibrium is reached when the atoms on each side have, on average, equal energies.

Collisions transfer energy from the warmer system to the cooler system as more-energetic atoms lose energy to less-energetic atoms.

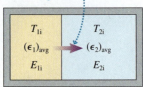

Thermal equilibrium occurs when the systems have the same average translational kinetic energy and thus the same temperature.

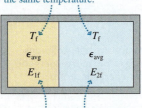

In general, the thermal energies E_{1f} and E_{2f} are *not* equal.

from which we can conclude

$$E_{1f} = \frac{N_1}{N_1 + N_2} E_{tot} = \frac{n_1}{n_1 + n_2} E_{tot}$$

$$E_{2f} = \frac{N_2}{N_1 + N_2} E_{tot} = \frac{n_2}{n_1 + n_2} E_{tot} \qquad (20.42)$$

where in the last step we used moles rather than molecules.

Notice that $E_{1f} + E_{2f} = E_{tot}$, verifying that energy has been conserved even while being redistributed between the systems.

No work is done on either system because the barrier has no macroscopic displacement, so the first law of thermodynamics is

$$Q_1 = \Delta E_1 = E_{1f} - E_{1i}$$

$$Q_2 = \Delta E_2 = E_{2f} - E_{2i} \qquad (20.43)$$

As a homework problem you can show that $Q_1 = -Q_2$, as required by energy conservation. That is, the heat lost by one system is gained by the other. $|Q_1|$ is the quantity of heat that is transferred from the warmer gas to the cooler gas during the thermal interaction.

NOTE In general, the equilibrium thermal energies of the system are *not* equal. That is, $E_{1f} \neq E_{2f}$. They will be equal only if $N_1 = N_2$. Equilibrium is reached when the average translational kinetic energies in the two systems are equal—that is, when $(\epsilon_1)_{avg} = (\epsilon_2)_{avg}$, not when $E_{1f} = E_{2f}$. The distinction is important. **FIGURE 20.14** summarizes these ideas.

EXAMPLE 20.6 | **A thermal interaction**

A sealed, insulated container has 2.0 g of helium at an initial temperature of 300 K on one side of a barrier and 10.0 g of argon at an initial temperature of 600 K on the other side.

a. How much heat energy is transferred, and in which direction?

b. What is the final temperature?

MODEL The systems start with different temperatures, so they are not in thermal equilibrium. Energy will be transferred via collisions from the argon to the helium until both systems have the same average molecular energy.

SOLVE a. Let the helium be system 1. Helium has molar mass $M_{mol} = 0.004$ kg/mol, so $n_1 = M/M_{mol} = 0.50$ mol. Similarly, argon has $M_{mol} = 0.040$ kg/mol, so $n_2 = 0.25$ mol. The initial thermal energies of the two monatomic gases are

$$E_{1i} = \tfrac{3}{2} n_1 R T_{1i} = 225R = 1870 \text{ J}$$

$$E_{2i} = \tfrac{3}{2} n_2 R T_{2i} = 225R = 1870 \text{ J}$$

The systems start with *equal* thermal energies, but they are not in thermal equilibrium. The total energy is $E_{tot} = 3740$ J. In equilibrium, this energy is distributed between the two systems as

$$E_{1f} = \frac{n_1}{n_1 + n_2} E_{tot} = \frac{0.50}{0.75} \times 3740 \text{ J} = 2493 \text{ J}$$

$$E_{2f} = \frac{n_2}{n_1 + n_2} E_{tot} = \frac{0.25}{0.75} \times 3740 \text{ J} = 1247 \text{ J}$$

The heat entering or leaving each system is

$$Q_1 = Q_{He} = E_{1f} - E_{1i} = 623 \text{ J}$$

$$Q_2 = Q_{Ar} = E_{2f} - E_{2i} = -623 \text{ J}$$

The helium and the argon interact thermally via collisions at the boundary, causing 623 J of heat to be transferred from the warmer argon to the cooler helium.

b. These are constant-volume processes, thus $Q = nC_V \Delta T$. $C_V = \tfrac{3}{2}R$ for monatomic gases, so the temperature changes are

$$\Delta T_{He} = \frac{Q_{He}}{\tfrac{3}{2} nR} = \frac{623 \text{ J}}{1.5(0.50 \text{ mol})(8.31 \text{ J/mol K})} = 100 \text{ K}$$

$$\Delta T_{Ar} = \frac{Q_{Ar}}{\tfrac{3}{2} nR} = \frac{-623 \text{ J}}{1.5(0.25 \text{ mol})(8.31 \text{ J/mol K})} = -200 \text{ K}$$

Both gases reach the common final temperature $T_f = 400$ K.

ASSESS $E_{1f} = 2E_{2f}$ because there are twice as many atoms in system 1.

The main idea of this section is that two systems reach a common final temperature not by magic or by a prearranged agreement but simply from the energy exchange of vast numbers of molecular collisions. Real interacting systems, of course, are separated by walls rather than our unrealistic thin membrane. As the systems interact, the energy is first transferred via collisions from system 1 into the wall and subsequently, as the cooler molecules collide with a warm wall, into system 2. That is, the energy transfer is $E_1 \rightarrow E_{wall} \rightarrow E_2$. This is still heat because the energy transfer is occurring via molecular collisions rather than mechanical motion.

STOP TO THINK 20.5   Systems A and B are interacting thermally. At this instant of time,

a. $T_A > T_B$
b. $T_A = T_B$
c. $T_A < T_B$

A	B
$N = 1000$	$N = 2000$
$\epsilon_{avg} = 1.0 \times 10^{-20}$ J	$\epsilon_{avg} = 0.5 \times 10^{-20}$ J
$E_{th} = 1.0 \times 10^{-17}$ J	$E_{th} = 1.0 \times 10^{-17}$ J

20.6 Irreversible Processes and the Second Law of Thermodynamics

The preceding section looked at the thermal interaction between a warm gas and a cold gas. Heat energy is transferred from the warm gas to the cold gas until they reach a common final temperature. But why isn't heat transferred from the cold gas to the warm gas, making the cold side colder and the warm side warmer? Such a process could still conserve energy, but it never happens. The transfer of heat energy from hot to cold is an example of an **irreversible process,** a process that can happen only in one direction.

Examples of irreversible processes abound. Stirring the cream in your coffee mixes the cream and coffee together. No amount of stirring ever unmixes them. If you shake a jar that has red marbles on the top and blue marbles on the bottom, the two colors are quickly mixed together. No amount of shaking ever separates them again. If you watched a movie of someone shaking a jar and saw the red and blue marbles separating, you would be certain that the movie was running backward. In fact, a reasonable definition of an irreversible process is one for which a backward-running movie shows a physically impossible process.

FIGURE 20.15a is a two-frame movie of a collision between two particles, perhaps two gas molecules. Suppose that sometime after the collision is over we could reach in and reverse the velocities of both particles. That is, replace vector $\vec{v}$ with vector $-\vec{v}$. Then, as in a movie playing backward, the collision would happen in reverse. This is the movie of **FIGURE 20.15b**.

FIGURE 20.15 Molecular collisions are reversible.

(a) Forward movie

Before: After:

(b) The backward movie is equally plausible.

Before: After:

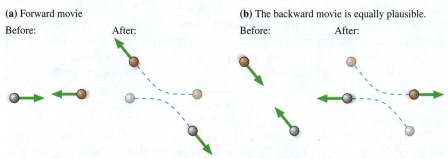

You cannot tell, just by looking at the two movies, which is really going forward and which is being played backward. Maybe Figure 20.15b was the original collision and Figure 20.15a is the backward version. Nothing in either collision looks wrong, and no measurements you might make on either would reveal any violations of Newton's laws. **Interactions between molecules are reversible processes.**

FIGURE 20.16 A car crash is irreversible.

(a) Forward movie

Before: After:

(b) The backward movie is physically impossible.

Before: After:

FIGURE 20.17 Two interacting systems. Balls are chosen at random and moved to the other box.

Balls are chosen at random and moved from one box to the other.

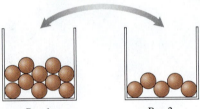

Box 1   Box 2
N_1 balls N_2 balls

Tossing all heads, while not impossible, is extremely unlikely, and the probability of doing so rapidly decreases as the number of coins increases.

Contrast this with the two-frame car crash movies in **FIGURE 20.16**. Past and future are clearly distinct in an irreversible process, and the backward movie of Figure 20.16b is obviously wrong. But what has been violated in the backward movie? To have the crumpled car spring away from the wall would not violate any laws of physics we have so far discovered. It would simply require transforming the thermal energy of the car and wall back into the macroscopic center-of-mass energy of the car as a whole.

The paradox stems from our assertion that macroscopic phenomena can be understood on the basis of microscopic molecular motions. If the microscopic motions are all reversible, how can the macroscopic phenomena end up being irreversible? If reversible collisions can cause heat to be transferred from hot to cold, why do they never cause heat to be transferred from cold to hot? There must be another law of physics preventing it. The law we seek must, in some sense, be able to distinguish the past from the future.

Which Way to Equilibrium?

Stated another way, how do two systems initially at different temperatures "know" which way to go to reach equilibrium? Perhaps an analogy will help.

FIGURE 20.17 shows two boxes, numbered 1 and 2, containing identical balls. Box 1 starts with more balls than box 2, so $N_{1i} > N_{2i}$. Once every second, one ball is chosen at random and moved to the other box. This is a reversible process because a ball can move from box 2 to box 1 just as easily as from box 1 to box 2. What do you expect to see if you return several hours later?

Because balls are chosen at random, and because $N_{1i} > N_{2i}$, it's initially more likely that a ball will move from box 1 to box 2 than from box 2 to box 1. Sometimes a ball will move "backward" from box 2 to box 1, but overall there's a net movement of balls from box 1 to box 2. The system will evolve until $N_1 \approx N_2$. This is a stable situation—equilibrium!—with an equal number of balls moving in both directions.

But couldn't it go the other way, with N_1 getting even larger while N_2 decreases? In principle, any possible arrangement of the balls is possible in the same way that any number of heads are possible if you throw N coins in the air and let them fall. If you throw four coins, the odds are 1 in 2^4, or 1 in 16, of getting four heads. With four balls, the odds are 1 in 16 that, at a randomly chosen instant of time, you would find $N_1 = 4$. You wouldn't find that to be terribly surprising.

With 10 balls, the probability that $N_1 = 10$ is $0.5^{10} \approx 1/1000$. With 100 balls, the probability that $N_1 = 100$ has dropped to $\approx 10^{-30}$. With 10^{20} balls, the odds of finding all of them, or even most of them, in one box are so staggeringly small that it's safe to say it will "never" happen. Although each transfer is reversible, **the statistics of large numbers make it overwhelmingly more likely that the system will evolve toward a state in which $N_1 \approx N_2$ than toward a state in which $N_1 > N_2$.**

The balls in our analogy represent energy. The total energy, like the total number of balls, is conserved, but molecular collisions can move energy between system 1 and system 2. Each collision is reversible, just as likely to transfer energy from 1 to 2 as from 2 to 1. But if $(\epsilon_{1i})_{avg} > (\epsilon_{2i})_{avg}$, and if we're dealing with two macroscopic systems where $N > 10^{20}$, then it's overwhelmingly likely that the net result of many, many collisions will be to transfer energy from system 1 to system 2 until $(\epsilon_{1f})_{avg} = (\epsilon_{2f})_{avg}$—in other words, for heat energy to be transferred from hot to cold.

The system reaches thermal equilibrium not by any plan or by outside intervention, but simply because **equilibrium is the *most probable* state in which to be**. It is *possible* that the system will move away from equilibrium, with heat moving from cold to hot, but remotely improbable in any realistic system. The consequence of a vast number of random events is that the system evolves in one direction, toward equilibrium, and not the other. **Reversible microscopic events lead to irreversible macroscopic behavior because some macroscopic states are vastly more probable than others.**

Order, Disorder, and Entropy

FIGURE 20.18 shows three different systems. At the top is a group of atoms arranged in a crystal-like lattice. This is a highly ordered and nonrandom system, with each atom's position precisely specified. Contrast this with the system on the bottom, where there is no order at all. The position of every atom was assigned entirely at random.

It is extremely improbable that the atoms in a container would *spontaneously* arrange themselves into the ordered pattern of the top picture. In a system of, say, 10^{20} atoms, the probability of this happening is similar to the probability that 10^{20} tossed coins will all be heads. We can safely say that it will never happen. By contrast, there are a vast number of arrangements like the one on the bottom that randomly fill the container.

The middle picture of Figure 20.18 is an in-between situation. This situation might arise as a solid melts. The positions of the atoms are clearly not completely random, so the system preserves some degree of order. This in-between situation is more likely to occur spontaneously than the highly ordered lattice on the top, but is less likely to occur than the completely random system on the bottom.

Scientists and engineers use a state variable called **entropy** to measure the probability that a macroscopic state will occur spontaneously. The ordered lattice, which has a very small probability of spontaneous occurrence, has a very low entropy. The entropy of the randomly filled container is high. The entropy of the middle picture is somewhere in between. It is often said that entropy measures the amount of *disorder* in a system. The entropy in Figure 20.18 increases as you move from the ordered system on the top to the disordered system on the bottom.

Similarly, two thermally interacting systems with different temperatures have a low entropy. These systems are ordered in the sense that the faster atoms are on one side of the barrier, the slower atoms on the other. The most random possible distribution of energy, and hence the least ordered system, corresponds to the situation where the two systems are in thermal equilibrium with equal temperatures. Entropy increases as two systems with initially different temperatures move toward equilibrium. Entropy would decrease if heat energy moved from cold to hot, making the hot system hotter and the cold system colder.

Entropy can be calculated, but we'll leave that to more advanced courses. For our purposes, the *concept* of entropy as a measure of the disorder in a system, or of the probability that a macroscopic state will occur, is more important than a numerical value.

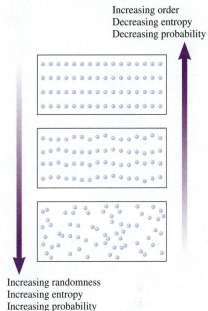

FIGURE 20.18 Ordered and disordered systems.

Increasing order
Decreasing entropy
Decreasing probability

Increasing randomness
Increasing entropy
Increasing probability

The Second Law of Thermodynamics

The fact that macroscopic systems evolve irreversibly toward equilibrium is a statement about nature that is not contained in any of the laws of physics we have encountered. It is, in fact, a new law of physics, one known as the **second law of thermodynamics.**

The formal statement of the second law of thermodynamics is given in terms of entropy:

Second law, formal statement The entropy of an isolated system (or group of systems) never decreases. The entropy either increases, until the system reaches equilibrium, or, if the system began in equilibrium, stays the same.

The qualifier "isolated" is most important. We can order the system by reaching in from the outside, perhaps using tiny tweezers to place the atoms in a lattice. Similarly, we can transfer heat from cold to hot by using a refrigerator. The second law is about what a system can or cannot do *spontaneously,* on its own, without outside intervention.

The second law of thermodynamics tells us that an isolated system evolves such that

- Order turns into disorder and randomness.
- Information is lost rather than gained.
- The system "runs down."

An isolated system never spontaneously generates order out of randomness. It is not that the system "knows" about order or randomness, but rather that there are vastly more states corresponding to randomness than there are corresponding to order. As collisions occur at the microscopic level, the laws of probability dictate that the system will, on average, move inexorably toward the most probable and thus most random macroscopic state.

The second law of thermodynamics is often stated in several equivalent but more informal versions. One of these, and the one most relevant to our discussion, is

> **Second law, informal statement #1** When two systems at different temperatures interact, heat energy is transferred spontaneously from the hotter to the colder system, never from the colder to the hotter.

This version of the second law will be used in Chapter 21 to understand the thermodynamics of engines.

The second law of thermodynamics is an independent statement about nature, separate from the first law. The first law is a precise statement about energy conservation. The second law, by contrast, is a *probabilistic* statement, based on the statistics of very large numbers. While it is conceivable that heat could spontaneously move from cold to hot, it will never occur in any realistic macroscopic system.

The irreversible evolution from less-likely macroscopic states to more-likely macroscopic states is what gives us a macroscopic direction of time. Stirring blends your coffee and cream; it never unmixes them. Friction causes an object to stop while increasing its thermal energy; the random atomic motions of thermal energy never spontaneously organize themselves into a macroscopic motion of the entire object. A plant in a sealed jar dies and decomposes to carbon and various gases; the gases and carbon never spontaneously assemble themselves into a flower. These are all examples of irreversible processes. They each show a clear direction of time, a distinct difference between past and future.

Thus another statement of the second law is

> **Second law, informal statement #2** The time direction in which the entropy of an isolated macroscopic system increases is "the future."

Establishing the "arrow of time" is one of the most profound implications of the second law of thermodynamics.

STOP TO THINK 20.6 Two identical boxes each contain 1,000,000 molecules. In box A, 750,000 molecules happen to be in the left half of the box while 250,000 are in the right half. In box B, 499,900 molecules happen to be in the left half of the box while 500,100 are in the right half. At this instant of time,

a. The entropy of box A is larger than the entropy of box B.
b. The entropy of box A is equal to the entropy of box B.
c. The entropy of box A is smaller than the entropy of box B.

SUMMARY

The goal of Chapter 20 has been to see how macroscopic properties depend on the motion of atoms.

GENERAL PRINCIPLES

The **micro/macro connection** relates the macroscopic properties of a system to the motion and collisions of its atoms and molecules.

The Equipartition Theorem

Tells us how collisions distribute the energy in the system. The energy stored in each mode of the system (each **degree of freedom**) is $\frac{1}{2}Nk_BT$ or, in terms of moles, $\frac{1}{2}nRT$.

The Second Law of Thermodynamics

Tells us how collisions move a system toward equilibrium. The entropy of an isolated system can only increase or, in equilibrium, stay the same.

- Order turns into disorder and randomness.
- Systems run down.
- Heat energy is transferred spontaneously from a hotter to a colder system, never from colder to hotter.

IMPORTANT CONCEPTS

Pressure is due to the force of the molecules colliding with the walls:

$$p = \frac{1}{3}\frac{N}{V}mv_{rms}^2 = \frac{2}{3}\frac{N}{V}\epsilon_{avg}$$

The **average translational kinetic energy** of a molecule is $\epsilon_{avg} = \frac{3}{2}k_BT$. The temperature of the gas $T = 2\epsilon_{avg}/3k_B$ measures the average translational kinetic energy.

Entropy measures the probability that a macroscopic state will occur or, equivalently, the amount of disorder in a system.

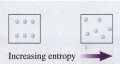

Increasing entropy

The **thermal energy** of a system is

E_{th} = translational kinetic energy + rotational kinetic energy + vibrational energy

- **Monatomic gas** $E_{th} = \frac{3}{2}Nk_BT = \frac{3}{2}nRT$
- **Diatomic gas** $E_{th} = \frac{5}{2}Nk_BT = \frac{5}{2}nRT$
- **Elemental solid** $E_{th} = 3Nk_BT = 3nRT$

Heat is energy transferred via collisions from more-energetic molecules on one side to less-energetic molecules on the other. Equilibrium is reached when $(\epsilon_1)_{avg} = (\epsilon_2)_{avg}$, which implies $T_{1f} = T_{2f}$.

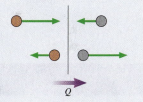

APPLICATIONS

The **root-mean-square speed** v_{rms} is the square root of the average of the squares of the molecular speeds:

$$v_{rms} = \sqrt{(v^2)_{avg}}$$

For molecules of mass m at temperature T, $v_{rms} = \sqrt{\dfrac{3k_BT}{m}}$

Molar specific heats can be predicted from the thermal energy because $\Delta E_{th} = nC\,\Delta T$.

- **Monatomic gas** $C_V = \frac{3}{2}R$
- **Diatomic gas** $C_V = \frac{5}{2}R$
- **Elemental solid** $C = 3R$

TERMS AND NOTATION

histogram	root-mean-square speed, v_{rms}	equipartition theorem	entropy
mean free path, λ	degrees of freedom	irreversible process	second law of thermodynamics

CONCEPTUAL QUESTIONS

1. Solids and liquids resist being compressed. They are not totally incompressible, but it takes large forces to compress them even slightly. If it is true that matter consists of atoms, what can you infer about the microscopic nature of solids and liquids from their incompressibility?

2. Gases, in contrast with solids and liquids, are very compressible. What can you infer from this observation about the microscopic nature of gases?

3. The density of air at STP is about $\frac{1}{1000}$ the density of water. How does the average distance between air molecules compare to the average distance between water molecules? Explain.

4. The mean free path of molecules in a gas is 200 nm.
   a. What will be the mean free path if the pressure is doubled while the temperature is held constant?
   b. What will be the mean free path if the absolute temperature is doubled while the pressure is held constant?

5. If the pressure of a gas is really due to the *random* collisions of molecules with the walls of the container, why do pressure gauges—even very sensitive ones—give perfectly steady readings? Shouldn't the gauge be continually jiggling and fluctuating? Explain.

6. Suppose you could suddenly increase the speed of every molecule in a gas by a factor of 2.
   a. Would the rms speed of the molecules increase by a factor of $2^{1/2}$, 2, or 2^2? Explain.
   b. Would the gas pressure increase by a factor of $2^{1/2}$, 2, or 2^2? Explain.

7. Suppose you could suddenly increase the speed of every molecule in a gas by a factor of 2.
   a. Would the temperature of the gas increase by a factor of $2^{1/2}$, 2, or 2^2? Explain.

   b. Would the molar specific heat at constant volume change? If so, by what factor? If not, why not?

8. The two containers of gas in FIGURE Q20.8 are in good thermal contact with each other but well insulated from the environment. They have been in contact for a long time and are in thermal equilibrium.
   a. Is v_{rms} of helium greater than, less than, or equal to v_{rms} of argon? Explain.
   b. Does the helium have more thermal energy, less thermal energy, or the same amount of thermal energy as the argon? Explain.

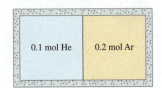

0.1 mol He   0.2 mol Ar

FIGURE Q20.8

9. Suppose you place an ice cube in a beaker of room-temperature water, then seal them in a rigid, well-insulated container. No energy can enter or leave the container.
   a. If you open the container an hour later, will you find a beaker of water slightly cooler than room temperature, or a large ice cube and some 100°C steam?
   b. Finding a large ice cube and some 100°C steam would not violate the first law of thermodynamics. $W = 0$ J and $Q = 0$ J because the container is sealed, and $\Delta E_{th} = 0$ J because the increase in thermal energy of the water molecules that became steam is offset by the decrease in thermal energy of the water molecules that turned to ice. Energy would be conserved, yet we never see an outcome like this. Why not?

EXERCISES AND PROBLEMS

Problems labeled ▨ integrate material from earlier chapters.

Exercises

Section 20.1 Molecular Speeds and Collisions

1. ‖ The number density of an ideal gas at STP is called the *Loschmidt number*. Calculate the Loschmidt number.

2. ‖ A 1.0 m × 1.0 m × 1.0 m cube of nitrogen gas is at 20°C and 1.0 atm. Estimate the number of molecules in the cube with a speed between 700 m/s and 1000 m/s.

3. ‖ At what pressure will the mean free path in room-temperature (20°C) nitrogen be 1.0 m?

4. ‖‖‖ Integrated circuits are manufactured in vacuum chambers in which the air pressure is 1.0×10^{-10} mm of Hg. What are (a) the number density and (b) the mean free path of a molecule? Assume $T = 20°C$.

5. ‖ The mean free path of a molecule in a gas is 300 nm. What will the mean free path be if the gas temperature is doubled at (a) constant volume and (b) constant pressure?

6. ‖ A lottery machine uses blowing air to keep 2000 Ping-Pong balls bouncing around inside a 1.0 m × 1.0 m × 1.0 m box. The diameter of a Ping-Pong ball is 3.0 cm. What is the mean free path between collisions? Give your answer in cm.

7. ‖ For a monatomic gas, what is the ratio of the volume per atom (V/N) to the volume *of* an atom when the mean free path is ten times the atomic diameter?

Section 20.2 Pressure in a Gas

8. ‖ The molecules in a six-particle gas have velocities

$$\vec{v}_1 = (20\hat{\imath} - 30\hat{\jmath}) \text{ m/s} \qquad \vec{v}_4 = 30\hat{\imath} \text{ m/s}$$
$$\vec{v}_2 = (40\hat{\imath} + 70\hat{\jmath}) \text{ m/s} \qquad \vec{v}_5 = (40\hat{\imath} - 40\hat{\jmath}) \text{ m/s}$$
$$\vec{v}_3 = (-80\hat{\imath} + 20\hat{\jmath}) \text{ m/s} \qquad \vec{v}_6 = (-50\hat{\imath} - 20\hat{\jmath}) \text{ m/s}$$

   Calculate (a) $\vec{v}_{avg}$, (b) v_{avg}, and (c) v_{rms}.

9. ‖ Eleven molecules have speeds 15, 16, 17, ... , 25 m/s. Calculate (a) v_{avg} and (b) v_{rms}.

10. | FIGURE EX20.10 is a histogram showing the speeds of the molecules in a very small gas. What are (a) the most probable speed, (b) the average speed, and (c) the rms speed?

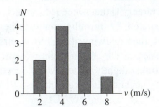

FIGURE EX20.10

11. ‖ The number density in a container of neon gas is 5.00×10^{25} m^{-3}. The atoms are moving with an rms speed of 660 m/s. What are (a) the temperature and (b) the pressure inside the container?

12. ‖‖ At 100°C the rms speed of nitrogen molecules is 576 m/s. Nitrogen at 100°C and a pressure of 2.0 atm is held in a container with a 10 cm × 10 cm square wall. Estimate the rate of molecular collisions (collisions/s) on this wall.

13. ‖ A cylinder contains gas at a pressure of 2.0 atm and a number density of 4.2×10^{25} m^{-3}. The rms speed of the atoms is 660 m/s. Identify the gas.

Section 20.3 Temperature

14. | A gas consists of a mixture of neon and argon. The rms speed of the neon atoms is 400 m/s. What is the rms speed of the argon atoms?

15. ‖ What are the rms speeds of (a) argon atoms and (b) hydrogen molecules at 800°C?

16. ‖ 1.5 m/s is a typical walking speed. At what temperature (in °C) would nitrogen molecules have an rms speed of 1.5 m/s?

17. ‖ At what temperature (in °C) do hydrogen molecules have the same rms speed as nitrogen molecules at 100°C?

18. | The rms speed of molecules in a gas is 600 m/s. What will be the rms speed if the gas pressure and volume are both halved?

19. ‖ By what factor does the rms speed of a molecule change if the temperature is increased from 10°C to 1000°C?

20. ‖ Atoms can be "cooled" to incredibly low temperatures by letting them interact with a laser beam. Various novel quantum phenomena appear at these temperatures. What is the rms speed of cesium atoms that have been cooled to a temperature of 100 nK?

21. | At STP, what is the total translational kinetic energy of the molecules in 1.0 mol of (a) hydrogen, (b) helium, and (c) oxygen?

22. | Suppose you double the temperature of a gas at constant volume. Do the following change? If so, by what factor?
   a. The average translational kinetic energy of a molecule.
   b. The rms speed of a molecule.
   c. The mean free path.

23. ‖ During a physics experiment, helium gas is cooled to a temperature of 10 K at a pressure of 0.10 atm. What are (a) the mean free path in the gas, (b) the rms speed of the atoms, and (c) the average energy per atom?

24. | What are (a) the average kinetic energy and (b) the rms speed of a proton in the center of the sun, where the temperature is 2.0×10^7 K?

25. ‖ The atmosphere of the sun consists mostly of hydrogen *atoms* (not molecules) at a temperature of 6000 K. What are (a) the average translational kinetic energy per atom and (b) the rms speed of the atoms?

Section 20.4 Thermal Energy and Specific Heat

26. ‖ A 10 g sample of neon gas has 1700 J of thermal energy. Estimate the average speed of a neon atom.

27. ‖ The rms speed of the atoms in a 2.0 g sample of helium gas is 700 m/s. What is the thermal energy of the gas?

28. ‖ A 6.0 m × 8.0 m × 3.0 m room contains air at 20°C. What is the room's thermal energy?

29. | The thermal energy of 1.0 mol of a substance is increased by 1.0 J. What is the temperature change if the system is (a) a monatomic gas, (b) a diatomic gas, and (c) a solid?

30. ‖ What is the thermal energy of 100 cm^3 of aluminum at 100°C?

31. | 1.0 mol of a monatomic gas interacts thermally with 1.0 mol of an elemental solid. The gas temperature decreases by 50°C at constant volume. What is the temperature change of the solid?

32. ‖ A cylinder of nitrogen gas has a volume of 15,000 cm^3 and a pressure of 100 atm.
   a. What is the thermal energy of this gas at room temperature (20°C)?
   b. What is the mean free path in the gas?
   c. The valve is opened and the gas is allowed to expand slowly and isothermally until it reaches a pressure of 1.0 atm. What is the change in the thermal energy of the gas?

33. ‖ A rigid container holds 0.20 g of hydrogen gas. How much heat is needed to change the temperature of the gas
   a. From 50 K to 100 K?
   b. From 250 K to 300 K?
   c. From 2250 K to 2300 K?

Section 20.5 Thermal Interactions and Heat

34. | 2.0 mol of monatomic gas A initially has 5000 J of thermal energy. It interacts with 3.0 mol of monatomic gas B, which initially has 8000 J of thermal energy.
   a. Which gas has the higher initial temperature?
   b. What is the final thermal energy of each gas?

35. ‖‖ 4.0 mol of monatomic gas A interacts with 3.0 mol of monatomic gas B. Gas A initially has 9000 J of thermal energy, but in the process of coming to thermal equilibrium it transfers 1000 J of heat energy to gas B. How much thermal energy did gas B have initially?

Section 20.6 Irreversible Processes and the Second Law of Thermodynamics

36. ‖ Two containers hold several balls. Once a second, one of the balls is chosen at random and switched to the other container. After a long time has passed, you record the number of balls in each container every second. In 10,000 s, you find 80 times when all the balls were in one container (either one) and the other container was empty.
   a. How many balls are there?
   b. What is the most likely number of balls to be found in one of the containers?

Problems

37. ‖ The pressure inside a tank of neon is 150 atm. The temperature is 25°C. On average, how many atomic diameters does a neon atom move between collisions?

38. ‖‖ From what height must an oxygen molecule fall in a vacuum so that its kinetic energy at the bottom equals the average energy of an oxygen molecule at 300 K?

39. ‖ Dust particles are $\approx 10\ \mu\text{m}$ in diameter. They are pulverized rock, with $\rho \approx 2500\ \text{kg/m}^3$. If you treat dust as an ideal gas, what is the rms speed of a dust particle at 20°C?

40. ‖ A mad engineer builds a cube, 2.5 m on a side, in which 6.2-cm-diameter rubber balls are constantly sent flying in random directions by vibrating walls. He will award a prize to anyone who can figure out how many balls are in the cube without entering it or taking out any of the balls. You decide to shoot 6.2-cm-diameter plastic balls into the cube, through a small hole, to see how far they get before colliding with a rubber ball. After many shots, you find they travel an average distance of 1.8 m. How many rubber balls do you think are in the cube?

41. ‖ Photons of light scatter off molecules, and the distance you can see through a gas is proportional to the mean free path of photons through the gas. Photons are not gas molecules, so the mean free path of a photon is not given by Equation 20.3, but its dependence on the number density of the gas and on the molecular radius is the same. Suppose you are in a smoggy city and can barely see buildings 500 m away.
    a. How far would you be able to see if all the molecules around you suddenly doubled in volume?
    b. How far would you be able to see if the temperature suddenly rose from 20°C to a blazing hot 1500°C with the pressure unchanged?

42. ‖ Interstellar space, far from any stars, is filled with a very low density of hydrogen atoms (H, not H_2). The number density is about 1 atom/cm^3 and the temperature is about 3 K.
    a. Estimate the pressure in interstellar space. Give your answer in Pa and in atm.
    b. What is the rms speed of the atoms?
    c. What is the edge length L of an $L \times L \times L$ cube of gas with 1.0 J of thermal energy?

43. ‖ You are watching a science fiction movie in which the hero shrinks down to the size of an atom and fights villains while jumping from air molecule to air molecule. In one scene, the hero's molecule is about to crash head-on into the molecule on which a villain is riding. The villain's molecule is initially 50 molecular radii away and, in the movie, it takes 3.5 s for the molecules to collide. Estimate the air temperature required for this to be possible. Assume the molecules are nitrogen molecules, each traveling at the rms speed. Is this a plausible temperature for air?

44. ‖‖ **CALC**
    a. Find an expression for the v_{rms} of gas molecules in terms of p, V, and the total mass of the gas M.
    b. A gas cylinder has a piston at one end that is moving outward at speed v_{piston} during an isobaric expansion of the gas. Find an expression for the *rate* at which v_{rms} is changing in terms of v_{piston}, the instantaneous value of v_{rms}, and the instantaneous value L of the length of the cylinder.
    c. A cylindrical sample chamber has a piston moving outward at 0.50 m/s during an isobaric expansion. The rms speed of the gas molecules is 450 m/s at the instant the chamber length is 1.5 m. At what rate is v_{rms} changing?

45. ‖ Equation 20.3 is the mean free path of a particle through a gas of identical particles of equal radius. An electron can be thought of as a point particle with zero radius.
    a. Find an expression for the mean free path of an electron through a gas.
    b. Electrons travel 3 km through the Stanford Linear Accelerator. In order for scattering losses to be negligible, the pressure inside the accelerator tube must be reduced to the point where the mean free path is at least 50 km. What is the maximum

possible pressure inside the accelerator tube, assuming $T = 20°C$? Give your answer in both Pa and atm.

46. ‖ Uranium has two naturally occurring isotopes. ^{238}U has a natural abundance of 99.3% and ^{235}U has an abundance of 0.7%. It is the rarer ^{235}U that is needed for nuclear reactors. The isotopes are separated by forming uranium hexafluoride, UF_6, which is a gas, then allowing it to diffuse through a series of porous membranes. $^{235}UF_6$ has a slightly larger rms speed than $^{238}UF_6$ and diffuses slightly faster. Many repetitions of this procedure gradually separate the two isotopes. What is the ratio of the rms speed of $^{235}UF_6$ to that of $^{238}UF_6$?

47. ‖ On earth, STP is based on the average atmospheric pressure at the surface and on a phase change of water that occurs at an easily produced temperature, being only slightly cooler than the average air temperature. The atmosphere of Venus is almost entirely carbon dioxide (CO_2), the pressure at the surface is a staggering 93 atm, and the average temperature is 470°C. Venusian scientists, if they existed, would certainly use the surface pressure as part of their definition of STP. To complete the definition, they would seek a phase change that occurs near the average temperature. Conveniently, the melting point of the element tellurium is 450°C. What are (a) the rms speed and (b) the mean free path of carbon dioxide molecules at Venusian STP based on this phase change in tellurium? The radius of a CO_2 molecule is 1.5×10^{-10} m.

48. ‖‖‖ 5.0×10^{23} nitrogen molecules collide with a 10 cm^2 wall each second. Assume that the molecules all travel with a speed of 400 m/s and strike the wall head-on. What is the pressure on the wall?

49. ‖‖‖ A 10 cm $\times$ 10 cm $\times$ 10 cm box contains 0.010 mol of nitrogen at 20°C. What is the rate of collisions (collisions/s) on one wall of the box?

50. ‖ **FIGURE P20.50** shows the thermal energy of 0.14 mol of gas as a function of temperature. What is C_V for this gas?

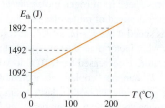

FIGURE P20.50

51. ‖ A 100 cm^3 box contains helium at a pressure of 2.0 atm and a temperature of 100°C. It is placed in thermal contact with a 200 cm^3 box containing argon at a pressure of 4.0 atm and a temperature of 400°C.
    a. What is the initial thermal energy of each gas?
    b. What is the final thermal energy of each gas?
    c. How much heat energy is transferred, and in which direction?
    d. What is the final temperature?
    e. What is the final pressure in each box?

52. ‖ 2.0 g of helium at an initial temperature of 300 K interacts thermally with 8.0 g of oxygen at an initial temperature of 600 K.
    a. What is the initial thermal energy of each gas?
    b. What is the final thermal energy of each gas?
    c. How much heat energy is transferred, and in which direction?
    d. What is the final temperature?

53. ‖ A gas of 1.0×10^{20} atoms or molecules has 1.0 J of thermal energy. Its molar specific heat at constant pressure is 20.8 J/mol K. What is the temperature of the gas?

54. ‖ Scientists studying the behavior of hydrogen at low temperatures need to lower the temperature of 0.50 mol of hydrogen gas from 300 K to 30 K. How much thermal energy must they remove from the gas?

55. ‖ A water molecule has its three atoms arranged in a "V" shape, so it has rotational kinetic energy around any of three mutually perpendicular axes. However, like diatomic molecules, its vibrational modes are not active at temperatures below 1000 K. What is the thermal energy of 2.0 mol of steam at a temperature of 160°C?

56. ‖ A monatomic gas and a diatomic gas have equal numbers of moles and equal temperatures. Both are heated at constant pressure until their volume doubles. What is the ratio $Q_{diatomic}/Q_{monatomic}$?

57. ‖ In the discussion following Equation 20.43 it was said that $Q_1 = -Q_2$. Prove that this is so.

58. ‖ A monatomic gas is adiabatically compressed to $\frac{1}{8}$ of its initial volume. Does each of the following quantities change? If so, does it increase or decrease, and by what factor? If not, why not?
    a. The rms speed.
    b. The mean free path.
    c. The thermal energy of the gas.
    d. The molar specific heat at constant volume.

59. ‖ n moles of a diatomic gas with $C_V = \frac{5}{2}R$ has initial pressure p_i and volume V_i. The gas undergoes a process in which the pressure is directly proportional to the volume until the rms speed of the molecules has doubled.
    a. Show this process on a pV diagram.
    b. How much heat does this process require? Give your answer in terms of n, p_i, and V_i.

60. ‖ The 2010 Nobel Prize in Physics was awarded for the discovery of graphene, a two-dimensional form of carbon in which the atoms form a two-dimensional crystal-lattice sheet only one atom thick. Predict the molar specific heat of graphene. Give your answer as a multiple of R.

61. ‖‖ The rms speed of the molecules in 1.0 g of hydrogen gas is 1800 m/s.
    a. What is the total translational kinetic energy of the gas molecules?
    b. What is the thermal energy of the gas?
    c. 500 J of work are done to compress the gas while, in the same process, 1200 J of heat energy are transferred from the gas to the environment. Afterward, what is the rms speed of the molecules?

62. ‖ At what temperature does the rms speed of (a) a nitrogen molecule and (b) a hydrogen molecule equal the escape speed from the earth's surface? (c) You'll find that these temperatures are very high, so you might think that the earth's gravity could easily contain both gases. But not all molecules move with v_{rms}. There is a distribution of speeds, and a small percentage of molecules have speeds several times v_{rms}. Bit by bit, a gas can slowly leak out of the atmosphere as its fastest molecules escape. A reasonable rule of thumb is that the earth's gravity can contain a gas only if the average translational kinetic energy per molecule is less than 1% of the kinetic energy needed to escape. Use this rule to show why the earth's atmosphere contains nitrogen but not hydrogen, even though hydrogen is the most abundant element in the universe.

63. ‖ n_1 moles of a monatomic gas and n_2 moles of a diatomic gas are mixed together in a container.
    a. Derive an expression for the molar specific heat at constant volume of the mixture.
    b. Show that your expression has the expected behavior if $n_1 \rightarrow 0$ or $n_2 \rightarrow 0$.

64. ‖ A 1.0 kg ball is at rest on the floor in a 2.0 m × 2.0 m × 2.0 m room of air at STP. Air is 80% nitrogen (N_2) and 20% oxygen (O_2) by volume.
    a. What is the thermal energy of the air in the room?
    b. What fraction of the thermal energy would have to be conveyed to the ball for it to be spontaneously launched to a height of 1.0 m?
    c. By how much would the air temperature have to decrease to launch the ball?
    d. Your answer to part c is so small as to be unnoticeable, yet this event never happens. Why not?

Challenge Problems

65. ‖‖‖ An experiment you're designing needs a gas with $\gamma = 1.50$. You recall from your physics class that no individual gas has this value, but it occurs to you that you could produce a gas with $\gamma = 1.50$ by mixing together a monatomic gas and a diatomic gas. What fraction of the molecules need to be monatomic?

66. ‖‖‖ Consider a container like that shown in Figure 20.12, with n_1 moles of a monatomic gas on one side and n_2 moles of a diatomic gas on the other. The monatomic gas has initial temperature T_{1i}. The diatomic gas has initial temperature T_{2i}.
    a. Show that the equilibrium thermal energies are

$$E_{1f} = \frac{3n_1}{3n_1 + 5n_2}(E_{1i} + E_{2i})$$

$$E_{2f} = \frac{5n_2}{3n_1 + 5n_2}(E_{1i} + E_{2i})$$

    b. Show that the equilibrium temperature is

$$T_f = \frac{3n_1 T_{1i} + 5n_2 T_{2i}}{3n_1 + 5n_2}$$

    c. 2.0 g of helium at an initial temperature of 300 K interacts thermally with 8.0 g of oxygen at an initial temperature of 600 K. What is the final temperature? How much heat energy is transferred, and in which direction?

21 Heat Engines and Refrigerators

This power plant is generating electricity by turning heat into work—but not very efficiently. The cooling towers dissipate roughly two-thirds of the fuel's energy into the air as "waste heat."

IN THIS CHAPTER, you will study the principles that govern heat engines and refrigerators.

What is a heat engine?

A heat engine is a device for transforming heat energy into useful work. Heat engines

- Follow a cyclical process that can be shown on a pV diagram or an energy-transfer diagram.
- Require not only a source of heat but also a source of cooling. These are called the hot reservoir and the cold reservoir.
- Are governed by the first and second laws of thermodynamics.

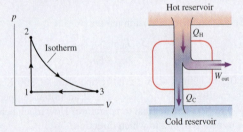

« **LOOKING BACK** Sections 19.2–19.4 and 19.7 Work, heat, the first law of thermodynamics, and the specific heats of gases

What is a refrigerator?

A refrigerator is any device—including air conditioners—in which external work is used to "pump energy uphill" from cold to hot. A refrigerator is essentially a heat engine running in reverse. A refrigerator's efficiency is given by its coefficient of performance.

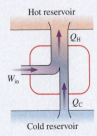

How is the efficiency of an engine determined?

How good is an engine at transforming heat energy into work? We'll define an engine's thermal efficiency as

$$\text{efficiency} = \frac{\text{work done}}{\text{heat required}}$$

Conservation of energy—the first law of thermodynamics—says that no engine can have an efficiency greater than one.

Is there a maximum possible efficiency?

Yes. And, surprisingly, it's set by the second law's prohibition of spontaneous heat flow from cold to hot. We'll find that a *perfectly reversible heat engine*—a Carnot engine—has the maximum efficiency allowed by the laws of thermodynamics. The Carnot efficiency depends only on the temperatures of the hot and cold reservoirs.

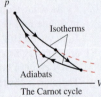

The Carnot cycle

« **LOOKING BACK** Section 20.6 The second law of thermodynamics

Why are heat engines important?

Modern society is powered by devices that transform fuel energy into useful work. Examples include engines for propelling cars and airplanes, power plants for generating electricity, and a wide variety of machines used in industry and manufacturing. These are all heat engines, and all are governed by the laws of thermodynamics. Maximizing efficiency—an important engineering and societal goal—requires a good understanding of the underlying physical principles.

21.1 Turning Heat into Work

Thermodynamics is the branch of physics that studies the transformation of energy. Many practical devices are designed to transform energy from one form, such as the heat from burning fuel, into another, such as work. Chapters 19 and 20 established two laws of thermodynamics that any such device must obey:

> **First law** Energy is conserved; that is, $\Delta E_{th} = W + Q$.
>
> **Second law** Most macroscopic processes are irreversible. In particular, heat energy is transferred spontaneously from a hotter system to a colder system but never from a colder system to a hotter system.

Our goal in this chapter is to discover what these two laws, especially the second law, imply about devices that turn heat into work. In particular:

- How does a practical device transform heat into work?
- What are the limitations and restrictions on these energy transformations?

Work Done by the System

The work W in the first law is the work done *on* the system by external forces from the environment. However, it makes more sense in "practical thermodynamics" to use the work done *by* the system. For example, you want to know how much useful work you can obtain from an expanding gas. The work done by the system is called W_s.

   Work done by the environment and work done by the system are not mutually exclusive. In fact, they are very simply related by $W_s = -W$. In **FIGURE 21.1**, force $\vec{F}_{gas}$ due to the gas pressure does work when the piston moves. This is W_s, the work done *by* the system. At the same time, some object in the environment, such as a piston rod, must be pushing inward with force $\vec{F}_{ext} = -\vec{F}_{gas}$ to keep the gas pressure from blowing the piston out. This force does the work W *on* the system, work that you've learned is the negative of the area under the pV curve of the process.

   Because the forces are equal but opposite, we see that

$$W_s = -W = \text{the area under the } pV \text{ curve} \qquad (21.1)$$

When a gas expands and pushes the piston out, transferring energy out of the system, we say "the system does work on the environment." While this may seem to imply that the environment is doing no work on the system, all the phrase means is that W_s is positive and W is negative.

   Similarly, "the environment does work on the system" means that $W > 0$ (energy is transferred into the system) and thus $W_s < 0$. Whether we use W or W_s is a matter of convenience. They are always opposite to each other rather than one being zero.

   The first law of thermodynamics $\Delta E_{th} = W + Q$ can be written in terms of W_s as

$$Q = W_s + \Delta E_{th} \qquad \text{(first law of thermodynamics)} \qquad (21.2)$$

Any energy transferred into a system as heat is either used to do work or stored within the system as an increased thermal energy.

Energy-Transfer Diagrams

Suppose you drop a hot rock into the ocean. Heat is transferred from the rock to the ocean until the rock and ocean are at the same temperature. Although the ocean warms up ever so slightly, ΔT_{ocean} is so small as to be completely insignificant. For all practical purposes, the ocean is infinite and unchangeable.

   An **energy reservoir** is an object or a part of the environment so large that **its temperature does not change** when heat is transferred between the system and the reservoir. A reservoir at a higher temperature than the system is called a *hot reservoir*.

FIGURE 21.1 Forces $\vec{F}_{gas}$ and $\vec{F}_{ext}$ both do work as the piston moves.

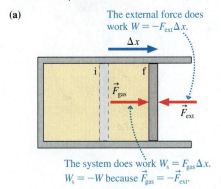

(a) The external force does work $W = -F_{ext}\Delta x$.

The system does work $W_s = F_{gas}\Delta x$. $W_s = -W$ because $\vec{F}_{gas} = -\vec{F}_{ext}$.

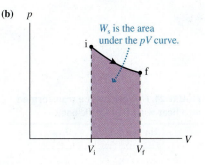

(b) W_s is the area under the pV curve.

FIGURE 21.2 Energy-transfer diagrams.

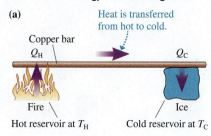

(a)

Heat is transferred from hot to cold.

Copper bar
Q_H Q_C

Fire   Ice

Hot reservoir at T_H Cold reservoir at T_C

(b)

Heat energy is transferred from a hot reservoir to a cold reservoir. Energy conservation requires $Q_C = Q_H$.

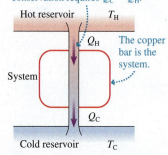

Hot reservoir T_H

Q_H The copper bar is the system.

System

Q_C

Cold reservoir T_C

(c)

Heat is never spontaneously transferred from a colder object to a hotter object.

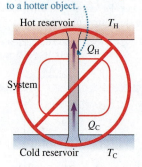

Hot reservoir T_H

Q_H

System

Q_C

Cold reservoir T_C

FIGURE 21.3 Work can be transformed into heat with 100% efficiency.

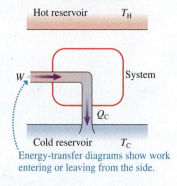

Hot reservoir T_H

W System

Q_C

Cold reservoir T_C

Energy-transfer diagrams show work entering or leaving from the side.

A vigorously burning flame is a hot reservoir for small objects placed in the flame. A reservoir at a lower temperature than the system is called a *cold reservoir*. The ocean is a cold reservoir for the hot rock. We will use T_H and T_C to designate the temperatures of the hot and cold reservoirs.

Hot and cold reservoirs are idealizations, in the same category as frictionless surfaces and massless strings. No real object can maintain a perfectly constant temperature as heat is transferred in or out. Even so, an object can be modeled as a reservoir if it is much larger than the system that thermally interacts with it.

Heat energy is transferred between a system and a reservoir if they have different temperatures. We will define

$$Q_H = \text{amount of heat transferred to or from a hot reservoir}$$

$$Q_C = \text{amount of heat transferred to or from a cold reservoir}$$

By definition, Q_H and Q_C are *positive* quantities. The direction of heat transfer, which determines the sign of Q in the first law, will always be clear as we deal with thermodynamic devices.

FIGURE 21.2a shows a heavy copper bar between a hot reservoir (at temperature T_H) and a cold reservoir (at temperature T_C). Heat Q_H is transferred from the hot reservoir into the copper and heat Q_C is transferred from the copper to the cold reservoir. **FIGURE 21.2b** is an **energy-transfer diagram** for this process. The hot reservoir is always drawn at the top, the cold reservoir at the bottom, and the system—the copper bar in this case—between them. Figure 21.2b shows heat Q_H being transferred into the system and Q_C being transferred out.

The first law of thermodynamics $Q = W_s + \Delta E_{th}$ refers to the *system*. Q is the *net* heat to the system, which, in this case, is $Q = Q_H - Q_C$. The copper bar does no work, so $W_s = 0$. The bar warms up when first placed between the two reservoirs, but it soon comes to a steady state where its temperature no longer changes. Then $\Delta E_{th} = 0$. Thus the first law tells us that $Q = Q_H - Q_C = 0$, from which we conclude that $Q_C = Q_H$.

In other words, all of the heat transferred into the hot end of the rod is subsequently transferred out of the cold end. This isn't surprising. After all, we know that heat is transferred spontaneously from a hotter object to a colder object. Even so, there has to be some *means* by which the heat energy gets from the hotter object to the colder. The copper bar provides a route for the energy transfer, and $Q_C = Q_H$ is the statement that energy is conserved as it moves through the bar.

Contrast Figure 21.2b with **FIGURE 21.2c**. Figure 21.2c shows a system in which heat is being transferred from the cold reservoir to the hot reservoir. The first law of thermodynamics is not violated, because $Q_H = Q_C$, but the second law is. If there were such a system, it would allow the spontaneous (i.e., with no outside input or assistance) transfer of heat from a colder object to a hotter object. The process of Figure 21.2c is forbidden by the second law of thermodynamics.

Work into Heat and Heat into Work

Turning work into heat is easy—just rub two objects together. Work from the friction force increases the objects' thermal energy and their temperature. Heat energy is then transferred from the warmer objects to the cooler environment. **FIGURE 21.3** is the energy-transfer diagram for this process. The conversion of work into heat is 100% efficient in that *all* the energy supplied to the system as work is ultimately transferred to the environment as heat. Notice that the objects have returned to their initial state at the end of this process, ready to repeat the process for as long as there's a source of motion.

The reverse—transforming heat into work—isn't so easy. Heat can be transformed into work in a one-time process, such as an isothermal expansion of a gas, but at the end the system is not restored to its initial state. To be practical, **a device that transforms heat into work must return to its initial state at the end of the process and be ready for continued use.** You want your car engine to turn over and over for as long as there's fuel.

Interestingly, no one has ever invented a "perfect engine" that transforms heat into work with 100% efficiency *and returns to its initial state* so that it can continue to do work as long as there is fuel. Of course, that such a device has not been invented is not a proof that it can't be done. We'll provide a proof shortly, but for now we'll make the hypothesis that the process of **FIGURE 21.4** is somehow forbidden.

Notice the asymmetry between Figures 21.3 and 21.4. The perfect transformation of work into heat is permitted, but the perfect transformation of heat into work is forbidden. This asymmetry parallels the asymmetry of the two processes in Figure 21.2. In fact, we'll soon see that the "perfect engine" of Figure 21.4 is forbidden for exactly the same reason: the second law of thermodynamics.

21.2 Heat Engines and Refrigerators

The steam generator at your local electric power plant works by boiling water to produce high-pressure steam that spins a turbine (which then spins a generator to produce electricity). That is, the steam pressure is doing work. The steam is then condensed to liquid water and pumped back to the boiler to start the process again. There are two crucial ideas here. First, the device works in a cycle, with the water returning to its initial conditions once a cycle. Second, heat is transferred to the water in the boiler, but heat is transferred *out* of the water in the condenser.

Car engines and steam generators are examples of what we call *heat engines*. A **heat engine** is any closed-cycle device that extracts heat Q_H from a hot reservoir, does useful work, and exhausts heat Q_C to a cold reservoir. A **closed-cycle device** is one that periodically *returns to its initial conditions*, repeating the same process over and over. That is, all state variables (pressure, temperature, thermal energy, and so on) return to their initial values once every cycle. Consequently, a heat engine can continue to do useful work for as long as it is attached to the reservoirs.

FIGURE 21.5 is the energy-transfer diagram of a heat engine. Unlike the forbidden "perfect engine" of Figure 21.4, a heat engine is connected to both a hot reservoir *and* a cold reservoir. You can think of a heat engine as "siphoning off" some of the heat that moves from the hot reservoir to the cold reservoir and transforming that heat into work—*some* of the heat, but not all.

Because the temperature and thermal energy of a heat engine return to their initial values at the end of each cycle, there is no *net* change in E_{th}:

$$(\Delta E_{th})_{net} = 0 \quad \text{(any heat engine, over one full cycle)} \quad (21.3)$$

Consequently, the first law of thermodynamics *for a full cycle* of a heat engine is $(\Delta E_{th})_{net} = Q - W_s = 0$.

Let's define W_{out} to be the useful work done *by* the heat engine *per cycle*. The first law applied to a heat engine is

$$W_{out} = Q_{net} = Q_H - Q_C \quad \text{(work per cycle done by a heat engine)} \quad (21.4)$$

This is just energy conservation. The energy-transfer diagram of Figure 21.5 is a pictorial representation of Equation 21.4.

NOTE Equations 21.3 and 21.4 apply only to a *full cycle* of the heat engine. They are *not* valid for any of the individual processes that make up a cycle.

For practical reasons, we would like an engine to do the maximum amount of work with the minimum amount of fuel. We can measure the performance of a heat engine in terms of its **thermal efficiency** η (lowercase Greek eta), defined as

$$\eta = \frac{W_{out}}{Q_H} = \frac{\text{what you get}}{\text{what you had to pay}} \quad (21.5)$$

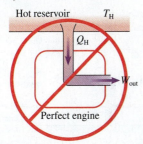

FIGURE 21.4 There are no perfect engines that turn heat into work with 100% efficiency.

Hot reservoir T_H
Q_H
W_{out}
Perfect engine

The steam turbine in a modern power plant is an enormous device. Expanding steam does work by spinning the turbine.

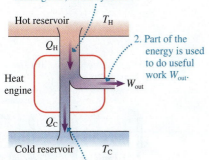

FIGURE 21.5 The energy-transfer diagram of a heat engine.

1. Heat energy Q_H is transferred from the hot reservoir (typically burning fuel) to the system.

Hot reservoir T_H
Q_H
2. Part of the energy is used to do useful work W_{out}.
Heat engine
W_{out}
Q_C
Cold reservoir T_C

3. The remaining energy $Q_C = Q_H - W_{out}$ is exhausted to the cold reservoir (cooling water or the air) as waste heat.

FIGURE 21.6 η is the fraction of heat energy that is transformed into useful work.

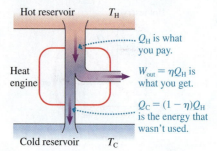

Using Equation 21.4 for W_{out}, we can also write the thermal efficiency as

$$\eta = 1 - \frac{Q_C}{Q_H} \qquad (21.6)$$

FIGURE 21.6 illustrates the idea of thermal efficiency.

A *perfect* heat engine would have $\eta_{perfect} = 1$. That is, it would be 100% efficient at converting heat from the hot reservoir (the burning fuel) into work. You can see from Equation 21.6 that a perfect engine would have no exhaust ($Q_C = 0$) and would not need a cold reservoir. Figure 21.4 has already suggested that there are no perfect heat engines, that an engine with $\eta = 1$ is impossible. A heat engine *must* exhaust **waste heat** to a cold reservoir. It is energy that was extracted from the hot reservoir but *not* transformed to useful work.

Practical heat engines, such as car engines and steam generators, have thermal efficiencies in the range $\eta \approx 0.1 - 0.5$. This is not large. Can a clever designer do better, or is this some kind of physical limitation?

> **STOP TO THINK 21.1** Rank in order, from largest to smallest, the work W_{out} performed by these four heat engines.

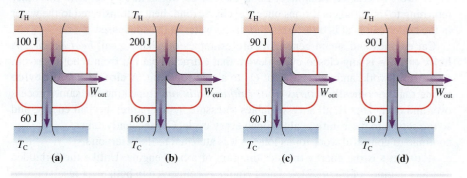

A Heat-Engine Example

To illustrate how these ideas actually work, **FIGURE 21.7** shows a simple engine that converts heat into the work of lifting mass M.

FIGURE 21.7 A simple heat engine transforms heat into work.

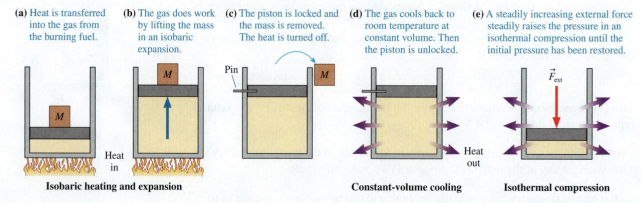

(a) Heat is transferred into the gas from the burning fuel.

(b) The gas does work by lifting the mass in an isobaric expansion.

(c) The piston is locked and the mass is removed. The heat is turned off.

(d) The gas cools back to room temperature at constant volume. Then the piston is unlocked.

(e) A steadily increasing external force steadily raises the pressure in an isothermal compression until the initial pressure has been restored.

Isobaric heating and expansion **Constant-volume cooling** **Isothermal compression**

The net effect of this multistep process is to convert some of the fuel's energy into the useful work of lifting the mass. There has been no net change in the gas, which has returned to its initial pressure, volume, and temperature at the end of step (e). We can start the whole process over again and continue lifting masses (doing work) as long as we have fuel.

FIGURE 21.8 shows the heat-engine process on a *pV* diagram. It is a *closed cycle* because the gas returns to its initial conditions. No work is done during the isochoric process, and, as you can see from the areas under the curve, the work done *by* the gas to lift the mass is greater than the work the environment must do *on* the gas to recompress it. Thus this heat engine, by burning fuel, does *net* work per cycle: $W_{net} = W_{lift} - W_{ext} = (W_s)_{1 \to 2} + (W_s)_{3 \to 1}$.

Notice that the cyclical process of Figure 21.8 involves two cooling processes in which heat is transferred *from* the gas to the environment. Heat energy is transferred from hotter objects to colder objects, so the system *must* be connected to a cold reservoir with $T_C < T_{gas}$ during these two processes. A key to understanding heat engines is that they require both a heat source (burning fuel) *and* a heat sink (cooling water, the air, or something at a lower temperature than the system).

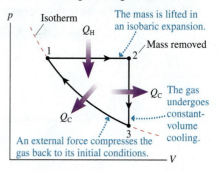

FIGURE 21.8 The closed-cycle *pV* diagram for the heat engine of Figure 21.7.

EXAMPLE 21.1 | Analyzing a heat engine I

Analyze the heat engine of FIGURE 21.9 to determine (a) the net work done per cycle, (b) the engine's thermal efficiency, and (c) the engine's power output if it runs at 600 rpm. Assume the gas is monatomic.

FIGURE 21.9 The heat engine of Example 21.1.

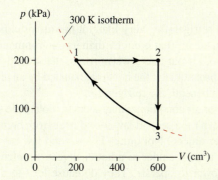

MODEL The gas follows a closed cycle consisting of three distinct processes, each of which was studied in Chapters 18 and 19. For each of the three we need to determine the work done and the heat transferred.

SOLVE To begin, we can use the initial conditions at state 1 and the ideal-gas law to determine the number of moles of gas:

$$n = \frac{p_1 V_1}{R T_1} = \frac{(200 \times 10^3 \text{ Pa})(2.0 \times 10^{-4} \text{ m}^3)}{(8.31 \text{ J/mol K})(300 \text{ K})} = 0.0160 \text{ mol}$$

Process $1 \to 2$: The work done *by* the gas in the isobaric expansion is

$$(W_s)_{12} = p \Delta V = (200 \times 10^3 \text{ Pa})\left((6.0 - 2.0) \times 10^{-4} \text{ m}^3\right) = 80 \text{ J}$$

We can use the ideal-gas law at constant pressure to find $T_2 = (V_2/V_1)T_1 = 3T_1 = 900 \text{ K}$. The heat transfer during a constant-pressure process is

$$Q_{12} = nC_P \Delta T$$

$$= (0.0160 \text{ mol})(20.8 \text{ J/mol K})(900 \text{ K} - 300 \text{ K}) = 200 \text{ J}$$

where we used $C_P = \frac{5}{2}R$ for a monatomic ideal gas.

Process $2 \to 3$: No work is done in an isochoric process, so $(W_s)_{23} = 0$. The temperature drops back to 300 K, so the heat transfer, with $C_V = \frac{3}{2}R$, is

$$Q_{23} = nC_V \Delta T$$

$$= (0.0160 \text{ mol})(12.5 \text{ J/mol K})(300 \text{ K} - 900 \text{ K}) = -120 \text{ J}$$

Process $3 \to 1$: The gas returns to its initial state with volume V_1. The work done *by* the gas during an isothermal process is

$$(W_s)_{31} = nRT \ln\!\left(\frac{V_1}{V_3}\right)$$

$$= (0.0160 \text{ mol})(8.31 \text{ J/mol K})(300 \text{ K}) \ln\!\left(\frac{1}{3}\right) = -44 \text{ J}$$

W_s is negative because the environment does work on the gas to compress it. An isothermal process has $\Delta E_{th} = 0$ and hence, from the first law,

$$Q_{31} = (W_s)_{31} = -44 \text{ J}$$

Q is negative because the gas must be cooled as it is compressed to keep the temperature constant.

a. The *net* work done by the engine during one cycle is

$$W_{out} = (W_s)_{12} + (W_s)_{23} + (W_s)_{31} = 36 \text{ J}$$

As a consistency check, notice that the net heat transfer is

$$Q_{net} = Q_{12} + Q_{23} + Q_{31} = 36 \text{ J}$$

Equation 21.4 told us that a heat engine *must* have $W_{out} = Q_{net}$, and we see that it does.

b. The efficiency depends not on the net heat transfer but on the heat Q_H transferred into the engine from the flame. Heat enters during process $1 \to 2$, where Q is positive, and exits during processes $2 \to 3$ and $3 \to 1$, where Q is negative. Thus

$$Q_H = Q_{12} = 200 \text{ J}$$

$$Q_C = |Q_{23}| + |Q_{31}| = 164 \text{ J}$$

Notice that $Q_H - Q_C = 36 \text{ J} = W_{out}$. In this heat engine, 200 J of heat from the hot reservoir does 36 J of useful work. Thus the thermal efficiency is

$$\eta = \frac{W_{out}}{Q_H} = \frac{36 \text{ J}}{200 \text{ J}} = 0.18 \text{ or } 18\%$$

This heat engine is far from being a perfect engine!

Continued

c. An engine running at 600 rpm goes through 10 cycles per second. The power output is the work done *per second:*

$$P_{out} = \text{(work per cycle)} \times \text{(cycles per second)}$$
$$= 360 \text{ J/s} = 360 \text{ W}$$

ASSESS Although we didn't need Q_{net}, verifying that $Q_{net} = W_{out}$ was a check of self-consistency. Heat-engine analysis requires many calculations and offers many opportunities to get signs wrong. However, there are a sufficient number of self-consistency checks so that you can almost always spot calculational errors *if you check for them.*

Let's think about this example a bit more before going on. We've said that a heat engine operates between a hot reservoir and a cold reservoir. Figure 21.9 doesn't explicitly show the reservoirs. Nonetheless, we know that heat is transferred from a hotter object to a colder object. Heat Q_H is transferred into the system during process $1 \rightarrow 2$ as the gas warms from 300 K to 900 K. For this to be true, the hot-reservoir temperature T_H must be ≥ 900 K. Likewise, heat Q_C is transferred from the system to the cold reservoir as the temperature drops from 900 K to 300 K in process $2 \rightarrow 3$. For this to be true, the cold-reservoir temperature T_C must be ≤ 300 K.

So, while we really don't know what the reservoirs are or their exact temperatures, we can say with certainty that the hot-reservoir temperature T_H must exceed the highest temperature reached by the system and the cold-reservoir temperature T_C must be less than the coldest system temperature.

Refrigerators

Your house or apartment has a refrigerator. Very likely it has an air conditioner. The purpose of these devices is to make air that is cooler than its environment even colder. The first does so by blowing hot air out into a warm room, the second by blowing it out to the hot outdoors. You've probably felt the hot air exhausted by an air conditioner compressor or coming out from beneath the refrigerator.

At first glance, a refrigerator or air conditioner may seem to violate the second law of thermodynamics. After all, doesn't the second law forbid heat from being transferred from a colder object to a hotter object? Not quite: The second law says that heat is not *spontaneously* transferred from a colder to a hotter object. A refrigerator or air conditioner requires electric power to operate. They do cause heat to be transferred from cold to hot, but the transfer is "assisted" rather than spontaneous.

A **refrigerator** is any closed-cycle device that uses external work W_{in} to remove heat Q_C from a cold reservoir and exhaust heat Q_H to a hot reservoir. **FIGURE 21.10** is the energy-transfer diagram of a refrigerator. The cold reservoir is the air inside the refrigerator or the air inside your house on a summer day. To keep the air cold, in the face of inevitable "heat leaks," the refrigerator or air conditioner compressor continuously removes heat from the cold reservoir and exhausts heat into the room or outdoors. You can think of a refrigerator as "pumping heat uphill," much as a water pump lifts water uphill.

Because a refrigerator, like a heat engine, is a cyclical device, $\Delta E_{th} = 0$. Conservation of energy requires

$$Q_H = Q_C + W_{in} \tag{21.7}$$

To move energy from a colder to a hotter reservoir, a refrigerator must exhaust *more* heat to the outside than it removes from the inside. This has significant implications for whether or not you can cool a room by leaving the refrigerator door open.

The thermal efficiency of a heat engine was defined as "what you get (useful work W_{out})" versus "what you had to pay (fuel to supply Q_H)." By analogy, we define the **coefficient of performance** K of a refrigerator to be

$$K = \frac{Q_C}{W_{in}} = \frac{\text{what you get}}{\text{what you had to pay}} \tag{21.8}$$

What you get, in this case, is the removal of heat from the cold reservoir. But you have to pay the electric company for the work needed to run the refrigerator. A better

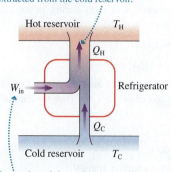

This air conditioner transfers heat energy *from* the cool indoors *to* the hot exterior.

FIGURE 21.10 The energy-transfer diagram of a refrigerator.

The amount of heat exhausted to the hot reservoir is larger than the amount of heat extracted from the cold reservoir.

External work is used to remove heat from a cold reservoir and exhaust heat to a hot reservoir.

refrigerator will require less work to remove a given amount of heat, thus having a larger coefficient of performance.

A perfect refrigerator would require no work ($W_{in} = 0$) and would have $K_{perfect} = \infty$. But if Figure 21.10 had no work input, it would look like Figure 21.2c. That device was forbidden by the second law of thermodynamics because, with no work input, heat would move *spontaneously* from cold to hot.

We noted in Chapter 20 that the second law of thermodynamics can be stated several different but equivalent ways. We can now give a third informal statement:

> **Second law, informal statement #3** There are no perfect refrigerators with coefficient of performance $K = \infty$.

Any real refrigerator or air conditioner *must* use work to move energy from the cold reservoir to the hot reservoir, hence $K < \infty$.

No Perfect Heat Engines

We hypothesized above that there are no perfect heat engines—that is, no heat engines like the one shown in Figure 21.4 with $Q_C = 0$ and $\eta = 1$. Now we're ready to prove this hypothesis. **FIGURE 21.11** shows a hot reservoir at temperature T_H and a cold reservoir at temperature T_C. An ordinary refrigerator, one that obeys all the laws of physics, is operating between these two reservoirs.

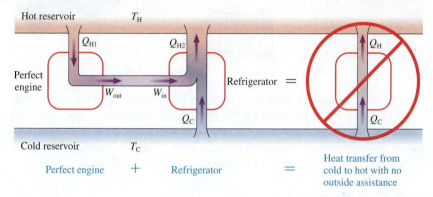

◀ **FIGURE 21.11** A perfect engine driving an ordinary refrigerator would be able to violate the second law of thermodynamics.

Suppose we had a perfect heat engine, one that takes in heat Q_H from the high-temperature reservoir and transforms that energy entirely into work W_{out}. If we had such a heat engine, we could use its output to provide the work input to the refrigerator. The two devices combined have no connection to the external world. That is, there's no net input or net output of work.

If we built a box around the heat engine and refrigerator, so that you couldn't see what was inside, the only thing you would observe is heat being transferred *with no outside assistance* from the cold reservoir to the hot reservoir. But a spontaneous or unassisted transfer of heat from a colder to a hotter object is exactly what the second law of thermodynamics forbids. Consequently, our assumption of a perfect heat engine must be wrong. Hence another statement of the second law of thermodynamics is:

> **Second law, informal statement #4** There are no perfect heat engines with efficiency $\eta = 1$.

Any real heat engine *must* exhaust waste heat Q_C to a cold reservoir.

STOP TO THINK 21.2 It's a hot day and your air conditioner is broken. Your roommate says, "Let's open the refrigerator door and cool this place off." Will this work?

a. Yes.   b. No.   c. It might, but it will depend on how hot the room is.

21.3 Ideal-Gas Heat Engines

We will focus on heat engines that use a gas as the *working substance*. The gasoline or diesel engine in your car is an engine that alternately compresses and expands a gaseous fuel-air mixture. A discussion of engines such as steam generators that rely on phase changes will be deferred to more advanced courses.

A gas heat engine can be represented by a closed-cycle trajectory in the *pV* diagram, such as the one shown in **FIGURE 21.12a**. This observation leads to an important geometric interpretation of the work done by the system during one full cycle. You learned in Section 21.1 that the work done *by* the system is the area under the curve of a *pV* trajectory. As **FIGURE 21.12b** shows, the net work done during a full cycle is

$$W_{out} = W_{expand} - |W_{compress}| = \text{area } inside \text{ the closed curve} \qquad (21.9)$$

FIGURE 21.12 The work W_{out} done by the system during one full cycle is the area enclosed within the curve.

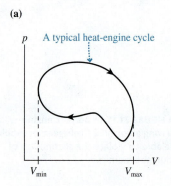

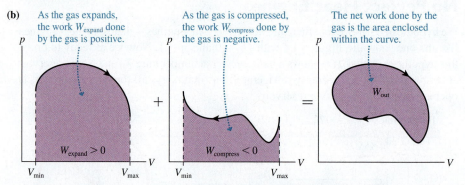

You can see that **the net work done by a gas heat engine during one full cycle is the area enclosed by the *pV* curve for the cycle.** A thermodynamic cycle with a larger enclosed area does more work than one with a smaller enclosed area. Notice that the gas must go around the *pV* trajectory in a *clockwise* direction for W_{out} to be positive. We'll see later that a refrigerator uses a counterclockwise (ccw) cycle.

Ideal-Gas Summary

We've learned a lot about ideal gases in the last three chapters. All gas processes obey the ideal-gas law $pV = nRT$ and the first law of thermodynamics $\Delta E_{th} = Q - W_s$. **TABLE 21.1** summarizes the results for specific gas processes. This table shows W_s, the work done *by* the system, so the signs are opposite those in Chapter 19.

TABLE 21.1 Summary of ideal-gas processes

Process	Gas law	Work W_s	Heat Q	Thermal energy
Isochoric	$p_i/T_i = p_f/T_f$	0	$nC_V \Delta T$	$\Delta E_{th} = Q$
Isobaric	$V_i/T_i = V_f/T_f$	$p \Delta V$	$nC_P \Delta T$	$\Delta E_{th} = Q - W_s$
Isothermal	$p_i V_i = p_f V_f$	$nRT \ln(V_f/V_i)$	$Q = W_s$	$\Delta E_{th} = 0$
		$pV \ln(V_f/V_i)$		
Adiabatic	$p_i V_i^\gamma = p_f V_f^\gamma$	$(p_f V_f - p_i V_i)/(1 - \gamma)$	0	$\Delta E_{th} = -W_s$
	$T_i V_i^{\gamma-1} = T_f V_f^{\gamma-1}$	$-nC_V \Delta T$		
Any	$p_i V_i/T_i = p_f V_f/T_f$	area under curve		$\Delta E_{th} = nC_V \Delta T$

There is one entry in this table that you haven't seen before. The expression

$$W_s = \frac{p_f V_f - p_i V_i}{1 - \gamma} \qquad \text{(work in an adiabatic process)} \qquad (21.10)$$

for the work done in an adiabatic process follows from writing $W_s = -\Delta E_{th} = -nC_V \Delta T$, which you learned in Chapter 19, then using $\Delta T = \Delta(pV)/nR$ and the definition of γ. The proof will be left for a homework problem.

You learned in Chapter 20 that the thermal energy of an ideal gas depends only on its temperature. **TABLE 21.2** lists the thermal energy, molar specific heats, and specific heat ratio $\gamma = C_P/C_V$ for monatomic and diatomic gases.

TABLE 21.2 Properties of monatomic and diatomic gases

	Monatomic	Diatomic
E_{th}	$\frac{3}{2}nRT$	$\frac{5}{2}nRT$
C_V	$\frac{3}{2}R$	$\frac{5}{2}R$
C_P	$\frac{5}{2}R$	$\frac{7}{2}R$
γ	$\frac{5}{3} = 1.67$	$\frac{7}{5} = 1.40$

A Strategy for Heat-Engine Problems

The engine of Example 21.1 was not a realistic heat engine, but it did illustrate the kinds of reasoning and computations involved in the analysis of a heat engine.

PROBLEM-SOLVING STRATEGY 21.1

Heat-engine problems

MODEL Identify each process in the cycle.

VISUALIZE Draw the pV diagram of the cycle.

SOLVE There are several steps in the mathematical analysis.

- Use the ideal-gas law to complete your knowledge of n, p, V, and T at one point in the cycle.
- Use the ideal-gas law and equations for specific gas processes to determine p, V, and T at the beginning and end of each process.
- Calculate Q, W_s, and ΔE_{th} for each process.
- Find W_{out} by adding W_s for each process in the cycle. If the geometry is simple, you can confirm this value by finding the area enclosed within the pV curve.
- Add just the positive values of Q to find Q_H.
- Verify that $(\Delta E_{th})_{net} = 0$. This is a self-consistency check to verify that you haven't made any mistakes.
- Calculate the thermal efficiency η and any other quantities you need to complete the solution.

ASSESS Is $(\Delta E_{th})_{net} = 0$? Do all the signs of W_s and Q make sense? Does η have a reasonable value? Have you answered the question?

EXAMPLE 21.2 | Analyzing a heat engine II

A heat engine with a diatomic gas as the working substance uses the closed cycle shown in **FIGURE 21.13**. How much work does this engine do per cycle, and what is its thermal efficiency?

FIGURE 21.13 The pV diagram for the heat engine of Example 21.2.

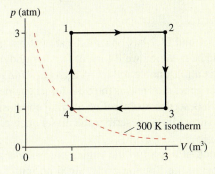

MODEL Processes $1 \rightarrow 2$ and $3 \rightarrow 4$ are isobaric. Processes $2 \rightarrow 3$ and $4 \rightarrow 1$ are isochoric.

VISUALIZE The pV diagram has already been drawn.

SOLVE We know the pressure, volume, and temperature at state 4. The number of moles of gas in the heat engine is

$$n = \frac{p_4 V_4}{RT_4} = \frac{(101{,}300 \text{ Pa})(1.0 \text{ m}^3)}{(8.31 \text{ J/mol K})(300 \text{ K})} = 40.6 \text{ mol}$$

$p/T = $ constant during an isochoric process and $V/T = $ constant during an isobaric process. These allow us to find that $T_1 = T_3 = 900$ K and $T_2 = 2700$ K. This completes our knowledge of the state variables at all four corners of the diagram.

Process $1 \rightarrow 2$ is an isobaric expansion, so

$$(W_s)_{12} = p\,\Delta V = (3.0 \times 101{,}300 \text{ Pa})(2.0 \text{ m}^3) = 6.08 \times 10^5 \text{ J}$$

Continued

where we converted the pressure to pascals. The heat transfer during an isobaric expansion is

$$Q_{12} = nC_P \Delta T = (40.6 \text{ mol})(29.1 \text{ J/mol K})(1800 \text{ K})$$
$$= 21.27 \times 10^5 \text{ J}$$

where $C_P = \frac{7}{2}R$ for a diatomic gas. Then, using the first law,

$$\Delta E_{12} = Q_{12} - (W_s)_{12} = 15.19 \times 10^5 \text{ J}$$

Process $2 \rightarrow 3$ is an isochoric process, so $(W_s)_{23} = 0$ and

$$\Delta E_{23} = Q_{23} = nC_V \Delta T = -15.19 \times 10^5 \text{ J}$$

Notice that ΔT is *negative*.

Process $3 \rightarrow 4$ is an isobaric compression. Now ΔV is negative, so

$$(W_s)_{34} = p \Delta V = -2.03 \times 10^5 \text{ J}$$

and

$$Q_{34} = nC_P \Delta T = -7.09 \times 10^5 \text{ J}$$

Then $\Delta E_{th} = Q_{34} - (W_s)_{34} = -5.06 \times 10^5 \text{ J}$.

Process $4 \rightarrow 1$ is another constant-volume process, so again $(W_s)_{41} = 0$ and

$$\Delta E_{41} = Q_{41} = nC_V \Delta T = 5.06 \times 10^5 \text{ J}$$

The results of all four processes are shown in **TABLE 21.3**. The net results for W_{out}, Q_{net}, and $(\Delta E_{th})_{net}$ are found by summing the columns. As expected, $W_{out} = Q_{net}$ and $(\Delta E_{th})_{net} = 0$.

TABLE 21.3 Energy transfers in Example 21.2. All energies $\times 10^5$ J

Process	W_s	Q	ΔE_{th}
$1 \rightarrow 2$	6.08	21.27	15.19
$2 \rightarrow 3$	0	−15.19	−15.19
$3 \rightarrow 4$	−2.03	−7.09	−5.06
$4 \rightarrow 1$	0	5.06	5.06
Net	4.05	4.05	0

The work done during one cycle is $W_{out} = 4.05 \times 10^5$ J. Heat enters the system from the hot reservoir during processes $1 \rightarrow 2$ and $4 \rightarrow 1$, where Q is positive. Summing these gives $Q_H = 26.33 \times 10^5$ J. Thus the thermal efficiency of this engine is

$$\eta = \frac{W_{out}}{Q_H} = \frac{4.05 \times 10^5 \text{ J}}{26.33 \times 10^5 \text{ J}} = 0.15 = 15\%$$

ASSESS The verification that $W_{out} = Q_{net}$ and $(\Delta E_{th})_{net} = 0$ gives us great confidence that we didn't make any calculational errors. This engine may not seem very efficient, but η is quite typical of many real engines.

We noted in Example 21.1 that a heat engine's hot-reservoir temperature T_H must exceed the highest temperature reached by the system and the cold-reservoir temperature T_C must be less than the coldest system temperature. Although we don't know what the reservoirs are in Example 21.2, we can be sure that $T_H > 2700$ K and $T_C < 300$ K.

STOP TO THINK 21.3 What is the thermal efficiency of this heat engine?

a. 0.10
b. 0.50
c. 0.25
d. 4
e. Can't tell without knowing Q_C

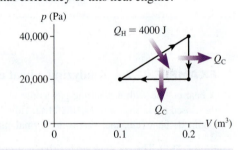

The Brayton Cycle

The heat engines of Examples 21.1 and 21.2 have been educational but not realistic. As an example of a more realistic heat engine we'll look at the thermodynamic cycle known as the *Brayton cycle*. It is a reasonable model of a *gas turbine engine*. Gas turbines are used for electric power generation and as the basis for jet engines in aircraft and rockets. The *Otto cycle*, which describes the gasoline internal combustion engine, and the *Diesel cycle*, which, not surprisingly, describes the diesel engine, will be the subject of homework problems.

FIGURE 21.14a is a schematic look at a gas turbine engine, and **FIGURE 21.14b** is the corresponding pV diagram. To begin the Brayton cycle, air at an initial pressure p_1 is rapidly compressed in a *compressor*. This is an *adiabatic process*, with $Q = 0$,

A jet engine uses a modified Brayton cycle.

because there is no time for heat to be exchanged with the surroundings. Recall that an adiabatic compression raises the temperature of a gas by doing work on it, not by heating it, so the air leaving the compressor is very hot.

The hot gas flows into a combustion chamber. Fuel is continuously admitted to the combustion chamber where it mixes with the hot gas and is ignited, transferring heat to the gas at constant pressure and raising the gas temperature yet further. The high-pressure gas then expands, spinning a turbine that does some form of useful work. This adiabatic expansion, with $Q = 0$, drops the temperature and pressure of the gas. The pressure at the end of the expansion through the turbine is back to p_1, but the gas is still quite hot. The gas completes the cycle by flowing through a device called a **heat exchanger** that transfers heat energy to a cooling fluid. Large power plants are often sited on rivers or oceans in order to use the water for the cooling fluid in the heat exchanger.

This thermodynamic cycle, called a Brayton cycle, has two adiabatic processes—the compression and the expansion through the turbine—plus a constant-pressure heating and a constant-pressure cooling. There's no heat transfer during the adiabatic processes. The hot-reservoir temperature must be $T_H \geq T_3$ for heat to be transferred into the gas during process $2 \rightarrow 3$. Similarly, the heat exchanger will remove heat from the gas only if $T_C \leq T_1$.

The thermal efficiency of any heat engine is

$$\eta = \frac{W_{out}}{Q_H} = 1 - \frac{Q_C}{Q_H}$$

Heat is transferred into the gas only during process $2 \rightarrow 3$. This is an isobaric process, so $Q_H = nC_P \Delta T = nC_P(T_3 - T_2)$. Similarly, heat is transferred out only during the isobaric process $4 \rightarrow 1$.

We have to be careful with signs. Q_{41} is negative because the temperature decreases, but Q_C was defined as the *amount* of heat exchanged with the cold reservoir, a positive quantity. Thus

$$Q_C = |Q_{41}| = |nC_P(T_1 - T_4)| = nC_P(T_4 - T_1) \qquad (21.11)$$

With these expressions for Q_H and Q_C, the thermal efficiency is

$$\eta_{Brayton} = 1 - \frac{T_4 - T_1}{T_3 - T_2} \qquad (21.12)$$

This expression isn't useful unless we compute all four temperatures. Fortunately, we can cast Equation 21.12 into a more useful form.

You learned in Chapter 19 that $pV^\gamma = $ constant during an adiabatic process, where $\gamma = C_P/C_V$ is the specific heat ratio. If we use $V = nRT/p$ from the ideal-gas law, $V^\gamma = (nR)^\gamma T^\gamma p^{-\gamma}$. $(nR)^\gamma$ is a constant, so we can write $pV^\gamma = $ constant as

$$p^{(1-\gamma)}T^\gamma = \text{constant} \qquad (21.13)$$

Equation 21.13 is a pressure-temperature relationship for an adiabatic process. Because $(T^\gamma)^{1/\gamma} = T$, we can simplify Equation 21.13 by raising both sides to the power $1/\gamma$. Doing so gives

$$p^{(1-\gamma)/\gamma}T = \text{constant} \qquad (21.14)$$

during an adiabatic process.

Process $1 \rightarrow 2$ is an adiabatic process; hence

$$p_1^{(1-\gamma)/\gamma}T_1 = p_2^{(1-\gamma)/\gamma}T_2 \qquad (21.15)$$

Isolating T_1 gives

$$T_1 = \frac{p_2^{(1-\gamma)/\gamma}}{p_1^{(1-\gamma)/\gamma}} T_2 = \left(\frac{p_2}{p_1}\right)^{(1-\gamma)/\gamma} T_2 = \left(\frac{p_{max}}{p_{min}}\right)^{(1-\gamma)/\gamma} T_2 \qquad (21.16)$$

If we define the **pressure ratio** r_p as $r_p = p_{max}/p_{min}$, then T_1 and T_2 are related by

$$T_1 = r_p^{(1-\gamma)/\gamma}T_2 \qquad (21.17)$$

FIGURE 21.14 A gas turbine engine follows a Brayton cycle.

(a)

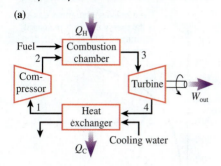

(b)

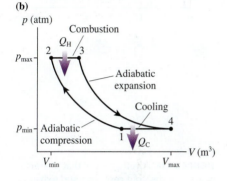

FIGURE 21.15 The efficiency of a Brayton cycle as a function of the pressure ratio r_p.

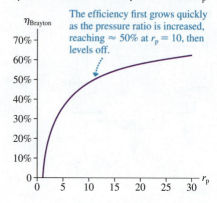

The efficiency first grows quickly as the pressure ratio is increased, reaching ≈ 50% at $r_p = 10$, then levels off.

The algebra of getting to Equation 21.17 was a bit tricky, but the final result is fairly simple. Process $3 \rightarrow 4$ is also an adiabatic process. The same reasoning leads to

$$T_4 = r_p^{(1-\gamma)/\gamma} T_3 \tag{21.18}$$

If we substitute these expressions for T_1 and T_4 into Equation 21.12, the efficiency is

$$\eta_B = 1 - \frac{T_4 - T_1}{T_3 - T_2} = 1 - \frac{r_p^{(1-\gamma)/\gamma} T_3 - r_p^{(1-\gamma)/\gamma} T_2}{T_3 - T_2} = 1 - \frac{r_p^{(1-\gamma)/\gamma}(T_3 - T_2)}{T_3 - T_2}$$

$$= 1 - r_p^{(1-\gamma)/\gamma}$$

Remarkably, all the temperatures cancel and we're left with an expression that depends only on the pressure ratio. Noting that $(1 - \gamma)$ is negative, we can make one final change and write

$$\eta_B = 1 - \frac{1}{r_p^{(\gamma-1)/\gamma}} \tag{21.19}$$

FIGURE 21.15 is a graph of the efficiency of the Brayton cycle as a function of the pressure ratio, assuming $\gamma = 1.40$ for a diatomic gas such as air.

21.4 Ideal-Gas Refrigerators

FIGURE 21.16 A refrigerator that extracts heat from the cold reservoir and exhausts heat to the hot reservoir.

(a)

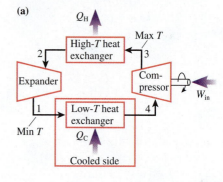

(b)

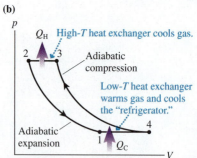

(c)

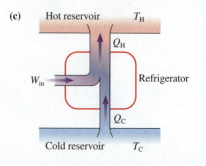

Suppose we were to operate a Brayton heat engine backward, going ccw rather than cw in the pV diagram. FIGURE 21.16a (which you should compare to Figure 21.14a) shows a device for doing this. FIGURE 21.16b is its pV diagram, and FIGURE 21.16c is the energy-transfer diagram. Starting from point 4, the gas is adiabatically compressed to increase its temperature and pressure. It then flows through a high-temperature heat exchanger where the gas *cools* at constant pressure from temperature T_3 to T_2. The gas then expands adiabatically, leaving it significantly colder at T_1 than it started at T_4. It completes the cycle by flowing through a low-temperature heat exchanger, where it *warms* back to its starting temperature.

Suppose that the low-temperature heat exchanger is a closed container of air surrounding a pipe through which the engine's cold gas is flowing. The heat-exchange process $1 \rightarrow 4$ *cools* the air in the container as it warms the gas flowing through the pipe. If you were to place eggs and milk inside this closed container, you would call it a refrigerator!

Going around a closed pV cycle in a ccw direction reverses the sign of W for each process in the cycle. Consequently, the area inside the curve of Figure 21.16b is W_{in}, the work done *on* the system. Here work is used to extract heat Q_C from the cold reservoir and exhaust a larger amount of heat $Q_H = Q_C + W_{in}$ to the hot reservoir. But where, in this situation, are the energy reservoirs?

Understanding a refrigerator is a little harder than understanding a heat engine. The key is to remember that **heat is always transferred from a hotter object to a colder object.** In particular,

- The gas in a refrigerator can extract heat Q_C *from* the cold reservoir only if the gas temperature is *lower* than the cold-reservoir temperature T_C. Heat energy is then transferred *from* the cold reservoir *into* the colder gas.
- The gas in a refrigerator can exhaust heat Q_H *to* the hot reservoir only if the gas temperature is *higher* than the hot-reservoir temperature T_H. Heat energy is then transferred *from* the warmer gas *into* the hot reservoir.

These two requirements place severe constraints on the thermodynamics of a refrigerator. Because there is no reservoir colder than T_C, the gas cannot reach a temperature lower than T_C by heat exchange. The gas in a refrigerator *must* use an adiabatic expansion $(Q = 0)$ to lower the temperature below T_C. Likewise, a gas refrigerator requires an adiabatic compression to raise the gas temperature above T_H.

EXAMPLE 21.3 | Analyzing a refrigerator

A refrigerator using helium gas operates on a reversed Brayton cycle with a pressure ratio of 5.0. Prior to compression, the gas occupies 100 cm³ at a pressure of 150 kPa and a temperature of −23°C. Its volume at the end of the expansion is 80 cm³. What are the refrigerator's coefficient of performance and its power input if it operates at 60 cycles per second?

MODEL The Brayton cycle has two adiabatic processes and two isobaric processes. The work per cycle needed to run the refrigerator is $W_{in} = Q_H - Q_C$; hence we can determine both the coefficient of performance and the power requirements from Q_H and Q_C. Heat energy is transferred only during the two isobaric processes.

VISUALIZE FIGURE 21.17 shows the pV cycle. We know from the pressure ratio of 5.0 that the maximum pressure is 750 kPa. Neither V_2 nor V_3 is known.

FIGURE 21.17 A Brayton-cycle refrigerator.

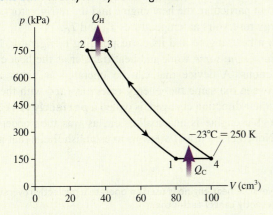

SOLVE To calculate heat we're going to need the temperatures at the four corners of the cycle. First, we can use the conditions of state 4 to find the number of moles of helium:

$$n = \frac{p_4 V_4}{RT_4} = 0.00722 \text{ mol}$$

Process $1 \rightarrow 4$ is isobaric; hence temperature T_1 is

$$T_1 = \frac{V_1}{V_4} T_4 = (0.80)(250 \text{ K}) = 200 \text{ K} = -73°C$$

With Equation 21.14 we found that the quantity $p^{(1-\gamma)/\gamma}T$ remains constant during an adiabatic process. Helium is a monatomic gas with $\gamma = \frac{5}{3}$, so $(1-\gamma)/\gamma = -\frac{2}{5} = -0.40$. For the adiabatic compression $4 \rightarrow 3$,

$$p_3^{-0.40} T_3 = p_4^{-0.40} T_4$$

Solving for T_3 gives

$$T_3 = \left(\frac{p_4}{p_3}\right)^{-0.40} T_4 = \left(\frac{1}{5}\right)^{-0.40} (250 \text{ K}) = 476 \text{ K} = 203°C$$

The same analysis applied to the $2 \rightarrow 1$ adiabatic expansion gives

$$T_2 = \left(\frac{p_1}{p_2}\right)^{-0.40} T_1 = \left(\frac{1}{5}\right)^{-0.40} (200 \text{ K}) = 381 \text{ K} = 108°C$$

Now we can use $C_P = \frac{5}{2}R = 20.8 \text{ J/mol K}$ for a monatomic gas to compute the heat transfers:

$$Q_H = |Q_{32}| = nC_P(T_3 - T_2)$$

$$= (0.00722 \text{ mol})(20.8 \text{ J/mol K})(95 \text{ K}) = 14.3 \text{ J}$$

$$Q_C = |Q_{14}| = nC_P(T_4 - T_1)$$

$$= (0.00722 \text{ mol})(20.8 \text{ J/mol K})(50 \text{ K}) = 7.5 \text{ J}$$

Thus the work *input* to the refrigerator is $W_{in} = Q_H - Q_C = 6.8 \text{ J}$. During each cycle, 6.8 J of work are done *on* the gas to extract 7.5 J of heat from the cold reservoir. Then 14.3 J of heat are exhausted into the hot reservoir.

The refrigerator's coefficient of performance is

$$K = \frac{Q_C}{W_{in}} = \frac{7.5 \text{ J}}{6.8 \text{ J}} = 1.1$$

The power input needed to run the refrigerator is

$$P_{in} = 6.8 \frac{\text{J}}{\text{cycle}} \times 60 \frac{\text{cycles}}{\text{s}} = 410 \frac{\text{J}}{\text{s}} = 410 \text{ W}$$

ASSESS These are fairly realistic values for a kitchen refrigerator. You pay your electric company for providing the work W_{in} that operates the refrigerator. The cold reservoir is the freezer compartment. The cold temperature T_C must be higher than T_4 ($T_C > -23°C$) in order for heat to be transferred *from* the cold reservoir *to* the gas. A typical freezer temperature is −15°C, so this condition is satisfied. The hot reservoir is the air in the room. The back and underside of a refrigerator have heat-exchanger coils where the hot gas, after compression, transfers heat to the air. The hot temperature T_H must be less than T_2 ($T_H < 108°C$) in order for heat to be transferred *from* the gas *to* the air. An air temperature ≈ 25°C under a refrigerator satisfies this condition.

STOP TO THINK 21.4 What, if anything, is wrong with this refrigerator?

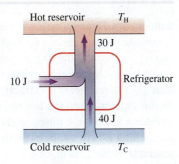

21.5 The Limits of Efficiency

Thermodynamics has its historical roots in the development of the steam engine and other machines of the early industrial revolution. Early steam engines, built on the basis of experience rather than scientific understanding, were not very efficient at converting fuel energy into work. The first major theoretical analysis of heat engines was published by the French engineer Sadi Carnot in 1824. The question that Carnot raised was: Can we make a heat engine whose thermal efficiency η approaches 1, or is there an upper limit η_{max} that cannot be exceeded? To frame the question more clearly, imagine we have a hot reservoir at temperature T_H and a cold reservoir at T_C. What is the most efficient heat engine (maximum η) that can operate between these two energy reservoirs? Similarly, what is the most efficient refrigerator (maximum K) that can operate between the two reservoirs?

We just saw that a refrigerator is, in some sense, a heat engine running backward. We might thus suspect that the most efficient heat engine is related to the most efficient refrigerator. Suppose we have a heat engine that we can turn into a refrigerator by reversing the direction of operation, thus changing the direction of the energy transfers, and with *no other changes*. In particular, the heat engine and the refrigerator operate between the same two energy reservoirs at temperatures T_H and T_C.

FIGURE 21.18a shows such a heat engine and its corresponding refrigerator. Notice that the refrigerator has *exactly the same* work and heat transfer as the heat engine, only in the opposite directions. A device that can be operated as either a heat engine or a refrigerator between the same two energy reservoirs and with the same energy transfers, with only their direction changed, is called a **perfectly reversible engine**. A perfectly reversible engine is an idealization, as was the concept of a perfectly elastic collision. Nonetheless, it will allow us to establish limits that no real engine can exceed.

FIGURE 21.18 If a perfectly reversible heat engine is used to operate a perfectly reversible refrigerator, the two devices exactly cancel each other.

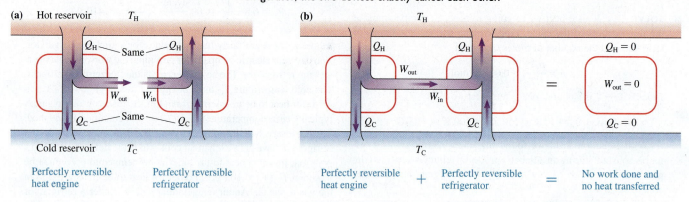

Suppose we have a perfectly reversible heat engine and a perfectly reversible refrigerator (the same device running backward) operating between a hot reservoir at temperature T_H and a cold reservoir at temperature T_C. Because the work W_{in} needed to operate the refrigerator is exactly the same as the useful work W_{out} done by the heat engine, we can use the heat engine, as shown in **FIGURE 21.18b**, to drive the refrigerator. The heat Q_C the engine exhausts to the cold reservoir is exactly the same as the heat Q_C the refrigerator extracts from the cold reservoir. Similarly, the heat Q_H the engine extracts from the hot reservoir matches the heat Q_H the refrigerator exhausts to the hot reservoir. Consequently, there is no net heat transfer in either direction. The refrigerator exactly replaces all the heat energy that had been transferred out of the hot reservoir by the heat engine.

You may want to compare the reasoning used here with the reasoning we used with Figure 21.11. There we tried to use the output of a "perfect" heat engine to run a refrigerator but did *not* succeed.

A Perfectly Reversible Engine Has Maximum Efficiency

Now we've arrived at the critical step in the reasoning. Suppose I claim to have a heat engine that can operate between temperatures T_H and T_C with *more* efficiency than a perfectly reversible engine. **FIGURE 21.19** shows the output of this heat engine operating the same perfectly reversible refrigerator that we used in Figure 21.18b.

FIGURE 21.19 A heat engine more efficient than a perfectly reversible engine could be used to violate the second law of thermodynamics.

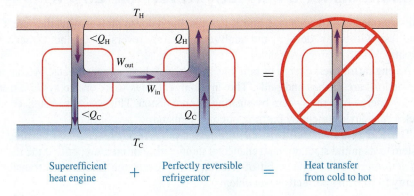

| Superefficient heat engine | + | Perfectly reversible refrigerator | = | Heat transfer from cold to hot |

Recall that the thermal efficiency and the work of a heat engine are

$$\eta = \frac{W_{out}}{Q_H} \quad \text{and} \quad W_{out} = Q_H - Q_C$$

If the new heat engine is more efficient than the perfectly reversible engine it replaces, it needs *less* heat Q_H from the hot reservoir to perform the *same* work W_{out}. If Q_H is less while W_{out} is the same, then Q_C must also be less. That is, the new heat engine exhausts less heat to the cold reservoir than does the perfectly reversible heat engine.

When this new heat engine drives the perfectly reversible refrigerator, the heat it exhausts to the cold reservoir is *less* than the heat extracted from the cold reservoir by the refrigerator. Similarly, this engine extracts *less* heat from the hot reservoir than the refrigerator exhausts. Thus the net result of using this superefficient heat engine to operate a perfectly reversible refrigerator is that heat is transferred from the cold reservoir to the hot reservoir *without outside assistance*.

But this can't happen. It would violate the second law of thermodynamics. Hence we have to conclude that no heat engine operating between reservoirs at temperatures T_H and T_C can be more efficient than a perfectly reversible engine. This very important conclusion is another version of the second law:

> **Second law, informal statement #5** No heat engine operating between reservoirs at temperatures T_H and T_C can be more efficient than a perfectly reversible engine operating between these temperatures.

The answer to our question "Is there a maximum η that cannot be exceeded?" is a clear Yes! The maximum possible efficiency η_{max} is that of a perfectly reversible engine. Because the perfectly reversible engine is an idealization, any real engine will have an efficiency less than η_{max}.

A similar argument shows that no refrigerator can be more efficient than a perfectly reversible refrigerator. If we had such a refrigerator, and if we ran it with the output of a perfectly reversible heat engine, we could transfer heat from cold to hot with no outside assistance. Thus:

> **Second law, informal statement #6** No refrigerator operating between reservoirs at temperatures T_H and T_C can have a coefficient of performance larger than that of a perfectly reversible refrigerator operating between these temperatures.

Conditions for a Perfectly Reversible Engine

This argument tells us that η_{max} and K_{max} exist, but it doesn't tell us what they are. Our final task will be to "design" and analyze a perfectly reversible engine. Under what conditions is an engine reversible?

An engine transfers energy by both mechanical and thermal interactions. Mechanical interactions are pushes and pulls. The environment does work on the system, transferring energy into the system by pushing in on a piston. The system transfers energy back to the environment by pushing out on the piston.

The energy transferred by a moving piston is perfectly reversible, returning the system to its initial state, with no change of temperature or pressure, only if the motion is *frictionless*. The slightest bit of friction will prevent the mechanical transfer of energy from being perfectly reversible.

The circumstances under which heat transfer can be *completely* reversed aren't quite so obvious. After all, Chapter 20 emphasized the *irreversible* nature of heat transfer. If objects A and B are in thermal contact, with $T_A > T_B$, then heat energy is transferred from A to B. But the second law of thermodynamics prohibits a heat transfer from B back to A. Heat transfer through a temperature *difference* is an irreversible process.

But suppose $T_A = T_B$. With no temperature difference, any heat that is transferred from A to B can, at a later time, be transferred from B back to A. This transfer wouldn't violate the second law, which prohibits only heat transfer from a colder object to a hotter object. Now you might object, and rightly so, that heat *can't* move from A to B if they are at the same temperature because heat, by definition, is the energy transferred between two objects at different temperatures.

This is true, but consider a heat engine in which, during one part of the cycle, $T_H = T_{engine} + dT$. That is, the hot reservoir is only infinitesimally hotter than the engine. Heat Q_H can be transferred to the heat engine, but *very slowly*. If you later, with a reversed cycle, try to make the heat move from the engine back to the hot reservoir, the second law will prevent you from doing so with perfect precision. But, because the temperature difference is infinitesimal, you'll be missing only an infinitesimal amount of heat. You can transfer heat reversibly in the limit $dT \rightarrow 0$ (making this an isothermal process), but you must be prepared to spend an infinite amount of time doing so. Similarly, the engine can reversibly exchange heat with a cold reservoir at temperature $T_C = T_{engine} - dT$.

Thus the thermal transfer of energy is reversible *if the heat is transferred infinitely slowly in an isothermal process*. This is an idealization, but so are completely frictionless processes. Nonetheless, we can now say that a perfectly reversible engine must use only two types of processes:

1. Frictionless mechanical interactions with no heat transfer ($Q = 0$), and
2. Thermal interactions in which heat is transferred in an isothermal process ($\Delta E_{th} = 0$).

Any engine that uses only these two types of processes is called a **Carnot engine.** A Carnot engine is a perfectly reversible engine; thus it has the maximum possible thermal efficiency η_{max} and, if operated as a refrigerator, the maximum possible coefficient of performance K_{max}.

21.6 The Carnot Cycle

The definition of a Carnot engine does not specify whether the engine's working substance is a gas or a liquid. It makes no difference. Our argument that a perfectly reversible engine is the most efficient possible heat engine depended only on the engine's reversibility. Consequently, **any Carnot engine operating between T_H and T_C must have exactly the same efficiency as any other Carnot engine operating between the same two energy reservoirs.** If we can determine the thermal efficiency of one Carnot engine, we'll know the efficiency of all Carnot engines. Because liquids and phase changes are complicated, we'll analyze a Carnot engine that uses an ideal gas.

Designing a Carnot Engine

The **Carnot cycle** is an ideal-gas cycle that consists of the two adiabatic processes $(Q = 0)$ and two isothermal processes $(\Delta E_{th} = 0)$ shown in **FIGURE 21.20.** These are the two types of processes allowed in a perfectly reversible gas engine. As a Carnot cycle operates,

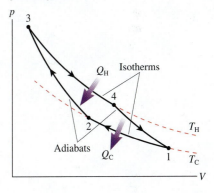

FIGURE 21.20 The Carnot cycle is perfectly reversible.

1. The gas is isothermally compressed while in thermal contact with the cold reservoir at temperature T_C. Heat energy $Q_C = |Q_{12}|$ is removed from the gas as it is compressed in order to keep the temperature constant. The compression must take place extremely slowly because there can be only an infinitesimal temperature difference between the gas and the reservoir.
2. The gas is adiabatically compressed while thermally isolated from the environment. This compression increases the gas temperature until it matches temperature T_H of the hot reservoir. No heat is transferred during this process.
3. After reaching maximum compression, the gas expands isothermally at temperature T_H. Heat $Q_H = Q_{34}$ is transferred from the hot reservoir into the gas as it expands in order to keep the temperature constant.
4. Finally, the gas expands adiabatically, with $Q = 0$, until the temperature decreases back to T_C.

Work is done in all four processes of the Carnot cycle, but heat is transferred only during the two isothermal processes.

The thermal efficiency of any heat engine is

$$\eta = \frac{W_{out}}{Q_H} = 1 - \frac{Q_C}{Q_H}$$

We can determine η_{Carnot} by finding the heat transfer in the two isothermal processes.

Process $1 \rightarrow 2$: Table 21.1 gives us the heat transfer in an isothermal process at temperature T_C:

$$Q_{12} = (W_s)_{12} = nRT_C \ln\left(\frac{V_2}{V_1}\right) = -nRT_C \ln\left(\frac{V_1}{V_2}\right) \tag{21.20}$$

$V_1 > V_2$, so the logarithm on the right is positive. Q_{12} is negative because heat is transferred out of the system, but Q_C is simply the *amount* of heat transferred to the cold reservoir:

$$Q_C = |Q_{12}| = nRT_C \ln\left(\frac{V_1}{V_2}\right) \tag{21.21}$$

Process $3 \rightarrow 4$: Similarly, the heat transferred in the isothermal expansion at temperature T_H is

$$Q_H = Q_{34} = (W_s)_{34} = nRT_H \ln\left(\frac{V_4}{V_3}\right) \tag{21.22}$$

Thus the thermal efficiency of the Carnot cycle is

$$\eta_{\text{Carnot}} = 1 - \frac{Q_C}{Q_H} = 1 - \frac{T_C \ln(V_1/V_2)}{T_H \ln(V_4/V_3)} \tag{21.23}$$

We can simplify this expression. During the two adiabatic processes,

$$T_C V_2^{\gamma-1} = T_H V_3^{\gamma-1} \quad \text{and} \quad T_C V_1^{\gamma-1} = T_H V_4^{\gamma-1} \tag{21.24}$$

An algebraic rearrangement gives

$$V_2 = V_3\left(\frac{T_H}{T_C}\right)^{1/(\gamma-1)} \quad \text{and} \quad V_1 = V_4\left(\frac{T_H}{T_C}\right)^{1/(\gamma-1)} \tag{21.25}$$

from which it follows that

$$\frac{V_1}{V_2} = \frac{V_4}{V_3} \tag{21.26}$$

Consequently, the two logarithms in Equation 21.23 cancel and we're left with the result that the thermal efficiency of a Carnot engine operating between a hot reservoir at temperature T_H and a cold reservoir at temperature T_C is

$$\eta_{\text{Carnot}} = 1 - \frac{T_C}{T_H} \quad \text{(Carnot thermal efficiency)} \tag{21.27}$$

This remarkably simple result, an efficiency that depends only on the ratio of the temperatures of the hot and cold reservoirs, is Carnot's legacy to thermodynamics.

NOTE Temperatures T_H and T_C are *absolute* temperatures.

EXAMPLE 21.4 | **A Carnot engine**

A Carnot engine is cooled by water at $T_C = 10°C$. What temperature must be maintained in the hot reservoir of the engine to have a thermal efficiency of 70%?

MODEL The efficiency of a Carnot engine depends only on the temperatures of the hot and cold reservoirs.

SOLVE The thermal efficiency $\eta_{\text{Carnot}} = 1 - T_C/T_H$ can be rearranged to give

$$T_H = \frac{T_C}{1 - \eta_{\text{Carnot}}} = 943 \text{ K} = 670°C$$

where we used $T_C = 283$ K.

ASSESS A "real" engine would need a higher temperature than this to provide 70% efficiency because no real engine will match the Carnot efficiency.

EXAMPLE 21.5 | **A real engine**

The heat engine of Example 21.2 had a highest temperature of 2700 K, a lowest temperature of 300 K, and a thermal efficiency of 15%. What is the efficiency of a Carnot engine operating between these two temperatures?

SOLVE The Carnot efficiency is

$$\eta_{\text{Carnot}} = 1 - \frac{T_C}{T_H} = 1 - \frac{300 \text{ K}}{2700 \text{ K}} = 0.89 = 89\%$$

ASSESS The thermodynamic cycle used in Example 21.2 doesn't come anywhere close to the Carnot efficiency.

The Maximum Efficiency

In Section 21.2 we tried to invent a perfect engine with $\eta = 1$ and $Q_C = 0$. We found that we could not do so without violating the second law, so no engine can have $\eta = 1$. However, that example didn't rule out an engine with $\eta = 0.9999$. Further analysis has now shown that no heat engine operating between energy reservoirs at temperatures T_H and T_C can be more efficient than a perfectly reversible engine operating between these temperatures.

We've now reached the endpoint of this line of reasoning by establishing an exact result for the thermal efficiency of a perfectly reversible engine, the Carnot engine. We can summarize our conclusions:

> **Second law, informal statement #7** No heat engine operating between energy reservoirs at temperatures T_H and T_C can exceed the Carnot efficiency
>
> $$\eta_{Carnot} = 1 - \frac{T_C}{T_H}$$

As Example 21.5 showed, real engines usually fall well short of the Carnot limit.

We also found that no refrigerator can exceed the coefficient of performance of a perfectly reversible refrigerator. We'll leave the proof as a homework problem, but an analysis very similar to that above shows that the coefficient of performance of a Carnot refrigerator is

$$K_{Carnot} = \frac{T_C}{T_H - T_C} \qquad \text{(Carnot coefficient of performance)} \qquad (21.28)$$

Thus we can state:

> **Second law, informal statement #8** No refrigerator operating between energy reservoirs at temperatures T_H and T_C can exceed the Carnot coefficient of performance
>
> $$K_{Carnot} = \frac{T_C}{T_H - T_C}$$

EXAMPLE 21.6 | **Brayton versus Carnot**

The Brayton-cycle refrigerator of Example 21.3 had coefficient of performance $K = 1.1$. Compare this to the limit set by the second law of thermodynamics.

SOLVE Example 21.3 found that the reservoir temperatures had to be $T_C \geq 250$ K and $T_H \leq 381$ K. A Carnot refrigerator operating between 250 K and 381 K has

$$K_{Carnot} = \frac{T_C}{T_H - T_C} = \frac{250 \text{ K}}{381 \text{ K} - 250 \text{ K}} = 1.9$$

ASSESS This is the minimum value of K_{Carnot}. It will be even higher if $T_C > 250$ K or $T_H < 381$ K. The coefficient of performance of the reasonably realistic refrigerator of Example 21.3 is less than 60% of the limiting value.

Statements #7 and #8 of the second law are a major result of this chapter, one with profound implications. The efficiency limit of a heat engine is set by the temperatures of the hot and cold reservoirs. High efficiency requires $T_C/T_H \ll 1$ and thus $T_H \gg T_C$. However, practical realities often prevent T_H from being significantly larger than T_C, in which case the engine cannot possibly have a large efficiency. This limit on the efficiency of heat engines is a consequence of the second law of thermodynamics.

EXAMPLE 21.7 | Generating electricity

An electric power plant boils water to produce high-pressure steam at 400°C. The high-pressure steam spins a turbine as it expands, then the turbine spins the generator. The steam is then condensed back to water in an ocean-cooled heat exchanger at 25°C. What is the *maximum* possible efficiency with which heat energy can be converted to electric energy?

MODEL The maximum possible efficiency is that of a Carnot engine operating between these temperatures.

SOLVE The Carnot efficiency depends on absolute temperatures, so we must use $T_H = 400°C = 673$ K and $T_C = 25°C = 298$ K. Then

$$\eta_{max} = 1 - \frac{298}{673} = 0.56 = 56\%$$

ASSESS This is an upper limit. Real coal-, oil-, gas-, and nuclear-heated steam generators actually operate at $\approx 35\%$ thermal efficiency, converting only about one-third of the fuel energy to electric energy while exhausting about two-thirds of the energy to the environment as waste heat. (The heat *source* has nothing to do with the efficiency. All it does is boil water.) Not much can be done to alter the low-temperature limit. The high-temperature limit is determined by the maximum temperature and pressure the boiler and turbine can withstand. The efficiency of electricity generation is far less than most people imagine, but it is an unavoidable consequence of the second law of thermodynamics.

A limit on the efficiency of heat engines was not expected. We are used to thinking in terms of energy conservation, so it comes as no surprise that we cannot make an engine with $\eta > 1$. But the limits arising from the second law were not anticipated, nor are they obvious. Nonetheless, they are a very real fact of life and a very real constraint on any practical device. No one has ever invented a machine that exceeds the second-law limits, and we have seen that the maximum efficiency for realistic engines is surprisingly low.

STOP TO THINK 21.5 Could this heat engine be built?

a. Yes.
b. No.
c. It's impossible to tell without knowing what kind of cycle it uses.

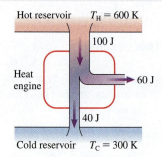

CHALLENGE EXAMPLE 21.8 | Calculating efficiency

A heat engine using a monatomic ideal gas goes through the following closed cycle:

- Isochoric heating until the pressure is doubled.
- Isothermal expansion until the pressure is restored to its initial value.
- Isobaric compression until the volume is restored to its initial value.

What is the thermal efficiency of this heat engine? What would be the thermal efficiency of a Carnot engine operating between the highest and lowest temperatures reached by this engine?

MODEL The cycle consists of three familiar processes; we'll need to analyze each. The amount of work and heat will depend on the quantity of gas, which we don't know, but efficiency is a work-to-heat ratio that is independent of the amount of gas.

VISUALIZE FIGURE 21.21 shows the cycle. The initial pressure, volume, and temperature are p, V, and T. The isochoric process increases the pressure to $2p$ and, because the ratio p/T is constant

FIGURE 21.21 The pV cycle of the heat engine.

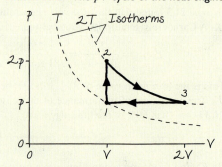

in an isochoric process, increases the temperature to $2T$. The isothermal expansion is along the $2T$ isotherm. The product pV is constant in an isothermal process, so the volume doubles to $2V$ as the pressure returns to p.

SOLVE We know, symbolically, the state variables at each corner of the pV diagram. That is sufficient for calculating W_s, Q, and ΔE_{th}.

Process $1 \rightarrow 2$: An isochoric process has $W_s = 0$ and

$$Q = \Delta E_{th} = nC_V \Delta T = \tfrac{3}{2}nRT$$

where we used $C_V = \tfrac{3}{2}R$ for a monatomic gas and $\Delta T = 2T - T = T$.

Process $2 \rightarrow 3$: An isothermal process has $\Delta E_{th} = 0$ and

$$Q = W_s = nR(2T)\ln\left(\frac{2V}{V}\right) = (2\ln 2)nRT$$

Here we used the Table 21.1 result for the work done in an isothermal process.

Process $3 \rightarrow 1$: The work done by the gas is the area under the curve, which is negative because $\Delta V = V - 2V = -V$ in the compression:

$$W_s = \text{area} = p\,\Delta V = -pV = -nRT$$

We used the ideal-gas law in the last step to express the result in terms of n and T. The heat transfer is also negative because $\Delta T = T - 2T = -T$:

$$Q = nC_P \Delta T = -\tfrac{5}{2}nRT$$

where we used $C_P = \tfrac{5}{2}R$ for a monatomic gas. Based on the first law, $\Delta E_{th} = Q - W_s = -\tfrac{3}{2}nRT$.

Summing over the three processes, we see that $(\Delta E_{th})_{net} = 0$, as expected, and

$$W_{out} = (2\ln 2 - 1)nRT$$

Heat energy is supplied to the gas ($Q > 0$) in processes $1 \rightarrow 2$ and $2 \rightarrow 3$, so

$$Q_H = (2\ln 2 + \tfrac{3}{2})nRT$$

Thus the thermal efficiency of this heat engine is

$$\eta = \frac{W_{out}}{Q_H} = \frac{(2\ln 2 - 1)nRT}{(2\ln 2 + \tfrac{3}{2})nRT} = 0.134 = 13.4\%$$

A Carnot engine would be able to operate between a high temperature $T_H = 2T$ and a low temperature $T_C = T$. Its efficiency would be

$$\eta_{Carnot} = 1 - \frac{T_C}{T_H} = 1 - \frac{T}{2T} = 0.500 = 50.0\%$$

ASSESS As we anticipated, the thermal efficiency depends on the *shape* of the pV cycle but not on the quantity of gas or even on the values of p, V, or T. The heat engine's 13.4% efficiency is considerably less than the 50% maximum possible efficiency set by the second law of thermodynamics.

SUMMARY

The goal of Chapter 21 has been to study the principles that govern heat engines and refrigerators.

GENERAL PRINCIPLES

Heat Engines

Devices that transform heat into work. They require two energy reservoirs at different temperatures.

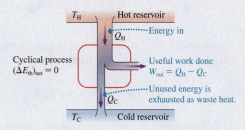

Thermal efficiency

$$\eta = \frac{W_{out}}{Q_H} = \frac{\text{what you get}}{\text{what you pay}}$$

Second-law limit:

$$\eta \le 1 - \frac{T_C}{T_H}$$

Refrigerators

Devices that use work to transfer heat from a colder object to a hotter object.

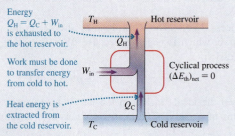

Coefficient of performance

$$K = \frac{Q_C}{W_{in}} = \frac{\text{what you get}}{\text{what you pay}}$$

Second-law limit:

$$K \le \frac{T_C}{T_H - T_C}$$

IMPORTANT CONCEPTS

A **perfectly reversible engine** (a **Carnot engine**) can be operated as either a heat engine or a refrigerator between the same two energy reservoirs by reversing the cycle and with no other changes.

- A **Carnot heat engine** has the maximum possible thermal efficiency of any heat engine operating between T_H and T_C:

$$\eta_{Carnot} = 1 - \frac{T_C}{T_H}$$

- A **Carnot refrigerator** has the maximum possible coefficient of performance of any refrigerator operating between T_H and T_C:

$$K_{Carnot} = \frac{T_C}{T_H - T_C}$$

The **Carnot cycle** for a gas engine consists of two isothermal processes and two adiabatic processes.

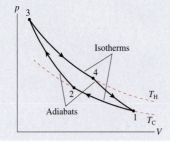

An **energy reservoir** is a part of the environment so large in comparison to the system that its temperature doesn't change as the system extracts heat energy from or exhausts heat energy to the reservoir. All heat engines and refrigerators operate between two energy reservoirs at different temperatures T_H and T_C.

The **work** W_s done *by* the system has the opposite sign to the work done *on* the system.

$$W_s = \text{area under } pV \text{ curve}$$

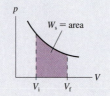

APPLICATIONS

To analyze a heat engine or refrigerator:

MODEL Identify each process in the cycle.

VISUALIZE Draw the pV diagram of the cycle.

SOLVE There are several steps:
- Determine p, V, and T at the beginning and end of each process.
- Calculate ΔE_{th}, W_s, and Q for each process.
- Determine W_{in} or W_{out}, Q_H, and Q_C.
- Calculate $\eta = W_{out}/Q_H$ or $K = Q_C/W_{in}$.

ASSESS Verify $(\Delta E_{th})_{net} = 0$. Check signs.

TERMS AND NOTATION

thermodynamics	closed-cycle device	coefficient of performance, K	Carnot engine
energy reservoir	thermal efficiency, η	heat exchanger	Carnot cycle
energy-transfer diagram	waste heat	pressure ratio, r_p	
heat engine	refrigerator	perfectly reversible engine	

CONCEPTUAL QUESTIONS

1. In going from i to f in each of the three processes of **FIGURE Q21.1**, is work done *by* the system ($W < 0$, $W_s > 0$), is work done *on* the system ($W > 0$, $W_s < 0$), or is *no* net work done?

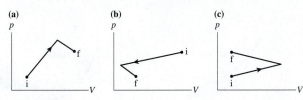

FIGURE Q21.1

2. Rank in order, from largest to smallest, the amount of work $(W_s)_1$ to $(W_s)_4$ done by the gas in each of the cycles shown in **FIGURE Q21.2**. Explain.

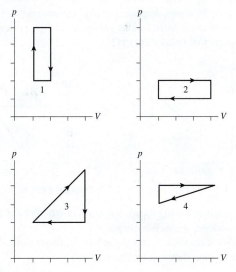

FIGURE Q21.2

3. Could you have a heat engine with $\eta > 1$? Explain.

4. **FIGURE Q21.4** shows the pV diagram of a heat engine. During which stage or stages is (a) heat added to the gas, (b) heat removed from the gas, (c) work done on the gas, and (d) work done by the gas?

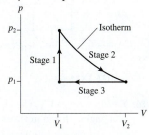

FIGURE Q21.4

5. Rank in order, from largest to smallest, the thermal efficiencies η_1 to η_4 of the four heat engines in **FIGURE Q21.5**. Explain.

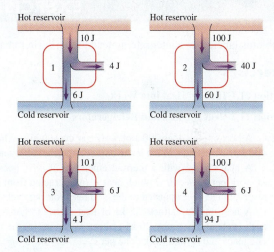

FIGURE Q21.5

6. **FIGURE Q21.6** shows the thermodynamic cycles of two heat engines. Which heat engine has the larger thermal efficiency? Or are they the same? Explain.

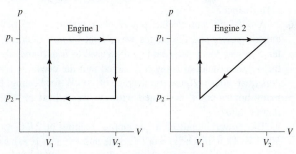

FIGURE Q21.6

7. A heat engine satisfies $W_{out} = Q_{net}$. Why is there no ΔE_{th} term in this relationship?

8. Do the energy-transfer diagrams in **FIGURE Q21.8** represent possible heat engines? If not, what is wrong?

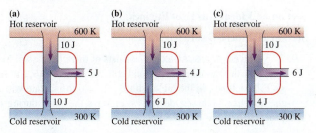

FIGURE Q21.8

9. Do the energy-transfer diagrams in **FIGURE Q21.9** represent possible refrigerators? If not, what is wrong?

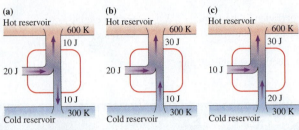

FIGURE Q21.9

10. It gets pretty hot in your apartment. In browsing the Internet, you find a company selling small "room air conditioners." You place the air conditioner on the floor, plug it in, and—the advertisement says—it will lower the room temperature up to 10°F. Should you order one? Explain.

11. The first and second laws of thermodynamics are sometimes stated as "You can't win" and "You can't even break even." Do these sayings accurately characterize the laws of thermodynamics as applied to heat engines? Why or why not?

EXERCISES AND PROBLEMS

Problems labeled ▮ integrate material from earlier chapters.

Exercises

Section 21.1 Turning Heat into Work

Section 21.2 Heat Engines and Refrigerators

1. ‖ A heat engine does 200 J of work per cycle while exhausting 400 J of waste heat. What is the engine's thermal efficiency?
2. | A heat engine with a thermal efficiency of 40% does 100 J of work per cycle. How much heat is (a) extracted from the hot reservoir and (b) exhausted to the cold reservoir per cycle?
3. | A heat engine extracts 55 kJ of heat from the hot reservoir each cycle and exhausts 40 kJ of heat. What are (a) the thermal efficiency and (b) the work done per cycle?
4. | 50 J of work are done per cycle on a refrigerator with a coefficient of performance of 4.0. How much heat is (a) extracted from the cold reservoir and (b) exhausted to the hot reservoir per cycle?
5. ‖ A refrigerator requires 200 J of work and exhausts 600 J of heat per cycle. What is the refrigerator's coefficient of performance?
6. ‖ A 32%-efficient electric power plant produces 900 MW of electric power and discharges waste heat into 20°C ocean water. Suppose the waste heat could be used to heat homes during the winter instead of being discharged into the ocean. A typical American house requires an average of 20 kW for heating. How many homes could be heated with the waste heat of this one power plant?
7. ‖ The power output of a car engine running at 2400 rpm is 500 kW. How much (a) work is done and (b) heat is exhausted per cycle if the engine's thermal efficiency is 20%? Give your answers in kJ.
8. ‖ 1.0 L of 20°C water is placed in a refrigerator. The refrigerator's motor must supply an extra 8.0 W of power to chill the water to 5°C in 1.0 h. What is the refrigerator's coefficient of performance?

Section 21.3 Ideal-Gas Heat Engines

Section 21.4 Ideal-Gas Refrigerators

9. ‖ The cycle of **FIGURE EX21.9** consists of three processes. Make a table with rows labeled A–C and columns labeled ΔE_{th}, W_s, and Q. Fill each box in the table with $+$, $-$, or 0 to indicate whether the quantity increases, decreases, or stays the same during that process.

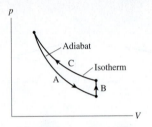

FIGURE EX21.9

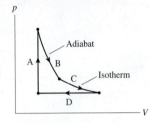

FIGURE EX21.10

10. ‖ The cycle of **FIGURE EX21.10** consists of four processes. Make a table with rows labeled A to D and columns labeled ΔE_{th}, W_s, and Q. Fill each box in the table with $+$, $-$, or 0 to indicate whether the quantity increases, decreases, or stays the same during that process.
11. ‖ How much work is done per cycle by a gas following the pV trajectory of **FIGURE EX21.11**?

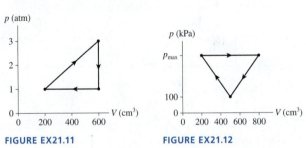

FIGURE EX21.11

FIGURE EX21.12

12. ‖ A gas following the pV trajectory of **FIGURE EX21.12** does 60 J of work per cycle. What is p_{max}?
13. ‖ What are (a) W_{out} and Q_H and (b) the thermal efficiency for the heat engine shown in **FIGURE EX21.13**?

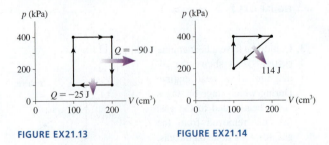

FIGURE EX21.13

FIGURE EX21.14

14. ‖ What are (a) W_{out} and Q_H and (b) the thermal efficiency for the heat engine shown in **FIGURE EX21.14**?

15. ‖ What are (a) the thermal efficiency and (b) the heat extracted from the hot reservoir for the heat engine shown in **FIGURE EX21.15**?

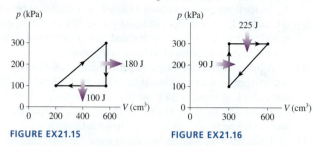

FIGURE EX21.15

FIGURE EX21.16

16. ‖ How much heat is exhausted to the cold reservoir by the heat engine shown in **FIGURE EX21.16**?

17. ‖ A heat engine uses a diatomic gas in a Brayton cycle. What is the engine's thermal efficiency if the gas volume is halved during the adiabatic compression?

18. | At what pressure ratio does a Brayton cycle using a monatomic gas have an efficiency of 50%?

19. ‖ The coefficient of performance of a refrigerator is 6.0. The refrigerator's compressor uses 115 W of electric power and is 95% efficient at converting electric power into work. What are (a) the rate at which heat energy is removed from inside the refrigerator and (b) the rate at which heat energy is exhausted into the room?

20. ‖ What are (a) the heat extracted from the cold reservoir and (b) the coefficient of performance for the refrigerator shown in **FIGURE EX21.20**?

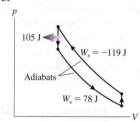

FIGURE EX21.20

21. ‖ An air conditioner removes 5.0×10^5 J/min of heat from a house and exhausts 8.0×10^5 J/min to the hot outdoors.
    a. How much power does the air conditioner's compressor require?
    b. What is the air conditioner's coefficient of performance?

Section 21.5 The Limits of Efficiency

Section 21.6 The Carnot Cycle

22. | Which, if any, of the heat engines in **FIGURE EX21.22** violate (a) the first law of thermodynamics or (b) the second law of thermodynamics? Explain.

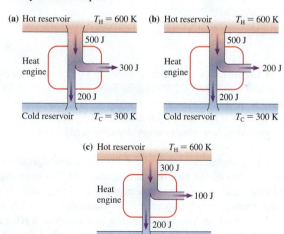

FIGURE EX21.22

23. | Which, if any, of the refrigerators in **FIGURE EX21.23** violate (a) the first law of thermodynamics or (b) the second law of thermodynamics? Explain.

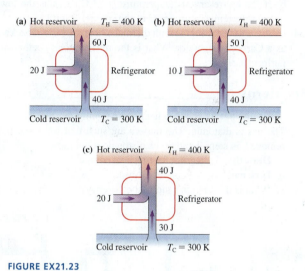

FIGURE EX21.23

24. ‖ At what cold-reservoir temperature (in °C) would a Carnot engine with a hot-reservoir temperature of 427°C have an efficiency of 60%?

25. ‖ A heat engine does 10 J of work and exhausts 15 J of waste heat during each cycle.
    a. What is the engine's thermal efficiency?
    b. If the cold-reservoir temperature is 20°C, what is the minimum possible temperature in °C of the hot reservoir?

26. ‖ a. A heat engine does 200 J of work per cycle while exhausting 600 J of heat to the cold reservoir. What is the engine's thermal efficiency?
    b. A Carnot engine with a hot-reservoir temperature of 400°C has the same thermal efficiency. What is the cold-reservoir temperature in °C?

27. | A Carnot engine operating between energy reservoirs at temperatures 300 K and 500 K produces a power output of 1000 W. What are (a) the thermal efficiency of this engine, (b) the rate of heat input, in W, and (c) the rate of heat output, in W?

28. ‖ A Carnot engine whose hot-reservoir temperature is 400°C has a thermal efficiency of 40%. By how many degrees should the temperature of the cold reservoir be decreased to raise the engine's efficiency to 60%?

29. ‖ The ideal gas in a Carnot engine extracts 1000 J of heat energy during the isothermal expansion at 300°C. How much heat energy is exhausted during the isothermal compression at 50°C?

30. ‖ A heat engine operating between energy reservoirs at 20°C and 600°C has 30% of the maximum possible efficiency. How much energy must this engine extract from the hot reservoir to do 1000 J of work?

31. ‖ A heat engine operating between a hot reservoir at 500°C and a cold reservoir at 0°C is 60% as efficient as a Carnot engine. If this heat engine and the Carnot engine do the same amount of work, what is the ratio $Q_H/(Q_H)_{Carnot}$?

32. ‖ A Carnot refrigerator operating between −20°C and +20°C extracts heat from the cold reservoir at the rate 200 J/s. What are (a) the coefficient of performance of this refrigerator, (b) the rate at which work is done on the refrigerator, and (c) the rate at which heat is exhausted to the hot side?

33. ‖ The coefficient of performance of a refrigerator is 5.0. The compressor uses 10 J of energy per cycle.
    a. How much heat energy is exhausted per cycle?
    b. If the hot-reservoir temperature is 27°C, what is the lowest possible temperature in °C of the cold reservoir?

34. ‖ A Carnot heat engine with thermal efficiency $\frac{1}{3}$ is run backward as a Carnot refrigerator. What is the refrigerator's coefficient of performance?

Problems

35. ‖ **FIGURE P21.35** shows a heat engine going through one cycle. The gas is diatomic. The masses are such that when the pin is removed, in steps 3 and 6, the piston does not move.
    a. Draw the pV diagram for this heat engine.
    b. How much work is done per cycle?
    c. What is this engine's thermal efficiency?

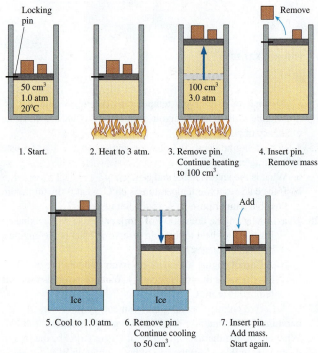

1. Start. 2. Heat to 3 atm. 3. Remove pin. Continue heating to 100 cm³. 4. Insert pin. Remove mass.

5. Cool to 1.0 atm. 6. Remove pin. Continue cooling to 50 cm³. 7. Insert pin. Add mass. Start again.

FIGURE P21.35

36. ‖ The engine that powers a crane burns fuel at a flame temperature of 2000°C. It is cooled by 20°C air. The crane lifts a 2000 kg steel girder 30 m upward. How much heat energy is transferred to the engine by burning fuel if the engine is 40% as efficient as a Carnot engine?

37. ‖ A heat engine with 50% of the Carnot efficiency operates between reservoirs at 20°C and 200°C. The engine inputs heat energy at an average rate of 63 W while compressing a spring 22 cm in 0.50 s. What is the spring constant?

38. ‖ Prove that the work done in an adiabatic process i → f is $W_s = (p_f V_f - p_i V_i)/(1 - \gamma)$.

39. ‖ A Carnot refrigerator operates between reservoirs at −20°C and 50°C in a 25°C room. The refrigerator is a 40 cm × 40 cm × 40 cm box. Five of the walls are perfect insulators, but the sixth is a 1.0-cm-thick piece of stainless steel. What electric power does the refrigerator require to maintain the inside temperature at −20°C?

40. ‖ Prove that the coefficient of performance of a Carnot refrigerator is $K_{\text{Carnot}} = T_C/(T_H - T_C)$.

41. ‖‖‖ An ideal refrigerator utilizes a Carnot cycle operating between 0°C and 25°C. To turn 10 kg of liquid water at 0°C into 10 kg of ice at 0°C, (a) how much heat is exhausted into the room and (b) how much energy must be supplied to the refrigerator?

42. ‖ A freezer with a coefficient of performance 30% that of a Carnot refrigerator keeps the inside temperature at −22°C in a 25°C room. 3.0 L of water at 20°C are placed in the freezer. How long does it take for the water to freeze if the freezer's compressor does work at the rate of 200 W while the water is freezing?

43. ‖ There has long been an interest in using the vast quantities of thermal energy in the oceans to run heat engines. A heat engine needs a temperature *difference*, a hot side and a cold side. Conveniently, the ocean surface waters are warmer than the deep ocean waters. Suppose you build a floating power plant in the tropics where the surface water temperature is ≈ 30°C. This would be the hot reservoir of the engine. For the cold reservoir, water would be pumped up from the ocean bottom where it is always ≈ 5°C. What is the maximum possible efficiency of such a power plant?

44. ‖ A Carnot heat engine operates between reservoirs at 182°C and 0°C. If the engine extracts 25 J of energy from the hot reservoir per cycle, how many cycles will it take to lift a 10 kg mass a height of 10 m?

45. ‖ A Carnot engine operates between temperatures of 5°C and 500°C. The output is used to run a Carnot refrigerator operating between −5°C and 25°C. How many joules of heat energy does the refrigerator exhaust into the room for each joule of heat energy used by the heat engine?

46. ‖ **FIGURE P21.46** shows a Carnot heat engine driving a Carnot refrigerator.
    a. Determine Q_2, Q_3, and Q_4.
    b. Is Q_3 greater than, less than, or equal to Q_1?
    c. Do these two devices, when operated together in this way, violate the second law?

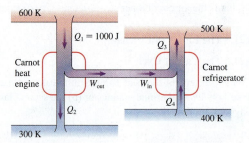

FIGURE P21.46

47. ‖‖‖ A Carnot heat engine and an ordinary refrigerator with coefficient of performance 2.00 operate between reservoirs at 350 K and 250 K. The work done by the Carnot heat engine drives the refrigerator. If the heat engine extracts 10.0 J of energy from the hot reservoir, how much energy does the refrigerator exhaust to the hot reservoir?

48. ‖ A heat engine running backward is called a refrigerator if its purpose is to extract heat from a cold reservoir. The same engine running backward is called a *heat pump* if its purpose is to exhaust warm air into the hot reservoir. Heat pumps are widely used for home heating. You can think of a heat pump as a refrigerator that is cooling the already cold outdoors and, with its exhaust heat Q_H, warming the indoors. Perhaps this seems a little silly, but

consider the following. Electricity can be directly used to heat a home by passing an electric current through a heating coil. This is a direct, 100% conversion of work to heat. That is, 15 kW of electric power (generated by doing work at the rate of 15 kJ/s at the power plant) produces heat energy inside the home at a rate of 15 kJ/s. Suppose that the neighbor's home has a heat pump with a coefficient of performance of 5.0, a realistic value. Note that "what you get" with a heat pump is heat delivered, Q_H, so a heat pump's coefficient of performance is defined as $K = Q_H/W_{in}$.

a. How much electric power (in kW) does the heat pump use to deliver 15 kJ/s of heat energy to the house?

b. An average price for electricity is about 40 MJ per dollar. A furnace or heat pump will run typically 250 hours per month during the winter. What does one month's heating cost in the home with a 15 kW electric heater and in the home of the neighbor who uses a heat pump?

49. ‖ A car's internal combustion engine can be modeled as a heat engine operating between a combustion temperature of 1500°C and an air temperature of 20°C with 30% of the Carnot efficiency. The heat of combustion of gasoline is 47 kJ/g. What mass of gasoline is burned to accelerate a 1500 kg car from rest to a speed of 30 m/s?

50. ‖ Consider a 1.0 MW power plant (this is the useful output in the form of electric energy) that operates between 30°C and 450°C at 65% of the Carnot efficiency. This is enough electric energy for about 750 homes. One way to use energy more efficiently would be to use the 30°C "waste" energy to heat the homes rather than releasing that heat energy into the environment. This is called *cogeneration,* and it is used in some parts of Europe but rarely in the United States. The average home uses 70 GJ of energy per year for heating. For estimating purposes, assume that all the power plant's exhaust energy can be transported to homes without loss and that home heating takes place at a steady rate for half a year each year. How many homes could be heated by the power plant?

51. ‖ A typical coal-fired power plant burns 300 metric tons of coal *every hour* to generate 750 MW of electricity. 1 metric ton = 1000 kg. The density of coal is 1500 kg/m³ and its heat of combustion is 28 MJ/kg. Assume that *all* heat is transferred from the fuel to the boiler and that *all* the work done in spinning the turbine is transformed into electric energy.

a. Suppose the coal is piled up in a 10 m × 10 m room. How tall must the pile be to operate the plant for one day?

b. What is the power plant's thermal efficiency?

52. ‖ A nuclear power plant generates 3000 MW of heat energy from nuclear reactions in the reactor's core. This energy is used to boil water and produce high-pressure steam at 300°C. The steam spins a turbine, which produces 1000 MW of electric power, then the steam is condensed and the water is cooled to 25°C before starting the cycle again.

a. What is the maximum possible thermal efficiency of the power plant?

b. What is the plant's actual efficiency?

c. Cooling water from a river flows through the condenser (the low-temperature heat exchanger) at the rate of 1.2×10^8 L/h (≈ 30 million gallons per hour). If the river water enters the condenser at 18°C, what is its exit temperature?

53. ‖ The electric output of a power plant is 750 MW. Cooling water flows through the power plant at the rate 1.0×10^8 L/h. The cooling water enters the plant at 16°C and exits at 27°C. What is the power plant's thermal efficiency?

54. ‖ Engineers testing the efficiency of an electric generator gradu-
CALC ally vary the temperature of the hot steam used to power it while leaving the temperature of the cooling water at a constant 20°C. They find that the generator's efficiency increases at a rate of 3.5×10^{-4} K⁻¹ at steam temperatures near 300°C. What is the ratio of the generator's efficiency to the efficiency of a Carnot engine?

55. ‖ A heat engine using 1.0 mol of a monatomic gas follows the cycle shown in FIGURE P21.55. 3750 J of heat energy is transferred to the gas during process $1 \rightarrow 2$.

a. Determine W_s, Q, and ΔE_{th} for each of the four processes in this cycle. Display your results in a table.

b. What is the thermal efficiency of this heat engine?

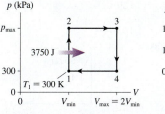

FIGURE P21.55

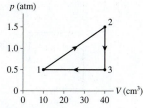

FIGURE P21.56

56. ‖ A heat engine using a diatomic gas follows the cycle shown in FIGURE P21.56. Its temperature at point 1 is 20°C.

a. Determine W_s, Q, and ΔE_{th} for each of the three processes in this cycle. Display your results in a table.

b. What is the thermal efficiency of this heat engine?

c. What is the power output of the engine if it runs at 500 rpm?

57. ‖ FIGURE P21.57 shows the cycle for a heat engine that uses a gas having $\gamma = 1.25$. The initial temperature is $T_1 = 300$ K, and this engine operates at 20 cycles per second.

a. What is the power output of the engine?

b. What is the engine's thermal efficiency?

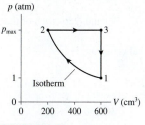

FIGURE P21.57

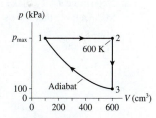

FIGURE P21.58

58. ‖ A heat engine using a monatomic gas follows the cycle shown in FIGURE P21.58.

a. Find W_s, Q, and ΔE_{th} for each process in the cycle. Display your results in a table.

b. What is the thermal efficiency of this heat engine?

59. ‖ A heat engine uses a di-
atomic gas that follows the pV cycle in FIGURE P21.59.

a. Determine the pressure, volume, and temperature at point 2.

b. Determine ΔE_{th}, W_s, and Q for each of the three processes. Put your results in a table for easy reading.

c. How much work does this engine do per cycle and what is its thermal efficiency?

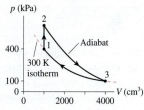

FIGURE P21.59

60. ‖ **FIGURE P21.60** is the pV diagram of Example 21.2, but now the device is operated in reverse.
   a. During which processes is heat transferred into the gas?
   b. Is this Q_H, heat extracted from a hot reservoir, or Q_C, heat extracted from a cold reservoir? Explain.
   c. Determine the values of Q_H and Q_C.
   Hint: The calculations have been done in Example 21.2 and do not need to be repeated. Instead, you need to determine which processes now contribute to Q_H and which to Q_C.
   d. Is the area inside the curve W_{in} or W_{out}? What is its value?
   e. The device is now being operated in a ccw cycle. Is it a refrigerator? Explain.

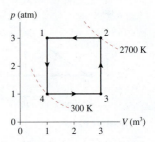

FIGURE P21.60

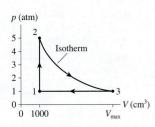

FIGURE P21.61

61. ‖ A heat engine using 120 mg of helium as the working substance follows the cycle shown in **FIGURE P21.61**.
   a. Determine the pressure, temperature, and volume of the gas at points 1, 2, and 3.
   b. What is the engine's thermal efficiency?
   c. What is the maximum possible efficiency of a heat engine that operates between T_{max} and T_{min}?
62. ‖ The heat engine shown in **FIGURE P21.62** uses 2.0 mol of a monatomic gas as the working substance.
   a. Determine T_1, T_2, and T_3.
   b. Make a table that shows ΔE_{th}, W_s, and Q for each of the three processes.
   c. What is the engine's thermal efficiency?

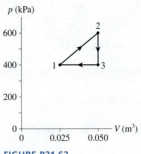

FIGURE P21.62

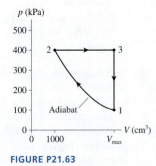

FIGURE P21.63

63. ‖ The heat engine shown in **FIGURE P21.63** uses 0.020 mol of a diatomic gas as the working substance.
   a. Determine T_1, T_2, and T_3.
   b. Make a table that shows ΔE_{th}, W_s, and Q for each of the three processes.
   c. What is the engine's thermal efficiency?
64. ‖ A heat engine with 0.20 mol of a monatomic ideal gas initially fills a 2000 cm³ cylinder at 600 K. The gas goes through the following closed cycle:

   ■ Isothermal expansion to 4000 cm³.
   ■ Isochoric cooling to 300 K.
   ■ Isothermal compression to 2000 cm³.
   ■ Isochoric heating to 600 K.
   How much work does this engine do per cycle and what is its thermal efficiency?

In Problems 65 through 68 you are given the equation(s) used to solve a problem. For each of these, you are to
   a. Write a realistic problem for which this is the correct equation(s).
   b. Finish the solution of the problem.
65. $0.80 = 1 - (0°C + 273)/(T_H + 273)$
66. $4.0 = Q_C/W_{in}$
   $Q_H = 100$ J
67. $0.20 = 1 - Q_C/Q_H$
   $W_{out} = Q_H - Q_C = 20$ J
68. 400 kJ $= \frac{1}{2}(p_{max} - 100$ kPa$)(3.0$ m³ $- 1.0$ m³$)$

Challenge Problems

69. ‖‖ 100 mL of water at 15°C is placed in the freezer compartment of a refrigerator with a coefficient of performance of 4.0. How much heat energy is exhausted into the room as the water is changed to ice at −15°C?
70. ‖‖ **FIGURE CP21.70** shows two insulated compartments separated by a thin wall. The left side contains 0.060 mol of helium at an initial temperature of 600 K and the right side contains 0.030 mol of helium at an initial temperature of 300 K. The compartment on the right is attached to a vertical cylinder, above which the air pressure is 1.0 atm. A 10-cm-diameter, 2.0 kg piston can slide without friction up and down the cylinder. Neither the cylinder diameter nor the volumes of the compartments are known.
   a. What is the final temperature?
   b. How much heat is transferred from the left side to the right side?
   c. How high is the piston lifted due to this heat transfer?
   d. What fraction of the heat is converted into work?

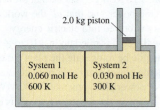

FIGURE CP21.70

71. ‖‖ A refrigerator using helium gas operates on the reversed cycle shown in **FIGURE CP21.71**. What are the refrigerator's (a) coefficient of performance and (b) power input if it operates at 60 cycles per second?

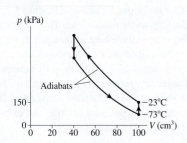

FIGURE CP21.71

72. ||| A heat engine using a diatomic ideal gas goes through the following closed cycle:
   - Isothermal compression until the volume is halved.
   - Isobaric expansion until the volume is restored to its initial value.
   - Isochoric cooling until the pressure is restored to its initial value.

   What are the thermal efficiencies of (a) this heat engine and (b) a Carnot engine operating between the highest and lowest temperatures reached by this engine?

73. ||| The gasoline engine in your car can be modeled as the Otto cycle shown in FIGURE CP21.73. A fuel-air mixture is sprayed into the cylinder at point 1, where the piston is at its farthest distance from the spark plug. This mixture is compressed as the piston moves toward the spark plug during the adiabatic *compression stroke*. The spark plug fires at point 2, releasing heat energy that had been stored in the gasoline. The fuel burns so quickly that the piston doesn't have time to move, so the heating is an isochoric process. The hot, high-pressure gas then pushes the piston outward during the *power stroke*. Finally, an exhaust value opens to allow the gas temperature and pressure to drop back to their initial values before starting the cycle over again.

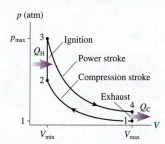

FIGURE CP21.73

a. Analyze the Otto cycle and show that the work done per cycle is

$$W_{out} = \frac{nR}{1 - \gamma}(T_2 - T_1 + T_4 - T_3)$$

b. Use the adiabatic connection between T_1 and T_2 and also between T_3 and T_4 to show that the thermal efficiency of the Otto cycle is

$$\eta = 1 - \frac{1}{r^{(\gamma - 1)}}$$

where $r = V_{max}/V_{min}$ is the engine's *compression ratio*.

c. Graph η versus r out to $r = 30$ for a diatomic gas.

74. ||| FIGURE CP21.74 shows the Diesel cycle. It is similar to the Otto cycle (see Problem 21.73), but there are two important differences. First, the fuel is not admitted until the air is fully compressed at point 2. Because of the high temperature at the end of an adiabatic compression, the fuel begins to burn spontaneously. (There are no spark plugs in a diesel engine!) Second, combustion takes place more slowly, with fuel continuing to be injected. This makes the ignition stage a constant-pressure process. The cycle shown, for one cylinder of a diesel engine, has a *displacement* $V_{max} - V_{min}$ of 1000 cm^3 and a compression ratio $r = V_{max}/V_{min} = 21$. These are typical values for a diesel truck. The engine operates with intake air ($\gamma = 1.40$) at 25°C and 1.0 atm pressure. The quantity of fuel injected into the cylinder has a heat of combustion of 1000 J.

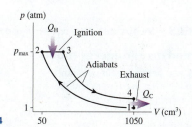

FIGURE CP21.74

a. Find p, V, and T at each of the four corners of the cycle. Display your results in a table.
b. What is the net work done by the cylinder during one full cycle?
c. What is the thermal efficiency of this engine?
d. What is the power output in kW and horsepower (1 hp = 746 W) of an eight-cylinder diesel engine running at 2400 rpm?

Thermodynamics

KEY FINDINGS   What are the overarching findings of Part V?

Thermodynamics is an expanded view of systems and energy.

- A system exchanges energy with its environment via both **work** and **heat.** These are energy transfers.
- In a **heat engine,** heat energy is transformed into useful work when a system follows a cyclical process.

- Some processes are **irreversible.** A reversible **Carnot engine** is a heat engine with the maximum possible efficiency.
- Many macroscopic thermal properties of materials can be understood in terms of the motions of atoms and molecules.

LAWS   What laws of physics govern thermodynamics?

First law of thermodynamics   Energy is conserved: $\Delta E_{th} = W + Q$.

Second law of thermodynamics   Heat is not spontaneously transferred from a colder object to a hotter object.

Ideal-gas law $pV = nRT$ or $pV = Nk_{B}T$

Equipartition theorem   The energy stored in each degree of freedom is $\frac{1}{2}Nk_{B}T$ or $\frac{1}{2}nRT$.

MODELS   What are the most important models of Part V?

Thermodynamic energy model

- **Work** and **heat** are energies transferred between the system and the environment.
  - Work is a mechanical interaction.
  - Heat is a thermal interaction.
- Transferring energy changes the system's thermal energy as given by the first law: $\Delta E_{th} = W + Q$.

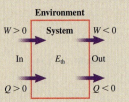

Ideal-gas model

- For low densities and temperatures not too close to the condensation point, all gases, regardless of composition, obey the ideal-gas law with the same value of the gas constant R.

Phases of matter

- Solid: Rigid, definite shape, nearly incompressible.
- Liquid: Fluid, takes shape of container, nearly incompressible.
- Gas: Non-interacting particles, highly compressible.

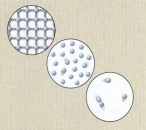

Carnot engine

- A perfectly reversible heat engine has the maximum possible efficiency of any heat engine operating between T_{H} and T_{C}.
- The efficiency depends only on the reservoir temperatures, not on any details of the engine: $\eta_{Carnot} = 1 - T_{C}/T_{H}$.

TOOLS   What are the most important tools introduced in Part V?

pV diagrams show

- States
- Processes

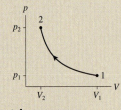

Four fundamental gas processes

- Isochoric: $\Delta V = 0$ and $W = 0$
- Isobaric: $\Delta p = 0$
- Isothermal: $\Delta T = 0$ and $\Delta E_{th} = 0$
- Adiabatic: $Q = 0$

Work in gas processes

- The work done *on* a gas is
$$W = -\int_{V_i}^{V_f} p\, dV$$
$$= -\text{area under the } pV \text{ curve}$$
- For a closed cycle, the work done *by* a gas is $W_s = $ area enclosed.

Heat and thermal energy

- Heat is energy transferred in a thermal process when there is a temperature difference.
- Thermal energy is the microscopic energy of moving atoms.
- Heat, thermal energy, and temperature are related but not the same.

Heat is transferred by

- Conduction
- Convection
- Radiation
- Evaporation

Heating and cooling

- The heat energy needed for a temperature change is $Q = mc\,\Delta T$ or $Q = nC\,\Delta T$.
- For thermally isolated systems
$$Q_{net} = Q_1 + Q_2 + \cdots = 0$$

Heat engines and refrigerators

Heat engines and refrigerators require both a **hot reservoir** and a **cold reservoir.**

- A heat engine transforms heat energy from a hot reservoir into work while exhausting energy to the cold reservoir.

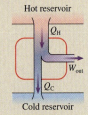

- A refrigerator uses external work to "pump" heat energy from a cold reservoir to the hot reservoir.
- A Carnot engine is the best possible engine or refrigerator.

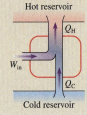

Mathematics Review

Algebra

Using exponents:
$$a^{-x} = \frac{1}{a^x} \qquad a^x a^y = a^{(x+y)} \qquad \frac{a^x}{a^y} = a^{(x-y)} \qquad (a^x)^y = a^{xy}$$

$$a^0 = 1 \qquad a^1 = a \qquad a^{1/n} = \sqrt[n]{a}$$

Fractions:
$$\left(\frac{a}{b}\right)\left(\frac{c}{d}\right) = \frac{ac}{bd} \qquad \frac{a/b}{c/d} = \frac{ad}{bc} \qquad \frac{1}{1/a} = a$$

Logarithms:
$$\text{If } a = e^x, \text{ then } \ln(a) = x \qquad \ln(e^x) = x \qquad e^{\ln(x)} = x$$

$$\ln(ab) = \ln(a) + \ln(b) \qquad \ln\left(\frac{a}{b}\right) = \ln(a) - \ln(b) \qquad \ln(a^n) = n\ln(a)$$

The expression $\ln(a + b)$ cannot be simplified.

Linear equations: The graph of the equation $y = ax + b$ is a straight line. a is the slope of the graph. b is the y-intercept.

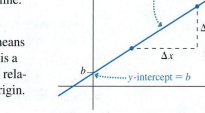

Proportionality: To say that y is proportional to x, written $y \propto x$, means that $y = ax$, where a is a constant. Proportionality is a special case of linearity. A graph of a proportional relationship is a straight line that passes through the origin. If $y \propto x$, then

$$\frac{y_1}{y_2} = \frac{x_1}{x_2}$$

Quadratic equation: The quadratic equation $ax^2 + bx + c = 0$ has the two solutions $x = \dfrac{-b \pm \sqrt{b^2 - 4ac}}{2a}$.

Geometry and Trigonometry

Area and volume:

Rectangle
$A = ab$

Rectangular box
$V = abc$

Triangle
$A = \frac{1}{2}ab$

Right circular cylinder
$V = \pi r^2 l$

Circle
$C = 2\pi r$
$A = \pi r^2$

Sphere
$A = 4\pi r^2$
$V = \frac{4}{3}\pi r^3$

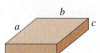

Arc length and angle: The angle θ in radians is defined as $\theta = s/r$.

The arc length that spans angle θ is $s = r\theta$.

$2\pi \text{ rad} = 360°$

Right triangle: Pythagorean theorem $c = \sqrt{a^2 + b^2}$ or $a^2 + b^2 = c^2$

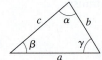

$$\sin\theta = \frac{b}{c} = \frac{\text{far side}}{\text{hypotenuse}} \qquad\qquad \theta = \sin^{-1}\left(\frac{b}{c}\right)$$

$$\cos\theta = \frac{a}{c} = \frac{\text{adjacent side}}{\text{hypotenuse}} \qquad\qquad \theta = \cos^{-1}\left(\frac{a}{c}\right)$$

$$\tan\theta = \frac{b}{a} = \frac{\text{far side}}{\text{adjacent side}} \qquad\qquad \theta = \tan^{-1}\left(\frac{b}{a}\right)$$

General triangle: $\alpha + \beta + \gamma = 180° = \pi \text{ rad}$

Law of cosines $c^2 = a^2 + b^2 - 2ab\cos\gamma$

Identities:

$$\tan\alpha = \frac{\sin\alpha}{\cos\alpha} \qquad\qquad \sin^2\alpha + \cos^2\alpha = 1$$

$$\sin(-\alpha) = -\sin\alpha \qquad\qquad \cos(-\alpha) = \cos\alpha$$

$$\sin(\alpha \pm \beta) = \sin\alpha\cos\beta \pm \cos\alpha\sin\beta \qquad\qquad \cos(\alpha \pm \beta) = \cos\alpha\cos\beta \mp \sin\alpha\sin\beta$$

$$\sin(2\alpha) = 2\sin\alpha\cos\alpha \qquad\qquad \cos(2\alpha) = \cos^2\alpha - \sin^2\alpha$$

$$\sin(\alpha \pm \pi/2) = \pm\cos\alpha \qquad\qquad \cos(\alpha \pm \pi/2) = \mp\sin\alpha$$

$$\sin(\alpha \pm \pi) = -\sin\alpha \qquad\qquad \cos(\alpha \pm \pi) = -\cos\alpha$$

Expansions and Approximations

Binomial expansion: $(1 + x)^n = 1 + nx + \dfrac{n(n-1)}{2}x^2 + \cdots$

Binomial approximation: $(1 + x)^n \approx 1 + nx \quad \text{if} \quad x \ll 1$

Trigonometric expansions: $\sin\alpha = \alpha - \dfrac{\alpha^3}{3!} + \dfrac{\alpha^5}{5!} - \dfrac{\alpha^7}{7!} + \cdots \quad \text{for } \alpha \text{ in rad}$

$\cos\alpha = 1 - \dfrac{\alpha^2}{2!} + \dfrac{\alpha^4}{4!} - \dfrac{\alpha^6}{6!} + \cdots \quad \text{for } \alpha \text{ in rad}$

Small-angle approximation: If $\alpha \ll 1$ rad, then $\sin\alpha \approx \tan\alpha \approx \alpha$ and $\cos\alpha \approx 1$.

The small-angle approximation is excellent for $\alpha < 5°$ (≈ 0.1 rad) and generally acceptable up to $\alpha \approx 10°$.

Calculus

The letters a and n represent constants in the following derivatives and integrals.

Derivatives

$$\frac{d}{dx}(a) = 0$$

$$\frac{d}{dx}(ax) = a$$

$$\frac{d}{dx}\left(\frac{a}{x}\right) = -\frac{a}{x^2}$$

$$\frac{d}{dx}(ax^n) = anx^{n-1}$$

$$\frac{d}{dx}\left(\ln(ax)\right) = \frac{1}{x}$$

$$\frac{d}{dx}(e^{ax}) = ae^{ax}$$

$$\frac{d}{dx}\left(\sin(ax)\right) = a\cos(ax)$$

$$\frac{d}{dx}\left(\cos(ax)\right) = -a\sin(ax)$$

Integrals

$$\int x\,dx = \frac{1}{2}x^2$$

$$\int x^2\,dx = \frac{1}{3}x^3$$

$$\int \frac{1}{x^2}\,dx = -\frac{1}{x}$$

$$\int x^n\,dx = \frac{x^{n+1}}{n+1} \qquad n \neq -1$$

$$\int \frac{dx}{x} = \ln x$$

$$\int \frac{dx}{a+x} = \ln(a+x)$$

$$\int \frac{x\,dx}{a+x} = x - a\ln(a+x)$$

$$\int \frac{dx}{\sqrt{x^2 \pm a^2}} = \ln\left(x + \sqrt{x^2 \pm a^2}\right)$$

$$\int \frac{x\,dx}{\sqrt{x^2 \pm a^2}} = \sqrt{x^2 \pm a^2}$$

$$\int \frac{dx}{x^2 + a^2} = \frac{1}{a}\tan^{-1}\left(\frac{x}{a}\right)$$

$$\int \frac{dx}{(x^2 + a^2)^2} = \frac{1}{2a^3}\tan^{-1}\left(\frac{x}{a}\right) + \frac{x}{2a^2(x^2 + a^2)}$$

$$\int \frac{dx}{(x^2 \pm a^2)^{3/2}} = \frac{\pm x}{a^2\sqrt{x^2 \pm a^2}}$$

$$\int \frac{x\,dx}{(x^2 \pm a^2)^{3/2}} = -\frac{1}{\sqrt{x^2 \pm a^2}}$$

$$\int e^{ax}\,dx = \frac{1}{a}e^{ax}$$

$$\int xe^{-x}\,dx = -(x+1)e^{-x}$$

$$\int x^2 e^{-x}\,dx = -(x^2 + 2x + 2)e^{-x}$$

$$\int \sin(ax)\,dx = -\frac{1}{a}\cos(ax)$$

$$\int \cos(ax)\,dx = \frac{1}{a}\sin(ax)$$

$$\int \sin^2(ax)\,dx = \frac{x}{2} - \frac{\sin(2ax)}{4a}$$

$$\int \cos^2(ax)\,dx = \frac{x}{2} + \frac{\sin(2ax)}{4a}$$

$$\int_0^\infty x^n e^{-ax}\,dx = \frac{n!}{a^{n+1}}$$

$$\int_0^\infty e^{-ax^2}\,dx = \frac{1}{2}\sqrt{\frac{\pi}{a}}$$

Periodic Table of Elements

Atomic number —— 27
Co —— Symbol
Atomic mass —— 58.9

Transition elements

Inner transition elements

An atomic mass in brackets is that of the longest-lived isotope of an element with no stable isotopes.

Period 1 2 3 4 5 6 7

Lanthanides 6
Actinides 7

1 H 1.0																	2 He 4.0
3 Li 6.9	4 Be 9.0											5 B 10.8	6 C 12.0	7 N 14.0	8 O 16.0	9 F 19.0	10 Ne 20.2
11 Na 23.0	12 Mg 24.3											13 Al 27.0	14 Si 28.1	15 P 31.0	16 S 32.1	17 Cl 35.5	18 Ar 39.9
19 K 39.1	20 Ca 40.1	21 Sc 45.0	22 Ti 47.9	23 V 50.9	24 Cr 52.0	25 Mn 54.9	26 Fe 55.8	27 Co 58.9	28 Ni 58.7	29 Cu 63.5	30 Zn 65.4	31 Ga 69.7	32 Ge 72.6	33 As 74.9	34 Se 79.0	35 Br 79.9	36 Kr 83.8
37 Rb 85.5	38 Sr 87.6	39 Y 88.9	40 Zr 91.2	41 Nb 92.9	42 Mo 95.9	43 Tc [98]	44 Ru 101.1	45 Rh 102.9	46 Pd 106.4	47 Ag 107.9	48 Cd 112.4	49 In 114.8	50 Sn 118.7	51 Sb 121.8	52 Te 127.6	53 I 126.9	54 Xe 131.3
55 Cs 132.9	56 Ba 137.3	71 Lu 175.0	72 Hf 178.5	73 Ta 180.9	74 W 183.9	75 Re 186.2	76 Os 190.2	77 Ir 192.2	78 Pt 195.1	79 Au 197.0	80 Hg 200.6	81 Tl 204.4	82 Pb 207.2	83 Bi 209.0	84 Po [209]	85 At [210]	86 Rn [222]
87 Fr [223]	88 Ra [226]	103 Lr [262]	104 Rf [265]	105 Db [268]	106 Sg [271]	107 Bh [272]	108 Hs [270]	109 Mt [276]	110 Ds [281]	111 Rg [280]	112 Cn [285]	113 [ ]	114 Fl [289]	115 [ ]	116 Lv [293]	117 [ ]	118 [ ]

57 La 138.9	58 Ce 140.1	59 Pr 140.9	60 Nd 144.2	61 Pm 144.9	62 Sm 150.4	63 Eu 152.0	64 Gd 157.3	65 Tb 158.9	66 Dy 162.5	67 Ho 164.9	68 Er 167.3	69 Tm 168.9	70 Yb 173.0
89 Ac [227]	90 Th 232.0	91 Pa 231.0	92 U 238.0	93 Np [237]	94 Pu [244]	95 Am [243]	96 Cm [247]	97 Bk [247]	98 Cf [251]	99 Es [252]	100 Fm [257]	101 Md [258]	102 No [259]

Answers

Answers to Stop to Think Questions and Odd-Numbered Exercises and Problems

Chapter 1

Stop to Think Questions

1. **B.** The images of B are farther apart, so it travels a larger distance than does A during the same intervals of time.
2. **a.** Dropped ball. **b.** Dust particle. **c.** Descending rocket.
3. **e.** The average velocity vector is found by connecting one dot in the motion diagram to the next.
4. **b.** $\vec{v}_2 = \vec{v}_1 + \Delta\vec{v}$, and $\Delta\vec{v}$ points in the direction of $\vec{a}$.

5. $d > c > b = a$.

Exercises and Problems

1.

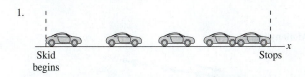

3.

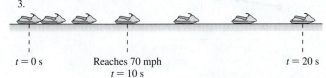

5.

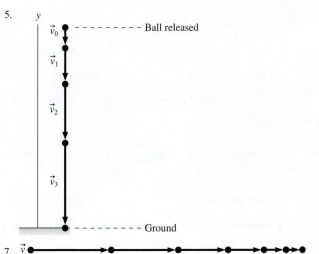

7.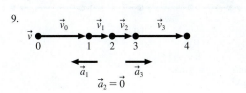

9.

11. a.

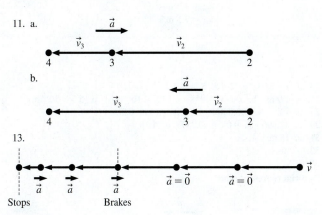

b.

13.

15.

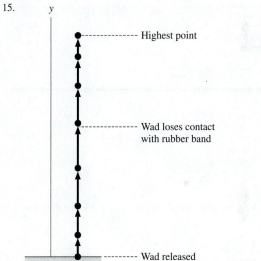

17.

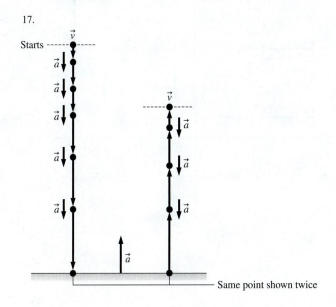

A-5

21.

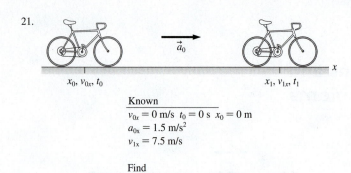

x_0, v_{0x}, t_0 x_1, v_{1x}, t_1

Known

$v_{0x} = 0$ m/s $t_0 = 0$ s $x_0 = 0$ m

$a_{0x} = 1.5$ m/s^2

$v_{1x} = 7.5$ m/s

Find

x_1

23. a. 1   b. 3   c. 2   d. 5
25. a. 1.9 m   b. 1.09×10^{14} s   c. 2.2×10^{-4} m/s   d. 5.7×10^{10} m^2
27. a. 7 m   d. 100,000 m   c. 30 m/s   d. 0.2 m
29. a. 32,590   b. 9.0   c. 0.237   d. 4.78
31. a. 15 m
33. 32 ms
35.

Pictorial representation

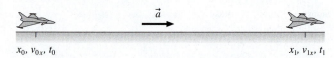

x_0, v_{0x}, t_0 x_1, v_{1x}, t_1

Motion diagram

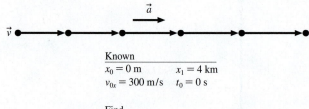

Known

$x_0 = 0$ m $x_1 = 4$ km

$v_{0x} = 300$ m/s $t_0 = 0$ s

Find

a_x

37.   **Pictorial representation**   **Motion diagram**

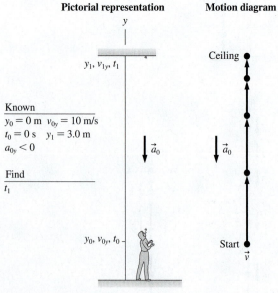

y_1, v_{1y}, t_1

Known

$y_0 = 0$ m $v_{0y} = 10$ m/s

$t_0 = 0$ s $y_1 = 3.0$ m

$a_{0y} < 0$

Find

t_1

y_0, v_{0y}, t_0

Ceiling

Start

39. **Pictorial representation**

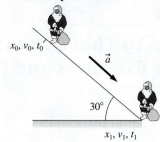

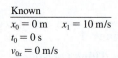

x_0, v_0, t_0

30°

x_1, v_1, t_1

Known

$x_0 = 0$ m $x_1 = 10$ m/s

$t_0 = 0$ s

$v_{0x} = 0$ m/s

Find

v_{1x}

Motion diagram

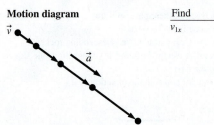

41. **Pictorial representation**

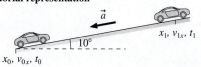

x_1, v_{1x}, t_1

10°

x_0, v_{0x}, t_0

Motion diagram

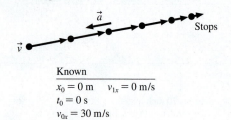

Stops

Known

$x_0 = 0$ m $v_{1x} = 0$ m/s

$t_0 = 0$ s

$v_{0x} = 30$ m/s

Find

x_1

43.

Pictorial representation

David $\vec{a}_D = \vec{0}$

x_{D0}, t_{D0}, v_{D0x} x_{D1}, t_{D1}, v_{D1x}

$\vec{a}_T$

Tina

x_{T0}, t_{T0}, v_{T0x} x_{T1}, t_{T1}, v_{T1x}

Motion diagram

$\vec{a}_D = \vec{0}$

$\vec{v}_D$

$\vec{a}_T$

$\vec{v}_T$

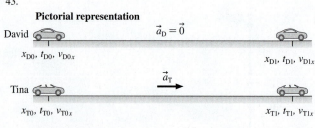
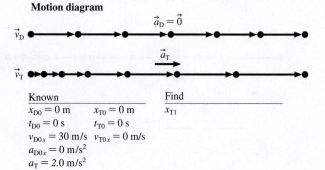

Known   **Find**

$x_{D0} = 0$ m $x_{T0} = 0$ m x_{T1}

$t_{D0} = 0$ s $t_{T0} = 0$ s

$v_{D0x} = 30$ m/s $v_{T0x} = 0$ m/s

$a_{D0x} = 0$ m/s^2

$a_T = 2.0$ m/s^2

49. a.

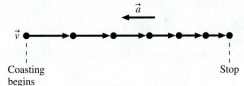

Coasting
begins   Stop

51. a.

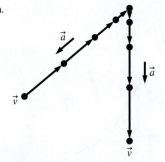

53. Smallest: 6.4×10^3 m^2, largest: 8.3×10^3 m^2
55. 2.9×10^{-4} m^3
57. a. 83.3 kg/m^3 b. 810 kg/m^3

Chapter 2

Stop to Think Questions

1. **d.** The particle starts with positive x and moves to negative x.
2. **c.** The velocity is the slope of the position graph. The slope is positive and constant until the position graph crosses the axis, then positive but decreasing, and finally zero when the position graph is horizontal.
3. **b.** A constant positive v_x corresponds to a linearly increasing x, starting from $x_i = -10$ m. The constant negative v_x then corresponds to a linearly decreasing x.
4. **a and b.** The velocity is constant while $a = 0$; it decreases linearly while a is negative. Graphs a, b, and c all have the same acceleration, but only graphs a and b have a positive initial velocity that represents a particle moving to the right.
5. **d.** The acceleration vector points downhill (negative s-direction) and has the constant value $-g \sin\theta$ throughout the motion.
6. **c.** Acceleration is the slope of the graph. The slope is zero at B. Although the graph is steepest at A, the slope at that point is negative, and so $a_A < a_B$. Only C has a positive slope, so $a_C > a_B$.

Exercises and Problems

1. a. Beth   b. 20 min
3. a. 48 mph   b. 50 mph
5. a. v_x (m/s)   b. None

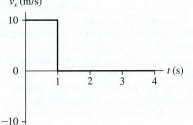

7. 8.0 cm

9. a_x (m/s^2)

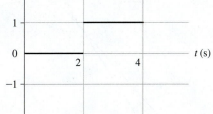

11. a. 8.0 m   b. 2.0 m/s   c. -2.0 m/s^2
13. a. 2.7 m/s^2 b. 1.3×10^2 m $= 4.3 \times 10^2$ feet
15. a. 360 d   b. 4.6×10^{15} m   c. 0.49
17. 2.8 m/s^2
19. 216 m
21. 3.2 s
23. 5.2 cm
25. 73 m
27. 265 m
29. a. 16 m/s   b. 31 m
31. 16 m/s
33. a. 21 m   b. 26 m/s   c. 24 m/s^2
35. a. 0 s and 3 s   b. 12 m and -18 m/s^2; -15 m and 18 m/s^2
37. a. -10 m/s   b. -20 m/s   c. 95 m/s
39. a. 2 s, 5 s   b. -3 m/s^2, 3 m/s^2
41. a. 20 s   b. 667 m
43.

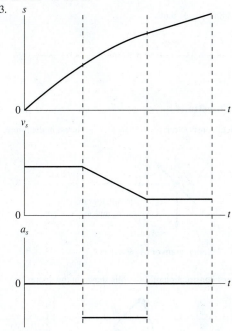

45.

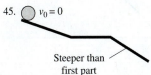

Steeper than
first part

47. a. Yes   b. 35 s   c. No
49. a. 5 m   b. 22 m/s
51. 5.7 m/s
53. Yes, 10 m
55. a. 54.8 km   b. 228 s
57. 19.7 m
59. 55 cm

61. a. 2.0 h b. 73 m
   c.   x (mi)

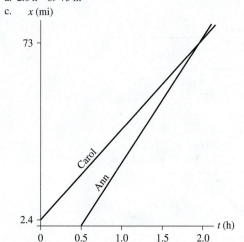

63. $v_f = \sqrt{2gh}$
65. 17 cm
67. 14 m/s
69. 0.36 m
71. 4.4 m/s^2
73. gh/d
77. c. 17.2 m/s
79. c. 750 m
81. a. 10 s b. 3.8 m/s^2 c. 5.6%
83. 12.5 m/s
85. 4500 m/s^2

Chapter 3

Stop to Think Questions

1. **c.** The graphical construction of $\vec{A}_1 + \vec{A}_2 + \vec{A}_3$ is shown in the figure.

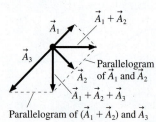

2. **a.** The graphical construction of $2\vec{A} - \vec{B}$ is shown in the figure.

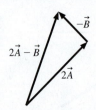

3. $C_x = -4$ cm, $C_y = 2$ cm.
4. **c.** Vector $\vec{C}$ points to the left and down, so both C_x and C_y are negative. C_x is in the numerator because it is the side opposite ϕ.

Exercises and Problems

1. a. $\vec{A} + \vec{B}$ b.

3. a. $E\sin\theta, -E\cos\theta$ b. $E\cos\phi, -E\sin\phi$
5. 8 m
7. a. 3.8 m/s, 6.5 m/s b. −1.3 m/s^2, 0.80 m/s^2 c. −30 N, 40 N
9. 100 m, west
11. a. 7.6, 67° b. 4.9 m/s^2, 66° c. 18 m/s, 38° d. 4.0 m, 34°
13. a. $1\hat{i} + 3\hat{j}$
   b.

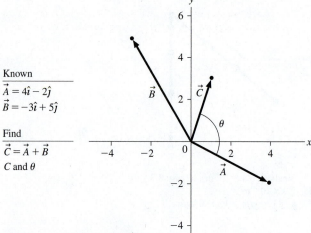

Known
$\vec{A} = 4\hat{i} - 2\hat{j}$
$\vec{B} = -3\hat{i} + 5\hat{j}$

Find
$\vec{C} = \vec{A} + \vec{B}$
C and θ

   c. 3.2, 72° above the +x-axis
15. a. $-1\hat{i} + 11\hat{j}$ c. 11, 5.2° ccw from the +y-axis
17. a. 2.8 b. 4.1 c. 6.1
19. $v_x = -50$ m/s, $v_y = -87$ m/s
21. $B = 2.2$ T, $\theta = 27°$
23. a. 0 m, 26 m, 160 m b. $\vec{v}(t) = (10\hat{i} + 8.0\hat{j})\,t$ m/s
   c. 0 m/s, 26 m/s, and 6 m/s
25. $\vec{C} = 0.8\hat{i} - 4.5\hat{j}$
27. $\vec{B} = \dfrac{1}{\sqrt{2}}\hat{i} + \dfrac{1}{\sqrt{2}}\hat{j}$
29. a. $\Delta\vec{r} = (1.7\text{ cm})\hat{i} - (3.0\text{ cm})\hat{j}$ b. $\Delta\vec{r} = 0.0$ cm
31. 90 m, 46° south of west
33. $v_x = 3.7$ m/s, $v_y = 2.6$ m/s
35. a. −3.4 m/s b. −9.4 m/s
37. 280 N, 350 N
39. 570 N, −380 N
41. 5.2 μm/s at 27° below the positive x-axis
43. 4.4 units at 83° below the negative x-axis
45. 7.3 N at 79° below the negative x-axis

Chapter 4

Stop to Think Questions

1. **c.** $v = 0$ requires both $v_x = 0$ and $v_y = 0$. Neither x nor y can be changing.
2. **d.** The parallel component of $\vec{a}$ is opposite $\vec{v}$ and will cause the particle to slow down. The perpendicular component of $\vec{a}$ will cause the particle to change direction downward.
3. **d.** A projectile's acceleration $\vec{a} = -g\hat{j}$ does not depend on its mass. The second marble has the same initial velocity and the same acceleration, so it follows the same trajectory and lands at the same position.
4. **f.** The plane's velocity relative to the helicopter is $\vec{v}_{PH} = \vec{v}_{PG} + \vec{v}_{GH} = \vec{v}_{PG} - \vec{v}_{HG}$, where G is the ground. The vector addition shows that $\vec{v}_{PH}$ is to the right and down with a magnitude greater than the 100 m/s of $\vec{v}_{PG}$.

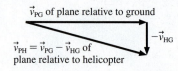

$\vec{v}_{PG}$ of plane relative to ground

$\vec{v}_{PH} = \vec{v}_{PG} - \vec{v}_{HG}$ of plane relative to helicopter

5. **b.** An initial cw rotation causes the particle's angular position to become increasingly negative. The speed drops to half after reversing direction, so the slope becomes positive and is half as steep as the initial slope. Turning through the same angle returns the particle to $\theta = 0°$.

6. $a_b > a_e > a_a = a_c > a_d$. Centripetal acceleration is v^2/r. Doubling r decreases a_r by a factor of 2. Doubling v increases a_r by a factor of 4. Reversing direction doesn't change a_r.

7. **c.** ω is negative because the rotation is cw. Because ω is negative but becoming *less* negative, the change $\Delta\omega$ is *positive*. So α is positive.

Exercises and Problems

3. C
5. E
7. 2.2 m/s²
9. $(2\hat{i} - 8\hat{j})$ m/s²
11. 19.6 m
13. 680 m
15. $r = 16.4$ m
17. $v_x = 27.7$ m/s
19. 30 s
21. a. 42° west of north   b. 45 s
23. 10 rev
25.

θ (rad) graph with y-axis values 0, 20, 40, 60, 80 and x-axis t (s) values 0, 2, 4, 6, 8. Line rises from 0 to about 78 at t=4, stays flat to t=6, then decreases to 60 at t=8.

27. a. 4.7 rad/s   b. 1.3 s
29. 680 km/h, 1040 mph
31. 43 m
33. a. 3.0×10^4 m/s   b. 2.0×10^{-7} rad/s   c. 6.0×10^{-3} m/s²
35. $v = 5.7$ m/s, $a_r = 108$ m/s²
37. a. 3.75 rad/s   b. 5.0 rad/s   c. 5.0 rad/s
39. 98 rpm
41. 47 rad/s²
43. 38 rev
45. $\left(\frac{1}{2}bt^2 + v_{0x}\right)\hat{i} + \left(e^{-ct} + v_{0y}\right)\hat{j}$
47. a. $\dfrac{v_0^2 \sin^2\theta}{2g}$ b. 14.4 m, 28.8 m, 43.2 m
49. a. 12 m/s   b. 0.90 m
51. Clears by 1.0 m
53. 7.4 m/s
55. a. 13 m/s   b. 48°
57. 470 m/s²
59. 4.8 m/s
61. a. 39 mi   b. 20 mph
63. No. The angular acceleration is negative.
65. a. 1.75×10^4 m/s²   b. 4.4×10^3 m/s²
67. 69 m/s at 21° with the vertical
69. a. -100 rad/s²   b. 50 rev
71. 0.75 rad/s²
73. a. $v = \sqrt{2\alpha \Delta\theta R}$ b. $a = 2\alpha \Delta\theta R$

75. 1940 rpm
77. 550 rpm
79. b. 30 m west
81. 34.3°
83. 3.8 m
85. 10°

Chapter 5

Stop to Think Questions

1. **c.**

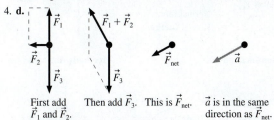

The y-component of $\vec{F}_3$ cancels the y-component of $\vec{F}_1$.

The x-component of $\vec{F}_3$ is to the left and larger than the x-component of $\vec{F}_2$.

2. **a, b, and d.** Friction and the normal force are the only contact forces. Nothing is touching the rock to provide a "force of the kick."

3. **b.** Acceleration is proportional to force, so doubling the number of rubber bands doubles the acceleration of the original object from 2 m/s² to 4 m/s². But acceleration is also inversely proportional to mass. Doubling the mass cuts the acceleration in half, back to 2 m/s².

4. **d.**

First add $\vec{F}_1$ and $\vec{F}_2$.   Then add $\vec{F}_3$. This is $\vec{F}_{net}$. $\vec{a}$ is in the same direction as $\vec{F}_{net}$.

5. **c.** The acceleration vector points downward as the elevator slows. $\vec{F}_{net}$ points in the same direction as $\vec{a}$, so $\vec{F}_{net}$ also points down. This will be true if the tension is less than the gravitational force: $T < F_G$.

Exercises and Problems

1. Gravity, normal force, static friction
3. Gravity, normal force, kinetic friction
5. Gravity, drag
7. a. 2.4 m/s²   b. 0.60 m/s²
9. $m_1 = 0.080$ kg, $m_3 = 0.50$ kg
11. 1.5 J
13. a. 4 m/s²   b. 2 m/s²
15. 25 kg
17.

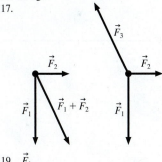

19.

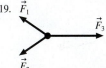

21.

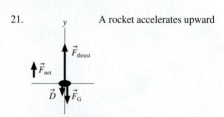

A rocket accelerates upward

23. **Force identification**   **Free-body diagram**

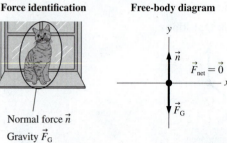

Normal force $\vec{n}$
Gravity $\vec{F}_G$

25. **Force identification**   **Free-body diagram**

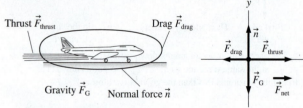

Gravity $\vec{F}_G$
Normal force $\vec{n}$
Kinetic friction $\vec{f}_k$

27. **Force identification**   **Free-body diagram**

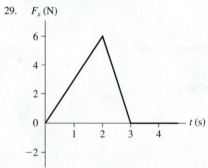

Thrust $\vec{F}_{thrust}$ Drag $\vec{F}_{drag}$
Gravity $\vec{F}_G$ Normal force $\vec{n}$

29. F_x (N)

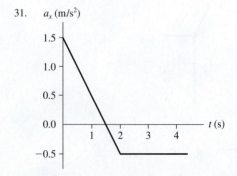

31. a_x (m/s^2)

33. a. 16 m/s^2 b. 4.0 m/s^2 c. 8.0 m/s^2 d. 32 m/s^2

35.

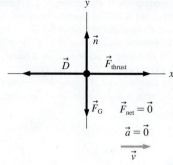

$$\vec{F}_{net} = \vec{0}$$
$$\vec{a} = \vec{0}$$
$$\vec{v}$$

37.

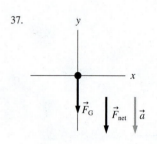

39.

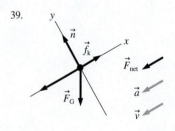

41. a.

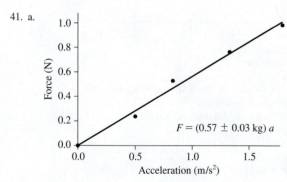

$F = (0.57 \pm 0.03 \text{ kg})\, a$

b. Yes. 0 m/s^2, 0 N   c. 57 kg

43.

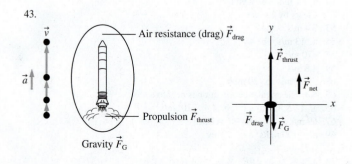

Air resistance (drag) $\vec{F}_{drag}$
Propulsion $\vec{F}_{thrust}$
Gravity $\vec{F}_G$

47.

49.

51.

53.

55.

57. a.

First contact

Loses contact

Magnified view of
ball in contact with
ground

Compressing

Expanding

Turning point

b.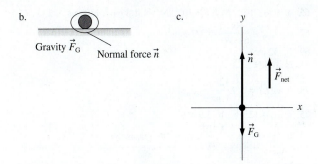

Gravity $\vec{F}_G$ Normal force $\vec{n}$

c.

Chapter 6

Stop to Think Questions

1. **a.** The lander is descending and slowing. The acceleration vector points upward, and so $\vec{F}_{net}$ points upward. This can be true only if the thrust has a larger magnitude than the weight.

2. **a.** You are descending and slowing, so your acceleration vector points upward and there is a net upward force on you. The floor pushes up against your feet harder than gravity pulls down.

3. $f_b > f_c = f_d = f_e > f_a$. Situations c, d, and e are all kinetic friction, which does not depend on either velocity or acceleration. Kinetic friction is smaller than the maximum static friction that is exerted in b. $f_a = 0$ because no friction is needed to keep the object at rest.

4. **d.** The ball is shot *down* at 30 m/s, so $v_{0y} = -30$ m/s. This exceeds the terminal speed, so the upward drag force is *larger* than the downward weight force. Thus the ball *slows down* even though it is "falling." It will slow until $v_y = -15$ m/s, the terminal velocity, then maintain that velocity.

Exercises and Problems

1. 94 N, 58° below the horizontal
3. 510 N
5. 160 N
7. a. 0.0036 N   b. 0.010 N
9. $a_x = 1.5$ m/s^2, $a_y = 0$ m/s^2
11. 0 m/s^2, 4 m/s
13. a. 490 N   b. 490 N   c. 740 N   d. 240 N
15. 40 s
17. a. 540 N   b. $m = 55$ kg; $mg = 210$ N
19. a. 7.8×10^2 N   b. 1.1 kN
21. a. 3.96 N   b. 2.32 N
23. 5.90×10^{-3} m/s^2
25. Yes
27. 9 m/s
29. 10,000 N
31. 2.6×10^3 m
33. 190,000 N
35. 140 m/s
37. 4.5 m/s
39. 6400 N, 4380 N
41. a. 59 N   b. 68°   c. 79 N
43. 59 N
45. a. $v(h) = \sqrt{2\left(\dfrac{F_{thrust}}{m} - g\right)h}$ b. 54 m/s
47. 2800 kg
49. a. 16.9 m/s   b. 229 m
51. a. 3.8 m   b. 7.0 m/s
53. $T_{max} = (M + m)\mu_s g$
55. a. Yes   b. Yes
57. Stay at rest
59. a. $-5g$ b. $3g$
61. a. 0 N   b. 220 N
63. a. $\dfrac{F_0 T}{m \, 2}$ b. $\dfrac{F_0 T^2}{m \, 3}$

65. a. 9.4×10^{-10} N, 5.7×10^{-13} N   b. 1.8 m/s^2, 130 m/s^2

67. a. $v_{\text{term}} = \dfrac{mg}{6\pi\eta R}$ b. 27 min

69. b. 32 m/s

71. c. 144 N

73. b. $v_x(L) = \sqrt{L\left(\dfrac{2F_0}{m} - \mu_0 g\right)}$

75. a. $a_x = (2x)g$ b. 3.9 m/s^2

77. b. $v_x(x) = v_0 - \dfrac{6\pi\eta R}{m}x$ c. Not reasonable

Chapter 7

Stop to Think Questions

1. **The crate's gravitational force and the normal force are incorrectly identified as an action/reaction pair.** The normal force should be paired with a downward force of the crate on the ground. Gravity is the pull of the entire earth, so $\vec{F}_G$ should be paired with a force pulling up on the entire earth.

2. **c.** Newton's third law says that the force of A on B is *equal* and opposite to the force of B on A. This is always true. The mass of the objects isn't relevant.

3. **b.** $F_{\text{B on H}} = F_{\text{H on B}}$ and $F_{\text{A on B}} = F_{\text{B on A}}$ because these are action/re-action pairs. Box B is slowing down and therefore must have a net force to the left. So from Newton's second law we also know that $F_{\text{H on B}} > F_{\text{A on B}}$.

4. **Equal to.** Each block is hanging in equilibrium, with no net force, so the upward tension force is mg.

5. **Less than.** Block B is *accelerating* downward, so the net force on B must point down. The only forces acting on B are the tension and gravity, so $T_{\text{S on B}} < (F_G)_B$.

Exercises and Problems

1. a.   **Interaction diagram**   b. The system is the soccer ball and bowling ball.

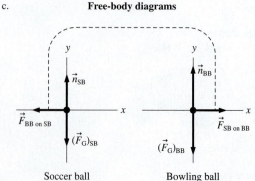

SB = Soccer ball
BB = Bowling ball
S = Surface
EE = Entire Earth

c.   **Free-body diagrams**

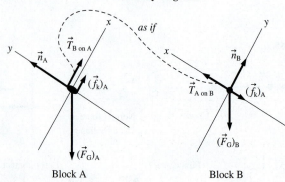

Soccer ball   Bowling ball

3. a.   **Interaction diagram**

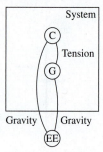

b. The system is the cable and the girder.

c.   **Free-body diagrams**

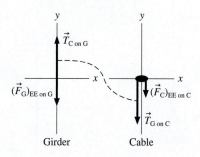

Girder   Cable

5. a.   **Interaction diagram**

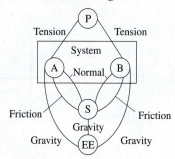

P = Pulley   S = Surface
A = Block A   B = Block B
EE = Entire Earth

b. The system is block A and block B.

c.   **Free-body diagrams**

Block A   Block B

7. $m_A = 12$ kg

9. a. 6 N   b. 10 N

11. 440 N

13. 42 m

15. 6.5 m/s^2

17. 590 N
19. 250 N
21. 9800 N
23. 67 N, 36°
25. 99 kg
27. a. 4 N   b. 3 N
29. No
31. a. 2800 N   b. 2800 N
33. Cheek, not forehead
35. 1.8 s
37. $T_1 = 100$ N, $T_2 = T_3 = T_5 = F = 50$ N, and $T_4 = 150$ N
39. a. 1.8 kg   b. 1.3 m/s^2
41. a. 0.67 m   b. Slides back down
43. a. 8.2×10^3 N   b. 4.8×10^2 N
45. 4.0×10^2 N
47. 2.4×10^{-6} N
49. $F = (m_1 + m_2)g \tan \theta$
53. 1.8 m/s^2
55. 2.8 m/s^2
57. 920 g

Chapter 8

Stop to Think Questions

1. **d.** The parallel component of $\vec{F}$ is opposite $\vec{v}$ and will cause the particle to slow down. The perpendicular component of $\vec{F}$ will cause the particle to change directions in a downward direction.
2. $T_d > T_b = T_e > T_c > T_a$. The center-directed force is $m\omega^2 r$. Changing r by a factor of 2 changes the tension by a factor of 2, but changing ω by a factor of 2 changes the tension by a factor of 4.
3. **b.** The car is moving in a circle, so there must be a net force toward the center of the circle. The circle is below the car, so the net force must point downward. This can be true only if $F_G > n$.
4. **c.** The ball does not have a "memory" of its previous motion. The velocity $\vec{v}$ is straight up at the instant the string breaks. The only force on the ball after the string breaks is the gravitational force, straight down. This is just like tossing a ball straight up.

Exercises and Problems

1. 39 m
3. c. 0.107°
5. 9.4 kN, static friction
7. a 3.9 m/s   b. 6.2 N
9. 2.01×10^{20} N
11. 45 s
13. $v = \sqrt{\dfrac{m_2 rg}{m_1}}$
15. 6.0×10^{-3} m/s^2
17. a. 24.0 h   b. 0.223 m/s^2 c. 0 N
19. 12 m/s
21. 20 m/s
23. a. 3.77 m/s, 0.95 m/s^2 b. 0.90   c. 1.1
25. a. 4.9 N   b. 2.9 N   c. 32 N
27. 1.6 s
29. a. 2.2×10^6 N   b. $-27°$
33. 0.034 N
35. 8.6 m
37. a. 165 m   b. Straight line
41. a. $\omega = \sqrt{\dfrac{g}{L\sin\theta}}$ b. 72 rpm
43. 0.79 N
45. a. 2.9 m/s   b. 14 N

47. a. $T = \dfrac{mgL}{\sqrt{L^2 - r^2}}$ b. $\omega = \sqrt{\dfrac{g}{\sqrt{L^2 - r^2}}}$ c. 5.0 N, 30 rpm
49. 24 rpm
51. a. 320 N, 1400 N   b. 5.7 s
53. 0. 38 N
55. a. $\sqrt{gL}$ b. 5.9 mph
57. 1.4 m to the right
59. 13 N
61. a. 1.90 m/s^2 at 21°   b. 15.7 m/s
63. a. 3.8 m/s   b. 19 m/s^2
65. b. 19.8 m/s
67. a. $\theta = \frac{1}{2}\tan^{-1}(mg/F)$ b. 11.5%
69. $\sqrt{gL}$

Chapter 9

Stop to Think Questions

1. **a.** Kinetic energy depends linearly on the mass but on the *square* of the velocity. A factor of 2 change in velocity is more significant than a factor of 2 change in mass.
2. **Positive.** The force (gravity) and the displacement are in the same direction. The rock gains kinetic energy.
3. **Positive.** Each puck experiences a force in the direction of motion, which is a positive amount of work. The net force is zero, but the total work is not zero because the pucks have different displacements.
4. **c.** The upward tension force is opposite the displacement, so it does negative work. The downward gravitational force is parallel to the displacement, so it does positive work.
5. **c.** $W = F(\Delta r)\cos\theta$. The 10 N force at 90° does no work at all. Because $\cos 60° = \frac{1}{2}$, the 8 N force does less work than the 6 N force.
6. **Zero.** The road does exert a forward force on the car, but the point of application does not move because the tires are not skidding on the road. The car's increasing kinetic energy is a transformation of chemical energy to kinetic energy, not a transfer of energy from the road to the car.
7. $k_a > k_b > k_c$. The spring constant is the slope of the force-versus-displacement graph.
8. **Zero.** The wall exerts a force on the right end of the spring, but the point of application does not move.
9. $P_b > P_a = P_c > P_d$. The work done is $mg\Delta y$, so the power output is $mg\Delta y/\Delta t$. Runner b does the same work as runner a, but in less time. The ratio $m/\Delta t$ is the same for runners a and c. Runner d does twice the work of a, but takes more than twice as long.

Exercises and Problems

1. The bullet
3. 2.0
5. a. 5.3 J   b. 0 J
7. 0 J
9. a 1.9 J   b. 26 m/s
11. a. -2 b. 0
13. 125°
15. a. -2.7 b. -20 c. 10
17. a. 0.20 J   b. 3.0 m/s
19. 1.7 kJ, 1.1 kJ, -2.0 kJ
21. 8.0 m/s, 10 m/s, 11 m/s
23. $\frac{1}{3}qd^3$
25. 38 N
27. a. 3.9×10^2 N/m   b. 17.5 cm
29. a. 49 N   b. 1450 N/m   c. 3.4 cm
31. 1360 m/s
33. 0.037
35. 540 J
37. a. $W_{net} = 176$ J   b. $P = 59$ W
41. Runner: $P_{avg} = 1.2$ kW; greyhound: $P_{avg} = 2.0$ kW

43. a. -9.8×10^4 J   b. 1.1×10^5 J   c. 1.0×10^4 J
45. 2.4 m/s
47. $2h$
49. a. 2.2 m/s   b. 0.0058
51. a. 2.9 J   b. 3.6%
53. a. $Gm_1m_2\left(\dfrac{x_2 - x_1}{x_1x_2}\right)$ b. 2.1×10^5 m/s
55. 19 m/s
57. 33 N/m
59. 3.7 m/s
61. 2.5×10^5 kg/s
63. 1.4 kW
65. 1.2×10^8 ly
71. 24 W

Chapter 10

Stop to Think Questions

1. $(U_G)_c > (U_G)_b = (U_G)_d > (U_G)_a$. Gravitational potential energy depends only on height, not on speed.
2. **b.** Potential energy depends only on the vertical displacement. At the elevation of the dashed line, both have gained the same gravitational potential energy, so both have lost the same kinetic energy.
3. $v_a = v_b = v_c = v_d$. Her increase in kinetic energy depends only on the vertical distance through which she falls, not on the shape of the slide.
4. **c.** Constant speed means no change of kinetic energy. But for motion on a slope, constant speed requires friction. All the gravitational potential energy is being transformed into thermal energy.
5. **c.** U_{Sp} depends on d^2. Doubling the compression increases U_{Sp} by a factor of 4. All potential energy is transformed into kinetic energy, so K increases by a factor of 4. But K depends on v^2, so v increases by only a factor of 2.
6. $x = 6$ m. From the graph, the particle's potential energy at $x = 1$ m is $U = 3$ J. Its total energy is thus $E = K + U = 4$ J. A TE line at 4 J crosses the PE curve at $x = 6$ m.
7. **e.** Force is the negative of the slope of the potential-energy diagram. At $x = 4$ m the potential energy has risen by 4 J over a distance of 2 m, so the slope is 2 J/m = 2 N.
8. **d.** The system is losing potential energy as the weight falls. It's gaining speed, so some U is tranformed into K. Energy is also being transferred out of the system to the environment via negative work done by the rope tension. The system is not isolated, so neither E_{mech} nor E_{sys} is conserved.

Exercises and Problems

1. a. $\Delta K_{tot} = 17$ J, $\Delta U_{int} = 0$ b. $\Delta K_{tot} = 12$ J, $\Delta U_{int} = 5$ J
3. $\Delta U = 2.9 \times 10^6$ J
5. a. 13 m/s   b. 14 m/s
7. 7.7 m/s
9. 1.4 m/s
11. 1.4 m/s
13. 81 m/s
15. 0.632 m
17. 2.0 m/s
19. 3.0 m/s
21. 9.7 J
23. $v_0/\sqrt{2}$
25. 3.5 m/s, 2.8 m/s, 4.5 m/s
27. a. 7.7 m/s   b. 10 m/s
29. 6.3 m/s
31. -60 N at $x = 1$ m, 15 N at $x = 4$ m
33. 2.5 N, 0.40 N, 0.16 N
35. a. 50 J   b. 50 J   c. 50 J, yes

37.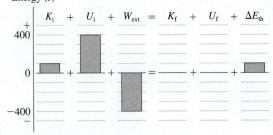

39. -1 J of work is done to the environment
41. 2.3 m/s
43. a. 2.2×10^4 N/m   b. 19 m/s
45. $\dfrac{5}{2}R$
47. 65 g
49. 0.54 m
51. a. 1.7 m/s   b. No
53. a. 0.51 m   b. 0.38 m
55. 1.37×10^8 m/s
57. a. $\dfrac{\pi}{3}$ and $\dfrac{2\pi}{3}$ b. $\dfrac{\pi}{3}$ is unstable, $\dfrac{2\pi}{3}$ is stable
59. a. 20 J, 40 J, 60 J, 70 J   b. 2.6 m
61. a. $-\pi B/L$ b. $-AL$ c. $-2AL + \pi B/L$
63. a. a^2b b. a^2b c. Yes
69. 80.4°
71. 11 m
73. 6.7 m

Chapter 11

Stop to Think Questions

1. **f.** The cart is initially moving in the negative x-direction, so $p_{ix} = -20$ kg m/s. After it bounces, $p_{fx} = -10$ kg m/s. Thus $\Delta p = (10 \text{ kg m/s}) - (-20 \text{ kg m/s}) = 30$ kg m/s.
2. **b.** The clay ball goes from $v_{ix} = v$ to $v_{fx} = 0$, so $J_{clay} = \Delta p_x = -mv$. The rubber ball rebounds, going from $v_{ix} = v$ to $v_{fx} = -v$ (same speed, opposite direction). Thus $J_{rubber} = \Delta p_x = -2mv$. The rubber ball has a larger momentum change, and this requires a larger impulse.
3. **Less than.** The ball's momentum $m_B v_B$ is the same in both cases. Momentum is conserved, so the *total* momentum is the same after both collisions. The ball that rebounds from C has *negative* momentum, so C must have a larger momentum than A.
4. **c.** Momentum conservation requires $(m_1 + m_2) \times v_f = m_1 v_1 + m_2 v_2$. Because $v_1 > v_2$, it must be that $(m_1 + m_2) \times v_f = m_1 v_1 + m_2 v_2 > m_1 v_2 + m_2 v_2 = (m_1 + m_2)v_2$. Thus $v_f > v_2$. Similarly, $v_2 < v_1$, so $(m_1 + m_2)v_f = m_1 v_1 + m_2 v_2 < m_1 v_1 + m_2 v_1 = (m_1 + m_2)v_1$. Thus $v_f < v_1$. The collision causes m_1 to slow down and m_2 to speed up.
5. **Right end.** The pieces started at rest, so the total momentum of the system is zero. It's an isolated system, so the total momentum after the explosion is still zero. The 6 g piece has momentum $6v$. The 4 g piece, with velocity $-2v$, has momentum $-8v$. The combined momentum of these two pieces is $-2v$. In order for P to be zero, the third piece must have a *positive* momentum $(+2v)$ and thus a positive velocity.
6. **e.** Momentum is conserved, so the total momentum of the two pieces must be the initial $2\,\hat{i}$ kg m/s. $\vec{p}_1 = 2\,\hat{j}$ kg m/s, so $\vec{p}_2$ has to be $(2\,\hat{i} - 2\,\hat{j})$ kg m/s.

Exercises and Problems

1. 75 m/s
3. 4.0 N s
5. 1.0×10^3 N

7. 80 m/s to the left

9. 3.0 m/s to the right

11. a. 1.5×10^4 N s   b. 30 s, 110 m/s

13. 6.0 m/s

15. 1.4 m/s

17. v_0

19. 2.0 mph

21. 7.6 cm in the direction Brutus was running

23. -1.69×10^7 m/s, 3.1×10^6 m/s

25. a. 2.6 m/s   b. 33 cm

27. a. 1.92 m/s   b. 1.90 m/s

29. 13 s

31. $(1\hat{i} + 5\hat{j})$ kg m/s

33. 45° north of east at 1.7 m/s

35. 1800 m/s

37. 1000 m/s

39. a. 6.4 m/s   b. $F_{avg} = 3.6 \times 10^2$ N $= 612(F_G)_B$

41. 9.3×10^2 N

43. $12t$ N

45. a. 6.7×10^{-8} m/s   b. $2.2 \times 10^{-10}\%$

47. $v_0/\sqrt{2}$, 45° east of north

49. a. $v_{bullet} = \dfrac{m + M}{m}\sqrt{2\mu_k g d}$ b. 4.4×10^2 m/s

51. 4.0×10^2 m

53. 0.021 m/s

55. 25.8 cm

57. a. $v_m = \dfrac{m + M}{m}\sqrt{5Rg}$ b. $v_m = \dfrac{1}{2}\dfrac{m + M}{m}\sqrt{5Rg}$

59. 14 m

61. 16 m/s

63. 7.9 m/s

65. a. 100 g ball: 5.3 m/s to the left; 200 g ball: 1.7 m/s to the right
   b. Both balls: 0.67 m/s to the left

67. 14.0 u

69. a. -1.4×10^{-22} kg m/s   b. and c. 1.4×10^{-22} kg m/s in the direction of the electron

71. 0.85 m/s, 72° below the $+x$-axis

73. 780 m/s, 1900 m/s, 4000 m/s

79. 5.7 m/s

81. 90 m/s

83. 8 bullets

Chapter 12

Stop to Think Questions

1. **b.** The center of the hitting end is *closer* to the center of mass, so it must have *more* mass.

2. **b.** Rotational kinetic energy depends on the angular velocity and the moment of inertia. Both have the same angular velocity, but a cylindrical shell has a larger moment of inertia than a solid cylinder with the same mass and radius.

3. $I_a > I_d > I_b > I_c$. The moment of inertia is smaller when the mass is more concentrated near the rotation axis.

4. $\tau_e > \tau_a = \tau_d > \tau_b > \tau_c$. The tangential component in e is larger than 2 N.

5. $\alpha_b > \alpha_a > \alpha_c = \alpha_d$. Angular acceleration is proportional to torque and inversely proportional to the moment of inertia. The moment of inertia depends on the *square* of the radius, so angular acceleration is proportional to F/r.

6. **c.** The scale exerts an upward force at half the distance from the pivot as the weight's downward force. To exert an equal but opposite torque for static equilibrium requires twice the force.

7. **d.** There is no net torque on the bucket + rain system, so the angular momentum is conserved. The addition of mass on the outer edge of the circle increases I, so ω must decrease. Mechanical energy is not conserved because the raindrop collisions are inelastic.

Exercises and Problems

1. a. 4.2×10^2 rad/s^2 b. 8.3 rev

3. a. 1.5 m/s   b. 13 rev

5. 4.7×10^6 m

7. 8.0 cm, 5.0 cm

9. 2.4 m/s

11. a. 0.032 kg m^2 b. 16 J

13. a. 0.063 m, 0.050 m   b. 0.0082 kg m^2

15. a. (0.060 m, 0.040 m)   b. 0.0020 kg m^2
   c. 0.0013 kg m^2

17. a. 6.9 kg m^2 b. 4.1 kg m^2

19. 50 N

21. −14.7 N

23. a. $\tau = 31.0$ N m   b. $\tau = 21.9$ N m

25. 0.50 rad/s

27. a. 1.75×10^{-3} N m   b. 50 rev

29. 1.4 m

31. $F_1 = 0.75$ kN, $F_2 = 1.0$ kN

33. a. 6.4×10^2 rpm   b. 40 m/s   c. 0 m/s

35. a. 88 rad/s   b. 2/7

37. a. $\vec{A} \times \vec{B} = (17$, out of the page)   b. $\vec{C} \times \vec{D} = \vec{0}$

39. a. $\vec{D} = n\hat{i}$, where n could be any real number
   b. $\vec{E} = 2\hat{j}$ c. $\vec{F} = 1\hat{k}$

41. $-(38$ N m$)\hat{k}$

43. $-0.025\hat{i}$ kg m^2/s

45. 91 rpm

47. 83 rpm

49. 1.8 J

51. $\dfrac{L^2}{4}\left(\dfrac{M}{3} + m_1 + \dfrac{m_2}{4}\right)$

53. $\dfrac{M}{3L}\left[(L - d)^3 + d^3\right]$

55. a. $I_{disk} = \frac{1}{2}M(R^2 + r^2)$ c. 1.4 m/s or 75%

57. 0.91 m

59. 15,300 N

61. Yes, $d_{max} = \dfrac{25}{24}L$

63. 800 N

65. a. $a = \dfrac{m_2 g}{m_1 + m_2}, T = \dfrac{m_1 m_2 g}{m_1 + m_2}$

   b. $a = \dfrac{m_2 g}{m_1 + m_2 + \frac{1}{2}m_p}, T_1 = \dfrac{m_1 m_2 g}{m_1 + m_2 + \frac{1}{2}m_p}, T_2 = \dfrac{m_2(m_1 + \frac{1}{2}m_p)g}{m_1 + m_2 + \frac{1}{2}m_p}$

67. 0.52 N

69. 4.3 m

71. $\frac{1}{3}\tan\theta$

73. a. $\sqrt{\dfrac{3}{4}Lg}$ b. $\sqrt{Lg}$

75. $h = 2.7(R - r)$

79. 50 rpm

81. 22 rpm

83. a. 137 km   b. 8.6×10^6 m/s

85. a. 2.9×10^{-5} N m   b. 7.0 m/s

87. a. kg/m^3 b. $\dfrac{12M}{L^3}$ c. $\dfrac{3}{20}ML^2$

89. $v_f = \dfrac{1}{5}v_0$ to the right

Chapter 13

Stop to Think Questions

1. **e.** The acceleration decreases inversely with the square of the distance. At height R_e, the distance from the center of the earth is $2R_e$.

2. **c.** Newton's third law requires $F_{1\text{ on }2} = F_{2\text{ on }1}$.

3. **b.** $g_{surface} = GM/R^2$. Because of the square, a radius twice as large balances a mass four times as large.

4. In absolute value, $U_e > U_a = U_b = U_d > U_c$. $|U_G|$ is proportional to $m_1 m_2 / r$.

5. **a.** T^2 is proportional to r^3, or T is proportional to $r^{3/2}$. $4^{3/2} = 8$.

Exercises and Problems

1. 6.00×10^{-4}
3. a. 6.7×10^{-9} N   b. 6.8×10^{-9}
5. 9.0 N
7. 12 cm
9. a. 1.62 m/s^2 b. 5.90×10^{-3} m/s^2
11. 1.35 m/s^2
13. $0.58R_e$
15. 39 m
17. 10 km/s
19. 4.3 km/s
21. a. 1.80×10^7 m   b. 9.41 km/s
23. 33 km/s
25. $M_p = 1.5 \times 10^{25}$ kg, $M_s = 5.2 \times 10^{30}$ kg
27. 2.9×10^9 m
29. 1.0×10^{26} kg
31. 1.44 km/s, 1.72×10^7 m
33. a. 7.0 m/s   b. 12 m/s
35. a. 1.3×10^{-6} N, 83° cw from the $+y$-axis
    b. 2.3×10^{27} N, 7.5° ccw from the $-y$-axis.
37. -1.48×10^{-7} J
39. 46 kg and 104 kg
41. $2000R_e$
43. 4.2×10^5 m
45. 1.6 s
47. a. 2.3 km/s   b. 11 km/s
49. $0.414R$
51. 6.7×10^8 J
53. 1.8×10^{25} J
55. a. 0.17 m/s   b. 1.0×10^{13} kg   c. 0.84 m/s
57. $T = \left[\dfrac{4\pi^2 r^3}{G} \dfrac{1}{(M + m/4)} \right]^{1/2}$
59. a. 6.3×10^4 m/s
    b. 1.3×10^{12} m/s^2
    c. 1.3×10^{12} N
    d. 6.29×10^{-4} s
    e. 1.5×10^6 m
61. 10 yr
63. a. 14,000 km   b. 2000 km
67. 3.0×10^4 m/s
69. a. 5.8×10^{22} kg   b. 1.3×10^6 m
71. b. $v_1 = 7730$ m/s, $v_1' = 10,160$ m/s
    c. 2.17×10^{10} J
    d. $v_2' = 1600$ m/s, $v_2 = 3070$ m/s
    e. 3.43×10^9 J
    f. 2.52×10^{10} J

Chapter 14

Stop to Think Questions

1. $\rho_a = \rho_b = \rho_c$. Density depends only on what the object is made of, not how big the pieces are.

2. **c.** These are all open tubes, so the liquid rises to the same height in all three despite their different shapes.

3. $F_b > F_a = F_c$. The masses in c do not add. The pressure underneath each of the two large pistons is mg/A^2, and the pressure under the small piston must be the same.

4. **b.** The weight of the displaced water equals the weight of the ice cube. When the ice cube melts and turns into water, that amount of water will exactly fill the volume that the ice cube is now displacing.

5. **1 cm^3/s out.** The fluid is incompressible, so the sum of what flows in must match the sum of what flows out. 13 cm^3/s is known to be flowing in, while 12 cm^3/s flows out. An additional 1 cm^3/s must flow out to achieve balance.

6. $h_b > h_d > h_c > h_a$. The liquid level is higher where the pressure is lower. The pressure is lower where the flow speed is higher. The flow speed is highest in the narrowest tube, zero in the open air.

Exercises and Problems

1. 50 mL
3. a. 8.1×10^2 kg/m^3 b. 8.4×10^2 kg/m^3
5. 2.4×10^3 kg
7. 38 cm
9. 1.2×10^5 Pa
11. 0.64 kN
13. 10.3 m
15. 3.2 m
17. 6.7×10^2 kg/m^3
19. 750 kg/m^3
21. 1.9 N
23. 2.49×10^3 kg/m^3
25. 1.1 kN
27. $1.27v_0$
29. 2.12 m/s
31. a. 0.38 N   b. 20 m/s
33. 1.0 cm
35. a. 5.1×10^7 Pa   b. -0.025 c. 1056 kg/m^3
37. 2.4 μm
39. a. 2.9×10^4 N   b. 30 players
41. 27 psi
43. 55 cm
45. a. Down by 3.0 cm   b. 3.5 cm
47. a. $F = \rho gDWL$ b. $F = \frac{1}{2}\rho gD^2 L$ c. 0.78 kN, 1.4 kN
49. 8.01%
51. $(\rho - \rho_1)/(\rho_2 - \rho_1)$
53. 0.16 kg
55. $m = 29$ g, $\rho = 1.1 \times 10^3$ kg/m^3
57. 5.5 m/s
59. 187 nm/s
61. 76 cm
63. 4.0 cm
65. a. Lower   b. 0.83 kPa   c. Blow out
67. a. $v_1 = 1.4 \times 10^2$ m/s, $v_2 = 5.8$ m/s   b. 4.5×10^{-3} m^3/s
69. a. $v_{hole}A_{hole} = \pi r^2 \sqrt{\dfrac{2gd}{1 - (r/R)^4}}$ b. 1.1 mm/min
71. 14 cm
73. $\dfrac{h}{l} = \left(1 - \dfrac{\rho_0}{\rho_f}\right)^{1/3}$
75. a. $\dfrac{2R^2 g}{9\eta}(\rho - \rho_f)$ b. $v_t\left(1 - e^{(-6t\pi\eta R/m)}\right)$ c. 3.1 s

Chapter 15

Stop to Think Questions

1. **c.** $v_{max} = 2\pi A/T$. Doubling A and T leaves v_{max} unchanged.

2. **d.** Think of circular motion. At 45°, the particle is in the first quadrant (positive x) and moving to the left (negative v_x).

3. **c > b > a = d.** Energy conservation $\frac{1}{2}kA^2 = \frac{1}{2}m(v_{max})^2$ gives $v_{max} = \sqrt{k/m}A$. k or m has to be increased or decreased by a factor of 4 to have the same effect as increasing or decreasing A by a factor of 2.

4. **c.** $v_x = 0$ because the slope of the position graph is zero. The negative value of x shows that the particle is left of the equilibrium position, so the restoring force is to the right.

5. **c.** The period of a pendulum does not depend on its mass.

6. $\tau_d > \tau_b = \tau_c > \tau_a$. The time constant is the time to decay to 37% of the initial height. The time constant is independent of the initial height.

Exercises and Problems

1. a. 3.3 s b. 0.30 Hz c. 1.9 rad/s d. 0.25 m e. 0.48 m/s
3. 2.27 ms
5. a. 20 cm b. 0.13 Hz c. 60°
7. a. $-\frac{2}{3}\pi$ rad b. 13.6 cm/s c. 15.7 cm/s
9.

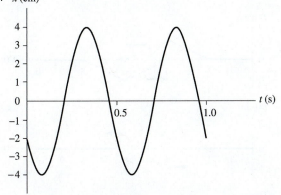

11. $x(t) = (4.0 \text{ cm})\cos\left[(8.0\pi \text{ rad/s})t - \frac{1}{2}\pi \text{ rad}\right]$
13. a. $\phi_0 = -\frac{2}{3}\pi$ rad, or $-120°$ b. $\phi = -\frac{2}{3}\pi$ rad, 0 rad, $\frac{2}{3}\pi$ rad, $\frac{4}{3}\pi$ rad
15. 5.48 N/m
17. a. 2.0 cm b. 0.63 s c. 5.0 N/m
    d. $-\frac{1}{4}\pi$ rad e. 14 cm/s
    f. 20 cm/s g. 1.00×10^{-3} J h. 1.5 cm/s
19. 0.13 s
21. a. 25 N/m b. 0.90 s c. 0.70 m/s
23. 3.5 Hz
25. 36 cm
27. 33 cm
29. 3.67 m/s²
31. 54 cm
33. 5.00 s
35. 250 N/m
37. 0.0955 s
41. 1.02 m/s
43. a. 55 kg b. 0.73 m/s
45. a. 3.2 Hz b. 7.1 cm c. 5.0 J
47. 1.6 Hz
49. 8.1×10^9 N/m²
51. a. 4.7×10^4 N/m b. 1.8 Hz
53. 5.0×10^2 m/s
55. a. 7.5 m b. 0.45 m/s
57. a. 0.84 s b. 7.1°
59. 8.7×10^{-2} kg m²
61. a. 2.0 Hz b. 1.2 cm
63. 7.9×10^{13} Hz
65. 5.86 m/s²
67. a. 9.5 N/m b. 0.010 kg/s
69. 25 s
71. 21 oscillations
75. $\sqrt{\dfrac{f_1^2 f_2^2}{f_1^2 + f_2^2}}$
77. $\dfrac{1}{2\pi}\sqrt{\dfrac{5g}{7R}}$

79. 1.8 Hz
81. 0.58 s

Chapter 16

Stop to Think Questions

1. **d and e.** The wave speed depends on properties of the medium, not on how you generate the wave. For a string, $v = \sqrt{T_s/\mu}$. Increasing the tension or decreasing the linear density (lighter string) will increase the wave speed.

2. **b.** The wave is traveling to the right at 2.0 m/s, so each point on the wave passes $x = 0$ m, the point of interest, 2.0 s before reaching $x = 4.0$ m. The graph has the same shape, but everything happens 2.0 s earlier.

3. **d.** The wavelength—the distance between two crests—is seen to be 10 m. The frequency is $f = v/\lambda = (50 \text{ m/s})/(10 \text{ m}) = 5$ Hz.

4. $n_c > n_a > n_b$. $\lambda = \lambda_{\text{vac}}/n$, so a shorter wavelength corresponds to a larger index of refraction.

5. **e.** A crest and an adjacent trough are separated by $\lambda/2$. This is a phase difference of π rad.

6. **c.** Any factor-of-2 change in intensity changes the sound intensity level by 3 dB. One trumpet is $\frac{1}{4}$ the original number, so the intensity has decreased by two factors of 2.

7. **c.** Zack hears a higher frequency as he and the source approach. Amy is moving with the source, so $f_{\text{Amy}} = f_0$.

Exercises and Problems

1. 141 m/s
3. 2.0 m
5.

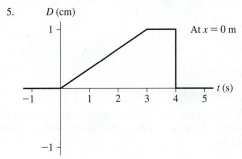

7.

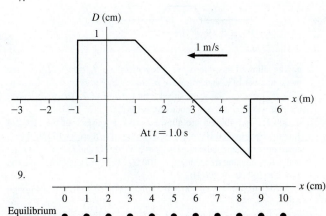

9.

11. a. 4.2 m b. 48 Hz
13. a. 20 Hz b. 2.3 m c. 46 m/s
15. $v = \sqrt{d/c}$
17. a. 3.2 mm b. 9.3×10^{10} Hz

19. a. 6.67×10^{14} Hz   b. 4.62×10^{14} Hz   c. 1.44
21. a. 10 GHz   b. 0.17 ms
23. a. 8.6×10^8 Hz   b. 23 cm
25. a. 1.88×10^8 m/s   b. 4.48×10^{14} Hz
27. a. 311 m/s   b. 361 m/s
29. 40 cm
31. 2.5 m
33. 3.4×10^{-6} J
35. a. 1.1×10^{-3} W/m^2 b. 1.1×10^{-7} J
37. 2700 W/m^2, 1400 W/m^2, 610 W/m^2
39. 5.0 W
41. a. 432 Hz   b. 429 Hz
43. 86 m/s
45. a. 0.80 m   b. $-\frac{\pi}{2}$ rad
    c. $D(x, t) = (2.0 \text{ mm}) \sin\left(2.5\pi x - 10\pi t - \frac{1}{2}\pi\right)$
47. 2.3 m, 1.7 m
49. 410 μs
51. 1230 km
53. 987 m/s
55. a. $-x$-direction   b. 12 m/s, 5.0 Hz, 2.6 rad/m   c. -1.5 cm
57. -19 m/s, 0 m/s, 19 m/s
59. 8
61. 9.4 m/s
63. 16 Hz and 2.0 cm
65. 4.0 m/s
67. a. 6.67×10^4 W   b. 8.5×10^{10} W/m^2
69. a. 3.10×10^9 W/m^2 b. 19 AU
71. 49 mW
73. 1.5
75. 19 m/s
79. Receding at 1.5×10^6 m/s
81. 0.07°C
83. 18 mJ
85. 29 s

Chapter 17

Stop to Think Questions

1. **c.** The figure shows the two waves at $t = 6$ s and their superposition. The superposition is the *point-by-point* addition of the displacements of the two individual waves.

2. **a.** The allowed standing-wave frequencies are $f_m = m(v/2L)$, so the mode number of a standing wave of frequency f is $m = 2Lf/v$. Quadrupling T_s increases the wave speed v by a factor of 2. The initial mode number was 2, so the new mode number is 1.

3. **b.** 300 Hz and 400 Hz are allowed standing waves, but they are not f_1 and f_2 because 400 Hz $\neq 2 \times 300$ Hz. Because there's a 100 Hz difference between them, these must be $f_3 = 3 \times 100$ Hz and $f_4 = 4 \times 100$ Hz, with a fundamental frequency $f_1 = 100$ Hz. Thus the second harmonic is $f_2 = 2 \times 100$ Hz $= 200$ Hz.

4. **c.** Shifting the top wave 0.5 m to the left aligns crest with crest and trough with trough.

5. **a.** $r_1 = 0.5\lambda$ and $r_2 = 2.5\lambda$, so $\Delta r = 2.0\lambda$. This is the condition for maximum constructive interference.

6. **Maximum constructive.** The path-length difference is $\Delta r = 1.0$ m $= \lambda$. For identical sources, interference is constructive when Δr is an integer multiple of λ.

7. **f.** The beat frequency is the difference between the two frequencies.

Exercises and Problems

1.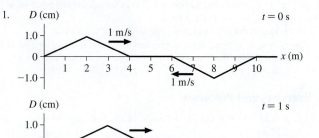
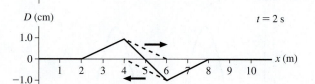
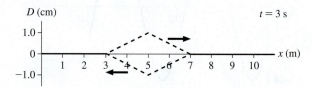
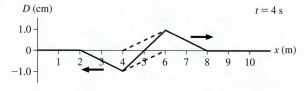
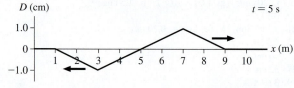
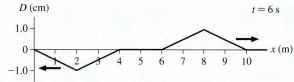
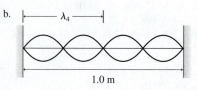

3. 4 s
5. 20 m/s
7. a. 8   b. 200 Hz
9. a. 12 Hz, 24 m/s   b.

11. 12 kg

13.

m	f_m (GHz)
7	10.5
8	12.0
9	13.5
10	15.0
11	16.5
12	18.0
13	19.5

15. 400 m/s
17. 35 cm
19. 1.9 Hz
21. 1.6 kHz
23. a. 25 cm   b. 25 cm
25. 216 nm
27. a. In phase   b.

m	r_1	r_2	Δr	C/D
P	3λ	4λ	λ	C
Q	$\frac{7}{2}\lambda$	2λ	$\frac{5}{2}\lambda$	D
R	$\frac{5}{2}\lambda$	$\frac{7}{2}\lambda$	λ	C

29. 5.0 m
31. Maximum destructive interference
33. 527 Hz
35. 1.255 cm
37. 5.4 m
39. 2.4×10^{-4} N
41. 260 N
43. 8.50 m/s^2
45. $T' = \dfrac{T_0}{4}$
47. 0.54 s
49. 11 g/m
51. 13.0 cm
53. 328 m/s
55. 26 cm, 56 cm, 85 cm
57. 450 N
59. 7.9 cm
61. a. 80 cm   b. $\dfrac{3\pi}{4}$ c. 0.77a
63. a. 473 nm   b. 406 nm, 568 nm   c. Reflected light is blue and trans-
    mitted light is yellowish green.
65. a. $\lambda_C = \dfrac{2.66d}{m + \frac{1}{2}}$, $m = 0, 1, 2, 3,\ldots$
    b. 692 nm and 415 nm; this gives a purplish color.
67. 679 nm, 428 nm
69. 170 Hz
71. a. a b. 1.0 m   c. 9
73. a. 5   b. 4.6 mm
75. 7.0 m/s
77. 4.0 cm, 35 cm, 65 cm
79. 2.0 kg
81. a. $\lambda_1 = 20.0$ m, $\lambda_2 = 10.0$ m, $\lambda_3 = 6.67$ m
    b. $v_1 = 5.59$ m/s, $v_2 = 3.95$ m/s, $v_3 = 3.22$ m/s
    c. $T_1 = 3.58$ s, $T_2 = 2.53$ s, $T_3 = 2.07$ s

Chapter 18

Stop to Think Questions

1. **d.** The pressure *decreases* by 20 kPa.
2. **a.** The number of atoms depends only on the number of moles, not the
   substance.
3. **a.** The step size on the Kelvin scale is the same as the step size on the
   Celsius scale. A *change* of 10°C is a *change* of 10 K.
4. **Increase.** When an object undergoes thermal expansion, all dimensions
   increase by the same percentage.
5. **a.** On the water phase diagram, you can see that for a pressure just slightly
   below the triple-point pressure, the solid/gas transition occurs at a higher
   temperature than does the solid/liquid transition at high pressures. This
   is not true for carbon dioxide.
6. **c.** $T = pV/nR$. Pressure and volume are the same, but n differs. The
   number of moles in mass M is $n = M/M_{\text{mol}}$. Helium, with the smaller
   molar mass, has a larger number of moles and thus a lower temperature.

7. **b.** The pressure is determined entirely by the weight of the piston
   pressing down. Changing the temperature changes the volume of the gas,
   but not its pressure.
8. **b.** The temperature decreases by a factor of 4 during the isochoric
   process, where $p_f/p_i = \frac{1}{4}$. The temperature then increases by a factor of 2
   during the isobaric expansion, where $V_f/V_i = 2$.

Exercises and Problems

1. 1900 cm^3
3. 8.3 cm
5. 1.1 mol
7. a. 6.02×10^{28} atoms/m^3 b. 3.28×10^{28} atoms/m^3
9. 2.7 cm
11. 184 K, 330 K
13. a. 171°Z   b. 671°C = 944 K
15. 12 mm
17. 1.68 L
19. a. 2   b. Unchanged
21. a. 0.050 m^3 b. 1.3 atm
23. 2.4×10^{22} molecules
25. 1.1×10^{15} particles/m^3
27. a. $T_2 = T_1$ b. $\dfrac{p_1}{2}$
29. a. 0.73 atm   b. 0.52 atm
31. a. 105,600 Pa   b. 42 cm
33. a. 93 cm^3
    b.

35. a. Isochoric   b. 1556°C, 337°C
37. a. Isothermal   b. −29°C   c. 133 cm^3
39.

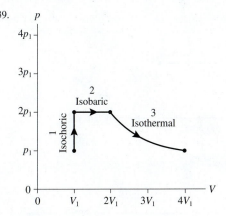

41. 0.228 nm
43. 333°C
45. a. 1.3×10^{-13} b. 1.2×10^{11} molecules
47. 1.0×10^{-16} atm
49. 10 g
51. a. 2.7 m   b. 11 atm
53. 35 psi
55. 93 cm^3
57. 24 cm
59. 1.8 cm
61. a. 880 kPa   b. $T_2 = 323$°C, $T_3 = -49$°C, $T_4 = 398$°C

63. −152°C

65. a. 4.0 atm, − 73°C

 b. p (atm)

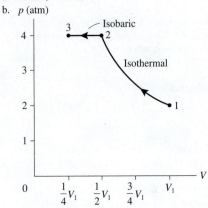

67. b. p (atm)

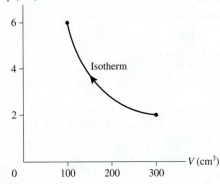

 c. 6 atm

69. b. p

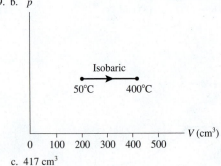

 c. 417 cm³

71. 1.8 g

73. 4.0 × 10⁵ Pa

Chapter 19

Stop to Think Questions

1. **a.** The piston does work W on the gas. There's no heat because of the insulation, and $\Delta E_{mech} = 0$ because the gas as a whole doesn't move. Thus $\Delta E_{th} = W > 0$. The work increases the system's thermal energy and thus raises its temperature.

2. **d.** $W_A = 0$ because A is an isochoric process. $W_B = W_{1\,to\,2} + W_{2\,to\,3}$. $W_{2\,to\,3} > W_{1\,to\,2}$ because there's more area under the curve, and $W_{2\,to\,3}$ is positive whereas $W_{1\,to\,2}$ is negative. Thus W_B is positive.

3. **b and e.** The temperature rises in d from doing work on the gas ($\Delta E_{th} = W$), not from heat. e involves heat because there is a tempera-

ture difference. The temperature of the gas doesn't change because the heat is used to do the work of lifting a weight.

4. **c.** The temperature increases so E_{th} must increase. W is negative in an expansion, so Q must be positive and larger than $|W|$.

5. **a.** A has a smaller specific heat and thus less thermal inertia. The temperature of A will change more than the temperature of B.

6. **a.** $W_A + Q_A = W_B + Q_B$. The area under process A is larger than the area under B, so W_A is *more negative* than W_B. Q_A has to be more positive than Q_B to maintain the equality.

7. **c.** Conduction, convection, and evaporation require matter. Only radiation transfers energy through the vacuum of space.

Exercises and Problems

1. −80 J
3. 200 cm³
5. 16.6 kJ
7.

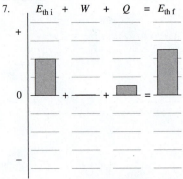

9.

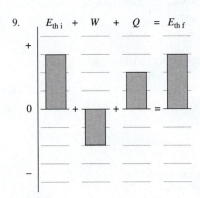

11. −700 J
13. a. 0 J   b. 3.4 kJ
15. 6.9 kJ
17. 1200 W
19. 0.55°C
21. 28°C
23. 73.5°C
25. 21%
27. a. 91 J   b. 140°C
29. a. 1.9 × 10⁻³ m³   b. 74°C
31. A: −1000 J, B: 1400 J
33. 3.14 × 10⁻⁴ m²
35. 5.8 s
37. 26 W
39. 8.7 h
41. 12 J/s
43. 87 min
45. 1.1 × 10⁵ L
47. 61 g
49. a. 5.5 kJ   b. 3.4 kJ

c. *p* (atm)

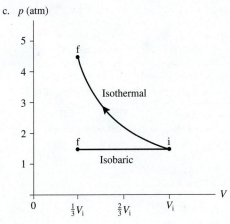

51. 5800 J
53. a. 250°C   b. 33 cm
55. a. 110 kPa   b. 24 cm
57. a. 4300 cm³, 610°C   b. 3000 J   c. 1.0 atm   d. −2200 J
    e. *p* (atm)

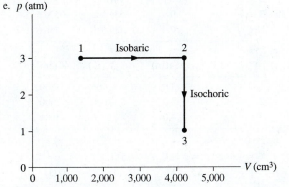

59. 15 atm
61. a. $T_{Af} = 300$ K, $V_{Af} = 2.5 \times 10^{-3}$ m³, $T_{Bf} = 220$ K, $V_{Bf} = 1.8 \times 10^{-3}$ m³
    b. *p* (atm)

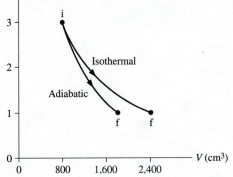

63. 28°C
65. a. 5500 K   b. 0 J   c. 54 kJ   d. 20
    e. *p* (atm)

67. 43°C, 109°F
69. 26°C
71. 12,000 K
73. −18°C
75. b. 217°C
77. −56°C
79. 36 J
81. 150 J

Chapter 20

Stop to Think Questions

1. $\lambda_B > \lambda_A = \lambda_C > \lambda_D$. Increasing the volume makes the gas less dense, so l increases. Increasing the radius makes the targets larger, so λ decreases. The mean free path doesn't depend on the atomic mass.
2. **c.** Each v^2 increases by a factor of 16 but, after averaging, v_{rms} takes the square root.
3. **e.** Temperature is proportional to the average energy. The energy of a gas molecule is kinetic, proportional to v^2. The average energy, and thus T, increases by 4^2.
4. **b.** The bead can slide along the wire (one degree of translational motion) and rotate around the wire (one degree of rotational motion).
5. **a.** Temperature measures the average translational kinetic energy *per molecule,* not the thermal energy of the entire system.
6. **c.** With 1,000,000 molecules, it's highly unlikely that 750,000 of them would spontaneously move into one side of the box. A state with a very small probability of occurrence has a very low entropy. Having an imbalance of only 100 out of 1,000,000 is well within what you might expect for random fluctuations. This is a highly probable situation and thus one of large entropy.

Exercises and Problems

1. $\dfrac{N}{V} = 2.69 \times 10^{25}$ m⁻³
3. 0.023 Pa
5. a. 300 nm   b. 600 nm
7. 84.8
9. a. 20.0 m/s   b. 20.2 m/s
11. a. 289 K   b. 200 kPa
13. Neon
15. a. 818 m/s   b. 3600 m/s
17. −246°C
19. 2.12
21. a. 3400 J   b. 3400 J   c. 3400 J
23. a. 310 nm   b. 250 m/s   c. 2.1×10^{-22} J
25. a. 1.24×10^{-19} J   b. 1.22×10^4 m/s
27. 490 J
29. a. 0.080°C   b. 0.048°C   c. 0.040°C
31. 25°C
33. a. 62 J   b. 100 J   c. 150 J
35. 5000 J
37. 61
39. 9.6×10^{-5} m/s
41. a. 310 m   b. 2900 m
43. 5.8×10^{-22} K
45. a. $\lambda_{electron} = \dfrac{1}{\sqrt{2}\pi(N/V)r^2}$ b. 1.82×10^{-6} Pa or 1.80×10^{-11} atm
47. a. 640 m/s   b. 2.6 nm
49. 8.9×10^{25} s⁻¹
51. a. $E_{helium\,i} = 30$ J, $E_{argon\,i} = 122$ J   b. $E_{helium\,f} = 47$ J, $E_{argon\,f} = 105$ J
    c. Helium gains 16.9 J and argon loses 16.9 J.   d. 307°C
    e. $p_{helium\,f} = 3.1$ atm, $p_{argon\,f} = 3.4$ atm
53. 482 K
55. 22 kJ

59. b. $\dfrac{15n + 3}{2} p_i V_i$

61. a. 1.6 kJ   b. 2.7 kJ   c. 1550 m/s

63. a. $C_V = \dfrac{(3n_1 + 5n_2)}{2(n_1 + n_2)} R$

65. $\frac{1}{2}$

Chapter 21

Stop to Think Questions

1. $W_d > W_a = W_b > W_c$. $W_{out} = Q_H - Q_C$.
2. **b.** Energy conservation requires $Q_H = Q_C + W_{in}$. The refrigerator will exhaust more heat out the back than it removes from the front. A refrigerator with an open door will heat the room rather than cool it.
3. **c.** W_{out} = area inside triangle = 1000 J. $\eta = W_{out}/Q_H = (1000\ \text{J})/(4000\ \text{J}) = 0.25$.
4. To conserve energy, the heat Q_H exhausted to the hot reservoir needs to be $Q_H = Q_C + W_{in} = 40\ \text{J} + 10\ \text{J} = 50\ \text{J}$, not 30 J. The numbers shown here would be appropriate to a heat engine if the energy-transfer arrows were all reversed.
5. **b.** The efficiency of this engine would be $\eta = W_{out}/Q_H = 0.6$. That exceeds the Carnot efficiency $\eta_{Carnot} = 1 - T_C/T_H = 0.5$, so it is not possible.

Exercises and Problems

1. 0.33
3. a 0.27   b. 15 kJ
5. 2.0
7. a. 13 kJ/cycle   b. 50 kJ
9.

	ΔE_{th}	W_s	Q
A	–	+	0
B	+	0	+
C	0	–	–

11. 40 J
13. a. 30 J, 0.15 kJ   b. 0.21
15. a. 0.13   b. 320 J
17. 0.24
19. a. 660 W   b. 760 W
21. a. 5.0 kW   b. 1.7
23. a. Engine c violates the first law.   b. Engine b violates the second law.
25. a. 40%   b. 215°C
27. a. 40%   b. 2500 W   c. 1500 W
29. 0.56 kJ
31. 1.7
33. a. 60 J   b. −23°C
35. a.

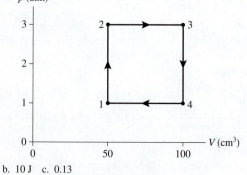

b. 10 J   c. 0.13

37. 250 N/m
39. 28 kW
41. a. 3.6×10^6 J   b. 3.0×10^5 J

43. 8.3%
45. 6.4 J
47. 8.57 J
49. 57 g
51. a. 48 m   b. 32%
53. 37%
55. a.

	W_s (J)	Q (kJ)	ΔE_{th} (kJ)
$1 \rightarrow 2$	0	3750	3750
$2 \rightarrow 3$	4990	12,480	7490
$3 \rightarrow 4$	0	−7500	−7500
$4 \rightarrow 1$	−2490	−6230	−3740
Net	2500	2500	0

b. 15%
57. a. 1.10 kW   b. 9.01%
59. a. 1000 cm³, 696 kPA, 522 K
b.

	ΔE_{th} (J)	W_s (J)	Q (J)
$1 \rightarrow 2$	741.1	0	741.1
$2 \rightarrow 3$	−741.1	741.1	0
$3 \rightarrow 1$	0	−554.5	−554.5
Net	0	186.6	186.6

c. 25%
61. a.

	p (atm)	T (K)	V (cm³)
1	1.0	406	1000
2	5.0	2030	1000
3	1.0	2030	5000

b. 29%   c. 80%
63. a. $T_1 = 1.6$ kK, $T_2 = 2.4$ kK, $T_3 = 6.5$ kK
b.

	ΔE_{th} (J)	W_s (J)	Q (J)
$1 \rightarrow 2$	327	−327	0
$2 \rightarrow 3$	1692	677	2369
$3 \rightarrow 1$	−2019	0	−2019
Net	0	350	350

65. b. $1.1 \times 10^{3\circ}$C
67. b. $Q_C = 80$ J
69. 5.3×10^4 J
71. a. 1.19   b. 227 W
73. c.

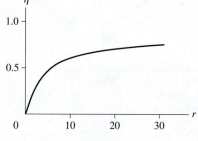

Credits

Cover and title page: Thomas Vogel/Getty Images, Inc.
Page **xv**: Stefano Garau/Shutterstock.

CHAPTER 1
Part I Opener Page **1**: Joe Davila/U.S. Air Force. Page **2**: Kai Pfaffenbach/Reuters Pictures. Page **3** Left: Jason Edwards/National Geographic/Alamy; Center Left: Loop Images Ltd/Alamy; Center Right: Richard Megna/Fundamental Photographs; Right: RooM the Agency/Alamy. Page **4**: MarchCattle/Shutterstock. Page **8**: Rangizzz/Shutterstock. Page **9**: US Air Force. Page **11**: Courtesy of Audi/ZUMAPRESS/Newscom. Page **20**: Blaj Gabriel/Shutterstock. Page **22** Top: U.S. Department of Commerce; Bottom: Bureau International des Poids et Mesures.

CHAPTER 2
Page **32**: Oleksiy Maksymenko/AGE Fotostock America Inc. Page **38**: STS-129 Crew/NASA. Page **49**: Jim Sugar/Corbis. Page **50**: S.Tuengler - Inafrica.de/Alamy.

CHAPTER 3
Page **65**: OJO Images Ltd/Alamy. Page **69**: Pincasso/Shutterstock.

CHAPTER 4
Page **80**: Igor Ivanko/Shutterstock. Page **85**: Tony Freeman/Photo Edit Inc. Page **87**: Richard Megna/Fundamental Photographs. Page **93**: Huntstock/Getty Images.

CHAPTER 5
Page **110**: Placzkowski Piotr/ChromeOrange/Newscom. Page **111** First: Tony Freeman/Photo Edit Inc.; Second: Dmitry Kalinovsky/Shutterstock; Third: Comstock/Getty Images; Fourth: Mike Dunning/Dorling Kindersley Ltd.; Fifth: Todd Taulman/Shutterstock; Sixth: Pearson Education, Inc. Page **122**: FStop Images GmbH/Alamy. Page **123**: Olivier Le Moal /Fotolia.

CHAPTER 6
Page **131**: Tom Tracy/Alamy. Page **132**: Akira Kaede/Getty Images. Page **140**: NASA. Page **145** Top: Pearson Education, Inc.; Bottom: Marc Xavier/Fotolia. Page **157**: Richard Megna/Fundamental Photograph.

CHAPTER 7
Page **159**: C. Kurt Holter/Shutterstock. Page **161**: Todd Taulman/Shutterstock.

CHAPTER 8
Page **182**: Bernd Mellmann/Alamy. Page **185**: Doug James/Alamy. Page **191**: STS-135 Crew/NASA.

CHAPTER 9
Part II Opener Page **205**: Foxbat/Shutterstock. Page **206**: EPA/Neil Munns/Alamy. Page **207** Left: Kyodo/Newscom; Center: John Foxx/Getty Images; Right: Stacey Bates/Shutterstock. Page **208** Top Left: Kimimasa Mayama/Reuters/Corbis; Bottom Left: Murat sarica/Getty Images; Top Right: Vibrant Image Studio/Shutterstock; Bottom Right: Galaxy Picture Library/Alamy. Page **220**: Paul Harris/Paul Harris BWP Media USA/Newscom.

CHAPTER 10
Page **231**: Im Perfect Lazybones/Shutterstock.

CHAPTER 11
Page **261**: Vincent St. Thomas/Alamy. Page **262**: Jupiter Images/Getty Images. Page **268**: bl1/ZUMA Press/Newscom. Page **274**: Richard Megna/Fundamental Photographs. Page **279**: Richard Megna/Fundamental Photographs.

CHAPTER 12
Part III Opener Page **293**: NASA. Page **294**: Atomic Imagery/Getty Images. Page **304**: Polka Dot Images/Getty Images. Page **311**: Karl Thomas/Robert Harding World Imagery/Alamy. Page **314** Top: Chris Nash/Getty Images; Bottom: Richard Megna/Fundamental Photographs. Page **323**: PCN Photography/Alamy.

CHAPTER 13
Page **336**: NASA. Page **338**: bilwissedition Ltd. & Co. KG/Alamy. Page **340**: Corbis. Page **343**: NASA Earth Observing System. Page **347**: NASA Headquarters.

CHAPTER 14
Page **357**: Oleksiy Maksymenko/Alamy. Page **365**: Joseph P. Sinnot/Fundamental Photographs. Page **371**: Louise Murray/Alamy. Page **373** Top: Alix/Science Source; Bottom: TECHART Automobildesign. Page **374**: Lisa Kyle Young/Getty Images. Page **379**: Rkriminger/Shutterstock.

CHAPTER 15
Part IV Opener Page **389**: AguaSonic Acoustics/Science Source. Page **390**: George Resch/Fundamental Photographs. Page **401**: Thomas D. Rossing. Page **408**: Balan Madhavan/Balan Madhavan/Alamy. Page **412**: Martin Bough/Fundamental Photographs. Page **417**: Dr. Harold G. Craighead, Professor.

CHAPTER 16
Page **420**: EpicStockMedia/Shutterstock. Page **421**: Olga Selyutina/Shutterstock. Page **422**: Adapted from PSSC Physics, 7th edition by Haber-Schaim, Dodge, Gardner, and Shore, Published by Kendall/Hunt, 1991. Page **425**: Aflo Co., Ltd./Alamy. Page **435**: B. Benoit/Science Source. Page **436**: David Parker/Science Source. Page **446**: NOAA/AP Images. Page **448**: Image courtesy of Space Telescope Science Institute and NASA.

CHAPTER 17
Page **455**: Stefano Garau/Shutterstock. Page **457**: Renn Sminkey/Creative Digital Vision. Page **458**: Prelinger Archives. Page **466**: Brian Atkinson/Alamy. Page **473**: Eric Schrader/Pearson Education, Inc. Page **475**: Richard Megna/Fundamental Photographs.

CHAPTER 18
Part V Opener Page **489**: Pablo Paul/Alamy. Page **490**: Toshi Sasaki/Getty Images. Page **493**: Richard Megna/Fundamental Photographs. Page **494**: Richard Megna/Fundamental Photographs. Page **499**: Sean White/Design Pics Inc/Alamy.

CHAPTER 19
Page **515**: Green Angel/Alamy. Page **517**: Alexey Dudoladov/Getty Images. Page **523**: Gordon Saunders/Shutterstock. Page **528**: Stocktrek Images/Getty Images. Page **537** Left: Cn Boon/Alamy; Center Left: Gary S. Settles/Jason Listak/Science Source; Center Right: Pascal Goetgheluck/Science Source; Right: Corbis. Page **538**: Sciencephotos/Alamy. Page **539**: Image courtesy of NASA.

CHAPTER 20
Page **548**: Wjarek/iStockphoto/Getty Images. Page **562**: Randy Knight.

CHAPTER 21
Page **570**: Michael Utech/Getty Images. Page **573**: Peter Bowater/Alamy Images. Page **576**: Teresa Olson/Shutterstock. Page **580**: Malcolm Fife/Getty Images.

Index

For users of the three-volume edition, pages 1–600 are in Volume 1, pages 601–1062 are in Volume 2, and pages 1021–1240 are in Volume 3.